PHYSICAL MODELS

PHYSICAL MODELS

Their historical and current use
in civil and building engineering design

Edited by Bill Addis

Ernst & Sohn
A Wiley Brand

Dr Bill Addis
Teddington
UK
bill.addis@cantab.net

Construction History Series/Edition Bautechnikgeschichte
edited by Karl-Eugen Kurrer and Werner Lorenz

Coverphotos (clockwise order):
Model, 1:30 scale, of the 15-storey Alexander Building, made by John Blume and Harry Hesselmeyer, 1932-33. (Image: John A. Blume Earthquake Engineering Center, reproduced by permission)

Soap-film model for the sports hall, Jeddah (end view, using parallel light). (Image: Courtesy of Institute of Lightweight Structures, Stuttgart)

Model test for a proposed monocable suspension bridge for a bridge over the Rhine at Emmerich, 1961. Designed by Fritz Leonhardt and tested at MPA Stuttgart. (Image: Südwestdeutsches Archiv für Architektur und Ingenieurbau, Karlsruhe; Fritz Leonhardt collection)

Water-bath model for Aldwyck Housing Group HQ, Houghton Regis, UK. (Image: Breathing Buildings)

Study of wall profiles in cinemas, 1930. (Image Archive, ETH Library Zurich)

University Library, Basel, 1964. 1:20 scale model in acrylic resin. (Image: Heinz Hossdorf)

1:37.5 scale celluloid model of the reinforced-concrete hangars (first version) designed by Pier Luigi Nervi, tested at the Politecnico di Milano, 1935-36. (Image: ISMES Historical Archive)

1:50 scale acoustic model for the concert hall in the Krakow Congress Centre, Poland. (Image: Ingarden and Ewy Architects)

1:50 scale model of the Thames Surge barrier to study the effectiveness of measures for protecting the river bed. (Image: Courtesy of HR Wallingford).

Library of Congress Card No.: applied for

British Library Cataloguing-in-Publication Data
A catalogue record for this book is available from the British Library.

Bibliographic information published by the Deutsche Nationalbibliothek
The Deutsche Nationalbibliothek lists this publication in the Deutsche Nationalbibliografie; detailed bibliographic data are available on the Internet at <http://dnb.d-nb.de>.

Print ISBN: 978-3-433-03257-2
ePDF ISBN: 978-3-433-60962-0
ePub ISBN: 978-3-433-60963-7
oBook ISBN: 978-3-433-60961-3

Typesetting SPi Global, Chennai, India
Printing mediaprint solutions GmbH, Paderborn

Printed in the Federal Republic of Germany
Printed on acid-free paper

Foreword of the series editors

Construction history has experienced amazing momentum over the past decades. It has become a highly vibrant, independent discipline attracting much attention through its international networks. Although research projects at national level focus on different themes, they are united through the knowledge that their diversity in terms of content and methods, and hence the associated synthesizing potential, are precisely the strengths that shape this new field of research. Construction history opens up new ways of understanding construction between engineering and architecture, between the history of building and history of art, between the history of technology and history of science.

Since Galileo's time, engineers and architects have been using physical models to build bridges between the conceptual design of engineering structures on the one hand and their detailed design on the other. As editor and one of the authors of the present volume of the *Construction History Series/Edition Bautechnikgeschichte*, Bill Addis has gathered together contributions by authors from very diverse backgrounds, different countries, to demonstrate the vital role that physical models play in the design of engineering structures. The authors' multi-method approach not only offers a fascinating and comprehensive insight into the historical development of building and civil engineering, but also enhances our perception of the changing relationship between experiment and theory in times of paradigm shifts in the aforementioned fields of historical research.

Karl-Eugen Kurrer and Werner Lorenz
Series editors

Foreword

The art of devising and capturing geometrical parameters is an essential part of architects' and engineers' daily working lives. In fact, no structures can be designed without it. The problem is that the individual components of many modern structures are subject to complicated internal stresses triggered by external loads from the wind, earthquakes and a range of other potentially complex phenomena. Building up a precise, scientifically rigorous picture of these loads and the internal stresses and deformations that arise as a result (such as the deflection of a bridge when a train travels over it) is certainly no easy task. Indeed, predicting the stresses and deformations experienced by structures and their components via purely theoretical means is, at best, only sufficient for anticipating general trends and cannot provide the precision necessary for reliable structural calculations. For this reason, both architects and engineers have long used models to help them with their plans. 'Models', in this context, are defined as physical or mathematical representations that demonstrate one or more specific properties of a structure. For example, a model of a load-bearing construction will display its structural and deformation characteristics, but not the acoustic qualities of the design in question.

Mathematical models of selected slices of reality stand out thanks to the high, comprehensive levels of precision they achieve in the predictions they produce. This is one of their great strengths. Of course, the accuracy of these predictions is dependent on the quality of each model's design. In comparison with their physical counterparts, mathematical models are also at a disadvantage when it comes to clarity and intelligibility: they typically give their results in two-dimensions and, as such, lack the potential for tactile exploration offered by three-dimensional representations. This tangibility is one of the main benefits of physical models and – alongside the consideration that they are relatively quick and easy to make – is precisely the reason why they continue to play such a central role in design processes in all branches of building and civil engineering today. Indeed, physical models are still used to grasp the key behavioural properties of a construction or engineering system at speed and to make initial optimisations before mathematical models are brought in to help generate the final result.

For hundreds of years, architects and engineers have depended on physical models for developing optimal constructions and determining the stresses and deformations experienced by entire structures and their constituent parts.

Since the late nineteenth century physical models have also played a major role in hydraulic, seismic, acoustic and wind engineering. In fact, for a long time, physical models have ranked alongside empirical knowledge as cornerstones of the construction industry, especially when new structures and novel materials were being included in designs. As a consequence, a huge range of different modelling techniques were employed from the days of the mason's lodges at Gothic building sites right up to the mid-twentieth century. In spite of their tremendous significance for the development of construction as a discipline overall, however, these creations have never been comprehensively described, classified, categorised or placed in a historical sequence. That is, at least, not until the publication of this book, which represents the very first time such a feat has ever been achieved. Even this simple fact alone is enough to make it a crucial, foundational text that will doubtless soon become an essential reference work for students and professionals in the fields of architecture, building and civil engineering alike.

Professor Werner Sobek
Institute for Lightweight Structures & Conceptual Design (ILEK)
University of Stuttgart

Preface

Both historical and current civil and building engineering are often considered in terms of the practical skills of using materials to construct artefacts, and the theoretical tools used to calculate and predict their engineering behaviour before construction begins. So often one hears talk of the theory and practice of engineering. Yet there is more to engineering than this.

Many books have dealt with the historical development of the practical aspects of construction – the history of canals, of dams, of bridges, of masonry structures, of iron construction. Many others have dealt with the history of engineering science and theory – the equations of fluid flow, the statics and equilibrium of structures, the elasticity of materials, the curious behaviour of soil, the reverberation and absorption of sound in a room.

I have long argued that there is a crucial third strand of engineering skill that is equally deserving of historical study – the skill of design. I have argued that design comprises two main activities or outcomes: to convey the designer's ideas to the people who will build an engineering artefact; and to provide the confidence that the proposed design will perform as wished by the client, and as intended by the designer [1].

Over centuries, the first outcome has been achieved by means of drawings, geometrically-faithful models at reduced scale, material data, design rules and codes of practice, and various types of performance specification specific to particular engineering disciplines.

The second outcome, providing the confidence to build, one might call it, has been achieved in simple ways such as following precedent, or learning from the experience of what was not successful. It has also been achieved in more sophisticated ways including making and testing a full-size prototype structure[1] or a reduced-scale model of an engineering construction, and the use of theories of engineering science. Today this might be described as reducing risk to an acceptable level.

The history of engineering design, then, comprises histories of the various ways that engineers have communicated their designs, and how they have provided

1 In UK English it is always awkward to have to refer to a full-size, or full-scale, or actual or real structure. In American English, the word 'prototype' is used, which is much easier. However, in UK English this word means making the first few examples of a product that later goes into batch- or mass-production. In this book the UK English terms are used, despite their awkwardness.

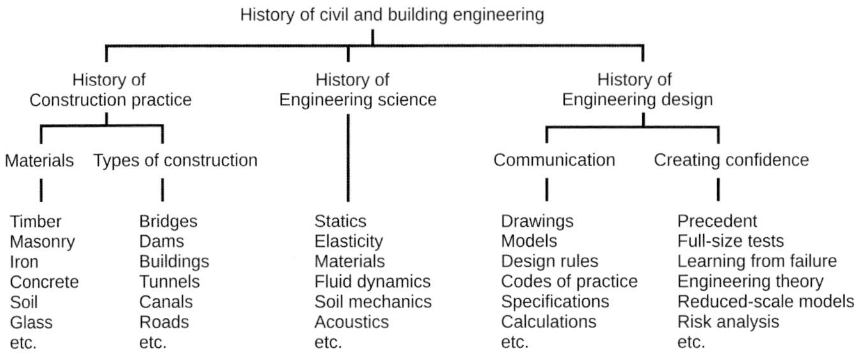

Figure 1 Diagram showing the scope of civil and building engineering history.

sufficient confidence in their designs to persuade clients to fund their projects and contractors to build them [2].

Using these ideas we can build up an overall picture of the history of civil and building engineering which also indicates where the subject of this book fits into the grand scheme (Figure 1). This diagram also provides an epistemological framework for engineering knowledge, applicable equally to the modern practice of civil and building engineering as to the history of the subject. For this reason, it also has consequences for the nature of progress in these fields. When considering the mechanism(s) by which progress is achieved, according to the idea that engineering is a matter of putting theory into practice, it is common to imply that progress occurs as a consequence of developments and progress in engineering science. However, this is patently wrong: there are no significant examples of progress in civil and building engineering construction that have arisen entirely as the result of progress in engineering science. Furthermore, there are countless examples of progress in engineering science that arose out of construction practice. Most progress has been symbiotic, with ideas and experience passing between the two with equal intensity in both directions. Reduced-scale models have often been the essential catalyst to the process, providing the only means by which the practical and theoretical worlds are brought together.

This book is devoted to the use of reduced-scale models, especially in the design process for civil and building engineering projects to help raise confidence in a proposed design. While the testing of full-scale prototypes is common in other engineering disciplines, the sheer size of construction projects generally prohibits full-scale testing. Faced with this constraint, making and testing a model is an intuitive thing to do, and surely goes back thousands of years for artefacts made from the traditional materials – timber, mud and masonry. One advantage of masonry construction is that the structural behaviour of a small model can be scaled up linearly to full size, and give reliable guidance. This is why masonry construction was able to make such dramatic progress from modest houses to the temples of Ancient Greece, the vaults and domes of Ancient Rome and then to the remarkable cathedrals of the Gothic era. However, most engineering phenomena cannot be scaled up linearly, and engineering theory is needed to transpose the results of small-scale tests to full-size behaviour. There was some understanding of this even in ancient Greece, and Vitruvius mentions that some aspects of model behaviour can be scaled up linearly, while others cannot (see Appendix A1).

This book has two main aims – to fill a gap in the history of construction by demonstrating the essential contribution to engineering progress made by physical models, and to give an overview of some uses of physical models in the twenty-first century. The greater part of the book looks at the history of using physical models and within this theme, the larger part is devoted to the use of models in structural and bridge engineering and some mechanical engineering fields such as pumping water. This partition largely reflects when and how models have been used in engineering design and also the availability of historical material. The first three sections look at mechanical and structural models from ancient times to around 1980, by which time their use was in decline as computers became more powerful and widely available. The next section deals with the use of models in engineering disciplines other than structural engineering – measuring the flow and forces associated with fluid flow in hydraulic engineering and wind tunnels, the loads and dynamic response cause by seismic events, the acoustical characteristics of buildings and the behaviour of soils under load. The final section takes a look at current practice in using physical models today in several branches of civil and building engineering.

The main focus of attention in the book is on physical models used by engineers to determine quantitative data – for example predicting wind loads on a building using a model in a wind tunnel. In German the word '*Messmodell*' is a convenient term that distinguishes this type of physical model from others that are merely mechanical or 'proof of concept' models or geometrically representative. Where appropriate, this book uses the term '*measurement model*' – a direct translation of the German word – for this purpose.

Despite being a large book, it has only scratched the surface of this enormous subject. Nearly every chapter would merit the more thorough attention of several doctoral students. I have not attempted to present the first instance of each model-testing technique, nor to cover every field of civil and building engineering that has made use of models, nor to look at the use of models in experimental science with purely scientific aims, nor to present the model-testing efforts of every country. The focus of the book has been on the use of models to inform engineering design, and it uses examples wherever it has been possible to find them. While aiming to provide an overview of the whole subject, I am aware of the unintentional biases that have pervaded my own researches due to the libraries I have been able to use, the relatively few languages that I speak, and the cultural filtering of information via the non-egalitarian Internet. I have done my best to overcome these challenges.

I would like to acknowledge the assistance I have been given in compiling the book – first, and most of all, from the authors, many of whom have squeezed the work required for their chapter into very tight work schedules. I would also like to thank the authors of Chapters 3, 4, 11, 13, 16 and 19 for the help they gave me in translating their contributions from their original languages. I would like to thank colleagues and library staff in several universities and the Institution of Structural Engineers in London for the help they gave me. I give special thanks to Annette Ruehlmann in the Institution of Civil Engineers in London, who found many sources and scanned many images for me. And finally, I give my heartfelt thanks to my partner, Martine Gowie, who has been very patient while I have written and compiled the book, and who has been such a great supporter of my project, in so many ways.

Before delving further into the book, it is worth shedding any idea that the model testing discussed in the book, especially since the mid-nineteenth century, is mainly a lot of (usually) men playing with toys. Even though two articles in a 1920s popular-science journal about model tests for the Boulder Dam were informative and 'serious', their titles portrayed a rather different image – 'Toys that save millions' and 'Toy dams to save lives'! The care and accuracy with which model tests were carried out was extraordinary – often measuring strains or deflections to a hundredth of a millimetre or better. They were no more 'playing' than when brain surgeon is at work. Nevertheless, even in the 1930s, there were engineers who scorned model testing – 'a vet would hardly be entrusted to operate on an elephant if he had gained his knowledge of anatomy from a mouse'. On the other hand, another engineer noted that you can learn a lot about the behaviour of dogs by observing puppies. It is to be hoped that this book will clarify matters.

Bill Addis
May 2020

References

1 Addis, W. (1990) *Structural Engineering – the Nature of Theory and Design.* Chichester, UK: Ellis Horwood.

2 Addis, Bill (2007) *Building: 3000 years of Design, Engineering and Construction.* London & New York: Phaidon.

Contents

Section A

Physical models from ancient times to the 1880s

The use of physical models can be traced back as far as there are books and other written documents about engineering and building. There is mention of using models in ancient Greek and Roman times in the design of military hardware. There is some evidence that the engineers of Gothic cathedrals used models, and indeed it is difficult to imagine how else they could have built such enormous, daring and successful structures. Some engineering models are known from the Renaissance and, based on the numerous sketches in his manuscripts, it is generally assumed that Leonardo da Vinci made and tested some models, ranging from flying machines to weapons to structures and canals; however nowhere does he say he made and used models, and no models attributable to Leonardo are known.

The opening chapter deals with a wide range of models that were produced in a largely non-scientific context in the sense that the models were used to demonstrate that something would work, and would often have been the end result of much development by means of trial and error. They also served to explain to builders, young craftsmen and potential clients how such structures and machines worked and could be manufactured. John Smeaton is identified as one of the first engineers who, in around 1850, conducted scientific tests with reduced-scale models (of a water wheel and windmill) to get data that would help improve the design of these two types of machine.

From the eighteenth century and the Age of the Enlightenment an interest in conducting real experiments finally replaced the classical Greek idea that questions about how the natural world functioned could be answered through logical reasoning alone. The activities of such scientific investigation were soon noticed by practising engineers – indeed many individuals were active in both fields.

The remaining chapters deal with models used in the design of the two areas of engineering that represented the cutting edge of technology from the seventeenth century onwards – masonry structures and bridges. In both cases models helped engineers explore how to push spans, material economy and safety towards their optimal limits. The 300-metre timber arch that Kulibin proposed in around 1770 to span the River Neva at St Petersburg is particularly interesting because it attracted the attention of Leonard Euler who first articulated the rules by which the results of model tests might be scaled up to a full-size bridge when the structural phenomena, such as buckling, could not be scaled up linearly.

The section concludes with the Britannia Bridge, a wrought-iron tubular bridge designed and built by William Fairbairn and Robert Stephenson in the late 1840s with main spans of 140 metres. This was about six times greater than any existing similar girder bridge. Such a step-change in the longest span was unprecedented and remains so. This mega-project required engineers to use all their ingenuity to take such a remarkable step into the unknown. It would have been impossible without the programme of testing reduced-scale models that was undertaken.

1

Models in civil engineering from ancient times to the Industrial Revolution

Dirk Bühler

1.1 Introduction

At the outset of this book, it will be useful to consider the difference between architectural and engineering models for the design and construction process. Architectural models have been documented since antiquity as showpieces or votive objects. In terms of content they represent architectural spaces, their creative design and sometimes the use of the enclosed or/and surrounding areas. Even if they may sometimes present elements related to the structural design, their main focus is still the design and shape of the space. Due to their attractive and often curious design and their use as showpieces and votive objects, many of them have been conserved until our times. It is only since the Renaissance, however, that construction-related (mechanical) models, began to be helpful not only for architectural, but also for structural design and hence this time marks the beginning of technical models as we know them today. These models for structural design fulfilled a new task in the design and planning process: they helped to visualize, explain, test or simply propose technical solutions for construction problems which could not be described or resolved using words or drawings alone. Their use was primarily experimental and less representational and were thus more ephemeral than their architectural counterparts. Their fundamentally practical use is probably one of the reasons why so few of them are still conserved, but also their use for experiments leading to failure was not always helpful for ensuring their conservation. If technical models were used for competitions, usually only the winning proposal has been preserved, not the discarded ones.

The different purposes for which these specific historical models were built comprise a large range, including: testing (new) mechanisms and designs, checking mechanical functioning and structural behaviour (stiffness, strength, rigidity and robustness), demonstrating the viability of new ideas, or new proposals to convince the clients to give their approval, exploring and confirming how the machines etc. should be constructed – the design of connections, the sequence of assembly, learning how to build full-size works, including apprentice 'masterpieces', and to demonstrate to visitors and potential clients how things

Physical Models: Their historical and current use in civil and building engineering design,
First Edition. Edited by Bill Addis.

worked. They have also been built for competitions and for patents, and because they were required as a part of construction contracts.

Many scientists and constructors described the use and the utility of physical models as tools for empirical cognition. Their manuscripts, together with contemporary publications from the times of Leonardo da Vinci (1452-1519), Galileo Galilei (1564-1642) and later, can be taken as a possible starting point. The writings of Leon Battista Alberti (1404-1472) and authors of many subsequent treatises on mechanics, structural design and architecture play an important role in the establishment of models as tools for testing and teaching construction techniques during the following centuries. During the Age of Enlightenment, associated with the emerging industrial revolution, the use of engineering models increased and documentation about their use improved. This process lasted during the early experimental phases of new technologies until the techniques became well-established – together with the testing of materials – in the practice and the education of engineers towards the end of the Industrial Revolution into the early decades of the nineteenth century.

Our understanding of the use of models for mechanical and construction solutions since the Renaissance has recently been growing more than ever before through the discovery of original drawings, manuscripts and publications, and also a growing number of surviving physical models have been discovered. Today we find the traces of them spread throughout archives, historical publications, museums and local collections, often still awaiting further research, thorough study and greater appreciation of their role in engineering.

1.2 Engineering models in classical times

The earliest evidence for the use of models in engineering design and construction is to be found in two manuals for military engineers (*Belopoeica* – On artillery) dating from around 250BC. One was written by Philon of Byzantium (c.280-c.220BC); the other was written by Heron of Alexandria (c.10-c.70AD) but he acknowledged what he wrote was largely taken from lost treatises by Ctesibius of Alexandria (c.300-222 BC) [1]. Both Philon and Heron discuss the use of experiments to determine the best dimensions to use when making artillery with a certain performance requirement (size of projectile and range). The power of a catapult was determined by the diameter (D, in *dactyls*. 1 dactyl = c.19.3 mm) of the elastic spring (twisted sinews). For a *euthytone* catapult, this diameter was calculated as one ninth of the length (L, in *dactyls*) of the metal-tipped wooden arrow – i.e. $D = L/9$. This design rule allowed the dimensions of a small catapult to be scaled up for use in building a larger one – a smaller one effectively served as a model for a larger one. No detail was given of the range of sizes for which this scaling process was valid [2]. Philon and Heron both give a more complicated relationship for the larger *palintone* catapult, able to throw stones weighing many tens of kilograms. The diameter D, in dactyls = $1.1 \times \sqrt[3]{(100M)}$, where M is the weight of the stone projectile in *minas* (1 mina = c.0.44 kg). Knowing the diameter of the spring, D, the design rules that Philon and Heron describe also give all the other dimensions of at respective catapults, such as the length of the catapult arms, the size of components of the timber frame, and so on. The dimension, D, called the *module*, is thus the single dimension needed to specify the entire design of a catapult.

The Roman engineer and architect Marcus Vitruvius Pollio (exact dates not known, c.70 B.C.-c.10 A.D.), who trained as a military engineer, included similar design rules for catapults in his book *De architectura*, compiled around 30-15 B.C. He cited both Philon and Ctesibius as his sources [3]. However, Vitruvius went a significant step beyond Philon and Ctesibius in giving the first clear description of using technical models in antiquity, again in the field of military engineering. In the tenth book of his *De architectura*, he recounts an exemplary story about the military engineers Diognetos from Rhodes, Callias from Aradus and Epimachos from Athens, who had to construct a military device for their King Demetrius. He discusses the advantages and disadvantages of both small-scale- and full-size models (Latin: *exemplaribus*) and notes that some dimensions in a model can be scaled up linearly to full size, while others cannot [4] (see Appendix A.1).

After the classical era, apart from some evidence of models for mill technology in the late Middle Ages, the earliest evidence of the existence of technical scale models, built in parallel with architectural ones, emerged in around 1400 [5]. One of the earliest documents refers to a competition for the design of a stone-saw for the builders of the cathedral of Milan in 1402, when a scale-model of the winner's proposition was built in wood in order 'to be able to try it out' [6].

1.3 The Master Builders of the Renaissance and their engineering models

1.3.1 Phillipo Brunelleschi and Florence cathedral

One of the oldest technical models is documented for the year 1418: it is Filippo Brunelleschi's (1377-1446) model of the cupola of Santa Maria del Fiore in Florence, which demonstrates an innovative bricklaying technique (It.: *Spinapesce*, Eng: herringbone) that permitted the cupola to be constructed without a supporting scaffold that was usually necessary for arch and dome constructions [7]. Although the scale of the model has not been documented, we know that he used regular bricks to build it and that the model could be examined from inside – indications of a large scale. With this proposition he had emerged as the winner of a contest seeking the best solution for the construction of the cupola. Ten models had been presented and their costs contribute an interesting fact to the practice of model making at this time: The range of their costs rises from 2 florin + 8 lira up to 127 florin + 510 lira: the latter was for Brunelleschi's model [8]. These costs illustrate the enormous range of quality that these models could offer, and their great economic value for the architects. Brunelleschi appreciated the use of models for both architectural and technical solutions and built many models of scaffolds and cranes [9]. However, his model from 1418 was not the only one he built for the Florence cupola: another, dating from around 1420, is still conserved in the *Museo dell'Opera di Santa Maria del Fiore* in Florence and illustrates their plainness. Recent measurements of this 1:60 scale model have proved at the same time the precision of its manufacture [10].

Models were also used for other important competitions, as for example in the one tendered for the construction of the crossing tower (It.: *Tiburio*) of Milan Cathedral in around 1490. Since the master builders of the fifteenth century were both architectural and structural designers, their treatises also attached great importance to the use of scale models as an additional resource for architectural and for technical planning and presentation. They could be helpful additions to drawings and written descriptions for their customers, the local approval authorities and the craftsmen during the building process. Filarete (Antonio di Pietro Averlino, 1400-1469), Leon Battista Alberti (1404-1472) and Francesco di Giorgio di Martini (1439-1501) explicitly mentioned them as essential means of architectural and technical representation and research in the planning process, and recommend checking the results of different means of presentation against one another. Martini also discusses modelling in the context of mill technology because drawings and written explanations were not able to explain technical contexts as well as mechanical models. Models also served as three dimensional plans for the reproduction of technical devices [11].

1.3.2 Leonardo da Vinci

In his book *Lives of the Most Excellent Painters, Sculptors, and Architects,* first published in 1550, Giorgio Vasari (1511-1574) praises, as the first among the many ingenious skills of Leonardo da Vinci (1452-1519), his creativity as a designer and model builder for 'tunnel boring machines, leverages, winches and screws for lifting heavy loads or to tow them away and [his proposals] in which way port basins can be cleaned or water can be elevated from the greatest depths by means of pumps' [12]. Leonardo himself was no less modest about his capacities as a model builder when he proposed, in his letter of application [13] to Lodovico il Moro (1452-1508) shortly before his arrival in Milan in 1482, presenting models of his inventions in Lodovico's park or any other space. Nevertheless, it is questionable whether Leonardo, with his own limited resources, really would have been able to afford the building of these expensive models to an acceptable quality. The only model we know he constructed was an architectural one, when he participated in the competition for constructing the lantern of the cathedral in Milan between 1487 and 1490. It was a wooden model, accompanied by a series of drawings [14]. We do not know whether his masterful drawings of machines and instruments for construction or his many sketches of testing devices for building materials and structural behaviour were based on real experiments using models, or not: some writers think they were [15].

It was only at the end of the nineteenth century, when Leonardo's technical drawings first appeared in facsimile, that engineering historians first began to interpret them by means of models. When the first generation of models, initially shown in 1939 in the *Palazzo dell'Arte* in Milan, had been lost because the ship transporting them to Tokyo in 1942 sank, a second generation of models was created in 1952 and exhibited from 1953 in the *Museo Nazionale della Scienza e della Tecnologia* in Milan in commemoration of the 500th anniversary of Leonardo's birth. However, the technical difficulties encountered when trying to construct accurate models of Leonardo's inventions from his drawings which, at first glance appeared perfect, suggest that his sketches were not really made for this purpose [16].

1.3.3 Andrea Palladio

In the second half of the sixteenth century, the use of models in architecture and civil engineering seems already to have been established to such an extent, that Andrea Palladio (1508-1580) does not mention them in this treatise *The four books of Architecture,* first published 1570 in Venice. Nevertheless, Palladio commissioned two models, paid for by the city council on 26 October 1569 [17], for the construction of the timber bridge over the Brenta river in Bassano di Grappa in order to support his carpenters who prepared and assembled the construction. Two years earlier, in 1567, his expert opinion had been in demand for the evaluation of a model for the cathedral in Brescia [18] which was surely not the only occasion when he had dealt with his own and other master builders' models as part of the general procedure in planning and building. There are further allusions in Palladio's and later in Claude Perrault's (1613-1688) writings that they used scale models of timber bridges for scaling dimensions for the real construction. More than a century later around 1750 the engineer Bartolomeo Ferracina (1692-1777) rebuilt Palladio's bridge at Bassano and manufactured a wooden model of a pile driving machine that he used on the site: this model is still conserved in the *Museo di Storia della Fisica* in Padua [19]. These facts suggest again the proliferation of models for engineering and construction purposes both for display and on the construction site, and the demand of models of hoisting devices, pumps, cranes, weirs, etc. for engineering and building competitions increased considerably towards the end of the seventeenth century.

1.3.4 Domenico Fontana and moving the Vatican obelisk

The most famous construction competition at this time was announced in 1585 by pope Sixtus V to find an engineer to undertake the moving of the so-called Vatican Obelisk. This announcement motivated many Italian and other European architects and engineers to present their proposals and ideas with plans, descriptions and models in a spectacular meeting with the pope and his administration in Rome. The winner of the contest was the engineer in charge of constructing St. Peter's Basilica, Domenico Fontana (1543-1607). His narrative of the project was published in 1590 entitled *Della Trasportatione dell'Obelisco Vaticano [...]* [20] and the first illustration in this work shows eight models that had been submitted in the competition, each with a short explanation. (Figure 1.1)

This illustration takes us to the place where the Vatican Obelisk (Fontana calls it *Guglia,* a spire) had stood since the Roman Emperor Caligula (reigned 37- 41 AD) had it erected in the central spine of the *Circus Gai et Neronis.* This site in 1586 was an open space adjacent to the old St. Peter's church just behind the

Figure 1.1 Fontana's illustration of nine models of methods for raising the Vatican Obelisk. His own is shown at the top of the page. (Source: [20] p. 7)

sacristy and apparently served as the place where the models were presented. In the background the construction of the new St. Peter's can be seen and from a cloud in the upper right corner Minerva (Pallas Athene), the Roman goddess of wisdom and crafts with a bundle of rods in her arms, with an olive branch emerging from one of them, observes the scene with curiosity. Fontana's model is identified by an 'A' and elevated towards heaven by two putti. Fontana presents his model at the top, drawing it larger than the others, and comments: 'A. The first one, which was accepted and executed which will be explained in this book'. The remaining models are all at a scale of around 1:10 (2.5 m tall, based on the size of the adjacent people) or 1:5 (5 m tall, based on the drawing of the full-size obelisk) and are positioned on the ground around the obelisk. They were all observed by two members of the Vatican's committee who are comparing what they see with the current literature. On the ground in front of them lies a wind rose, which serves to highlight the scientific standing of the models. Fontana comments on them as follows: 'B. The second, which was to support the spire just with slight ropes on the top. C. The third, which wants to balance [the spire] on a half wheel. D. The forth, which wants to establish [the spire] using only wedges. E. The fifth, which wants to lower [the spire] using screws, and to carry it inclined, as they say, half in the air. F. The sixth, which wants to establish and lower [the spire] with a single bar in the form of a steelyard. G. The seventh, which consists of a half wheel, upon which they want to lower [the spire] tooth by tooth, like the buckets of a millwheel. H. The eighth, which wants to establish, lower, and pull [the spire] by the power of screws.' In order to elevate his own solution – as the putti do in the drawing- the alternative proposals are shown as not very feasible and their description rather dismissive.

Fontana's systematic and well-organized procedure for calculating the physical geometry and the loads involved, drawing elevations, and resolving the transport and placement of the obelisk, establishes him as perhaps the first engineer to work in a modern scientific manner. The purpose of his book is not only an early-published presentation of using engineering scale models and a documentation of the methods he employed, but also a science-based instruction for further tasks and future engineers.

One more scientist of the sixteenth century cannot be omitted: Galileo Galilei (1564-1642), came to fame not only by his 'Discorsi…' and his telescope, but also thanks to his mechanical experiments with models. He received a privilege based on a model of a water-lifting device that he presented in Venice in 1594 [21]. Only a few decades later Robert Hooke (1635-1703) and Christopher Wren (1632-1723) used, for the first time recorded, a hanging chain model (catenary) to help define the optimal structural form for the cupola of London's St. Paul's Cathedral (see Section 3.2.1).

1.3.5 The Fleischbrücke: a case study in technology transfer

The close trading relations between Northern Italy and Southern Germany and the visits to Venice, Florence, Milan, and elsewhere by German master builders such as Elias Holl (Augsburg, 1573-1646) and Joseph Furttenbach (Ulm, 1591-1667), stimulated an intense technology transfer between Italy and Germany in the late Renaissance. A striking example is the case of the construction of the Fleischbrücke in Nuremberg in 1598. There is no doubt that the master builders in Nuremberg knew about the construction of the Rialto Bridge in Venice. The winning design of a spectacular competition in 1551 was famous for its innovative construction, and was built by Antonio da Ponte (1512-1597) between 1588 and 1591. This was just a few years before the planning of the Fleischbrücke began. The similarity between the two bridges concerning their structural challenges and their solutions is clear: both are designed to cross the river with a single flat, segmental arch (Rialto Bridge, 28.8 m span and Fleischbrücke, 27 m); both required careful design of the timber centring, and both had their foundations in very marshy ground [22]. Although it is clear that there has been some technical exchange from Venice to Nurenburg, a study of the original documents in Venice and Nuremberg do not reveal a direct influence by the Venetian designers on the German bridge. It seems most likely, therefore, that there was a personal exchange of experience [23], and the strongest evidence for this is the existence of a model.

Wolf Jakob Stromer (1561-1614), who was Nuremberg's chief master builder between 1595 and 1598, commissioned and supervised the planning and building of the Fleischbrücke. His descendants today still own a curious model of the Rialto Bridge which had recently been completed (constructed 1588-1591). The model is not dated but was already in Nuremberg when the Fleischbrücke was being built. However, as it is not mentioned in either the Nuremberg or the Italian archives [24], it cannot have been used for the design or construction of the Fleischbrücke. Nevertheless it is a remarkable piece that is conserved in its original wooden case in the private archives of the Stromer family in the castle of Grünsberg near Nuremberg. The painted wooden model, without a declared scale (65.5 x 23.4 x 29.1 cm, approximately 1:70) represents the complete Rialto Bridge, fixed on a two-piece baseplate, including abutments and centring. [25] The model consists of fifteen pieces and can be disassembled (upside down) into these pieces: first the buildings on top of the bridge can be removed, and then the arch below and the centring; finally, the whole model can be separated into its individual pieces. As a special feature there is a drawer in one abutment which shows a cross section through the foundation of the bridge. This three-dimensional model

Figure 1.2 Model of the Rialto Bridge, conserved in the Stromer family archives, Grünsberg Castle, near Nuremberg. (Image: [26])

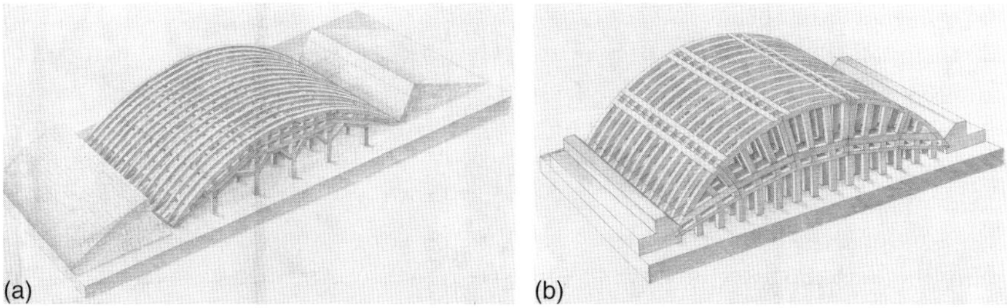

Figure 1.3 Two prints, dated 1595-1603, showing models of designs for the timber centring for the Fleischbrücke, Nuremberg. (a) Probably the adopted solution. (b) An alternative, heavier solution. (Image: Courtesy of Staatsarchiv Nürnberg. Stromer-Archiv [28])

demonstrates beautifully the main challenges of the original design for the bridge and an approach to their solution: it was clearly intended to be an instructive device. (Figure 1.2)

The documentation of the Fleischbrücke revealed the existence of two other models produced at the time when the bridge was under construction: both models are lost, but two drawings of them, and related documents still survive (Figure 1.3). The models show two possible solutions for the timber centring used for the erection of the main arch. They were probably made to find the best possible construction of this essential and difficult element. It is likely that the final solution is that shown in Figure 1.3a [27].

1.4 The first collections of engineering models

The increasing number of technical models produced during Renaissance stimulated local rulers, building authorities and academies, from the second half of the sixteenth century, to collect these objects for the purpose of documentation in their archives. Architects and engineers, who were gradually joining the emerging bourgeois elite in society, also began to assemble collections of models in their Cabinets of curiosities (*Wunderkammer*) [29].

For instance, already in 1532, the city council of Nuremberg recognized the value of their building models, including models of mills and particular construction elements, and brought them together in a new secure space in the town hall, specifically designed to house the collection, having previously been dispersed in the armoury and several other locations [30]. Shortly after its installation, the town hall of Nuremberg presented a temporary exhibition of mechanical models in 1569, which may have been the first of its kind [31]. Half a century later, in 1620, the city council of Augsburg established a model chamber in its new town hall that would later become well known (see Section 1.3.1).

The German architect Joseph Furttenbach (1591-1667) in Ulm mentioned in the foreword of his treatise *Architectura Civilis*, published in 1628, that an architect must be capable, not only of drawing plans, but also of producing explicitly 'scale models made of timber' of his constructions [32]. In later years Furttenbach installed a *Kunstkammer* (cabinet of curiosities) in his own house in Ulm and, in

1641, published another treatise, *Architectura Privata,* dealing particularly with the design of this house and the collection presented there, illustrated with additional explanatory plates showing models of bridges, construction equipment and water-raising devices [33]. Amongst the many other objects, he included models of different kinds of mills and also important waterworks, all of them listed and described in his book. He mentioned proudly that his models included also a cutaway model ('one shoe high') of the rope winch used by Fontana in 1586 for the translocation of the Vatican's obelisk, and improved by Galilei in 1617. We do not know if he really met Galilei in his *Grand Tour* of Italy between 1606 and 1617, but he claims to have received the model of an endless spindle or winch from Galileo's own hands, which was, of course, displayed in his collection.

Thanks to a catalogue, published in 1683 and distributed not only in France, but also in Britain and Germany and maybe elsewhere, we learn about one of the first public exhibitions of mechanical models [34]. The author of the catalogue is not explicitly mentioned but it was most likely to have been Jean Baptiste Picot (dates unknown), quartermaster during the reign of Louis XIV, who was in charge of organizing the exhibition under the patronage of the then superintendent of buildings and later Grand Master of Ceremonies, Jean Jules Armand Colbert, Marquis de Blainville (1663-1704) [35].

As Colbert and Picot adhered to a rigorous experimental rationalism, this exhibition was intended to encourage engineers and inventors to use models to present their inventions, rather than relying only on drawings or written descriptions, and probably also to promote technical inventions within France [36]. The exhibition consisted of twenty-one models with indications of their scale, manufactured by diverse workshops. Twelve of these models were explained and illustrated with plates in the catalogue. Most were of mill machinery such as timber mills, drilling equipment for cannons, and numerous water-lifting devices. Some of the models represented machinery already published in treatises by authors such as François Beroald (Lyon 1578), Georg Andreas Böckler (Nuremberg 1601) and Salomon de Caus (Frankfurt am Main 1615) in an improved and technically more convincing version. Eight of the models were by two French inventors: of Jean Baptiste Picot, the curator of the exhibition and author of the catalogue, and M. Tarragon, a somewhat unknown artisan. Without going into further details of such models, it can be confirmed that, by the end of the seventeenth century, the interest in and use of engineering models had become ubiquitous and common knowledge in Europe, and their utility motivated more and more engineers to build them in order to confirm the correct function of their inventions, as well as to promote their acceptance and use.

1.4.1 Elias Holl and the Augsburg model chamber

The Free Imperial City of Augsburg has collected design models and models of record of important building projects since the early Renaissance. In 1620, Elias Holl (1573-1646), Augsburg's most famous architect and designer of its new town hall (built 1609-20), turned the room above the Golden Hall into a dedicated model chamber. The Augsburg collection was one of the first to focus not only on architecture, but also on what would come to be known as civil engineering. This original collection was augmented by additional engineering models donated later by the St. Anna grammar school, a protestant educational institution with firm humanistic and scientific aspirations. The school was founded in 1531, and located in the courtyards of the St. Anna Carmelite monastery, which was dissolved in 1534. Between 1613 and 1615 the school built an additional building, designed by Elias Holl, alongside the old one; both are still remaining. Scientific instruments and mechanical models were collected there for educational purposes.

Another well-known collection of models was later incorporated into the city's model chamber. These were models of the hydraulic engineering works of Caspar Walter (1701-1769) which he had installed in the water towers of Augsburg (see Section 1.3.2). The model chamber in Augsburg later incorporated several bridge models from the former Building Museum, which was part of the local Building School, founded in 1893 in the former monastery of the Holy Cross. Today, 126 models are listed in the inventory of the municipal model chamber including fourteen models of bridges, eight models of machinery and construction equipment, twenty-five of waterworks such as pumping stations and water towers, sixteen of mills, seven of roof trusses and one of a scaffold. Two models are of double and triple helix staircases, seven represent the city gates and eight more show the towers. The model chamber also has eight models related to Holl's historic, surviving town hall: the old town hall (1515-1516); a second Italian model (1609); a Palladian model (1611); model I by Holl (1614); a three-gable model by Holl (1614); model II by Holl (1614); a roof crossing, model III by Holl (1614/15); a Venetian model, model IV by Holl (1614/15); and a model of the floor structure for the town hall [37]. There is also a model of a boat and six miscellaneous models with mainly architectural content.

One of the oldest models in the model chamber represents the *Lueg-ins-Land-Turm*, an observation tower and part of the city wall. The model was made in 1514 by the carpenter and sculptor Adolf Daucher (c.1460-1523/24) to present the architectural and structural design of the tower (Figure 1.4).

Perhaps the most spectacular models of Augsburg's architecture are two models of the entire city, the older one dating from 1563 (Figure 1.5) and the other from the seventeenth century.

Even if, as we have already seen, Elias Holl's idea was not really new, nevertheless, his particular selection of models for his collection makes it so very remarkable. In his time, Holl, a traditional type of master builder for those days, was appreciated more for his technical and engineering achievements than for his architectural designs, which are what are generally admired today [40]. Son of a master builder family for two generations and a master mason by training, Holl was appointed as the city's 'master builder' in 1602 and took this opportunity to rebuild the flourishing city of Augsburg. His oeuvre ranges from civil and religious architecture to engineering projects such as bridges, canals and mills. In his personal family chronicle, it is with some pride that Holl himself mentions his successful solutions to structural and technical problems rather than his architectural designs [41]. Among the many drawings and written documents

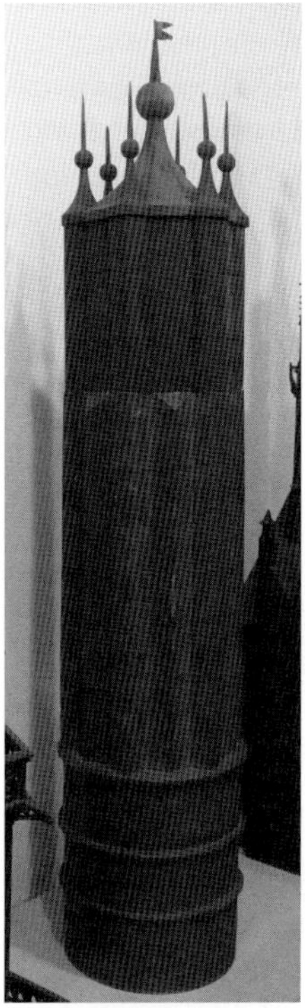

Figure 1.4 Model of the 'Lueg-ins-Land Tower' by Adolf Daucher manufactured in 1514. Modellkammer in the Maximilianmuseum in Augsburg. (Image: Dirk Bühler [38])

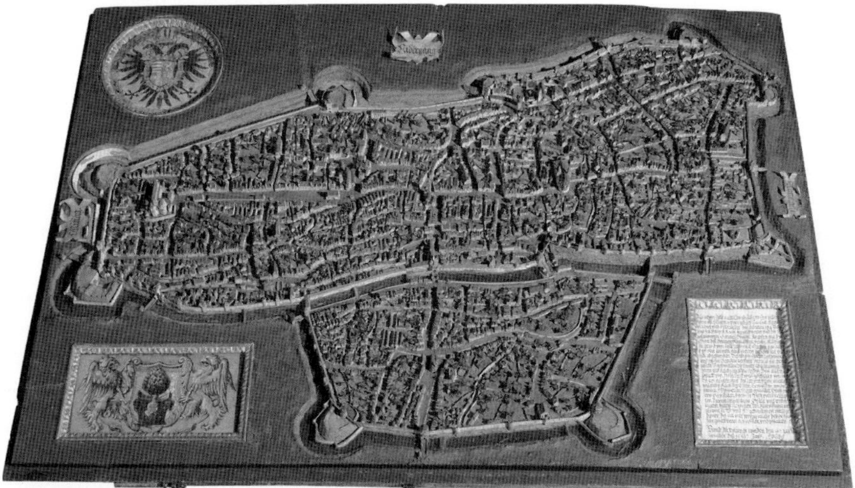

Figure 1.5 Model of the city of Augsburg made in 1563 by the master printer Rogel after three years of survey and wood-carving work. (Image: Courtesy of Kunstsammlungen und Museen Augsburg [39])

still conserved in the city and other archives there is an illustrated manuscript of 209 folios, entitled 'Master Holl's Book of Drawings.' It deals formally with 'geometry and the art of measurement', and contains not only a compound treatise on geometry, surveying and the related instruments, but also precise indications about building materials, their composition and practical application, together with tables of their measures and weights [42]. This valuable document, probably intended for later publication, is a reliable and valuable source for deepening our knowledge of Renaissance construction methods. Based on his built legacy, his written and drawn documents, and contemporary sources, we know that Holl was constantly engaged in designing innovative construction equipment and collected models of both his architectural and engineering achievements.

After Elias Holl had received the commission to design and build the new city hall, and before the old Gothic building could be demolished, the precious bell hanging in the old town hall had to be relocated to the nearby Perlach Tower between 1614 and 1615. After confirming the stability of the tower's structure, Holl proposed an additional roofed space on top of the tower to accommodate the bell. For this task he planned a scaffold surrounding the tower and a special crane. This scaffold was a self-supporting structure independent of the tower's wall and without the use of clamps, thus protecting the façade from damage, as Holl explained in his chronicle [43]. This technically demanding task is represented by one his most important models - still conserved - in the model chamber, together with his extraordinary architectural models of the town hall. Holl mentions that he delivered several models to the city council at that time, but without further specification of them we do not know if the scaffold model was made before or after the construction and hence we cannot be sure about its use. Other models from Elias Holl's time include a lifting device for the gentle handling of cut stones (Figure 1.6), a crane with a worm gear, moveable in two directions, and an automatic ram. Later models include the scaffold used for the vaulted ceiling of the St. Anna grammar school (1747), the architecture and the timber roof truss of the Schrannenhalle, and a covered market hall close to St. Moritz church (1750).

1.4.2 Caspar Walter and hydraulic engineering in Augsburg

The second great contribution to Augsburg's model collection relates to hydraulic engineering. Founded by the Romans in the first century BC, during the reign of Emperor Augustus, on a mound surrounded by the rivers Lech and Wertach, Augsburg has always had a plentiful supply of fresh water, but the careful management and distribution of running water was not common until the late middle ages. In 1412, the city council built the first pumping station to feed the newly-installed public fountains. From that time, the city had at its disposal a double hydraulic infrastructure: a grid of canals providing the mills with hydropower, and a network of wooden water pipes fed by the pumping station at the 'Rotes Tor', one of the main city gates. With the establishment of its first waterworks, Augsburg became an international exemplar for urban water supply, and the city soon installed a collection of models and explanatory panels in the new water towers for interested travellers who came to learn from the 'Brunnenmeister' (master of the waterworks).

Figure 1.6 Model of an innovative lifting device (17th century), Modellkammer in the Maximilianmuseum in Augsburg. (Image: Dirk Bühler [44])

Caspar Walter (1701-1769) had been Brunnenmeister in Augsburg since 1741 and, in 1754, he published *Hydraulica Augustana* in which he described the three central water towers, the well houses of the waterworks with all their technical installations, and expanded the model collection displayed in the main tower [45]. Walter's description begins on the third floor of the main tower, where several crankshaft models are presented. On display on the fourth floor are models of weirs, dams and sluice gates suitable for installation in canals and rivers. Other models illustrate the use of stone, bricks and timber in the construction of basins, gates and pit lining. There are even models of a canal bridge, a boat used when stabilising river banks, and several roof frameworks. Visitors to the fifth floor see a number of text panels, five models of Augsburg's famous pumping stations, a model waterwheel driving three different types of machinery, as well as water towers, well houses, and a mill driven either by muscle power or hydraulic power. Also on display, with a description, is a model of the sluices and weirs along the branch of the Lech River known as the 'Hochablass' which conducts water into Augsburg's canal system. The exhibition culminates on the sixth floor, with three large models of pumping stations. Different types of transmission shaft show additional practical uses of hydropower. The presentation is completed by a series of panels, showing the water distribution system and all its technical installations. The seventh floor – where a small stairway leads to the viewing platform on top of the tower – contains the water tank. The models that Walter describes were made to present the technical achievements of Augsburg's water infrastructure, and for educational purposes in his workshop – many of the models were masterpieces made by his apprentices.

The distribution of fresh water to the public fountains and households in Augsburg created a need for bored timber pipes for the supply system. Such pipes – straight tree trunks up to four metres in length that were drilled using a borer with an iron cutting head – had been well-known since Roman times and used in many places, even until the end of the nineteenth century. The making of these pipes posed two main challenges for the workers. The first was how to keep the cutting head, which was usually fixed to the end of a wooden or iron rod, straight while steadily drilling through the centre of the trunk. The second challenge was to provide enough human or mechanical power to carry out the drilling, which required considerable energy. In this context, the functional model of an eighteenth-century timber pipe-drilling machine (Figure 1.7) deserves particular consideration because it illustrates clearly how the power transmission provided by the waterwheel, and the labour-saving gears, are combined into a horizontal, water-powered drilling machine to perform the task. This practical machine pulls and fixes the trunks in position so that they can be drilled easily. Aside from a few drawings published in artisan treatises, this is the only three-dimensional representation of this kind of machinery. The model can be operated with a crank handle to move the waterwheel.

Two of the twenty-five hydraulically-powered machines are outstanding examples [47, 48]. The first is an eighteenth-century model of a pumping station, formerly exhibited in the main tower of the Augsburg waterworks and described by Walter. The model represents two water pumps, each consisting of three hydraulic cylinders. The devices were operated by a waterwheel placed between them. Water pipes conduct the water to the tank on top of the tower (not represented in this model) (Figure 1.8). The main difference between the two pumping devices lies in the power transmission between the waterwheel and the pumps: in one model, the transmission is performed by a cumbersome swinging beam; in the other, a more modern transmission is achieved by means of a cranked shaft.

Figure 1.7 Model of a boring machine, driven by a waterwheel, for making timber water conduits, Maximilianmuseum in Augsburg (Image: Dirk Bühler [46])

Figure 1.8 Models of two water pumps made by Caspar Walter in 1754, Maximilianmuseum in Augsburg (Kunstsammlungen und Museen Augsburg, Inv.-Nr. 9753) (Image: Dirk Bühler)

The model thus displays the transmission technologies used at the turn of the sixteenth and seventeenth centuries (the swinging beam) and the more modern transmission, introduced during the eighteenth century (the cranked shaft).

The second remarkable model demonstrates several possible uses of hydraulic power. As the brass label on the base plate states, it was made by the carpenter Andreas Seuffert in 1758. It shows a central waterwheel surrounded by seven mills with sixteen different types of gear. The waterwheel can be turned by a hand crank to demonstrate the operating mode of mills and gears. This sophisticated and functional teaching model was used by its manufacturer for presentations at various courts. The mills and their corresponding mechanical devices include a flour mill, fulling mill, paper mill and an oil mill on one side of the waterwheel, and sawing, hammering and polishing mills on the other side. This well-constructed model is able to convey the state-of-the-art milling techniques in the mid-eighteenth century, and is a precious technical document of its time.

Caspar Walter was not only an important hydraulic engineer and *Brunnenmeister*; he had also trained as a master carpenter and bridge builder. In his treatise on bridge construction [49], published in 1766, he described all types of bridge and their construction details, and includes some discussion on models. For the first time he described how he deduced from a load test on a wooden model at a scale of 1:20, the dimensions and load capacity of the real bridge, using direct proportion. However, Walter was mistaken in this interpretation of the results of a model test. Leonard Euler (1707-1783) read Walter's publication and explained the mistake before going on to discuss, for the first time, the non-linear scaling of bending behaviour in models of structures (see Chapter 4).

Finally, there is an interesting insight into professional relationships in hydraulic engineering in the eighteenth century. When the pumping station for the great fountain in the ducal gardens of the Nymphenburg Castle near Munich had to be replaced in 1767, the Bavarian prince-elector Maximilian III

Joseph (1745-1777) did not consult the famous hydraulic engineer of the nearby Imperial City of Augsburg. Refusing his bourgeois ambience, he looked out in his own aristocratic circles for a French pump engineer, successfully trained in the ducal gardens of Lunéville, considered as a second garden of Versailles. He contracted Jacques Faye Poitevin (life dates unknown), a travelling pump engineer, who worked between 1767 and 1786, most of his time successfully, in Nymphenburg [50]. Documents in the archives indicate that Poitevin was required to build a model of his pump to demonstrate how it functioned, unfortunately not conserved or described in detail.

1.5 Models in the Age of Enlightenment

1.5.1 Cabinets of curiosities

From the beginning of the eighteenth century an innovative approach to mathematics, physics, mechanics and engineering, based on rational experiments, grew and flourished among scientists and engineers. This new approach enhanced the status and prospects of existing scientific academies and promoted the foundation of new ones. A central activity of the academies was the scientific experiment, realized by demonstrations and by means of models; this practice also led to larger collections of original equipment and models used for representing the behaviour of full-size reality as well as for illustration, documentation and education. At the same time, more and more private amateurs began collecting mechanical devices and scale models for their Cabinets of mechanical curiosities as a symbol of their wealth and learning [51]. Perhaps the best-recorded *Cabinet de curiosités* was created in Paris in the 1730s by the amateur scientist Joseph Bonnier de la Mosson (1702-1744) and comprised many thousands of objects. The largest of the nine rooms was devoted to mechanical devices and included models of machines powered by clockwork, a coin press, various machine tools, a machine to desalinate seawater, another for desilting ports, a machine to transplant trees, a water pump motor, machines to help on construction sites (one from Fontana), machines from various mills (for pressing sugar cane, paper, grain, etc.). There were also many pump models, several cranes, a model of a bridge that could be built quickly, Archimedean screws, a horse-powered grain threshing and winnowing machine, and two dynamic model paddle-boats [52]. Meanwhile, in Britain, the collection of Archibald Campbell, the first Earl of Islay and third Duke of Argyll (1682-1761) became famous for the variety and quantity of objects, which included models of improvements for agricultural machinery. Both collections were sold and scattered after the death of their owners. Also in Britain, King George III (1738-1820) established his own collection starting in 1760 with the appointment of George Adams (1750-1795) as his Mathematical Instrument Maker and initiator of his collection. This collection contained engineering models too, and is largely conserved today in the Science Museum in London.

The first scholastic Cabinet of Physics was created for the first time in Europe in 1675 at the University of Leiden after experiments were introduced into science education. The scientific exchange stimulated the creation of similar collections, mainly in the second half of the eighteenth century, in Stockholm and Uppsala, Florence and Padua, Lisbon and Coimbra, Vienna and Munich, Prague, Oxford and London.

The collection of the University of Oxford includes a number of significant objects from civil engineering. The original Museum was built between 1679 and 1683 in order to host the collection of Elias Ashmole (1617-1692) and developed constantly to a considerable size, mainly after 1714 when demonstrations began to accompany the lectures. The collection today is embodied in the University Museum. A catalogue from 1790 lists a model of a machine for breaking pieces of wood, models of cranes, pile drivers, pumps and mills and even 'Smeaton's pulley with a three legged stand' [53]. Instrument and model makers such as Benjamin Cole (1695-1766) and Edward Nairne (1726-1806), scientists such as Jean Desaguliers (1683-1744) and Stephen Demainbray (1710-1782) left their legacy to today's University Museum in Oxford and the Science Museum in London. It was Demainbray who designed models of water lifting machines, waterwheels, pile drivers and other helpful construction devices in order to use them for his famous lecture-demonstrations.

Another remarkable collection of models encompassing civil engineering applications was established in the Netherlands by Pieter Teyler van der Hulst (1702-1778). After his death, the Teyler Foundation decided, in 1779, to construct a Museum for his collection. When the Museum opened in 1784 in Haarlem, it presented not only usual mechanical demonstrations, but also devices of particular relevance to typical Dutch hydraulic construction problems including hoists, cranes, pile-drivers, polder mills, a saw mill, a bucket dredger, a water-lift and a lock gate.

By the end of the eighteenth century engineers and constructors realized that models represented a marvellous means for promoting their products. The completion of the Iron Bridge at Coalbrookdale, built between 1777 and 1779 and opened to traffic in 1781, was the first cast-iron bridge in Europe. As such, it was considered one of the wonders of the age and was promoted as a spectacle. Only a few years after the opening, in 1787, the ironmaster and constructor of the bridge, Abraham Darby III (1750-1789), ordered a model of his exemplary bridge to be assembled by his ironworks foreman and patternmaker Thomas Gregory (dates unknown). The intention was to donate it to the Royal Society in order to promote his engineering work and his enterprise in its famous collection [54]. The model was built at an impressive scale of 1:24 (with a span of 1.2 m) and presented to the Society in 1787, where Darby proudly received a gold medal for his much-appreciated gift. After some repairs and mounting in a glass display case, the model was added to the collection of the Society, and is still conserved in the Science Museum in London [55].

1.5.2 The models of Hans Ulrich Grubenmann

In October 1755 the carpenter Hans Ulrich Grubenmann (1709-1783) presented a model of his proposed bridge over the river Rhine to the city council of Schaffhausen in Switzerland [56], planning to construct a single span of 119 metres, the largest timber arch ever built up to that time. After several previous requests in larger cities in Germany and inquiries within the scientific community, the city council decided in August 1755 to invite Grubenmann to present his proposal, even though he was neither the scientist nor the engineer they had been looking for, rather 'just a carpenter'. Grubenmann was already well-known in Switzerland for his timber roof structures and bridges. It was said that, when the councillors doubted the stability of the construction, Grubenmann placed himself on top of the model with the famous words: 'If the model supports me, the real bridge will support several carriages' [57]. On the one hand, this suggests that the model could have been made for an empirical test of its strength and stability, in which case Grubenmann would have been one of the first constructors to use a model for such a practical test or demonstration. On the other hand, this episode obviously did not take into account that there can be great differences between the structural behaviour of the model with respect to the real construction, a problem which already had been mentioned by Vitruvius and already in the mid-eighteenth century, was understood as the 'state of the art' in engineering. Nevertheless, the explanation of his structural design finally convinced the councillors to accept one essential change: the bridge should be built with an additional pier in the centre of the span. Recent examinations of the structure suggest that the decision to build an additional pier was not so much due to mistrust of the stability of the construction, but rather due to the additional timber needed for the larger span [58]. The construction of the bridge was completed only three years later and contributed to Hans Ulrich Grubenmann and his family achieving international fame.

Since Schaffhausen was at a crossroads for international travellers, many visitors started to write about this and other bridges by the Grubenmann family. Mathematicians, architects and engineers commented on the construction, copied his plans and published and distributed them in Germany, France and Britain. One of the most famous was Sir John Soane (1753-1837) who visited Schaffhausen in 1780 and studied the bridge thoroughly, copied the plans, completed them with his own drawings and used these documents for his lectures as examples for other architects and engineers [59]. In 1833 he set up a museum in his home in London to present his collection of architectural, archaeological and structural curiosities to his clients and students. Apart from his important collection of Canelatto's paintings (including bridges) he also displayed cork models of some buildings in Pompeii and, a wooden model of a pile driving machine, dating from around 1754 that Soane had purchased in 1801 [60]. Drawings of Grubenmann's (and other) wooden bridges are still conserved in the museum's archives [61]. Even the renowned French engineer Jean Rodolphe Perronet (1708-1794) collected drawings and probably a model of a Grubenmann bridge [62].

Today there exist six models of Grubenmann bridges dating from their time of construction, of which four are confirmed as being made by Grubenmann: the Schaffhausen Bridge (around 1755); the Wettingen Bridge (around 1765); the bridge in Stein (around 1760); and the so called 'Trogen Bridge' (around 1745-1755, real site unknown) (Figure 1.9) (see Section 39.2.1). These models are unique remains of the original bridges. Grubenmann made them mainly to work out their complex construction and to instruct his own workers.

Figure 1.9 Original model of the 'Trogen bridge' (around 1745-1755), Grubenmannmuseum, Teufen, Switzerland. (Image: Dirk Bühler)

Along with various drawings and publications, Grubenmann's models served to raise awareness of his constructions throughout Europe. For example, they inspired others such as Konrad Altherr (unknown dates) from Chur and Johann Konrad Langenegger (1749-1818) from Gais in Switzerland, who built a model for a 300-metre timber arch bridge for a competition to cross the River Foyle in Londonderry, Ireland [63, 64] (see Section 4.2.2). The history of this design competition is told in different ways. The prize was probably awarded originally in 1771 by Frederick Hervey, Lord Bishop of Derry (1730-1803) for the design of a timber bridge that should span the river, '827 shoe-lengths wide', in 'one or two Schwibbogen' (i.e. spans) [65]. In 1770, inspired by the fame of the Grubenmann brothers' bridges, Hervey and the young architect Michael Shanahan (c.1731-1811) visited the Alpine countries and northern Italy and had engravings of various timber bridges drawn, presumably to determine a realistic basis for the construction of his timber bridge planned in Derry [66]. According to some reports the winner of the design competition was Johannes Grubenmann, although most report that the winner was Altherr. In 1772, he and two colleagues travelled from Appenzell in Switzerland to Londonderry with a 19-foot (5.8 m) long model of the bridge. An account in the Gentleman's Magazine noted that the journey that took them about five months, that the model comprised 11,734 separate parts and about 4000 screws, and that the bridge had 62 windows on each side [67, 68]. News of this competition and the model travelled to Russia where its description in the St. Petersburg Vedomosti (business report) corresponded to that given in the Gentleman's Magazine in 1772 [67] and its construction was attributed to Konrad Altherr. Remarkably, this model survives in the Science Museum in London (see Section 39.2.1). Altherr and Langenegger built two models later in the 1770s to promote Swiss timber bridge construction in St. Petersburg, Vienna and London.

1.5.3 The models of John Smeaton

John Smeaton (1724-1792) was a truly remarkable engineer and only came to civil engineering at the age of 29. By this time he had learned much in his father's workshop at home, had worked for four years as a maker of scientific instruments and, at the age of 28, had been elected as a Fellow of the Royal Society.

During the first half of the eighteenth century, energy efficiency was an issue that captured the interest of the scientific community across Europe, in particular, the relative efficiency of different types of generators of mechanical power – waterwheels and windmills. At this time, many eminent scientists were gaining great confidence in using theoretical science to justify claims about how the real world functioned, and water power was no exception. The French scientist, Antoine Parent (1666-1716), for example had stated in 1704, that an undershot waterwheel had a maximum efficiency of about 15% and that this was achieved when the rim speed was one third that of the water stream. Despite their errors, such views were widely accepted, for example by Bernard Forest de Bélidor (1698-1761) in his *Architecture Hydraulique* (1737), who went on to state that an undershot wheel was six times more efficient than the overshot type. John Theophilus Desaguliers FRS (1683-1744), on the other hand, claimed that the overshot wheel was ten times more efficient than the undershot type. Such statements had clearly been made without the robust evidence of experimental investigation. It was Smeaton who would tackle this question and resolve the 'monstrous disagreement' as Thomas Telford later called it.

Smeaton undertook his experiments on models of waterwheels (about 60 cm diameter, a scale of approx. 1:8) (Figure 1.10) and windmills (about 1.6 m diameter, a scale of approx. 1:5) (see Figure 24.1) in 1752-1753. From his model tests on waterwheels, among many conclusions, he found that overshot wheels were about twice as efficient as undershot wheels (66% compared to 30%) thus refuting the assertions by many of the scientific community [69]. Based on the understanding he gained from these experiments, he went on to work on over fifty water-powered mills and a several windmills.

Smeaton was well aware of the dangers inherent in scaling up the results of model tests to full size, and withheld publication of the results until 1759, by which time he had verified many of the design guidelines that followed from his experiments; he had designed and constructed several waterwheels, and carried out tests on full-size windmills, in which he changed the shape and angle of attack of the sails:

> 'What I have to communicate on the subject was originally deduced from experiments made on working models, which I look upon as the best means of obtaining the outlines in mechanical enquiries. But in this case it is very necessary to distinguish the circumstances in which a model differs from a machine in large; otherwise a model is more apt to lead us from the truth than towards it. Hence the common observation, that a thing may do very well in a model, that will not answer in large. And indeed, tho' the utmost circumspection be used in this way, the best structure of machines cannot be fully ascertained, but by making trials with them, when made of their proper size. It is for this reason, that, tho' the models referred to, and the greatest part of the following experiments, were made in the years 1752 and 1753 yet I deferred offering them to the Society, till I had an opportunity of putting the deductions made therefrom in real practice, in a variety of cases, and for various purposes; so as to be able to assure the Society, that I have found them to answer' [70].

Figure 1.10 John Smeaton's model waterwheel used to compare the efficiency of an overshot and undershot wheel (Source: [70] Plate 4)

Smeaton's greatest project was the construction of a stone lighthouse upon the Eddystone rock, 20 km south of Plymouth. He received the commission in 1756 when he was just 32 years old, and it was his first civil engineering project. In order to design the foundation of the stone structure on the irregular-shaped rock, Smeaton made two models of the rock, one before cutting it to shape (Figure 1.11), and the other after. This enabled him not only to fully understand the geometrical nature of the interface, but also to get a feeling (in his hands) of the effectiveness of the bond created by the cut stones interlocking with the recesses in the rock. He went on to explain why he placed such importance in making physical models:

> 'It seeming therefore to be a first principle, to cut the rock as *little as we could help*; and for this end, to humour its irregularities as far as we could, so as to get a firm fixing for our work ; on this account it appeared necessary, as the first step to be taken (from the dimensions already obtained, and by the methods already specified) to construct a *complete Model* of

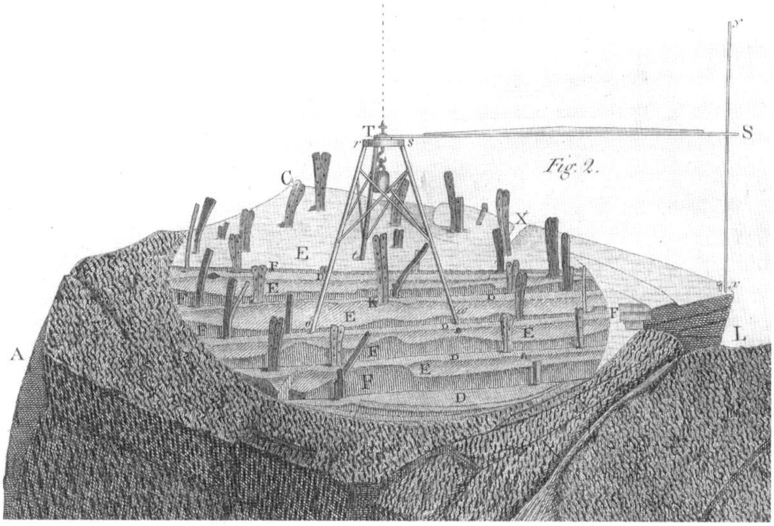

Figure 1.11 Smeaton's drawing of the model he made of the rough-cut Eddystone rock, used to determine how it would need to be cut precisely to receive the first five courses of cut stones forming the base of the lighthouse. (Source: [71] Plate 7)

> the rock, in the condition I found it : which being done, a second model might *then* be formed, shewing to what the rock was to be reduced, [...] These models I determined should be the work of *my own hands* ; and this I foresaw, must in its own nature be a work of *Time.*
> [...] I had also a further reason for undertaking this part of the work myself; which those who shall peruse this account for the sake of information, may not be displeased to know. I have always found in subjects of mechanical invention and investigation, that I can seldom form an original idea so complete, but that by laying it down in its proper dimensions on paper, I could very much mature and improve it ; and where the subject is attended with intricacy, it is in a greater degree necessary : but in reducing this to a *Solid,* as is the case in making a model, still further corrections and advantages will often present themselves, that did not appear upon *Paper* : and this in a much more eminent degree when the *Solid* is produced from the drawing by the artist's *own* hand, than by the hand of another ; and still further improvements will occur, by going again over the *detail,* in constructing the work itself at large. Therefore [...] I determined, from the paper materials that I had brought from Plymouth; as well as those I carried thither; at once to construct the models above-mentioned myself' [71] (spelling, italics and punctuation as original).

The cutting and placing of the stones for the lighthouse was supervised by Josias Jessop (c.1715-1761), father of the eminent canal builder, William Jessop. During the winter of 1756-1757 Jessop completed the drawings of each of the sixty-six courses of the structure and wooden models of the most important courses of masonry at a scale of about 1:16 (50 cm diameter), including Course VII, the first full course of interlocking stones, with the eighteen stone cubes serving as shear connectors between this course and the ones above and below [72].

Before reaching the detailed design and construction stages of the lighthouse, Smeaton had made much use of models in 'selling' his novel method of construction in the first place, among others to King George II. Later he made models to help to raise finance for the project and then to communicate to the backers of the project the changes he was proposing as the final scheme was developed. Smeaton continued to use the 1:48 scale model of the final design to publicise the structure, which had attracted great public interest. Indeed: 'so great was the curiosity of the public, at the completion of the present building, in the year 1759; and such were the numbers, by the intercession and recommendation of friends, that for some years after, flocked daily to see the model; that, to avoid having the whole of my time consumed in satisfying their curiosity, I found myself under a necessity of deputing Mrs. Smeaton, to shew and explain the model' [73].

1.6 Final remarks

This overview of early engineering models, from antiquity until the beginning of the industrial revolution, describes the enormous efforts engineers made to resolve mechanical and structural problems using physical models, and the obstacles they had to overcome to obtain useful results. In early times the focus of employing these models was mainly on military devices. Master builders, many of them coming from a military training, soon recognized the utility of engineering models for the design of civil and religious constructions and, since at least the beginning of the fifteenth century, have continued to benefit from this approach. The main uses of models were for devising machines that reduced manual workload, optimizing the efficiency of mechanical devices, determining the bearing capacities of structures, and resolving construction problems. Gradually, physical findings helped the engineers to perfect their methods and to create common rules for improved structural models.

The chapter has presented a base upon which were built the great advances in using models during and after the industrial revolution. It provides a vital point of departure for the appreciation and validation of the later development of physical models. In order to deepen these traces of the past and to improve their visibility and recognition in the public eye, the discovery, conservation and presentation of engineering models need to become a focus for future research into the history of engineering (see Chapter 39).

References

1 Marsden, E.W. (1971) *Greek and Roman Artillery: Technical treatises.* Oxford: Oxford University Press.
2 Marsden, E.W. (1969) *Greek and Roman Artillery: Historical development.* Oxford: Oxford University Press.

3 Marcus Vitruvius Pollio, (c.30-15 BC) *De architectura.* Book 10, Chapters X, XI.
4 Marcus Vitruvius Pollio, (c.30-15 BC) *De architectura.* Book 10, Chapter XVI, Paragraphs 3-6.
5 Popplow, Marcus (2011) Lemma 'Technisches Modell'. In: *Enzyklopädie der Neuzeit:* Vol. 13, column 298. Stuttgart: Friedrich Jaeger.
6 Popplow, Marcus (2002) *Models of Machines: A 'Missing Link' Between Early Modern Engineering and Mechanics?* Berlin: Max-Planck-Institut für Wissenschaftsgeschichte, Preprint 225, 5.
7 Lepik, Andres (1995) Das Architekturmodell der frühen Renaissance. In: *Architekturmodelle der Renaissance – Die Harmonie des Bauens von Alberti bis Michelangelo* (ed. Bernd Evers), 13. München / New York: Prestel.
8 Lepik, Andres (1994) *Das Architekturmodell in Italien 1335-1550.* Worms: Wernersche Verlagsgesellschaft. 60ff and 148.
9 Saalmann, Howard (1980) *Filippo Brunelleschi and the Cupola of S. Maria del Fiore.* London: Zwemmer.
10 Evers, Bernd (Ed.) (1995) *Architekturmodelle der Renaissance – Die Harmonie des Bauens von Alberti bis Michelangelo.* München / New York: Prestel. (Kat.-Nr. 82) p. 264.
11 Popplow, Marcus (2002) *Models of Machines: A 'Missing Link' Between Early Modern Engineering and Mechanics?* Berlin: Max-Planck-Institut für Wissenschaftsgeschichte, Preprint 225: 5, 7.
12 Vasari, Giorgio (1974) *Lebensläufe der berühmtesten Maler, Bildhauer und Architekten.* Zürich: Manesse Verlag. 315-316.
13 Leonardo. *Codex Atlanticus.* Biblioteca Ambrosiana, Milan.
14 Giorgione, Claudio (2009) *Leonardo da Vinci – the models collection.* Milan: Museo Nazionale della Scienza e della Tecnologia Leonardo da Vinci. 138-139.
15 Giorgione, Claudio (2009) *Leonardo da Vinci – the models collection.* Milan: Museo Nazionale della Scienza e della Tecnologia Leonardo da Vinci. 178-183, 235-239.
16 Giorgione, Claudio (2013) Leonardo interpretieren – von der Zeichnung zum Modell. In: *Leonardo da Vinci: Vorbild Natur – Zeichnungen und Modelle, Katalog zur gleichnamigen Ausstellung im Deutschen Museum vom 11.10.2013-03.08.2014* (ed. Patrick Boucheron, Claudio Giorgione & Dirk Bühler), 49-53. München: Hirmer Verlag.
17 Puppi, Lionello (2000) *Andrea Palladio - Das Gesamtwerk.* Stuttgart/ München: Deutsche Verlags-Anstalt. 389.
18 Puppi, Lionello (2000) *Andrea Palladio - Das Gesamtwerk.* Stuttgart/ München: Deutsche Verlags-Anstalt. 383.
19 Maggi, Angelo & Navone, Nicola (Eds.) (2003) *John Soane and the wooden bridges of Switzerland. Architecture and the culture of technology from Palladio to the Grubenmanns.* Mendrisio & London: Archivio del Moderno, Accademia di architettura, Mendriso. 112-113.
20 Fontana, Domenico (1590) *Della Trasportatione dell'Obelisco Vaticano…* Roma: Domenico Basa. https://www.e-rara.ch/zut/content/titleinfo/23759 (6/9/2019)

21 Popplow, Marcus (2002) *Models of Machines: A 'Missing Link' Between Early Modern Engineering and Mechanics?* Berlin: Max-Planck-Institut für Wissenschaftsgeschichte, Preprint 225. 10.

22 Kaiser, Christiane (2005) *Die Fleischbrücke in Nürnberg – 1596-1598.* Dissertation an der Fakultät für Architektur, Bauingenieurwesen und Stadtplanung an der Brandenburgischen Technischen Hochschule, Cottbus. Vol. 1, 80.

23 Kaiser, Christiane (2005) *Die Fleischbrücke in Nürnberg – 1596-1598.* Dissertation an der Fakultät für Architektur, Bauingenieurwesen und Stadtplanung an der Brandenburgischen Technischen Hochschule, Cottbus. Vol. 1, 95-96.

24 Kaiser, Christiane (2005) *Die Fleischbrücke in Nürnberg – 1596-1598.* Dissertation an der Fakultät für Architektur, Bauingenieurwesen und Stadtplanung an der Brandenburgischen Technischen Hochschule, Cottbus. Vol. 1, 71.

25 Reuter, Hans & Berckenhagen, Ekhart (1994) *Deutsche Architekturmodelle. Projekthilfe zwischen 1500 und 1900.* Berlin: Deutscher Verein für Kunstwissenschaft. 113.

26 Heiner Grillenberger. https://commons.wikimedia.org/w/index.php?curid=5864877 Creative Commons Attribution-Share Alike 3.0 Unported license. (18/9/2019).

27 Kaiser, Christiane (2005) *Die Fleischbrücke in Nürnberg – 1596-1598.* Dissertation an der Fakultät für Architektur, Bauingenieurwesen und Stadtplanung an der Brandenburgischen Technischen Hochschule, Cottbus. Vol. 1, 108 and Vol. 2, 30.

28 Staatsarchiv Nürnberg. Stromer-Archiv. B 15 (BMB1) folios 41 and 41 verso.

29 Popplow, Marcus (2002) *Models of Machines: A 'Missing Link' Between Early Modern Engineering and Mechanics?* Berlin: Max-Planck-Institut für Wissenschaftsgeschichte, Preprint 225. 11.

30 Bischoff, Franz & Helmberger, Werner (1993) *Rom über die Alpen tragen. Fürsten sammeln antike Architektur.* Landshut/Ergolding: Arcos-Verlag. 41.

31 Feldhaus, Franz M. (1914) *Die Technik der Vorzeit, der geschichtlichen Zeit, und der Naturvölker.* Leipzig / Berlin W. Engelmann. Lemma: Ausstellungen, column 45. The author mentions as source the review *Neues Universum* No.2, 1881, p. 98, but remarks that he could not confirm the fact with the authorities of Nuremberg.

32 Furttenbach, Joseph (1628) *Architectura Civilis.* Ulm: Johann Saurn. Vorwort, 8.

33 Furttenbach, Joseph (1641) *Architectura Privata.* Augsburg: Johann Schultes the Younger. 37-38, 40.

34 Picot, Jean Baptiste (1683) *Explication des modèles des machines et forces mouvantes, que l'on expose à Paris dans la rue de la Harpe, vis-à-vis Saint Cosme.* Paris: Guillery.

35 Birembaut, Arthur (1967) L'exposition de modèles de machines à Paris, en 1683. *Revue d'histoire des sciences et leurs applications* 20 (2): 141-158. 153.

36 Birembaut, Arthur (1967) L'exposition de modèles de machines à Paris, en 1683. *Revue d'histoire des sciences et leurs applications* 20 (2): 141-158. 145 and 155.

37 Baer, W., Kruft, H.-W. & Roeck, B. (eds.) (1985) *Elias Holl (1573-1646) und das Augsburger Rathaus.* Regensburg: Friedrich Pustet.

38 Kunstsammlungen und Museen Augsburg, Inv.-Nr. 3448.

39 Kunstsammlungen und Museen Augsburg. Inv.-Nr. 3818.

40 Roeck, Bernd (2004) *Elias Holl – Ein Architekt der Renaissance*. Regensburg: Pustet. 24.

41 Roeck, Bernd (2004) *Elias Holl – Ein Architekt der Renaissance*. Regensburg: Pustet. 94.

42 Haberstock, Eva (2016) *Der Augsburger Stadtwerkmeister Elias Holl (1573-1646) Werksverzeichnis*. Petersberg (Germany): Imhof Verlag. 111-136 (transcript), 439-471 (facsimile).

43 Haberstock, Eva (2016) *Der Augsburger Stadtwerkmeister Elias Holl (1573-1646) Werksverzeichnis*. Petersberg (Germany): Imhof Verlag. 69 (fol. 30v and 31r).

44 Kunstsammlungen und Museen Augsburg, Inv.-Nr. 9740.

45 Walter, Caspar (1754) *Hydraulica Augustana*. Augsburg: Detleffsen. 22.

46 Kunstsammlungen und Museen Augsburg, Inv.-Nr. 9616.

47 Bühler, Dirk (2015) Historical Models of Civil Engineering in Collections in Augsburg and Munich. In: *Proceedings of the fifth International Congress on Construction History – Chicago (Ill), June 2015* (Eds Brian Bowen, Donald Friedman, Thomas Leslie & John Ochsendorf), Vol. 1, 275-282. Chicago.

48 Mair, Raimund (2017) *Die städtische Wunderkammer. Die hydrotechnischen Architekturmodelle in der Modellkammer des Maximilianmuseums in Augsburg*. Masterarbeit an der Fakultät für Architektur der Leopold-Franzens-Universität Innsbruck. (unpublished)

49 Walter, Caspar (1766) *Brücken-Bau, oder Anweisung, wie allerley Arten von Brücken, sowohl von Holz als Steinen, nach den besten Regeln der Zimmerkunst dauerhaft anzulegen sind*. Augsburg: Gebrüder Veith.

50 Bühler, Dirk (2017) Technik für Fontänen. Die Pumpenanlage von 1767 im Nymphenburger Schlosspark. In: *Augsburg und die Wasserwirtschaft – Studien zur Nominierung für das UNESCO-Welterbe im internationalen Vergleich* (ed. Stadt Augsburg), 132-135. Augsburg: Context-Verlag.

51 Fox, Celina (2009) *The Arts of Industry in the Age of Enlightenment*. New Haven & London: Yale University Press. 148-160.

52 https://fr.wikipedia.org/wiki/Cabinet_de_curiosit%C3%A9s_de_Joseph_Bonnier_de_La_Mosson (consulted on 6/9/2019).

53 Fox, Celina (2009) *The Arts of Industry in the Age of Enlightenment*. New Haven & London: Yale University Press. 154.

54 Fox, Celina (2009) *The Arts of Industry in the Age of Enlightenment*. New Haven & London: Yale University Press. 162-163.

55 Object-Nr. 1882-29.

56 Killer, Joseph (1942) *Die Werke der Baumeister Grubenmann*. PhD Thesis. Zürich: ETH. 22-24. https://www.research-collection.ethz.ch/handle/20.500.11850/133364 (6/9/2019).

57 Killer, Joseph (1942) *Die Werke der Baumeister Grubenmann*. PhD Thesis. Zürich: ETH. 24.

58 https://zeughausteufen.ch/grubenmann-museum/bruecken/ (18/9/2019).

59 Maggi, Angelo & Navone, Nicola (Eds.) (2003) *John Soane and the wooden bridges of Switzerland. Architecture and the culture of technology from Palladio to the Grubenmanns.* Mendrisio & London: Archivio del Moderno, Accademia di architettura, Mendriso. 11 and 31 ff.

60 Maggi, Angelo & Navone, Nicola (Eds.) (2003) *John Soane and the wooden bridges of Switzerland. Architecture and the culture of technology from Palladio to the Grubenmanns.* Mendrisio & London: Archivio del Moderno, Accademia di architettura, Mendriso. 111-112.

61 Maggi, Angelo & Navone, Nicola (Eds.) (2003) *John Soane and the wooden bridges of Switzerland. Architecture and the culture of technology from Palladio to the Grubenmanns.* Mendrisio & London: Archivio del Moderno, Accademia di architettura, Mendriso. 90-111.

62 Maggi, Angelo & Navone, Nicola (Eds.) (2003) *John Soane and the wooden bridges of Switzerland. Architecture and the culture of technology from Palladio to the Grubenmanns.* Mendrisio & London: Archivio del Moderno, Accademia di architettura, Mendriso. 36.

63 Maggi, Angelo & Navone, Nicola (Eds.) (2003) *John Soane and the wooden bridges of Switzerland. Architecture and the culture of technology from Palladio to the Grubenmanns.* Mendrisio & London: Archivio del Moderno, Accademia di architettura, Mendriso. 42.

64 Nägeli, A. (1932) Ein appenzellischer Baumeister, Joh. Konrad Langenegger, 1749-1818. *Appenzeller Kalender* 211: 5 pages (no page numbers). http://doi.org/10.5169/seals-374897 (6/9/2019).

65 Killer, Josef (1998) *Die Werke der Baumeister Grubenmann.* Dietikon: Baufachverlag. 57-58.

66 Ruffinière du Prey, Pierre de la (1979) Eighteenth-century English sources for a history of Swiss wooden bridges. *Zeitschrift für Schweizerische Archäologie und Kunstgeschichte* 36: 51-63.

67 Anon (1772) Account of a model for a bridge in Ireland. *Gentleman's Magazine* XLII: 399-400.

68 Maggi, Angelo & Navone, Nicola (Eds.) (2003) *John Soane and the wooden bridges of Switzerland. Architecture and the culture of technology from Palladio to the Grubenmanns.* Mendrisio & London: Archivio del Moderno, Accademia di architettura, Mendriso. 40-42.

69 Smith, N.A.F. (1981) Scientific work. In: *John Smeaton FRS* (ed. A.W. Skempton), 35-58. London: Thomas Telford.

70 Smeaton, J. (1760) An experimental Enquiry concerning the natural Powers of Water and Wind to turn Mills, and other Machines, depending on a circular Motion. *Phil. Trans.* 51 (for 1759): 100-174. 101.

71 Smeaton, J. (1791) *A Narrative of the Building and a Description of the Construction of the Edystone Lighthouse.* London. 70.

72 https://www.nms.ac.uk/explore-our-collections/collection-search-results/?item_id=180908 (20/8/2019).

73 Smeaton, J. (1791) *A Narrative of the Building and a Description of the Construction of the Edystone Lighthouse.* London. 7.

2

Block models of the masonry arch and vault

Santiago Huerta

2.1 The beginnings of arch construction

The process of masonry building consists mainly in piling up blocks to obtain a stable structure (*struere* in Latin means 'to pile up'). The stability of such structures depends on geometry and not on size; if a small model made of blocks is stable, the full-size masonry building will also be safe, a truth that has been rigorously proven using modern masonry theory [1]. However, this property of masonry construction had been well known by master builders since ancient times: building any arch or vault larger than a previous one was, in effect, based on an earlier successful test of a reduced-scale model.

From the small vaulted tombs in Mesopotamia more than 5000 years ago, passing through the tomb of the Egyptian King Ramesses II, the Roman *thermae* and Byzantine churches, Romanesque and Gothic cathedrals, the domes at Santa Maria del Fiore in Florence and St. Peter's in Rome, up until the end of masonry architecture in the early twentieth century, the problem has remained the same: to design and build a certain form in stable equilibrium with gravity loads, in which all the stones or bricks are working in compression. There is no unique solution to this problem and the history of architecture shows countless examples of different equilibrated geometries.

The evolution of such stable geometries was slow at the beginning. Until around 2000 BC, the span of vaults remained less than 2 metres. The great variety of trials revealed by archaeological excavations is remarkable, yet there seems to be no steady improvement or convergence on a certain geometry: rather there were a great many trials [2]. After many examples of 'good' arches (catenary), typically with radial joints, there later appeared many 'false' vaults with horizontal courses of stones (corbelled).

After about 2000 B.C. spans suddenly increased by a factor of about four, in the barrel vaults of the granaries of the Ramesseum span 4-5 m, and both the design and the erection of the vault became more difficult. The geometry, the profile of the arch or vault, had be precisely defined *before construction commenced* which is, in fact, a good way of defining what is meant by engineering design. Temporary centring had to be built, or some other way to control the

Physical Models: Their historical and current use in civil and building engineering design,
First Edition. Edited by Bill Addis.

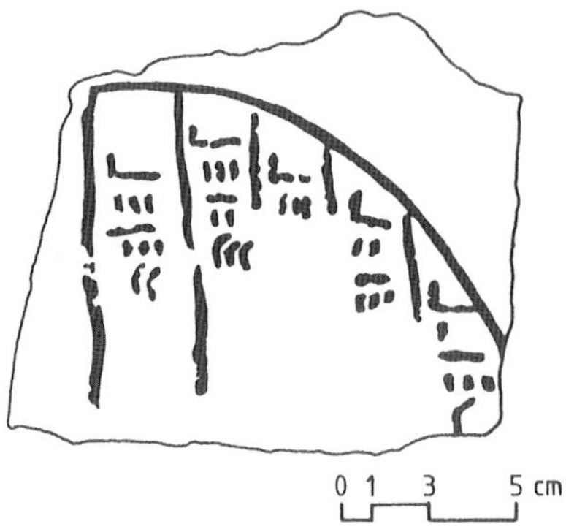

Figure 2.1 Egyptian scale drawing of an arch, Third Dynasty, ca. 2600 BC. Ordinates in cubits, palms, and digits. (From left to right: 3 cubits, 3 hands, 2 digits; 3 cubits 2 hands, 3 digits, etc.). (Source: [3])

curvature had to be devised. Eventually, simple geometric forms derived from the circle, became widespread. Semi-circular or oval arches were used in Ancient Egypt from which time small-scale representations of intended forms have survived as plans or rudimentary blueprints. The oldest of these scaled drawings to be discovered was in Egypt and, rather surprisingly, it uses a coordinate system to define the form of a vault [3] (Figure 2.1). In other examples, the form of the vault was incised in the stone so that the masons could fabricate templates to control the form during construction.

The idea of scale drawings can thus be traced back to the third millennium BC and monolithic models have survived which would have served the purpose of portraying appearance. It is highly probable that builders from that time also made models using small blocks to resolve construction issues. These could be structural, as small block models faithfully represent full-size behaviour, or questions of stereotomy, to determine the shapes of individual stones in a structure. However, no trace of such block models have survived from antiquity.

2.2 The use of block models from 1400 to 1700

It is evident that mediaeval master builders knew that the stability of a masonry structure was independent of size since the structural design rules which were passed down through the masons' guilds were all based on simple proportion. For example, the width of the buttress was a certain fraction of the span. These rules could be expressed either arithmetically or geometrically. A late-Gothic rule from Germany gave the thickness of a wall as one tenth of the span of the vault, and the thickness of the buttress was an additional two times the thickness of the wall [4]. A geometric design rule, dating from the fourteenth century, was first recorded in print in the 1640s [5] and later given the name 'Blondel's rule' after the French architect who published it in the 1670s [6] (Figure 2.2). It was widely published in stone cutting manuals of the eighteenth century, throughout the nineteenth century and was still to be found even in the mid-twentieth century [7].

While it is highly probably that most builders of Gothic cathedrals used small models to resolve structural and construction issues related to masonry construction, the earliest documentary evidence of their use was in the late-fourteenth

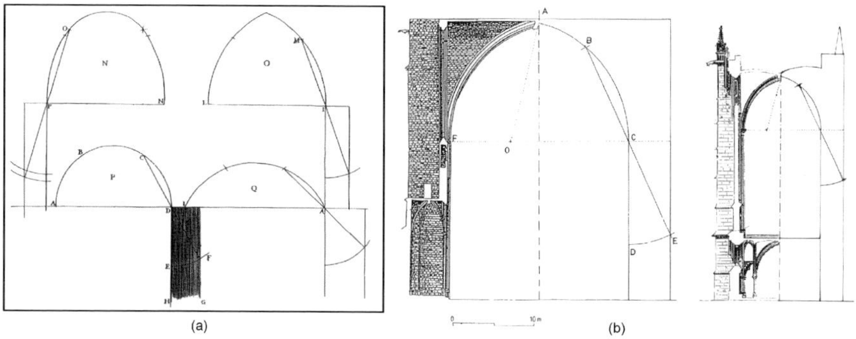

(a) (b)

Figure 2.2 (a) Gothic rule for buttress design (Source: [5]); (b) Application to the rule to the cathedral of Girona, left, and the Sainte-Chapelle, Paris, right. (Source: [4] p. 148)

century. For the construction of the Basilica of San Petronio in Bologna: '[The architect] Antonio di Vincenzo … made a model of brick and plaster in 1390 on an enormous scale. Its length was 18.71 metres; one could walk into it, and it seemed at the time that later generations would have no difficulties if they only adhered to this model' [8]. While this model would have served as a three-dimensional drawing for the masons, 'Such a model can be used with confidence to check the stability of the whole or of any part of the structure, since questions of stability, depending as they do on individual proportions, that is, on geometrical shape, can be checked at any scale' [9]. To ensure the correct geometry of cut stones, it seems that Brunelleschi made many models using clay, wax, wood and even large turnips, like those that 'are sold in in the market in springtime' [10].

A few years later Brunelleschi built many models, at different scales, of different materials and for different purposes during his work of the cupola of Florence cathedral. Most impressive of all was one made of brick and mortar:

> 'Near the end of 1418, four masons … had constructed Filippo's model, without [formwork], in about ninety days. … The model … showed the proposed cupola and its support on a scale of about 1:12. … The reported amount of lime [forty-nine horse loads] provided cement for ca. 150 tons of brick masonry [for the model and a wall]. … Estimating that the model weighed 100 to 120 tons, and that it stood on masonry piers ca. 7 feet high to provide access, we find that the Tambour probably had some 11 to 12 feet clear space from wall to wall' [11].

Another source suggests that 'The model made by Brunelleschi for the cupola of Santa Maria del Fiore, built in 1417, consisted of 6,000 bricks of regular size, and had a span of 10-12 *braccia* (5.8-7 m) [12].

During the final stages of the construction of Milan cathedral in the late fifteenth century, there was great uncertainty regarding the construction of the *tiburio* on top of the partly built cathedral, not least regarding the challenge faced by main architect, Bramante, in introducing a modern Renaissance design to a city still conservatively wedded to the Gothic style. To help resolve the issue an expertise was organised. As well as the main architect Donato Bramante

(1444-1514), there were both Leonardo da Vinci and Francesco di Giorgio (1439-1502) and several other experts. At least ten models were made and used to facilitate discussion about conformity with the rest of the building, lightness, beauty and, most significantly, strength. Although the models were probably of wood, there was discussion of stability, the actions of buttresses and flying buttresses, and the impact of the *tiburio* on the structure below. The models were clearly central to the method of dealing with issues of design and construction and, as Bramante concluded, 'if all the competitors came together, one could settle in less than an hour on a new model that would unite the advantages of all the others' [13].

2.2.1 Leonardo da Vinci

Leonardo da Vinci (1452-1519) was the first engineer we know who apparently used small models of arches to study their mechanical behaviour and he goes on to suggest practical guidance. While there is no evidence that he passed this knowledge on to masons, he would certainly have used it himself, for example when advising on the stability of designs for the *tiburio* at Milan Cathedral mentioned above. It is more than likely that practising masons already knew what he concluded, however it is remarkable that he gained his understanding of masonry construction using models and reason, much as a modern engineer would, and not after years of practical experience as a mason would.

Leonardo's numerous references to the statics and behaviour of arches and vaults are dispersed among his several Codices, and his drawings show great ingenuity. However, his 'medieval' mechanics prevented him from elaborating a theory to calculate the thrust of arches or to determine 'correct' forms of arches. Nevertheless, he arrived at some valid conclusions that were later demonstrated to be true. For example, he stated explicitly that the condition for an arch to support a large load at the crown is that it must be possible to draw a straight line through the masonry (Figure 2.3a). Leonardo explains: 'An arch will not break if the chord of the outer arch does not touch the inner arch. This is manifest by experience, because whenever the chord *aon* of the outer arch *nra* approaches the inner arch *xby*, the arch will be weak, and it will be weak in proportion as the inner arch passes beyond the chord'; and referring to Figure 2.3b, Leonardo says:

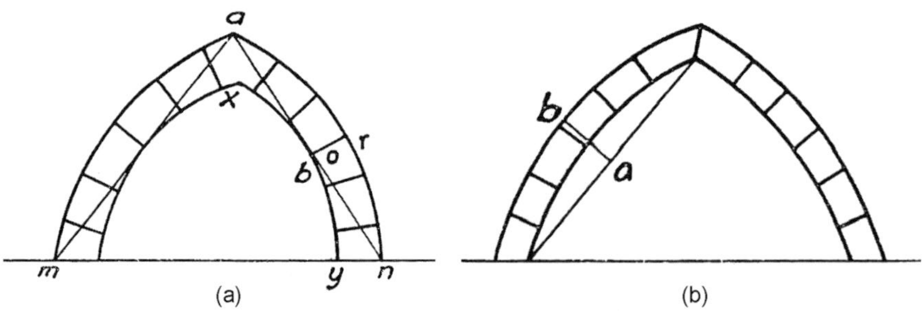

Figure 2.3 Leonardo's drawings on arch mechanics: (a) Rule for an arch that will not collapse. (b) Drawing showing the weakest part in a side-loaded arch. (Source: [14] p. 71)

'When an arch is loaded only on one side, the thrust will press on the top of the other side and be transmitted to the spring of the arch on that side; and it will break at a point halfway between its two extremities, where it is farthest from the chord' [14].

These statements would not surprise any medieval or modern mason or, indeed, a modern architect or engineer with a good grasp of masonry behaviour. In his first sentence, Leonardo was referring to a point load on a pointed arch, and it was well known by medieval masons that pointed arches could resist great loads at the crown. The 'rule' that, if it is possible to trace two straight lines within the thickness of the arch, then it can carry a very large load, could be a medieval rule or Leonardo's own idea; it is impossible to know. His second sentence is saying that if the load is applied not on the crown but on one side, then the arch becomes much weaker and will collapse. Again, this was common knowledge. What is uncommon is the interest which Leonardo showed in the problem, and which appears many times in his manuscripts: how do arches collapse and what methods can prevent this collapse?

A separate question is whether Leonardo actually made tests on model arches. Clifford Truesdell in his long essay on the 'Mechanics of Leonardo da Vinci' [15] doubted that Leonardo did experiments. Rather, he was mainly recording, with extraordinary drawings, what was common knowledge. He clearly had a remarkable imagination and ability to conduct 'thought experiments'. For example, in the section of Leonardo's collected sketches headed 'The Nature of the Arch', is an experiment whose purpose is clear: 'I here ask what weight will be needed to counterpoise and resist the tendency of each of these arches to give way' [16] (Figure 2.4.) Qualitatively they are correct (insofar as they were correctly redrawn) as the outward thrust of the pointed arch is less than the thrust of the semi-circular or segmental arches. This was well known by masons. Was Leonardo expressing his thoughts? Or, were the model arches actually built? Or, he may have constructed one model and speculated on other experiments.

The Madrid Codex, rediscovered only in the 1960s, contains new information that reveal Leonardo's understanding of arch mechanics; for the first time, he shows the correct mechanism of collapse of masonry arches [17]. The upper sketch shows the collapse of a semi-circular arch under a point load at the crown. Five hinges form: one under the load, two in the haunches and two at the feet of the arch. Leonardo draws the internal cords, perhaps trying to show the

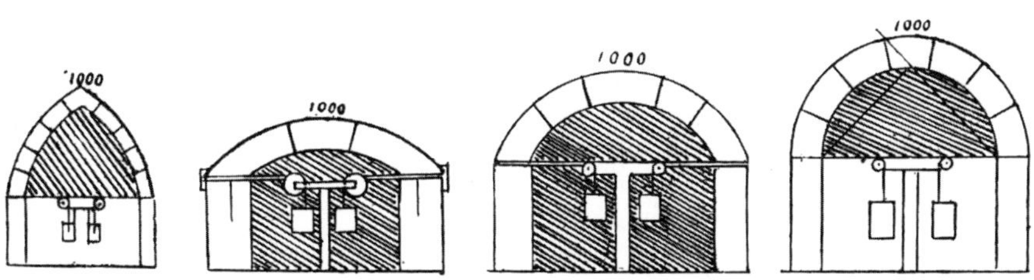

Figure 2.4 Leonardo's proposed models to compare the thrusts of different arches. (Source: [16])

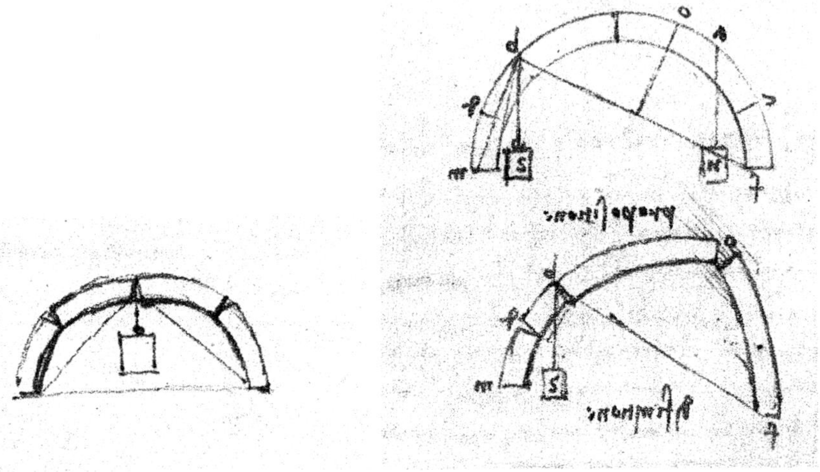

Figure 2.5 Leonardo's sketches for the collapse of semi-circular arches under point loads. (Source: [17])

transmission of the load to the ground, or trying to determine the position of the hinge, or both (Figure 2.5).

In the lower two sketches, the same semi-circular arch is tested for an eccentric load. There is no explanation in the manuscript and the test is subject to several interpretations. It appears that in the upper figure, Leonardo first places two point loads, approximately at the mid-distance between springing and crown, marked S and N. The arch is equilibrated. Then, after removing the right load N, the arch collapses forming four hinges. In the upper figure with the undeformed arch, he traces two straight lines, as before, joining the point of application of the left load, S, to the feet of the arch. The arch must be composed of individual voussoirs. The marked lines probably indicate the location of cracks. The drawing below shows the precise moment of collapse. Four collapse hinges have formed (including the right foot of the arch), converting the arch in a four-bar mechanism, just as illustrated using modern masonry theory [18] (Figure 2.6). Leonardo's drawing is particularly interesting in how the hinge directly beneath the applied load is shown: there is both rotation and sliding between the pieces of masonry. This could have occurred in practice, and does not follow directly from

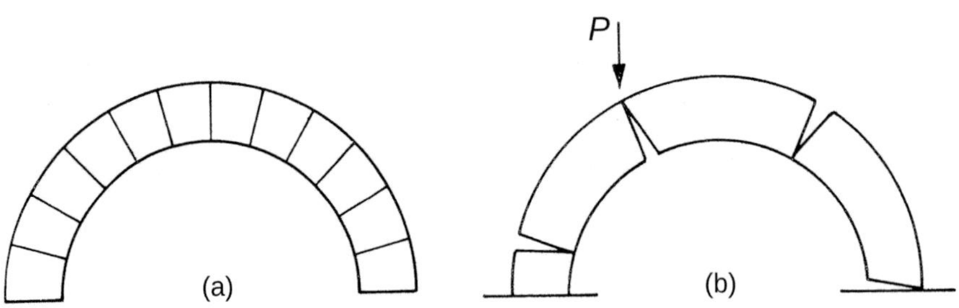

Figure 2.6 Collapse mechanism for a masonry arch. (Source: [18])

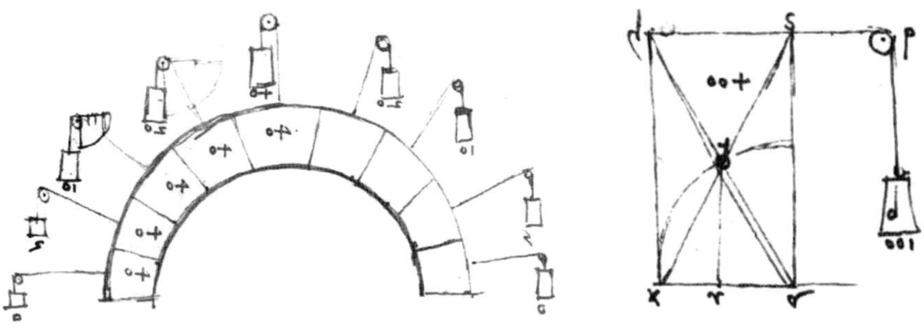

Figure 2.7 Possible experiments with arches by Leonardo da Vinci. (Source: [19])

a thought experiment, which suggests that Leonardo may indeed have performed such an experiment.

Two other sketches from the Madrid Codex are also illuminating. Leonardo proposes how the action of the individual voussoirs in an arch could be studied (Figure 2.7a); he proposes an experiment to determine the horizontal force needed to overturn a rectangular buttress [19] (Figure 2.7b). The value of 100 units shown, is correct for a buttress of the proportions indicated, with a weight of 400 units. A detailed analysis of all Leonardo's drawings and texts on masonry structures and, indeed, bending structures, still has to be undertaken.

2.2.2 Robert Hooke and Christopher Wren

Robert Hooke is well known for establishing the law that an inverted catenary defines the shape of a stable arch – 'As hangs the flexible line, so but inverted will stand the rigid arch' (see Section 3.1.1). Most probably Robert Hooke arrived to his crucial idea by conducting practical experiments – indeed he was the curator of experiments at the Royal Society in London. It is known that he presented some models to the Royal Society around 1670 to demonstrate his theory of arches, but their nature is unknown. Besides being a scientist, Hooke worked as an architect and collaborated closely with Christopher Wren in the reconstruction of the City of London after the great fire of 1666; he would thus have been in contact with masons and must have known something of their thinking about the behaviour of arches and vaults.

When designing St Paul's Cathedral, Wren clearly took advice from Hooke and there exists a sectional drawing of one design for the dome that shows an inverted catenary overlaying the dome, indicating that a hanging chain was somehow used to check the proposed shape (see Section 3.2.1). Apart from this, there is considerable evidence of many models that were built for the construction of the cathedral – as would have been usual for any such large building. Of eighty-four references to 'model' in the monthly accounts for the cathedral, many of these were references to what we could now call drawings, and many were to wooden models. However thirty-one were models made of stone or wood and stone. Several of the payments made to masons mention vaulting or the dome:

'Jan 1691 "Modell for 1/4 part Dome" taking "86 days 2 Masons"
May 1691 "part of a Modell in small stones for part of the Dome";
May 1691 "little Model for the Vaulting of the Choire";
Oct 1693 "23 days Mason making a modell for ye Cupilo upon the Vaulting of the side Isles at 2s 6d & 3 days Carver about the same":
Nov 1693 "19 1/2 days Mason making a 2nd Modell of a Cupilo over brick vaulting of Side Isles…Carver 1 1/2 days"
Jun 1694 "18 days Mason making a large Modell of 1/8th of the Great Dome at 2s 8d per day"
Dec 1695 "for 22 days 1 Mason. Making a Modell for the pedament of the N. Portico"
Dec 1695 "Alter and add to the Modell of ye Legg of the Dome"
May 1696 "Making a Model for ye Intabliture on ye inside of the Great Dome"
Jun 1696 "making another Modell for the Intabliture on the Inside of the Great Dome" ' [20].

We will never know exactly what these models were, or why they were built, but it is more than likely that they must have been to investigate issues of structure or of stereotomy. However, some of them must have had a considerable size and could have served to test the equilibrium of the design, as in the case of San Petronio.

Although Hooke tried to conceal his idea of the stable arch as an inverted catenary, two decades later, in 1695, the Scottish mathematician David Gregory (1661-1708) in a paper on the mathematical form of the catenary stated crucially: 'none but the *catenaria* is the figure of a true legitimate arch … And when an arch of any other figure is supported, it is because in its thickness some *catenaria* is included' [21]. This statement opened the way to an experimental check of the stability of arches and vaulted structures. However, in Britain, mathematicians and engineers adhered rigidly to the assumption that a catenary, located in the intrados, is the only path for the internal forces. In fact, it appears that the no-friction hypothesis of block voussoirs came to Britain only at the beginning of the nineteenth century (see Section 2.4.2), probably under French influence. In any case, there is no evidence during the whole of the eighteenth century that any kind of tests were made on arches in Britain.

2.3 Block models in eighteenth- and nineteenth-century France

The eighteenth century witnessed an explosion of scientific study in France during the Enlightenment. This included much activity related to engineering and in particular, to the behaviour and construction of arches. Although much of the work was theoretical, it was often supported by scientific experiments and, furthermore, because much of the work was undertaken by practising engineers in military academies or, from 1747 at the *École Nationale des Ponts et Chaussées* in Paris, the results of the research were often quickly able to influence live projects.

2.3.1 Henri Gautier, 1717

Henri Gautier (1660-1737) was an architect, engineer and inspector of roads and bridges and wrote the first treatise on bridges, *Traité des ponts* in 1716 [22]. A year later he published his *Dissertation* on the depth of abutments and piles, and on the thickness and thrust of arches [23]. He also presented the results of some experiments on a model arch made of wooden voussoirs that he carried out to convince himself regarding his ideas about the thrust of arches and vaults. These seem to be the first published experiments made using models to investigate the behaviour of arches.

The model arch was a quarter circle with a thickness of about 1/15 of the span of the full arch, a ratio derived from Alberti's recommendations (Figure 2.8). He probably thought that, because the arch was symmetric, he could safely work with half an arch and draw conclusions for the complete arch; as we shall see, this is incorrect. The demi-arch was made of wooden voussoirs loaded with blocks numbered 1 to 12 also made of wood, set without mortar, in dry contact. He described the construction of the model: first, the demi-arch was built; then he placed blocks 1 to 9 touching the extrados; and finally, he placed blocks 10 to 13. The arch stood. Then, Gautier proceeded to remove, one by one, the blocks from the top, i.e. 9, 8, 7, 6, 5, 13, 4, 12 and 3. The arch still stood and there now remained only blocks 1, 2, 10 and 11. However, Gautier then remarks, when he tried to remove block 11, the arch collapsed.

Gautier extracted several consequences from his experiment. First, that it is good to make the arches with a solid extrados and intrados. Second, that since the voussoirs were set dry, they maintain their positions only by virtue of their *coupe*, that is, their form. Third, that it is good to construct in this way, first setting the stones dry and then pouring in the mortar, without packing the poor joints with 'les calles, les coins de bois & autres vilaines choses', which the Ancients (Romans) had never done. And, fourth, he noted that, if a vertical line CK is drawn through the springing of the arch C, then the voussoirs to the right of CK (six and one half, though not mentioned explicitly) should be in equilibrium

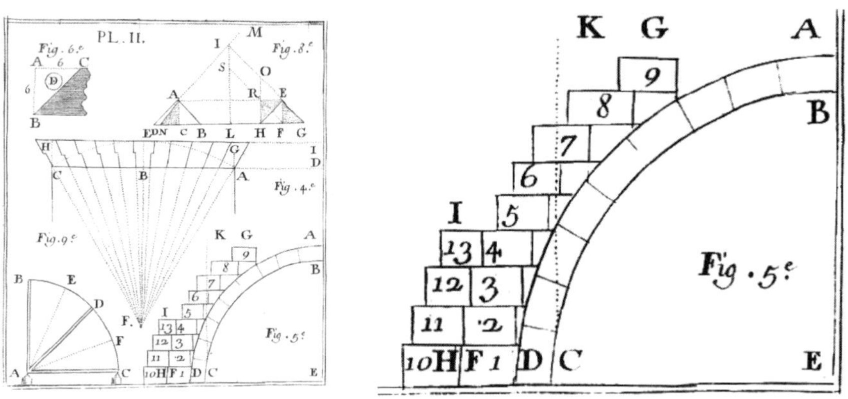

Figure 2.8 (a) Plate from the *Dissertation* showing the model arch.
(b) Detail of the model arch. (Source: [22])

with the voussoirs and blocks to the left, which form the abutment (the two and a half voussoirs and four blocks that remained, before the removal of block 11 and the subsequent collapse). Gautier interpreted this as showing that equilibrium is achieved by making equal the weights on either side of the springing, which is, of course, incorrect.

A modern analysis of Gautier's arch using the Safe Theorem of Limit Analysis shows that it would have been stable without any blocks to act as abutments. Indeed, the limiting thickness of a complete semi-circular arch is about 1/18 of the span and Gautier's demi arch would be stable with a thickness of about 1/24 of the complete span.

Why then did Gautier's arch collapse even when it had an abutment? Anyone who has worked with block models of arches near their limit will know how difficult it is. The slightest imperfection or tremor causes collapse. It is likely that this happened to Gautier. In any case, his experiment confirmed the traditions of masonry construction, that it is important to equilibrate the arch with weight on the haunches, and it is better to cut the stones well and not pack uneven joints with wedges. Gautier was attempting to combine mechanics with experience, one of the main objectives of his *Dissertation* [24] but his work, perhaps unsurprisingly, had little influence and was not cited in later literature.

2.3.2 Augustin Danyzy, 1732

Augustin Danyzy (1698-1777) was a little-known professor of mathematics and hydrography in Montpellier who has passed into the history of structural engineering as being the first to perform systematic tests on models of arches that correctly demonstrate that they collapse by forming cracks which open up to become hinges. Danyzy presented his experiments in a Public Assembly to the Société Royale des Sciences de Montpellier on 27 February 1732. A summary with a plate showing the tests was published and preserved in the Library of the Société, but they became widely known only after their importance was recognised by Amédée Frézier (1682-1773), author of the most influential treatise on stereotomy of the eighteenth century, who included them in his book in 1739 [25].

Danyzy undertook his experiments to resolve the inconsistency between two well-known theories concerning the collapse of a masonry arch or barrel vault. Philippe de La Hire (1640-1718) had observed that when a barrel vault collapses due to the weakness of the buttresses, it breaks along a line about half way between the crown and the springings [26]. His theory, published in 1712, supposed the upper part of a cracked vault (between L and N in Figure 2.9) was a monolith which pushed downwards and outwards on the joints of rupture, which formed by the breaking of the monolithic sections of the vault. He did not fix the point of rupture or give a physical explanation of the thrust. Bélidor simplified this theory in 1729 and gave a physical interpretation: the outward thrust was caused by the upper monolith acting as a wedge, sliding downwards without friction, and forcing the two abutments of the vault apart.

Danyzy, by contrast, argued that no sliding took place, and that the failure would occur due to the opening of joints between voussoirs, eventually forming a hinge. He developed this theory of hinge formation, based on statical

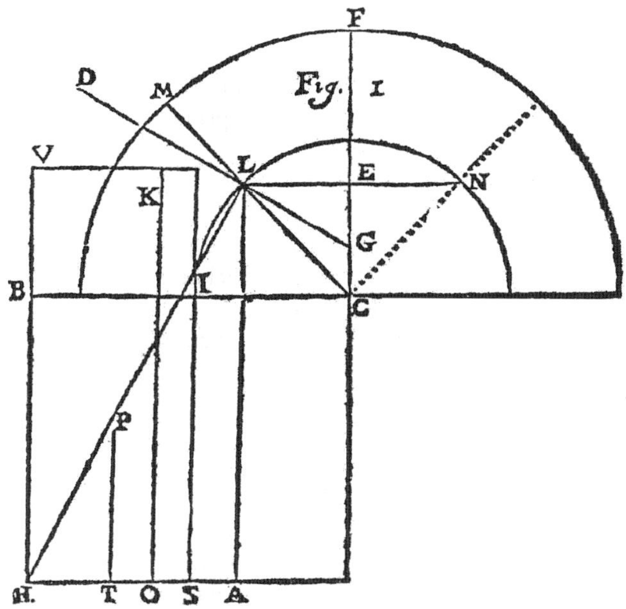

Figure 2.9 Vault theory of La Hire. (Source: [26])

considerations and undertook four experiments on model arches made with plaster voussoirs to prove his ideas. He found that the voussoirs moved not by sliding but by forming hinges: 'The movement that must result cannot take place except by cracks in the arch opening up at the haunches, in the extrados and at the crown in the intrados [...] When a vault collapses, it is not by means of a force that makes the voussoirs slide over each other, like a wedge, but on the contrary: because they cannot slide due to friction, they are forced to rotate about the edges of the joint as if around a fixed point' [27]. These results confirmed his ideas, which he also used to make a calculation of the thrust of a vault by considering the equilibrium of the mechanism, just before collapse (Danyzy's Fig. 1 in Figure 2.10). In fact, just a few years earlier, the engineer and scientist Pierre Couplet (c.1670-1743) had published papers on the thrust and collapse of arches and vaults which, like Danyzy, were based on the formation of hinges to create a collapse mechanism, although it seems that, in 1732, Danyzy was not aware of this work.

Nevertheless, despite these convincing demonstrations, which could have been repeated by anyone, it was the theory of La Hire and Bélidor that prevailed among engineers throughout the eighteenth century and even into the early nineteenth century.

2.3.3 François-Michel Lecreulx, 1774

In the second half of the eighteenth century the French engineer Jean-Rodolphe Perronet (1708-1794) initiated a revolution in bridge design. He introduced two radical changes: reducing drastically the width of the piers, to facilitate the flow of the water, and using arches with a much lower rise to improve access onto

Figure 2.10 Tests made by Danyzy on small plaster voussoir arches. (Source: [27])

the bridge. Never before had arch bridges been built with such extremely low rise-to-span ratios, between 1/9 and 1/18 rather than 1/2 for a semi-circular arch.

One of the first bridges of this type was the Fouchard Bridge, over the river Thouet at Saumur in north-western France, designed in 1772 by Jean-Baptiste de Voglie (1723-1777) who worked under Perronet. It consists of three equal arches, with spans of 26 m and a rise of 1/10 of the span. These low-rise segmental arches posed new problems, both for their construction and the use of arch theory to design them. In particular, although the thrust of the arch could be computed approximately using La Hire's theory (assuming joints of rupture at the springings), the abutments could no longer be considered as monolithic, as was the usual practice. The risk of some kind of fracture was evident and, with such uncertainty, the sensible thing was to make some model tests.

The engineer François-Michel Lecreulx (1729-1812), responsible for constructing the bridge, planned the appropriate tests carefully and executed them in 1774. The models were made at a scale of 1:16 of tufa stone, although Lecreulx stated that stone of any density could be used because the stability was a matter of geometric conditions, not forces. He recorded and discussed the results of the six '*expériences*' in a detailed Memoir which remained unpublished, but accessible, in the library of the *École Nationale des Ponts et Chaussées.* The drawings and

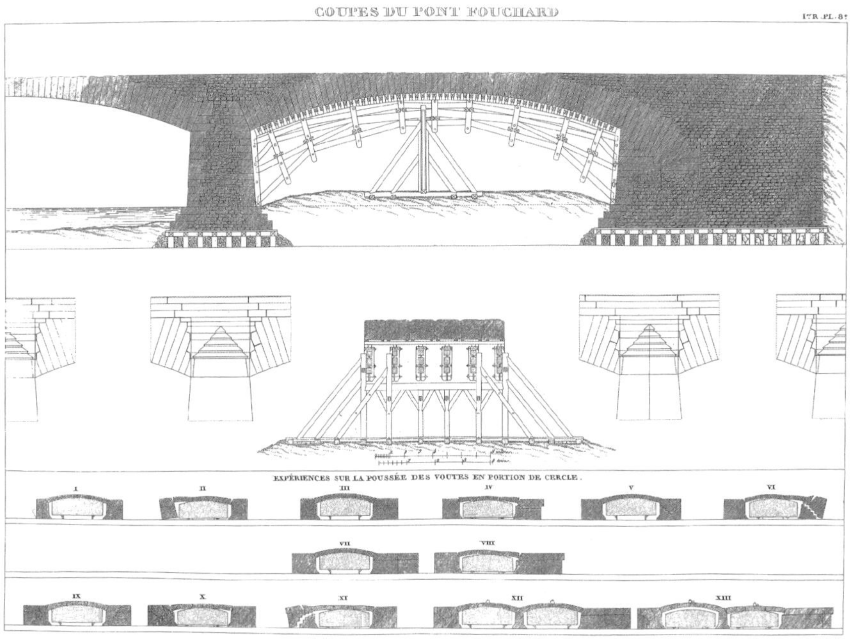

Figure 2.11 The Fouchard Bridge, Saumur. Below, the tests made by Lecreulx in 1774. (Source: [28])

part of the Journal of Lecreulx (but not the text of the Memoir) were published in 1823 [28] (Figure 2.11). The full memoir was published only recently, in 2015 [29].

Lecreulx's experiments were hardly known by later authors on arch bridge design, with the exception of the English engineer Edward Cresy who included the drawings of the tests in his Encyclopaedia of Civil Engineering [30]. The tests demonstrated a high risk of failure of the abutments by sliding. As a result, the arch was prolonged inside the abutment, and its width increased above the usual dimensions (1/3 of the span) to more than half the span. The foundation, built on wooden piles, carried the thrust from the arch into a large mass of soil, which further resisted the tendency for the masonry abutment to slide at its base.

2.3.4 Louis-Charles Boistard, 1797

The bridge over the River Loing at Nemours, a little south of Paris, was designed by Perronet in 1771. Louis-Charles Boistard (1763-1823) was appointed to supervise the construction of the foundations and the arches of the bridge between 1794 and 1797. After the decentring of the arches, work was halted until 1803 and the bridge was completed in 1804.

The bridge was an audacious design, with a rise-to-span ratio of 1/15. Boistard, as many other French engineers of his time, was searching for a theory of arches which was not based on the unrealistic assumption of no friction between the voussoirs. In 1797, he undertook a series of systematic tests on the collapse of arches using models of great size – with a span of 2.60 m. Boistard wrote a memoir describing minutely the experiments which remained unpublished in the Library of the *École Nationale de Ponts et Chaussées*; he later published all his memoirs relating to the construction of the bridge (1822) [31]. In his memoir, Boistard criticized the no-friction theory of La Hire-Bélidor and cited Couplet and Prony, but he does not cite Danyzy or Coulomb. He was aware of some model tests made by Gauthey using small wooden voussoir arches, but he considered that more and better tests were needed. Boistard carried out a total of twenty-two experiments on arch models and drew careful conclusions from the results [32].

All the models spanned 8 French feet (2.60 m) and the voussoirs had a breadth of 8 inches (21.6 cm) and a uniform thickness of 4 inches (10.8 cm), i.e. 1/24 of the span, following, Boistard said, the rule of Perronet. The voussoirs consisted of two bricks, one face polished and the other rough. The two bricks were placed in a wooden mould, with the appropriate angle between the bricks and with the polished surfaces facing outwards. The mould left a small gap between the rough interior faces of the bricks (no more than a knife thickness) and plaster was poured into the gap. In this way, it was easy to produce many identical voussoirs. The bricks could be reused by separating them with a chisel. The centring was formed with two planks separated by 6 inches. The arch was built up from the springings on both sides, placing the voussoirs in dry contact. To ensure the arch was closed precisely, the central keystones were made in situ. The weight of each voussoir was 3.8 kg and the angle between the joints was 3° 45′, corresponding to forty-eight voussoirs in the whole arch (Figure 2.12).

Immediately after decentring the semi-circular arch with a thickness to span ratio of 1/24, is unstable. Couplet had found that the limiting thickness of such an arch was little less than 1/18.8 of the span (modern calculations give 1/18.6) [33]. The arch settled and the upper part rested on the centring (Fig.1 in Figure 2.12). To correct this, Boistard restrained the arch using two cords tensioned by weights of 100 lbs. The arch now showed the typical pattern of cracks in the crown and the haunches (Fig. 2 in Figure 2.12). Boistard states that he then increased the tension in the cords to 300 lbs, so that the crown rose a little and no cracks were visible in either the intrados or extrados (although the drawing shows small cracks in the extrados of the crown) (Fig. 3 in Figure 2.12). Finally, the perfectly circular arch was loaded with bricks on the extrados to form a horizontal surface, and the position of the cracks and hinges can be seen (Fig. 4 in Figure 2.12). Boistard also tested four simple variants of the arch form: oval arches, with rises of one third and one quarter of the span, and segmental arches, with rises of one quarter and one eighth of the span. In every case, he repeated the observations for different loads (Figure 2.13). Finally, he carried out tests on a model having the profile of the bridge of Nemours (1/15), and on a flat arch (*plate bande*) (Figure 2.14). Boistard recorded minutely all the movements and the aperture of the cracks and also registered the weights of the individual voussoirs (which were always numbered) and of the complete arches.

Having proved that sliding between voussoirs never occurred, and that the collapse required the creation of a four-bar mechanism, he went on to develop a way to calculate the thrust of the arches at full scale, following the same approach that Gauthey had also developed [34].

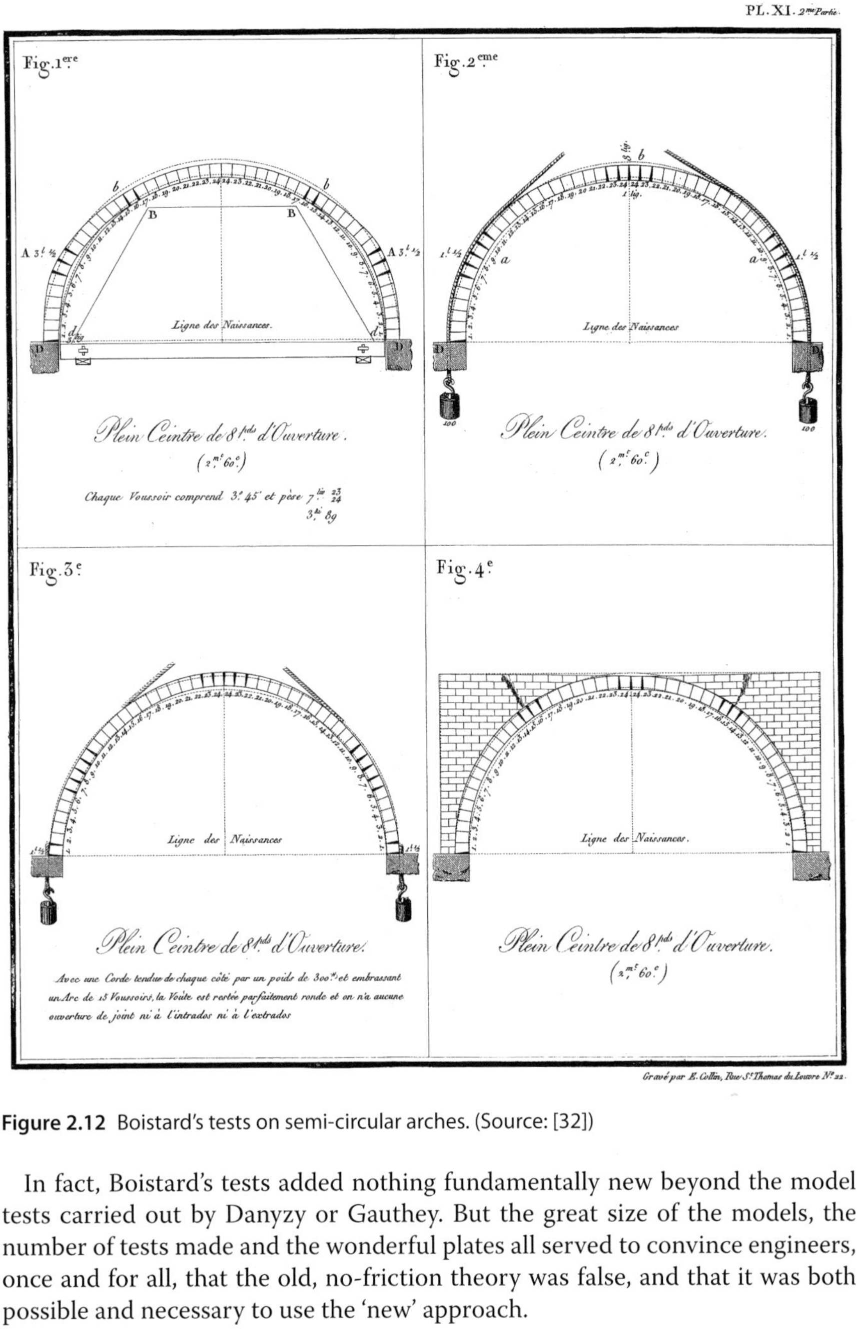

Figure 2.12 Boistard's tests on semi-circular arches. (Source: [32])

In fact, Boistard's tests added nothing fundamentally new beyond the model tests carried out by Danyzy or Gauthey. But the great size of the models, the number of tests made and the wonderful plates all served to convince engineers, once and for all, that the old, no-friction theory was false, and that it was both possible and necessary to use the 'new' approach.

2.3.5 Émiland-Marie Gauthey, 1798

The safety of the designs for the new church of St. Geneviève in Paris (French Panthéon) by Jacques-Germain Soufflot (1713-1780) in the 1750s aroused

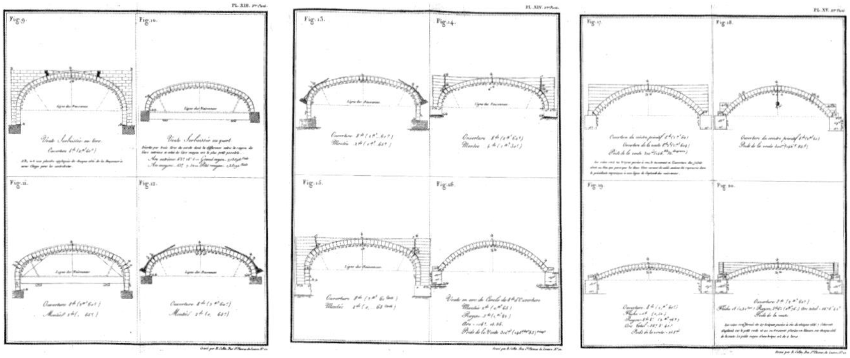

Figure 2.13 Boistard's tests in different oval and segmental arches. (Source: [32])

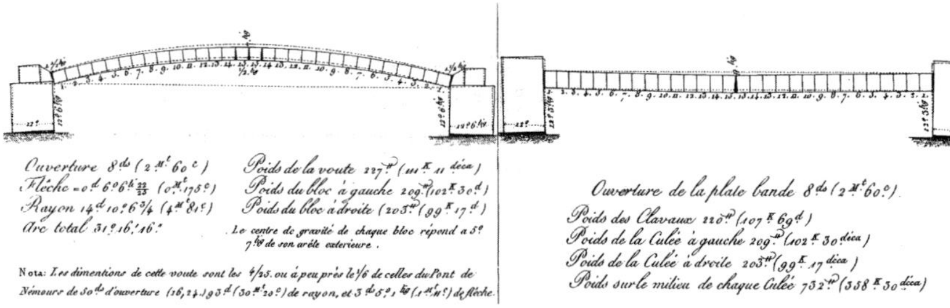

Figure 2.14 Boistard's tests on a model with the profile of the bridge of Nemours, left, and a flat arch. (Source: [32])

a heated debate between engineers and architects. Several memoirs were published and, of course, in the background was how to design the vaults and the dome. Pierre Patte (1723-1814) defended the traditional approach using geometrical rules [35]. Émiland-Marie Gauthey (1732-1806) an eminent bridge engineer, supported Soufflot's designs and wrote two memoirs on the subject of arch and vault behaviour. In the first [36] he used the widely accepted wedge, no-friction theory of La Hire and Bélidor. But in the second memoir, twenty seven years later [37], he had changed his views on the matter, and used Couplet's approach assuming no sliding of voussoirs and collapse by the formation of a four-bar mechanisms. Having stated that La Hire's method does not give an accurate thickness for the foot of the vaults, he notes that vaults collapse by breaking into at least four parts, something which he learned having undertaken many experiments at large and small scale [38].

In the memoir he gives no details about these tests. However, he describes them briefly in his *Traité des ponts* published posthumously by Navier where he dedicates a whole chapter to arch theory. There, after a brief review of previous theories (La Hire, Parent, Couplet, Danyzy, Coulomb and Boistard) he describes his own theory which, he says was confirmed by *'expériences directes'* – direct experimentation, using wooden models:

'We have had circular vaults built: segmental vaults a one-third and one-quarter lower than a semi-circle, and a semi-circle. Their opening was 65 centimetres; the voussoirs, 27 millimetres deep, were made of wood and cut with precision. We have sought to upset the balance between the thrust of the upper parts and the stability of the lower parts, either by reducing the thickness of the abutments or by loading the top of the vault, and we have constantly noticed that the failure tends to occur under the circumstances we have just explained [the formation of hinges]' [39].

2.3.6 Jean-Baptiste Rondelet, 1797-1813

The French architect and engineer Jean-Baptiste Rondelet (1743-1829) also made tests on model arches with the objective of calculating the width of their buttresses. He dedicated a whole chapter of 187 pages in the third volume of his *Traité* to this subject [40]. Rondelet carried out a large number of tests with models: 'an infinity of experiments made using fifty-four models of vaults of different forms and different extrados'. He was looking for a formula that would give the thickness of the buttress for any form of arch, and believed he had found one: '... from the experiments we have just mentioned, and a large number of others that it would take too long to report, the results of which are these, we have established a general formula for all kinds of barrel vaults, with the same thickness, regardless of their shape' [41]. He arrive at an expression which involves a square root and some geometrical parameters. In the 'demonstration' of his formula, his reasoning aims to be scientific, but is defective and, in many parts, incoherent (besides which his formula is not dimensionally correct). Eventually, he gives an entirely geometric procedure to obtain the depth of the buttress, claiming that it gives the same results. Rondelet was still operating within the tradition of empirical rules; however, he wanted to appear rigorous and scientific.

Nevertheless, he produced perhaps the largest set of arch and vault models ever made, all of which were made of stone (Figure 2.15). He first made a 'great model', with a semi-circular arch of 36 ¼ inch (98 cm) span and thickness 3 inches; the height of the buttress was 40 ¼ inches (1.09 m). The arch was divided in two equal voussoirs with the joint at 45°. His formula gave a buttress thickness of 5.8 inches, but the model needed 6.25 inches to stand. Rondelet attributed the difference to the imperfections of the model (Fig. 11 in Figure 2.15). These experiments convinced him that the collapse was always by the division of the arch in several parts, with the voussoirs rotating as a mechanism, acting as levers [42].

Rondelet then built a series of sixteen smaller models all with the same span (9 inches or 24 cm), thickness (¾ inch or 2 cm) and buttress height (10 inches or 27 cm), but with different shapes of arch and extrados in order to compare their thrusts and the way they behave [43] (Figure 2.16).

From these model tests, he obtained the depth of the buttress that caused collapse and compared it with his arithmetic and geometric formulae. He claimed a good agreement. In fact, it is rather exasperating to read this part of Rondelet's treatise. For over two hundred pages he struggles to obtain a design rule based on the results of his '*expériences*', but using no statical theory. At least, this constitutes a warning to anyone who hopes to build any theory using only 'the facts' – a

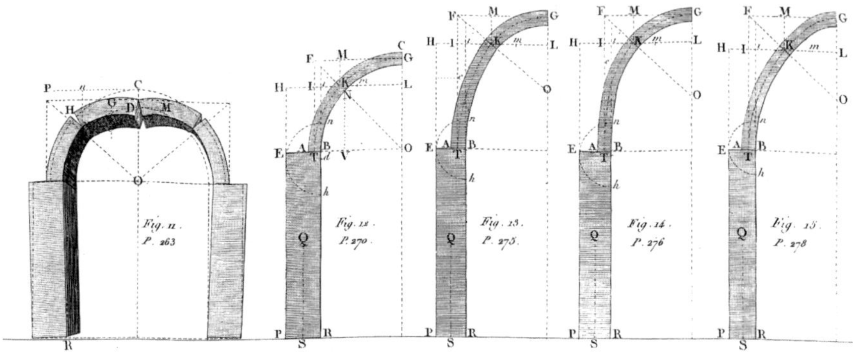

Figure 2.15 Rondelet's stone models of arches. The model in Fig.11 (left) had a span of 98 cm with buttresses 109 cm high. (Source: [40])

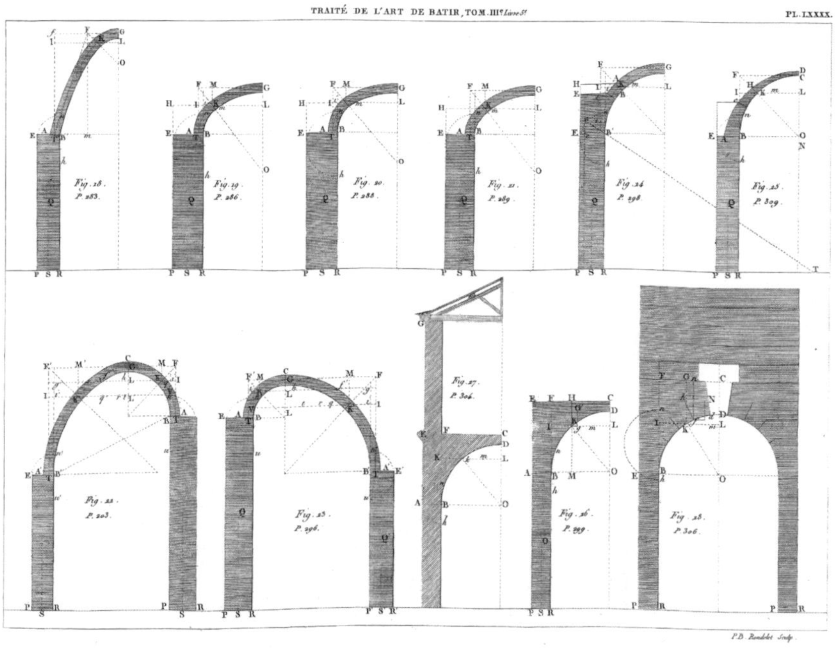

Figure 2.16 Stone models of arches of different forms tested by Rondelet. All the arches had a span of 24 cm and thickness of 2 cm. (Source: [40])

warning, indeed, that is as relevant today as it was then. The tradition of empirical rules was based on generations of masons building and making critical observations; a completely different process with no pretentions.

The rules derived by Rondelet for abutments were copied time and again throughout the nineteenth century in numerous construction handbooks all across Europe (the *Traité* was translated to German and Italian, and large parts of it appeared in English in Joseph Gwilt's *Encyclopaedia of Architecture* published in 1842). The rules were, in many cases, unsafe: that they were ignored by practical builders, is a sign that a reliable feeling for safe construction still existed in the nineteenth century.

Rondelet also conducted some model tests on masonry domes, but the conclusions he drew from his tests, and his knowledge of basic statics were manifestly faulty. To help understand the stability of domes when preparing his study of the structure of the Panthéon in Paris, he made a model of a hemispherical dome of uniform thickness, which was divided by four vertical planes [44] (Figure 2.17a). He studied two different means of supporting the dome: a continuous circular wall the same thickness as the dome, and eight, equally spaced columns. In both cases, said Rondelet, the dome stood. He argued that the model could be divided into two halves by a vertical plane, and that each half would be stable and, therefore, one half dome did not thrust on its opposing half. Hence, he concluded that domes do not thrust outwards against their supports. This last assertion is false, and it unnerved Gauthey. A demi-dome of sufficient thickness can, indeed, be stable, exerting the thrusts, not radially, but parallel to the front arch of the demi-dome [45]; a masonry dome, however, made of material with no tensile strength, must exert a lateral thrust.

Gauthey, who saw Rondelet's model, noted first that, although the lower part of the model was divided into eight slices, the upper part was divided into only four slices, and this would considerably change the conditions relating to the stability of the dome; as a result, it was stabilised sufficiently by the curved wall of the tambour. Secondly, in the case where the dome was supported on monolithic

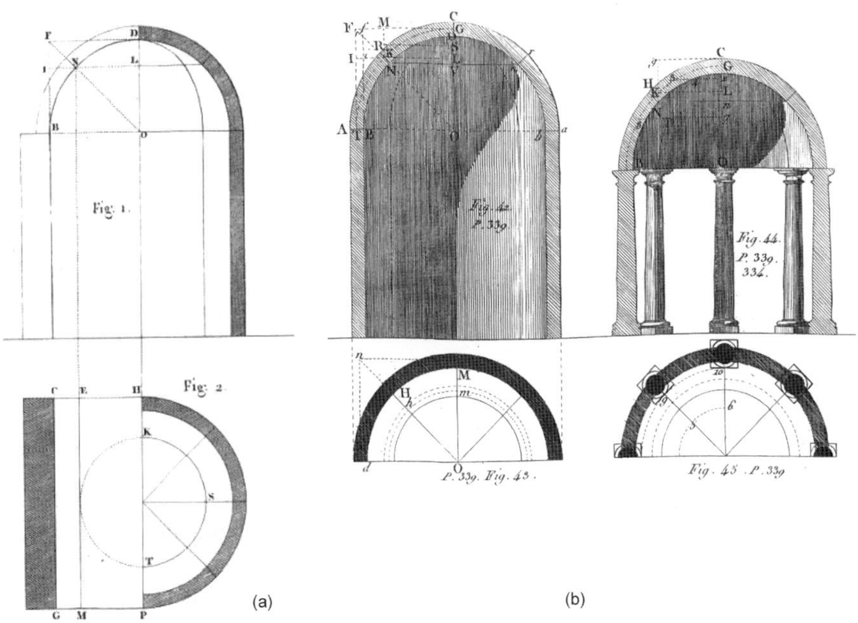

Figure 2.17 Rondelet's stone models of domes. Model of vaults divided in blocks to test their stability. Left to right: (a) cloister vault on walls or columns (Source: [44]); (b) hemispherical dome on wall and columns. (Source: [40])

columns, as well as the columns being monolithic rather than a number of stone drums, each slice of vault was also supported between two columns, so this monolithic slice was effectively acting as a tie between the adjacent columns [46]. Rondelet also discussed the stability of domes in his *Traité* making reference to these same model tests [47]. After a new analysis of the results, and despite Gauthey's comments, he still maintained that domes do not exert any outward thrust.

2.3.7 Louis Vicat, 1832

The French engineer Louis Vicat (1786-1861) made several models in 1832 to study the problem in the masonry piers which served to attach the cables of the suspension bridges. He said that '… l'absence de données précises sur le mode de destruction que les cables tendent à provoquer à l'égard des massifs auxquels ils s'attachent ou sur lesquels ils s'appuient … j'ai cru devoir interroguer l'expérience en opérant sur des piliers construits à l'échelle 1/10' [48] (Figure 2.18). The models were made using adobe bricks ('briques crues') with no mortar. Vicat made several tests varying the mode of anchorage of the cable and the detail of the crown support on the masonry. He noted the appearance of cracks and which order it occurred. Eventually he pulled until the collapse occurred remarking that the force was 10% less than calculated. He attributed this diminution to the fracture and separation of part of the pile. However he gave no hint about how he made the calculations.

2.3.8 Édouard-Henri-François Méry, 1828 and 1840

The French engineer Édouard-Henri-François Méry (1805-1866) is remembered in the history of structural analysis for just one work, his memoir on the '*courbes de pression*', lines of thrust, published in 1840 [49]. The method of studying the stability of arches with the aid of lines of thrust is still known in France as 'la méthode de Méry'. In fact, he had developed his theory more than a decade earlier. In

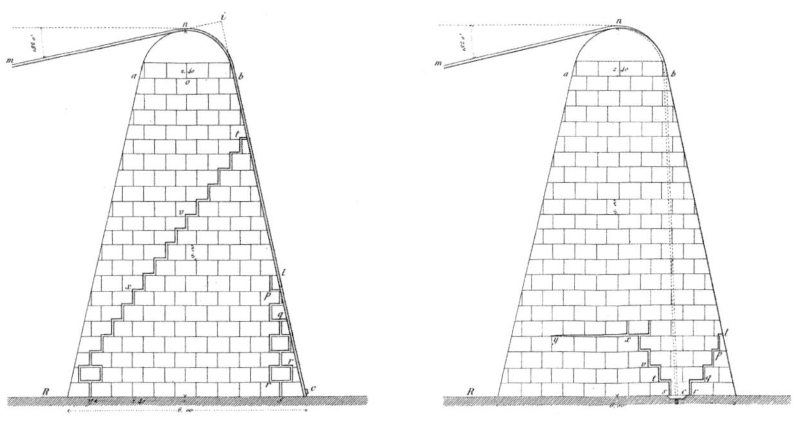

Figure 2.18 Tests on 1:10 scale models of masonry piers for a suspension bridge anchor. The models, made of adobe bricks without mortar, were 90 cm tall with a base 60 cm wide. (Source: [48])

1828, before finishing his career as an engineer in 1830, he was sent for service to the port of Brest. At that time, the vaults of the Clermont-Tonnerre hospital were under construction and there were some doubts about their safety. The engineer director of the work commissioned Méry to undertake a study of the stability of these vaults, with calculations and some full-scale tests. In 1829, Méry wrote a long report for the director in Brest, which was not published at the time.

The report by Méry has extraordinary interest and represents an enormous step in the understanding of arch behaviour. The full report was eventually published by V. Chevallier in 1862 in the *Mémorial des Travaux Hydrauliques de la Marine* [50]. In it Méry develops the theory of lines of thrust, and used it to interpret the results of his tests on the vaults. He made three tests on pointed vaults with a span of 8.1 m and height of 2.8 m. The first test was for the design shown in his Figure 1 in Figure 2.19, in which the line of thrust has been drawn. The vault was built of rubble masonry, *möellons,* and collapsed shortly after decentring. Méry documented minutely the cracks and movements and, comparing these observations with his calculations, (Figure 2 in Figure 2.19), he concluded that the geometry was too near to the *l'équilibre mathématique* and, being built with rubble, the collapse was expected. Three more tests were made on vaults with a slightly modified geometry, to preclude collapse: the radius of the intrados was reduced and the thickness increased. This time the lines of thrust were comfortably within the masonry (Figure 3 in Figure 2.19). The supports were on wooden rollers, permitting horizontal displacement, and were tied together with iron ties. The ties incorporated a dynamometer to measure the outward thrust. Méry observed the influence of movements of the supports on the outward thrust, and obtained a fairly good agreement with his calculations.

According to Chevallier, Méry continued his theoretical work and, by 1837, he had the manuscript of a memoir ready for publication. It has recently

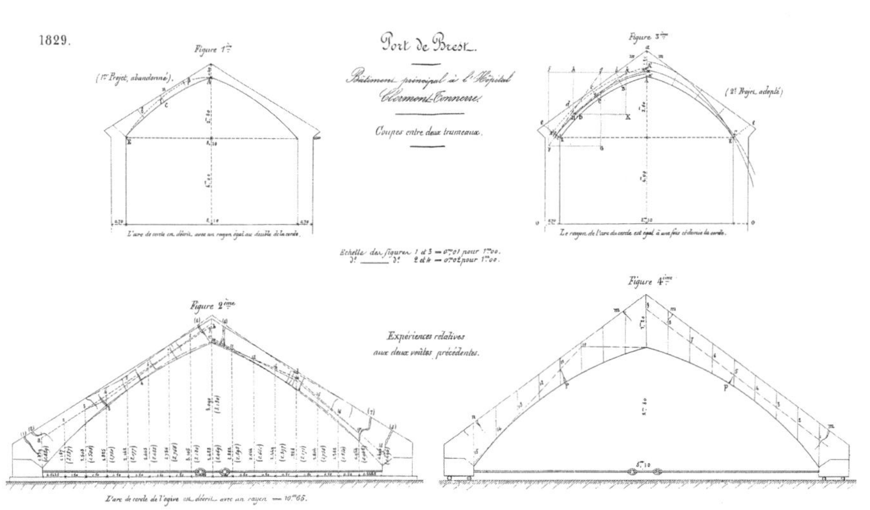

Figure 2.19 Méry's tests on full-scale masonry arches, with a span of 8.1 m and height of 2.8 m, in 1828. (Source: [50])

been discovered that Méry contacted the eminent French theoretician, Adhémar Barré de Saint-Venant (1797-1886), asking him for comments on his manuscripts [51], and the final form of the memoir, published in 1840, owes much to Saint-Venant's advice. Méry used Boistard's tests to check the validity of his new theory of 'courbes de pression'. He reproduced Boistard's tests, drawing the line of thrust in the active part, and, on the right side, he drew the corresponding limit arch that he had calculated (Figure 2.20). He found a perfect agreement between his theory and Boistard tests. It is perhaps surprising that he did not include in the memoir any plates showing the tests he had made in Brest in 1828.

No more tests on model arches were made in France. The next experiments, with arches of real size, were made in the late nineteenth century to demonstrate their 'elastic' behaviour.

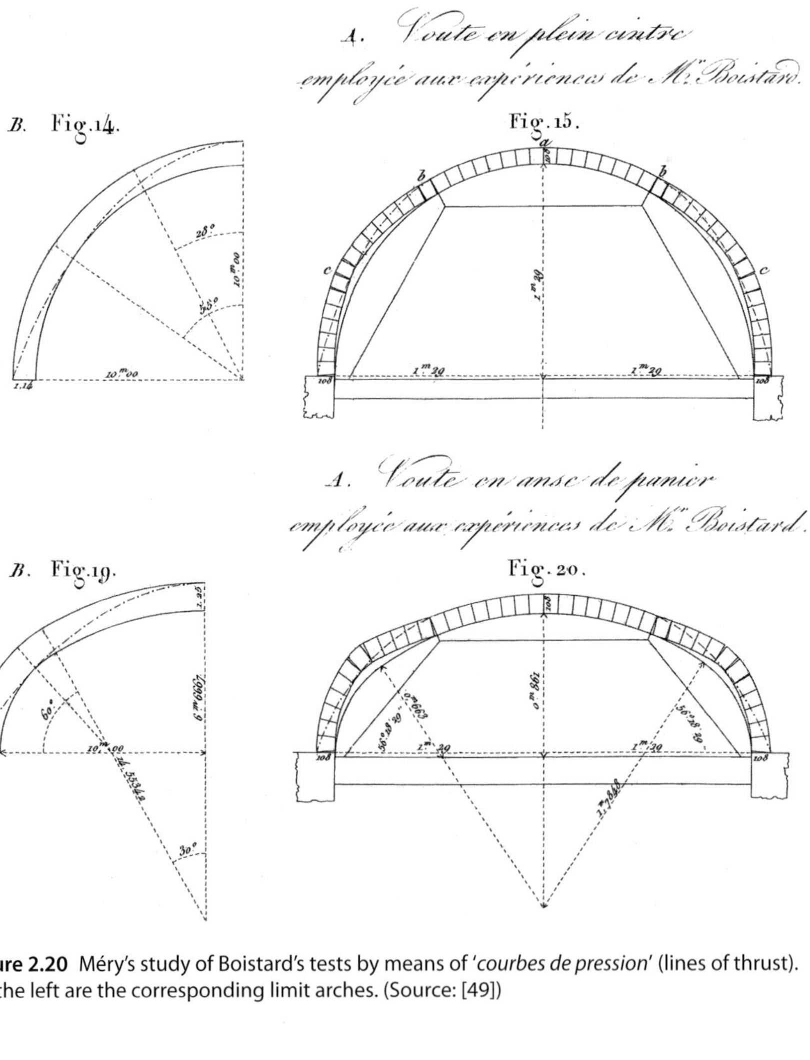

Figure 2.20 Méry's study of Boistard's tests by means of '*courbes de pression*' (lines of thrust). On the left are the corresponding limit arches. (Source: [49])

2.4 Block models in nineteenth-century Britain

After the pioneering work of Robert Hooke and David Gregory in the early eighteenth century, there was little interest in exploring the design of arches in eighteenth-century Britain, although the work undertaken in France did not go unnoticed. Road and canal bridges were generally of modest size and could be reliably designed and constructed using familiar and standard designs. Matters began to change at the beginning of the nineteenth century when cast iron was first used for large bridges. As there was no body of experience upon which to draw, engineers had to draw on every possible means to create the necessary confidence to design and construct arch bridges of cast iron.

2.4.1 John Robison, 1801

The first British engineer to make a serious criticism of the hanging-chord theory was the Scottish scientist and engineer John Robison (1739-1805), professor of Natural Philosophy at the University of Edinburgh. In his article 'Arch' for the Supplement to the Third Edition of the Encyclopaedia Britannica, after presenting the equilibration theory and referring to Hooke's hanging chain, he made a series of observations which would have a great influence in later times [52]. His main contention was that the equilibrated arch was in a state of unstable, 'tottering' equilibrium, as any applied force will cause collapse. To explain his ideas about arch behaviour Robison used a model of a hanging arch formed by magnets (Figure 6 in Figure 2.21) (it is not clear if he ever build such a model; he may have used it as an image to think about the problem). He then says that, when the hanging shape is inverted and: 'set up as an arch, if a small weight be laid on any part of it, it will bring the whole to the ground'. Robison then noted that, if the contact is through a surface instead of a point, the arch will possess some degree of stability: 'if the same weight be laid on the same part of the arch now constructed with flat joints, it will be sustained, if the new curve of equilibration still passes through the touching surfaces' and he added: 'the longer the joints are, the greater will be the stability of the arch, or that it will require a greater force to break it down' [53].

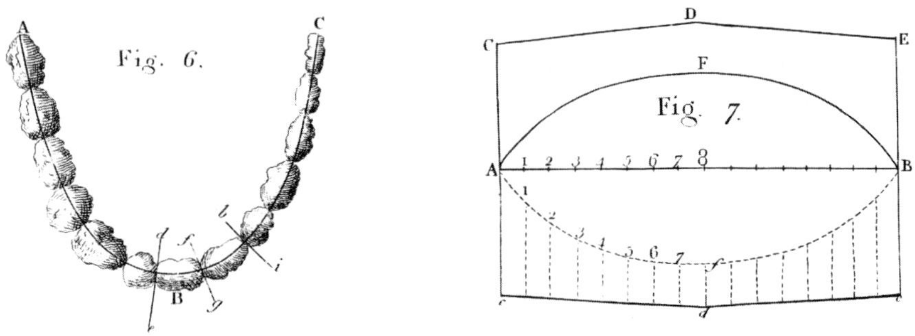

Figure 2.21 Robison's illustrations of the catenary. Fig. 6, a catenary made with magnets; Fig. 7, a hanging model for designing an arch bridge. (Source: [52])

Here, for the first time, the curve of equilibration is allowed to move within the masonry of the arch. Though Robison does not mention him, this is an explanation of Gregory's assertion quoted above.

Robison explains in great detail how to design a bridge of 60 feet span, 21 feet rise and thickness of 2 feet 9 inches, by means of a hanging model (Fig.7 in Figure 2.21). He also calculated the shape of the curve and found a difference of only 2 inches between the results of the model test and the calculation: 'This was very agreeable information; for it showed us that the first curve, formed by about two hours labour . . . would have been sufficiently exact for the purpose, being in no place three inches from the accurate curve, and therefore far within the joints of the intended arch stones' [54]. He stressed that the process could be followed by any mason, though there are very few examples indicating that it was later used in Britain (see Section 3.3.3).

Later in the article, Robison refers to the 'Inutility of the common theory of equilibration' and notes some of the most obvious deviations from reality. First he stresses the need to study the way arches collapse: 'The general facts which occur in the failure of old arches are highly instructive, and deserve the most careful attention of the engineer; for it is in this state that their defects, and the process of nature in their destruction, are most distinctly seen'. He then remarks, correctly, that 'The way in which circular arches commonly fail, is by the sinking of the crown and the rising of the flanks', and concludes that 'a very great majority of these facts are irreconcilable to the theory' [55]. He continues 'arches very rarely fail where their load differs most remarkably from that which this theory allows. Semi-circular arches have stood the power of ages, as may be seen in the bridges of ancient Rome'. Robison describes minutely the process of collapse of an actual bridge and eventually decided to build small model arches made of chalk, though he gives no dimensions, 'and subjected them to many trials, in hopes of collecting some general laws of the internal workings of arches which finally produce their downfall' [56] (Figure 2.22).

When loading the arch at A, he always found that 'the arch always broke at some place B' apparently obtained drawing a straight line within the thickness of the arch.' Robison was unable to explain the actual mechanism of collapse - though the dotted lines suggest a four-bar mechanism. Previously he had stated that: '... it will appear that an arch, when exposed to a great overload on the crown (or indeed in any part), divides, of itself, into a number of parts, each of which contains as many arch stones as can be pierced (so to speak) by one straight line'. He also insisted on the non-vertical action of the fill over the haunches of the arches, which differed from the usual hypothesis that the action was vertical. Finally, he

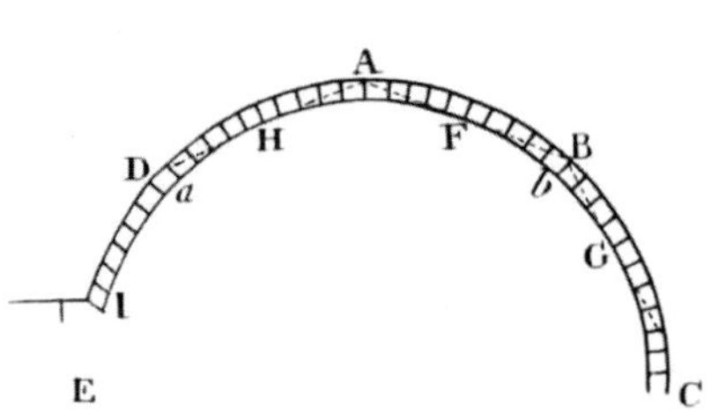

Figure 2.22 Robison's model arch made with blocks of chalk to study the failure of arches. (Source: [52])

commented briefly on the importance of friction, acting as a kind of cohesion to maintain the stones in their places. Robison did not formulate a coherent theory of masonry, but his criticisms shook the foundations of the equilibration theory and showed the way for later English engineers.

2.4.2 Thomas Young, 1807-1824

One particular project at the turn of the century stimulated considerable interest among mathematicians and practising engineers in Britain. It was the design by Thomas Telford (1757-1834) and James Douglass for a 600 foot cast-iron arch bridge [57]. They submitted their proposal for consideration in the competition held in 1800 for a new bridge over the River Thames in London as part of a wide-ranging inquiry for improving the Port of London (Figure 2.23).

The design was greeted with widespread approval, but 'considerable doubts were expressed of the feasibility of the proposal which was at that period so novel and so bold'. A Select Committee was formed to inquire into its viability and a series of twenty one questions were posed and sent to seventeen of the most eminent men of science and engineering in the country. The responses were a great disappointment to the Committee: 'The answers which were given were singularly humiliating to the pride of philosophy: they were not only altogether at variance with each other, but in every instance incomplete and unsatisfactory' [58]. Eventually the design was rejected. However, the whole episode provoked an awakening in the interest in arch theory in England which finally led to the abandoning of the equilibration theory.

Two of the questions posed by the Select Committee are of particular interest to the role of models in engineering design:

> 'VIII. Is it necessary or advisable to have a model made of the proposed bridge, or any part of it, in cast iron? If so, what are the objects to which the experiments should be directed; to the equilibration only, or to the cohesion of the several parts, or to both united, as they will occur in the intended bridge?

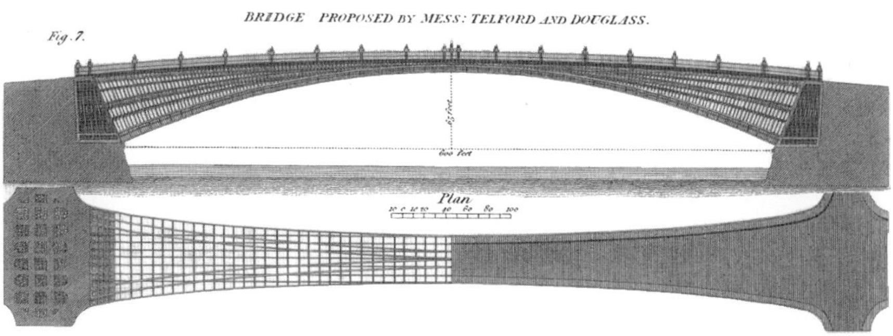

Figure 2.23 The design by Thomas Telford and James Douglass for a single cast-iron arch over the Thames in London, with a span of 600 feet, 1800. (Source: [57])

IX. Of what size ought the model to be made, and what relative proportions will experiments, made on the model, bear to the bridge when executed?' [59].

Before considering Thomas Young's responses to these questions, made about five years later, it is worth remarking on the views about arch behaviour by one of the seventeen men who were consulted for the inquiry. (All eleven responses to these questions are reprinted in full in Appendix A4). George Atwood (1745-1807) was an eminent mathematician and physicist, and published two books (1801 and 1804) in which he defended the old theory of masonry arches which assumed no friction between the voussoirs which had been assumed to be correct for a century. To validate this long-held theory, he built and tested some elaborate and expensive models made with voussoirs of polished brass. One of the models corresponded precisely to Telford's design for his 600 foot arch [60] (Figure 2.24). The models were so obvious as to arouse surprise at the time and money expended on them. Atwood's memoir is both difficult and indigestible, and an anonymous reviewer of the book in the British Critic commented: 'Now it certainly appears to us … that Mr. Atwood's method is extremely circuitous and his mathematical process unnecessarily complex and intricate' [61]. Nevertheless, they demonstrated that a theory which was considered correct for a century, despite being discredited by eminent engineers, did not disappear all of a sudden.

Thomas Young (1773-1829) was, at this time, Professor of Natural Philosophy at London's Royal Institution and between 1801 and 1803 he gave 91 lectures on a wide range of subjects, including the properties of materials and the behaviour of arches. Young had studied Robison's ideas with care and expounded his own criticism of the traditional theory of arches, reprinted as the article 'Architecture and Carpentry' in his Lectures on Natural Philosophy. To illustrate his lecture he made several models, the first two of which were models of catenarian and polygonal arches (Fig.152 and Fig.153 in Figure 2.25). He remarked that 'a slight degree of friction is sufficient for retaining in equilibrium an arch formed by the inversion of a chain of beads' [62] which suggests that he actually was able to invert these models. The last model (Fig.154 in Figure 2.25) represented Robison's proposal discussed above (Figure 2.21).

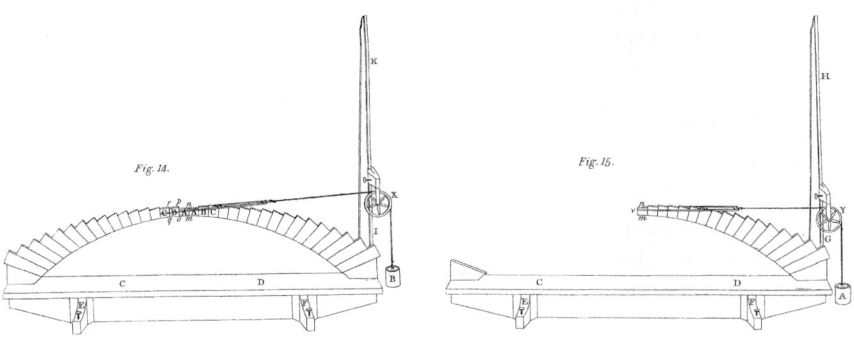

Figure 2.24 Attwood's 'no-friction' model of an arch made with blocks of polished brass. (Source: [60])

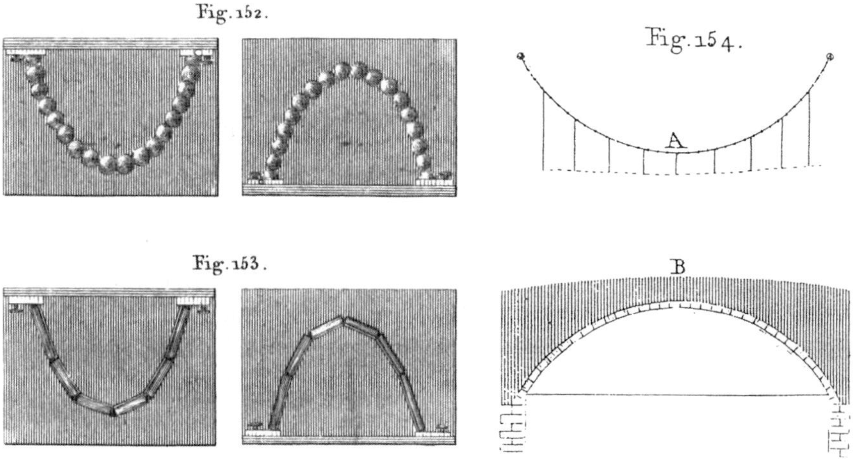

Figure 2.25 Arch models used by Thomas Young in his lecture on Architecture and Carpentry, 1807. (Source: [62] Plate XI)

Eventually, Young elaborated a complete and correct theory of arches in his article 'Bridge' written for the Encyclopaedia Britannica [63]. However, the place of publication and the extreme difficulty of his explanations precluded his work from having later influence. Here, it is of greatest interest to look at how he answered the two questions about the use of models (VIII and IX) which had been posed by the Select Committee in 1801 with regard to Telford's design for a 600-foot cast-iron arch. (Young's responses to these questions are reprinted in full in Appendix A4).

Young is very clear about the matter: hanging models will enable the stability of the arch to be verified. However, if the model is used to study the effect of 'the cohesion or connection of the parts' (the strength) the results will be 'extremely uncertain'. He observed that 'the form and proportion of the joints could scarcely be imitated with sufficient accuracy, and since the strength of some of the parts concerned would vary as the thickness simply, and that of others as the square or cube of the thickness, it would be more difficult to argue from the strength of the model upon that of the bridge, than to calculate the whole from still more elementary experiments'. Young went on to recommend undertaking experiments 'on the force required to crush a block of the substance employed' and explained how such an experiment should be carried out: 'the parts corresponding to the blocks of the arch should be formed of their proper thickness and length, and connected with each other and with the abutments by a short joint or hinge in the middle of each, allowing room for a slight degree of angular motion only ... [and] if the curve underwent no material alteration by the suspension, we should be sure that the calculation was sufficiently correct'. If this is not the case, 'the arrangement of the materials might be altered'. He, then, makes a suggestion to ease the use of the model: '... the investigation might be facilitated by allowing the joints or hinges connecting the block to slide a little along their surfaces, within such limits as would be allowable, without too great a reduction of the powers of resistance of the blocks' [64]. Young provided no drawing, but the

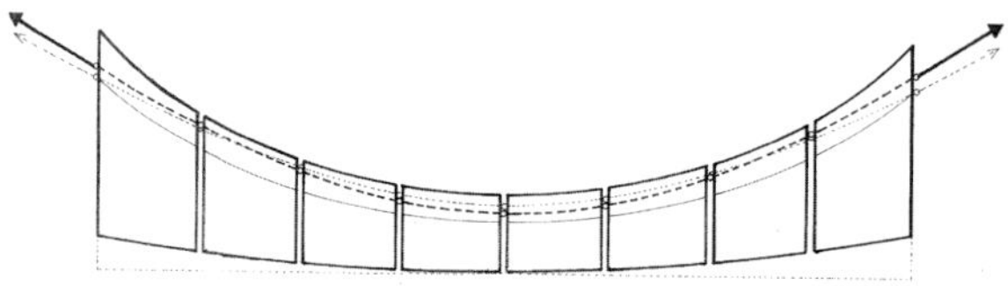

Figure 2.26 Hypothetical reconstruction of Young's hanging-block model. The hinges can move vertically within the ring of the arch, materializing different curves of equilibrium. (Source: [63])

text may be interpreted as a hanging block model in which the size of the blocks corresponds to their respective weights, and the hinges are located within the section of the arch and would allow a vertical displacement within it (Figure 2.26).

As for the size of the model, he stated that it 'is of little importance, and it would be unsafe to calculate the strength of the bridge from any general comparison with that of the model'. Of particular interest is that he cites Euler's essay (in Latin) on testing models of bridges [65], discussed in Chapter 4 of this book and reproduced in full, in English, in Appendix A3. With characteristic disdain, Young commented that, though 'relating expressly to the mode of judging the strength of a bridge from a model ... it contains only an elementary calculation, applicable to ropes and simple levers, and by no means comprehending all the circumstances that require to be considered in the structure of an arch' [66].

2.4.3 William Bland, 1836-1839

In February 1836 a series of eleven articles was begun in Loudon's Architectural Magazine, signed by William Bland, Esq., a land owner from Kent (of whom nothing else is known), with the title 'Experimental essays on the principles of construction in arches, piers and buttresses'. These essays were later collected in a book which had enormous success and went through seven editions, the last in 1912 [67]. The author seems to have been an amateur, interested in bridges and cathedrals, and, after perusing some works (he cites Charles Hutton, George Atwood and Samuel Ware) he felt dissatisfied and decided to investigate the matter himself. He proceeded in a completely experimental way, making and testing hundreds of models (Figure 2.27).

As a curiosity, one of his more 'scientific' essays is worth closer scrutiny. He began (as did Leonardo) trying to measure the forces acting in each voussoir (cf. Figure 2.9b). For this purpose, he built an apparatus on which to place arches (the box in Figure 2.27), with pulleys and hanging weights to measure the horizontal and vertical components of the thrust. The diagram showing the horizontal component was drawn afterwards. Of course, the horizontal component should have been uniform, as all the forces are vertical, but Bland obtained a different distribution for the semi-circular arch. For the parabolic arch his diagram is almost correct. Bland makes no comment on this. In summary, Bland simply recorded the 'data': dimensions and collapse loads. He did not describe the manner of the collapse, nor did he arrive at any sensible theory, except tracing some rather arbitrary straight lines. While he had no influence on the development of arch theory,

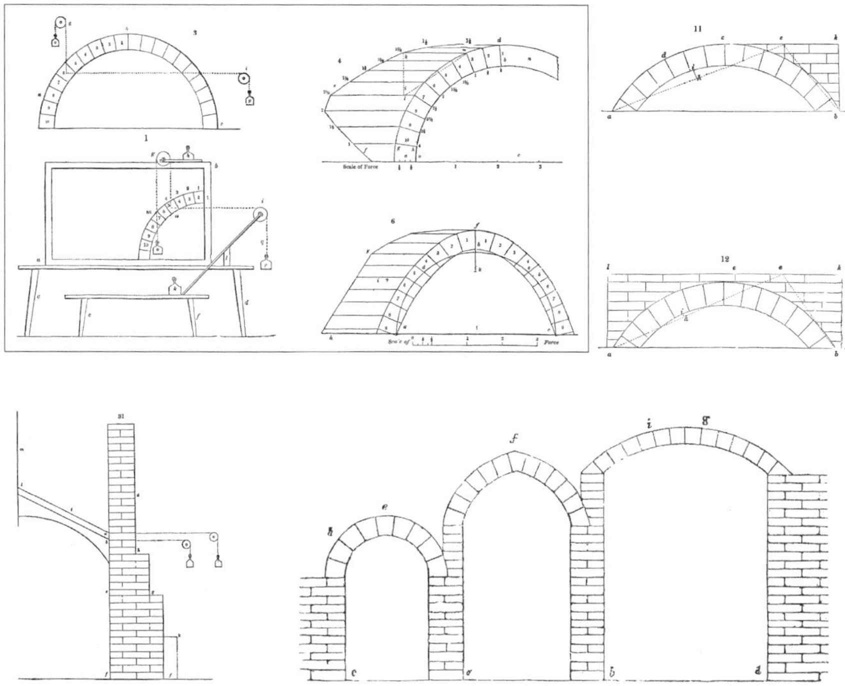

Figure 2.27 Some tests on block models made by Bland. (Source: [67])

the work is worth mentioning as it demonstrates that, as was also the case with Rondelet, making a large number of tests is no guarantee to arrive to a useful theory. Nevertheless, given the success of his book, it would seem likely that it was consulted by many builders of small arches for buildings and bridges.

2.4.4 Henry Moseley, 1833-1837

With the exception of Thomas Young, Robison's observations about collapse were ignored in England during the following twenty years. For example, in the first edition of his *Elementary Mechanics*, published in 1819, the eminent scientist William Whewell (1794-1866) still advocated the equilibration theory. Only in the second edition of 1828 did he admit the effect of friction and a possible deviation of the 'line of pressure' from the centre line of the arch. He cited the tests on small arches by Robison and accepted his hypothesis that the collapse is explained by tracing straight lines within the thickness of the arch. Similar observations were included in other manuals by Cambridge scholars in the 1830s [68]. However, a deep grasp of the problem, such as Young had shown two decades earlier, was still lacking.

It was Henry Moseley (1801-1872), Professor of Natural and Experimental Philosophy and Astronomy at King's College, London between 1831 and 1844, who established a firm theoretical foundation for the behaviour of arches.

Moseley defined, for the first time, with mathematical precision, the concept of 'line of thrust'. He seems to have been the first English engineer writing about arches who had a deep knowledge of the French theories (he cited Coulomb, Navier, Lamé and Clapeyron, Poncelet etc.), and he reconciled his theory of thrust lines with Coulomb's theory of collapse. He handled the matter in an entirely mathematical way, but he made many references to the experiments on model arches known to him.

2.4.5 William Henry Barlow, 1846

The theory of thrust lines immediately became an essential tool to understand the behaviour of arches. Moseley had demonstrated that the study of the thrust lines enabled the range of the thrusts to be defined, between a maximum and a minimum, and enabled the cracks observed in model tests or real arches to be interpreted. However, Moseley's purely mathematical approach was not suitable for use by practising builders, architects or engineers.

William Henry Barlow (1812-1902) published a memoir in 1846 [69] with the aim of making comprehensible Moseley's theory of thrust lines ('lines of resistance') and of verifying the two conditions for the stability of an arch: 1) the line of thrust must be inside the cross-section of arch, i.e. that all the forces are compressive; 2) the inclination of the force should not exceed certain limits, defined by the angle of friction of the materials. He also wanted to stress the importance of a property which he considered crucial: in arches subject only to the force of gravity (vertical forces), the horizontal component of the thrust remains uniform along the arch. To demonstrate the validity of his theory, Barlow used a set of very ingenious models of arches. He thus developed a simple semi-graphical method (some quantities must be calculated analytically) to draw the thrust lines. Barlow then applied his method to draw thrust lines in some typical arch forms, and, hence, determine the limits to their thickness. Finally, he applied his method to the analysis and design of buttresses. He presented his paper and demonstrated his models at a meeting of the Institution of Civil Engineers held on 3rd February 1846, presided over by John Rennie and with Isambard Brunel and Robert Stephenson in the audience; a lively discussion followed.

Barlow's first model was a voussoir arch with the joints made curved so that the contact is always at only one point (Figure 2.28). Barlow correctly noted that: 'If the original form of the arch be such that the line of resistance passes through the points of contact, no motion will arise among the voussoirs, on removing the centre; but if the arch be a segment of a circle, or any other form which does not coincide with the line of resistance, the voussoirs will take up a new position, the curved surfaces of the voussoirs rolling on each other' [70]. If another load is now applied, the arch will move by the voussoirs rotating slightly, and adopt a new position of equilibrium. If, now, the load is removed, the arch will return to its original geometry.

Barlow remarks: 'In this experiment it is obvious, that the pressure must be transmitted through the points of contact' and a line passing through all the points of contact is the line of thrust. Barlow claimed that the model also 'affords

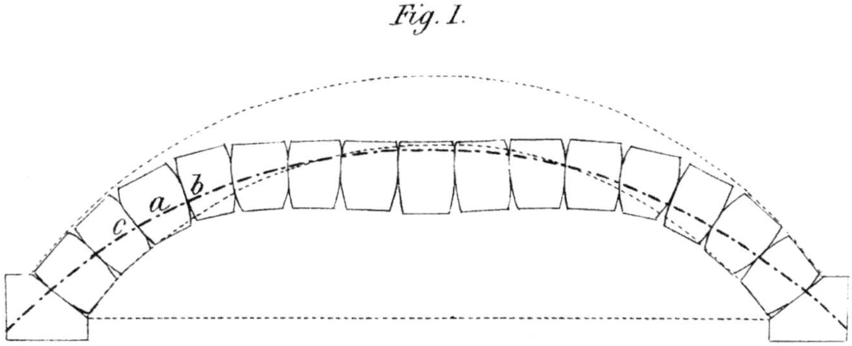

Figure 2.28 Voussoir arch of curved joints. The points of contact allow the thrust line to be seen. (Source: [69])

a practical proof, that this line is the curve of equal horizontal thrust'. The equilibrium of the voussoir marked 'a' in Figure 2.28, is only possible if the forces acting in points c and b have the same horizontal component. Of course, this is correct only for vertical loads. If the same model is subjected to loads slightly inclined to the vertical, the voussoirs will again move again slightly and a new position of equilibrium will be found, but now the horizontal components of the forces will vary along the arch. The model served also to verify the two conditions mentioned above: the thrust line being inside the arch and no sliding between the voussoirs.

Barlow, then, showed a model which demonstrated a crucial fact: if the thickness of the arch is greater than the one which just contains the thrust line, then, 'it is obvious, from the nature of the construction, that more than one such curve will be contained in it, and if the theory advanced is correct, the arch ought to be capable of being supported in any one of these curves' [71]. The arch of Fig.9.a in Figure 2.29a consists of six voussoirs, separated at each joint by four small pieces of wood, each of which could be removed by hand. Barlow draws first, using his graphical method, three possible thrust lines: the minimum thrust *abc*, the maximum thrust *df* and a third, which approaches the middle line of the arch, *ghi*. Then, considering the first *abc*, it is possible to remove the pieces which the line does not cut and equilibrium is maintained. He performs the same operation with the other thrust lines with the same result. Figure 2.29b shows the results or removing the correct pieces for each case (The idea for this drawing was taken from a lecture by Prof. Heyman). However, it could be seen also, that if the pieces removed are so placed that no thrust line could pass through them, then the arch collapses.

Two lines represent the limits of the horizontal thrust: minimum thrust, line *abc*, and maximum thrust, line *df*. The test makes it clear that, between these limits, there is an infinite number of possible lines of thrust, all achieving a different state of equilibrium for the arch. The problem arises (although Barlow did not mention it) as to which one is the 'actual' or 'true' line of thrust of the arch.

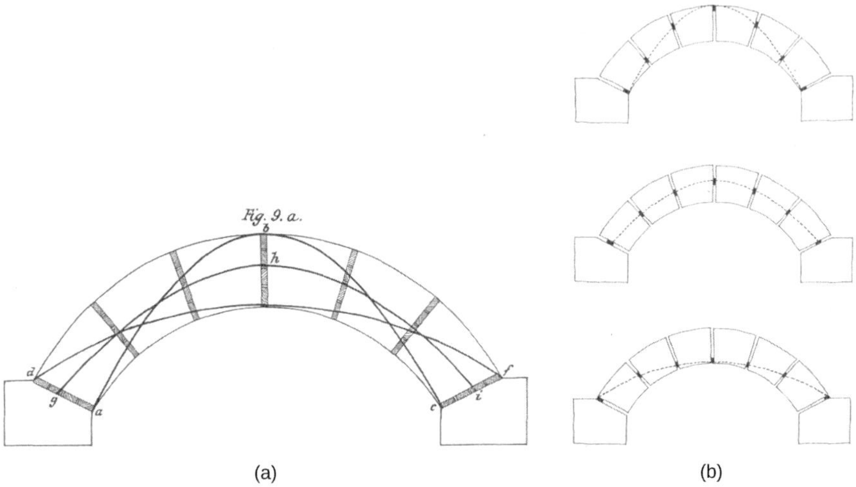

Figure 2.29 (a) Barlow's arch model with small removable pieces of wood forming the joints (Source: [69]). (b) Three lines of thrust made visible by removing pieces of wood in the joints between voussoirs. (Source: [4] p. 51)

2.4.6 Henry Charles Fleeming Jenkin, 1876

Henry Charles Fleeming Jenkin (1833-1885), Professor of Engineering at the University of Edinburgh, was invited by the editors of the Encyclopaedia Britannica to write the article on 'Bridges', for the new (9th) edition [72]. He accepted with the condition that he would 'arrange the matter as to produce a short treatise which might serve as a text-book for one portion of the course of lectures delivered by him in the University of Edinburgh' and it was published in the same year [73]. The article gives a good state-of-the-art review of the theory, design and construction of bridges in the 1870s.

Jenkin discussed how a three-voussoir arch with curved joints would find its state of equilibrium and argued that, since an arch with more voussoirs could be considered as a number of three-voussoir arches, the same mechanism would apply. He demonstrated this with a model arch with eight wooden voussoirs, whose curved joint surfaces were roughened to prevent sliding, to demonstrate how the thrust line would move according to the imposed loads. When a weight was placed at the crown, it descended a little, but the line of thrust, 'the curve passing through the points of contact', rose at the crown and descended at the haunches as a result of rotations of the blocks (Figure 2.30a). If two weights were placed symmetrically on the haunches, the voussoirs rearrange in a different manner. The line of thrust rises slightly at the crown, under the loads it descends, and at the haunches it rises (Figure 2.30b). Finally, if a weight is placed on only one of the haunches, the line of thrust rises at this haunch and descend at the other (Figure 2.30c).

Jenkin commented on the analogy with a hanging model: 'The general character of the curve passing through the points of contact may be easily conceived by thinking of a string similarly loaded and inverted', and he went on to point out that

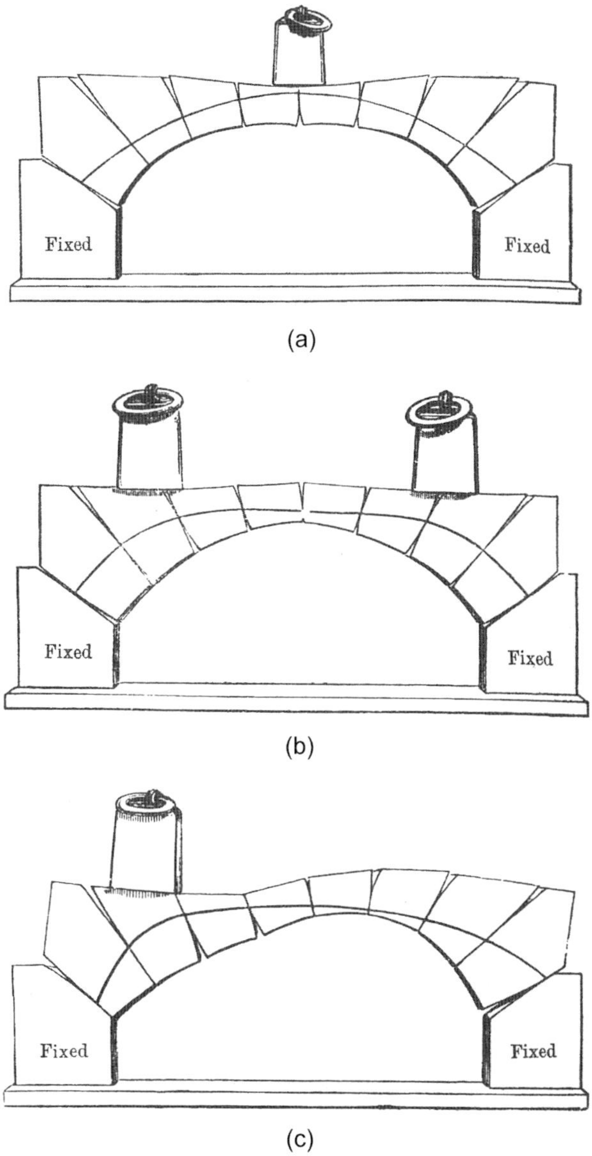

Figure 2.30 Load tests on an arch with voussoirs with curved joints, to show the line of thrust. The figures were drawn by Jenkin from photographs he took of the actual tests). (Source: [72])

'the length of the curve will not be constant in the arch, whereas it must lie constant with any given chain' [74]. Jenkin then compared the behaviour of the arch with curved joints with a real arch bridge with flat bedding-joints. In this case the voussoirs cannot rotate to adapt themselves to the thrusts, but the moments which modify the position of the thrusts arise from unequal compression at the joints. Therefore, 'an equilibrated polygon cutting the joints at various distances from the centre is therefore as correct an indication of the actual forces present in a practical arch with flat joints as in the model with curved joints'. The arch with plane joints 'is in stable equilibrium, and adapts itself to new distributions of load for precisely the same reasons as the model with curved joints', and Jenkin concluded with a crucial statement: 'the incipient yielding of an arch under loads will produce a reaction at the abutments suited to keep the whole ring in equilibrium, provided only an equilibrated polygon can be drawn, cutting the joints within the ring at suitable angles' [75]. This constitutes perhaps the first reasoned exposition of the Safe Theorem of modern Limit Analysis applied to masonry structures.

2.5 Block models in the late nineteenth and twentieth centuries

2.5.1 Modelling the elastic behaviour of masonry arches

Before the 1870s arches were generally divided into 'rigid' arches, made of voussoirs, and continuous 'elastic' arches, made of timber or iron. The way that forces are transmitted from one stone to another was considered as an entirely different problem from the transmission of forces through a continuous elastic arch. Yet, as early as 1852, the French engineer Jean-Victor Poncelet (1788-1867) had suggested that a comprehensive theory of arches must consider the elastic properties of the masonry. He affirmed that 'the theory of vaults and the theory of naturally curved elastic solids constitute only one [theory]' [76] and commented that there was a lack of tests on the elastic properties of masonry.

Tests made to study the elastic behaviour masonry arches had to be made at full size, since the elastic movements are very small, maybe two orders of magnitude smaller than the movements due to the yielding of abutments. Perhaps the first engineer to propose the elastic analysis of masonry arches was the Spanish engineer Eduardo Saavedra (1829-1912). He considered masses of masonry as elastic bodies, and arches as curved, continuous and elastic bars. In 1866 he built and tested a masonry voussoir arch with a span of 19 m and thickness of 25 cm (Figure 2.31). Saavedra measured the movements of the arch and compared them with the results of his detailed elastic calculations based on an arch encastred in the abutments. Saavedra considered the experiment as a conclusive proof of the validity of elastic analysis: 'It is demonstrated that the stone arches obey the laws of elasticity, to the point of being possible to calculate its descent and deformation provided that, by some observations, the coefficients of elasticity that correspond to the different classes and thicknesses of mortar could be fixed' [77]. However, some doubt must be cast on Saavedra's conclusion. One of the abutments suffered a displacement of 37 mm just before the complete decentring and it is not clear how he could have taken this into account; such a movement would have produced enormous changes in the internal forces of an encastred elastic arch.

Around the same time, in 1865-1866, a great '*arche d'essai*' was built and tested in the quarry of Souppes, near Nemours south of Paris [78]. The arch had the same dimensions as a proposed new bridge (which was never built) over the Seine

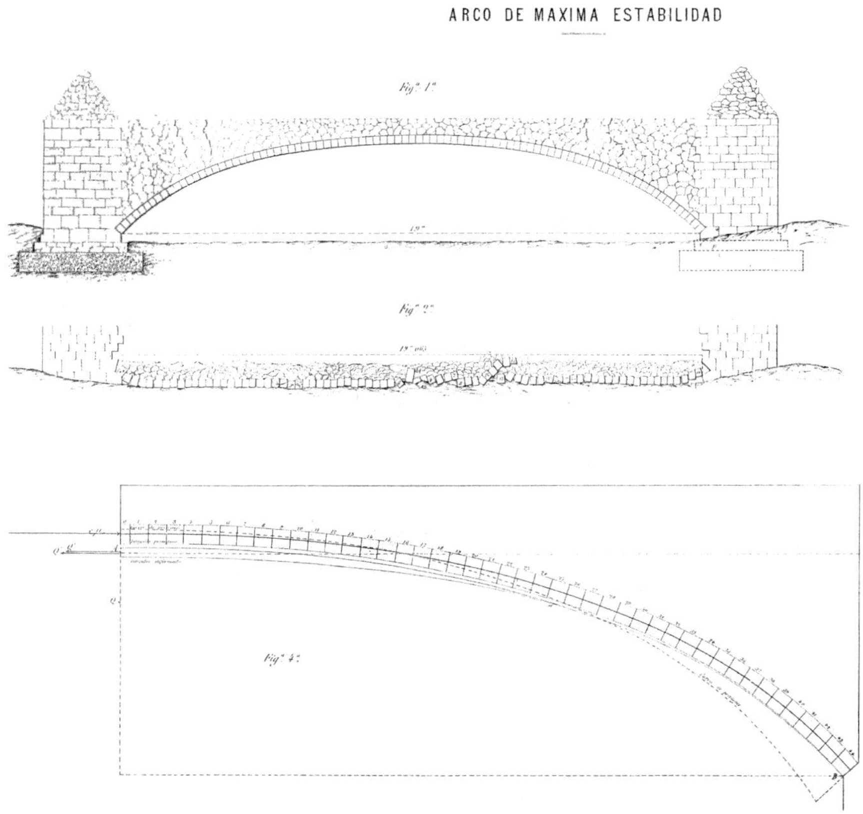

Figure 2.31 The first test of a masonry arch to investigate the application of elastic theory, by Eduardo Saavedra, 1866. (Source: [77])

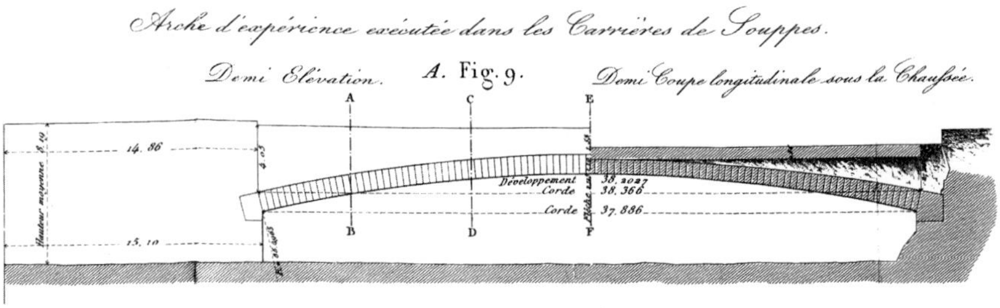

Figure 2.32 Test of a full-size masonry arch in the quarry of Souppes. (Source: [78])

in Paris at the western end of the Île de la cite, parallel to the Pont Neuf. It had a span of 37.88 m and a height of 2.125 m, a rise-to-span ratio of 1/18 (Figure 2.32). The thickness at the keystone had to be just 0.80 m to allow ships to pass beneath. The stones of Souppes had a crushing strength of 40 N/mm^2. Two years later the arch was demolished, but during this time several tests were made on both the vault and the buttresses [79].

As the sophistication of and confidence in elastic analysis grew in the 1870s, especially for iron bridge structures, so engineers tried to apply it to masonry structures. For example, in order to determine precisely where the line of thrust would lie within a masonry arch or vault, it would be necessary to take account of everything that might influence its position. The German engineer Emil Winkler (1835-1888), for example, defined the 'normal' state for the arch as having rigid centring, constant temperature, uniform setting of the mortar and rigid, unmovable, abutments [80]. An elastic analysis of this 'normal' arch would give the actual position of the thrust line in the real arch. This could then be modified by considering perturbations (*Störungen*) to this normal state – a lack of rigidity in the centring, variations in temperature, differences in the setting of the mortar, and the yielding of the abutments. These intellectual contortions to justify what appears unjustifiable, by one of the greatest engineers and professors of his day, shows the enormous pressure that the ideas of elasticity imposed on ways of thinking towards the end of the nineteenth century. This is understandable given the development of elastic theory, a mathematical monument on which the best minds of the time worked and which has the (implicit) promise of eliminating all uncertainties from the structural calculations.

At about this time, Friedrich Engesser (1848-1931), a professor at the engineering school in Karlesruhe, undertook a series of tests on small concrete vaults to test the suitability of the material for such structures. They had a span of 3.5 m and a thickness of about 10 cm, and he used elastic theory to predict their elastic behaviour and collapse loads. The actual collapse loads were much higher than those predicted, leading him to conclude that: 'For the load cases studied the behaviour at collapse is completely different from the behaviour in the elastic range. Besides, the collapse results are more economical than those obtained by the elastic theory' [81]. It appears that this is the first time that such a comparison was made, and would remain unique until the development of plastic analysis in the twentieth century.

Doubts remained among bridge engineers about the appropriateness of elastic analysis to a something so inelastic as a masonry arch. Eventually in the 1890s, the Association of Austrian and Engineers promoted a large series of tests on small-scale and full-size arches to resolve the matter once and for all. In particular, five arches of 23 m span made of stone, brick, mass concrete, reinforced concrete (Monier system) and steel, were tested to collapse. The displacements, both in the elastic and plastic zones, were minutely recorded. Laboratory tests were also made on sample specimens of the five materials. The tests and subsequent analysis were gathered in a 200-page publication with numerous plates [82]. The panel of experts (Auschuss) comprised twenty-one of the most notable Austrian engineers. Even today, these represent the most ambitious set of structural tests ever made on arches.

The results were interpreted as a demonstration of the validity of elastic analysis for masonry arches. This is curious; after an initial elastic behaviour, with very small deformations, the behaviour became manifestly non-elastic, and the collapse loads were much higher than those calculated as the limit of the elastic behaviour. However, the tests were made to confirm the validity of elastic analysis, and the data concerning the non-elastic behaviour was ignored.

Although the panel had in their hands data suitable to formulate a plastic theory of arches, the frame of mind was limited to elastic theory and, indeed, the tests were used to validate the use of elastic analysis. The episode is further evidence for the need to have a theory with which to interpret test results. In any case, the Austrian tests were considered incontrovertible proof of the validity of elastic analysis of masonry arches, and all later contributions to this subject aimed only to simplify the elastic analysis.

2.5.2 Alfred Pippard, 1936-1938

There was a renewed interest in the behaviour of masonry bridges in the 1930s, due to the increase in loads from road traffic. Old bridges had to be checked and the earlier doubts about the usefulness of elastic analysis of masonry arches were aroused again. By then, the old theories of arches, developed before the era of elastic analysis, had been completely forgotten. Responding to the need to assess existing bridge structures in 1936, Alfred Pippard (1891-1969) carried out a series of tests at Imperial College, London on models of voussoir arches [83]. The first model arch consisted of fifteen steel voussoirs, and had a span of 4 feet (1.22 m) and rise of 1 foot (30.5 cm) (Figure 2.33). The voussoirs were 3 inches (7.5 cm) deep, i.e. 1/16 of the span, with a width of 1.5 inches (3.8 cm). The arch could have either pinned ends or a skew-back in the abutments. The right-hand support could slide horizontally and the tie included an extensometer to measure the horizontal thrust of the arch.

The results of the model tests were compared with others made on a continuous steel arch of the same dimensions and, also, with calculations using elastic theory and it was noted that the line of thrust should fall everywhere within

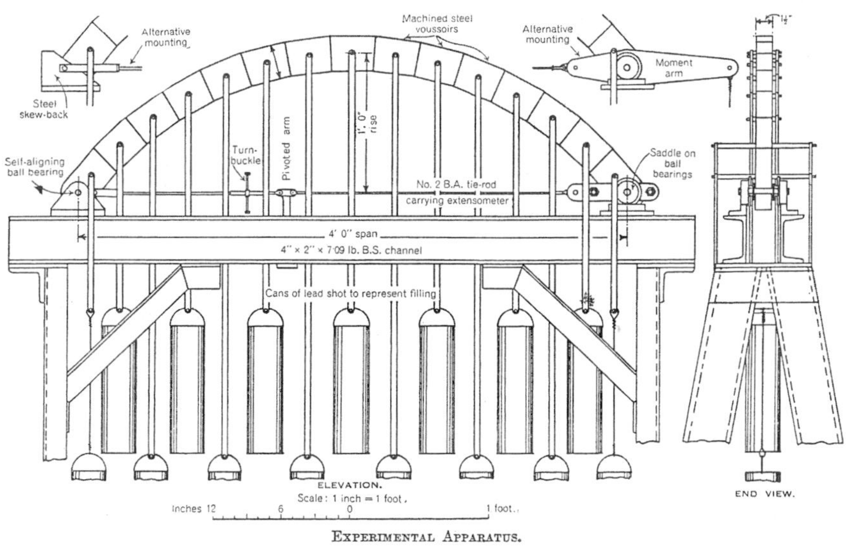

Figure 2.33 Voussoir arch model made with steel voussoirs and a span of 1.2 m. (Source: [83])

the middle-third of the section. However, being statically indeterminate, the structure could only be analyzed by considering its elastic behaviour. The theory of Castigliano was considered more trustworthy than simple elastic analysis, as it considers the reduction of the section of the arch when the thrust line passes outside the middle third [84]. The authors commented that 'the process is lengthy but rational, and constitutes the first scientific approach to the question' [85]. However, they also noted that Castigliano assumed rigid abutments, so that their span must remain invariable. They considered this assumption 'very doubtful'. The tests recorded the opening of joints between the voussoirs when the span increased or decreased. Collapse load tests were made and it was discovered – or rather re-discovered – that this occurs when a sufficient number of joints open to form an unstable mechanism. No relationship was established between the opening of joints and possible lines of thrusts.

The second series of experiments were made two years later, on a larger model with a span of 10 feet (3.05 m) and similar proportions [86] (Figure 2.34). The apparatus was also similar, but this time the rigid springings were on concrete skew-backs. Concrete voussoirs, made with either limestone or granite aggregate were used to reduce cost. The voussoirs were mounted on a steel centring with thin mortar joints (both lime mortar and Portland cement mortar were used). As the point load on a certain voussoir was increased, the appearance of cracks was noted at every step. The test continued until collapse occurred by formation of the fourth hinge, but complete collapse was prevented by the centring beneath. On removing the point load the arch recovered its original geometry. Twenty-six separate tests were made.

In this second series of tests the opening of the joints (formation of hinges) was related to the calculated thrust lines due to each point load (Figure 2.35). The middle-third of the arch is drawn to relate the appearance of the cracks to the paths of the lines of thrust. It became apparent that, in many cases, no opening of joints were observed even though the line of thrust passed outside the middle-third. The mechanism of collapse was correctly identified and a simple trial method was developed to calculate the position of the critical cracks.

The authors concluded that in an uncracked arch, with the thrust line within the middle-third, elastic analysis was the correct tool to calculate the thrust of the arch and the internal stresses. They claimed that the tensile strength and adhesion of a good mortar will prevent the opening of the cracks when the line of thrust passes outside the middle third. They suggested that the middle-half of the section could be considered a safe region for the line of thrust, and emphasized that the opening of a single joint does not imply that the arch was unsafe. In these tests they ignored the possible movements of the abutments, although they mentioned that any small movement would have a dramatic effect on the internal state of the arch. The elastic lines of thrust for the first crack (obtained using the energy method) were drawn (Figure 2.35a). Eventually the same mechanism of collapse was identified that had already been described in the eighteenth century (Figure 2.35b) and a trial and error procedure was developed for obtaining the value of the collapse load.

Pippard included a summary of these experiments in a chapter on 'The voussoir arch' in the second edition of his book *The Analysis of Engineering Structures* [87].

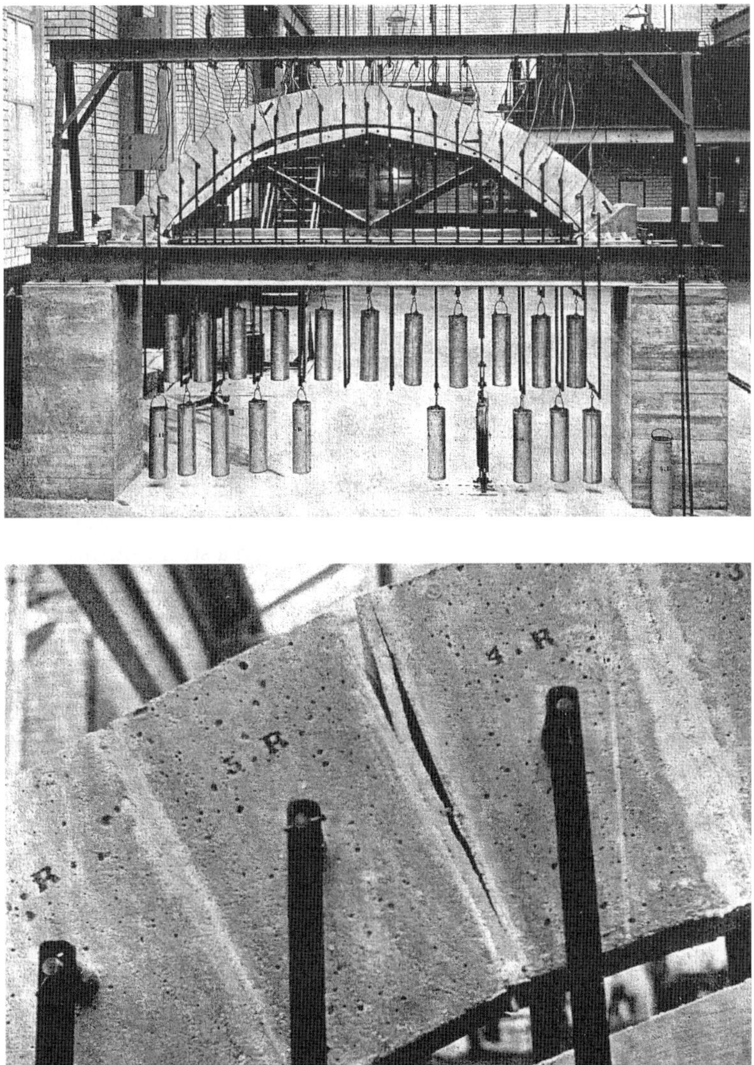

Figure 2.34 Model arch of concrete voussoirs of 3.05 m span. (a) Arch resting on centring after collapse; (b) detail of a hinge ('pinning') between two voussoirs. (Source: [86])

A historical note was also included which reviewed previous theories of the arch and related experiments. Castigliano's approach was considered preferable. Later, Pippard described a simple apparatus to test small arches of 1.5 foot (45 cm) span and similar proportions in his book on *The Experimental Study of Structures* [88], and proposed an elastic check on existing bridges based on a two-hinged parabolic arch. Finally, responding to a particular need after the war, Pippard proposed an elastic method to estimate the safe loads that existing arch bridges could carry [89].

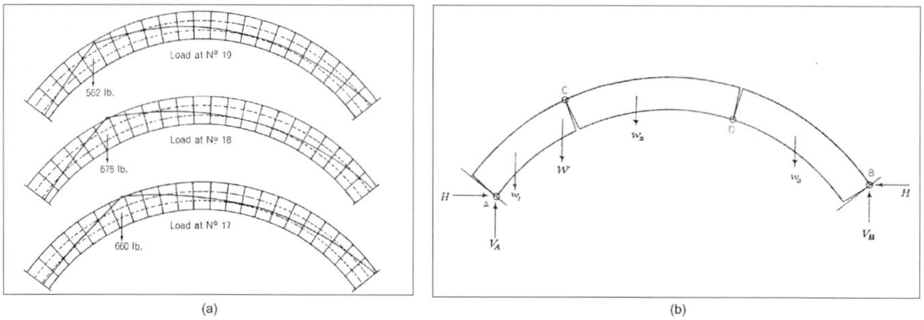

Figure 2.35 Series of tests on an arch of voussoirs of limestone concrete are represented. (a) Lines of thrust for first cracking; (b) Mechanism of collapse. (Source: [86])

2.5.3 Epilogue – the 'plastic' theory of masonry structures

Since the 1960s there has been a revolution in understanding the behaviour of masonry structures; and curiously, perhaps, it evolved out of another revolution that occurred in relation to understanding the behaviour of steel structures. Faced with the irrational and manifestly incorrect assumption that the strength of steel structures depends on elastic behaviour criteria of steel, the plastic theory of steel frames was developed in the 1950s and 1960s. The first book presenting this work and gathering all the experimental evidence on the subject, was published in 1956 [90]. Plastic theory is founded on the observation that collapse occurs only when a mechanism is created by the formation of 'hinges' in which all the steel at a cross section has been stressed beyond its elastic limit. Once sufficient hinges have formed to create a mechanism, a structure can collapse. Crucially, a minimum load can be identified that would create sufficient hinges to cause collapse – the Safe Theorem' of modern plastic theory or Limit Analysis.

Having helped develop plastic theory for steel structures in the 1950s and 1960s, Jacques Heyman (b.1925) turned his interest to masonry structures, stimulated in part by the experimental work of Pippard in the 1930s and 1940s. Being aware that the elastic analysis of masonry structures was unable to predict collapse loads, he saw the parallel with steel structures and set about adapting the plastic theory developed for steel, for use in masonry structures. Effectively, Heyman saw the collapse mechanisms of steel structures as a model for the collapse of masonry structures. To make this intellectual step, it was only necessary to make three assumptions about masonry structures: the material should have infinite compressive strength, and zero tensile strength (i.e. no cohesion between stones), and there should be no sliding at the joints between stones. This allows failure mechanisms to be identified, and the condition for a hinge to form is simply that the notional line of thrust should not pass outside the cross section of the structure (a column, buttress, arch or vault). This observation mirrors the findings of the earlier experimenters who developed the use of block models, from Leonardo to Pippard. Heyman published his seminal paper in 1966, formulating the theory of the masonry arch within the broader framework of modern limit state analysis [91].

The possibility of cracks opening up to form hinges is what makes masonry structures 'plastic' – as long as fewer hinges form than would allow collapse, a masonry structure is able to adapt to any small variation of the boundary conditions, a phenomenon known to all conservers of masonry structures which are demonstrably able tolerate the opening of joints, and even the formation of cracks, without the overall safety of the structure being threatened. It is thus impossible to predict, or to 'know', the precise internal conditions of a masonry structure, and it can be shown that there is thus no 'true' or 'actual' line of thrust. As long as one line of thrust can be drawn within the masonry of a structure, then it is safe. Heyman went on to address the safety of masonry arches [92, 93] and to develop a method for assessing the strength of an existing masonry structure [94].

Given the long traditions of masonry construction, it is not surprising that developing the plastic theory of masonry structures took Heyman back to the days before the explosion of elastic theory in the late nineteenth century that had clearly shown its inadequacy in dealing with masonry structures. He discovered the wealth of historical experimentation and theoretical work that have been summarised in this chapter, and committed many of these discoveries to print [95, 96, 97, 98].

Ironically, the 'old theory' is now seen to be essentially correct (could it have been otherwise when it was used with success during more than a century?). The use of elastic theory to predict the behaviour of masonry structures is now discredited, yet it is still sustained by computer packages designed to obtain a single solution. No matter the complexity or sophistication of the program (linear elastic, non-linear elastic plastic, of rigid blocks, etc.) the structural analyst with a deep understanding of modern 'plastic' masonry theory will look with great scepticism at any analysis which claims to have identified a unique solution for the safe state of real masonry structure [99].

This chapter has presented an overview of using block models of arches to develop an understanding of arch behaviour, and to develop new and better ways of designing arches and vaults. It has also shown that experimenting without an underlying theoretical model is useless, and that the ideas in a researcher's mind could make him or her blind to the most obvious observations – of course, stones do not slide; of course, the hanging chain moves by the action of point loads; of course, it is impossible to predict or observe the actual behaviour of a real structure; of course, masonry is essentially discontinuous and non-elastic. History, in this case, has shown how blind we can be to what is evident and how difficult it can be to change an acquired frame of mind.

References

1 Heyman, Jacques (1982) *The Masonry Arch*. Chichester: Ellis Horwood.
2 Besenval, Roland (1984) *Technologie de la voûte dans l'Orient Ancien*. Paris: Editions Recherche sur les Civilisations.
3 Heisel, J. P. (1993) *Antike Bauzeichnungen*. Darmstadt: Wissenschaftliche Buchgesellschaft.

4 Huerta, S. (2004) *Arcos, bóvedas y cúpulas. Geometría y equilibrio en el cálculo tradicional de estructuras de fábrica.* Madrid: Instituto Juan de Herrera. 154.

5 Derand, François (1643) *L'architecture des voûtes ou l'art des traits et coupe des voûtes.* Paris: Sebastian Cramoisy.

6 Huerta, S. (2004) *Arcos, bóvedas y cúpulas. Geometría y equilibrio en el cálculo tradicional de estructuras de fábrica.* Madrid: Instituto Juan de Herrera. 142.

7 Cassinello, F. (1964) *Arcos de ladrillo,* Madrid: Consejo Superior de Investigaciones scientificas.

8 Frankl, Paul (1960) *The Gothic: Literary sources and interpretations through eight centuries.* Princeton: Princeton University Press. 299.

9 Heyman, Jacques (1995) *The Stone Skeleton: Structural Engineering of Masonry Architecture.* Cambridge: University Press. 26.

10 Baldinucci, Filippo (1812) *Vita di Filippo de ser Brunellescho.* Firenze: Niccolò Carli. 333.

11 Prager, Frank D. & Scaglia, Gustina (1970) *Brunelleschi: Studies of his Technology and Inventions.* Mass: MIT Press. 30-31.

12 Saalman, Howard (1980) *Filippo Brunelleschi: The Cupola of. Santa Maria del Fiore.* London: Zwemmer. 62.

13 Frankl, Paul (1960) *The Gothic: Literary sources and interpretations through eight centuries.* Princeton: Princeton University Press. 263.

14 Parsons, Barclay (1939) *Engineers and engineering in the Renaissance.* Baltimore: Williams & Wilkins. 71.

15 Truesdell, Clifford A. (1968) The mechanics of Leonardo da Vinci. In: *Essays in the History of Mechanics,* 1-82. Berlin: Springer.

16 Richter, Jean Paul (1883) *The literary works of Leonardo da Vinci* (2 Vols). London: Samson Low, Marston, Searle and Rivington. Vol.2, 86-94 (§779-788).

17 Leonardo da Vinci, Madrid Codex, Biblioteca Nacional de España, Mss. 8937. fol. 134v.

18 Heyman, Jacques (1995) *The Stone Skeleton: Structural Engineering of Masonry Architecture.* Cambridge: University Press.

19 Leonardo da Vinci. Madrid Codex, Biblioteca Nacional de España, Biblioteca Nacional de España, Mss. 8937. fols. 142v, 143r.

20 Campbell, James (2012) The First Complete List of All the Models Made for the Construction of St Paul's Cathedral, London 1675-1720. In: *Nuts and Bolts of Construction History* (eds. R. Carvais, A. Guillerme, V. Negre, & J. Sakarovitch) Vol.1, 173-183. Paris: Picard.

21 Gregory, David (1697) Catenaria. *Philosophical Transactions* (231) 637-652.

22 Gautier, Hubert (1716) *Traité des Ponts. Disstertation sur l'épaisseur des ponts* … Paris: Cailleau.

23 Gauthey, Emiland-Marie (1798) *Dissertation sur les dégradations survenues aux piliers du dome de Panthéon Français, et sur les moyens d'y remedier.* Paris: Perronneau Imprimeur.

24 Heyman, Jacques (1972) *Coulomb's Memoir on Statics: An Essay in the History of Civil Engineering.* Cambridge: Cambridge University Press. 124, 129.

25 Frézier, Amedée François (1737-39) *La théorie et la pratique de la coupe des pierres et des bois ...* Strassbourg/Paris: Charles-Antoine Jombert.

26 La Hire, Philippe de (1712) Sur la construction des voûtes dans les edifices. *Mémoire de l'Académie Royale des Science de Paris.* 70-78.

27 Danyzy, Augustin (1732) Méthode générale pour déterminer la résistance qu'il faut opposer à la poussée des voûtes. In: *Histoire de la Société Royale des Sciences: établie à Montpellier* (1788, 2: 40-53). Lyon. 43.

28 Bruyère, Louis (1823) Ponts en pierre. In: *Études relatives à l'art des constructions* (2 Vols). Paris: Bance Ainé. 1-16, Plates I-XVI. 10-12.

29 Huerta, S. (2015) Designing by "Expérience": Lecreulx models tests for the design of the abutments for the Bridge of Fouchard. In: *Masonry structures: Between mechanics and architecture* (ed. Danila Aita, Orietta Pedemonte & Kim Williams), 21-47. Basel: Birkhäuser.

30 Cresy, Edward (1847) *Encyclopaedia of Civil Engineering* (2 Vols). London: Longman. Vol II, 1499.

31 Boistard, Louis-Charles (1822) *Recueil d'expériences et d'observations faites sur différens travaux exécutés pour la construction du pont de Nemours ...* Paris: Fuegueray. Expériences sur la stabilité des voûtes: 95-134.

32 Boistard, Louis-Charles (1810) Expériences sur la stabilité des voûtes. In: *Recueil de divers memoires extraits de la Bibliothèque Impériale des Ponts et Chaussées, à l'usage de MM. les Ingénieurs* (ed. P.C. Lesage), Vol.2, 171-217, Pl. XI-XVI. Paris: Didot.

33 Couplet, P. (1730) Seconde partie de l'examen de la pousée des voûtes. *Mémoires de l'Académie Royale des Sciences Paris.* 117-141, Plates 6-7.

34 Gauthey, Emiland-Marie (1798) *Dissertation sur les dégradations survenues aux piliers du dome de Panthéon Français, et sur les moyens d'y remedier.* Paris: Perronneau Imprimeur.

35 Patte, Pierre (1770) *Memoire sur la construction de la coupole projectée pour couvronner la nouvelle église de Saint Geneniève a Paris...* Amsterdam: Gueffier.

36 Gauthey, Emiland-Marie (1771) *Mémoire sur lápplication des principes de la Mèchanique à la construction des voûtes et des domes, dans lequel on examine le Problême proposé par M. Patte, relativement à la construction de la Coupole de l'Eglise Saint-Genenieve de Paris.* Dijon: Louis Nicolas Frantin.

37 Gauthey, Emiland-Marie (1798) *Dissertation sur les dégradations survenues aux piliers du dome de Panthéon Français, et sur les moyens d'y remedier.* Paris: Perronneau Imprimeur.

38 Gauthey, Emiland-Marie (1798) *Dissertation sur les dégradations survenues aux piliers du dome de Panthéon Français, et sur les moyens d'y remedier.* Paris: Perronneau Imprimeur. 111.

39 Gauthey, Emiland-Marie (1809-1813) *Traité de la construction des ponts... Publié par M. Navier.* Paris: Firmin Didot. Vol.I, 313.

40 Rondelet, J. (1814) *Traité théorique et pratique de l'art de bâtir.* 2nd ed. Vol. III. Paris: Chez l'auteur.

41 Rondelet, J. (1814) *Traité théorique et pratique de l'art de bâtir.* 2nd ed. Vol. III. Paris: Chez l'auteur. 260.

42 Rondelet, J. (1814) *Traité théorique et pratique de l'art de bâtir.* 2nd ed. Vol. III. Paris: Chez l'auteur. 258.
43 Rondelet, J. (1814) *Traité théorique et pratique de l'art de bâtir.* 2nd ed. Vol. III. Paris: Chez l'auteur. 270.
44 Rondelet, J. (1797) *Mémoire historique sur le dôme du Panthéon Français.* Paris: Chez Du Pont. 59-60.
45 Heyman, Jacques (1995) *The Stone Skeleton: Structural Engineering of Masonry Architecture.* Cambridge: University Press. 43.
46 Gauthey, Emiland-Marie (1798) *Dissertation sur les dégradations survenues aux piliers du dome de Panthéon Français, et sur les moyens d'y remedier.* Paris: Perronneau Imprimeur. 109-110.
47 Rondelet, J. (1814) *Traité théorique et pratique de l'art de bâtir.* 2nd ed. Vol. III. Paris: Chez l'auteur. 339 ff.
48 Vicat, L.-J. (1832) Influence du mode d'attache des chaines sur la résistance des ponts suspendus. *Annales des Ponts et Chaussées.* 394-397, Plate 26. 394.
49 Méry, E. [Édouard-Henry-François] (1840) Mémoire sur l'equilibre des voûtes en berceau. *Annales des Ponts et Chaussées.* 50-70, Plates 133-134.
50 Méry, E. (1862) Mémoire sur l'équilibre des diverses voûtes projetées pour l'hôpital Clermont-Tonnerre, à Brest. *Mémorial des Travaux Hydrauliques de la Marine.* 73-116, Plates 10-14.
51 Foce, Federico (2010) A pearl from the history of structural mechanics. Méry, Saint-Venant and the théorie des voûtes. In: *Mechanics and Architecture: Between Epistéme and Téchne* (ed. A. Sinopoli), 125-154. Roma: Edizioni di Storia e Letteratura.
52 Robison, John (1803) Article on 'Arch'. In: *Supplement to the third edition of the Encyclopaedia Britannica* (2 Vols) (ed. George Gleig), Vol.1, 14-30, Plates 1-4. Edinburgh: Thomas Bonar.
53 Robison, John (1803) Article on 'Arch'. In: *Supplement to the third edition of the Encyclopaedia Britannica* (2 Vols) (ed. George Gleig), Vol.1, 24. Edinburgh: Thomas Bonar.
54 Robison, John (1803) Article on 'Arch'. In: *Supplement to the third edition of the Encyclopaedia Britannica* (2 Vols) (ed. George Gleig), Vol.1, 25. Edinburgh: Thomas Bonar.
55 Robison, John (1803) Article on 'Arch'. In: *Supplement to the third edition of the Encyclopaedia Britannica* (2 Vols) (ed. George Gleig), Vol.1, 28. Edinburgh: Thomas Bonar.
56 Robison, John (1803) Article on 'Arch'. In: *Supplement to the third edition of the Encyclopaedia Britannica* (2 Vols) (ed. George Gleig), Vol.1, 30. Edinburgh: Thomas Bonar.
57 Young, Thomas (1824) Article on 'Bridge'. In: *Supplement to the fourth, fifth and sixth editions of the Encyclopaedia Britannica.* Vol. 2: 497-520, Plates 42-44. (Written in 1816). Edinburgh: Archibald Constable.
58 Peacock, George. (1855) *Life of Thomas Young.* London: John Murray. 442.
59 REPORT (1801) from Select Committee upon further Measures for Improvement of The Port of LONDON. In: *Reports from the Committees of the House of Commons, Vol.XIV (1803),* 604-635. London: HMSO. 604-605.

60 Atwood, George (1804) *A Supplement to a tract, entitled 'A treatise on the construction and properties of arches', Published in the year 1801: Part II.* London: Bulmer. Plate 4.
61 Anon. (1804) A Dissertation on the Construction and Properties of Arches. By G. Atwood. *British Critic* 23: 6-14.
62 Young, Thomas. (1807) On Architecture and carpentry. In: *A Course of Lectures on Natural Philosophy and the Mechanical Arts* (2 Vols). 156-171, Plates XI-XIII. London: Joseph Johnson. 161. (Reprint: Bristol, Thoemmes Press, 2002.)
63 Huerta, S. (2005) Thomas Young's theory of the arch: His analysis of Telford's design for an iron arch of 600 feet span. In: *Essays in the history of the theory of structures, in honour of Jacques Heyman* (ed. S. Huerta), 189-233. Madrid: Instituto Juan de Herrera, CEHOPU.
64 Young, Thomas (1824) Article on 'Bridge'. In: *Supplement to the fourth, fifth and sixth editions of the Encyclopaedia Britannica.* Vol. 2: 497-520, Plates 42-44. (Written in 1816). Edinburgh: Archibald Constable. 513.
65 Euler, Leonhard (1776) Regula facilis pro dijudicanda firmitate pontis aliusve corporis similis ex cognita firmitate moduli. *Novi commentarii academiae scientiarum Petropolitanae* 20: 271-85, Tab. II.
66 Young, Thomas (1824) Article on 'Bridge'. In: *Supplement to the fourth, fifth and sixth editions of the Encyclopaedia Britannica.* Vol. 2: 497-520, plates. 42-44. (Written in 1816). Edinburgh: Archibald Constable. 514.
67 Bland, William (1839) *Experimental essays on the principles of construction in arches, piers and buttresses.* London: Weale.
68 Pratt, J.H. (1836) *The mathematical principles of mechanical philosophy.* Cambridge: University Press.
69 Barlow, William Henry (1846) On the existence (practically) of the line of equal Horizontal Thrust in Arches, and the mode of determining it by Geometrical Construction. *Minutes and Proceedings of the Institution of Civil Engineers* 5: 162-172. Discussion: 172-182.
70 Barlow, William Henry (1846) On the existence (practically) of the line of equal Horizontal Thrust in Arches, and the mode of determining it by Geometrical Construction. *Minutes and Proceedings of the Institution of Civil Engineers* 5: 162-182. 164.
71 Barlow, William Henry (1846) On the existence (practically) of the line of equal Horizontal Thrust in Arches, and the mode of determining it by Geometrical Construction. *Minutes and Proceedings of the Institution of Civil Engineers* 5: 162-172. 168.
72 Fleeming-Jenkin, H.C. (1876) Article on Bridges. In: *Encyclopaedia Britannica.* 9th ed. (Vol.2: 284-341) Edinburgh: Adam and Charles Black.
73 Fleeming-Jenkin, H.C. (1876) *Bridges, an elementary treatise of their construction and history.* Edinburgh: Adam and Charles Black.
74 Fleeming-Jenkin, H.C. (1876) Article on Bridges. In: *Encyclopaedia Britannica.* 9th ed. (Vol.2: 284-341) Edinburgh: Adam and Charles Black. 308.

75 Fleeming-Jenkin, H.C. (1876) Article on Bridges. In: *Encyclopaedia Britannica*. 9th ed. (Vol.2: 284-341) Edinburgh: Adam and Charles Black. 309.

76 Poncelet, J.V. (1852) Examen critique et historique des principales théories ou solutions concernant l'équilibre des voûtes. *Comptes Rendus de l'Académie des Sciences* 35: 494-502, 532-540 & 577-587. 586.

77 Saavedra, Eduardo (1866) Experimento sobre los arcos de maxima estabilidad. *Revista de obras públicas* 14: 13-21.

78 Féline-Romany, E.J. (1866) Notice sur la construction d'une arche d'essai très-surbaissée. *Annales des ponts et chaussées.* 12 (129): 10-17.

79 De Lagrené, Henri Melchior (1868) Note sur l'arche d'essai des carrières de Souppes. *Annales des ponts et chaussées* 16 (191): 130-146.

80 Winkler, Emil (1879/1880) Die Lage der Stützlinie im Gewölbe. *Deutsche Bauzeitung* 13: 117-119, 127-130 & 14: 58-60.

81 Engesser, Friedrich (1880) Über die Lage der Stützlinie in Gewölben. *Deutsche Bauzeitung* 14: 184-186, 210, 243.

82 Oesterreichische Ingenieur- und Architekten-Vereines (1895) *Bericht des Gewölbe-Auschusses.* Wien: Im Selbstverlage des Vereines.

83 Pippard, A.J.S., Tranter, E. & Chitty, L. (1936/1937) The mechanics of the voüssoir arch. *Journal of the Institution of Civil Engineers* 4: 281-306 and 6:4-26 (discussion).

84 Castigliano, C. A. P. (1919) *The theory of equilibrium of elastic systems and its applications.* Reprint (1966). New York: Dover.

85 Pippard, A.J.S., Tranter, E. & Chitty, L. (1936/1937) The mechanics of the voussoir arch. *Journal of the Institution of Civil Engineers* 4: 281-306. 283.

86 Pippard, A.J.S., Ashby, R.J. (1938) An experimental study of the voussoir arch. *Journal of the Institution of Civil Engineers* 10: 383-404.

87 Pippard, A.J.S. & Baker J.F. (1943) *The analysis of engineering structures.* 2nd Ed. London: Edward Arnold.

88 Pippard, A.J.S. (1947) *The Experimental Study of Structures.* London: Edward Arnold.

89 Pippard, A.J.S. (1948) The Approximate Estimation of Safe Loads on Masonry Bridges. In: *The civil engineer in war: A symposium of papers on war-time engineering problems,* 365-372. London: Institution of Civil Engineers.

90 Baker, J.F., Horne, M.R. & Heyman, J. (1956) *The steel skeleton. Vol. 2: Plastic behaviour and design.* Cambridge: University Press.

91 Heyman, Jacques (1966) The Stone Skeleton. *International Journal of Solids and Structures* 2: 249-279.

92 Heyman, Jacques (1969) The Safety of Masonry Arches. *International Journal of Mechanical Sciences* 11: 363-385.

93 Heyman, Jacques (1982) *The Masonry Arch.* Chichester: Ellis Horwood.

94 Heyman, Jacques (1980) The estimation of the strength of masonry arches. *Proceedings of the Institution of Civil Engineers* 69 Part 2: 921-937.

95 Heyman, Jacques (1972) *Coulomb's Memoir on Statics: An Essay in the History of Civil Engineering.* Cambridge: Cambridge University Press.

96 Heyman, Jacques (1998) *Structural analysis: a historical approach.* Cambridge: Cambridge University Press.

97 Heyman, Jacques (1996) *Arches, Vaults and Buttresses: Masonry Structures and their Engineering*. Aldershot: Variorum.
98 Heyman, Jacques (2016) *Geometry and Mechanics of Historic Structures.* Madrid: Instituto Juan de Herrera.
99 Huerta, S. (2004) *Arcos, bóvedas y cúpulas. Geometría y equilibrio en el cálculo tradicional de estructuras de fábrica*. Madrid: Instituto Juan de Herrera.

3

The catenary and the line of thrust as a means for shaping arches and vaults

Rainer Graefe

> *'To draw a catenary with the help of calculation is one of the most annoying, and to generate it manually, one of the most pleasant and easy tasks to perform'* (Joh. E. Silberschlag, 1772 [1]).

3.1 Introduction

The chain or cord, hanging from both ends, is a construction that automatically takes up the correct shape under its own weight; this curve is called the *catenary*. Many constructions in both nature and technology follow this principle, which is so familiar to all of us that we hardly notice it (Figure 3.1). We encounter it in nature, for example, in the thread constructions of spiders and insects, in hanging lianas in tropical forests, and even in the spine of the sloth' [2]. In technology, we find it in power lines, cable cars and suspension bridges.

The theme of this essay is an aspect of construction whose neglect Eduardo Torroja already complained about: 'The nineteenth and twentieth centuries have produced such astounding mechanical advances in the structural field that ontological study of resistant morphology has been overshadowed and bypassed' [3]. In the historical developments we address here, *form* is of central importance as a constructive element: we follow the centuries of efforts in theory and practice resulting in constructively conclusive vault forms and, in parallel, the emergence of a new type of architectural model. The suspended model makes it possible, for the first time, to determine statically correct building forms and it has another significant novelty: the shapes are created in a forming process over which the designer has only limited influence. We will be travelling down a side road in building history where, however, we will meet many well-known buildings.

3.1.1 Robert Hooke's theory

Most of the vaults whose constructional boldness we still admire today, had already long-since been constructed when the question of the statically-best-possible shape for a vault was first raised. At a meeting of the Royal Society

Physical Models: Their historical and current use in civil and building engineering design,
First Edition. Edited by Bill Addis.

Figure 3.1 The catenary. (Image: Rainer Graefe)

in London in 1670, Robert Hooke (1635-1703) put this question for the first time, and answered it. The minutes of the meeting made a clear note that this was the first time that this question has been asked: Hooke 'brought in this problem in architecture – the basis of the pillars and the altitude of the arch being given, to find out the right figure of this arch' [4]. Hooke's answer was that brick arches are correctly formed when their curve follows the catenary (Figure 3.2). Just as only tensile forces occur in the equilibrium figure of a chain suspended at both

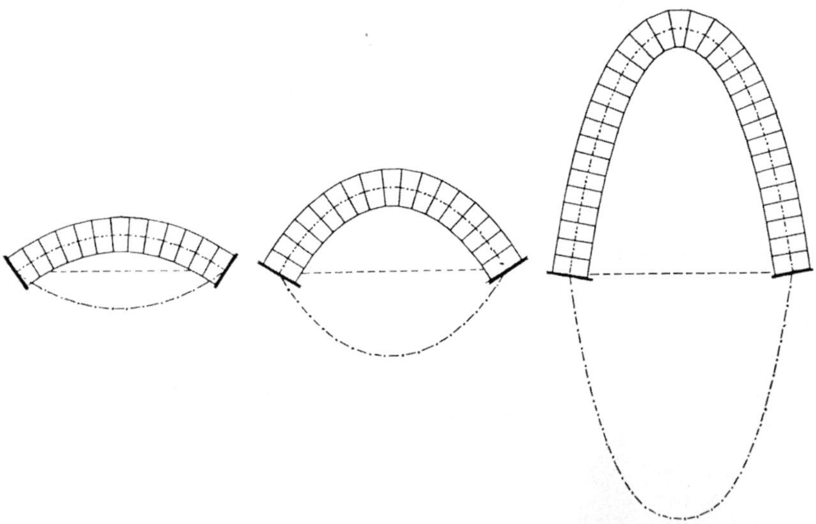

Figure 3.2 Robert Hooke: *ut pendet continuum flexile, sic stabit contiguum rigidum inversum* (As hangs the flexible line, so but inverted will stand the rigid arch). Drawing by Frei Otto of brick arches in the form of inverted catenaries. (Image: Courtesy of the Institute of Lightweight Structures, Stuttgart)

ends, only axial compressive forces occur in arcs following the inverted – upside down – catenary. By the trick of reversing the suspension form, arch forms are created in which no bending forces occur under their own weight.

3.1.2 Early arch forms

Although Hooke's principle of generating forms was recognised only later, arches and vaults whose shape largely corresponded to the inverted catenary had been built very early on. Indeed, perhaps they are the oldest vaults of all. However, they were not, of course, created by inverting a catenary. Rather, in the course of time these arch forms evolved themselves quasi independently, in an empirical process of formation that could almost be compared to wavering around the catenary shape [5, 6]. In the case of barrel vaults, for example, this applies to the ancient-oriental brick vaulting technique used in the thirteenth century BC for the construction of the warehouses beside the tomb of Ramses II. The vaults, made of air-dried bricks, were built without centring as inclined arches arranged one next to the other, with the first arches built leaning against a transverse end wall. With this type of construction, the vault was in a critical state during its formation, in which its stability turned out to be a test of strength, which must have had a regulating effect on the design: failure and success automatically led to the most favourable form of the vault [7, 8]. This form of construction reached its climax almost two millennia later at the Sassanid Royal Palace at Ctesiphon, whose imposing barrel vault spans an incredible 25.80 metres (Figure 3.3a). Even today, Nubian craftsmen perform vaults in the same way [9] (Figure 3.3b). From equally early times, by the same learning processes, it is likely that correspondingly shaped clay domes were created, as they are still built in numerous variants in countries with a tradition of clay construction techniques. And, since domes span in two directions – the meridian and radial directions – there is a wider range of curved shapes in which only compressive forces occur.

3.1.3 Circular and spherical shapes

In contrast to these early vault forms, derived from construction considerations, there were also, since Roman antiquity, vaults determined by other formal

(a) (b)

Figure 3.3 (a) Sassanid Royal Palace at Ctesiphon, Iraq. Photograph from 1864. (Image: [10]) (b) Nubian craftsmen building barrel vaults. (Source: [9])

considerations. Since that time, vaults have been formed almost exclusively using circles, segments of circles, and parts of spheres. Whatever the reasons for this, it is clear that statical considerations were not the first priority [11, 12]. Such geometrically-derived curves often deviate from the statically-correct curve to such an extent that they threaten to bulge or crack in certain areas. In such cases the danger was avoided, not by deviating from the geometric form, but by different kinds of corrections: loads were redistributed, for example, by placing additional material on the lower regions of the vault and constructing the upper regions using a particularly light construction (e.g. the Pantheon in Rome). Alternatively, for a different shape of vault, by adding additional deadweight onto the crown of the vault, for example, the cathedral of Palma de Mallorca [13, 14] and the bridge in Martorell [15, 16] (Figure 3.4). The endangered vault area could also be supported (e.g. the Florentine Baptistery) or held together with tie rods (e.g. St. Peter in Rome). Or, even more simply, the endangered lower part a hemispherical dome could be omitted (e.g. the dome of Hagia Sophia in Istanbul).

These solutions which made it possible to construct the desired circular and spherical shapes, and which are often highly interesting in terms of design and construction, are generally the result of long periods of development. A remarkable exception is the construction history of the bridge in Pontypridd in Wales, where such a learning process took place in a very short timescale. The master builder William Edwards (1717-1789) built a stone bridge over the Taff 1746. He received £500 for the work with the contractual condition that the bridge must last for at least seven years. Two years later, the three bridge arches were destroyed by a flood. Edwards decided to bridge the river with a single arch of almost 43 m span. When this second bridge was almost finished, the falsework collapsed and he then erected a third bridge with a lighter arch on stronger falsework. After the falsework had been removed, the balustrade was bricked up and backfill was added into the spandrels beneath the carriageway. During this work, the vertex of the arch of the bridge was forced upwards by the weight of the ends of the arch and the bridge collapsed. Apparently, the arch had the wrong shape to bear its own weight. In 1754 the stubborn master bricklayer built the fourth

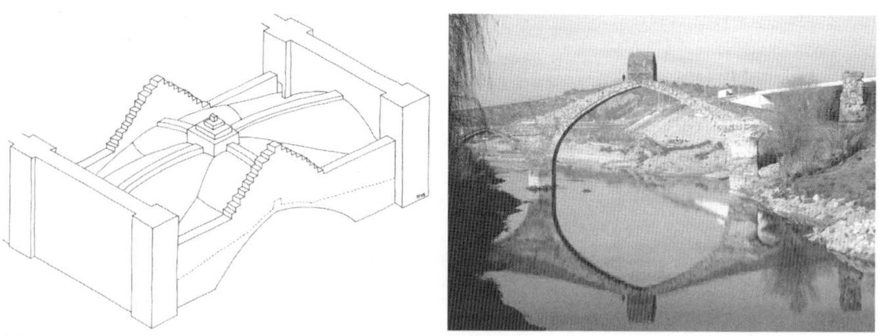

(a) (b)

Figure 3.4 (a) Vault of the Cathedral of Palma de Mallorca. (Image: Paula Fuentes). (b) Martorell bridge, near Barcelona, Spain. (Image: Rainer Graefe)

Figure 3.5 William Edwards' bridge at Pontypridd, Wales, 1754. (Source: [20])

and final bridge, which still stands today (Figure 3.5), with the following modifications: the weight of the crown of the arch was increased, the weight of the spandrels was reduced by introducing three cylindrical recesses, and a lighter backfilling was used in the spandrels [17, 18, 19]. Edwards changed many things, but not the circular shape of the bridge arch. This series of bridges provides a rare example of undertaking and learning from what were effectively four full-scale model tests!

3.1.4 The special shape of the catenary

With his discovery, Hooke added a new arch form, developed from statical conditions, to the traditional repertoire of forms, which ranged from pointed arches to basket arches. This first step towards a scientifically-based design of structures meant a departure from the ideal of perfect geometric form. Before pursuing the consequences of this innovation for the further development of arches by means of examples from building practice and theory [21, 22], it should still be mentioned that, before Hooke, others had already occupied themselves with hanging cords and chains. In 1586 Simon Stevin (1548-1620) published his investigations into the composition and decomposition of forces. He had explored his ideas with the help of small hanging models made of cords loaded with weights 'checked in practice and found to agree with theory' [23] (Figure 3.6). At about the same time the architect, painter and sculptor Bernardo Buontalenti (c.1536-1608) had also noticed the particular shape of the curve of a hanging chain. The outline of a vase of lapis lazuli designed by him and made by Jacopo Bilivert in 1583 reproduces exactly the catenary shape whose form is accentuated by means of a carrying chain attached to the handles of vase [24].

Galileo, however, is generally regarded as the real discoverer of the catenary. In his *Discorsi e Dimonstrazioni Matematiche*, published in 1638, Galileo mentions the hanging chain as an aid to drawing parabolas. He then states that deep catenaries deviate noticeably from the parabolic form, while those with an end inclination of 45° fit the parabola 'almost exactly'. These observations directed the interest of later researchers to the special form of the catenary [25].

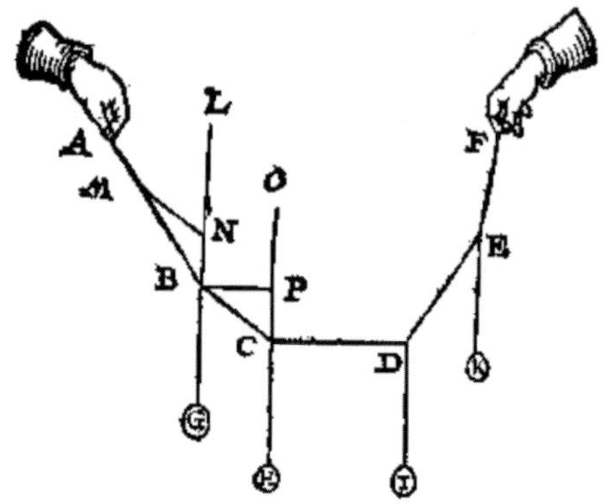

Figure 3.6 Suspended weights in equilibrium, Simon Stevin (Source: [23])

But now back to Hooke's discovery of the correspondence of catenary and the line of thrust. Hooke, who was repeatedly involved in disputes regarding the priority of his inventions, considered this discovery so important that he hesitated to disclose it to the public. At the meeting of the Royal Society of 1670, mentioned at the beginning of this chapter, he found himself demonstrating his solutions only to the President. Asked again at a meeting in the following year, he referred to this demonstration. It was not until 1675 that he decided to publish it, but even then as a riddle in the form of an anagram. Only after Hooke's death, did Richard Waller find a solution to the anagram in 1705: '*ut pendet continuum flexile, sic stabit inversum rigidum*' (as the flexible line hangs, so, but inverted, will the rigid (arch) stand) [26, 27].

In the meantime, however, other scientists had also come to the same conclusion. In 1691 the equation of the catenary was found by Jacob Bernoulli, Leibniz and Huygens. La Hire and Parent were the first physicists to study the equilibrium conditions of arches as a mathematical-statical task. In 1697 David Gregory (1661-1708) summarized their findings to the effect that the axis of a theoretically correct arch had to be formed according to the line of an inverted catenary: 'none but the catenaria is the figure of a true legitimate arch, or fornix. *And when an arch of any other figure is supported, it is because in its thickness some catenaria is included.* Neither would it be sustained if it were very thin, and composed of slippery parts... For the force, which in the chain draws inwards, in an arch equal to the chain drives outwards' [28].

3.2 The early use of the catenary in construction

3.2.1 Hooke and Wren – theory and construction practice

In the years 1705 to 1708 Christopher Wren (1632-1723) completed the dome of St. Paul's Cathedral. This dome is the first vault construction whose design takes into account the new findings about the catenary [29, 30] (Figure 3.7). After a large number of preliminary designs, Wren had only determined the final form of the dome in 1703 or 1704, at a time when the publication by Gregory – who, by the way, was Wren's successor as professor of astronomy at Oxford – must

Figure 3.7 St. Paul's Cathedral, London. (Image: Rainer Graefe)

have long been known to him. It is also certain that Hooke's new arch theory had been familiar to him from much earlier [31]. As curator of the Royal Society and City Surveyor, Hooke worked closely with Wren, with whom he was also a private friend [32]. Numerous notes in Hooke's diary since November 1672 testify to the lively interest he took in Wren's lengthy design work for St. Paul's. On 5 June 1675, Hooke noted, 'Sir Ch. Wren (...) was making up of my principle about arches and altered his module [of St. Paul's] by it.' [33]

A few weeks earlier, the construction of the new St. Paul building had begun. Unfortunately, we do not know which module changes were involved. Further diary entries show that Hooke also advised Wren on the design of the dome construction in the following years: 'I spoke with him (...) about Bow and Colums' (11 Oct. 1678). 'Walked with Sir Chr. Wren in the Park, told him of double vaulting Paules with cramps between' (4 Sept. 1679).

Further proof that Wren also included the new catenary theory in his considerations in the following years is provided by a draft sketch with two dome sections (Figure 3.8). Instead of the double dome mentioned by Hooke, they now show three domes on top of each other. A line of thrust is drawn in both cross-sections, which extends from the junction between the lantern and the dome down to the foundation of the outer supporting pillar. As can easily be checked, these lines of thrust are exact catenaries. They have obviously been drawn using a hanging cord or chain. From the form of the outer dome, which is reminiscent of the domes of St. Peter's in Rome and the Cathedral of St Louis des Invalides in Paris, this drawing belongs to a group of sketches that were probably made in the second half of the 1680s [34]. From the structure of the inner dome, it leads towards the final group of design drawings and to the final design.

The dome actually constructed, with a span of 33 m, represents a peculiar combination of ingenious constructional solutions and illusionistic effects. The outer dome, which seems to carry the heavy lantern, is in reality a light wooden

Figure 3.8 Sketch by Christopher Wren showing the line of a hanging chain within the masonry of the dome of St. Paul's Cathedral, London (c. 1680) (Source: [12])

construction (Figure 3.9). It is partly supported on the conical, 45 cm thick brick dome that is hidden between the outer and inner domes. The stone lantern weighing approx. 850 tons is placed on this cone. At its base the cone is held together by an iron tie ring. The tambour, on which the three domes sit, spreads the loads downwards according to the flow of forces.

The design of the two masonry domes makes it clear that Wren had dealt with the new thrust-line theory. The curve of a catenary (more precisely, of a catenoid) is followed by the inner dome, which, with the exception of the pressure ring of its apex opening, is loaded only under its own weight. The conical central dome is straight-sided, but the masonry is thick enough to accommodate the slightly curving line of thrust arising from its own self weight and the heavy weight of the lantern. We do not know how Wren came to this complex shape in detail [36]. It is hard to imagine that he could have developed it without hanging experiments. In any case, he did not like the result. Even earlier design drawings with similar dome shapes already reveal Wren's efforts to remove them from view both inside and out.

3.2.2 Hidden or on view?

Wren's invention of the triple dome with a steep 'working dome' hidden between separate visible domes was later imitated in other prominent large domes: St. Genévième in Paris, St Isaac's Cathedral in St. Petersburg, San Gaudenzio in Novara and the Capitol in Washington. This distinction between the visible

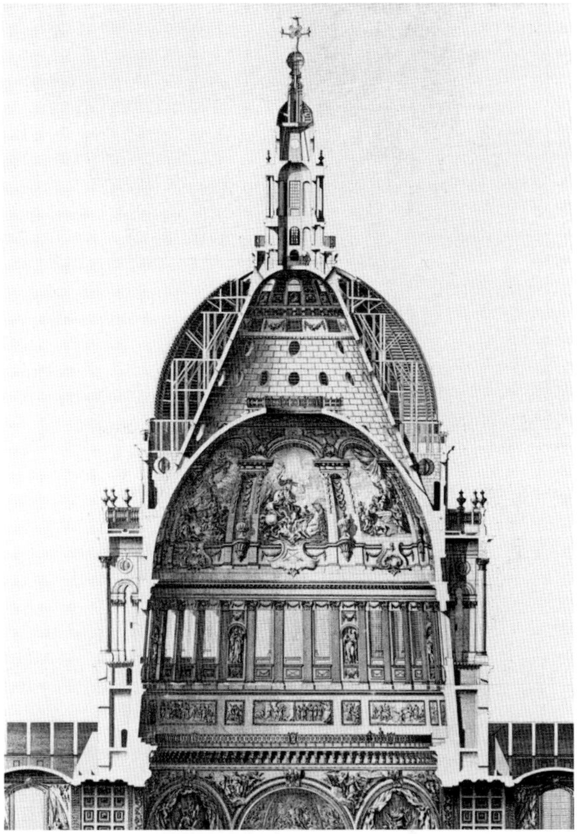

Figure 3.9 St Paul's cathedral, London. Cross section. (Source: [35])

architectural form (outer and inner dome) and the hidden structural form (cone) was, of course, not new in St Paul's. However, it was to become characteristic of later development in which statically-correct building forms can be determined and achieved ever more precisely, without the need to decide to present them as an architectural form.

One interesting example is the gateway of a building (1776) in the Bavarian monastery of Thierhaupten. The elegant basket arch, which seems to support the façade wall, in reality, only supports itself (Figure 3.10a). The weight of the wall is taken by a relieving arch above, which was originally concealed behind the plaster. For his statically-determined parabolic form, the master builder would have found role models in the pattern of cracks in distressed masonry walls: when old walls settle, parabolic arches can develop and stabilize the masonry structure again – an example of an autonomous self-repairing process (Figure 3.10b).

While the hiding of the unattractive structural forms has long been the rule, there have also been exceptions. For a timber bridge designed in the 'Chinese style' by William Halfpenny (1723-1755) in 1755, he deliberately contrasted the semi-circular truss of the supporting arch with the truss forming the walkway and railings had the shape of an inverted catenary: 'the Floor of the Bridge with the Cornice which covers the Ends of the Joists, and Rails adjoining thereto and

(a) (b)

Figure 3.10 (a) Parabolic relieving arch above a basket arch over a gateway. (b) A wall supported by arching between two rigid supports. (Images: Rainer Graefe)

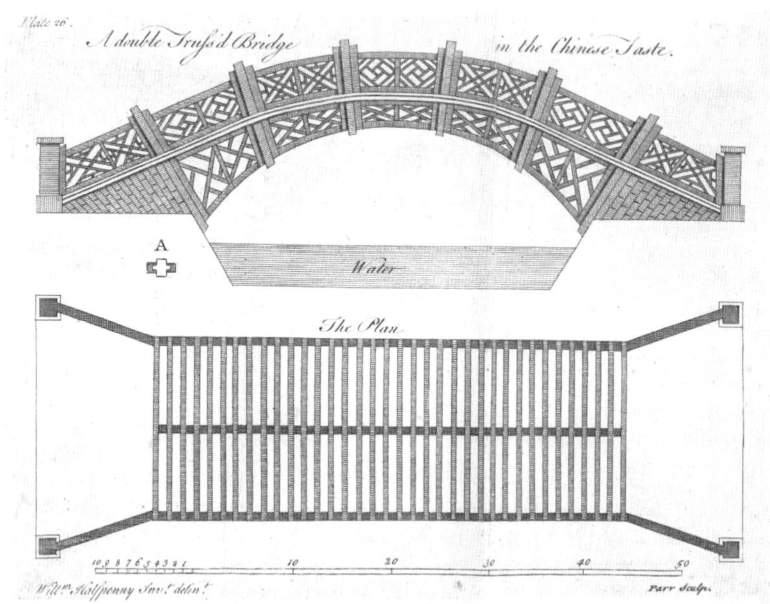

Figure 3.11 Timber bridge whose floor follows 'the curve of a *Cathanarian Arch*, which is created by the swagging of a Chain', William Halfpenny, 1755. (Source: [37])

upper Rail are answerable to the Curve of a *Cathanarian Arch* [i.e. a Catenary Arch], which is created by the swagging of a Chain'. (Figure 3.11) This conspicuous design was not conditioned by construction issues. Halfpenny chose the form of the catenary precisely because it did not correspond to aesthetic norms: a strange, exotic arched form that was in keeping with the desire for 'variety' in garden art at the time [37].

3.2.3 Poleni's hanging experiment for St. Peter's Basilica, Rome

Giovanni Poleni (1683-1761) provided the earliest description of making a suspension model to determine the load-bearing behaviour of a masonry vault in his well-known work on the dome of St. Peter's in his memoire of 1748 [38].The dome had developed a large number of radial cracks and the consequential outward movement had caused many cracks in the drum of the cupola. To analyse the causes of this cracking, Poleni first illustrates the statically correct shape of an arch made of spheres, which remain in stable equilibrium forming a catenary (Figure 3.12). Jean-Baptiste Rondelet later actually tried to build such an arch with small balls. Of forty attempts he was successful on at least two occasions and he correctly noted that achieving it even once would have been enough for him to confirm the correctness of the catenary theory [39]. Heinrich Hübsch also made an experiment similar to Rondelet's using disc-shaped stones from a board game (nine-men's morris), which is much simpler [40, 41].

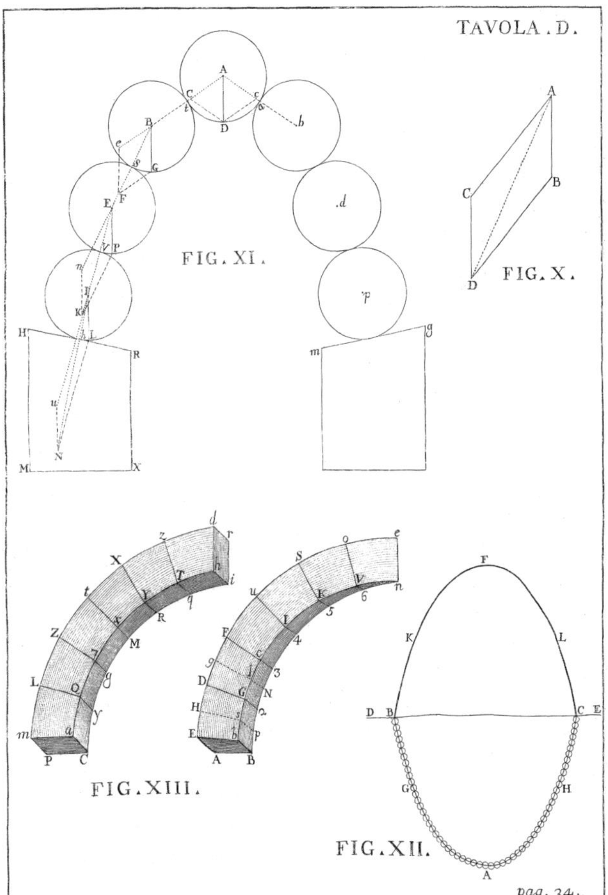

Figure 3.12 Poleni's underlying notions of the ideal dome. (Source: [38])

Poleni's hanging model test served to check the stability of the St. Peter's dome, the cracks in which had given cause for concern. Poleni first divided a vertical segment of the dome into sections, determined their respective centres of gravity and connected them to form a '*curva di gravità*'. The hanging chain consisted of two halves with 16 wire links each. Their lower ends were connected by a light thread, the length of which corresponded to the diameter of the apex of the dome. A lead ball was attached to each chain link, the weight of which corresponded to that of the dome section in question; an additional weight was added to the lowest link to correspond to the weight of the lantern section (Figure 3.13). The chains of his model should have coincided with the curve of the dome if it had the right shape and stability. In the lower part of the dome, the approximate catenary thus obtained passed close to the inner surface, while near the top of the dome it coincided with the centre-of-gravity line. Poleni rightly concluded from this outcome that the dome was wrongly shaped at the bottom, namely too steep, and therefore had a tendency to spread outwards. At his suggestion, the dome was stabilised with six new iron hoops, which were in addition to the three hoops built

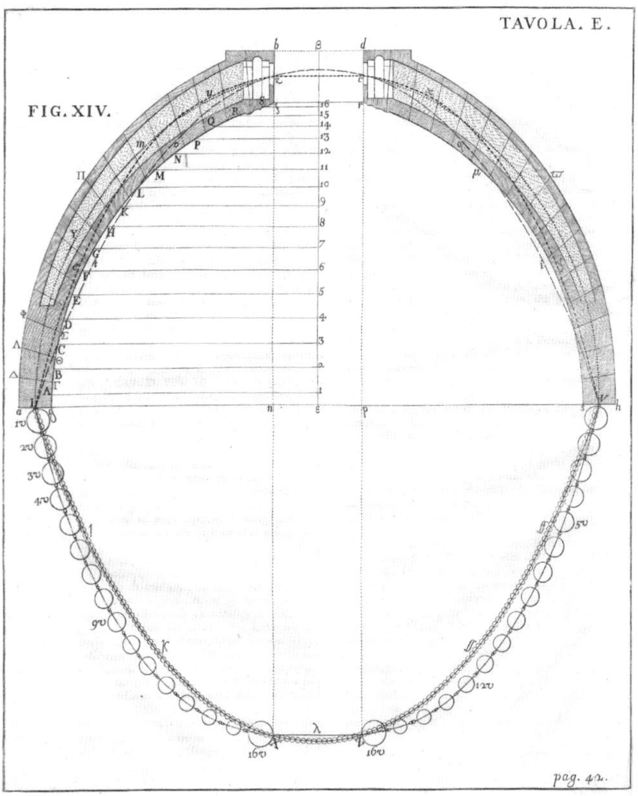

Figure 3.13 Poleni's hanging model of the dome with balls representing the different weights of the voussoirs. This confirmed the line of thrust lay within the thickness of the dome, which was thus stable. In the lower region of the dome, the lines of thrust come too close to the intrados, which was the cause of the crack formation. (Source: [38])

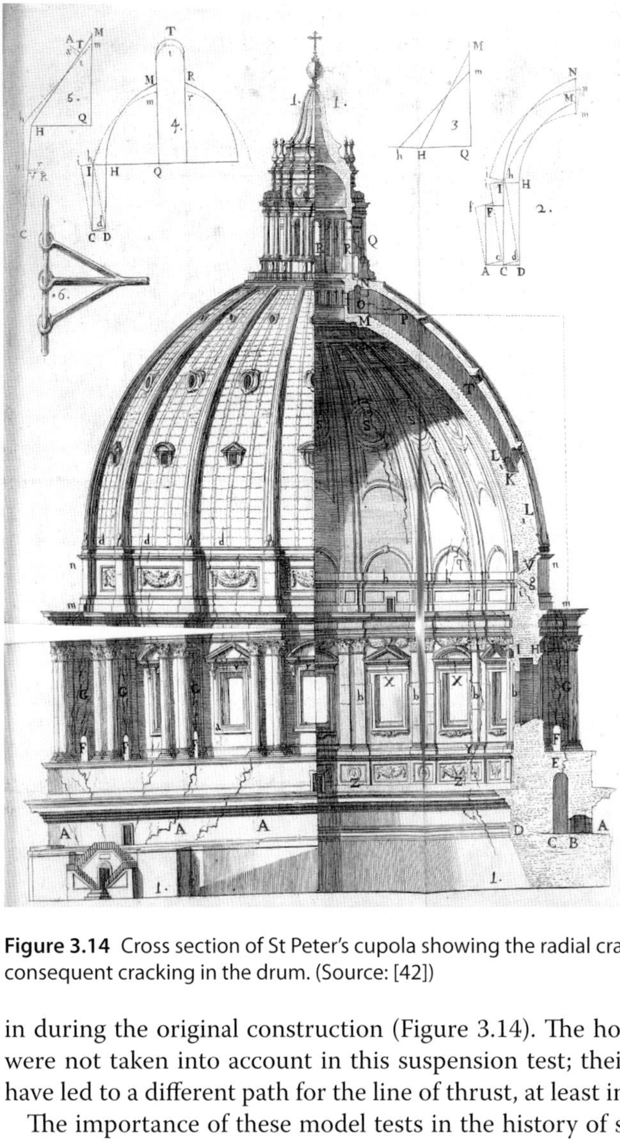

Figure 3.14 Cross section of St Peter's cupola showing the radial cracks in the dome and consequent cracking in the drum. (Source: [42])

in during the original construction (Figure 3.14). The hoop forces in the dome were not taken into account in this suspension test; their consideration would have led to a different path for the line of thrust, at least in the upper dome area.

The importance of these model tests in the history of structural engineering, together with the earlier analytical studies undertaken by Le Seur, Jacquier and Boscovich in 1743 [42], cannot be understated. It is widely acknowledged that they mark the real beginning of modern structural science and its use by design engineers.

3.2.4 Kulibin's Bridge over the River Neva

Another remarkable series of similar experiments undertaken by the Russian inventor, mechanic, watchmaker and optician Ivan Petrovic Kulibin (1735-1818) have not shared the same acclaim and have gone largely unnoticed in Western literature. Kulibin was appointed by Catherine the Great in 1769 as head of the

Figure 3.15 Kulibin's design for a 300 m timber arch bridge across the River Neva in St. Petersburg, 1772. (Source: [43])

mechanical workshop of the St. Petersburg Academy of Sciences. In 1772 he proposed a design for a timber bridge which was to cross the River Neva in St. Petersburg with a single arch of 300 m span (Figure 3.15) (see Chapter 4). His design included several revolutionary innovations. The bridge construction consisted of lattice arch trusses and two massive arches composed of beams. The form of these arches was an inverted catenary. Kulibin determined its shape – apparently without knowledge of Poleni's model – using strings with weights attached, representing the weights of the construction elements. Using these simple, innovative string models, he also determined the magnitude of the horizontal thrust exerted by the bridge on its abutments.

A commission set up by the Academy, which was said to have included both Leonhard Euler and Daniel Bernoulli, examined the project and gave a positive recommendation. In 1776 Kulibin built an experimental model of the bridge at a scale of 1:10, i.e. with a span of 30 m, and he used a method of interpreting the model statics at full scale which he had developed (50 years before Cauchy's similitude law). The model bridge, allegedly in the presence of Euler, was loaded with twelve times its own weight. The load tests were completed after 28 days. The model was relocated as a public attraction, a pedestrian bridge, in the new Tavrichesky (Tauride) Garden in St. Petersburg where it survived for several decades [44] With the construction of this experimental model, Kulibin was the first engineer to propose a structural design whose form was undoubtedly determined using a hanging model. Kulibin's design became well-known abroad and he proposed a revised version to cross the Thames in London with a 200 m span, but he received no response.

3.2.5 Silberschlag's hanging models

Johann Esaias Silberschlag (1721-1791), hydraulic engineer to Frederick the Great in Prussia, also interested himself in the catenary at around the same time as Kulibin. In 1772 he published a still-readable, entertaining treatise which included some discussion about the subject: 'In recent times the question arose as to what kind of curve it would be that would give an equal compression throughout an arch? Galilaeus, a man born to invention, brought the catenary into consideration, and gave to the learned world the means to determine the nature of this line using an algebraic equation. He died, and with him died the desire to calculate the catenary. Jacob Bernoulli, a scholar who has propagated his deep insights into algebra to his gender, revived interest in this line from the dead, as it were, he believed to find in it the long-desired line of equal tension, and in the end one could see that one was not mistaken in this matter. Thus was the fate of the catenary' [45].

From various remarks made by Silberschlag it becomes clear that the particular difficulties they had at the time in understanding the laws governing the formation of catenaries: 'Others believe that the catenary of heavy chains deviates very much from the line of lighter chains, and I believe that this deviation is due to the unequal tension of the chain links [46]. This uncertainty was demonstrated, for example, by the sectional drawing of the vault of King's College Chapel, Cambridge published in 1842, showing an unloaded chain attached: 'The catenarian curve exhibits the direction of the thrust of the vault, which falls within the base' [47].

Silberschlag proposed three-dimensional net-shaped hanging models for the design of net vaults. That was new! However, it remained only an idea; Silberschlag has obviously not yet carried out model tests. The threads of the model would have represented the ribs of the vault. The orientation and distances of each thread to the flat surface mounted above would have been measured and transferred to the actual ribbed construction. With regard to the usability of such shaped vaults, however, Silberschlag imposed certain limitations. In the case of high nave vaults, the catenary form could be used, because the vault shape could not be clearly read from below. Furthermore, due to its constructional advantages, it was also acceptable for bridge construction. However, it was not acceptable 'in magnificent architecture' since, for whatever proportion of height to span was chosen, the eye finds little regularity in the form: it lacks, in one word, 'beauty' [48]. In fact, already in 1739, Amédée-François Frézier (1682-1773) had discarded the forms of barrel vaults based on an inverted catenary as unattractive, even though this statically derived curve would ensure that the line of thrust would remain within the thickness of the vault, especially in the case of pointed vaults [49].

3.2.6 Rondelet's dome for Sainte Geneviève

The dome of Sainte Geneviève (the Panthéon) in Paris, conceived in the 1750s by Jacques-Germain Soufflot (1713-1780), was completed posthumously by Jean-Baptiste Rondelet (1743-1829) in 1790 with substantial structural modifications. This building marks the end of the era of monumental vaulted domes. It was the first significant vaulted structure whose design was based on calculations involving both statics and the strength of materials. It is of incomparable boldness and never again would a masonry dome be constructed so lightly (Figures 3.16 and 3.17).

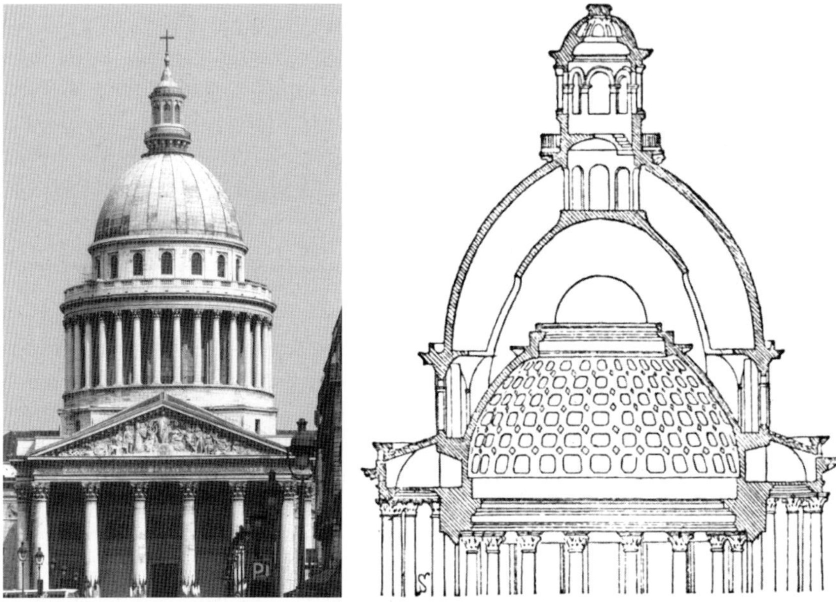

(a) (b)

Figure 3.16 Dome of Sainte Geneviève, Paris, designed by Soufflot and Rondelet, 1790. (a) Exterior view. (Image: Eda Schaur). (b) Cross section. (Source: [50])

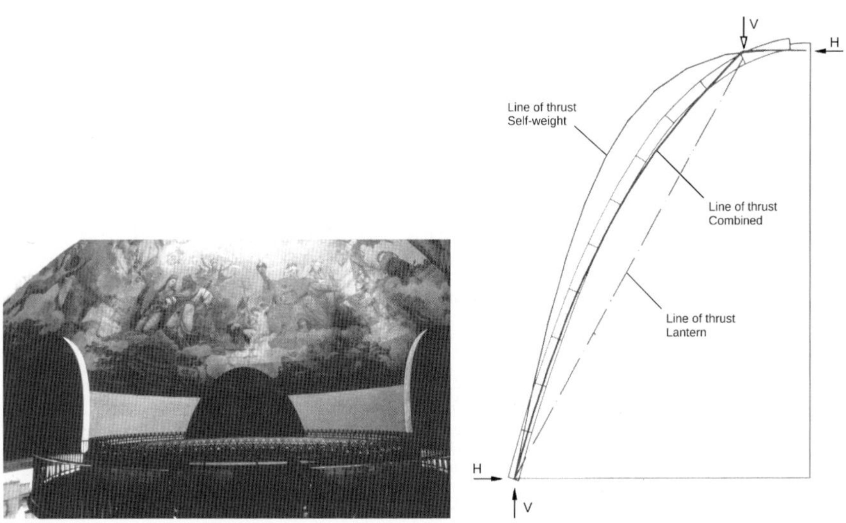

(a) (b)

Figure 3.17 Dome of Sainte Geneviève, Paris. (a) The thin, middle, load-bearing dome. (Image: Eda Schaur) (b) Lines of thrust in the middle, load-bearing dome. (Source: [54] relabelled)

Both in appearance and structural form, the dome construction is based on St. Paul's. Soufflot had obtained exact plans of Wren's building before he began his design [51]. Here again, three domes are set on top of each other, but in the Panthéon, the outer dome, like the other two, is made of stone and self-supporting. Blind arches on the inside reduce the weight of the outer dome. The middle, more stable dome, serves as the supporting structure for the lantern tower, whose weight is thereby distributed to arcades in the substructure. This dome has four large arch openings. Rondelet states that this dome was shaped according to the catenary (which he had studied), since it had to bear a considerable weight [52]. The outline silhouette of the dome follows exactly the curve of a hanging chain which, in fact, is rather curious since the form does not take into account either the spatial distribution of loads from the dome, or the unequal load distribution within the dome caused by the large arch openings, or the significant point load imposed by the lantern. Rondelet therefore does not seem to have fully understood the purpose of inverting the catenary to determine the line of thrust. The inversion does indeed enable the optimal vault forms to be determined precisely, since the statical equilibrium diagram of the hanging chains (and the corresponding vaulted form) alters as changes to the weight distribution lead to corresponding changes in the hanging shape (and the line of thrust). Compared to the inner dome of St. Paul's, whose design Rondelet criticized by the way [53], the shape of his intermediate dome shows an inferior structural understanding [54].

Rondelet repeatedly expressed strange views on the load-bearing behaviour of vaults and was directly reproached by Navier: 'The notions presented by this architect, who has always maintained that domed vaults do not have a horizontal thrust, do not conform with the truth' [55]. Thus we have the memorable case that the dome for Ste. Geneviève, a masterpiece of vaulting art, was designed and built with an obviously poor understanding of form and load-bearing behaviour. It is also clear that Rondelet used a hanging chain or cord model to determine only the contour of the intermediate dome, and did not carry out further hanging experiments to inform the structural design.

3.3 Catenary models in the early nineteenth century

3.3.1 Models of upright catenary arches

Uncertainties regarding the design and dimensioning of vaulted structures were only gradually overcome in building practice during the following decades. Numerous model tests had already been carried out in order to better understand structural behaviour. As early as 1732, for example, Augustin Danyzy (1698-1777) had carried out load tests on small models of barrel vaults up to collapse at the Academy in Montpellier. Émiland-Marie Gauthey (1732-1806) and Rondelet later carried out similar experiments. Louis-Charles Boistard (1763-1823), French director of road construction, reported in 1800 on experiments 'which were more large-scale and done with care. The vaults were 2.27 m wide and made of bricks sanded with sandstone.' In Britain, George Atwood (1745-1807) used 'well-polished copper' voussoirs in similar experiments the following year [56].

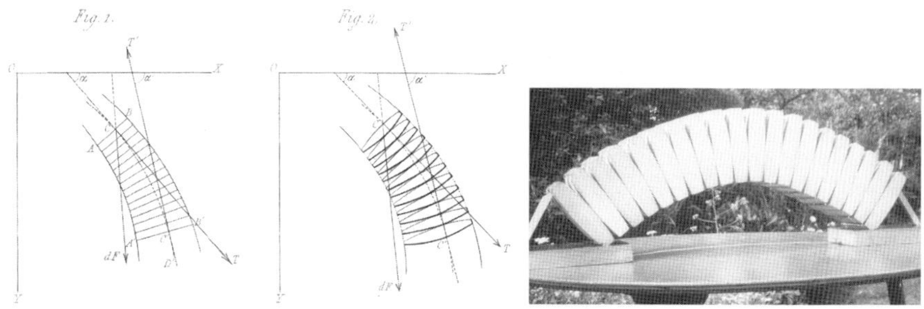

(a) (b)

Figure 3.18 (a) Diagrams by Yvon-Villarceau explaining the self-equilibrating arch. (Source: [57]). (b) A 'standing chain' (*Stehende Kette*) or 'wobbly arch'. Model by Frei Otto. (Image: Courtesy of Institute of Lightweight Structures, Stuttgart)

By contrast, the vertical link chain published by Antoine-Joseph Yvon-Villarceau (1813-1883) in 1844 was a fundamentally new type of model. It allowed the line of thrust of an arch to be determined experimentally for different load cases – an amazing effect [57] (see Section 2.4.6). The voussoirs, in the form of circular segments, roll over each other. The contact points between the voussoirs mark the line of thrust (or the inverted catenary) (Figure 3.18a). Their position changes according to the imposed loads and changes the shape of the line of thrust which they indicate: a self-forming arch shape analogous to the freely hanging chain. This type of model is sometimes called a 'wobbly arch'. Frei Otto developed a similar arch model in 1952 (Figure 3.18b), in which the voussoirs are secured against slipping by threads.

3.3.2 Rejection of the catenary as an arch form

The mathematician Franz Gerstner (1756-1832) described the situation in 1798 aptly: 'With the advances that the mechanical arts has made in a century, with the help of higher mathematics, construction suffered the undeserved fate of falling far short of the other arts. Leibniz and Bernoulli had also scattered some seeds which matured to perfection in declaring the catenary to be the most correct form for vaults. Moreover, the fact that this line could not be reconciled with the good taste of the elderly meant that master builders did not know how to make any advantageous use of it, especially because it seemed to address more the centres of gravity of vaults than the centrings [58].

Several authors appeared as decisive opponents of the catenary theory. Atwood, for example, in 1801, was of the view (paraphrased later by Bartholomew), 'that the current opinion, that a true arch of equilibrium is an inverted catenary, is neither founded on mathematical nor on experimental investigation' [59]. In the following year, the engineer Carl Friedrich Meerwein (1737-1810) declared himself unable to accept Silberschlag's explanations based on the correspondence of the catenary and line of thrust. The catenary is neither structurally advantageous nor 'more pleasing to the eye'. 'Here it will clearly be seen that the advantages of catenaries, while not entirely based on imagination alone, are certainly not of such significance that it would be worth the trouble to disfigure a vault [60]. The circular arc is the best shape for an arch and, where it is too high, an ellipse is the best alternative [61]. When Meerwein designed a 150-metre arch bridge across the Neva at St. Petersburg, he, of course, chose a segment of a circle [62]. Like Kulibin, Meerwein proposed building a 1:10 scale model to test his design which, he suggested, could later be used as a footbridge [63].

As late as 1817, Sebastian von Maillard, in his 'Mechanics of the Vault, dealt with in its entirety' referring to the catenary, noted tersely: 'this shape is simply unworkable' and makes no further comment [64]. It also remained unclear to the proponents of the catenary which position the corresponding line of thrust would take up within the cross section of an arch. In other words how far could the shape of the arch deviate from the statically, most-favourable shape or what cross section would the arch need to have to ensure the line of thrust would remain in a safe zone. These uncertainties were only eliminated around 1860 with the investigations of Rankine and Culmann who defined as stable, those vaults in which the line of thrust passes within in the middle third of the cross-section [65].

Catenary arches and vaults were not accepted as architectural forms for a long time. In 1772, Silberschlag had recommended that the catenary should be used only high up in church vaults where the shape of the vault was not clearly visible from below. David Gilly (1748-1808) recommended in 1797 their use only in basements and undercrofts. The undercroft vaults which Gilly describes are, strictly speaking, not really shaped according to the catenary. They are translational shells generated by two perpendicular circular segments to form sail vaults (Figure 3.19). Nevertheless, they are remarkable in our context: 'Such a vault

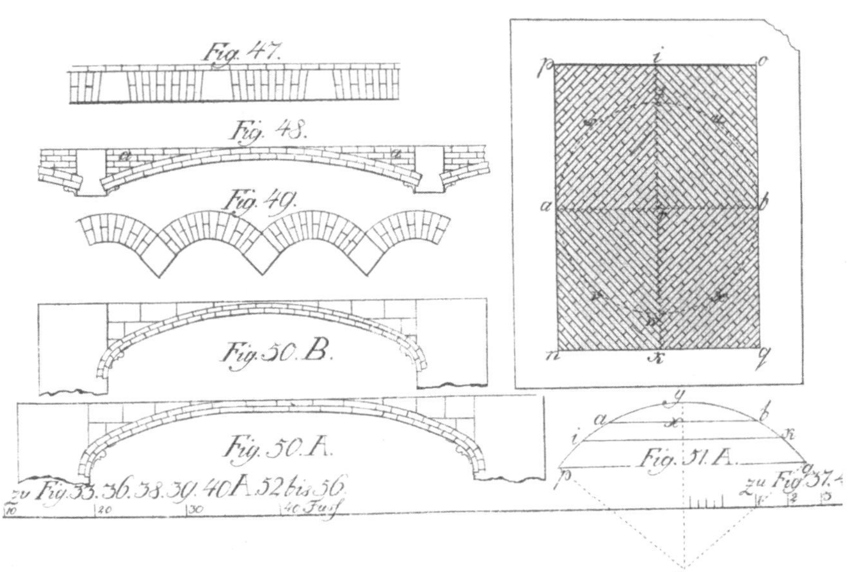

Figure 3.19 Various vaults for basements and undercrofts proposed by David Gilly. (Source: [66])

would have the inverted form of a flaccid cloth attached at the four corners and hanging down [66]. Gilly imagines the curvature of the vault as the inversion of a hanging surface. This was an entirely new and original thought which was only rediscovered 150 years later in the three-dimensional form-finding models of Heinz Isler (see Section 21.2.3).

3.3.3 Hanging models and the design of arch bridges

In 1801 the Scottish engineering John Robison (1735-1809) proposed making a suspension model for an arched bridge which took into account the load of the backfill and the roadway [67]. Thomas Young showed corresponding models in his lectures in 1807 and in 1824 described a hanging-block model of the sections of a bridge arch connected by hinges [68].

In 1839 the English engineer John Millington (1779-1868) built a suspension model for an arch bridge with the weights of dead load and superimposed load. At regular intervals there were pieces of chain hanging from the chain arch, which simulated the loads. Millington describes a similar experiment that the John Robison had carried out in Edinburgh on a large scale: the arch had a span of 60 feet (18.29 m) and a height of 21 feet (6.40 m). The initial assumption was an elliptical arc, from which the chain curve deviated greatly. This chain curve was also calculated. The result hardly deviated from the shape determined in the model test. 'This result was highly satisfactory, since it showed that the first curve formed by about two hours labour, on a supposition considerably different from the truth, would have been sufficiently exact for the purpose, since it did not vary in any place three inches from the accurate curve, and was, therefore, far within the joints of the intended arch stones. Therefore this process which any intelligent workman, though ignorant of mathematical science may go through with little trouble, will give a very proper form for an arch subject to any conditions.' [69] Millington recommended Robinson's form-finding procedures for practical use. It is 'so just and so easy in its practical operation' [70].

Some years later, between 1845 and 1850, the railway company 'Wilts, Somerset and Weymouth Railway' built several bridges in England which were designed using hanging models made with chains. These were 'skew bridges and bridges with inclined carriageways and spans up to 66 feet' (20 m), which crossed railway lines. Gustav Meyer, from whom this communication originates, was himself the designer of one of these bridges, a brick road bridge with an incline of 1 in 8 (12.5%) gradient, which he designed in the following way: 'After first making an approximate sketch of the vault with the roadway above on a large scale, the length of the test chain and the position of its suspension points were chosen accordingly. After this, the weights to be attached were determined in such a way that they represented the load of the spandrel fill, while the chain itself represented the weight of the vault' [71] (Figure 3.20).

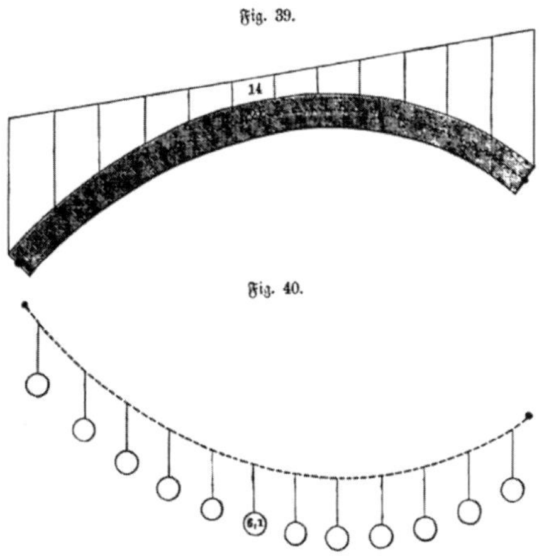

Figure 3.20 Hanging chain used to design an inclined brick arch bridge over a railway (Source: [71])

3.4 Hanging models for architectural designs

3.4.1 Heinrich Hübsch

Heinrich Hübsch (1794-1863) was the first architect routinely to use two-dimensional hanging models in the design of his churches; yet he did not shape his buildings externally according to forms determined experimentally. In 1838 Hübsch published the 'Method for the determination of the required strengths of arches and Abutments for each genus and composition of arches by means of a graphical method', he had developed five years earlier [72, 73]. Several hints indicate that he had studied various treatises on vault theory, but precise references are still not known. As mentioned above, Hübsch first demonstrated the principle of inverting the catenary using the round flat stones of a board game. He then explained that, not only the line of thrust of the arch itself but also that of the abutments must be determined and, furthermore, that the 'common' catenary does not correspond to the line of thrust of most arch constructions because these are only seldom of the uniform thickness and weight throughout. He rejects the mathematical determination of the catenary as too complicated for construction practice. It is far too time-consuming in the design process, in which architectural-artistic and construction aspects have to be coordinated step by step: 'This difficulty tortured me for a long time, until I finally came up with a graphical process, which is not a strictly geometric construction, but so simple as to be infallible. I (...) make use of this method (which, to my knowledge, is not yet mentioned anywhere else) with great advantage in my designs for vaulted buildings. I find here a safe means to design the profile of a building in a very short time in such a way that the task can be solved with the minimum of material and with the greatest possible protection of space' [74].

Hübsch hung two-dimensional hanging models in front of upside-down sectional drawings of his projects. Small loops were tied to the cords of the models at regular intervals. Weights were hung in the loops that corresponded to the weights of the respective sections of the vaulting. With the help of these simple hanging models he modified and improved the shapes of his designs. He even simulated certain load cases up to collapse, which he considered to have occurred when the catenary passed outside the section of any part of the vaulted structure (vault, column and abutment). The detailed descriptions of individual tests prove the effectiveness of this new form-finding technique. In his 'Investigations of various vaulted forms and structures with the aid of suspension models', Hübsch demonstrates how masonry loads at the foot of the vault can be redirected towards the vertical, enabling slimmer constructions, more highly-loaded vaults or generally higher constructions [72] (Figure 3.21a).

Hübsch designed several churches using his method of finding forms – the Bishop's cathedral in Rottenburg am Neckar, a church in Bulach near Karlsruhe, a Catholic and a Protestant church in Karlsruhe, and a church in Bietigheim. For the first building, St. Cyriakus' Church in Bulach, begun in 1834, he first erected an experimental building on a scale of 1:2 and carried out a series of load tests – which was also a novelty (Figures 3.21b and 3.22). The experimental building was about 50 cm in front of a building from whose windows the load tests could be checked. They showed that the main arch could carry up to one third of its own weight when uniformly distributed. More precise ideas of the loads that would have been necessary to achieve realistic results in this model experiment were still lacking. The observation of the experiment, however, obviously enabled Hübsch to arrive at a correct assessment with the help of his feel for structural behaviour and his practical professional experience. In addition, the load assumptions for the suspension model had been simplified: the string passing through the vaulting profile was weighted with equal weights throughout (and the strings within the pier profiles were not weighted at all due to the slenderness of the piers). Hübsch was thus given, as he emphasised, a safety reserve for the stability of the construction due to the additional weight of the brickwork of the

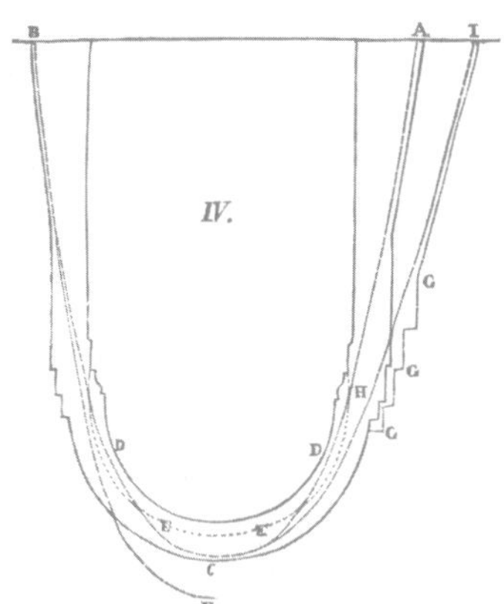

(a)

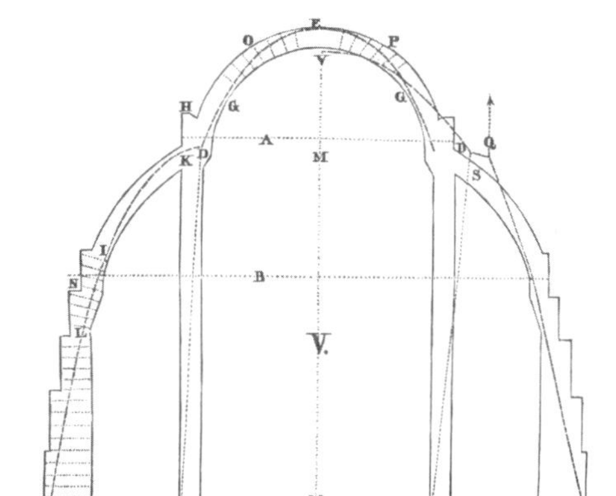

(b)

Figure 3.21 (a) Investigation of arch forms for different load cases, suspension tests by Heinrich Hübsch, 1838. (b) Church at Bulach. Cross section of a 1:2 scale model made by Hübsch for load tests. (Source: [72] Figs IV, V)

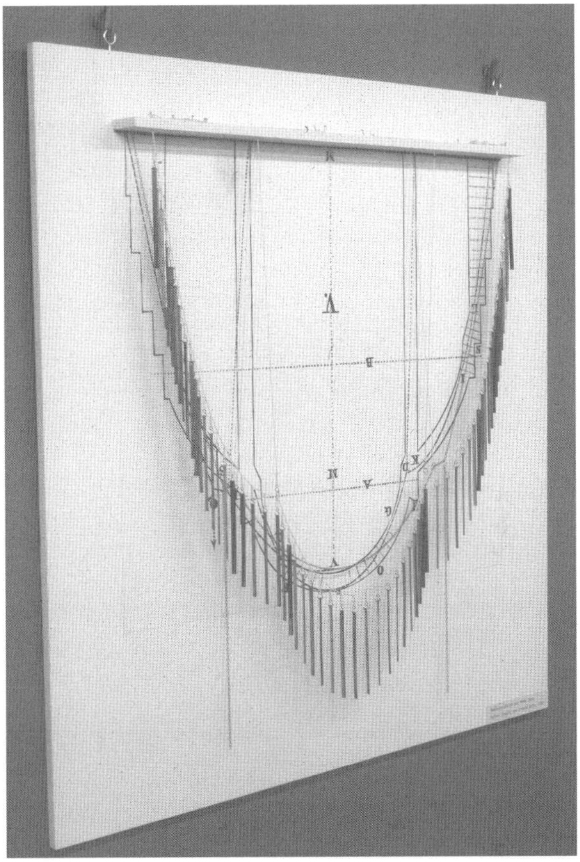

Figure 3.22 Church at Bulach. Two-dimensional hanging model after Hübsch. Reconstruction by Rainer Graefe and Arnold Walz, 1983. (Image: Rainer Graefe).

abutments and that of the roof above, which were not taken into account in the hanging model test [72].

For the construction of the Bulach church, Hübsch used a system of flat transverse barrel vault rising to the centre [75] in order to reduce the outward thrust on the walls (Figure 3.23). The arches thicken towards the apex so that the line of thrust does not come too close to the surface of the arch; for the same reason stepped piers are placed at the foot of the outer wall (Figure 3.24). The arches and the brickwork supporting the main roof beams are hollow to reduce weight. Nevertheless, Hübsch's new design procedure had no effect on the appearance of his buildings. The right to draw such formal conclusions was reserved for others. Heinrich Hübsch's design method and its implementation in building practice attracted some attention. As late as 1903, Otto Warth dealt extensively with the form-finding methods of the Karlsruhe architect in his book 'Constructions in stone' [76].

3.4.2 Carl-Anton Henschel

In 1862 Gotthilf Hagen (1797-1884) pointed out that Hübsch 'had not only used his hanging models for designing simple arches' but also 'used the

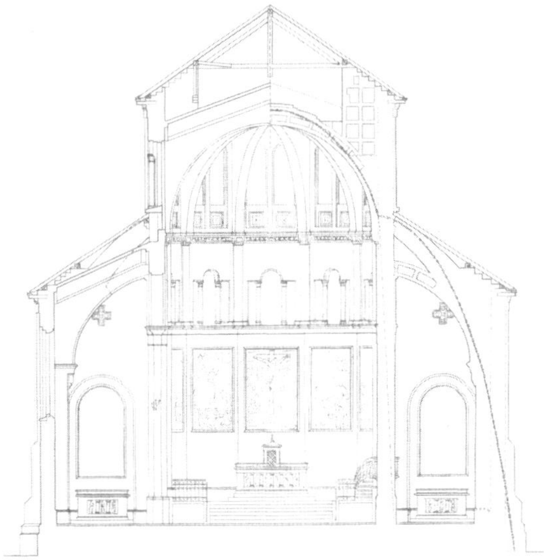

Figure 3.23 St Cyriakus Catholic Church at Bulach near Karlesruhe, completed 1837. Cross section showing the line of the hanging chain used to determine the form of the vaults and the width of the buttresses. (Source: [72]) Plate 7)

(a)

(b)

Figure 3.24 St Cyriakus Catholic Church at Bulach. (a) Exterior view. Contemporary lithograph. (b) Interior view. (Image: Rainer Graefe)

same method to balance out the geometry of complex vaulted roofs and buttresses in large Basilica-like churches, and then to determine the masses of the individual parts' [77]. Hagen added: 'This is the path Hübsch took, and Ober-Bergrath-Henschel from Kassel, probably followed even more exhaustively when he wanted to avoid seeing the dome covering his foundry collapse for a second time' [78]. Hagen was here referring to the Henschel

(a)

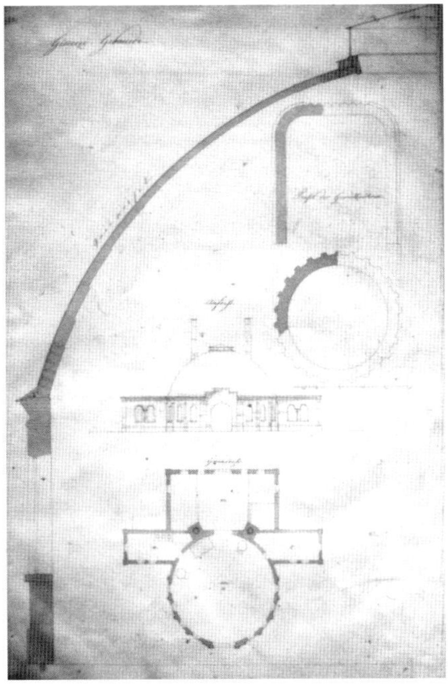

(b)

Figure 3.25 The Henschel foundry in Kassel. (a) A contemporary aquatint by J.H. Martens. (Source: [81]). (b) Ground plan and elevation of the foundry; section of the dome, preliminary design, 1837. (Source: [81])

foundry in Kassel. The old foundry was destroyed by fire in 1836 and a new circular building was completed in 1839 (Figure 3.25a). The design came from Carl-Anton Henschel (1780-1861), an industrial pioneer, visionary and inventor and founder of Maschinenfabrik Henschel. However, the construction of the dome over the foundry, with an inner diameter of 16 m, was only successful at the second attempt, after Henschel had used a hanging model to determine the statically-correct thickness and shape for the dome (Figure 3.25b). The dome was made using hollow clay pots to make it as light as possible (Figure 3.26). Hagen's remark that Henschel's procedure was 'probably followed even more exhaustively' than Hübsch's elaborate model experiments suggests that Henschel may well have used a three-dimensional model instead of the two-dimensional models to which Hübsch limited himself; we will probably never know [79].

Henschel did not describe his design process. In 1850, however, the versatile inventor and entrepreneur published 'On the Aesthetics of Higher Architecture' in which a passage in the text reveals that he had dealt with hanging structures in connection with vaulting: 'A suspended fabric usually shows such pleasing forms because the hanging folds are correct in all their variations and therefore appear predictable and pleasing to the eye. The insightful person feels how the law of gravity, which binds together the world's bodies, prevails here as well, is also

(a) (b)

Figure 3.26 The Henschel foundry today, after refurbishment. (a) Exterior view. (b) Interior view showing the hollow clay pots used to reduce the weight. (Images: Jos Tomlow)

at work in this case, and the weight of the smallest thread contributes to shaping the whole form, while the upholsterer may well often believe that he has created the fold. We look up at the boldest arched vault with calm, only if we keep in mind every part of it, in a position where it is able to support its neighbour' [80]. Henschel thus regards the self-forming shape of a hanging fabric as a quasi-natural process and relates it to correspondingly-formed shapes of vaults. This is a remarkably early move into three dimensions.

The foundry dome, shaped as a thin pointed arch, is built up in horizontal ring courses of hollow clay pots. Such so-called *tubi fittili* were used as lightweight construction technique in Roman antiquity and in Byzantine vaults and reused in the nineteenth century [81]. This building is the only one of the industrial ensemble to have survived the bombardments of World War II and is now used as the auditorium of Kassel University. We are thus lucky that the first two buildings were designed using hanging models – the Bulacher Kirche (1837) and the Kasseler Gießhaus (1839) – have both survived.

3.4.3 The catenary architecture of Wilhelm Tappe

Wilhelm Tappe (1769-1823) made the earliest attempt to bring the catenary form into architectural design. Tappe was a member of the architectural movement represented by David Gilly, Gentz, Catel and others, which emphasized more the practical-social aspects of building than the established aesthetic-historical dogmas [82]. In his 'presentation of a new highly fire-resistant construction technique requiring very little wood' [83] published between 1818 and 1821, Tappe suggested founding a new German architectural style in which all buildings were

(a) (b)

Figure 3.27 Wilhelm Tappe's vision for an architecture for masonry buildings based on the catenary. (a) The basic unit – a hut made of clay, which served as a dwelling for the simple, impecunious man. (b) Tappe's vision for a townscape. (Source: [83])

arched according to the form of the catenary (Figure 3.27). The basic unit of this development was to be a hut made of clay, which served as a dwelling for the simple, impecunious man. In addition, Tappe called for a return to the German people's own art of building, the Gothic, which was also exemplary because of its use of statically-favourable structural forms. However, from a construction point of view, this represented only a first step in the right direction. Gothic churches were still 'so surrounded by disfiguring buttresses (...) that they appear like sick majesties, supported and carried by servants' [84]. Four decades later Jakob Burckhardt would repeat this criticism of Gothic architecture almost literally: 'The exterior [of the whole church] is a fragmented [...]. The building shows its crutches on all sides' [85]. Later, Antoni Gaudí would also refer to Gothic buttressing as 'crutches'.

If vaults are correctly shaped, which transfer their load directly to the ground rather than to vertical walls, such auxiliary constructions should be superfluous after the stage: 'Let your eye get used to the large arch, the legs of which are founded in solid earth without supports and resistances' [86]. Tappe saw hardly any limits to the efficiency of such vault constructions: 'Give me the means and I'll vault you over a county with a dome vault and put the capital on its crown' [87].

In his search for the best vault form, Tappe correctly recognized that the catenary line only gives the most favourable line of thrust for barrel vaults of the constant thickness throughout. However, he then wrongly concluded that hanging chains, which (as in his designs) become thinner and lighter towards the centre, take the forms of half ellipses. In fact, there may be a near coincidence of catenaries and ellipses around the vertex. However, the ends of half ellipses run tangential to the vertical, while hanging chains always carry a horizontal inward component of the tensile force, and vaults of the same shape always exert a horizontal outward thrust.

Tappe was therefore mistaken in his assumption that he would be using an optimal vault shape in his designs. Apparently, he had not checked his designs with hanging models. It is interesting, however, that his early attempt to elevate the constructional requirements, from which the form of his vaults were developed, to an artistic principle. Tappe explained that statically-correct forms of construction are beautiful because they are natural, or rather, are the result of the same forming processes as the objects of living nature. Nature creates forms whose beauty is based on their constructional practicality. Architecture should also develop building forms in which the dualism of practicality and beauty is abandoned.

Tappe's radical constructional functionalism met with a lack of understanding among his contemporaries. When he wanted to build a church in his new style for the Protestant congregation in Lüdenscheid in 1822 – albeit with a comparatively conventional design – Schinkel prevented the building with the longest and most unfriendly expert opinion he is known to have given [88]. Criticism of Tappe's projects in the Leipziger Literatur-Zeitung was milder: 'If he had stopped at the simple hut intended for poor peasants, and not gone beyond this, then he would more likely have received …' [89].

3.4.4 Vaulted stairs

Vaults in the form of catenaries have been built in large numbers in Catalonia since the middle of the nineteenth century. The triumphal march of this vaulted form took place on a side stage: in the stairwells. In 1841 Manuel Fornes y Gurrea published the simple, effective method of creating the forms of stair vaults by inverting the curves of ropes [90] (Figure 3.28a). The method is obviously older, and the origin unclear [91]. In any case, in Barcelona the stairwells of the buildings of the 'Eixample' and Modernisme are almost all constructed in this way. The optimized arch shape and the efficient Catalan masonry technique with flat bricks and quick-setting mortar enable the construction of frighteningly thin, stable vaults (Figure 3.28b). Even today, some bricklayers still master this vaulting technique. Rafael Guastavino, master builder from Barcelona and architect of numerous impressive vaults in North American cities, also exported the Catalan method of forming stair vaults with the aid of hanging ropes and patented the method in the United States in 1886 [91, 92] (Figure 3.29).

3.4.5 Aesthetics of the catenary

In the middle of the nineteenth century in England 'a wide discussion of the catenary principle applied to Gothic churches' took place which addressed not only an understanding of historical building forms, but also about design aids for contemporary architecture [95]. Edward Garbett uses the image of a chain with

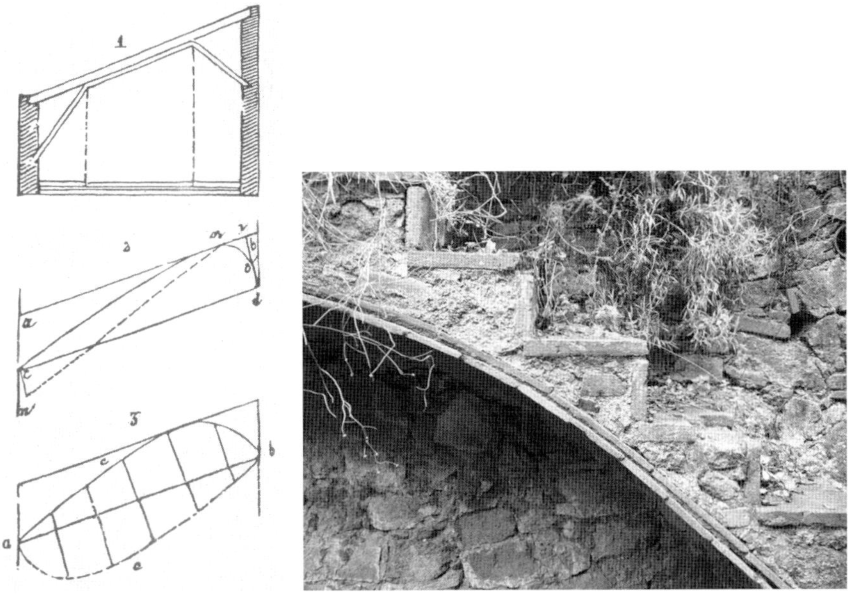

(a) (b)

Figure 3.28 (a) Fornes y Gurrea's method for generating the catenary arch for an inclined plane or staircase. (Source: [90]. (b) Remains of a vaulted staircase in Barcelona. (Image: Hubertus Pöppinghaus)

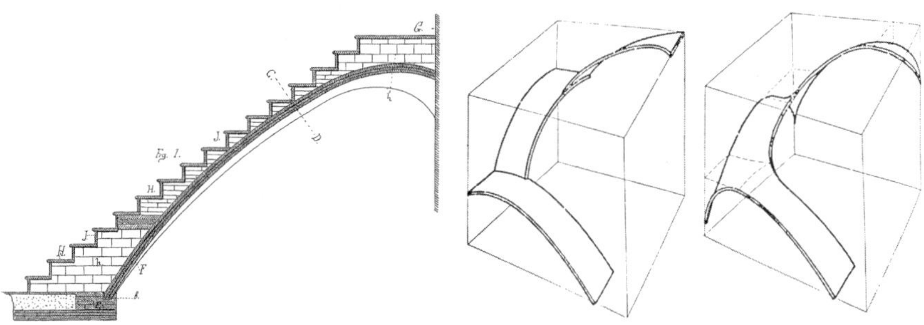

(a) (b)

Figure 3.29 (a) Staircase patented by Guastavino (Source: [93]). (b) Ways of constructing a three-flight staircase. (Source: [94])

a street lamp hanging from it, to illustrate the shape and load-bearing behaviour of Gothic pointed and flying buttress arches: 'This property of arches (by which each pressure concentrated on a point calls for a cusp at that point, and each cusp calls for a concentration of pressure on it) may be shown by the "catenary", which becomes an inverted Gothic arch whenever a weight is suspended from one link. Hooke's discovery, '*ut pendet continuum flexile, sic stabit contiguum rigidum inversum*', is a motto never to be forgotten in Gothic building. A French street lamp, or a spider's web, may thus teach the architect important lessons' [96]. Garbett expressed the assumption that the equilibrium figures of the boldest Gothic vaults had already been determined with hanging chains. In his opinion, every pointed arch that is statically correct is also beautiful. As already with Tappe, Hübsch or Henschel, the idea of the naturalness of self-forming hanging forms, and arch or vaulted forms derived from them, can also be found with Garbett, for example when he cites the spider's web as a natural construction of catenaries.

In 1862, Gotthilf Hagen, the 'old master of German hydraulic engineering' and professor at the Building School in Berlin [97], seems to have been the first after Tappe who was positively disposed to vaults which are formed according to the catenary. He rebuffed any objections to such arch forms: 'That something should lie in these peculiarities of the arch form which would contradict the requirements of aesthetics is probably not to be assumed. On the contrary, one should think that the curve of the arch and its swelling should give a more pleasing appearance than if it were bounded by two concentric arcs of a circle' [98]. The arch shape of the catenary is thus more appealing than that of the pointed arch.

Hagen recommends again and with detailed information to hang two-dimensional hanging models in front of upside-down sectional drawings: 'Now one fastens a silken thread at the ends of the arch and loads it with different attached weights until the thread follows the centre line of the arch. These weights then denote the load on the arch that must be chosen in order to achieve equilibrium of the form which has been chosen' [99]. It is in this context that he refers, as mentioned above, to the corresponding design procedures of Heinrich Hübsch and Carl-Anton Henschel.

3.5 Hanging models become three-dimensional

Despite being widely recommended in construction publications during the second half of the nineteenth century, hanging models were relatively little used by architects or engineers in construction design and practice. Even as late as 1908, Karl Esselborn was recommending the use of simple two-dimensional hanging models as aids to design: 'This method is recommended as a means of verifying the correctness of a design for 'vaulting'; the drawing is turned upside down and the catenary is laid over it' [100]. To some extent, the opportunities for using physical models were reduced by the growing use of graphical statics, a fundamentally identical but *theoretical* procedure, which had grown considerably in popularity from the 1850s because of its convenience, great versatility and, indeed, lower cost. It became the method of choice for both masonry (compression) structures and the trusses used widely in roof structures and railway bridges. However, graphical statics was only possible in two dimensions and unable to help architects and engineers who began thinking and working more in three dimensions towards the end of the nineteenth century [101]. Physical models were able to fill this gap.

3.5.1 Three-dimensional net models by Karl Mohrmann

In 1890, the architect Karl Mohrmann (1857-1927) recommended that three-dimensional hanging models could be used for form-finding and for the design of complex vaults:

> 'An extension of graphic statics to deal with spatial network forms is, in most cases, not so easy, since the diagrams can only be drawn on a flat plane. (...) For special cases in practice, it might actually be advisable to make small net models (perhaps at 1/10 of natural size) from ropes or threads, the production of which is not at all as difficult, as experiments by the author have shown. If it is only required to determine the location of the nodes of the net, this would be possible by combining the coincident forces to which they are subjected in an appropriate manner. The threads running from node to node would then appear as taut straight lines. If one wanted to determine in this way the theoretically most correct curvature of the ribs (...), then one would have to divide up the rope into an appropriate number of individual lengths corresponding to each rib, and to apply a force to each part' [102].

Unfortunately, at present, no more information is known of Mohrmann's experiments on three-dimensional net models. His legacy in the family's personal archive in Freiburg (im Breisgau) has not yet been edited [103].

3.5.2 The Reichstag Project by Fritz Gösling

A few years later Antoni Gaudí began building his legendary hanging model for the church of Colònia Güell. But before we look at it, a project of an outsider should be considered that already pre-formulated some of the constructional and formal solutions of Antoni Gaudí's architecture. In 1871, the competition for the construction of the German Reichstag took place in Berlin. Among the 101 submissions, one project caused a sensation [104] (Figure 3.30). It was by Fritz Gösling (1837-1899), architect and brickworks owner in Bad Pyrmont, and a pupil of the Hanoverian neo-Gothicist, Conrad Wilhelm Hase (1818-1902). The design by this completely unknown, provincial architect with the forms of an organic Gothic met with broad and vehement rejection. Pointed, helmet-like roofs are staggered above horizontally storeys, rising towards the centre, and surmounted by a giant horsetail tower shooting into the sky. The eccentric formal language was alienating. The design was regarded as bizarre, crazy, and even pathologically hostile.

It is generally not understood that the project represents nothing less than an attempt to develop a fundamentally new architecture from optimized forms of construction. Gösling used shapes in which there are only compressive forces, namely straight-line building structures as inversions of hanging net structures. 'The line of thrust is the tension line, the rope or the cable' [104]. That was a step that went far beyond any form-finding techniques previously described – an

Figure 3.30 Fritz Gösling's scheme for the new German Reichstag building, 1871. (Source: [104])

enormous, complex of buildings that should be developed, down to the last detail, by means of a hanging model.

In drawings, Gösling depicted the reversal of the catenary to the line of thrust (with an exact reproduction of the catenary curve), as well as changes in the curve shape where the loading changed (Figure 3.31). However, if his sectional drawing of the building (Figure 3.32) is turned upside down, it becomes clear that this structure does not really hang, that not all the shapes have been determined by means of a hanging model. The reason is simple: a hanging model of the entire giant building would have far exceeded the practical possibilities open to Gösling. Even so, one major part of the building, the central main hall with its 30 m span arched ceiling, was actually designed with the help of a thread model. Gösling depicted it in a modest drawing ('axial picture') without scale in which the hanging structure is reproduced, already inverted (Figure 3.33a). A central ring is supported by eight radial threads, and four parallel transverse threads, all with the same height, span between adjacent radial threads. They hang under

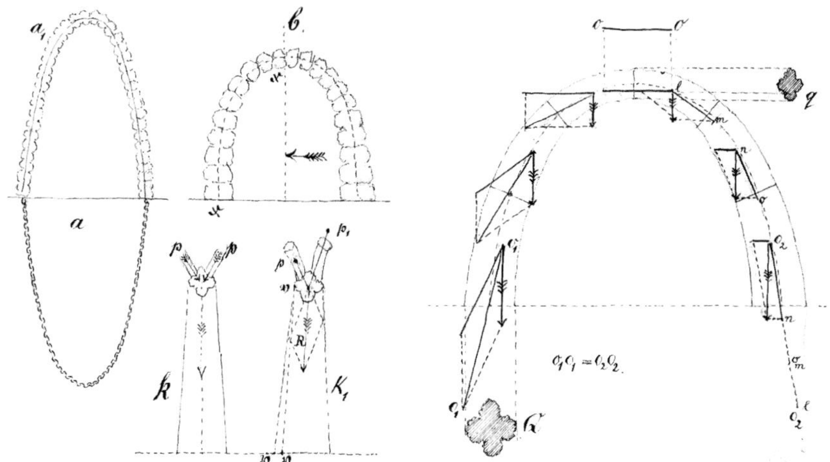

(a) (b)

Figure 3.31 German Reichstag design by Gösling. (a) Arch design based on an inverted catenary. (b) Statics of the catenary arch. (Source: [104])

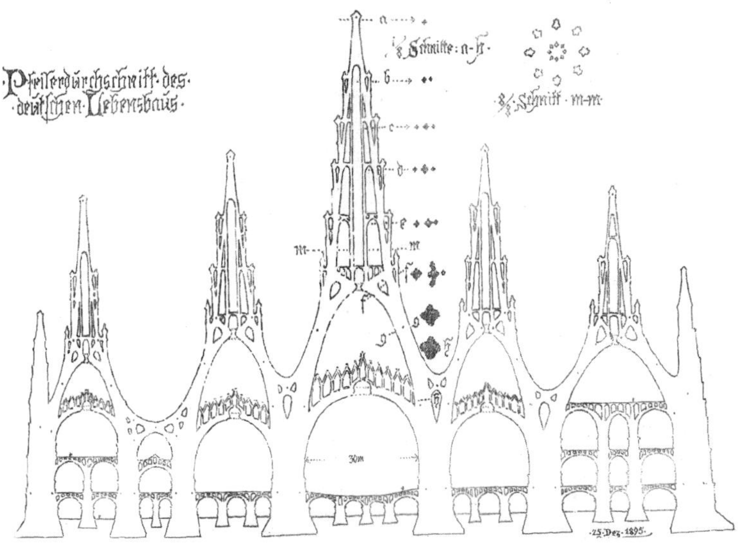

Figure 3.32 German Reichstag design by Gösling, 1871. Section. (Source: [104])

their own weight, and are connected to each other at the apex by square meshes. The drawing shows the structural realization: a ribbed dome with a central compression ring (Figure 3.33b). A later view of the hall ceiling from below shows a modified arch-rib system (Figure 3.34a). This is the earliest reproduction of a spatial hanging model known to me. Gösling seeks to develop the construction details by analogy to construction forms derived from living nature, above

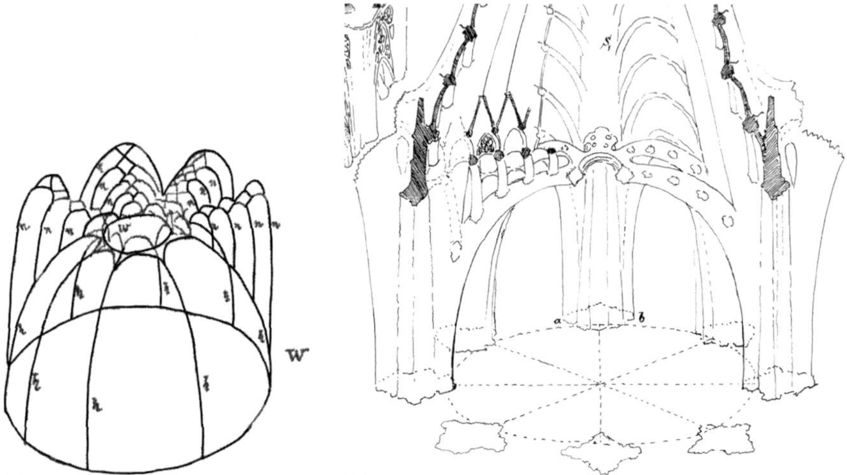

(a) (b)

Figure 3.33 (a) The thread model, shown inverted, used to design the central main hall. (b) The central main hall with a 30 m span, designed with the help of a thread model. (Source: [104])

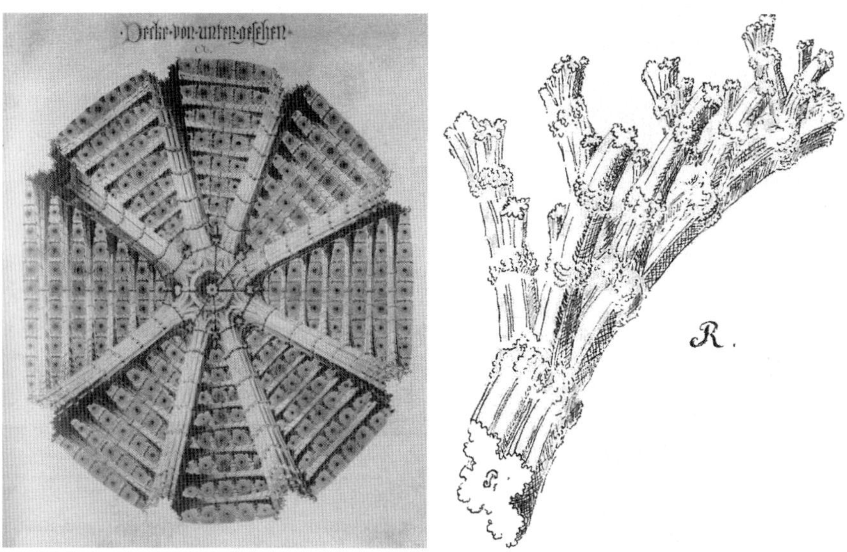

(a) (b)

Figure 3.34 (a) Gösling's modified vault design with a crown replacing the compression ring. (b) An example of Gösling's structural elements deriving their forms from the stem and branches of plants. (Source: [104])

all from the stem and branch structures of plants (Figure 3.34b). Gösling was striving for a new, quasi-natural unity of structural form and architectural form, by optimising the vaulted structure created using self-forming hanging structures, and by developing a harmonised family of building details borrowed from vegetation.

3.5.3 The legendary model of Colònia Güell built by Antoni Gaudí

Between 1898 and 1908 the Catalan architect Antoni Gaudí (1852-1926) built his famous hanging model for the church of Colònia Güell in Barcelona. It is not known how the idea for this model was born. The underlying principle of inverting the catenary had been discussed countless times in the relevant literature of the eighteenth and nineteenth centuries and had long been known in Barcelona [105]. Gaudí, however, did not know that hanging models had already been made in the past. He believed he was the first to use this 'system of equilibrated structure' [106].

What Gösling dreamed of, Gaudí realized. For the first time, he made a model which reproduced the loadbearing structure of an entire, complex building (Figures 3.35). He used threads to represent the supports, arches, vault ribs and wall surfaces (or the lines of thrust within them), and he attached wooden discs to make clear the diameters of the various components (Figures 3.36). The self-weight of the individual components were simulated by bags filled with lead shot suspended from the threads. The 1:10 scale model was almost four metres

Figure 3.35 Church of the Colònia Güell. Photograph of Gaudí's original model in the workshop adjacent to the construction site, c.1903. (Image: Courtesy of Archivo Casa Amatler)

Figure 3.36 Photograph of Gaudí's original model in the workshop adjacent to the construction site, c.1903. (Image: Courtesy of Archivo FotográficoTerrassa)

high and six metres long and was constructed in a hut right next to the church construction site. The delicate equilibrium shape of the model could only be changed and improved in a holistic process. In an elaborate interplay between model tests and establishing the self-weight of the components, the weights of the individual bags were determined ever more precisely. Finally, the positions of the key locations in the model were precisely measured and used to set out the structure in situ. For the lengthy development work on the model, Gaudí was assisted by the architects Francesco Berenguer and Joseph Canaleta and the engineer Eduard Goetz.

Gaudí's main objective was as unusual as the design process. He ventured to develop a new architecture based on experimentally-determined and optimized structural form. The structurally-correct form became the architectural form. In modern times, this was one of the earliest concrete attempts to create a new unity of structural form and artistic form. This goal, however, only partially captures the revolutionary nature of the design. The new unity of structure and form is not arbitrary: rather, it develops through a process of shaping, the inherent dynamic of which permits no arbitrary intervention, only cautious controlling assistance. With the construction of the hanging model, Gaudí entered into a process which constrained his influence as a designer, requiring him to hold back as a designer and to allow the building forms to develop autonomously.

The church of Colònia Güell served as an experimental building. Revolutionary innovations were created – the inclined columns angled to the direction of the thrust, the branched column, the wall in the form of a folded structure, and the vaulted surfaces in the form of hyperbolic paraboloids. The innovations that emerged were immediately used again in other projects, especially in the designs for the Sagrada Família, which were revised over and over again. Gaudí himself called the church of Colònia Güell a 'monumental model' for his large church.

Construction of the church of Colònia Güell began in 1908. The complex structural forms required an immense expenditure. A six-year construction period was needed to complete the crypt and the staircase leading up to the church itself. Then the construction work came to a standstill due to lack of money, and the building remained a torso. It is regrettable that it was this very building – a key work by Gaudí and a landmark of modernism – could not be realized in its entirety. Free from the pressure of time, from pre-determined specifications (as with the Sagrada Familia) and from the need to make a grand gesture, this design had been able to mature, and it must be regarded as one of Gaudí's most convincing works in terms of construction and design.

Nothing is known of any design drawings such as would normally be made. The hanging model is the overall design. Gaudí created an elaborate photographic documentation of the design which fortunately is largely preserved. In order to illustrate the design of the exterior and interior of the church for the client, Gaudí created a number of views by painted over photographs of the hanging model (Figure 3.37). The hanging model and almost all documents relating to the church

(a) (b)

Figure 3.37 Two views of the Colònia Güell painted by Gaudí over photographs of the model. (Images: Courtesy of Archivo Casa Amatler) were destroyed in the Spanish Civil War; so the overall design of the church had to be regarded as lost.

All efforts to reconstruct this important church design must begin with Gaudí's hanging model. The many years of research I and my colleagues undertook cannot be described in detail here. However, some highlights of the reconstruction processes, in which the hanging model played a key role, can be given. The initiator was Harald Szeemann. He commissioned a reconstruction of Gaudí's hanging model for his exhibition *Der Hang zum Gesamtkunstwerk* in Zurich in 1983. This reconstruction was carried out by myself together with Jos Tomlow, Arnold Walz and a team of students at Frei Otto's Institute for Lightweight Structures in the University of Stuttgart, in cooperation with the Gaudí Group of Delft Technical University [107, 108, 109] (Figure 3.38). By reconstructing the model at a scale of 1:15, compared to the original 1:10, it was possible to capture almost every part of the thread model and to reproduce approximately all shapes and proportions of the original. After an exhibition tour around half the world, the model is now exhibited in the Museum of the Sagrada Família.

Of course, the hanging model did not represent the real architecture of the church but rather, the statical system of the load paths running through the columns, arches and walls. In a following research project, we therefore decided to reconstruct the design of church itself. [110]. The aim of our research was to provide as complete, precise and authentic a picture as possible of the overall design, with the help of all the information and sources available, the unique complexity of which surpasses even that of the Sagrada Família, which is by no means simple.

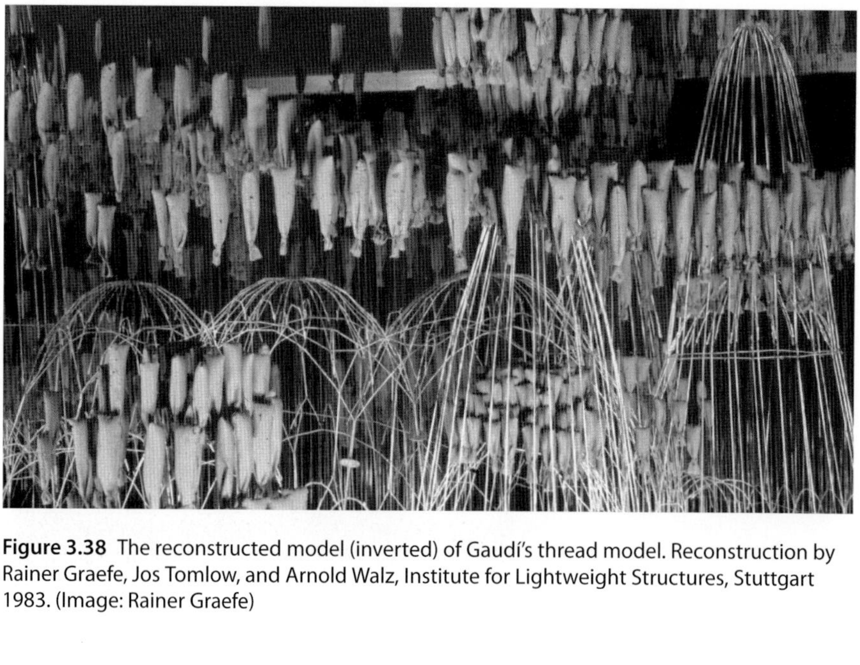

Figure 3.38 The reconstructed model (inverted) of Gaudí's thread model. Reconstruction by Rainer Graefe, Jos Tomlow, and Arnold Walz, Institute for Lightweight Structures, Stuttgart 1983. (Image: Rainer Graefe)

Numerous architecture students from Innsbruck and Moscow were involved in these studies. Together, we first carried out an event as a playful homage to Gaudí and his little church in 1998. Above the existing crypt, at full size (37m high), we erected the front façade and the main tower as an installation which consisted of 200 large, helium-filled balloons, representing an upside-down version of Gaudí's hanging model [111] (Figure 3.39).

Using a photogrammetric evaluation of the historical model photographs, a breakthrough was made enabling a much more precise form analysis of the original. Based on this survey, a 1:25 model of the complete church could be produced using a CNC milling machine. Since 2008, together with other models and exhibits from our project, this has become part of a permanent exhibition at the Diocesan Museum in Barcelona [112] (Figure 3.40).

Further recent investigations have enabled the church design to be captured even more precisely and completely, updated in the 3D model (Figure 3.41) and reinterpreted [113, 114, 115]. Analysis of the different construction phases of the model, which reflect different planning phases, enabled new and surprising insights to be gained about Gaudí's original design process. The reconstruction also included additions and changes that further complicate the project, including the addition of two side portals to the main portal, which Gaudí had to carry out without imposing any changes to the overall equilibrium of the hanging model.

Figure 3.39 Helium-filled balloons creating at full size, over Gaudì's crypt, the form of the front façade and main tower. Creation by the Institute for Building History, University of Innsbruck led by Rainer Graefe. (Image: Rainer Graefe)

(a) (b)

Figure 3.40 Church of the Colònia Güell. (a) Computer-generated model of the reconstructed thread model. (Image: Gaudi-Projekt Innsbruck/Moskau: R. Graefe, L. Demjanov, A. Lyubimkin, K. Lyubimkina, M. Moser). (b) 3D digital model reconstruction of the building. (Image: Gaudi-Projekt Innsbruck/Moskau: R. Graefe, L. Demjanov, A. Lyubimkin, K. Lyubimkina). (Images: Archiv für Baukunst Innsbruck)

3.6 Epilogue

Gaudí's huge, complex hanging model was undoubtedly the culmination of development in using hanging models in building design. We end our reflections with a more modest example. While Gaudí was still working on his great model, the Krefeld architect Karl Buschhüter (1872-1956) invented a supposedly new shape of arch. He conceived his 'Fallbogen', the shape of which, as the name suggests, was to be created by inverting a hanging arc, in 1903. It was published in 1905 on a poster for an exhibition at the Kaiser-Wilhelm Museum in Krefeld, and first used it for the construction of the Dürerheim community building in Krefeld [116]. Buschhüter illustrated (Figure 3.42) the development of the 'coverings of openings', from the Greek architrave, through the Roman round arch and Gothic pointed arch, to his Fallbogen, the only one that needed no abutments:

> 'The dilettantish statics in all parts of old buildings (...) was overcome with one blow with the appearance of my new forms of expression. The momentum that sends waves the world over, into which all the forces join and snuggle, the forces pushing outwards and forcing inwards were precisely suspended in power and counter-power, was also the reason for the arch, my arch, which did not push outwards and did not push inwards. It didn't need the clumsy Gothic buttress...' [117]

Buschhüter believed that his Fallbogen would initiate a new development in architecture. Yet he completely overlooked the fact that similar arches had long

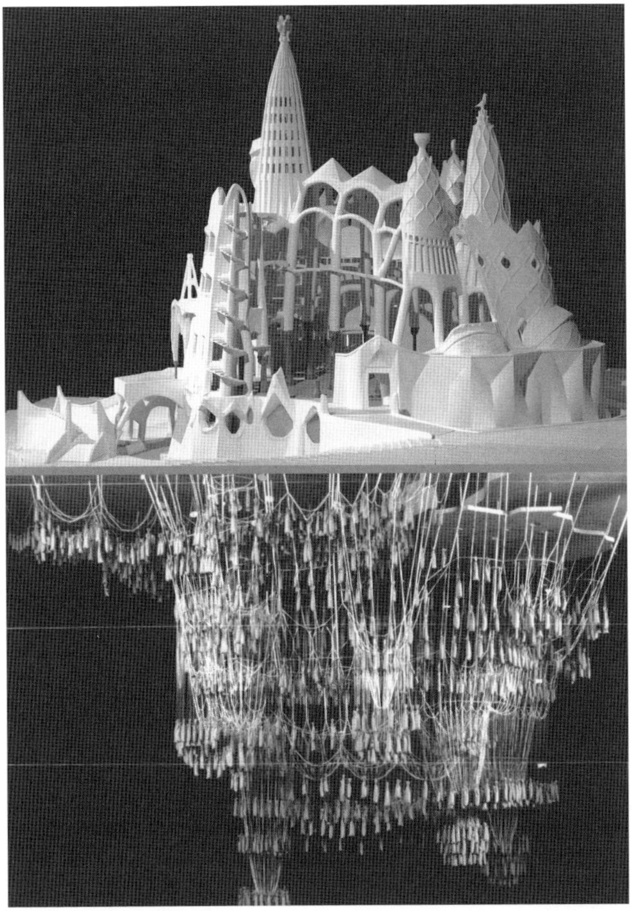

Figure 3.41 Reconstruction of the Church of the Colònia Güell: photomontage. Top: 1:25 scale model of the church produced using a CNC milling machine, based on the computer model. (Gaudi-Projekt Innsbruck/Moskau: R. Graefe, T. Grabner and team) (Image: Rainer Graefe). Bottom: 1:25 scale reconstruction of Gaudi's thread model by R. Graefe, J. Tomlow and A. Walz, Institute for Lightweight Structures, Stuttgart. 1983. (Image: Arnold Walz)

Figure 3.42 Karl Buschüter's drawing to demonstrate that his 'Fallbogen' is the only structural form that requires no abutments. (Source: [117])

been fashionable forms throughout Art Nouveau architecture. In the history of construction, Buschhüter was, for the time being, the last to develop new designs from the catenary. Only half a century later would Heinz Isler and Frei Otto open up new possibilities for finding forms with their hanging models, and open up fascinating new worlds of form for architecture.

References

1 Silberschlag, Joh. E. (1772/73) *Ausführliche Abhandlung der Hydrotechnik oder des Wasserbaus.* Leipzig:Fritsch. 261, §682.

2 Thompson, d'Arcy Wentworth (1945) *On growth and form.* 2nd Ed. Cambridge: University Press. 1007.

3 Torroja, Eduardo (1967) *Philosophy of structures.* Berkeley & Los Angeles: University of California Press. 1.

4 Hooke, Robert (1676) *A description of helioscopes and other instruments.* London: John & Martyn, Printer to the Royal Society. 31.

5 Koßmann, Bernhard (1908) *Lehrbuch des Hochbaues.* Leipzig: Karl Esselborn. Vol.1, 144.

6 Cowan, H.J. (1978) *The Master Builders. A History of Structural and Environmental Design from Ancient Egypt to the Nineteenth Century.* New York: Wiley. 104ff.

7 Hart, Franz (1965) *Kunst und Technik der Wölbung.* München: Callwey. 23, 26 ff.

8 Mainstone, Rowland J. (1975) *Developments in Structural Form.* Cambridge, Mass.: MIT Press. 75.

9 Fathy, Hassan (1973) *Architecture for the Poor.* Chicago: University of Chicago Press. 6-11.

10 https://commons.wikimedia.org/wiki/File:Ctesiphon-ruin_1864.jpg. Public domain. (19/9/2019)

11 Koßmann, Bernhard (1908) *Lehrbuch des Hochbaues.* Leipzig: Karl Esselborn. Vol.1, 144.

12 Mainstone, Rowland J. (1973) Intuition and the Springs of Structural Invention. In: *Structures implicit and explicit.* (ed. J. Bryan & R. Sauer), 47-67. Philadelphia: University of Pennsylvania Press. 64. Reprinted in Mainstone, Roland J. (1999) *Structure in Architecture: History, Design and Innovation.* Aldershot: Ashgate Variorum.

13 Rubió i Bellver, Joan (1912) Conferencia acerca de los conceptos organicos, mecanicos y constructivos de la Cathedral de Mallorca. *Annario de la Asociacion de Arquitectos de Cataluña*, 87-140.

14 Huerta, Santiago, (2019) The Cathedral of Palma de Mallorca: a Wonder of Equiliobrium. Rubió i Bellver's Equilibrium Analysis. In: *The Art of Vaulting: Design and Construction in the Mediterranean Gothic* (ed. Paula Fuentes & Anka Wunderwald), 165-202. Basel: Birkhäuser.

15 Hertwig, A. (1941) Die Entwicklung der Statik der Baukonstruktion im 19. Jahrhundert, *Technikgeschichte* 30, 82-98. 91.

16 Séjourné, Paul (1913) *Grandes Voutes.* Bourges: Tardy-Pigelet et fils. Vol. 3, 313-314.
17 Millington, John (1839) *Elements of Civil Engineering.* Philadelphia: Smith & Palmer. 652f.
18 Séjourné, Paul (1913) *Grandes Voutes.* Bourges: Tardy-Pigelet et fils. Vol. 3, 26ff.
19 Heyman, Jacques, (1982) *The Masonry Arch.* Chichester: Ellis Horwood. 9f.
20 Varitek. https://commons.wikimedia.org/wiki/File:The_Old_Bridge,_Pontypridd.jpg Creative Commons Attribution-Share Alike 3.0 Unported license. (19/9/2019)
21 Graefe R. (1986) Zur Formgebung von Bögen und Gewölben. *Architectura, Zeitschrift für Geschichte der Baukunst* 1: 50-67.
22 Bassegoda Nonell J. (1986) El arco de festón. *Memorias de la Real Academia de Ciencias y Artes de Barcelona,* Epoca 3, Núm. 847, XLVI (20) 505-560.
23 Stevin, Simon (1586) *De Beghinselen der Weeghconst.* Leyden (Reprint Amsterdam 1955), 183.
24 Cresti, Carlo (1980) Arti decorative: cultura e strumenti di progetto. In: *Firenze e la Toscana dei Medici nell'Europa del Cinquecento. Il Potere e lo Spazio. La Scena del Principe.* Firenze: Electa. 185, 187.
25 Galileo Galilei (1638) *Dialogues Concerning Two New Sciences.* Ed. by Antonio Favaro, 1914; reprinted 1954 New York: Dover Publications. 290-293.
26 Waller, Richard (1705) *The Posthumous Papers of Robert Hooke.* London: Sam Smith & Benj. Walford (Printers to the Royal Society). xxi.
27 Kurrer, Karl-Eugen (2018) *The History of the Theory of Structures: Searching for Equilibrium.* 2nd Ed. Berlin: Ernst und Sohn. 212 ff.
28 Gregory, David (1697) 'Catenaria', *Philosophical Transactions* (231) 637-652; Translation from Latin in Ware, Samuel (1809) *A Treatise of the properties of arches, and their abutment piers.* London: Taylor & Longman. 5-7.
29 Graefe R. (1986) Zur Formgebung von Bögen und Gewölben. *Architectura, Zeitschrift für Geschichte der Baukunst* 1: 50-67. 54-56.
30 Heyman, Jacques (2014) Wren's domes. *Proceedings of the First Construction History Society Conference,* Queen's College, University of Cambridge, April 2014, ed. J.W. Campbell et al., Cambridge: Construction History Society. 209-218.
31 Sekler, Edward F. (1956) *Wren and his Place in European Architecture.* London: Faber. 130 Note 1.
32 Heyman, Jacques (2003) Wren, Hooke and Partners. *Proceedings of the First International Congress on Construction History.* Madrid: Instituto Juan de Herrera. 3-9.
33 Robinson, H.W. & Adams, W. (1935) *The Diary of Robert Hooke 1672-1680.* London: Taylor & Francis.
34 Sekler, Edward F. (1956) *Wren and his Place in European Architecture.* London: Faber. 128.
35 Engraving by Samuel Wale and John Gwynn (1755).
36 Heyman, Jacques (1998) Hooke's cubico-parabolical conoid. *Notes and Records of the Royal Society of London* (52) 1, 39-50. 42.

37 Halfpenny, William (1755) *Rural Architecture in the Chinese Taste*. London: Robert Sayer. 6, Plate XXVI.
38 Poleni, Giovanni (1748) *Memorie istoriche della gran cupola del Tempio Vaticano*. Padua.
39 Rondelet, J. (1814) *Traité théorique et pratique de l'art de bâtir*. Vol.2 (Book 3). Paris. 137ff.
40 Hübsch, Heinrich (1838) *Bau-Werke*. Karlsruhe. 42, Footnote.
41 Eytelwein, J. A. (1808) *Handbuch der Statik fester Körper. Mit vorzüglicher Rücksicht auf ihre Anwendung in der Architektur*. Berlin. §95 (Vol.1, 131ff) and §415 (Vol.2, 201).
42 Le Seur, T., Jacquier, F. & Boscovich, R.G. (1743) *Riflessioni de padri Tommaso Le Seur, Francesco Jacquier deel' Ordine de' Minimi, e Ruggiero Giuseppe Boscovich della Compagnia di Gesù, sopra alcune difficoltà spettanti i danni, e risarcimenti della cupola di S. Pietro*. Rome.
43 Bozherianov I.N. (1902) = Божеряновъ И. Н. (1902) Невский проспект. Культурно-исторический очеркъ двухвековой жизни Петербурга. Томъ. 2. СПБ, А.И. Вильборг. 251.
44 Bernsteyn, S.A. (1957) = Бернштейн, С.А. (1957) Очерки по истории строительном механики. Москва Государственное издательсво литературы по строительству и архитектуре. A German translation of Ch.4, 99-108 is available on line from Stuttgart University. https://elib.uni-stuttgart.de/bitstream/11682/6255/1/Ue_371_Bernstejn.pdf (24/10/2019)
45 Silberschlag, Joh. E. (1772/73) *Ausführliche Abhandlung der Hydrotechnik oder des Wasserbaus*. Leipzig: Fritsch. 260ff, §680.
46 Silberschlag, Joh. E. (1772/73) *Ausführliche Abhandlung der Hydrotechnik oder des Wasserbaus*. Leipzig: Fritsch. 261, §681.
47 Gwilt, Joseph (1867) *The Encyclopedia of Architecture*. London (Reprint New York 1982). 1004, Fig. 1306.
48 Silberschlag, Joh. E. (1772/73) *Ausführliche Abhandlung der Hydrotechnik oder des Wasserbaus*. Leipzig: Fritsch. 265, §687.
49 Straub, Hans (1975) *Die Geschichte der Bauingenieurkunst*. 3rd Ed. Basel: Birkhäuser. 182.
50 Chabat, Pierre (1881) *Dictionnaire des termes employés dans la construction*. Paris: Ve A. Morel. Vol 2: 233.
51 Petzet, Michael, (1961) *Soufflots Sainte Geneviève*. Berlin: Walter de Gruyter. 113.
52 Rondelet, J. (1835) *Theoretisch-praktische Anleitung zur Kunst zu bauen*, Vol.4. Leipzig/Darmstadt. 434.
53 Rondelet, J. (1835) *Theoretisch-praktische Anleitung zur Kunst zu bauen*, Vol.4. Leipzig/Darmstadt. 426.
54 Stocker, Jürgen (1987) *Zum Tragverhalten der Zwischenkuppel von St. Geneviève (Pantheon) in Paris*. Diplomarbeit Institut für Baustatik, Universität Stuttgart (Juli)
55 Navier, Claude Louis (1826) *Résumé des Lecons données à l'École des ponts et chaussées*. Paris. 152.
56 Navier, Claude Louis (Ed. By G. Westphal & A. Foeppl) (1881) *Mechanik der Baukunst*. Hannover. 171-173.

57 Yvon-Villarceau, Antoine-Joseph (1844) Équilibre des voutes en berceaux cylindriques. *Revue Générale de l'Architecture et des Travaux Publics, Paris.* 5: 58-82, Plate 4.

58 Gerstner, Franz (1789) *Einleitung in die statische Baukunst.* Prag.Cited in Meyer, Alfred Gotthold (1907) *Eisenbauten.* Esslingen: Paul Neff. 38.

59 Bartholomew, A. (1840) *Specifications for Practical Architecture, preceded by an essay on the decline of excellence in the structure and in the science of modern English buildings; with the proposal of remedies for those defects.* London: John Williams. Art. 92.

60 Meerwein, Carl Friedrich (1798) Über die Stärke der Gewölbebögen. In: *Sammlung nützlicher Aufsätze und Nachrichten, die Baukunst betreffend.* Berlin: Part 1. 74-80, 146.

61 Meerwein, Carl Friedrich (1802) *Beytrag zur richtigen Beurtheilung der Eigenschaften und der Wirkungen der Gewölbe (...).* Frankfurt am Main: Guilhauman. 193ff.

62 Meerwein, Carl Friedrich (1802) *Beytrag zur richtigen Beurtheilung der Eigenschaften und der Wirkungen der Gewölbe (...).* Frankfurt am Main: Guilhauman. 90ff.

63 Meerwein, Carl Friedrich (1802) *Beytrag zur richtigen Beurtheilung der Eigenschaften und der Wirkungen der Gewölbe (...).* Frankfurt am Main: Guilhauman. 132.

64 von Maillard, Sebastian (1817) *Die Mechanik der Gewölbe in ihrem ganzen Umfange abgehandelt.* Pest. ix.

65 Hertwig, A. (1941) Die Entwicklung der Statik der Baukonstruktion im 19. Jahrhundert. *Technikgeschichte* 30, 82-98. 85.

66 Gilly, David (1797) *Handbuch der Landbaukunst* 5th Ed. Part 2. Braunschweig. 73-75 and Fig. 51 c.

67 Huerta, Santiago (2005) Thomas Young's theory of the arch. His analysis of Telford's design for an iron arch of 600 feet over the Thames in London. In: *Essays in the history of the theory of structures* (ed. S. Huerta), 189-233. Madrid: Instituto Juan de Herrera.

68 Huerta, Santiago (2005) Thomas Young's theory of the arch. His analysis of Telford's design for an iron arch of 600 feet over the Thames in London. In: *Essays in the history of the theory of structures* (ed. S. Huerta), 189-233. Madrid: Instituto Juan de Herrera. 222-223 and Figure 13. (Reconstruction by Huerta).

69 Millington, John (1839) *Elements of Civil Engineering.* Philadelphia: Dobson. 644ff.

70 Millington, John (1839) *Elements of Civil Engineering.* Philadelphia: Dobson. 657.

71 Meyer, Gustav (1862) Ueber Englische Eisenbahnbrücken. *Zeitschrift des Architekten- und Ingenieur-Vereins für das Königreich Hannover* VIII (3-4): Columns 281-232. Column 320, Figures 39 & 40.

72 Hübsch, Heinrich (1838) *Bau-Werke.* Carlsruhe. 40-53, Figs IV, VI, VII, VIII, IX, Plate 7.

73 Graefe R. (1983) Heinrich Hübsch als Konstrukteur. In: *Heinrich Hübsch 1795-1863. Der grosse badische Baumeister der Romantik. Ausstellungskatalog* (ed. Stadtarchiv Karlsruhe and Institut für Baugeschichte der Universität. Karlsruhe), 184-189.

74 Hübsch, Heinrich (1838) *Bau-Werke*. Carlsruhe. 44.

75 Göricke, Joachim (1974) *Die Kirchenbauten des Architekten Heinrich Hübsch*. (Dissertation) Karlsruhe. Note 142.

76 Warth, Otto (1903) *Die Konstruktionen in Stein*. (7th Ed.) Leipzig: J.M. Gebhardt. 300 ff.

77 Hagen, G.H.L. (1862) *Über Form und Stärke gewölbter Bögen und Kuppeln*. Berlin: Ernst & Korn. 6.

78 Hagen, G.H.L. (1862) *Über Form und Stärke gewölbter Bögen und Kuppeln*. Berlin: Ernst & Korn. 25.

79 Tomlow, Jos (1993) Die Kuppel des Gießhauses der Firma Henschel in Kassel (1837) – Ein frühe Anwendung des Entwurfsverfahrens mit Hängemodellen. *Architectura, Zeitschrift für Geschichte der Baukunst* 23 (2): 151-172.

80 Tomlow, Jos (1993) Die Kuppel des Gießhauses der Firma Henschel in Kassel (1837) – Ein frühe Anwendung des Entwurfsverfahrens mit Hängemodellen. *Architectura, Zeitschrift für Geschichte der Baukunst* 23 (2): 151-172. 172 citing Henschel, Carl-Anton (1850) *Zur Aesthetik der höheren Baukunst*. Cassel. 38.

81 Tomlow, Jos (1993) Die Kuppel des Gießhauses der Firma Henschel in Kassel (1837) – Ein frühe Anwendung des Entwurfsverfahrens mit Hängemodellen. *Architectura, Zeitschrift für Geschichte der Baukunst* 23 (2): 151-172. 162ff.

82 Habicht, V. C. (1937) Architekturtheorie. In: *Reallexikon zur deutschen Kunstgeschichte*. Vol.1. Stuttgart. Sp. 988 ff.

83 Tappe, Wilhelm (1818-21) *Darstellung einer neuen, äußerst wenig Holz erfordernden und höchst feuersicheren Bauart*. Essen. Eight booklets.

84 Tappe, Wilhelm (1818-21) *Darstellung einer neuen, äußerst wenig Holz erfordernden und höchst feuersicheren Bauart*. Essen. Booklet 5, 10.

85 Siebert, Irmgard (ed.) (1992) *Jacob Burckhardt: Ästhetik der Bildenden Kunst. Text der Vorlesung 'Zur Einleitung in die Aesthetik der bildenden Kunst' aufgrund der Handschriften (1863)*. Darmstadt: Wissenschaftliche Buchgesellschaft. 47.

86 Tappe, Wilhelm (1818-21) *Darstellung einer neuen, äußerst wenig Holz erfordernden und höchst feuersicheren Bauart*. Essen. Booklet 4, 10.

87 Tappe, Wilhelm (1818-21) *Darstellung einer neuen, äußerst wenig Holz erfordernden und höchst feuersicheren Bauart*. Essen. Booklet 5, 11.

88 Schreiner, Ludwig (1970) Wilhelm Tappe (1769-1823), ein Architekturtheoretiker des 19. Jahrhunderts. *Niederdeutsche Beiträge zur Kunstgeschichte* (IX) 195-234. p. 214

89 Anon (1822) *Leipziger Literatur-Zeitung*. Nr. 37, 11 Feb. Columns 293-294.

90 Fornés y Gurrea, Manuel (1841) *Observaciones sobre la práctica del arte de Edificar*. Valencia: Imprenta de Cabrerizo. 21.

91 Ochsendorf, J. (2010) *Guastavino Vaulting. The Art of Structural Tile.* New York: Princeton Architectural Press. 21-33.

92 Huerta, Santiago (ed.) (2006) *Estritos sobre la construccíon cohesiva y su funcíon en arquitectura: Rafael Guastavino.* Madrid: Instituto Juan de Herrera.

93 US Patent 336,047, 9 February 1886. R. Guastavino: Fire proof building.

94 Gulli, R. & Mochi, G. (1995) *Bóvedas Tabicadas – architettura e costruzione.* Rome: Edilstampa.

95 Mackenzie, Frederick (1840) *Observations on the Construction of the Roof of King's College Chapel.* London: Weale. 17-18.

96 Garbett, E.L. (1850) *Rudimentary Treatise on the Principles of Design in Architecture, as deducible from Nature and exemplified in the Works of the Greek and Gothic Architects.* London: John Weale. 191.

97 Straub, Hans (1975) *Die Geschichte der Bauingenieurkunst.* 3rd Ed. Basel: Birkhäuser. 186.

98 Hagen, G.H.L. (1862) *Über Form und Stärke gewölbter Bögen und Kuppeln.* Berlin: Ernst & Korn. Cited in Hertwig, A. (1941) Die Entwicklung der Statik der Baukonstruktion im 19. Jahrhundert. *Technikgeschichte* 30, 82-98. 85.

99 Hagen, G.H.L. (1862) *Über Form und Stärke gewölbter Bögen und Kuppeln.* Berlin: Ernst & Korn.

100 Esselborn, Karl (1908) *Lehrbuch des Hochbaues.* Leipzig: Engelmann. Vol.1, 105, Figs 146-148.

101 Addis, Bill (2007) *Building: 3000 years of design, engineering and construction.* New York: Phaidon. 438-446.

102 Ungewitter, G. (1890) *Lehrbuch der gotischen Konstruktionen, neu bearbeitet von K. Mohrmann.* Leipzig: Wiegel. Vol.1, 45-46.

103 Personal communication to the author from Günther Kokkelink.

104 Graefe, Rainer (2012) 'Natürliche' Formbildungsprozesse in der Neugotik – Friedrich W. H. Gösling (1837-1899) In: *Bautechnik des Historismus* (ed. U. Hassler & C. Rauhut), 144-159. München: Hirmer.

105 Bassegoda Nonell J. (1986) El arco de festón. *Memorias de la Real Academia de Ciencias y Artes de Barcelona* Epoca 3, Núm. 847, XLVI (20): 505-560. 46.

106 Martinell, César (1975) *Gaudí: His Life. His Theories. His Work.* Barcelona: Editorial Blume. 112-113.

107 Graefe, R. & Tomlow, J. (1983) Antoni Gaudí i Cornet. In: *Der Hang zum Gesamtkunstwerk.* Aarau: Sauerländer (Catalogue of exhibition in the Kunsthaus, Zürich). 215-220.

108 Graefe, R., Tomlow, J. and Walz A. (1989) Ein verschollenes Modell und seine Rekonstruktion. *Bauwelt* 15, 568-573.

109 Tomlow, J. (1989) *Das Modell. Antoni Gaudís Hängemodell und seine Rekonstruktion. IL34.* Stuttgart: Institut für leichte Flächentragwerke.

110 Dieses Projekt wurde unter meiner Leitung und in enger Zusammenarbeit mit Prof. Leonid Demjanov (MIGAIK Moskau) am Innsbrucker Institut für Baugeschichte durchgeführt

111 Graefe, R. (1999) Fliegende Kirche. Eine Installation über Antoni Gaudís Krypta in der Colonia Güell. *Baukultur* (6) 28-30.

112 Graefe, Rainer, (2008) Els dos temples de Gaudí. In: *El primer temple de Gaudí. La recuperación d'una obra mestra.* Exhibition Catalogue. Barcelona: Museu Diocesà de Barcelona.

113 Graefe, Rainer (2009) Reconstruction of Gaudí's Church of the Colonia Güell. *Proceedings of the Third International Congress on Construction History*, Cottbus. Vol.2, 729-736.

114 Graefe, Rainer (2010) Gaudís Hängemodell für die Kirche der Colonia Güell und seine Umsetzung im Bau. In: *Erhalten historisch bedeutsamer Bauwerke. Festschrift Fritz Wenzel.* (ed. Rainer Barthel), 135-141. München: Technische Uni München Fachbereich Architektur, Lehrstuhl f. Tragwerksplanung.

115 Graefe, Rainer (2010) Rekonstruktion des Entwurfs von Antoni Gaudís kleiner Kirche – ein Arbeitsbericht. In: *Bauforschung. Zur Rekonstruktion des Wissens.* (ed. Uta Hassler), 46-53. Zürich: Vdf Hochschulverlag.

116 Walfried, Pohl & Schwanke, Hans-Peter (1987) *Der Krefelder Architekt Karl Buschhüter 1872-1956.* Krefeld: Stadtarchiv Krefeld. 365, Note 911.

117 Walfried, Pohl & Schwanke, Hans-Peter (1987) *Der Krefelder Architekt Karl Buschhüter 1872-1956.* Krefeld: Stadtarchiv Krefeld. 367.

4

Leonhard Euler and the model tests for a 300-metre timber arch bridge in St. Petersburg

Andreas Kahlow

4.1 Timber bridges and material experiments in the eighteenth century

4.1.1 Timber bridges in the mid-eighteenth century

The bridge over the Rhine at Schaffhausen, built by the Grubenmann brothers in 1758, suddenly made Switzerland a pioneer in bridge construction in the eighteenth century. There were signs that, in addition to the familiar simple truss structures with hangers, there was a growing tendency to use trussed-arch construction, employing polygonally-arranged timber members. The consequent transition to an arch structure took place in 1766 with the completion of the bridge over the River Limmat at Wettingen [1] (Figure 4.1). The beams of the timber arch were connected to each other by interlocking or sawtooth splices and bolting to form a shear-resistant composite beam [2].

Simultaneously with the transition to larger spans, there was an increase in the use of load tests on small-scale models of what were often elaborately-fabricated bridges. Reports about experiments on structural models go back a long way; perhaps the first example to be reported in print was that described in 1561 by Philibert de l'Orme for his timber plank, arched roof construction. In his home, he tested two models of his invention, one made of 225 separate pieces, the other 263 pieces, applying loads using ropes and two turnbuckles. The arches deformed by less than a half a finger's width before he stopped the test for fear of damaging the floor of his room [4].

In the mid-eighteenth century, however, model tests were usually undertaken to make decisions for or against a particular type of bridge construction. With the increasing complexity of construction, they also formed a necessary preliminary stage in developing precise construction techniques, as in the case of the Grubenmann brothers [5]. Only rarely were books written by the carpenters themselves; the book *Brückenbau* by Caspar Walter (1701-1769), from Augsburg in southern Germany, became widely known after its publication in 1766 [6]. It provided relatively precise information about the experiments he made on bridge models. He reports on two models that he tested, made at a scale of 1:20 for bridges with spans

Physical Models: Their historical and current use in civil and building engineering design,
First Edition. Edited by Bill Addis.

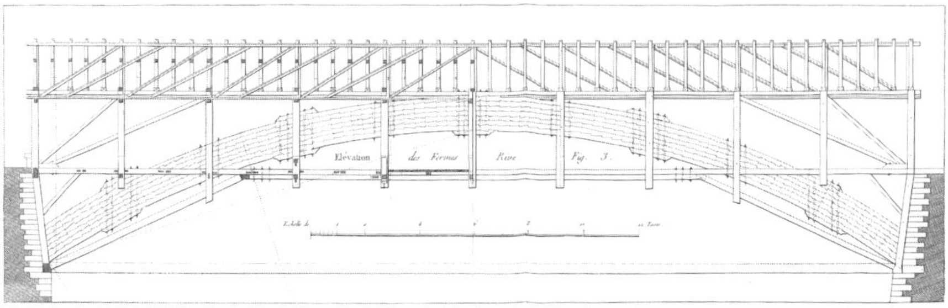

Figure 4.1 Wettingen Bridge (1766). Built by Hans Ulrich Grubenmann, his brother Johannes Grubenmann and his son (also named Johannes Grubenmann). (Source: [3])

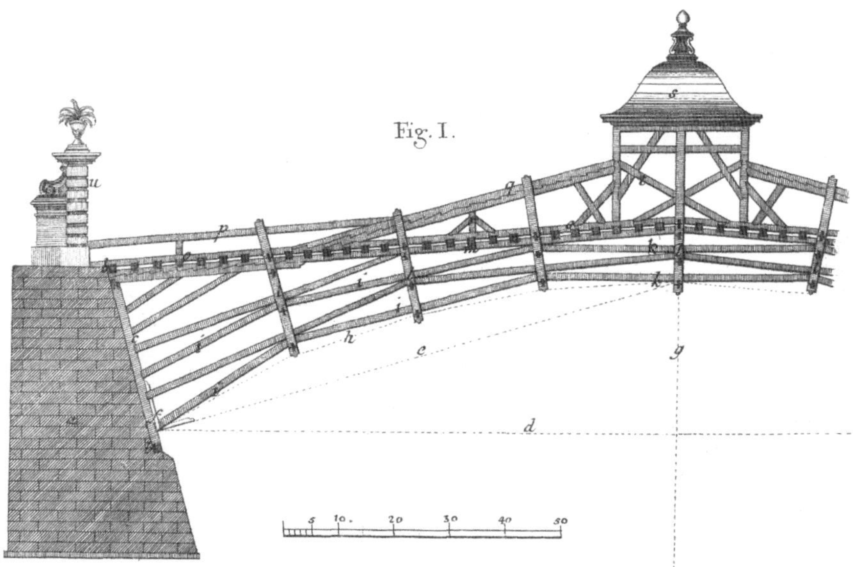

Figure 4.2 A bridge of span 62.8 m tested by Caspar Walter by applying loads to a 1:20 scale model. (Source: [6] p. 26-27, Tab. XVII)

of 200 and 300 Rhenanian feet (62.8 & 94.2 m) (Figures 4.2 and 4.3). The model of the smaller bridge, made from 1/2 inch (13 mm) thick timber sections, was loaded with 6 quintals and 6 persons, which Caspar Walter equates to 15 quintals (1 quintal = 1 centner = one hundredweight = 50 kg). Since the timber in the full-size bridge is 10 inches (26.2 cm) thick, it would 'carry 300 centners for sure'. The second, larger model of a 300-foot span withstood a load of 10 quintals and 10 people, equivalent to a load of 25 quintals. Also here the scaling up from the model to the real bridge was in direct proportion to the linear scale factor, from ½ inch to 10 inches, i.e. 20-fold, from 25 hundredweight test load to a load capacity of 500 hundredweight (25 tonnes) for the full-scale bridge, which was

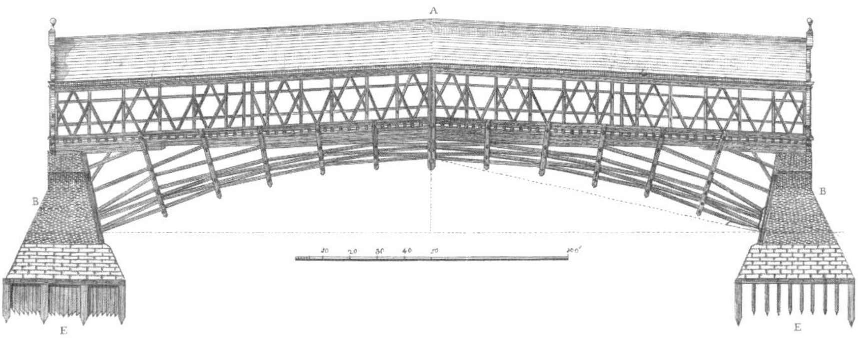

Figure 4.3 A bridge of span 94.2 m tested by Caspar Walter by applying loads to a 1:20 scale model. (Source: [6] p. 28, Tab. XVIII)

an overestimate. With regard to the construction of this second bridge, Walter stresses that the 'timbers ... form a vault'. Caspar Walter presents the transfer of the vault principle from masonry bridge construction to timber bridge construction as something new [7]. However, in contrast to Leonhard Christoph Sturm (1669-1719) [8] and others, Walter does not take into account the effect of the dead weight increasing as the cube of the scale factor, well-known since Galileo, which leads to a relative reduction of its load-bearing capacity with the increasing size of a beam-shaped load-bearing structure [9].

4.1.2 Experimental testing of material properties

In addition to practitioners such as the Grubenmann brothers and Caspar Walter, a number of mathematicians, architects and engineers were involved in model experiments in the eighteenth century. Most of them knew the Galilean beam theory, but they were not carpenters. The purpose of experiments was aimed at different knowledge objectives from those of the practitioners. The geometrical and physical modelling had priority: the aim was not to test the load-bearing capacity of a particular construction, but to gain knowledge about the behaviour of materials and to establish general rules for the dimensioning of components. The works of Pieter van Musschenbroek (1692-1761) and Bernard Forest de Bélidor (1697-1761), both published in 1729, were especially influential.

While Musschenbroek deals with the cohesive properties of solid bodies in the section on *Cohaerentia Corporum firmorum* in his book [10], Bélidor compiles results for the bending strength of beams, a fundamental element in the practice of timber construction. Both try to start out from the scientific results achieved so far, and to investigate the forces involved by 'geometrization', but they do not yet establish relationships describing the state of equilibrium. Bélidor's experiments confirm the quadratic increase in strength with a linear increase in the depth of a beam. He examines the cases of the simple supports, and also those of beams

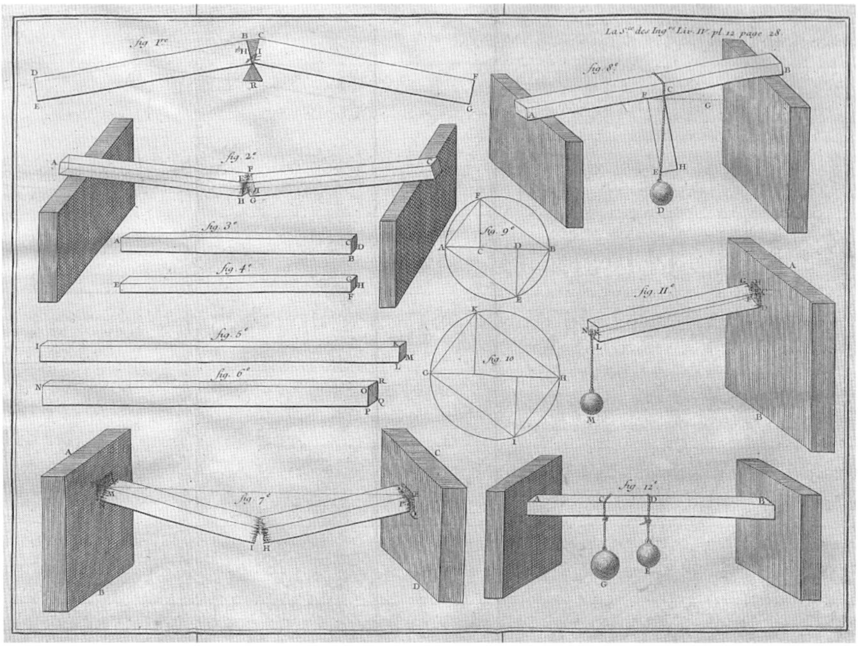

Figure 4.4 Illustrations by Bélidor of tests to determine the strength of beams. (Source: [13])

encastred at both ends in masonry (Figure 4.4). However, there is no analytical consideration of the effects of force acting on the individual fibres, so that he uses equations of simple proportion on the basis of the test results. In so doing, he hopes to give practical help more well-founded rules than a catalogue of beam dimensions that have been verified by testing. Bélidor reported the outcome of most of the tests he conducted in the Arsenal at La Fère [11, 12].

Musschenbroek, who is concerned with the general rules of *Cohaerentia* [14], is even more thorough in his approach. He distinguishes between the *Cohaerentia absoluta* and *Cohaeretia respectiva* and presents various ways of applying loads to test pieces, indeed the same ways that Galilei had already partly considered. Among them is a rod loaded in tension, the cantilever, the beam on two supports and a strut loaded in compression (Figures 4.5, 4.6, 4.7). Subsequently, definitions, propositions and experiments are presented in the *Caput secundum* [15]. From his experiments on the buckling strength of struts, Musschenbroek found that the load causing the buckling of the strut is inversely proportional to the square of the strut length [16], nearly 30 years before Euler's famous paper on buckling in 1757 (see Section 4.5.1). Musschenbroek was also the first to distinguish the elastic buckling from the later fracture of the rod [17]. Musschenbroek also describes the fracture behaviour of various simple bodies (Figure 4.8) and reports the different strengths he has determined experimentally. All the experiments were carried out up to fracture – at that time the transition between elastic and plastic behaviour was not addressed.

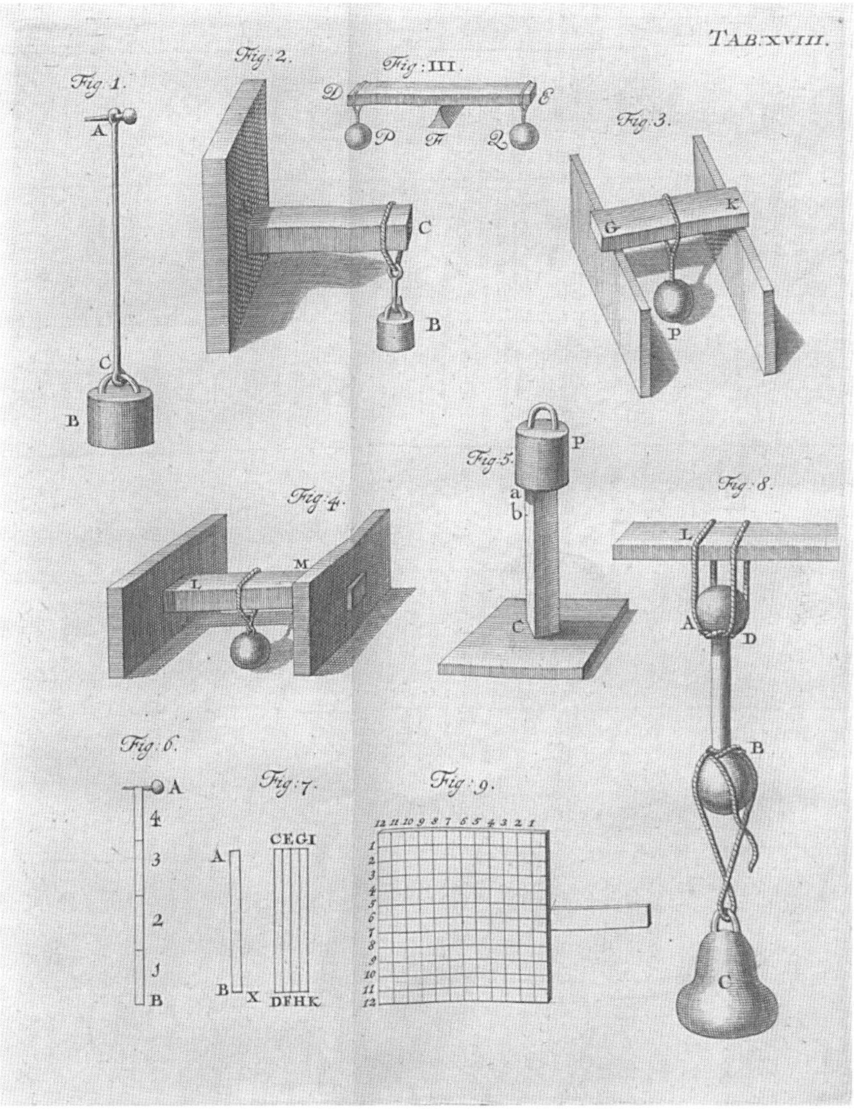

Figure 4.5 Illustration by Musschenbroek of tests to determine the tensile strength of materials. (Source: [10] Plate XVIII)

4.2 Proposals for a bridge across the River Neva in St Petersburg

4.2.1 The first design proposals by Ivan Kulibin

During the last third of the eighteenth century, the question of a bridge over the Great Neva in St. Petersburg, to connect Vasily Island with the left bank of the Neva River, was discussed more and more seriously. Until then, pontoon

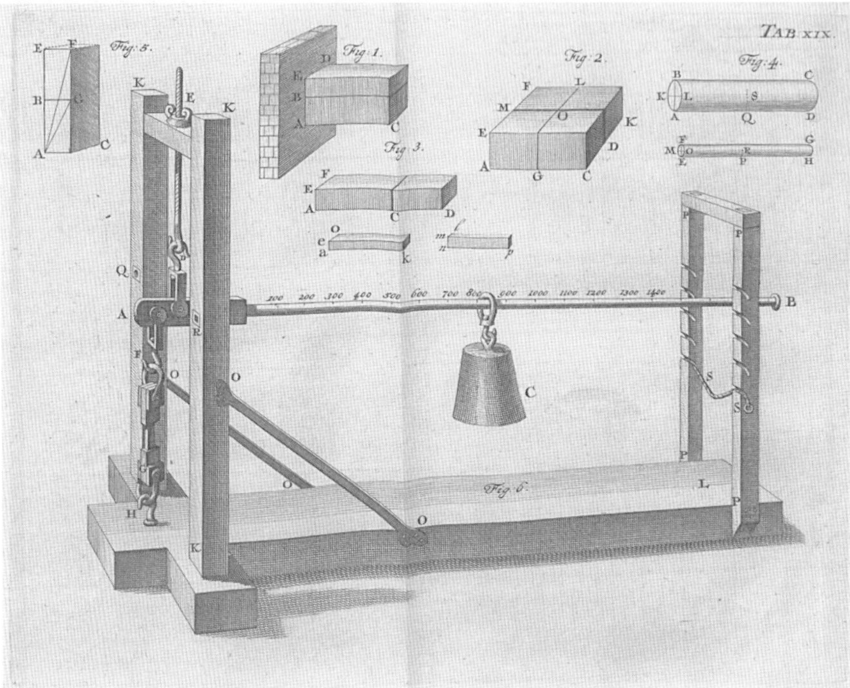

Figure 4.6 Illustration by Musschenbroek of his apparatus for determining the tensile strength of materials. (Source: [10] Plate XIX)

bridges had been used to establish a transport link. In 1771, Ivan Petrovich Kulibin (1735-1818), the head of the mechanical workshops of the St. Petersburg Academy of Sciences, presented an extremely daring project to the Academy members: the construction of a timber bridge with a single arch. The span of the arch would be 140 Sashen (a Sashen, or Russian fathom = 2.134 m), indicating the timber arch would have had a span of almost 300 metres.

The earliest document preserved in this context is one by Kulibin dated 9 December, 1772 (old calendar – Russia adopted the Gregorian calendar in 1918). It was first published in 1953 together with other documents [18] in the course of the processing of the complete collection of Kulibin's documents in the archive of the St. Petersburg Academy of Sciences [19]. An evaluation of Kulibin's work, especially on his bridge project, was undertaken in 1936 by B.V. Jakubovskij [20]. In particular, N.R. Raskin has devoted himself from the 1950s to the 1980s to many publications evaluating Kulibin's documents [21, 22, 23].

The document dating from 1772 is a request that Kulibin made to Alexej Reshevski, the Deputy Director of the St. Petersburg Academy. Kulibin describes in it that he had already tackled the project of a bridge over the Great Neva shortly after his arrival in St Petersburg in 1769. From his drawings a first model was made, 4 Sashen (8.5 m) long, consisting of a tube formed by vertical and horizontal trusses. He presented this model to the members of the Academy in 1771, although no mention of this is made in the minutes of the proceedings of

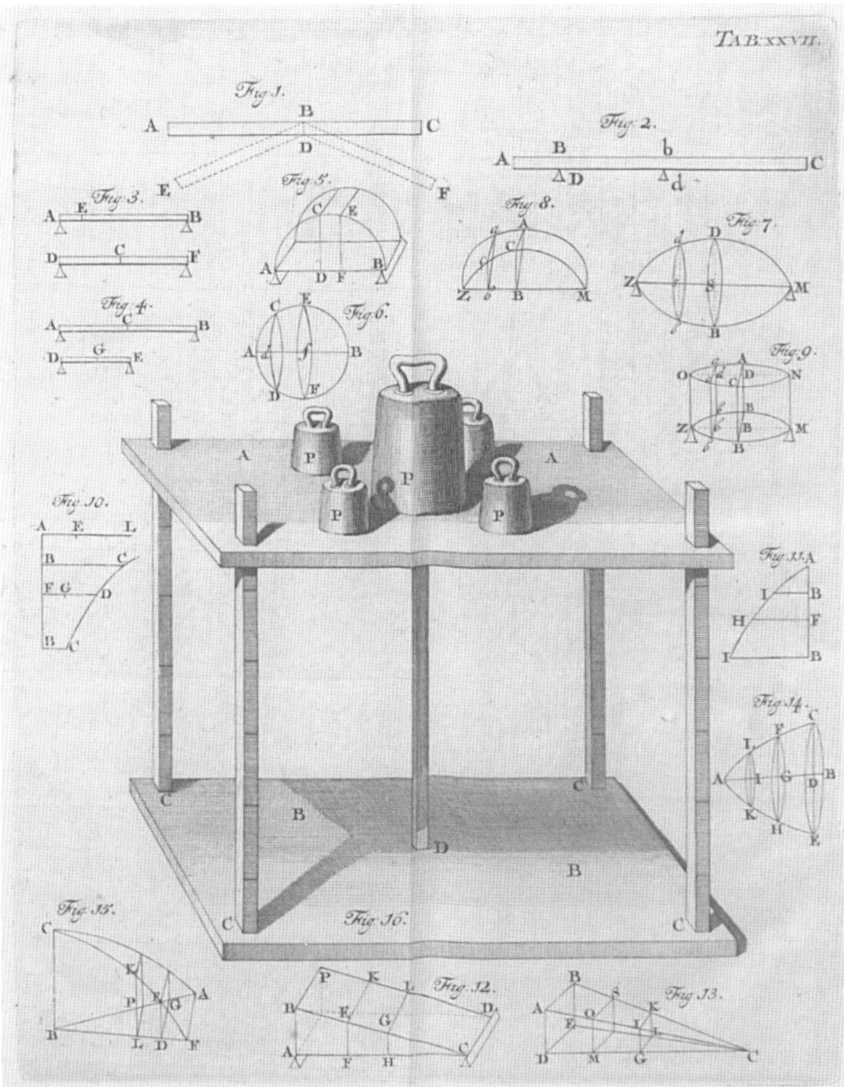

Figure 4.7 Illustration by Musschenbroek of his apparatus for determining the buckling strength of struts. (Source: [10] Plate XXI)

the Academy in St. Petersburg. Kulibin himself, however, goes on to mention in his written request that, since this presentation in 1771, he had not only understood the effect of the dead weight through many experiments, but had also found a method to achieve a correct scaling up of the loads measured in a model to the full size bridge that would be built [24]. From Kulibin's remarks it seems likely that he may have been informed about the experiments published by Bélidor and Musschenbroek through his contact with scientists in the Academy. The fact that reference was made to these books during the subsequent tests of bridge proposals in St. Petersburg and, in particular, that Leonhard Euler made

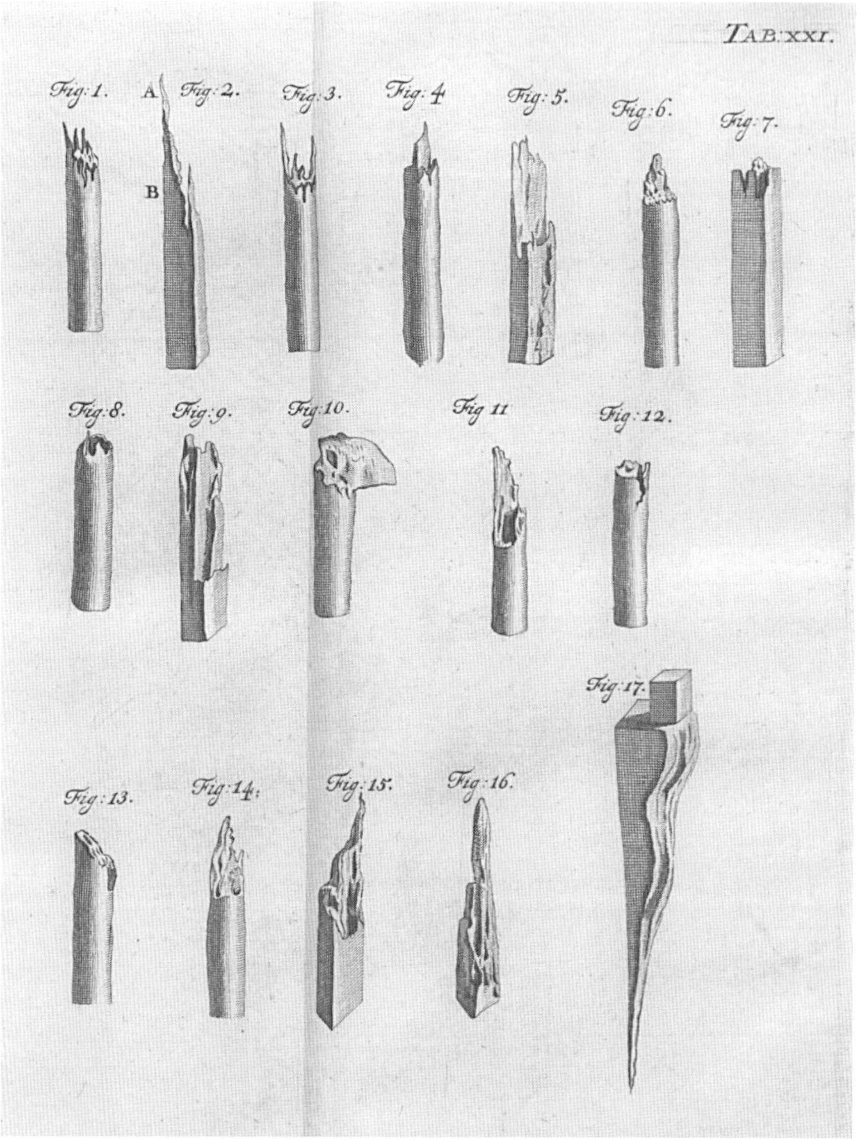

Figure 4.8 Illustration by Musschenbroek showing how specimens of different materials and shapes ruptured when loaded in tension. (Source: [10] Plate XXVII)

reference to them, points towards the later outcome of events. Furthermore, Kulibin was probably inspired by an existing bridge model dating from 1762, attributed to a Swiss bridge builder named Aldon which still survives and is displayed in the Central Museum for Railways in the Russian Federation in St. Petersburg. The length of the model is given as 2 1/3 Sashen (4.98 m), the width as 1 3/4 feet (53 cm), and the height as 2 feet (61 cm). The scale of the model is 1:60, indicating a full-size bridge with a span of 298 metres (Figures 4.9 and 4.10).

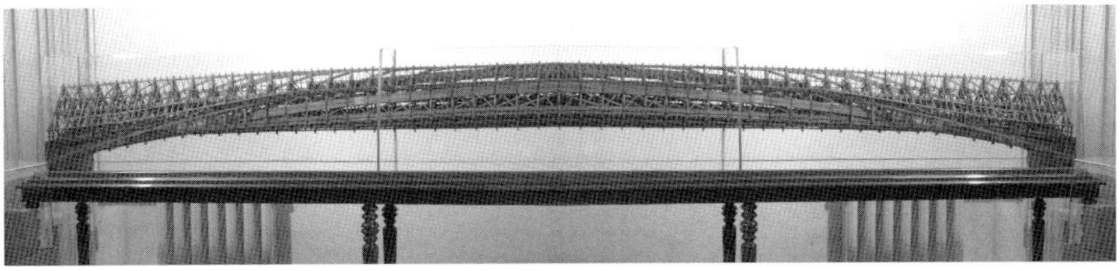

Figure 4.9 Original 1:60 scale model of a timber arch bridge across the Neva, made in 1762 by a Swiss bridge builder named Aldon (picture mirrored in the centre). (Image: Central Museum for Railways in the Russian Federation, St. Petersburg. = Центральный музей железнодорожного транспорта Российской Федераций, Ст. Петербург (ЦМЖТ)).

Figure 4.10 Original 1:60 scale model of a timber arch bridge across the Neva, made in 1762 by a Swiss bridge builder named Aldon. Detail showing scarfed joints. (Image: Andreas Kahlow)

In the description of the model collection in 1862 it is stated that the collection also contains unrealized proposals of bridges over the great Neva that have emerged over the last hundred years: 'they give insight into proposals for dealing with the technically important question of bridging over a tidal river, with a poor quality river bed, low banks, a fast flowing current and large ice drifts. The series [of models] begins with a project (1762) by the Swiss Aldon, long attributed to the mechanic and autodidact Kulibin' [25]. The author continues:

> 'In his project, the author proposes to build the arch in such a way that it spans the entire width of the river, Between the main ribs of the arch, two carriageways are planned, each with a pavement for pedestrians. The

destructive effect of the wind is prevented by side struts. The outside of the arch is protected by a canopy.
The model [by Aldon] also allows a better understanding of the remarkable project of the well-known Russian mechanic Kulibin, who proposed building a timber arch bridge with a span of 140 Sashen over the Neva River and at the place where the Isaakevski Bridge was located. The model for the project was built in 1776. The difference between the Aldon system and the Kulibin system is that the rise in the model [by Aldon] is less than in Kulibin's model, and that some of the iron brackets have a different shape. The system of arches [by Kulibin] is remarkable and, with some success, perhaps applicable to larger non-loadbearing arches, but difficult to execute in an arch spanning 140 Sashen. In any case, Kulibin was not the first to deal with the question of a permanent bridging of the Neva, which is still a magnificent task to solve today. Finally, it should be noted that the model in question [by Aldon] was kept for a long time in the Art Chamber of the St. Petersburg Academy of Sciences and was incorrectly presented to the public as a model of Kulibin. The latter model had been built on a massive scale of 1/10 of the real size, and has not existed for a long time' [26].

In his written application of December 1772, Kulibin presents to the Academy a revised bridge project, together with the associated explanations and drawings. Kulibin indicates he wants to have a model built, not just for the sake of a better visual understanding, but specifically in order to carry out load tests. In his application, he formulates a similitude rule for determining the load to be applied to the model in which he assumes a linear decrease in the relative load-bearing capacity: in summary – if the bridge is n times larger than the model, the model must be tested with a dead load n times larger than its own weight.

Kulibin demonstrates this rule mathematically using the example of his planned Neva Bridge, which will span 140 Sashen. Based on its likely construction, the full-size bridge would weigh 237,568 pudi (3870 tonnes; 1 *pood* (plural *pudi*) ≈ 16.3 kg), so that a 1:16 scale model bridge would weigh 237,568/(16^3) = 58 pudi (945 kg). To achieve correct similitude with the real bridge, the model bridge would need to be loaded with a dead load n-times its own weight, i.e. 58 x 16 = 928 pudi (15.13 tonnes). Likewise, for a 1:8 scale model, weighing 237,568/(8^3) = 464 pudi (7563 kg), it would have to be loaded with a dead load of 464 x 8 = 3712 pudi (60.5 tonnes) [27]. In modern understanding this is correct for a beam on two supports and, depending from its form and design, an arch can bear even more.

In his application, Kulibin proposed a grid construction that was as light as possible in order to avoid collapse due to its own weight given such a great span. Since the drawing that accompanied the application has been lost, two later drawings, the first probably by Kulibin, are shown here (Figures 4.11 and 4.12). There exists, in the Hermitage Museum in St Petersburg, an undated model made by Kulibin of one of his designs for his bridge; at present it lies in storage and is in need of some conservation work.

Figure 4.11 Drawing of the final variant of Kulibin's bridge project, probably in his own hand, 1770s. (Source: [28])

Figure 4.12 General view of Kulibin's timber arch bridge over the Neva. Engraving. (Source: [29] p. 251)

4.2.2 The bridge design competition of 1771

The idea of choosing a compression-loaded arch as the basic structural form had gradually gained acceptance in Switzerland in the mid-eighteenth century, as the Limmat Bridge shows (Section 4.1.1). The early wood model by the Swiss bridge builder Aldon illustrates this trend [30]. Kulibin's 1772 project was further stimulated by the news arriving in St Petersburg that the Royal Society of London had awarded a prize to the Swiss bridge builder Konrad Altherr for the best design for a bridge capable of spanning the River Foyle in Londonderry, Ireland without intermediate supports (see Section 1.5.2). The 1:48 scale model of this timber trussed-arch bridge, 5.8 m long, represented a bridge spanning 900 feet (275 m)

[31]. It is interesting that Nicolas Fuss wrote to Daniel Bernoulli (1700-1782) in 1777 [70] that he had seen this model in Basel before he travelled to St. Petersburg in 1773. It was, he said, made by a man from Appenzell named Schwarzenbach (possibly a mistake by Fuss) and was destined for Londonderry, and later was displayed in London (see Section 4.4.1). Remarkably, this model survives in the Science Museum, London [32] (see Section 39.2.1).

It is surprising that the name of the client for this competition penetrated as far as Russia. It is even more surprising, however, that Kulibin combined the problem of increasing scale, which had been known since Galilei and had already been dealt by Sturm, Musschenbroek and Bélidor [8, 10, 11], with the practical skills of a model maker. The example of the Grubenmann brothers shows the astonishing achievements that were made at that time through experience and intuition. Certainly, the transfer of the idea of a compression-loaded arch from masonry construction into timber construction was within the reach of someone with practical experience of structures. Kulibin, however, arrived at his conclusions through systematic experimental design, supported by calculations, and arising directly from his approach as an engineer working at the Academy of Sciences of Saint Petersburg.

Finally, some further observations are needed in order to complement the calculation of the rules for loading the model bridge that Kulibin described in his manuscript and illustrated with some numerical examples. One of these concerns his finding the correct form of the arch bridge. In the appendix to his application of 9th December 1772, Kulibin described how he intended to determine the shape of his structure. Using thin 'Dutch' thread stretched between two supports, he determined the line of thrust for the structure with the aid of Solotniks (weights of 4.26 grams). Kulibin thus demonstrated that he was equally aware that the outward thrust on the supports of the arch depended on the distribution of weight along the length of the bridge [33]. After getting the financial means of 1000 Rubles, Kulibin started the construction of his third and final bridge model on 1 May 1775. It had been completed by 1 October 1776 [34].

4.2.3 Kulibin's experimental model from 1776

The aim of Kulibin's experiments before the completion of his final model 1776 was not only to determine the form of an ideal arch shape by hanging weights on a rope. Kulibin was also concerned to achieve a sensible distribution of the weight of the arch.

Based on his form-finding experiments between 1772 and 1775, Kulibin reduced the weight of the centre section of the arch in order to reduce the horizontal thrust at the supports. Thus he developed a construction that varied in weight along the length of the arch – with a wider and heavier cross section near the abutments, and a smaller and lighter cross-section towards the centre of the arch (Figure 4.13).

In Kulibin's only printed publication which appeared in 1799, 13 years after the completion of his final test-model in 1776, there is a description of the experiments he carried out to determine the arch thrust, illustrated by a drawing of the experimental machine ('Ispytatelnaja Maschina') which he constructed for these

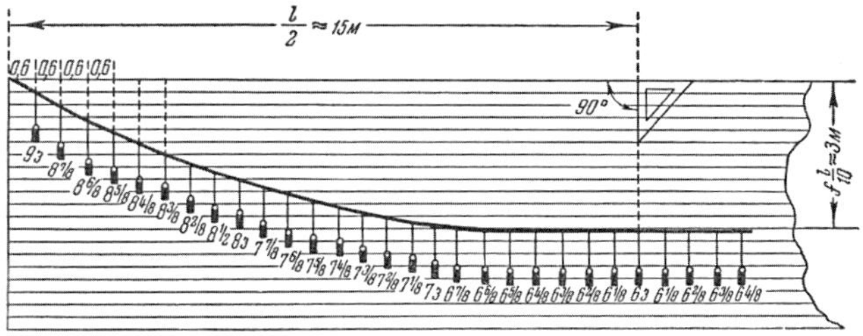

Figure 4.13 Reconstruction of a hanging rope model used by Kulibin to determine the axis of the arch for the final test model built in 1775-1776. (Source: [35])

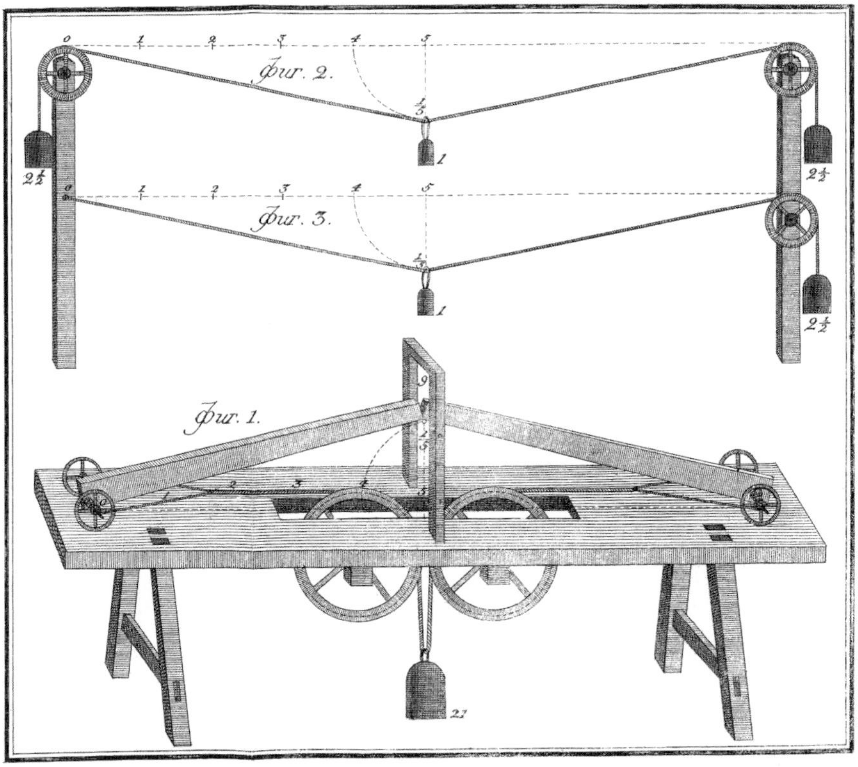

Figure 4.14 Kulibins experimental machine (*Ispytatelnaja Maschina*). (Source: [36] Plate 2)

experiments (Figure 4.14). He shows two struts which taper towards the centre. In the upper part of his illustration, Kulibin illustrates similarity between his three-hinged (or three-pin) arch and the equivalent, inverted system using ropes loaded in tension which clearly demonstrates his train of thought. With the help of this apparatus, he derives the relationship for the horizontal thrust acting on a three-hinged arch.

A weight of 21 pounds (8.6 kg) hanging from the central two ropes, giving a horizontal force at each support of 10½ pounds (4.3 kg), is needed to establish equilibrium with the struts weighing 9 pounds (3.7 kg) inclined at an angle of 1:5.

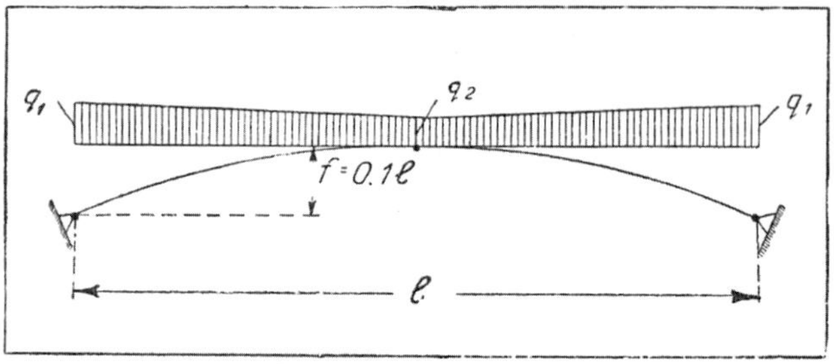

Figure 4.15 Three-hinged arch with the load distribution described by Kulibin. (Source: [38] Fig. 19)

Kulibin writes that these ratios can be transferred in proportion to a construction forty times larger: i.e. if each strut weighs 9 pudi (147 kg), there would be a horizontal thrust of 10½ pudi (171 kg) at each support [37].

Jakubowski has compared the forces Kulibin measured experimentally with those calculated using statics in the modern way, and demonstrated that the results agree exactly – for a dead load of 330,000 pudi, the horizontal thrust on the supports would be 385,000 pudi. Using Kulibin's specification that the weight of the centre parts of the bridge to the weight of the side parts should be in the ratio of 2 : 3 [37], Jakubowskij set the ratio of the unit lengths q2 : q1 as 2 : 3 [38] (Figure 4.15).

The horizontal reaction from the supports is created in Kulibin's model using ropes, pulleys and a single weight. For single load, P, acting on a three-hinged arch, a span, l, and rise, f, this horizontal reaction would be $\mathrm{H} = Pl/4f$, which Kulibin formulates verbally. Using his apparatus, he came to the conclusion that a laterally displaced vertical load, P, would lead to lower values of H. On the basis of this insight he optimised his design for the bridge by reducing the cross-section of his lattice construction towards the centre of the span in order to make it lighter in this region [39].

Of course Kulibin did not use the term 'three-hinged arch', nor did he establish equations for the equilibrium of forces and moments as we would today. His apparatus nevertheless shows that he understood that his timber arch worked as a three-hinged structure. Instead of establishing equations for equilibrium, he determined the values by experiment and obtained correct results. Kulibin's immediate goal was not to establish general laws, but rather to optimize the construction and determine the magnitude of the horizontal forces acting on the supports. Nevertheless, in a certain way, he anticipated basic insights of modern structural design [40].

The physical modelling carried out by Kulibin formed the starting point of Euler's derivation of his 'simple rule for assessing the strength of bridges'.

4.3 Euler's 'Simple rule' for assessing the strength of bridges

Inspired by Kulibin's work, Leonhard Euler (1707-1783) presented a paper to the St. Petersburg Academy of Sciences on 25 September 1775 on a 'Simple rule to determine the strength of a bridge or similar structure, on the basis of the known strength of a model' [41] (see Appendix A3). Although the inversion of the hanging rope to give the shape of a stable arch had become known above all through the work of Hooke and Wren on the dome of St. Paul's Cathedral, the idea had previous history, such as the use of a hanging cord by Bartolomeo Ammannati in 1569 to help find the form of the Ponte Santa Trinità in Florence [42].

Euler develops the problem in the form of relational equations, as was customary at the time. Following general considerations on the strength of ropes in §5, he first defines their strength by their breaking load when hanging vertically: two ropes or wires of the same material and different cross-sections, *cc* and CC, break under their own weight when their length L is the same. Their strengths are in the ratio L*cc*:LCC, in other words, *cc*:CC.

Then in §6 there follows a consideration of similar inclined ropes, just as Kulibin describes them. They consist of the same material and are loaded in the same way. The ropes or wires have different cross-sections, *cc* and CC, and different dimensions *abc* and ABC (Figure 4.16). Euler assumes that the strengths behave like the cross-sections according to the total loads – the weight of the ropes and the imposed load. (To help understanding, the equations below are numbered, unlike in Euler's original):

$$\mathrm{V} : v = \mathrm{CC} : cc \tag{4.1}$$

In §8, the total loads shortly before breaking are indicated by V and *v*, the respective loads to be carried by Q and *q*, and the self-weights of the ropes by P and *p*. Then Euler establishes a relation between the geometrically similar entities. The self-weights and volumes are in the proportion:

$$\mathrm{P} : p = \mathrm{ACC} : acc \tag{4.2}$$

In §8 the bridge rule is stated as follows. The total loads, the sum of self-weight and live load, are given by:

$$\mathrm{V} = \mathrm{P} + \mathrm{Q} \text{ and } \mathrm{v} = p + q \tag{4.3}$$

Hence, using equation (4.1), $\mathrm{V} = (\mathrm{CC}/cc) * v$, Euler concludes that:

$$\mathrm{P} + \mathrm{Q} = (\mathrm{CC}/cc)^{*} (p + q) \tag{4.4}$$

and, using equation (4.2), $\mathrm{P} = \mathrm{ACC}/acc * p$, and equation (4.4), he derives the following expression for the load capacity of the full-size bridge as:

$$\mathrm{Q} = \mathrm{CC}/cc^{*} (p + q) - \mathrm{ACC}/acc^{*} p$$

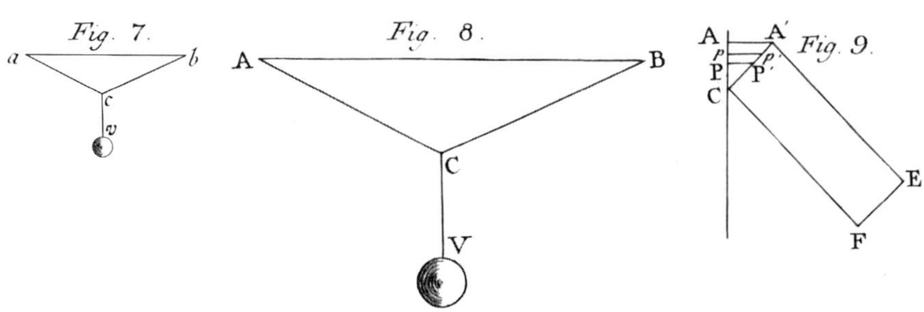

Figure 4.16 Figures 7, 8 and 9 from Euler's memorandum. (Source: [41]; see Appendix A3)

or, as he prefers to express it, using the geometric scale factor $A/a = n$:

$$Q = CC/cc\,[(p + q) - A/a * p].$$

Thus, the full size bridge can carry its own weight if, in the model test, (p + q) > (A/a * p).

Using this result, Euler concludes in §9, that a rope bridge spanning 1000 feet over the Neva can be built if, for example, a reduced-scale model with a span of 20 feet length, or geometrical scale factor of A/a = 50, can carry 49 times the self-weight of the model. For the full-size bridge to carry an additional imposed load, Q, the model must be subjected to a correspondingly higher load.

After considering a rope construction subjected to tensile loads as the inverse of an arch, Euler tries to take into account bending in the model for the case of a cantilever (encastered) beam. He considers the case of a beam rotating around the lower edge, where he (wrongly) assumes the neutral fibre is located (Figure 4.16). The bending moment carried by a beam clamped on one side is today known to be proportional $bh^2/6$ (where b = beam width, and h = beam height). In §11 Euler, however, determined that the bending moment in a beam to be proportional to $bh^2/3$ (written by Euler as *bcc*/3) due to the wrongly-positioned neutral fibre.

However, this incorrect factor does not play a role if only the *ratio* of a small beam to a large beam is considered. In the following paragraphs, §12 to §16, Euler works with simple relational equations and derives the same 'bridge rule' as before.

Two beams with lengths *a* and A resist the imposed bending moments (*a* * *v*) and (A * V) in the ratio of ⅓(*bcc*) to ⅓(BCC). Since the (incorrect) factor of ⅓ is common to both, the following relationship can be stated:

$$v : V = \frac{bcc}{a} : \frac{BCC}{A}$$

Since, from Equation 4.3, $v = (p + q)$, and $V = (P + Q)$, this can be expressed as:

$$(p + q) : (P + Q) = \frac{bcc}{a} : \frac{BCC}{A}$$

From this, it follows that:

$$(P + Q) = \frac{aBCC}{Abcc}(p + q)$$

Since the ratio of self-weights, P/*p*, is the same as the ratio of the volumes (ABC/*abc*), P can be substituted by *p**ABC/*abc* with the result that, the load capacity of the full-size bridge, Q, is given by:

$$Q = \frac{aBCC}{Abcc}(p + q) - \frac{ABC}{abc}p \tag{4.5}$$

Expressing the quantities in terms of the scale factor, *n*, then A = *na*, B = *nb* and C = *nc*, and Equation 4.5 can be written as:

$$Q = \frac{a.nb.nc.nc}{na.bcc}(p + q) - \frac{na.nb.nc}{abc}p$$

or

$$Q = n^2(p + q - np)$$

This also leads to the rule that the load $(p + q)$ applied to the model must be n times the self-weight of the model.

In §16, Euler gives the example of a bridge whose full size is 30 times larger than the model. Accordingly, the model must be loaded with 29 times its own weight in order that loaded by a total of 30 times its own self-weight, as in the previous case.

With these derivations, Euler had given practical rules for model testing which confirmed the results of Kulibin.

In the following paragraphs, §17 to §21, Euler deals with a disproportionate increase in the depth of the beam; in the transfer of results from the model test to the full-size bridge, the load-bearing capacity can thus be increased. Euler uses the example of the previous bridge to illustrate how an additional load capacity of the full-size bridge of 3600 times the self-weight of the model, can be accommodated. He shows it could be achieved by increasing the depth of the beam by a factor of 32.595, rather than the scale factor of 30 for all the linear dimensions. Euler then gives a table in which he shows the additional load capacity of a full-size bridge that can be achieved for various increases in the depth of the beam.

In his final two paragraphs, §22 and §23, Euler also proposes measures against the effect of wind, and is design for the bridge included cross-bracing (in plan). He observes that, since the wind load is proportional to the surface area, then, if the model is exposed to 'an extremely strong wind' then the full size bridge would be able to resist the same wind force. He also recommends that the resistance to wind loads can be increased by making the beams wider, as long as this increased self-weight can be tolerated, or the resistance could be increased without compromising the load capacity of the bridge by making the bridge wider towards the supports so that it can act as a horizontal arch.

4.4 Competition between rival designs for the Neva Bridge

4.4.1 The models of Kulibin and de Ribas

In May 1775 Ivan Kulibin began with the construction of a bridge model on a scale of 1:10, which had the enormous span of 30 metres. Its construction cost amounted to 3524 roubles, it was completed in October 1776 [43].

However, at this time, a serious competitor for the construction of a Neva bridge emerged in the form of Major José de Ribas (1749-1801), a Spanish officer in Russian service. De Ribas came from Barcelona and while serving as an officer in a Neapolitan infantry regiment, was recommended to enter the Russian service as captain of the cadet corps, arriving in St. Petersburg in 1772. He soon joined the circle of Catherine II's favourites [44].

De Ribas worked faster than Kulibin and took advantage of his close relations to the court. He presented his first model to Tsarina Catherine II (Catherine the Great, 1729-1796) while Kulibin was still constructing his model. The consequences of involving the Court were immediate: on 22 February 1776, Sergei Domashnev, the director of the Academy of Sciences declared that, according to the wishes of the Tsarina, the Academy would have to undertake and evaluation of the de Ribas bridge model. Domashnev appointed a commission of Academy members to visit the model in the building which housed the cadet corps. The commission consisted of Leonhard Euler, his son Johann Albrecht Euler (1734-1800), academician and his father's secretary, Semyon Kirillovich Kotelnikov (1723-1806), Stepan Yakovlevich Rumovsky (1734-1812) and the two associates of the Academy, Michael E. Golovin (an assistant of Euler, 1756-1790) and Nicolas Fuss (1755-1826) [45]. Nicolas Fuss (also Nikolaus Fuß and Nikolai Fuss) was a mathematician and a pupil of Daniel Bernoulli. In 1773, aged 18, he travelled from Basel to St Petersburg to become the secretary and assistant of Euler whose eyesight had been deteriorating since 1766; he later married Euler's granddaughter [46].

The minutes of the Academy's meetings indicate that Kulibin's project would, from this time on, play only the role of an alternative design to the one proposed by de Ribas. Thus, Fuss and Golovin were asked to prepare the description and design of the 'other' bridge, i.e. the Neva bridge proposed by Kulibin, for examination by the Commission. Golovin first had to translate Kulibin's Russian text into Latin. It was stated that his model would still take some time to complete. The German scientist Wolfgang Ludwig Krafft (1743-1814) and the Swedish scientist Anders Lexell (1740-1784) were also involved in this assessment, indeed Krafft described Kulibin's calculation as inaccurate but sufficient to estimate the total weight of the bridge [47]. Like Euler, many of the people involved came from Switzerland and therefore also knew the bridges built by Swiss carpenters.

On 27 February 1776, the model of de Ribas' design for the bridge, which had been made at his own expense, was tested in the Land Cadet Corps building in the presence of the Academy Commission [48]. Its length was 1/30 of the width of the Neva, which is given as 151 *toises* or 294.2 metres. The load test demonstrated that it could not even carry a load of one quarter of that required. With explicit reference to Euler's loading rule, the report of the test dealt mathematically with the possibility of using lighter wood from a fir tree instead of the oak intended by de Ribas. The specific weights and strengths of these woods were taken from Bélidor's *Science des Ingénieurs* [49]. Since, according to the calculation, this would lead to a substantial improvement in the ratio of self-weight to load capacity, de Ribas welcomed the proposal to build a new model made from the less dense wood, and also to increase the rise of the arch in the bridge.

The test of the new model of the design by de Ribas took place eight months later, on 15 November 1776 [50]. This model was made at a scale of 1:20, and so had to be loaded with 20 times its own self-weight. For this purpose, de Ribas drew up a table with the weights of the components of his model multiplied by a factor of 20 [51]. The load test was apparently successful. First a timber-frame scaffold was constructed on which bricks were placed in layers. The test continued until the load was 23 times the self-weight of the model which it was ended because the model threatened to buckle. Thus, the required load of 20 times the self-weight was exceeded, and the live load due to traffic on the bridge could also be considered.

4.4.2 The Academy is required to judge between the designs

The following session of the Academy, however, on 18 November 1776, did not end with a conclusion [52]. Internal papers of the academic session reveal that the discussion about the test led to a scandal: the secretary, Johann Euler, read out a report of the load test, which was heavily criticized. Objections were made to the procedure used by de Ribas for loading the model: the point of contention for the academicians was the timber-frame scaffolding, which was used by de Ribas to provide a stable base for the bricks that were placed upon it. No agreement could be reached during the session, as to whether the timber-frame scaffold structure carried part of the load intended for the model bridge, or not. Finally, it was agreed that a report would be finalised only after a further load test had been carried out under modified conditions. The scaffold frame should be removed and only loose layers of timber planks should be used as a base for the bricks. At this point, Academician Roumovski became so concerned that the report criticising the test by de Ribas would be received with considerable disfavour by the Court, that he asked to be released from signing the report [53]. In the following academic session on 25th November, a letter from Field Marshal Prince Golitzin was discussed in which he urged for a second time the report of the test on 15 November 1776 should be finished. The academy members decided that a new test should be carried out on the following day 'to end this affair and make their rapport to the Academy' [54].

The various unprinted versions of the Report of the Bridge Commission, which contain numerous annotations, give a vivid picture of the test procedure carried out on 26 November 1776:

'But since one cannot be too cautious in research of this importance, and since it seems that the apparatus of the beams & ropes also carried part of the burden by itself, the Academicians thought it necessary to go to the noble Corps des Cadèts de Terre on the 26th of this month for the second time, to proceed in a different way by immediately loading the model with the burden it must support according to the precepts of the Calculation.

The model having therefore been stripped of the beam assembly, which covered it at the first examination, it was first loaded with simple boards, on which Major de Ribas wanted to place two more rows (which together already weighed around 70 pudi (1140 kg)), before putting the rest of the weight in bricks. But as the people Major de Ribas had used to store these boards had neglected to observe all the precautions to ensure that they rest vertically on the model, the model suffered an oblique shock, which bent it in different places and even though they tried to straighten it to a straight line, it was so much so damaged, that it disintegrated completely, even before it was loaded with bricks' [55].

This attempted load test was unsuccessful and left the Commission faced with a dilemma: the formal requirements of the Euler Bridge Rule had previously been met in two tests carried out on 15 and 26 November, yet the first test was judged successful, while the following one had not. The Academicians sought appropriate wording that would not offend de Ribas, and would find reasons why the overall results of the tests were judged as inadequate. On the one hand, there was talk of the 'unexpected catastrophe' of the failure of the bridge model; on the other hand, the 'high esteem' of Major de Ribas, who had wanted to overcome the 'countless difficulties in the execution of such a worthwhile project for the good of mankind', was not denied [56]. The test of the de Ribas model had become an extremely unpleasant affair for the Academy, mainly because its own competence had been called into question. De Ribas accused the Academy of incompetence in requiring that his model should be tested for a second time, and then not being able to agree on assessing and explaining the outcomes.

Since it was obvious that a generalized rule for scaling up results obtained from the model to the full size structure was not sufficient for the evaluation of a structure, and that an analysis of the individual construction elements was necessary, it became clear that the issue of buckling strength had to be formally considered. As there are no drawings of the model of the bridge designed by Major de Ribas, it is not now possible to comment on the exact nature of the buckling observed, or which particular compression members were so endangered.

4.5 Further work by Euler

4.5.1 Euler's work on column buckling of 1776

At the height of the bridge discussion, on 16 December 1776, Euler presented to the Academy his paper on the 'Determination of the loads that columns are able to support' [57, 58]. Three weeks have passed since the load test on the model of the bridge designed by de Ribas, and eleven days remain until the Kulibin's model bridge is due to be tested.

In this work Euler refers to his earlier investigation *Sur la force de colonnes* of 1757, in which he had already dealt in a general way with the question of buckling [59]. His first, famous work on this problem *Methodus inveniendi...* dates back to 1744 [60]. Starting from the differential equation of the bending curve, Euler had determined the eigenvalue for the case of a buckling load; the stiffness had been expressed with the symbol E*kk* (EI in today's notation). In his new work he was interested in determining the buckling load of a column under its own weight. He first explains the difference between a cantilever arm deflection and a buckling deflection, in general.

First, Euler treats the case of the cantilever arm more closely, although it is shown standing upright and clamped at the base (Figure 4.17). In the case of a horizontal load, F, Euler derives the differential equation of the bending curve. In today's nomenclature, as follows:

$$v'' = -\frac{\mathrm{F}(a-x)}{\mathrm{EI}}$$

$$v' = -\left(\frac{\mathrm{F}}{\mathrm{EI}}\right) \times \left(ax - \frac{x^2}{2}\right) + \mathrm{C}_1$$

$$v = -\left(\frac{\mathrm{F}}{\mathrm{EI}}\right) \times \left(\frac{ax^2}{2} - \frac{x^3}{6}\right) + \mathrm{C}_1 x + \mathrm{C}_2$$

At the clamping point

$$v(0) = 0 \text{ i.e. } C_2 = 0$$
$$v'(0) = 0 \text{ i.e. } C_1 = 0$$

Thus, the maximum deflection, i.e. at position α, is

$$v(a) = -\left(\frac{F}{EI}\right) \times \left(\frac{aa^2}{2} - \frac{a^3}{6}\right)$$
$$= -\left(\frac{F}{EI}\right) \times \left(\frac{a^3}{3}\right)$$

Or, rearranged to give an expression for EI,

$$EI = -\left(\frac{F}{v(a)}\right) \times \left(\frac{a^3}{3}\right)$$

Euler designates the second moment of area, I, as kk and the maximum deflection $v(a)$ as α. Since Euler has used a coordinate system with deflection positive, $v(a) = \alpha$ has no minus sign, i.e. in Euler's notation:

$$Ekk = F\frac{a^3}{3\alpha}$$

Euler himself did not carry out any deflection experiments (time was pressing), but used Musschenbroek's then well-known experiments on struts made of wood, which had already been published in 1729. From them he could determine the stiffness Ekk, since force, deflection and dimensions of the test strut were known. Now that Euler has the stiffness EI (or Ekk: *'rigor absolutus'*) of a strut, he can come to his buckling problem (Figure 4.17). He sets up the equation for the elastic strut mounted so that it can rotate at both ends and determines the critical load (today referred to as the second Euler case) as:

$$O = \pi\pi\frac{Ekk}{aa}$$

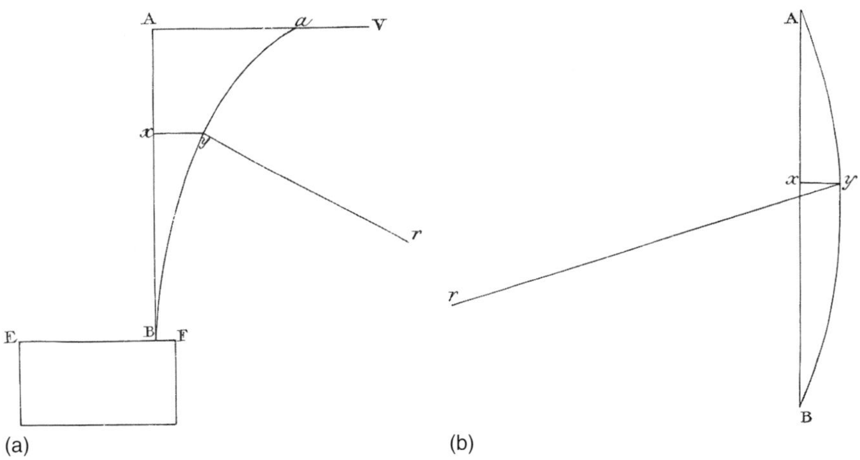

Figure 4.17 (a) Euler's diagram of a cantilever arm, encastered at its base. (Source: [58] Figure 5). (b) Euler's diagram of a column with position-fixed, pinned ends. (Source: [58] Figure 6)

Since the aim of his work was to determine the influence of the dead load on buckling, Euler proceeds next to determine the proportion that the material contributes to the total stiffness.

4.5.2 Definition of the modulus of elasticity

Euler develops an interesting idea which, in the history of theoretical mechanics, is regarded as the birth of the idea of 'modulus of elasticity'. Neither Euler nor anyone else had previously been able to separate the material-elastic contribution to stiffness from its shape-dependent contribution, i.e. distinguishing E and I. The result was that the stiffness of the material was not a function of the shape-dependent part of the stiffness. Euler achieved this by defining a column 'modulus' that embodies this material constant. Using the fact that the height of a mercury column or the height of a water column is a known measure of pressure, Euler turns the situation around mentally and represents a tractive force by a hanging column.

He considers a hanging rod loaded vertically by a weight, P, equal to the weight of a cylinder, length h, of the same cross-section as the rod (Figure 4.18). For a certain material there is a characteristic length (h), which causes an extension of the rod, Δl, for example a doubling of its original length. Since this takes place independently of the cross-section of the test rod, h characterizes the property of the material to lengthen more or less under load: its elasticity. One can easily

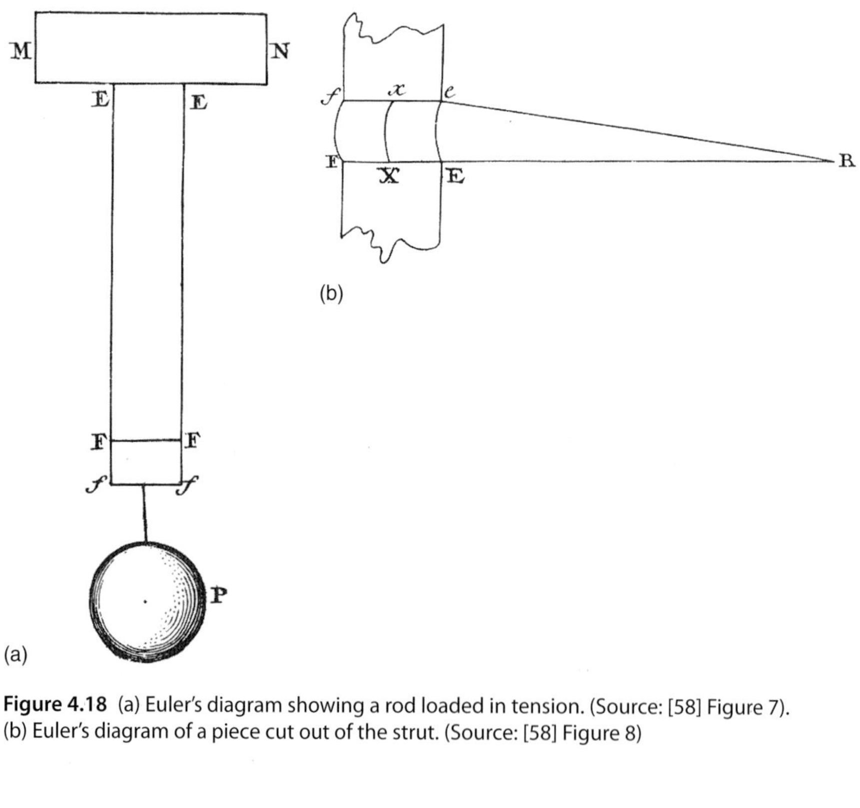

Figure 4.18 (a) Euler's diagram showing a rod loaded in tension. (Source: [58] Figure 7). (b) Euler's diagram of a piece cut out of the strut. (Source: [58] Figure 8)

identify Euler's $P = h\,\Delta l$ with today's equivalent, $\sigma = E * \varepsilon$. Euler's h is the constant of proportionality relating stress and strain, and the strain is given by $\varepsilon = \Delta l\,/\,l$. Euler developed here the *modulus of elasticity*, as a material constant, h, independent from the diameter of the hanging column and its length.

4.5.3 Calculation of the modulus of elasticity *h* from the data of Musschenbroek's buckling tests

Euler now derives the geometric, shape-dependent relationships by considering an infinitesimal element of the previously described strut or column with position-fixed, pinned ends (Figure 4.18). His aim is to express the strain, on the one hand, in terms of the modulus of elasticity, h, developed on the basis of the thought experiment described previously, in which the suspended column extends under its own weight; and, on the other hand, in terms of the experimentally-determined stiffness Ekk. The resulting equation establishes a relationship between h and Ekk that allows h to be determined numerically.

In a piece cut out of the strut in Figure 4.17, but with deformation in the opposite direction, the distance Ee is given as ds. The differential extension of the fibre at distance x from the right edge of the strut or column is xds/r (Euler again places the neutral axis, i.e. the unstretched fibre, at the edge of the strut). Euler formulates an integral for the deformation of the strut using his modulus, h, when the bending moment is buckling the strut:

$$\frac{ch}{r}\int xx\,dx = \frac{1}{3}\,\frac{chx^3}{r} = \frac{b^3c\,h}{3\,r}$$

With x = b and y = c as the dimensions of the cross-section of a prismatic strut, Euler assumed the neutral axis on the edge of the cross-section. Instead of what we would say today:

$$\mathrm{EI} = \mathrm{E}\,\frac{b^3c}{12}$$

Euler writes: $\mathrm{E}kk = \frac{1}{3}\,b^3c\,.\,h$ where h is given by $h = \dfrac{\mathrm{E}}{\rho g}$.

Now Euler compares Ekk given by his geometrical approach, $Ekk = h\dfrac{b^3c}{3}$ with Ekk given by Musschenbroek's experiments for critical load, O,

$$\mathrm{O} = \pi\pi\,\frac{\mathrm{E}kk}{a^2} \text{ or } \mathrm{E}kk = \frac{\mathrm{O}a^2}{\pi\pi}$$

Thus, $h\,\dfrac{b^3c}{3} = \dfrac{\mathrm{O}a^2}{\pi\pi}$

and for a square cross-section, with $c = b$,

$$h = \frac{\mathrm{O}\,.\,a^2}{\left(\dfrac{\pi\pi\,.\,b^4}{3}\right)}$$

Which differs from the correct expression by a factor of four:

$$\mathrm{E} = \frac{\mathrm{F}\,.\,l^2}{\left(\dfrac{\pi\pi\,.\,b^4}{12}\right)}$$

By considering the results of Musschenbroek's experiments, Euler calculates a value of the elastic modulus, $h = 2{,}774{,}980$ feet, regarding which he believes that he does not need to 'shy away' from the facts gained by experience [61]. Euler's modulus of elasticity is four times lower than it should be, due to his value of I being four times larger. If this is corrected, it corresponds quite well to today's values [62, 63]. Another source of inaccuracy results from the fact that Musschenbroek's buckling experiments were taken up as far as breaking (*'dirupta fuit a pondere 226 lib.'*) [64].

With regard to the problem of a critical self-weight mentioned at the beginning, Euler comes to the paradoxical conclusion that there is no length, l, at which a column would break due to its own self-weight, which gave rise to two further papers by Euler in 1778 [65, 66]. The particular difficulty of the three listed works was the numerically complex calculation of the critical self-weight [67].

The correspondence between D. Bernoulli and N. Fuss also shows that the question of buckling was by no means settled with this result from Euler. Fuss reports in detail on the experiments undertaken on Kulibin's bridge project in St. Petersburg and answers Bernoulli's questions [68, 69, 70], and Bernoulli asks whether Kulibin could not carry out experiments to confirm Euler's buckling theory [71].

In spite of all this, because of their highly-theoretical nature, Euler's considerations of buckling in his paper of 16 December 1776 *Determinatio onerum, quae columnae gestare valent* were not able to be used to support a negative verdict regarding the design of de Ribas' bridge. The need for justification continued.

4.6 The practical solution to the academic question: Kulibin's experiment

The dilemma of the Bridge Commission to assess the tests of de Ribas' model correctly was resolved, apart from a few further theoretical considerations, eleven days after the reading of Euler's paper. The load tests of Kulibin's model proved to be clearly superior to those of de Ribas. The 14 Sashen (30 metre) long model of Kulibin's lattice arch bridge had been built between May 1775 and October 1776. It was too large to erect in an enclosed space, and so, on 27 December 1776, the members of the St. Petersburg Academy named in the bridge affair left the Academy building to test the model in the open air. To the great satisfaction of the members of the Audit Committee, the model withstood the tenfold burden required by Euler's rule. Due to its very light construction, its self-weight was only 330 pudi (5379 kg), and it carried not only the required 2970 pudi, but even 550 pudi more (in total, 3520 pudi = 57,376 kg) [72]. In his publication of 1799 Kulibin gives the even higher number of 570 pudi (in total, 3540 pudi = 57,702 kg). Even after sustaining this load continuously for a month, it showed no signs of deformation. With this result, de Ribas' project was finally out of the running, and the face of the Academy was saved by a successful assessment of the outcome of a model test based on the 'bridge rule' [36].

Sixteen years later, in 1793, Kulibin was required – despite his protests – to alter and reconstruct his model in the gardens of the Taurian Palace in St Petersburg

[73] where it remained an attraction for every visitor to the city for many years. The full-size bridge he proposed to cross the Neva, however, did not come to fruition. One decisive issue was the practical question of the durability of the timber under the harsh weather conditions, which would allow a service life of only about 30 years. No less important were aesthetic concerns about the bridge and, of course, the costs would have been enormous – an entire forest would have had to be cleared to build the bridge, a point that Fuss had already noted [70].

The questions that arose about the St. Petersburg bridge became widely known through the many visitors from other countries during the time of Catherine the Great. This is shown by the design for a bridge over the Neva by the French engineer Jean-Rodolphe Perronet in 1778. The proposal was for a conventional stone bridge with seven arches [74]. A bascule bridge in the centre span to allow the passage of ships anticipates the design that was realized later in the Nikolaewskij bridge (today the Blagoveshenskij bridge), which was equipped with a similar bascule device. This bridge was eventually constructed in 1843-1850 as a cast-iron structure with arches spanning up to 47 metres [75].

4.7 Final remarks

It is hardly surprising that Kulibin's activities have received little attention outside Russia, especially since he wrote only one work – his Russian-language publication of 1799 [36]. Nevertheless, some news of his work did percolate abroad through second-hand reports such as one that gave the dimensions of the bridge as 960 shoe-lengths long and 160 shoe-lengths high [76].

Kulibin was clearly aware of the transition in timber bridge construction in the 1760s from simple truss structures to a shear-resistant arch construction, as exemplified in the model by the Swiss bridge builder Aldon that had been available in St. Petersburg since 1762. However, Kulibin's bridge design went far beyond the shear-resistant, saw-tooth connections and dowelling of layers in a deep beam used by the Swiss bridge builders. In using a lattice arch construction, Kulibin created a completely new construction method, and he even published some construction details for his bridge in 1799 (Figures 4.19 and 4.20).

The model experiments that Kulibin carried out were extraordinary. On the one hand, it was based on the exact determination of the line of thrust and an optimization of the weight distribution of the construction. On the other hand, he built an apparatus to determine the horizontal thrust in a model conceived as a three-hinged arch. Kulibin results from his 'experimental machine' were taken up by Euler in the reformulation of his 'bridge rule'. Thirdly, Kulibin derived a model rule experimentally that correctly scaled up the measurements made on a reduced-scale model to the full-size structure.

While transferring the idea of a compression-loaded arch from masonry construction to timber construction was within the capabilities of craftsmen's experience, Kulibin's position as a mechanic in the Academy gave rise to an approach involving the systematic planning of tests and subsequent calculations that enabled him to go far beyond what could be achieved using a craftsman's knowledge and experience.

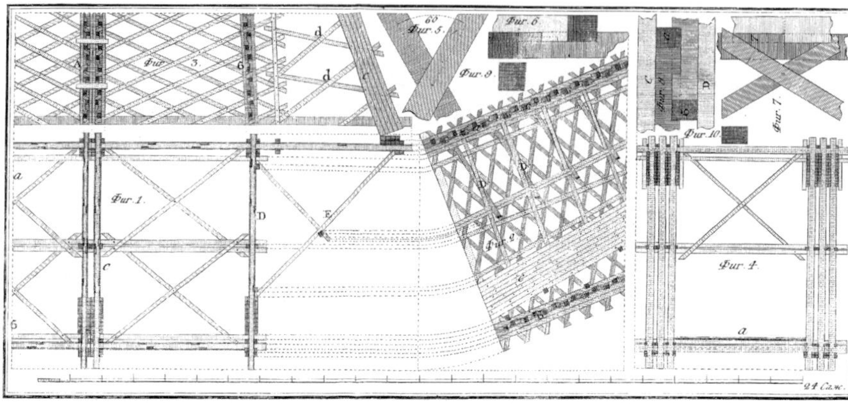

Figure 4.19 Construction details of the lattice truss from Kulibin's publication of 1799. (Source: [36] Plate 1)

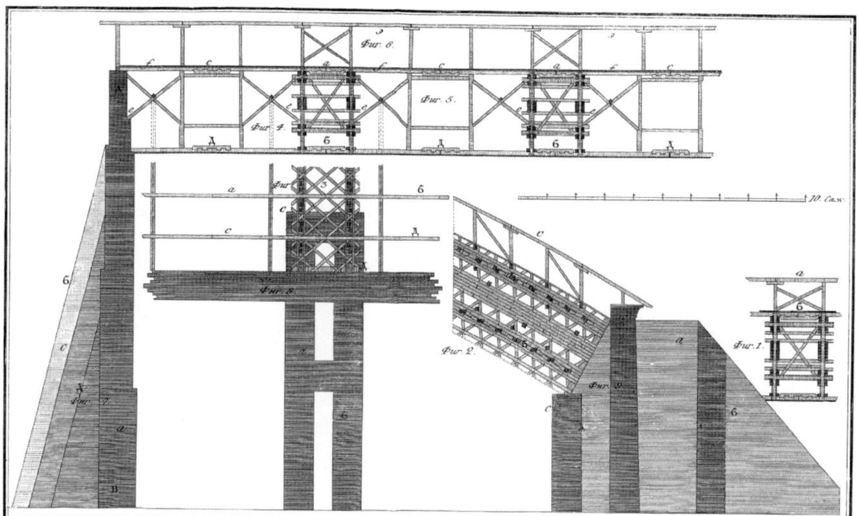

Figure 4.20 Construction details of the cross section and springing from Kulibin's publication of 1799. (Source: [36] Plate 3)

On the occasion of the Kulibin Bridge Project in 1775, Leonhard Euler developed his paper 'Regula facilis…' (E480), dedicated entirely to practical design questions and how to scale up the results from tests on a reduced-scale model to a full-size bridge. Beginning with the work 'Determinatio onerum…' (E508) Euler then created a series of three works on the buckling of columns, from December 1776 onwards.

The special and interesting aspect of Euler's first work (E508) is the idea of separating the material and geometric properties of the column and deriving a 'modulus of elasticity' from Musschenbroek's experiments, with which theoretical calculations about the buckling of columns could be combined with empirical, numerical values.

While the history of Euler-Kulibins' bridge rule and the success of Kulibins' model experiments are well-known in Russian-speaking countries, it has been largely overlooked that Euler's work on buckling was intimately linked to the bridge experiments in St. Petersburg.

The failure of de Ribas' tests gave Euler the motivation to investigate the buckling of columns under their own weight, building on his earlier work of 1744 and 1757. This assumed that the self-weight case would prove to be insignificant; the outcome, however, was Euler's theory of buckling. Euler thus continued along the path begun by Musschenbroek to move from understanding a whole construction to the consideration of the behaviour of individual components, using theoretical means.

The impact of Euler's papers seems, therefore, to have been more profound than previously recognised. For example, his work was directly continued in a *Mémoire* by Nicolas Fuss from 1778, in which the forces acting at the nodes of statically-determinate trusses are considered [77]. In Britain, the Scottish physicist John Robison (1739-1805) knew and developed Euler's work. Robison had visited St Petersburg from 1772-1773 and taught mathematics to the cadets at the Naval Academy in Kronstadt. He was later the author of numerous articles for the third edition of Encyclopaedia Britannica in 1797 [78, 79]. The English physicist Thomas Young (1773-1829) was also familiar with Euler's work and mentioned the *Regula facilis* ... paper in his discussion of the value of using a model to test Thomas Telford's 1801 proposal for a 180 metre cast-iron arch over the River Thames, in London (see Appendix A4). And in Germany, Johann Eytelwein (1764-1848) related his own work on strength of materials to the Euler-Bernoulli theory of elasticity and also used Euler in his treatment of 'reactive strength' (in buckling) [80].

For the introduction of the concept of elasticity into the world of engineering science, however, it was Thomas Young who has a particularly significant role, since he published the elastic moduli of several materials in tabular form in the article 'Carpentry' in the supplement to the 4th, 5th and 6th editions of the Encyclopaedia Britannica published in 1824, although probably written some years earlier [81].

Euler's theory of buckling, although constantly present in the scientific literature and after a long period of purely theoretical significance, was only brought into structural design procedures towards the end of the nineteenth century in relation to the failure of steel structures.

Abbreviations and citation methods

AN = Archive of the Academy of Sciences of the (former) USSR, now St. Petersburg. Fund 1, Op. 2 contains protocol sheets or protocol booklets with supplements of the academic meetings.

AN F.1, Op. 2-1776, No.10 Documents relating to the model experiments of de Ribas.
AN F.1, Op. 2-1777, No.1 Documents relating the model experiments of Kulibin.

The works of Leonhard Euler are quoted from the edition of the Leohardi Opera Omnia, for example:

Euler, Leonhard: Determinatio onerum, quae columnae gestare valent (E508). *Acta academiae scientiarum Petropolitanae* (1778:I),1780, 121-145 Leonhardi Opera Omnia sub auspisis Societatis Scientarum Naturalium Helveticae. Series secunda. Opera Mechanica et astronomica. Volume septemdecimum. Commentationes Mechanicae ad Theoriam Machinarum Pertinentes. Bernae 1982, Orell Füssli Turici, 232-251.

Abbreviated for the first quote:

Euler, Leonhard: Determinatio onerum, quae columnae gestare valent (E508). *Acta academiae scientiarum Petropolitanae* (1778:I),1780, 121-145 EO (1982) 2,17. 232-251.

When further quoting:

Euler, L (1982) Determinatio onerum (E508). EO (1982) 2.17. 232-251.

References

1 Killer, Josef (1998) *Die Werke der Baumeister Grubenmann.* Dietikon: Baufachverlag. 42.
2 Killer, Josef (1998) *Die Werke der Baumeister Grubenmann.* Dietikon: Baufachverlag. 27, 42-44.
3 Krafft, J.-Ch. (1820) *Plans coupes et élévations de divers productions de l'Art de la Charpente.* 2nd Ed. Mannheim: Artaria et Fontaine. Plate 28.
4 De l'Orme, P. (1561) *Nouvelles inventions pour bien bastir et a petits fraiz.* Paris: Federic Morel. Book 2, 38-39.
5 Killer, Josef (1998) *Die Werke der Baumeister Grubenmann.* Dietikon: Baufachverlag. 36.
6 Walter, Caspar (1766) *Brücken-Bau, oder Anweisung, wie allerley Arten von Brücken, sowohl von Holz als Steinen, nach den besten Regeln der Zimmermannskunst dauerhaft anzulegen sind.* Augsburg: Veith.
7 Walter, Caspar (1766) *Brücken-Bau, oder Anweisung, wie allerley Arten von Brücken, sowohl von Holz als Steinen, nach den besten Regeln der Zimmermannskunst dauerhaft anzulegen sind.* Augsburg: Veith. 27.
8 Sturm, Leonard Christoph (1726) *Gründlicher Unterricht [...] von Heng- und Sprengwercken.* Stockholm und Leipzig: Johann Heinrich Rußworm. 51.
9 Holzer, Stefan (2010) Hölzerne Bogenbrücken und Modellstatik von Perrault bis Euler. *Bautechnik* 87 (3): 158-170.
10 Musschenbroek, Pieter van (1729) Introductio ad Cohaerentiam Corporum firmorum. In: *Physicae experimentales et geometricae,* 421-672. Leyden: Luchtmans.

11 Bélidor, B.F. de (1729) *La Science des Ingénieurs dans la conduite des travaux de fortification et d'architecteure civile.* Paris: Claude Jombert. Livre IV, 16.

12 Tardini, Chiara (2014) *Towards structural mechanics through wooden bridges in France (1716-1841).* London: Springer. 4-7.

13 Bélidor, B. F. de (1729) *La Science des Ingénieurs dans la conduite des travaux de fortification et d'architecteure civile.* Paris: Claude Jombert. Livre IV. Pl. 12.

14 Musschenbroek, Pieter van (1729) Introductio ad Cohaerentiam Corporum firmorum. In: *Physicae experimentales et geometricae.* 421-672, Leyden: Luchtmans. 429.

15 Musschenbroek, Pieter van (1729) Introductio ad Cohaerentiam Corporum firmorum. In: *Physicae experimentales et geometricae.* 421-672, Leyden: Luchtmans. 465-466.

16 Musschenbroek, Pieter van (1729) Introductio ad Cohaerentiam Corporum firmorum. In: *Physicae experimentales et geometricae.* 421-672, Leyden: Luchtmans. 660.

17 Truesdell, Clifford (1960) The Rational Mechanics of flexible or elastic bodies. 1638-1788. (Introduction to *Leonhardi Euleri opera omnia*, vol X, XI, seriei secundae). In: *Leonhardi Euleri Opera Omnia* II, 11-2. Turici: Orell Füssli. 153.

18 Kulibin (1772) = Кулибин И.П. (1772) Прошение на имя вице-директора Академии Наук А. А. Ржевского об изготовлеии модели деревянного арочного моста. 1772, декабря 9. Архив Академии наук СССР Ф. 296, оп. 1, No.38. л.л. 1-9 об. In: Raskin N. M. & Malkevitch B. A. (1953) = Раскин, Н.М. и Малькевич, Б. А. (1953) Рукописные материалы И.П. Кулибина в архиве академии наук СССР. Академии наук СССР. Труды Архива. Выпуск 11. Москва-Ленинград: Издательство Академии наук СССР.151-160.

19 Raskin, N.M. & Malkevitch, B.A. (1953) = Раскин, Н.М. и Малькевич, Б. А. (1953) Рукописные материалы И.П. Кулибина в архиве академии наук СССР. Академии наук СССР. Труды Архива. Выпуск 11. Москва-Ленинград: Издательство Академии Наук СССР.

20 Jakubovskij, B.V. (1936) = Якубовский Б.В. (1936) Проекты Мостов И. П. Кулибина. Архив истории науки и техники. Выпуск 8. Москва-Ленинград: Издательство Академии Наук СССР. 191-252.

21 Raskin, N.M. (2007) Euler and I.P. Kulibin. In: *Euler and Modern Science* (eds N.N. Bogoljubov, G.K. Mikhaĭlov & A.P. Youshkevich), 317-333. Originally published in Russia as *Razvitie ideĭ Leonharda Eulera i sovremennaya nauka,* Izdatel'stvo Nauka 1988. (translated from Russian by Robert Burns) USA: The mathematical Association of America. 317.

22 Raskin, N.M. (1958) = Раскин Н.М. (1958) Вопросы Техники у Ейлера (russ.). In: *Sammelband der zu Ehren des 250. Geburtstages Leonhard Eulers der Akademie der Wissenschaften der UdSSR vorgelegten Abhandlungen* (eds M.A. Lavrentjev, A.P. Youchkevich & A.T. Grigorian), 499-554. Moskau: Verlag der Akademie der Wissenschaften der UdSSR.

23 Raskin, N.M. (1960) = Раскин Н.М. (1962) Иван Петрович Кулибин 1735-1818. Москва-Ленинград: Издательство Академии Наук СССР.

24 Raskin, N.M. & Malkevitch, B.A. (1953) = Раскин, Н.М. и Малькевич, Б. А. (1953) Рукописные материалы И.П. Кулибина. 151.

25 Sokolov N. (1862) = Соколовъ Н. (1862) Описание моделй Музеума института Корпуса Инженеровъ Путей Сообщенія. Ст. Петербургъ. III-IV.

26 Sokolov N. (1862) = Соколовъ Н. (1862) Описание моделй Музеума института Корпуса Инженеровъ Путей Сообщенія. Ст. Петербургъ. 26-27.

27 Raskin, N.M. & Malkevitch, B.A. (1953) = Раскин, Н.М. и Малькевич, Б. А. (1953) Рукописные материалы И.П. Кулибина. 154-156.

28 Kulibin. Архив Академии наук СССР Ф. 296, оп. 1, No.134. л.1. In: Raskin N. M. and Malkevitch B. A. (1953). Рукописные материалы И.П. Кулибина. Деревянный арочны мост через р. В. Неву (обчий вид). Описание No.16. 37. 507.

29 Bozherianov I.N. (1902) = Божеряновъ И. Н. (1902) Невский проспект. Културно-исторический очеркъ двухвековой жизни Петербурга. Томъ. 2. СПБ, А.И. Вильборг. 251.

30 Jakubovskij, B.V. (1936) = Якубовский Б. В. (1936) Проекты Мостов И. П. Кулибина. Архив истории науки и техники. Выпуск 8. Москва-Ленинград: Издательство Академии Наук СССР. 191-252. 241.

31 St. Peterburgskije Vedomosti (4 Mai 1772) = Санкт-Петербургские Ведомости No.36, пятницу Маія 4 дня 1772 года. 5.

32 Maggi, A. & Navone, N. (eds.) (2003) *John Soane and the wooden bridges of Switzerland. Architecture and the culture of technology from Palladio to the Grubenmanns.* Mendrisio, Accademia di architettura Mendrisio, Università della Svizzera italiana. 42.

33 Raskin N.M. & Malkevitch B.A. (1953). = Раскин, Н.М. и Малькевич, Б. А. (1953) Рукописные материалы И.П. Кулибина. 159.

34 Kulibin (1772). In: Raskin, N.M. & Malkevitch, B.A. (1953) = Раскин, Н.М. и Малькевич, Б. А. (1953) Рукописные материалы И.П. Кулибина. 223.

35 Bolschakov, V.V. (1954) = Большаков, В.В. (1954) Краткий исторический обзор развития деревянных конструкций в СССР. Туды по истории техники. Выпуск 8. Издательство Академии наук СССР. Москва 1954, 37-63. 47.

36 Kulibin I. (1799) = Кулибин, И. (1799) Описание представленнаго на чертежѣ моста, простирающагося изъ одной дуги на 140 саженяхъ, изобрѣтеннаго Механикомъ Иваномъ Кулибинымъ, съ разными вычисленіями состоящихъ въ немъ тяжестей по разстоянію и другихъ обширныхъ Зданіи. Ст. Петербургъ. Republished in Raskin N.M. & Malkevitch B.A. (1953) = Раскин, Н.М. и Малькевич, Б. А. (1953) Рукописные материалы И.П. Кулибина 163-178.

37 Kulibin (1799). In: Raskin, N.M. & Malkevitch, B.A. (1953) = Раскин, Н.М. и Малькевич, Б. А. (1953) Рукописные материалы И.П. Кулибина. 170-172.

38 Jakubovskij, B.V. (1936) = Якубовский Б. В. (1936) Проекты Мостов И. П. Кулибина. Архив истории науки и техники. Выпуск 8. Москва-Ленинград: Издательство Академии Наук СССР. 227-228.

39 Kulibin (1772). In: Raskin, N.M. & Malkevitch, B.A. (1953) = Раскин, Н.М. и Малькевич, Б. А. (1953) Рукописные материалы И.П. Кулибина. 159-160.

40 Bernstein, S.A. (1957) = Бернштейн, С.А. (1957) Очерки по истории строительном механики. Москва: Государственное издательсво литературы по строительству и архитектуре. 99-108. 103-104.

41 Euler, L. *Regula facilis pro diiudicanda firmitate pontis aliusve corporis similis, ex cognita firmitate moduli.* Read on 25. Sept. 1775. In: *Novi commentarii academiae scientiarum Petropolitanae* 20 (1775), 1776, 271-285. Also in Euler L. (1982) EO II,17 (E480). 220-231.

42 Falter, H., Kahlow, A. & Kurrer, K.-E. (2001) Vom geometrischen Denken zum statisch-konstruktiven Brückenentwurf. *Bautechnik* 78 (12): 889-902.

43 Raskin, N.M. & Malkevitch, B.A (1953) = Раскин, Н.М. и Малькевич, Б. А. (1953) Рукописные материалы И.П. Кулибина. 161-162.

44 Raskin, N.M. (1958) 542. Also Русский биографический словарь, С. Петербургъ 1913, 16, 168-173.

45 Procès-verbaux (1900) *Procès-verbaux des séances de l'Académie impériale des Sciences depuis sa fondation jusqu'à 1803.* Tom III. 1771-1785. St Petersburg. 228-229 (Meeting of 22 February 1776).

46 Stäckel, P. & Ahrens, W. (1908), *Der Briefwechsel zwischen C.G.J. Jakobi und P.H. von Fuss über die Herausgabe der Werke Leonhard Eulers.* Leipzig: Teubner. 1.

47 Procès-verbaux (1900) *Procès-verbaux des séances de l'Académie impériale des Sciences depuis sa fondation jusqu'à 1803.* Tom III. 1771-1785. St Petersburg. 229 (meeting of 22 February 1776). cf.: Minutes of the Academy of February 1776 In: AN: F.1, Op. 2-1776, No.2, sheet 1.

48 Procès-verbaux (1900) *Procès-verbaux des séances de l'Académie impériale des Sciences depuis sa fondation jusqu'à 1803.* Tom III. 1771-1785. St Petersburg. 230-232.

49 Bélidor, B.F. de (1729) *La Science des Ingénieurs dans la conduite des travaux de fortification et d'architecteure civile.* Paris: Claude Jombert. Livre. III, ch. 6, 25.

50 Procès-verbaux (1900) *Procès-verbaux des séances de l'Académie impériale des Sciences depuis sa fondation jusqu'à 1803.* Tom III. 1771-1785. St Petersburg. 265-266.

51 AN: F.1, Op. 2-1776, No.10, Sheet 6: de Ribas: Calcul du poids du Modelle du Pont et de celui qu'il doit porter. St. Petersburg, 15. November 1776. Manuscript.

52 Procès-verbaux (1900) *Procès-verbaux des séances de l'Académie impériale des Sciences depuis sa fondation jusqu'à 1803.* Tom III. 1771-1785. St Petersburg. 267.

53 AN: F.1, Op. 2-1776, No.10, Sheets 11-12: Letter from J.A. Euler with a request for comments on the draft report.

54 Procès-verbaux (1900) *Procès-verbaux des séances de l'Académie impériale des Sciences depuis sa fondation jusqu'à 1803.* Tom III. 1771-1785. St Petersburg. 268.

55 AN: F.1, Op. 2-1776, No.10, Sheet 21.

56 AN: F.1, Op. 2-1776, No.10, Sheet 18 Comments on the draft report; Sheets 29-30 Remarks by Krafft concerning the experiments of de Ribas.

57 Procès-verbaux (1900) *Procès-verbaux des séances de l'Académie impériale des Sciences depuis sa fondation jusqu'à 1803.* Tom III. 1771-1785. St Petersburg. 270.

58 Euler, Leonhard (1778) *Determinatio onerum, quae columnae gestare valent* (E508). *Acta academiae scientiarum Petropolitanae* (1778:I),1780, 121-145. EO II,17 (1982) 232-251.

59 Euler, Leonhard (1757) *Sur la force des Colonnes* (E238). *Mémoires de l'académie des sciences de Berlin* 13 (1757), 1759, 252-282. EO (1982) II,17, 89-118.

60 Euler, L.: EO I,24. (1952) *Methodis inveniendi lineas curvas. Addimentum I. De Curvis elasticis.* Bern: Orell Füssli Turin. 255.

61 Euler, Leonhard: *Determinatio onerum* (E508). EO II,17 (1982) 232-251. 245.

62 Truesdell, Clifford (1960) The Rational Mechanics of flexible or elastic bodies: 1638-1788. *Leonhardi Euleri Opera Omnia II*, 11-2. Turici: Orell Füssli. 404.

63 Nowak, Bernd (2008) Der Eulersche Knickstab aus der Sicht eines Ingenieurs. In: *Leonhard Euler 1707-1783. Mathematiker – Mechaniker – Physiker* (eds G. Biegel, A. Klein & Th. Sonar), 271-291. Braunschweig: Braunschweigisches Landesmuseum. 282.

64 Euler, L.: EO II,17. 244.

65 Euler, L. (1780) *Examen insignis paradoxi in theoria columnarum occurrentis.* (E509). *Acta academiae scientiarum Petropolitanae* (1778:I),1780, 146-162. EO II,17(1982) 252-265.

66 Euler L. (1780) *De altitudine columnarum sub proprio pondere corruentum* (E510). *Acta academiae scientiarum Petropolitanae* (1778:I),1780, 163-193. EO II,17 (1982) 266-293.

67 Nowak, Bernd (1981) *Die historische Entwicklung des Knickstabproblems und dessen Behandlung in den Stahlbaunormen.* Dissertation. Darmstadt: Technische Hochschule Darmstadt. 13-18.

68 Letter N. Fuss to D. Bernoulli, St. Petersburg 17(6) January 1777. EO IV A, 3.2 (2016) Commercium epistolicum. Basel: Birkhäuser. 1007-1008.

69 Letter D. Bernoulli to N. Fuss, Basel, 18(7) June 1777, 1010-1011 EO IV A, 3.2 (2016) Commercium epistolicum. Basel: Birkhäuser. 1010-1011.

70 Letter N. Fuss to D. Bernoulli, St. Petersburg 25(14) July 1777. EO IV A, 3.2 (2016) Commercium epistolicum. Basel: Birkhäuser. 1015-1019.

71 Letter D. Bernoulli to N. Fuss, Basel 18 March 1778. EO IV A, 3.2 (2016) Commercium epistolicum. Basel: Birkhäuser. 1027.

72 AN: F.1, Op.2-1777, No.1, Sheets 3-4 Examen d'un model de pont d'une seul arche construit de bois et inventé par le Méchanicien Koulibin. Draft report by Jean A. Euler 6. January 1777.

73 Raskin, N.M. & Malkevitch, B.A. (1953) = Раскин, Н.М. и Малькевич, Б. А. (1953) Рукописные материалы И.П. Кулибина. 162-163.

74 Perronet, Jean-Rodolphe (1987) *Construire des Ponts au XVIIIe Siècle. L'Oevre de J.R. Perronet.* [Reproduction of the original *Oeuvres de M. Perronet* (1788). Paris: Didot]. Paris: Press de L'école nationale des ponts et chaussées. 187-193.

75 Bunin, M.N. (1986) = Бунин, М.Н. (1986) Мосты Ленинграда. Ленинград: Стройиздат 1986. 123.

76 Anon (1818) *Über eine neue Konstruktionsmethode von hölzernen Brücken mit 100 und mehrschuhigen Joch-Öffnungen.* Kunst- und Gewerb-Blatt des polytechnischen Vereins im König-Reiche Bayern 4 Nr. 2. Leipzig und Frankfurt a.M.: München: Zellersches Commissions Magazin. 25.

77 Fuss, Nicolas (1780) *Varia problema circa statum aequilibrii trabium compactilium oneratarum, earumque vires et pressionem contra anterides. Acta academiae scientiarum Petropolitanae* (1778 I.) 194-215.

78 Timoshenko, S.P. (1953) *History of Strength of Materials,* New York: Dover Publications. 98-99.

79 Young, Th. (1855) Life of Robison. In: *Miscellaneous Works of the late Thomas Young.* (ed. George Peacock), Vol II, 505-517. London: John Murray. 509-510.

80 Eytelwein, J.A. (1808) *Handbuch der Statik fester Körper* vol. II. Berlin: Realschulbuchhandlung. 409-424.

81 Young, Th. (1824) Carpentry. In: *Supplement to 4th, 5th and 6th Editions of Encyclopaedia Britannica.* Vol 2, 621-646. Edinburgh: Constable. 624. (Part reprinted in *Miscellaneous Works of the late Thomas Young* (ed. George Peacock), Vol II, 248-261. London: John Murray. 256.)

5

The use of models in early nineteenth-century British suspension bridge design

Denis Smith

5.1 Suspension bridges in the early nineteenth century

In the nineteenth century the suspension bridge became a prominent feature of bridge-building on both sides of the Atlantic, especially in Great Britain, France and the United States. In Great Britain the first half of the century promised more, both in terms of theoretical ideas and practical techniques, than was to be sustained after about 1850 by which time the suspension bridge played a declining role in the British bridge builders' repertoire. France was a prolific builder of suspended bridges in the 1830s and 1840s but the effort petered out after 1850 following the collapse of the Pont de la Basse-Chaine at Angers which killed two hundred and twenty-three people. By comparison, in the United States suspension bridge building went from strength to strength in the hands of such able practitioners as Charles Eller and John Roebling, although precisely why American technology succeeded so fully where its European counterpart failed so utterly remains unexplained.

This chapter is concerned only with early suspension bridges in Great Britain and with but one, and a very distinctive, facet at that. For the first two-thirds of the nineteenth-century British bridge building in general was characterized by the use of models as an aid to design and as a basis for evaluating new ideas. This is not the place to examine the technical, scientific or even the psychological reasons as to why model techniques endeared themselves so thoroughly to British engineers but it does seem that this was the most distinctive contribution to bridge designing techniques of British origin. The characteristic French approach was through mathematical analysis while in the United States the prevailing philosophy was essentially that of 'trial-and-error'.

Perhaps it is no accident that the demise of the suspension bridge in Britain coincided with a rapid stagnation of the use of models (although one should not overlook the significance of a shift in the application of scale-model techniques typified by the work of Robert Manning, William Froude, Osborne Reynolds [hydraulics engineers] and Horatio Phillips [aviation pioneer]). Nevertheless for half a century suspension bridge design by means of models was a matter of

Physical Models: Their historical and current use in civil and building engineering design,
First Edition. Edited by Bill Addis.

interest to some of the most eminent civil engineers (and some of the least) in Britain, and what follows is an assessment of their work.

The suspension bridge has several advantages over the traditional masonry arch structure. Its relatively light-weight construction made large spans possible, and the structural system being above the deck left unobstructed space below. This was an important factor leading to its adoption in cases where headroom was of prime importance. As a tension structure problems of lateral instability, later to be encountered in the compression zones of metal girders, did not arise. Under the action of dead load only tension structures are inherently stable. However, the price paid for structures with such favourable strength/weight ratios is lack of stiffness and it was undoubtedly the suspension bridge which first made designers face the distinction between *strength* and *stiffness*. The calculation of the necessary sizes of suspension links and hanger rods was, after all, an easy matter after the materials strength experiments of Telford, Barlow, and others had established the tensile properties of wrought iron. But to predict the response of the structure to *live* load was a more complicated matter. This first became a problem with the dynamic effects of wind loading on the earlier road bridges and was the cause of several bridge and pier failures. Methods of making the deck sufficiently stiff to maintain the chain geometry under the passage of moving loads were increasingly discussed and became the prime consideration in the design of railway suspension bridges.

British model experimental work in this field in the nineteenth century can be divided into two phases. Firstly, the exploitation of wrought iron in suspension bridges capable of carrying vehicular traffic. This was in response to the growth of planned road systems in the early part of the century. In this period, model experimenters were preoccupied with the properties of the suspension chain itself. Models were used to help resolve problems arising from variations in geometry, link material, and cross-sectional area of the chain. Road bridge models were concerned, quantitatively, with *strength*, and the *stiffening* effect of the road deck was considered only in a qualitative manner. Secondly, models were resorted to by designers in their attempts to adapt the suspension bridge to railway purposes. Here the preoccupation was with overall *stiffness*. Models were built to investigate the effect of combining a girder with a suspension chain. These were amongst the last model experiments recorded in Britain in the nineteenth century. The models that were made to help achieve all these objectives will be discussed here in chronological order.

5.2 The proposed suspension bridge at Runcorn

The earliest reference to a model test in connection with suspension bridge design relates to the proposed bridge over the Mersey estuary at Runcorn undertaken by Thomas Telford (1757-1834) in 1814. A petition to Parliament was drawn up in 1813 to bring in a Bill to establish the River Weaver Fishery Company, 'to preserve the fry in the said River Weaver, and in certain parts of the River Mersey ... and also to erect a Bridge over the River Mersey at or near a place called Runcorn Gap'. Telford was approached during 1814 and in

September a newspaper said 'A magnificent bridge is intended to be erected over the Mersey at Runcorn, which unite Lancaster and Chester ... A plan worthy of this great object, has been produced by Mr. Telford'. Some years later Telford said 'the crossing of the River Mersey at Runcorn has always been the most important consideration whenever shortening the communication between London and Liverpool has come under discussion. In the year 1814 I had my attention particularly drawn to the subject, and was employed to investigate it'. Few details of his work in 1814 survive and the principal source is Telford's Report to the Runcorn Bridge Committee of 1817. However, three of his pocket notebooks survive in the library of the Institution of Civil Engineers, one of which is titled, in his own hand, 'Runcorn Bridge 1814'. This contains river survey notes at Runcorn and details of various iron chains seen in his travels, recording the size of links and the loads carried. In addition, two pages marked 'model' give details of wire joints and incremental loading up to 3,033 lbs. This does not tell us much of the model except that the tests were made on 8 July. Another part of the notebook mentions 'Experiments at Brunton's' made from 5 April, and these he described in his Report to the Committee three years later:

> 'On considering the matter maturely, it occurred to me, that on the suspending principle, chains are not the most perfect means to be employed, but that the metal should be kept, as far as practicable, in straight lines, and also have few joinings. In order to proceed with due caution upon this principle, I made during the summer of 1814 above 200 experiments upon malleable iron, of from one twentieth to one half inch diameter, and on lengths varying from 31 to 900 feet. These experiments were made perpendicularly, horizontally, and with different degrees of curvature [1]'.

These tests on the tensile strength of wrought iron were made 'both in regard to elementary parts, and also when combined, partly by welding, and partly by jointing in a model'. Those relating to the 'elementary parts' were carried out with a hydrostatic press at 'Mr. Brunton's patent chain-cable manufactory, Commercial-Road, London' [2]. The tests made with 'different degrees of curvature' were conducted with the link-polygonal apparatus shown in Figure 5.1 [3]. The tests sought to establish the suitability of the material for suspension bridges and he subsequently proceeded with a design and a model – but whether the model was made at Brunton's works is not clear. Telford was assisted by James Rendel (1799-1856) who, 'At an early age ... went to London, and obtained an engagement, as a surveyor, under ... Mr. Telford, by whom he was employed

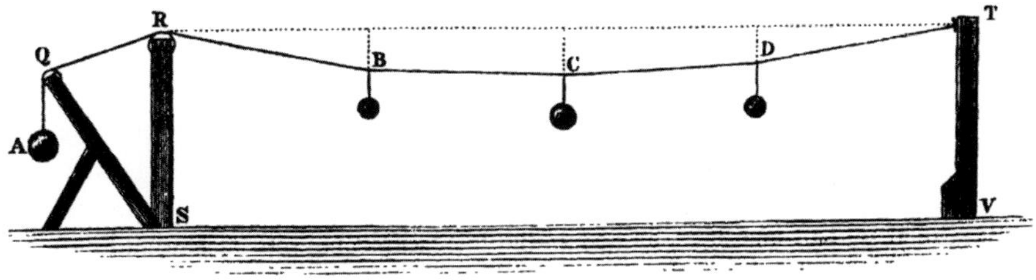

Figure 5.1 Telford's wire test apparatus. (Source: [2] p. 555)

on the surveys and experiments for the proposed suspension bridge across the Mersey at Runcorn' [4].

Telford's crucial statement about the model test is the prime source of evidence:

> 'I constructed a model of 50 feet in length, and one twelve hundredth part of the strength of the intended bridge, and although composed of wire only, and of course, without proper Joints or Braces, it bore 3000 lbs. without any symptom of its parts being deranged' [5].

Several questions arise from this; firstly that of the geometric scale of the model. Telford's design was for a bridge having a central span of 1,000 feet together with two side spans of 500 feet each – making it a bridge of unprecedented span. When he said 'a model of 50 feet in length' we cannot be certain if this included all three spans; although W. A. Provis (1792-1870) was more specific saying 'a model of the central opening was constructed' [6] and this is certainly more likely and logical. If so, the geometric scale factor was 50/1000 = 1/20 or 20:1 so far as the span was concerned, but unfortunately we have no further details of the model. Secondly, as Telford loaded this model it indicates that it was not merely of the visual type, and indeed we know that another, smaller, model was made. Telford made another model of this bridge, presumably for this purpose, to a scale of 20 feet to 1 inch (120:1) and in 1906, 'this interesting memento' was still in existence in a building belonging to the Shropshire Union Canals at Ellesmere, Shropshire [7]. Lacking details of the model chains it is difficult to understand what Telford meant by describing the model as being 'one twelve hundredth part of the strength of the intended bridge'. In using the word *strength* we may reasonably assume that he is concerned with the breaking *stress* of a member subjected to pure tension, in which case a square law relates the tensile stress to the geometric scale factor. If we assume Telford's span scale factor to be 1/20 and that this factor was also used to scale down the dimension of the proposed chain then the model would have been 1/400 (i.e. $1/20^2$) the strength of the full-sized structure. The chains of this model were, however, 'composed of wire only' and 'without proper joints' and unless the geometric scale factor for these was $1/\sqrt{1200} = 1/34.6$ it is difficult to add anything to Telford's statement relating the strength of his model to that of the full-sized structure. The model was loaded up to c. 3,000 lbs, 'without any symptom of its parts being deranged'; in other words, it was not tested to destruction, and indeed Provis added that it 'would have carried considerably more without being injured' [6]. After the tests, Telford said: 'This far I proceeded in 1814. I was perfectly convinced of the practicability of the scheme, and am now satisfied I went further than necessary in the quantity of material, and also as to sundry precautions in forming the plan, but I was willing to err on the safe side' [5]. Telford's design for the chains exist; they were to consist of '16 iron cables, each formed of 36 square half-inch iron bars; and of the segments of cylinders proper for forming them into one immense cylindrical iron cable, which was to be nearly half a mile long (including the fixings on shore), and about 4¼ inches diameter' [8, 9].

The bridge Committee met several times in the spring of 1817 and Telford was again consulted to help them decide on various proposals. He was convinced that a suspension bridge was the only possible solution and reminded them that he had 'already prepared a Design and Model', and offered his Report gratuitously. Telford mentioned only one other scheme submitted:

> 'It is that produced by CAPTAIN BROWN only which corresponds to the principle adopted by me in 1814; that is to say, Suspension by malleable Charcoal Iron, preserved as nearly as possible in straight lines . . . His perpendicular Suspenders, Cross Ties, and diagonal braces, correspond precisely with mine. His main suspending lines of the upper Curve, instead of being composed of a number of flexible rods, as in mine, consist each of one Rod or Bar of malleable Iron united by Forelocks. From four of these it was proposed that the whole structure should be suspended' [5].

It is interesting to note Telford's preference for *charcoal* smelted iron. Captain (later Sir) Samuel Brown's (1776-1852) interest in the applications of wrought iron chains led to his patent of 1817, relating to suspension bridges. During 1808 he had made a series of experiments on the 'comparative strength of bolts and bars and chains of different descriptions' at his works on the Isle-of-Dogs in London. His catenary design involved replacing ordinary chain links with long bars having forged eyes at their ends, connected by pins in shear, reducing both the weight of the chain and the number of connections. After the 1808 experiments he made 'drawings and calculations of the strength of bridges of suspension' but was too busy to test the design experimentally 'on a scale commensurate to its importance' until 1813:

> 'when I constructed a bridge of straight bars for this purpose on my own premises ... The span or extent of this bridge is one hundred and five feet, and although the whole of the ironwork weighs only thirty-seven hundred weight, it has supported loaded carts and carriages of various descriptions' [10].

No further details of this small-scale bridge have emerged but we may reasonably assume that its chain configuration accorded with his recommendation that they should 'rise or have a versed sine, in the proportion of twenty feet in a thousand feet of length'. This gives a very shallow curve producing a large horizontal force component on the tops of the supporting towers. Brown claimed a favourable strength/weight ratio for his bridge and it is worth noting that he never used the word *model* in describing it. Both Telford and John Rennie (1761-1821) visited Brown's works and use the model in describing the bridge. Mentioning it to the Runcorn Bridge Committee in 1817 Telford said 'Your Solicitor and I examined this model and drove a Hackney Coach over it' [5]. John Rennie, when questioned on iron suspension bridges in 1819, said 'the only thing of that kind that I have seen is a model that was made by Captain Brown ... over which I was drawn in a carriage, and found myself perfectly safe and easy'. In addition, he noticed 'very little vibration'.

Telford and Brown met in London and exchanged ideas on suspension bridge design. Both considered the Runcorn scheme practicable but 'although the Committee reported favourably upon the scheme, and subscribed liberally to a guarantee fund, nothing further was done' [11]. A thousand-foot span bridge was obviously ahead of its time and John Farey (1791-1851) (under the pseudonym John Nicholson) said: 'The size, and I may say, the daring character of this project frightened the capitalists to whom it was proposed; but it had, at least, the advantage of drawing public attention to this new species of construction' [12]. Other bridge and tunnel schemes were subsequently proposed but the Mersey was not bridged at Runcorn Gap until the transporter bridge was built in the first decade of the twentieth century [13].

5.3 The proposed suspension bridge at Montrose

The next suspension bridge design containing a brief reference to a model is that of George Buchanan – a civil engineer born in Montrose and a lecturer on mechanics at the School of Arts, Edinburgh. In 1824 he published a 'Report on the present State of the Wooden Bridge at Montrose, and the practicability of Erecting a Suspended Bridge of Iron in its stead' [14]. The Report contains full details of the proposed bridge over the Esk with a single span of 420 feet and a versed sine of 60 feet – much steeper than that advocated by Samuel Brown. Buchanan was particularly concerned with achieving adequate *stiffness* in his bridge and said: 'Besides that motion which takes place in suspended roadways from side to side, we must equally guard against any vertical undulation to which such a roadway may be subject' [15]. He attempted to achieve this by having vertical longitudinal iron plates beneath the decking, using the side railings, and multi-strand chains comprising 1½ inch diameter rods 15 feet long. In addition, between the vertical suspender rods he proposed to:

> 'extend a similar series of rods horizontally ... so that these horizontal rods, interweaving with those which hang vertically, the whole space directly under the chains, and between them and the roadway, will be filled up by a kind of net or wire work, on a large scale, and which, without interrupting the view, will yet communicate a remarkable firmness and steadiness to the roadway, the reason of which may be easily shown, and has indeed been experienced in the model which I have constructed' [15].

This is the only reference to the model and we therefore have no knowledge of the scale factor, whether a load test was made, or indeed whether this was anything more than a visual model. Even a small-scale model would demonstrate a qualitative increase in stiffness, which is all he is claiming for his chain configuration. Buchanan's bridge was not built and the first suspension bridge at Montrose was that of Samuel Brown built in 1829 and destroyed in 1838.

5.4 The Menai suspension bridge – chain geometry model

There is no evidence to suggest that Telford made a structural load test on a model as a preliminary to his Menai Bridge design. He would have gained valuable information relating to *strength* from his Runcorn model and tests on wrought iron a few years earlier, indeed these results were repeated in evidence to the Select Committee which recommended the building of the Menai Bridge. Nevertheless, he did make a quarter-scale model in 1823 to help resolve problems of the chain geometry. W. A. Provis, Telford's resident engineer on the bridge, said:

> 'In order to get the lengths of the ordinates or vertical rods which were to suspend the roadway from the main chains, a model was made of a single chain to a scale of one-fourth of the real one, and a set of boards on edge, curved to the intended rise of the roadway, were placed below it. Between these the perpendiculars were measured at every joint of the chain, which became the standards for making the rods of the vertical suspenders' [16].

The model has a chord line (or span) of 142 feet 6 inches, a versed sine of 10 feet 9 inches, and the chain comprised links 2 feet 6¾ inches long. Provis published a 'Table showing the lengths of the vertical suspending rods and their horizontal distances from each other'. In this measuring exercise no load was applied to the chain to simulate the decking and hence the 'intended rise of the roadway' did not take into account change of chain profile due to its weight. However, any differences due to this interaction could most likely be accommodated by screw thread and nut adjustment at the ends of the suspender rods. Provis explained the reason for using this type of model very succinctly when he said; 'It is true their ordinates might have been determined by calculation, but with a practical man an experiment is always more simple and satisfactory than theoretical deductions' [16]. Whereas Charles Drewry (1805-1881), in his Memoir of 1832 commented:

> 'In reality, those lengths cannot be calculated with perfect accuracy, because ... the chains of a bridge are neither a pure catenary nor a pure parabola, and their curve varies in every case; hence, when it is possible, the lengths of the vertical rods should be determined by experiment' [17].

Nothing could illustrate more clearly the type of dilemma leading engineers to use a model. Opinions differed as to whether calculation could provide the required data, and even Provis (a practical man) who felt that it could, nevertheless felt it 'more simple and satisfactory' to use a model. The expense of a model was justified on the grounds of the reassurance, or peace of mind, which it bought.

5.5 Wood as a material for suspension chains

An episode worthy of brief mention was the model test made by Drewry in order to investigate an alternative material for the chains of suspension bridges. In

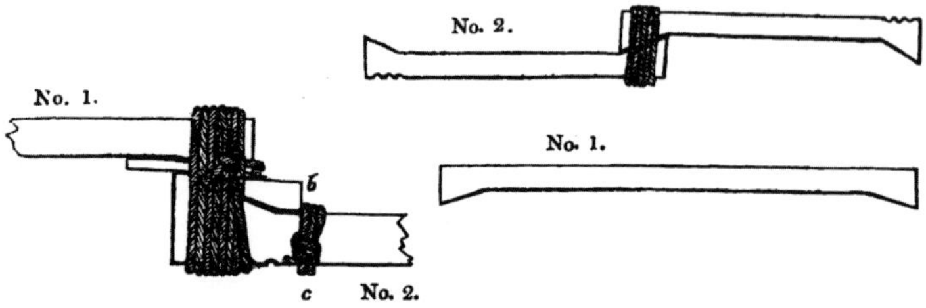

Figure 5.2 Two methods of linking bars of wood in the chain of Drewry's model suspension bridge. (Source: [19])

his well-known Memoir of 1832 he made out a case for using wood in place of wrought iron for chain links. He admitted it would not normally be desirable to seek a substitute for iron, but for bridges 'erected in remote places, particularly abroad, where iron and mechanics are not easily or cheaply procured', and for lightly loaded structures, wood might be a useful material:

> 'Wood, used as a material for making the chains of a suspension bridge, is in some respects superior, in others inferior, to iron. The advantages are, that, weight for weight, its absolute strength is greater than that of iron, and that it is worked more readily and with fewer tools' [18].

He proposed joining the wooden links by means of binding with rope the tapered ends (Figure 5.2) and it was to test the efficiency of this method that he made his experiment. As he put it:

> 'The author has tried this mode of fastening in a model 26 1/4 feet span ... The bars of wood were 14 inches long, out and out. Of two kinds, viz. No. 1 of good pine, 3/4 inch square in the middle, with solid wedge ends 1 1/8 inches deep, and 1 1/2 long. Some of the bars No. 1 were bound together with seven turns of common cord, of barely 3/16 of an inch diameter; others with iron wire: some had notches cut in the back, to receive the cord; others were left quite smooth. The bars (No. 2) were of very common fir, 1 in. deep by 3/4 in. in the middle, with solid wedge ends 2 inches deep and 2 inches long' [19].

The middle 8 or 10 feet of the span was made of the wooden bars, the remainder being iron wires (see Figure 5.3). The results of his experiments are given in Table 5.1. The model was loaded with both static and dynamic loads and it is obvious from experiments 1, 2, and 8 that he ignored the self-weight of the model in

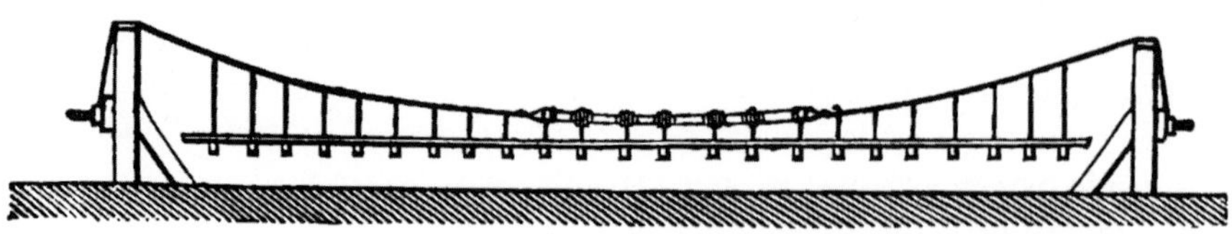

Figure 5.3 Drewry's model with timber 'chains' in the central portion of the 26 1/4 foot span. (Source: [19])

Table 5.1 Experiments on wood chains (span 26 ¼ feet) by Drewry.

Experiment	Deflection (Inches)	Load (dead weight)	Height (h) through which 130 lbs weight falls on the Platform (Inches)	Momentum of the falling weight 130 x $\sqrt{(2gh)}$ (Lbs)	Remarks
I	26	None	18 inches	1274	No injury
II		None	33 inches	1726	One of the wood-bars torn in two
III	26	52 lbs	34 inches	2735	No injury done
IV	22	52 lbs	30 inches	1645	One of the wood-bars torn in two
V	21	52 lbs	34 inches	1735	No injury
VI	26	4 persons. c. 580 lbs	None	-	While the weight remained steady, nothing gave way. On swaying up and down, one of the wires broke.
VII	26	Same, with additional 52 lbs = 632 lbs	None	-	While at rest, the bridge bore the weight. One person jumping broke one of the iron wires.
VIII	26	none	32 inches	1697	One of the wood-bars was torn in two pieces. The fastenings remained sound.

(Source: [19] p. 158)

his 'Dead Weight'. He calculated the momentum of the dropped weight correctly, although he mistakenly equated this with the force (load) imposed by the falling weight on the model deck, and expressed the result in 'lbs'.

e.g. for experiment 1:

$$\text{Mass} = 130 \text{ lbs}; g == 32.08 \text{ ft/sec}^2; \text{Distance fallen by weight} = 1.5 \text{ ft.}$$

$$\text{Momentum} = \text{m x v} = \text{m}\sqrt{(2\text{ gh})} = 130 \text{ x} \sqrt{(2 \text{ x } 32.08 \text{ x } 1.5)} = 1{,}274 \text{ lbs.}$$

Drewry made no attempt to calculate an equivalent static load and combine it with the dead load but merely uses the relative momentum in each case for the purposes of comparison. In experiments 6 and 7 he introduces four people as loading who obligingly acted as both dead and live load. There is no evidence that Drewry's work on wooden chains had much impact even on military engineers who might have been expected to use such bridges for temporary structures.

It is interesting to note that at the same time that Drewry was discussing wood as a substitute for wrought iron, a speaker at the summer meeting of the British Association in Oxford said:

'I am very desirous of calling the attention of the scientific world, and particularly of Civil engineers, to the serious consideration of the question, – How far ought iron to be hereafter used for suspension-bridges, since it is ascertained that a steel bridge can be built of equal strength, and superior durability, with one third or one fourth of the weight of an iron one, and at a much less expense, ... The only doubt of this being practicable on the large scale, arises from the circumstance, that in this country iron is made with mineral coal, but in Styria with charcoal of wood' [20].

This echoes Telford's expressed preference for charcoal-smelted iron and the speaker added; 'I have calculated that upwards of 1000 feet span may with confidence be depended on.' This was Telford's planned span for the avant-garde Runcorn Bridge but at that time the largest span built in Britain was 579 feet in the Menai bridge.

5.6 James Dredge and the taper chain controversy

In the 1830s a proposal to span the Avon at Bath with a suspension bridge led to an interesting discussion on chain design. The common chain form at that time was that of Telford and Brown as discussed above. They comprised an equal number of links between the pins giving the chain a uniform cross-sectional area throughout the span. In James Dredge's (1794-1863) designs the area of the chain decreased progressively from the points of suspension towards the centre of the span. Although it can be argued that the idea of matching the area of the chain to the tension was not Dredge's, his name became inextricably identified with the idea. A typical Dredge bridge is shown in Figure 5.4.

The Victoria Bridge Company of Bath invited designs for a bridge in the centre of the city and two local men were the principal contestants – James Dredge and Thomas Motley. Dredge's design was chosen and the Victoria Bridge was built in 1836 on his *Taper chain* principle and is indeed still standing. The span was 150 feet, versed sine 25 feet, weight of chains 5 tons, with a platform width

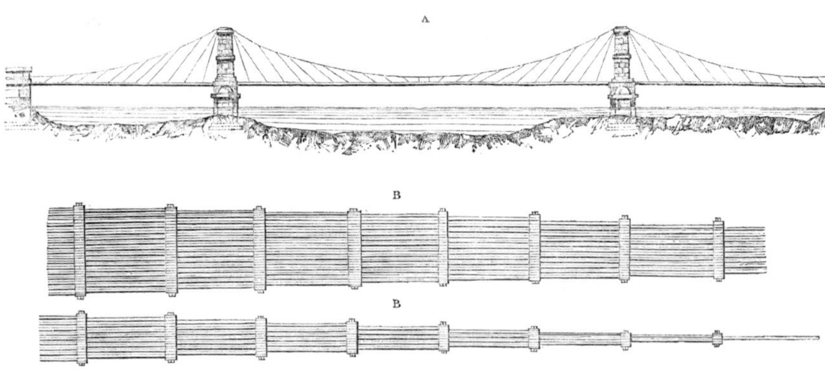

Figure 5.4 A typical taper chain bridge of James Dredge. (Source: [21])

of 18 feet. Motley, extremely bitter about the rejection of his scheme, entered on a long exchange with Dredge in the *Bath Journal*. Dredge claimed a superior strength/weight ratio for his system and made experimental models to demonstrate its performance. He was a great propagandist and the discussion extended for nearly ten years in the technical press and elsewhere. How this west-country brewer became interested in bridges is not clear. We know that he submitted a design in the Clifton Bridge competition of 1829-30, and had erected a bridge over the Leven at Balloch Ferry in 1832.

In 1836 Dredge took out a patent for 'Certain improvements in the construction of suspension chains for bridges, viaducts, aqueducts, and other purposes'. His improvements comprised two distinct parts, namely; 'the construction of a chain ... which shall diminish in breadth and weight as it recedes from the point or points of suspension; also the construction of a bridge or other suspended erection by means of such chains in connection with inclined suspension rods supporting the platform' [22]. Inclined suspender rods had previously been proposed by Poyet in France and the combination of these with the taper chain in Dredge's model tests helps confuse his argument. When the patent was enrolled the Victoria bridge was under construction and the loyal local press considered the structure 'well worthy the attention of the scientific [community]' and 'equally suitable for a large as a narrow span and the certainty of strength insured to the satisfaction of any one who may be able to know that two and two make four'. Not everyone had such faith in Dredge bridges. The first reference to models in connection with the design appears in a letter from the disappointed Motley to the bridge Proprietors in July 1836:

> 'You are most of you aware that the project of the bridge originated entirely with me, and that I have been more or less engaged for more than 12 months endeavouring to get the object effected, and that I have incurred considerable expense in journies to Bristol, Marlboro', &c. &c. to produce what was deemed needful information, and also that a heavy expense has been incurred for models, drawings and experiments, with which all parties appeared satisfied except James Dredge, brewer, and Isaac Wilson, dentist, to whose conduct I unquestionably attribute the rejection of my plan' [23].

This gives an insight into the way in which bridge design competitions were managed at that time. No further information regarding Motley's design work has materialized, but that he was thinking in terms of a load test on a model we know from a letter written a week later urging the Proprietors to:

> 'direct some proper *disinterested* party to make a model upon J.D.'s first plan, beginning with eight bars and ending with two, or his second plan of beginning with ten bars and ending with three, and another model on my plan, to contain exactly the same quantity of material, each party to have the superintendence of his own plan, then appoint a time for trial, and I will engage that my model shall sustain at least double the weight which J .D.'s will do' [24].

The confident Motley added a postscript saying; 'If you object to the expence (two or three pounds) I will propose that J.D.'s model may be loaded till it breaks, and I will then arrange the same materials and make it sustain double the weight.' In the subsequent argument it was never suggested that a model should represent, to scale, a particular bridge so the question of *similarity* does not arise. He merely proposed using two models to adjudicate the rival claims of the two protagonists. But he was not even comparing like with like. In a Motley bridge the roadway was supported by a series of inclined parallel rods connecting the deck to the towers without a catenary. They were, in fact, not suspension bridges but highly statically indeterminate stayed-girder bridges. He built a bridge of this type over the Avon at Twerton of 120 foot span which was described as being 'Superior to the common suspension bridge, in that it is more firm, and experiences much less friction, owing to the absence of vibration' [25]. After this, Motley the 'Grocer, Tea Dealer, and Bridge Designer' withdraws and we must return to the subsequent work of James Dredge.

No evidence suggests that Dredge used a model test to justify his designs for either the Clifton or Victoria bridges, and it was not until the beginning of 1838 that he began his series of much-publicized tests. Considerable discussion followed the communication of his results to the press – invariably in a letter from Dredge. His experiments compared the strength/weight ratio (unlike Motley he did not specifically consider *stiffness*) of parallel and taper chain models of the same span/versed sine ratio, each containing 9 oz of iron wire (exclusive of hanging rods). The first trial was made in Bath on 2 January 1838 with two models of 4 feet 6 inches span, 6 inch deflection, and a platform 2 feet wide. His amusing results were given as follows – the parallel chain model 'broke down when loaded with six sacks of horse beans' and the taper (or *mathematical*) chain model 'Bore six sacks of horse beans, 7 sacks of malt, 2 cwt. cast iron, and 11 men, all this did not break it down' [26]. Considering the loading material it is to be assumed that the experiment was carried out in Dredge's brewery using materials to hand – but how, in addition to the inanimate load, he got eleven men on to the platform of so small a model is difficult to imagine. This test was witnessed by 'Messrs. Worsams from London, Ball, Cambridge, and others of Bath'. These unscientific results were presumably sufficiently dramatic to demonstrate something in favour of Dredge's scheme but this must surely be discounted as a serious, well-conducted, experiment.

In the following fortnight three further tests were made on models of the same size in Bristol, 'in the presence of several eminent, commercial, scientific, and literary men, with attendant engineers to superintend the uniform loading of the model bridges' [27]. This suggests the exercise of greater care in these tests, the result of which are given in Table 5.2. No additional information on the loading technique has been found but perhaps the consistently large difference in favour of the taper-chain models (approximately 2½ : 1) renders a detailed study unnecessary. For the experiment of 6 January (A) he commented that 'both broke on adding the subsequent 1 cwt.' which does not suggest small increments in the 'uniform loading', although the load carried is given in quarters and pounds. This trial was witnessed by 'Messrs. Protheroe, Guppy, and others'. In the experiments of 13 January (B and C) the models were both broken on adding ½ cwt. revealing

Table 5.2 The model tests of James Dredge in Bristol and London.

	Test		A	B	C	D	E
	Date		6 Jan 1838	13 Jan 1838	13 Jan 1838	19 May 1840	13 Jun 1840
	Location		Bristol	Bristol	Bristol	London (Royal Gallery of Science)	London (College of Civil Engineers)
	Span	Feet-inches	4'6"	4'6"	4'6"	5'8½"	
	Versed sine	Inches	6"	6"	6"	8½"	
PARALLEL CHAIN	Weight of iron	Ounces				6½	
	Suspension rods					Vertical	
	Load sustained	Cwt (Lbs)	13.97 (1,565)	13.0 (1,456)	23.5 (2,632)	7 persons (1,015)	9 persons (1,305)
TAPER CHAIN	Weight of iron	Ounces				7	
	Suspension rods					Oblique	Oblique
	Load sustained	Cwt (Lbs)	34.47 (3,861)	33.0 (3,696)	61.15 (6849)	11 persons (1,595) + dynamic load	18 persons (2,610)
	Notes		Both broke on adding the subsequent 1 cwt.	Each broke on adding the subsequent ½ cwt.	Each broke with another ½ cwt.		Steelyard used to measure load. Cords used for suspender rods

(Source: [56] p. 183)

inconsistency in the loading method. The models for experiment C were 'constructed by Mr. Cross of Bristol, unknown to Mr. Dredge, in order to prove that all was fair in the former trials'. These performed considerably better than those made by Dredge to the identical specification and we may therefore assume that Cross was the better craftsman. These tests were witnessed by 'Messrs. Acrarnan, Daniels, Hillhouse, and many others of the first merchants of Bristol, Dr. Waldren and many others of Bath'. These model tests were made at a time when research was not undertaken behind the closed doors of a laboratory and Dredge's meticulous lists of those present lend the events an atmosphere of theatre. He made an important statement which may be deemed as renouncing any claim to have considered *similarity* between model and full-sized structure: 'It now remains for scientific men to demonstrate why the same result should not take place in the largest chain piers and bridges' [28].

In the early part of 1840 Dredge came under the influential patronage of Lord Western. On behalf of Dredge he wrote a long letter to the Prime Minister, Lord Melbourne, which was published in The Times in April 1840, and subsequently widely discussed [29]. On Thursday 30 April he said 'On Saturday last I introduced Mr. Dredge, the inventor of the improvement in bridge-building, to the Marquis of Northampton, at his evening assembly; He brought his models and drawings with him and they attracted great attention and admiration.' This was, presumably, a mere *divertissement* on a social occasion but, as the preamble to Lord Western's letter noted, more importantly:

> 'On the Monday morning following a trial of strength of the old system of building relatively to Mr. Dredge's, with two models of bridges, each formed out of the same weight of iron wire, namely six ounces, was made in the Marquis's garden, the Marquis and Lord Compton and others present. The model upon the old system was first loaded. It bore 18 half-hundred weights, and with the next half broke down. The model formed upon Mr. Dredge's principle sustained 34 half-hundreds, when upon the addition of another half-hundred, the wooden structure on which the chains were hung (answering to the towers of masonry, which form the fulcrum of suspension bridges) gave way altogether, and the bridge came down, but the chains were unbroken' [29].

No further details have been found, but comparing the weight of the chain with other Dredge models, the span must have been between 5 and 6 feet. Western enthusiastically embraced Dredge's principle and publicly claimed that 'Mr. Dredge can build bridges on his plan at one-third the expense of the present method; he requires only one-third the quantity of iron, and his bridge will be stronger and freer from vibration and pendulous motion.' He described the model tests at Bristol (giving the results a very wide audience) and mentioned the 'prudent caution' of clients:

> 'Mr. Dredge's principle of suspension bridge building completely overthrows the theory and practice of a Telford, a Brunel, whose experience and talents we are bound highly to respect . . . can we then be surprised

that the public should evince some fear, and some reluctance, hastily to adopt Mr. Dredge's novel principle or theory, in substitution of that which has been so long acted upon?' [29].

He went on to say 'No human was ever exempt from error, and Messrs. Telford, Brunel, and others, must not be considered to be infallible.' But his main plea as Dredge's advocate was that he should be allowed to repeat his model experiments before the Prime Minister, the Menai Bridge Commissioners, or 'any of the most eminent engineers'. In addition he claimed that Dredge would repair the storm-damaged Menai Bridge, built by Telford in 1826, at his own expense, and that the *superfluous* iron of the bridge would pay him and give a balance in favour of the Government. This dramatic claim, appearing in The Times on the last day of April I 840, must have initiated the experiments made in London during May and June 1840. What information is available on these is contained in Table 2 (experiments D and E). That of 19th May was made at the Royal Gallery of Science, Adelaide Street in London, with two models somewhat larger than those used at Bath and Bristol:

> 'The first experiment was with a model constructed with two chains on the ordinary principle, each consisting of 3 wires laid parallel to each other ... to which by the aid of vertical wires ... a platform of wood was suspended, this platform was loaded with 7 full grown persons, and upon the eighth getting on, it broke down. The wire chains were fractured at the point of suspension. The weight of the wire in this model was 6 ½ ounces. The next experiment was with two wire chains consisting of six wires at the point of suspension, and diminishing off to one in the centre ... these chains supported by suspension wires placed obliquely ... the platform which was loaded with 11 persons, without producing any fracture, until one or two of the party stamped on the platform, when it broke down, the fracture taking place at the junction of the oblique wires with the chain of suspension. The weight of the wire in this model was only 6 ounces' [30].

Here, as at Bath, there was no precision in loading and this was obviously undertaken as a dramatic public-relations exercise. The main point of interest is that not only are the parallel and taper chains compared, but the effect of vertical and inclined suspender rods. In a letter of August 1840 Dredge claimed his experiments had been made 'before the President of the Royal Society, and other noble lords' and this must have been at the Royal Gallery of Science. In the following month he made his last experiments at the College of Civil Engineers, Gordon House, Kentish Town in London. He again resorted to a human load, and the parallel chain model 'gave way when only nine persons were standing still upon it, while eighteen persons standing on the tapered chain, had to jump before they could produce a fracture, the momentum of which must have been equivalent to a very considerable addition of weight' [31]. A more interesting experiment was made to demonstrate the relative performance of vertical and inclined suspender rods and was shown:

> 'in a model constructed for that purpose. This model consisted of the chains, with a spring-beam, or steel-yard at the centre, a flexible piece of timber for the roadway, and cords for the suspension rods, which could be arranged either vertically, as in chain bridges of the old construction – or diverging from the centre as in Mr. Dredge's improvement. When the cords were arranged vertically and the roadway pressed down, the strain ... was shown by the action of the spring, which, when Mr. Dredge pressed down the roadway with both his hands, indicated a weight of 30 or 40 lbs. straining on the centre. When the suspension cords were arranged obliquely ... the whole chain was called into action; and although the roadway was pressed down with equal or even greater force than before, the spring steel-yard remained quiescent' [31].

The steel-yard is the first reference to Dredge using a measuring instrument in conjunction with a model but the loading method was again hopelessly imprecise. No details of span, versed sine, or platform width were recorded for this experiment but it would seem to have been only a small demonstration model. The year 1840 saw the most extensive discussion of Dredge's ideas, no doubt initiated by Lord Westem's letter to *The Times*.

During 1841 the debate about Dredge's system continued and five modest footbridges were built to his design in Regent's Park. In July a journal published an article by *Common Sense* on suspension bridges which extolled the virtue of Dredge bridges but had reservations about data which could only be obtained by experiment:

> 'Experiments in such matters are not always to be trusted. They are usually made by means of models vastly smaller than the structures that are to be erected. Independently of this, it is difficult to get them done by the proper parties, or in the proper manner. The experiments of engineers are but too frequently made, not for the purpose of obtaining such results might turn out, but in order, if possible, to obtain some result previously desired by the experimenter; and in such a case, but small reliance can be placed of them' [32].

Common sense indeed. The mathematical reasoning associated with the taper-chain principle is not strictly within the scope of this paper, but it is important to note that it did receive attention. In 1841 William Turnbull published a treatise which attempted 'To trace the principle which inclined Mr. Dredge to adopt the tapering chain and the oblique suspending rods, and to prove mathematically that the principle thus adopted is strictly in accordance with the maxims of accurate mechanics' [33]. During 1843 Dredge himself published his own analysis using both mathematics and graphical statics [34, 35] (Figure 5.5). His mathematics was quite sophisticated and having made fundamental assumptions, he showed himself adept at handling equations of statics in developing his argument.

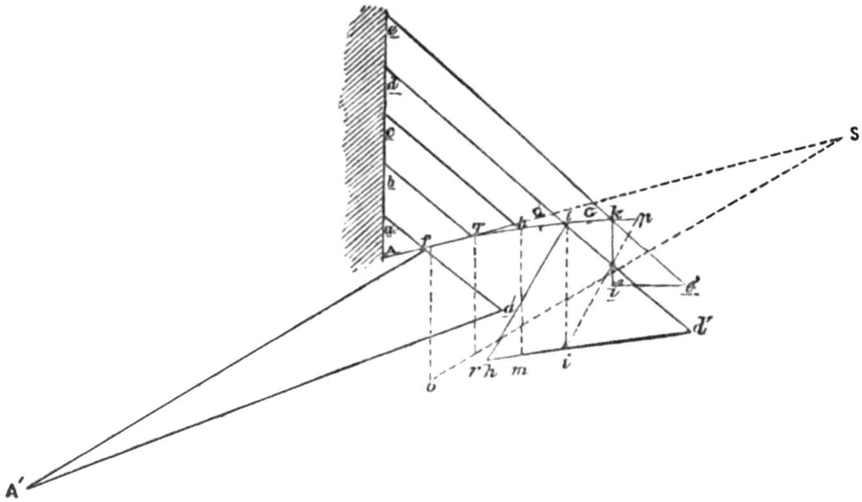

Figure 5.5 Graphical statics calculation (1843) by James Dredge of forces in Clive's proposed cable-stayed bridge (Figure 5.7a). (Source: [34] 463)

5.6.1 The Dredge and Clive correspondence

During 1843 yet another resident of Bath entered into the controversy with Dredge and used models to justify his claims. This was the relatively brief exchange between Dredge and a Mr J.H. Clive. Like Motley, Clive was not discussing a suspension bridge but a stayed-girder bridge. His suspender rods were grouped in clusters and radiated from several points on the masonry towers (Figure 5.6) [36]. He is at least quite clear about this distinction:

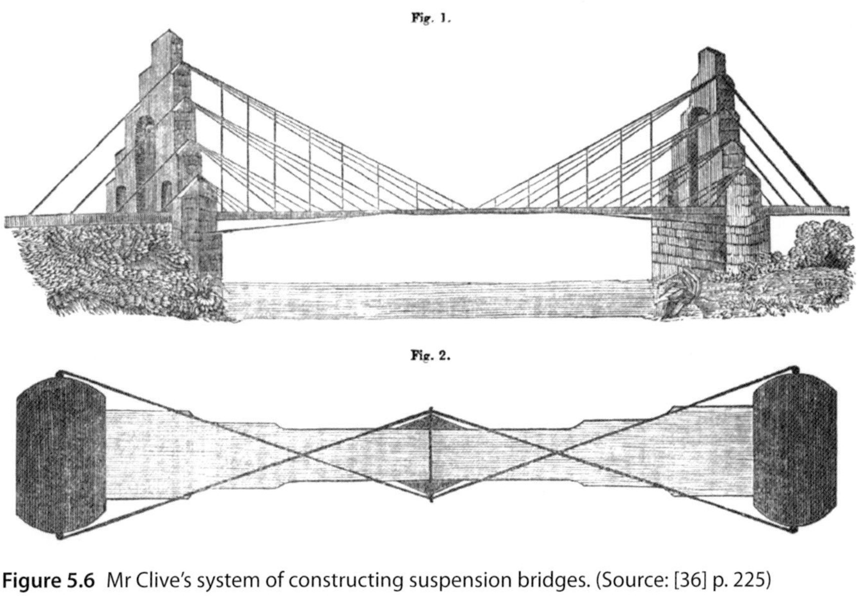

Figure 5.6 Mr Clive's system of constructing suspension bridges. (Source: [36] p. 225)

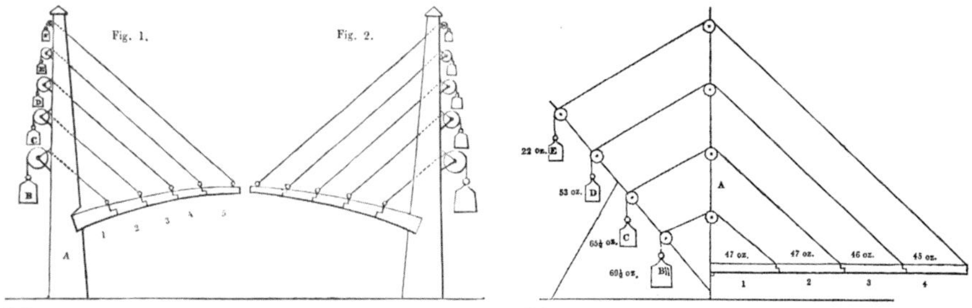

(a) (b)

Figure 5.7 (a) Model of a proposed cable-stayed bridge tested by Clive. (Source: [38] 407). (b) Model of the proposed cable-stayed bridge, rebuilt to meet the criticism of James Dredge, tested by Clive. (Source: [40] 57)

> 'On considering my models and their indications, it appears to me, that the main chain or bearing, though one of the first and most naturally occurring modes of making a suspension bridge, is by no means the best ... My models prove that the rods alone are equal to carrying the platform and load ... therefore all the weight in the main chains which exceeds the difference between inclined and perpendicular rods is worse then useless' [37].

Clive had emerged in May 1843 when a letter of his was published in the *Mechanic's Magazine* [38] where he stated that he had seen both Dredge's and Motley's bridges over the Avon at Bath. He favoured Motley's system 'on account of its being free from any swinging motion, which chains can never be', and to test his ideas he 'constructed a rough model' (Figure 5.7a). His drawing of it indicates an increasing tension in the parallel inclined suspender rods from the top to the bottom of the tower. He ended his letter with a challenge: 'If there be any fallacy in this reasoning of mine, for there is none in the model, I shall be most glad to have it pointed out' [38]. Dredge obliged in a letter a fortnight later; 'If a bracket were constructed in the manner he has shown ... with the platform curved ... the strain in the several tension cords would not vary as indicated by the weights' [39]. This drew a reply from Clive some six weeks later:

> 'Feeling assured that Mr. Dredge ... had assumed an erroneous value for the relative forces brought into action, and thereby arrived at wrong conclusions ... I reconstructed a model of the half bridge with four loose blocks or platform compartments, just ledged on each other (which I shall be happy to show him any day) in order to furnish such a near approach as a model would give' [40].

Clive's rebuilt model is shown in Figure 5.7b. The subject, as presented by Clive, is confused by his individual blocks, 'just ledged on each other'. These introduce discontinuities into the deck making it incapable of sustaining bending at the junctions, rendering the system statically determinate and capable of analysis by simple mathematics. However, if such a calculation is made by taking moments about the junctions only a poor correlation with the experimental results is obtained. It is further confused by Clive's statement that the blocks were 'not of uniform texture and thickness, so as to be of equal weights in all parts' [40]. However, he admitted that in a real bridge he had no intention of letting the load 'pass over a parcel of loose blocks, merely kept on balance, but presupposed that they were to be suitably secured' [41].

Clive admitted that his model experiments were 'only intended loosely to show one mode of making a projection in imitation of the horizontal bough of a tree, which is, in my opinion, an excellent example for study to all suspension bridge builders' [41]. No evidence suggests that Clive ever built a suspension bridge and we may assume that he entered the discussion in a philosophical manner, and it must be admitted that he did not shed much light on the subject.

5.6.2 The Ballee Khal bridge – Calcutta

Dredge's fame, and indeed the controversy, spread to the outposts of the British Empire when his principle was adopted by the Royal Corps of Engineers in India. This was its first use by the military and discussion began when a Dredge bridge in Calcutta collapsed during construction in 1845 [42]. The erection of the Bailee Khal bridge was noted in the British technical press in October 1844 [43] in an article quoting a letter dated July 1844 to James Dredge from Captain Goodwyn, R.E. of Calcutta. Goodwyn superintended the design and erection of the ill-fated bridge which had a span of 250 feet, a versed sine of 26 feet, and a deck 18 feet wide. He claimed to have carefully checked the preliminary design:

> 'With the assistance of a very able and first rate mathematician here, I have studied the theory of these bridges most thoroughly; and the model I have made 22 feet long, and 4 feet width of platform, is on so large a scale, that I have been able to test it in every possible way, and it has withstood the utmost efforts to derange its parts. The Governor-General, and all the scientific people here, have perfectly satisfied themselves of the efficiency of the system, and all these proofs, with my models, assure me that the theory is correct' [44].

We know nothing of either the mathematician or the theory studied but we may be sure that the model mentioned was not a scale representation of the bridge. After the collapse, a three-man committee of military engineers was directed to 'investigate the circumstances of the failure ... and to report on the expediency of reconstructing it on the same principles as before'. Goodwyn said the cause of failure was not one of principle but lack of care during construction. Just before completion of the decking the overseer in charge, during Goodwyn's absence, placed a concentrated load of 40 tons at one end 'in innocent ignorance of the mischief he was doing' causing the horizontal iron edge-beams to buckle. The Committee accepted Goodwyn's analysis and, after making further experiments, decided that with modifications the bridge could safely be rebuilt. In describing the experiments, the Committee made an interesting comment on the role of model tests in general:

> 'Useful, invaluable as models are, for rendering a particular mode of construction intelligible, all practical Engineers, as all mathematicians who

are practically as well as theoretically acquainted with mechanics, know that *mere models* often lead to the most fallacious conclusions; and this from its being generally assumed (often without a shadow of satisfactory proof) that the dimensions of particular parts of the fabric only require to be increased in direct proportion to one of their linear dimensions, as for instance, in the case of bridges, directly in proportion to the spans; whereas extended or *full-scaled* experiments may be found to prove that the dimensions of some of the parts ought to be increased in some higher power than either the squares or cubes of the spans' [45].

This acknowledges the importance of the visual model and the caution expressed almost paraphrases the words of John Smeaton (1724-92) in 1759:

> 'It is very necessary to distinguish the circumstances in which a model differs from a machine in large; otherwise a model is more apt to lead us from the truth than towards it. Hence the common observation, that a thing may do very well in a model, that will not answer in large. And indeed, though the utmost circumspection be used in this way, the best structure of machines cannot be fully ascertained, but by making trials with them, when made of their proper size' [46].

It also reveals an awareness of the *similarity* problem, albeit rather vaguely. Experiments were made in India between October 1845 and August 1846 with the limited objective of testing the 'theory of a system based on the "resolution of forces" '. Although this was a general analytical model, scaling laws were involved. The first experiment was 'on full scale as regards heights and distances, but formed of material 1/220 of the strength of the real bridge' It is not clear just what is meant by *strength* in this case; whether a material with a lower tensile breaking stress was used, or whether, in addition a scaling law related the cross-sectional dimensions of model and prototype. However, in the third experiment 'the points of suspension, the lengths of the rods and beam, heights and distances, being to a full scale ... the sectional area of the iron was 1/196 part of reality'. This obviously indicates a linear scale factor for the cross-sectional dimensions of $1/\sqrt{196} = 1/14$, making the self-weight of the model members also 1/196 of that of the prototype. The model was allegedly made to investigate *forces* but how these were measured is not described and the reason for reducing the cross-sectional area is not clear. Even though Dredge regarded the quantity of material in the failed bridge to be 'greater than was indispensably requisite', the new bridge was rebuilt, as a result of these experiments, with more material. The whole investigation was confused and involved both model tests and mathematical analysis. The Committee remarked that 'the eminent mathematician who attempted to investigate the principles of Mr. Dredge's mode of bridge construction appears only *partially* to have succeeded' [47]. One mathematician showed that suspender-rod tension increased uniformly towards the centre whilst Turnbull arrived at exactly the reverse result [48]. Indeed, Turnbull added that 'the successful application of the principle to practice must in a great measure depend on the sagacity and skill of the Engineer by whom the fabric is raised' [49]. The Committee expressed 'regret that the profound science which has been applied to its investigation has led to such an indefinite conclusion' [50].

Other Dredge bridges were built, and more proposed, by the Royal Engineers in India at this time. Major Abbott of the Bengal Engineers visited a Dredge bridge of 120 feet span near Meerut (about 800 miles from Calcutta) and in a letter to Calcutta said:

> 'On being called upon to report on a project for bridging the Jumna, with spans of 500 feet, constructed on Mr. Dredge's system I no longer delayed to inquire more fully into the soundness of the principle ... If I am troubling you with views that have been already set forth, you must attribute it to the distance which separates me, as well as many others of my brother officers from the scientific world. I am utterly destitute of the means of experimenting in support of my theory. I have referred the subject to Calcutta, where there are models, and on a large scale, but as yet I have received no reply' [51].

News of Goodwyn's model tests had obviously reached remote parts and the whole story is an interesting example of the way in which field military engineers undertook research at this time. The army's use of Dredge bridges in India ended with a catastrophe. Writing in October 1846, Goodwyn (now Major) said, 'A bridge on Mr. Dredge's principle, which was put up a short time since across the Kubudduck River, near Jessore, has fallen.' On testing, the bridge was found to be weak even though '*guaranteed* by Mr. Dredge to stand any traffic' and before its use could be restricted, 'some grand festival took place on the water, to witness which, a crowd of natives rushed on the bridge, nearly filling the platform, and after a few minutes it gave way, and drowned nearly one hundred and fifty of them' [52].

Although the Bailee Khal bridge committee, writing of Dredge's system, felt that 'the Civil Engineers of Europe have shown a marked disinclination to having any concern with its mathematical merits' one Englishman at least expressed strong views. After the news of the Ballee Khal bridge failure reached Britain an irate letter was published in the *Civil Engineer and Architect's Journal*:

> 'Any new plan for a suspension bridge unaccompanied by its correct mathematical theory is, in my opinion, unworthy of confidence. I do not dispute the fact of Mr. Dredge's having constructed bridges with a smaller quantity of material ... but I do question *their safety* ... We see here a laudable desire to take every precaution to ensure success – the theory is studied, a model is made, 1/12 of the size of the bridge, yet somehow theory, common sense, and experience all fail, and the blame is laid on the contractor; in my opinion there was another weighty reason' [53].

The writer accused Dredge of supporting his theory 'in face of its palpable absurdities' and charged him to 'state distinctly what is given and what is to be found by theory, and also what theory it was that the Indian engineer studied'.

Dredge, and his use of models in bridge design have been dealt with here at some length, not so much because he extended model testing techniques (that he certainly did not do) but largely because of his considerable influence on the *discussion* of suspension bridge design. A study of the literature of the period confirms this. It is true that Dredge began his rather dramatic model experiments at a time when a young, rapidly growing, technical press must have been looking for *copy*, and he almost certainly received a greater share of attention than he deserved. Nevertheless many Dredge bridges were built. The first of 1832 in Scotland was followed by many others in England, Ireland, India, and some, during 1851, in Jamaica. It was during 1851 that the design and manufacture of the bridges was put on a formal footing, although not many were built after that date. Messrs Charles D. Young & Co., with extensive iron works in Edinburgh, published a pamphlet stating that they had 'completed arrangements with Messrs. Dredge and [Robert] Stephenson C.E., London, whereby they become the Manufacturers and Contractors for the Patent Taper Suspension Bridge invented and patented by Mr. Dredge' but that 'the entire engineering, and the immediate superintendence of the erection of all such structures devolve upon Messrs. Dredge and Stephenson' [54].

5.7 Final remarks

The real justification for Dredge's place in this account of structural model testing is his role in stimulating discussion on suspension bridge design – *the* subject for experiment in the decade 1830-40. The engineering world divided itself into two camps – either *for* or *against* the taper chain principle. Dredge's greatest contribution to our topic was to provoke argument and keep alive the idea that a model (however crude) could be part of an engineer's analytical equipment. This he did superbly well in the years before model studies undertaken in the mid-1840s during the development of the huge tubular girders used by Robert Stephenson (1803-1859) and William Fairbairn (1789-1874) for the railway bridges across the Conwy Estuary and the Menai Strait in North Wales (see Chapter 6).

By the 1850s all major new bridges in Britain were built for rail traffic. Suspension bridges are inherently too flexible to carry the great weights of locomotives without suffering excessive deflection. Robert Stevenson had built a suspension bridge in 1830 on an extension to the Darlington to Stockton railway over the River Tees but it had to be replaced by a girder bridge in 1844. With the benefit of hindsight, Stevenson summed up his experience to a Parliamentary commission in 1849 as follows:

> 'I do not think that a railway bridge could be made on suspension principles; we have one at Stockton, which I replaced by one of these iron-girder bridges, and it was fearful when the engine went on to it ... when the engine and train went over the first time, there was a wave before the engine of something like 2 feet, just like a carpet' [55].

Nevertheless, the debate about suspension bridges was rekindled after John Roebling had completed his suspension bridge at the Niagara Falls in 1855. This two-level, rail and road bridge achieved the necessary rigidity by means of a 24-foot deep iron deck truss and by cable stays beneath the deck tied to anchors on the river bank. In Britain, several practising and research engineers conducted many experiments with models during the decade 1857-67 to determine the degree of rigidity that could be achieved by means of a stiff deck [56]. Although this work did not influence the design of any bridges, one researcher, George Biddell Airy (1801-1892), did use his work to propose a practical design method:

> 'For planning the tube or girder of such a bridge, the rule founded on known constructions is very simple. The Engineer will decide mentally on the amount of deflection that he will tolerate in the bridge as the effect of a given load. He will then, by the rules of engineering, compute the section of the tube or girder which will permit that deflection in a bridge of ⅓ of the length in question or less; and this will be the section to be used for his long bridge' [57].

Here we have a model used, in the classic circumstances of analytical complexity, to provide the basis for an approximate design method to achieve a pre-determined degree of stiffness. It was the last model used in nineteenth-century suspension bridge design in Britain.

Acknowledgements

This chapter was prepared by Bill Addis, after the untimely death of Denis Smith in November 2017. The use of models in the nineteenth century was the theme of his doctorate [58] and he published two seminal papers on the subject. The material for this chapter is drawn from his paper 'Models in British Suspension Bridge Design', published in *History of Technology*, 1977, 2: 169-214, edited by Rupert Hall, published by Mansell Publishing Ltd. This chapter consists of about 60% of the original paper, largely unedited, with minor changes to the beginning and end to create a stand-alone piece, and with many of the minor references removed. The material for this chapter is used by permission of Bloomsbury Publishing Plc. It is used also with the permission of Denis's widow, Lyn. A short tribute to Denis can be seen on the website of the Panel of Historical Engineering Works, of which Denis was a founder member.

https://www.ice.org.uk/ICEDevelopmentWebPortal/media/Documents/About%20Us/PHEW-Newsletter-June-2018.pdf

References

1 *Parl. Pap.* (1817) Report of the Select Committee, appointed to consider the most practicable and expedient mode of effecting the proposed BRIDGE AT

RUNCORN, submitted to the General Committee on the 8th. April 1817. Mr. Telford's Report. (I.C.E. Lib. ref: Tract 4to, Vol. 13). 11.

2 Rickman, J. (Ed.) (1838) *Life of Thomas Telford (written by himself).* London: Hansard and Sons. 542.

3 Rickman, J. (Ed.) (1838) *Life of Thomas Telford (written by himself).* London: Hansard and Sons. 555.

4 Anon. (1857) Obituary for James Meadows Rendel (1799-1856). *Proc. I.C.E.* 16: 133-142. 133.

5 *Parl. Pap.* (1817) Report of the Select Committee, appointed to consider the most practicable and expedient mode of effecting the proposed BRIDGE AT RUNCORN, submitted to the General Committee on the 8th. April 1817. Mr. Telford's Report. (I.C.E. Lib. ref: Tract 4to, Vol. 13). 13.

6 Provis, W.A. (1828) *An historical and descriptive account of the Suspension bridge constructed over the Menai Strait.* London: Ibotson and Palmer. 16.

7 Webster, John James (1906) The Widnes and Runcorn Transporter Bridge. *Proc. I.C.E.* 165: 87-118. 88 footnote.

8 Drewry, Charles (1832) *A Memoir on Suspension Bridges.* London: Longman. 14.

9 Rickman, J. (Ed.) (1838) *Life of Thomas Telford (written by himself).* London: Hansard and Sons. 550.

10 *Patent No. 4137,* 10 July 1817: Invention or Improvement in the Construction of a Bridge, by the Formation and Uniting of its Component Parts in a Manner not hitherto Practised. 4.

11 Webster, John James (1906) The Widnes and Runcorn Transporter Bridge. *Proc. I.C.E.* 165: 87-118. 87.

12 Nicholson, John (1853) *The Operative Mechanic and British Machinist.* 4th. Ed. London: Knight & Lacey. 805.

13 Webster, John James (1906) The Widnes and Runcorn Transporter Bridge. *Proc. I.C.E.* 165: 87-118.

14 Buchanan, George (1824) Report on the present State of the Wooden Bridge at Montrose, and the practicability of Erecting a Suspended Bridge of Iron in its stead. *Edinburgh Philosophical Journal* 11: 140-156 and 267-281.

15 Buchanan, George (1824) Report on the present State of the Wooden Bridge at Montrose, and the practicability of Erecting a Suspended Bridge of Iron in its stead. *Edinburgh Philosophical Journal* 11: 140-156 and 267-281. 268.

16 Provis, W.A. (1828) *An historical and descriptive account of the Suspension bridge constructed over the Menai Strait.* London: Ibotson and Palmer. 48.

17 Drewry, Charles (1832) *A Memoir on Suspension Bridges.* London: Longman. 160.

18 Drewry, Charles (1832) *A Memoir on Suspension Bridges.* London: Longman. 154.

19 Drewry, Charles (1832) *A Memoir on Suspension Bridges.* London: Longman. 156.

20 Hawkins, James J. (1832) On the Steel Suspension Bridge recently built over on arm of the Danube at Vienna; and on the mode by which the exceeding tough steel employed was manufactured in Styria, at a small advance upon

the cost of Iron. *Report of the British Association for the Advancement of Science (Oxford meeting).* 608-610.

21 *The Surveyor, Engineer, and Architect,* June 1840. 100.

22 *Patent No. 7120* dated 17 June 1836.

23 Motley, T. (1836a) Open letter in *Bath Journal.* Monday 18 July. 2.

24 Motley, T. (1836b) Open letter in *Bath Journal.* Monday 25 July. 2.

25 Anon. (1838) *British Association Report, Newcastle Meeting.* 157.

26 Dredge, J. (1838b) Letter in *Mechanic's Magazine* 30 (794) 27 October: 53-55. 54.

27 Dredge, J. (1838a) Letter in *Mechanic's Magazine* 28 (763) 24 March: 422-423. 422.

28 Dredge, J. (1838a) Letter in *Mechanic's Magazine* 28 (763) 24 March: 422-423. 423.

29 Lord Western (1840a) Letter to Lord Melbourne. *The Times* Thursday 30 April: 6. Republished in full, with illustrations and editorial appendix in Lord Weston (1840b). Letter to Lord Melbourne. *Civil Engineer and Architect's Journal* 3 (33) 1 June: 193-194, Plate X. Republished in abridged form in Lord Western (1840c). *Mechanic's Magazine* 32 (874) 9th May: 706-708.

30 Editor's appendix to Lord Weston (1840b). Letter to Lord Melbourne. *Civil Engineer and Architect's Journal* 3 (33) 1 June: 193-194, Plate X.

31 Dredge, James (1840) *Mechanic's Magazine* 33 (884) 18 July: 122-123. 122.

32 Anon. (1841) Common Sense on Suspension Bridges. *The Surveyor, Engineer, and Architect* 2 (1 July). 131.

33 Turnbull, William (1841) *The Mathematical Principle of Dredge's Suspension Bridge.* London: John Weale.

34 Dredge, James (1843b) Mathematical Demonstration of the Principles of Dredge's Patent iron bridges, by the Inventor. *Mechanic's Magazine* 39 (1023) 18 March: 213-215, and (1024) 25 March: 234-239.

35 Dredge, James (1843c) Principles of suspension bridges. *Mechanic's Magazine* 39 (1034) 3 June: 463-465.

36 Clive, J. H. (1843d) Mr Clive's system of constructing suspension bridges. *Mechanics Magazine,* 39 (1050) 23 September: 225-227.

37 Clive, J. H. (1843c) Letter in *Mechanic's Magazine* 39 (1038) 19 August: 133-134.

38 Clive, J. H. (1843a) Letter in *Mechanic's Magazine* 38 (1032) 20 May: 406-407.

39 Dredge, James (1843a) Letter in *Mechanic's Magazine* 38 (1034) 3 June: 463-465. 463.

40 Clive, J. H. (1843b) Letter in *Mechanic's Magazine* 39 (1041) 22 July: 57-58. 57

41 Clive, J. H. (1843b) Letter in *Mechanic's Magazine* 39 (1041) 22 July: 57-58. 58.

42 Goodwyn, H. (1847) The Taper Chain Tension Bridge at Bailee: Khat, near Calcutta, in its renewed form after the Failure in June 1845. *Professional Papers of the Royal Corps of Engineers* IX: 83-144.

43 Anon. (1844) Description of a Suspension Bridge on Mr. Dredge's principle, erected over the Bailee Khal, for the Indian Government, from the designs, and under the superintendence of Captain Goodwyn. *Mechanic's Magazine* 41 (1106) 19 October: 257-261.

44 Anon. (1844) Description of a Suspension Bridge on Mr. Dredge's principle, erected over the Bailee Khal, for the Indian Government, from the designs, and under the superintendence of Captain Goodwyn. *Mechanic's Magazine* 41 (1106) 19 October: 257-261. 261.

45 Goodwyn, H. (1847) The Taper Chain Tension Bridge at Bailee: Khat, near Calcutta, in its renewed form after the Failure in June 1845. *Professional Papers of the Royal Corps of Engineers* IX: 83-144. 121.

46 Smeaton, John (1759) An experimental Enquiry concerning the natural Powers of Water and Wind to turn Mills, and other Machines, depending on a circular Motion. *Philosophical Transactions* 51: 100-174. 100-101.

47 Goodwyn, H. (1847) The Taper Chain Tension Bridge at Bailee: Khat, near Calcutta, in its renewed form after the Failure in June 1845. *Professional Papers of the Royal Corps of Engineers* IX: 83-144. 114.

48 Turnbull, William (1841) *The Mathematical Principle of Dredge's Suspension Bridge.* London: John Weale.

49 Turnbull, William (1841) *The Mathematical Principle of Dredge's Suspension Bridge.* London: John Weale. 35.

50 Goodwyn, H. (1847) The Taper Chain Tension Bridge at Bailee: Khat, near Calcutta, in its renewed form after the Failure in June 1845. *Professional Papers of the Royal Corps of Engineers* IX: 83-144. 115.

51 Abbott, Major F. (1844) An Inquiry into the Principle of Mr. Dredge's Tension Bridge. *Professional Papers of the Royal Corps of Engineers,* Vol. I, 2nd ed., pp. 24-6. Letter from Abbott, to Captain Denison, p. 24.

52 Goodwyn, H. (1847) The Taper Chain Tension Bridge at Bailee: Khat, near Calcutta, in its renewed form after the Failure in June 1845. *Professional Papers of the Royal Corps of Engineers* IX: 83-144. 138.

53 Bashforth, F. (1845) Letter on Dredge's Suspension Bridges, *Civil Engineer and Architect's Journal* 8 October: 313-314. 313.

54 Anon. (1851) *Description of Suspension Bridges on Dredge's Patent taper principle.* Pamphlet, I.C.E. Tract, Vol. 69. London and Edinburgh: Charles D. Young. 5.

55 *Parl. Pap.* (1849) Report of the Commissioners appointed to inquire into the Application of Iron to Railway Structures. Robert Stephenson giving evidence on 16 March 1848. London: HMSO. 340, §936.

56 Smith, Denis (1977) Models in British Suspension Bridge Design. *History of Technology* 2: 169-214.

57 Airy, George Biddell (1867) On the Use of the Suspension Bridge with Stiffened Roadway, for Railway and other Bridges of Great Span. *Proc. I.C.E.* 26: 258-264.

58 Smith, Denis (1976) *The role of structural models in the design of British bridges – 1800-1870.* PhD Thesis. London: Imperial College of Science and Technology. Available on line at https://spiral.imperial.ac.uk/bitstream/10044/1/22788/2/Smith-DP-1977-PhD-Thesis.pdf (24/10/2019)

6

Models used during the design of the Conway and Britannia tubular bridges

Bill Addis

6.1 The Chester to Holyhead railway

The railway line between Chester and Holyhead in Anglesey was planned in the early 1840s to provide the final link between England and Ireland. The route faced two major bridging challenges – the estuary of the River Conway, about 600 metres wide, and the Menai Strait, about 300 metres wide, between mainland Wales and the island of Anglesey. The Menai Strait was deemed by the British Government to be of great military importance and so the bridge would have to provide at least 30 m clearance above high water, making an arch bridge unsuitable. These various constraints led to proposals for an entirely unprecedented type of bridge with hollow rectangular, wrought-iron tubes. At Conway, a single span of 122 m was built; at the Menai Strait, the high clearance required four spans – two of 140 m and two of 70 m (Figure 6.1). To put these spans in context, the longest spans of iron bridges in the early 1840s were 20-25 metres. The long span of the Britannia Bridge was thus about six times longer than previous bridges, the greatest single technical leap in the history of bridge building. The story of the construction of these two bridges is one of the greatest in engineering history [1, 2, 3, 4, 5]. The enormity of the achievement was well recognised at the time of their construction and there were many accounts of the projects, most importantly by Edwin Clark (1814-1894) who was the superintending engineer for the Britannia Bridge [6], William Fairbairn (1789-1874) who conducted the many experiments on the iron construction [7], and George Drysdale Dempsey (1815-1859), a railway engineer [8].

Design work began on the two bridges in 1844 and a large number of model tests were undertaken, in three phases, between mid-1845 and mid-1847. Assembly of the first girder for the Conway Bridge was begun April 1847 and was always intended to be the pilot, with the aim of solving problems ahead of the construction of the later spans. A train ran through the first Conway tube in April 1848 and through the first Britannia Bridge tube in March 1850. The whole Britannia bridge was finally opened in October 1850.

Physical Models: Their historical and current use in civil and building engineering design,
First Edition. Edited by Bill Addis.

Figure 6.1 Britannia Bridge across the Menai Strait in North Wales. (Source: [6] Plate VI)

6.2 Bridge engineering around 1840

Major railway bridges at this time were generally arches made of masonry or cast iron. The first cast-iron beam or girder bridges were built in the 1830s and their span was limited to about 10 m due to the low tensile strength of cast iron, and the difficulty of casting longer beams. To overcome these limitations, several engineers began making bridges of two or three girders bolted together and strengthened with wrought-iron ties performing a function similar to the steel reinforcement in a reinforced-concrete beam. Sadly, the most famous of these bridges, designed by Robert Stephenson (1803-1859), was the Dee Bridge, with 30 metre spans, which collapsed in 1847 [9]. Although this collapse led quickly to the demise of cast-iron girder bridges, there were already many engineers exploring the possibilities of wrought iron, which is much stronger in tension, notably, William Fairbairn. Since the early 1830s Fairbairn had been making ships of wrought iron, and by 1840 had built several around 60 m long. The board of the Menai bridge company was said to have been particularly impressed at the structural properties of a wrought-iron tube when one of their number witnessed an accident that befell a 60 m long iron ship that became stranded as the tide receded, with one end resting on a jetty and the other on the sea bed [10].

At the time, the established design approach for an iron ship was to make the wrought-iron tube strong enough to span, as a beam, between the crests of two waves, and when supported on a wave crest at mid-span. The theory of beams and bending was, by this time, well understood and both the strength and deflection of such a structure could be calculated. However, there was no theoretical understanding of the buckling of thin iron plates in compression, which was likely to be the failure mechanism of a long, slender beam. It was this particular lack of knowledge that led to the decision, early in the design of the Conway and Britannia bridges, to undertake tests on models [11, 12].

6.3 The first set of model tests – Fairbairn in Millwall

Robert Stephenson began work as engineer-in-chief for the two bridges in 1844 and his first task was to decide on the form of the bridges. His initial idea seems to have been a suspension bridge with a 'deck' consisting of a wrought iron tube or circular or elliptical cross section to provide sufficient stiffness to the structure. Stephenson was familiar with the difficulties facing a suspension bridge for rail traffic having seen the problems with the insufficiently stiff deck of the River Tees bridge on the extension to his father's Stockton to Darlington railway. However, the idea soon transformed into a tubular bridge supported, at least during construction, by supplementary chains. With regard to using an iron tube, Stephenson was reported as saying, in May 1845 that 'I feel perfectly confident; but, with a view to remove any doubt upon that point, I feel it necessary to make a series of experiments: not that it will convince me more than I am at present, but that it shall convey confidence to the Board of Directors under whom I am acting; not that I have any doubt in my own mind of it' [13]. In 1848, with the benefit of hindsight and with less need to reassure people, he was perhaps a little more honest in recalling that, with the consent of the railway directors, he had instituted 'a very laborious, and elaborate, and expensive series of experiments, in order, most thoroughly, to test experimentally the theory I had formed, and also to add suggestions for its full development' [14]. Knowing of his expertise with wrought iron, Stephenson invited William Fairbairn to undertake the first series of model tests beginning in July 1845 at his engineering works at Millwall in London.

It is interesting that, at this stage, there was not yet a precise aim of the tests. As Edwin Clark explained:

> 'The experiments [...] were not, at first, specific in their object. It was necessary to determine what kind of information was required, rather than to pursue any definite course, and to ascertain generally in what manner tubes might be expected to fail, and to what extent their strength might be modified by form. Although the square or rectangular section was first proposed, round or elliptical tubes appeared to offer considerable advantages. If suspended in chains, they had equal rigidity in every direction, and small tendency to change of form; they were simple of construction; and if the vertical portions of the circle or ellipse could be securely retained in shape, the top and bottom were well adapted to resist extension and compression; but the most important advantage was the small resistance they offered to the violent winds, to which their great height and peculiarly unsheltered situation would expose them; and on these accounts they were suggested instead of the square tube. It is difficult to retrace the steps by which any design is perfected; and it is remarkable that the first conceptions are frequently returned to, and discovered to have been correct. In the present instance this was eminently the case, and in the course of experiment the round tube was superseded by the elliptical, which in its turn merged into the original square' [15].

Between July and October 1845 thirty-seven models of different sizes and cross sections were tested. The load was applied at mid span, either on top of the tube, or from a hole in the bottom of the tube (Figure 6.2).

In August, Fairbairn invited his colleague Eaton Hodgkinson (1789-1861) to observe the test and offer advice and help with the bending theory needed to interpret the results. Fairbairn and Hodgkinson had collaborated in the 1820s to determine the most efficient shape for a cast iron beam, leading to the first

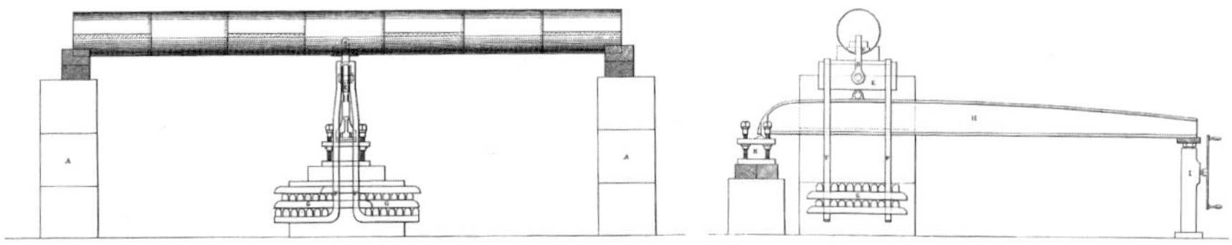

(a) (b)

Figure 6.2 Apparatus used for the experiments with the sheet iron tubes and beams. (a) Elevation; (b) Section. (Source: [6] Plate 1 facing p. 85, Plate 2 facing p. 86)

conclusive demonstration of the superiority of the I-section, albeit an asymmetric I, because of the different strengths of cast iron in tension and compression [16].

The results of different tests were normalised by calculating the value of the constant, c, in the equation $W = \frac{a\,d}{l}.c$ where W is the breaking load; *a*, the area of iron in the cross section; *d*, the diameter or depth of the tube; and *l*, the span of the girder. i.e. $c = \frac{W\ l}{a\ d}$.

A total of 34 models with spans of between 4.5 m and 9.5 m were loaded to failure, and the results recorded, with illustrations showing the nature of the failures (Figure 6.3):

> 'The [twelve] models [of cylindrical tubes] varied in length from 15 to 31 feet, and in diameter from 12 to 24 inches, bearing at the centre from 1 to 6 tons.
> … The [six] oval tubes were from 17 to 24 feet long, varying in depth from 12 to 22 inches, and in width from 7 to 14 inches, and broke with weights varying from 1 to 8 tons, suspended from the centre.
> … The [sixteen] rectangular tubes first constructed varied in length from 17 feet 6 inches to 24 feet, and in depth from 1/15th to 1/20th of the length, or from 8 inches to 18¼ inches, the thickest plates used being about ¼ inch thick. They failed by weights placed at the centre, varying from a ton and a half to about 10 tons. … Some of the models were not of uniform thickness, the top and bottom differing considerably.
> … Some experiments [were also carried out on] three solid wrought-iron T-girders, about ten feet long. They failed by bending over laterally without

TABLE III.

Preliminary Experiments on the Transverse Strength of Rectangular Tubes of riveted Boiler-Plate.

No. of Experiment.	Date.	Length.		Weight.		Sectional Area.			Weight applied.	Deflection.	Permanent Sets.	Remarks.
		Total.	Between Supports.	Total.	Between Supports.	Top.	Sides.	Bottom.				
		Feet.	Feet.	Tons.	Tons.	Inches.	Inches.	Inches.	Tons.	Inches.	Inches.	
20	July 31	18·5	17·5	·0902	·0853	1·074	1·432	1·074	·419	·17		Shackle hung across top of tube.
									·919	·55		
									1·419	·96	·1	Failed by compression, the top doubling up and the sides bulging out.
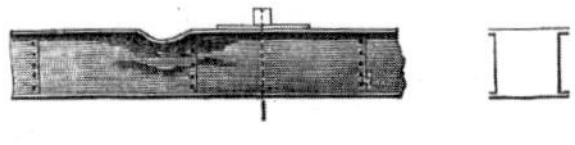						Depth 9·6 inches. Breadth 9·6 —			1·679	1·1		

Figure 6.3 Preliminary experiments on the Transverse Strengths of rectangular Tubes of riveted Boiler-Plate. (Source: [6] Table III, p. 118)

fracture of the flanges. On account of the ductility of wrought iron, square bars, or beams, of this material invariably become useless or destroyed by bending under transverse strain, and not by tearing asunder, as in tubes, where the distance of the material from the neutral axis ensures great stiffness and prevents much flexure without fracture' [17].

Once the rectangular cross section had been chosen, Fairbairn reported on experiments to determine the best cross section of the top and bottom flanges, for example a 4.8 m span model with 'a corrugated top, forming two longitudinal cavities along the whole of its top side' [18] (Figure 6.4). In later tests on the larger model (9 m span) the upper flange was strengthened and stiffened using plates of cast iron to prevent failure by buckling of the top flange. Additional strengthening measures were taken (a) by adding more iron plate to the tension flange and (b) stiffening the side walls to prevent buckling. Earc of these measures increased the strength of the girder by a factor of about two (Figure 6.5).

From these tests Stephenson was able to conclude that:

- Wrought-iron plate is better at resisting tension than resisting compression (due to bucking).
- The rectangular section was stronger than the circular or elliptical sections, for the same amount of material.
- The rigidity and strength of a tube are best obtained by 'throwing the greatest thickness of material into the upper side'.

FIG. 27.

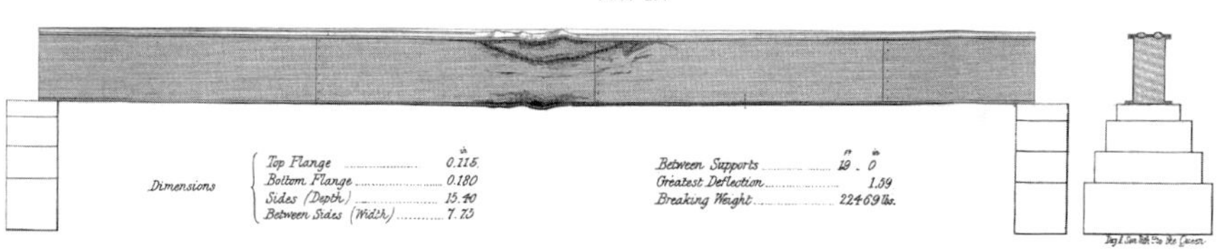

Figure 6.4 Experiments on sheet iron tubes and beams. 4.8 m span. (Source: [18])

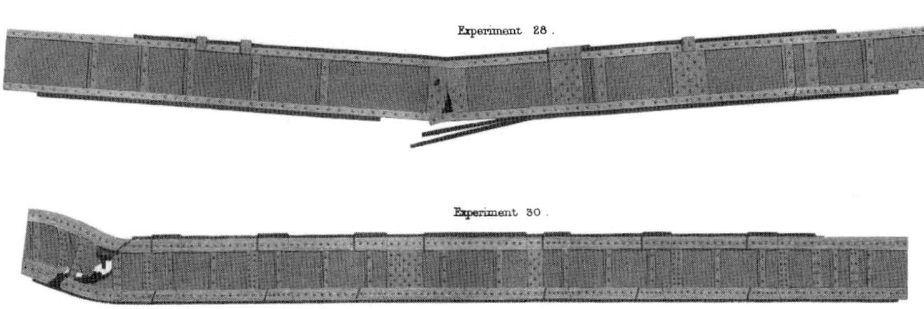

Figure 6.5 Load tests on a 9-metre span model. Cast-iron plate was attached to the upper flange to prevent failure by buckling in compression. (a) Experiment 28. Additional wrought iron riveted to the bottom flange increased the strength by a factor of two. Failure was in tension in the lower flange by tearing the wrought-iron plate and shearing of rivets.
(b) Experiment 30. Additional wrought iron riveted to the side walls to prevent buckling increased the strength by a factor of two. Failure was in shear near one support. (Source: [6] facing p. 416)

Fairbairn also noted the beneficial effect of forming corrugations in the top surface of the beam to increase the resistance to buckling.

From the tests on rectangular tubes, Stephenson deduced that a full-size tube would be able to carry a total central load of 1100 tons. Subtracting the weight of the girder of 747 tons gave a load capacity of 253 tons point load, or 506 tons uniformly distributed load. While adequate, it was felt that this did not give a sufficient margin of safety. Stephenson believed that the necessary capacity could be achieved by undertaking more experiments to refine the geometry of the cross section and he commissioned Hodgkinson to undertake these.

Already on the evidence of this first set of tests, Fairbairn believed that the suspension chains could be dispensed with after they had been used during the construction (if required). Hodgkinson on the other hand, believed to chains would be essential, or at least a useful auxiliary to help provide the necessary strength and rigidity of the girders. Stephenson was undecided, recommending to await the outcome of the next set of tests.

6.4 The second set of model tests – Hodgkinson in Manchester

Eaton Hodgkinson was a research engineer, not a consulting engineer or contractor. He had a different approach to experimentation to Stephenson and Fairbairn. He was more interested in finding general rules for predicting the behaviour of tubes of different sizes than finding a practical and timely design for the bridge. He commissioned eight models from Fairbairn's Manchester works which he tested in January 1846. He set out the aims clearly:

> 'To ascertain how far this value of *f* [the stress in the iron] would be affected by changing the thickness of the metal, the other dimensions of the tube being the same.
> To obtain the strength of tubes precisely similar to other tubes fixed on, – but proportionately less than the former in all their dimensions, as length, breadth, depth, and thickness, – in order to enable us to reason, as to strength from one size to another, with more certainty than hitherto, as mentioned before. Another object, not far pursued, was to seek for the proper proportion of metal in the top and bottom of the tube. Much more is required in this direction' [19].

He tested three sizes of rectangular-section beam, 30, 7½ and 3¾ feet long, keeping the width and depth in proportion; only the thickness of the iron sheet was changed – for example, the 30-foot beam, made with plate 0.5, 0.25 and 0.125 inches thick, collapse at loads of 57.5, 22.75 and 7.74 tons, respectively.

He was not able to make precise sense of the test results, despite mentioning using Navier's bending theories and his own, published some years earlier. He supposed that breaking stresses in the iron could only be determined by experiment, and not by calculation and that, hence, more tests would be needed. He mentioned specifically the anomalies in the results caused by the riveted connections between the iron plates. He also noted that the strength of the middle

30-foot beam was increased from 22.75 to 32.53 tons when the damaged top was replaced by thicker plate, which he had calculated, and that the new beam failed simultaneously on the top and bottom faces. Nevertheless, he failed to find a formula that would allow scaling of the model tests up to full size. His only conclusion was that, as a consequence of the thinner plate reducing the strength of the beam significantly, this confirmed his view that the suspension chains were necessary, to avoid the increased weight of a girder made with thicker iron plate.

6.5 The third series of tests, on the large model – Fairbairn in Manchester

Following the preliminary sets of experiments:

> 'The necessity for further experiments was obvious. Everybody had some doubts and fears to be overcome as soon as these details became known; dismal warnings came in on all hands, suggesting every imaginable apprehension, and Mr. Stephenson appeared at times disheartened when he withdrew, as was his daily custom, to give instructions on the subject, and to deliberate on the weighty difficulties that had to he encountered in his undertaking. Very few are aware of the painful anxiety that falls to the lot of the engineer in circumstances of such deep responsibility; he can be satisfied with no uncertainty or doubt, – and what other foundations were possible?' [20]

Hodgkinson recommended that the design should be modified to ensure that the girder would fail simultaneously at the top, in compression, and at the bottom, in tension. To achieve this he recommended the top plate be replaced by a series of circular tubes that would be more effective at resisting compression. He was commissioned to undertake a series of tests on the compression strengths of model columns of wrought iron, of both circular and rectangular section, in order to determine the minimum weight to be effective.

It was agreed that a much larger model of the girder should be tested, made as precisely as possible to a scale of one sixth. The model of the girder proposed for the Britannia Bridge would be 75 feet long, 4½ feet deep and 2⅔ feet wide (22.9 x 1.4 x 0.81 m); it weighed about 5¼ tons. It was noted that, in accordance with the square-cube law, the area of iron in any component in the actual bridge would be 36 times greater than in the model, and the self-weight would be 316 times greater. The load to be applied to the model was calculated accordingly. Hodgkinson's recommendation to use circular cells in the top of the girder was modified for practical reasons to be a row of six rectangular cells, each 6 x 6½ inches. The bottom of the girder was made with flat plates. An angle section in the form of an arch, was riveted to the each side wall of the girder to provide some additional stiffness to this vulnerable part.

In mid-July 1846 the first of ten tests were carried out [21]. The deflection of the centre of the girder were measured as it was loaded up to collapse at 35.5 tons, with a deflection of 4.4 inches. The collapse was on the bottom, tension side, due to failure of some riveted connections. The failed sections of plate were

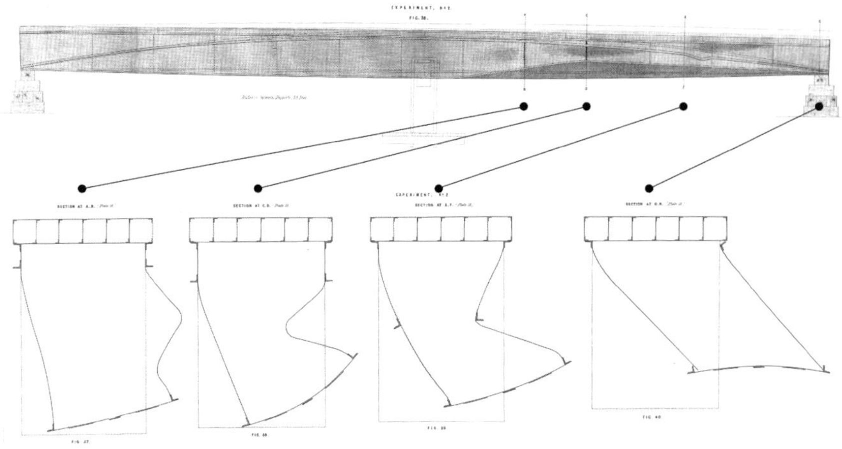

Figure 6.6 (a) Failure in the second test by sway buckling of the side walls. (b) Deformed shape at four sections on the right-hand half of the girder. (Source: [7] Plates XVIII & XIX)

repaired and additional plates – two strips 6 inches wide by five sixteenths of an inch thick – were added to the central 40 feet of the girder. The girder failed at 43.3 tons, with a deflection of 4.25 inches. This time, the failure was due to buckling of the flat vertical side walls of the tube (Figure 6.6).

After this failure, it was also noted that:

> 'A further and interesting proof of the position of the strain in the sides was afforded by the condition of the paint, which, being less elastic than the metal, did not follow the extending particles of iron, but peeled away in long flakes along the line of maximum tension, and parallel to its direction.*

This is an early observation of what was developed in the twentieth century as the brittle lacquer technique for measuring surface strains (Figure 6.7) (see Section 13.4.6).

Following this failure, 33 vertical stiffening pillars of angle section were riveted to the each side wall inside the tube, at 2 foot centres. This improved model

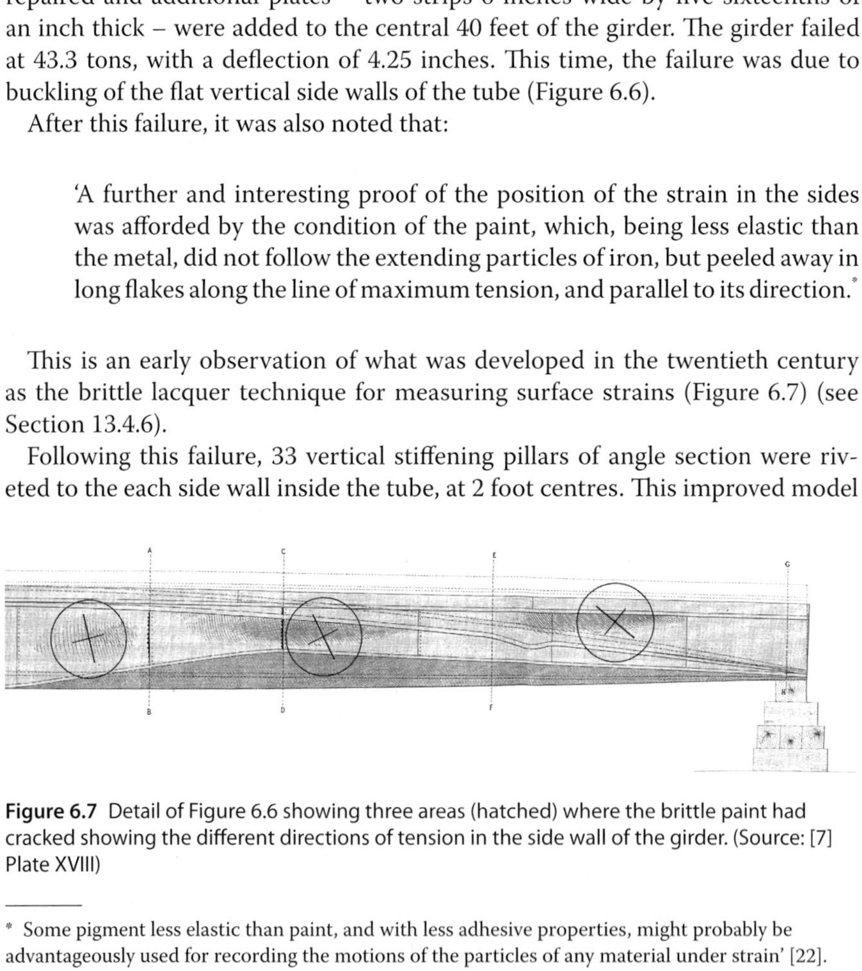

Figure 6.7 Detail of Figure 6.6 showing three areas (hatched) where the brittle paint had cracked showing the different directions of tension in the side wall of the girder. (Source: [7] Plate XVIII)

* Some pigment less elastic than paint, and with less adhesive properties, might probably be advantageously used for recording the motions of the particles of any material under strain' [22].

carried a load of 54.2 tons when the lower half of the girder failed in tension when several riveted joints pulled apart.

Three further tests were conducted between October and December 1846, and after each failure, as well as repairing the damage, plates of iron were riveted onto the girder to provide additional strength where it was needed.

At the start of the fifth test in December it was decided that a new method of measuring the deflection of the girder was needed. Previously this had been done using as a datum a wire stretched between the ends of the beam but this was found to be unreliable, especially when it was windy, and difficult to achieve the accuracy of one-hundredth of an inch. A simple optical method was devised in which a card attached to the centre of the beam was raised to be aligned with cards fixed at each end.

One of the experiments in this fifth test was performed on the girder lying on its side in order to assess the strength and stiffness under wind loads. It was also noted that a person could easily set the whole model vibrating by repeatedly pushing on the model. There was anxiety that similar vibrations could be excited in the actual bridge girder by the wind. It was considered that applying a load of 12 tons would be an appropriate test, under which the deflection was 3.21 inches.

The final test on the 1:6 scale model was carried out in April 1847 and was by then too late to influence the design of the actual girders. However, it was noted that this was the only time when the model girder failed in compression, which proved the great effectiveness of the square tubes. With the final additional strengthening plates riveted to the girder the weight of the model by this time had grown to nearly 7 tons, an increase of just over a ton. However, this model sustained a load of 192,892 pounds at failure – more than twice the strength of the model in the first test – a remarkable increase given the addition of only about 20% more iron. Thus, by modifying the model after each failure or collapse, and rebuilding it in a way that would avoid the failure just observed, Fairbairn increased the strength of the 22.9 m span girder from 35.5 tons to 86.1 tons (Figure 6.8).

Model Tube.

No. of experiments.	Distance between supports.	Depth of tube.	Width of tube.	Thickness of plate in inches. Top.	Bottom.	Ultimate deflection.	Breaking weight.	Remarks.
	ft. in.	ft. in.	ft. in.	area.	area.		lbs.	
1	75 0	4 6	2 11	24·02	8·80	......	79,578	Broke by tension.
2	75 0	4 6	2 11	24·02	12·80	......	97,102	Twisted over.
3	75 0	4 6	2 11	24·02	12·80	......	126,128	By tension.
4	75 0	4 6	2 11	24·02	17·80	......	148,129	By tension.
5	75 0	4 6	2 11	24·02	22·45	......	129,009	Weight left suspended. Tube turned on its side.
6	75 0	4 6	2 11	......	......	......	26,781	Experiment discontinued.
7	75 0	4 6	2 11	24·02	22·45	......	135,255	Weight left suspended for nine days.
8	75 0	4 6	2 11	24·02	22·45	......	135,255	Broke by tension.
9	75 0	4 6	2 11	24·02	22·45	......	154,452	Broke by tension.
10	75 0	4 6	2 11	24·02	22·45	......	192,892	Crushed on the top.

Figure 6.8 Table showing the strengths of ten 1:6 scale models, each rebuilt more strongly than the previous, failed design. (Source: [7] p. 268)

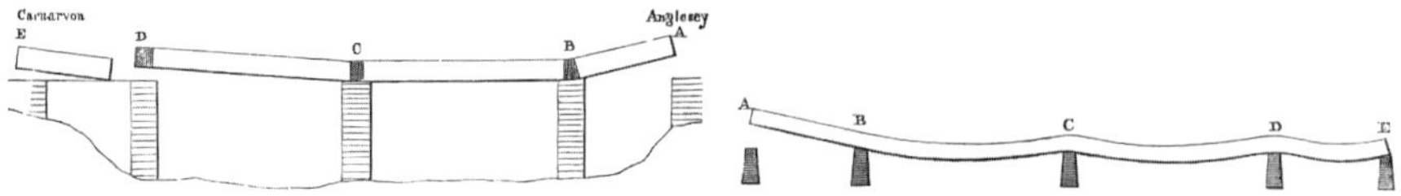

(a) (b)

Figure 6.9 (a) Sequence of raising and connecting the four spans of the bridge. (Source: [6] p. 704) (b) Deflected shape of the girder before lowering the left-hand span onto its support. (Source: [6] p. 764)

One final model, of an entirely different type, was described by Clark. When constructed, the beams of the Britannia Bridge were made continuous over the three intermediate supports. This was achieved by raising one end of a beam while riveting it to the next, and repeating this at the next two supports (Figure 6.9). The reason for this was to reduce the maximum bending moment in the centre of the two long spans. It also made the beam carry a hogging bending moment over the supports, which is why the final girders of the Britannia Bridge had rectangular cells at the bottom of the cross section, as well as at the top, to resist the compression forces (Figure 6.10). In order to determine the shape of this continuous beam, and the location of the points of contraflexure, Edwin Clark made a simple model of wood. It was made of red Memel deal, ½ inch square (12.5 mm) and 33 feet long. The deflected shape was measured four times, with each face of the wooden beam in turn uppermost, and an average taken (Figure 6.11).

The curvature shown in this diagram indicated the relative magnitudes of the bending moments at the intermediate supports and was the basis of calculation of the required depth of the rectangular girder which varied from 30 feet at mid-span, to 27 feet at the intermediate piers, and 23 feet at the ends [23].

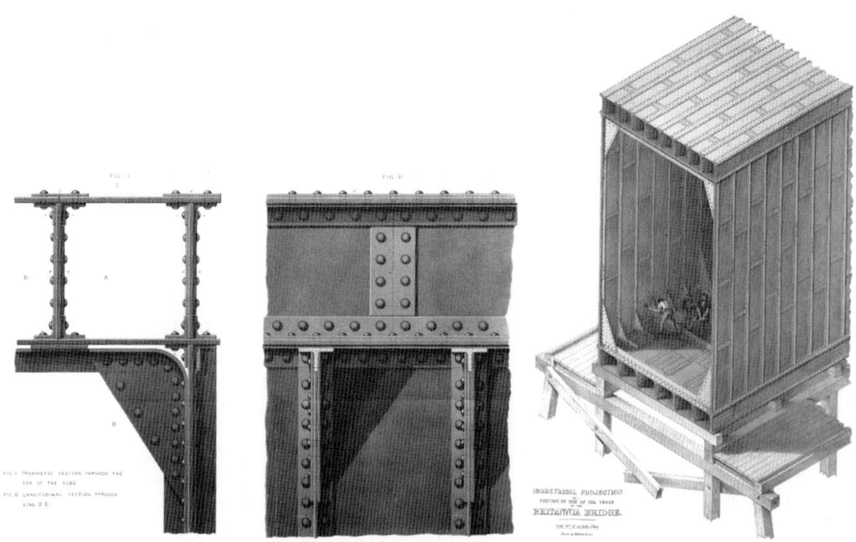

(a) (b)

Figure 6.10 (a) Detail of a riveted, wrought-iron cell in the girder. (Source: [7] Plate VI). (b) The arrangement of the cells in the top and bottom faces of the girder. (Source: [6] Plate 13)

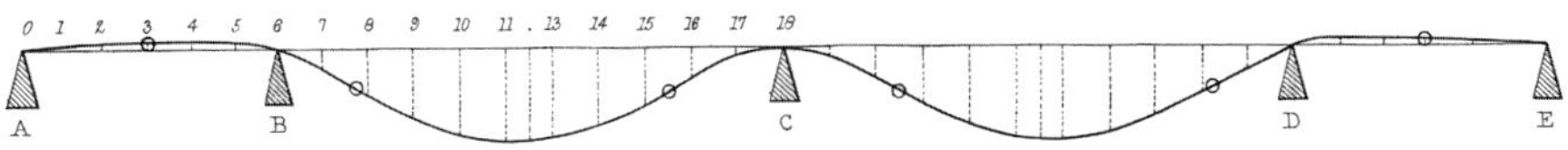

Figure 6.11 Deflected shape of the wooden model beam, 10 m x 12.5 mm x 12.5 mm, resting on five supports. (Source: [6] facing p. 462)

Clark went on to calculate the reactions on each support using the elastic formulae for a continuous beam. This was, of course, erroneous since the system is highly redundant, and an elastic solution is meaningless.

Although not strictly model tests, as they were conducted on the full-size bridge girder, two further experiments are noteworthy. The first was carried out on 27 February 1848. It was a full scale load test on the first of the tubes for the single span Conway Bridge. This prototype contained all the various strengthening measure that had been discovered during the extensive model-testing programme, including cellular construction in the top and bottom faces and vertical stiffening ribs in the walls of the girder (Figure 6.10). The load test consisted of applying loads of 95, 154, 201 and 301 tons distributed over centred lengths of 21, 32, 36 and 58 metres, respectively, on the 122 metre span [24]. The measured deflections corresponded to what had been predicted. The deflection under self-weight was 20 cm and 28 cm with the imposed load of 301 tons (Figure 6.12). It was also noted that a severe gale caused the girder to oscillate with an amplitude of 6 mm.

As mentioned earlier, one of the 1:6 scale models was loaded when on its side, to determine the stiffness of the girder under wind loads. Although it was recognised that the girders would also suffer movement due to the heating effect of the sun, this was not studied using a model. While the first completed girder was lying on the sea shore before being floated out to be placed on its supports, a great many measurements were taken to monitor how the girder moved due to the heating effect of the sun and due to wind [25].

A self-registering thermometer was located at each end of the tube, recording temperatures over a 24-hour cycle. It was found that the longitudinal expansion of the main tubes, between -3.3°C and +24.4°C was 162 mm. This movement was accommodated by the roller bearings. To record the movement of the centre of the tube an ingenious device was built. The observations were first taken using

No. of experi-ments.	Distance between supports.	Depth.	Width.	Deflec-tion.	Weight.	Remarks.
	ft. in.	ft. in.	ft. in.	in.	tons.	
1	400 0	25 6	15 0	7·91	0	The weight of the tube, 1300 tons, gave a deflection of nearly 8 inches, and 95 tons left in the inside of the tube for four hours, increased the deflection from 9·02 to 9·25 =·23 inch. This weight was continued for seventeen hours, with an increase of deflection ·10 inch. After this 301 tons, exclusive of the weight of the tubes, were laid on, when the experiment was discontinued.
2	400 0	25 6	15 0	9·02	95	
3	400 0	25 6	15 0	9·50	154	
4	400 0	25 6	15 0	10·50	201	
5	400 0	25 6	15 0	10·95	301	

Figure 6.12 Results of the load-deflection test on the full-size girder, later forming the first girder of the Conway Bridge. (Source: [24])

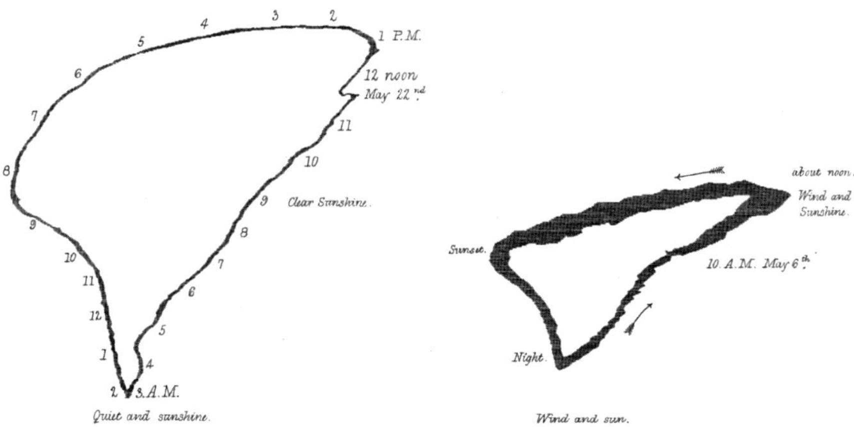

Figure 6.13 Daily horizontal and vertical motion of the tubes (Source: [6] facing p. 717)

a self-acting clockwork arrangement, which slowly elevated a board in a vertical plane, while the motions of the tube were delineated upon it by a pencil mounted on an arm which was fixed to the tube. This revealed the significant difference between the movement of the faces in the sun and the shade. A second apparatus was built in which the pencil was pressed against the board by a spring, and the rise and fall, and the lateral motions of the tube were traced on the board. The result was a closed-loop trace showing horizontal and vertical movements during a 24-hour cycle (Figure 6.13). Movements of up to 63 mm in each direction were recorded and the trace gave an instant impression of the weather – the larger the area drawn, the warmer the day and the thicker the line, the windier it was. The degree of movement was also affected by rain and even by clouds passing in front of the sun. The conclusion of the experiment was that the deformations were not of danger to the structure.

6.6 Discussion of *similarity*

The initial premise regarding similarity between the model and the actual structure was based on simple bending theory, namely that the strength of similar beams varies as the square, and the weight as the cube of their lineal dimensions. Also, if a represents the sectional area at the centre, d the depth, and l the length of any beam in inches, and C be a constant derived from a similar beam, then the weight W applied at the centre that will cause the beam to break, is given by $W = \frac{a\,d}{l}.C$ [26]. Curiously, the dimensions of C are given as tons. The reason is the W is measured in tons, and the quantities a, d and l are treated as numbers with no dimensions.

There is much discussion of 'similarity' when evaluating the results of the model tests [27], but it does not really go much further than the simple rules cited above. The reasons for this lack of clarity are probably that the discussion wanders between ultimate strength and elastic behaviour, and when considering strength, there were no theoretical models for the crucial buckling of the iron plates in the upper face of the girder. There were also ambiguities arising from the different contributions made by imposed load and self-weight, and how these should be scaled up, and also the fact that models were loaded with a central point load, rather than a uniformly distributed load. Fairbairn, Hodgkinson and Edwin Clark had slightly different ideas about similarity – Hodgkinson was always seeking general rules that could be applied to girders of any scale, as an extension of classical bending theory. Fairbairn and Clark, however, were seeking useful answers to the particular problem of their bridge girders in North Wales. Their different views are discussed in detail by Denis Smith [28].

6.7 Final remarks

The scale of the Britannia and Conway box-girder bridges so far exceeded anything that had been built previously that it was essential that the engineers undertake some type of model tests, given the inadequacy of current theory in dealing with a complex tubular section, and with the buckling of thin plates. They were mainly concerned to determine the strength of the beam, although they also provided a basis for calculating the camber that would need to be built in to the full-size girder to ensure that the deck was horizontal when subject to its self-weight. Although the tests were quantitative, in recording load-deflection behaviour and ultimate strength, the main contribution of the model tests, both the preliminary ones as small scale, and the later ones at one-sixth scale, was qualitative. The demonstrated the superiority of the rectangular cross section; they demonstrated the efficacy of small cells or tubes in the compression flange of the girder at resisting buckling; they exposed the vulnerability of the side walls to sway buckling and they proved the efficacy of vertical stiffeners in preventing this failure; they verified how best to arrange the material in the section to ensure that the top and bottom failed simultaneously, respectively in compression and tension.

In summary, the first tests established that the rectangular cross section was the most economical to achieve a desired strength. Subsequent tests identified the various ways in which a tubular beam could fail and, hence, the ways in which these modes of failure could be avoided. The effectiveness of these remedies were then verified by further tests. A tension failure in the bottom flange was prevented by increasing the thickness of the iron plates and by improving the quality of the riveted connections between plates. A buckling failure in the top flange was prevented by making it as a series of square tubes that were more resistant to buckling than a simple plate. A sway failure of the rectangular section was prevented by using corner bracing. Buckling of the vertical walls of the girder was prevented by adding vertical stiffening ribs at the connections between adjacent iron plates. A shear failure at the ends of the girder was prevented by increasing the thickness of the walls of the girder and by the addition of the vertical stiffening ribs. And finally, the stresses in the flanges of the girder were reduced by connecting the four spans of the bridge over the intermediate supports to create a continuous beam.

This series of tests undoubtedly mark the beginning of modern structural model testing although, curiously little similar testing was undertaken until the twentieth century. The reason for this was the rapid development of elastic bending theory and, from the early 1870s, the ability to deal with redundant structures. It is interesting that Benjamin Baker and John Fowler did not undertake model tests during their design of the Forth Railway Bridge in the 1880s, which was clearly a structure on an entirely unprecedented scale. The next time that structural model testing would be used to a great extent was for the design of the Boulder Dam in Colorado in the 1930s (Chapter #).

Two small anecdotes bring this remarkable story of engineering achievement to an end. First was that, after the completion of the preliminary tests, several of the small rectangular model tubes were used as chimneys for the rivet-furnaces at the works on the banks of the Menai Strait [29] (Figure 6.14).

The second story dates from 1869, twenty years after the bridges were open, and reflects the intense international interest that the project aroused, especially regarding the use of models. The French engineer Édouard Phillips, mindful of the fact that Fairbairn's load tests on the models involved a self-weight that was an inaccurate model of the self-weight of the full-size girder, together with a point load applied at the centre of the span, suggested, for the first time, using a centrifuge to increase the effective force of gravity acting on the model, to give a

Figure 6.14 Rivet-furnaces at the works by the Menai Strait with chimneys made with reclaimed models from the preliminary model tests. (Source: [6] Plate 3. p84.)

more faithful representation of the full-size girder, 430 m long with a total weight of 4,740 tonnes. He proposed a centrifuge with a model at 1:50 scale, rotating about a vertical axis at a radius of 2 m and a speed of 2½ revolutions per second (equivalent to 50g). The proposed model of 8.6 m overall length and a weight of 38 kg, would have correctly simulated the uniformly distributed load representing the self-weight of the entire bridge structure, and, if subjected to speeds of 3½ or 4⅓ revs/s, could simulate overstressing by a factor of 2 or 3. He repeated the calculations for the Conway Bridge, for models at scales of 1:20 and 1:50. He concluded: 'It is reasonable to believe that this would be a useful way of clarifying in advance the question of the possibility or impossibility of this project' (see Section 27.2.1) [30].

References

1 Chilton J.C. (2000) The Britannia and Conwy Tubular Bridges of Robert Stephenson. *Proceedings of IASS Symposium: Bridging Large Spans from Antiquity to the Present, Istanbul.* 193-204.

2 Ryall, M. & Stephenson, J.R. (1999) Britannia bridge: From concept to construction. *Civil Engineering* 132 (2): 132-143.

3 Bailey, Michael (2003) *Robert Stephenson – the Eminent Engineer.* Aldershot: Ashgate. 318-335.

4 Byrom, Richard (2017) *William Fairbairn – The Experimental Engineer.* Market Drayton: Railway & Canal Historical Society.

5 Beckett, Derek (1984) *Stephensons' Britain.* Newton Abbot: David & Charles. 131-151, 218-226.

6 Clark, Edwin (1850) *The Britannia and Conway Tubular Bridges, with General Inquires on Beams and on the Properties of Materials Used in Construction.* London: Day. https://www.e-rara.ch/doi/10.3931/e-rara-29676 (4/2/2019)

7 Fairbairn, William (1849) *An Account of the Construction of the Britannia and Conway Tubular Bridges, with a Complete History of their Progress* [...]. London: Weale. https://www.e-rara.ch/doi/10.3931/e-rara-29365 (4/2/2019)

8 Dempsey, George Drysdale (1850) *Rudimentary Treatise on Tubular and Other Iron Girder Bridges: Particularly Describing the Britannia and Conway Tubular Bridges* [...]. London: Weale. 92-132. https://archive.org/details/tubularandother00unkngoog (4/2/2019)

9 Bailey, Michael (2003) *Robert Stephenson – the Eminent Engineer.* Aldershot: Ashgate. 309-317.

10 Clark, Edwin (1850) *The Britannia and Conway Tubular Bridges, with General Inquires on Beams and on the Properties of Materials Used in Construction.* London: Day. 31-32.

11 Smith, D.P. (1977) *The role of structural models in the design of British bridges – 1800-1870.* PhD Thesis. London: Imperial College. 103-130. https://spiral.imperial.ac.uk/bitstream/10044/1/22788/2/Smith-DP-1977-PhD-Thesis.pdf (4/2/2019)

12 Smith, Denis (1976/67) Structural Model Testing and the Design of British Railway Bridges in the Nineteenth Century. Transactions of the Newcomen Society 48: 73-90. 74-79.

13 Clark, Edwin (1850) *The Britannia and Conway Tubular Bridges, with General Inquires on Beams and on the Properties of Materials Used in Construction.* London: Day. 60.

14 Fairbairn, William (1849) *An Account of the Construction of the Britannia and Conway Tubular Bridges, with a Complete History of their Progress* [...]. London: Weale. 175.

15 Clark, Edwin (1850) *The Britannia and Conway Tubular Bridges, with General Inquires on Beams and on the Properties of Materials Used in Construction.* London: Day. 83.

16 Swailes, T. (1996) 19th century cast-iron beams: their design, manufacture and reliability. *Proceedings of the Institution of Civil Engineers – Civil Engineering* 114 (1): 25-35.

17 Clark, Edwin (1850) *The Britannia and Conway Tubular Bridges, with General Inquires on Beams and on the Properties of Materials Used in Construction.* London: Day. 85, 94, 116, 126.

18 Fairbairn, William (1849) *An Account of the Construction of the Britannia and Conway Tubular Bridges, with a Complete History of their Progress* [...]. London: Weale. 246-247, Plate XIV, Fig. 27.

19 Clark, Edwin (1850) *The Britannia and Conway Tubular Bridges, with General Inquires on Beams and on the Properties of Materials Used in Construction.* London: Day. 152.

20 Clark, Edwin (1850) *The Britannia and Conway Tubular Bridges, with General Inquires on Beams and on the Properties of Materials Used in Construction.* London: Day. 156.

21 Fairbairn, William (1849) *An Account of the Construction of the Britannia and Conway Tubular Bridges, with a Complete History of their Progress* [...]. London: Weale. 250-265.

22 Clark, Edwin (1850) *The Britannia and Conway Tubular Bridges, with General Inquires on Beams and on the Properties of Materials Used in Construction.* London: Day. 166.

23 Fairbairn, William (1849) *An Account of the Construction of the Britannia and Conway Tubular Bridges, with a Complete History of their Progress* [...]. London: Weale. 184.

24 Fairbairn, William (1849) *An Account of the Construction of the Britannia and Conway Tubular Bridges, with a Complete History of their Progress* [...]. London: Weale. 265-267.

25 Clark, Edwin (1850) *The Britannia and Conway Tubular Bridges, with General Inquires on Beams and on the Properties of Materials Used in Construction.* London: Day. 713-720.

26 Clark, Edwin (1850) *The Britannia and Conway Tubular Bridges, with General Inquires on Beams and on the Properties of Materials Used in Construction.* London: Day. 93, 107, 112.

27 Clark, Edwin (1850) *The Britannia and Conway Tubular Bridges, with General Inquires on Beams and on the Properties of Materials Used in Construction.* London: Day. 184-205, 419-422, 788-789.

28 Smith, D.P. (1977) *The role of structural models in the design of British bridges -1800-1870.* PhD Thesis. London: Imperial College. Chapter 4, Similarity, 201-208. https://spiral.imperial.ac.uk/bitstream/10044/1/22788/2/Smith-DP-1977-PhD-Thesis.pdf (4/2/2019)

29 Clark, Edwin (1850) *The Britannia and Conway Tubular Bridges, with General Inquires on Beams and on the Properties of Materials Used in Construction.* London: Day. 84, Plate III.

30 Phillips, E. (1869) De l'équilibre des solides élastiques semblables. *CR Acad. Sci., Paris* 68: 75-79. 79.

Section B

Physical models used in structural design, 1890s-1930s

After some occasional examples of using measurement models before this time, the modern art of scientific testing of reduced-scale models was firmly established in the 1860s with the tests by William Froude on the resistance of ships moving through water (see Section 13.2). The idea was soon adopted in open-water hydraulic engineering and in the 1890s began to impact on civil engineering in the design of dams. As observed throughout the book, model testing was of crucial importance in mega-projects that took engineering into new territory, so to speak, and for which there was no relevant precedent. The plan to raise the first Aswan Dam, opened in 1902, by five metres was led by another engineering giant, Benjamin Baker, designer of the mighty Forth Bridge. Not only did he undertake small-scale model tests made of gelatine to test the plan to raise the dam, his work stimulated the seminal studies by Wilson and Gore that brought dam design firmly into the twentieth century.

This work was followed by another mega-project, the Boulder Dam, completed in 1936, the scale of which, compared to previous dams, is astonishing. It is fair to say that this project stimulated more progress in experimenting with measurement models that any other in history, in both structural engineering and hydraulic engineering. In particular, it stimulated the use of measurement models for reinforced-concrete structures in Germany, Italy and Spain.

Franz Dischinger used models made of both reinforced concrete and steel sheet in developing the Zeiss-Dywidag shells that achieved unprecedented spans and ratios of span to thickness. In Italy, the early testing of a model shell dam in 1926 led to the establishment, by Arturo Danusso, of the world's first research laboratory devoted to model testing, in which models of several aircraft hangars design by Pier Luigi Nervi were tested in the 1930s. In Spain, Eduardo Torroja used measurement models for the design of several of his remarkable shells in the mid-1930s.

The late-1920s saw the first use of photoelastic stress analysis for civil engineering projects and its use grew steadily during the 1930s and again after the war, until the 1970s. During this time, many universities and independent research laboratories, such as the Chrysler Corporation and Rolls Royce Aeroengines had their own photoelastic laboratories. Although its expense generally limited its use to mechanical engineering, for example gear teeth, and engineering in the aerospace industry, for example, the design of compressor and turbine discs, it was also used successfully in studying the flow of stresses in dams, tunnel walls and riveted steel-plate structures. Despite its limited use in construction projects, photoelasticity was familiar to all mechanical and structural engineers because it was able to give a unique view of stresses acting *inside* loaded structures and remains the only technique that allows a direct measurement of stress rather than deflection or strain.

7

The use of models to inform the structural design of dams, 1890s-1930s

Mike Chrimes

7.1 Introduction – the British experimental approach in the nineteenth century

A characteristic of British engineering development in the late eighteenth and nineteenth centuries was a healthy scepticism of engineering science and the associated, if not consequent, extensive use of material and component testing, and the use of physical models to develop and verify structural designs. Around 1800 this was seen famously in the development of the use of cast iron for bridges and building frames, and persisted through the use of structural wrought iron and steel later in the century [1, 2] as well as in less well-known examples associated with retaining wall design [3]. Two engineers who were firmly imbued in this approach were (Professor) William Cawthorn Unwin (1838-1933) and (Sir) Benjamin Baker (1840-1907) [4], the former through his early association with the great millwright and constructional iron fabricator William Fairbairn (1789-1874), and the latter from his early experience in ironworks, and decades spent in materials testing associated with his advocacy of the use of structural steel as well as investigations into wind pressure and (reinforced) concrete. In both cases their approach was associated with full awareness of advances in engineering science and the use of analytical methods, generally based on elastic theory, as the basis for engineering design. Around 1900 the two came together to play a crucial part in considering the use of physical models to investigate stresses in masonry dams.

7.2 Masonry dam design before 1900

Despite a long tradition of using masonry in hydraulic structures [5, 6, 7], nineteenth-century British and Irish engineers rarely designed large masonry dams in the British Isles, generally relying instead on earth embankment dams with puddle clay cores, for impounding water for the water supply of towns. This was a tradition established in the eighteenth century for ornamental lakes in landscaped gardens and canal reservoirs, as well as the design and construction

Physical Models: Their historical and current use in civil and building engineering design,
First Edition. Edited by Bill Addis.

methods developed empirically through the observation of performance of such structures, and the impact of some notable failures [6, 8]. Although some engineers would have been aware of a tradition of masonry dam design in continental Europe, notably in Spain, and the design methods being adopted in France for more 'rational' masonry dam profiles [9, 10, 11], it was only the British and Irish engineers working overseas, generally in the British colonies and most notably in the Indian Empire, who had much experience of masonry dams at that time [12].

It was in this context, and reflecting lack of experience of designing with the French methods, that Professor William John Macquorn Rankine (1820-1872) of Glasgow University was consulted about the design of a masonry dam for Mumbai (Bombay) water supply. Rankine's views, which included a critique of the French approach, were widely disseminated through their publication in *The Engineer*, one of the leading British engineering journals of the time, and included his classic 'middle third rule' according to which resultant force should be kept within the middle third of all horizontal joints or planes, in order that no tensile forces should be allowed to develop [13]. He proved that the ideal section to satisfy the condition that there would be no tension with the reservoir full or empty, was a simple right-angled triangle, with the width of the base equal to the depth of water divided by the square root of the specific gravity of the masonry. Recognising that the maximum stresses at the faces would act on planes perpendicular to those faces, and not on horizontal planes, he concentrated on stresses in the downstream face as it was the most inclined, and the maximum stress there would exceed the vertical component. Unable to calculate this point of maximum stress, he calculated the vertical stresses on horizontal planes using the trapezium law, and used a scale of limiting vertical pressure, decreasing as the slope of the face increased towards the dam's base. This meant the maximum stresses would remain constant at different heights above the base of the dam (Figure 7.1). His conclusions were widely accepted in the broader engineering community, although some French engineers ignored elements of his advice in their pursuit of more economic profiles. Henri-Joseph Bouvier (1828-1898) [14] developed a method for calculating the point of maximum stress – of little help if other factors indicated by Rankine were disregarded. Most late nineteenth-century US and German engineering textbooks reflected Rankine's approach. In Germany Otto Intze (1843-1904) developed a well-known design taking account of uplift, as well as a sound theoretical approach [15].

Rankine's intervention had little impact on general dam practice in the British Isles where embankment dams remained the norm, and only a handful of masonry dams were built – rare examples being: Abbeystead, in Lancashire (1851, Robert Rawlinson; 1881, James Mansergh), Vyrnwy, in north-Wales (1890, Thomas Hawksley and George Deacon), Thirlmere, in the English Lake District (1894, G.H. Hill) and Caban-coch in mid-Wales (1904, James Mansergh). Of these, Vyrnwy was the most important for its early consideration of uplift. This was in part due to reports of the Habra dam failure in Algeria (1881), and the ongoing problems of the Bouzey dam near Epinal in France [16]. The latter dam failed in April 1895, with a number of British engineers commentating on the mechanism of failure, including Unwin, who passed on his report to Baker, who was then finalising the design of the Aswan Dam on the Nile and who remarked

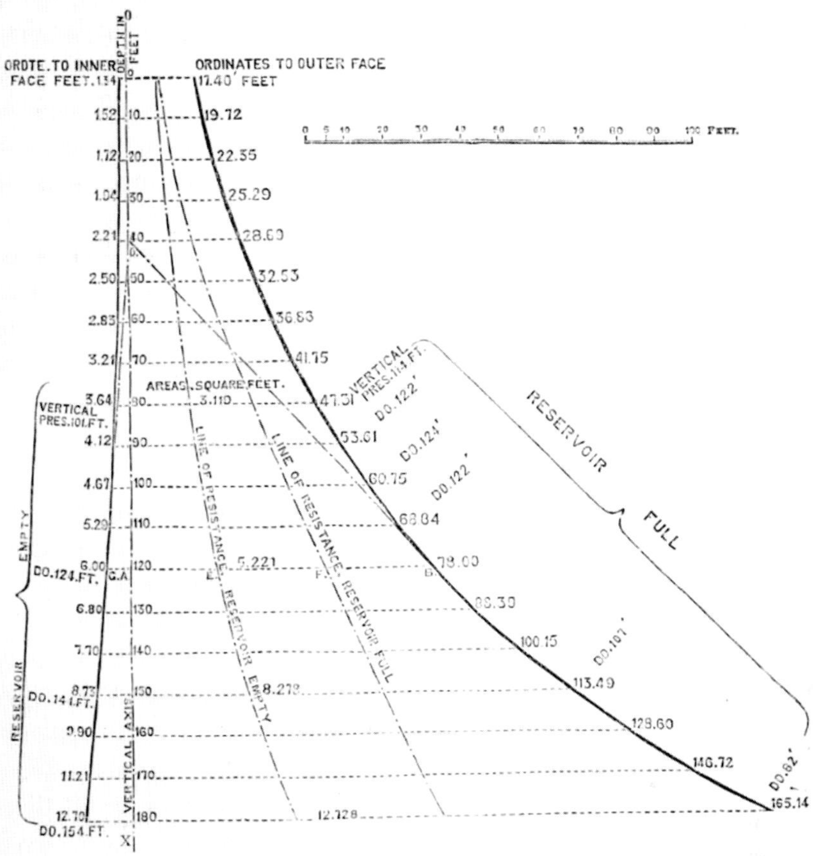

Figure 7.1 Diagram illustrating Rankine's preferred profile from his Report on the design and construction of masonry dams, 1872. (Source: [13])

'It must have been an awful jerry-built affair' [17, 18]. Among the many French commentators the most significant was Maurice Lévy (1838-1910) [19, 20, 21], widely followed not only in France but also in Italy and Spain, and the failure had a conservative impact on dam design across those countries for some time.

The relative lack of interest in masonry dam design in the British Isles after Rankine's work was published was in contrast to the work of British engineers overseas, which included arch, buttress and gravity types. In India the Periyar scheme was one of the world's engineering wonders of the 1890s, while in Australia several daring masonry dams were built [22]. Interestingly, Betaloo dam in South Australia was probably the first, possibly the only dam to be designed strictly using Rankine's methods.

Masonry dam design, despite simplifications suggested by Rankine and others, and the use of graphical methods espoused by Unwin, was a laborious business. The dam was divided into a number of sections by vertical planes perpendicular to its length. Each section was then analysed to ensure it satisfied, independently of its neighbours, the criteria of stability enunciated by Rankine and Unwin. The major issues, long since identified, were cracks opening up in the water face, crushing developing near the air face leading to overturning, and sliding at the base or along horizontal planes within the structure. In a practical sense, many of these problems could be avoided by careful construction and large factors of safety, as they had been for centuries.

7.3 The Aswan Dam

7.3.1 The genesis of the First Aswan Dam scheme

From 1883 some experienced India-based hydraulic engineers were transferred to Egypt following the effective takeover of the state by Britain, to tackle problems with the existing irrigation and river control structures. A masterly series of reports followed, addressing the control of the Nile floodwaters and how a dam might enable perennial irrigation in Egypt, with consequent increases in agricultural production [23, 24]. Amongst other recommendations William Willcocks (1852-1932), previously involved in the design of the Betwa (Parichha) dam in India, based on Rankine's methods, designed a masonry dam, curved in plan, for a site at Aswan/Assouan, which had the advantage of a supply of granite masonry close by [25]. It contained a large number of sluiced openings to allow flood waters to continue north and also retain water for later irrigation-use down river.

The scheme was expensive and, partly because it would potentially flood the important ancient temple complex at Philae in the tourist season, the proposal was referred to an international commission on which Baker, President-elect of the Institution of Civil Engineers (ICE), sat. Partly due to his influence the dam's plan was changed to a straight one but, more significantly, the height of the intended dam was reduced, with a view to subsequent heightening being possible, if required. The revised scheme was still an engineering wonder, c.30 metres high and 1950 metres long, with 180 sluices in 18 groups of 10, permitting 360 metres width of flow when open (Figure 7.2) [26].

Perhaps surprisingly, the Bouzey failure had little impact on the design at Aswan as it was soon demonstrated it had been designed neglecting analytical orthodoxy, compounded by poor construction. Instead, the great challenges at Aswan were addressed by considering how to ensure the overall line of thrust lay within the middle third at every level even though the dam had a large number of openings for the sluices. This meant it was not a uniform solid wall, and the differing height of sluices at different points made it even more difficult to determine typical cross-sections. Simplification was required [27]. To achieve this, it was assumed that, with regard to the masonry down to the sill of the lowest sluices, the five metres of masonry between the sills took all the pressure of water over the whole length of the sluice and masonry, i.e. five metres of masonry took the pressure of seven metres of water. Below the level of the lowest sluices, the lengths of water and masonry were equal and the weight of all the masonry above the sluices was considered effective; it was also calculated that the water level might be raised three metres more than originally planned. With the profile adopted, the maximum compressive stress on the air face, with the reservoir full, was calculated at 4.35 kg/cm^2, adequately below the commission's limit of 5 kg/cm^2, while, with the dam empty, the maximum stress on the water

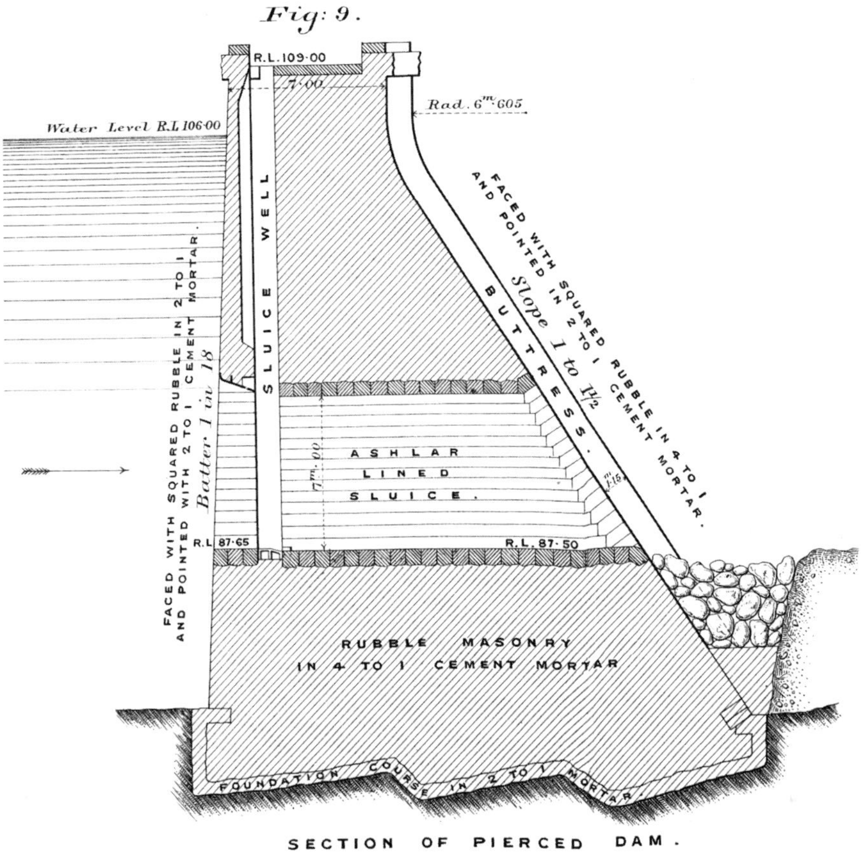

Figure 7.2 Section of the Aswan dam as designed by Baker. (Source: [27] Plate 2, Figure 9)

face was 6.32 kg/cm^2. The structure was of granite with Portland cement mortar, and was built by Sir John Aird (1833-1911).

7.3.2 Raising the Aswan dam and the stability controversy

Despite Willcocks' criticism of the straight design, to an extent vindicated by weak granite in the river bed and springs affecting construction progress and costs, and of the reduced height, the built dam proved a great success in improving agricultural production. As a consequence, in 1904, Baker was asked to consider its heightening. Put simply, raising the dam by five metres would provide more than two and a half times the existing storage capacity; whatever problems it might present at Philae, this proposal seemed more than justifiable in economic terms. As chance would have it, this request coincided with the publication of a paper by Professor Karl Pearson and L.W. Atcherley of University College, London 'On some disregarded points in the stability of masonry dams' [28].

This paper challenged the existing orthodoxies of gravity dam design, claiming that while such a dam might satisfy the middle third rule in horizontal planes, there would be dangerous tensile stresses on vertical planes, especially near the tail or base corner of the air face. With a treatment based on classic elastic theory, the paper considered Rankine's work, which it considered unsatisfactory, and the work of Maurice Lévy, although the research was commenced before reading his three papers [19, 20, 21]. A major point of their investigation was the magnitude and distribution of shear stresses within a dam and their work comprised not only extensive mathematical analysis but also the use of wooden models (Figure 7.3). One model consisted of vertical strips of wood that resisted overturning by the friction between the strips, representing vertical shear forces; the other consisted of horizontal pieces of wood that again resisted overturning by friction between the strips, and now representing horizontal shear forces. Compared to what followed, these were very crude models indeed.

The investigation assumed that the traditional approach, as reflected in Rankine's work, was erroneous, and the models were rather intended to demonstrate that the development of tensile forces in vertical sections was a major concern. The 'vertical-sliced' model is on the left, and the 'horizontal-sliced' model is on the right in Figure 7.3. In the horizontal model, water pressure was simulated by a cord attached to a stiff lath, one third of the way up, which bore on the ends of the horizontal strips through two longitudinal strips of India rubber. The cord was adjusted to be perpendicular to the battered front of the dam, with the aim that the force be applied at the centre of pressure in the manner of actual water pressure. The cord was carried over a vertical pulley and attached to a bucket which could be gradually filled with steel balls or shot to represent the water pressure. For the vertical model, the force in the cord was applied to the first courses, including the battered front. One problem with the models was that the coefficient of friction between the wooden strips varied; attempts were made to bring it into closer alignment with the coefficient for masonry, using sand paper and tissue paper, but without significant modification, the strips began to move under relatively low loads. However, the researchers were relatively content that the models demonstrated that the horizontally stratified model sheared either at its base or in the first three or four layers, while the vertical model saw the opening of tensile cracks and then sheared at its tail, although it proved difficult to monitor the failure as it happened. In a sense, the researchers had 'proved' what they already believed, i.e. that their theory showed that 'the vertical and not the horizontal sections are the critical sections, and that a dam gives first by tension of the vertical sections near the tail' [29].

(a)

(b)

Figure 7.3 Atcherley and Pearson's model dams made of wood. The left model consists of vertical slices of wood; the right model consists of horizontal slices of wood. (a) unloaded; (b) shown collapsing due to excessive 'water pressure'. (Source: [28] Plate 1, Figures (i) & (ii))

Professor Karl Pearson (1857-1936) was a widely-respected elastician and the research had to be taken seriously. Baker drew the attention of the engineering community to it an ICE meeting in 1904 and also to the attention of his colleagues in Egypt [30]. There followed nearly four years of model testing, mathematical studies and intellectual dispute, which ultimately largely discredited Pearson's concerns but, in the process, demonstrated the value of properly conducted investigations using physical models.

Baker encouraged Pearson and his colleagues to pursue their work as he recognised the limitations of existing dam theory, but he himself was relatively soon assured that the major problems concerning the raising of a dam with operational sluices on Aswan's scale, were practical ones. Of particular importance were the thermal stresses in the masonry, given that a dam of such length in granite, exposed to the large temperature changes at the location, could expand/contract by 30 cm, and it was estimated it would take two years before it could be safely assumed that the newly-built dam would have cooled to the same thermal gradient as the existing structure.

Baker designed a separate structure, attached to the existing dam by 8 ft (2.42 m) steel bearing rods driven into both structures, leaving a space filled with clean rock chips (Figure 7.4). Performance could be monitored and the

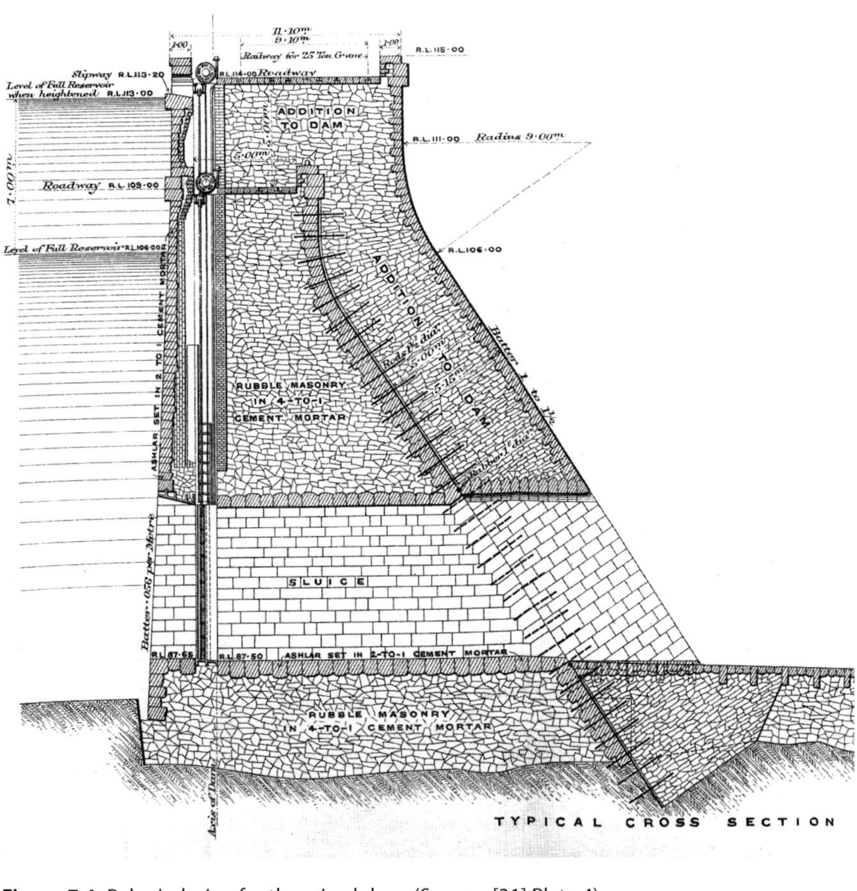

Figure 7.4 Baker's design for the raised dam. (Source: [31] Plate 4)

space then grouted up when the engineers were satisfied that thermal gradients could be reconciled between the two structures, and details of the construction sequence were drawn up accordingly. Recognising the weak and fissured nature of the underlying granite, and the need to ensure the dam was capable of dealing with possible tensile stresses at the downstream toe, work was initially carried out on constructing a downstream apron; slow progress on this work was responsible for delays in the main work to heighten the dam. The works were eventually completed after d the end of the dispute about the best analytical approach and after Baker's death [31].

For engineers actively involved in dam design, the indisputable fact that many masonry dams had stood in apparent defiance of Pearson and Atcherley's finding was enough to discredit their thinking. Some argued that they were mistaken in not seeing a masonry dam and its foundations as a continuum, and that any analysis had to recognise this. Unwin wrote a series of articles in *Engineering* in 1905 considering, in particular, the shear stress distribution at the base of a dam, and identifying errors in Pearson and Atcherley's reasoning [32, 33, 34].

Baker himself had used a jelly model when considering the original design to investigate overturning of the proposed trapezoidal cross-section, and the distribution of stresses in the dam. He used this model, made of 'ordinary jelly' and probably about 200 mm high, to corroborate the design calculations that had been made assuming a dam and its base behaved as a homogenous elastic body. As soon as he learned of the problem raised by the two 'mathematicians', he again used a stiff (gelatine) model to illustrate the stress patterns (Figure 7.5).

Baker thus satisfied himself, as Unwin had argued, that there was a more even distribution of shear tresses across the base than had been suggested by Pearson

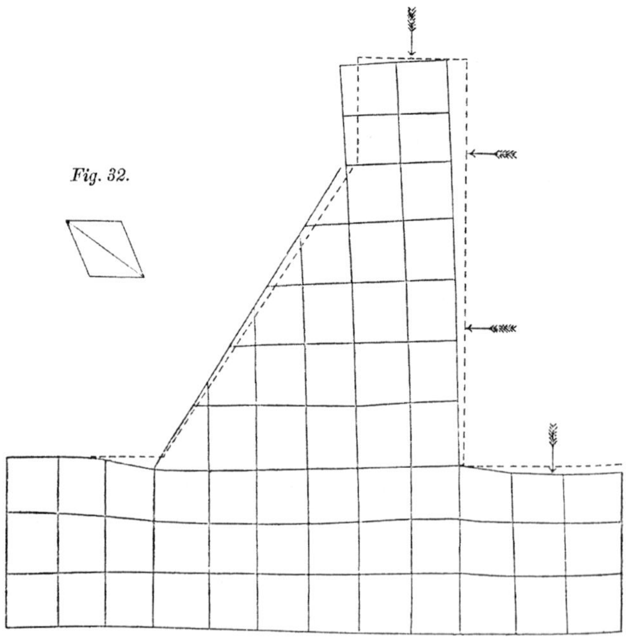

Figure 7.5 Baker's diagram showing deformation of the gelatine model under load. (Source: [30], p. 122)

and Atcherley. Hence, the tensile stresses were unlikely to exceed the generally accepted 3 tons/ft^2 (2.93 kg/cm^2) rather than reach the 10 tons/ft^2 (9.8 kg/cm^2) suggested by Atcherley and Pearson. However, he also found that loads from the dam were perhaps likely to continue into the underlying rock deeper than generally assumed, making the design of the dam foundation particularly important. When in Egypt Baker had a dozen further such models made 'for the engineers to experiment with, in different ways, including cutting out portions in front to represent rock scoured away by water rushing through the sluices. It was found that if rock was cut away in front of the dam the stresses on the dam were appreciably altered. After 3 or 4 hours' trial with the models, everybody had agreed with him that raising the dam was not such a simple problem as it seemed at first' [35].

As Baker wrestled with a practical solution to raising the dam, he encouraged others to reconcile mathematical analysis of dam stresses with practical dam design through the use of physical models.

7.4 Responses to Baker's call for the use of models for the Aswan dam

7.4.1 The introduction of India-rubber models

The complexity of the analysis required to model the stresses in dams mathematically encouraged the use of physical models, but critics readily found fault with the crude apparatus of Pearson and Atcherley, and the unsuitability of jelly with its lack of set. Baker recognised this and suggested the use of India rubber, which had been used by Arnulph Mallock (1851-1933) (a nephew of William Froude) for firearm experiments. An experienced experimental scientist, Mallock had worked with Baker before on several projects including the Forth Railway Bridge and the Central London Railway [36, 37]. He had used rubber to investigate a number of problems [38, 39, 40], and carried out some tests for Baker in the autumn of 1904. These used thin sheets of India rubber placed horizontally and supported on ball bearings. Lines were drawn upon them before any load was applied, and after the strain was applied, the same series of lines was ruled on the model. When the strain was removed, the two sets of lines were compared and the displacement caused by the stress was calculated. Mallock applied separate forces to represent the hydrostatic load and the weight of the dam [41].

While the results generally vindicated Rankine's approach, they did not really help in investigating the stresses apparently identified by Pearson and Atcherley. To give Baker more time to consider all the issues, a postponement of the heightening work in Aswan was announced in March 1905. This further ignited the debate over Pearson and Atcherley's findings. The experienced British dam engineer E.P. Hill asked in a letter to *Engineering* 'is the dam or the theory to give way?' [42].

7.4.2 The mathematicians respond

Pearson, in the face of bitter criticism from experienced practitioners and some academics, notably Unwin, carried out further work, described at a Royal Society soirée on 9 May 1906 [43] and another report was issued in 1907 with

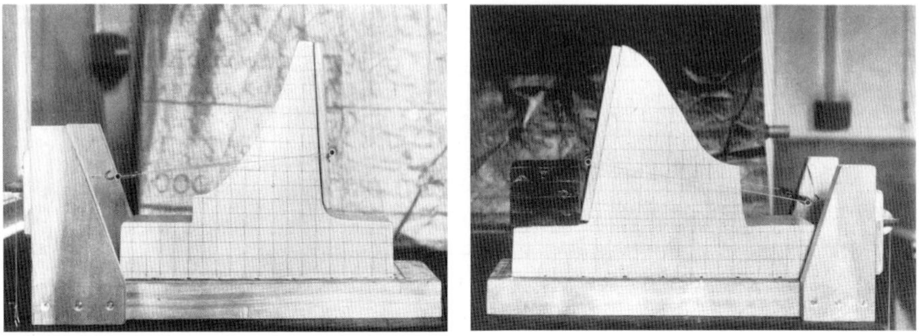

(a) (b)

Figure 7.6 Pollard and Pearson's gelatine-based models. (a) Aswan Dam – unloaded condition. (b) Vyrnwy Dam – overloaded condition. (Source: [44], Plate 1)

A.F. Campbell Pollard (1877-1948): 'An Experimental study of the stresses in masonry dams' [44]. This considered in particular the examples of the Vyrnwy and Aswan dams. With regard to the latter, a maximum compressive stress of 6.3 kg/cm^2 was acceptable to the Aswan dam designers, but a maximum tensile strength of 2.4 kg/cm^2 left further cause for concern. However, the model material – water gelatine and, latter, a moderately stiff glycerine gelatine jelly – was again open to objection, despite the use of such a material having been suggested by Baker himself (Figure 7.6).

The editor of *Engineering*, William Henry Maw (1838 1924), a future President of the ICE, joined the chorus of critics objecting to the use of 'plastic jelly' to study elastic behaviour, while recognising the quality of the mathematical portion of the work [45]. The experimenters themselves recognised the problems of conducting the model experiments and their cost, and suggested better results might have been obtained with larger models – 6 x 5 feet (1.8 x 1.5 m), rather than the 18 x 14 inch (45 x 35 cm) model actually used, calling on the ICE to fund the work.

7.5 The ICE debate in January 1908

As already noted, Baker, in general agreement with Unwin, had decided that the simplified assumptions necessary in any mathematical analysis were of little help in solving his practical problems: 'there were numerous disturbing influences arising from changes of temperature, continuances of workmanship, and other causes, which would entirely upset any reasoning based on the assumptions of the dam being a perfect elastic solid at constant temperature' [46]. He had, however, pointed out how a more satisfactory physical model might be made using an elastic material, India rubber. In the year after his death three papers were presented on 31 January 1908 at the ICE, which collectively laid the matter to rest. Of these, the most significant was by John Sigismund Wilson (1875-1955) and William Gore [47].

7.5.1 Wilson and Gore's studies with mechanical models

Wilson had been part of Baker's team working on the original Aswan Dam while Gore was an assistant to the leading water engineer G.F. Deacon, jointly responsible for the Vyrnwy masonry dam. The pair followed Unwin's argument about the importance of the distribution of shear stresses and Baker's suggestion of using a model made of India rubber. They tested a model of a cross section of the Aswan dam, subjecting it to pressures that could be measured and analysed. They combined this work with visits to European dams and studying the international literature on the design and construction of masonry dams. The model and testing apparatus were much more sophisticated than that used by Baker or Pearson's team. To replicate the effect of gravity loading on a triangular dam, weights were hung from cross-bars passing through the model dam. In this way the specific gravity of the dam could be increased to a value sufficient to give measurable strains while retaining an accurate ratio of water pressure to the weight of the dam.

Work, undoubtedly influenced by Mallock, began in May 1905, and was presented at an ICE *conversazione* in July; preliminary results were published in *Engineering* in August 1905 [48]. The first rubber model represented a dam of triangular section, with a base two thirds of the height, 4 x 7 inches and 1 inch thick (100 x 178 x 25 mm) thick). It rested on a beam secured to a trestle to which guide pulleys were attached (Figures 7.7 and 7.8).

The dam was divided into rectangular and triangular elements, and a hole was made through the centre of gravity of each section through which a small-diameter tube was inserted. Through these tubes pins were placed from which weights could be hung. The tubes were used to enable the lines to be drawn after they had been inserted and any local distortion resulting from their placement would not affect the straightness of these lines (Figure 7.9). The length of the pins was varied so that the weights, when hung, did not interfere with each other. A layer of shot or small steel balls, was placed between the base of the model and its support, so that the support could move freely along its base and would not offer appreciable resistance to the 'water pressure'.

The 'water pressure' acting on the dam face was divided into six sections of equal depth, with the forces distributed over the section of the face by six bearing plates (Figure 7.9b), with the forces acting along the resultants of the pressure on each section, and carried over pulleys. A line of six tube holes with loaded pins were inserted through the base, so that the whole of the horizontal thrust could be taken up by known forces which, if equal to the forces representing water pressure, ensured the model was in horizontal equilibrium. By varying the load on the six points in the base blocks, the intensity of shear could vary between the lower rectangular or base part of the model representing the foundation, and the upper part representing the dam subjected to water pressure. This distribution of load along the line through the base could be adjusted to follow a straight or curved line according to how the six points of application were arranged. The deformations were photographed by a fixed, full-plate camera, effectively doubling the dimensions. Wilson and Gore were able to show the distribution of shear and vertical and horizontal pressures on vertical and horizontal planes, and some

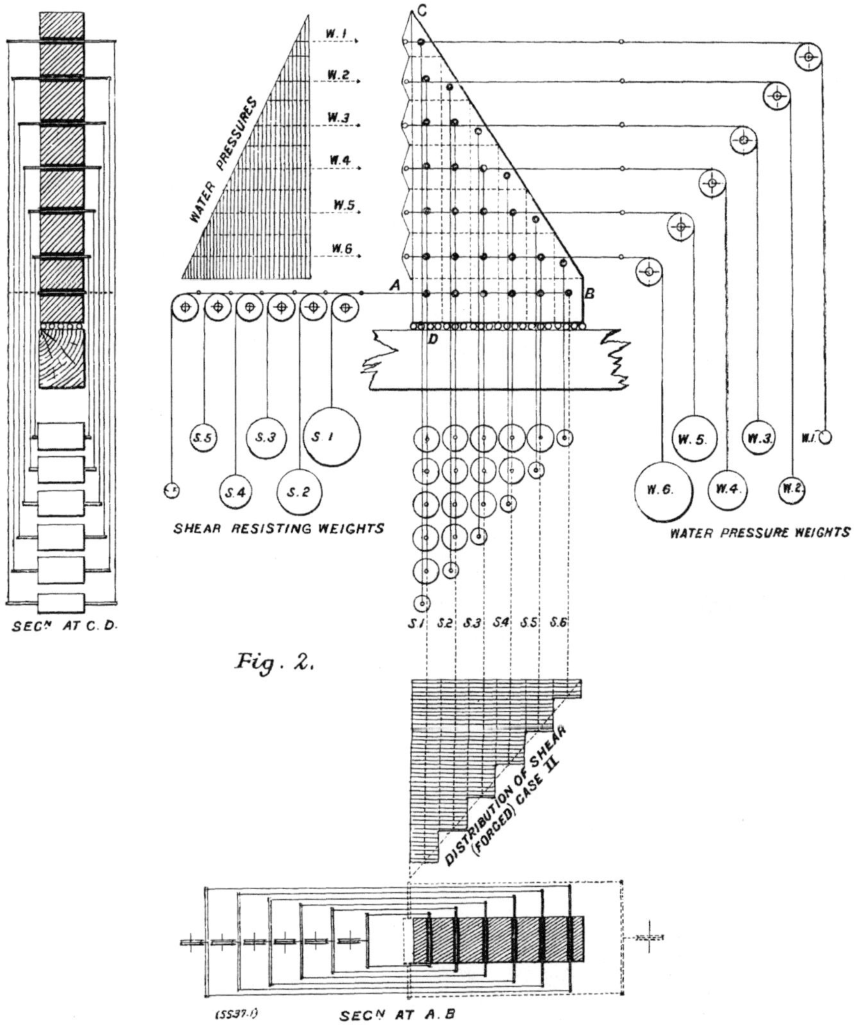

Figure 7.7 Drawing of Wilson and Gore's first model test rig. (Source: [48] Figure 2)

tensile stresses on the (unanchored) downstream, and upstream toes; they also demonstrated that the approach was successful and would merit further work.

At this stage, the results were summarily rejected by Pearson and, in their 1907 memoir, Pearson and Pollard criticised the results derived by Wilson and Gore as inapplicable to 'real' dams as the experiments did not allow for the horizontal bending that occurred in practice, i.e. the model failed to take into account the effects transmitted from the main structure on either side of the section being studied. When criticised by the editor of *Engineering*, Pearson claimed, mistakenly as he subsequently acknowledged, following a debate in *Nature*, that the error introduced by using the model in checking stress distribution was of the order of 30%.

Figure 7.8 Photograph of Wilson and Gore's first model test rig. (Source: [48] Figure 1)

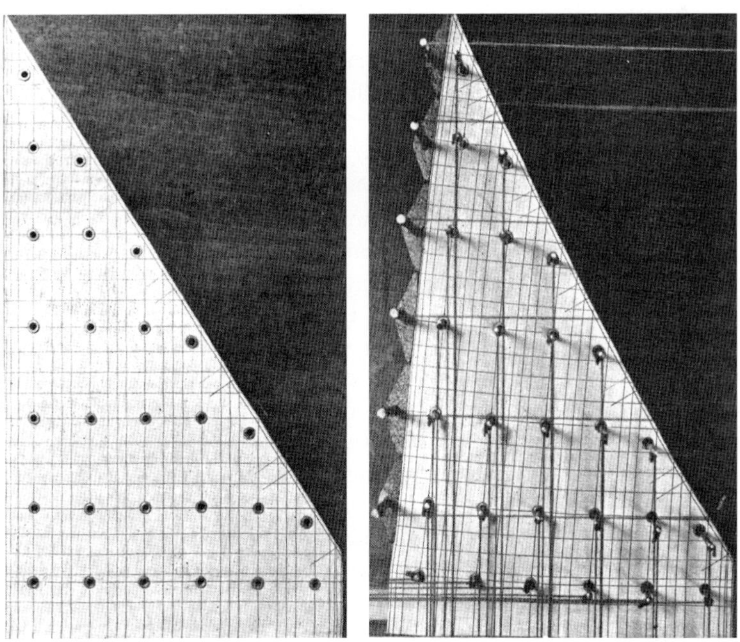

(a) (b)

Figure 7.9 Wilson and Gore's first model (a) marked up with lined grid. (b) under load showing pins and bearing plates. (Source: [48] Figures 3 & 4)

Figure 7.10 Wilson and Gore's second test rig. (Source: [49], p. 15, Figure 17)

Wilson and Gore noted that the curves of stress distribution obtained with the Pollard and Pearson models were very irregular and thus difficult to interpret, suggesting that all the academics' problems with conducting the experiments had not been overcome, although they did indicate the development of serious tensile stresses at the upstream toe of the dams.

Wilson and Gore persisted with the India-rubber models, using equipment available at Wilson's home in Hampstead, and India-rubber models of dam sections with various profiles about 150 mm high and 25 mm thick [49] (Figure 7.10). The loads were again applied mechanically using a similar system of weights, fishing line and pulleys, and the deformations were recorded photographically and measured to an accuracy of one thousandth of an inch (0.025 mm).

The models were shaped to adopt the desired profile when fully strained and therefore an allowance had to be made in the initial unstrained shape. In this context, the properties of India rubber proved invaluable; when only slightly strained, it was found to have an almost linear stress-strain relationship. The pressures on vertical and horizontal planes, and the maximum and minimum, or principal, stresses were found and the ellipses of stress at various points were calculated from the strains using the equations of continuum mechanics (Figure 7.11). With regard to shear, the measured tangents of the angles of distortion, plotted at a suitable scale, gave the shear stress. All the strains were measured on photographic negatives taken of the model's relative positions when strained and

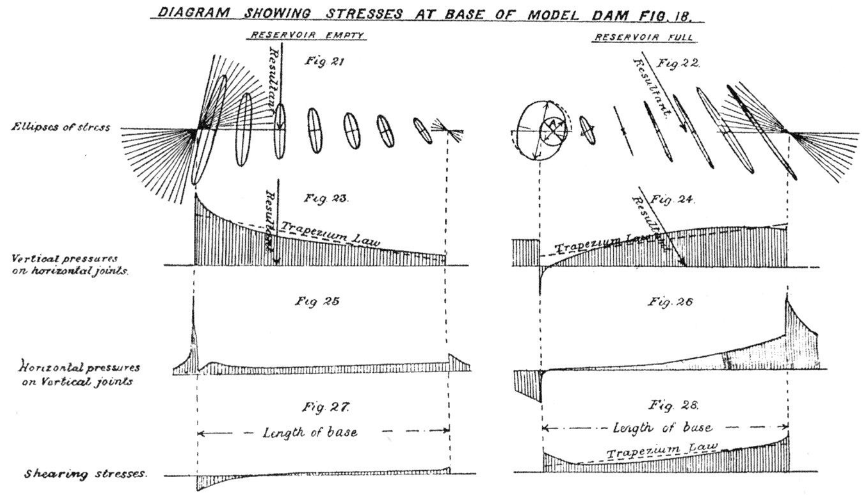

Figure 7.11 Wilson and Gore's diagram of ellipses of stresses (see Figure 7.13). (Source: [54])

unstrained. The model, including the foundation, was 30.5 cm wide and the height of the dam was 17 cm (Figure 7.12). While not indicating the stresses, it showed where the greatest strains develop, viz. near the upstream toe where both the dam and foundation are stretched, while at the downstream toe they are compressed. The published paper contained curves of stress distribution over the bulk

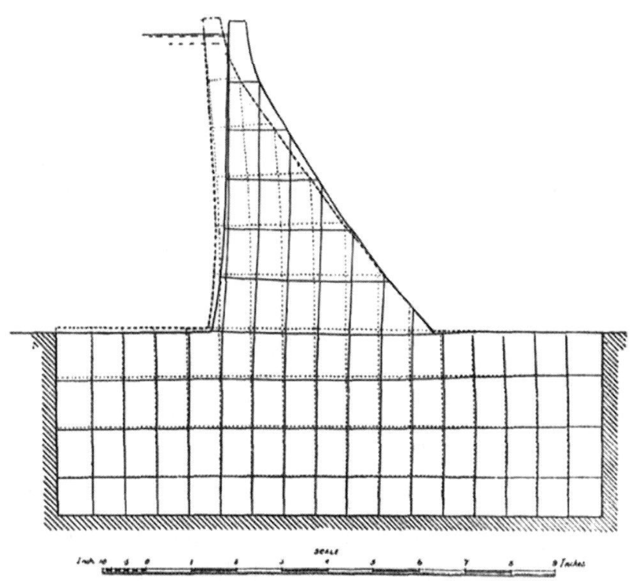

Figure 7.12 Drawing of deformations in Wilson and Gore's rubber model. The broken lines show the model section with the principal lines drawing thereon when unstrained, and the full lines show the model section with the principal lines drawing thereon when fully strained with both 'gravity-' and 'water-loads'. (Source: [54])

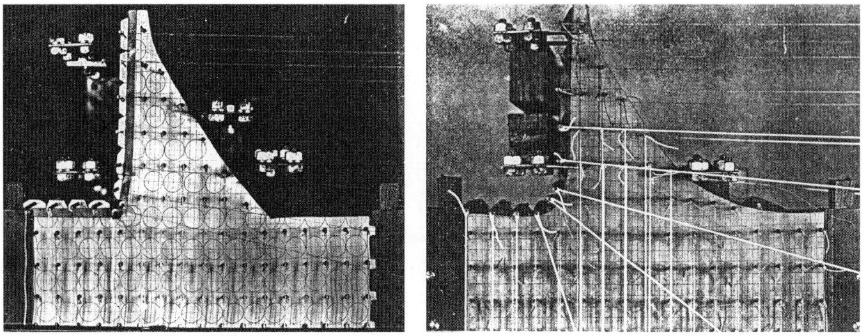

(a) (b)

Figure 7.13 Wilson and Gore's models which are preserved in London's Science Museum. (see Section 39.2.1) (a) Model with sharp corners between the dam and the foundations. (b) Model with rounded corners between the dam and the foundations. (Source: [54])

of the sectional area of the models. Although the models were less suitable for the reservoir-empty case, reasonable measurements were also obtained for that situation. The results demonstrated that the stresses at the base of a dam were different from both those derived by the traditional trapezium method, and other analytical methods hitherto suggested.

The major findings were that, in the case of a full reservoir, there is a gradual transition in the direction of the maximum principal stresses from perpendicular to the surface of the water face, to parallel with the downstream face at that face. The minimum principal stresses are compression near the downstream face, and pass through zero near the upstream face and become tensions at the face itself. Extraordinary stresses were found at the toes (Figure 7.13). In the model with rounded toes, these stresses were so reduced, that it was clear that sharp angles should be avoided at these points. To reduce the tension significantly it would be necessary to make significant, and unnecessary, changes to designs. These tests took into account the scaling factors for both stiffness and Poisson's ratio and, as before, displayed the results as ellipses of stress, thus indicating the lines of principal stress. Their investigations were designed to take account of the classic method, as espoused by Rankine, Lévy's mathematics analysis, and Unwin's graphical approach post-Bouzey, as well as the findings of Atcherley, Pearson and, later, Pollard. In one of their model tests they looked specifically at Bouvier's allowable stress design method. Almost all those present at the ICE discussion praised their work for its help in resolving the main issues.

7.5.2 Ottley and Brightmore's studies with Plasticine models

Alongside Wilson and Gore's paper another was presented by Sir John Ottley (1841-1931), of the India Public Works Department, and Arthur William Brightmore (1865-1927) of the Royal Indian Engineering College, Coopers Hill Brightmore had worked on both the Vyrnwy and Elan projects, while Ottley was an experienced irrigation engineer before being transferred to the Royal Indian Engineering College. In 1909 they reported on their work done with Plasticine models of dam sections [50]. The model was 750 mm high with a base of 650 mm and length of 300 mm and loaded using an India-rubber bag filled with water. Movement of the dam was observed through glass plates at each end of (but not touching) the model. Both the glass plates and the ends of the model were marked with an orthogonal grid of lines at 50 mm spacing so that their relative movement could be measured. Plasticine had only been patented in 1899 and, while the model material was again criticised for its lack of elastic behaviour, the researchers had found that, when loaded at low stresses for a long period of time, the Plasticine fractured perpendicular to the tensile stresses, giving information about internal stresses in the dam section. The water pressure load was applied for over a month and displacements of the originally vertical and horizontal lines were indicated by horizontal and vertical figures on a diagram (Figure 7.14). The crack at the inner toe began to appear after seven days.

Fig: 2.

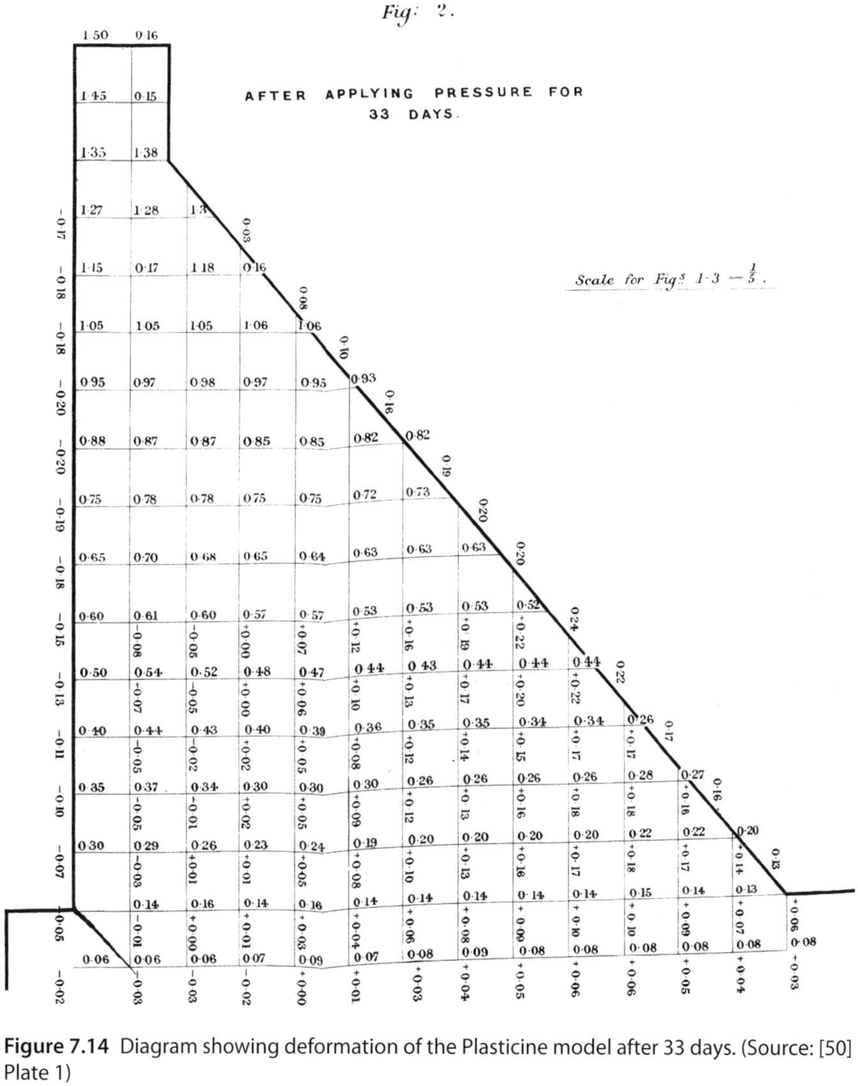

Figure 7.14 Diagram showing deformation of the Plasticine model after 33 days. (Source: [50] Plate 1)

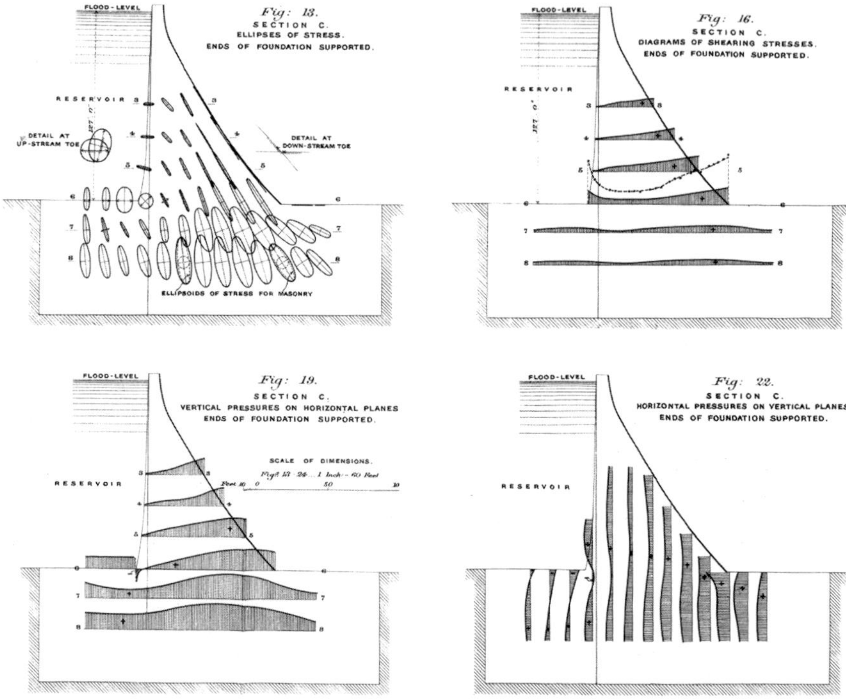

Figure 7.15 Ottley & Brightmore's diagrams showing the stress field in the loaded model of the dam. (Ellipses of stress; Shear stresses; Vertical stresses; Horizontal stresses). (Source: [50] Plate 3)

The tests yielded comprehensive data about the distribution of horizontal, vertical and shear stresses both within the dam and at its base (Figure 7.15). Overall conclusions were that a masonry dam designed on the assumption that stresses on the base varied uniformly, would have smaller stresses than those estimated by calculation, that there could be no tension near the downstream toe, and that there would be tension on certain planes other than the horizontal near the upstream toe.

7.5.3 Further contributions to the debate

Another paper was presented at the ICE by E.P. Hill, of the leading firm of water engineers G.H. Hill & Sons, which dealt with an analytical approach that Unwin claimed added nothing to the graphical method he had used when rebutting Atcherley and Pearson's ideas at the beginning of the debate [51]. All the contributors to this 1908 debate agreed there was some tension near the heel of the base section where it joined the foundation, and that nowhere else in the (traditional) dam profile was tension possible. None of the contributors to the discussion thought this tension was of significance. By May 1908 the matter was regarded as closed, and the 'mathematicians' dam' had finally been vanquished [52].

The findings were ably summarised by the US Engineer and foremost dam authority, Edward Wegmann:

> 'Tensile stresses which may have large local magnitude occur at the upstream toe, tensile stresses occur in no other part of the dam owing to fixity of the base, and changes in water pressure at the base stresses in foundations distributed almost uniformly, for joints above foundation the usual assumption of linear distribution of normal stress in horizontal plane is approximately correct, and overestimates maximum intensity of stress, the maximum compressive stresses on the downstream face occur on planes normal to the face, near the base the maximum compressive stresses on horizontal planes do not exceed those calculated on the assumption of linear distribution of normal stress.' [53].

Gore and Wilson themselves summarised the issues involved in a paper published in 1909 [54]. The value of physical models was well recognised by the Egyptian Irrigation service who went on to use such an approach not only for structural models but also in developing hydraulic models during the following decade, borrowing from advances in aerodynamic modelling [55].

7.6 After Wilson and Gore

7.6.1 The use of physical models for designing dams between the Wars

The dispute over the 'mathematicians' and engineers' dams', and the key role played by the Gore-Wilson rubber model was one of the great intellectual disputes of early twentieth century engineering. It was a discussion that illustrates a frequent tension between academic engineering scientists and practising engineers. It demonstrated the complexity of calculating stresses in masonry dams, and the difficulties in constructing physical models that provided meaningful help in resolving the associated practical design issues. After an understandable hiatus caused by the First World War, engineers designing major dams continued to develop both their analytical tools and physical modelling techniques. For fairly obvious economic reasons, as predicted by Baker, this research was largely concerned with developments in reinforced-concrete rather than stone masonry dams [56].

In the early 1920s the U.S. Bureau of Reclamation began a huge programme of research into the design of reinforced-concrete arch and gravity dams [57] (see Chapter 8). This work, supported in part by the Engineering Foundation and American Society of Civil Engineers, drew much from the Wilson-Gore debate and led directly to new modelling techniques using stiff rubber, cardboard and celluloid, as well as models made of micro-concrete using aggregate from the same sources that would be used in actual dams.

Parallel research on the design of arch dams in Europe began in the mid-1920s. One of the pioneers was Professor Camillo Guidi (1853-1941) at the Polytechnic of Turin who undertook a model test of a cylindrical shell dam in 1926 [58, 59] (see Section 10.1). In France, the government had set up the *Office national des recherches scientifiques et industrielles et des inventions* in 1922 and soon created a department for testing structures and reduced-scale models of dams, made with plaster and loaded using mercury. In 1926, in their retirement, Augustin Mesnager (1862-1933) and J. Veyrier, a colleague at the *Ecole Nationale des Ponts et Chaussées* in Paris carried out model tests on a design by André Coyne (1891-1960) for the 70-metre high Marèges arch dam on the River Dordogne [60, 61, 62]. The 1:100 scale model was made of plaster with a strength 1/7.4 that of concrete, and using mercury (density 13.6) to apply the load. It was loaded up to collapse, verifying that the safety factor for the dam was four. Coyne went on to become the most significant designer of dams during this period, and was ultimately responsible for sixty-one large dams in France and twenty-six elsewhere [63].

By the early-1930s, model testing of both the structure [64] and the hydraulics [65] of dams had become the norm, mainly in order to give added confidence to the theoretical calculations which were both complex (the calculation of large dam could take up to a year) and still not considered trustworthy. Also at this time, a new technique was added to the model-testing repertoire – photoelastic stress analysis (see Chapter 12).

7.6.2 Physical and numerical models – post-war developments

The masonry dam controversy, largely resolved by the use of Gore and Wilson's model approach, was in part stimulated by the complexity of mathematical models and analysis. The mathematical approach was radically strengthened by the development of the relaxation method by R.V. Southwell (1888-1970) from the mid-1930s. Professor A.J.S. Pippard (1891-1969), among others, had been inspired by the Wilson-Gore controversy early in his academic career to forge links between the use of physical models and theoretical calculations [66]. In the 1940s at Imperial College, London, Pippard encouraged the young O.C. Zienkiewicz (1921-2009) to apply his numerical methods to revisiting the problem of stress distribution in gravity dams, investigating triangular, rectangular and Wegmann profiles, on the basis of perfect elasticity – an assumption sought by Baker and carried out by Wilson and Gore. He found that, for a full reservoir, the maximum compressive stresses do not exceed those indicated by the traditional approach, while the most severe compressive and shear stresses occur on the downstream face where the structure is subjected to a state of uniaxial compression. He also found that compliance with the middle-third rule does not always eliminate tension, as tension develops below ground level on the upstream side of the dam when the reservoir is full, but tension does not develop elsewhere. In the upper portion of the dam, the classical theory and elastic theories coincide and, for the reservoir-empty condition, the traditional approach is adequate [67]. This work by Zienkiewicz provided a remarkable vindication of Wilson and Gore's experimental work (Figure 7.16) [66]. It also proved the starting point for rapid advances in numerical modelling and,

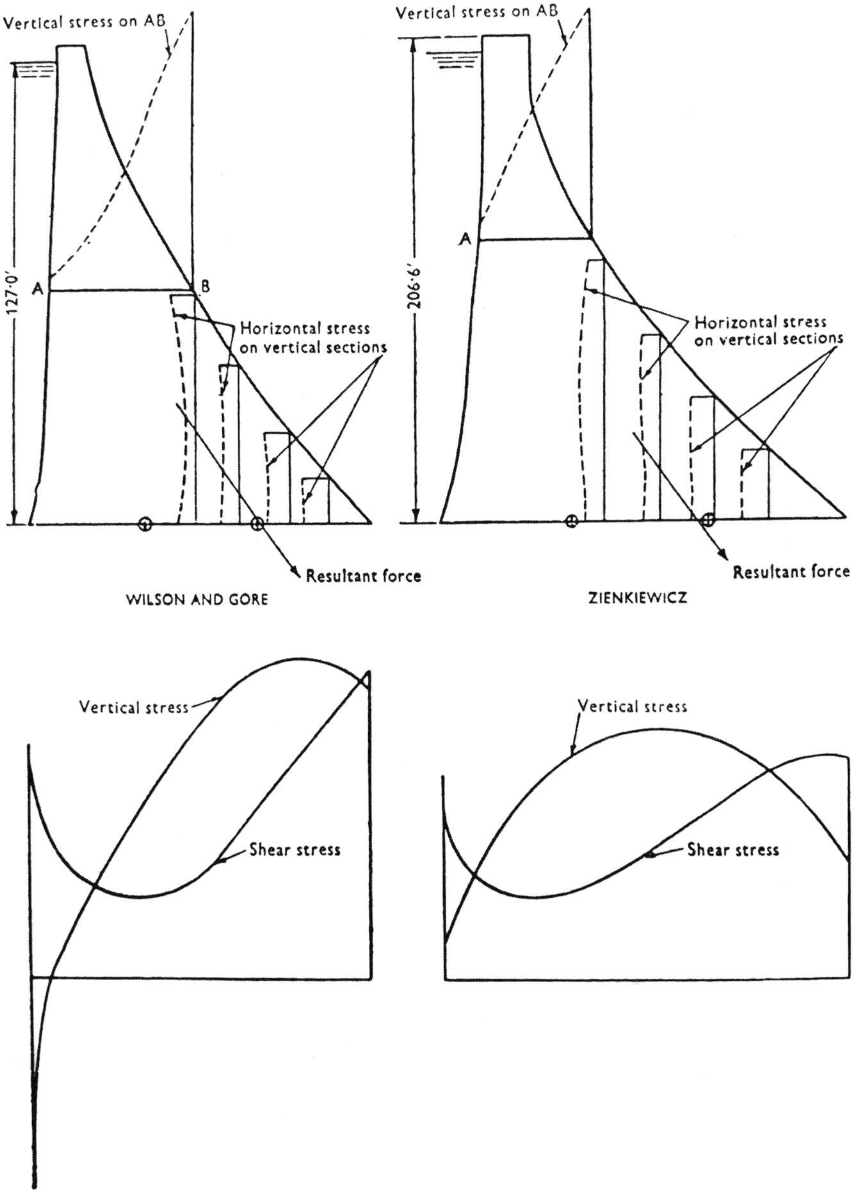

Figure 7.16 Professor Pippard's comparison of the results from model tests by Wilson and Gore and the analysis by Zienkiewicz. (Source: [66], p. 278)

ultimately, the establishment of finite-element techniques, largely pioneered by Zienkiewicz.

The methods developed by Zienkiewicz were first vindicated in the design of the Dokan Dam in Iraq by UK consultants Binnie and Partners with support from Imperial College, London on both mathematical and experimental analyses [68].

India-rubber models were again used, alongside the theoretical techniques based on the relaxation method [69]. By the mid-1960s, as reflected in the Arch Dams Symposium organised by the Institution of Civil Engineers in 1968, numerical models had gained the ascendancy. Nevertheless, in Britain at least, this came only after a major research effort, sponsored by the ICE's Research Committee, in which physical models had also played an important part [70].

7.7 Conclusions

In 1905-1907 Wilson and Gore demonstrated that carefully-constructed physical models in a suitable material could help resolve issues regarding stress distributions in masonry dams that were too complex to resolve satisfactorily using the available mathematical analysis. They were also able to demonstrate that the 'classical theory' used to design many masonry dams during the previous fifty years was a safe approach, and that concerns raised particularly about shear stresses near the toe were alarmist, provided due care was taken in the design and construction of such dams. Not only were their findings vindicated by later mathematical modelling, but many large dams built over the next sixty years used physical models to enable careful consideration of stresses, both physically and analytically. This debate is worthy of study by any engineering student embarking on a career where mathematical models are now the norm as a reminder that, as Baker had pointed out, an 'engineer's dam' is infinitely more complex than any 'mathematician's dam' [52].

Further reading

Biographical information on most British engineers mentioned in this chapter can be found in the *Biographical Dictionary of Civil Engineers of Great Britain and Ireland* (edited by A.W. Skempton et al., 3 vols). 2002-2014. London: ICE Publishing.

Acknowledgements

I am grateful to the late Dr Norman Smith, formerly of Imperial College, London, for his guidance. I also thank the library staff of the Institution of Civil Engineers for their help in finding many of the archive documents and printed material used in preparing this chapter.

References

1 Smith, D. (1976/77) Structural model testing and the design of British railway bridges in the nineteenth century. *Trans. Newcomen Society* 48: 73-90.
2 Smith, D. (1977) The Use of models in nineteenth century British suspension bridge design. *History of Technology.* 2: 169-214.

3 Skempton, A.W. (1981/82) Landmarks in early soil mechanics. In: *The measurement, selection and use of design parameters in geotechnical engineering: Proceedings 7th European Conference on Soil Mechanics and Foundation Engineering, Brighton, 1979*. Vol. 5: 1-26. London: British Geotechnical Society.
4 Skempton, A.W. et al. (eds.) (2002-2014) *Biographical Dictionary of civil engineers of Great Britain and Ireland*. 3 vols. London: ICE Publishing.
5 Binnie, G.M (1981) *Early Victorian water engineers*. London: Thomas Telford.
6 Binnie, G M (1987) *Early dam builders in Britain*. London: Thomas Telford.
7 Skempton, A.W. (ed.) (1981) *John Smeaton, FRS*. London: Thomas Telford.
8 Skempton, A.W. (1990) Historical development of British embankment dams to 1960. In: *Clay barriers for embankment dams: Proceedings of the conference organised by ICE, London, 1989*. 15-52. London: Thomas Telford.
9 De Sazilly, J.A.T. (1853) Note sur un type de profil d'égale résistance proposé pour les murs de réservoirs d'eau. *Annales des ponts et chaussées* 3rd series VI: 191-222.
10 Graeff, A. (1866) Rapport sur la forme et le mode de construction du barrage du gouffre d'Enfer, sur le Furens, et des grands barrages en général. *Annales des ponts et chaussées* 4th series XII (2): 184-211.
11 Delocre, E. (1866) Mémoire sur la forme du profil à adopter pour les grands barrages en maconnerie des réservoirs. *Annales des ponts et chaussées* 4th series XII: 212-272.
12 Chrimes, M.M. (2009) Ahead of the game – masonry dam design in the British colonies 1800-1900. *Dams and reservoirs* 19 (2): 55-66; 19 (4): 171-183.
13 Rankine, W.J.M. (1872) Report on the design and construction of masonry dams. *The Engineer* 33: 1-2.
14 Bouvier, H.J.M. (1875) Calculs de résistance des grands barrages en maçonnerie. *Annales des ponts et chaussées* 5th series X: 173-205.
15 Konig, W. (2006) Der Ingenieur als Politiker: Otto Intze. *Technikgeschichte* 73 (1): 27-46.
16 Smith, N.A.F. (1994) The Failure of the Bouzey dam in 1895. *Construction History* 10: 47-65.
17 Unwin, W.C. (1896) The failure of the masonry dam at Bouzey France. *Cassiers magazine* 11: 9-19.
18 Baker, Benjamin (1897) Letter to W.C. Unwin on Bouzey dam. Quoted in Smith, N.A.F. (1994) The Failure of the Bouzey dam in 1895. *Construction history* 10: 47-65. 53.
19 Lévy, M. (1895) Quelques considérations sur la construction des grands barrages. *Comptes rendus hebomaires … Academie des Sciences* 121: 288-300.
20 Lévy, M. (1898) Sur la légitimé de la règle dite 'du trapèze' dans l'étude de la résistance des barrages en maçonnerie', *Comptes rendus … Académie des Sciences* 126: 1235-1240.
21 Lévy, M. (1898) Sur l'equilibre élastique d'un barrage en maçonnerie à section triangulaire. *Comptes rendus … Académie des Sciences* 127: 10-15.
22 Chrimes, M.M. (2009) Ahead of the game – masonry dam design in the British colonies 1800-1900. *Dams and reservoirs* 19 (2): 55-66; 19 (4): 171-183.

23 Willcocks, W. (1894) *Perennial irrigation*. Cairo: Ministry of Public Works.
24 Garstin, W. (1901) *Report as to irrigation projects on the upper Nile*. Cairo: Ministry of Public Works.
25 Willcocks, W. (1901) *The Nile Reservoir at Assuan and after*. London: Willcocks.
26 Smith, N.A.F. (2002) *The Centenary of the Aswan dam 1902-2002*. London: ICE.
27 Fitzmaurice, Maurice (1903) The Nile Reservoir, Assuan. *Min. Procs. ICE*. 152: 71-107.
28 Atcherley, L.W. & Pearson, K. (1904) On some disregarded points in the stability of masonry dams. *Drapers Company Research memoir, Technical series, II*.
29 Atcherley, L.W. & Pearson, K. (1904) On some disregarded points in the stability of masonry dams. *Drapers Company Research memoir, Technical series, II*. 21.
30 Baker, B. (1904) Discussion on Coolgardie water supply. *Min. Procs. ICE*. 162: 120-126.
31 Macdonald, Murdoch (1913) The Aswan dam: thickening and heightening. *Min. Procs. ICE*. 194: 249-277.
32 Unwin, W.C. (1905) Note on the theory of unsymmetrical masonry dams. *Engineering* 79 (21 April): 513-514.
33 Unwin, W.C. (1905) Further note on the theory of unsymmetrically masonry dams. *Engineering* 79 (12 May): 593-594.
34 Unwin, W.C. (1905) On the Shearing stress in masonry dams. *Engineering*. 79 (30 June): 825-826.
35 Baker, B. (1904) Discussion on Coolgardie water supply. *Min. Procs. ICE*. 162: 120-126. 124.
36 Anon. (1933) *Obituary Notices of Fellows of the Royal Society* 1 (2): 95-100. https://doi.org/10.1098/rsbm.1933.0003 (25/10/2019)
37 Mallock, A. (1902) *Vibrations produced by the working of traffic on the Central London Railway. Report to the Board of Trade, 1901*. London: HMSO. Cmnd 951.
38 Mallock, A. (1891) On the instability of inflated rubber tubes and balloons when distorted by fluid pressure. *Procs Royal Society* 49: 458-463.
39 Mallock, A. (1897) On model experiments. *Engineering* 94: 771.
40 Mallock, A. (1901) Vibrations of rifle barrels. *Procs. Royal Society* 68: 327.
41 Mallock, A. (1908) Discussion on stresses in masonry dams. *Min. Procs. ICE*. 172: 171-172.
42 Smith, N.A.F. (2002) *The Centenary of the Aswan dam 1902-2002*. London: ICE. 38.
43 Anon. (1906) *Engineering*. 82 (22 June): 791.
44 Pearson, K. & Pollard, A.F.C. (1907) An experimental study of the stresses in masonry dams. *Drapers Company Research memoirs, Technical Series, V.*
45 Anon. (1907) The Stresses in masonry dams. *Engineering* 84 (13 Sept.): 368.
46 Baker, B. (1904) Discussion on Coolgardie water supply. *Min. Procs. ICE*. 162: 120-126.

47 Wilson, J.S. & Gore, W. (1908) Stresses in dams: an experimental investigation by means of India rubber models. *Min. Procs. ICE* 172: 107-133; discussion 172: 143-194.
48 Wilson, J.S. & Gore, W. (1905) Stresses in dams: an experimental method of investigating the distribution of stresses in dams. *Engineering* 80 (4 Aug.): 134-135.
49 Wilson, J.S. (1953) *Stresses in dams: inception of the tests by Wilson and Gore.* Typescript. Copy at ICE TFV 598.
50 Ottley, J.W. & Brightmore, A.W. (1908) Experimental investigations of the stresses in masonry dams subjected to water pressure. *Min. Procs. ICE.* 172: 89-106.
51 Hill, E.P. (1908) Stresses in masonry dams. *Min. Procs. ICE.* 172: 134.
52 Anon. (1908) The Stresses in masonry dams. *Engineering* 85 (22 May): 692.
53 Wegmann, E. (1922) *The design and construction of dams.* 7th ed. New York: Wiley.
54 Gore, W. & Wilson, J.S. (1909) The design of high masonry dams: A Review of the history of its development. *The Central* VI (17, April): 1-18.
55 Hurst, H.E. & Watt, D.A.F. (1924) The motion of water through sluices and scale models of the Assuan Dam. *Min. Procs. ICE.* 218: 72-112.
56 Bruggemann, D.A., Hollock, K.J. & Sims, G.P. (2001) Historic concrete in dams. In: *Historic concrete: background to appraisal* (eds R.J.M. Sutherland et al.), 343-358. London: Thomas Telford.
57 Billington, D.P, Jackson, D.C. & Melosi, M.V. (2005) *The history of large federal dams: planning, design and construction in the era of big dams.* Denver: Bureau of Reclamation.
58 Guidi, C. (1921) *Statica delle dighe per laghi artificiali.* Torino: Bona.
59 Guidi, C. (1927) Etudes expérimentales sur les constructions en beton armé (Translated by A. Paris) *Bulletin technique de la Suisse romande* 53: 181-187; 201-206.
60 Mesnager (Augustin) & Veyrier (J) (1927) Détermination de la résistance d'un ouvrage sur un modèle réduit. *Comptes rendus des séances de l'académie des sciences* 184 (Jan-June): 1026-1027.
61 Mesnager, A. & Veyrier, J. (1928) Les barrages réservoirs à voûtes et à charge fractionnée. *Annales des Travaux Publics de Belgiques,* 219-257.
62 Mesnager, Augustin (1928) Les barrages hydrauliques. *La Houille Blanche* 27 (213) (July-Aug): 97-105.
63 Bordes, Jean-Louis & Tardieu, Bernard Tardieu (2015) André Coyne, de la Dordogne au Zambèze, la passion de construire. *Bulletin SABIX* 57: 8–53. http://journals.openedition.org/sabix/1452 (8/5/2019)
64 Bažant, Z. (1936) Le développement du calcul des barrages arqués. *Report of the IABSE Congress.* 1115-1147.
65 Bourgeat, R. (1933) Essais sur modèles réduits des conditions d'établissements des grands ouvrages hydrauliques. In: *Actes du 1er Congrès des Grands Barrages, Stockholm,* 17-46. Paris: Commission Internationale des Grands Barrages.
66 Pippard, A.J.S (1950) The Functions of engineering research in the university. Unwin Memorial Lecture, ICE. *ICE Journal* 33 (4): 265-284.

67 Zienkiewicz, O.C. (1947) The stress distribution in gravity dams. *ICE Journal* 27 (3): 244-271.

68 Twort, A.C. (1959) The problem of stresses in concrete dam design. *New Scientist* 5 (119, 26 Feb): 454-456.

69 Allen, D.N. de G. et al. (1956) The Experimental and mathematical analysis of arch dams with special reference to Dokan. *ICE Proceedings* 5 (3, May): 198-244; Discussion: 244-258; 412-434.

70 Sparkes, S.R. et al. (1968) The use of models in the analysis of arch dams. In: *Arch dams: a review of British research and development. Proceedings of the symposium held at the Institution ... 20-21 March 1968*. 117-128.

8

Models used during the design of the Boulder Dam

Bill Addis

8.1 The US Bureau of Reclamation dams: 1925-1940

Two factors contributed to a huge transformation in the use of models in the 1920s. In 1902 the western states of the USA formed the Reclamation Service (later the Bureau of Reclamation) to prevent the terrible damage caused by regular flooding. The Bureau was given an enormous budget for research into the construction of dams, both for flood management and the generation of hydroelectric power. One major research theme, from about 1920, was to develop reinforced concrete arch dams as an alternative to conventional cantilever gravity dams. In essence, these were thin concrete shells, not too remote from the concrete shell roofs that were being developed in several European countries from around 1900.

For the first arch dams built in the early years of the twentieth century, the theoretical analysis of the dam treated the shell as a cylinder. However, this was far from ideal since, in a typical V-shaped canyon, at each higher level in the shell, the arch elements had increasingly longer spans, and there were significant discontinuities between adjacent arch elements. Despite various attempts at resolving these difficulties, it became widely agreed that the mathematical models being used were still not adequately representative of the real structure and, indeed, were becoming more and more abstract and distanced from reality [1]; a particularly harsh critic was the eminent structural engineering academic, Hardy Cross (1885-1959) [2].

8.2 Preliminary model studies

8.2.1 Stevenson Creek Experimental Dam

In 1923 a proposal was made to the Bureau's board to build an experimental arch dam to investigate thoroughly the behaviour of a concrete arch (shell) dam. Construction of the test dam at Stevenson Creek, California began in 1925 at the site of a disused miners' dam, and was completed in summer 1926. Deflections of

Physical Models: Their historical and current use in civil and building engineering design,
First Edition. Edited by Bill Addis.

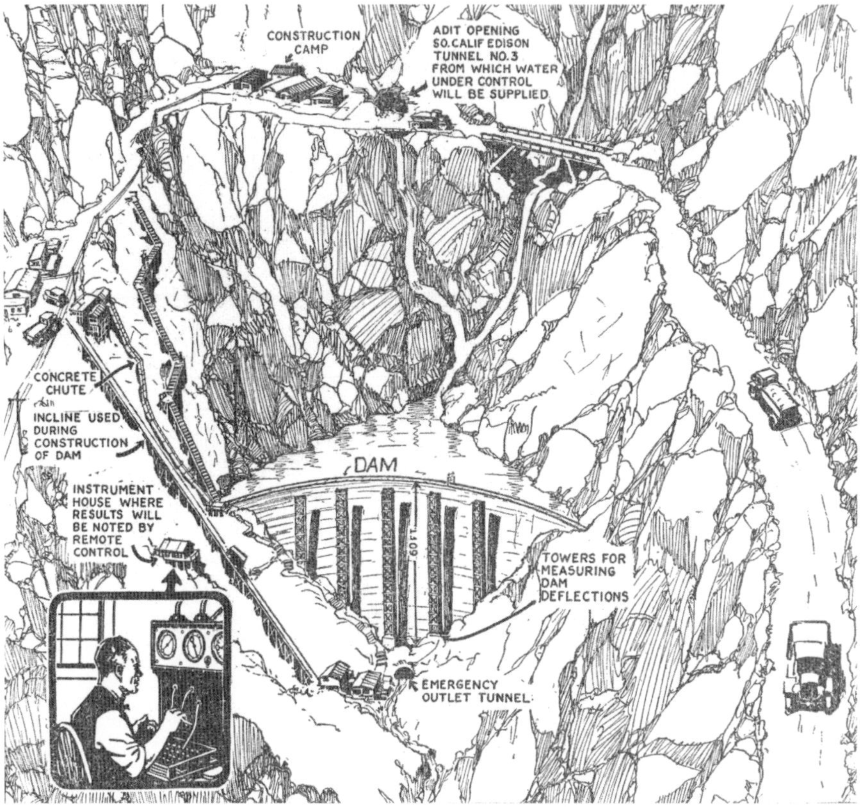

Figure 8.1 Artist's impression of the Stevenson Creek Experimental Dam, 1926. (Image: Courtesy of the Huntington Library, San Marino, CA. Southern California Edison Photographs and Negatives. No. p16003coll2_5463_full)

the thin shell were measured from the access scaffolding on the downstream face of the dam. Internal strains within the concrete were measured using electric carbon-pile telemeters, which had recently been developed by Roy Carlson (1900-1990), a physicist working on the project. These were embedded in the concrete during construction and their output was monitored remotely, via 'sensor wires', in a hut above the dam (Figure 8.1). In addition, internal temperatures of the concrete were monitored in order to be able to separate the influences of temperature-induced loads from physical loads. The dam was 18.3 m high and about 25 m long, with a constant radius of 30.5 m. Its thickness reduced from 2.3 m at the base to just 600 mm at half height, and continued with this thickness to the crest of the dam. Such a slim profile was unprecedented (Figure 8.2).

Meanwhile, measurements of stress and strain taken on other dams were not as convincing as had been hoped, as it proved difficult to interpret the data that had been collected. In December 1925 it was therefore decided that a programme of testing reduced-scale models should be carried out in parallel with the tests on the Stevenson Creek dam. The main model of the Stevenson Creek dam was made of concrete and built at 1:12 scale, 1.5 m high, 3.6 m long, and reducing

Figure 8.2 Stevenson Creek Experimental Dam in October 1926. (Image: Courtesy of the Huntington Library, San Marino, CA. Southern California Edison Photographs and Negatives. No. p16003coll2_22295_full)

from 19 cm at the base to just 5 cm thick at the top of the dam [3, 4]. One of the research team at the University of Colorado in Boulder was Dr Fredrik Vogt (1892-1970), an engineer who had previously carried out tests on model dams made of stiff rubber in his native country of Norway. His contribution to the Stevenson Dam model was aptly celebrated in the popular scientific press under the heading 'Toy dams to save lives' [5]. The concrete was made using rock aggregate from the site of the real dam, and the model was loaded using mercury in a rubber bag. Testing began in 1926 and, after a full programme of elastic tests, the main model was loaded to destruction in December 1928. Measurements taken from the model and the full-size dam were in close agreement and were used to check and improve the mathematical model used for the theoretical analysis. The model tests allowed the researchers to draw the following principal conclusions:

- A properly constructed small-scale model can be relied upon to produce strains and deformations similar to its prototype;
- Mercury is a satisfactory medium for producing model testing loads;
- The trial load method of analysis gives accurate results for a thin arch dam (The 'trial load method' involved reconciling the deflections of the shell calculated in two ways, one assuming the dam consisted of a series of vertical cantilevers, the other as a series of horizontal arches.)

In parallel with the Vogt's model tests on the Stevenson Creek Experimental Dam, George E. Beggs (1883-1939) at Princeton University undertook his own study of the dam using a 1:40 scale celluloid model of the cross section, and employing his patented 'deformeter' to measure deflections (see Section 13.5) [6].

8.2.2 Gibson Dam

The experience gained on the Stevenson Creek Experimental Dam was carried over to the design of the arch-gravity high Gibson Dam in Montana, which is 60 m high. A 1:68 scale model was tested in 1929 [7]. It was 88 cm high, 4.1 m long at the crest with a thickness that reduced from 39 cm at the base to 6.7 cm at the top. Like the Stevenson Creek model, it was constructed in concrete and loaded using mercury in a rubber bag. The Gibson model investigations included a temperature test as well. The temperature of the model was first raised by running hot water over the faces; when a fairly uniform high temperature was obtained the model was allowed to cool, after which the temperature was further lowered by running ice water over the faces. Continuous observations of radial deflection and temperature were taken throughout the temperature cycle [8]. These model studies corresponded well with theoretical calculations and measurements of the actual dam, and gave further confidence in the trial-load theoretical calculations and the use of reduced-scale model tests for the design of large arch dams.

8.3 Boulder Dam – structural model studies

The use of model testing for the Stevenson Creek and Gibson dam served as preparation for the largest dam of all – the enormous Boulder Dam on the Arizona/Nevada border (now the Hoover Dam) (1930-1936). It is an arch gravity dam – 221 m high, 200 m thick at the base, 12 m thick and 379 m long at the crest. At its completion in 1936 it was the largest dam in the world. The hydraulic works extend about 700 metres up- and downstream from the dam. Given the unprecedented size of the project, the design engineers for both the structure of the dam and the hydraulics believed it necessary to undertake model studies to complement and check their theoretical calculations [9, 10] (Figure 8.3).

8.3.1 The 1:240 scale plaster-Celite model

The purpose of the main structural model was to confirm calculations obtained using the trial-load method of analysis which was the best available at the time [11, 12]. This would be done by measuring surface stresses on the up- and downstream faces of the dam, and measuring deflections of the faces of the dam and radial and tangential deformations of the crest, when subject to a hydrostatic loads.

The very size of the dam posed new challenges for the programme of model testing. First of all, based on the experience of the Gibson model, it was clear that a model made of concrete, even using mercury to load the model, would be too

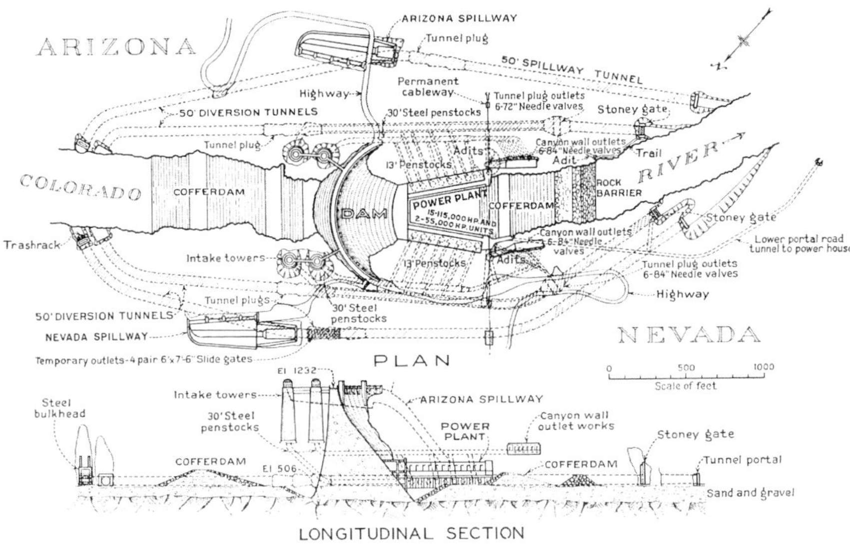

Figure 8.3 General scheme for the Boulder Dam (Source: [9] p. iv)

stiff to give deformations that could be measured to sufficient accuracy. A new model material had to be developed, with a lower modulus of elasticity but with other qualities necessary to ensure similarity with the real dam, and suitable to use for a model test. Thirteen different mixtures of materials were considered before a suitable material was found – a mixture of normal builder's plaster and a diatomaceous earth, a soft, sedimentary rock, abundant in Colorado and sold under the trade name Celite.

The model was 90 cm high, 82 cm thick at the base, reducing to 6.7 cm at the crest. The model was cast in layers 6 cm deep, which were allowed to cure for seven to ten days, and thermocouples were embedded in the model for measuring internal temperatures. The load was applied using mercury in a rubber bag that was placed in the 20 mm gap between the model dam and a fixed wooden form, that had been used for casting the model (Figure 8.4). Uniform loads could also be applied using water under mains pressure in the rubber bag. Construction of the model and test rig began in December 1930 and the first tests were undertaken on 7th July 1931.

While the model remained unchanged, different arrangements for measurement were used according to the purpose of different tests. For example, dial gauges sensitive to 0.0025 mm were used to measure the radial movement of the crest of the model dam (Figure 8.5). All rods used for mounting the gauges were made of Invar to avoid temperature effects.

After the early tests it was decided it would be important to take account of hydrostatic loads acting on the rock sides of the canyon which supported the dam. Additional loading bags for mercury were made and fitted to the model. A different arrangement of clock gauges was used to measure the tangential movement of the crest of the dam, due, in part to movement of the canyon sides (Figure 8.6).

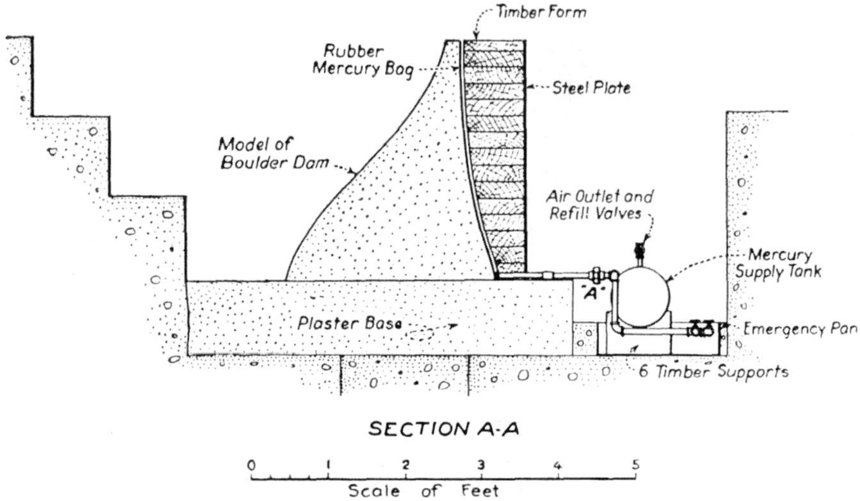

Figure 8.4 Section of the model showing the wooden form and the apparatus for introducing the mercury from below. (Source: [9] Fig.70)

Figure 8.5 Arrangement of gauges for measurement of radial deflections of the crest on the downstream face. (Source: [9] Fig.72)

In order to measure cantilever strains on the downstream face, dial gauges were mounted between pairs of anchors (Figure 8.7). Again, tests were undertaken with and without loads applied to the canyon walls.

A further series of measurements were taken to determine the change of slope of the dam face. These were made using an optical lever system – a narrow beam of light was directed to a mirror mounted on the model, and its reflection

Figure 8.6 Arrangement of gauges for measurement of tangential deflections of the crest on the downstream face. (Source: [9] Fig.79)

Figure 8.7 Arrangement of gauges set to measure cantilever strains on the downstream face. (Source: [9] Fig.95)

projected onto a scale about 3.8 m from the model. This system was sensitive to a change in slope of the dam surface of 0.000033 radians (7 seconds of arc).

During earlier tests it was noticed that for the first few tests each day, the model displayed a small plastic flow (an equivalent of creep or visco-elasticity) and so, in December 1931, some 'flow tests' were begun. The load was applied and measurements taken every 15 minutes for the first few hours, then every 30 minutes for the rest of the first day, and then left until all movement had ceased after eight days. The deflection of the top of the dam increased from 0.125 to 0.136 mm (8.75%) during these eight days, while near the bottom, it was from 0.0165 to 0.0254 mm (53.9%). It was also noted the deformation of the lower part of the dam was slower than in the upper part due to the lower levels of stress. The load was then removed and it was found that the upper part of the dam returned to its original position after four days, while the lower portion maintained its deformation for rather longer. These tests allowed an appropriate correction to be applied to measurements taken in other tests.

The many thousands of measurement data were presented in both tabular and graphical form to facilitate understanding of the behaviour of the dam (Figure 8.8).

Finally, a series of temperature tests were undertaken, beginning in July 1932, to determine the model deflections and the internal temperature distribution caused by temperature differentials between the upstream and downstream faces of the dam – a situation that would arise both during construction and in service, when the downstream face would be subjected to strong sunlight, while the upstream face was at the temperature of the reservoir water. In fact, it proved impossible to use water to heat and cool the surfaces of the model, as had been done for the Gibson dam model, as the model material, plaster-Celite, was vulnerable to damage by water. Also, it was found to be impractical to control temperatures on each face of the dam independently, so it was decided to determine the temperature response of the dam by imposing three temperatures on the whole model, and monitoring its dynamic temperature response during the transition between them.

The model dam was encased in a shell with ducts to allow air cooled with carbon dioxide from dry ice at -2°C to be circulated through the jacket for a week until the internal temperature and movement had stabilised. A uniform temperature

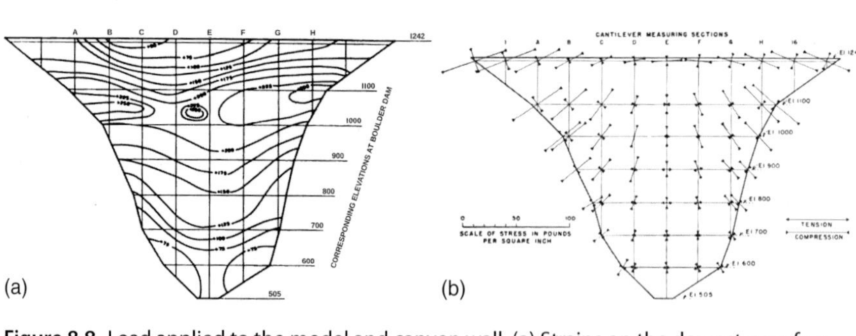

Figure 8.8 Load applied to the model and canyon wall. (a) Strains on the downstream face. (Source: [9] Fig.99). (b) Principal stresses (isostatic lines) on the downstream face. (Source: [9] Fig.112)

Figure 8.9 Model dam encased in a jacket, ready for connection of pipes to introduce cooling and heating gases, with clock gauges in position for monitoring movement of the downstream face of the dam. (Source: [9] Fig.128)

distribution was achieved by means of cardboard vanes inside the jacket to direct the air flow across the whole structure. The system was then returned to room temperature over a period of a week. The temperature at ten points in the centre line of the dam, distributed across the full height, was monitored every hour by thermocouples. The same procedure was then followed while the model was heated to about 40°C by two 550 W heaters blowing air through the jacket. Movement of the dam surface was measured using clock gauges moved by Invar rods which passed through small holes in the jacket, sealed with cotton wool to prevent the escape of the cooling/heating gases, allowing the dam to be free to move without hindrance (Figure 8.9).

The results of the hundreds of tests were numerous and complex. In summary, they provided quantitative data regarding the movement of the dam under hydrostatic and temperature loading, and gave values for the principal stresses on the downstream face of the dam. Most importantly, they supported the results of the trial-load analysis of the dam, as illustrated by the close agreement between deflections of the dam determined by the model tests and the trial-load analysis (Figure 8.10).

During the test programme with the plaster-Celite model two things became clear. The results indicated that a modification to the original design was needed – the arches needed to be thickened at the junction with the canyon wall in roughly the middle third of the height of the dam. Also, it was decided that more information was needed about the internal strains and stress in the dam, especially where the arches in the dam were too thick to be considered as a 'thin'

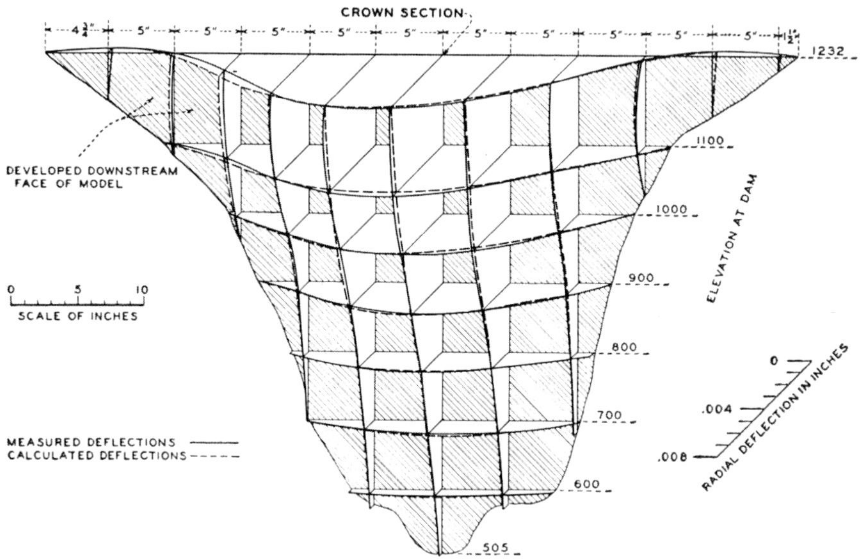

Figure 8.10 Comparison of measured and calculated deflections of the downstream face of the dam. (Source: [9] Fig.162)

arch. The three further series of model tests were commissioned. The first comprised two two-dimensional models: one was of a vertical cantilever section of the dam; the other was a two-dimensional model of a horizontal arch section of the dam, just above half height, where the trial-load analysis had determined that a large proportion of the load (76%) was carried by arch action, rather than cantilever action. The second series of tests was carried out using a new three-dimensional model of the dam, made with a more flexible material than the plaster-Celite model.

8.3.2 The 1:240 scale, 2-D plaster-Celite cantilever model

The tests on the cantilever model were undertaken by a research student at the University of Colorado in collaboration with the Boulder Dam design team [13]. He followed the same basic approach that had been taken by Wilson and Gore some twenty years earlier (see Chapter 7). It was modified, however, in light of the fact that Wilson and Gore had taken no account of the self-weight of the dam, which would be significant for a high dam, and that the deformations were large in relation to the geometry of the model, meaning that the change in shape would significantly affect the measured strains.

A more sophisticated test was planned, and tried out on an early design for another dam that was being planned, the Grand Coulee Dam. The section model was 7.5 cm thick, 53 cm high, and rested on a foundation slab 46 cm deep and 178 cm long. It was made of the same plaster-Celite mix used for the Boulder Dam model. The hydrostatic load was applied using mercury contained in a rubber bag held in contact with the upstream face. Dead loads were applied in the same manner that Wilson and Gore had used for live loads (see Section 7.5.1) – weights were suspended from pins inserted into the model located at the centres of gravity of a number of imaginary rectangles and trapezoids. Stain measurements made using a number of different hand-operated mechanical strain gauges proved inconsistent and a more accurate method was sought. Finally, a Tuckerman optical strain gauge was adapted for the model and, for the first time, a grid of lines arranged in the form of a rosette was used to provide a systematic measurement approach, using basic elastic theory (Mohr's circle of stress) to determine the magnitude and direction of the principal stresses (isostatic lines).

Having proved the technique with the Coulee Dam model, the cantilever model of the Boulder Dam was constructed. The model, made of plaster-Celite, was 7.5 cm thick, 92 cm high on a foundation slab 91 cm deep and 2.67 m long. Both the self-weight and the live load on the upstream face were applied mechanically using weights, wires and pulleys (Figure 8.11). The live load on the foundation due to water in the reservoir, and tail water on the downstream side, were applied by means of mercury in a rubber bag whose pressure could be adjusted by raising the mercury supply reservoir.

The first measurements recorded the deformation of the cantilever and of the foundations both under dead load and under full live load plus dead load (Figure 8.12). The clock gauges were able to measure the deflections of up to 0.25 mm to an accuracy of 1% (0.0025 mm). The deflections are graphically shown when plotted at an exaggerated scale (Figure 8.13).

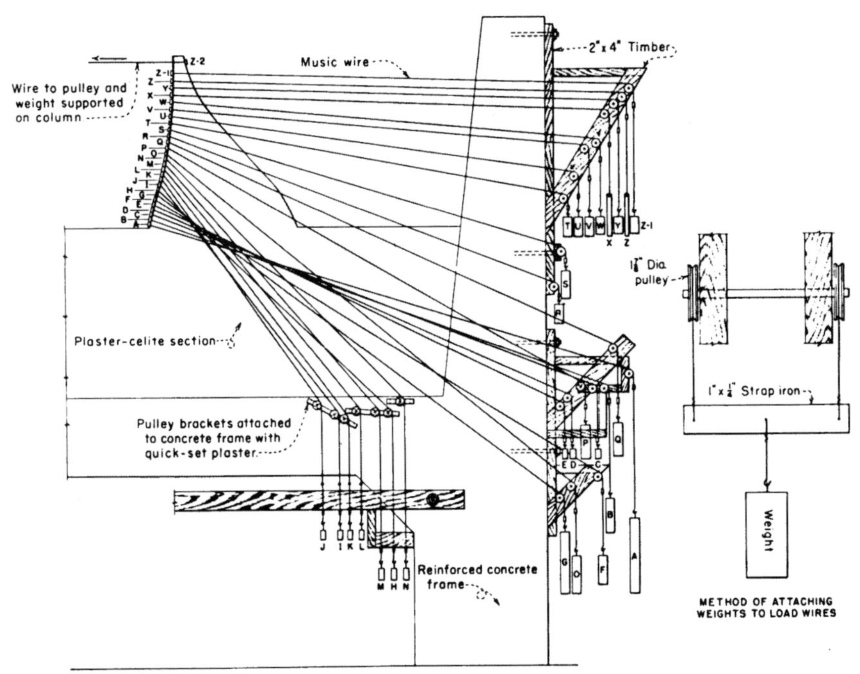

Figure 8.11 Apparatus for applying live loads. (The similar system for applying dead loads is omitted for clarity). (Source: [13] Fig.10)

Figure 8.12 Gauge installation for measuring cantilever and foundation deflections. (Source: [13] Fig.12)

Internal strains in the cantilever and foundations were measured using strain gauges. Where it was possible to clamp the strain gauges to the model, at the boundaries, Huggenberger tensometers were used (see Figure 13.8). Tuckerman optical strain gauges were used for the interior areas, as they can easily be mounted on the surface. The movement in an optical strain gauge causes a lozenge-shaped glass mirror to turn and hence rotate a reflected beam of light, which is viewed on a screen (Figure 8.14).

As had been developed in the Coulee Dam model, the gauges were deployed in a rosette arrangement, and read with the aid of a telescope, seen at the left of the photograph (Figure 8.15). The principal stresses inside the dam were calculated from the strain measurements (Figure 8.16). There was generally a good

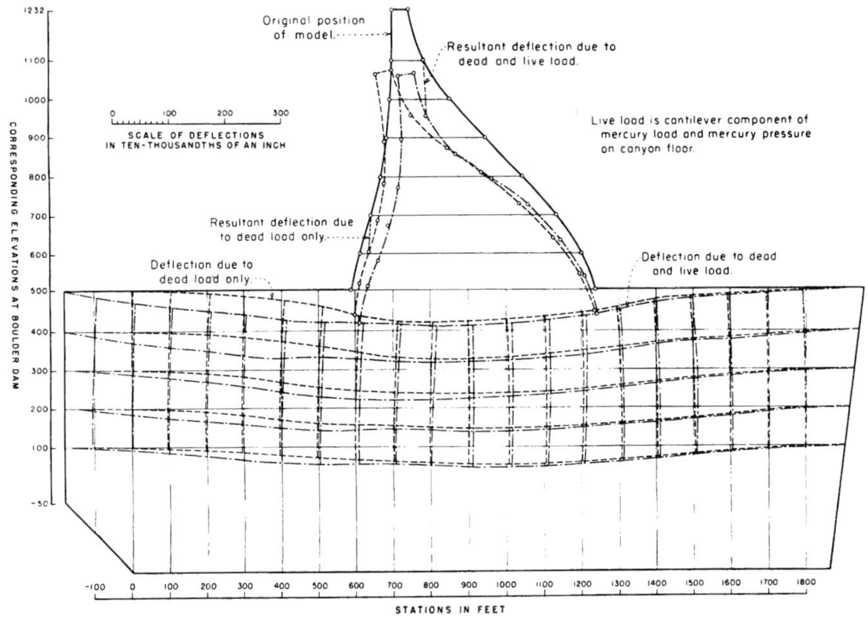

Figure 8.13 Deformations of the cantilever and foundations under no load, dead load and live load plus dead load. (Source: [13] Fig.14)

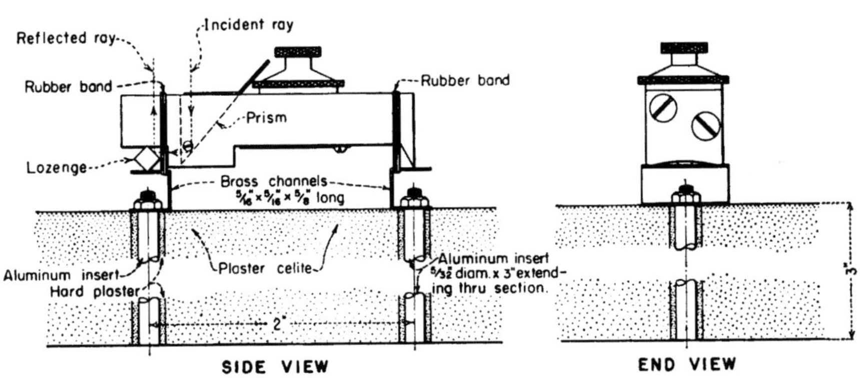

Figure 8.14 Tuckerman-type optical strain gauge mounted on the model. (Source: [13] Fig.15)

correspondence between the measured stresses and those calculated using the trial-load method, except in areas of high stress concentration where it was evident that the model material had undergone some plastic deformation.

8.3.3 The 1:120 scale, 2-D plaster-Celite arch model

Like the cantilever tests, the main purpose of the arch tests was to investigate the internal stresses in the dam [14]. This was particularly important where the thickness of the dam was too large for it to be considered as a simple arch; rather it was subject to a combination of arch and bending action. The section chosen was

Figure 8.15 Model set up for application of mercury loads on the upstream edge of the section, and along the foundation surfaces. (Source: [13] Fig.17)

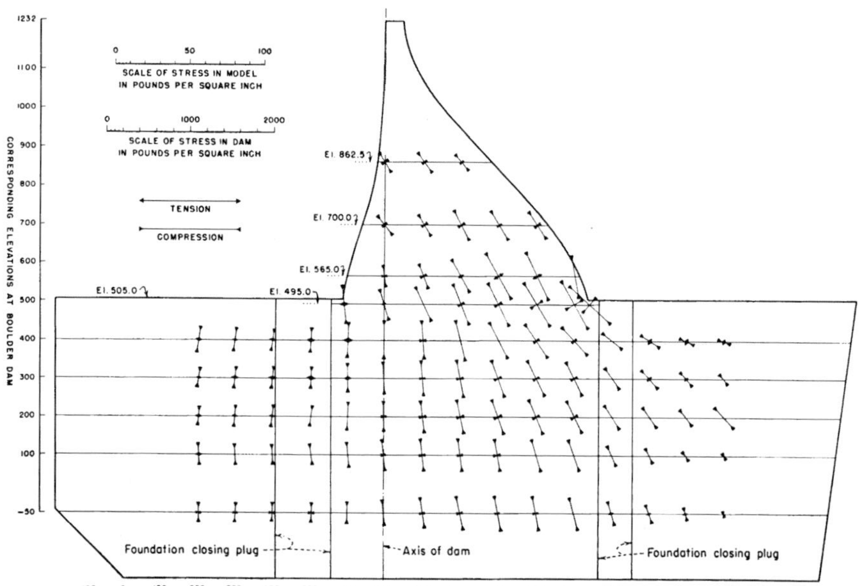

Figure 8.16 Principal stresses in the cantilever and foundations under live load plus dead load. (Source: [13] Fig.20)

about half way up the dam where the thickness was large in relation to the span, and where the span between canyon walls was still quite large. The arch model was again made with plaster-Celite, but at half the scale as the cantilever model, i.e. 1:120. The arch was 15 cm thick, 67 cm deep and spanning approximately 1.37 m between abutments (the canyon walls).

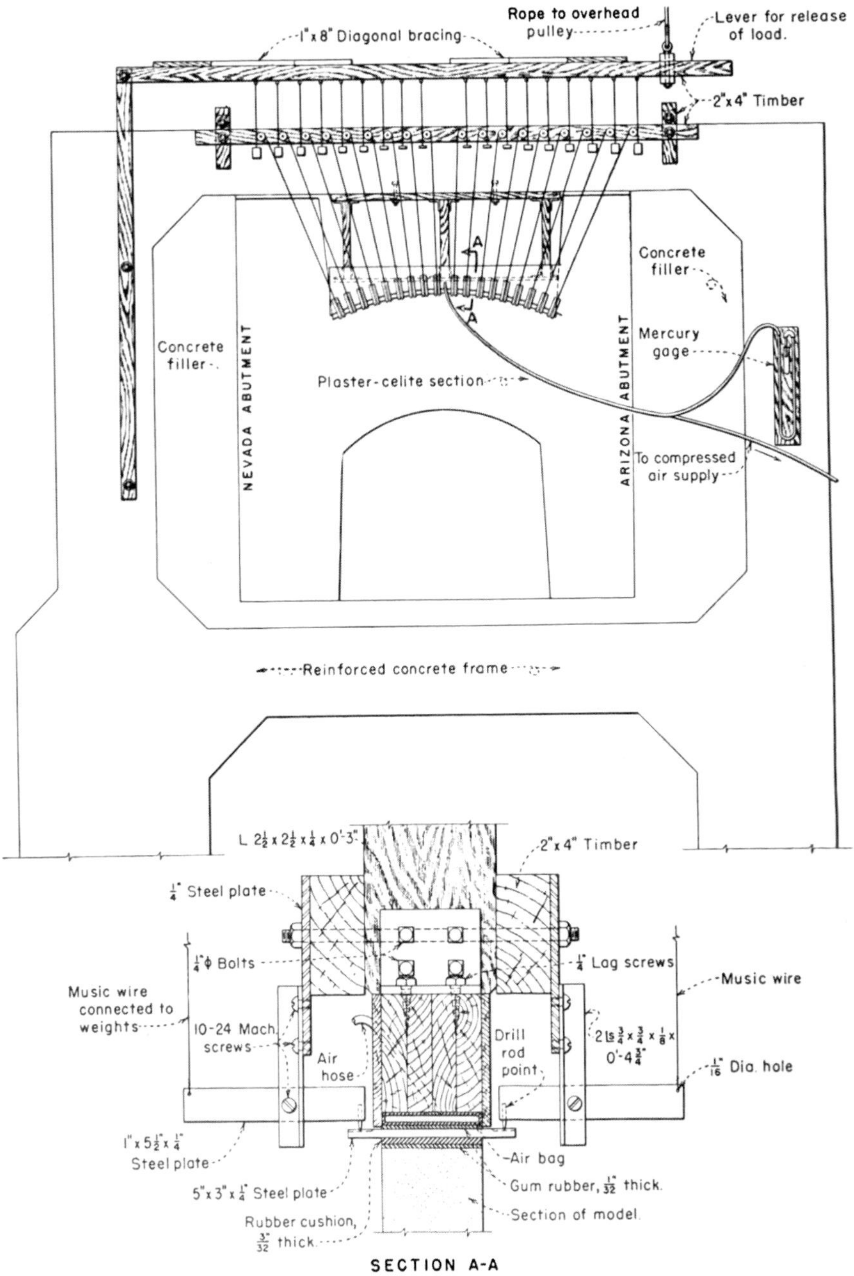

Figure 8.17 Apparatus for applying loads to the extrados of the arch. The triangular loads are applied mechanically, by means of weights and levers; the uniform radial load is applied pneumatically, by means of air in a rubber bladder. (Source: [13] Fig.49)

The loading system for the arch was particularly ingenious. As with the cantilever model, the load acting on the section was derived from both the trial-load analysis and the tests on the full model. These identified two components to the internal strains within the section – one was the strains due to a constant radial load equivalent to pure arch action; the other was the strains due to bending which varied as the shear force in a beam – from zero at the centre to a maximum at each side, equivalent to a triangular load on each half of the dam section.

The triangular loads were applied using small weights, wires, pulleys and levers. The wire from each weight passed over a pulley, down to a lever, pivoted in a frame at the level of the extrados of the arch (Figures 8.17 and 8.18). This reversed the upward pull in the wire, to a downward, radial push, via steel transfer plates and a layer of rubber, onto the extrados (see Cross section A-A in Figure 8.17). The whole frame could be lifted by a large lever to allow the entire triangular load to be applied or removed at will.

The constant load on the arch due to arch action, was applied by means of compressed air inside a thin rubber bladder placed between the wooden frame and

Figure 8.18 Arrangement of gauges for the radial deflection tests. (Source: [13] Fig.60)

the steel transfer plates on the extrados of the arch (see Cross section A-A in Figure 8.17). This load could be applied uniformly by increasing the air pressure in the bladder.

Although the first series of tests yielded a useful set of results for both radial and tangential deflections of the arch, they also exposed a weakness in the design of the arch. The sharp angle between the extrados of the arch on the upstream face and the abutment caused a considerable stress concentration at the root of the notch. This led to a fracture at both ends of the model arch, and it was realised that this would also be a weakness in the full-size dam. The model was repaired and a smoothly-curved fillet introduced to replace the sharp notch and reduce the intensity of the stress concentration (Figure 8.19). It was also found that the intricate system of applying the loads made it difficult to observe the deflections of the structure. In the later series of tests, a simplified loading system was used to facilitate the measurement of deflections and surface strains on the model.

The tests on the second model produced useful data about the deflections of the dam as well as the pattern and magnitude of the internal principal stresses (Figures 8.20 and 8.21).

Figure 8.19 Modified model arch, with smooth fillets between the extrados and abutments. Three optical strain gauges are mounted for taking measurements. The full grid of rosettes can also be seen. (Source: [13] Fig.74)

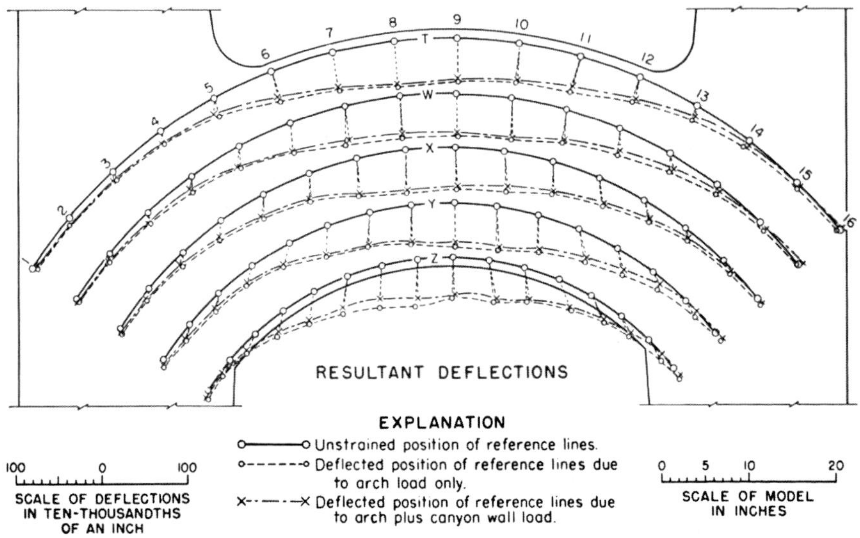

Figure 8.20 Radial and tangential deflections of the arch. (Source: [13] Fig.71)

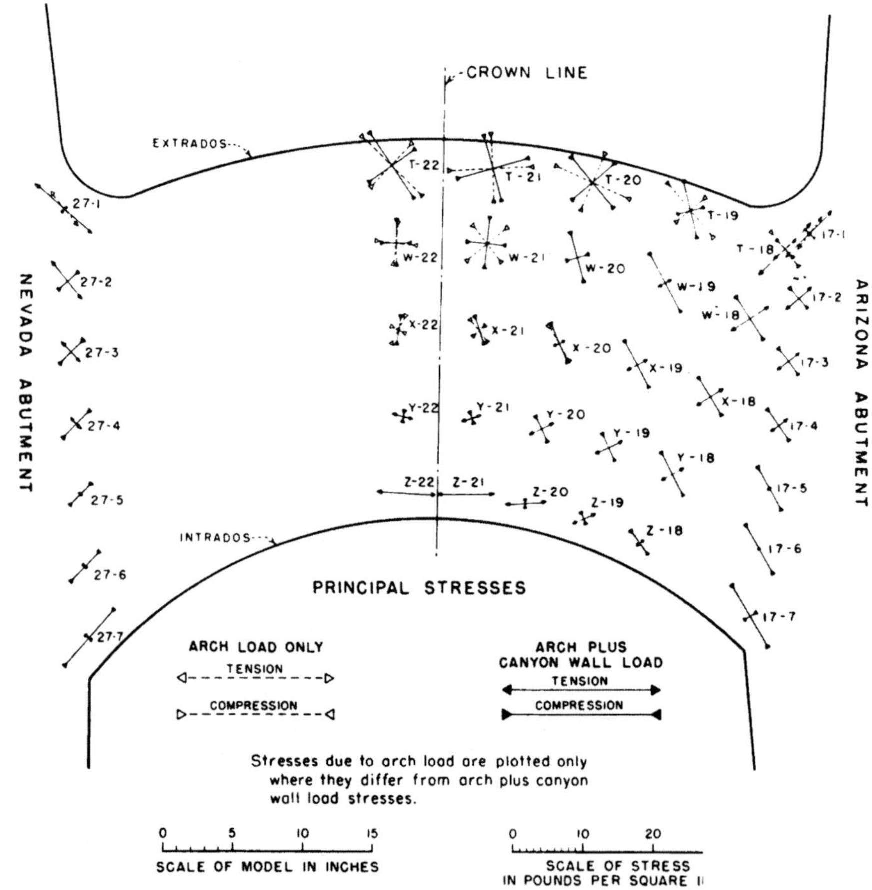

Figure 8.21 Principal stresses (isostatic lines) in the arch. (Source: [13] Fig.75)

The tests revealed the unanticipated result that the strains and deflections in the canyon walls were as large as in the concrete dam itself, and also that this deformation of the canyon walls had a significant effect on the stress distribution within the dam itself. These observations highlighted the need to take this into consideration during the detail design. A comparison of the measured strains and those calculated using the trial-load method reveals good correspondence for some places in the dam, but poor correspondence in either magnitude or direction in other replaces. This was generally explained by the difficulty of modelling a real structure accurately in either a mathematical model or a physical model. Nevertheless, the use of plaster-Celite models was advocated as a useful design tool when used in conjunction with a theoretical analysis.

8.3.4 The 1:180 scale 3-D rubber-litharge model

Following the results of the full plaster-Celite models and the two-dimensional cantilever and arch models, it was decided to test a second three-dimensional model, in particular including the fillet in the transition zone between the dam and the canyon abutments [15].

Rubber-litharge (rubber made stiffer using lead oxide) was chosen as the material for the second full model of the dam because it had a lower modulus of elasticity that plaster-Celite and it was resistant to damage by water. This meant that the model could be loaded using water, and deflections were still large enough to be measured easily. In particular, this meant that the surface stresses could be measured on both the downstream face of the dam and also on the upstream face, using gauges that would work under water. However, the material had disadvantages, particularly a higher Poisson's ratio than that of concrete and the plaster-Celite, and also greater visco-elasticity. These differences in similitude between the model material and actual material were compensated for using appropriate scaling factors.

18,000 kilograms of the rubber-litharge was delivered in 15, 30 and 60 cm squares, all 2.5 cm thick. Much of this was used in the many tests to establish its mechanical properties. The model dam was made by cutting squares to the approximate shape and building up the dam, gluing each layer to the one below, and finally finishing the dam to the correct shape. The model was 1.23 m high, 1.10 m thick at the base, reducing to 7.6 cm at the crest (Figure 8.22).

In addition to using dial gauges to measure deflection of 55 points in the downstream face of the model, measurements were also taken from the downstream side, at three points at the mid-thickness of the dam, and three points 4.2 cm from the upstream face of the dam (at 16, 33 and 67 cm from the base) using aluminium rods anchored inside the rubber model.

As with the plaster-Celite model, measurements were also taken of the vertical and tangential movement of the crest of the dam using clock gauges, and changes in slope of the downstream face of the dam, using an optical lever system. Tests were also undertaken with rubber-litharge canyon walls to determine the effect on the shape of the dam due to movement of the rock foundations.

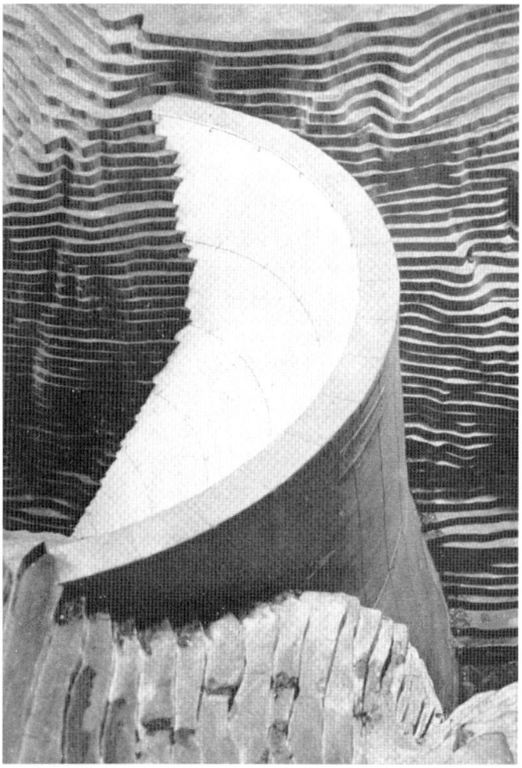

Figure 8.22 The completed rubber-litharge model. (Source: [9] Fig.187)

A series of tests was also undertaken to determine the surface strains on both faces of the dam. On the upstream side, a mechanical strain-gauge rosette was used that would work under water. Each of the three gauges, mounted at 0°, 90° and 135°, had a gauge length of 75 mm and was located in holes pre-drilled using a jig. The lever arm magnified the strain by about 30 times and the indicator hands were restrained by light elastic bands. Each gauge was individually calibrated before use and measured strains to an accuracy of 0.0025 mm (Figures 8.23 and 8.24).

On the downstream side, small hand-operated strain gauges were used to measure strains in four directions over a gauge length of 50 mm, between countersunk holes in the heads of tacks, located using a template, and driven into the rubber to form rosettes (Figures 8.25 and 8.26).

As with the plaster-Celite model, further tests were carried out to determine the visco-elastic ('flow') characteristics of the model and its response to extreme changes of temperature. At ten days, the flow tests were of longer duration that for the plaster model because of the material properties, and the deformation of the dam was large enough to require that water to be added to the reservoir. Unlike the temperature tests on the plaster-Celite model, the tests on the rubber-litharge

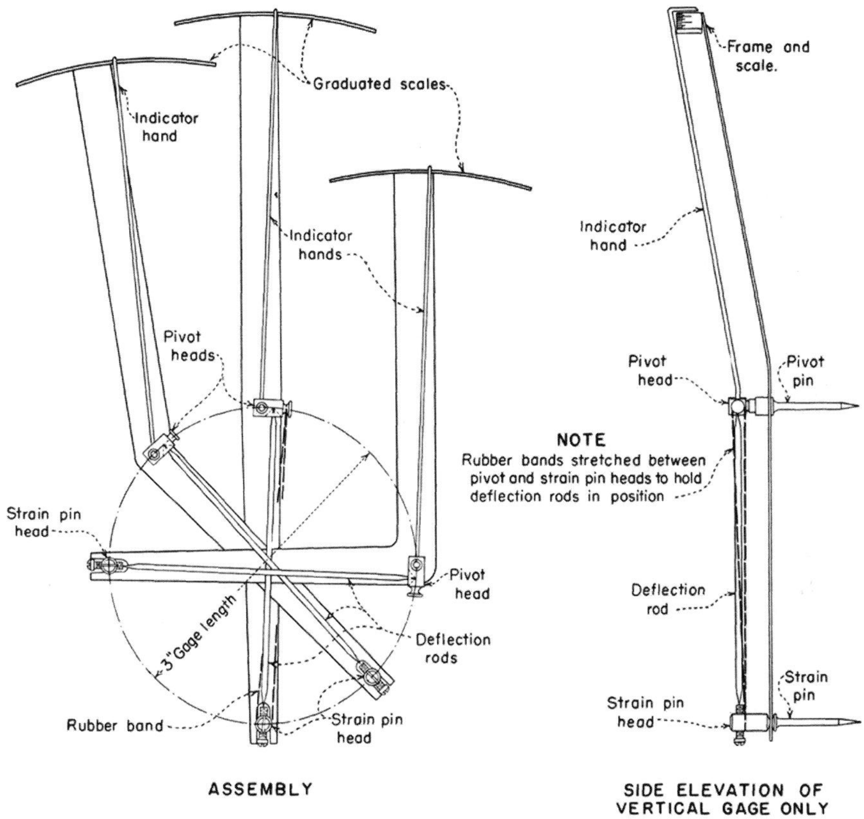

Figure 8.23 Mechanical strain-gauge rosette for the upstream face. (Source: [9] Fig.207)

model could be conducted with better a correspondence to the real dam – water could be used to impose loads on the upstream face and it was also possible to heat and cool only the upstream face of the dam, using warm and cold water. Each test lasted twelve days to allow temperatures to stabilise.

Finally, strain measurements were taken as the model was disassembled, in order to quantify the contribution to strain measurements made under load, due to the self-weight of the dam itself. This allowed the contributions to deformations and principal stress diagrams made only by live loads to be assessed and used to correct other data. In fact, the correspondence between the calculated and measured results was not as close as for the plaster-Celite, probably due to the variable properties of the rubber-litharge and the lack of homogeneity of the model material, due to its availability only in 25 mm thick sheet.

With this series of tests, the model testing programme was ended as the designers now felt they understood well enough the behaviour of the dam and were more confident in their use of the theoretical analyses. The final design could be made.

Figure 8.24 Strain-gauge rosettes fixed to the upstream face of 1:180 scale model in stiff rubber. (Source: [9] Fig.210)

Figure 8.25 Hand-operated strain gauge for use on the downstream face. (Source: [9] Fig.211)

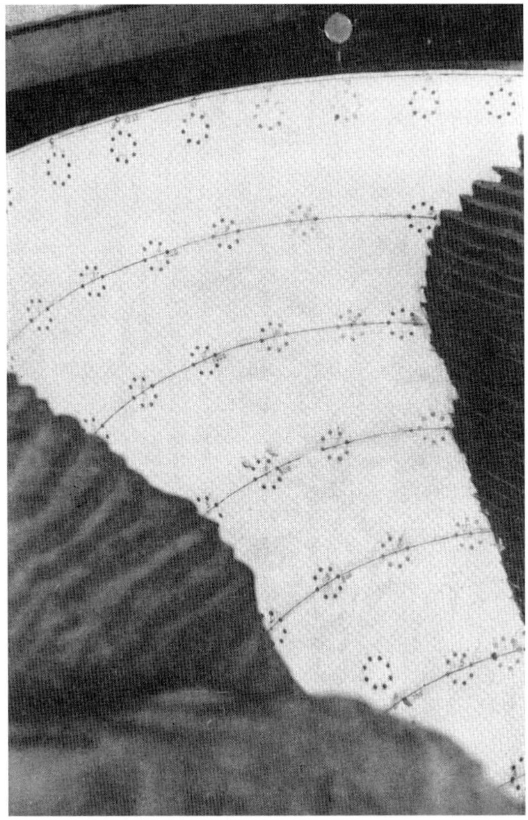

Figure 8.26 Locations for placing the hand-operated strain gauges on the downstream face to create the equivalent of strain-gauge rosettes. (Source: [9] Fig.212)

8.4 Boulder Dam – hydraulic model studies

As with the structure of this dam of unprecedented size, it was also felt that the hydraulic system was so different in scale from previous experience that theoretical calculations could not be used with sufficient confidence to generate an effective and safe design. The design team decided it was necessary to conduct a number of model experiments, drawing on the growing expertise in this field of research in the 1920s (see Chapter 23).

8.4.1 Model studies for design of the spillways

The two spillways, which act as a safety overflow should heavy rainfall threaten to exceed the capacity of the reservoir, are very large. Each one is about 200 m long, 46 m wide and 52 m deep; the volume capacity is equal to the Niagara Falls – 5,660 m^3/s – and, due to the greater height, embody more than three times the energy. The theoretical speed of the water flowing out of the 15-metre diameter spillway tunnel is 190 kph. Nothing on this scale had been built before the Boulder Dam.

The theoretical design was well under way by 1932 when it was decided that it would be prudent to undertake a programme of model studies in parallel. Various model studies relevant to spillways had already been undertaken in the Germany, Britain and the USA, but it not on this scale. The studies were therefore undertaken mainly to provide added confidence to the theoretical design proposals and, in particular, to compare the results of model studies carried out at different scales [16].

The first study, at the University of Colorado, was on a 'glory hole' spillway design, for the original design flow of 2,830 m^3/s – a 70 m diameter intake would feed a 15 m diameter, 200 m vertical pipe. The 1:60 scale model had an intake 1.15 m in diameter made of paraffin wax; the 25 cm diameter vertical pipe and 90° bend into the horizontal discharge pipe was made of transparent Pyralin, a celluloid plastic, so that the water flow could be studied. This design was not successful as the pipe repeatedly choked in a pulsating fashion. Minor improvements were achieved by ensuring a more uniform entry into the intake, but further work was abandoned.

A side-channel spillway design, with a free crest, was adopted and a 1:60 model was built to test the flow into the spillway tunnels however, before these tests were complete, it was decided to double the design capacity to 5,660 m^3/s. A second 1:60 model was built incorporating a Stoney gate (rising gate) entry to manage the flood water. Although this system proved unsatisfactory, these tests provided useful guidance for constructing a 1:20 scale model at the Bureau of Reclamation laboratories at Montrose which had a drum-gate system for managing the incoming water. Tests on many forms of the spillway were undertaken to improve the flow through the portal into the spillway tunnel. These included changes to the slope of the spillway floor, a curved wall at the upstream end, improvements to the portals well as a number of modifications to the floor to reduce spiral movement of the water (Figure 8.27). The results of these initial tests on many variations to

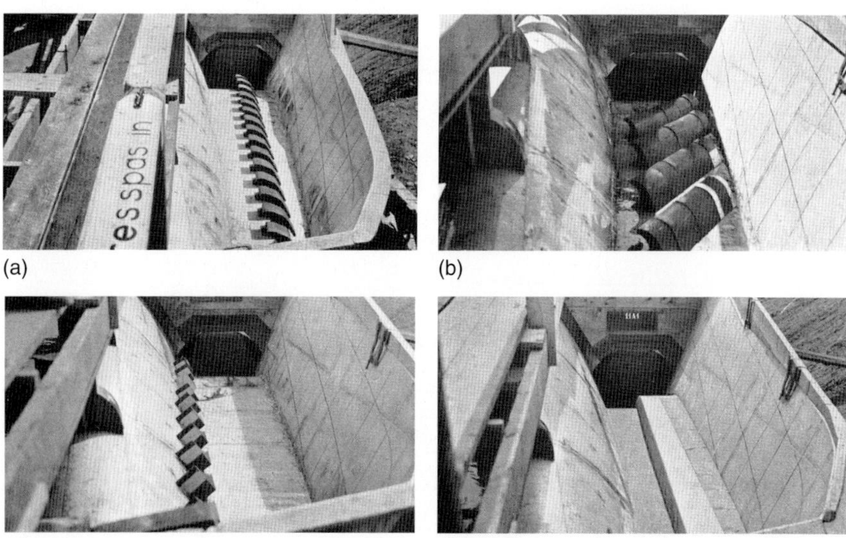

Figure 8.27 1:20 scale model of the spillway. Alternative means were tested for reducing spiral movement of the water flowing in from the left side of the model over the spillway crest. (a) Cylindrically-faced dentated baffles on centre line of the channel. (b) Cylindrically-faced baffles inclined downstream. (c) Vanes on the spillway crest. (d) Trough on the crest side of the channel floor. (Source: [16] Figs 42, 43, 45, 46)

(a) (b) (c)

Figure 8.28 1:20 scale model of the spillway with equivalent of full flow (56 m^3/s). (a) Original mid-stage design. (b) As (a) with curved end wall to spillway. (c) As (a) with curved end wall and dentated baffles. (Source: [16] Fig.65, Fig.66B, Fig.67)

the shape of the spillway at 1:60 scale led to a new model of a revised, intermediate design at 1:20 scale, which was also tested in various alternative versions (Figure 8.28).

Following the tests on many versions of the intermediate design, new models of the final design were made at three scales in order to test the correspondence in the observed behaviour of the water flow at different scales. Since there would be no opportunity to make measurements on the full-size dam, this was considered to be the best way of justifying an interpolation from small scale to full size (Figure 8.29).

Overall, the many tests provided strong evidence that tests on reduced-scale models were an effective and reliable tool in the design of unprecedented structures. They led directly to a number of significant design changes which would not have been suggested by the theoretical analysis, for which the model tests provided a quick and cheap means of checking the effects of design changes. The model tests also demonstrated that, while some design changes, such as baffles on the spillway floor, might lead to improved performance, such improvements would not be worth the additional construction costs. The studies on models at different scales also led the engineers to conclude that on a similar future project, it would be more cost- and time-effective to undertake most of the preliminary

(a) (b) (c)

Figure 8.29 Models of the final design with water flow at 60% of maximum. (a) Model scale 1:100. (b) Model scale 1:60. (c) Model scale 1:20. (Source: [16] Fig.86) tests at small scale (1:100) and only use larger scale, more expensive models when major issues had been resolved.

8.4.2 Model studies for design of the penstocks and intake tower

The success of the model tests in the design process for the spillways led the design team to undertake model tests on several other aspects of the hydraulic design, even though the tests were begun too late to be able to lead to major design changes. The first series of tests, which comprised some 25,000 observations, concerned the losses at the junction between the 9.1 m diameter main header pipe and the 4 m diameter turbine penstocks, at an angle of 105° [17]. First of all, qualitative observations were carried out using a 1:36 scale model of a junction made of transparent Pyralin. The concern was that eddies forming immediately after the junction would lead to a reduction in flow. The extent of the zone affected by eddies was identified by injecting a coloured die into the water at several locations in the small diameter pipe. Following these visual tests, quantitative studies were made using a 1:36 scale model made of steel pipe. The results were found to agree well with the theoretical calculations and no significant losses were found. Further studies were made on a 90° junction, giving results which could be compared directly with similar tests carried out by Professor Thoma at the Hydraulic Institute of the Technical University in Munich, albeit that these were made with models six times smaller. From the differences between models of different sizes, it was concluded that the size was significant, as viscous effects became more significant at smaller scales. The tests also repeated the Munich tests looking at the effect of introducing a conical transition between the two pipes of different diameters, and this was found to improve flow by about a third – which justified the use of such cones that had already been included in the Boulder Dam design.

Extensive tests were made using a 1:64 scale model of the intake tower (Figure 8.30) and the penstock (Figure 8.31) to determine the various hydraulic losses, in particular due to the filter that covered the intake to prevent rubbish from the reservoir being ingested. The results agreed well with the theoretical calculations for the model, and they were used to give approximate values for the full-size dam using hydraulic scaling factors.

To supplement the results of the 1:64 scale model tests, it was considered useful to study the flow of water through the intakes into the feed pipe to the penstocks, and the method chosen was one familiar to those studying the flow of water through earth dams [18, 19]. The flow net of such water movement maps the loss of hydrostatic pressure through a dam as water is forced to move along the hydraulic gradient. Such a flow net displays the same characteristics as the fall of voltage in an electric field, with the electric current moving along the potential gradient [20]. A similar analogue technique was used for tests on the intake towers.

A radial section of the intake tower, 90 cm long, made of wood (horizontal, at the bottom of Figure 8.32), was placed in a shallow tray containing salt solution. The positive electrode, a strip of copper, was fixed at the source (top edge of the tray in Figure 8.32) and at two negative electrodes were fixed inside the intake, immediately beyond each of the two inlet gates at the bottom of the tower

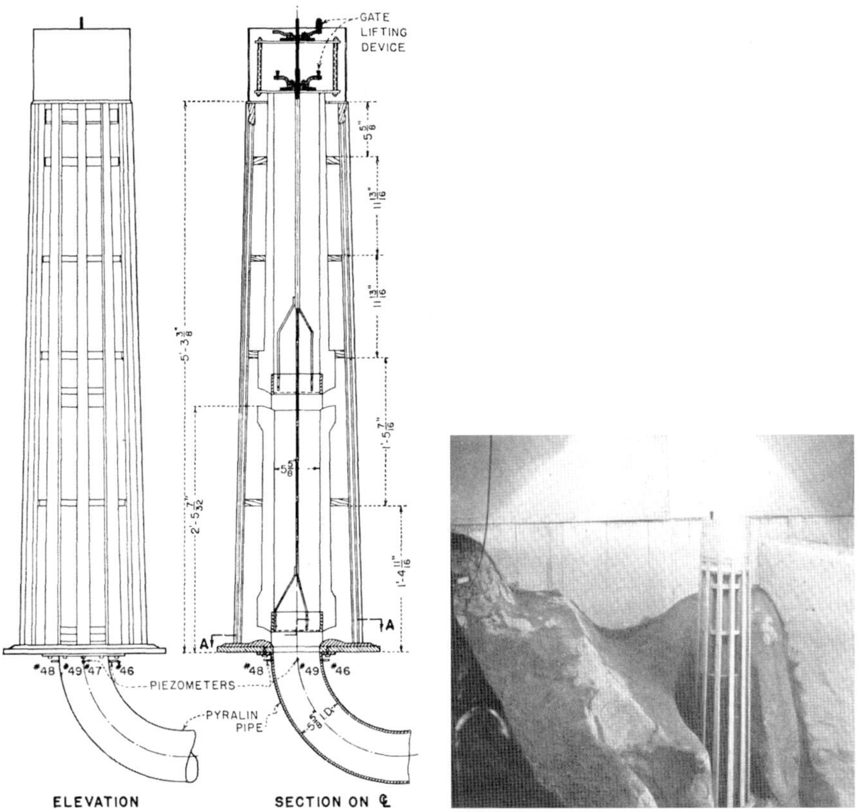

(a) (b)

Figure 8.30 1:64 scale model of the intake tower, 1.6 m tall (a) Cross section, showing mechanism for opening the two inlet gates. (b) The model tower in situ. (Source: [17] Fig.43, Fig.42C)

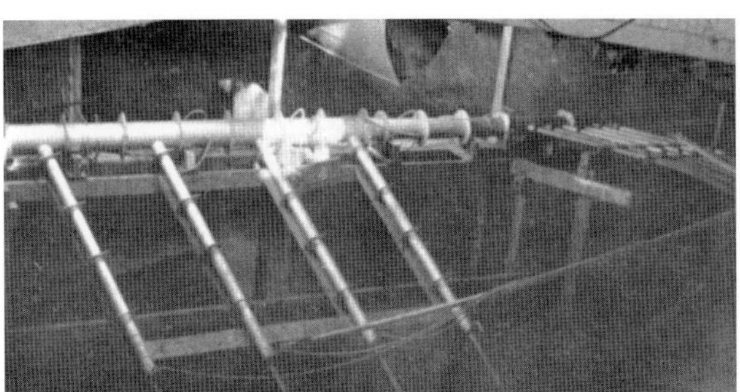

Figure 8.31 1:64 scale model of the penstock for delivering water to the turbines. (Source: [17] Fig.56)

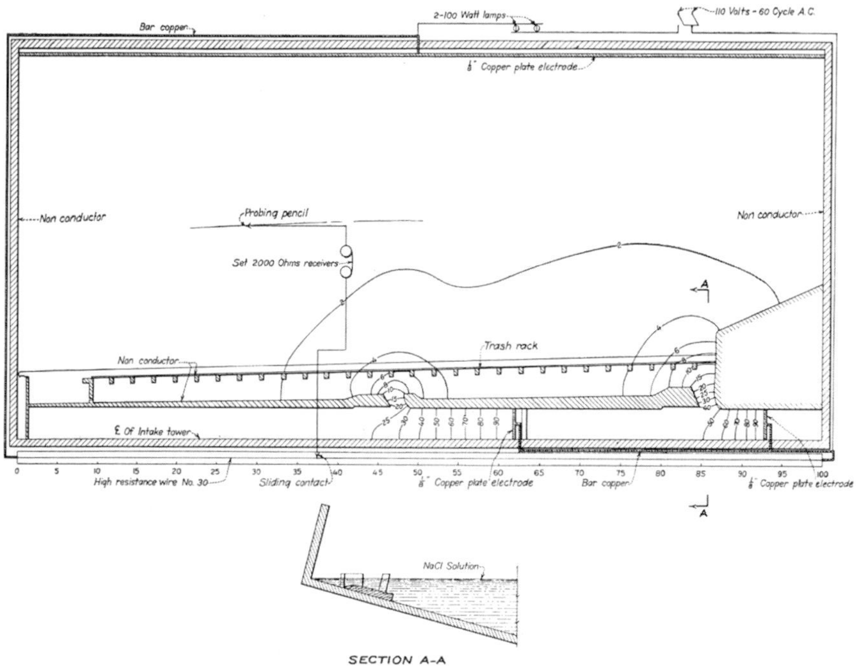

Figure 8.32 Plan view of the electrical analogy model of the intake tower. (Source: [17] Fig.62)

and half way up (bottom right of the tray in Figure 8.32). The third dimension of the water flow was reproduced by tilting the tray so that the depth of the electrolyte increased from zero at the centre line of the tower, towards the positive electrode (Section A-A in Figure 8.32). The potential falls gradually through the electrolyte between the positive and negative electrodes, and an electrical probe, feeding into a Wheatstone bridge, was used to plot the lines of equipotential. In Figure 8.33, the equipotential lines are shown for the situation where both inlet gates are open. From the pattern of equipotential lines, the orthogonal flow net was constructed and, in turn, the velocities and quantities flow were calculated. The flow net shown in Figure 8.33 is for the case where a filter was in place for the lower half of the tower only, and only the lower inlet gate was open. However, the electrical analogue model did not, it seems, influence the design of the Boulder Dam, but it confirmed the assumptions made in the theoretical calculations, and also proved the methodology to be useful for hydraulic studies.

8.4.3 Model studies for design of the outlet works

A further series of hydraulic model studies were undertaken to investigate the outflow of water into the river, and the needle valves used to manage this flow. Water from the 7.6 m diameter penstock is split into six pipes containing the 1.8 m diameter needle valves for controlling the flow, each of which needs to be able to carry up to 100 m^3/s at speeds up to 190 kph. After the needle valves, the flow is recombined into a 15 m diameter duct taking the water to the outlet portal

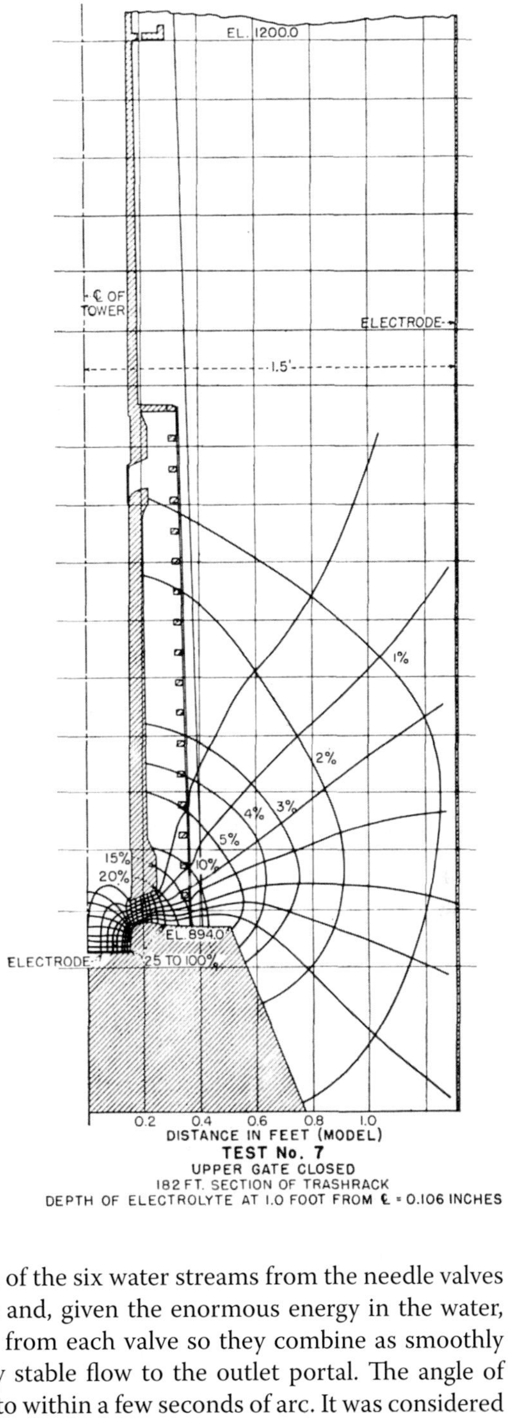

Figure 8.33 Flow net into the intake tower for the case with a filter in place for the lower half of the tower only, and with only the lower gate open. (Source: [17] Fig.65)

(Figure 8.34). The recombination of the six water streams from the needle valves is a particularly delicate process and, given the enormous energy in the water, it is essential to direct the flows from each valve so they combine as smoothly as possible to ensure a relatively stable flow to the outlet portal. The angle of these valves has to be controlled to within a few seconds of arc. It was considered essential to study the behaviour of these water streams using physical models to check the theoretical calculations.

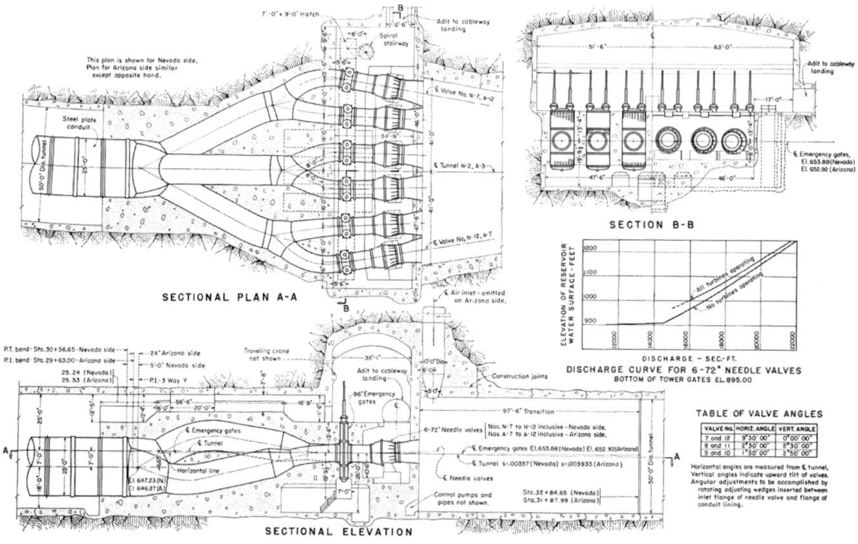

Figure 8.34 Tunnel-plug outlet works. The water from the 7.6 m penstock is split into six pipes of 1.8 m diameter that can be regulated using needle valves, and then recombined into a 15 m diameter duct taking the water to the outlet portal. (Source: [17] Fig.69)

Figure 8.35 1:20 scale model of outlet works. (a) View of the model. (b) Action in the outlet tunnel with the original orientation of needle valves. (c) Action with the revised orientation of needle valves. (Source: [17] Fig.71)

Models were made at three different scales. Tests using the first model, made of transparent Pyrolin at a scale of 1:106.2, showed that the initial design produced far too much turbulence. After eight tests with different angles, the horizontal and vertical angles of the valves were finally changed by between two seconds of arc and two degrees, which gave a much less turbulent flow. So successful was this study that it was repeated at a scale of 1:20 to refine the geometry even more carefully. Fourteen different arrangements of the valves were studied and the valve angles finally optimised to the nearest second or two of arc, giving a much improved performance compared to the original design (Figure 8.35).

No less important than the performance of the valves when all were fully open, were the various combinations of valves that would give the best performance when fewer than six were open. Studies were undertaken using the 1:20 model to find the combinations of valve operation that gave the most stable flows. For

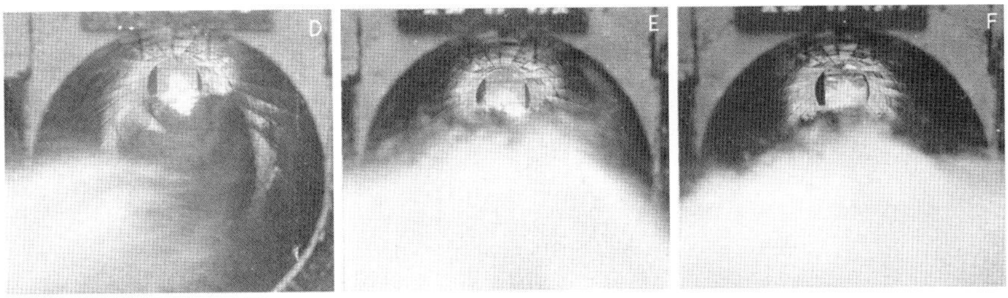

Figure 8.36 1:20 scale model of outlet works. Action in tunnel for different combinations of valves. (a) Valves 1 and 2. (b) Valves 1, 2, 3 and 4. (c) Centre four valves (2, 3, 4, 5). (Source: [17] Fig.72)

example it was found that using the centre four valves gave much better results than valves 1 and 2 or valves 1, 2, 3 and 4 (Figure 8.36).

Following these optimisation studies, the results obtained with the 1:106.2 and 1:20 scale models were checked with a new, 1:60 scale model which was made as a more faithful representation of the full-size valves and pipes, for example including the manifold and emergency gates. A further optimisation of the valve angles was undertaken, this time also taking account of the forces exerted by the pipes on their supports.

The final stage of the hydraulic path through the dam was the outlet from the various portals into the river below the dam. The classic view of the downstream face of the dam shows the nature and magnitude of the problems (Figure 8.37). On each side of the canyon there could be up to four different streams of water entering the river – the Powerhouse water exiting from the turbines, the water from the Canyon Wall and Tunnel Plug outlets from the two penstocks and, in the event of excess water in the reservoir, up to 5660 m^3/s from the spillway.

The model, built at a scale of 1:150, was 8.5 m by 5.8 m (Figure 8.38). Thirty nine different combinations of the four possible sources of water on each side were studied, for three different initial levels of the tailwater in the river, and measurements were made at two points – one in the upper section, the other in the lower section. The observations were made both of the measured changes in water level in each case and, in a more qualitative way, the nature and severity of the erosion of the river bed (Figure 8.39).

8.5 Final remarks

The Boulder Dam was probably the world's largest construction project in the first half of the twentieth century and the unprecedented size forced engineers to push back the boundaries of their knowledge. Just as with the Britannia Bridge in the 1840s, physical models played a significant role in this effort; indeed, no project had made more use of physical model studies. In the field of physical model testing in structures and hydraulics, the project served as a unique catalyst for developing old methods and establishing new ones. Even before the 15-volume report of the project was published in 1938-1940, new developments

Figure 8.37 Hoover Dam during jet flow gate testing, June 1998. (Source: Bureau of Reclamation. No. C45-300-21094) (Image: [21])

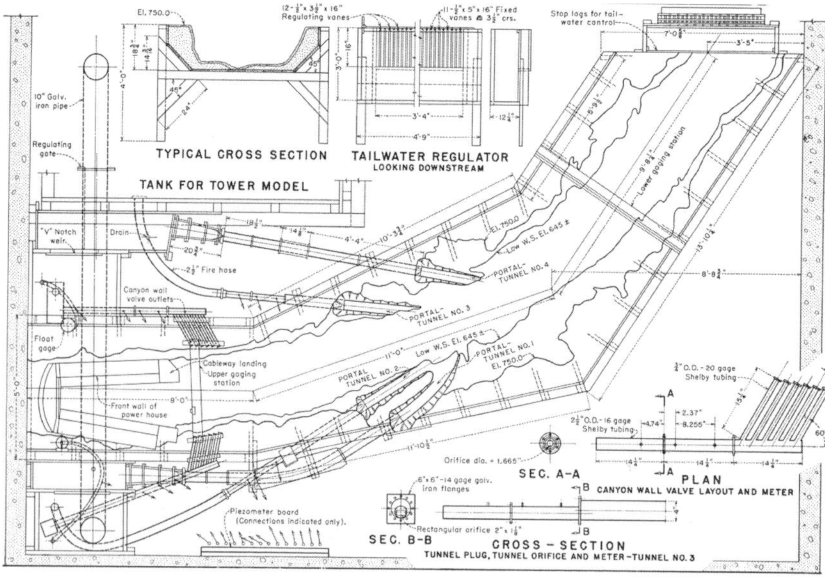

Figure 8.38 Plan and details of the 1:150 scale river model. (Source: [17] Fig.79)

Figure 8.39 1:150 scale model of the river. (a) Water from the Powerhouse and the Canyon Wall outlets. (b) River bed erosion caused by water from Spillways, the Canyon Wall outlets and the Tunnel Plug outlets (i.e. almost the maximum possible). (Source: [17] Fig.83C, Fig.87D)

had been circulated in the industry throughout the world via journal and conference papers. Despite the undoubted success of the model studies, the reports make it clear that things could have gone better. Many of the studies were only given the go-ahead when it would be too late for the outcome to recommend fundamental changes to the designs, for example, concerning the penstock and intake tower model tests:

> 'In no instance did the studies described in this report lead to drastic changes in design such as resulted from the model studies of spillways. All minor changes indicated in the penstock junction and the intake tower studies were not made because the design and construction had progressed to the stage where further changes were not economically feasible. However, results of the studies gave substantial evidence that the adopted design were satisfactory' [22].

This may have been largely due to the overconfidence in engineering theory that was normal at that time among design engineers and the academics who often supported them. Elsewhere, the authors of another report pointed to other influences: 'Expediting Boulder Dam project as an unemployment relief measure made necessary a decision on the general type of spillway without extensive tests on this model' [23]. Nevertheless, the introductory words to one of the reports gives a fair overview of the great contribution to both theoretical design methods and to the use of physical models:

> 'The preparation of plans for Boulder Dam involved the development of many new methods of theoretical analysis and masonry dam design, as well as the application of many existing theories to problems of greater magnitude than those for which they had been developed. In verifying the fundamental accuracy of new and existing theories the use of models took an important part, in both hydraulic and structural problems. In contrast to the hydraulic models, which provided direct empirical data, the principal function of the structural models was to furnish a check on analytical methods of design [24].'

Acknowledgements

I am grateful to Professor Yves Rammer of the Université Libre de Bruxelles who made high-quality scans of most of the images from original copies of the Boulder Dam reports.

References

1 Bureau of Reclamation (2008) *History Essays from the Centennial Symposium.* Denver, Colorado: Bureau of Reclamation. 258-263.
2 Billington, D.P., Jackson, D.C. & Melosi, M.V. (2005) *The history of large federal dams: Planning, design, and construction in the era of big dams.* Denver, Colorado: Bureau of Reclamation. 75-79.
3 Savage, J.L. & Houk, I.E. (1931) Checking Arch Dam Designs with Models: Results of tests on small-scale models of the Stevenson Creek and Gibson Arch Dams. *Civil Engineering* 1 (8): 695-699.
4 Bureau of Reclamation (1939) Model tests of Boulder Dam. *Boulder Canyon Project, Final reports. Part V (Technical Investigations), Bulletin 3.* Denver, Colorado: Bureau of Reclamation. 22.
5 Stuart, A.A. (1929) Toy dams to save lives. *Popular Science Monthly* 115 (2) (August): 47.
6 Beggs, G.E. (1928) Test of a Celluloid Model of the Stevenson Creek Dam. *Arch Dam Investigation.* ASCE.
7 Savage, J.L. & Houk, I.E. (1931) Checking Arch Dam Designs with Models: Results of tests on small-scale models of the Stevenson Creek and Gibson Arch Dams. *Civil Engineering* 1 (8): 695-699.
8 Bureau of Reclamation (1939) Model tests of Boulder Dam. *Boulder Canyon Project, Final reports. Part V (Technical Investigations), Bulletin 3.* Denver, Colorado: Bureau of Reclamation. 22-23. All the final reports are available online at https://catalog.hathitrust.org/Record/001959384 (consulted 26/10/2019)
9 Bureau of Reclamation (1939) Model tests of Boulder Dam. *Boulder Canyon Project, Final reports. Part V (Technical Investigations), Bulletin 3.* Denver, Colorado: Bureau of Reclamation.
10 Rogers, J. David (2010) Hoover Dam: Evolution of the Dam's Design. In: *Hoover Dam 75th Anniversary History Symposium: Proceedings of the Hoover Dam 75th Anniversary History Symposium, October 2010, Las Vegas, Nevada* (eds. Richard L. Wiltshire, David R. Gilbert & Jerry R. Rogers), 85-123. Reston, Virginia: American Society of Civil Engineers. 115-118.
11 Bureau of Reclamation (1939) Model tests of Boulder Dam. *Boulder Canyon Project, Final reports. Part V (Technical Investigations), Bulletin 3.* Denver, Colorado: Bureau of Reclamation. 225-248.
12 Bureau of Reclamation (1938) Trial load method of analyzing arch dams. *Boulder Canyon Project, Final reports. Part V (Technical Investigations), Bulletin 1.* Denver, Colorado: Bureau of Reclamation.

13 Bureau of Reclamation (1940) Model tests of arch and cantilever elements. *Boulder Canyon Project, Final reports. Part V (Technical Investigations), Bulletin 6.* Denver, Colorado: Bureau of Reclamation.

14 Bureau of Reclamation (1940) Model tests of arch and cantilever elements. *Boulder Canyon Project, Final reports. Part V (Technical Investigations), Bulletin 6.* Denver, Colorado: Bureau of Reclamation.

15 Bureau of Reclamation (1939) Model tests of Boulder Dam. *Boulder Canyon Project, Final reports. Part V (Technical Investigations), Bulletin 3.* Denver, Colorado: Bureau of Reclamation.

16 Bureau of Reclamation (1938) Model studies of spillways. *Boulder Canyon Project, Final reports. Part VI (Hydraulic Investigations), Bulletin 1.* Denver, Colorado: Bureau of Reclamation.

17 Bureau of Reclamation (1938) Model studies of penstocks and outlet works. *Boulder Canyon Project, Final reports. Part VI (Hydraulic Investigations), Bulletin 2.* Denver, Colorado: Bureau of Reclamation.

18 Zangar, Carl N. (1953) *Theory and problems of water percolation.* Engineering Monograph No.8. Denver, Colorado: US Bureau of Reclamation. https://www.usbr.gov/tsc/techreferences/hydraulics_lab/pubs/EM/EM08.pdf (12/05/2019)

19 Bureau of Reclamation (1938) Model studies of penstocks and outlet works. *Boulder Canyon Project, Final reports. Part VI (Hydraulic Investigations), Bulletin 2.* Denver, Colorado: Bureau of Reclamation. 107-118.

20 Lane, E.W., Campbell, F.B. & Price, W.H. (1934) The Flow Net and the Electrical Analogy. *Civil Engineering* 4: 510-514.

21 Bureau of Reclamation. https://commons.wikimedia.org/wiki/File:Hoover_Dam_1_(4307418516).jpg

22 Bureau of Reclamation (1938) Model studies of penstocks and outlet works. *Boulder Canyon Project, Final reports. Part VI (Hydraulic Investigations), Bulletin 2.* Denver, Colorado: Bureau of Reclamation. 14.

23 Bureau of Reclamation (1938) Model studies of spillways. *Boulder Canyon Project, Final reports. Part VI (Hydraulic Investigations), Bulletin 1.* Denver, Colorado: Bureau of Reclamation. 8.

24 Bureau of Reclamation (1939) Model tests of Boulder Dam. *Boulder Canyon Project, Final reports. Part V (Technical Investigations), Bulletin 3.* Denver, Colorado: Bureau of Reclamation. 9.

9

The role of models in the early development of Zeiss-Dywidag shells

Roland May

9.1 Zeiss-Dywidag shells: an absence of empiricism?

The building firm Dyckerhoff & Widmann (Dywidag – abbreviation for Dyckerhoff & Widmann Aktien-Gesellschaft) was one of the significant pioneers of reinforced-concrete construction in Germany. Internationally, the company attracted attention for the first time with the dome of the Centennial Hall in Breslau (today Wrocław), whose span of 65 m marked a new world record for reinforced-concrete structures. A few years later, the company substantially increased its international reputation when it started to devote itself to the systematic development of thin concrete-shell construction. The *System Zeiss-Dywidag*, initiated together with the optics manufacturing company Carl Zeiss, would become pivotal for the worldwide dissemination of thin concrete-shell construction in the second quarter of the twentieth century [1].

Franz Dischinger (1887-1953) and Ulrich Finsterwalder (1897-1988), the Dywidag engineers who played a key role in this process, occupy a prominent place in the history of shell construction. Initially, contemporaries had marvelled at their complex calculation methods, some of which were guarded like state secrets and seemed like 'a sort of transcendental mystery only divested by the German school' [2]. But once the secrets were decoded, criticism began to grow, and nowadays the names Dischinger and Finsterwalder rarely evoke enthusiasm.

A key reason for this remarkably cool treatment may be found in Denis P. Gardner's statement that civil engineers such as Torroja, Maillart, Candela and Isler 'stressed the pleasing aesthetics of thin shells', while 'the Germans emphasized numeric calculations that would bolster their thin-shell plans' [3]. Gardner's allegation has to be seen in the larger context of a smouldering conflict between the deductive method based on abstract statical calculations and the idea of a 'statical sense' based on practical experience and observations contained in inductive approaches. And Dischinger, according to Jörg Schlaich, in some way represented the former attitude in Germany: 'Dischinger was, so to speak, the opponent of Mörsch [...]. Mörsch, the pragmatist, the empiricist, the sensitive man and Dischinger, the rational man, the logical, systematic thinker' [4].

Physical Models: Their historical and current use in civil and building engineering design,
First Edition. Edited by Bill Addis.

A number of commentators, including Anton Tedesko and David P. Billington, categorized Dischinger – along with Wilhelm Flügge and Eric Reissner – as an 'engineer-analyst' of reinforced-concrete construction from the 'strong theoretical German tradition'. In contrast, they saw personalities such as Robert Maillart, Pier Luigi Nervi and Eugène Freyssinet as creative 'engineer-designers' [5]. However, although they also ascribed Ulrich Finsterwalder to the latter group, the image of 'tireless number crunching' [6] detached from all creative and practical construction questions, became increasingly associated, internationally, with the Dywidag shell pioneers. To this day, this accusation often still sticks to Zeiss-Dywidag shell constructions.

There has been some ambiguity in attitudes to the 'German tradition' of theoretical analysis. On the one hand it has been viewed critically [7], while on the other, the importance of model testing – and thus of empiricism – in the development of the Zeiss-Dywidag shells has also been acknowledged [8]. Of course, the two views are not incompatible. As Lutz Schöne has observed, the abstract theoretical work of the Dywidag engineers was an 'apparatus of analysis and sensitive training' without which a 'constructive-empirical approach' would hardly have been possible; cylindrical shells, for example, whose load-bearing behaviour contradicted any experience in nature, had only become 'conceivable for engineers as a construction element' through theory [9]. However, before a practical implementation could be considered, the knowledge gained deductively would have required 'empirical verification in a downscaled experiment' [10]. According to Billington and Hines, both in Germany and, subsequently, in the case of the American Zeiss-Dywidag licensee Roberts & Schaefer, the 'basic pattern for innovation' consisted of 'four overlapping steps':

> 'First, designers performed simple calculations based on the membrane theory in order to determine basic dimensions for proposed shells. Second, they constructed models and structures based on these calculations to verify their assumptions about the proposed shell's behaviour. Third, they developed a more exact bending theory of thin-shell behaviour based on qualitative observations and numerical results from such tests. Finally, both Dyckerhoff & Widmann in Germany, and Roberts & Schaefer in America, monitored actual structures and applied their observations to future designs' [11].

9.2 The Jena models: domes

In fact, the early phase of Zeiss-Dywidag shell construction contributed more to the development of structural engineering than using model tests to check complex theories for their practical suitability. It built substantial bridges between the abstract theories of applied mechanics and the practice-oriented calculation methods of civil engineers. Hence, Karl-Eugen Kurrer defines this early phase as both a springboard for the so-called 'invention phase of theory of structures' and as a milestone for the 'second developmental stage of industry-type science [...] in which the technological side of construction prepares itself to have a significant influence on theory' [12].

It was the impact of Walther Bauersfeld (1879-1959) that was decisive for the creation of this new path. Bauersfeld had studied mechanical engineering at the Royal Institute of Technology in Charlottenburg where he had also earned his doctorate in 1904 with a dissertation on the regulation of water turbines. Only a year later he was working as a design engineer for the Carl Zeiss company in Jena. After a brief interlude at the Research Institute for Aeronautical Navigation in Berlin, he joined Zeiss in 1908 as a member of the management board. Under his direction, numerous inventions and patents were developed which, in addition to precision mechanics, were anchored in various fields of optics, photogrammetry and cinematography.

In 1919, Bauersfeld took on the task of developing a new type of projection planetarium for the *Deutsches Museum* in Munich, an ongoing project on the brink of failure. Three years later, this project was nearing completion and an experimental building was needed to test the projection apparatus. For this purpose, Bauersfeld developed a spatial latticework system based on triangulation of a spherical surface that could produce a structure with an almost perfect hemispherical shape. This structure consisted of 3840 rigidly-connected steel bars with 51 different lengths and, despite its low weight, proved to be remarkably stiff and resistant.

To find a weather-resistant mantle for the projection dome, advice was sought from August Mergler, chief engineer at Dywidag's Nuremberg branch which was regularly contracted by Zeiss. Mergler suggested encasing the core steel structure with a thin layer of concrete using the novel shotcreting technique. He also recognized the potential of the method to 'open a quite new way for roofing over large halls' [13]. Thereupon, at the end of 1922, Franz Dischinger, chief engineer at the Dywidag headquarters in Biebrich near Wiesbaden, was sent to Jena to explore jointly with Bauersfeld the possibilities of the new construction method.

Subsequently, an experimental planetarium was erected on the roof of building B 11 at the Zeiss main factory in Jena during the winter of 1922/23 (Figure 9.1). While of special interest because it was the first Zeiss-Dywidag shell, there also arises, in the current context, the question as to whether it can also be regarded as a model. With respect to the planned use, this issue can be affirmed, despite the fact that the Jena structure, with a diameter of 16 m, exceeded the dimensions of the planned Munich project by 6 m. Concerning the structural design, it is less clear how to classify the building because Bauersfeld's research interest had originally not been directed at questions of structural design. In the end, with regard to both this aspect and the acoustic problems of such circular buildings, the experimental structure acquired the character of a model only later in the story.

Following this pioneering construction, Bauersfeld and Dischinger jointly sought to explore the potentials of the new construction system with a cupola above a new building at the glassworks Schott & Genossen, a subsidiary of Carl Zeiss, also located in Jena. With a diameter of 40 m, the dome spanned much farther than the experimental planetarium and also had a much lower rise-to-span ratio of 1:5. Although the shell thickness was almost doubled to

Figure 9.1 Experimental planetarium, Jena, 1922/23. The Zeiss steel bar latticework can still be perceived under the concrete layer. (Source: [14])

6 cm, critical questions arose with regard to buckling safety both for the steel latticework and also for the final reinforced-concrete shell. To answer these questions, the field of applied mechanics offered Robert Zoelly's formula for the buckling resistance of spherical shells against uniformly distributed and radially directed external pressure [15]. Especially with regard to the impact of point loads, Bauersfeld was keen to make sure that he also had his own theoretical justification before construction began. For this purpose, he hired the young, mathematically talented civil engineer Josef Geckeler, who had received his doctorate studying under Sebastian Finsterwalder in Munich.

In retrospect, Bauersfeld stressed that the methods 'of calculating the strains in shell constructions differ largely from the methods used in architectural work'; for a mechanical engineer, however, they 'were not new' [16]. In fact, his future son-in-law, Geckeler, tackled the task under Bauersfeld's guidance more like a fundamental research scientist than a civil engineer interested in findings that could be put to practical use. His starting point was the approach already used for the experimental planetarium which was based on earlier ideas of Johann Wilhelm Schwedler (1823-1894), viz.: Not to determine the stresses in the lattice shell with the usual methods for iron trusses, but to use a homogeneous metal sheet as a substitute in the calculation. On the basis of this idea, it was reasonable to verify the theoretical investigations through the use of models made with metal sheets.

One of Geckeler's recorded models dealt with the consequences of temperature stresses in the shallow dome. Geckeler chose a semi-ellipsoid pressed from brass sheet with a base diameter of 44 cm. It had a mostly uniform thickness of 0.9 mm in the area of the edge stresses, but then increased to 1.07 mm towards the crown due to the manufacturing method (Figure 9.2). This model was exposed to internal pressure and the resulting elastic displacements of individual points on the shell, normal to the surface, were measured using 37 Zeiss dial gauges

Figure 9.2 Semi-ellipsoid brass sheet model for the Schott dome with Zeiss dial gauges, c.1924. (Source: [17])

mounted on twelve brackets. Since the intention was to achieve results of general validity, the shape of the semi-ellipsoid only approximated that of the planned dome. Due to the special importance of this shape of the vessel, a major-to-minor axis ratio of 2:1 was chosen. Notwithstanding the irregularities of the model in terms of wall thickness, supports and general shape, the measured results generally confirmed an approximate theoretical solution developed by Geckeler.

Particularly critical for the shallow Schott dome was the danger of buckling in the support area. Geckeler investigated this problem using a whole series of models. Once again, he used semi ellipsoids pressed from brass sheet, which this time were joined together to form rotational ellipsoids. Also in this case, the tests confirmed Geckeler's approximate solution, although variations in shape, wall thickness and material properties, which were difficult to avoid, led to a number of deviations. In particular, the expected undulation often did not form along the entire circumference, as a strong fold arose at the weakest point (Figure 9.3). Geckeler's experiments, which also included measurements on vessel models with a diameter of 30 cm and a sheet thickness of 0.28 mm, brought several new insights. Specifically, it was found that the critical values calculated using Zoelly's buckling formulae were in some cases overestimates by a factor of four.

In contrast to previous studies by Ernst Meissner and his students at ETH Zurich, Geckeler's investigations aimed – despite their high degree of idealisation – at approximate solutions for practical applications. For this work he soon received support from his friend Ulrich Finsterwalder, who had joined Dyckerhoff & Widmann at the beginning of December 1923 and was sent directly to Jena as their site engineer. Owing to their time-consuming systematic approach, many of the results came too late for the Schott dome. For example, it was not foreseen that around the turn of the year 1923/1924 waves would appear in the lattice in

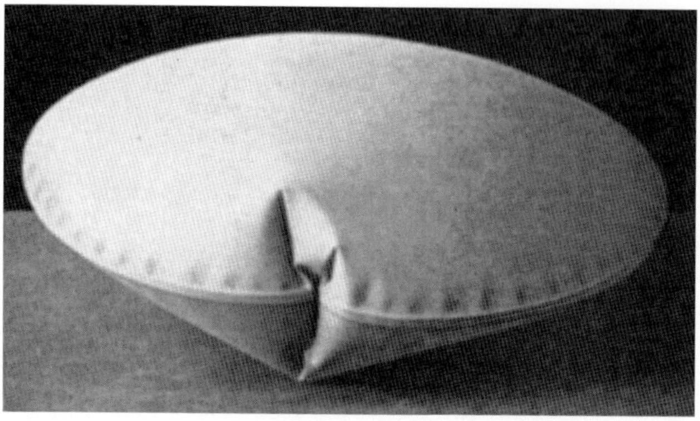

Figure 9.3 Ellipsoid brass sheet model with kink fold across the equator caused by internal overpressure, c.1924. (Source: [18])

the region of the support while the structure was being assembled. Moreover, due to the lack of a theoretical basis, a 'civil engineer's solution' by Dischinger had to be applied that prevented bending in the support area by means of a transition curvature.

9.3 The Jena models: cylindrical shells

Models used in the development of the cylindrical (barrel) shells, which entered new territory in many ways, were much more important than those used for the rotational domes. According to Lutz Schöne, a 'design doctrine' (*Konstruktionslehre*) was developed in the interplay between theories derived from the appropriate differential equations and experiments undertaken to verify them, which can be regarded as the 'prime example' for the 'Second Modernism in Engineering' [19].

The driving force behind this development was apparently Franz Dischinger. After he had studied civil engineering at the *Technische Hochschule,* Karlsruhe (Karlsruhe Polytechnic), Dischinger had been working at Dywidag's headquarters in Biebrich near Wiesbaden since 1913. In 1921, he was appointed Chief Engineer and took responsibility for large-scale bridges and pneumatic caisson foundations. In November 1922, Dischinger was sent to Jena to investigate the potential of thin-shell construction. He was thus looking for possibilities to open up a much broader field of application by adapting shells to structures with rectangular ground plans. Dischinger's first proposals were aimed at virtually cutting the rotational domes at the sides, and stiffening them along the cut edges, because he assumed that the loads would concentrate in the four corners and be transferred vertically. However, he was unsuccessful in this attempt to calculate the behaviour of such doubly-curved structures because of the mathematical complexity.

This led Dischinger in spring 1923 to transfer the concept to shells loaded in uniaxial bending. Equilibrium of the forces was to be ensured only through stresses in the plane of the shell, whereby its thinness should prevent the

Figure 9.4 Model of a cylindrical steel-lattice shell by August Föppl, c.1895. (Source: [21])

occurrence of significant bending stresses. Bauersfeld initially rejected this idea, but Josef Geckeler reported that a model tested by August Föppl as early as 1891 at the *Technische Hochschule,* Munich had followed the same line of thinking. Föppl had described for the first time the surface-active load-bearing capacity of a simply curved structure that was stiffened at each end [20]. Directly after his appointment as professor of applied mechanics, he had arranged for a model of a cylindrical steel-lattice shell with a base area of 2.80 x 3.80 m to be constructed at the *Technische Hochschule,* Munich (Figure 9.4). Föppl had reported in detail the various load tests carried out with this model, as well as tests on a small experimental building [21]. However, his work remained unnoticed beyond narrow expert circles, and thus should be seen as paradigmatically representing the entirety of early research on shell theories, which developed almost independently of actual construction activities.

By 1923, this situation had now fundamentally changed. As Dischinger considered the work on the cylindrical shells to be 'of highest importance' [22], Geckeler initially attempted to base his calculation on a framework corresponding to the structure of the Zeiss latticework that was embedded in the concrete. However, like Föppl with his 'cylindrical lattice shell' (*Tonnenflechtwerk*) in around 1895, Geckeler too failed to derive general laws from the individual bar forces. More promising was Bauersfeld's approach to replace the latticework once more by an imagined continuous sheet-metal surface. His preserved notes bear witness to how Bauersfeld now committed himself to the development of a first membrane theory for cylindrical shells with circular as well as elliptical cross-sections [23]. Of particular interest to him were the stresses and elastic deformations of the loaded cylindrical shell. His theoretical investigations confirmed Dischinger's original idea that the displacements perpendicular to the surface were only slightly greater than in the tangential direction, as long as the shell was not too large in the cylinder axis direction.

As with the Schott dome, Bauersfeld carried out tests in the autumn of 1923 with a brass-sheet model to verify his calculations. That model was a vessel of elliptical cross-section with a ratio of the semi-axes of $\sqrt{2}:1$ (Figure 9.5). It was stiffened by rigid end walls that were anchored against each other. Once this model was exposed to internal pressure by means of a liquid, Bauersfeld was

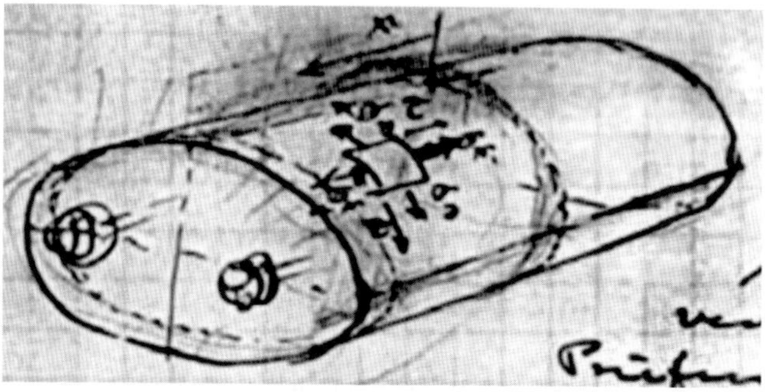

Figure 9.5 Brass-sheet model vessel with elliptical cross-section, sketch by Walther Bauersfeld, c.1923. (Source: [24])

astonished to find that the maximum deformations perpendicular to the sheet metal skin were indeed almost equal to those along the surface.

Now it was at least approximately proven that a cylindrical shell stiffened by walls had a spatial load-bearing capacity similar to that of shell domes, differing fundamentally from traditional cylindrical vaults. In their ongoing collaboration, Bauersfeld and Dischinger realized that bending moments caused by effects at the shell edges could be minimized by choosing an appropriate cross-section curve. Cross sections that were elevated above the catenary curve and had vertical end tangents proved to be the optimum. On 22 January 1924, the joint work resulted in the filing of a patent application for a 'purlin-free reinforced-concrete barrel roof' by the Carl Zeiss company [25].

On the very same day, Bauersfeld also requested permission from the City of Jena to change the plans for Building 23, already under construction at the Zeiss South Works in Jena [26]. Its roof, which had already been planned in the shape of a barrel, was now to be constructed according to the new ideas. Like the domes, this cylindrical shell consisted of a Zeiss latticework that was later embedded in concrete. Due to the lack of experience and the remarkable dimensions of the experimental construction, stiffening arches were placed at relatively close spacing for reasons of security. Thus, the structural capabilities of long barrel shells, which had already been identified, were by no means yet fully exploited (Figure 9.6).

It was Ulrich Finsterwalder, son of the renowned Munich geometry professor Sebastian Finsterwalder, who would shape the further development of cylindrical shells. Coming straight from university, he had joined Dyckerhoff & Widmann a few weeks before the patent application. At the suggestion of his friend Josef Geckeler, Finsterwalder had already devoted the final phase of his civil engineering studies at the *Technische Hochschule*, Munich, to working at the chair of Ludwig Föppl, who had followed his father August in 1922. In the course of analysing stiffened tubes for various load cases, he thereby – almost simultaneously with Bauersfeld – had also already developed his first membrane theory of the cylindrical shell [28].

The problem now was that such a membrane state could only occur if strict support conditions were fulfilled, a conclusion that resulted inevitably from

Figure 9.6 Construction works at Building 23 of the Zeiss South Works in Jena, 1924. The Zeiss latticework is mounted between the formwork for the stiffening arches. (Source: [27])

the differential equations. Particularly problematic were the tension members required to absorb the shear forces along the edges of the shell. Their tensile forces increased greatly with the distances between the stiffening elements but, at the same time, membrane theory demanded that the tensile members should be rigid and weightless. Of course, this was physically impossible, so that certain bending stresses near the edges had to be accepted.

It was precisely when Finsterwalder arrived at Jena that the question concerning both the quality and the quantity of these bending stresses had to be answered. The wavelengths of the bending disturbances were significantly larger in the case of the cylindrical shells than in the case of the rotational shells that were being investigated in parallel. As a result, the additional stresses were not limited to the outermost edge zones and therefore had to be tracked across the entire roof surface. Even though Finsterwalder proved himself to be 'very clever in mathematical calculations' [29], the computational work turned out to be far too complex for practical application. Under Dischinger's guidance, Finsterwalder therefore developed new sheet-metal models to determine approximate values for the bending moments flowing across the cross-section. Unfortunately, up to now, no information about the appearance of these models or the exact arrangement of the experiments carried out in the course of 1924 has been traced. Probably again with the help of Zeiss dial gauges, Finsterwalder apparently succeeded in proving that the bending moments in this case also had, primarily, the character of edge disturbances [30]. In addition, there are indications that, in this context, he had already investigated the possibility of using shallow circular segments with edge beams on both sides, in which slight bending moments were tolerated [31].

9.4 Basic research in Biebrich

After about a year, the cooperation in Jena, described by Bauersfeld as 'an excellent example of team work' [32], ended. Bauersfeld returned to his original tasks at Zeiss and Finsterwalder relocated to Biebrich, where he continued his work under Dischinger's direction at Dywidag in a department set up specifically to work on shells. In addition to the practical design of various shell domes for planetaria, the basic research work from Jena was now continued with experiments that were closer to practical conditions.

For example, Dywidag erected a shipbuilding hall at its own shipyard in Neuss which also served as a model (Figure 9.7). The cylindrical shell, some 4 cm thick and around 30 m long in total, spanned about 12 m in the transverse direction. Presumably it was divided into two sections in the longitudinal direction by means of stiffening walls. According to a document apparently written by Finsterwalder, the deflection of the crown line during decentring was only 0.4 mm. This was taken as clear proof that 'the arched roof stretched between rigid constructions does not carry the loads as a slab but almost exclusively as a shell' [33].

At the same time, in the summer of 1925, at least two model shells in reinforced concrete were erected on the Dywidag premises in Biebrich, of which one is documented only by photographs. This was a cylindrical shell with a length of about 4 m and possibly only 1 cm thick, mounted on four supports with stiffening walls at both ends [35] (Figure 9.8). Its design looks like a direct physical realisation of

Figure 9.7 Shipbuilding hall at the Dywidag shipyard in Neuss on the River Rhine near Düsseldorf, 1925. (Source: [34])

Figure 9.8 Dischinger and Finsterwalder seated on a model cylindrical shell in reinforced concrete at the Dywidag premises in Biebrich, 1925. (Source: [36])

Figure 9.9 Schematic drawing of a Zeiss-Dywidag cylindrical shell, c.1927. (Source: [37])

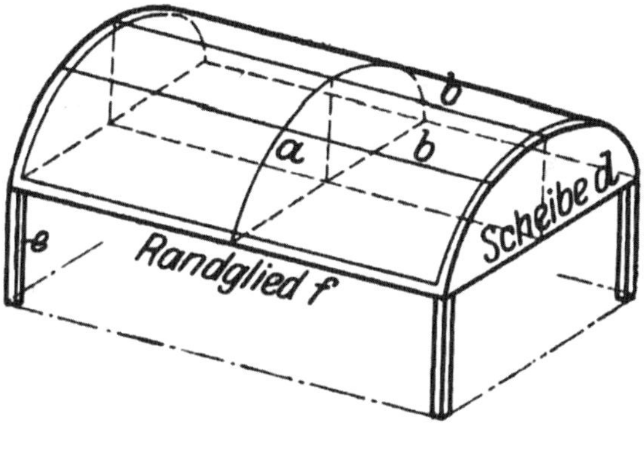

a schematic drawing that was subsequently published several times to explain the Zeiss-Dywidag barrel roof (Figure 9.9).

The second experimental structure was stiffened by two internal walls, 6 m apart. On both sides there was a cantilevered section extending 2 m beyond the supports. The 4 m wide shell had edge members 8 x 10 cm thick, and its ends were flanged to avoid buckling phenomena (Figure 9.10). Concreting of a Zeiss latticework was now omitted. Instead, the 1.5 cm thin shell was reinforced with a conventional steel mesh with 3 mm diameter wire at 8 cm centres, which was supplemented by wires to carry the highest tensile stresses at an oblique angle.

Figure 9.10 Second model cylindrical shell in reinforced concrete at Biebrich, 1925. Load test with sandbags weighing a total of 18 tonnes. (Source: [38])

Various load tests were carried out on both models with the help of sacks filled with sand. The first model was subjected to an area load of about 4 tonnes as well as a concentrated load of 3.2 tonnes in the central crown area. In the second model, the total load was increased to 20 tonnes (2 tonnes dead weight + 18 tonnes superimposed load) under the supervision of the local building inspectorate, corresponding to a considerable 500 kg per square metre (5 kN/m^2) on the plan area. No hairline cracks were observed and the deflection was hardly measurable. This again confirmed the theory that the transmission of forces to the stiffening elements was almost exclusively effected by tensile and compressive stresses in the shell surface, i.e. by membrane action. In addition, load tests on the boundaries showed that the dead weight of the edge members caused only insignificant bending moments.

Based on their theoretical preliminary work, the Dywidag engineers were certain that the same stresses would continue to occur on a larger scale if the load per square metre remained the same. Therefore, after about three years of research and development, the company ventured into the public eye. In the spring of 1926, the first roof with Zeiss-Dywidag barrel shells in a public arena was built for the 'Dywidaghalle' at the exceptionally well-attended GeSoLei exhibition in Düsseldorf (Figure 9.11). However, the superstructure, presumably developed mainly by Finsterwalder, revealed some of the fundamental problems of the initial concept. For example, concreting of the vertical sections of the adjacent shells proved to be extremely complicated. In addition, deep valleys were formed between the shells, which carried the danger of forming snow traps. Problems also arose in producing the complex elliptical geometry of the Zeiss latticework, which now served only to shape a conventional timber formwork.

During the same year, a further test program with sheet-metal models was launched in Biebrich. Based on the knowledge gained in 1924, Finsterwalder investigated the possibilities of cylindrical shells with cross-sections made up of shallow circular segments. Shells with semi-elliptical cross-sections were capable of absorbing the vault thrust mainly by membrane stresses. In the case of flatter cross-sections, on the other hand, a very high vault thrust was expected that could not be carried by the edge beams. Nevertheless, Finsterwalder assumed

Figure 9.11 Dywidaghalle at the GeSoLei exhibition in Düsseldorf, 1926. In the foreground can be seen that here the Zeiss latticework was used only to support a timber formwork. (Source: [39])

that, for the case of a flexible edge beam, the shell would behave like a 2-web tee beam with a curved slab. In this case, it would continue to transfer the thrust almost exclusively by means of membrane stresses to the stiffening elements. The type and dimension of the bending stresses arising during this process were, however, unclear to him.

For his experiments, Finsterwalder used two types of models (*Versuchskörper*) with lens-shaped cross-sections, which were prepared using metal sheets of 1 mm and 1.5 mm thickness (Figure 9.12). Unfortunately, Finsterwalder and Dischinger were rather vague in describing both the models and the investigations that were carried out. In addition, Anton Tedesko's papers include depictions of the models with dimensions slightly different from the published ones [40]. In one model type, which was hardly described at all, two elements with segmented circular cross-sections were firmly connected in order to simulate, as Lutz Schöne suspected, 'a circular cylindrical shell which typically is vertically supported along its boundaries by an edge member' [41]. The circular segments of the second model type had flanged edges. Furthermore, 3 mm gaps between the two elements were left, and sealed with Plastoline. The aim was to simulate an edge support to the shell that was flexible in the transverse direction. All models were subjected to external pressure by evacuating the internal void. With the aid of Zeiss precision dial gauges, the deformations were measured and the value of the bending stresses occurring in the shells could be estimated.

On this basis, Finsterwalder developed an early theoretical approximate solution, according to which even the new shell type, when correctly shaped, developed only minor bending stresses. A decisive role in this case was played by the edge beams, whose main stiffness had to be normal to the plane of the surface of

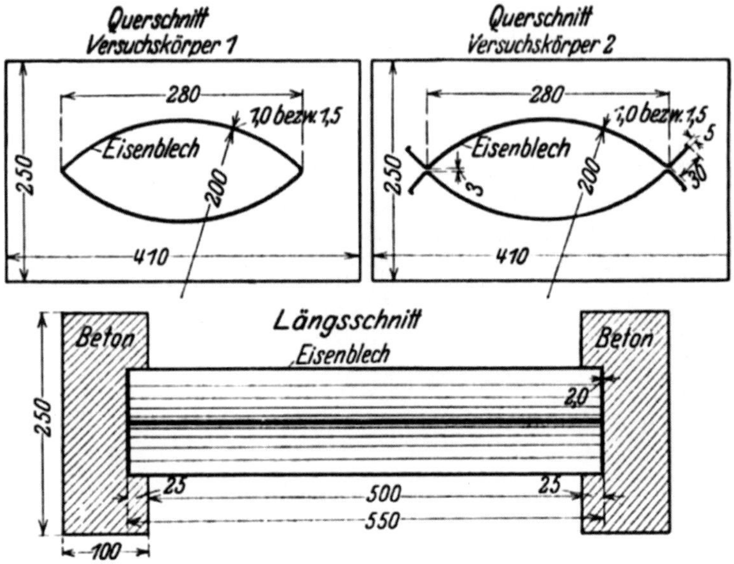

Figure 9.12 Metal-sheet models with lens-shaped cross-sections made of circular segments, c.1926. (Source: [42])

the shell and whose height ensured that the overall cross-section was still elevated in relation to the catenary curve. By neglecting bending moments in the direction of the generatrix, the associated transverse forces and the torsion moments, Finsterwalder subsequently created a calculation method which made it possible to calculate approximately shell roofs with varying cross-sections in the form of circular segments and with any elastic support of the edges. This approximate solution was central to his dissertation submitted to the *Technische Hochschule*, Munich, in early 1930 and also formed the basis for the calculation of numerous shells with segmented circular cross-section developed afterwards by Dywidag and other designers, including Eduardo Torroja.

9.5 Models of large-scale buildings

Finsterwalder's new solution made it possible for engineers to respond to many functional and architectural requirements. For the first large-scale building, however, his findings came too late, as Dywidag had already been awarded the contract for the wholesale market hall in Frankfurt am Main at the end of 1926. The execution of the hall, with 15 cylindrical shells, each 36.9 m long, was to be carried out together with the Frankfurt-based competitor Wayss & Freytag. Acting as a consultant to Wayss & Freytag, Emil Mörsch inspected the Biebrich models and on this occasion received an 'hour-long talk about the theory and the experiments carried out' from Dischinger [43]. Mörsch, however, responded laconically: 'If you build this on a large scale, then it will collapse' [44]. This scepticism, in addition to concerns by the building authorities, was probably responsible for the

Figure 9.13 Reinforced-concrete model for the Frankfurt wholesale market hall, 1927. (Source: [46])

request to test a model specifically designed for the Frankfurt case. Erected in the spring of 1927, northeast of the already-prepared construction site, this experimental structure opened up a new phase for the Zeiss-Dywidag model testing program. While earlier models were mainly to support the formation of general theories, they were now specifically oriented towards planned buildings.

The Frankfurt model had a length of 12.3 m, a vault span of 4.7 m with a rise of 2 m, and a shell thickness of 2.7 cm (Figure 9.13). Including even the arrangement of the steel reinforcement, it formed a realistic replica of the planned structure at the large scale of 1:3. Geometrical differences existed, however, in the substructure, and the multi-layer reinforcement in the actual structure was modelled by overlapping nets of steel inlays. Precise measurements were taken during the various load tests carried out under the direction of the assessors Heinrich Spangenberg and Adolf Kleinlogel. A paper by Kleinlogel, whose assistant Kálman Hajnal-Kónyi was responsible for supervising the experiments, can be regarded as the most comprehensive description of a Zeiss-Dywidag model shell and the associated tests [45].

The load tests revealed some awkward aspects of model experiments. As already mentioned, it was assumed that the stresses in the shell were dependent on the respective surface loads and the ratio of shell thickness to curvature, without regard to absolute dimensions. The snow and wind loads to be simulated could therefore be applied at their full-scale values by means of sandbags. In addition, however, loads corresponding to twice the dead weight of the model shell had to be applied because its own weight per unit area was one third that of the actual shell's weight per unit area. Furthermore, additional loads were applied to simulate the plasterwork and insulation boards that had not been built into the model. The edge members were subjected in turn to additional line loads which, in contrast to the distributed loads, had to be reduced in the model ratio in order to produce the same stresses as in the planned structure. Since the edge beam of the model weighed only one ninth of the full-size structure per unit length, additional loads had to be applied that were twice the dead weight. Finally, the load resulting from the adjacent skylights had to be applied in addition, in a 1:3 model ratio.

In accordance with these specifications, on 11 April 1927, after only nine days of curing, the full load of the official load case for dead weight and snow, totalling 313 kg per square metre on the ground plan area, was applied. Subsequently, the surcharge was increased by a further 3 tonnes, corresponding to 52 kg/m^2 or an increase of 50% to the snow loading. After 10 days, the sandbags were removed. It was found that the small deformations of the crown did not disappear completely after unloading, a phenomenon which Kleinlogel attributed to temperature influences. A few days later, a second load test was carried out until the first hairline cracks appeared in the edge beams. This time, the shell was first loaded on one side and then over the entire surface with up to three times the design snow load. Subsequently, the theoretical maximum load of 586 kg/m^2 was imposed on the model, which was then gradually increased up to 964 kg/m^2. The maximum deflection, however, reached only 1/3500 of the length of the shell.

Finally, the worries regarding the execution of the full-size structure, with cylindrical shells only 7 cm thick, disappeared, and the Frankfurt wholesale market hall marked the breakthrough for Zeiss-Dywidag shell construction (Figure 9.14). Furthermore, the shells in the market were built with a circular cross-section for the first time – apparently due to an unauthorized decision by Finsterwalder [47]. Unfortunately, it is not clear from the descriptions whether this was also already the case with the model.

In his discussion of the Frankfurt model, Adolf Kleinlogel concluded that buckling safety was now 'practically guaranteed' [49], although this had only been proven theoretically through an approximate solution. The specific problem of buckling resistance had also been considered only marginally in the previous models [50]. This was to change during the preparations for the

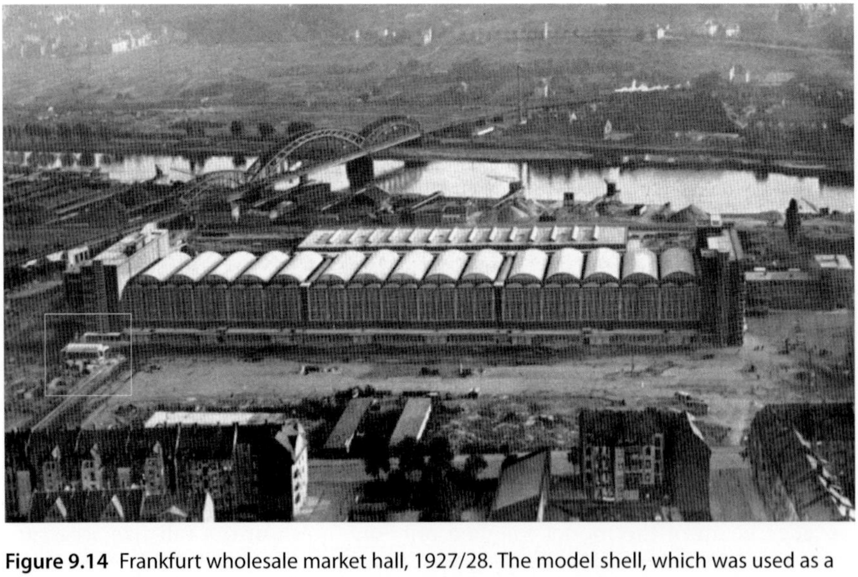

Figure 9.14 Frankfurt wholesale market hall, 1927/28. The model shell, which was used as a bicycle shed after the tests, can be seen at the left end of the building. (Source: [48])

Figure 9.15 Architectural model of the Leipzig wholesale market hall, 1927. Only two of the three planned domes were actually built. (Source: [51])

construction of the Leipzig wholesale market hall, which was largely in the hands of Franz Dischinger and Hubert Ritter. Dywidag had won the contract for this building in the summer of 1927 with a design that included the world's largest reinforced-concrete domes at that time (Figure 9.15). In the shape of an octagonal cloister vault, these domes were made up of the segments of four intersecting barrel shells. At the ridge junctions between these sections, stiffening ribs were inserted, whose function corresponded to that of the stiffening walls in cylindrical shells. Counterparts to their self-supporting shell edges were here the impost areas of the cupola segments, each of them more than 27 m long. The dimensions of the segments, which were designed for a thickness of 9 cm, and had only slight curvature, caused concerns about their buckling safety.

During the structural design, the Dywidag engineers had considered a strict Robert von Mises solution for the buckling of stiffened cylinders, which resulted in a safety factor of around six. Yet the solution was verified for only parallel stiffeners and for spacings between them that were considerably larger than the radius of curvature. Since von Mises' equations were dependent only on the basic proportions, it was decided, for safety's sake, to carry out a test on a 1:60 scale model (Figures 9.16 and 9.17). Unlike the planned cupolas, the model was mounted upside down and consisted of a sheet-steel dome with a wall thickness of 1.5 mm, which was anchored in a reinforced concrete block. It was surrounded by a slightly larger second dome and water was pumped into the void between, until buckling occurred in the inner dome. The test confirmed the expected safety against buckling. However, buckling waves, whose dimensions were measured by dial gauges, had already formed before failure. This unexpected effect was problematic for the real domes in reinforced concrete because of their

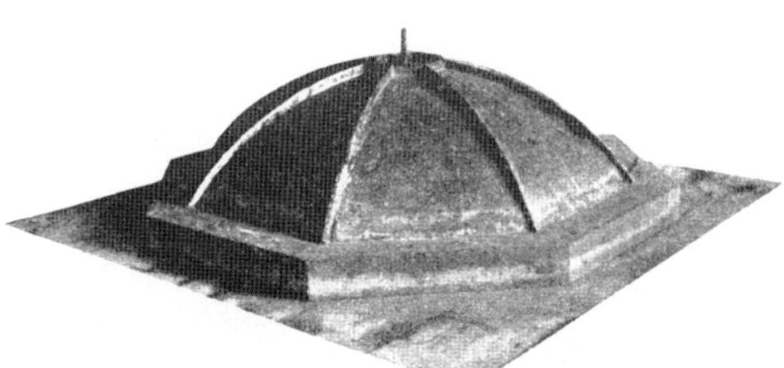

Figure 9.16 1:60 scale model for the dome of the Leipzig wholesale market hall, 1927. (Source: [52])

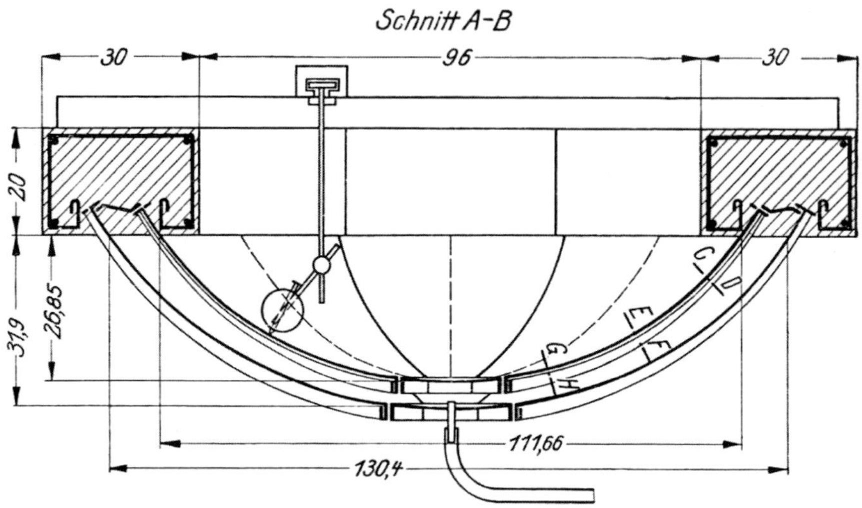

Figure 9.17 Cross-section of the 1:60 scale model and test rig for the Leipzig dome. The model was made with steel sheet 1.5 mm thick and loaded in compression by hydrostatic pressure. (Source: [53])

inferior relationship between permissible stress and Young's modulus. It was therefore decided to insert additional stiffening ribs in the middle of the shell segments.

Besides the model dome, another structural model was employed, which was only indirectly related to the shell structure. It aimed to simulate the main lines of force in the cellar ceiling that acted as a tie between the eight inclined columns of the domes. The model was made of a thick sheet of rubber measuring about 10 x 10 cm, which was pulled radially outwards at eight points (Figure 9.18). The test served as an important aid for the subsequent reinforcement layout.

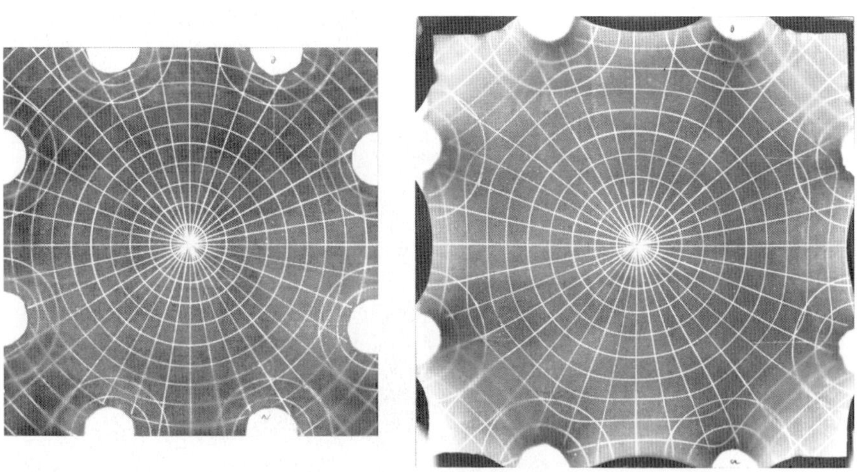

Figure 9.18 1:600 scale (approx.) rubber-sheet model of the cellar ceiling of the Leipzig wholesale market hall, 1927. State before (left) and during (right) the application of radial tensile loads. (Source: [54])

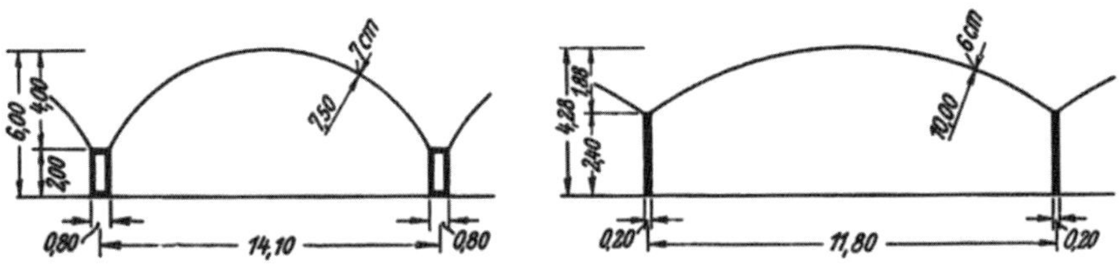

Figure 9.19 Cross-sections of the Frankfurt (left) and Budapest (right) wholesale market halls. (Source: [55])

A combination of several models was also used during the design of another major project with cylindrical shells – the wholesale market hall for Budapest. Designed from 1929, the building's fundamental disposition was clearly related to that of the Frankfurt Hall. The latter had been created in the early days of Finsterwalder's intensive examination of the cylindrical shell with edge beams, while the Budapest hall represented, in a certain sense, the culmination of this structural form (Figure 9.19). Nevertheless, some questions still remained unanswered. In particular, the safety against buckling in the direction of the generatrix for the segmented circular cylindrical shells, which had been increased to a length of 41 m in Budapest, was by no means adequately demonstrated using theory. Also, possible superimposition effects had to be investigated. To evaluate the buckling behaviour, another double-shell steel-sheet model was created, this time at a scale of 1:30 (Figure 9.20). As in the Leipzig model, the load on the shell was applied by

Figure 9.20 1:30 scale steel-sheet model for the Budapest wholesale market hall during a load test, c.1930. The double-shell model (at the top of the photograph) was subjected to internal water pressure and the edge beams were loaded by suspended weights. (Source: [56])

pressurised water, while the load on the edge beams was simulated by suspended weights. A fourfold safety factor against buckling was deduced from the tests.

In a further experiment, the edge beams were examined in detail. According to Finsterwalder's theory, they required a large height because – just as in the case of the elliptical shells – the cross-section of the overall system should be elevated in relation to the line of thrust. At the same time, the web width should ideally be only twice the thickness of the shell at the junction with the edge beam. This resulted in the unusually slender dimensions of 2.40 x 0.20 m for the girder which led to two rather unorthodox solutions in the reinforcement concept: on the one hand, an exceptionally large number of tension reinforcement layers had to be arranged one above the other; on the other hand, the rebar stirrups for shear reinforcement were much thinner than usual. These unconventional arrangements were tested for practicality before construction began using an equally unusual model. Since it was not wanted to replicate a complete shell, they designed a T-girder with a span of 6.35 m. Its geometry was not strictly proportional to the later design, but had been optimised for almost identical stress ratios. This beam was successively loaded with up to 3.75 times the normal area load (Figure 9.21). The resulting crack pattern largely confirmed the earlier expectations.

Another model was created for the canopy roof planned to run along the entire south-western longitudinal front. The roof was conceived as a 6.10 metre-deep cantilever shell structure working in three-dimensions. In the transverse direction, it spanned between the main hall's gallery girder and an outer edge beam only 40 cm in height; in the longitudinal direction, it stretched continuously over transverse girders that protruded from the columns of the main hall. However, the test engineer, Gábor Kazinczy, found the statical calculations unconvincing. And so Kazinczy, the pioneer of the ultimate-load method, ordered the making of a model in reinforced concrete measuring about two thirds of the full size,

Figure 9.21 Test beam for the Budapest wholesale market hall during load test, c.1930. (Source: [57])

for which unfortunately no image is known. This model was successively loaded with bricks until it collapsed. The truss construction collapsed at 2.8 times the theoretical full load, while in the shell itself, no hairline cracks had occurred.

9.6 The 'white temple' of Biebrich

The last of the structural models from the early days of the Zeiss-Dywidag shells to be discussed here was also developed in connection with the project for a wholesale market hall. In this case it was intended to be roofed with doubly-curved synclastic translational shells, stiffened by vertical frames. As mentioned above, Dischinger had brought this shell type into play as early as the spring of 1923, when he sought a way to exploit the structural and statical characteristics of the Jena test planetarium for covering rectangular ground plans. Already at that time, Dischinger had presumed that such a structure could transfer its loads with very low membrane stresses towards the corners. But he had only been able to calculate the statically-indeterminate structure approximately as a space framework (Figure 9.22).

By 1929, Dischinger had achieved an important breakthrough. In the context of preparing a paper for a prize competition of the Building Academy (*Akademie des Bauwesens*) he was able to solve the differential equations at least for spherical shells with rectangular ground plans and thus to provide a rough statement on the stress distribution [59]. During 1930 he had finally solved the problem of the simultaneous differential equations, so that the doubly-curved translational shell became calculable [60]. That same year, Dywidag developed a proposal for a wholesale market hall in Dresden in which five shell domes with spans of 37.5 x 39.7 m were each arranged in a row over a rectangular ground plan. The design met with an enthusiastic response and, after minor revisions, Dywidag was progressing towards being awarded the contract by the municipal committees in May 1931 (Figure 9.23). A few weeks later Dywidag erected a structural model in reinforced concrete on its premises in Biebrich under Dischinger's supervision. In contrast to the other Zeiss-Dywidag models of those years, some documentation

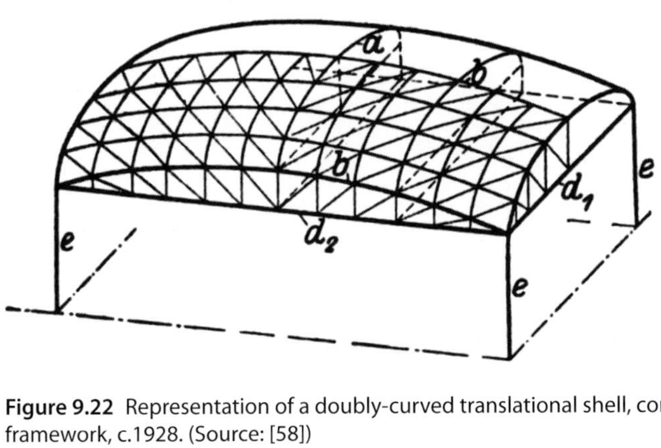

Figure 9.22 Representation of a doubly-curved translational shell, converted into a space framework, c.1928. (Source: [58])

Figure 9.23 Architectural model of the Dresden wholesale market hall (detail), c.1931. A total of five domes arranged one behind the other were supposed to cover the hall space. (Source: [62])

of the tests has been preserved [61]. Thus, the model as well as the load tests can be reconstructed quite well.

The model was not built exactly to scale. It was approximately one fifth of the dimensions of the shells intended for Dresden and did not adopt their slightly oblong shape, as it was erected over a square ground plan with sides of 7.32 m. The structure was thus essentially a trimmed rotational shell, but the scaffolding corresponded to a translational shell, as all the trusses of the falsework had the same shape (Figure 9.24). As in the case of the two cylindrical shell models that had been erected at the same site six years earlier, a wire mesh with wire of 3 mm diameter was chosen for the reinforcement, so that a shell thickness of only 1.5 cm could once again be achieved. Only in the edges were supplementary reinforcing bars added to carry the tensile stresses expected there, and for practical reasons the shell was thickened to 2.5 cm in the area of transition with the 10 cm

Figure 9.24 Assembly of the formwork for the reinforced-concrete model, approximately 1:5 scale, of a shell dome on four supports on the Dywidag premises in Biebrich, 1931. The scaffolding reveals the model's inherent character of a translational shell. (Source: [63])

thick stiffening walls. In the crown region, a circular area with a diameter of 1 m remained unreinforced and was surrounded by a compression ring.

For the tests, the load intensity was chosen in such a way that the model shell was subjected to the same membrane stresses as would be the case with the large-scale version. First, the cupola was loaded over its entire surface with sandbags. In several steps, the loads were raised to 15 tonnes, which, including the dead weight, corresponded to a total load of 316 kg per square metre of ground area. Extensive measurements were carried out for each load condition and the deformations that occurred were recorded. Subsequently, the same procedure was repeated for a load applied to half the plan of the shell. Finally, the shell was subjected to a crowd of 50 persons. The model survived all tests without any cracking. Relative to the corner points of the shell, the load of 15 tonnes resulted in the apex sinking only 2.05 mm. At the same time, the centres of the stiffening walls bent inwards horizontally by 0.94 mm in relation to their ends, which confirmed the assumption that no thrust was transmitted to them by the shell.

It is little known that the Biebrich model shell has survived the course of time. Used as a storage shed for many years and translocated several times since 1975, the model, deprived of its pillars, is now temporarily stored on the premises of Dyckerhoff GmbH in Wiesbaden-Amöneburg (Figure 9.25). An initiative was launched in 2015 to try to get it rebuilt in Wiesbaden. The model is truly remarkable, not only for being one of the few preserved experimental structures

Figure 9.25 Provisional storage of the model shell on the premises of the Dyckerhoff cement plant in Wiesbaden, photographed in 2016. (Image: Knut Stegmann)

Figure 9.26 An icon of construction history. Thirty-nine Dyckerhoff & Widmann employees demonstrate the load-bearing capacity of the 1.5 cm thick model shell, 1931. Finsterwalder (4th from left) and Dischinger (to his right) also form part of the 'human load'. (Source: [64])

in reinforced-concrete construction, it also testifies that model tests could be used as spectacular advertising for the efficiency and trustworthiness of a new product. A photograph that proved particularly effective to this end showed not the relatively modest applied loads used in the test, but rather, thirty-nine Dywidag employees gathered on the shell (Figure 9.26). Its promotional effect was further enhanced by the fact that the newly developed *Dyckerhoff Weiß* cement from the sister company Dyckerhoff & Söhne was used here for the first time. Its quartzite aggregate gave the structure an especially pleasing, light appearance.

In the end, the Biebrich model of 1931 formed part of a long tradition of attention-grabbing showpieces by the emerging reinforced-concrete industry. But it was precisely here that the achievement was particularly enduring: Dresden's new wholesale market hall was never built due to the global economic crisis, and today only a few specialists know about this spectacular project. In contrast, the photograph of the project model has been in continuous use as an icon of the international history of shell construction.

9.7 Outlook and conclusion

With the Biebrich model, which was later even called the 'white temple' [65], the Zeiss-Dywidag model-making reached its climax. At the same time, the iconic experimental building can also be seen as a certain end-point of a phase that lasted for nearly a decade during which the development of effective theoretical approximation methods and empirical verification went hand in hand.

The numerous model experiments had generally confirmed the correctness of the shell pioneers' theories and this seems to have led to a certain 'overestimation of calculation methods and theory' [66]. In the summer of 1933 Dischinger left Dywidag to become a professor at the *Technische Hochschule* in Berlin. Only a short time later, the company ventured to build two aircraft hangars with long barrel shells in Cottbus. They had a totally novel asymmetric cross-section, and their dimensions of up to 40 x 35 m had never been approached before. Nevertheless, this considerable step appears not to have been accompanied by model tests. In the spring of 1934, one of the two buildings collapsed. Whether Dywidag reacted to this catastrophe by returning to model tests in addition to the hasty introduction of stiffening ribs and monitoring measures, has yet to be investigated. It is known only that Anton Tedesko, who worked for the US licensee Roberts & Schaefer, carried out numerous tests on completed shell structures as well as on sheet-metal models [67]. However, a model comparable with the Biebrich concrete models was apparently first created in the USA only in 1950 in Harvey, Ill., when a scale model of a 'ribless' shell roof was tested (Figure 9.27).

In fact, Franz Dischinger had earlier been involved in developing the underlying theoretical calculations for this project. Once again, empirical model testing was primarily used to verify theoretical deductions. The close link to the abstract practices of applied mechanics, which can be traced back to Bauersfeld's influences, and was so typical for Zeiss-Dywidag model making, was again demonstrated here. At the same time, it also symbolised the virtually unstoppable advance of the 'abstract art of analytical statics', whose consequences were extensively criticized by Kurt Leibbrand as early as 1959:

Figure 9.27 Reinforced concrete model of a 'ribless' shell roof in Harvey, Ill., 1950. With a gentle irony Anton Tedesko referred in this photo to the inspiring effect of the earlier Zeiss-Dywidag model experiments. (Source: [68])

> 'The new calculation methods meant great progress, but the close relationship between form and statics of a structure was loosened. Gradually there was a separation between static calculation and creative design. The auxiliary science of statics was now carried out independently by specialists for whom questions of building materials, constructive necessities and practical experience receded' [69].

Already in the 1940s there were intensive discussions about a 'widening gap between the theoretical analyst and the practical designer' [70] in civil engineering. Billington's evaluation of the 'German School', mentioned at the beginning of this chapter, also lies within this tradition. However, at the same time, he emphasized Dischinger's 'propensity for combining analysis, structural testing and construction' [71], which also characterized Finsterwalder's method. Precisely because of their ability to combine 'systematic basic research and the tenacious development of mathematical theories for calculating their constructions' with practical application, Leonhard Fünfschilling saw the two engineers as both the 'grand theorists and practitioners of shell construction' [72]. The Zeiss-Dywidag models, which have been presented comprehensively here for the first time, played a decisive role in their bridge-building between theory and practice.

References

1 May, Roland (2015) Shell Sellers. The International Dissemination Of The Zeiss-Dywidag System, 1923-1939. In: *Proceedings of the 5th International Congress on Construction History, June 2015, Chicago, Illinois* (ed. Brian Bowen, et al.), Vol. 2, 557-565. Chicago: Construction History Society of America.

2 Piz[z]etti, Giulio (1958) Some Considerations of the Past and the Future of Reinforced Concrete. *Journal of the Boston Society of Civil Engineers* 45 (2): 151-160. 155.

3 Gardner, Denis P. (2004) *Minnesota Treasures: Stories Behind the State's Historic Places*. St. Paul: Minnesota Historical Society Press. 102.

4 Dechau, Wilfried, et al. (2014) Rückblick und Ausschau auf das Berufsbild des Bauingenieurs: Ein Interview mit Jörg Schlaich aus dem Jahr 1999. Herrn Professor Dr.-Ing. Drs. h.c. Jörg Schlaich zur Vollendung seines 80. Lebensjahres gewidmet. *Beton- und Stahlbetonbau* 109 (11): 838-849. 839.

5 Tedesko, Anton & Billington, David P. (1982) The Engineer's Personality and the Influence It Has on His Work – An Historical Perspective. *Concrete International: Design & Construction* 4 (12): 20-26.

6 Gardner, Denis P. (2004) *Minnesota Treasures: Stories Behind the State's Historic Places*. St. Paul: Minnesota Historical Society Press. 103.

7 Billington, David P. (1983) *The Tower and the Bridge. The New Art of Structural Engineering*. New York: Basic Books. 174-175.

8 Billington, David P. & Hines, Eric M. (1997) Anton Tedesko, Model Research and the Introduction of Thin Shells into the United States. In: *Proceedings*

of the International Conference on Experimental Model Research and Testing of Thin-Walled Structures, Prague in September 1997 (ed. Miloš Drdácký & Teoman Peköz), 21-30. Telč: Aristokrat.

9 Schöne, Lutz (2011) *Eisenbetonschalen zwischen 1898 und 1928*. Aachen: Shaker. 116.

10 Schöne, Lutz (2011) *Eisenbetonschalen zwischen 1898 und 1928*. Aachen: Shaker. 128.

11 Billington, David P. & Hines, Eric M. (1997) Anton Tedesko, Model Research and the Introduction of Thin Shells into the United States. In: *Proceedings of the International Conference on Experimental Model Research and Testing of Thin-Walled Structures, Prague in September 1997* (ed. Miloš Drdácký & Teoman Peköz), 21-30. Telč: Aristokrat. 21.

12 Kurrer, Karl-Eugen (2018) *The History of the Theory of Structures. Searching for Equilibrium*. 2nd ed., Berlin: Ernst & Sohn. 733.

13 Bauersfeld, Walter (1957) Projection planetarium and shell construction. *The Chartered Mechanical Engineer* 4 (12): 509-516. 514.

14 Photo (n.d.) Universitätsarchiv der TU Berlin, 423 Nachlass Franz Dischinger, no. 118.

15 Zoelly, Robert (1915) *Über ein Knickungsproblem an der Kugelschale*. PhD diss., ETH Zürich.

16 Bauersfeld, Walter (1957) Projection planetarium and shell construction. *The Chartered Mechanical Engineer* 4 (12): 509-516. 514.

17 Geckeler, Josef (1926) *Über die Festigkeit achsensymmetrischer Schalen*. Berlin: VDI-Verlag. 48.

18 Geckeler, Josef W. (1928) Elastostatik. In: *Mechanik der elastischen Körper*. (ed. Richard Grammel). Berlin: Springer. 304.

19 Schöne, Lutz (2011) *Eisenbetonschalen zwischen 1898 und 1928*. Aachen: Shaker. 150.

20 Föppl, August (1891) Über das Flechtwerk. *Schweizerische Bauzeitung* 17 (16): 95-96.

21 Föppl, August (1896) Belastungsversuche an einem Tonnenflechtwerkdache. *Mittheilungen aus dem Mechanisch-Technischen Laboratorium der Königlichen Technischen Hochschule München* (24): 41-49 and pl. 2. Pl. 2.

22 Bauersfeld, Walther (1928) Letter to Ulrich Finsterwalder, 4 June 1928, ZEISS Archiv, BACZ, no. 22853.

23 Bauersfeld, Walther (c.1923) Konstruktion für cylindrisches Hallendach (manuscript): 330-566, ZEISS Archiv, BACZ, no. 26027.

24 Bauersfeld, Walther (c.1923) Kuppelkonstruktion für das Projektionsplanetarium (Part IV) (manuscript): 567, ZEISS Archiv, BACZ, no. 26028.

25 Firma Carl Zeiss (1926) Pfettenloses Eisenbeton-Tonnendach, German Patent 431,629, filed 22 January 1924, and issued 8 July 1926.

26 Kurze, Bertram (2006) *Industriearchitektur eines Weltunternehmens. Carl Zeiss 1880-1945*. Erfurt: Thüringisches Landesamt für Denkmalpflege und Archäologie. 125.

27 Photo (n.d.) Universitätsarchiv der TU Berlin, 423 Nachlass Franz Dischinger, no. 112.

28 Rüsch, Hubert (1973) Ulrich Finsterwalder zu seinem fünfzigsten Dienstjubiläum. Sein Lebensweg als Mensch und Ingenieur. In: *Festschrift Ulrich Finsterwalder – 50 Jahre für Dywidag* (ed. Dyckerhoff & Widmann AG: 9-16. Karlsruhe: G. Braun. 10.

29 Bauersfeld, Walter (1957) Projection planetarium and shell construction. *The Chartered Mechanical Engineer* 4 (12): 509-516. 516.

30 Pucher, Adolf (1943) Die Schalenbauweise im Bauwesen. *VDI – Zeitschrift des Vereines Deutscher Ingenieure,* 87 (17/18): 251-260, here 252.

31 Finsterwalder, U [lrich] (1933) Die querversteiften zylindrischen Schalengewölbe mit kreissegmentförmigem Querschnitt. *Ingenieur-Archiv* 4 (1): 43-65. 43.

32 Bauersfeld, Walter (1957) Projection planetarium and shell construction. *The Chartered Mechanical Engineer* 4 (12): 509-516. 514.

33 N.N. [Finsterwalder, Ulrich ?] (1926?) Theorie der zylindrischen Schalengewölbe, typescript, ZEISS Archiv, BACZ, no. 26028.

34 Photo album (c.1929) Archiv des Deutschen Museums, FA 010, no. 329.

35 Dischinger, Franz (1952) *Die Schalenbauweise.* Typescript, Universitätsarchiv der TU Berlin, 423 Nachlass Franz Dischinger, no. 75.

36 Photo (n.d.) Universitätsarchiv der TU Berlin, 423 Nachlass Franz Dischinger, no. 112.

37 Dischinger, Fr[anz] & Finsterwalder, U[lrich] (1928) Eisenbeton-Schalendächer System Zeiss-Dywidag, *Der Bauingenieur* 9 (44): 807-812, (45): 823-827, and (46): 842-846. 809.

38 Dischinger, F. (1928) *Schalen und Rippenkuppeln.* In: *Handbuch für Eisenbetonbau* 12, pt. 2 (ed. F. Emperger), 151-371. Berlin: Ernst & Sohn. 258.

39 N.N. (1927) *Schalen-Gewölbe D.R.P. nach dem System Zeiss-Dywidag,* s.l.: [Dyckerhoff & Widmann AG], unpag.

40 Billington, David P. & Hines, Eric M. (1997) Anton Tedesko, Model Research and the Introduction of Thin Shells into the United States. In: *Proceedings of the International Conference on Experimental Model Research and Testing of Thin-Walled Structures, Prague in September 1997* (ed. Miloš Drdácký & Teoman Peköz), 21-30. Telč: Aristokrat. Footnote 7.

41 Schöne, Lutz (2011) *Eisenbetonschalen zwischen 1898 und 1928.* Aachen: Shaker. 171.

42 Dischinger, Fr[anz] & Finsterwalder, U[lrich] (1928) Eisenbeton-Schalendächer System Zeiss-Dywidag. *Der Bauingenieur* 9 (44): 807-812, (45): 823-827, and (46): 842-846. 823.

43 Dischinger, Franz (1950) Weitgespannte Tragwerke. In: *Deutscher Beton-Verein. Ansprachen und Vorträge zum 50jährigen Bestehen*: 390-449. Herborn: Beck. 449.

44 Dischinger, Franz (1952) *Die Schalenbauweise.* Typescript, Universitätsarchiv der TU Berlin, 423 Nachlass Franz Dischinger, no. 75.

45 Kleinlogel, Adolf (1928) Die Schalengewölbe der Großmarkthalle Frankfurt a. M. *Beton und Eisen* 27 (1): 11-16, and (2): 25–28.

46 Dischinger, F. (1928) *Schalen und Rippenkuppeln.* In: *Handbuch für Eisenbetonbau* 12, pt. 2 (ed. F. Emperger), 151-371. Berlin: Ernst & Sohn. 258.

47 May, Roland (2012) Shell Wars: Franz Dischinger and Ulrich Finsterwalder. In: *Nuts & Bolts of Construction History* (ed. Robert Carvais, et al.), Vol. 3, 133-141. Paris: Picard.

48 Ernährungsamt & Hochbauamt, Frankfurt a. M. (eds.) (1928) *Die neue Grossmarkthalle in Frankfurt a. M. Zur Eröffnung am 25. Oktober 1928*. Frankfurt a. M.: Englert und Schlosser. 4.

49 Kleinlogel, Adolf (1928) Die Schalengewölbe der Großmarkthalle Frankfurt a. M. *Beton und Eisen* 27 (1): 11-16, and (2): 25-28. 28.

50 Cf. Herzog, Max (1996) Die Beulfestigkeit von Stahlbetonschalen: Theorie, Versuche, Schadenfälle. *Beton- und Stahlbetonbau* 91 (5): 115-121, (6): 141-145, and (7): 164-169.

51 Dischinger, Franz (1930) Großmarkthalle Leipzig. In: *Bericht über die XXXII. Haupt-Versammlung des Deutschen Beton-Vereins (E.V.) am 7., 8. und 9. März 1929*, s.l.: Deutscher Beton-Verein: 167-192. Plate following p. 174.

52 Dischinger, Franz & Rüsch, Hubert (1929) Die Großmarkthalle in Leipzig, ein neues Kuppelsystem, zusammengesetzt aus Zeiss-Dywidag-Schalengewölben. *Beton und Eisen* 28 (18): 325-329, (19): 341-346, (23): 422-429, and (24): 437-442. 328.

53 Dischinger, Franz & Rüsch, Hubert (1929) Die Großmarkthalle in Leipzig, ein neues Kuppelsystem, zusammengesetzt aus Zeiss-Dywidag-Schalengewölben. *Beton und Eisen* 28 (18): 325-329, (19): 341-346, (23): 422-429, and (24): 437-442. 328.

54 [Rüsch, Hubert] (1927) Großmarkthalle Leipzig. Statische Berechnung Teil 9: Zugring der Kellerdecke, manuscript (copy), 17 September 1927, archive of the Amt für Bauordnung und Denkmalpflege der Stadt Leipzig, file Zwickauer Str. 40-44.

55 Dischinger, Franz & Finsterwalder, Ulrich (1932) Die weitere Entwicklung der Schalenbauweise 'Zeiss-Dywidag'. *Beton und Eisen* 31 (7/8): 101-108, (10): 149-155, (11): 165-170, (12): 181-184, (14): 213-220, (15): 229-235, (16): 245-247, and (23): 371. 213.

56 Dischinger, Franz & Finsterwalder, Ulrich (1932) Die weitere Entwicklung der Schalenbauweise 'Zeiss-Dywidag'. *Beton und Eisen* 31 (7/8): 101-108, (10): 149-155, (11): 165-170, (12): 181-184, (14): 213-220, (15): 229-235, (16): 245-247, and (23): 371. 218.

57 Dischinger, Franz & Finsterwalder, Ulrich (1932) Die weitere Entwicklung der Schalenbauweise 'Zeiss-Dywidag'. *Beton und Eisen* 31 (7/8): 101-108, (10): 149-155, (11): 165-170, (12): 181-184, (14): 213-220, (15): 229-235, (16): 245-247, and (23): 371. 216.

58 Dischinger, Fr[anz] & Finsterwalder, U[lrich] (1928), Eisenbeton-Schalendächer System Zeiss-Dywidag. *Der Bauingenieur* 9 (44): 807-812, (45): 823-827, and (46): 842-846. 809.

59 [Dischinger, Franz] (1929/30) Anlage 5. Die Anwendung der Kugelschale zur Ueberdachung von Hallen und Kuppelbauten mit rechteckigen und vieleckigen Grundrissen. Typescript, ZEISS Archiv, BACZ, Nr. 8344.

60 [Dischinger, Franz] (c.1947) Schalen und Kuppeln aus Stahlbeton. Typescript: 26. Universitätsarchiv der TU Berlin, 423 Nachlass Franz Dischinger, no. 65.

61 Gra[udenz?] & Wa[gner?] (1932) Schalenbauweise. Versuche mit Vierecks-kuppeln. Typescript, 21 January 1932, Universitätsarchiv der TU Berlin, 423 Nachlass Franz Dischinger, no. 42.

62 Büttner, Oskar & Hampe, Erhard (1977) *Bauwerk, Tragwerk, Tragstruktur*. Stuttgart: Hatje. 243

63 Photo (n.d.) Universitätsarchiv der TU Berlin, 423 Nachlass Franz Dischinger, no. 118.

64 Photo (n.d.) Universitätsarchiv der TU Berlin, 423 Nachlass Franz Dischinger, no. 118.

65 Thode, D. (1974) Die Rettung des 'weißen Tempels'. *Die Bauwirtschaft* 28 (51/52): 2066-2067.

66 Leibbrand, K[urt] (1959) Wende des Bauingenieurwesens? *Österreichische Ingenieur-Zeitschrift* 2 (6): 205-207. 206.

67 Tedesko, Anton (1949) Letter to John Kalinka, 30 November 1949, Princeton University, Library, Dept. of Rare Books and Special Collections, Anton Tedesko Papers, C1478, box 13, folder 43.

68 Hines, Eric M. & Billington, David P. (2004) Anton Tedesko and the Introduction of Thin Shell Concrete Roofs in the United States. *Journal of Structural Engineering* 130 (11): 1639-1650. 1648.

69 Leibbrand, K[urt] (1959) Wende des Bauingenieurwesens? *Österreichische Ingenieur-Zeitschrift* 2 (6): 205-207.

70 Budgen, William Edward James (1944) *Modern Computation Methods and Office Practice in the Design of Statically Indeterminate Structures* (The Institution of Civil Engineers, Structural and Engineering Division, Session 1943-44, Structural Paper No. 6) London: The Institution of Civil Engineers. 4.

71 Hines, Eric M. & Billington, David P. (2004) Anton Tedesko and the Introduction of Thin Shell Concrete Roofs in the United States. *Journal of Structural Engineering* 130 (11): 1639-1650. 1642.

72 Fünfschilling, Leonhard (1967) Günter Günschel: Große Konstrukteure I [review]. *Das Werk* 54 (7): 461-462. 461.

10

Model testing of structures in pre-war Italy: the School of Arturo Danusso

Mario Alberto Chiorino and Gabriele Neri

10.1 The beginning of model testing in Italy

In Italy, experimentation on reduced-scale structural models started quietly during the 1920s. The first well-known cases were probably developed at the Polytechnic of Turin. One of these was the model of a dam composed of a cylindrical concrete thin shell, loaded with hydrostatic pressure in order to test the statical behaviour of the shell [1]. The model was built and tested in July 1926 by Camillo Guidi (1853-1941) – the first professor who gave university lectures about reinforced concrete, in 1900 [2] – at the *Mostra Internazionale di Edilizia* (International Building Exhibition) in Turin [3]. The structure supporting the model and the water tank for loading was 5 m high, 8.1 m wide and 5.9 metres deep (Figure 10.1). Although not a model of an actual dam, it represented a typical dam of its type, about 40 m high. The scale was, therefore, about 1:8.

The model of the dam itself was 4.4 m high, 2.61 m radius, and 16 cm thick, fixed at the bottom and with an unrestrained, semi-circular upper edge. The water was held in the semi-circular annulus between the shell dam and the outer wall (Figure 10.2). The deflection of the shell was measured using seven radial struts, at three heights above the base, each fitted with a gauge to show the change of length, magnified by 150 times. The deflection of the shell varied between 0.06 and 0.26 mm. To ensure comparability of results, Guidi was particularly attentive to measuring the internal temperature of the dam at six places, as well as the temperatures of the water and ambient air, and made the measurements of deflection quickly in order to eliminate the effects of temperature changes. His key observation was that the maximum deflection of the shell was about six times less than the theoretical predictions.

Shortly after Guidi's experiment, Gustavo Colonnetti (1886-1968), also a professor at the Polytechnic of Turin, used photoelastic stress analysis (see Chapter 12) to study the behaviour of dam elements, followed by other researches carried out by his assistant G.M. Pugno [4].

Physical Models: Their historical and current use in civil and building engineering design,
First Edition. Edited by Bill Addis.

Figure 10.1 The structure supporting the model of a semi-cylindrical shell dam, water tank for loading the shell, and measuring equipment. Camillo Guidi, 1926. (Source: [1])

10.2 The *Laboratorio Prove Modelli e Costruzioni* at the Polytechnic of Milan

10.2.1 Arturo Danusso and Guido Oberti

Systematic research using model testing began in Italy only in 1930, especially thanks to the impetus of Arturo Danusso (1880-1968), one of the fathers of Italian engineering in the last century [5, 6]. Born in a small town in Piedmont in 1880, he became familiar with the theory and practice of reinforced concrete at the very beginning of the century, initially during his time at university in Turin with Camillo Guidi and later, from 1903 to 1915, during his working experience within the building company of Antonio Porcheddu (1860-1937), one of the first agents in Italy for the system of François Hennebique (1842-1921) [7].

Danusso was seduced by the French entrepreneur's approach to structural design, based on few theoretical principles merged with considerable practical experience. In the following years, through several academic and scientific achievements [8], Danusso never forgot his debt to Hennebique. He tried to overcome the limitations of traditional theory which, constrained by the assumption of perfect elasticity, could not adapt to the real behaviour of reinforced concrete. On the one hand, this path led Danusso into the theoretical investigation of plastic deformations [9]; on the other, he would try to find answers through empirical experimentation, making use of reduced-scale models tested in a laboratory [6]. In his words:

> 'Building means working in reality, and if reality – as we discover every day from physical experience – is highly complex and interrelated, [...] science,

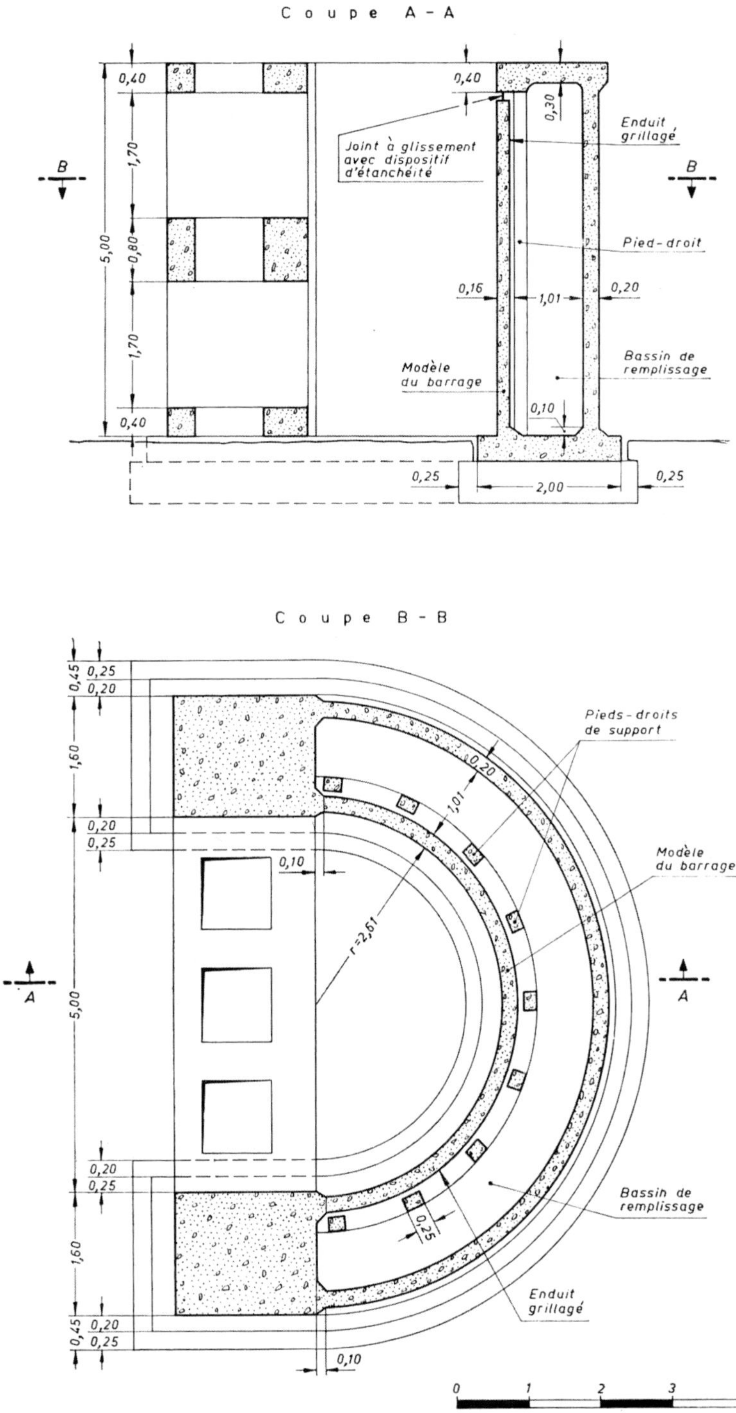

Figure 10.2 Model of a semi-cylindrical shell dam, by Camillo Guidi. 1926. (Source: [1])

by the poverty of its instruments, is forced to proceed by dint of limitations and simplifications, leaving an abyss between it and reality' [10].

And so, while Eduardo Torroja was carrying out his pioneering tests in 1930 in Spain [11, 12], Arturo Danusso founded the *Laboratorio Prove Modelli e Costruzioni* (Model and structures testing laboratory), annexed to the *Istituto di Scienza delle Costruzioni* (Institute of Construction Science) of the Royal Polytechnic of Milan, with the aim of developing experimental research using reduced-scale models. The creation of this small laboratory was made possible thanks to generous grants by Italcementi – one of the leading groups in the field of cement in Italy – and other companies. Following a short trial period while rooms were adapted and equipment and measurement instruments were installed, the photoelastic stress analysis section officially started its activity in 1933, followed, in 1935, by space dedicated to large scale model testing, a total of about 400 m^2 of floor space in total [13] (Figures 10.3 and 10.4). The Laboratory's scientific purposes were the following:

> 'a) experimental investigations on existing or specifically built structures, in order to determine their real statical behaviour and potentially compare it with that deduced by ordinary calculations; b) experimental studies of structures, carried out in a laboratory, working on reduced-scale models,

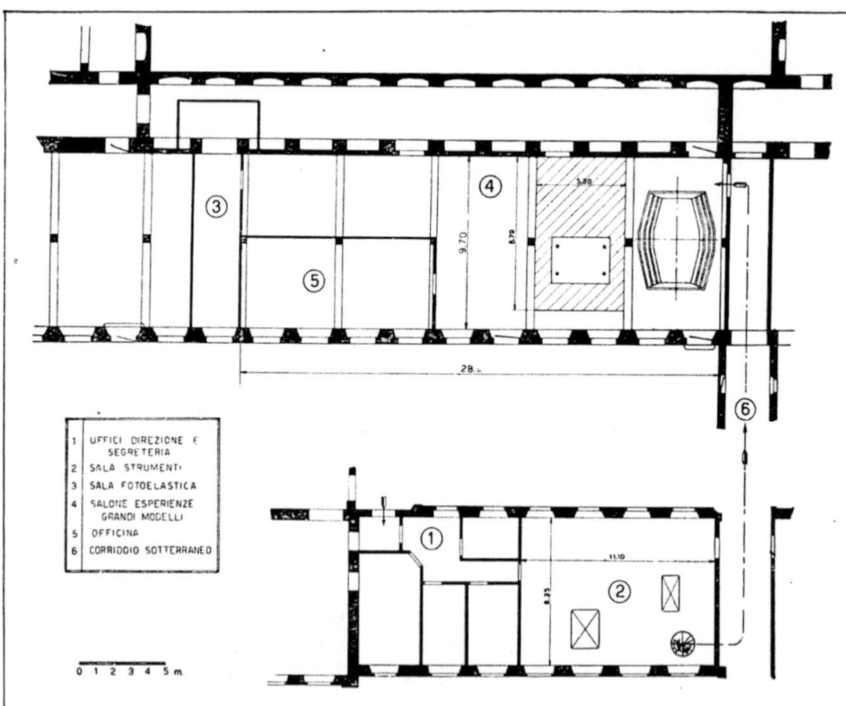

Figure 10.3 Plan of the *Prove Modelli e Costruzioni* Laboratory in 1941. 1 – Office and secretariat; 2 – Instrument room; 3 – Photoelastic laboratory; 4 – Experimental room for large models; 5 – Office; 6 – Underground corridor. (Source: [13])

Figure 10.4 Universal machine for testing of structural elements in the *Prove Modelli e Costruzioni*. (Source: [13])

> using instruments to determine global effects (...) as well as extensometers or special methods (photoelasticity) that allow investigation, within the elastic domain, of local deformations and stresses that are generated in models by loads or external forces. Those researches allow investigation of the stress situation on entire structures, or at particular locations, with an approximation to reality that could hardly be achieved using other methods; c) investigations on elasticity, plasticity and time effects, (...); d) investigations on dynamic actions (...)' [13].

Alongside Danusso, another brilliant researcher stood out: the young engineer and mathematician Guido Oberti (1907-2004), who later became one of the most renowned experts in the field of structural modelling [14, 15, 16, 17]. In 1931 he was hired as assistant to the Chair of Construction Science at the Polytechnic of Milan, held by Danusso. In addition to the above-mentioned research activities, the Milan Laboratory also had didactic purposes. From experimental investigations a large number of pictures was collected, that were useful for the Institute's courses. From 1936, Guido Oberti began organizing special exercises – mainly on photoelastic stress analysis – for students.

10.2.2 Models for concrete dam design in the 1930s

The construction sector that really drove the research by Danusso and his team was hydroelectric engineering. Structural modelling made it possible to optimize the sections of very complex projects, and so save on the total quantities of materials used. These savings were particularly marked on large-scale projects such as dams which used huge quantities of concrete. Moreover, the particular geometry used for dams was found to be more readily reproducible by a physical model at a reduced scale rather than by mathematical formulae.

Following the well-known model tests carried out in the USA – e.g. for the Stevenson Creek experimental dam (1925-32) and the Boulder arch-gravity dam (1930-35) [18] (see Chapter 8) –, Danusso and Oberti started a series of important experiments on dams using photoelastic stress analysis [19] (Figure 10.5).

The photoelastic study for the Santa Giustina dam, 140 m high, carried out in 1935-38, was fundamental to examining and understanding the behaviour of thick-arch dams. Tests were conducted assuming the continuous elastic arch was built not on rigid rock, as was commonly assumed, but on elastic rock, with a modulus equal to that of the concrete. Many arches were studied, with circular profiles on the up-stream and down-stream sides, both with constant and variable thicknesses and with different subtended angles. As Oberti stated: 'The analysis

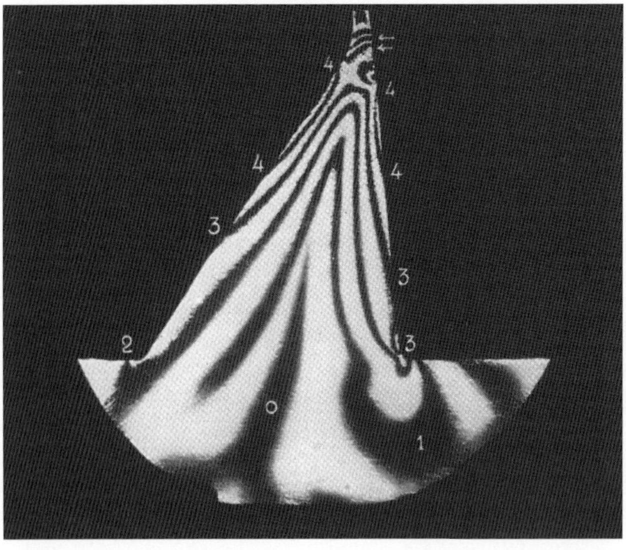

Figure 10.5 Photoelastic model analysis of a gravity dam subject to the action of ice at the *Prove Modelli e Costruzioni* Laboratory. (Source: [13])

of the results and their comparison with those of the elastic computation made it possible to clarify the statical behaviour of the structure and to reduce the initially planned volume for the lower half of the dam' [20].

Three-dimensional tests on gravity dams were carried out for the first time in 1936 for the design of the maximum-height buttress of the straight Scais dam, near Sondrio in the Italian Alps. Results were interesting for many reasons: first, to check the approximations made in the calculations which assumed a linear variation of the vertical stresses; secondly, because a new approach was achieved in the construction, which was developed in two stages by fixing the profile of the construction joint according to the isostatic lines obtained using the model [20].

Among the other tests carried out by Oberti we can mention those on a 120-metre high buttress adopted for dams at Lake Ancipa (Sicily) and Lake Bau-Muggeris (Sardinia). He used a celluloid physical model (linear scale, $\lambda = 100$) tested under hydrostatic load, imposed using mercury (13.5 times denser than water), and using a great number of electric resistance strain-gauges to determine the principal stresses and isostatic lines both on external and inner faces: 'The results, compared with those of conventional calculation, bore out a satisfactory agreement and made it possible to determine the pattern of stresses acting on the foundations' [20].

More important and extensive were the experiments carried out on three-dimensional models for arch dams. Oberti did the first investigation on the Rocchetta arch dam (70 m high), east of Genoa in 1937-38, using a 1:40 scale model made of gypsum-Celite which was tested both for elastic behaviour and for failure. Tests showed a satisfactory behaviour of the dam and – for the first time – emphasized 'how single curvature arches react to the loads acting on them, even beyond the elastic range, and made it possible to evaluate the statical resources of arch dams (even if slender and thin) and the high overall safety factor. Moreover, these results confirmed those obtained for the Stevenson Creek dam in tests confined to the elastic range, investigating high bending stresses acting on vertical elements (cantilevers). Even taking into account the balancing effect of the dead weight, the idea of designing double-curvature, rather than single-curvature, arch dams arose spontaneously' [21].

These experiments were useful for the investigations for the Osiglietta dam (1935-37) to the west of Genoa, which had to close off a 200 m wide and 75 m high gorge. Tests were planned to compare a single curvature classical type and a double-curvature type, using celluloid models and mercury to simulate the hydraulic load within the elastic range. They demonstrated that the doubly-curved dam reduced tensile stresses along the cantilevers, increasing the contribution of those elements just by reason of their arch shape. Other structural behaviour demonstrated by models led to a complete modification of the boundary conditions of the dam: rather than being encastred, it would bear upon a standard pulvino foundation embedded in the rock face, through a continuous perimeter joint [21].

10.3 The encounter between Arturo Danusso and Pier Luigi Nervi

In 1935 there took place an important encounter in the history of Italian model testing. At that time, the engineer Pier Luigi Nervi (1891-1979) – who had just become well known thanks to his design for the Berta Stadium in Florence [22, 23, 24] – was trying to define the structure of a series of reinforced-concrete aircraft and seaplane hangars, but he was not, in fact, able to determine analytically the complex interplay of forces acting on the highly statically-indeterminate structural system he had devised. Rather than calculated, the shape of the vault had been designed 'by intuition': 'I have to clarify – Nervi later admitted – that my hangars had been defined through approximate calculations derived from simplified hypotheses, on which basis it would be really imprudent to proceed with the real construction' [25]. So, he turned to Danusso and his team for help.

The complex roof – a shell composed of a double-lattice structure of reinforced-concrete ribs on a rectangular plan (110 x 36 m) – was thus reproduced at the Polytechnic of Milan as a 1:37.5 scale celluloid model (Figure 10.6). Various tests were carried out on this model to analyse its elastic behaviour under a dead load and overloads by hanging calibrated weights on the structure to simulate real loads. The dead loads acting on the structure were the self-weight of the reinforced-concrete framework and the self-weight of the roof, evaluated at 100 kg/m^2.

Loads were applied to the model using calibrated weights, hung by wires fixed to the nodes of the lattice. The lateral nodes were more heavily-loaded than the central ones owing to the slope of the vault. As live load, only wind pressure on the main doors, at an incident angle of 10° to the horizontal, was simulated. Strain-gauges on the upper and lower edges of the various beams, allowed the

Figure 10.6 1:37.5 scale celluloid model of the reinforced-concrete hangars (first version) designed by Pier Luigi Nervi, tested at the Politecnico di Milano, 1935-36. (Image: ISMES Historical Archive)

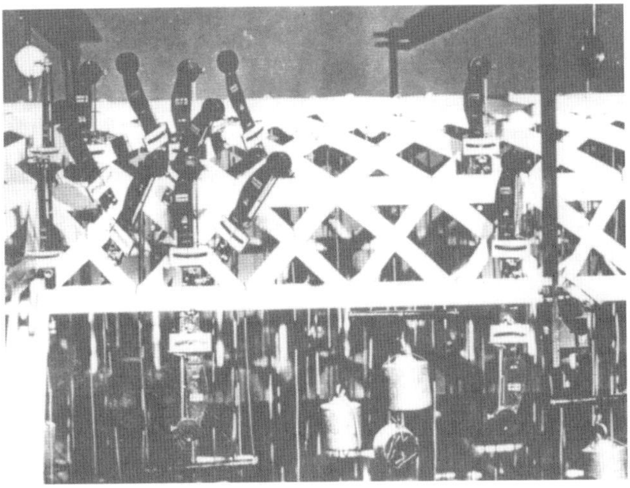

Figure 10.7 1:37.5 scale celluloid model of the reinforced-concrete hangars. Detail showing Huggenberger Tensometers (strain gauges) in place (see Section 13.4.2). (Image: ISMES Historical Archive)

axial and bending actions on the most important ribs to be studied [26, 27, 28] (Figure 10.7).

The model was tested only in the elastic range under its own self-weight and wind loads. In these experiments the model thus became the simple analogue substitute for a calculation of the system of elastic stresses, a calculation which nowadays can easily be made by means of digital numerical models. The problem of a more realistic assessment of the margin of safety thus remained open, as happens every time a structure is modelled as fully elastic.

The main reason for this choice was undoubtedly due to the large dimensions of the hangar vault, and to the unusual slenderness of the crossed ribs of which it was formed. If these extreme characteristics had been taken together it would have required excessive dimensions for models designed to simulate the real behaviour of the reinforced-concrete structure beyond the elastic range up to collapse, at the same time suffering the distortions caused by the effects of scale. Albeit within these limits, the observations drawn from the tests allowed Nervi to obtain information useful for the fine-tuning of the initial project, in particular with regard to the design of the large horizontal bracing beam, as observed so incisively by Danusso and Oberti: 'Here, too, the calculation – like matter for Dante – may be unresponsive, deaf. Not the model, however: in fact the model was a precious counsellor of useful modifications' [13].

It should be emphasised, however, that the bold, innovative structural design developed by Nervi, despite the minor changes prompted by the tests, was generally perfectly adequate – proof of his remarkable capacity for structural design and statical intuition, and his mastery of construction. This was a constant that characterized the results of all the experiments carried out on the works of the great architect-engineer. Experimental testing, conceived as an assessment of the

Figure 10.8 1:37.5 scale celluloid model of the reinforced-concrete hangars (second version) designed by Pier Luigi Nervi, tested at the Politecnico di Milano, 1938-39. (Image: ISMES Historical Archive)

original structural design, was, in effect, a substantial validation, providing indications for improvements in the dimensioning of single components that brought little alteration to the initial outline. The measures implemented by Nervi at the conclusion of the construction work during the decentring of the two hangars built using this structural scheme at Orvieto (1935-38) offered further confirmation of this.

A second model, at the same scale, was built and tested in 1938-39 for the six hangars to be built at Orvieto, Orbetello and Torre del Lago, conceived in this case with a symmetrical structural system and improved in accordance with the results of the tests performed on the first model (Figure 10.8). The deflections under load were determined in this second series of model tests, and later confirmed by *in-situ* measurements of the finished structure, once the thermal effects, which were not negligible, had been taken into account [29].

10.4 Model analysis and structural intuition in the work of Pier Luigi Nervi

These investigations discussed above revealed to Nervi the power of experimental model analysis and, moreover, confirmed the particular vision of engineering and structural design that he had expressed in many writings from 1931 [30, 31]. Rejecting the idea that engineering was a science, built by means of perfectly-describable phenomena and exact rules, Nervi criticized the mathematical hegemony in the curriculum of engineering schools, considering it to be the cause of the lack of creativity and, most of all, the cause of the loss of the designer's essential skill: *structural intuition*. From the moment that, in a statical problem, some data such as external forces, internal elasticity and the real materials' resistance turn out to be 'essentially unknowable', Nervi asked himself: 'what is the real value of those numbers, reached using formulae which treat accurately things which are not accurate, unless they are to give the order of magnitude of highly-approximate results, to be interpreted with personal judgement and in which the judging criteria are only feeling and intuition, the only faculties capable of evaluating things which are not measured with a ruler and scale?' [32]. From this reasoning there followed one of his most famous statements: 'in measuring such variability in the choice of these coefficients, in the estimation of non-measurable factors, we see the real essence of engineering; but in this light this is no longer 'science', but becomes 'art'' [32].

Matured over more than twenty years of experience on construction sites, this thesis was derived from his observation of the poor correspondence between theory and reality, especially in relation to the technique of reinforced concrete. Faced with an extraordinary but hardly decodable material, which often expresses unpredictable features, Nervi developed the firm belief that a strictly mathematic thinking could be counter-productive, and that it should be contrasted with the development of an intuitive way of thinking, in order to learn to *think* more than to *calculate*.

From this perspective, clear analogies can be found with the thought of Arturo Danusso, especially about the role of intuition. For instance, in his lecture *Intuition and science in concrete,* given in 1952, Danusso [33] later recalled, as often done by his colleague, the role of this 'property' in the major architectural masterpieces of the past and in the value that it fulfilled and still fulfils. In fact, the first statement of that 'delicate, and often mysterious psychological process in which the approach to the truth matures' – which corresponds to scientific research and, broadly speaking, to structural design – is essentially based on the action of intuition which 'suggests, orientates and directs', and the subsequent phases of logical determination and verification follow from this. So, intuition is not something 'taken out' of science, but a determining faculty, capable of merging men's patrimony of feelings and observations, 'placing oneself with his sensibility in front of nature, to probe the depth in which it disguises its mysteries' [34]. As was the case with Nervi so, also, Danusso's view resulted in a careful assessment of the role of calculations – 'aimed at analyzing more than synthesizing' – in an engineer's study course, often leading him to conclusions which are similar to his colleagues: 'For these reasons I agree with the people who want the suppression of the exclusivity of mathematics in today's introductory courses, providing an alternative view which gradually improves in each year, first from an intuitive point of view and then from scientific elaboration' [35].

Considering this common theoretical background, one can understand Nervi's gradual involvement in experimental model analysis in the following years, which led him to test many of his projects and even to be appointed President of the ISMES when Danusso retired (see Chapter 15). Moreover, besides its practical applications, experimental stress analysis was – in Nervi's mind – of great didactic importance. 'No theoretical derivation – Nervi wrote – will ever clarify the behaviour of a structure so much as an experimental investigation, nor can any other procedure check our theoretical calculations more efficiently. It is, therefore, to be hoped that colleges, designers and builders will appreciate the importance of model analysis, from which we may derive improvements to our teaching methods, checks on further avenues for our theoretical research and exact solutions to our actual problems' [36].

10.5 Further experimental model studies by Nervi and Oberti

From this auspicious beginning, enabling the hangars to be constructed safely, Nervi made use of the Laboratory at the Polytechnic of Milan to test many other structures, either as designer or as consultant through his construction company, Nervi & Bartoli.

10.5.1 The monumental arch for the 1942 Universal Exhibition in Rome

An unusual case study was the investigation for the monumental arch conceived for the Universal Exhibition of 1942 (E42) in Rome (Figure 10.9). In 1938, Nervi – who had already designed some structures for the same exhibition

Figure 10.9 Manifesto for the 1942 Universal Exhibition in Rome, 1939. (Source: ISMES Historical Archive)

[37] – was commissioned, through his building company 'Nervi & Bartoli', to study how to realize the Roman Arch that Mussolini had dreamt as the symbol of the E42. The daring structure, designed by architect Adalberto Libera with engineer Vittorio Di Berardino, was a non-reinforced concrete semi-circular arch with a span of 200 m and a rise of 103,65 m, with an elliptical section gradually narrowing towards the top. Actually, the problem was not to determine its statical behaviour, but rather how to build it [38].

On 13 December 1938, Nervi submitted two solutions – one made out of non-reinforced concrete elements; the other in reinforced concrete – but at the same time the need for experimental tests emerged. Therefore, in February 1939 the Nervi & Bartoli was commissioned to test the arch using reduced-scale models, in collaboration with the Institute of Construction Science of the Polytechnic of Milan and, in particular, with Arturo Danusso and Di Berardino (Figures 10.10 and 10.11).

The arch was going to be tested in Milan by Danusso's team using a 1:200 scale model, but suddenly the President of the E42 decided to abandon concrete (both reinforced and not) and to examine a new version made of aluminium alloys. Nervi's contribution to this project stopped here, but experimental investigations continued for the new design, which considered a 300 (or more)

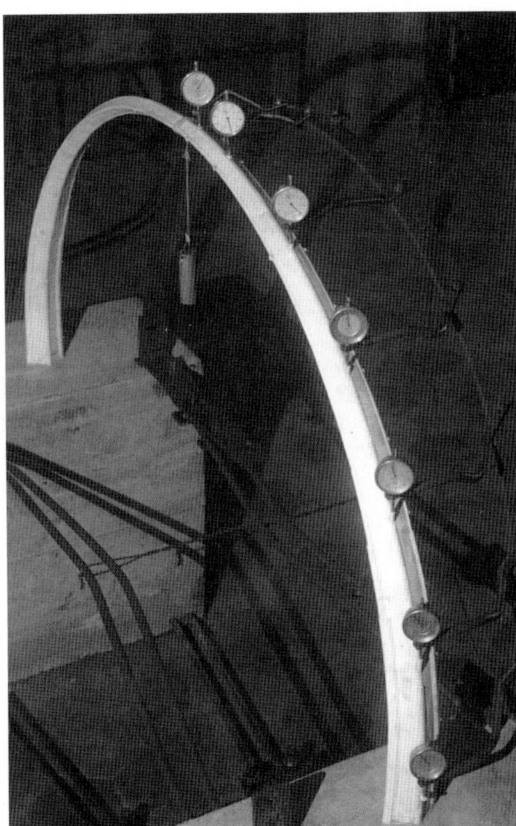

Figure 10.10 1:200 scale model of the E42 arch, tested at the Politecnico di Milano, c. 1939. (Image: ISMES Historical Archive)

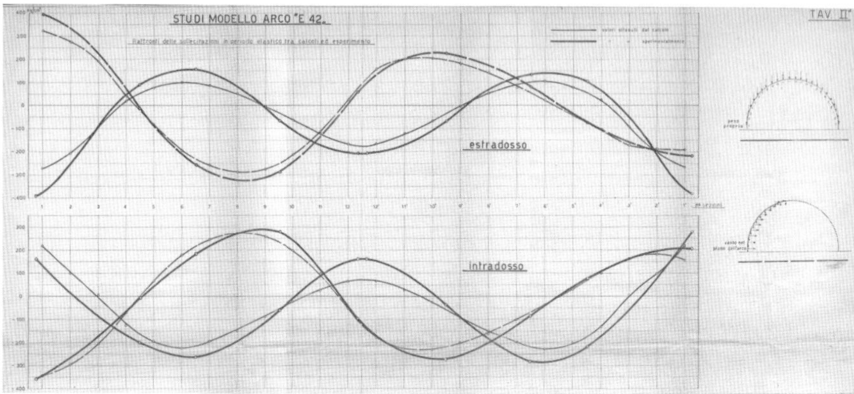

Figure 10.11 Diagram of stresses in the model of the E42 arch, tested at the Politecnico di Milano. (Image: ISMES Historical Archive)

m span. In August 1939 Aristide Giannelli (1888-1970), professor in Rome [39] who was hired as consultant, suggested new tests [40]. In the following months, experimental tests were carried out in different places: in Guidonia (Rome), with the collaboration of the Minister of the Air Force for wind testing; and with the Institute of Construction Science in Rome and of Danusso's Laboratory.

In 1940, Danusso assisted the 'S.A. Lavorazione Leghe Leggere' (the Building Company commissioned to realize the arch) in the checking of the new version of the arch. Initially, a reduced scale model made of aluminium alloy was planned, but later a celluloid model ($\lambda = 200$) was built. The latter was tested for dead loads and wind effects; then a series of tests was carried out until failure with progressively increasing vertical and horizontal loads applied. Despite all these efforts, the monumental arch was not built as the E42 exhibition was cancelled because of WWII.

10.5.2 Pavilion for the 1947 Milan Trade Fair

A major experimental study was carried out in the laboratory of Milan Polytechnic during this period for a pavilion at the Milan Trade Fair on behalf of Pier Luigi Nervi, though it was not constructed until just after the war (Figure 10.12). Nervi asked Danusso and Oberti to test the ferrocement elements for the semi-circular pavilion. Ferrocement was one of Nervi's most famous and important patents. He started developing applications of this material – which consists of a thin layer of cement mortar reinforced with closely spaced layers of small-diameter steel-wire mesh – before WWII and first tested it for a small warehouse, the Magliana Pavilion, built in Rome in 1945 which can be considered as a 1:1 model test of a ferrocement structure [41]. These experiments, as well as the semi-circular pavilion in Milan, were fundamental for the realization of Hall B for the Exhibition Palace in Turin (1947-48), one of Nervi's most famous works [42].

Figure 10.12 Semi-circular pavilion for the Milan Trade Fair, 1947, designed by Pier Luigi Nervi. (Image: CSAC, Parma)

The structural system of the pavilion canopy at Milan consisted of a tubular-steel frame, capable of supporting a wire-mesh, stretched between the frames themselves. This mesh was then impregnated with a special cement mortar. The 'ferrocement' structure thus obtained was sufficiently stiff to form, at the same time, the covering panels and the load-bearing frame. This system, given the characteristics of ferrocement, was waterproof in itself. Each individual bay covered an area of 64 m^2, with an average span of 16 m and width of 4 m. The internal surface of the canopy – the main decorative element of the pavilion – corresponded precisely to a conical surface. The latter was described by a straight-line generatrix rotated about a point outside the pavilion and lower than the canopy itself. It thus described an undulating movement of pre-determined shape [43].

Oberti carried out tests on a 1:5 scale model of three bays of the canopy, constructed in 12 days at the Polytechnic of Milan using the same material as the final structure, with some minor adjustments (Figure 10.13). Oberti concluded that 'The tests: (1) proved that the thrust of steel struts in the skylight had minimal impact on the stability of the main structure due to their very low stiffness relative to the whole envelope, and therefore their number was reduced in the final structure; (2) highlighted the significant lateral support provided by the continuous covering structure; this required that a strong horizontal

Figure 10.13 1:5 scale model of three bays for the semi-circular pavilion for the Milan Trade Fair, tested at the Politecnico di Milano, 1947. (Source: [43])

member had to be incorporated so that, in the event of breakage of the skylight, lateral displacements could be avoided that would have led to a functional weakening of the entire structure; (3) showed a reasonable local safety margin, almost uniform for equally-loaded points. The weakest section was between the continuous covered area (the bays at the extreme ends) and the area containing the skylights. This local zone was strengthened in the final structure by means of a suitable increase in reinforcement' [44]. Thanks to Oberti's investigations, the canopy and skylight were safely constructed.

10.5.3 Civic Centre for Tucumán, Argentina

Also during the 1940s, Oberti and Nervi worked together on a project in Argentina: the Civic Centre of Tucumán (also known as the University City Management Centre), designed by the Argentinian architect, Horacio Caminos (1914-1990) for the new local University [45] (Figure 10.14). It consisted of the repetition of huge triangular structural elements, with 20 m sides, made of conical, reinforced-concrete thin shells. These elements, whose orientation alternated (convex/concave), stood on columns, 20 m high. Nervi – presumably thanks to Oberti, who was part of an Italian team of professors in touch with the new University [45] – was entrusted with the static analysis of the structure, while Oberti was asked to undertake the experimental tests in 1949. Thus, the structural element was reproduced in a 1:25 scale ferrocement model, using a special cement mortar, reinforced with layers of a one-centimetre mesh of steel wires (0.4 mm diameter), arranged with a density that corresponded to the anticipated stress levels (Figure 10.15). In Oberti's words, 'The model worked very well as clever computer, and made it possible to appraise the stress state of the structure in the linear elastic range, after the first settling load cycles had been carried out. No visible crack appeared during or after the tests up to three times the expected maximum applied loads that would have led to maximum

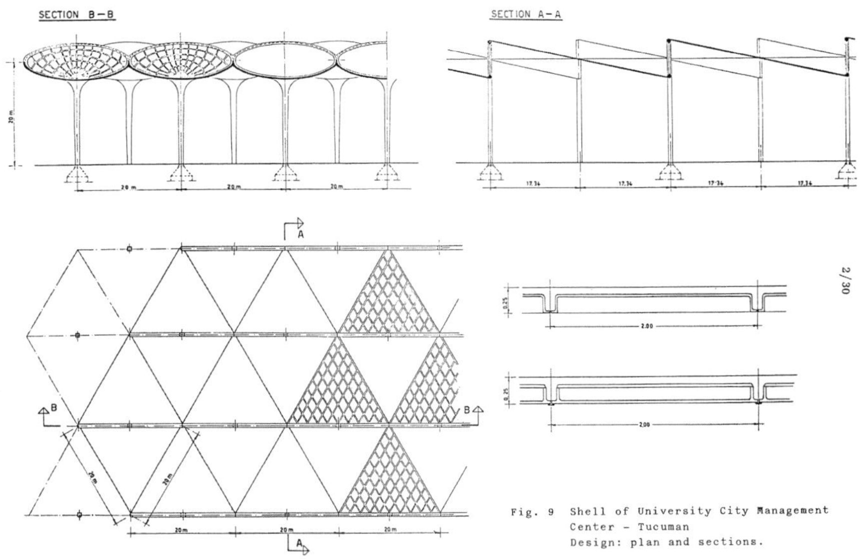

Figure 10.14 Plan and sections of the Civic Centre of Tucumán, Argentina in 1949. (Source: [43])

Figure 10.15 1:25 scale model of a structural element of the Civic Centre of Tucumán, tested at the Politecnico di Milano in 1949. (Source: [43])

tensile stresses of 50-60 kg/cm^2 in the full size structure [44]. Unfortunately, even though Nervi had already defined the construction method to be used, the project was not realized.

10.6 Final remarks

During the 1920s and 1930s a wide range of model-testing techniques and measuring instrumentation were developed in Italy, for example at laboratories in Turin, Rome and Milan. These were used mainly for the design of important innovative structures such as large dams, spatial structures and thin shells, made with relatively new construction materials including reinforced concrete and ferrocement. The laboratory set-up and led by Arturo Danusso at the Polytechnic of Milan soon became Italy's leading model-testing institution. Danusso and his assistant Guido Oberti were gifted experimentalists who used tests on large-scale and, later, small-scale models and, from 1936, photoelastic stress analysis to verify design calculations and to improve engineering designs. In 1935, Danusso and Oberti worked with Nervi for the first time and, as a designer of structures, he was able to exploit the benefits of model tests to the maximum. Several prominent projects on which they collaborated closely helped to establish an international reputation for the Milan laboratory. After WWII, the three engineers continued to collaborate on many innovative designs for which experimentation became, in an exceptional way, an extremely refined conjunction of art and science. Danusso and Oberti were also involved in setting up the Experimental Institute for Models and Structures (ISMES) in Bergamo, led by Danusso, whose activities are discussed in Chapter 15.

Further reading

Many further references on physical models not included in this chapter can be found in several key works by the authors, especially:

Chiorino, M.A., Sabia D. & Bruno, L. (2003) Structural Models: Historical Notes and New Frontiers. *Proceedings International Seminar Eduardo Torroja: from the Philosophy of Structures to the Art and Science of Building, Politecnico di Torino, November 2000* (eds F. Levi, M.A. Chiorino & C. Bertolini Cestari), 120-159. Milano: FrancoAngeli.

Chiorino, M.A. (2010) Reduced scale mechanical models in 20th century structural architecture. The case study of Pier Luigi Nervi. In: *Geometría y proporción en las estructuras, Geometry and proportion in structures. Simposio en homenaje a Ricardo Aroca* (ed. Santiago Huerta), 209-225. Madrid: Universidad Politécnica de Madrid.

Chiorino, M.A. (2010) Experimentation in the Work of Pier Luigi Nervi. In: *Pier Luigi Nervi. Architecture as Challenge* (eds Carlo Olmo & Cristiana Chiorino), 61-83. Cinisello Balsamo (MI): Silvana Editoriale.

Neri, G. (2010) En miniature: Die Modelle des I.S.M.E.S. in Bergamo. *Bauwelt* 101 (May): 20-27.

Neri, G. (2013) Pier Luigi Nervi's Reduced Scale Models for HP Shells. *Journal of the IASS* 176-177 (September): 149-158.

Neri, G. (2014) *Capolavori in miniatura. Pier Luigi Nervi e la modellazione strutturale.* Mendrisio: Mendrisio Academy Press / Cinisello Balsamo: Silvana Editoriale. https://doc.rero.ch/record/288862/files/Capolavori_in_miniatura.pdf (29/9/2019)

Neri, G. (2017) Pier Luigi Nervi al Politecnico di Milano. I modelli delle aviorimesse in cemento armato. In: *Pier Luigi Nervi. Il modello come strumento di progetto e costruzione* (ed. Giulio Barazzetta), 25-33. Macerata: Quodlibet.

References

1 Guidi, C. (1927) Etudes expérimentales sur des constructions en béton armé. *Bulletin technique de la Suisse romande* 53 (15): 181-187; 53 (17): 201-206; 53 (18): 213-215.

2 Guidi, C. (1903) *Le costruzioni in 'béton' armato. Conferenze tenute nel maggio 1900 nella R. Scuola d' Applicazione per gl' Ingegneri in Torino.* Torino.

3 Oberti, G. (1956) La sperimentazione su modelli strutturali in Italia. In: *I modelli nella tecnica. Atti del Convegno di Venezia, 1-4 ottobre 1955, vol. I,* 35-55. Roma: Accademia Nazionale dei Lincei. 35.

4 Oberti, G. (1956) La sperimentazione su modelli strutturali in Italia. In: *I modelli nella tecnica. Atti del Convegno di Venezia, 1-4 ottobre 1955, vol. I,* 35-55. Roma: Accademia Nazionale dei Lincei.

5 Chiorino M.A. (2010) Experimentation in the Work of Pier Luigi Nervi. In: *Pier Luigi Nervi. Architecture as Challenge* (eds Carlo Olmo & Cristiana Chiorino), 61-83. Cinisello Balsamo (MI): Silvana Editoriale.

6 Neri, G. (2014) *Capolavori in miniatura. Pier Luigi Nervi e la modellazione strutturale.* Mendrisio: Mendrisio Academy Press / Cinisello Balsamo: Silvana Editoriale.

7 Porcheddu, G.A. (1910) *Elenco dei lavori eseguiti in calcestruzzo armato sistema Hennebique.* Torino: Società Porcheddu Ing. G.A.

8 Locatelli, P. (ed.) (1978) *La scienza e lo spirito negli scritti di Arturo Danusso.* Brescia: Morcelliana.

9 Danusso, A. (1934) Le autotensioni. Spunti teorici ed applicazioni pratiche. In: *Rendiconti del seminario matematico e fisico di Milano,* VIII: 217-246.

10 Danusso (1937) Scienza delle costruzioni. Premesse e concetti fondamentali. In: *Scienza delle costruzioni,* Milano: Ravezzani Editore. 1-8: 7. Republished in: Locatelli, P. (ed.) (1978) *La scienza e lo spirito negli scritti di Arturo Danusso.* Brescia: Morcelliana.

11 Levi, F., Chiorino M.A. & Bertolini Cestari C. (2003) *Proceedings of International Seminar Eduardo Torroja: from the philosophy of structures to the art and science of building, Politecnico di Torino, November 2000.* Milano: Franco Angeli.

12 Neri, G. (2014) *Capolavori in miniatura. Pier Luigi Nervi e la modellazione strutturale.* Mendrisio: Mendrisio Academy Press / Cinisello Balsamo: Silvana Editoriale. 26-31.

13 Danusso, A. & Oberti, G. (1941) Il Laboratorio 'Prove modelli e costruzioni' dell'Istituto di Scienza delle Costruzioni del R.° Politecnico di Milano. *Il Cemento Armato – Le Industrie del Cemento.* 5: 1-5.
14 Chiorino M.A. (2010) Experimentation in the Work of Pier Luigi Nervi. In: *Pier Luigi Nervi. Architecture as Challenge* (eds Carlo Olmo & Cristiana Chiorino), 61-83. Cinisello Balsamo (MI): Silvana Editoriale.
15 Chiorino M.A. and Neri, G. (2011) I legami tra Pier Luigi Nervi e la cultura politecnica milanese e torinese: una stagione d'oro del'ingegneria e architettura italiana del Novecento. In: *Pier Luigi Nervi: Torino, la committenza industriale, le culture architettoniche e politecniche italiane* (ed. Sergio Pace), 131-143. Cinisello Balsamo: Silvana Editoriale.
16 Neri, G. (2014) *Capolavori in miniatura. Pier Luigi Nervi e la modellazione strutturale.* Mendrisio: Mendrisio Academy Press / Cinisello Balsamo: Silvana Editoriale. 35, 51-53.
17 Piga, C. (1996) *Storia dei modelli: dal tempio di Salomone alla realtà virtuale.* Seriate (BG, Italy): ISMES. 119, 145-162.
18 Oberti, G. (1980) The Development of Physical Models in the Design of Plain and Reinforced Concrete Structures. *L'Industria Italiana del Cemento* 9: 659-690. 663.
19 Danusso, A. & Oberti, G. (1932) *Indagini sperimentali sulle Costruzioni – La fotoelasticità.* Milano: Libreria Editrice Politecnica.
20 Oberti, G. (1980) The Development of Physical Models in the Design of Plain and Reinforced Concrete Structures. *L'Industria Italiana del Cemento* 9: 659-690. 664.
21 Oberti, G. (1980) The Development of Physical Models in the Design of Plain and Reinforced Concrete Structures. *L'Industria Italiana del Cemento* 9: 659-690. 665.
22 Abraham, G. (1933) Le stade Giovanni Berta, à Florence. *La Technique des Travaux.* 2 (febbraio): 93-101.
23 Olmo, C. & Chiorino, C. (2010) *Pier Luigi Nervi. Architecture as Challenge.* Cinisello Balsamo: Silvana Editoriale.
24 Neri, G. (ed.) (2014) *Pier Luigi Nervi: Ingegneria, architettura, costruzione. Scritti scelti 1922-1971.* Novara: Città Studi Edizioni. 5-20.
25 Nervi, P.L. (1968) Contributo italiano nel campo della sperimentazione su modelli. In: *Atti del XV Convegno Nazionale degli Ingegneri Italiani, Politecnico di Milano, 15-17 novembre 1968.* Milan: De Silvestri. 3-6.
26 Oberti, G. (1980) The Development of Physical Models in the Design of Plain and Reinforced Concrete Structures. In: *L'Industria Italiana del Cemento.* Special issue devoted to 'The Progress of Concrete in Italy'. 129-160. Rome: AITEC.
27 Neri, G. (2014) *Capolavori in miniatura. Pier Luigi Nervi e la modellazione strutturale.* Mendrisio: Mendrisio Academy Press / Cinisello Balsamo: Silvana Editoriale. 37-43.
28 Chiorino M.A. (2010) Experimentation in the Work of Pier Luigi Nervi. In: *Pier Luigi Nervi. Architecture as Challenge* (eds Carlo Olmo & Cristiana Chiorino), 61-83. Cinisello Balsamo (MI): Silvana Editoriale.

29 Oberti, G. (1980) The Development of Physical Models in the Design of Plain and Reinforced Concrete Structures. *L'Industria Italiana del Cemento* 9: 659-690. 674-675.

30 Nervi, P.L. (1931) Scienza o arte dell'ingegnere? *L'Ingegnere: Rivista Tecnica del Sindacato Nazionale Fascista Ingegneri* 7: 473-474.

31 Neri, G. (ed.) (2014) *Pier Luigi Nervi: Ingegneria, architettura, costruzione. Scritti scelti 1922-1971.* Novara: Città Studi Edizioni.

32 Nervi, P.L. (1931) *Scienza o arte dell'ingegnere? L'Ingegnere: Rivista Tecnica del Sindacato Nazionale Fascista Ingegneri* 7: 473-474. 474.

33 Danusso, A. (1952) Intuito e scienza nel cemento armato. In: *Il Corso di Perfezionamento per le costruzioni in cemento armato – Rendiconti e Pubblicazioni.* 53-62. Milano: Editrice Politecnica Tamburini. Reprinted in Locatello (1978): 79-91.

34 Danusso, A. (1952) Intuito e scienza nel cemento armato. In: *Il Corso di Perfezionamento per le costruzioni in cemento armato – Rendiconti e Pubblicazioni.* 54.

35 Danusso (1946) Sulla riforma dei politecnici (September). 1-8: 7. Republished in: Locatelli, P. (ed.) (1978) *La scienza e lo spirito negli scritti di Arturo Danusso.* Brescia: Morcelliana.

36 Nervi, P.L. (1956) *Structures.* New York: F.W. Dodge Corporation. 93.

37 Greco, C. (2008) *Pier Luigi Nervi. Dai primi brevetti al Palazzo delle Esposizioni di Torino 1917-1948.* Lucerna: Quart Edizioni.

38 Nervi, P.L. (1942) Un arco monumentale in conglomerato non armato. *Casabella-Costruzioni* 176 (Agosto): 23-25. 23.

39 Giannelli, A. (1936) *Lezioni di cemento armato dettate al R. Istituto Superiore di Ingegneria di Roma dettate al R. Istituto Superiore di Ingegneria di Roma, I anno 1935-36.* Roma: R. Istituto Superiore di Ingegneria.

40 La Torre, A. (1987) Arco monumentale. In: *E42. Utopia e scenario del regime* (eds M. Calvesi, E. Guidoni, & S. Lux), 467-470. Venezia: Marsilio.

41 Greco, C. (1994) The Ferro-Cemento Experimental Storehouse by P.L. Nervi, Rome, 1945. In: *Proceedings of Docomomo Third International Conference, Barcelona, 16-19 settembre 1994.* Barcelona.

42 Olmo, C. & Chiorino, C. (2010) *Pier Luigi Nervi. Architecture as Challenge.* Cinisello Balsamo: Silvana Editoriale.

43 Oberti, G. (1981) Structural design and testing, by means of models, of some special constructions (using ferro-cement). *Preprint of papers in Proceedings of International Symposium on Ferrocement held at ISMES, Bergamo 22-24 July 1981; 2nd Session. RILEM-ISMES* (eds G. Oberti and S. P. Shah). Paper 2.3. 138.

44 Oberti, G. (1981) Structural design and testing, by means of models, of some special constructions (using ferro-cement). *Preprint of papers in Proceedings of International Symposium on Ferrocement held at ISMES, Bergamo 22-24 July 1981; 2nd Session. RILEM-ISMES* (eds G. Oberti and S. P. Shah). Paper 2.3.

45 Neri, G. (2014) *Capolavori in miniatura. Pier Luigi Nervi e la modellazione strutturale.* Mendrisio: Mendrisio Academy Press / Cinisello Balsamo: Silvana Editoriale. 52-53.

11

Eduardo Torroja and his use of models up to 1936

Joaquín Antuña

11.1 Introduction*

Between 1931 and 1935, the engineer Carlos Fernández Casado (1905-1988) published several articles in the *Revista de obras Publicas* under the generic title of 'Teoría del arco' [1]. In the second of them, in September 1931, he classified the methods that were available at that time for obtaining quantitatively the forces in structures. He mentioned three types of procedure – analytical, graphical and experimental and, among the latter, he included four types: tests on 'equivalent reduced-size structures', tests on 'reduced-scale models', photoelastic stress analysis (see Chapter 12) and the experimental interpretation of the Beggs method using his deformeter (see Section 13.5). (An 'equivalent reduced-size structure' is one in which mechanical properties, such as absolute and relative stiffnesses, are similar, while precise geometric similitude is not maintained.) In the following months, until January 1933, he published six more articles under the same title showing the experimental results obtained using equivalent reduced-size models and reduced-scale models of arches (bridges and arch dams) obtained in several countries. No reference was made in the articles to experiments carried out in Spain.

A few years earlier, Eduardo Torroja (1899-1961), working for the company Hidrocivil, carried out the Tempul aqueduct project. The crossing of the Guadalete River was resolved using a cantilever construction to eliminate two piers (Figure 11.1). As Torroja wrote: 'This is a simple project consisting of a series of straight sections of 20 m span, plus a span of 60 m over the main riverbed, consisting of two cantilever sections of 20 m and a central section of similar span, supported on the ends of these' [2]. The cantilevers were made up

* In this chapter, and the later one on Torroja's post-war work (Chapter 16) source documents about the his projects can be consulted at the Archivo Eduardo Torroja deposited in the *Centro de Estudios Históricos de Obras Públicas y Urbanismo* (CEHOPU) http://www.cehopu.cedex.es/etm/indices/obraindx.htm . References to his projects contain two numbers; one is the number of the project file in Toroja's Oficina Técnica and the other is the number given in the Archivo Eduardo Torroja in CEHOPU where the content can be consulted or viewed online. For each project, references are given to addresses where conserved documents can be consulted.

Physical Models: Their historical and current use in civil and building engineering design,
First Edition. Edited by Bill Addis.

Figure 11.1 The Tempul aqueduct, 1927. (Image: AET, CEDEX-CEHOPU)

of two 20-metre-long sections supported by concrete-encased cables supported on a frame placed on 6 m high towers built on piles. The ties consisted of two cables capable of supporting 7800 kN, that is, about four times the maximum load to which they would be subjected. In order to balance the cantilevers, a counterweight was provided at the ends. The description of the project dates from May 1925 and the contract for the works was awarded in August 1925. The manoeuvre to lift the heads of the towers in order to tighten the ties was carried out in October 1926. The contract specified a load test in which the lower slab was loaded with a uniform load of 6 kN/m^2. The work was completed in May 1927 [3].

To carry out the operation of loading the ties by jacking up the saddles from the towers, Torroja considered it necessary to instrument the structure in order to measure the movements that occurred: the vertical, horizontal and torsional displacements of the end of the cantilevers (Figure 11.2). The vertical displacement of the frame needed to separate the aqueduct channel from the formwork was measured, which was 400 mm above the towers to achieve 50 mm at the

Figure 11.2 The Tempul aqueduct during tests to determine the rise of the saddle over the tower to lift the channel clear of the formwork. (Image: AET, CEDEX-CEHOPU)

end of the cantilever. By measuring of the extension of the cable end (20 mm) when applying the test overload, a value of Young's modulus for the composite cable was calculated as 210 kN/mm^2, which, being the same value as steel, indicated that, once the cables were encased in concrete and firmly fixed at each end, the concrete was moving together with the steel. The effect of the temperature variation was also checked, and discovered to be different for the ties and the spans.

It was, without a doubt, a novel project, for which there was no precedent and, although the solution was similar to that used by Freyssinet on the Laon bridge in the same year, it is unlikely that Torroja knew of it [4]. He encountered at least three difficulties: the first was to select the right material for the ties, the second was to devise a suitable construction process and, finally, the need to check that the calculated stresses were actually occurring. Each tie consisted of twisted wire cables fixed to the cantilever at each end, passing continuously over the saddle. In this way, using jacks, the whole section could be raised, allowing the cables to take load until the end of the cantilevered section was raised and separated from the formwork. The ends of the tie rods were grouted in place after the formwork had been removed. When describing the work, Torroja paused to relate the process of applying the load, with which he was able to study the elastic properties of the material forming the tie, as much as the construction of the structure as a whole. With this project, almost the first of his professional career, Torroja was able to demonstrate that the experimentation in the work itself made it possible to verify the expected results. This would become a strategy that he used throughout his career. This work marks the beginning of a manner of working in which highly original works would follow one another, for which there were neither precedent nor sufficiently reliable tools of analysis. For that reason, he would regularly use experimentation as a means of verifying that the degree of safety of a structure was sufficient. However, experimentation on the finished structure only allows the expected results to be verified. The work is already finished and the modifications that can be made at a reasonable cost are few. Experimental methods in laboratories that were beginning to be used at the conception stage in a number of countries were a viable and readily available alternative. At an affordable cost, the proposed structure could be studied and, if necessary, appropriate modifications and corrections could be proposed.

In the following years, Torroja began conducting tests on models, combining them with tests on the finished works. In order to carry out both tasks, a highly-specialized infrastructure was required, which could be specified in three needs:

- precision instruments to make the necessary measurements, both on site and in the models;
- specialised technical staff capable of handling the instruments with precision;
- craftsmen capable of making the models.

This chapter will discuss the works carried out by Torroja before 1936, for which there are data regarding experimental procedures that were used to verify the structure. Included are works for which tests were undertaken on equivalent reduced-size models, as well as others in which the tests were carried out on full-scale models or on site. This gives a clear idea of what can be obtained from each type of test, and its economic implications. In structures where the main uncertainty is the construction process itself (e.g. the grandstand roofs of the Madrid hippodrome, see Section 11.2.3), the construction of the full-scale model is what best provided the necessary information. In cases where the method of analysis used involved a certain degree of uncertainty (e.g. the roof of the *Frontón Recoletos* see Section 11.3.3), the model test made it possible to confirm the results obtained from analysis. For this chapter, the works are organised into two large groups, those for which trials were carried out on full-scale models, and those for which reduced-scale models were used.

11.2 Tests on full-scale models

Initially, Torroja's engineering activity was focused on public works projects. Those were the years when Rafael Benjumea was Minister of Public Works and there was a significant increase in the contracting of works by the administration. However, this situation ended in 1928 and so Torroja had to look for a new job. He found it in the *Ciudad Universitaria* (University City) in Madrid, the construction of which was just beginning [5]. This new job determined his professional future. In fact, the Technical Office of Eduardo Torroja kept 135 files for projects at the *Ciudad Universitaria,* from the first in 1929 until the last in 1961, out of a total of 920. In this new working environment, Torroja began collaborating with a series of architects on the designs of new buildings. Of all of them, for the purpose that interests us here, the collaboration with Manuel Sánchez Arcas (1897-1970) was fundamental. In the *Ciudad Universitaria* they collaborated on the clinical hospital and the thermal power station.

11.2.1 The *Hospital Clínico* in the University City in Madrid

The *Hospital Clínico* was a conventional concrete building [6], in which two singular constructions were proposed. These were two novel designs for which there were few precedents in Spain. On the one hand, the floor slabs without beams on the balconies and, on the other, the skylight roof of the operating theatres [7] (see Section 11.3.1).

The solid reinforced-concrete flat slabs on separate supports had already been used in several previous works and been known since at least 1905 [8]. In Spain Torroja had used this system before, for the roof of a water supply tank in the Tetouan in Morocco [9]. The danger of the columns punching through the thin slab was resolved by enlarging the section of the column at the upper end to form a capital. The column-slab connection was achieved by distributing the reinforcement in several directions and on both faces of the slab. In the case of the Tetouan tank, the support was reinforced by increasing its thickness at the upper end and with a top screed of the slab in the form of a cone above the column. The slab reinforcement at the support was arranged in four layers, a group of orthogonal bars arranged in the direction of the support grid and, superimposed, another grid rotated by 45°.

The cantilever slab at the *Hospital Clinico*, square on plan, rests on three supports, two of them joined by a beam with a greater depth than the slab on two of its sides (Figure 11.3). Part of the slab is cantilevered and also rests on a separate support. In this section the bending moments were defined and the reinforcement could be determined. The rest of the slab rests on the beam arranged diagonally between the other two supports and on both sides. In this part, Torroja had no way of knowing the distribution of the bending moments and, therefore, of deciding the distribution of the reinforcement. The usual solution today would be to have a homogeneous reinforcement with bars arranged orthogonally on both sides, to which the necessary extra reinforcement is added. It was not possible to obtain the 'exact' distribution of the bending moments to decide the precise placement of the reinforcement, and to check the chosen layout a test was carried out on a full-scale model (Figure 11.4). It was built on site and, although no details of the test results are preserved, a scheme was published describing the qualitative movements of the slab due to the vertical load. This part of the building was seriously damaged during the years of the civil war, in which it was the front line of combat for two and a half years. At the end of the conflict the balconies were rebuilt following the same design.

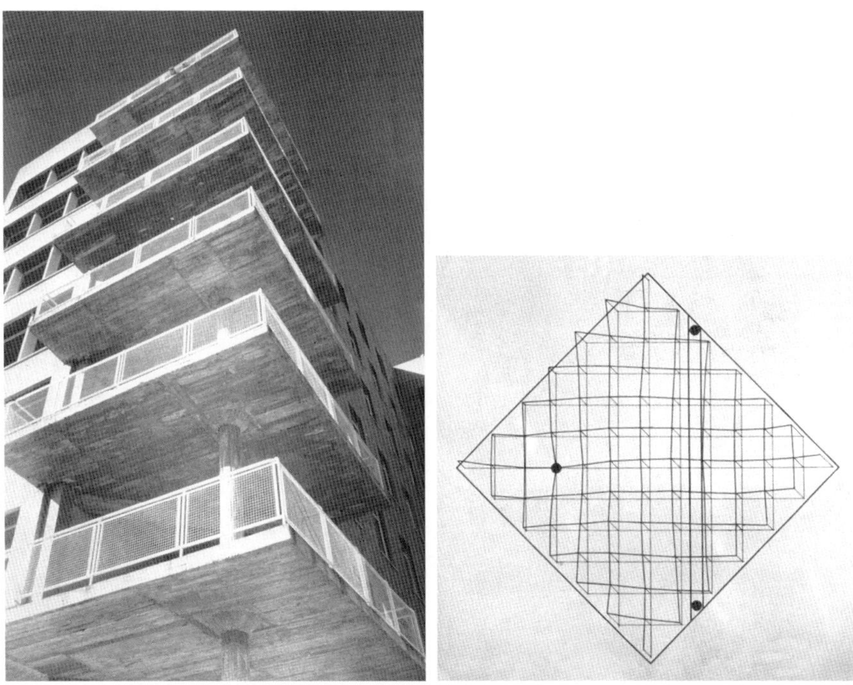

Figure 11.3 The Hospital Clínico. (a) The completed building. (b) Deflections of the slab. (Images: AET, CEDEX-CEHOPU)

Figure 11.4 The cantilever slab section of the Clinical Hospital. (a) During construction when the slab reinforcement. (b) Load-testing the slab. (Images: AET, CEDEX-CEHOPU)

11.2.2 Courtyard roof at the *Escuela elemental de trabajo* (Elementary Technical School), Madrid

In 1934, Torroja, along with other engineers and architects, collaborated in the foundation of a private institution, the *Instituto Técnico de la Construcción y Edificación*. One of the objectives was to disseminate in Spain the most advanced techniques in the field of construction. At the time, the reinforced-concrete thin shell was the technique which offered most possibilities. Franz Dischinger's invention was barely 7 years old and his achievements had been published in various journals and international conferences (see Chapter 9). In addition to the durability and fire resistance of the concrete, a thin shell was also very light: it was possible to build shells just 5 cm thick, with a self-weight of just 1.20 kN/m^2.

The extension of the school was a work by the architect Manuel Martínez Chumillas and the aim of the project was to make a roof in the courtyard of an existing building. The dimensions of the room were 22.0 x 8.0 m. A conventional solution, with a reinforced concrete reticular structure of beams, columns and a concrete slab, would mean an average of 20 cm of concrete per square metre, in addition to the columns and their foundations. The self-weight of the floor slab would be around 5 kN/m^2. As an added drawback, the foundation for the work was about 4.0 m below the level of the courtyard where the roof should be made [10].

Compared to the conventional solution, with high self-weight, supports and an expensive foundation, Torroja proposed the construction of a barrel shell with an elliptical section (major axis, 8 m; minor axis, 5 m) supported on edge beams with variable depths. The self-weight was reduced by almost a quarter, the required formwork area was reduced by half, and both the columns and the foundation could be eliminated. Indeed, the total weight of the roof was only around 280 kN and, suspended from four ties, it could be hung from the building's own façade. The 70 kN weight of each tie increased the total weight of the façade by only about 5%. Taking into account the weight of the slabs, the impact of the weight of the deck would be even less.

The solution was clearly competitive: it weighed less, required less material than a conventional solution and, in addition, the columns and foundation with their corresponding excavation were eliminated. The economic advantage was

Figure 11.5 Courtyard canopy at the Elementary Technical School, Madrid. (a) Preparation for the load test, showing projecting reinforcement for construction of the stiffening rib, should it prove necessary. (b) Load testing of the canopy. (Images: AET, CEDEX-CEHOPU)

unquestionable. The only difficulty was its novelty; it was the first construction of this type to be carried out in Spain. Dischinger had already built similar ones successfully, several years earlier, and had published the theoretical analysis and the results of the trials he carried out. Torroja analysed his shell following the procedure proposed by Dischinger. Despite this, the extreme slenderness of the shell could lead to fears of instability problems. In the documentation for the project that is conserved, it can be seen that the proposed shell had, in addition to the edge beams at the ends of the shell, two other stiffening ribs of similar shape added at the third points of the span. In the images of the construction of the roof that are conserved, one can see the reinforcement below the timber frame awaiting the construction of the stiffening ribs (Figure 11.5a). However, before the ribs were constructed, a load test was carried out (Figure 11.5b). Although the results of the test are not preserved, the fact that the planned ribs were not built, suggests that the stiffness of the shell was sufficient and the ribs were found to be superfluous.

This decision was significant. With the shell in place, the additional cost of casting the two ribs would have been negligible; yet, nevertheless, Torroja chose not to construct them and was thus able to show the shell without ribs covering the 22 m span. It was an authentic declaration of intent on the part of Torroja: he wanted to build thin shells without the need for stiffening ribs. This episode demonstrates the potential for load tests to be used to make design decisions about the structure itself, even when they are carried out on the nearly-finished work.

11.2.3 Roof of the grandstands of the Zarzuela Hippodrome, Madrid

In the same year that the roof of the elementary school was built, 1934, the architects Carlos Arniches and Martín Domínguez, with whom Eduardo Torroja collaborated, won the competition to build the *Hipódromo de la Zarzuela*, the race course grandstand, in Madrid. Agustín Aguirre, director of the company in charge of the construction, has claimed that it was he who proposed the construction of a model of an element of the canopy at full scale, a decision justified by his desire to check the strength of the roof deck [11]. However, another reason, probably more important, was the need to define the shape of

Figure 11.6 Canopy of the *Hipódromo de la Zarzuela*, Madrid. (a) Load test of one bay of the shell. (b) The shell after being loaded to collapse. (Images: AET, CEDEX-CEHOPU)

the roof deck and the arrangement of the reinforcement. The detailed drawings of the roof element and the reinforcement layout are dated June 1935 [12].

In this case, it was a rupture test that was undertaken to confirm the predicted bearing capacity; in fact the collapse load was several times higher than expected [13]. However, the trial revealed the weakness of the corners at the end of the cantilever, and their excessive flexibility. In the photographs of the collapse that are preserved, one can see the excessive deformation of these edges, as well as the breakage that occurs on the upper part of the shell where the steel reinforcement bars change direction (Figure 11.6). Nevertheless, no action was taken and the design was not modified. Only after the civil war, when proposals were made to repair the damage suffered by the roof, were the corners of the end modules of the canopies removed and re-made in their initial form, but reinforced above with ribs.

11.3 Tests on equivalent reduced-size models

Once a project was completed, the carrying out of experiments and load tests was not difficult and the cost was low. It was enough to apply, in specified ways, a load of a value equal to the maximum expected load, and to measure the observed deflections. The displacements were typically several tens of millimetres and could be measured easily. Tests on the finished construction enabled a large amount of information to be obtained at a low cost. However, in return, all that could be obtained was confirmation of the anticipated results. If the tests revealed the need to modify the design, the cost of improvements was high, since the work was finished.

Although the use of models for catapults for firing rocks or bolts has been documented since ancient times (see Section 1.2), the use of structural models only allowed conclusions to be drawn about structural stability. At least since Galileo, it was clear that strength and stiffness could not be extrapolated from small scale to larger sizes. Fernández Casado mentions the results of several tests on arches [1] and Dischinger's tests on domes and cylindrical vaults carried out at reduced scale (see Chapter 9) were already known when Torroja started working at the *Hospital Clínico*. In order to obtain significant results using model tests, it is necessary to have measuring instruments with the appropriate precision. Therefore, to undertake model testing meant also the need to develop the necessary measuring devices.

With the growing use of reinforced-concrete, the possibility of constructing structures with a high degree of statical redundancy was opened up. However, the capacity for theoretical analysis was limited and only simplified methods had been developed for analysing structures such as planar frames. To take advantage of the possibilities of the new material, it was necessary to find new techniques to validate proposals for new forms. In these circumstances, the construction and testing of reduced-scale models was a route that offered great possibilities. In the mid-1930s, Pier Luigi Nervi was beginning to experiment along similar lines (see Chapter 10).

11.3.1 The roofs of operating theatres at the *Hospital Clínico* in the University City in Madrid

The construction of the new medical faculty was at the heart of the University City project. Following the example of some American universities, one of the requirements was to build a University Hospital associated with the Faculty of Medicine, whose activity was a fundamental part of the training. That need led to the design of the new operating theatres like small amphitheatres in which students and researchers could witness the operations. For this purpose, rooms with a circular floor were designed, in which operations were carried out in the centre with tiered seating arranged around the sides. The roof of the theatre was supported on sixteen columns arranged at the vertices of a regular polygon, 21.9 m in diameter (Figure 11.7).

A central skylight provided good lighting and ventilation. It had an octagonal prismatic body, 11.1 m in diameter and 4 m high, that protruded above the roof plane. The structure consisted of the skylight supports arranged at the vertices of the octagonal prism joined to two horizontal beams that connected the lower part of these supports with two corresponding columns in the outer perimeter. The supports of the cylindrical skylight were joined together by two octagonal rings, at the top and the base of the lantern. The mechanical scheme of the roof can be understood as a central, rigid octagonal cylinder, from which 16 beams cantilever out to the perimeter supports. Thus, the bending stresses in these 16 beams would be determined by dividing the total weight of the lantern structure by 16 and multiplying it by the span of the cantilever beams. However the elastic analysis of what would happen in the rest of the structure was not easy to solve.

To study this structure Torroja undertook experiments with an equivalent reduced-size model (Figure 11.8). While no data from the tests have survived, a photograph of the model shows that it was an equivalent reduced structural model, not one made to a precise scale reflecting the sizes of the individual members and connections in the built structure.

When thc roof had been built, a series of load/deflection tests were undertaken on the final structure for comparison with the model tests. There survives a collection of drawings on which deflections and rotations of the vertices in the

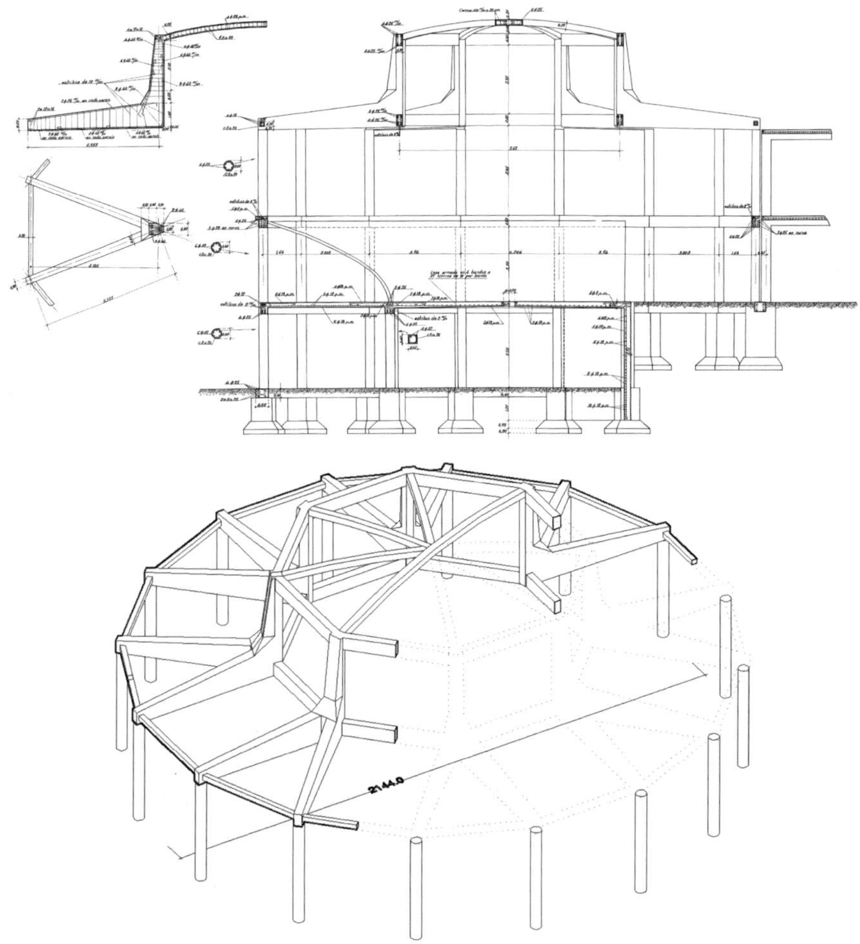

Figure 11.7 Roof of operating theatre at the *Hospital Clinico*. (a) Cross section. (Image: Archivo Ciudad Universitaria). (b) Isometric view. (Image: the author)

structure are noted, although there are no indications of what type of load was applied, nor whether they were the results of the tests on the model or those of the load test on the completed structure. Several roofs of this type were made, at both the University Hospital and the Faculty of Medicine, which are still in use today.

11.3.2 The market hall in Algeciras, Spain

In 1932, Manuel Sánchez Arcas and Eduardo Torroja began working on a project to build a covered market in the city of Algeciras in Andalucía. At the same time, two large markets were being built in Madrid: the fish market in Puerta de Toledo, and the fruit and vegetable market in Plaza de Legazpi. The structure of the latter was conventional – reinforced-concrete beams and columns supporting a grid of secondary beams and a slab about 10 cm thick.

Figure 11.8 Roof of operating theatre at the *Hospital Clinico*. (a) Equivalent reduced-size model of the roof (scale approx. 1:35). (b) The roof under construction. (Images: AET, CEDEX-CEHOPU)

In Algeciras, the daily market was held in a roughly square *plaza*, where the stalls were set up outdoors. The aim of the project was to build a roof to shade both traders and customers from the sun. The economic aspect was crucial and obtaining an economic solution was a priority objective. Market buildings had already become established as highly suitable for covering with reinforced-concrete thin shells, so it is not surprising that Manuel Sánchez Arcas and Eduardo Torroja decided to use a shell in Algeciras [14].

The 40 m diameter dome for the Schott Company in Jena, designed by Walter Bauersfeld and Franz Dischinger had been completed in 1925. It was a spherical cap supported on columns at less than 5.0 m centres with a deep circumferential beam between them. Its structure provided great rigidity, both horizontally and vertically, and it could be assumed that the support of the spherical cap was continuous. With these conditions of support, it could be considered that the edge of the dome did not move, and its behaviour could be taken to be that of a rigid membrane, in which bending could be neglected and only the stresses in the plane tangential to the shell need be considered.

The roof proposed by Arcas and Torroja presented a great novelty in comparison to the dome at Jena. The shell is supported on eight columns, joined at their tops by eight steel tie rods, linked to form a complete ring (Figure 11.9). These were encased in concrete when the roof was complete. To stiffen the edges of the dome between the columns, there are eight sections of radial cylindrical vault. In fact, the engineer Ildefonso Sanchez del Río (1898-1980) had resolved the edges of the domed roofs of several water tanks (1926-1928) in a similar way, though at a smaller size [15]. At Algeciras, Torroja had created an entirely original type of roof.

In his theoretical analysis of the roof, Torroja shows in a note about the project that it is possible to obtain the stress state of a regular sphere. However,

Figure 11.9 Market hall in Algeciras, Andalucía, 1934. (a) Shell under construction, showing the perimeter ties before encasement in concrete. (b) Interior view. (Images: AET, CEDEX-CEHOPU)

he confesses the impossibility of doing so with the dome section that he was proposing for his project, in which the dome is supported at only eight points, and the edge has the irregularity of the intersection of two cylindrical shells.

In spite of this difficulty, the analysis of the roof was carried out, and has been conserved; there are two parts. In the first, the shell is considered to behave like a rigid membrane and the calculation yields a value for the outward thrust and its direction. With this value, the necessary cross section of the tie rods was decided. The second set of calculations studied the effects of temperature variation in the tie rods that would cause an increase in length and a corresponding separation of the columns, causing bending in the shell. To complete the analysis, Torroja stated that it would be necessary to carry out tests on a model (Figure 11.10).

The only reference to the tests is in a document by the contractor Agustín Aguirre. We do not know exactly what material was used for its construction, nor the results obtained, although it is likely that a scaled-down form of reinforced concrete was used (micro-concrete). The most notable result of the trial was the sudden collapse of the dome, having been left under load for a few hours.

Figure 11.10 1:10 (estimate) scale model of Algeciras market hall roof, 4.76 m in diameter. (Image: AET, CEDEX-CEHOPU)

It was discovered that the failure had been caused by the fracture of one of the bars that formed the tie beam connecting the heads of the eight columns.

At least this conclusion allowed an important modification to be made to the design. Although the overall design was not modified substantially, changes were made to the composition of the tie rods and to the planned process for decentring the shell and introducing loads into the tie rods. On the one hand, the test was able to reproduce the proper behaviour of the dome but, on the other, it exposed the great sensitivity of the strength and stability of the dome to any defect in the tie rods.

In general, in singular works such as this one, both the definition of the project and the construction process are usually closely related. In order to reduce the effects of the elongation of the tie rods when it started to carry loads, a detailed loading procedure was defined in which the dome was separated from the formwork by tensioning the circumferential reinforcement. The initial plan described in the project report was to generate this tension by hanging weights from the centre of each of the tie rods. This was changed for the actual construction in which the tension was applied using screwed connectors in the tie rods.

The proposed procedure was imprecise and difficult to control. The initial proposal from December 1933 with 6 bars (of which the diameter is not indicated in the drawing, but a note by Torroja suggests they would be 50 mm in diameter, total area 118 cm^2), was replaced in June 1934 by a similar solution formed by 16 bars of 30 mm in diameter (total area 113 cm^2) with threads at the ends [16] (Figure 11.11). In this way, a progressive and controlled load could be applied to each bar.

In this project, the test on the equivalent reduced-size model made it possible to correct the arrangement of the tie rods and suggested an improvement to the construction method, which undoubtedly facilitated the construction and improved the final result. The project was successfully completed in 1935 and is still in use today. In 1997 the building was refurbished, during which the tie rods in the tension ring were renewed [17].

11.3.3 The *Frontón Recoletos*, Madrid

The *Frontón Recoletos* was a remarkable building for many reasons, and one of these was the sophisticated and complex model of the structure that was made and tested prior to the construction of the roof. The most outstanding and novel part of the building was the enormous roof (32.5 x 55.0 m), whose design was defined in September 1935. In the same month construction of the building began and, at the same time, tests were carried out on the model. Decentring of the shell roof was carried out between 18 and 20 January 1936. In less than twelve months, in addition to the rest of the project, the type of roof was devised and its entirely original form was decided upon, which was then precisely defined and analysed, and a reduced-scale model of the shell roof was built and tested which allowed some adjustments to be incorporated into the project, and construction was completed. A remarkable achievement indeed [18].

The building had a rectangular plan, covering the playing area for the Basque game of *pelota*, with two levels of seating along one of the long sides, and with

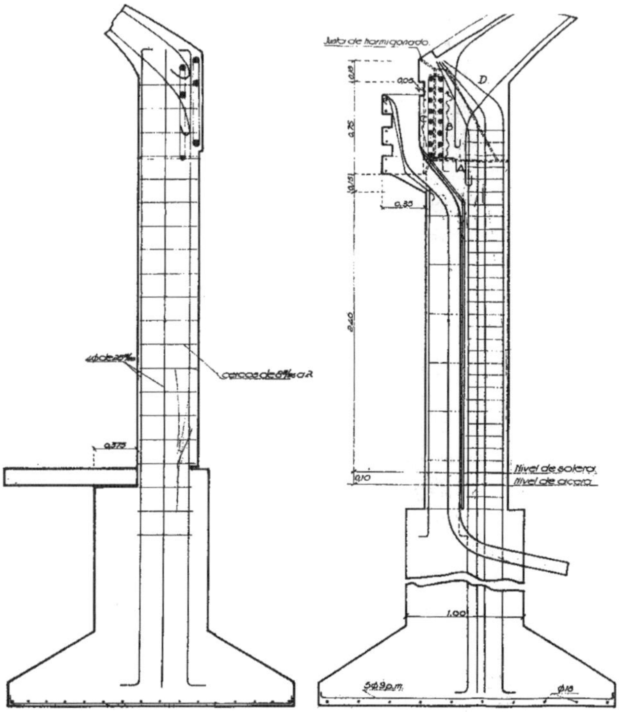

Figure 11.11 Algeciras market. Section of a column showing the original scheme (left) with six tie rods and the final scheme (right) with sixteen tie rods. (Image: AET, CEDEX-CEHOPU)

the necessary accesses (Figure 11.12). The most significant element consists of the unique roof design that protects the enclosure. The roof is a cylindrical vault of reinforced concrete with two unequal lobes, oriented in the longitudinal direction, supported at each end by the end walls, and by the external walls on the long sides. The shell was generally 90 mm thick, and large parts of the continuous shell were replaced by a triangulated lattice made of concrete struts with a larger section than the thickness of the sheet itself, to introducing daylight.

The size of the vault was more than twice that of two other cylindrical vaults that Torroja had made and tested. It was also one third larger than the shell roof designed by Ulrich Finsterwalder for the market hall in Budapest (1931), for which he used a new method of analysis for this type of structure in which, in addition to membrane stresses, bending stresses were also taken into account.

To analyse his roof, Torroja used the calculation procedure developed by Finsterwalder, although adopting important simplifications, such as that of assuming the continuous roof surface, without considering the irregularity of the triangulations, in both the shape and the variation of loads due to self-weight, and without representing precisely the way in which the shell was actually supported along the long perimeter walls (on many short columns). In order to check both the collapse load of the roof and the deflections due to the various loads, a 1:10 scale model was built using reinforced mortar (model concrete) and

Figure 11.12 The interior of the Frontón Recoletos, Madrid, 1935. Engineer: Eduado Torroja; architect: Secundino Zuazo. (Image: AET, CEDEX-CEHOPU)

tested. It was therefore a model which could be used to test the elastic behaviour of the roof as well as, finally, determining the ultimate load causing collapse.

11.3.3.1 Description of the model

The realization of the model was complex, both because of the difficulty of applying the loads, and because of the need to obtain data about the strains on the inner surface of the shell. The main objective was to check the results for the displacements and rotations obtained from the theoretical analysis. In order to do this, the aim was to obtain the displacements relative to the central section of the vault for the different load combinations that had been analysed: self-weight, self-weight plus snow, self-weight plus wind, and self-weight plus wind and more snow.

On the other hand, previous experience gained while testing the dome of the Algeciras market had shown the serious danger of collapse of the model, with the possibility of serious accidents to the staff conducting the tests. For this reason, it was carefully planned how best to carry out the test without needing to go underneath the model during the load tests. In addition, the test apparatus was equipped with an appropriate safety mechanism to prevent the whole model from collapsing to the ground in the event of a local failure or the breaking of cables applying the loads.

The model of the roof was made with reinforced mortar: the compressive strength of the mortar was measured as 21 N/mm^2, and the steel reinforcement wires had a tensile strength of 380 N/mm^2, with an elastic limit of 220 N/mm^2. The model reproduced to scale the shape of the roof as well as that of the walls that supported the ends of the shell, and the supports for the long edges of the shells. The latter were elevated on walls, so that there was enough headroom under the roof to work on the model (Figure 11.13). In addition to the roof itself, the model had to have three further elements: a fixed reference platform from which measurements of movement could be made, a system for applying the various loads, and a system for measuring deflections and bending. Under the

Figure 11.13 1:10 scale model of the Frontón Recoletos, 3.5 m wide, 3 m high. Interior view when awaiting installation of the loading apparatus. (Image: AET, CEDEX-CEHOPU)

roof and resting directly on the same floor as the model, there was a platform that served both as a base on which to place the square plates used as datum surfaces from which to make the measurements, and as a means of access to the interior of the model for assembly and maintenance of the whole. The key internal features of the model are shown in Figure 11.14.

The system for applying the loads was complex and had at least three noteworthy qualities. First, it was decided to simulate the uniformly distributed loads by a number of point loads located at a maximum distance of 200 mm from each other. For this reason, the roof was provided with holes for fixing a plate to which the cables applying the loads could be anchored (Figure 11.15). Second, the load application system was organized so that if any of cables applying the loads were to break, and the load no longer applied to one of the point loads, the entire set of loads acting on the same cross-section of the shell would be automatically unloaded. This would avoid abrupt imbalances that could damage the model due to application of asymmetric loads that, in reality, would never occur. Finally, the loading system as able to apply the loads slowly and, in an emergency, unload the structure instantly. The loading system would also automatically unload the model if a vertical deformation of more than 10 mm were exceeded. In this way, the model could be inspected at the first moment when failure began, and its causes could be found.

11.3.3.2 Description of the test, application of loads

Three types of load were applied, self-weight, wind action and snow action. To represent the self-weight, a reinforced concrete slab and a beam of the same material were used, located in the lower part of the model. To represent the wind, counterweights outside the model were used, and the action of the snow was simulated by placing sand directly on the model.

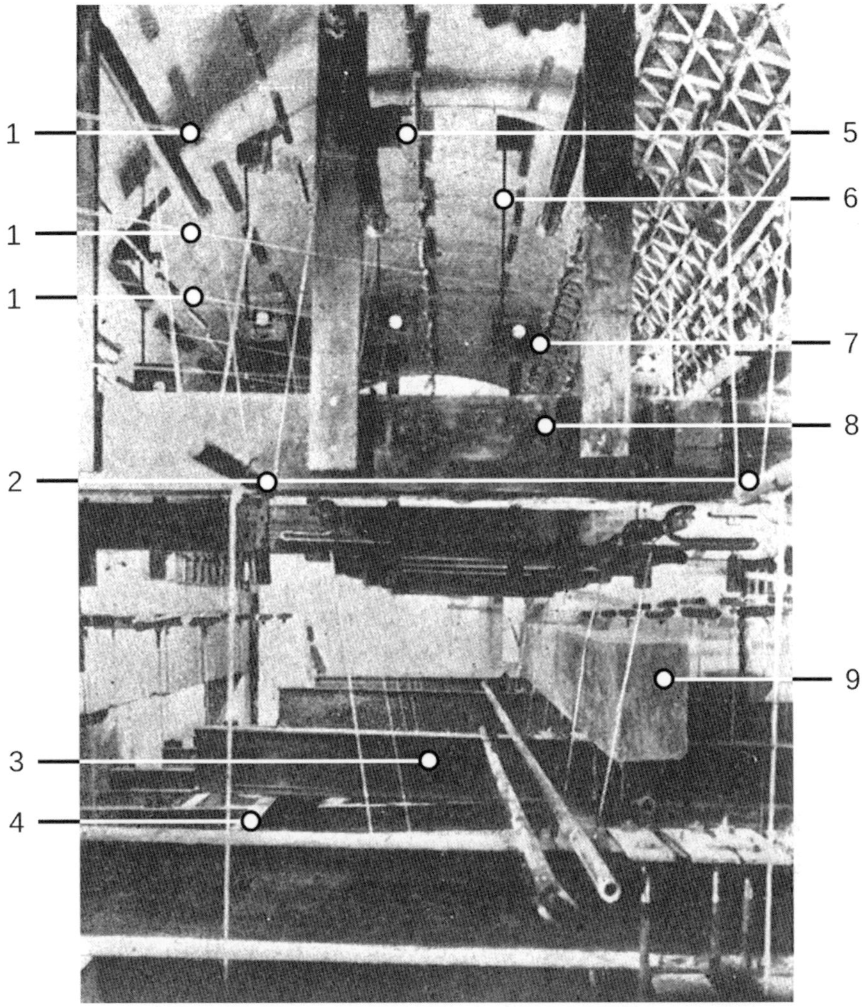

Figure 11.14 Interior of the model. 1: Cables used to apply the wind loads radially. 2: Load distribution device. 3: Load-distribution beam type D. 4: Load-distribution beam type B. 5: Plate fixed to the vault at the point where the movements are to be measured. 6: Needle and plummet used as a reference to measure movements. 7: Square surface used as a reference for taking measurements. 8: Fixed platform on which the reference plates are placed. 9: Concrete beam used as a counterweight. (Image: AET, CEDEX-CEHOPU).

A detailed description of the test facility has not survived. However, from the documentation that does exist it can be seen that, to apply the loads, a series of beams (type A) were suspended by four points by means of two cables anchored to the vault with an intermediate structure. In the centre of beam A is supported a longitudinal beam (type B) and, on this, another transverse beam (type D). Finally, from this beam another longitudinal beam (G) was suspended (Figure 11.16).

Using hydraulic jacks, the counterweight, formed by the concrete slab (E) and the concrete beam (C), joined together, could be lifted off the ground and be

Figure 11.15 Exterior of the model. 1: Frames that support the extreme edges of the shell. 2: Microscopes that enable internal measurements to be taken of the movements of the average profile of the shell. 3: Frame supporting the cables used to apply the negative wind load (suction). 4: Cable fixed to the shell to apply the suction loads. 5: Supports for the fixed inner platform from which reference measurements were taken. This platform supports the model independently. (Image: AET, CEDEX-CEHOPU).

suspended from the longitudinal beam G. At that instant, the entire weight was acting on the model. To remove the load from the model, it was thus only necessary to release the jacks.

On the other hand, if one of the cables broke, the beam, A, suspended from it was without one of its supports, so that beam B was also without support. We do not know the length of the B beams and it cannot be deduced from the images that are preserved, but supposing that it rests on only two A beams, when a cable breaks, this B beam would also lose one of its two supports. In this way, all the burdens applied in those cross-sections would cease to act.

The application of the wind action had two effects, one of pressure acting horizontally on the windward faces of the vaults, and another of suction acting radially on the rest of the shell. The first effect was achieved with weights that were put in place before the test (WP). The suction effect was achieved with a device similar to the application of self-weight, but acting outside the model (WS). Finally, the snow action was applied by spreading sand over the vault. The complete model shell had a surface area of 1782 m^2. The self-weight was applied

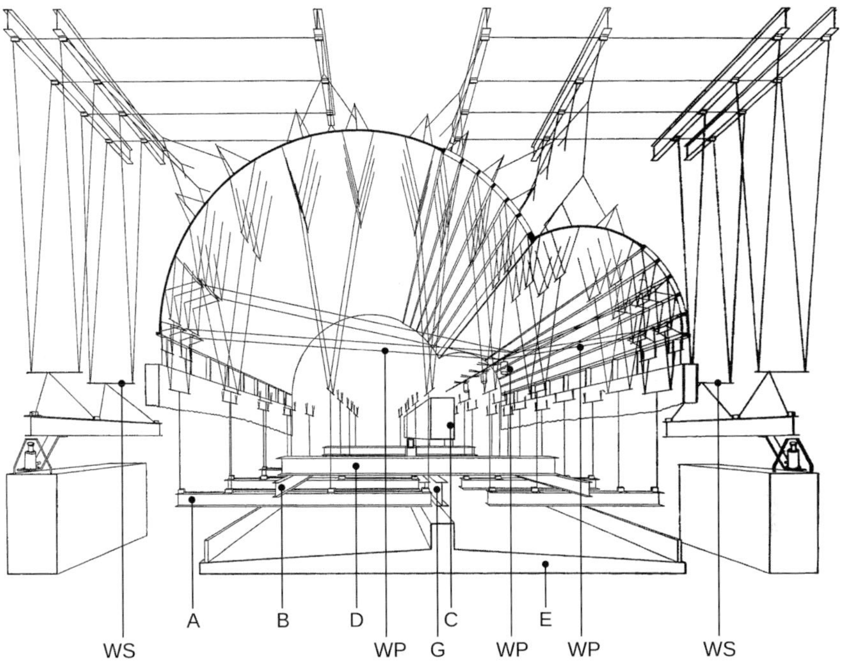

Figure 11.16 Drawing of the model showing arrangement of cables, distribution system and loads. (Image: AET, CEDEX-CEHOPU).

to the model by means of 1024 cables fixed to the shell, and the wind load was applied by a further 1024 cables.

The model was tested between 10 November and 2 December 1935, when construction had already started on site. Different load combinations were performed over several days, including a load test to observe the effects of a temperature variation of 35°. The tests made it possible to obtain the deformation of the middle cross-section for different load combinations. When these results were compared with those of the theoretical analysis, a good correspondence was found (Figure 11.17). In addition, another series of qualitative conclusions were obtained that led to slight modifications to the shape and structural details of the vault. Cracks were observed in the upper ends of the first four diagonals of the triangular lattice windows of the roof light, which led to the reinforcement of these ribs being increased to almost twice as much as originally planned. In addition, the transition zone where these struts joined the main shell was lengthened, and made thicker. In the model a longitudinal fissure was also observed in the crown of the large lobe of the vault, which was later also found in the completed structure.

The experimental work did not end with the tests on the model. It continued after construction with the observation of the movements of the shell during the decentring of the vault, and during various load tests carried out on the completed structure. The test rig remained in the Frontón building until it was dismantled in the first weeks of the civil war in the summer of 1936.

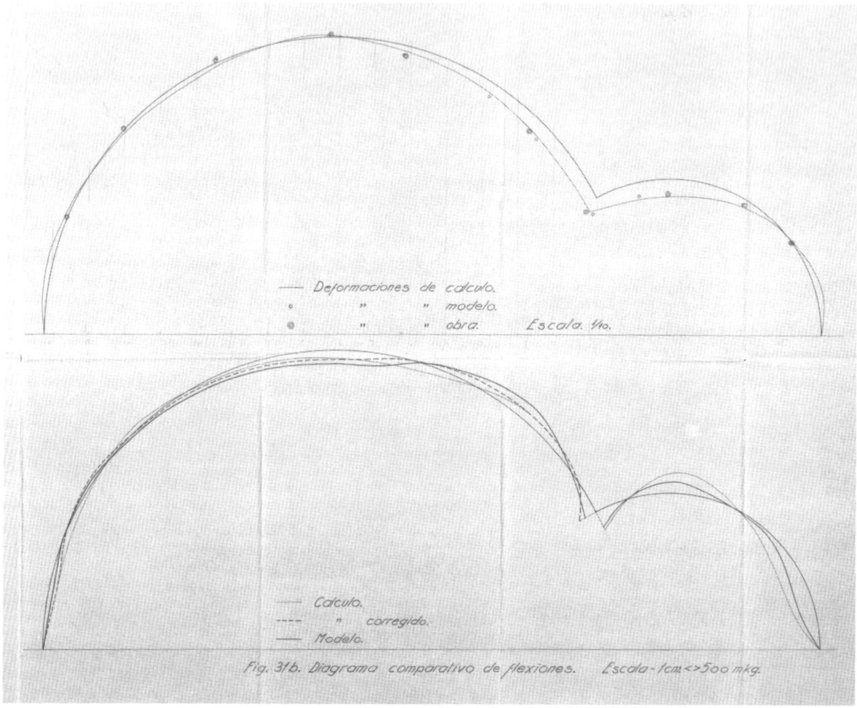

Figure 11.17 Comparison of deformations of the shell: calculation, model and actual structure (above); calculation, modified calculation and model (below). (Image: AET, CEDEX-CEHOPU)

11.4 Final remarks

The monitoring devices placed in the vault of the Frontón allowed permanent observation of the long-term behaviour of the roof. As the culmination of the process of studying a project, permanent monitoring of the work would make it possible to confirm definitively the predicted behaviour, and would make it possible to detect unforeseen and little-known long-term effects such as creep, little studied up to that moment in real structures. Furthermore, in this case, it could be done at a reduced cost, since the instrumentation was already installed in the structure and it was only necessary to take periodic readings.

The Frontón project allowed Torroja to use various techniques of experimentation in the study of the structure, adding to those he had already carried out previously (model testing and monitoring the work during the construction process), the possibility of collecting data on the actual performance of the built work.

In order to carry out the necessary work, as well as to manufacture the necessary measuring instruments, Torroja had promoted the creation of the company ICON, in whose facilities the model of the Frontón was made and which was also in charge of the manufacture of the necessary measuring instruments, as well as their installation and data collection. This initiative was the beginning of an experimental research laboratory under construction that was interrupted by the civil war. The company ICON continued to operate and supplied measuring devices that were used in the following years.

His previous experience allowed Torroja to identify the difficulties and risks of working with models, as well as the need to define test protocols that were simple to apply and that allowed the greatest amount of information to be obtained.

At the same time, Torroja's intention to unite professional practice with research and the dissemination of the results obtained is evident. Together with other professionals, among others the architect Manuel Sánchez Arcas with whom he had worked on the Clinical Hospital and the thermal power station at the University City, and on the dome of the market in Algeciras, he participated in the organization of the *Instituto Técnico de la Construcción y Edificación*. Subsequently, his participation in the formation of the company ICON is a proof of that intention to connect the activities of research and professional practice with the dissemination and education of technicians and operators.

Although the situation in Spain after the war was radically different, Torroja maintained his intention to promote experimental research in the world of construction, linked to professional practice. Therefore, some of the work of his technical office would always involve unconventional solutions and the object of research, with the intention of exploring new construction materials or techniques. In his future work he would often combine his role as a designer, researcher, teacher and promoter of professional and research organisations. Some of these activities will be covered in Chapter 16.

Further reading

A complete bibliography on Eduardo Torroja can be consulted in:

Antuña, J. (2003) *Las estructuras de edificación de Eduardo Torroja*. PhD Thesis. Universidad Politécnica de Madrid. http://oa.upm.es/1348/

The Torroja Archive can be consulted at http://www.cehopu.cedex.es/etm/indices/obraindx.htm

Torroja, Eduardo (1957) *Razón y Ser de los tipos estructurales*. Madrid: Instituto Técnico de la Construcción y del Cemento.

Torroja, Eduardo (1958) *Philosophy of Structures*. New York: F.W. Dodge Corporation.

Torroja, Eduardo (1958) *The structures of Eduardo Torroja: An autobiography of engineering accomplishment*. New York: F.W. Dodge Corporation.

Torroja, Eduardo (1999) *Las estructuras de Eduardo Torroja*. Madrid: Centro de Estudios y Experimentación de Obras Publicas.

References

1 Fernández Casado, C. (1931) Teoría del Arco. *Revista de Obras Públicas* 79 (2580): 354-358. http://ropdigital.ciccp.es/pdf/publico/1931/1931_tomoI_2580_02.pdf (11/11/2019)

2 Torroja, E. (1927) Acueducto-sifón sobre el río Guadalete. *Revista de Obras Públicas* 75 (2477): 193-195.

3 Eduardo Torroja Archive. File No. 002 *Acueducto-sifón sobre el río Guadalete en Jerez de la Frontera.* http://www.cehopu.cedex.es/etm/obras/ETM-002.htm (28/05/2020)

4 Fernández Ordoñez, J. A. (1978) *Eugène Freyssinet.* Barcelona: 2C Ediciones.

5 Chias, P. (1986) *La Ciudad Universitaria de Madrid: Génesis y Realización.* Madrid: Editorial Complutense.

6 See https://archivoshistoria.com/la-batalla-de-ciudad-universitaria/ (29/9/2019) and http://www.madrid1936.es/universitaria/minas.html (29/9/2019)

7 Antuña, J. (2003) *Las estructuras de edificación de Eduardo Torroja.* PhD Thesis. Universidad Politécnica de Madrid. http://oa.upm.es/1348/ (29/9/2019). 47-84.

8 Kierdorf, A. (2006) Early mushrooms slab construction in Switzerland, Russia and the E.S.A. – A study in parallel technological development. In: *Proceedings of the Second International Congress on Construction History,* Vol 2, 1793-1807. Cambridge.

9 Eduardo Torroja Archive. File No. 017 *Depósito de agua para el abastecimiento de Tetuán (Marruecos).* http://www.cehopu.cedex.es/etm/obras/ETM-016.htm (28/5/2020).

10 Eduardo Torroja Archive. File No. 275 *Estructura de un cobertizo para Escuela Elemental del Trabajo.* http://www.cehopu.cedex.es/etm/obras/ETM-123.htm (28/5/2020)

11 Arredondo, F. et al. (1977) *La obra de Eduardo Torroja.* Madrid: Instituto de España.

12 Eduardo Torroja Archive. File No. 246 *Proyecto del nuevo Hipódromo en Madrid.* http://www.cehopu.cedex.es/etm/obras/ETM-115.htm (28/5/2020)

13 Eduardo Torroja Archive. File No. 246 *Proyecto del nuevo Hipódromo en Madrid.* http://www.cehopu.cedex.es/etm/pictindx/ETM-115.htm (28/5/2020)

14 Eduardo Torroja Archive. File No. 158 *Gran Mercado de Algeciras.* http://www.cehopu.cedex.es/etm/obras/ETM-093.htm (28/5/2020)

15 Cassinello, P. (2013) The evolution of concrete shells: innovations by Ildefonso Sánchez del Río. *Informes de la Construcción* 65 (530) (Apirl-June): 147-154.

16 Eduardo Torroja Archive. File No. 158 *Gran Mercado de Algeciras.* http://www.cehopu.cedex.es/etm/pdf/planos/158.302.pdf (28/5/2020)

17 http://www.torrojaingenieria.es/HTML/TrabDetalle.php?Proyecto=142&texto1=Ver%20en%20el%20mapa (29/9/2019)

18 Torroja, E. (1942) Comprobación y comportamiento de una estructura laminar. *Memorias de la Real Academia de Ciencias Exactas, Físicas y Naturales (Madrid)* III: 11-12.

12

Photoelastic stress analysis

Bill Addis

12.1 The principles of photoelastic stress analysis

12.1.1 Photoelasticity

Photoelastic stress analysis depends upon the property of certain transparent materials to change their refractive index in proportion to the internal stresses. This property is known as double refraction or birefringence. At a particular point in a two-dimensional stress field in an isotropic material, the stress will be defined by Mohr's circle of stress in which two orthogonal directions are of particular significance – the directions at which the difference between the principal stresses, (σ_1-σ_3), is a maximum and a minimum. Light passing through the material will be slowed down most in the direction of maximum (σ_1-σ_3) and least in the orthogonal direction. If the incident light is polarised then the effect of the stressed material is to rotate the plane of polarisation. Since the rotation depends on the wavelength of the incident light, it is usual to use monochromatic light source, normally a sodium light (wavelength 5893 Å). If the plane of polarisation is rotated by 90°, then no light is transmitted at that point and it appears black. In a stress field, all the points at which the rotation is 90° will join to form a line, or 'fringe', which is called an *isoclinic*. Similarly for all the points at which the plane of polarisation is rotated by 270°, 450°, 630° and so on. The level of stress difference which causes a rotation of 90° depends on the thickness of the material sample and the birefringent characteristic of the material itself – the unit fringe value, measured in N/mm^2, which is determined by empirical testing [1].

A test piece under load thus appears with a pattern of lines (Figure 12.1). A number of positions in such a pattern are of particular importance: at the boundary of model the principal stress perpendicular to the edge must be zero, and so the magnitude of the orthogonal principal stress, parallel to the edge, can be determined. In this way a model can be calibrated.

The apparatus used for viewing a model under test is called a polariscope. It consists of a light source, a collimator lens to create a beam of parallel light, a sheet of polaroid to polarise the light, the *polariser*, the model under test, and a second sheet of polaroid, the *analyser*, and, as necessary, a camera to record the image for

Physical Models: Their historical and current use in civil and building engineering design,
First Edition. Edited by Bill Addis.

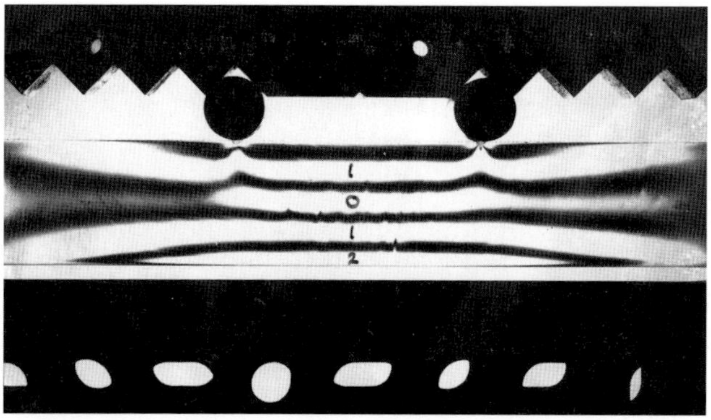

Figure 12.1 A beam loaded in bending. (Image: Bill Addis)

Figure 12.2 A polariscope. (Image: Sharples Stress Engineers)

later analysis. Typically the field of view is 20-25 cm which places a limit on the size of model that can be tested (Figure 12.2). Although photoelastic fringes are clearly visible using plane-polarised light, the different intensities of the orthogonal components transmitted through the model can render the fringes unclear. This problem is avoided by circularly polarising the light so that it is transformed to a certain degree of elliptical polarisation. The fringes thus generated are much sharper and enable more accurate quantitative measurements to be made [1, 2].

Figure 12.3 A beam with a notch, loading in bending, showing the effect of the stress concentration. (Image: Bill Addis)

12.1.2 The basic photoelastic image

The image visible through the polaroid analyser of the polariscope is a series of lines that indicate the isoclinics – lines connecting points with the same value of difference between the principal stresses, (σ_1-σ_3). In a monochromatic image, the closer the lines are spaced, the greater the internal stresses. If the model is loaded and unloaded using one's hand, there is an immediate visual feedback between applied load and induced stresses. In a model with notches, holes or sudden changes of cross-section, there is also an immediate visual feedback of the location and magnitude of any stress concentrations (Figure 12.3). The decreased spacing / increased density of lines is dramatic and, with some experience, such images can help a designer understand how to avoid stress concentrations with no need for calculations. Although white light is not used for quantitative analysis, it is nevertheless worth noting the brightly-coloured photoelastic image can be a thing of great beauty [3].

12.1.3 Analysing the results of a photoelastic model test

A basic photoelastic stress analysis is used to generate three types of result. The first is a quick comparison of the relative stress at key points in a test piece, simply by counting the number of fringes – whole numbers and fractions of a fringe – starting at the zero fringe (Figure 12.4). This example shows how

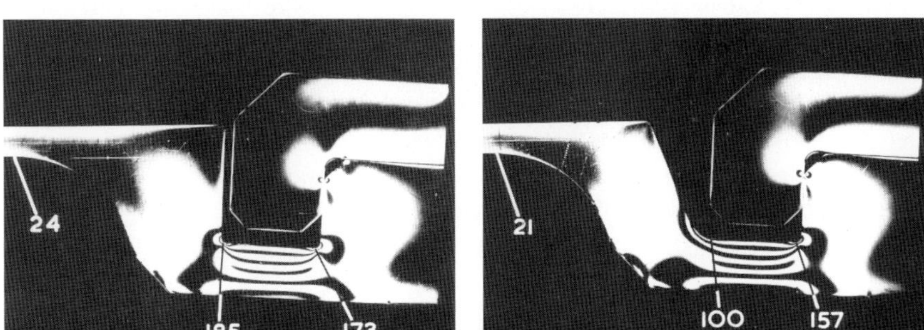

Figure 12.4 Photoelastic models of a connector under (horizontal) tensile load (a) Original design with a peak stress of 4.5 fringes (normalised to 185). (b) Revised design, with increased radius at the root of the cut-back, with a peak stress of 2.4 fringes (normalised to 100). (Image: Bill Addis)

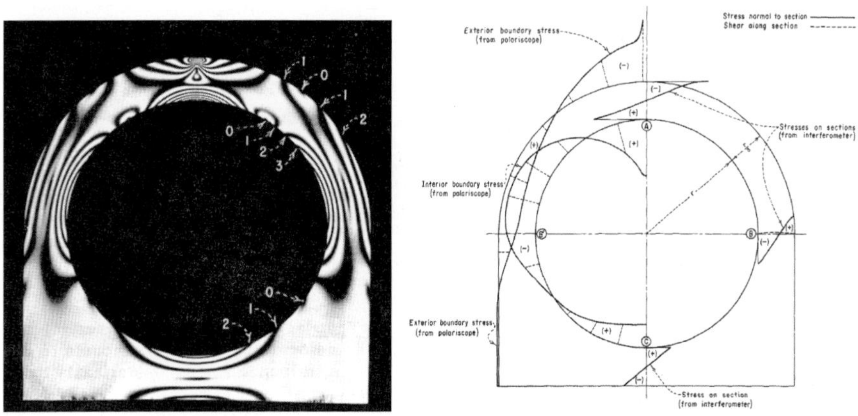

Figure 12.5 Model test of a reinforced-concrete conduit under application of a central, vertical point load. (a) The photoelastic model showing where fringes intersect the boundary. (b) A radial plot showing the compressive and tensile boundary stresses. (Source: [1])

increasing the root radius in a connector loaded in tension reduced the level of stress concentration by 45%.

The second type of result from a photoelastic study is the tangential stress at any point on the boundary of the model. This is derived from the unit fringe value of the material and the 'order' of the fringe at a particular point – first, second, etc. and intermediate values. Knowing the material properties, such as Young's modulus and Poisson's ratio of the model material and actual material, the effects of scale can be taken into account (Figure 12.5).

The third outcome from a photoelastic stress analysis enables the lines of principal stress inside the structure to be determined. The model is loaded, and the polaroid analyser is rotated to give a suitable view of the isoclinic lines. This image is projected onto tracing paper, and a suitable isoclinic line is drawn onto the paper, and labelled 0°. This line joins all the points at which the lines of principal stress are orientated at 0° (to the arbitrary datum). The polaroid analyser is rotated by 15°, and the new key isoclinic line is traced onto the paper, and the process repeated until a rotation of 180° has been completed (Figure 12.6a). Over this series of twelve isoclinic lines, an orthogonal grid of lines is drawn over each isoclinic line, orientated at the angle corresponding to that isoclinic (Figure 12.6b). Finally, a series of orthogonal curved lines are drawn over the grids, indicating the directions of the tensile stresses and, intersecting at right angles, the compressive stresses (Figure 12.6c). These are the lines of principal stress, or isostatic lines.

12.1.4 Photoelastic materials

A photoelastic material has two essential requirements – it must be transparent and have the property of birefringence. After that, the properties make a material more or less convenient to use for particular tests – a reasonably high stiffness so that deflections are not too large; very low creep or viscoelasticity;

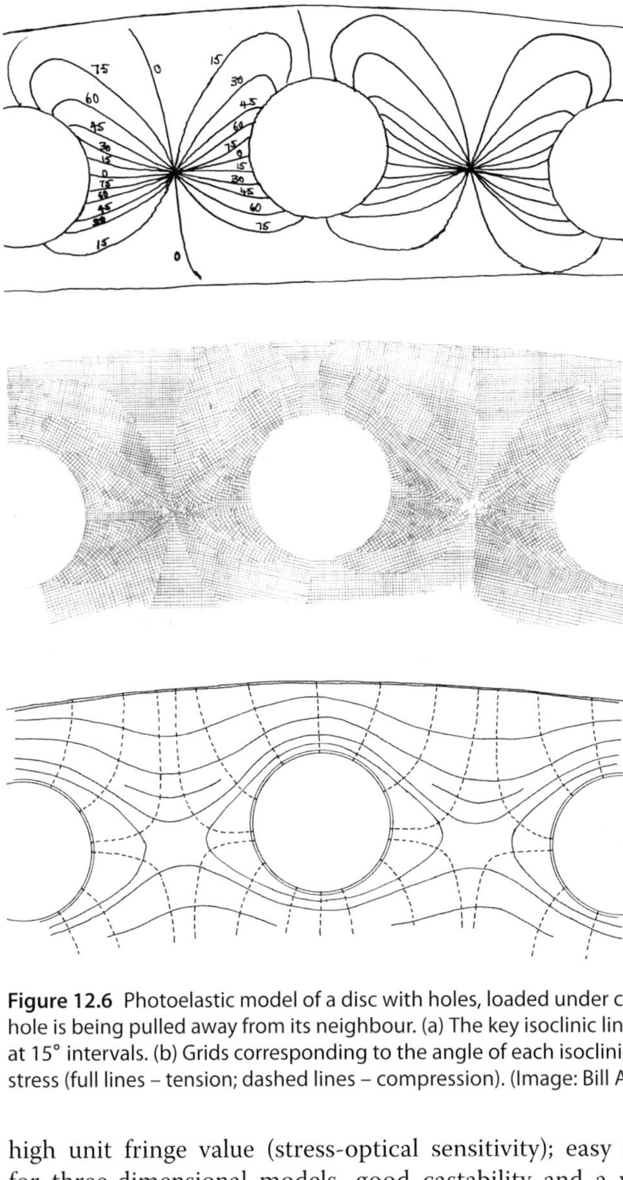

Figure 12.6 Photoelastic model of a disc with holes, loaded under centrifugal load, i.e. each hole is being pulled away from its neighbour. (a) The key isoclinic line, drawn with the analyser at 15° intervals. (b) Grids corresponding to the angle of each isoclinic. (c) The lines of principal stress (full lines – tension; dashed lines – compression). (Image: Bill Addis)

high unit fringe value (stress-optical sensitivity); easy to cut to shape; and, for three-dimensional models, good castability and a well-defined softening temperature to facilitate stress freezing. Some examples are given in Table 12.1. Glass is included because it was the first material used for photoelastic analysis. For two-dimensional model studies, the materials were generally available in 6-7 mm thick sheet that could be cut to the size and shape required. Complex 2-D models could be made by joining several pieces using solvent glues. For three-dimensional photoelastic studies a limiting factor with the first material used – *Bakelite* – was that it was available only in sheets 30 x 15 x 2.5 cm thick. *Fosterite*, developed in 1947, had the advantage of being available in sheets 7 cm thick and in cylinders 75 cm long and 15-20 cm in diameter [4]. From

Table 12.1 Properties of some common photoelastic materials

Material	Temperature	Young's modulus kN/mm^2	Tensile strength N/mm^2	Unit fringe value N/mm^2 per fringe
Plate glass	20°	62.0	-	21.93
Celluloid	20°	2.1	48	1.55
Acrylic resin / Perspex	20°	3.1	48	5.90
Bakelite BT-61-893	20°	4.2	117	0.63
	114°	0.008	1	0.01
Fosterite	70°	0.016	7	0.03
Epoxy resin / Araldite CT200	20°	2.9	72	0.41
	100°	0.01	-	0.01

such blocks, quite large components could be modelled. From around 1960 a number of epoxy resins, including Araldite CT200, became available which revolutionised photoelastic studies both because of its suitability for stress freezing and because it could be easily cast into large shapes, thus reducing the amount of machining necessary to make complex three-dimensional models, such as the turbine disc of an aero-engine, 50 cm in diameter.

Although photoelastic stress analysis makes use only of solid birefringent materials, some liquids display birefringent properties. A rare example in which this phenomenon was used was for studies into the collapse of the Tacoma Narrows Bridge. Using a form of bentonite, a polariscope was used to make visible the different velocities of fluid flowing past the bridge deck and generating eddies in its wake (see Figure 24.31).

12.2 History of photoelastic stress analysis

12.2.1 Birefringence, or double refraction

The phenomenon of birefringence or double refraction in glass was first explored in detail in 1815 by the Scottish physicist David Brewster (1781-1868) [5, 6]. Brewster illustrated the fringes created when a beam made of glass, 150 x 38 x 7 mm thick, was subjected to bending (Figure 12.7). Although his interest was mainly in the optics, he recognised the potential use of photoelastic models for studying the behaviour of structures, referring particularly to models used by his friend John Robison (see Section 2.4.1):

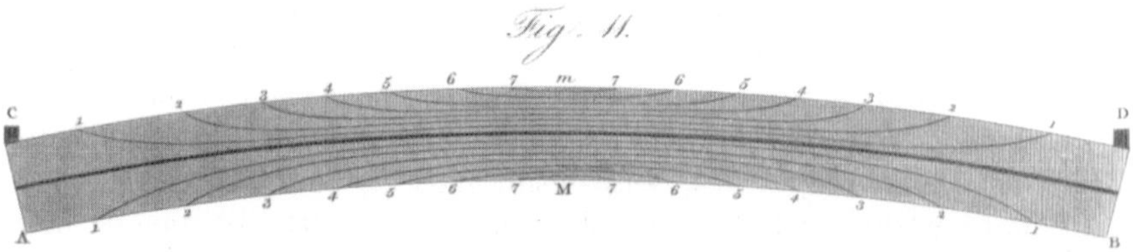

Figure 12.7 Fringes observed by Brewster in a glass beam loaded in bending. (Source: [7])

> 'There is one practical application of these views which is particularly deserving of notice. In order to observe the manner in which stone arches yield to a superincumbent pressure, Dr. Robison executed several models in chalk, and deduced many general laws relative to the internal forces by which they were crushed. If the arch stones of models are made of glass, or any other simply refracting substance, such as gum copal, &c. the intensity and direction of all the forces which are excited by a superincumbent load in different parts of the arch, will be rendered visible by exposing the model to polarised light. If different degrees of roughness are given to the touching surfaces of the glass voussoirs, the results may be observed for any degree of friction at the joints. The intensity and direction of the compressing and dilating forces which are excited in loaded framings of carpentry, may be rendered visible in a similar manner' [7].

Brewster was misleading in saying that 'the intensity and direction of all the forces ... will be rendered visible', and he was, indeed, aware that the fringes did not correspond to lines of force. Despite Brewster's conviction that photoelasticity would be a useful tool for analysing forces and stresses in structures, nothing of direct application to real structures happened until the early twentieth century. During the intervening years the technique was studied by engineering scientists interested in the theory of elasticity: it provided them with an experimental means of corroborating their theories of elastic behaviour and stress distributions inside solid objects, in particular the internal distribution of stresses and the position of the neutral axis in a beam loaded in bending. The theoretical basis for the photoelastic technique was established by Franz Neumann (1798-1895) who demonstrated the correspondence between internal strains and the visible fringes [8] and in 1891 the English engineer Charles Carus-Wilson (1860-1942) was the first to use photoelastic analysis to plot the lines of principle stress in a beam [9].

12.2.2 The beginnings of photoelastic analysis in construction

In the first decade of the twentieth century, the French engineer and director of the laboratories at the *École des Ponts et Chaussées* in Paris, Augustin Mesnager (1862-1933), began to develop practical uses for the technique using models made of glass. In his most important paper, in 1913, he covered the theoretical basis of photoelasticity and discussed the work by Ernest Coker on stress concentrations [10]. He then reported on his own study of a 95-metre arch bridge in reinforced concrete, over the River Rhône at la Balme. Given the very high span-to-rise ratio (10.5), the designers invited Mesnager to undertake photoelastic studies to confirm their calculations. Mesnager had wanted to make his models of Xylonite or Celluloid, two early transparent plastics, because they were easy to cut to shape and to glue pieces together, and had a stiffness thirty-five times less than glass, which generated more fringes than an equivalent model in glass. However, being unable to find sheets of the plastic that were sufficiently uniform and transparent, he resorted to using glass, gluing pieces together with a solution of gelatine and then carefully baking the model. The models, with a span of about 28 cm (i.e. about 1:340 scale), were used to measure surface stresses which were then compared with those found by calculation, for both

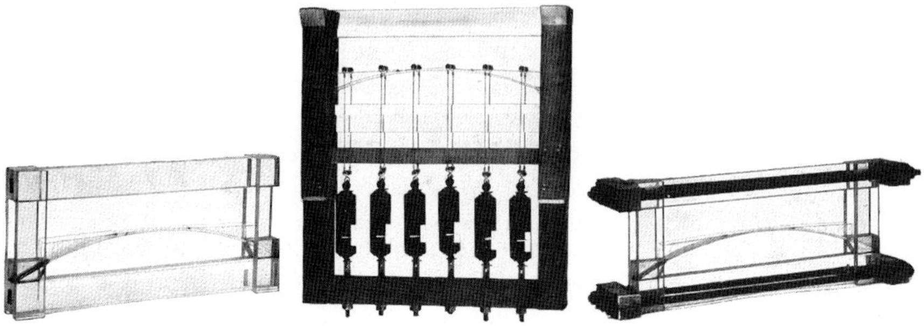

Figure 12.8 Mesnager's 28 cm long glass model for photoelastic analysis of a 95 metre span reinforced-concrete bridge, 1913. (a) The model in its supporting frame. (b) The model subject to six point loads. (c) Compression applied horizontally to the model to reproduce the effects of thermal expansion. (Source: [10])

vertical loads and the effects of thermal expansion (Figure 12.8). The greatest difference between the calculated and measured values was found to be 14.7%.

In around 1910, Ernest Coker (1869-1946) began his research on photoelasticity at University College, London with a study of bending and the position of the neutral axis using models made from sheets of Xylonite about 5 mm thick [11]. He followed this in 1911 with a study of stress concentrations that occurred around rivet holes in narrow steel plates loaded in tension [12]. He found that stresses could be raised by up to 300%, which were rather higher than values calculated using the current theories of elasticity, and thus of direct use to the designers of steel ships and bridges. Coker developed this work in a paper to the Institution of Civil Engineers in 1919, this time focusing on the stress concentrations associated with the radius of the transition at an abrupt change in section in a tension member – the tighter the radius, the higher the stress concentration (Figure 12.9) [13].

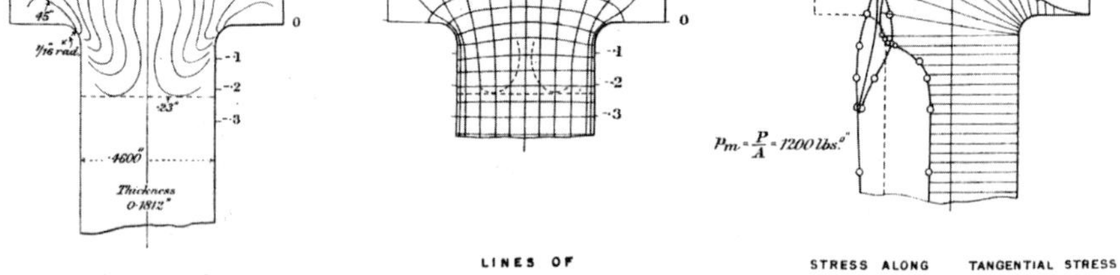

Figure 12.9 Photoelastic measurements of the stress distribution in tension members used in the testing of materials. Demonstration of the stress concentration caused by a small radius at a change of cross-section. (Source: [13] (reconfigured))

In 1925 Coker presented a paper to the Royal Institute of British Architects [14] focusing on the stress concentrations that may arise at the changes of cross-section in the footings of buildings, and at the corners of rectangular windows in a masonry wall. Although his quantitative results for stress concentrations must be treated with scepticism, given the large radii he assumed at the corners in the masonry construction, the general idea of stress flowing through building elements and concentrating where cross-sections change would have been useful lessons. In 1930, Coker returned to the Institution of Civil Engineers, to present the results of his studies for a reinforced-concrete two-pin parabolic arch with rigid connections between the arch, piers and deck (Figure 12.10) [15]. The model, made from Xylonite sheet 5.2 mm thick, had a span of 51 cm and a rise of 12.7 cm. Being larger than the field of view, the polariscope had to be moved relative to the model as setting up the test rig was a lengthy process. Three people were involved in carrying out the each test – one was responsible for ensuring the load was correctly applied; another took the reading of the surface strain at a certain point by counting the number of fringes relative to the zero fringe at the neutral axis; and the third person converted the fringe reading to a corresponding stress which was immediately plotted onto an outline drawing

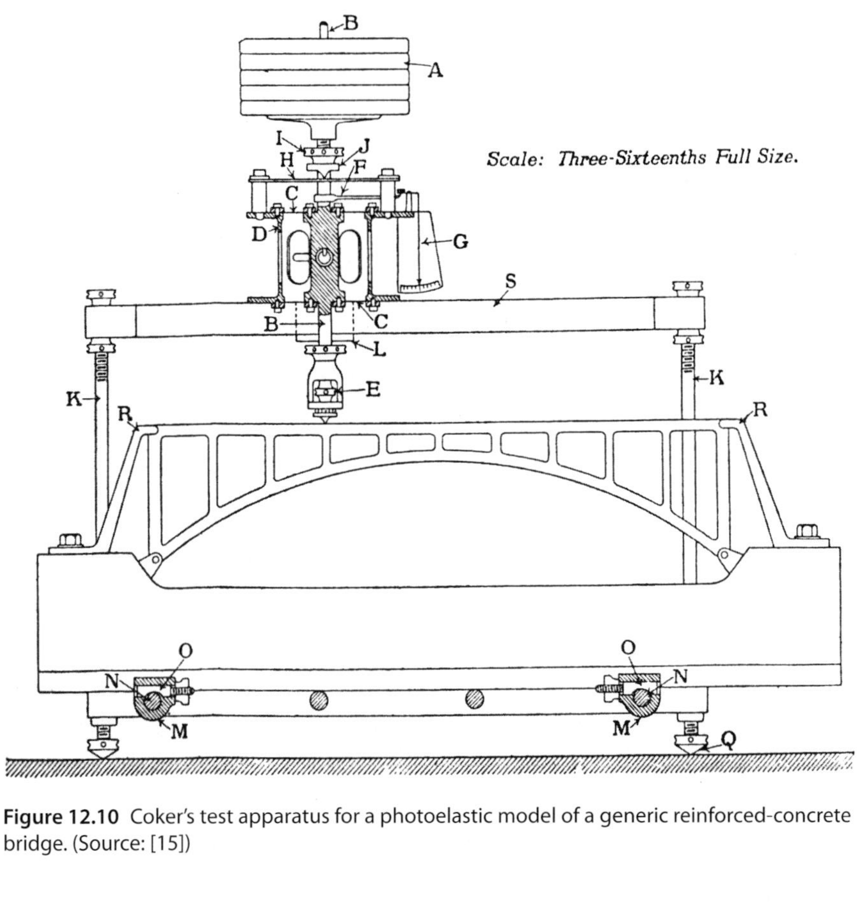

Figure 12.10 Coker's test apparatus for a photoelastic model of a generic reinforced-concrete bridge. (Source: [15])

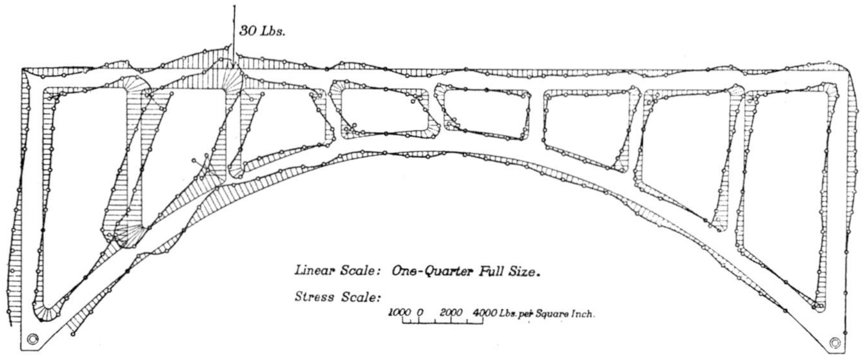

Figure 12.11 Graphical summary of surface stresses for a point load of 13.6 kg applied at the quarter point of a generic reinforced-concrete arch bridge. Compression stress is plotted outward from the boundary, and tension stress inward. (Source: [15])

of the bridge section (Figure 12.11). Depending on the structure under test, this would incorporate the material stiffness appropriate to reinforced concrete or steel. Knowing the surface stresses, it was then possible to identify points of contraflexure and the bending moments along each member. A great strength of this experimental technique was, according to Coker, that it was sufficiently accurate and very much quicker than undertaking hand calculations, especially for highly-redundant structures.

Coker was an academic engineer and, unlike other experimenters at the time, his models were generic, leading only to general conclusions for practising engineers and architects, albeit important ones, which were aimed more at publicising the technique rather than informing the design of particular structures. Nevertheless, the 1931 book on the subject that Coker co-authored with his colleague Louis Filon (1875-1937) became a classic for several decades in the English-speaking world [16].

12.2.3 The growth of photoelastic stress analysis after 1930

From around 1930 universities and other research institutions in most western countries set up laboratories for photoelastic stress analysis (probably several hundred), and the researchers began publishing many hundreds of papers on the subject – both on improvements and new developments to the technique itself, and on particular projects. The leaders of many of these institutions also published their own manuals on the technique, with new editions appearing right up to the 1970s. The following are the best known of these authors:

- Arturo Danusso (1880-1968) and Guido Oberti (1907-2003) at the Polytechnic of Milan (1932) [17]
- Ernest Coker and Louis Filon at University College, London (1931) [16]
- Gustav Mesmer (1905-1981) at the Aerodynamic Institute in Aachen (1939) [18]
- Henri le Boiteux and Robert Boussard at *L'École supérieure de physique et de chimie industrielles* (l'ESPCI) in Paris (1940) [19]

- Max Frocht (1894-1974) at the Carnegie Institute of Technology and, from 1946, the Illinois Institute of Technology in Chicago (1941, 1948) [20]
- Leonardo Villena at the *Instituto Técnico de la Construcción* in Madrid (1943) [21]
- Albert Pirard (1910-) at the University of Liège (1947) [22]
- Herbert Jessop and Francis Harris at University College, London (1949) [23]
- Ludwig Föppl (1887-1976) at the Technical University in Munich (1950) [24]
- Helmut Wolf at the photoelastic laboratory of Siemens, in Mülheim, Germany, whose book includes over 100 references to publications on photoelasticity between 1814 and 1959 (1961) [25]
- Albrecht Kuske at the Technical University of Klausthal in Germany (1951) [26]
- Roland Heywood at the photoelastic laboratory of Rolls Royce Aero-engines and later the Royal Aircraft Establishment in England (1952) [27]

Of these manuals only *Praktische Spannungsoptik* by Ludwig Föppl and Ernst Mönch [28] gave particular examples of studies in the field of construction engineering, including the flow of stresses through foundations, the reinforced-concrete cross-bracing structure in a 14-storey steel-frame apartment block, a cantilever dam, and the design of reinforced-concrete elements including glass-fibre reinforcement.

Other well-known institutions that undertook photoelastic stress analysis, for both scientific and project-related studies, included:

- The laboratories at the *École Nationale des Ponts et Chaussées*, Paris (Augustin Mesnager, V. Tesař). [29, 30]
- The Swiss Federal Institute of Technology (ETH, Zurich) and Federal Materials Testing Station (EMPA), (Robert Baud) [31, 32, 33]
- Materials Testing Institute at the University of Stuttgart (Otto Graf Institut) [34]
- The Westinghouse Research Laboratories, Pittsburgh (Milton Leven)
- The University of Columbia (Raymond Mindlin)
- The Institute of Physical and Chemical Research, and Science University of Tokyo (Z. Tuzi, M. Nisida) [35, 36, 37]
- The Laboratory of the Institute of Optics in Paris [38]
- The Engineering Institute of the Chrysler Corporation, Detroit (Charles Lipson)
- The photoelastic laboratory at Rolls Royce Aero-engines, Derby (Spooner and Heywood)
- The National Physical Laboratory, Teddington, UK (Brown & Hickson) [39]
- *Laboratório de Engenharia Civil, Ministério das Obras Públicas*, Lisbon [40]
- United States Department of the Interior Bureau of Reclamation, Denver Colorado [1].

In the 1930s and 1940s photoelastic stress analysis was widely used for mechanical engineering studies, especially for the design of gear teeth, moving components in machinery and internal combustion engines, and various mechanical connections in which stress concentrations were identified to be of great importance. It is no exaggeration to say that the single, most-important contribution of photoelasticity to engineering was in raising the general awareness of the stress concentrations caused by small radii, for example at the corners of holes in objects under stress, and at the root of screw threads and gear teeth [41]. The technique provided valuable, visible and quantitative evidence at a time when the brittle fracture of metal components was of growing concern, especially when associated with metal fatigue [42]. The revelations about stress concentrations led to an immediate change – indeed a revolution – in the detail design of such components, either by increasing the radius of curvature or, even better, by including a gradual transition between two tangents.

In civil engineering, photoelastic studies were undertaken mainly when there was one of two areas of uncertainty: highly redundant, reinforced-concrete arch and truss bridges in which bending was of particular concern; and reinforced-concrete structures with particularly thick sections, such as an underground conduit, and both cantilever and arch dams, for which theoretical tools were known to be unreliable [43, 44, 45]. In both cases, the photoelastic technique was unrivalled in being able to determine the lines of principal stress through a thick section and, using 3-dimensional photoelastic techniques, to determine absolute levels of internal stress (see Section 12.4.3).

The industry which developed the most sophisticated photoelastic techniques, and benefited most from their results, was the aerospace industry – particularly for aero-engines. Apart from employing many components to the very limit of their strength, just as in internal combustion engines, there was the added requirement to reduce weight to an absolute minimum. It is no coincidence that the development of the modern strain gauge also took place in the aircraft industry, during the same period – early 1930s to early 1950s – that photoelastic techniques were being developed (see Section 13.4.5) [46, 47].

12.3 Technical developments in photoelastic stress analysis

12.3.1 Three-dimensional model analysis – stress freezing

The two-dimensional photoelastic technique, using models cut from a sheet of material, was suitable for many structures that could be represented by a simple cross-section, such as an arch bridge or a cantilever dam. However, it was already becoming easier to analyse stresses and bending moments in such structures using elastic theory. The theoretical analysis of large reinforced-concrete sections was, at the time, effectively impossible. With the development of a new type of Bakelite in the mid-1930s, photoelastic stress analysis was able to provide a solution. The mechanical properties of this Bakelite were dependent on temperature. It was possible to apply loads to a model, such as piston engine connecting rod or an arch dam, at a high temperature – typically about 115° – and then to cool the model, under load, to room temperature and lock in or freeze the strains/stresses in the model. The model could then be cut up into appropriate slices, about 2 mm thick, and the locked-in stresses analysed in the usual way. This technique of 'stress-freezing' was developed in the late 1920s [48] and first described in detail by George Oppel (1902-?) at the Technical University in Munich [49, 50], and Miklós Hetényi (1906-1984) [51] at Westinghouse Research Laboratories. An example of stress freezing is given in Section 12.4.3. An unusual application of the stress-freezing technique was also undertaken in the field of centrifuge studies, for example the stress distribution around excavations in deep mines (see Section 27.2.1).

12.3.2 Three-dimensional model analysis – scattered light method

In 1939 a second technique of three-dimensional photoelastic analysis was developed by researchers at the Langley Memorial Aeronautical Laboratory (now part of NASA) [52, 53]. It is based on the physical phenomenon by which a beam of light is scattered in all directions transverse to its path when it passes through a material. If the incident light source is unpolarized, then the scattering will be of equal intensity in all directions transverse to the beam, and the scattered light will be plane-polarized. If, however, the material is birefringent and the incident light beam is plane-polarized, then the scattered light will be resolved into two orthogonal components moving at different velocities through the material. The apparent amplitude of these components will depend on the point of observation in the plane normal to the direction of the light beam. Since the two components travel at different velocities, they will alternately reinforce and weaken the intensity along their scattered light beams. If the two components have equal amplitudes, they will alternately double the intensity and cancel each other when they are in and out of phase. Thus, when the scattered light is observed in the plane of polarization, or perpendicular to it, alternate light and dark fringes will be visible. And, given the birefringence of the material, these fringes will be in proportion to the internal stress differences (σ_1-σ_2), (σ_1-σ_3) and (σ_2-σ_3). The effect of one of the stresses can be removed by moving the viewing point to give a maximum number of fringes, and the remaining orthogonal stress differences can be calculated. As with two-dimensional photoelasticity, the situation at a free boundary is significant since the stress normal to the boundary is zero. At this point in a three-dimensional model the other two stresses lie in the plane tangential to the boundary. Thus, if the direction of the light is made parallel to this tangent plane, and the model rotated until the maximum and minimum fringe spacings occur, these spacings will give values for the separate principal stresses [1].

Although the experimental procedure was more complex than two-dimensional photoelasticity, and it required the construction of special apparatus, it had the advantage over the stress freezing method, that the model could be used many times to create a full three-dimensional picture of an object. With the development of lasers in the 1960s, this technique was widely used in the aerospace and mechanical engineering fields, and for analysing stress in geometrically complex bones such as the ankle joint and the jawbone. Its expense precluded widespread use in the civil engineering field.

12.3.3 The photoelastic interferometer

The weakness of conventional photoelasticity was that it was only possible to determine absolute levels of strain and stress at the boundary of the model, where one principal stress is zero. This limitation was overcome in the late 1920s by Henry Favre (1901-1966) working at the photoelasticity laboratory at the Swiss Federal Institute of Technology (ETH) in Zurich [54, 55]. Using an interferometer, it is possible to compute the absolute retardation of the orthogonal polarised light rays passing through a birefringent material by comparing the light beam with a reference beam that passes around the specimen being studied. Using this process, the information obtained gives a complete picture of the state of stress at any desired point in the structure since each individual principal stress and its direction are determined. As with the scattered light technique, its complexity and expense meant it was little used for civil engineering studies. Although Favre studied the stresses in encastered walls subject to eccentric vertical loads in his doctoral research [56], the technique was used little in civil engineering outside the university research community.

12.3.4 Photoelastic analysis using birefringent coatings

If a birefringent coating is applied to the reflective surface of an object and illuminated with a beam of polarized light, then isoclinic fringes can be observed on the surface when viewed through an analyser sheet of polariod. The light passes through the birefringent material twice between polariser and analyser and the difference in principal stresses at the interface between the coating and the object can be determined. This idea was originally proposed by Augustin Mesnager in 1930 but could not be exploited at the time because the materials available were insufficiently birefringent to be measured reliably [57]. When expoxy resins became available in the 1950s, the idea was brought to reality by researchers in the French aerospace industry who referred to the technique as a 'photoelastic strain gauge' [58]. It is interesting that Max Frocht commented that many users of coatings seemed to overlook that fact that true values for the difference between principal stresses were only possible when the surface being studied was flat and noted that many researchers had overlooked this constraint [59]. The main advantage of the technique was that it could be used with components and materials that were not satisfactorily represented using a homogeneous model material such as the normal two-dimensional models made from sheets of Araldite. It was, for example, used to study materials such as ferroconcrete, timber reinforced with carbon-fibre reinforced plastic and honeycomb sandwich structures. Such studies were generally undertaken in research institutes and not used directly to inform the design of civil or building engineering structures.

12.4 Some case studies

12.4.1 Studies of the Vierendeel girder (1936)

The first girder bridge carrying the name of its inventor Arthur Vierendeel (1852-1940) was constructed in 1902 using the calculation method he had developed in the 1890s. The girder bridge had a period of popularity in Belgium during the 1930s, many of which were designed by Vierendeel's students at

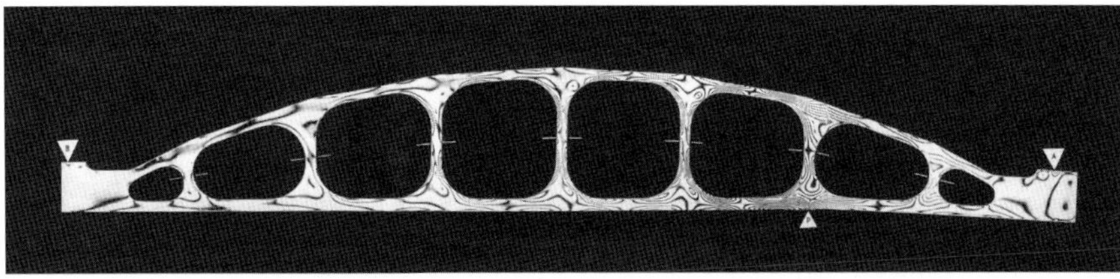

Figure 12.12 Photoelastic model used to determine the points of contraflexure in a Vierendeel bridge girder. (Image: © BATir Department, *Université Libre de Bruxelles*. By courtesy of Bernard Espion)

the Catholic University of Louvain. Vierendeel's method of calculation was very complex and its reliability was not precisely known. In the mid-1930s, Professor Luis-Charles Baes (1883-1961), at the *Université Libre de Bruxelles*, began a profound study to develop a simplified calculation procedure and to test its accuracy and reliability using photoelastic models. He made models of Vierendeel girders in many different forms, with different proportions between the thicknesses of the upper flange, lower flange and vertical members. His particular point of interest was the precise location of the points of contraflexure in the vertical members since this governs the precise stiffness of the structure in response to point loads. His work left us with some of the most beautiful photoelastic images of all time (Figure 12.12) [60].

In his study of a pylon constructed using the Vierendeel form, Baes compared different types of fixity of the two feet of the structure and found that, while all the horizontal members featured a point of contraflexure (at their mid-points), this was not the case for the (nearly) vertical members (Figure 12.13) [61]. Baes noted that this demonstrated the inaccuracy of a method of calculation based on the assumption that all members would have a point of contraflexure.

12.4.2 Oleftal Dam (1956-1959)

In the early years of reinforced-concrete dams it was realised that the cost could be significantly reduced if the dam were not built as a solid structure, as if constructed of masonry, but rather as a curved shell. A large single shell was only possible in a steep-sided narrow valley, as for the Boulder Dam (see Chapter 8). For a wide valley an alternative was a series of vertical or slightly inclined barrel shells, supported by a series of vertical, triangular cantilevers. A model of such a shell was tested by Guido Oberti in Milan in 1926 (see Section 10.1). The first such dam in Germany, completed in 1925, was built at Linachtal, 30 km east of Freiburg in Baden-Württemberg, and is now a listed monument [62]. A similar design was planned in the 1950s for a dam at Oleftal, 50 km south west of Cologne, and would need to be twice as long, and twice as high. The design for Oleftal differed from Linachtal as each barrel shell was supported by its own two cantilevers, rather than each shell sharing a cantilever with its neighbour. The main area of uncertainty lay in the transition between a structure that was clearly a thin shell, at the foot of the dam, to a section that became progressively more like a beam than a shell and, being stiffer, would

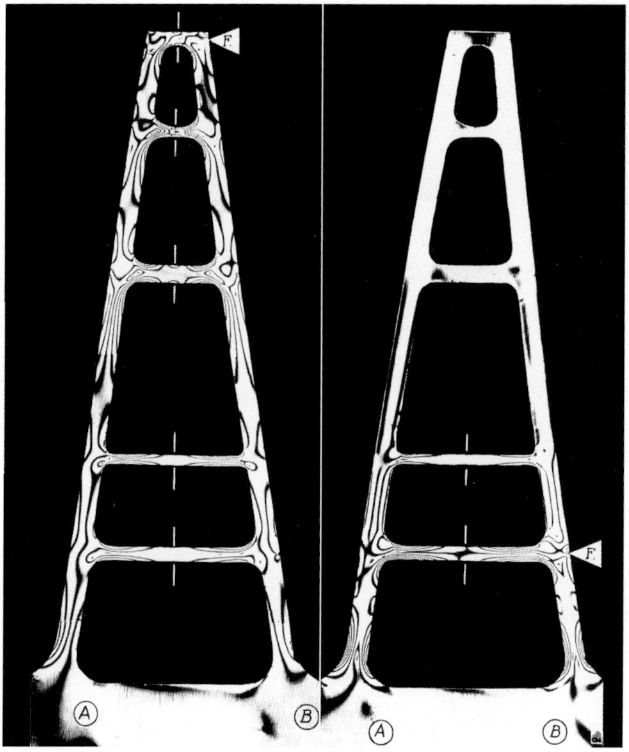

Figure 12.13 Photoelastic model used to determine the points of contraflexure in a Vierendeel pylon girder. (a) Loaded at the top of the tower. (b) Loaded at the top of the first bay. (Source: [61] Courtesy of Ghent University Library)

attract a greater share of the load. The designers wanted to study this changing distribution of stresses. The photoelastic investigation was undertaken in the Otto Graf Institute at the University of Stuttgart. The models were mounted in a test rig in which the load due to water pressure was applied using bolts [63] (Figure 12.14).

The measurements of stress in the edges of the section-models revealed the location and magnitude of tensile stresses at different depths and thicknesses of the arch/beam (Figure 12.15). The construction of the lines of principal stress revealed three singularities in the stress field where the principal stresses are of equal magnitude and both compressive (Figure 12.16).

During the construction of the dam some vertical cracks appeared in the triangular diaphragm walls which were later found to be due to the rock on which the foundations were constructed being much stiffer than expected. As the concrete was thus not able to shrink freely as it cured, a crack formed, starting at the base and rising to about two thirds the height of the wall (about 40 metres). A scheme was immediately devised for strengthening the diaphragm walls by building in some post-tensioned elements. A photoelastic study was

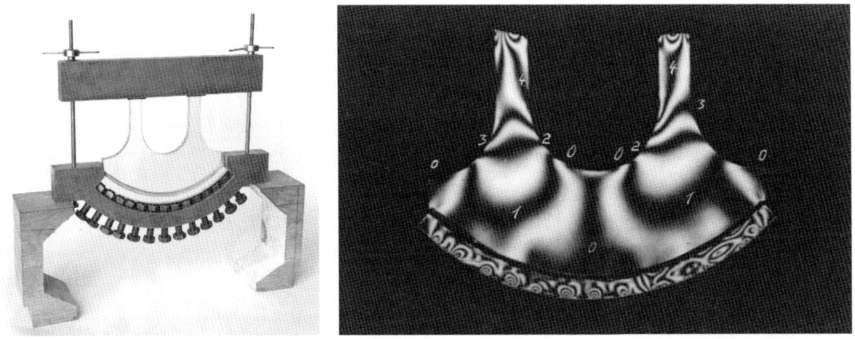

Figure 12.14 1:100 scale photoelastic model of a section of the shell. (a) The loading rig for the model. (b) Pattern of fringes in the loaded model, with the order of fringe identified. (Image: University Archive, Stuttgart 33/1/1699)

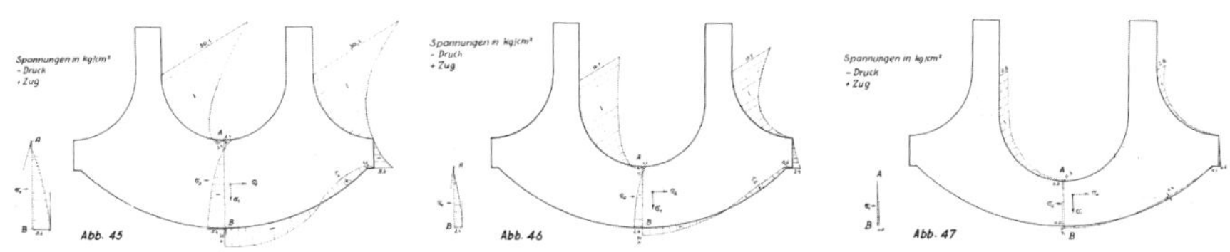

Figure 12.15 Diagrams showing surface stresses in section-models of the dam at three depths (Tension positive; compression negative). (a) Model III 5.0 m (b) Model II 24.3 m (c) Model I 57.1 m. (Images: University Archive, Stuttgart 33/1/1699)

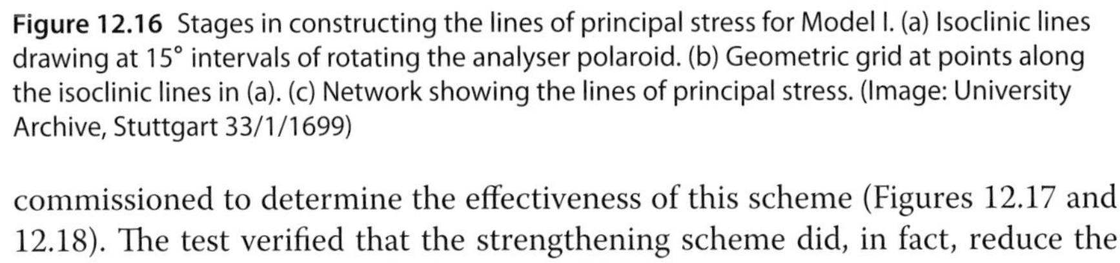

Figure 12.16 Stages in constructing the lines of principal stress for Model I. (a) Isoclinic lines drawing at 15° intervals of rotating the analyser polaroid. (b) Geometric grid at points along the isoclinic lines in (a). (c) Network showing the lines of principal stress. (Image: University Archive, Stuttgart 33/1/1699)

commissioned to determine the effectiveness of this scheme (Figures 12.17 and 12.18). The test verified that the strengthening scheme did, in fact, reduce the tension stresses inside the diaphragm wall [64].

12.4.3 3D photoelastic study for a strong road base, using stress-freezing (1960s)

During the 1960s a number of photoelastic studies were undertaken at the Otto Graf Institute regarding designs for strong road bases, able to carry wheeled and

Figure 12.17 The 1:200 scale photoelastic model of one barrel shell and the triangular diaphragm wall with a vertical crack. (a) The 27 cm high model of Araldite was cast roughly to size and then machined to shape. (b) The model loaded using bolts and springs to simulate the post-tensioning of the added elements. (Image: University Archive, Stuttgart 33/1/1698)

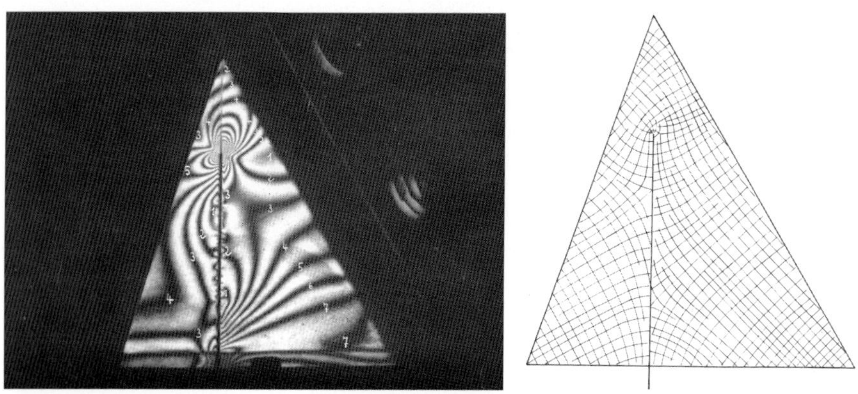

Figure 12.18 (a) The photoelastic model under load (loaded from right to left), showing fringes. (b) Lines of principal stress derived from the photoelastic study. (Image: University Archive, Stuttgart)

tracked vehicles weighing up to 50 tonnes. The main purpose of the tests was to determine the maximum shear stresses that would be generated inside the road base [65]. A 1:25 scale photoelastic model made of Araldite was loaded in an oven at about 80°. After a period to allow the system to stabilise, the oven was turned off and the loaded model allowed to cool slowly to room temperature. The model was then cut into a number of orthogonal slices about 3 mm thick to enable the lines of principal stress to be determined in three dimensions (Figures 12.19, 12.20, 12.21).

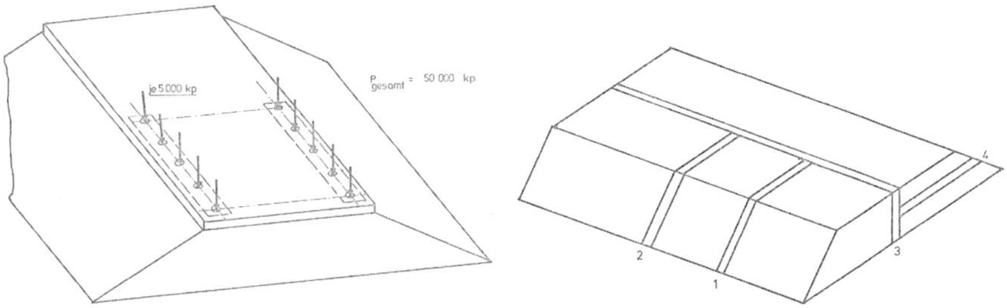

Figure 12.19 Details of the photoelastic model of the road base. (a) Drawing showing the loads applied. (b) Drawing showing how the three-dimensional model was cut after stress freezing to create 3 mm thick slices for two-dimensional photoelastic analysis. (Image: University Archive, Stuttgart 33/1/1761)

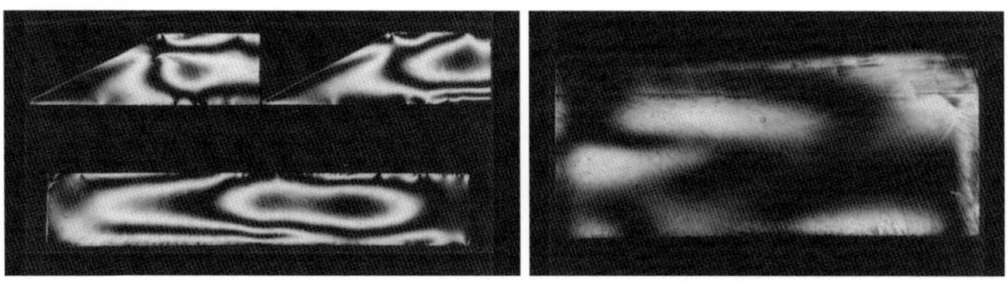

Figure 12.20 Fringes observed in the 3 mm thick slices taken from the solid model of the road base. (a) Vertical slices 1, 2 and 3. (b) Horizontal slice 4. (Image: University Archive, Stuttgart 33/1/1761)

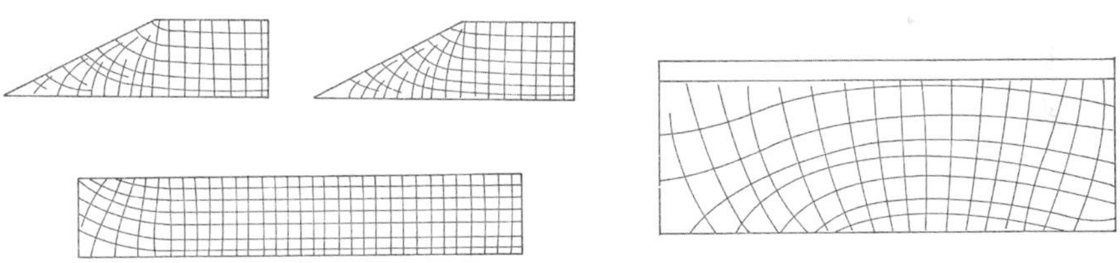

Figure 12.21 Lines of principal stress in each slice, determined from the photoelastic model test. (Image: University Archive, Stuttgart 33/1/1761)

12.5 Conclusion

There is no technique for making visible the forces or stresses inside an object – but photoelasticity is the next best thing. To those who used it regularly it gave a direct sense of internal two-dimensional stresses, and, rather less directly, it could give the same for three-dimensional stress fields. As one author wrote in his report 'Photoelastic investigations can provide a qualitatively correct and clear picture of stress states that would be very difficult or often impossible to calculate. However, it is in the nature of things that the quantitative evaluation contains a certain inaccuracy, [around] +/- 10 % for the present experiment' [66].

Being rather costly, photoelastic stress analysis was never adopted as widely for construction projects as it was for mechanical engineering, especially in the aerospace industry. Nevertheless, it was technique familiar to all experimental engineers from the 1930s to the 1970s, as it was included in all books on experimental stress analysis and could be carried out in most research laboratories.

As a main-stream activity, photoelasticity died out in the early 1980s as finite-element computer software became available. The chair in Photoelastic analysis at the Technical University in Munich was ended in 1984 [67]. The last major symposium on the subject was held in Japan in 1986 [68].

Acknowledgements

I am grateful for the assistance of Christiane Weber for finding and making available archive material for the case studies from the University of Stuttgart.

References

1 Moody, W.T. & Phillips, H.B. (1955/1967) *Photoelastic and experimental analog procedures*. Engineering Monograph No. 23. Denver, Colorado: United States Department of the Interior Bureau of Reclamation. https://www.usbr.gov/tsc/techreferences/hydraulics_lab/pubs/EM/EM23.pdf (14/07/2019)

2 https://tiedemann-instruments.de/en/photoelasticity/principles/ (15/07/2019)

3 https://classes.mst.edu/civeng2200/lessons/12/photoelasticity/index.html (14/7/2019)

4 Frocht, M.M. (1969) The growth and present state of photoelasticity. In: *The selected scientific papers of M.M. Frocht* (ed. M.M. Leven), 1-23. Oxford: Pergamon Press. (Contains over 200 references)

5 Brewster, D. (1816) On the communication of the structure of doubly refracting crystals to glass, muriate of soda, fluor spar and other substances, by mechanical compression and dilatation. *Phil Trans Roy Soc London*: 156-178.

6 Smith, D.P. (1977) *The role of structural models in the design of British bridges – 1800-1870*. PhD Thesis. London: Imperial College. 152-161.

7 Brewster, D. (1816) On the communication of the structure of doubly refracting crystals to glass, muriate of soda, fluor spar and other substances, by mechanical compression and dilatation. *Phil Trans Roy Soc London*: 160.

8 Neumann, F.E. (1841) Üiber die Gesetze der Doppelbrechung des Lichtes in komprimierten oder ungleichformig erwarmten unkrystallinischen Körpern. In: *Franz Neumanns gesammelte Werke* (1912) (eds C. Neumann, W. Voigt & A. Ivangerin), Vol. 3: 5-254. Leipzig: Teubner.

9 Carus-Wilson, C.A. (1891) On the influence of surface loading on the flexure of beams. *Phil. Mag*. Series V, 32 (199): 481-503.

10 Mesnager, A. (1913) Détermination complète sur un modèle réduit des tensions qui se produiront dans un ouvrage. *Annales des Ponts et Chaussées* 4th Trimestre: 133-186.

11 Coker, E.G. (1910) The optical determination of stress. *Phil. Mag.* Series VI, 20 (118): 740-751.

12 Coker, E.G (1911) The determination, by photoelastic methods, of the distribution of stress in plates of variable section, with some applications to ships' plating. *Transactions of the Institution of Naval Architects (London)* 53 (Part 1): 265-296.

13 Coker, E.G. (1919) Photoelastic measurements of the stress distribution in tension members used in the testing of materials. *Minutes of the Proceedings of the Institution of Civil Engineers* 208 (Part 2): 362-401.

14 Coker, E.G. (1926) Some problems in the construction of buildings considered experimentally. *Journal of the Royal Institute of British Architects* Series 3, 33 (7): 209-225.

15 Coker, E.G. (1930) Some experimental methods and apparatus for determining the stresses in bridges and framed structures. *Minutes of the Proceedings of the Institution of Civil Engineers* 229 (Part 1): 33-53. Discussion, 229 (Part 1): 54-71. Correspondence, 229 (Part 1): 72-76.

16 Coker, E.G. & Filon, L.N.G. (1931) *A treatise on photoelasticity.* Cambridge University Press.

17 Danusso, A. & Oberti, G. (1932) *Indagini sperimentali sulle Costruzioni – La fotoelasticità.* Milano: Libreria Editrice Politecnica.

18 Mesmer, G. (1939) *Spannungsoptik.* Berlin: Springer.

19 Boiteux, H. Le & Boussard, R. (1940) *Élasticité et photoélasticimétrie.* Paris: Hermann.

20 Frocht, M.M. (1941, 1948) *Photoelasticity* vols 1,2 New York: Wiley.

21 Villena, L. (1943) *Fotoelasticidad.* Madrid: Anales del Instituto Técnico de la Construcción y Edificación.

22 Pirard, A. (1947) *Photoélasticité.* Paris: Dunod / Liège: Vaillant-Carmanne.

23 Jessop, H.T. & Harris, F.C. (1949) *Photoelasticity: Principles and Methods.* London: Cleaver-Hume Press.

24 Föppl, L. & Mönch. E. (1950) *Praktische Spannungsoptik.* Berlin: Springer.

25 Wolf, H. (1961) *Spannungsoptik.* Berlin: Springer.

26 Kuske, A. (1951) *Verfahren der Spannungsoptik.* Düsseldorf: Verein Deutscher Ingenieure.

27 Heywood, R.B. (1952) *Designing by photoelasticity.* London: Chapman & Hall.

28 Föppl, L. & Mönch. E. (1950) *Praktische Spannungsoptik.* Berlin: Springer.

29 Tesař, V. (1932) La Photoélasticimétrie et ses applications dans les constructions aéronautiques. *La Science Aérienne* (Paris) 372 (Sept.-Oct.).

30 Fernandez Casado, C. (1932) Fotoelasticimetría – El laboratorio de la Escuela de 'Ponts et chaussées', Paris. *Revista de Obras Públicas* LXXX (2590): 55-61. http://ropdigital.ciccp.es/directorio_articulos.php?anio_ini=1930&anio_fin=1939&anio=1932&numero_revista=2590 (13/7/2019)

31 Baud, R.V. (1932) Technische Methoden photoelasticher Forschung *Schweizerische Bauzeitung* 100 (1): 1-4, 100 (2): 15-20.

32 Tank, F. (1934) Die Tätigkeit des photo-elastischen Laboratoriums der Eidg. Technischen Hochschule 1927 bis 1933. *Schweizerische Bauzeitung* 104 (5): 45-48.

33 Tank, F., Baud, R.V. & Schiltknecht, E. (1937) Die neuen Einrichtungen des Photoelastischen Laboratoriums an der Eidg. Techn. Hochschule und an der Eidg. Materialprüfungsanstalt. *Schweizerische Bauzeitung* 109 (21): 249-252.

34 https://www.ub.uni-stuttgart.de/archiv/Projekte/Projekt-FMPA/index.html (13/7/2019)

35 Tuzi, Z. (1929) On the development of experimental methods in photoelasticity. *Sci. Pap. Inst. Phys. Chem. Research Japan* 12, 21-69.

36 Tuzi, Z. (1963) Recent activity in photoelasticity in Japan. In: *Photoelasticity – Proceedings of the International Symposium held at Illinois Institute of Technology, Chicago, October 1961* (ed. M.M. Frocht), 59-74. London/New York: Pergamon.

37 Nisida, M. (1986) Photoelasticity in Japan – a survey. In: *Proceedings of the International Symposium on Photoelasticity, Tokyo* (ed M. Nisida), 1-8. Tokyo: Springer.

38 Fernandez Casado, C. (1932) Fotoelasticimetría – El laboratorio del Instituto de Optica de Paris. *Revista de Obras Públicas* LXXX (2602): 353-355. http://ropdigital.ciccp.es/directorio_articulos.php?anio_ini=1930&anio_fin=1939&anio=1932&numero_revista=2602 (13/7/2019)

39 Brown, A.F.C. & Hickson, V.M. (1951) Photoelastic Laboratory at the National Physical Laboratory. *Engineering* 171 (4454): 701-704.

40 Rocha M. & Borges, F. (1952) Photoelasticity applied to structural design. *IABSE Congress Report (Cambridge)* 4: 353-369.

41 Dolan, T.J. & Broghamer, E.L. (1942) A photoelastic study of stresses in gear tooth fillets. Engineering Experiment Station Bulletin series No. 335. *University of Illinois Bulletin* XXXIX (31) (24th March). https://core.ac.uk/download/pdf/4814501.pdf (13/7/2019)

42 Gordon, J. E. (1976) *The new science of strong materials.* 2nd Ed. Harmondsworth, UK: Penguin.

43 Föppl, L. (1938) Die Spannungsoptik im Dienste des Bauingenieurs. *Der Bauingenieur* 19: 341-345.

44 Mönch, E. (1986) A historical survey of the development of photoelasticity in Germany, especially in Munich. In: *Proceedings of the International Symposium on Photoelasticity, Tokyo* (ed M. Nisida), 1-8. Tokyo: Springer.

45 Rocha M. & Borges, F. (1952) Photoelasticity applied to structural design. *IABSE Congress Report (Cambridge)* 4: 353-369.

46 Tesař, V. (1932) La Photoélasticimétrie et ses applications dans les constructions aéronautiques. *La Science Aérienne* (Paris) 372 (Sept.-Oct.).

47 Heywood, R.B. (1952) *Designing by photoelasticity.* London: Chapman & Hall.

48 Frocht, M.M. (1969) The growth and present state of photoelasticity. In: (1969) *The selected scientific papers of M.M. Frocht* (ed. M.M. Leven), 1-23. Oxford: Pergamon Press.

49 Oppel, G. (1936) Polarisationsoptische Untersuchung raumlicher Spannungs- und Dehnungszustande. *Forschung Geb Ing Wesen* 7 (5): 240-248.

50 Oppel, G. (1937) *The photoelastic investigation of three-dimensional stress and strain conditions.* (Translation of Oppel 1936). Technical Memorandum No. 824. Washington: National Advisory Committee for Aeronautics.

51 Hetenyi, M. (1938) The fundamentals of three-dimensional photoelasticity. *J Appl Mech* 5: 149-155.

52 Weller, R. (1939) A new method for photoelasticity in three-dimensions. *Jnl. Appl. Phys.* 10 (4): 266.

53 Weller, R. & Bussey, J.K. (1939) *Photoelastic analysis of three-dimensional stress systems using scattered light.* Technical Memorandum No. 737. Washington: National Advisory Committee for Aeronautics.

54 Favre, H. (1927) Méthode purement optique de détermination des tension intérieurs se produisant dans les constructions. *Schweizerische Bauzeitung* 90 (23): 291-294; 90 (24): 307-310.

55 Favre, H. (1929) *Sur une nouvelle méthode optique de détermination des tensions intérieures.* Dissertation, ETH Zurich. https://www.research-collection.ethz.ch/handle/20.500.11850/133570

56 Favre, H. (1929) *Sur une nouvelle méthode optique de détermination des tensions intérieures.* Dissertation, ETH Zurich. https://www.research-collection.ethz.ch/handle/20.500.11850/133570

57 Mesnager, A. (1930) Sur la détermination optique des tensions intérieures dans les solides à trois dimensions. *Comptes Rendus Académie des sciences, Paris* 190 (1e semestre, 22): 1249-1250.

58 Fleury, R. & Zandman, F. (1954) Jauge d'effort photoélastique, *Comptes Rendus Académie des sciences, Paris* 238 (1e semestre, 15): 1559-1561.

59 Frocht, M.M. (1969) The growth and present state of photoelasticity. In: (1969) *The selected scientific papers of M.M. Frocht* (ed. M.M. Leven), 1-23. Oxford: Pergamon Press.

60 Baes, L. (1936) La Poutre Vierendeel. Généralisation de la méthode de calcul par ouverture des mailles par sectionnement d'une des membrures. *L'Ossature Métallique* 5 (10): 447-477. https://lib.ugent.be/viewer/collection/COLL-RUG01-000013457#?c=0&m=42&s=0&cv=18&r=0&xywh=-2112%2C-1%2C7116%2C3307 (18/7/2019)

61 Baes, L. (1937) Poutres Vierendeel en porte-à-faux. Application de la méthode de calcul par ouverture des mailles par sectionnement d'une des membrures. 3ème mémoire. *L'Ossature Métallique* 6 (9): 427-443. https://lib.ugent.be/viewer/collection/COLL-RUG01-000013457#?c=0&m=52&s=0&cv=26&r=0&xywh=-2112%2C-1%2C7116%2C3307 (18/7/2019)

62 https://goo.gl/maps/1Lv7GQCVeHJtHtFWA (17/07/2019)

63 Sammlung von Prüfberichten (Gotterbarm): *Spannungsoptische Untersuchungen des wasserseitigen Kopfes der Olef-Staumauer vom 7.2.1956.* Archives of Otto Graf Institute, University of Stuttgart. Archive No. 33/1/1699.

64 Sammlung von Prüfberichten zu spannungsoptischen Untersuchungen: *Prüfungsberichte zu Modellversuchen Oleftalsperre vom 25.3.1960.* Archives of Otto Graf Institute, University of Stuttgart. Archive No. 33/1/1698.

65 *Prüfungsbericht zu Spannungsoptischen Untersuchungen über die 'Beanspruchung der Straße durch Schwerstfahrzeuge und Kettenfahrzeuge' vom 15/3/1971.* Archives of Otto Graf Institute, University of Stuttgart. Archive No. 33/1/1761.

66 Sammlung von Prüfberichten (Gotterbarm): *Spannungsoptische Untersuchungen des wasserseitigen Kopfes der Olef-Staumauer vom 7.2.1956.* Archives of Otto Graf Institute, University of Stuttgart. Archive No. 33/1/1699.

67 Mönch, E. (1986) A historical survey of the development of photoelasticity in Germany, especially in Munich. In: *Proceedings of the International Symposium on Photoelasticity, Tokyo* (ed M. Nisida), 1-8. Tokyo: Springer.

68 Nisida, N. & Kawada, K. (eds) (1986) *Photoelasticity: Proceedings of the International Symposium on Photoelasticity, Tokyo, 1986.* Tokyo: Springer-Verlag.

Section C

Physical models used in structural design, 1940s to 1980s

The use of measurement models in structural engineering reached a peak in the 1950s which continued until the mid-1970s when computers came onto the scene. One reason for this was the development of the foil electrical-resistance strain gauge in the aircraft industry, which became widely available soon after the end of World War II. This era also saw the development of many new materials with which to make reduced-scale models, including many plastics such as Perspex and various types of micro-concrete. During this time the question of how to achieve similitude between the behaviour of reduced-scale models and full-size structures, although first used by Froude in about 1870, and further developed in the 1920s for use in wind tunnels, now became more widely understood and applied to structural model testing, especially for long-span bridges and reinforced-concrete structures. This dramatically increased the accuracy with which models could predict the behaviour of real structures.

This period saw the establishment and thriving of many model-testing laboratories throughout the world, as well as the emergence of a few individuals who made their own unique contributions to the art of model testing. In this section we have chapters about work undertaken in the laboratories of the University of Stuttgart, the *Istituto Sperimentale Modelli e Strutture* (ISMES) in Bergamo, the *Laboratorio Central de Ensayo de Materiales de Construcción* in Madrid, whose director from 1940 was Eduardo Torroja, the Cement and Concrete Association in the UK, the Institute for Lightweight Structures in Stuttgart – in particular, the use of soap-film models – under the leadership of Frei Otto, and also the laboratories of several notable individuals – Heinz Hossdorf and Heinz Isler in Switzerland and Sergio Musmeci in Italy.

The final chapter in this section describes the use of models by Ove Arup and Partners and Frei Otto during the design of the Multihalle built in 1975 for the Bundesgartenschau in Mannheim, one of the most remarkable buildings constructed in the twentieth century. The design of a lightweight timber gridshell to span a space 60 x 85 m was unprecedented. As with other structures exploring uncharted territory, it was made possible only through the use of measurement models, in conjunction with computers which, at that time, still had very limited power.

13

Structural modelling technique

Bernard Espion and Bill Addis

13.1 Introduction

As we saw in Chapter 1, there is evidence that engineering models have been used for at least two millennia. When planning a large construction of any type, it is surely intuitive to make a small-scale version. It is also intuitive, though wrongly so, to assume that for every case, everything can be scaled up linearly. As it happens, this is the case for masonry constructions which rely only on gravity and the geometry of the structure for their stability. This would have meant that builders could have relied on scaling up the results of model tests, and we have some evidence that this was done (see Chapter 2). But even without evidence, it is surely inconceivable that they did not building models of arches, flying buttresses, vaults and domes.

Even in Roman times it appears that engineers realised that there were some phenomena that could not be scaled-up linearly (see Appendix A1). The first author to address the issue in print was Galileo, in 1638 (see Appendix A2). Nevertheless, even after that time, there were engineers who drew wrong conclusions from tests on models, for example Caspar Walter in 1766 (see Section 4.1.1). To a large extent the issue was resolved by Euler in his paper of 1776 entitled 'A simple rule to determine the strength of a bridge or similar structure, on the basis of the known strength of a model' (see Appendix 3 and Section 4.3). However, Euler's paper was written in Latin, and was not widely known, especially among engineers.

By the mid-nineteenth century, Euler's approach had been largely adopted, in that many formulae describing the behaviour of engineering structures were now known and, with a formula, the results could be scaled. For example the strength of a beam can be expressed using modern bending theory in terms of its width, depth, length and the failure stress of the material – $S = \dfrac{2bd^2\sigma}{3L}$. Before the twentieth century it was common to express such formulae for the common materials and shapes of beam in the form – $S = \dfrac{kbd^2}{L}$ for a rectangular timber beam. Countless experiments were undertaken to determine the value of k for various materials and shapes of beam. Generally speaking, the aim of many

Physical Models: Their historical and current use in civil and building engineering design,
First Edition. Edited by Bill Addis.

experiments was first to establish the 'correct' formula for a certain phenomenon, and then to determine values for the empirical constant k which allowed the formula to be applied to a particular case. Thus, from his experiments on the buckling strength of struts, Musschenbroek found that the load causing the buckling of the strut is inversely proportional to the square of the strut length nearly 30 years before Euler's famous paper on buckling in 1757. Musschenbroek's formula was useful and scalable, without needed to know Euler's theoretical evaluation of the constant. Thus, there was much intelligent use of scaling calculations from the mid-nineteenth century, including the Britannia Bridge (see Section 6.6) and more general discussions (see Appendices A5 and A6).

However, this approach to scaling and the idea of similarity between model and full-size structure, was limited to phenomenon that could be reduced to relatively simple and deterministic formulae. In order to deal with more complex engineering phenomena for which 'formulae' were not known, an entirely different approach was needed, and this was developed first in the field of fluid mechanics.

13.2 Dimensionless numbers and similitude

13.2.1 The beginning – fluid dynamics in the nineteenth century

In the mid-nineteenth century a number of scientists dealing with the flow of fluids created *dimensionless numbers* (see Section 13.2.2) associated with various phenomena in the field of fluid dynamics. The best known of these is, perhaps, Reynolds Number (Re), posthumously named after the British physicist Osborne Reynolds (1842-1912) and used by him from around 1880. (In fact the ratio had first been used by George Stokes in 1851.) Reynolds Number represents a ratio of the key quantities that relate to the transition between smooth laminar flow and turbulent flow of a fluid, for example in an open channel or over an aircraft wing. The relevant quantities are: the density of the fluid (ρ kg/m^3); the velocity of the fluid with respect to the object (u m/s); a characteristic linear dimension (L m); the dynamic viscosity of the fluid (μ kg/m·s) and the kinematic viscosity of the fluid (ν m^2/s). Reynolds Number (Re) is defined as:

$$Re = \frac{\rho u L}{\mu} = \frac{uL}{\nu}$$

The significance of a dimensionless number is that if a small-scale model is made and tested in a manner that keeps the value of the number the same as is found in the full-size prototype, then the behaviour of the model can be reliably scaled up to full-size.

Other dimensionless numbers were soon found useful for other aspects of fluid flow; one of these was the Froude Number.

William Froude (1810-1879) was a marine engineer who became engaged in the search for the most efficient shape of a ship hull. This became an issue for ship designers in the 1860s when steam engines were first used to power ships. In order to determine the size of engine that a new ship would require and the quantity of coal that would be needed for a voyage, before the ship was built, it was necessary to know the resistance the ship would encounter when moving through water. Initially, it was assumed that the resistance came only from friction between the hull of the ship and the water, and it was believed that testing a model of a ship's hull might enable the resistance of the full-size ship to be determined. Calculations based on this assumption, however, were found to be seriously wrong and the results of such tests proved to be 'embarrassing and expensive mistakes'.

Froude had been working for Isambard Brunel's Great Western Railway company, constructing the line from London to the west of England. At the time, such employment was not typical for someone who had studied mathematics and physics at Oxford University. It turned out to be particularly fortunate for Brunel when both Froude and John Scott Russell (1808-1882), a graduate of Glasgow University, joined Brunel's design team for the Great Eastern in 1853. This ship was to be the largest the world had ever seen – at over 200 metres long and 26 metres wide, she was nearly twice as long and five times the displacement of the next largest. Froude joined the team, both to address the question as to how much power would be needed to take her across the Atlantic, and also to study how the ship would roll in high seas – his analysis of the rolling of ships was still in use in the second half of the twentieth century. Froude's work on the resistance of the hull led him to the conclusion that friction was not the only cause; for the first time he realised that the surface wave created by a ship also impeded its movement [1]. This insight led him to believe, against the opinions of all others, that models of ships' hulls might be successfully used to predict the resistance to movement of full-size ships. In 1867 Froude tested models of two different hull shapes in three sizes, 0.9, 1.8 and 2.7 metres long (Figure 13.1). He found that they generated similar wave patterns when towed at speeds proportional to the square of the length of the models. He was thus able to conclude that the wave resistance of the model and a full-size ship would be proportional if the ratio $V/\sqrt{gL}$ for the model and the prototype had the same value. This ratio, now known as the Froude Number, is a dimensionless number, which is effectively to say that it is independent of the scale of the model.

Froude's model tests had been done in the open water of the River Dart in South Devon, hardly ideal for precise measurements. Nevertheless, their success persuaded the British Admiralty to grant Froude £2000 to build the world's first towing tank near his house in Torquay. It was here, in 1871, that he conducted the first model tests that influenced the design of a new ship, HMS Greyhound. The

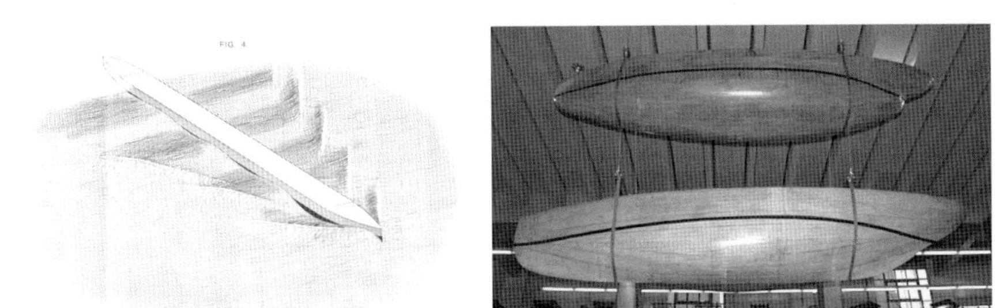

(a) (b)

Figure 13.1 (a) Froude's drawing of the two types of wave formed by a vessel passing through water. (Image: [1]) (b) Froude's model hulls, 2.7 metres long – the Swan (above) and the Raven (below). Despite being broader, the Swan offered lower resistance the motion. (Image: [2])

sea trials a year later showed Froude's design method to be a complete success. His name has been immortalised in the Froude Number, Fr, $= V/\sqrt{gL}$.

Froude's work laid the basis for one of the most powerful tools available to the engineer. As long as a designer can find a suitable dimensionless number, it would be possible to use a scale model to simulate accurately the behaviour of a full-size prototype. Froude's method was soon being used to great effect in wind tunnels in the early years of the twentieth century and for many other engineering problems involving fluid flow. Its use in structural engineering began in the 1950s.

13.2.2 Dimensionless numbers for structural model testing

There are two common ways to give the dimensions of a physical variable. The Mass System uses the dimensions of mass (M), length (L) and time (T); the Force System, which is more convenient in structural engineering, uses force (F), length (L) and time (T). For example:

Quantity	Dimensions using the Mass system	Dimensions using the Force system
Length	$[L]$	$[L]$
Mass	$[M]$	$[FL^{-1}T^{-2}]$
Force	$[MLT^{-2}]$	[F]
Modulus of elasticity	$[ML^{-1}T^{-2}]$	$[FL^{-2}]$

The dimensions of some common quantities relating to structural behaviour (using the Force system, and with temperature denoted by θ) are as follows ([1] indicates no *FLT* dimensions):

Quantity	Dimensions using the Force system
Length	$[L]$
Mass	$[FL^{-1}T^{-2}]$
Force	$[F]$
Angle	[1]
Stress & pressure	$[FL^{-2}]$
Strain	[1]
Moment	$[ML]$
2nd moment of area	$[L^4]$
Modulus of elasticity	$[FL^{-2}]$
Poisson's ratio	[1]
Linear coefficient of thermal expansion	$[\theta^{-1}]$

In order that the behaviour of a model can be scaled up to full size, it is necessary to find a dimensionless number that relates all the variables of a structural system.

For a problem in elastic theory, the following physical quantities are adopted as the significant ones: modulus of elasticity E, length l, and Poisson's ratio ν, for the structural member, and the quantities load P, reaction R, stress σ, deflection δ and strain ε. The dimensional matrix for these variables is as follows:

	l	E	ν	P	R	σ	δ	ε
F	0	1	0	1	1	1	0	0
L	1	−2	0	0	0	−2	1	0

By inspection of the various coefficients we can identify the following dimensionless numbers:

$$\pi_1 = \frac{\sigma}{E} \qquad \pi_2 = \frac{l}{\delta} \qquad \pi_3 = \frac{P}{R} \qquad \pi_4 = \nu \qquad \pi_5 = \varepsilon \qquad \pi_6 = \frac{P}{El^2}$$

It is, of course, possible to find other dimensionless numbers, but these can all be expressed as the multiplication of two or more of the six above.

In 1914 the American mathematician Edgar Buckingham (1867-1940) discovered a theorem that relates the number of dimensionless numbers (π) for a system, the number of physical variables (r) and the number of independent primary dimensions (n). He proved that (r – n) = π. In the example above, this would be (8 – 2) = 6. In other words, for this system, only six (different) dimensionless numbers need be found. This is known as Buckingham's π-(or pi-)theorem [3, 4]. (In fact Buckingham's theorem had been found earlier and independently by several mathematicians, but he managed to get his name attached!) A fuller discussion of dimensionless numbers, scaling and similitude are given in the standard books on structural model testing (see Section 13.3.1) and in Bridgman's classic work [5].

The example of a simple structural problem will serve to illustrate the use of dimensionless numbers to determine scale factors – a cantilever loaded at the free end (Figure 13.2).

The vertical deflection, y, of the free end of the cantilever in Figure 13.2 depends on four variables – W, l, I and E. The dimensional matrix for these variables is as follows:

	y	W	l	I	E
F	0	1	0	0	1
L	1	0	1	4	−2

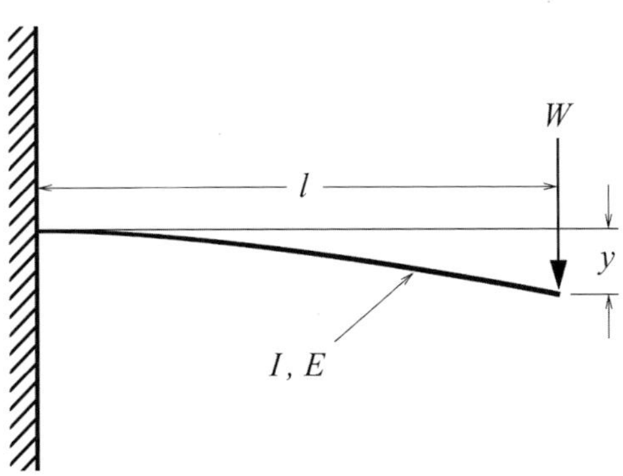

Figure 13.2 A cantilever of length l, second moment of area I, stiffness E, with a load W and deflection y.

As there are five variables and two independent primary dimensions (F and L), from Buckingham's π-theorem, there will be $(5-2)=3$ dimensionless quantities. By inspection, it can be seen that they are:

$$\pi_1 = \frac{y}{l} \quad \pi_2 = \frac{W}{El^2} \quad \pi_3 = \frac{I}{l^4}$$

Hence, where subscript p denotes values for the full-size structure (or prototype), subscript m denotes values for the model, and the scaling factor is given by λ, then:

$$\text{for} \quad \pi_1 \quad \frac{y_p}{l_p} = \frac{y_m}{l_m}$$

$$\text{or} \quad \frac{y_p}{y_m} = \frac{l_p}{l_m}$$

$$\text{and} \quad \lambda_y = \lambda_l$$

i.e. the scale factor for deflection (λ_y) is the same as for the geometric scale factor (λ_l) (for deformation similarity).

$$\text{Similarly for} \quad \pi_2 \quad \frac{W_p}{E_p l_p^{\,2}} = \frac{W_m}{E_m l_m^{\,2}}$$

$$\text{or} \quad \lambda_W = \lambda_E \lambda_l^2$$

i.e. if the geometric scale factor (λ_l) is chosen (say 20), the scale factors for W (λ_W) and E (λ_E) must be in the ratio of 400:1 (for loading similarity).

$$\text{and also for} \quad \pi_3 \quad \frac{I_p}{l_p^{\,4}} = \frac{I_m}{l_m^{\,4}}$$

$$\text{or} \quad \lambda_I = \lambda_l^{\,4}$$

i.e. if the geometric scale factor (λ_l) is chosen, the scale factor for the second moment of area (λ_I) must be the fourth root of the geometric scale factor (for geometrical similarity).

Further examples of scaling calculations for more complex structures can be found in the standard books on structural model testing (see Section 13.3.1). Some of the treatment of scale factors for the Mannheim Multihalle are given in Section 22.6.1.

13.3 Experimental stress analysis using measurement models

13.3.1 Manuals on experimental stress analysis using measurement models

After the pioneering work in the 1920s and 1930s, the technique of model testing had become a standard tool available to designers of complex or unprecedented structures ranging from concrete shell roofs, to bridges and dams. The 1950s saw further developments in structural model testing, in particular, the use of scaling factors based on dimensional analysis.

During the 1950s and 1960s, many universities and research institutes built laboratories for model testing, both for pure research and for commercial application on major structures. At that time the classic reference work was by Hetenyi [6]. The first text books devoted specifically to the art and science of structural model-testing appeared in the late 1960s and early 1970s, such as the classic works by Henry Cowan (et al.) based in Sydney [7], Heinz Hossdorf based in Basel [8, 9, 10], Robert Müller based in Stuttgart [11] and Emanuele Fumagalli, based at the *Istituto Sperimentale Modelli e Strutture* (ISMES) in Bergamo, Italy [12]. Many others followed, notably the classic work on Structural modelling and Experimental Techniques by Sabnis, Harris, White and Mirza in 1983 [13, 14]. Ironically, the mid-1970s also saw the beginning of the decline in structural model-testing for actual construction projects as computers and computer software became more and more useful and reliable for structural design.

This chapter presents an overview of some of the techniques of structural model testing and their development up to the 1970s when they reached their zenith.

13.3.2 Structural model-testing procedure

The level of detail included in most of the manuals on structural model testing is impressive in its depth and comprehensiveness, and can easily be followed up. By way of an overview of the whole process, many authors gave a series of guidelines to would-be model testers. The following is typical, taken from the book by Harris and Sabnis [13], beginning with the wise advice to be clear about why you are doing the test!

1. Define the scope of the problem, deciding what is needed from the model and what is not needed.
2. Specify similitude requirements for geometry, materials, loading and interpretation of results. Pay particular attention to those similitude requirements that cannot be met, such as the desired equality of Poisson's ratio for concrete and plastics when doing elastic modelling of shell and slab structures.
3. Decide on the size of model and required level of reliability or accuracy. What size model should be used consistent with the accuracy? If ±30% is adequate for design purposes, then an attempt to achieve ±10% accuracy is wasted effort and time. Typical scales are as follows:

Type of structure	Elastic models	Strength models
Shell roof	1:200 to 1:50	1:30 to 1:10
Highway bridge	1:25	1:20 to 1:4
Reactor vessel	1:100 to 1:50	1:20 to 1:4
Beam-slab structures	1:25	1:10 to 1:4
Dams	1:400	1:75
Wind effects	1:300 to 1:50	n/a

4. Select model materials with proper attention to steps 1, 2 and 3 above.
5. Plan the fabrication phase in consultation with the technicians who will be constructing the model, and follow the fabrication activities closely. This can be a frustrating part of modelling because it is often quite time-consuming.
6. Select instrumentation and recording equipment for strains, displacements, forces and other quantities. This step must be closely coordinated with steps 4 and 5, particularly if embedded strain gages are to be used in concrete models. Special strain gauges and other equipment must be ordered well in advance of the actual time of usage.
7. Design and prepare the loading equipment; new systems should be thoroughly checked out and calibrated before use on a model.
8. Observe the response of the model during loading, taking complete notes and photographic records of the behaviour. Do not rush through a test, and never leave anything to memory. Some investigators use tape or video recorders in this phase to record detailed comments on load history, cracking development, instability modes and other information that may be difficult to describe numerically. Approximate calculations should be done before the experiment to estimate expected levels of response. Equilibrium checks should be done on results obtained early in the test.

13.4 The measurement of strain

13.4.1 Extensometry

A construction, a structure, must be designed to have adequate in-service behaviour in relation to its function, as well as sufficient bearing capacity to withstand the loads that will be applied to it during its lifetime. Adequate in-service behaviour most often means limiting the deflection under permanent load and variations in deflection under moving loads to certain defined values, either absolute or within a certain range. The notion that comes into play here is that of the rigidity (stiffness) of the structure. The question of adequate load-bearing capacity is of a different nature: it underlies the introduction of quasi-philosophical concepts relating to safety – buildings must have a 'sufficient' reserve of load-bearing capacity, quantified by a 'factor of safety', with regard to ruin, collapse and failure.

The precise prediction of the collapse load of a real construction by calculation is complicated because it involves non-linear physical phenomena which are complex to model, either in relation to the behaviour of the materials (plasticity, cracking, creep, fatigue, ...), or in relation to the large displacements that occur as the collapse approaches (various instabilities, buckling, ...). The history of the evolution of structural analysis and structural design methods since the beginning of the nineteenth century testifies to the search for ever more sophisticated and 'enriched' models to predict as accurately as possible the real load-bearing capacity of structures of increasing complexity. This enrichment of computational models by taking into account non-linear phenomena in computational assumptions, which began to accelerate in the 1920s and 1930s, quickly led to equations, or systems of mathematical equations, that could not be solved analytically. Only with the introduction of computers, gradually starting in the 1960s, did it become possible to consider predicting in a relatively reliable way, through numerical modelling, the real load-bearing capacity of buildings.

Since the estimation by analytical calculation of the actual load-bearing capacity of buildings was inaccessible, structural designers, certainly until the 1960s, and sometimes even much later, based the design of structures on a different concept: the calculated stresses had to remain, in service, below certain 'permissible' values, fractions of the intrinsic strength of the materials. Typically, for example, this meant that for a mild-steel structure with a yield strength of 240 MPa, the in-service stresses should not exceed, say, 120 MPa. The basic calculation hypothesis was that the construction behaves, under service loads, according to the laws of elasticity theory, a theoretical corpus that has gradually been enriched since the beginning of the nineteenth century. However, it should be stressed that this safety philosophy (elastic dimensioning using 'admissible stresses') gives no reliable indication of the safety of a structure in relation to its actual load-bearing capacity, which is why this approach has been gradually abandoned since the 1960s, particularly in view of the major achievements of the theory of plasticity applied to buildings.

To return to the 'elastic, permissible stress' dimensioning of constructions, it was still necessary to have realistic mathematical models for stress analysis in constructions. As soon as the constructions were innovative, or somewhat complex, the calculation models were too poor to deduce by calculation (or graphically), in a safe or realistic way, the displacements and stresses. As a result, the justification of a construction, particularly a pioneering construction, has always involved applying a test load to prove its safety under the service load conditions. Sometimes, when a prototype has been built, a load test might even be taken up to collapse. The control of deflections and stresses calculated during the loading test of actual constructions is historically one of the sources of development of model instrumentation, and particularly of *extensometry*, the measurement of deformations. The other source of development of extensometers is the testing of materials in the laboratory in order to characterize their behavioural laws under different types of stress (traction, compression, torsion ...).

The use of small-scale physical structural models is also part of this approach to overcome the impossibility of understanding the structural functioning of complex constructions through calculation, either because there was no basic mathematical model describing it, or because the solution of the equations would have been impossible with the means available at the time. The small size of physical structural models required the development of specific extensometers, different from those used for the monitoring of full-size constructions.

The principle that has governed the design of buildings for at least 150 years is the dimensioning to achieve 'admissible stresses'. Concerning the reasons for wanting to collect experimental knowledge of stresses in actual structures or in reduced-scale models, it should first be recalled that 'stress' is not a scalar concept, but a tensor quantity. In general, one must speak of a 'state of stress' at a point in a construction. The state of stress can be uniaxial (tension-compression), as in elements of a truss whose nodes are perfectly articulated (rarely achieved in practice), or biaxial, as on the surface of a structure such as a shell, or triaxial, as inside a massive structure such as a gravity-arch dam.

Let us consider the important case of the state of bi-axial stress (which includes the uni-axial case). At any point on a surface of a construction, or a flat model, there are two perpendicular directions in the plane of the construction, or in a plane tangential to the surface, in which the normal stress components are respectively maximum and minimum: these directions are called the principal directions, and the associated values of the normal stress components are called the principal stresses (σ_1, σ_2). In planes whose normals are aligned with the principal directions, stresses act only as normal stress components, with no tangential stress components (shear stress). In the intermediate plane (at 45° to the directions of the principal stresses) the maximum shear value, τ_{max} is given by $(\sigma_1 - \sigma_2)/2$). The isostatic lines of a stress field are the trajectories of the principal stress directions: a knowledge of them is important to optimize the shapes of the constructions and, in concrete constructions, to arrange the reinforcement rationally.

It should also be recalled that while knowledge of the state of stress at any point in a construction or model is the purpose of experimental stress analysis, it is not directly accessible because it is impossible to measure stress components experimentally. On the other hand, it is possible to measure deformation or strain components and, in particular, normal deformation components: the measurement of strains is the object of extensometry. Having determined the main strain components by means of extensometers, it is possible, by making a hypothesis about the behaviour law of the material of which the structure or model is made, to deduce the stress state. Most often, this hypothesis is that of a linear-elastic and isotropic behaviour (Hooke's law), requiring only two mechanical characteristics to be introduced: the Young's modulus, E, and Poisson's ratio for the material.

Assuming linear-elastic and isotropic behaviour for the material, the principal directions in the strain field coincide with the principal directions in the stress field. They should therefore also be determined. Generally speaking, in the bi-axial case, to know the state of deformations at a point fully, it is important to measure normal deformations in three directions to deduce the principal strains (ε_1, ε_2), and the principal directions. These three measurements are typically made in three directions separated by 45°, or three directions separated by 120°. The arrangement of the extensometers in these three directions are called a rectangular or triangular rosette. For the tests on the model of the Boulder Dam, a strain-gauge particularly suitable for low-modulus materials was used – the Three-component Templin Strain Gauge was used [15] (see Section 8.3.4) (Figure 13.3). The two principal strain components are derived from the three measurements from the strain gauge rosette using two equations. Before the advent of computers this was often done by means of a graphical construction using Mohr's circle.

It is therefore through extensometry that the stresses are (indirectly) accessed. Many physical principles have been used to develop extensometers. Before reviewing some of them, it is useful to recall some orders of magnitude. Assuming that we want to evaluate a stress of 10 N/mm^2 (about 1 kg/mm^2) in an iron

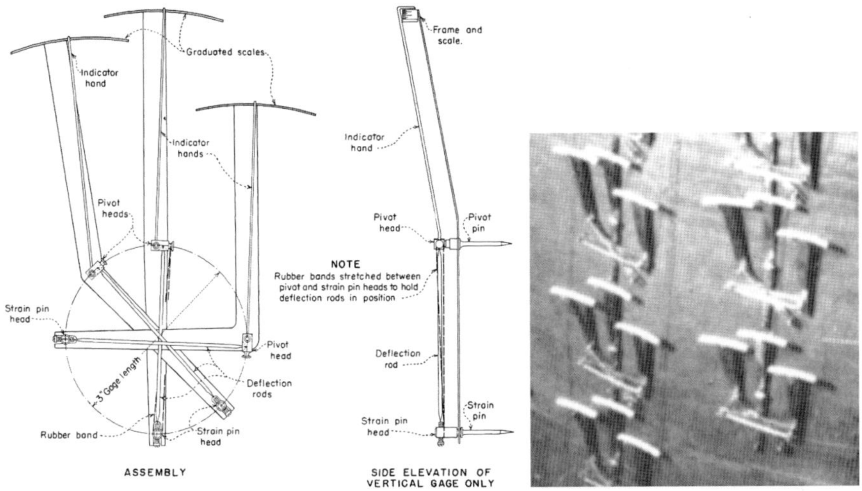

(a) (b)

Figure 13.3 Mechanical strain-gauge rosette used in the model studies for the Boulder Dam. (a) Plan and elevation. (b) Gauges mounted in situ. (Source: Chapter 8, endnote [9])

or steel structure with a modulus of elasticity of 200 kN/mm^2, we must therefore be able to measure a strain of 0.00005 or 50 microstrain units ($\mu\varepsilon$), or a change in length of 50 microns per metre, 5 microns per decimetre or 0.5 microns per cm. This immediately sets the challenge to be met according to the gauge length to obtain sufficient sensitivity. It should be added that a temperature variation of 1°C causes iron or steel to change length by 0.00001 (10 $\mu\varepsilon$).

The history of the development of extensometry techniques tells the story of the development of sensors:

- more and more sensitive and precise;
- yet, whose gauge length is as short as possible, which is contradictory;
- without human intervention for the measurement and allowing remote measurement;
- robust and increasingly lighter;
- avoiding disturbance of the stress field to be studied;
- providing stable (drift-free) and repetitive measurements over time;
- insensitive to parasitic phenomena (electromagnetic waves, temperature variations, relative humidity);
- cheap!

While some authors have traced the historical development of many extensometers whose names are now completely forgotten [16, 17], the purpose of this chapter is not to give an exhaustive history. Rather it is to review various extensometers and measuring techniques developed during the first half of the twentieth century, that were successful, particularly for the study of reduced-scale structural models. Many of them are still in common use today.

13.4.2 Mechanical strain measurement

By mechanical means, and to obtain highly sensitive readings for the measurement of deformation on relatively short gauge lengths, it is necessary to use levers in order to provide a strong amplification of the elongation of the measurement base. This requires an extremely fine mechanical construction to limit play and friction in the joints which are common sources of inaccurate and unreliable measurements.

One of the first extensometers of this type, which was quite widely used, was developed by Augustin Mesnager (1862-1933) in around 1900 [18, 19, 20]. The gauge length was 50 mm and the amplification factor, achieved by a three-lever mechanism, was often given as 2000 [18, 21, 22], although some gauges had lower factors, for example 1050 [23], and 940 for a Mesnager gauge in the laboratory of *Université Libre de Bruxelles.* The reading of deformation is made on a drum recorder, where a needle displacement of 5 mm corresponds to a stress variation of 10 N/mm^2 for steel [24] (Figure 13.4). However, at 6 kg, the Mesnager extensometer was a cumbersome and heavy instrument, intended more for measuring deformations in the structural elements of actual bridges than for models.

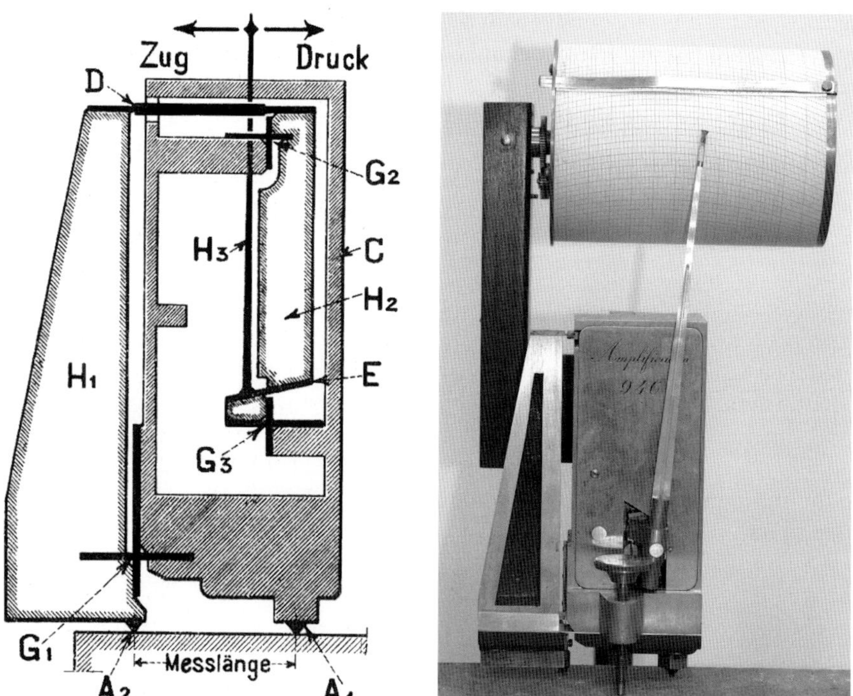

(a) (b)

Figure 13.4 Mesnager extensometer. (a) Cross section. Gauge length A1-A2. Pivots at G1, G2 and G3. Indicator pointer H3. Tension to the left; compression to the right. (Source: [18]) (b) An extensometer with an amplification of 946, including the drum recorder. (Image: Bernard Espion. Collection BATIr Dept ULB)

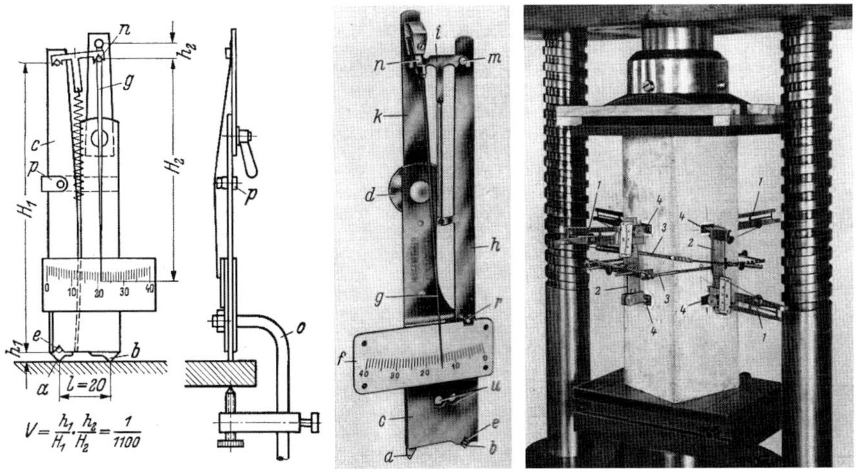

Figure 13.5 Okhuizen extensometer. Gauge length – ab; lower pivot – e; upper pivot – n; pointer – g. (Source: [20])

In 1902, the Dutchman Dirk Okhuizen designed a much more compact and lightweight lever extensometer with a 1100-fold amplification for a 20 mm gauge length [25, 26] (Figure 13.5). It could be fitted with electromagnets to clamp the gauge to the test piece, and adapted for a longer gauge length [27]. Since the 1870s, many mechanical extensometers were developed with the primary purpose of determining stresses in metal bridges. In the 1920s, a major research programme on this subject was carried out at the ETH, Zurich under the aegis of the Technical Commission of the Association of Swiss Bridge and Steelworkers [28, 29, 30]. Many measurements using Mesnager, Okhuizen and other strain gauges were made on bridges in service [31], but it is particularly worth mentioning here that the researcher Theophil Wyss used Okhuizen strain gauges for experimental analysis of the state of stress in gusset plates at structural nodes in a reduced-scale model of a truss bridge with a 6 metre span [32] (Figure 13.6).

In the same year in which Wyss submitted his thesis at the ETH Zurich, Arnold Huggenberger (1895-1982), an assistant to the eminent Slovak engineer Aurel Stodola (1859-1942), finished his doctoral studies at the ETH, Zurich devoted to the strength of steam turbine components [33]. He then began a prolific and world-renowned career in the development of instruments for stress analysis and instrumentation for the monitoring of structures, especially dams [34]. For his studies for the *Schweizerischen Vereins von Dampfkessel-Besitzern* (Swiss Steam Boiler Owners Association) Huggenberger developed the mechanical extensometer, known as the 'Tensometer', which made him famous [35, 36]. The Huggenberger Type 'A' Tensometer (Figures 13.7, 13.8a) with a 20 mm gauge length (reducible to 10 mm, or extendible), with an amplification factor of 1200, was, for many years, the workhorse of the 'Tenso' brand, registered in 1933. The Huggenberger Tensometer became available in many less-expensive versions, or adapted to particular purposes such as the *Biegungsverzerrungsmesser* which could measure changes in the curvature of a surface due to bending (Figure 13.8b). This was the case for the Type B Tensometer which was claimed

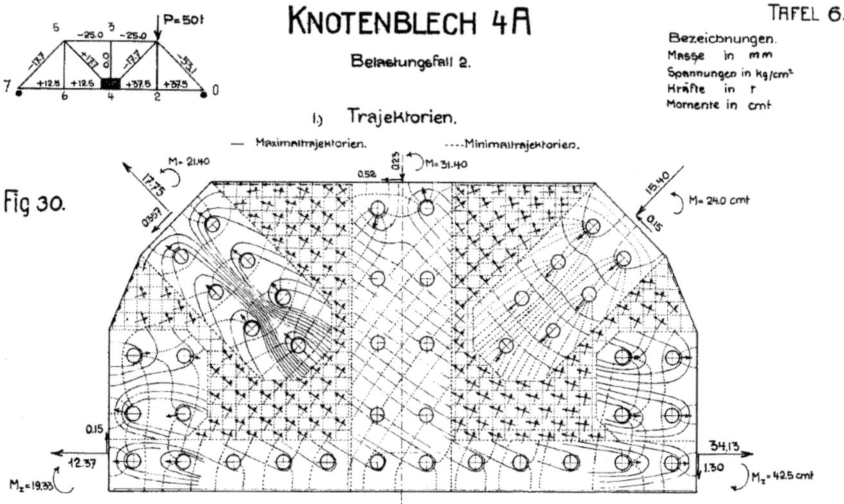

Figure 13.6 Isostatic lines in a gusset plate of an eccentrically loaded steel truss, plotted using strain readings obtained with Okhuizen extensometers. (Source: [32])

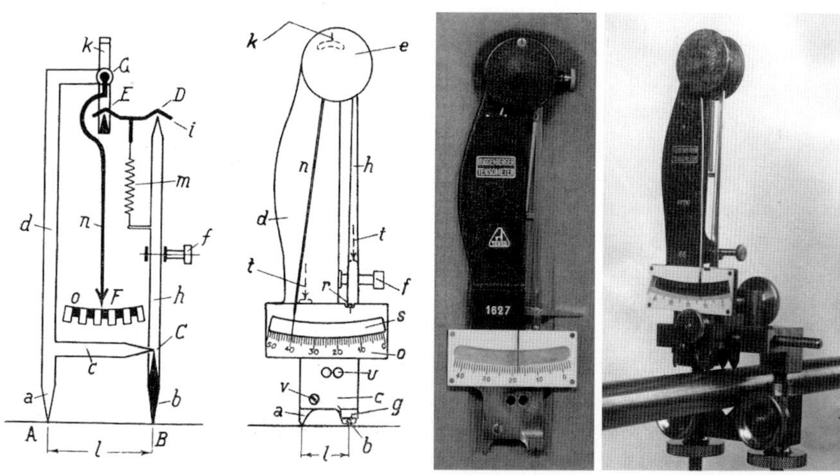

Figure 13.7 Huggenberger-A tensometer. Height – 16.5 cm; weight – 70 g. (a) Principle of the mechanism and cross-section. (Source: [40]) (b) Tensometer in its storage box. (Image: Bernard Espion. Collection BATIr Dept ULB) (c) Tensometer mounted on a tube. (Image: Collection BATIr Dept ULB)

as an adaptation of the Okhuizen model [37, 38]. It was cheaper than the Type A and required less care to use, but was also less precise with an amplification factor of 1000, and was intended more for outdoor applications or for use on concrete structures.

The first publications on the Huggenberger Tensometer appeared in 1927 [39] and, as with resistance strain gauges two decades later, the early applications of

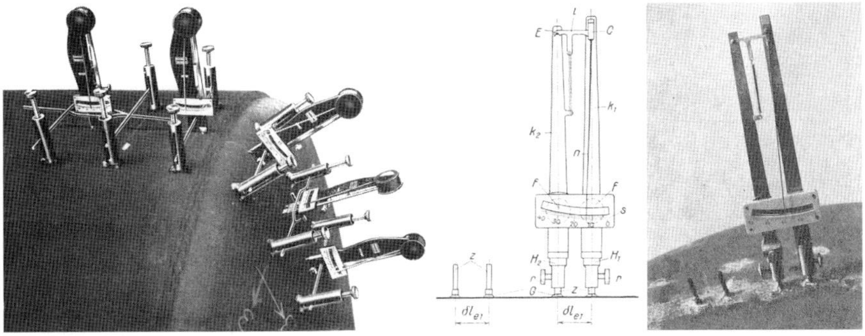

(a) (b)

Figure 13.8 (a) Huggenberger-A tensometers mounted on a pressure vessel. (Source: [40]). (b) Huggenberger Biegungsverzerrungsmesser (Bending deformation meter): Cross section and the device attached to pegs welded to the surface of a pressure vessel. (Source: [36])

the Huggenberger Tensometer were in mechanical engineering industries and, in particular, in aircraft construction [40, 41, 42]. Huggenberger Tensometers quickly found their vocation in civil engineering laboratories and detailed studies regarding its operation were published [43, 44]. Tests on a reduced-scale (probably 1:10) Vierendeel girder were undertaken in 1931 by Daniel Rosenthal (1900-1989) at the *Université Libre de Bruxelles* as part of a major study on the effectiveness and reliability of electric-arc welding [45] (Figure 13.9).

In the 1930s, the Huggenberger Tensometer was sold in the USA through the Baldwin Southwark division of the Baldwin Works [46]. They can also be seen in use in photographs of the early experimental studies of structural models in

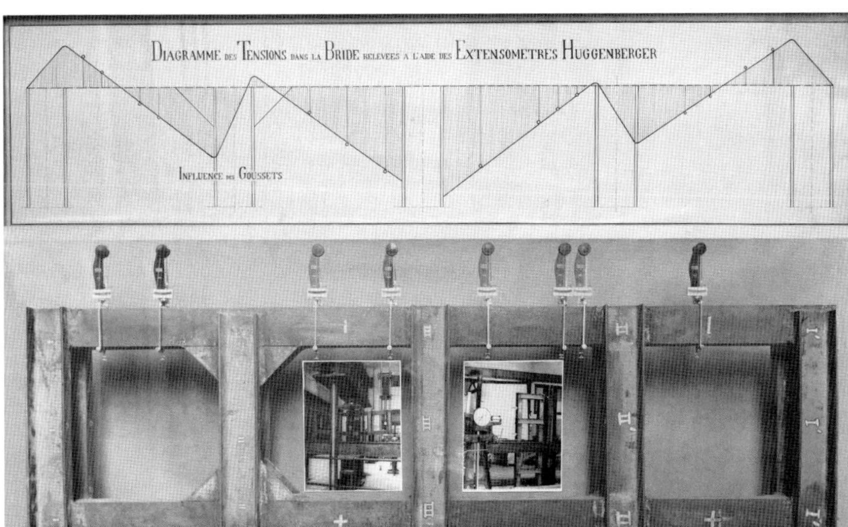

Figure 13.9 A 1:10 scale model of a welded Vierendeel girder with Huggenberger Tensometers, studied by Daniel Rosenthal at ULB in 1931. (Image: Bernard Espion. Collection BATIr Dept ULB)

the USA, such as for the design of the towers of the Hudson River Bridge in New York [47] and in Italy, such as for Nervi's designs for seaplane hangars at Orvietto undertaken at the *Politecnico di Milano* in 1935 (see Figure 10.7). The Huggenberger Tensometer achieved rare longevity, long after the widespread introduction of electrical-resistance strain gauges: they were still being used at ISMES in Bergamo in the 1960s, for example to study the model of Morandi's bridge in Genoa (see Figure 15.22b) and until the early 1980s by EMPA in Zurich [48].

The appearance of the Huggenberger Tensometer was a significant breakthrough in experimental stress analysis in the 1920s and 1930s. However, it was costly to use because each measuring point had to be equipped with an expensive, though reusable instrument, held in position by clamps or electromagnets. All measurements needed to be carried out individually on each instrument by an operator who had to be able to read the position of the fine pointer on a millimetre scale. Furthermore, the strain readings then had to be processed to give the desired stresses or forces.

13.4.3 Acoustic strain gauges

Anyone who has played a stringed instrument, piano or harpsichord is aware that the vibration frequency of a stretched string depends on its tension. In 1868, Wilfred Airy, son of Sir George Biddel Airy (1801-1892), the Astronomer Royal, used this principle to study the behaviour of a bowstring truss bridge, stiffened by wire cross bracing. He built and tested a model to determine the distribution of forces in the different elements of the bracing structure. The model had a span of 1.83 metres and a rise of 30 cm. The bow was made of steel, 9.5 x 9.5 mm, and the bowstring (the horizontal tie) comprised two strips of oak, 19 x 9.5 mm, placed 16 mm apart. All the ties were made of steel wire, 0.4 mm in diameter. For each of the ties, which are in principle in tension, the experimenter compared the pitch of the musical note made when a tie was plucked with the note produced by a wire stretched by a known load in an apparatus next to the bridge model [49] (Figure 13.10). The load in the apparatus was adjusted until the two notes were of the same pitch. This ingenious method was a very precise way to determine directly the force in the ties of the model. The experiment was then repeated for twelve different load cases.

The idea of using the vibration of a wire as the principle for extensometric measurement was first proposed in mathematical terms, in 1889, by Pietro Cardini [50]. However this early paper seems not to have been noticed by the several inventors of 'acoustic' extensometers that began to appear in the 1920s. An acoustic extensometer is basically a stretched wire, whose ends are fixed to two points on the structure being studied, which create the gauge length. If this gauge length is extended due to axial deformation, which is what we want to measure, the tension in the wire increases and so also will its vibration frequency. The relationship between the parameters is as follows: the vibration frequency (N) of a wire, of specific weight ρ, and length L, subject to tension T, is given by:

$$N = \frac{1}{2L}\sqrt{\frac{T}{\rho}}$$

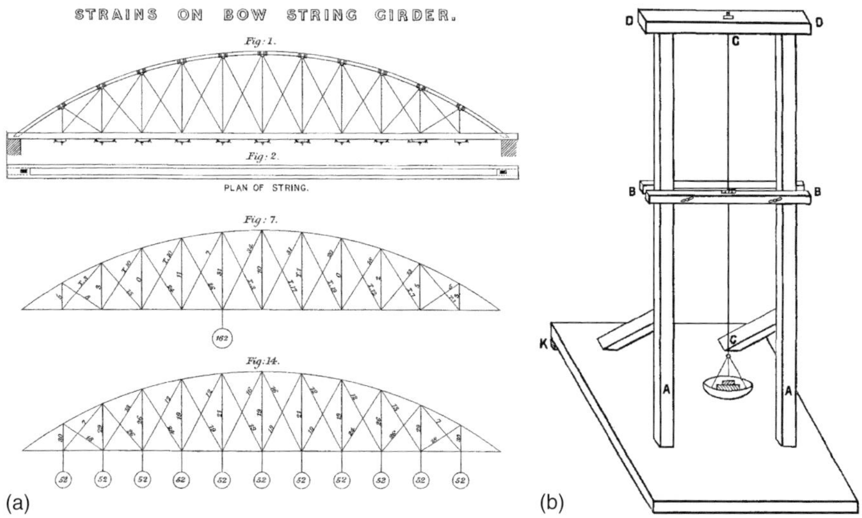

Figure 13.10 Model tests to determine the forces in the members of a bow-string truss bridge. (a) The 1.83 metre span model and two of the loading cases. (b) The apparatus to reproduce the musical pitch in a wire of the same length as a tie in the model bridge. The length of the wire was adjusted by moving the support B. The weight in the load pan gave the force in the tie. (Source: [49])

To determine the force in the wire it is only necessary to measure its vibration frequency. It was the development of practical triode-valve amplifiers in the 1910s that made it possible to realize the principle of the vibrating wire to produce an acoustic extensometer [51]. The wire is remotely excited by means of an electromagnet activated by a small current, and the vibrations are detected by means of an amplifier at a central listening station in the same electric circuit. Since the electrical cable connecting the extensometer to the measuring station only has to transmit the frequency of a signal, and not absolute changes to current intensity or voltage, very long cables can be used, and the measurements are insensitive to many potentially interfering effects such as humidity or disturbing electromagnetic fields. The acoustic extensometers developed in the 1920s were the first extensometers based on an electrical principle allowing remote measurement and were both very sensitive and precise. They were developed, apparently independently, in Germany, the USSR and France.

In Germany, Otto Schaefer republished the equations and theoretical principles in 1919 [52, 53] and developed the equipment with the firm Maihak from Hamburg. The patent number DE544023 *Akustischer Dehnungsmesser* was submitted on 1 February 1928 and registered on 12 February 1932. The first promotional publications were published in 1928 [54, 55] and then in 1930 [56]. A detailed study on the functioning of the Schaefer-Maihak extensometer was published in Belgium in 1932 [57] and Professor Ferdinand Campus (1894-1983) used extensometers of this type to measure stresses in a large steel structure under construction at the University of Liège in 1933 [58]. At the same time, Professor John Baker (1901-1985) began a large-scale research programme devoted

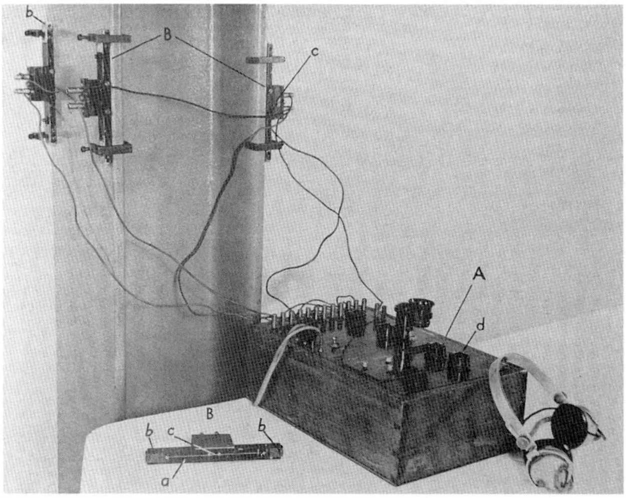

Figure 13.11 Three Maihak acoustic strain gauges mounted on a steel column. The gauge on the table shows the 12 cm long wire, a, adjacent to the electromagnet, c, used to excite vibration in the wire. (Source: [59])

to the design of steel structures which led, twenty years later, to the acceptance of their design according to the plastic theory. For the tests conducted in around 1934 on these steel structures, which were effectively reduced-scale models of a full-size steel frame, at the Building Research Station near London, Baker used Maihak acoustic strain gauges from Germany (Figure 13.11). The principle was the same as in Airy's apparatus. A standard wire in the central receiver was vibrated and adjusted so the two notes were the same, simply by eliminating interference beats between the notes produced in the two wires, and the tension in the wire, i.e. the strain measured by the strain gauge, was read off from the graduated micrometer screw adjustment. Baker recounts, at some length, the amusing story leading up to the research team's adoption of these gauges [59].

In France, the development of acoustic extensometry was due to the *Ponts et Chaussées* engineer André Coyne (1891-1960), who began his work on the subject in around 1925 [60]. He probably started developing extensometers around 1928 when he was appointed head of the special development service for the Upper Dordogne and identified a particular need for extensometers in the concrete of large dams for monitoring purposes. Coyne ignored the work of Schaefer-Maihak in Germany and developed his ideas based on studies of vibrating string sensors by two French railway engineers, Sabouret and Legrain some forty years earlier – probably Victor Sabouret (1851-1938) and Gustave Legrain (1864-1904). Coyne also recalled that Ferdinand Arnodin (1845-1924), the builder of suspension and transporter bridges, adjusted the tension of the cables according to their 'sound' [61]. André Coyne and the firm of Henry-Lepaute filed French patent number 721,629 on 19 August 1931 entitled 'Method and devices for measuring deformations of solids and in particular internal deformations of concrete'; addition number 41,835 to patent 721,629 was requested on 30 April 1932. Coyne first used his acoustic strain gauges from around 1934 for

(a)

(b)

Figure 13.12 Tests under way on the 1:2 scale model / prototype shell at Dreux, France designed by Bernard Laffaille, 1933. (a) Operator taking readings from the 'Coyne sound tubes' (Telemac acoustic strain gauges). (b) Cables leading from the gauges embedded in the concrete shell to the monitoring station. (Images: Fonds Laffaille. SIAF/Cité de l'architecture et du patrimoine/Archives d'architecture du XXe siècle. Dossier 140/6)

monitoring full-size structures including the 90-metre high Marèges arch dam [62, 63]. A pioneering use of the 'Coyne sound tubes', though not mentioned by Coyne himself, was made by Bernard Laffaille to determine the stresses in the thin concrete shells for the experimental prototype structure he made at Dreux in 1933 [64] (Figure 13.12). This structure, which is in fact a 1:2 scale model of a roof for the 'Caquot' type aircraft hangar, is sometimes cited as the ancestor of thin hyperbolic-paraboloid shells, although they are actually conoid shells.

In the USSR, vibrating-wire sensors had been developed and used since 1926, and a report of this work was published in English in 1934 [65].

Acoustic strain gauges are mainly found in two forms: either to be embedded in concrete, or to be attached to the surface of the structure to be studied by means of anchors or clamps. They are renowned for their reliability for long-term measurements and have been used mainly for the monitoring of full-scale structures. Their use for model study was reserved for relatively large models, such as the one by Laffaille in 1933. Although the minimum length of the vibrating wire is only about 5 cm, the extensometer itself has a certain tensile/compressive rigidity. Depending on the means of attaching the vibrating wire to the structure, its length and the specific rigidity of the structure, it is possible that the extensometer itself may disturb the deformation of the structure being studied. The Telemac company, created by Coyne in 1947 to market his acoustic strain gauges still exists and other companies now also make them. The vibrating wire principle is widely used today in sensors for geotechnical applications (pressure cells, etc.).

13.4.4 Electrical-resistance strain gauges

The first electrical resistance strain gauges were not based on the same principle as modern strain gauges. They made use of pieces of carbon, whose electrical resistance varies with the pressure exerted on it. This is the same principle upon which the carbon microphone was based and which was used in all telephones

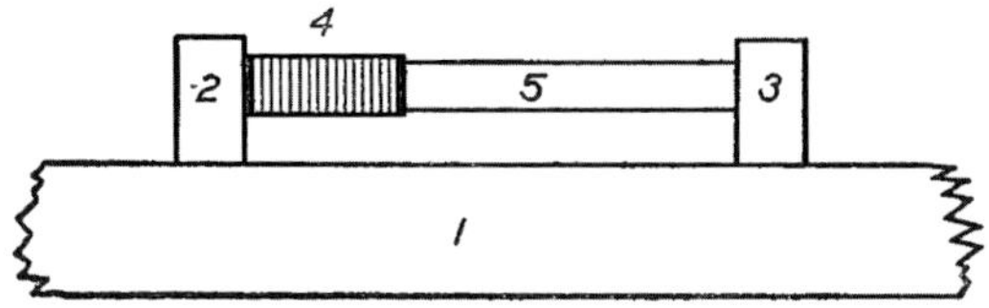

Figure 13.13 (a) A carbon-pile 'strain telemeter'. Supports 2 and 3 are fixed to the structural element under test, 1. The actuator rod, 5, pushes on the pile of carbon discs, 4. (Source: [67])

until around 1960; it was called 'carbon-pile' technology. This device was developed into a means of measuring strains in the early 1920s when they were known as 'strain telemeters' [66, 67]. A carbon-pile telemeter had a stack of 50-60 carbon discs, each about 0.5 mm thick (Figure 13.13). A strain of 0.025 mm would typically cause a change of resistance of about 1 ohm which was measured using a Wheatstone bridge (see Figure 13.16).

The strain-resistance curve for carbon is not linear, a problem that was overcome by using two pre-stressed resistors – when the load on one increases, the load on the other decreases, and the combined change of resistance is nearly linear. A strain gauge was about 20 cm long and could be clamped to the structural member under test (Figure 13.14). The great advantage of these strain gauges was, of course, that that they could be read from a great distance, and readings from many gauges could be brought to a single place to be recorded; the wires could be 30 metres long.

In the early 1920s these strain gauges were used to determine the forces and strains in the structural members of rigid airships and the members of railway bridge trusses for which both static and dynamic readings were recorded using an oscillograph [67] (Figure 13.15). A variant of the strain gauge was developed to measure the forces in the wires of aircraft wings (when aircraft wings were made with struts and ties to stiffen the wings of a biplane). This strain gauge worked on the same principle as the Montreal micrometer that Frei Otto developed to measure forces in cablenet structures in the 1970s (see Figure 14.15).

A further development of the electric carbon-pile telemeter was made in 1925 by Roy Carlson (1900-1990), a physicist who had been employed to make and record measurements on the Stevenson Creek Experimental Dam. His telemeter could be embedded in the concrete when the dam was constructed and deliver internal strain readings via 'sensor wires' to the monitoring station in a hut above

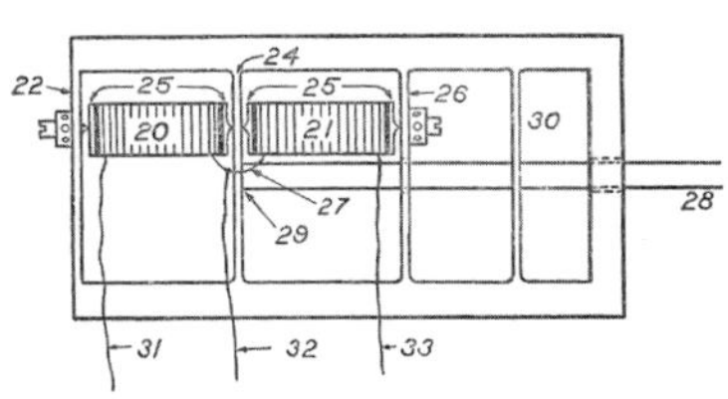

(a)

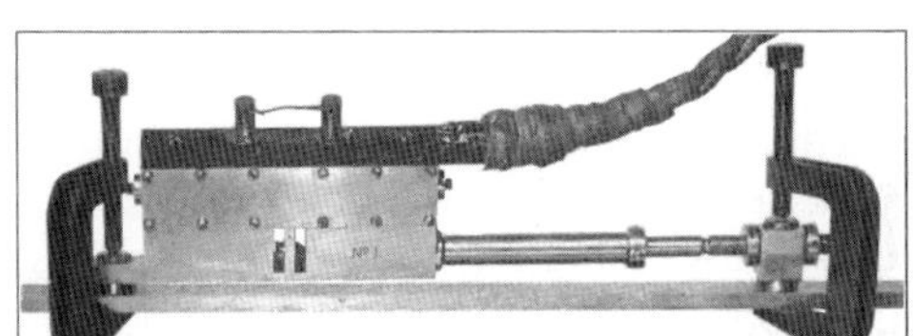

(b)

Figure 13.14 (a) Double-resistor carbon-pile telemeter (strain gauge) which eliminated the non-linearity of the strain-resistance curve of carbon. The actuator (28) pushes or pulls on the thin element (24) between the two carbon piles (20 and 21). (b) A device, 20 cm long, clamped to a structural member under test. (Source: [67])

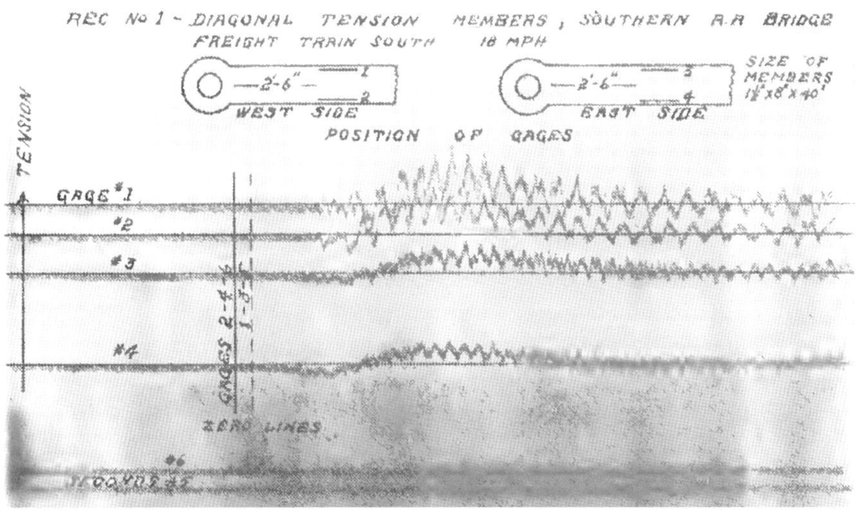

Figure 13.15 Records of strains produced in two ties of a truss bridge when a freight train crossed the bridge at 29 kph. Two carbon-pile gauges were fitted to a tie on each side of the bridge, about 75 cm from the end. The strains were recorded using an oscillograph. (Source: [67])

the dam [68, 69] (see Section 8.2.1). Further developments in carbon-pile telemeters were made in the early 1930s [70, 71].

A new approach to studying strains on the surface of metals developed in the mid-1930s, namely to attach a material to the surface that would respond to strain. One early idea, developed in 1934 by Greer Ellis (1910-1997), was a brittle lacquer called 'Stresscoat'. When strained, cracks form perpendicular to the direction of tensile strain (see Section 13.4.6). Another, in 1936, was to paint a layer of a graphite-emulsion called Aquadag onto the surface. When strained, the electrical resistance changed and could be measured using a Wheatstone bridge. Although it was very difficult to control the thickness and the gauge length, this method was used widely, especially for measuring dynamic strains on aircraft propellers and moving machinery. Also in 1936, Charles M. Kearns (b.1915), who worked for Hamilton Standard, a firm manufacturing aircraft propellers, made the first bonded electrical-resistance strain gauge to measure strains on rotating propeller blades. He ground flat a conventional carbon-composite electrical resistor and mounted it onto an insulating strip which he cemented to the blade. He made the electrical connections to the Wheatstone bridge by means of brushes and rings similar to those used in electric motors. The electrical properties of the carbon-composite were very variable and also changed with temperature and time. This made them unsuitable for measuring static or slowly-changing strains, but were highly satisfactory for dynamic strains. His bonded-carbon strain gauges were widely used in the aircraft industry and led directly to a dramatic fall in the numbers of propeller failures, from several dozen per year in 1937 to zero in 1940. Kearns' breakthrough was quickly followed by the development of bonded-wire strain gauges [72, 73].

13.4.5 The bonded electrical-resistance strain gauge

With regard to the measurement of strains in model structures, the most advanced strain-measuring device appeared at the end of the 1930s and has been continuously improved since then: the bonded electrical-resistance strain gauge, whose use goes far beyond experimental stress analysis. For example, nowadays, almost any weighing operation, whether to weigh a letter or the axle-load of a lorry, is carried out by means of a transducer whose basic principle is the measurement of strain components by means of strain gauges on the surface of a sensitive element such as a sheet loaded in bending, a cylinder loaded axially, or a thin-walled tube subjected to internal pressure.

The physical principle of the electrical-resistance strain gauge is based on the fact that a wire subjected to strain has its electrical resistance, R, changed. This physical property was first reported in 1856 by William Thomson (later Lord Kelvin) (1824-1927) [74]. When limited to first-order deformations, within a certain range and without significant temperature variation, the relative variation of resistance of a metal wire of resistance, R, is proportional to the deformation applied, ε, and is given by Equation (13.1).

$$\frac{\Delta R}{R} = K\varepsilon \tag{13.1}$$

The proportionality factor, K, is called the *gauge factor*. The variation in the resistance, ΔR, is very small for typical elastic deformations. It is measured using an electrical circuit called the Wheatstone bridge [75]. The Wheatstone bridge is an idealized scheme that neglects some factors that influence resistance (Figure 13.16). The principle of the measurement is as follows: the bridge is powered by a voltage source, V, between terminals D and C. If the bridge is initially balanced, we have the relationship in Equation (13.2):

$$R_1 \cdot R_3 = R_2 \cdot R_4 \tag{13.2}$$

If the bridge arms each undergo a variation in resistance, the imbalance of the bridge results in a voltage variation, ΔE, between A and B (Equation (13.3)). This imbalance, both at the time of Thomson and at the beginning of the development

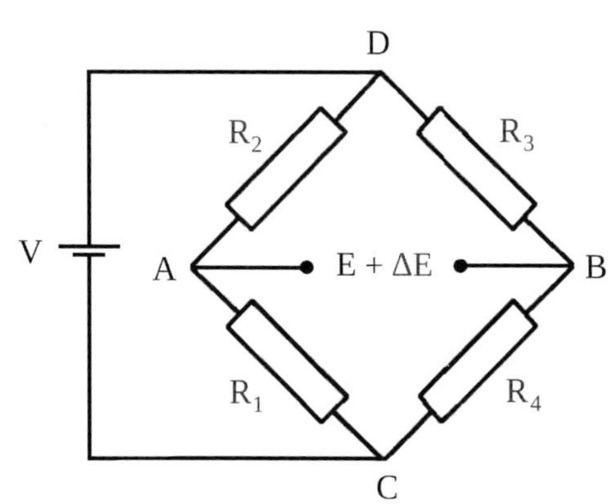

Figure 13.16 A Wheatstone bridge.

of strain gauge, was measured by means of a galvanometer.

$$\frac{\Delta E}{V} = \frac{1}{4}\left[\frac{\Delta R_3}{R_3} + \frac{\Delta R_1}{R_1} - \frac{\Delta R_2}{R_2} - \frac{\Delta R_4}{R_4}\right] \tag{13.3}$$

If the resistance of only one arm of the bridge varies – for example D-B (R_3) – the arrangement is called a 'quarter-bridge' circuit. If the resistances of two adjacent arms of the bridge – for example D-B (R_3) and D-A (R_2) – vary, the arrangement is called a 'half-bridge' circuit. If the resistances of all four bridge arms vary, it is a 'full-bridge' circuit.

The history of the appearance and early use of bonded resistance strain gauges is well-documented [76]. It has been reported in detail by Peter Stein (1928-2016) [72, 73, 77, 78] who collected the testimonies of all the pioneers and actors in this adventure, notably on the occasion of various meetings organized in 1988-1989 to commemorate the fiftieth anniversary of the first appearance of the wire strain gauge. The autobiographical account by Frank Tatnell is a primary source [46]. An interesting complementary historical perspective from the German viewpoint is provided by Keil [79, 80] and Rohrbach [81].

In 1936, Donald Clark (1906-1976) and Gottfried Dätwyler (1906-1976) were conducting research at Caltech (Pasadena, CA) on the properties of metals under impact. The electrical engineer in charge of the laboratory, Edward E. Simmons, Jr. (1911-2004), designed a specific sensor to mount on the impact-testing apparatus. This sensor is the ancestor of all gauge sensors: Simmons suggested bonding a fine, insulated wire to the surface of a prismatic bar to act as a force-measuring transducer [78]. The material used was a Constantan wire (alloy 60Cu–40Ni, K≈2) with a diameter of 0.003 in (.075 mm). Simmons developed all the electrical instrumentation, but it was Dätwyler who glued the wire to the bar, Simmons having never glued a wire or a resistance gauge in his life. This team did not realize at all the potential of this innovation, and the report of the research which describes the results achieved from the measurements, only mentions the technique for acquiring the measurements incidentally and was not published until 1939.

Since 1935, Arthur C. Ruge (1905-2000) had been conducting research in the MIT Civil Engineering Department in the field of seismic engineering (see Section 25.2.3). In particular, in the years 1937-1939, his collaborator J. Hans Meier (b.1913) was undertaking tests on models of water towers subjected to seismic loading on a shake table that Ruge had developed himself [82]. The measurement of the stresses generated in the model was a real headache until, on 4 April 1938, Ruge thought of sticking thin wires, 0.001 in (.025 mm) in diameter, made of the alloy *Elinvar* (52Fe–36Ni–12Ce, K=3.5), to the surface of the model and measuring the change in resistance by means of a galvanometer (Figure 13.17). Ruge immediately realized the potential of this innovation and approached the MIT Patent Committee to get their support. With hindsight, their response in March 1939 was, for the MIT, perhaps the wrong one:

> 'It is the general policy of the Committee [...] to pay attention primarily to matters which may prove to be of major importance and, while this development is interesting, the Committee does not feel that the commercial use is likely to be of major importance.

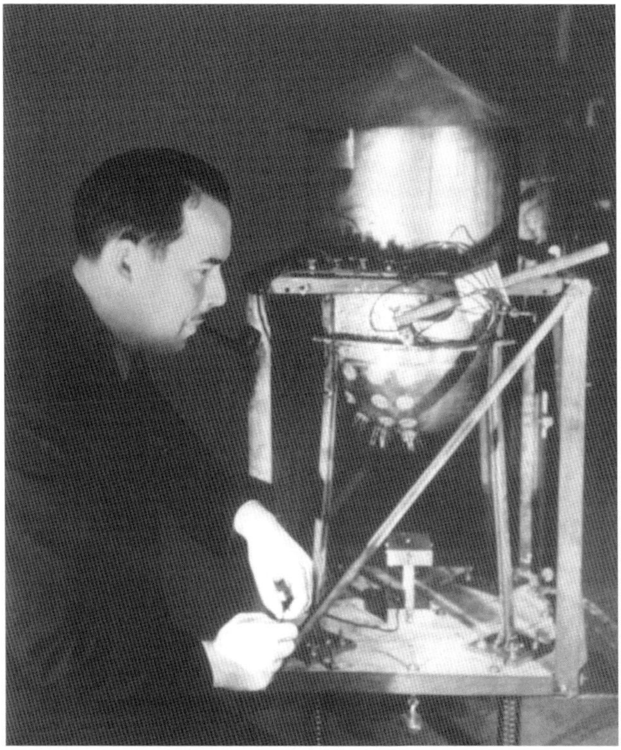

Figure 13.17 Arthur Ruge engaged in using his bonded electrical-resistance strain gauges for studying the effects of seismic loads on water towers in 1938. (Source: [80])

> Accordingly, the Committee has voted that any rights that the Institute may have in this invention should be waived in your favor. This leaves you free to treat the invention entirely as a personal matter' [79].

In 1939, Arthur Ruge joined forces with Alfred V. de Forest (1888-1945) from the Mechanical Engineering Department of MIT, who supported him in his research and was in contact with the firm Baldwin Southwark, which marketed, among other things, testing machines and various types of extensometers. Simmons' earlier discovery in 1936 is now known, and it was Baldwin's lawyer who wrote the US patent that wrote the US patent 2,292,549 (filed 23 Feb 1940) on behalf of Simmons, allowing Ruge take the application patents (filed 16 Sept 1939) (Figure 13.18). The technical development was carried out by Ruge and de Forest, and the product – the electrical-resistance wire strain gauge – came to be known as SR-4 (S for Simmons, R for Ruge and '4' for the four people involved in the invention: Simmons, Ruge, Clark and de Forest). Very quickly, the invention interested aircraft manufacturers on the west coast, but the first industrial batch of 50,000 gauges was not sold by Baldwin until 1941. The strain gauge immediately became an essential tool in the development of the aeronautics, automotive and armament industries in the USA during the Second World War.

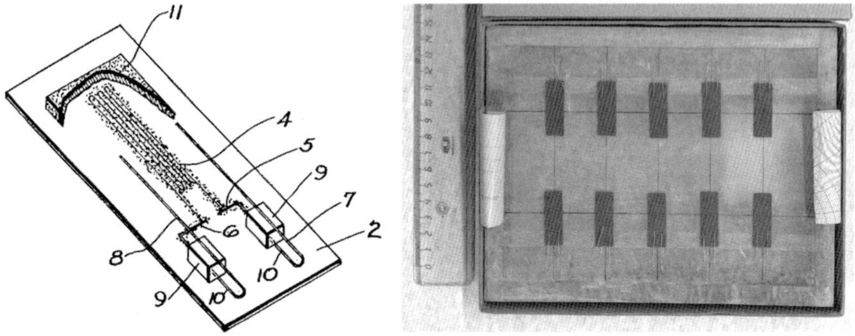

(a) (b)

Figure 13.18 The SR-4 stain gauge. (a) Drawing from Ruge's patent for the bonded wire strain gauge. [83] (b) A box of ten SR-4 strain gauges. (Image: Bernard Espion. Collection BATIr Dept ULB).

It is remarkable to note here that the gauge originated from the need to perform dynamic measurements (at Caltech and MIT) and model measurements (at MIT)! In the immediate post-war period, Baldwin Hamilton's SR–4 gauge obviously had no competition; production in Europe did not start until the late 1940s and early 1950s. Its main application (by volume) was probably more for the manufacture of sensors (transducers) to measure forces, torques and pressures than in experimental stress analysis or strain measurement on models. For these latter applications, the electrical-resistance wire strain gauge was a significant advance over the Huggenberger tensometer, even though the latter was still in common use in laboratories for at least two further decades.

The advantages of the electrical-resistance wire gauge over the mechanical extensometer included:

- very small size (a few mm) allowing very local strains to be measured;
- the possibility of measuring the deformation on a surface with any inclination;
- virtually zero weight;
- the possibility of carrying out the measurement remotely using an electrical signal;
- very low cost.

But the resistance wire gauge is not as obvious to use as a mechanical extensometer: it requires a very specific know-how, which strain-gauge manufacturers still provide nowadays through training. All the difficulties in using strain gauges lie in electrical problems, because the variations in resistance to be measured are very small, and many sources of errors can affect the measurements.

Nowadays, and it was already of the same order of magnitude in the 1940s, the electrical resistance gauges commonly used in extensometry have a resistance of 120 ohm and a gauge factor $K \approx 2$, although the current flowing through the gauge heats it up by the Joule effect, which locally heats the surface and affects the strain measurement. The supply voltage, V, of the Wheatstone bridge (see Equation (13.3)), should be as high as possible to achieve the best sensitivity, but balanced against the adverse heating effects of a higher current. For materials that do not conduct heat well, such as plastics or concrete, a supply voltage of less than 1 V is required. For metals (steel, aluminium), voltages in the range of 5-10 V can be used. With these values, we can see immediately from Equation (13.3) that the order of magnitude of the potential difference to be measured in the elastic range, cannot be measured with a voltmeter! For a long time, until the 1970s, it was necessary to supply the Wheatstone bridge with an alternating-current carrier frequency; this remains the preferred option for high-precision measurements even though, nowadays, DC power supplies and amplifiers are available.

Sources of measurement error with strain gauges are 'nested' everywhere. The most obvious, and generally a problem in mechanical extensometry, is the inclusion in the measured deformations of variations arising from the ambient temperature during the measurement period, which can extend over several days when studying a model. This effect can be eliminated by adopting a half-bridge circuit incorporating a control gauge (dummy gauge), in which the 'active' gauge is R_3 and the 'passive' (dummy) gauge is R_2. The R_2 control gauge is bonded to a freely expandable sample made of the same material as the structure to be studied and subject to the same thermal variations.

Among other electrical problems which can strongly affect strain-gauge measurements to the point of rendering them meaningless, it is worth mentioning:

- Electrical insulation problems between the gauge and the structure to be studied, to the surface of which it is glued;
- The effect of moisture which can affect both the insulation of the gauge and the measuring cables; this has always severely limited the use of strain gauges use outside the laboratory or over long periods of time;
- Contact resistances in the switches that are widely-used in experimental stress analysis when the model or structural element to be studied has, as is often the case, many measuring points and only one measuring instrument is being used with a multi-way channel selector.

A subsequent development in electrical-resistance strain gauges of significant importance for stress analysis, was the invention of the foil strain gauge by Peter Scott-Jackson (1916-2000) in 1952, while working for the British aircraft manufacturer Saunders-Roe Ltd [84]. This used a manufacturing method developed for printed-circuit boards making the gauge even lighter and cheaper than the wire gauges which they quickly replaced (Figure 13.19). Other significant developments since the 1970s have been in the electrical and electronic aspects of amplification, measurement and data collection. This means, in particular, that the connection of a gauge to the measuring instrumentation is made with three wires, not two, and that all the gauge sensors, which operate in a full-bridge circuit with four (or even eight) gauges, require a six-wire circuit and no longer only four.

In 1975, thirty-five years after the launch of the SR-4, Vaughan identified at least thirty manufacturers of strain gauges, and this list was certainly not exhaustive [86]. The strain gauge had become not only the strain measurement sensor used in experimental stress analysis, on real structures or models, but also a sensor increasingly present in everyday life [87].

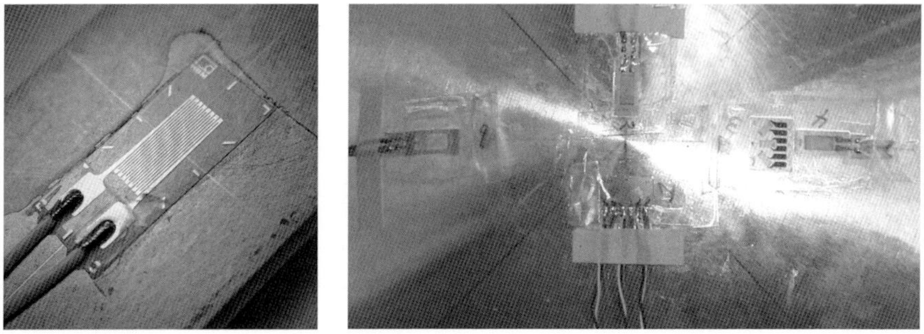

(a) (b)

Figure 13.19 (a) A single foil strain gauge. (Image: [85]). (b) Five foil strain gauges mounted on a metal surface. Single gauges – left, top and right; rosettes – bottom and right. (Image: Bill Addis)

13.4.6 Some miscellaneous measuring techniques

Evidence of man's ingenuity and creativity is to be found in the astonishing number of ways that have been devised to measure the movement of an object subject to loads; here is not the place to explore this field of endeavour in any depth. It is, however worth drawing attention to some of the techniques mentioned in other chapters of this book and a few more besides.

Someone must have invented the dial or clock gauge, but their name is lost to us. This is not the case for the inventor of that elegant means of reading a scale to great accuracy – the Vernier scale. It was invented in 1631 by the French mathematician Pierre Vernier (1580-1637). Its appearance stimulated the English astronomer William Gascoigne (1612-1644) in around 1638 to invent the micrometer to improve his readings of star positions made with his telescope. It became the measuring instrument we know today only in the 1840s. The 1630s, by the way, was also the decade when the slide rule was becoming known among scientists, although it was first adopted by engineers around 1800 when the firm Boulton and Watt developed a more accurate and useful slide rule and sold it widely as the Soho Slide Rule, named after the site of their factory in Birmingham.

In order to achieve a greater accuracy when reading mechanical strain gauges, a number of so-called 'optical strain gauges' were developed. These involved causing the movement in a strain gauge to rotate a mirror and this movement could be measured to great accuracy by measuring the movement of a beam of light reflected by the mirror. One example is the Tuckerman-type optical strain gauge that was used by the engineers studying the models of the Boulder Dam (see Figure 8.14), and which was often also referred to as an 'optical lever' [6].

Photography has served engineers well as a means of recording the instantaneous positions of structures and, indeed, fluids and sound waves passing through air. The engineers in Frei Otto's Institute of Lightweight Structures in Stuttgart made much use of photography, using photogrammetry to survey complex 3-D shapes (see Figure 14.14) and double exposure photographs to measure the deformation of complex 3-D cablenet structures (see Figure 14.16).

They also devised a miniature device to measure the tension in the many dozens of wires in models of cablenet structures, known as the Montreal micrometer after the name of the project (Expo'67) for which it was perfected (see Figure 14.15).

Another optical technique that has been much used in the field of mechanical engineering and in research laboratories is the *moiré* method [88]. This technique is based on the patterns that are formed when two grid patterns are moved relative to one another. It was named after the moiré fabric in which this effect is observed, and which can also be seen when two layers of net curtain overlap. Depending on the spacing of the grid lines, the movement of the pattern can be ten or a hundred times greater than the movement of the grid itself. This visually-striking phenomenon has not been widely used in structural model testing but it has been applied by Bill Harvey to create a simple tell-tale that will allow movements on full-size structures such as masonry bridges to be monitored and recorded [89].

Another strain-measurement method that has been widely used in the mechanical and aeronautical engineering industries, but has also been used in structural model testing, is the brittle-lacquer technique. The technique has its origin in a phenomenon that was well known to engineers of wrought iron and steel bridges in the nineteenth and early twentieth centuries. It had often been observed that the brittle coating of ferric oxide that forms on the surface of iron and steel during rolling and annealing will crack or flake off when the metal is greatly stressed – indeed such cracks could be highlighted using whitewash. The engineers testing models of the wrought-iron girders for the Conway and Britannia bridges in 1846 noticed similar cracking in the paint that had been applied to prevent rusting (see Figure 6.7).

The technique was developed as a tool in experimental stress analysis in the early 1930s by Otto Dietrich and Ernst Lehr in Germany [90]. A few years later, Greer Ellis (1910-1997) produced an improved material that was easy to apply and gave more consistent results than other materials. He found that carbon disulfide was the only satisfactory solvent for a rosin-based coating. He called it 'Stresscoat' and it was successfully marketed by the Magnaflux Corporation which had been founded by Alfred V. de Forest [91, 92]. It is still available today, among many other brands.

The principle is simple. A lacquer is applied by painting or spraying and allowed to harden by drying in air or by the application of heat to form an even, brittle layer 0.1-0.2 mm thick. When the structural component is loaded, the surface is strained and the lacquer forms a series of parallel cracks perpendicular to the direction of tensile strain (Figure 13.20). The cracks can be made more evident using a die. The most visible outcome is the direction of the tensile stresses, which allows the directions of the isostatic lines (lines of principal stress) on the surface to be mapped [93]. The technique was used for this purpose in 1948 by Carlos Bonito for the model test of the shell roof of the Xerallo church in Spain (see Figure 16.1b). The region of cracking of the lacquer also indicates the extent over which the strain was greater than that needed to crack the lacquer. It is also possible to derive quantitative information about the strains, the degree of strain being indicated by the spacing of the cracks. Since the strain at which a crack forms

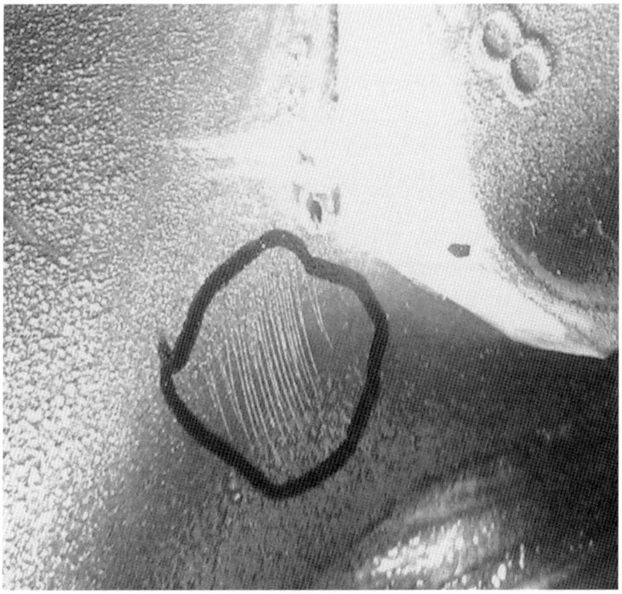

Figure 13.20 Brittle-lacquer coating. The region of cracking is visible on the surface of a structural component under load indicating the direction, intensity and extent of tension stresses. (Image: Courtesy of Stresscoat)

depends on the strength of the lacquer and its thickness, it is usually necessary to prepare a test strip of metal which is strained using a known load to calibrate the crack formation.

13.5 The Beggs Deformeter

The Beggs 'Deformeter' is by no means a mechanical extensometer in the sense defined in Section 13.4.2. It is a mechanical device used in an experimental method of analysis of statically-indeterminate structures. It was patented in 1922 by Professor George Erle Beggs (1883-1939) of Princeton University [94] and was used extensively in the following decade, particularly among concrete bridge designers.

It is worth recalling the context of the analysis of structures at the time of this invention. In the early 1920s, the calculation of redundant structures could only be approached analytically by the force (stress) method, which involves solving a system of equations for displacement or deformation compatibility, whose number is equal to the degree of statical indeterminacy. Reinforced-concrete construction, however, particularly favoured the construction of continuous structures, leading to flat structures with rigid nodes, whose rigorous analytical study was tedious, or even impossible by analytical means as soon as the degree of indeterminacy exceeded three.

The Beggs method was an experimental technique for the analysis of planar structures of this type, based on the Maxwell-Mohr reciprocal displacements theorem applied to the search for the line of influence of forces in redundant systems [95, 96]. As a typical case, consider a continuous multi-span viaduct. Suppose we look for the influence line of the bending moment to the right of a section of the deck under the effect of a unit vertical force moving from one end of the bridge deck to the other. If we have this influence line, it is easy to determine how to arrange the load to have the maximum effect in the section considered and to calculate this maximum value.

Analytically, the solution to this problem is as tedious, if not more so, as the direct search for the distribution of forces throughout the redundant system for a given load: it consists in imagining, in the section under consideration, a cut that releases the force whose line of influence is sought, and in applying to the faces of the cut (on both sides of the cut) a deformation combined with this force, which produces an overall deformation of the structure. This deformation, in the direction of the unit force, is the influence line of the force considered. Since the deformation is applied as a certain strain, imposed by the standard plugs, it is necessary to do one calculation of an influence line in order to determine the effective scale factor for the test. For the influence line of a bending moment, it is therefore necessary to introduce an internal joint (hinge) and to apply a relative rotation (angular discontinuity) in the section considered.

As Beggs reported in his patent [94] and in two papers published in 1922 and 1923 [97, 98], he proposed to do this analysis of influence lines very easily and economically using flat models of the structure, made with sheets of either cardboard or celluloid a few millimetres thick. A Beggs deformeter is a mechanical device that enables the application by means of gauge plugs, to the faces of a cut in the model, of imposed displacements associated with either a normal force, a shear force or a bending moment (Figures 13.21, 13.22, 13.23). These imposed, relative displacements are very small and so are the displacements of the measuring points of the structure: they are therefore measured by means of graduated microscopes. Of course, as many models have to be created as there are different sections to study.

Beggs patented his invention because he clearly wanted to make a commercial profit from it. He therefore disseminated his ideas widely to promote his invention, not only in scientific papers but also in a popular magazine [99]. In Europe, he personally presented his method to the ETH, Zurich on 7 July 1923 [100] and in Paris, Augustin Mesnager presented it on his behalf to the Academy of Sciences on 26 March 1923 [101]. Beggs returned to Europe in 1926 and presented his method at the Second International Congress of Applied Mechanics at the ETH Zurich on 14 September 1926. Kann [102] and Hofacker [100] synthesized in German the two papers by Beggs [97, 98] whose figures they faithfully reproduced. A 'kit' for marketing of Beggs Deformeters included six pairs of clamps and various accessories designed to support the model on a table by minimizing friction [103, 104] (Figure 13.24).

Beggs' first papers [97, 98, 103] are devoted to a presentation of his method: the examples he considered are 'academic', and the measurements are compared with theoretical solutions, which allowed Beggs to claim an accuracy of around 1% for the results obtained using his method. Later, independent studies speak rather of errors of between 5% [105] and 8% [106].

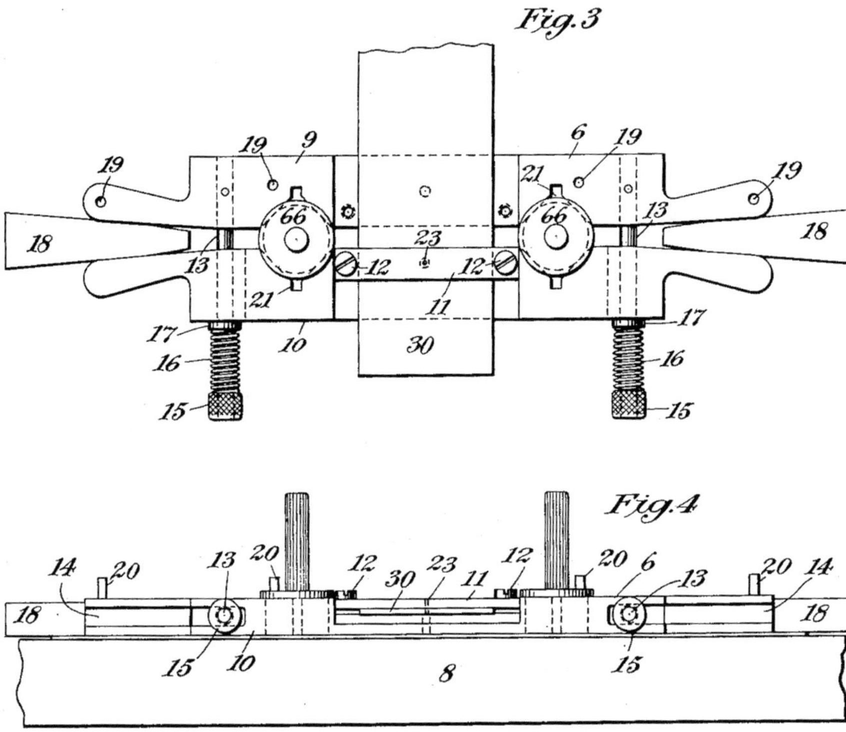

Figure 13.21 Drawings in Beggs' patent showing the Deformeter with which known strains can be introduced at a cut in the model. Plugs are inserted at position 66, circular of equal size to introduce compression, of unequal size to introduce bending, and of rectangular section to introduce shear. (Source: [94])

The first 'practical' case studied was undoubtedly in Switzerland, in 1926: the application of the method to the Grandfey railway viaduct over the Sarine River near Fribourg. This viaduct, with seven concrete arch spans, the largest opening of which is 43 m, was, at the time, the largest railway bridge under construction in Switzerland. The study of a single arch using the Beggs method was carried out by the Swiss Federal Railways (SBB) bridge office [107]. Robert Maillart is cited as the consulting engineer for this bridge, but his involvement in the design appears to have been quite limited [108]. Beggs immediately used this work on the Grandfey Viaduct to further promote his method by publishing a photo of the complete bridge and a model [109]. These same illustrations were reproduced in a popular science article that reviewed a number of different techniques of model studies under the heading 'Toys that save millions' [110].

The second noteworthy application of the Beggs method was an analysis commissioned from Beggs for the study of the arches of the Arlington Memorial Bridge over the Potomac River in Washington, a large concrete bridge, comprising eight arches with spans between 50 and 55 m [111].

As Beggs Deformeter gauges became commercialized, they were used by a number of researchers and engineers working in design offices or in administrations. Some examples of their use are:

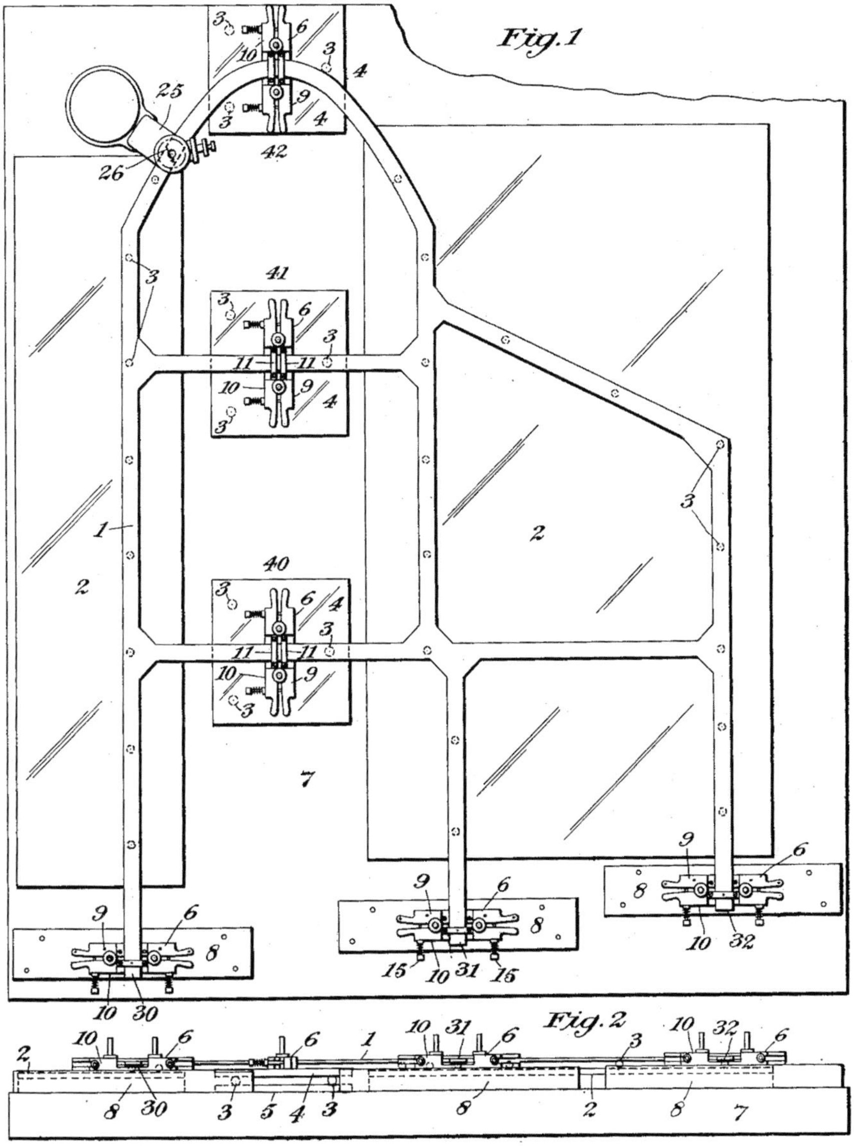

Figure 13.22 Drawing from Beggs' patent showing the arrangement of Deformeters for testing a celluloid or cardboard model of a structural frame. (Source: [94])

- the optimization of variable-inertia, reinforced-concrete frame bridges in its parkways by Arthur G. Hayden, consulting engineer of the Westchester County Park Commission (N.Y.) [109, 112];
- a study of the effect of gussets at the column-beam intersection in reinforced-concrete portal frames by Frank E. Richart and Rex L. Brown, of the University of Illinois at Urbana-Champaign [113];

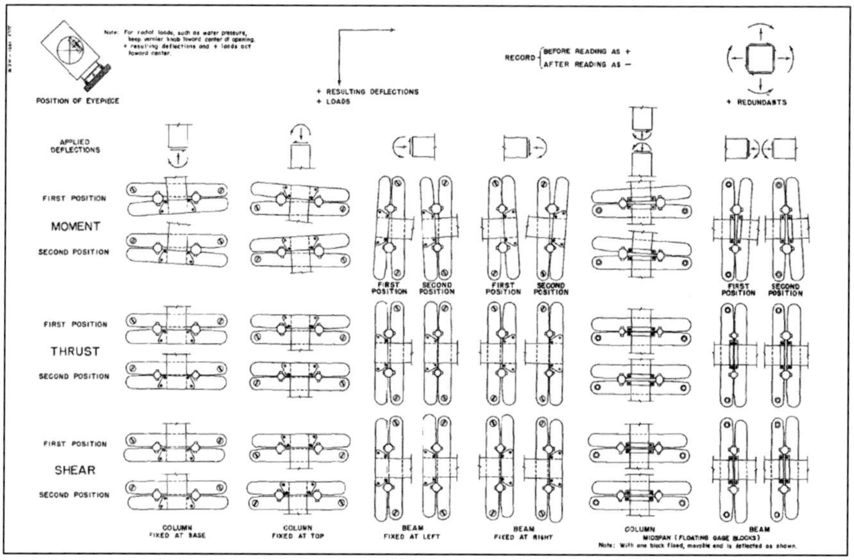

Figure 13.23 Drawing showing the different arrangements of the gauge plugs to generate bending moment, compression and shear. (Source: [127])

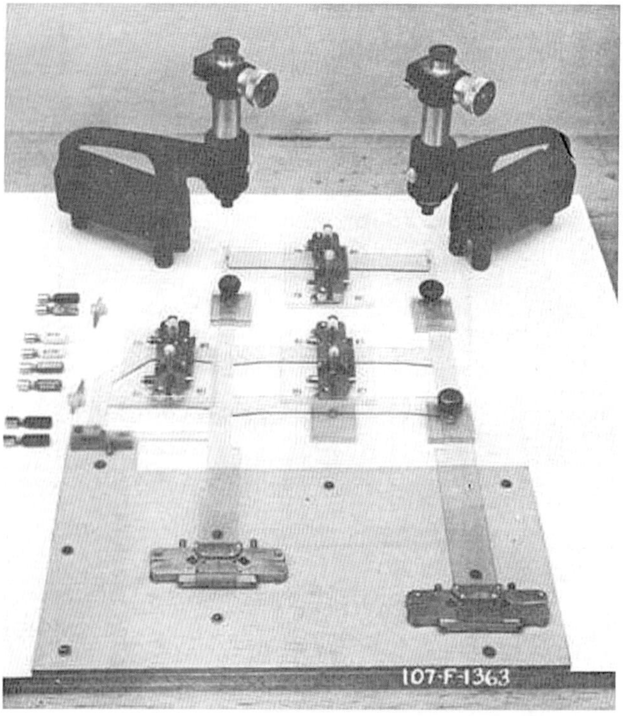

Figure 13.24 Typical arrangement for undertaking model tests on a celluloid model using Beggs' Deformeter. The photograph shows two Deformeters in the foreground, three Deformeters mounted on the model in the centre, seven plugs at the left and two microscopes at the back. (Source: [127])

- the study of a concrete bridge with three arches of 45 m span over the Yadkin River between Albermale and Mount Gilead (N.C.) by J.T. Thompson and Albin L. Gemeny, engineers from the US Bureau of Public Roads; a large model of the whole bridge (all three arches), 2.54 m long, was studied at John Hopkins University [114];
- the study, in 1927, of a bridge with seven arches – the largest of which was 39 m wide – over the Ashtabula River by D. Henry Overman of the Bureau of Bridges of the Ohio Department of Highways; a Beggs model of one arch is illustrated in Hampton et al. [115];
- the study of a concrete bridge with five arches of 59 m span on the White River Bridge (Lake Taneycomo) between Branson and Hollister (Mo). This was a record span for a bridge of this type at the time. Since the structural analysis of the highly redundant structure involved solving some 270 equations, the engineers felt it wise to undertake a model test using Beggs' method. The 4.3 metre model of the full bridge (all five arches) was studied in 1931 by engineers from the Missouri Department of Highways Bridge Office [116] (Figure 13.25);
- an evaluation of the accuracy of the Beggs method by K. Schaechterle (1879-1971), by comparing the results obtained on a Beggs model with an analytical calculation, for a rail bridge consisting of a single concrete arch with a span of 27.4 m, built in 1927 near Weikersheim [117];

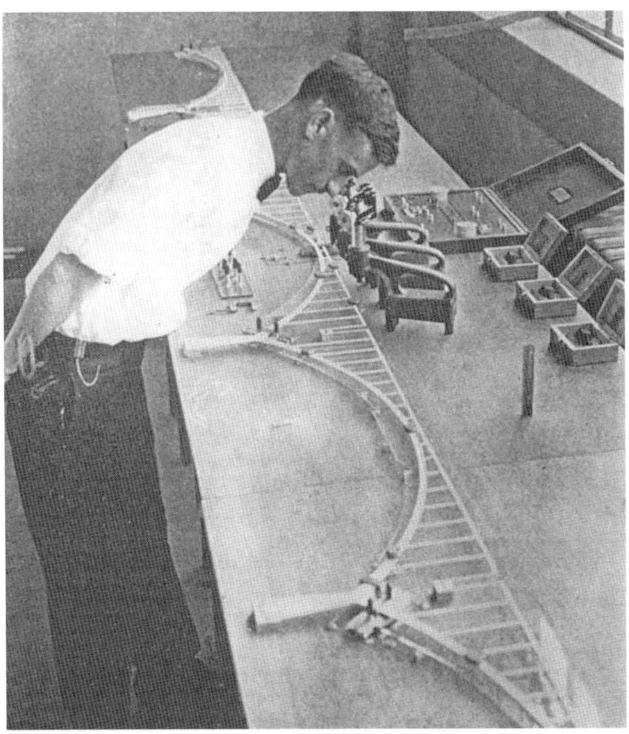

Figure 13.25 A 4.3-metre long celluloid model of the White River Bridge being tested using Beggs Deformeters. Several thousand separate measurements were taken. (Source: [122])

- a study by Beggs himself, at the request of a design office, of a mat foundation, 28.7 x 20.4 m in plan, in the form of a two-storey high Vierendeel structure, for an eleven-storey building housing a telephone exchange in Albany (N.Y.) [118];
- a study by Conde B. McCullough (1887-1946), the eminent concrete bridge builder of the Oregon Coast Highway 101, for his iconic bridge over the Rogue River comprising seven 70-metre spans, for which a three-arch Beggs model of the bridge was tested [119];
- a study by G. S. Richardson in 1932 for the George Westinghouse Bridge in East Pittsburgh (Pa) with five concrete arches spanning between 60 and 140 metres. A Beggs model was made for the largest arch which, with a span of 140 m, was the largest concrete arch in the USA [120].
- a study by K. Hofacker (1897-1991) using a Beggs model of the Fürstenland Bridge near St. Gallen in Switzerland, a 134-metre concrete arch built between 1938 and 1941 [121].

It can be seen that most of the published case studies concern large continuous concrete viaducts consisting of a deck resting on arches via piers embedded in the deck and in the arches, thus forming a spandrel. For these large concrete bridges of a new typology, the question arose concerning the interaction between the deck and the arches in a structure made monolithic by the rigid connections; such an interaction was not taken into account using a classical approach that considers the arches and the deck separately. An ASCE Special Committee on Concrete and Concrete Arches, of which Beggs was a member, was active between 1923 and 1934 [123].

At the First International Congress of Concrete and Reinforced Concrete, held in Liège in 1930, Gustave Magnel (1889-1955) proposed an improvement to the Beggs deformeter and presented the 'micro-influenciometer' that he had developed in Ghent [105]. He used it to establish the influence lines of the forces in the

Figure 13.26 Magnel's 'microinfluencemètre' applied to the analysis of a Vierendeel girder. (Source: [124])

Vierendeel girders published in his book [124] (Figure 13.26). It is worth noting here that the study of Vierendeel girders was explicitly mentioned by Beggs as an interesting field of application for his method [109]. Brebera briefly reports the results of another analysis of a Vierendeel bridge using a Beggs method adapted by Blazek in Prague [125, 126].

In 1930 Hardy Cross (1885-1959) published his moment distribution method for analysing flat redundant frame structures by means of an iterative, rapidly-converging method of solving the equations of equilibrium of the displacement method. The method was adopted very quickly and spread widely during the 1930s. This method enabled many of the structures which Beggs had been considering in the 1920s to be solved analytically. Nevertheless, the Beggs method remained a practical solution for many problems, including structures composed of non-prismatic elements such as the curved elements of arch bridges and elements with varying second moments of inertia, structures with substantial haunches and those whose solution converges only very slowly using the Hardy-Cross method, as is the case for Vierendeel girders.

After the Second World War, many papers present the Beggs method as one of several methods for model-based structural analysis; perhaps the best of these is by Moody and Phillips [127]. However, there were rather few publications of projects for which the Beggs method had been used. As exceptions, there should be noted:

- A parametric study by the US Bureau of Reclamation, of the distribution of stresses in buried pipes for nine different forms of cross-section [128];
- The use of a Beggs model during the design of the Ash Street Bridge on Interstate 93, Londonderry (N.H.) in 1962 by R.J. Prowse of the New Hampshire Department of Public Works and Highways: 'the bridge was the first known example of all-welded steel, rigid-frame technology used along the interstate Highway System' [129].

Beggs, who died in 1939, is also known to have carried out model studies for the Stevenson Creek Experimental Dam (see Section 8.2.1), the Mount Hope Bridge in Rhode Island [130], the San Francisco-Oakland Bay Bridge [131, 132] and the piers for the Golden Gate Bridge [133]. He also undertook many other studies using the Maxwell-Mohr theorem principle for 3-D structures such as skew bridges, but these were not direct applications of the 'Beggs Method' as described above.

In fact, the Beggs Deformeter was not the only device of this kind developed in the 1920s and 1930s, based on the Maxwell-Mohr reciprocal theorem. Among others there were Kann's method [134], Gottschalk's 'Continostat' [135], Rieckhof's 'Nupubest' (Nullpunkt Bestimmung) [136, 137], Eney's 'Large displacement deformeter' [138, 139] and Ruge and Schmidt's 'moment indicator' [140]. However, while these were no doubt intellectually satisfying and probably useful to convey understanding, they were not used by practising engineers on live projects. Indeed, in 1930, Magnel strongly condemned on theoretical grounds all the methods based on the reciprocal theorem that relied on measurement of large displacements [105].

In conclusion, it should be noted that the published case studies of applications of the Beggs method, with a few exceptions, do not allow us to assess to what extent the use of a Beggs model actually contributed to the 'design process' for the structures studied. Rather, we get the feeling that, often, the model study served mainly to raise the engineer's confidence in a design derived using analytical techniques. In the case of parametric theoretical studies, such as those by Magnel [124] or Philips and Allen [128], a Beggs model can rather be considered as an analogue computing machine.

13.6 Concluding remarks

The development of modern structural modelling technique depended on developments in two separate strands of engineering – the ability to conduct tests in a scientific manner, which evolved from the scientific revolution in the Age of Enlightenment; and the ability to scale up of results of model tests to the full-size structure. The former skill was already well-developed by 1750. The latter was only achieved in the interwar period, following the lead taken in fluid mechanics and aerodynamics during the period from about 1890 to 1925. Structural modelling technique reached its full maturity in the 1960s and 1970s, ironically, at the same time that computers began to take the lead in modelling the real world, at least in the field of structural engineering. As we shall see, while physical modelling in structural engineering is now rare, it is still of paramount importance in other fields of civil and building engineering.

References

1 Froude, W. (1877) Experiments upon the effect produced on the wave-making resistance of ships by lengths of parallel middle body. *Transactions of the Institution of Naval Architects* XVIII: 77-87; discussion, 87-97.
2 William M. Connolley. Licensed under the Creative Commons Attribution-Share Alike 3.0 Unported license. https://commons.wikimedia.org/wiki/File:Boat_models_by_William_Froude.JPG (23/11/2019)
3 Buckingham, E. (1914) On Physically Similar Systems: Illustrations of the Use of Dimensional Equations. *Phys. Rev.* 4 (4): 345-376.
4 Buckingham, E. (1915) Model Experiments and the Form of Empirical Equations. *Trans. ASME* 37: 263-296.
5 Bridgman, P. W. (1922) *Dimensional analysis.* New Haven: Yale University Press.
6 Hetenyi, M. (ed.) (1950) *Handbook of experimental stress analysis.* New York: Wiley.
7 Cowan, H.J., Gero, J.S., Ding, G.D. & Muncy, R.W. (1968) *Models in Architecture.* Amsterdam: Elsevier.
8 Hossdorf, Heinz (1971) *Modellstatik.* Wiesbaden and Berlin: Bauverlag GMBH.

9 Hossdorf, Heinz (1972) *Modelos reducidos: metodo de calculo* (trans. Carlos Benito Hernández). Madrid: Instituto Eduardo Torroja.

10 Hossdorf, Heinz (1974) *Model Analysis of Structures* (trans. C. van Amerongen). New York: Van Nostrand Reinhold.

11 Müller, Robert K. (1971) *Handbuch der Modellstatik*. Berlin: Springer Verlag

12 Fumagalli, Emanuele (1973) *Statical and Geomechnical Models.* Springer-Verlag/Wien.

13 Sabnis G.M., Harris H.G., White R.N. & Mirza MS. (1983) *Structural Modeling and Experimental Techniques.* Prentice-Hall: Englewood Cliffs, NJ.

14 Harris, Harry G. & Sabnis, Gajanan (1999) *Structural Modeling and Experimental Techniques.* 2nd ed. Boca Raton (Florida): CRC Press.

15 Hetenyi, M. (ed.) (1950) *Handbook of experimental stress analysis.* New York: Wiley. 401-403.

16 Lehr E. (1940) Meßverfahren und Meßeinrichtungen für Dehnungsmessungen. In: *Prüf- und Meßeinrichtungen. Handbuch der Werkstoffprüfung)* (ed. E. Siebel), 463-571. Berlin: Springer.

17 Donnell, L.H. & Savage, W.T. (1950) Mechanical gages and extensometers. In: *Handbook of experimental stress analysis* (ed. M. Hetenyi), 72-117. New York: Wiley.

18 Mesnager, A. (1903) Deux appareils nouveaux pour la mesure des déformations élastiques. *Annales des Ponts et Chaussées* 73ème année, 8ème série, Tome XI: 212-245.

19 Bühler, A. & Meyer, A. (1924) *Beschreibung von Apparaten zur Untersuchung von* (2. Auflage). Bern: Schweizerischen Bundesbahnen. 26-28.

20 Huggenberger, A.U. & Schwaigerer, S. (1958) Meßverfahren und Meßeinrichtungen für Verformungsmessungen. In: *Prüf- und Meßeinrichtungen. Handbuch der Werkstoffprüfung* (Zweite Auflage) (eds E. Siebel & N. Ludwig), 371-518. Berlin: Springer. 403.

21 Lehr E. (1940) Meßverfahren und Meßeinrichtungen für Dehnungsmessungen. In: *Prüf- und Meßeinrichtungen. Handbuch der Werkstoffprüfung* (ed. E. Siebel), 463-571. Berlin: Springer.

22 Huggenberger, A.U. & Schwaigerer, S. (1958) Meßverfahren und Meßeinrichtungen für Verformungsmessungen. In: *Prüf- und Meßeinrichtungen. Handbuch der Werkstoffprüfung* (Zweite Auflage) (eds E. Siebel & N. Ludwig), 371-518. Berlin: Springer.

23 Bühler, A. & Meyer, A. (1924) *Beschreibung von Apparaten zur Untersuchung von eisernen und massive Bauwerken* (2. Auflage). Bern: Schweizerischen Bundesbahnen.

24 Mesnager, A. (1903) Deux appareils nouveaux pour la mesure des déformations élastiques. *Annales des Ponts et Chaussées* 73ème année, 8ème série, Tome XI: 212-245.

25 Cantieni, A. (1983) *Dynamic load tests on highway bridges in Switzerland. 60 Years Experience of EMPA.* Bericht 211. Dübendorf: EMPA.

26 Huggenberger, A.U. & Schwaigerer, S. (1958) Meßverfahren und Meßeinrichtungen für Verformungsmessungen. In: *Prüf- und Meßeinrichtungen. Handbuch der Werkstoffprüfung* (Zweite Auflage) (eds E. Siebel & N. Ludwig), 371-518. Berlin: Springer. 403.

27 Bühler, A. & Meyer, A. (1924) *Beschreibung von Apparaten zur Untersuchung von eisernen und massive Bauwerken* (2. Auflage). Bern: Schweizerischen Bundesbahnen. 9-17.

28 Dommer, A. (1920) Commission technique de l'Association des constructeurs suisses de ponts et de charpentes métalliques. *Bulletin technique de la Suisse romande* 46 (19): 223-224.

29 Roš, M. (1921) Technische Kommission des Verbandes Schweizer. Brücken- und Eisenhochbau- Fabriken. *Schweizerische Bauzeitung.* 78 (15): 185-187.

30 Roš, M. (1922-1923) Nebenspannungen infolge vernieteter Knotenpunkt-Verbindungen eiserner Fachwerk-Brücken. *Schweizerische Bauzeitung* 80 (15): 168-171; 80 (16): 178-181; 80 (17): 187-189; 81 (4): 42-45; 81 (5): 49-51.

31 Bühler, A. & Meyer, A. (1924) *Beschreibung von Apparaten zur Untersuchung von eisernen und massive Bauwerken* (2. Auflage). Bern: Schweizerische Bundesbahnen.

32 Wyss, Th. (1923) *Beitrag zur Spannungsuntersuchung an Knotenblechen eiserner Fachwerke.* Doctoral Thesis, Nr. 283, ETH Zürich.

33 Huggenberger, A. (1926) *Festigkeit halbkreisförmiger Platten und Dampfturbinen-Leiträder.* Doctoral Thesis, Nr. 361 (Submitted 1923), ETH Zürich.

34 Huggenberger, A. (1951) *Talsperren-Meßtechnik. Meßverfahren, Instrumente und Apparate für die Prüfung der Bauwerke in Massenbeton.* Berlin: Springer-Verlag.

35 Huggenberger, A. (1926) Meßverfahren des Schweizerische Vereins von Dampfkesselbesitzern usw. In: *Bericht der V. Tagung des allgem. Verbandes der deutschen Dampfk.- Überw.-Vereine,* 116 ff, Figs 1, 4. Berlin: Verlag des Vereins Deutscher Ingenieure.

36 Huggenberger, A. (1928) Über die günstigste Gestalt des vollen, gewölbten Bodens zylindrischer Kesseltrommeln gleicher Dicke und ihre Festigkeitsberechnung. *Schweizerische Bauzeitung* 91 (17): 203-208; (18): 217-221.

37 Huggenberger, A.U. & Schwaigerer, S. (1958) Meßverfahren und Meßeinrichtungen für Verformungsmessungen. In: *Prüf- und Meßeinrichtungen. Handbuch der Werkstoffprüfung* (Zweite Auflage) (eds E. Siebel & N. Ludwig), 371-518. Berlin: Springer. 404.

38 Cantieni, A. (1983) *Dynamic load tests on highway bridges in Switzerland. 60 Years Experience of EMPA.* Bericht 211. Dübendorf: EMPA.

39 Zindel, Georges (1927) Das Tensometer von Huggenberger. *Schweizerische Bauzeitung* 90 (19): 250-251.

40 Baes, L. & Vandeperre, L. (1929) L'extensomètre Huggenberger, *Bulletin de la Société belge des Ingénieurs et Industriels* IX (3) and (10). Offprint, 128 pages.

41 Huggenberger, A. (1930) Festigkeitsuntersuchung im Luftfahrzeugbau. *Schweizerische Bauzeitung* 95 (3): 31-34; (4): 51-56.

42 Ivanow, G. (1932) Technique d'emploi des extensomètres dans les constructions aéronautiques. *L'Aéronautique* No. 157: 187-191.

43 Baes, L. & Vandeperre, L. (1929) L'extensomètre Huggenberger, *Bulletin de la Société belge des Ingénieurs et Industriels* IX (3) and (10). Offprint, 128 pages.

44 Vose, R.W. (1934) Characteristics of the Huggenberger Tensometer. *Proceedings Am. Soc. Testing Mats* 34 (2): 862-873; discussion, 874-876.

45 Rosenthal, D. (1932) Etude expérimentale sur modèle réduit d'une poutre Vierendeel soudée. *PV de la 1ère séance de l'ABEM tenue le 3 février 1932, N° 1bis. 1932.* 21-30.

46 Tatnall, Frank G. (1958) *Tatnall on testing: An autobiographical account of adventures under 13 vice presidents.* Ohio: American Society of Metals.

47 Moisseiff, L.S. (1928) Designing the Tower of the Hudson River Bridge. *Engineering News-Record* 100 (21): 819-822.

48 Cantieni, A. (1983) *Dynamic load tests on highway bridges in Switzerland. 60 Years Experience of EMPA.* Bericht 211. Dübendorf: EMPA. 7.

49 Airy, W. (1868) On the experimental determination of the strains of the suspension ties of a bowstring girder, *Minutes of the Proceedings of the ICE* 27: 443-447. Discussion 27: 448-453.

50 Cardini, Pietro (1889) Metodo acustico per la misura di piccoli allungamenti, e determinazione dei moduli di elasticità. *Atti della Reale Accademia dei Lincei* Series 4, *Vol.* 5: 892-899.

51 Coyne, A. (1938) Quelques résultats d'auscultation sonore sur les ouvrages en béton, béton armé ou metal. *Annales de l'Institut Technique du Bâtiment et des Travaux Publics* 3ème année, no 4: 33-56. 34.

52 Schaefer, Otto (1919) Die schwingende Saite als Dehnungsmesser. *V.D.I. Zeitschrift* 63 (41): 1008-1009.

53 Schaefer, Otto (1920) Die Schwingungen biegungssteifer Saiten. *Annalen der Physik* 367 (10): 156-164.

54 Anon. (1928) Akustischer Ferndehnungsmesser. *Zeitschrift des V.D.I.* 72 (49): 1810.

55 Bacharach Industrial Instrument Company, (1928) New Maihak acoustical Strain Gage. *Instruments* 1 (5): 251-252.

56 Hoffmann, H. (1930) Genaue Dehnungs- und Spannungsmessungen an Eisenkonstruktionen und Steinbauten. *Der Bauingenieur* 18: 312-315.

57 Campus, F. & Dantinne, R. (1932) Le télé-extensomètre acoustique du Dr Schaefer. *Bulletin de la Société belge des Ingénieurs et Industriels,* No 10. Offprint, 58 pages.

58 Campus, F. (1933) La charpente métallique de l'Institut de chimie et de métallurgie du Val-Benoît. Partie 2. *Revue Universelle des Mines* 8éme série, 9 (6): 146-157.

59 Baker, J.F. (1954) *The Steel Skeleton: Vol.1 Elastic Behaviour and Design.* Cambridge: Cambridge University Press. 24-26.

60 Coyne, A. (1938) Quelques résultats d'auscultation sonore sur les ouvrages en béton, béton armé ou metal. *Annales de l'Institut Technique du Bâtiment et des Travaux Publics* 3ème année, no 4: 33-56. 55.

61 Coyne, A. (1932) Nouveau procédé d'auscultation des ouvrages en béton ou béton armé, et notamment des barrages. *Le Génie Civil* 100 (9): 225.

62 Coyne, A. (1936) Mesure acoustique des allongements: application à l'étude de points singuliers dans les constructions. In: *Rapport du 2ème congrès de l'AIPC (Berlin-Munich)*, Vol. 2: 580-586.
63 Coyne, A. (1938) Quelques résultats d'auscultation sonore sur les ouvrages en béton, béton armé ou metal. *Annales de l'Institut Technique du Bâtiment et des Travaux Publics* 3ème année, no 4: 33-56. 55.
64 Laffaille, B. (1935) Mémoire sur l'étude générale de surfaces gauches minces. *Mémoires de l'Association Internationale des Ponts et Charpentes (Zürich)* 3: 295-332.
65 Davidenkoff, N. (1934) The Vibrating Wire method of Measuring Deformation. *Proceedings ASTM* 34 (2): 847-860.
66 Peters, O.S. & Johnston, R.S. (1923) New developments in electric telemeters. *Proc. ASTM* 23 (Part II): 592-601.
67 McCollum, B. & Peters, O.S. (1924) A New Electrical Telemeter. *U.S. Bureau of Standards Technologic Papers* 17 (247, 4 Jan): 737-777.
68 Davis, R.E. & Carlson, R.W. (1932) The electrical strain meter and its use in measuring internal strains. *ASTM Proceedings* 32: 793-801.
69 Rogers, J. David (2010) Hoover Dam: Evolution of the Dam's Design. In: *Hoover Dam 75th Anniversary History Symposium: Proceedings of the Hoover Dam 75th Anniversary History Symposium, October 2010, Las Vegas, Nevada* (eds. Richard L. Wiltshire, David R. Gilbert & Jerry R. Rogers), 85-123. Reston, Virginia: American Society of Civil Engineers. 115-118. https://web.mst.edu/~rogersda/hoover_dam/Rogers-HooverDam-Pt1.pdf (21/11/2019)
70 Eaton, E.C. (1931) Resistance Strain Gage Measures Stresses in Concrete. *Engineering News-Record* 107 (15 Oct): 615-616.
71 Carlson, R.W. (1935) Five years' improvement of the elastic-wire strain meter. *Engineering News-Record* 114 (16 May): 696-697.
72 Stein, P.K. (1990) A brief history of bonded resistance strain gages from conception to commercialization. *Experimental Techniques* 14 (5), 13-19.
73 Stein, P.K. (2006) 1936-A banner year for strain gages and experimental stress analysis: an historical perspective. *Experimental Techniques* 30 (1): 23-41.
74 Thomson, W. (1856) On the Electro-dynamic Qualities of Metals. *Philosophical Transactions of the Royal Society of London* 146 (3): 649-751.
75 Wheatstone, C. (1843) An Account of Several New Instruments and Processes for Determining the Constants of a Voltaic Circuit. *Philosophical Transactions of the Royal Society of London* 133 (2): 303-327.
76 Dohrenwend C.O. & Mehaffey, W.R. (1950) Electrical-resistance gages and circuit theory. In: *Handbook of experimental stress analysis* (Ed. M. Hetényi), 160-237. New York: Wiley.
77 Stein, P.K. (2000) A tribute to Arthur C. Ruge. *Experimental Techniques* 24 (4): 15-18, 47-49; 24 (5): 19-22, 43-44; 24 (6): 23-24.
78 Stein, P.K. (2004) Edward E. Simmons, Jr. The invention of the bonded resistance wire strain gage. *Experimental Techniques* 28 (5): 11, 49-51.
79 Keil, S. (1988) On the strain gage's 50th Jubilee – a review of its evolution and 33 years strain gage production at Darmstadt. *HBM Reports in Applied Measurements* 4 (2): 39-48. (19/11/2019)

80 Keil, S. (2017) *Technology and Practical Use of Strain Gages: with Particular Consideration of Stress Analysis Using Strain Gages*. Berlin: Ernst & Sohn. 1-16. https://application.wiley-vch.de/books/sample/343303138X_c01.pdf (24/11/2019)

81 Rohrbach, Chr. & Keil, S. (1988) 50 Years Strain Gages and Brittle Coatings: The German Point of View. *Preprints of the IMEKO World Congress in Houston, Texas, October* 1988. 49-76.

82 https://ascelibrary.org/action/showImage?doi=10.1061/9780784410714&iName=master.img-003.jpg&w=360&h=450 (23/11/2019)

83 US Patent No. 2,340,146, Strain Gauge. A.C. Ruge, Filed 2 Oct 1941. Granted 25 Jan 1944.

84 Stein, P.K. (2001) Strain gage history and the end of the twentieth century. *Experimental Techniques* 25 (2): 15-16.

85 Cristian V. Licensed under the Creative Commons Attribution 4.0 International license. https://commons.wikimedia.org/wiki/File:Strain_gauge_-.jpg (25/11/2019)

86 Vaughan, J. (1975) *Strain Measurement*. Nærum, Denmark: Brüel & Kjaer.

87 Nielsen, Howard H. (1985) *From Locomotives to Strain Gages: the true story of a tortuous conversion from smokestack to high-technology industry*. New York: Vantage Press Inc.

88 Low I.A.B. (1964) *The Use of Moiré Fringes in Analysing Strains around Mine Openings*. PhD Thesis. Imperial College, London. https://spiral.imperial.ac.uk/handle/10044/1/16870 (22/11/2019)

89 http://www.moiretelltales.com/ (22/11/2019)

90 Dietrich, O. & Lehr, E. (1932) Dehnungslinienverfahren ein Mittel zur Bestimmung der für die Bruchsicherheit bei Wechselbeanspruchung massgebenden Spannungsverteilung. *Zeitschrift des Vereins deutscher Ingenieure* 76 (41): 973-982.

91 De Forest, A.V. & Ellis, Greer (1940) Brittle Lacquers as an Aid to Stress Analysis. *Journal of Aeronautical Science* 7: 205-208.

92 De Forest, A.V., Ellis, Greer & Stern, F.B. (1942) Brittle Coatings for Quantitative Strain Measurements. *Journal of Applied Mechanics* 9 (4, Dec): 184-188.

93 Durelli, A.J. (1942) Experimental Determination of Isostatic Lines. *Journal of Applied Mechanics* 9 (4, Dec): 155-160.

94 Beggs, G.E. (1922) Method of determining stresses in structures, US Patent 1,551,282, filed 30.03.1922, granted 25.08.1925.

95 Maxwell, James Clerk (1864) On reciprocal figures and diagrams of forces. *The London, Edinburgh, and Dublin Philosophical Magazine and Journal of Science* 27: 182, 250-261.

96 Mohr, O.C. (1874) Beitrag zur Theorie des Fachwerks. *Zeitschrift des Architekten und Ingenieur-Vereins zu Hannover* 20: 509ff and 21: 17ff.

97 Beggs, G.E. (1922) An accurate mechanical solution of statically indeterminate structures by use of paper models and special gages *Proc. American Concrete Institute* 18 (2): 58-78.

98 Beggs, G.E. (1923) Design of elastic structures from paper models. *Proc. American Concrete Institute* 19 (1): 53-65.

99 Anon. (1922) Designs Concrete Arches with Microscope. *Popular Science Monthly* 101 (5, November): 47.

100 Hofacker, Karl (1926) Mechano-statische Untersuchungen hochgradig statisch unbestimmter Tragsysteme. *Schweizerische Bauzeitung* 87(12): 153-157; 87(13): 232-234; 87(15): 191-193; 87(16): 211-212; 87(26): 321-323.

101 Beggs, G.E. (1923) Solution précise au moyen de modèles en papier des problèmes statiquement indéterminés. Note présentée par M. Mesnager. *Comptes Rendus hebdomadaires de l'Académie des Sciences (Paris)* Tome 176 (26 March): 885-886. *Le Génie Civil* 82(14): 336.

102 Kann, F. (1923) Amerikanische Beiträge zur Statik der hochgradig unbestimmten Systeme. Eine genaue mechanische Lösung mittels Modelle aus Pappe von Prof. Beggs in Princeton N.-J. *Beton und Eisen* 22 (21): 260-264.

103 Beggs, G.E. (1925) Discussion of 'Design of a multiple-arch system and permissible simplifications' by A.C. Janni. *Trans. ASCE* 88: 1208-1230.

104 Anon. (1928) The Beggs Deformeter, *Engineering (London)* 126 (13 July): 31-35.

105 Magnel, G. (1930) Recherche expérimentale des lignes d'influence relatives aux constructions hyperstatiques planes: preprint of paper (II-7). In: *Premier Congrès International du Béton et du Béton Armé*, Vol.1, 147-156. Liège, Belgium: La Technique des Travaux.

106 McCullough, Conde B. & Thayer, Edward S. (1931) *Elastic Arch Bridges*, New York: J. Wiley and Sons, 282-300.

107 Bühler, A. (1927) Zur Anwendung des Modellverfahrens von Prof. Beggs, *Beton und Eisen* 26 (16): 304-306.

108 Billington, D.P. (1997) *Robert Maillart. Builder, Designer, and Artist.* Cambridge: Cambridge University Press. 126-129 and notes 111, 112, 113, page 278.

109 Beggs, G.E. (1927) The use of models in the solution of indeterminate structures. *Journal of Franklin Institute* 203 (3): 375-386.

110 Armagnac, A.P. (1928) Toys that save millions, *Popular Science Monthly* 112 (2, February): 17-18, 128.

111 Anon. (1928) The Beggs Deformeter, *Engineering (London)* 126 (13 July): 31-35.

112 Hayden, Arthur G., Clarke, G.D. & Beggs, G.E. (1931) *The Rigid frame bridge.* London: Chapman and Hall; New York: J. Wiley & Sons. (2nd ed, 1940).

113 Richart, F.E. (1928) Tests of the effect of brackets in reinforced concrete rigid frames; Appendix II: Supplementary tests of paper models. *Bureau of Standards Journal of Research* Vol.1 (RP-9): 190-251; 250-253.

114 Thomson, J.T. (1928) Model Analysis of a Reinforced Concrete Arch. *Public Roads* (U. S. Department of Agriculture) 9 (11): 218-220.

115 Hampton, R.A., Trebellas, Ch. & the Hardlines Design Company (2012) *Historic Context Report for the Ohio Department of Highways*, Bureau of Bridges 1911-1945, OH DOT: 19-20.

116 Austin, D.C. (2009) *Documentation of Historic Bridge over White River (Lake Taneycomo) at Branson, MoDOT Bridge No. J0705R*, MoDOT Job No. J8P0764, Missouri Department of Transportation.

117 Schaechterle, K. (1930) Anleitung zur Ermittlung der inneren Kräfte von beliebig belasteten, statisch unbestimmten Tragwerken mit Hilfe des Beggs'schen Gerätes. *Beton und Eisen* 29 (22): 406-410.
118 Click, G.W. (1930) Rigid Rectangular Frame Foundation for Albany Telephone Building. *Engineering News-Record* 105 (27 November): 836-838.
119 McCullough, Conde B. & Edward S. Thayer (1931) *Elastic Arch Bridges.* New York: J. Wiley and Sons. 282-300.
120 McCullough, C.B. (1940-1941) Modern design and construction practice for wide-span arches in U.S.A., *IABSE Publications,* Vol. 6, 189-209.
121 Hofacker, K. (1948) Fürstenlandbrücke bei St.Gallen (Schweiz). *Report of the 3rd IABSE Congress (Liège),* 421-428.
122 Missouri State Highway Commission (1932) *Eighth Biennial Report of the State Highway Commission of Missouri, for the Period Ending December 1, 1932.* 237.
123 Wilson, W.M. (1937-1938) Tests of reinforced concrete arch bridges. *IABSE Publications* 5: 375-422.
124 Magnel, G. (1934) *Le calcul pratique des poutres Vierendeel.* Ghent: Van Rysselberghe & Rombaut.
125 Brebera, A. (1932) Addenda, *Report of the 1st IABSE Congress* (Paris, 1932) 709-715.
126 Brebera, A. (1936) Quelques exemples de constructions soudées réalisées en Tchécoslovaquie. *Report of the 2nd IABSE Congress* (Berlin-Munich, 1936), 425-436.
127 Moody, W.T. & Phillips, H.B. (1955, 1967) Photoelastic and experimental analog procedures. *US Bureau of Reclamation Engineering Monograph,* No. 23 (originally published 1955). 35-48.
128 Phillips, H.B. & Allen, I.E. (1952, 1968) Beggs deformeter stress analysis of single-barrel conduits. *US Bureau of Reclamation Engineering Monograph,* No. 14 (originally published 1952). 57-58.
129 Garvin, J., Casella, R. & The Preservation Company (2010) *New Hampshire Historical Property Documentation Ash Street Bridge (Robert J. Prowse Memorial Bridge),* NH State No. 650. 22-28.
130 Beggs, G.E., Timby, E.K. & Birdsall, B. (1932) Suspension Bridge Stresses Determined by Model. *Engineering News-Record* 108 (9 June): 828-832.
131 Beggs, G.E., Davis, P.E. & Davis, H.E. (1934) Tests on structural models of proposed San Francisco Oakland suspension bridge. *Engineering* 138 (July 6): 4-6.
132 Kurrer, K.-E. (2018) *The history of the theory of structures.* Berlin: Ernst & Sohn. 967-968.
133 Morey, C.R., Willis, C.H., Greene, A.M. (1939) Obituary of George Erle Beggs. *The Princeton Alumni Weekly* 40: 304-305.
134 Kann, F. (1924) Beitrag zur Lösung statisch unberstimmter Systeme durch Messung. *Beton und Eisen* 20: 10.
135 Otto Gottschalk (1926) Mechanical Calculation of Elastic Systems. *Journal of the Franklin Institute* 202 (1): 61-88.

136 Carot Paul & Rieckhof, Christian (1931) *La statique expérimentale des constructions continues: Étude et resolution, au moyen de modèles réduits des systèmes hyperstatiques*. Paris: Constructeur de ciment armé.

137 Toussaint, Daniel (2017) *Les outils de simulation: Le Nupubest de Christian Rieckhof*. Linealis.org. http://www.linealis.org/wp-content/uploads/2017/03/Nupubestw.pdf (22/11/2019)

138 Eney, W.J. (1939) New Deformeter Apparatus. *Engineering News-Record* 122 (7): 16, 221.

139 Eney, W.J. (1953) A Large Displacement Deformeter Apparatus for Stress Analysis with Elastic Models. In: *Steel Rigid Frames Manual: Design & construction* (ed. M.P. Korn), 67-76. Ann Arbour, Michigan: Edwards.

140 Ruge, A.C. & Schmidt, E.O. (1939) Mechanical structural analysisby the moment indicator. *Trans. ASCE* 104; also *Proc. ASCE* 65 (1): 161-170; 65 (6): 1037-1040.

14

Physical modelling at the University of Stuttgart

Christiane Weber

14.1 The Materials Testing Institute at the Technical University of Stuttgart

The Materials Testing Institute (*Materialprüfungsanstalt, MPA*) was established in 1884 at the Polytechnic High School in Stuttgart – later the Technical University of Stuttgart – by Carl Bach (1847-1931). In 1927 the Civil Engineering section was renamed the *Institut für Bauforschung und Materialprüfung des Bauwesens* (Institute of Building Materials and Materials Testing of Civil Engineering) and Otto Graf (1881-1956) was appointed head of the Institute which role he held until 1950. Under his leadership the Institute became one of the most important institutions of its kind in Germany [1]. In 1936 Otto Graf was appointed Professor for materials testing.

After the rise to power of the National Socialist Party in Germany the construction of a national road and motorway network was launched on a larger scale. In this context, the testing of construction materials at MPA Stuttgart received increasing interest and attention, as Otto Graf had become an acknowledged expert on concrete and reinforced concrete. It was therefore no coincidence that Graf played an important role in the process of road construction in National Socialist Germany. The Materials Testing Institute was awarded research contracts from the *Forschungsgesellschaft für das Straßenwesen* (Highway Research Institute) which had been founded in 1934 by Fritz Todt (1891-1942) to support the construction of highways with practical research contracts. Todt was appointed to the newly created position of *Generalinspektor für das deutsche Straßenwesen* (Inspector General of the German Road Administration) by Adolf Hitler himself. As well as Todt and Graf, the Stuttgart bridge engineers Fritz Leonhardt (1909-1999) and Karl Schaechterle (1879-1971) played an important role in developing the National Socialist road system. Karl Schaechterle was Councilman of Bridges and Railroad Construction within the *Reichsautobahndirektion* (Agency for Motorway Construction) in Stuttgart. He rose rapidly within the *Reichsautobahn* administration and in 1935, with his vast experience in building bridges, was appointed to the directorate of the *Reichsautobahn*. He then moved to Berlin and took with him his collaborator Fritz Leonhardt [2].

Physical Models: Their historical and current use in civil and building engineering design,
First Edition. Edited by Bill Addis.

14.1.1 Model-testing of bridges for the *Reichsautobahn*

In 1937 the young engineer Fritz Leonhadt was appointed by Fritz Todt as head of the construction management for the Rhine Bridge near Cologne-Rodenkirchen. This bridge was designed by the Stuttgart architect Paul Bonatz (1877-1956). When arriving in Cologne, Fritz Leonhardt immediately changed the basic design which was to be the first major suspension bridge in the German road system. Leonhardt modified the bridge to be a self-anchored suspension bridge, a typology that he had been able to study thoroughly during his year-long stay in the USA [3]. To establish this new method of construction in Germany, numerous experiments had to be undertaken. These tests were carried out by the MPA Stuttgart. As well as undertaking material tests on wire rope technology, a measurement model (scale 1:100) consisting of steel wire ropes and steel profiles had been built by MPA to explore the deformation under load as well as length changes under the influence of temperature (Figure 14.1). Leonhardt published his experience on this project in cooperation with Karl Schaechtlerle in three papers in *Die Bautechnik* in 1941 [4, 5].

According to Leonhardt's autobiography, it had been Karl Schaechterle who was responsible for introducing model testing into the Technical University, Stuttgart [6]. Since the ninteenth century, models had served to determine the distribution of forces in a structure. Using scaling factors, the readings taken on a model are used to specify the dimensions of the full-size structure. This appraoch to designing load-bearing structures only became established in the twentieth century, following advances in the theory of similitude which is an important prerequisite. The scientific understanding of scale effects enabled the engineers to deduce real forces from the data gained from the model. Measurement technology and material science provided the conditions needed to use physical models in this way. Since 1934 Karl Schaechterle had carried out model tests for arched bridges in Stuttgart, using methods developed in the USA. In the years that followed, structural model analysis conducted at the Materials Testing Institute underwent significant development by Schaechterle's colleague, Fritz Leonhardt. The tests on models by the MPA Stuttgart, carried out in the context of the construction of the bridge near Cologne-Rodenkirchen, set a new standard [7].

These experiments were of vital importance when, while the works at the Cologne bridge were well under way, the design of a large suspension bridge spanning the Elbe at Hamburg was begun [8]. Since the suspension bridge would be part of the German national road network, the design of a new bridge was

Figure 14.1 1:100 scale measurement model of the, Rhine Bridge near Cologne-Rodenkirchen by Fritz Leonhardt, 1939. Tested at the MPA Stuttgart. (Image: University Archive, Stuttgart 33/1/B11817)

Figure 14.2 1:125 scale measurement model for a proposed suspension bridge spanning the Elbe at Hamburg, 1940. Designed by Fritz Leonhardt and tested at MPA Stuttgart. (Image: University Archive, Stuttgart 33/1/1242 – BA 6295)

placed under the authority of Fritz Todt in 1937. The chosen design by the MAN company followed a sketch by Hitler himself, but the bridge experts Karl Schaechterle and Fritz Leonhardt did not approve this project. They considered the stone pylons to be too massive and too heavy. With the support of Fritz Todt, they therefore proposed an alternative bridge design in steel. In January 1940 Fritz Todt commissioned Karl Schaechterle to undertake a research project [9]. A working committee was appointed to initiate, conduct and interprete experiments on physical models. Representing the Technical University, Stuttgart, Professors Otto Graf and Hermann Maier-Leibnitz (1885-1962) were involved, and Karl Schaechterle and Fritz Leonhardt represented the directorate of the *Reichsautobahn*. In addition to cable tests, mechano-static model tests on suspension structures were carried out. The MPA Stuttgart was commissioned to build two measurement models at the same 1:100 geometric scale as the Cologne-Rodenkirchen bridge model. However, to achieve the necessary constant temperature conditions, the models had to be built in the basement of the MPA building where – due to lack of space – the scale had to be modified to 1:125 (Figure 14.2).

The evaluation of the experiments was carried out by Professor Hermann Maier-Leibnitz at Stuttgart, who published interim reports in the journal *Die Bautechnik* in 1941/1942, bearing the title: 'Basic information about model measurements made to determine the deformations and stresses in self-anchored suspension bridges' [10].

14.1.2 Model tests for a dome proposed for the new main station in Munich

Alongside the showpiece projects for the German construction industry in the new motorway network, were the so-called Nazi *Führerstadtplanungen*, city plans, directed by Hitler himself [11]. These masterplans for Berlin, Nuremberg, Hamburg, Munich and Linz offered a rich body of work for civil engineers. The Materials Testing Institute and the Stuttgart networks were involved in the plan for Hamburg and then the plan for Munich. In May 1939, the firm August

Klönne with consulting architect Paul Bonatz (1877-1956) (with both of whom Leonhardt had worked at Cologne-Rodenkirchen) won the competition for the construction of a steel domed hall for a new Central Station in Munich [12]. This was the largest single project of the city planning activities for Munich by the architect Hermann Giesler (1898-1987). Hitler had appointed Giesler on 21 December 1938 as *Generalbaurat für die Hauptstadt der Bewegung* (General Director of Buildings for the Capital of the Movement). After winning the competition, Paul Bonatz was nominated head of the Planning Bureau for the New Central Station and the companies August Klönne and Krupp established the *Stahlbau-Gemeinschaft Klönne-Krupp, München.* At the end of September 1939, Giesler turned to Fritz Todt with a request for 'the design of the steel cupola [...] to have Professor Bonatz assisted by *Regierungsbaumeister* Dr.-Ing. Fritz Leonhardt as consulting engineer'. Todt agreed, remarking that 'Herr Leonhardt is one of the best bridge designers and likely the best structural engineer among the staff of the Chief Building Management'. Leonhardt would help with the design 'alongside his official function' of construction supervision for the suspension bridge at Rodenkirchen. Todt also arranged to provide Leonhardt with 'sufficient staff to limit his own work to supervision'.

In Giesler's plans, the new station was at the end of a monumental avenue in Munich between the Monument to the Party and the New Central Station, thereby being the defining element in remodeling the city as the Capital of the Movement, just as Albert Speer's (1905-1981) *Volkshalle* (Hall of the People) was to be the key element in the plan for Berlin. By lowering the railway tracks, the streets would be led around the station as an express-way interchange, while the dome above had to cover both the tracks and platforms. As the station was a technical building, Hitler had decided it should be a steel dome with a span of more than 265 m to demonstrate the skill of the German steel construction engineers. A sketch by Hitler dated 22 March 1939, served as the basis for the competition for the Munich Central Station – a flat dome resting on a ring of supporting buildings. A circular ribbon window and a lantern would light the interior of the giant cupola. Hitler specifically wanted a distinction between the Munich Central Station as a 'monument of our century's technology', and Speer's dome for the *Volkshalle* in Berlin, which was to be nearly hemispherical, spanning 250 m and made of granite.

The winning competition by Klönne-Bonatz envisaged a twelve-ribbed dome over the tracks and platforms. The giant cupola, sketched by Hitler, would be covered with a folded sheet metal roof to a design by Bonatz. A perspective rendering of the dome, dated 9 November 1939, was drawn by Bonatz. Since the construction of the steel dome was to demonstrate German engineering know-how, Fritz Leonhardt was allowed to choose the technical materials: 'Aluminum for the profiled dome roofing, aluminum for the windows and other elements, glass and glass mosaic for the translucent and wall surfaces'. Leonhardt planned to span the 265 metres with a ribbed dome in steel. Between the ribs with a box profile (2.5 × 2.0 m) was a double roof, externally of folded, and internally of corrugated aluminum sheets, reinforced by six horizontal ring beams and lattice rafters (0.6 × 4.9 m). Aluminum was chosen to reduce the deadweight to be carried by the steel structure. The calculations for the structural system, consisting of 384 equations

Figure 14.3 1:50 scale measurement model for the new main station in Munich, 1940, designed by Fritz Leonhardt and tested by MPA Stuttgart. (Image: University Archive, Stuttgart 33/1/776)

and 284 unknowns, were a challenge to the engineers and mathematicians, and meant work for many months. The structural behaviour was also examined with the help of structural models, for which Leonhardt commissioned a 1:50 model 'with riveted hollow sections of thin sheet metal to scale' (Figure 14.3).

The MPA Stuttgart built this six metre-high model in a building constructed for the purpose in Munich. With the help of this model the effects of the self-weight of the roof and the ring beam, and the live load of snow and wind could be studied (Figure 14.4). Unfortunately, the model was destroyed by a bombing raid on 25 April 1944 and only photographs have been preserved in the University Archive in Stuttgart. The structural calculations by Leonhardt and his assistants in 1942 are also preserved and prove that, in the middle of the war, the detail design was proceeding at full speed, and technical questions, such as compensation for thermal influences, were being studied.

The engineers were intrigued by the construction of a steel cupola with a diameter of 265 m in the 1940s. In fact, 200 m spans would only be realized some twenty years later. In retrospect, Leonhardt said that he and his team 'conducted megalomaniac planning, at least surmising, that it would not be put into practice' [13].

At the end of the Second World War only Otto Graf, the head of MPA and professor at the Technical University, was reprimanded. He was suspended from his position by the American occupying power in 1945, accused of the mistreatment of foreign workers employed at the MPA during the war. Nevertheless, as early as 1948 he was rehabilitated by a German Tribunal – so called

Figure 14.4 Detailed model of the compression ring for the dome, under load, 1940. Tested by the MPA Stuttgart. (Image: University Archive Stuttgart 33/4/133)

Spruchkammerverfahren – and classified as *Mitläufer*, which meant that he could continue to work as head of the Materials Testing Institute and remain a professor and civil servant. One can assume that this decision was supported by the German construction industry. Engineers like Otto Graf, Karl Schaechterle and Fritz Leonhardt were desparately needed for the reconstruction in post-war Germany [14].

On the subject of model testing, it can be noted that it was indirectly encouraged by National Socialist construction activities, especially the suspension bridges in Cologne and Hamburg and the dome in Munich, for which substantial programmes of model testing were carried out. By the substantial financial support provided for these experiments in National Socialist construction, model testing became an established method in Stuttgart.

14.2 Model testing after the Second World War

After the war, the MPA applied the model testing techniques to optimizing the construction of both suspension- and cable-stayed bridges, for example the cable-stayed *Nordbrücke* in Düsseldorf built in the years 1952 to 1957 to designs by the engineering firm Leonhardt + Andrä [15]. The engineers at MPA also conducted a series of model tests for the monocable suspension bridge which

Figure 14.5 Model test for a proposed monocable suspension bridge for a bridge over the Rhine at Emmerich, 1961. Designed by Fritz Leonhardt and tested at MPA Stuttgart. (Image: *Südwestdeutsches Archiv für Architektur und Ingenieurbau*, Karlsruhe; Fritz Leonhardt collection)

Fritz Leonhardt proposed in 1961 for a bridge across the Rhine near Emmerich, which was not built [16] (Figure 14.5).

14.2.1 The *Institut für Spannungsoptik und Modellmessungen* (Institute for Photoelasticity and Model Measurement) at the Faculty of Construction

The continuity of the model testing techniques was largely due to the staff who remained employed at the Technical University, Stuttgart. In the following years, model analysis was deployed as the most effcient and most economical method of determing the behaviour of large or unusual structures. In his autobiographical memoirs Fritz Leonhardt reports that, for his project work as a consulting engineer, he commissioned research work in the field of steel construction at the *Materialprüfungsanstalt* in Stuttgart, in particular at the *Institut für Bauforschung und Materialprüfungen des Bauwesens*, renamed in 1953 the *Otto-Graf-Institut*.

In 1954 the Chair of *Konstruktiver Ingenieurbau A und Statik A*, Karl Deiniger (1896-1956), reported that the questions posed by prestressed concrete could not be answered mathematically. Hence Deiniger had established in 1953 the *Institut für Spannungsoptik und Modellmessungen* [17]. The Institute carried out photoelastic stress analysis using transparent isotropic acrylic materials such as

Plexiglas, Dekorit or Trolon [18], and *Modelstatik* (model statics) was offered as a supplementary taught subject to *Statik* at the Faculty of Civil Engineering.

When Fritz Leonhardt was appointed in 1958 as Deinigers successor to the Chair for *Massivbau* (Concrete Construction) at the Technical University, Stuttgart, the scope of the Chair was redesigned and a separate Chair for *Baustatik* (Building statics) was created. Hence the faculty decided that the Chair for *Massivbau* should no longer be the only director responsible for the *Institut für Spannungsoptik und Modellmessungen* (Institute for Photoelasticity and Model Measurements). From 1957, the Institute became available to the entire Faculty of Construction. Fritz Leonhardt and Friedrich Bornscheuer (1917-2015), who was appointed Professor for *Elastizitätslehre und Baustatik* (Theory of Elasticicity and Statics) in the same year as Leonhardt, negotiated in their appointment agreements a considerable sum of money for technical apparatus, measuring equipment and new staff for the *Institut für Spannungsoptik und Modellmessungen* [19]. The Institute was supervised by a Board of Trustees (*Kuratorium*) consisting of the five Chairs of Civil Engineering – *Technische Mechanik, Baustatik, Massivbau, Stahlbau, Wasserbau* (mechanics, statics, concrete construction, steel construction and hydraulic engineering). At the request of the Ministry, the distinction between the *Otto-Graf-Institut* and the *Institut für Spannungsoptik und Modellmessungen* was defined such that the *Institut für Spannungsoptik und Modellmessungen* had to concentrate on fundamental research, while commissions from the industry were handled by the *Otto-Graf-Institut*. To represent the interests of the *Otto-Graf-Institut* when arbitration was required, professor Friedrich Tölke (1901-1992), the director of the *Otto-Graf-Institut*, was entitled to vote with the Board of Trustees [20].

In 1960, the physicist and mathematician Robert Müller (1925-2020) joined the *Institut für Spannungsoptik und Modellmessungen* and he was appointed *Akadmischer Rat* in 1963 [21]. In 1965, Müller was granted responsibility for *Modellstatik* [22] and, some time before July 1967, he became director of the renamed *Institut für Modellstatik* (Institute for Structural Model Analysis) which was the only independent university institute of its kind in Germany [23].

In model tests that were used to determine load capacities or breaking loads, it was usual to use the same material as in the actual structure – for example concrete. For these so called *Realmodelle*, a suitable micro-concrete was developed, whereby the Stuttgart scientists built on experiments carried out by their English, Italian (Bergamo) and Portuguese (Lisbon) colleagues. Since these models were very large, a new building for the institute was erected in around 1977 [24]. Models tested at the *Institute für Modellstatik* included the shell structures for the platforms of the television communication towers erected by the engineering firm Leonhardt + Andrä in numerous German cities [25]. A special focus of the institute's activities was the use of plastic models. These were used both for photoelastic stress analysis (see Chapter 12) and also to explore the elastic behaviour of structures using electrical-resistance strain gauges [26].

14.2.2 Model tests for the roof of the Alster swimming baths, Hamburg

One particular model from this era was especially remarkable, and has been preserved to this day: the shell roof over the Alster swimming baths in Hamburg.

The project was based on a 1956 design by the Hamburg architects Niessen & Störmer and is striking on account of its expressively projecting pre-stressed concrete shell. It was only completed in 1973. The engineering firm Leonhardt + Andrä was responsible for the structural design of the construction comprising two hyperbolic paraboloids [27]. Jörg Schlaich (b.1934) was the engineer responsible within the company. The calculation of the distribution of stresses in the overhanging pre-stressed concrete shells using the 1960s methods was very complex and therefore costly. The engineers therefore chose to determine the distribution of stresses using experiments on a reduced-scale model. Initial plans envisaged the physical model in micro-concrete. With this particular material, the shells would have needed to be at least one centimetre thick on account of the pre-stressing reinforcement, which meant that the micro-concrete model would have been too big [28]. Not least due to the complexity of the formwork and the far higher costs involved, micro-concrete was therefore rejected. Instead, the engineers at the *Institut für Modellstatik* built an elastic model made of Perspex (Figure 14.6).

The model comprised easy-to-make individual parts, which were then stuck together using adhesives to form a single piece. All the parts had to be geometrically similar and at the same scale. In his publications, Müller propagated this type of plastic model, whose elasticity quotient must correspond to the actual structure's material. For manufacturing reasons, the thickness of the shell for the model could not be less than 3 mm. Shells 8 cm thick were envisaged for the structure itself. This resulted in a scale of 1:26.67. At this scale the maximum

Figure 14.6 1:26.67 scale Perspex model for the shell roof over the Alster swimming baths in Hamburg tested by the Institute for Structural Model Analysis, late 1960s [27]. (Image: *Südwestdeutsches Archiv für Architektur und Ingenieurbau*, Karlsruhe; Fritz Leonhardt collection)

dimension of the model was four metres. To achieve sufficient height for the weights and load equipment, the model was placed on three concrete pedestals.

The physical model was used to test the behaviour under dead loads, snow and loads resulting from the movement of the foundations, as well as localised wind loads at the highest points. Strains, bending moments, and deflections of the boundary supports and the shells were all measured using the model. In the model test the dead load and snow load were applied by means of several small individual loads. To this end, holes two millimetres in diameter were drilled in the shell and the boundary supports for hanging the weights. The individual lead weights were secured by means of rubber rings, such that the application and removal of the loads was gradual, not sudden. A pneumatic lifting platform was used to raise and lower the lead weights (Figure 14.7).

In order to simulate load scenarios caused by the supports shifting and deforming, the supports were mounted on roller bearings in channels to constrain their movement, and moved by weights [29]. The strains were measured using total of 317 strain-gauge rosettes and 26 foil strain gauges. Only through the use of an automatic data-logging system, which transferred the measured data onto a punched tape for input into a computer, was it possible to make the study of a model with so many measuring points economical.

As far as we know, this particular model is the last plastic model to have come from the former Institute for Structural Model Analysis in Stuttgart. It was

Figure 14.7 1:26.67 scale Perspex model for the shell roof over the Alster swimming baths in Hamburg under load, late 1960s. (Source: [27] Fig. 15).

reassembled in the 1980s, without the measuring devices and horizontal cross bracing, and put on display in the foyer of the Faculty of Structural Engineering and Environmental Sciences at Stuttgart University. Since then, the number of small defects has increased, splices have become brittle and the Perspex shells have been broken by mechanical impact. To date, its large size has prevented it from becoming part of a museum collection.

14.3 The *Institut für Leichte Flächentragwerke*, University of Stuttgart

Although Frei Otto (1925-2015) was trained as an architect, he had a major influence on structural engineering in the second half of the twentieth century, especially in the field of lightweight structures. His engagement with this theme began in the 1950s and he opened his *Entwicklungsstätte für den Leichtbau* (Development Laboratory for Lightweight Construction) in Berlin in 1958. In 1964, when Fritz Leonhardt was Dean of the Department of Construction at the Technical University, Stuttgart, he succeeded in establishing the *Institut für Leichte Flächentragwerke* (Institute for Lightweight Structures, IL) especially for Frei Otto. The following decades of Frei Otto's work in Stuttgart saw a wealth of models being built, reflecting Frei Otto's passion for model making, which were greatly admired by his peers. Since it was nearly impossible to draw with precision many of the forms of structures that he developed, he needed the models as a medium 'to materialize the idea'. Models were thus employed at all phases of the design process: presentation models, form-finding models and finally, measurement models with which the forces within structures could be determined [30]. The forty or so publications by IL attest to the Institute's extensive commitment to structural model analysis and to the collaboration between IL and the *Institut für Modellstatik* under Robert Müller and, especially, the *Institut für Anwendungen der Geodäsie im Bauwesen* (Institute of Applications of Geodesy to Engineering/IAGB) under the supervision of Klaus Linkwitz (1927-2017). Apart from his collaboration with German Engineers such as Fritz Leonhardt and Jörg Schlaich, Otto had particular success working with the English structural engineers, Ted Happold (1930-1996), Ian Liddlell (b.1938) and Michael Dickson (1944-2018), initially while working in the Structures 3 Group at Ove Arup and Partners and, after 1976, at Buro Happold [31, 32].

14.3.1 Form-finding models with soap films and soap bubbles

One area in which Frei Otto worked extensively from 1958 onwards was with soap-film models (see Chapter 19) [33]. His starting point here was his realization that films made with soap solutions invariably and, of their own accord, form the smallest possible surface within a given frame, taking on a form with equal surface tension in all directions. This enabled him to generate minimal surfaces whose forms served as models for numerous tent structures realized in the course of the 1950s and 1960s. It was thanks to his soap-film experiments that he was able to develop the cable loops in cable net structures which began in 1965 with the IL pavilion in Stuttgart, were used later for the German pavilion at Expo '67 in

Montreal, and for the roofs of the Olympic stadium in Munich in 1972. In 1963, Frei Otto built a *Minimalwegegerät* (minimum-path apparatus) in Berlin as part of his series of experiments on minimal networks [34]. The machine underwent further development at the IL in Stuttgart, where it became the minimum-surface apparatus or soap-film apparatus used for photographing the short-lived forms of soap-films. Using a technique devised by engineers at the Institute for Geodesic Applications in Civil Enginering (IAGB) headed by Klaus Linkwitz, it was possible to obtain precise photogrammetric images of soap-films. An optical bench was added to the soap-film apparatus to project a grid of parallel light beams onto the minimum surface creating contours on the soap film that could be made visible and recorded by cameras (see Figure 19.4). In order to increase the life of the soap films, the apparatus was further adapted in 1973 so that the soap-film could be formed within a climate chamber with high humidity to reduce evaporation of liquid in the film (see Figure 19.3). Finally, a wind tunnel was added to enable the testing of soap-film forms exposed to wind. The research team also studied minimal soap-film surfaces created by internal air pressure to create a variety of pneumatic structures [35]. These air-dome structures brought Frei Otto very close to his goal of roofing-over large areas with minimum effort and expense and of creating climatic domes. In contrast to the tent structures, however, very few of these ideas were ever implemented.

14.3.2 The IL Pavilion, University of Stuttgart

At the university campus in Stuttgart-Vahingen an inconspicuous tent construction can still be found: the IL-pavilion (Figure 14.8). It was originally designed

Figure 14.8 IL Pavilion, University of Stuttgart. (Image: [37])

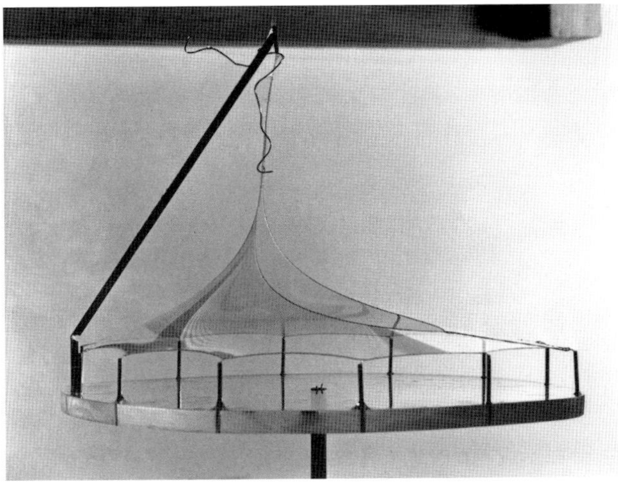

Figure 14.9 Soap-film model used to create the basic form for the final cablenet structure for the Institute of Lightweight Structures pavilion. (Image: Courtesy of ILEK, University of Stuttgart)

and erected as a life-size proof-of-concept prototype for the German Pavilion at Expo '67 in Montreal. The experimental building was built to test the prestressing and the method of construction and assembly for the cablenet, and it also provided a live project on which to test the various ways of using reduced-scale models in the design proocess. It was completed in 1966 and two years later was relocated on the campus to become the permanent home for Frei Otto's Institute for Lightweight Structures. Today it still houses the successor to IL, the Institute for the Design and Construction of Lightweight Building (*Institut für Leichtbau Entwerfen und Konstruieren* (ILEK)) [36].

The initial form of the cablenet tent was determined using a soap-film model. It was based on the simple idea of lifting a circular bubble from its centre point which, of course, is not possible. The boundary was created by a series of threads to create a twelve-sided perimeter, and the centre was lifted using a teardrop loop (Figure 14.9). The geometry of the soap-film model was recreated as a cablenet model using steel wire (Figure 14.10). The correct tension in the wires was set approximately using adjustments at the end connections, and then refined using a screw in a three-point tensioning device fixed to the wire itself, similar to the one shown on the right in Figure 14.14.

The overall change in geometry, and the changes of tension in the individual wires due to snow load, were measured by hanging small weights from the nodes of the net and recording the deformation of the net using double-exposure photographs of the model (Figure 14.11).

14.3.3 The German pavilion at Montreal.

As Frei Otto planned his tent and tensile structures with increasingly large spans, he had to rely more and more on model analysis. The first extensive use of measurement models was during the design phase of the German pavilion at Expo '67

Figure 14.10 Making the model of the cablenet envelope for the Institute of Lightweight Structures pavilion. (Image: Courtesy of ILEK, University of Stuttgart)

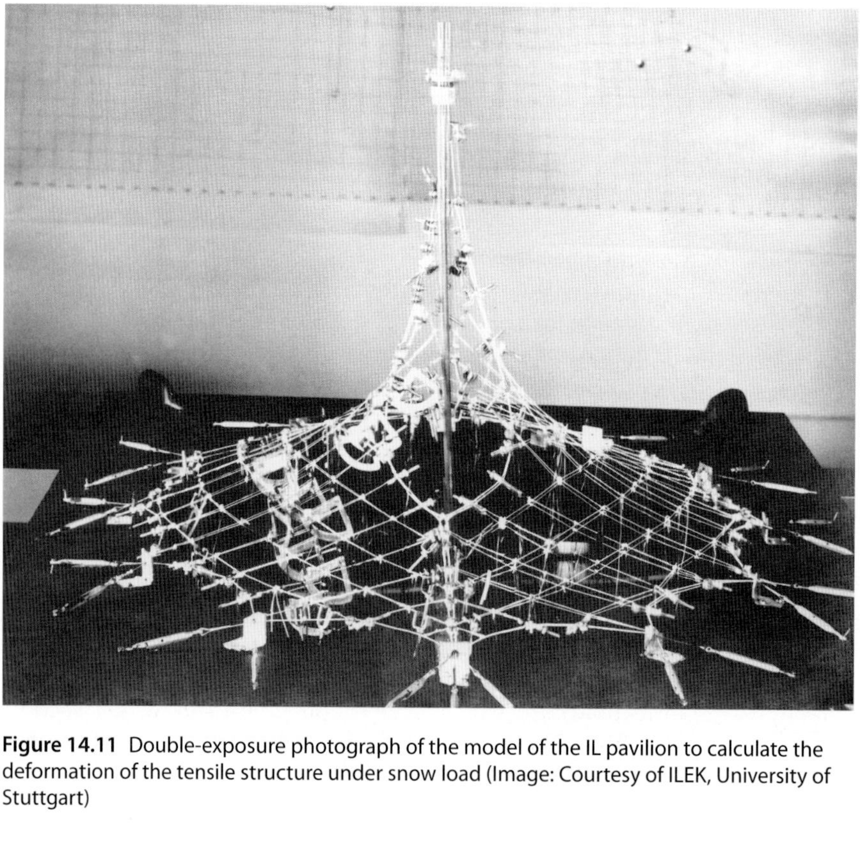

Figure 14.11 Double-exposure photograph of the model of the IL pavilion to calculate the deformation of the tensile structure under snow load (Image: Courtesy of ILEK, University of Stuttgart)

Figure 14.12 1:75 scale measurement model of the German pavilion at Expo '67 in Montreal. (Image: Courtesy of ILEK, Universitof of Stuttgart)

in Montreal. For this project, Frei Otto's Institute for Lightweight Structures built a measurement model at a scale of 1:75, based on the principle of geometric and statical similarity. The model net was made using 2 mm diameter stainless-steel wires at 27 mm spacing. Each wire represented every fourth cable in the full-size structure. (Figure 14.12).

One fundamental difficulty presented by Frei Otto's lightweight, load-bearing membrane structures was how to capture their complex shapes in graphical and digital form. The spatially complex geometries are difficult to represent using the standard representational techniques of floor plan and section. Since the forms were based on models, the challenge was to translate these into two-dimensional drawings, in order, for example, to produce cutting patterns for the membranes. A measuring table designed specifically for this stage in the design process was built at the Institute for Lightweight Structures. Using this apparatus, the xyz-coordinates of the models could be taken by means of plumb lines. The measurements could be displayed directly and exported onto punched paper tape for input into a computer. A computer-assisted process for the evaluation of the measurements was introduced in 1966. Under the leadership of Klaus Linkwitz, the IAGB at the Technical University, Stuttgart also created a computer program that could translate the data recorded on punched tape directly to produce drawings, which made it possible to draw any sections of the tent surface. This *Zuse-Graphomat* output device was one of the first computer-based drawing systems used in the construction industry, developed by the computer pioneer Konrad Zuse (1910-1995) [38] and itself displayed as an exhibit in the German pavilion in Montreal. Another measurement technique,

Figure 14.13 Apparatus for measuring the form of the model for the German pavillion at Montreal using photogrammetry. Images were taken by a camera mounted on the guide rail at the top, from two positions a known distance apart. (Source: Courtesy of ILEK, University of Stuttgart)

first used for the Expo '67 pavilion, was photogrammetry. The 1:75 scale model was measured using photogrammetry by taking photos from above at specific distances, running along a guide rail (Figure 14.13). This non-contact method had also been used for the soap-film experiments, as the delicate forms were liable to burst on the slightest contact.

The tensions in the individual wires in the model were measured using a small gauge specially developed for this project and always referred to as 'Montreal micrometers'. It was mounted on the wire at three points and worked on the same principle as the three-point tensioning device [39] (Figure 14.14). Overall changes to the geometry and changes to tensions in the wires under snow load were measured using the double-exposure photographic technique developed for the IL building (Figure 14.15).

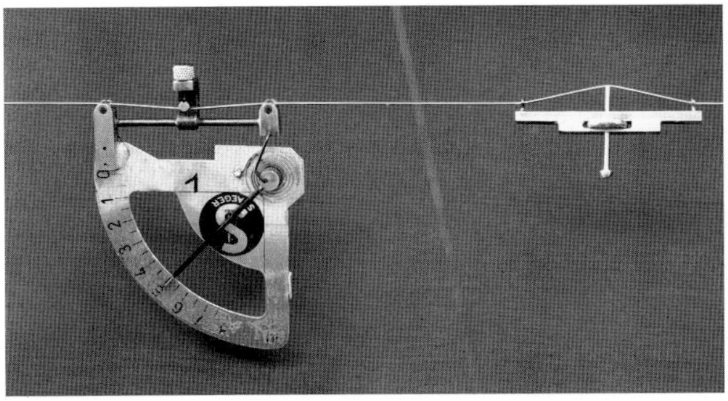

Figure 14.14 Left: The Montreal micrometer: a small gauge mounted at three points to determine the level of prestress in the cables. Right: a device for increasing the tension in a wire. (Image: Courtesy of ILEK, University of Stuttgart)

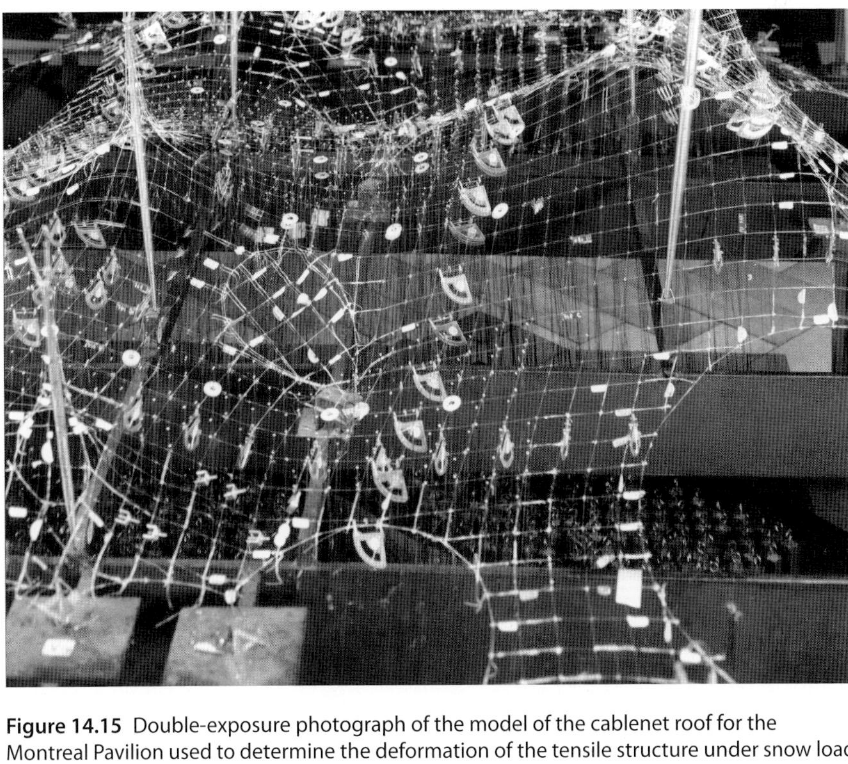

Figure 14.15 Double-exposure photograph of the model of the cablenet roof for the Montreal Pavilion used to determine the deformation of the tensile structure under snow load. (Image: Courtesy of ILEK, University of Stuttgart)

14.3.4 The cablenet roofs for the Munich Olympic Games

Following the success of the Montreal Pavilion, a number of cablenet structures by Frei Otto, based on a scheme-design by Günter Behnisch (1922-2010), were chosen for several venues at the 1972 Munich Olympic games (Figure 14.16).

Figure 14.16 The cablenet roofs of the Olympic Games sports venues, Munich. Top: the Olympic Stadium. Lower right: Sports hall; Lower left: Swimming pool. (Image: [40])

A large number of form-finding experiments and structural model analyses were undertaken at IL for the Olympic Games structures. These included a soap-film model to determine the minimal surface. In order to increase the durability of the soap film, soap solution was fed through funnel tubes in the support ponts to replenish liquid lost by evaporation [41] (Figure 14.17a). Based

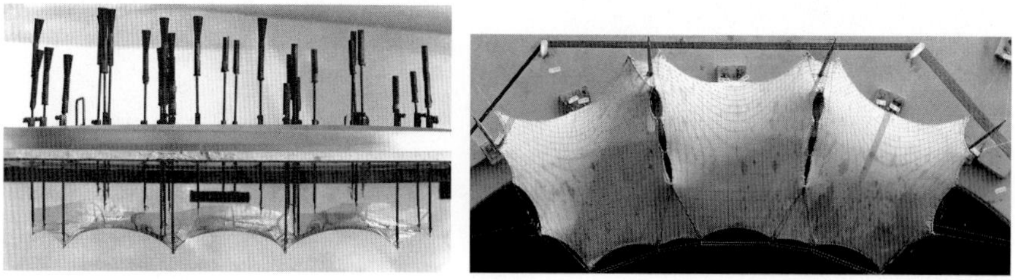

(a) (b)

Figure 14.17 Form-finding models for the canopy of the Olympic Stadium, Munich. (a) Soap-film model to determine the minimal surface. (b) Tulle model to refine the form using an elastic material, and enable the contours to be mapped using projected shadow lines. (Image: Courtesy of ILEK, University of Stuttgart) on the minimal surface, a model was made using tulle to refine the geometry using an elastic material, and to enable the contours of the complex surface to be mapped using projected shadow lines, and converted to a digital model that was used to plan the cutting patterns for the acrylic sheets used to clad the cablenet structure (Figure 14.17b).

The task of IL comprised 'producing measurement models for all building parts, including for the measurement of forces and distortion'. The team of engineers at Leonhardt + Andrä, headed by Jörg Schlaich, worked intensly with the models and 'exerted a certain control function during the model construction phase' [42]. Based on the geometric specifications of the numerous form-finding models, physical models were built at a scale of 1:125. This scale resulted from the size of the measuring table at IL, the foundations of which are still visible today inside the IL pavilion. The largest of the models was for the stadiuim canopy, measuring 1.9 x 4 metres. The tension in the wires was measured using an improved version of the Montreal micrometer gauges (Figure 14.18).

By means of photogrammetry, the *Institut für Anwendungen der Geodäsie im Bauwesen* (IAGB) determined the pattern of the roof skin and the coordinates needed for the subsequent digital calculation. Working together with the the *Institut für Modellstatik*, mechanical strain-gauge and photogrammetric measurements were made simultaneously to avoid loading the highly-sensitive models too often (Figure 14.19). As in Montreal, the cutting patterns for the Munich roofs were determined in close collaboration with the IL and the IAGB. Since the measurement models for the various nets at the Olympic Park were not sufficiently precise to define the dimensions for manaufacturing the plastic sheet cladding, Leonhardt + Andrä involved the Institute for the Statics and Dynamics of Aerospace Engineering at the Technical University, Stuttgart, headed by John

Figure 14.18 1:125 scale model of the cable net roof for the Munich Olympic stadium. (Image: Courtesy of ILEK, University of Stuttgart)

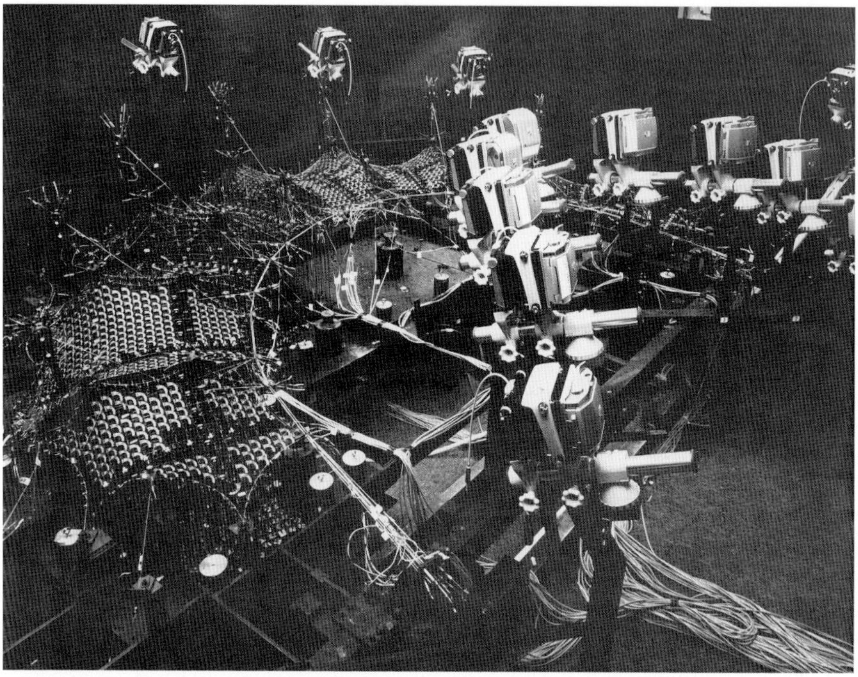

Figure 14.19 1:125 scale model of cable net roof for the Munich Olympic stadium, showing cameras in position to take double-exposure photographs of the unloaded/loaded model. (Image: Courtesy of ILEK, University of Stuttgart)

Argyris (1913-2004). As a pioneer of the Finite Element Method, Argyris was able to calculate the cutting pattern for the sports-hall roof mathematically, using the data from the IAGB [43].

14.3.5 Hanging models

Frei Otto and his team also took a different approach to optimizing the form of load-bearing structures using various forms of hanging models. In this case, researchers were able to consult previous studies, such as those by Heinrich Hübsch (1795-1863) and Antoni Gaudí (1852-1926) (see Chapter 3). These structures find their own shape by the physical phenomenon that a hanging mesh will always take on the form corresponding to the line of (compressive) thrust of the inverted structure [44]. Applying this law of physics, Frei Otto began experimenting in 1946 with hanging cloths [45] and, later, with a wide range of hanging chains and meshes. In the early 1970s, for example, a hanging mesh made of machine-made hooks and eyes was used for optimizing the grid shell for a government building that was to be built in Saudi Arabia, although it was never actually realized in this form [46]. Perhaps the most spectacular of these hanging models was a replica made at IL of the hanging model constructed by Gaudí to determine the forms for the church at the Colonia Güell (see Section 3.5.3) [47].

A large hanging model was used in the design of the Mannheim Multihalle for the *Bundesgartenschau* (Federal horticultural show) in 1975, designed by architect Carlfried Mutschler (1926-1999) and the Atelier Frei Otto in Warmbronn, working with Ove Arup (see Chapter 22). In order to define precisely the form of the shell, after measuring the pre-design model, a large hanging model – today preserved in the DAM collection – was built in Frei Otto's studio in Warmbronn. The grid of the hanging model, at a scale of 1:98.5, was created by hand with 15-millimetre-long chain links and rings with a diameter of 2.5 millimetres (see Figure 22.5). This model was measured photogrammetrically by the IAGB under Klaus Linkwitz, from which a digital, spatial model was produced. The data, created with a CDC 6600 mainframe computer, could be drawn and distributed using a so-called *Zeichenautomat* (drawing machine) which served as basis for the detail design of the structure, including the determination of the grid pattern, and the lengths of all timber laths of the grid and all the edge details [48, 49].

14.3.6 The sports hall in Jeddah

One of the last large projects undertaken at IL which made considerable use of models was the King Abdul Aziz University Sports Hall in Jeddah (1979-1981). As with many projects, a number of different models were used during the design development, each with a different purpose. To explore the geometry of a prestressed, doubly-curved cablenet, form-finding experiments were undertaken with (non-elastic) soap films to determine the minimum surface shape. This work enabled a suspended-chain model to be built to introduce gravity forces into the form-finding process and, finally, a model was made using fine polyester fabric to define the surface geometry that would be formed by an elastic material [50, 51] (see Figures 19.12, 19.13, 19.14, 19.15, 19.16). Measurements on this model were processed by computer to generate the cutting patterns for the membrane in the final tent, which is 27 metres high and covers an area of 9,500 m^2.

14.4 The decline of physical model testing in Stuttgart

Fritz Leonhardt was among the first to recognize the potential of computer-aided measuring processes and, as early as 1961, he launched the company for *Rechneranwendungen im Bauwesen* (RIB) together with Volker Hahn (1923-2011) and Friedrich Wilhelm Bornscheuer. It was this company that brought the first Finite Element Method program (FEM) to the market in 1972. In the following years, the development of ever faster and cheaper computers made them a more-economical modelling technique than physical model testing. Robert Müller, head of the Institute for Model Analysis at the Technical University, Stuttgart, saw this coming in his *Handbuch zur Modellstatik* [52] in 1971. When Müller retired in 1993, the institute he had headed was dissolved and the professorship renamed Information Processing in Engineering Design, marking the end of the era of model analysis in structural engineering in Stuttgart [53].

References

1 Weber, Christiane & Ziegler, Volker (2011) Otto Graf (1881-1956) und die Baustoffprüfung an der Technischen Hochschule Stuttgart. *Beton- und Stahlbetonbau* 9: 594-603.

2 Weber, Christiane (2011) *Fritz Leonhardt: Leichtbau – eine Forderung unserer Zeit. Zur Einführung baukonstruktiver Innovationen im Leichtbau in den 1930er und 1940er Jahren.* Doctoral thesis. Karlsruhe: KIT Scientific Publshing. http://dx.doi.org/10.5445/KSP/1000025243. 39-46.

3 Weber, Christiane & Voormann, Friedmar (2009) Fritz Leonhardt. Erste Bauten und Projekte. In: *Fritz Leonhardt 1909-1999. Die Kunst des Konstruierens/The Art of Engineering* (Eds. Joachim Kleinmanns & Christiane Weber), 30-47. Stuttgart/London: Menges.

4 Leonhardt, Fritz & Schaechterle, Karl (1940) Hängebrücken I. *Die Bautechnik* 18 (33): 377-386.

5 Leonhardt, Fritz & Schaechterle, Karl (1941) Hängebrücken II & III. *Die Bautechnik* 19 (7): 73-82, 19 (12/13): 125-133, 200.

6 Leonhardt, Fritz (1984) *Baumeister in einer umwälzenden Zeit. Erinnerungen,* Stuttgart: DVA. 78.

7 Materialprüfungsanstalt Stuttgart (1939) *Prüfbericht. Antragsteller Direktion der Reichsautobahnen, Berlin. Modelluntersuchungen zur Rheinbrücke Köln-Rodenkirchen. 3. Bericht mit den Ergebnissen der Untersuchungen an der Modellbrücke III. Fortsetzung unserer Berichte vom 22. April und 30. Juni 1938.* 15. Januar 1939. (Universitätsarchiv Stuttgart, 33/1/1240 and 33/1/1241).

8 Weber, Christiane (2011) *Fritz Leonhardt: Leichtbau – eine Forderung unserer Zeit. Zur Einführung baukonstruktiver Innovationen im Leichtbau in den 1930er und 1940er Jahren.* Karlsruhe: KIT Scientific Publshing. http://dx.doi.org/10.5445/KSP/1000025243. 87-104.

9 Materialprüfungsanstalt Stuttgart (1940) *Niederschrift der Besprechung des Prüfungsausschusses für die Elbehochbrücke Hamburg,* 6. April 1940 (Universitätsarchiv Stuttgart 33/1/743). 4.

10 Maier-Leibnitz, Hermann (1941/1942) Grundsätzliches über Modellmessungen zur Klarlegung der Formänderungen und Spannungen von verankerten Hängebrücken. *Die Bautechnik* 46/47: 508-515; 48: 522-526; 53/54: 573-583; 53/54: 457-463; 54/56: 486-490.

11 Weber, Christiane (2011) *Fritz Leonhardt: Leichtbau – eine Forderung unserer Zeit. Zur Einführung baukonstruktiver Innovationen im Leichtbau in den 1930er und 1940er Jahren.* Karlsruhe: KIT Scientific Publshing. http://dx.doi.org/10.5445/KSP/1000025243. 115-134.

12 Weber, Christiane (2009) Fritz Leonhardt's contribution to the construction of the new main station of Munich (1939-1942) In: *Proceedings of the Third International Congress on Construction History* (eds K.-E. Kurrer, W. Lorenz & V. Wetzk), 1485-1492. Cottbus.

13 Leonhardt, Fritz (1984) *Baumeister in einer umwälzenden Zeit. Erinnerungen.* Stuttgart: DVA. 91-92.

14 Weber, Christiane & Ziegler, Volker (2012) The construction material testing at the MPA Stuttgart during the Third Reich. In: *Nuts and Bolts of Construction History: Proceedings of the 4th International Congress on Construction History, Paris 3-7 July* (2 Vols) (eds Robert Carvais, André Guillerme, Valérie Nègre & Joël Sakarovitch), Vol 2: 297-304. Paris: Picard.

15 Materialprüfungsanstalt Stuttgart (1952) *Modellversuche für die Nordbrücke in Düsseldorf*. Reports from 20.10.1952 and 8.1.1953. (Universitätsarchiv Stuttgart, 33/1/800).

16 Kleinmanns, Joachim; Weber, Christiane (2009) *Fritz Leonhardt* 1909-1999. *Die Kunst des Konstruierens/The Art of Engineering*. Stuttgart/London: MENGES. 210.

17 Abteilung Bauingenieur und Vermessungswesen, *Stellungsnahme bezüglich des Arbeitsgebiets des Instituts für Spannungsoptik und Modellmessungen und seine Abgrenzung gegenüber dem Otto Graf Institut*, 25. November 1957 (Universität Stuttgart, (Akte der) Zentralen Verwaltung, Az. 7712.17-0212-0 / 216), 1.

18 Technische Hochschule Stuttgart (ed.) (1954) *Die Technische Hochschule Stuttgart 1954. Bericht zum 125jährigen Bestehen*. Stuttgart. 94-97.

19 Kultusministerium Baden-Württemberg, *Berufungsvereinbarung Prof. Leonhardt*, Stuttgart, den 23. April 1958 (Universitätsarchiv Stuttgart/Personalakte Prof. Dr.-Ing. Fritz Leonhardt).

20 Abteilung Bauingenieur und Vermessungswesen, *Stellungsnahme bezüglich des Arbeitsgebiets des Instituts für Spannungsoptik und Modellmessungen und seine Abgrenzung gegenüber dem Otto Graf Institut*, 25. November 1957 (Universität Stuttgart, (Akte der) Zentralen Verwaltung, Az. 7712.17-0212-0 / 216), 2.

21 Technische Hochschule Stuttgart. *Mitteilungen des Rektoramts 1/103 1948/1967*: Mitteilungsblatt THS Nr. 79, Stuttgart, den 15. Januar 1963 (Universitätsarchiv Stuttgart/Hb UA).

22 Technische Hochschule Stuttgart. *Mitteilungen des Rektoramts 1/103 1948/1967*: Mitteilungsblatt THS Nr. 90, Stuttgart, den 8. März 1965 (Universitätsarchiv Stuttgart/Hb UA).

23 Müller, Robert (1969) Modellstatik an der Universität Stuttgart. *Südwestdeutsche Monatsschrift für Kultur, Wirtschaft und Reisen* 19: 101-102.

24 Universität Stuttgart (1972) *Personal- und Vorlesungsverezeichnis Wintersemester 1977/78*. 126. Address Institut für Modellstatik, Keplerstr. 11 and Pfaffenwaldring 7.

25 Leonhardt, Fritz (1984) *Baumeister in einer umwälzenden Zeit. Erinnerungen*, Stuttgart: DVA. 246.

26 Müller, Robert (1969) Modellstatik an der Universität Stuttgart. *Südwestdeutsche Monatsschrift für Kultur, Wirtschaft und Reisen* 19: 101-102.

27 Kayser, Rolf, Lehmitz, Karl, Leonhardt, Fritz, Müller, Robert K., Schlaich, Jörg & Voßbein, Hartmut (1970) Das Hyperschalendach des Hallenbades Hamburg Sechslingspforte; Teil I: Entwurf und Tragverhalten (Leonhardt/Schlaich). *Beton- und Stahlbetonbau* 9: 207-215; Teil II: Modelluntersuchung (Müller/Kayser). *Beton- und Stahlbetonbau* 10: 245-249; Teil III: Bauausführung (Voßbein/Lehmitz). *Beton- und Stahlbetonbau* 11: 261-264.

28 Kayser, Rolf (1970) Das Hyperschalendach des Hallenbades Hamburg Sechslingspforte. *Die Bautechnik* 65: 245-247.
29 Kayser, Rolf & Müller, Robert K. (1970) Das Hyperschalendach des Hallenbades Hamburg Sechslingspforte. *Beton- und Stahlbetonbau.* part II Modelluntersuchung. 10 (1970): 245-249.
30 Weber, Christiane (2014) 'Ich traue der reinen Rechnung nicht, wenn diese nicht am Modell überprüft werden kann' (I cannot trust the pure calculation, if this cannot be verified with a model). Frei Otto and model testing. In: *Proceedings of the first Conference of the Construction History Society, Queens' College, Cambridge* (ed. J. W. P. Campbell, et. al.), 445-453.
31 Brensing, Christian (2005) Frei Otto und Ove Arup. Szenen einer gegenseitigen Inspiration. In: *Leicht Bauen – Natürlich Gestalten. Frei Otto – Das Gesamtwerk* (ed. Winfried Nerdinger), 102-109. Basel: Birkhäuser.
32 Dickson, Michael (2005) Frei Otto und Ted Happold. 1967-1996 und darüber hinaus. In: *Leicht Bauen – Natürlich Gestalten. Frei Otto – Das Gesamtwerk* (ed. Winfried Nerdinger), 111-123. Basel: Birkhäuser.
33 Bach, Klaus, Burkhardt, Berthold & Otto, Frei (1987) Seifenblasen: Forming bubbles. *Mitteilungen des Instituts für leichte Flächentragwerke.* IL 18. Universität Stuttgart.
34 Burkhardt, Berthold (1969) Minimalnetze: Minimal nets. *Mitteilungen des Instituts für leichte Flächentragwerke.* IL 1. Universität Stuttgart. 12-14.
35 Roland, Conrad (1965) *Frei Otto – Spannweiten. Ideen und Versuche zum Leichtbau.* Frankfurt/Berlin: Ullstein. 84-105.
36 Weber, Christiane (2011) Der Deutsche Pavillon auf der Weltausstellung 1967 in Montréal. In: *Rolf Gutbrod. Bauten in den Boomjahren der 1960er* (ed. Klaus Jan Philipp), 68-83. Salzburg: Müry Salzmann.
37 Kamahele. Licensed under the Creative Commons Attribution-Share Alike 3.0 Germany license. https://commons.wikimedia.org/wiki/File:0136-Stuttgart-Otto.jpg (2/10/2019)
38 http://www.horst-zuse.homepage.t-online.de/z64.html (30/10/2019)
39 Otto, Frei (1967) *Protokoll über die Arbeiten des Instituts für Leichte Flächentragwerke* (Private collection of Bertold Burkhardt, Braunschweig).
40 Dave Morris. Licensed under the Creative Commons Attribution 2.0 Generic license. https://commons.wikimedia.org/wiki/File:Olympic_park_12.jpg (1/10/2019)
41 Bach, Klaus, Burkhardt, Berthold & Otto, Frei (1987) Seifenblasen: Forming bubbles. *Mitteilungen des Instituts für leichte Flächentragwerke* IL 18. Universität Stuttgart. 106-107; 348-351.
42 Otto, Frei (1975) Netze in Natur und Technik: Nets in nature and technics. *Mitteilungen des Instituts für leichte Flächentragwerke.* IL 8. Universität Stuttgart. 275.
43 Spieker, Elisabeth (2009) Die Planung des Olympiadachs in München. Fritz Leonhardts Mitwirkung und Impulse. In: *Fritz Leonhardt 1909-1999. Die Kunst des Konstruierens/The Art of Engineering* (eds. Joachim Kleinmanns & Christiane Weber), 118-125. Stuttgart/London: Menges.
44 Graefe, Rainer (1974) Gewölbe und Kuppeln, Schalen und Stabtragwerke, Gitterschalen. Beispiele aus der Baugeschichte. In: Gitterschalen: *Mitteilungen des*

Instituts für leichte Flächentragwerke. IL 10 (ed. Frei Otto), 10-25. Universität Stuttgart.

45 Otto, Frei & Trostel, Rudolf (1962) *Zugbeanspruchte Konstruktionen,* Vol. 1. Frankfurt/Berlin: Ullstein. 22.

46 Otto, Frei (ed.) (1974) Gitterschalen. *Mitteilungen des Instituts für leichte Flächentragwerke.* IL 10. Universität Stuttgart. 294.

47 Tomlow, Jos (1989) Das Modell: Antoni Gaudis Hängemodell und seine Rekonstruktion. Neue Erkenntnisse zum Entwurf für die Kirche der Colonia Güell. *Mitteilungen des Instituts für leichte Flächentragwerke.* IL 34. Universität Stuttgart.

48 Otto, Frei (1978) Multihalle Mannheim. *Mitteilungen des Instituts für leichte Flächentragwerke.* IL 13. Universität Stuttgart. 33.

49 Weber, Christiane (2012) Frei Otto – Experimentelle Modelle. In: *Das Architekturmodell. Werkzeug, Fetisch, Kleine Utopie* (eds Peter Cachola Schmal & Oliver Elser), 174-195. Zürich: Scheidegger & Spiess.

50 Bach, Klaus, Burkhardt, Berthold & Otto, Frei (1987) Seifenblasen: Forming bubbles. *Mitteilungen des Instituts für leichte Flächentragwerke.* IL 18. Universität Stuttgart. 108-110.

51 Walker, Derek & Addis, Bill (1998) *Happold. The Confidence to Build.* London: Taylor & Francis. 64-65.

52 Müller, Robert K. (1971) *Handbuch der Modellstatik.* Berlin/Heidelberg/New York: Springer. 8.

53 Anon (1993) *Senatsprotokoll der Universität Stuttgart, November 3, 1993.* (Universitätsarchiv Stuttgart). 18.

15

Model testing of structures in post-war Italy. The activity of ISMES, 1951-1974

Mario Alberto Chiorino and Gabriele Neri

15.1 *Istituto Sperimentale Modelli e Strutture* (ISMES)

15.1.1 The founding of ISMES

In Italy, research on model testing has always been connected to the development of hydroelectric power, which was, through the design of large dams, the primary source of huge structural and, therefore, economic challenges. From 1945, the necessity of putting back in order a run-down and exhausted country at the end of WWII, provided the conditions for a difficult but confident recovery. Post-war Italy was facing the need for a broad reconstruction plan, in which renewed infrastructure and energy networks were essential. Within this context, several companies became more and more interested in the work carried out by the eminent professor Arturo Danusso (1880-1968) and his team at the Polytechnic of Milan since the 1930s, focusing especially on model testing of concrete dams and other structures (see Chapter 10).

Ambitious plans for new hydroelectric plants had been already proposed in 1943, when the fall of the dictatorship wrongly heralded the idea of an imminent end to the war. One of the first initiatives proposed was the use of the Piave river basin, beginning with a new large plant at Pieve di Cadore, near Belluno. In 1945-46 the design of a large concrete dam began again. Due to the complexity of such a structure, an arch-gravity dam, Carlo Semenza (1893-1961), Technical Director of the *Società Adriatica di Elettricità* (SADE, the Adriatic Energy Corporation), asked for an intensive series of experimental analyses to be carried out under Danusso's supervision.

After completion of the dam's final design, a large concrete model (scale factor, $\lambda = 40$) was needed in order to check and optimize its structure. In this case, the laboratory at the Polytechnic of Milan was unsuitable, for various reasons. On the one hand, the size of the model was too big for its small room and equipment; on the other, university rules would have led to excessive costs [1]. These factors shed light on the intrinsic limitations of the Milan laboratory which was too small to deal with the work of the large companies that were rebuilding the country. The expansion of model testing therefore had to be done outside the University, in a new and independent institute.

Physical Models: Their historical and current use in civil and building engineering design,
First Edition. Edited by Bill Addis.

This situation led to the creation, in 1946, of a new laboratory – the *Istituto Sperimentale Applicazioni in Calcestruzzo* (ISAC, Experimental Institute for Concrete Applications) – in Bergamo, financed by SADE, Italcementi (the cement group which had already supported the laboratory at the Polytechnic of Milan in the 1930s) and other companies. Thus, in 1947-48, the 10-metre long model of the Piave dam was built and tested by Danusso's pupil Guido Oberti, followed by a second model. The first study played a decisive part in the final selection of the arch-gravity type instead of the gravity type, which had also been investigated. It exerted a direct influence on the acceptance of this kind of dam. The physical models showed how the results deduced from mathematical models were too schematic: thanks to the new investigations at ISAC the project was optimized, reducing the volume of the dam by around 20 percent, and bringing associated economic benefits. At the end of the tests, SADE claimed to have saved one billion Italian lira (around €30 m today) in comparison with the original project [2].

The success of the models for the Piave dam attracted many large Italian companies. This allure brought in a large amount of funding which enabled the creation of a permanent experimental laboratory. On 6 September 1951, the *Istituto Sperimentale Modelli e Strutture* (ISMES; Experimental Institute for Models and Structures) was founded as a joint-stock company, involving Italcementi; several electricity companies including Edison, Romana Elettricità, Valdarno, SADE, SIP, Falck, Montecatini, Meridionale Elettricità; and building companies such as Lodigiani, Torno and Girola. The principal stakeholders were Italcementi (30%) and Edison (20%) [3]. Danusso was appointed President and Oberti the Director of the Institute. ISMES was based in Bergamo, in a new building designed by architect Carlo Barbieri, a former assistant of Oberti in Milan. The latter highlighted some of the Institute's distinctive traits:

'In order to fully appreciate the function performed by the ISMES [...], it should be noted that in Italy most of the study and research work in the field of civil engineering is done by University Institutes. Their work is largely theoretical and, in the field of experimental research, is confined chiefly to material testing, due also to the limited facilities and staff available for this work. ISMES is a private corporation established for the purpose of solving, with adequate financial means and freedom of movement (which, by virtue of their own structures, University Institutes do not possess), specific structural problems arising from the needs of designers and builders. Therefore, ISMES carries out a technical-scientific activity which is complementary to the activity which is carried out, or should be developed, in the Italian University Research Laboratories' [4].

15.1.2 The facilities and fields of research at ISMES

ISMES' test facilities were located in a group of buildings, enlarged in various steps. They included special instruments and equipment to carry out static and dynamic tests up to failure on large-scale models or full-size elements of any type of structure. After several years of setting up, the buildings for static tests consisted of two large reinforced-concrete structures capable of containing models or structural elements to be tested. One of these structures was shaped

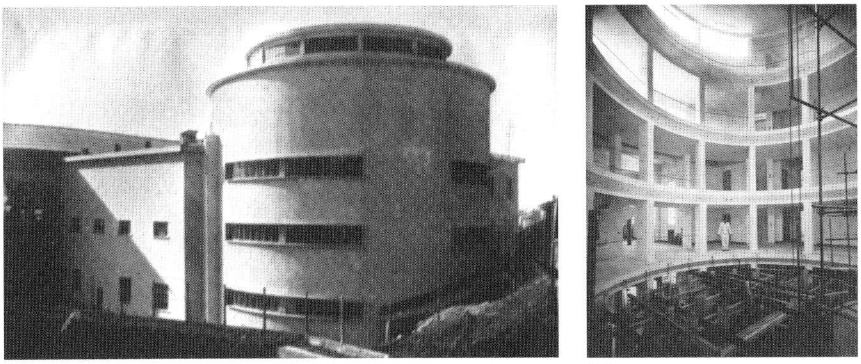

(a) (b)

Figure 15.1 The four-story reinforced-concrete cylindrical tower at ISMES, Bergamo, 1953. (a) Exterior view. (b) Interior view. (Images: ISMES Historical Archive)

like a two-story rectangular tank, 10 x 5 x 3(h) m, built in 1945 for the Piave dam tests and then enlarged in 1952 to become the favourite site for testing dams. The other was a four-story reinforced-concrete cylindrical tower (9.80 m diameter and 16 m high), built in 1952, which permitted experimental research on tall structures (Figure 15.1). Thanks to its rigid frame, composed of 24 reinforced-concrete columns and a solid foundation system, it could also bear strong horizontal loads.

A large rectangular hall (210 m^2) equipped with a crane was specifically available for testing prefabricated and even very heavy elements and structures and, in general, large-scale models of structures resting or anchored on a level floor [5]. Some years later, a vast hall was equipped to host fixed facilities for dynamic tests, capable of reproducing various types of earthquake acting on model dams and structural elements. This second hall housed also a large compression-testing machine with a capacity of 2,000 tonnes, particularly suited for tests on large structural elements and specimens [5].

ISMES was especially devoted to research on the materials to be used in the preparation of models and the laboratory for testing materials was provided with a compression-testing machine. Among other facilities at the beginning of the 1960s, ISMES provided a physics and chemistry laboratory, an electrical laboratory, a photoelastic laboratory, a soil mechanics laboratory, a hall for model preparation, a vast range of equipment for precision measurement of both strains and deflections, as well as a considerable variety of loading equipment (hydraulic jacks of various diameters, spring dynamometers of various capacities, etc.) [6].

In its first decades, ISMES pursued four main fields of research: research on materials, especially those required for model making; preparation and testing of structural models under both static and dynamic loading; testing of structural models and real structures; and the development of measuring equipment. Among these fields, the second one is particularly interesting to explore. Since the model had to reproduce the full-size structure very closely, the construction was carried out using an accurate process. For a reinforced-concrete structure, for example, firstly, a trial model in plaster or wood was made, to the same scale of that foreseen for the experimental model. This trial model would serve to check dimensions and, if necessary, to improve the actual shape. It was then used in the manufacture of the formwork for the experimental model and to test methods of applying the desired loading and arranging the measuring equipment. Later, the experimental model was built with great care. As the models were built using delicate materials in which the binding element was mixed with aggregates of various origins: sand, pumice, Celite, bentonite, barite and rubber, it was in many cases necessary to protect the models with impermeable coatings to retard drying and reduce shrinkage. Before casting the model, the attachments for increasing the dead load, whenever such an increase was required by similitude considerations, were inserted throughout the entire body of the future model. The completed and cured model was then carefully tested in the elastic range and then incrementally loaded to destruction, in order to obtain the total factor of safety of the structure [6]. 1953 saw the publication of the first issue of the journal *I Quaderni ISMES* which reported the activities of ISMES until 1974 (Figure 15.2).

Figure 15.2 Front cover of the magazine *I Quaderni ISMES*, No. 1, September 1953. (Source: ISMES Historical Archive)

15.2 ISMES' activity 1951-1961

During the 1950s, model testing at ISMES was mainly divided into two distinct experimental procedures. The first one investigated elastic problems by means of small models of entire structures (mainly dams but also buildings, bridges, etc.), typically 1:25 to 1:150, or part of them, acting as a 'calculating machine' for stresses by measuring strains. This method presumes that the boundary conditions of the problem should already be well known and that the assumptions postulated by theory are accepted (homogeneity and isotropy of material, continuity and linear stress-strain behaviour). It provided an elastic solution to the problem, which could be compared with the theoretical results. As Guido Oberti observed: 'It has, therefore, principally the purpose of confirming the method followed and the procedures adopted in the calculations themselves' [7].

The second method was intended to reproduce the prototype – more frequently for buildings or structural elements, but also for dams – with a large-scale model to be tested up to the failure. Usually, the scale of these models was between 1:1 and 1:5 for single structural elements (buttresses, beams, slabs, reinforced-concrete shells, reinforced-concrete pipes, etc.) and around 1:10 – 1:15 for entire buildings, bridges and other structures. This meant that more parameters were introduced into the problem, such as the anisotropy of the structure and its foundations, the elasto-plasticity of the materials, and discontinuities. It led to a broader and more complete understanding of the real statical behaviour of a dam or building. As Oberti noted: 'We search for a better agreement with the prototype rather than for a confirmation of calculations' [7].

15.2.1 Large dams

The size of ISMES activity in its first years is well summarised in two tables published to celebrate its first decade (1951-1961) [8]. The first reports the models of dams, the main source of income for the Institute: seventy-four models, related to fifty-nine different projects (Figure 15.3). The second table lists all the models and prototypes for other structures: eighty-nine in total. Besides the total number of models, which is significant, their geographic distribution is particularly impressive.

Up to 1955, there were only models relating to Italian structures. From the following year, a consistent number of foreign commissions began: the first was the Drossen dam in Austria (1956); then the Swiss Reno di Lei dam (1957) (Figure 15.4) and, from 1958, models for dams in Japan, Yugoslavia, Germany, Bulgaria, Mexico, Iran, Argentina, Spain, France and even Australia, some 16,000 km from Bergamo. Less international, on the other hand, is the list for other structures, with the exception of a 1:24 scale model for a circular concrete roof at the School of Design of the State College of North Carolina, USA [9]. This was designed in 1958 by Horacio Caminos, the same Argentinian architect who had designed the Civic Centre of Tucumán, whose structure had been studied by Oberti using model tests in 1949 at the *Politecnico di Milano* (see Section 10.5.3).

Figure 15.3 1:50 scale model of the Beauregard Dam, in Valle d'Aosta in Italy (c. 1951). (Image: ISMES Historical Archive)

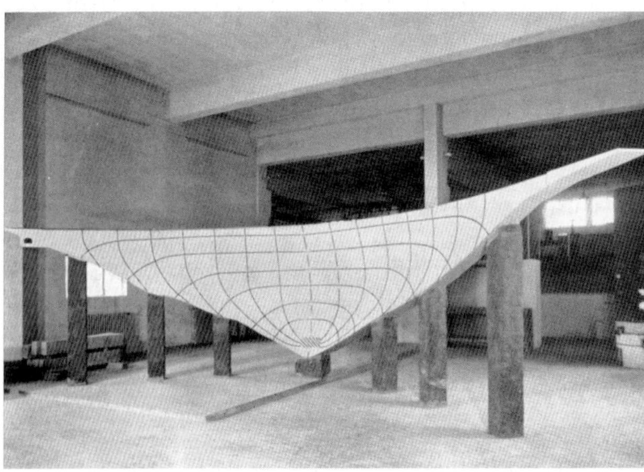

Figure 15.4 1:66.6 scale model of the Valle di Lei Dam, Switzerland, 1957. (Image: ISMES Historical Archive)

The reasons for this success were many. For instance, at the end of the 1950s, Bulgaria – in the oppressive context of the Cold War – resorted to ISMES, following Moscow's suggestion, 'among the astonishment and incredulity of ministerial and diplomatic authorities, in order to solve some serious problems regarding dams under design or construction' [10]. On the other hand, the failure of the Malpasset (Fréjus) dam in 1959, which claimed 421 victims, persuaded the World Bank to ask for compulsory model tests at ISMES for all dam projects that the bank helped to finance.

Among the many investigations for dams we can mention the most important models. The two models for the Piave dam (also known as Pieve di Cadore

Figure 15.5 1:40 scale model of the Pieve di Cadore Dam, 1946. (Image: ISMES Historical Archive)

dam) (Figure 15.5), with a maximum height of 112 m above the river bed, and a length of 323 m) were particularly important, because they marked the beginning of ISMES' investigations on large-scale models in 1946. Besides the large saving of money that they enabled, also interesting were the comparisons with computations that these models made possible. As Guido Oberti remarked, 'Comparing the experimental elastic curve of cantilevers with the curve that results by computing the dam as a portion of vertical rotation solid, clamped at the toe into a yielding foundation (classical computation); the agreement, for the whole central section, may be considered satisfactory. A marked difference occurs towards the upper part, where experimental elastic curves bend towards the upstream direction, showing the marked stiffness of the crest arch' [11].

The 265 metre high, double-curvature arch dam at Vajont, which became tragically famous in 1963 (see Section 15.3), was tested at different stages. Two models had been made in the early 1940s to study the behaviour of the lower section of the dam for a height of some 100 m, and also to test for the effect of the weight of the upper section, though not part of the model. A third model, built in 1947-48, was tested under hydrostatic pressure, in order to check and modify its shape. Later, when SADE decided to raise the dam up to the final height, ISMES was asked to build and test a large-scale model in 1957-58 with a linear scale factor, $\lambda = 35$; and loading scale factor, ζ (force per unit area of prototype / force per unit area of model) = 9 (Figure 15.6). Another special so-called 'geomechanical' model was tested to evaluate, through failure tests, the safety factor, which was always very high and typically greater than sevevn [12, 13].

The Forte Buso dam (maximum height, 110 m; length, 275 m), tested in 1951-52, was the first significantly asymmetric structure checked using models. It allowed the dam's design to be improved and a better distribution of stresses to be achieved, especially at the plug. Two models were made, with: $\lambda = 50$; $\zeta = 4.35$; i_x (stiffness of rock / stiffness of concrete) = 1.9.

Figure 15.6 1:35 scale model of the 265 m high Vajont Dam, 1957. (Image: ISMES Historical Archive)

Another interesting arch-gravity dam was the Reno di Lei dam, at the Swiss-Italian border. It had the maximum length ever achieved (605 m) and a height of about 140 m. Two models were built in 1957: (a) $\lambda = 66.6$; $\zeta = 8.59$; $i_x = 1.9$; (b) $\lambda = 66.6$; $\zeta = 6.04$; $i_x = 1.9$. They made it possible to determine the stress conditions under water pressure and dead-load action. The second model was used, for the first time, to check thermal behaviour in order to find the total deflections produced by seasonal heat cycles and the resulting temperature distribution within the dam [14].

The Kurobe dam in Japan, 184 m high, 385 m long, was tested between 1958 and 1961 with two models that permitted the solving of complex problems due to the inadequate and uneven strength of abutments and foundations. The first model had the following scale factors: $\lambda = 90$; $\zeta = 3.85$. For the first time, it also reproduced areas of the foundation crossed by faults and subject to dead loads, to make it possible to study the stability of the dam-mountain unit as a whole, by introducing all the most important factors [15]. A second model was made at a scale of $\lambda = 75$.

In 1965 ISMES tested models for the Hongrin dam, near Montreux in Switzerland, modelled with $\lambda = 55$. The dam consisted of two arch dams in series, closing off two valleys and supported at the centre by a large pillar of stratified rock, designed by the Swiss engineer Henri Gicot [16].

The model tests for the Tolmezzo-Ambiesta double-curvature arch-dam (height 56 m; length 118 m), began in 1955, and marked the beginning of dynamic studies of models of large dams situated in seismic areas. A model ($\lambda = 50$; $\zeta = 3.3$; $i_x = 2.3$) showed the behaviour of the structure under hydrostatic load. In 1956 three more models were tested ($\lambda = 75$; $\zeta = 3.85$; $i_x = 1.86$) to check the effect produced both by normal and exceptional earthquakes and to determine the degree of safety of the structure. The problem was especially interesting because of the considerable downstream slope of the dam, which was adopted for better stability under seismic action. The Ambiesta dam withstood the 1976, magnitude 6.5, earthquake at Friuli less than 20 km distant [17, 18, 19].

15.2.2 High-rise buildings

One of the most complex models produced during the 1950s at ISMES was for the reinforced-concrete frame of the Pirelli Tower in Milan, the 135-metre tall skyscraper designed by the architect Gio Ponti, whose structural design was entrusted to Pier Luigi Nervi and Arturo Danusso (1955-56) (Figure 15.7). Nervi had met Danusso at his Polytechnic in Milan in 1935, when he needed to verify the structural design for his reinforced-concrete hangars built for the Royal Air Force at various Italian sites (see Section 10.3).

For this building the large spaces and the innovative equipment available at ISMES allowed Oberti to create a model at large scale ($\lambda = 15$), around 9 metres tall, cast in a micro-concrete of pumice-stone and cement mortar, and resting on a layer of rubber to simulate the deformability of the soil. The chosen scale, combined with the similarity between the material used for the model and that in the actual building, also enabled him to test the model beyond service conditions up to failure.

The model reproduced the basic lines of the real structure, with the following schematizations: reproduction of every second floor, with a slab 4.2 cm thick having the same average moments of inertia; simplifications in the end walls and walls of the vertical cores, retaining similar sections and moments of inertia. Reinforcement was reproduced using netting and wire reinforcement.

The construction of the model started in March 1955 in ISMES' four-story cylindrical tower (Figure 15.8). First of all, the formwork was placed; then reinforcement was laid and the cement mortar was poured. After removal of the formwork, surfaces were coated with waterproof paint to prevent evaporation and shrinkage effects, and to ensure homogeneous curing conditions. The model was completed in about two months and two further months were needed for curing.

Figure 15.7 The Pirelli Tower, Milan, 1956-60. Engineers, Arturo Danusso and Pier Luigi Nervi; Architect, Gio Ponti. (Image: [20])

(a) (b)

Figure 15.8 (a) and (b) 1:15 scale micro-concrete model of the Pirelli tower, during construction, 1955. (Images: ISMES Historical Archive)

Figure 15.9 1:15 scale micro-concrete model of the Pirelli tower, 1956. (Image: ISMES Historical Archive)

As pictures show, the model of the Pirelli tower was outstanding for its size – being one of the biggest models of an entire structure ever built – and complexity; it can, therefore, be seen as one of the symbols of the golden age of Italian structural engineering (Figure 15.9).

The tests focused on the static behaviour of the whole structure and the interaction and connection between its various parts. As ISMES' engineers reported, of special interest were the tests dealing with wind action, carried out by using both symmetrical and asymmetric loads. Tests also determined the natural periods of vibration due to bending and torsional oscillations [21, 22]. The application of dead loading to the columns and floors, as well as of arbitrarily-oriented wind thrusts, was done by means of the steel framework.

The first series of tests fully confirmed the validity of the structural configuration that Nervi had designed, this time together with his now elderly friend Danusso, to give solidity and statical strength to the elegant form designed by Ponti. The variants introduced in the project and in a second series of tests,

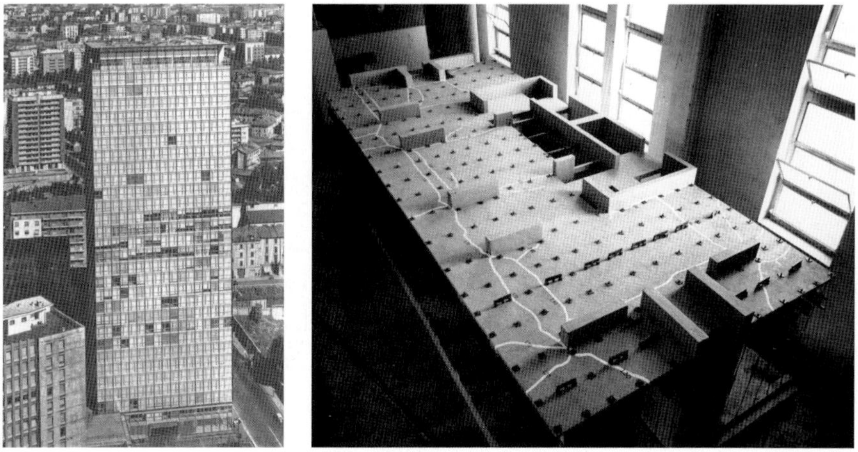

(a) (b)

Figure 15.10 (a) Torre Galfa, Milan, 1959. (Image: [24]) (b) 1:5 scale model of typical floor of the Galfa Tower, 1958. (Image: ISMES Historical Archive)

following the results of the first series, were minimal, aimed at obtaining an increase in the flexural rigidity of the main vertical cantilever – formed of pairs of large dovetailed pylons – through the insertion of architraves joining the pylons themselves in the upper two thirds of their height, and also at improving the torsional behaviour which was affected by the eccentric position of the lift shafts. The model, brought to collapse by proportionally increasing the two basic loads (dead load and wind pressure), revealed unexpected strength capacities in the elastoplastic range [23].

During the same period, two other Milanese skyscrapers – icons of the so-called 'Economic Boom' that characterized the end of the 1950s – were checked at ISMES. The 109 metre Galfa Tower (Figure 15.10a) was designed by the architect Melchiorre Bega and, among the structural engineers, Arturo Danusso. A 1:5 model of a typical-floor was made and tested (Figure 15.10b). The 106 metre Velasca Tower (Figure 15.11a) was designed by BBPR architects working with Danusso as structural engineer. ISMES' engineers built models ($\lambda = 2$) of three different elements of the columns with different percentages of reinforcement, and a typical-floor ($\lambda = 5$). All these elements were tested up to failure (Figure 15.11b).

In 1962, ISMES carried out tests on a celluloid model at small scale (1:52.8) for another tower: the 145-metre Stock Exchange Tower in Place Victoria, Montreal – at the time, the highest building in reinforced concrete in the world. This building was the pride of Italian engineering and enterprise, designed by Luigi Moretti with structural design by Nervi and constructed by the Roman firm, Sogene. The tests focused above all on the structural behaviour in the elastic field, to study the dynamic response to the action of wind and earthquakes [26] (Figure 15.12).

These were followed by tests for a series of other prestigious buildings around the world for which Nervi acted either as engineer or consultant. Studying the

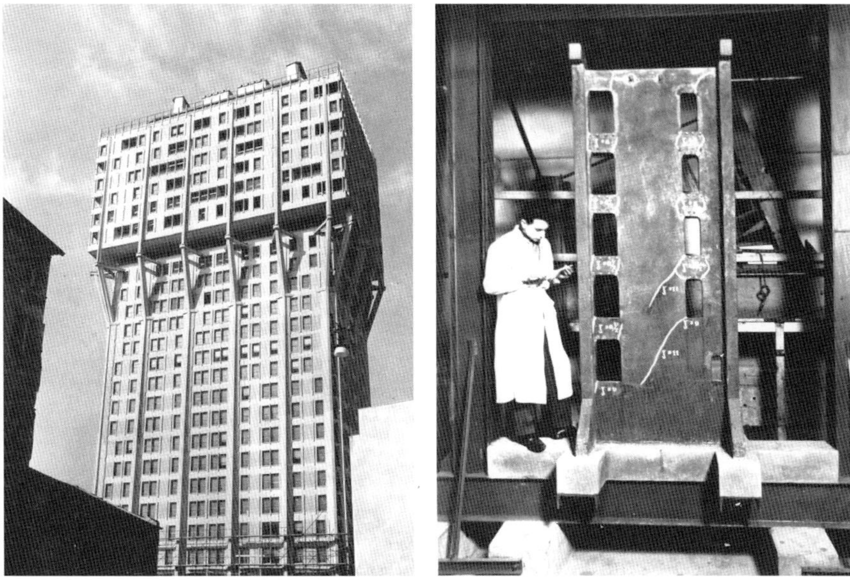

(a) (b)

Figure 15.11 (a) Torre Velasca, Milan 1958. (Image: [25]). (b) 1:2 scale model of a wind brace of the Torre Velasca, tested to destruction, 1956. (Image: ISMES Historical Archive)

dynamic response of structures had become the norm since the mid-1950s. For structures and buildings whose dimensions were not excessive, it became usual to produce separate models: a medium-scale model (typically 1:40 to 1:50) was made with an elastic material for the dynamic tests; and a large-scale model (typically 1:15 for a whole building, or between 1:5 and 1:10 for a building element) was made with a material which, in terms of similitude, reflected the stiffness and strength of concrete, for tests beyond service conditions up to the failure.

In the 1960s, instead of being simplified in the conventional manner as a pseudo-static load, the action of the wind began to be assessed experimentally by means of model tests in wind tunnels. This led to developing a line of research for medium-sized buildings which had already begun in the 1930s, especially in the United States, for skyscrapers such as the Empire State Building. To undertake such tests, a further series of models were created at small scale (1:100). Nervi and Oberti took this path at an early date, using the wind tunnel at the Polytechnic of Turin, a choice made by Oberti, who taught at the Polytechnic and enjoyed close relations with the school of Carlo Ferrari, since ISMES lacked a facility of this type. Models of the St. Mary's Cathedral in San Francisco and the Norfolk Scope Arena, Virginia (both 1:100 for wind-tunnel tests) are among the few ISMES models still surviving [27] (see Chapter 39).

With respect to the results of these tests used by Nervi and Oberti with their customary brilliance in the process of verifying the safety of constructions, it should be pointed out that a new wind-tunnel technique was being developed during this period, in particular by Canadian engineers. Now in common use,

Figure 15.12 1:52.8 scale celluloid model of the Stock Exchange Tower in Montreal, 1962. (Image: ISMES Historical Archive)

Boundary Layer Wind Tunnels recreate the atmospheric boundary layer, characterized by conditions of turbulent flow which has a more severe impact on buildings than the smooth airflow used in Turin (see Chapter 32).

15.2.3 *I modelli nella tecnica* (1955): an important conference for structural modelling in Italy

While the model of the Pirelli tower was being tested at ISMES in October 1955, an important conference took place in Venice, entitled 'I modelli nella tecnica' (Models in technology). Promoted by the prestigious *Accademia Nazionale dei Lincei* of Rome, it was supported, not by chance, by SADE on the occasion of its Fiftieth anniversary. The President of SADE, Count Vittorio Cini, remarked: 'SADE believes it has something to claim in this field. Its technicians have widely contributed to the practice of modelling. Almost all its buildings were studied, examined and checked using models' [28]. This was definitely true; SADE supported the foundation of ISMES and, up to 1955, it had already commissioned nineteen experimental investigations at the Institute.

The exceptional achievement of the Venice conference was its international dimension: for the first time, in Italy, some of the most eminent specialists in structural modelling from Europe and United States, gathered together. This allowed a fruitful debate about different methods and perspectives. Furthermore, the overall scope was not limited to a structural testing. The conference embraced many other engineering disciplines including acoustic and photometric models, aerodynamic and electrical models, seismic and hydraulic models, and many more. The proceedings of the congress probably represent the best collection of papers about the full scope of engineering modelling ever published [29].

The session dedicated to structural models opened the conference, with an impressive speech given by Arturo Danusso, who remarked on the importance of the occasion. The other participants were Professor Lydik S. Jacobsen (Stanford University, USA), engineer Gaston Réméniéras (Electricité de France), Professor Alfred Stucky (Ecole Polytechnique Lausanne, CH), Professor Eduardo Torroja (University of Madrid), Professor Friedrich Tölke (University of Stuttgart), Professor Manuel Coelho Mendes da Rocha (Director of the National Laboratory of Civil Engineering in Lisbon, LNEC) and, finally, Professor Guido Oberti, on behalf of the Polytechnic of Turin and ISMES. Danusso summarised the current situation:

> 'After a century of investigating the statics of structures – investigations that got a major impulse from the observed deformability of materials of any kind, and confirmation, at first in metal structures and then in structures of reinforced concrete; after the trust and distrust about the value of the theory of elasticity as the real representation of the actual behaviour of structures; after the significant recourse to experiments on materials in order to get, so to speak, the grammatical grounding of the language spoken by nature itself; finally, the idea of the recourse to structural models has come – initially in a shy manner, then strengthening itself – to suggest the syntax of this language and how to express itself in elegant style (...)' [28].

Danusso also remarked, with clear disappointment, how a 'real awareness' of the function of structural models was still not adequately recognized, due to a sort of 'discord', common among scholars, between 'the synthetic vision of the intuition which looks at the phenomenon in its whole physical objectivity, and the analytic vision of science which, within its idealised boundaries, controls, specifies and asserts' [28]. From this erroneous contrast, we obtain two antithetical positions; thus 'some, linked by tradition to this frame of mind, find in calculations the support for logical perfection, relying on it and being wary of intuition as a banana skin to slip on, towards empiricism; while others keep intuition in the forefront as an inspiring instrument to be fed and checked scientifically through experimentation' [30]. The latter, who indeed were the supporters of using structural models, did not have an easy time since they had to face a wall made of preconceived ideas that led to a generalized refusal of any expense concerning models.

Danusso would express a similar mix of dissatisfaction and pride some years later, on the occasion of the tenth anniversary of ISMES:

> 'Ten years of existence in modest but growing prosperity is a highly significant achievement for a non-profit-making Institute that had set out to offer the benefit of experimental research and testing to all those who bear the burden of responsibility in construction work. This is all the more so when one considers how remote the idea of such a benefit, or rather of its necessity in the construction economy, was and still is, in current opinion. It would serve no purpose to remind designers of the number of limitations which are imposed upon theory in order to reduce ordinary problems to a plan of calculation and hence its resulting impaired validity. This, however, can for the most part be avoided through experimentation. It would be even more superfluous to insist upon the prospective client, in the hope of convincing him, that the addition of a small expenditure for a model (half a percent, or less, of the total cost of the structure) may greatly contribute to the reduction of the two hundred percent minimum risk coverage with which the designer must insure himself in setting the dimensions of the structure, and thereby its cost, when using regular calculation procedures. One cannot reasonably claim that this should be taken care of by the Construction Company, since the latter is forced to hold expenses to their lowest limit due to competition and the restrictions included in the terms of the contract. The Construction Company could do so in the case of a fixed fee contract and full freedom in adopting structural changes. Such cases, however, are infrequent and would, in regular business practice, involve an advance assurance of a profit; such assurance, even if plausible, is in contrast with the necessary reserve of the experimenter. He who would doubt our statement, wondering at it, ought to be reminded that the imposing budget of public works ignores the item of structural models' [31].

The main exception to this practice concerned large dams. Danusso proudly drew attention to the role of ISMES in the promotion of modelling in this field, and underlined how – thanks to the support of enlightened clients – models were elevated 'from the status of calculating machines to that of representatives, as far as possible true ones, of the resistance process, above all in the range which surpasses the elastic limit up to the point of failure' [32]. So, models became capable of revealing not only the behaviour of structures in normal conditions, but also 'the reserve of strength that protects that range and represents the greatest hidden treasure of the structure' [32]. Again, in Danusso's own words:

'The models must therefore be enlarged, their materials are to be carefully selected and the boundary conditions, as well as the constructional procedures of the prototype, must be accurately reproduced; the models then are to be fitted with a great number of instruments and patiently tested. Model engineering thus becomes more difficult and must constantly be improved. One must meditate anew day-by-day, punctiliously and without conceit, haughtiness or mental crystallization, on the two basic pivots upon which our work hinges: the degree of similitude attainable between model and prototype, and calibration of the measuring apparatus. The model, like the theory, is also a representative scheme, even though an improved one, but it is not the prototype and, hence, it has its own defects.

The engineer endeavours to reduce these defects, though he is certain that he will never quite succeed in eliminating them. But he does not resent it because he knows that all his procedures are but games of wise approximation and that this wisdom includes the best of his art. The measuring instruments, too, have their caprice and whims; we call them so because our ignorance is unable to grasp the fineness and the endless variety of factors operating in Nature. In this field our workers, from the graduate technicians to the labourers, perform miracles of intuition; and ultimately the instruments surrender to them like puppies to the owner who treats them well. Thus, they impart valuable intuitive lessons to us who are charged with inspiring and directing their work' [32].

Danusso's words are clear: they reveal the spirit and philosophy that the great Professor instilled in his Institute, conceived as a 'clinic of structures' where the latter were patiently tested before their encounter with the real world. Despite the scepticism of certain scholars and building actors, the Venice conference demonstrated the existence of an international network comprising Institutes and research laboratories for structural modelling.

15.3 1964-1974. Changes for ISMES: new guidance, new horizons

If the 1950s can be considered, for ISMES, the decade of Arturo Danusso, then the 1960s were ruled by Pier Luigi Nervi who was appointed its President in 1964, the year after Danusso had suffered a serious illness that prevented him from keeping his work going.

Nervi's election was not a surprise. He was considered internationally to be one of the leading contemporary engineers and architects and had been closely allied to Danusso's School since the 1930s, when he started testing his buildings using models. Thanks to their multiple roles as engineers, designers, scientists and consultants, Danusso, Nervi and Oberti formed a robust team, in which information, knowledge and commissions were shared. Moreover, like Danusso, Nervi was also linked to several national and international institutions and organizations, such as the *Consiglio Nazionale delle Ricerche* (CNR, Italian National Research Council) and the *Comité européen du béton* (CEB), which were useful for expanding ISMES relationships.

During the 1950s, most of the tests for dams had been commissioned by the companies which were managing the Italian electricity network, region by region: SADE for Veneto and part of Emilia; Edison for Lombardy, Emilia and Liguria; etc. Following what was happening in other countries such as France and Britain, the nationalization of the electricity industry took place even in Italy. In autumn 1962, the government proclaimed the nationalization of all the companies and enterprises working on production, trade, distribution and transportation of electrical power, as well as all the related fields.

The newly constituted ENEL (*Ente Nazionale Energia Elettrica* – National Authority for Electric Energy) absorbed all the 1,270 Italian companies dealing with electricity, in order to have a shared technical, administrative and logistical organization. Among these companies there were also the main shareholders of ISMES: thus, ENEL suddenly became the majority shareholder of the Institute with around 55%. ENEL, through its Central Directorate of Studies and Research, took over the financing plans and the development of applied research in accordance with organic programs of long duration. As some witnesses remarked, the new board was composed of valuable engineers, but they were not linked to the specific field of modelling and thus they had a propensity for diversifying ISMES' activities [1]. Initially this condition caused some worries, but later the new direction turned out to be favourable for the future of ISMES, thanks to the huge assets of the new company – at that time, ENEL was the third largest energy provider in the world.

Almost at the same time, a tragic event shook the Italian world hydroelectricity. On 9 October 1963, a massive landslide occurred on Monte Toc in the Piave Valley, where the Vajont Dam had been built in 1957-60. The landslide caused a mega-wave of 50 million cubic metres which over-topped the dam (265 m high), leading to the flooding and complete destruction of several towns and villages in the valley below and 1,917 casualties. However, as is well-known, the behaviour of the Vajont Dam itself after the landslide was outstanding: the dam held and disaster was caused by the resulting wave, not by failure of the structure.

This disaster had profound consequences for national energy policy. Even if the catastrophe was not the only cause (we must add the depletion of hydroelectric sources), it can be taken as a stimulus for the progressive shifting of ISMES' activities towards new fields of interest. The Vajont hecatomb led to all ongoing works on new hydroelectric plants being stopped, which sharply reduced the requests for model testing of dams in Italy. More requests arrived from abroad, but they were not enough to guarantee commercial survival.

The Vajont disaster also highlighted the need to make a realistic mechanical model of the rock strata beneath the site of a new dam – the possible collapse of a dam is related directly to the yielding of the abutments. For the Grančarevo Dam in Yugoslavia (now Montenegro), a geotechnical survey was undertaken to determine the orientation of the rock strata, their thickness and the existence of local faults. Using this data a model was built using bricks to replicate this rock structure. The precise placing of the bricks enabled three or sometimes four systems of joints in different bedding layers to be reproduced [33] (Figure 15.13). It was also recognised that when a discontinuity penetrates deep into the rock, the stability of the abutments depends on the angle of friction between the surfaces in contact. If the angle of friction is similar in magnitude to the internal friction in the rock, a sliding failure is unlikely since a discontinuity does not represent a significant weakness. When the angle of friction between the discontinuity surfaces is less than the angle of internal angle of friction of the rock, then the fractures will propagate. This was the situation that was observed in the behaviour of the model of the rock beneath the Ca' Selva Dam in Italy [34] (Figure 15.14).

This situation led ENEL to extend the horizons of ISMES' scientific research to include dynamic analysis, geotechnics, rock mechanics and analyses of foundations, as well as the design of hydro-electric power plants located in rock caves [35]. In essence, a transfer from pure model-testing specialization and activity, to applied research in general in the structural and geotechnical fields [36].

Figure 15.13 1:125 scale model of the rock strata beneath the proposed Grančarevo Dam, Yugoslavia (now Montenegro) (c.1963). The bricks were placed to replicate the orientation of the strata. (Image: ISMES Historical Archive)

Figure 15.14 1:125 scale model of the Ca' Selva Dam, North of Venice (c.1960). Fractures were due to sliding of layers at discontinuities in the rock. (Image: ISMES Historical Archive)

Given this situation – both uncertain and variable – Nervi's arrival brought useful benefits for the development of structural modelling at ISMES. Cleverly, he succeeded in bringing together his company's strengths with those of ISMES, and began to provide the Institute with a great number of projects designed or managed by Studio Nervi (Nervi's office) and Nervi & Bartoli (Nervi's building company). Nervi was thus a many-sided figure: he played the role of President of a prestigious research Institute; he was an entrepreneur, scientist and structural designer [27, 37, 38, 39, 40, 41]. His advantages were many. On one hand, his role at ISMES was a guarantee for his clients, reassured by the fact that the engineer's structural ingenuity was supported by one of the best research laboratories in the world. On the other, ISMES helped Nervi to solve complex structural challenges, like those presented below.

In addition to Guido Oberti, who at this point was one of the most important experts in the field of structural modelling, another brilliant engineer came to the forefront: Enzo Lauletta (1927-1971), who became co-director of ISMES from 1965, and who contributed to spreading ISMES expertise throughout the world through his investigations and scientific papers [42, 43, 44]. Lauletta's contribution to model testing was particularly related to aseismic design. He participated in the 2nd World Conference of Earthquake Engineering in Tokyo (July 1960), together with Guido Oberti [45] who had already contributed to the

Figure 15.15 1:40 scale, elastic resin model of the Parque Central building in Caracas, Venezuela, tested on a shaking table to study soil-structure interaction in the foundations (c.1969). (Image: ISMES Historical Archive)

San Francisco/Berkeley World Conference on Earthquake Engineering Research in 1956 [4].

Italy had not been at the forefront in this field, but ISMES research gave an important impulse, gradually, from the mid-1950s. In 1954-1955, the Institute, as already mentioned, was equipped with facilities for dynamic tests, capable of reproducing the action of various types of earthquake on model dames and structural elements. The experiments on the Ambiesta arch dam were the first in this field at ISMES in 1955-1956.

It should also be mentioned that Arturo Danusso had been one of the first engineers in Italy to deal with aseismic design, at the beginning of the twentieth century. In 1909 he won the international competition of ideas and projects for anti-seismic construction held after the earthquake that had hit Messina and Reggio Calabria the year before [46]. In the following years, he promoted many initiatives and research in this field, in particular with Aldo Castoldi (b.1939) [47, 48] (Figure 15.15). This growing interest led to the founding of the *International Center of Earthquake Engineering* in Milan [49], which offered courses on earthquake engineering for graduate students, with the support of CNR and UNESCO, from 1967 to 1969.

15.4 Nervi's projects at ISMES

The dynamic tests on a model of the Stock Exchange Tower in Montreal in 1962 were followed by those carried out in 1964-1965 for another prestigious building in North America: St. Mary's Cathedral in San Francisco, designed by Nervi as structural consultant together with the architect Pietro Belluschi [27, 40, 50]. In this project, a model at small scale (1:100) (Figure 15.16) was

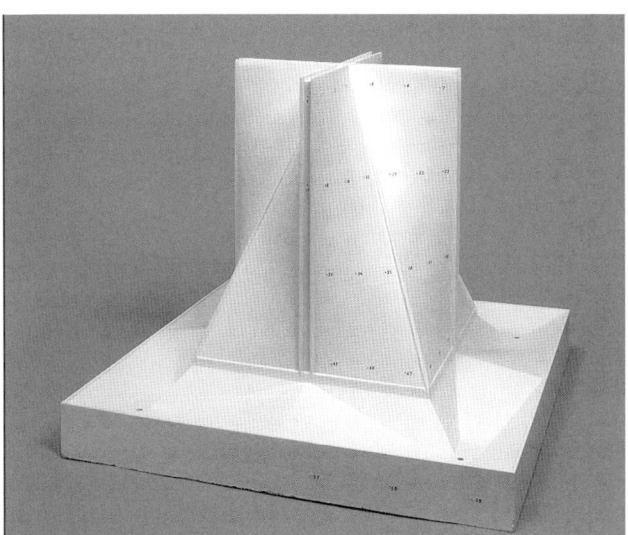

Figure 15.16 1:100 scale wind-tunnel model of St. Mary's Cathedral, San Francisco, 1964. (Image: Politecnico di Torino, Area Bibliotecaria e Museale)

used for the wind-tunnel tests performed at the Polytechnic of Turin (see Section 15.2.2). This was followed by two medium-scale (1:40 and 1:36.89) resin models (Figure 15.17) for static and dynamic tests in the elastic domain, with special attention to seismic tests due to the building's location. Finally a large-scale (1:15) model in micro-concrete (Figure 15.18) was tested up to failure. The validity of the beautiful geometrical form resulting from the union of eight segments of hyperbolic paraboloids, designed by Belluschi in collaboration with Eduardo Catalano, was substantially confirmed, despite Nervi's initial fears. The model tests also furnished useful indications, both for the correct dimensions and the pattern of the stiffening ribs that form the intrados of the dome. Finally, the first series of tests identified the need for significant strengthening of the supporting structures of the cupola in order to carry its self-weight and horizontal seismic action; the modifications were confirmed in the second series of tests. The results of the model tests were accompanied by checks using numerical models made in California by the engineering studio of Leonard Robinson responsible for the executing the project, and by the review board of Californian experts responsible for the control of the project [27].

It is interesting to compare this San Francisco Cathedral with St. Mary's Cathedral in Tokyo, designed by Kenzo Tange and completed a few years earlier in 1964, with a roof that was also formed by the union of eight hyperbolic paraboloid shells. The structural engineer, Yoshikatsu Tsuboi, placed his confidence for all the safety checks, including the seismic assessments, in analytical modelling in the elastic field. He used a small-scale (1:100) experimental model only for elastic testing to verify the results of the analytical modelling [51].

In the following years, other important design and consultancy commissions were awarded to Nervi around the world, especially in North America. In most of these cases, Nervi opted once again for experimentation on models in addition to analytical modelling. The tests were generally conducted at ISMES, again under the guidance of Oberti and his team of researchers. Two exceptions were the Australia Square Tower (180 m) and the MLC Center (240 m), two buildings in reinforced concrete built in Sydney and designed by architect Harry Seidler, for which Nervi was responsible for the structural design. The elastic models – in resin at medium scale (1:30) for the first building and in methacrylate at small scale (1:95) for the second – were produced and tested at the University of New South Wales (1964-1965) and the University of Sydney (1972) respectively.

In the second half of the 1960s, ISMES hosted experiments on models for a number of important projects in the USA and Europe. The SCOPE Cultural and Convention Center, Norfolk, Virginia, USA (1967) has a magnificent ribbed dome with a diameter of 135 m., representing a larger version of Nervi's roofs at the Palazzetto dello Sport and the Palazzo dello Sport in Rome. Nervi used a model at 1:100 scale for the wind-tunnel tests at the Polytechnic of Turin (Figure 15.19a) and an elastic model in resin, at 1:50 scale, for static and dynamic tests (Figure 15.19b) [27].

These experiments were followed in 1968-1969 by model tests on the hyperbolic paraboloid element for the roof of Newark International Airport, New Jersey. Static tests were carried out on a micro-concrete model at very large scale (1:6.6) of one of the umbrella-type elements [52] (Figure 15.20). Nervi was also

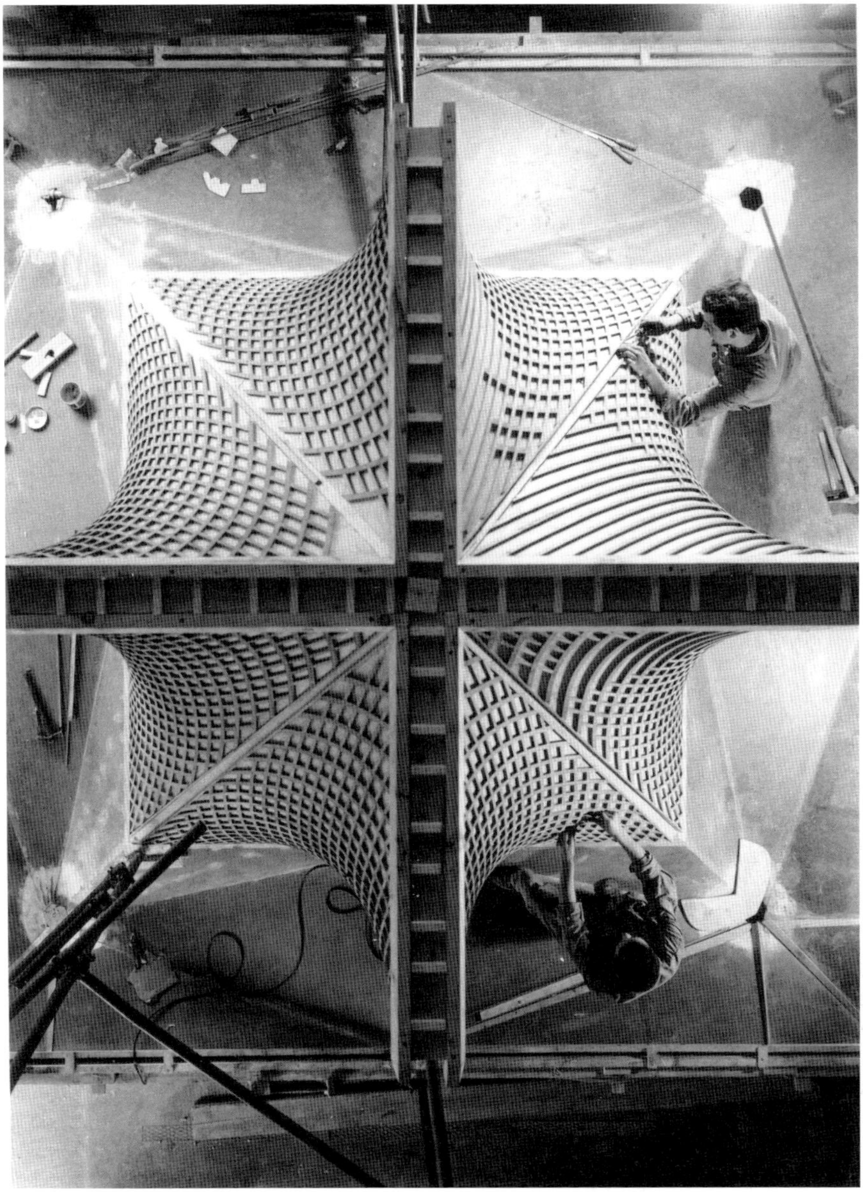

Figure 15.17 1:15 scale reinforced micro-concrete model of St. Mary's Cathedral, San Francisco, during construction, 1964. (Image: ISMES Historical Archive)

consultant engineer for the B.I.T. offices in Geneva for which he commissioned dynamic tests on two elastic models in resin at 1:40 scale. The last model made for a project by Nervi was an elastic model at 1:50 scale of the Rupert C. Thompson Arena at Dartmouth College (New Hampshire, USA), tested in Bergamo in 1970-1971.

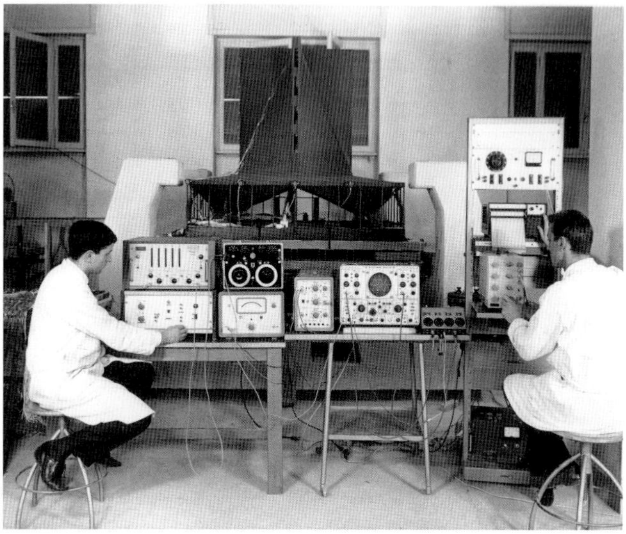

Figure 15.18 1:36.89 scale epoxy resin model of St. Mary's Cathedral, San Francisco, during dynamic testing on a shake table, 1965. (Image: ISMES Historical Archive)

(a) (b)

Figure 15.19 SCOPE Cultural and Convention Center, Norfolk, Virginia, USA, 1967. (a) 1:100 scale model for wind-tunnel tests. (Image: Politecnico di Torino, Area Bibliotecaria e Museale). (b) 1:50 scale model made of resin for statical and dynamic load testing. (Image: ISMES Historical Archive)

15.5 Other structures tested at ISMES

Many other structures were tested at ISMES in the 1950s and 1960s, including reinforced-concrete pipes and penstocks, some of which were designed by Nervi [1], motorway tunnels which led to the use of 'geomechanical' models from 1961, foundation structures for turbo-generators, flumes for irrigation, decks for motorway bridges, and nuclear reactor pressure vessels in prestressed concrete (1966-1972) on behalf of ENEL.

Figure 15.20 1:6.6 scale model of a hyperbolic paraboloid element of Newark Airport, NJ, 1968. (Image: ISMES Historical Archive)

A particularly interesting case study is that of the Mole Antonelliana in Turin, the well-known masonry building (164 m-high) designed by Alessandro Antonelli, begun in 1862. After the collapse of the original masonry spire in 1951, tests were undertaken in 1954 to study the action of wind in the wind tunnel of the Polytechnic of Turin in smooth layer conditions, using a 1:50 scale model of the new steel-core spire, as in the case of the skyscrapers. At ISMES, a series of static tests were carried out in 1955 on a 1:4 scale model of the reinforced-concrete structure for the Mole Antonelliana designed by professors Giuseppe Albenga (1882-1957) and Danusso for the connection between the new spire and the existing masonry cupola [53] (Figure 15.21).

A number of investigations on the design of large span bridges took place at ISMES in the 1950s and 1960s. The most important were those carried out for bridges designed by Riccardo Morandi (1902-1989) and Sergio Musmeci (1926-1981), which are among the masterpieces of post-war Italian engineering.

Figure 15.21 1:50 scale model of the connecting structure between new spire and the existing masonry cupola of the Mole Antonelliana, being tested at ISMES in 1955. (Image: ISMES Historical Archive)

Morandi asked for ISMES' help in 1958 in order to test the deck of his Lake Maracaibo Bridge, in Venezuela, spanning 8.7 km from shore to shore (1957-1962) [54]. The cable-stayed bridge was made in reinforced and prestressed concrete with five main spans, each of 235 m, and towers 92 m high, providing a clearance of 46 m for shipping. A celluloid elastic model ($\lambda = 28$) of the deck of the bridge was made and tested (Figure 15.22a). Every detail of the structure was reproduced, except for the steel prestressing cables. In 1962 static and dynamic tests were also carried out for Morandi's similar Polcevera Bridge in Genoa (1960-1965), with four cable-stayed spans (maximum 208 m), one of which sadly collapsed in August 2018. One pier of the bridge was tested using a ($\lambda = 50$) celluloid model (Figure 15.22b). The following year, a series of statical tests were made on the deck of a railway bridge in Venice also using a celluloid elastic model ($\lambda = 16.6$).

The investigations for the bridge across the Basento River near Potenza (1967-1975), a daring structure designed by Sergio Musmeci, were more complex and varied. Its structure consists of 4 spans, each of 70 m, formed with a continuous thin shell (30 cm thickness), supporting the bridge superstructure at several points and resting on the ground at other points [55]. For his bridge, Musmeci carried out tests on several different reduced-scale models: after initial investigations with soap-film models, he used a neoprene model to determine the shape of the structure, through a form-finding process. Later, a third model was made in methacrylate ($\lambda = 100$) reproducing 2 of the spans, according to

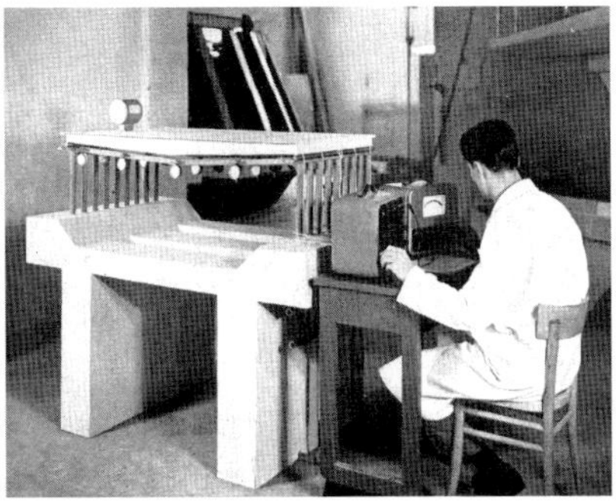

(a)

(b)

Figure 15.22 (a) Maracaibo Bridge, Venezuela, 1958. Engineer, Riccardo Morandi. 1:28 scale celluloid elastic model of section of the bridge deck. (b) Polcevera Bridge, Genoa, 1961. Engineer, Riccardo Morandi. 1:50 scale celluloid elastic model of the tower and the stayed portion of the deck. Several Huggenberger Tensometers are visible. (Images: ISMES Historical Archive)

Figure 15.23 Basento Bridge, Italy. 1:10 scale reinforced-concrete model, tested at ISMES. (Image: ISMES Historical Archive)

the design shaped after the neoprene model. This third model, built at the *Laboratorio di Ricerche su Modelli* (Laboratory for Model Researches) at the Faculty of Engineering in Rome, was tested in the elastic range in order to check the theoretical hypothesis. ISMES was consulted for the fourth model, also reproducing 2 spans of the bridge. This model was made of micro-reinforced concrete at a scale of 1:10 and was 14 m long, with a thickness of only 3 cm (Figure 15.23). Musmeci described it as 'an egg shell gripped in a vice' [56]. The tests, which were carried out up to failure, led to a number of changes to the final design, and the model was also used to check the construction process before work started on site. Further details of Musmeci's model tests are covered in Chapter 20.

The Zarate-Brazo Largo rail and road bridge in Argentina, by Italian engineer Fabrizio de Miranda, has a central span of 330 m and a railway that is to one side of the bridge deck [57]. Despite using the best available tools for analysis and design, the engineers decided it was important to establish during the

Figure 15.24 1:33 scale model of the Zarate-Brazo Largo rail and road bridge, Argentina, for statical and dynamic tests at ISMES, c.1971. (Image: ISMES Historical Archive)

design stage, the reliability of design assumptions and calculations by testing a 1:33 structural model (Figure 15.24) under both statical and dynamic loading. Several design issues required further information that could only be obtained from suitable experimental work; these included the torsional stiffness of the highly indeterminate bridge deck, its interaction with the cables, their fatigue behaviour and the influence of secondary cables under the action of eccentric loads from the high-speed railway, and response of the bridge under impact loads. Although the model could not be constructed as geometrically faithful in every way, the bending and axial rigidity of structural members were scaled correctly. The discrepancies between the model test results and the analytical calculations for vertical deflections were found to be between 5 and 15%, and for cable stresses, between 5 and 20% which were considered sufficiently close to give confidence in the design methods proposed for this unique bridge [58].

15.6 From physical to virtual models

In 1974, Pier Luigi Nervi passed the torch to Guido Oberti, who became President of ISMES. This handover, due to Nervi's age (he was 82), marked a profound change in the Institute's policies, and there were significant changes in the world of structural engineering and modelling.

In 1969, after a long disease, Danusso passed away and just two years later, while ISMES was celebrating its twentieth anniversary, Enzo Lauletta died prematurely at the age of 44. At the same time, the change of course that had begun with the nationalization of the electricity industry in 1962 became more and more evident. During the 1970s, research on structural models were flanked by work in broader sectors, already mentioned above: geomechanical models; activity in the dynamic field; geophysical prospecting and Finite Element field [59]. The arrival of the Finite Element Method (FEM), in particular, was indicative of how things were changing. Anticipating modern model updating procedures and techniques, engineer Roberto Riccioni explained the introduction of FEM at ISMES in the following words:

'The importance and astonishing development of the finite element method in the past few years are substantially due to its adaptability to a wide range of different problems: two- and three-dimensional structural analysis, rock mechanics, seepage flow, fluid velocity fields, heat conduction, and so on. This is proved by the increasing number of international congresses and symposia on the theoretical development of this method and its applications. The already amply-demonstrated importance of this type of numerical analysis suggested in 1969 to ENEL and ISMES a close cooperation in this field for both research and practical purposes.

At ISMES this new activity, consisting in associating the conventional physical model with a mathematical one, finds a particularly ideal complement and an inexhaustible wealth of experimental comparisons.

The mathematical model should be looked upon from this angle since, because of its greater general applicability and lower cost, it can serve as a sure guide in constructing more accurate physical models as well as in equipping and more accurately locating the measuring instruments. Moreover, in an ideal feed-back, it can receive a series of test data indispensable for its development and refinement' [60].

In these early days of computing, physical and virtual models were often used together, with the output of one feeding into the other. This was done, for instance, during the design of the Sydney Opera House [61], and in the work of the Swiss engineer Heinz Hossdorf (1925-2006), who combined results from reduced scale models with computer analysis [62] (see Section 18.4). In Guido Oberti's mind, computer and FEM had a role similar to that of elastic models, which worked as a 'calculating machine', and they were not able to achieve what the 'true' structural models could: 'The superiority of the experimental method over the analytical one is unquestionable when the structure's behaviour has to be checked beyond the elastic range when determining its factor of safety' [63]. Up to the 1970s, in the field of thin shell analysis, for instance, computer programs were based on linear elastic behaviour [64].

Despite the validity of Oberti's observations, as a matter of fact, such distinction faded. Even if the new instruments and methods had their limitations, they were more adaptable and flexible in comparison with physical modelling. More than its actual effectiveness, physical modelling of structures suffered from its intrinsic laboriousness and slowness. Construction of these 'masterpieces in miniature' had to follow a unvarying course: the preparation of drawings, appropriately scaled and schematized; the choice of materials and the investigation of their mechanical properties; the construction of the model using specially-crafted formwork or moulds and its curing and finishing; the design and installation of the loading and measuring equipment; and, finally, conducting the actual tests which entailed a further phase of verification, with checks for errors, and often the repetition of the whole procedure. At ISMES, the conduct of these operations had attained extraordinary levels of excellence, thanks to the expertise of its engineers and the progress of electronics, but some limitations appeared virtually unsurmountable. Apart from the costs – dependent on the size of the model, the materials used, the geometric complexity and the price of labour – the physicality of the model entailed some laborious procedures when changing the geometric and constructional parameters defined at the conception stage, particularly as they often lagged behind the corrections made during the design phases.

In some sectors the physical model, much more expensive and less versatile than the virtual one, had already had its day. As Enzo Lauletta had stated in around 1970: 'models are over, killed by computers' [65]. However, looking at the subsequent development of modelling techniques, Lauletta's statement has been shown to be not quite true, as demonstrated in other chapters in this volume. Nevertheless, at that stage, the arrival of digital models, as well as the related software and calculation methods, opened up new horizons, provoking a profound shift in this particular sector of engineering and the end of an extraordinary era for structural modelling.

15.7 Concluding remarks

The art of building, with all the complexity of the formal and structural solutions that extraordinary designers such as Nervi and Torroja developed in what was an unrepeatable period of architectural engineering was, during that period and in that particular phase of the development of scientific method, largely indebted to the art of experimentation. Perhaps justifying Oberti's frequent citation, with unconcealed satisfaction, of the saying attributed by Vasari to Michelangelo: 'The most blessed monies that are spent by those who would build are on models'. A debt towards experimentation and a passionate, creative confidence in its almost magical qualities that the words of Nervi express so well at the end of the chapter *Scienza o arte del costruire?* ('Building: Art or Science?') in the book of the same name: 'From these brief considerations we can see what progress can be brought to the whole art of building by the broad development of experimental research ... Experimental research opens up the confines of construction, almost without limits. Bold structural inventions which only ten years ago, in the face of the inability of theoretical models to obtain a sufficiently accurate quantitative verification, would have been impossible to execute, can now, or rather in the future, be faced on the basis of the results of experimental research' [66].

In the fertile and long-lasting cooperation between Oberti and Nervi, experimentation in itself became, in a very exceptional way, an extremely refined conjugation of art and science: in the conception and construction of models and selection of proper materials, in applying the loads and all external actions, in obtaining and interpreting the results. Keen attention is paid in all these steps, on the one hand, to the significant constraints imposed by the theory of similitude and the related scale effects and, on the other hand, to the significance of the outputs in terms of contributions to the conceptual and assessment stages of the specific structural project, finding in each case the best and most intelligent solutions in the frame of an intense dialectic interaction between the art and science of testing and the art and science of building.

The first of the authors of this chapter, who had the privilege during his education to be instructed by Guido Oberti, could directly appreciate such a wide vision of experimentation, and his passion for these very special conjugations between art and science.

Further reading

Many further references on physical models not included in this chapter can be found in several key works by the authors, especially:

Chiorino, M.A., Sabia D. & Bruno, L. (2003) Structural Models: Historical Notes and New Frontiers. *Proceedings International Seminar Eduardo Torroja: from the Philosophy of Structures to the Art and Science of Building, Politecnico di Torino, November 2000* (eds F. Levi, M.A. Chiorino & C. Bertolini Cestari), 120-159. Milano: FrancoAngeli.

Chiorino, M.A. (2010) Reduced scale mechanical models in 20th century structural architecture. The case study of Pier Luigi Nervi. In: *Geometría y proporción en las estructuras, Geometry and proportion in structures. Simposio en homenaje a Ricardo Aroca* (ed. Santiago Huerta), 209-225. Madrid: Universidad Politécnica de Madrid.

Chiorino, M.A. (2010) Experimentation in the Work of Pier Luigi Nervi. In: *Pier Luigi Nervi. Architecture as Challenge* (eds Carlo Olmo & Cristiana Chiorino), 61-83. Cinisello Balsamo (MI): Silvana Editoriale.

Neri, G. (2010) En miniature: Die Modelle des I.S.M.E.S. in Bergamo. *Bauwelt* 101 (May): 20-27.

Neri, G. (2013) Pier Luigi Nervi's Reduced Scale Models for HP Shells. *Journal of the IASS 176-177* (September): 149-158.

Neri, G. (2014) *Capolavori in miniatura. Pier Luigi Nervi e la modellazione strutturale.* Mendrisio: Mendrisio Academy Press / Cinisello Balsamo: Silvana Editoriale. https://doc.rero.ch/record/288862/files/Capolavori_in_miniatura.pdf (1/10/2019).

Neri, G. (2017) Pier Luigi Nervi al Politecnico di Milano. I modelli delle aviorimesse in cemento armato. In: *Pier Luigi Nervi: Il modello come strumento di progetto e costruzione* (ed. G. Barazzetta), 25-33. Macerata: Quodlibet.

References

1 Neri, G. (2014) *Capolavori in miniatura. Pier Luigi Nervi e la modellazione strutturale.* Mendrisio: Mendrisio Academy Press / Cinisello Balsamo: Silvana Editoriale.

2 ISMES (1961) *ISMES 1951-1961.* Bergamo: Istituto Sperimentale Modelli e Strutture. x.

3 Bocca, G. (1993) *ISMES, quarant'anni.* Bergamo: ISMES. 4-5.

4 Oberti, G. (1956) Development of aseismic design and construction in Italy by means of research on large model test. *I Quaderni ISMES* 6: 1-11. 3-4.

5 ISMES (1961) *ISMES 1951-1961.* Bergamo: Istituto Sperimentale Modelli e Strutture. xii.
6 ISMES (1961) *ISMES 1951-1961.* Bergamo: Istituto Sperimentale Modelli e Strutture.
7 Oberti, G. (1958) The Aid of Models in the Study of Static Behaviour of Large Dams in Concrete Built in Successive Stages. *I Quaderni ISMES* 10: 1-25. 2.
8 ISMES (1961) *ISMES 1951-1961.* Bergamo: Istituto Sperimentale Modelli e Strutture. 28-31.
9 Oberti, G. (1980) L'evoluzione dei modelli fisici impiegati nello studio delle strutture in calcestruzzo semplice ed armato. *L'industria italiana del Cemento* (9): 659-690. 679.
10 Fumagalli, E. *ISMES. Origins and development.* http://www.asim.it/ismes/Ismes_Uk/1.htm (1/10/2019).
11 Oberti, G. (1981) Structural design and testing, by means of models, of some special constructions (using ferro-cement). *Preprint of papers in Proceedings of International Symposium on Ferrocement held at ISMES, Bergamo 22-24 July 1981; 2nd Session. RILEM-ISMES* (eds G. Oberti & S.P. Shah), Paper 2.3. 139.
12 Oberti, G. (1981) Structural design and testing, by means of models, of some special constructions (using ferro-cement). *Preprint of papers in Proceedings of International Symposium on Ferrocement held at ISMES, Bergamo 22-24 July 1981; 2nd Session. RILEM-ISMES* (eds G. Oberti & S.P. Shah), Paper 2.3. 138.
13 Oberti, G., Lauletta, E. & Sanmartino, R. (1964) Esperienze termiche sul modello della diga del Vajont. In: *Memorias de las IX Jornadas Sudamericanas de Ingenieria Estructural – Montevideo 1964, 20-25 April 1964.* Publicación n. 25: 491-494. Montevideo: Instituto de Estatica.
14 ISMES (1961) *ISMES 1951-1961.* Bergamo: Istituto Sperimentale Modelli e Strutture. 78.
15 ISMES (1961) *ISMES 1951-1961.* Bergamo: Istituto Sperimentale Modelli e Strutture. 88.
16 Oberti, G. (1981) Structural design and testing, by means of models, of some special constructions (using ferro-cement). *Preprint of papers in Proceedings of International Symposium on Ferrocement held at ISMES, Bergamo 22-24 July 1981; 2nd Session. RILEM-ISMES* (eds G. Oberti & S.P. Shah), Paper 2.3. 143.
17 Oberti, G. (1956) La sperimentazione su modelli strutturali in Italia. In: *I modelli nella tecnica. Atti del Convegno di Venezia, 1-4 ottobre 1955, vol. I.* 35-55. Roma: Accademia Nazionale dei Lincei.
18 Oberti, G. (1958) The aid of models in the study of static behaviour of large dams in concrete built in successive stages. *I Quaderni ISMES* 10: 1-25.
19 Oberti, G. & Lauletta (1962) Dynamic tests on models of structures. (Paper presented at the 2nd World Conference of Earthquake Engineering, Tokyo, July 1960). *I Quaderni ISMES* 19: 1-14.
20 Paolo Monti. Licensed under the Creative Commons Attribution-Share Alike 4.0 International License. https://commons.wikimedia.org/wiki/File:Paolo_Monti_-_Servizio_fotografico_-_BEIC_6338549.jpg (1/10/2019).

21 ISMES (1961) *ISMES 1951-1961.* Bergamo: Istituto Sperimentale Modelli e Strutture. 60.

22 Rowe, R.A. (1957) *Instituto Sperimentale Modelli e Strutture, Bergamo, Italy.* Technical Report 351. London: Cement & Concrete Association.

23 Oberti, G. (1980) La modellazione strutturale. In: *Pier Luigi Nervi e la sua opera, incontro di studio organizzato dal Comitato Premio Ingersoll Rand Italia* (Ed. Silvio Ceccato), 11. Milano: Boroni.

24 Albertomos. Licensed under the Creative Commons Attribution-Share Alike 4.0 International License. https://commons.wikimedia.org/wiki/File:MI-Milano-1965-grattacielo-Galfa.jpg (1/10/2019).

25 Paolo Monti. Licensed under the Creative Commons Attribution-Share Alike 4.0 International License. https://commons.wikimedia.org/wiki/File:Paolo_Monti_-_Servizio_fotografico_-_BEIC_6338811.jpg (1/10/2019).

26 Lauletta, E. (1965) Theoretical Considerations and Experimental Research on the Behaviour of Tall Buildings during Earthquakes. *I Quaderni ISMES* 28: 3-15.

27 Chiorino M.A. (2010) Experimentation in the work of Pier Luigi Nervi. In: *Pier Luigi Nervi. Architecture as Challenge* (eds C. Olmo & C. Chiorino), 61-83. Cinisello Balsamo (MI): Silvana Editoriale.

28 Accademia Nazionale dei Lincei (1956) *I modelli nella Tecnica, Atti del Convegno di Venezia del 1-4 ottobre 1955* (2 Vols). Roma: Accademia Nazionale dei Lincei. Vol.1: 1.

29 Accademia Nazionale dei Lincei (1956) *I modelli nella Tecnica, Atti del Convegno di Venezia del 1-4 ottobre 1955* (2 Vols). Roma: Accademia Nazionale dei Lincei.

30 Accademia Nazionale dei Lincei (1956) *I modelli nella Tecnica, Atti del Convegno di Venezia del 1-4 ottobre 1955* (2 Vols). Roma: Accademia Nazionale dei Lincei. Vol.1: 2.

31 ISMES (1961) *ISMES 1951-1961.* Bergamo: Istituto Sperimentale Modelli e Strutture. vii.

32 ISMES (1961) *ISMES 1951-1961.* Bergamo: Istituto Sperimentale Modelli e Strutture. ix.

33 ISMES (1971) *ISMES 1961-1971.* Bergamo: Istituto Sperimentale Modelli e Strutture. 30.

34 ISMES (1971) *ISMES 1961-1971.* Bergamo: Istituto Sperimentale Modelli e Strutture. 33.

35 ISMES (1971) *ISMES 1961-1971.* Bergamo: Istituto Sperimentale Modelli e Strutture. 39-48.

36 Bocca, G. (1993) *ISMES, quarant'anni.* Bergamo: ISMES. 36.

37 Gargiani, R. & Bologna, A. (2016) *The Rhetoric of Pier Luigi Nervi. Concrete and Ferrocement Forms.* Lausanne: EPFL Press.

38 Greco, C. (2008) Pier Luigi Nervi. *Dai primi brevetti al Palazzo delle Esposizioni di Torino 1917-1948.* Luzern: Quart.

39 Neri, G. (ed.) (2014) *Pier Luigi Nervi, Ingegneria, architettura, costruzione. Scritti scelti 1922-1971.* Novara: Città Studi Edizioni.

40 Neri, G. (2018) The Design of Saint Mary's Cathedral, San Francisco (1963-71) In: *Modernism and American mid-century sacred architecture* (ed. Anat Geva), 93-112. London: Routledge.
41 Olmo, C. & Chiorino, C. (2010) *Pier Luigi Nervi. Architecture as Challenge.* Cinisello Balsamo (MI): Silvana Editoriale.
42 Lauletta, E. (1955) Communication sur deux appareils de mesure pour modèles statiques de barrages. *I Quaderni ISMES* 4: 61-74.
43 Lauletta, E. (1965) Theoretical Considerations and Experimental Research on the Behaviour of Tall Buildings during Earthquakes. *I Quaderni ISMES* 28: 3-15.
44 Lauletta, E. (1964) Thermoelastic test on arch dam models. In: *Proceedings of Symposium on Concrete Dam Models, 14-19 October 1963.* 4.1-4.16. Lisbon: LNEC.
45 Oberti, G. & Lauletta (1962) Dynamic tests on models of structures. (Paper presented at the 2nd World Conference of Earthquake Engineering, Tokyo, July 1960). *I Quaderni ISMES* 19: 1-14.
46 Locatelli, P. (ed.) (1978) *La scienza e lo spirito negli scritti di Arturo Danusso.* Brescia: Morcelliana.
47 Castoldi, A. (1973) New techniques of model investigation of the seismic behaviour of large structures. *I Quaderni ISMES* 56: 1-22.
48 Castoldi, A. & Casirate, M. (1976) *Experimental Techniques for the Dynamic Analysis of Complex Structures.* Report No. 74. Bergamo: Istituto Sperimentale Modelli e Strutture.
49 Nervi, P.L. (1970) La ricerca sperimentale nel campo costruttivo. In: *Rapporto sulla ricerca in Italia atti del Convegno di studio I problemi e lo stato della ricerca in Italia svoltosi il 25-27 maggio in Palazzo Torlonia a Roma,* 153 ff. Roma: Istituto Accademico di Roma.
50 Chiorino, M.A. (2010) Reduced scale mechanical models in 20th century structural architecture. The case study of Pier Luigi Nervi. In: *Geometría y proporción en las estructuras, Geometry and proportion in structures. Simposio en homenaje a Ricardo Aroca* (ed. Santiago Huerta), 209-225. Madrid: Universidad Politécnica de Madrid.
51 Tsuboi & Nasukawa (1966) Tsuboi, Y. & Nasukawa, R. (1966) Hyperbolic paraboloidal shell structure for Tokyo Roman Catholic Cathedral. *Bulletin of IASS* 28: 37-52.
52 Neri, G. (2013) Pier Luigi Nervi's reduced scale models for HP shells. *Journal of the IASS* 54 (176-177, September): 149-158.
53 Oberti, G. (1981) Structural design and testing, by means of models, of some special constructions (using ferro-cement). *Preprint of papers in Proceedings of International Symposium on Ferrocement held at ISMES, Bergamo 22-24 July 1981; 2nd Session. RILEM-ISMES* (eds G. Oberti & S.P. Shah), Paper 2.3.
54 Morandi, R. (1980) Strutture strallate in cemento armato / Concrete cable-stayed structures. *L'Industria italiana del cemento* 10: 887-909.
55 Musmeci, S. (1977) Ponte sul Basento a Potenza. *L'Industria italiana del cemento* XLVII (2, February): 77-98.
56 Musmeci, S. (1977) Ponte sul Basento a Potenza. *L'Industria italiana del cemento* XLVII (2, February): 77-98. 86.

57 Harris, H.G. & Sabnis, G.M. (1999) *Structural modeling and experimental techniques.* 2nd Ed. Boca Raton: CRC Press. Sections 10.2.2.1 & 11.5.1.2.
58 Baglietto, E., Casirati, M., Castoldi, A., Demiranda, F. & Sammartino, R. (1976). *Mathematical and Structural Models of Zarate-Brazo Largo Bridges.* Report No. 85. Bergamo: Instituto Sperimentale Modelli e Strutture.
59 ISMES (1971) *ISMES 1961-1971.* Bergamo: Istituto Sperimentale Modelli e Strutture. 53-58.
60 ISMES (1971) *ISMES 1961-1971.* Bergamo: Istituto Sperimentale Modelli e Strutture. 53.
61 Taffs, D. (2006) Computers and the Opera House: pioneering a new technology. In: *Building a Masterpiece: The Sydney Opera House* (ed. A. Watson), 84-101. Sydney: Powerhouse Publishing.
62 Hossdorf, H. (1974) *Model Analysis of Structures.* (trans. C. van Amerongen). New York: Van Nostrand Reinhold.
63 Oberti, G. & Castoldi (1973) Oberti, G. & Castoldi, A. (1973) New trends in model research on large structures. (Paper presented at the 9th Congress of International Association for Bridge and Structural Engineering (IABSE), Amsterdam, May 1972). *I Quaderni ISMES* 53: 1-11.
64 Scordelis, A.C. (1979-80) General analysis using discretizing methods. *Bulletin of IASS* 71/72 (December): 67-72. 70.
65 Bocca, G. (1993) *ISMES, quarant'anni.* Bergamo: ISMES. 35.
66 Nervi, P.L. (1945) *Scienza o arte del costruire? Caratteristiche e possibilità del cemento armato.* Roma: Edizioni della Bussola. New edition (2014). Novara: Città Studi Edizioni. 33.

16

Eduardo Torroja and his use of models from 1939

Joaquín Atuña

16.1 Introduction

In 1939, after the end of the civil war, Torroja's activity diversified. In addition to continuing with the activity of his consulting engineering office, he assumed other responsibilities. One was to serve as director of the *Instituto de la Construcción y la Edificación*, the private organization in whose foundation he had participated in 1934; another was to join the *Escuela Especial de Ingenieros de Caminos* as a lecturer. The *Instituto de la Construcción y la Edificación* was integrated into the structure of the *Consejo Superior de Investigaciones Científicas*, within which the country's research activity was organized. Later it merged with the *Instituto del Cemento* and its field of activities expanded to become the *Instituto Técnico de la Construcción y el Cemento* (Technical Institute of Construction and Cement). While serving as director of this organisation, Torroja commissioned several investigations and tests, some of them on reduced-scale models that we will see later. Already part of the faculty of the *Escuela de Ingenieros* (School of Engineers), he was appointed, in November 1940, director of the *Laboratorio Central de Ensayo de Materiales de Construcción* (LCEMC, Central Laboratory for Testing Construction Materials), a position from which he promoted the reorganization of the centre and where he developed his activities in testing reduced models during the following 20 years.

16.2 The Central Laboratory for Testing Construction Materials (LCEMC)

The Laboratory associated to the School of Civil Engineering had been in operation since 1899 and its purpose was to study the properties of the materials, mainly their strength, responding to requests from both the state and private companies and individuals. According to the founding regulations of 1898, the sole purpose of the Laboratory was to study the properties of engineering materials, mainly their strength, but changes in 1926 allowed it to include a collaboration between teaching and research work.

Physical Models: Their historical and current use in civil and building engineering design,
First Edition. Edited by Bill Addis.

Although Torroja carried out his activities in many different centres, he always kept his Consulting Engineering office in operation and many of the works he carried out were registered in the archive of the office itself. This is the case of the first tasks he carried out as director of the Laboratory that have been registered as works of his office. As already indicated in Chapter 11, the *Archivo Eduardo Torroja* (AET, Eduardo Torroja Archive) [1] which keeps the files of Torroja's Consulting Engineering office, is located in the *Centro de Estudios Históricos de Obras Públicas y Urbanismo* (CEHOPU) [2]. In one of the entries of the AET [3] there are several files related to the Laboratory. In the first of these, Torroja proposes a new organization in which several laboratories are created, establishes the personnel required and proposes the construction of a new building [4]. One of the new sections created was dedicated to the testing of reduced-scale models [5]. The new building was inaugurated in 1943.

In addition to having adequate space to carry out the laboratory's new activities, it needed a team of specialists in a number of fields, as well as guaranteed financing. With regard to model testing, in addition to having technicians capable of evaluating the results, it was necessary to have specialists who could build the models and their instrumentation and carry out their loading and the data collection.

16.2.1 The new organization and staff

In the existing Laboratory, the position of head of the Laboratory had been considered a temporary occupation, a preliminary step in developing a professional career. It was a position with a small salary and poor career prospects. Torroja's intention was to create a large facility, with support from industry and the state, which would initiate its own problems to work on, and for which they could earn part of the income needed for its operation. This would create a working group of well-paid managers, practitioners and support staff who could be offered a career in which to take responsibility and recognition. The annual budget for the new laboratory was 500,000 pesetas (well over €1 million today).

Among the people who were part of that initial team in the mid-1940s were Leonardo Villena (1917-2015), Carlos Benito (1915-2001), and Florencio del Pozo (1924-1993). Villena was a graduate in physical sciences and specialist in photoelastic stress analysis. He later published a manual on photoelasticity, in 1943 [6], directed the construction of the new equipment for photoelastic stress analysis for the Laboratory [7] and wrote several publications on the projects carried out using this technique in the Laboratory [8, 9, 10, 11]. Benito was the chief engineer of the model-testing laboratory and director of the Laboratory from 1961; the majority of the Laboratory's files are signed by him. Del Pozo, also a civil engineer, became secretary and director of the Laboratory.

16.2.2 The objectives of the new laboratory

The main objective of the new laboratory was to serve external clients, both private individuals and the state administration itself. In the new organization, however, the variety of materials studied was increased, as well as the material properties that could be tested. Several specific laboratories were created, and new test procedures were incorporated into the work of the laboratory. With the new means available, it was possible to study complete structures in the model laboratory and not be limited only to the materials – in a new materials testing laboratory it was possible to study works in different phases of their construction. Thus, in addition to the initial objective of describing the properties of the materials, there was the possibility of studying constructions of all kinds, either in the design phase or at any time during the construction process.

In addition to external assignments, Torroja also commissioned several research projects in his own laboratory in a range of different fields, including 'Photoelastic methods' [12], 'Cross-section of reinforced concrete aqueducts' [13], 'Investigations on the strength ratio between cylindrical and cube specimens' [14], 'Study to establish a method for determining stresses in pieces of concrete' [15] and 'Reduced-scale model of ENASA' [16].

Torroja conceived the new Central Laboratory as an organization that could have economic independence, since it could be financed with the work it carried out, but in which it could also commission research projects, since it had the qualified and experienced personnel to carry them out – technicians and craftsmen capable of building reduced-scale models with the necessary precision and with the necessary measuring equipment to complete the data collection.

16.2.3 Types of tests performed

In most of the model tests carried out in the LCEMC, the objective was to check an already-defined design for a project. Importantly, although both elastic and collapse tests were performed, in no case did the results of the tests replace the theoretical analysis of the structure. Other trials could be counted as exploratory or qualitative where the aim was to observe the possibilities of a new material; for example, tests with gelatines to be used as a photoelastic material, testing parts assembled with photoelastic gelatine, 'Isostatic studies using brittle lacquers', and the 'Study of flat models' in which a composite of gelatine and sand was studied. Tests were also used to explore the possibilities of a new test procedure, such as using photoelastic stress analysis to study flat grillages, and 'Determining the sum of principal stresses by electrical analogy of the points of a flow net. Other tests were carried out to confirm the behaviour of a new type of structure, in which a shell was formed with several prefabricated elements, is studied [17] (see Section 16.4.4).

While Torroja was director of the Laboratory, around 11,000 projects were carried out. The great majority consisted of tests of the strength of specimens of diverse materials, while only a small part consisted in tests on reduced-scale models. The aim of these tests was usually to confirm the strength and stiffness of a design that had already been carried out. However, in some cases, the experiments were carried out to test a new calculation method [16] while in others, experiments were undertaken to study the methods and techniques by which the testing of scale models was carried out [17].

16.2.4 Works carried out

When Torroja began to direct the LCEMC, Spain was impoverished by three years of war, and little means were available. On the other hand, modelling research was a new discipline, which in many cases forced the design of new procedures and materials to be used, or new applications of existing ones. In the following sections we review the different types of tests that were carried out in the model laboratory up to 1961 when Torroja died, including the techniques used and various innovations. The first results of reduced-scale model tests date back to 1945 and were in the field of photoelastic stress analysis.

16.3 Photoelastic stress analysis

In the LCEMC there had been a photoelasticity installation since 1932 with an optical field of 8 cm diameter built according to the scheme proposed by the French engineer Mesnager (see Section 12.2.2). For the new model testing department, the company ICON (*Investigaciones de la Construcción S. A.*), under the direction of Leonardo Villena, built a new polariscope with a simpler operation than in existing installations and with an optical field of 35 cm diameter [7]. In addition to having a new facility that simplified the analysis procedure, various materials were studied to make the transparent models, in order to simplify the tests and increase their field of action. For example, various reports cover studies on photoelastic research using gelatine models, and a 'Study of pieces assembled in photoelastic gelatine' focusing on mixtures with different gelatine contents between 20.7% 'and 23.5%', and models made with clarified gelatine, reinforced with vulcanized rubber [18].

16.3.1 Three-dimensional problems: grillages

One of the first investigations promoted by Torroja was the photoelastic study of grillages [19]. This project demonstrated the possibility of using photoelasticity, a technique developed to study elasticity in plane structures, to address a three-dimensional problem. The objective was achieved by constructing a grillage in which the joists in the direction perpendicular to that of observation were slightly staggered and placed at different heights. The transverse joists, in the direction of the observation, were arranged with a slight inclination. In this way, the behaviour of the two beams arranged in one direction could be measured using a single observation. The model was made with artificial resin. Two configurations were tested, one in which the beams overlapped and the other in which both beams were in the same plane. In the second case, the influence of the torsional stiffness of the beams could also be evaluated.

The results of the trials differed slightly from the analytical results, but considering the simplifications required to perform the analysis, it was concluded that the experimental results would give a better approximation to reality. On the other hand, the photoelastic method could be used to study complex grillages to which the simplified analytical procedure could not be applied. Finally, he mentioned the possibility of carrying out new experiments to study problems of grillages in order to take into account the contribution of the upper slab that joins the ribs of the grillage, since it is possible to build monolithic ribbed slabs. With the proposed system it would be possible to obtain the tensile or compressive stresses in the upper face of the slab and the distribution of stresses in the longitudinal beams.

This work succeeded in demonstrating that with the available photoelasticity apparatus it was possible to study complex three-dimensional grillage arrangements, for which a numerical analysis was not possible. However, no practical application of this procedure has been found in Torroja's archives.

16.3.2 Three-dimensional problems: stresses in solid materials

The intention to extend the scope of photoelasticity to three-dimensional problems led to experiments using the technique of stress freezing. The research focused on finding a material that could be deformed by applying a load to it when hot, cooling it to freeze the stress, and then machining it to obtain cross-sections without altering its internal state of stress (see Section 12.3.1). Several materials were studied, finding that the mixture formed by glycerine and water at 50% was the one that presented the most suitable properties. This material was used to study a problem whose analytical solution was solved: the aim was to determine the maximum tangential stress values on the axis of a semi-defined elastic solid uniformly loaded on a circular area of its surface. The results obtained experimentally differed by less than 4% from the theoretical results [20].

In addition to its own research, photoelasticity was used in several of the Laboratory's commissioned projects. In 1946, photoelastic studies using reduced-scale models were made on stiffeners for thin-web beams, tunnels, a buttress dam and the inclined faces of a dam [21]. In 1954 the stresses in a flat portico were studied using a birefringent material in which one of the tests included the horizontal prestressing of the lintel [22]. In 1960, two models were made for the 'Photoelastic test of an underground power station' and various pipelines. One model, made of Plexiglas, was used to study the stresses at the edges of a flat gantry subjected to five load combinations; the other model, to test the effects of boring holes for the pipes, was made of Decorite [23].

16.4 Tests on reduced-scale physical models

During Torroja's twenty years in charge of the Laboratory, almost eleven thousand tests of different types were carried out, of which about sixty were on reduced-scale models. Considering that the minimum time for a model test was three months, the activity of the model laboratory was constant during the whole of that period. In this section we describe, in chronological order, some of the models tested. In each case, both the model itself and the technique used are described if they presented any novelty, either in the materials used or in the procedure followed.

16.4.1 Initial investigations

Between 1946 and 1949 several files were opened for tests on model structures, including a 'Study of a hyperbolic paraboloid shell' [24], a 'Study of a viaduct over the river Miño' (1947) [25] and 'Tests on a reinforced-mortar vault' (1949) [26]. A file dated 1952 reported tests to find a model material with equal strengths in tension and compression, greater than 200 gr/cm^2 (0.02 N/mm^2) with linear behaviour up to fracture, and a consistency dry enough to be applied with a spatula [27]. On the one hand, they were looking for a material with which to build models to test both in the elastic range and up to collapse. For the former, it was a matter of finding a material with low Young's modulus that was inexpensive and could be easily shaped. In order to make models for fracture, the right ratio of aggregates and mortar was needed. On the other hand, it was necessary to make appropriate measuring devices. In the case of elastic models, in addition to accuracy, the devices had to be light enough not to distort the measurements when attached to the model.

A material called *Litargel* was developed in the laboratory to meet these specifications. It consisted of a mixture of litharge (lead monoxide), gelatine, glycerine and water. It had several properties that made it useful for elastic testing. It melted at a temperature of 50° C, so that it could be injected into a mould with the shape of the model. It had a linear-elastic behaviour over a range of stresses up to five times those that the model would experience during the test, which guaranteed its perfect elasticity. On the other hand, its Young's modulus was low enough that deformations could be measured with sufficient precision. The Young's modulus was temperature sensitive and so, to avoid errors in interpreting the results, a procedure was devised in which a number of test pieces were made at the same time as constructing the model. These specimens were tested under compressive load at the same time (and temperature) as the model, so that the exact value of Young's modulus was known at the time data was collected in the test.

To measure strains, some very precise and lightweight instruments were created, including an electrical resistance extensometer which was much lighter and more sensitive than a normal extensometer. It was usual to fix these extensometers to a model by means of a strip of paper, but for a low-modulus model material, it was found that the high stiffness of the paper affected the measured results. And so two of the extensometers were glued to the beam of a small frame whose legs rested on the surface of the model. Each extensometer, with a gauge length of 30 mm, weighed only nine grams. Thus the weight of the gauges and their stiffness were negligibly small compared to the loads and the stiffness of the model [28].

The study of the hyperbolic paraboloid shell was closed in February 1948 without results, because no deformations could be measured due to irregularities in the shape of the model. Nevertheless, it was noted that the tests helped to improve the testing technique. Carlos Benito published a description of some of the works with reduced models that were carried out in the laboratory and in which the experiment and results from these initial works were condensed and which, once established, were used regularly in subsequent tests on different models [28]. Among the tests mentioned by Benito is that of the shell roof for the church of Xerallo in the Pyrenees [29] in which he describes the frame and

(a) (b)

Figure 16.1 Model of the roof of the church of Xerallo (a) Model mounted in support structure that allowed it to be turned 180°. (b) Model with strain gauges in position. (Images: LCEMC)

bell mounting they devised to support the model, made of litargel, to enable it to be rotated through 180° (Figure 16.1a). The photograph showing the lightweight strain gauges in position (Figure 16.1b) also shows a curved line on the face of the model shell. This line was produced using the brittle-lacquer technique. First the shell was coated with a lacquer and, once dry, it was turned over so that it was then subject to the action of its own self-weight. When deformed, the lacquer cracked, producing isostatic lines that were made visible using powdered lead sulphide which, through the cracks, combined with the lead oxide of the litargel to form black lead sulphide.

16.4.2 Innovation in shell structures

As an alternative to the construction of continuous shells, Torroja began to study the possibility of replacing the continuous shell with a mesh of steel bars located in the plane of the shell [30]. In the project that he carried out in his Technical Office for a building for the National Automotive Company (ENASA) several solutions were studied [31]. The ensemble consisted of a series of sections of cylindrical shells arranged parallel to the generatrices for which various forms of section – an elliptical, cycloidal or cylindrical arch – were studied. In all cases a part of the concrete vault was replaced by a triangular lattice structure, also made of reinforced concrete.

One of the options consisted of a cylindrical shell of struts made with rolled-steel profiles, a novel solution for which there was no calibrated method of analysis. Two approaches were proposed – a model test and, in addition, a new method of analysis. The test would serve both to verify the proposed dimensions and to validate the analysis procedure. The 9.0 metre long model was built and tested in the laboratory and consisted of a triangulated cylindrical shell formed by double-T profiles, built to scale of 1:5 [32]. The load assumed for the structure was 1.15 kN/m^2, and was the same for the model. The loading procedure consisted of suspending counterweights from the model and floating them in water tanks; the load was applied by lowering the water level in the tank to allow careful control of the process. In order to consider the asymmetrical action of snow, 0.65 kN/m^2 was imposed on one half of the shell (Figures 16.2 and 16.3).

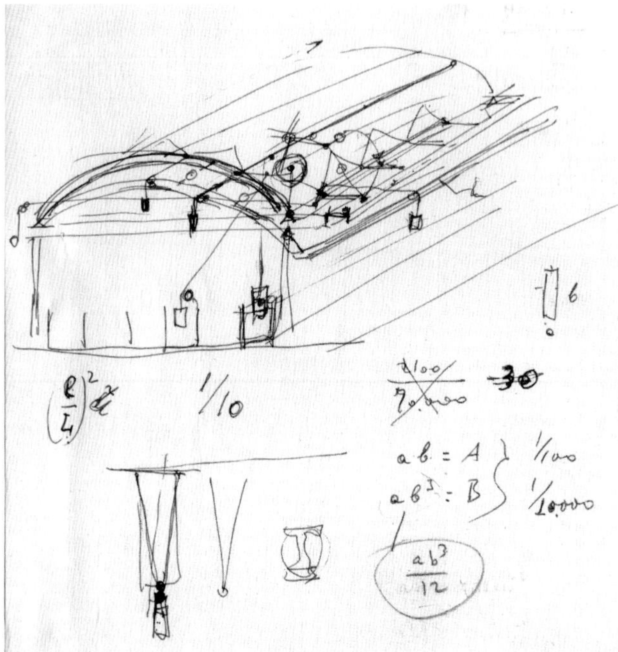

Figure 16.2 Original sketch of the apparatus for testing the cylindrical shell proposed for the ENASA project. (Source: AET, CEDEX-CEHOPU [16])

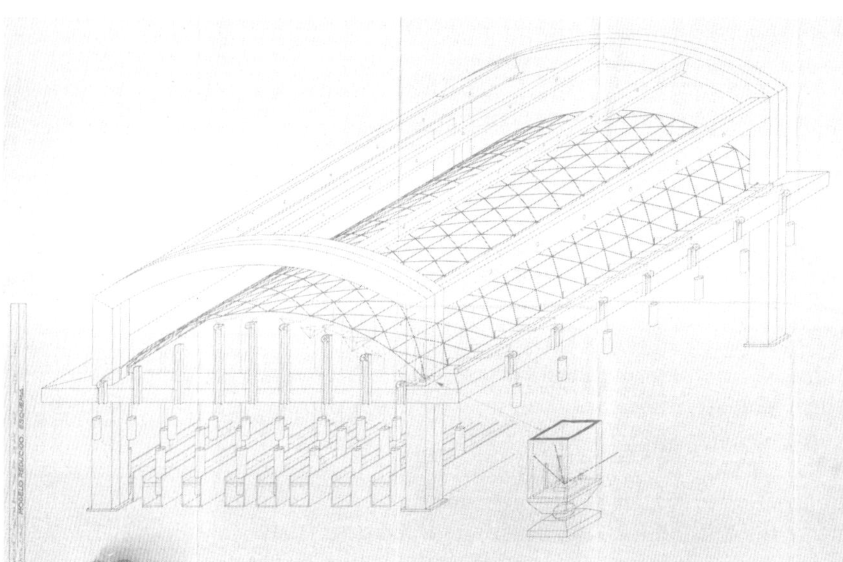

Figure 16.3 Drawing of the apparatus for testing the 1:5 scale model (9 m long) of a cylindrical shell proposed for the ENASA project. (Source: AET, CEDEX-CEHOPU [16])

The roof studied for ENASA was never built, but the proposed analysis procedure was published in a monograph of the *Instituto Técnico de la Construcción y del Cemento* [33, 34] and, several years later, in 1955, this type of structure was used for the roof of the test building in the new headquarters of the *Instituto Técnico de la Construcción* [35].

16.4.3 The roof of the church of Saints Felix and Régula, Zurich

With the experience of these previous experiments, work began in March 1949 on model tests commissioned by engineer E. Schubiger for the design and construction of a roof for the church of Saints Felix and Régula in Zurich [36]. The elliptical dome had a plan in the form of a 'Cassini oval' (major axis, 24 m; minor axis 17 m) with a rise of only 1.6 m. Given this very low rise and low curvature, a key aim of the model tests was to study the possibility of failure by buckling. The tie rod for the shell was incorporated in the perimeter beam which is supported on 32 point-supports arranged symmetrically. Three identical models of the shell vault were constructed at a scale of 1:10. A total of eight loading tests were carried out, as well as studies of the impact of the degree of pre-tensioning in the perimeter tie rod. The loading was carefully controlled by using steel cylinders floating in a tank of water, whose level was lowered to apply the load (Figure 16.4).

The 1:10 scale model was 2.4 m by 1.7 m and had a thickness of 8 mm. It was made with reinforced mortar in which the reinforcement anticipated in the final structure was reproduced using 0.7 mm diameter steel wires. In the perimeter, the shell increased in thickness and was reinforced with 7 wires of 2 mm diameter. The tension rod in the perimeter of the model was an 8 mm diameter steel tie, which was fixed to the shell at the points where it rested on the 32 supports. The connection was made by means of a device that was adjustable in the radial direction to allow the tension of the tie rod to be regulated. Where the circumferential reinforcement of the shell was connected to the vertical support it incorporated both the device for tensioning the tie rod and the means for allowing the inclination of the support to change (Figure 16.5).

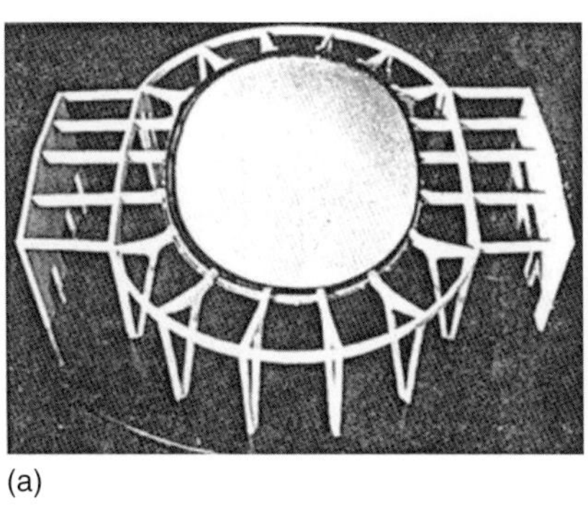
(a)

(b)

Figure 16.4 Model for the roof of the church of Saints Felix and Régula, Zurich, 1949. (a) Architectural model conserved at the LCEMC (b) 1:10 scale model showing clock gauges used to measure the deflection of the shell. Steel loading cylinders, floating in a water tank, are visible beneath the model. (Images: LCEMC)

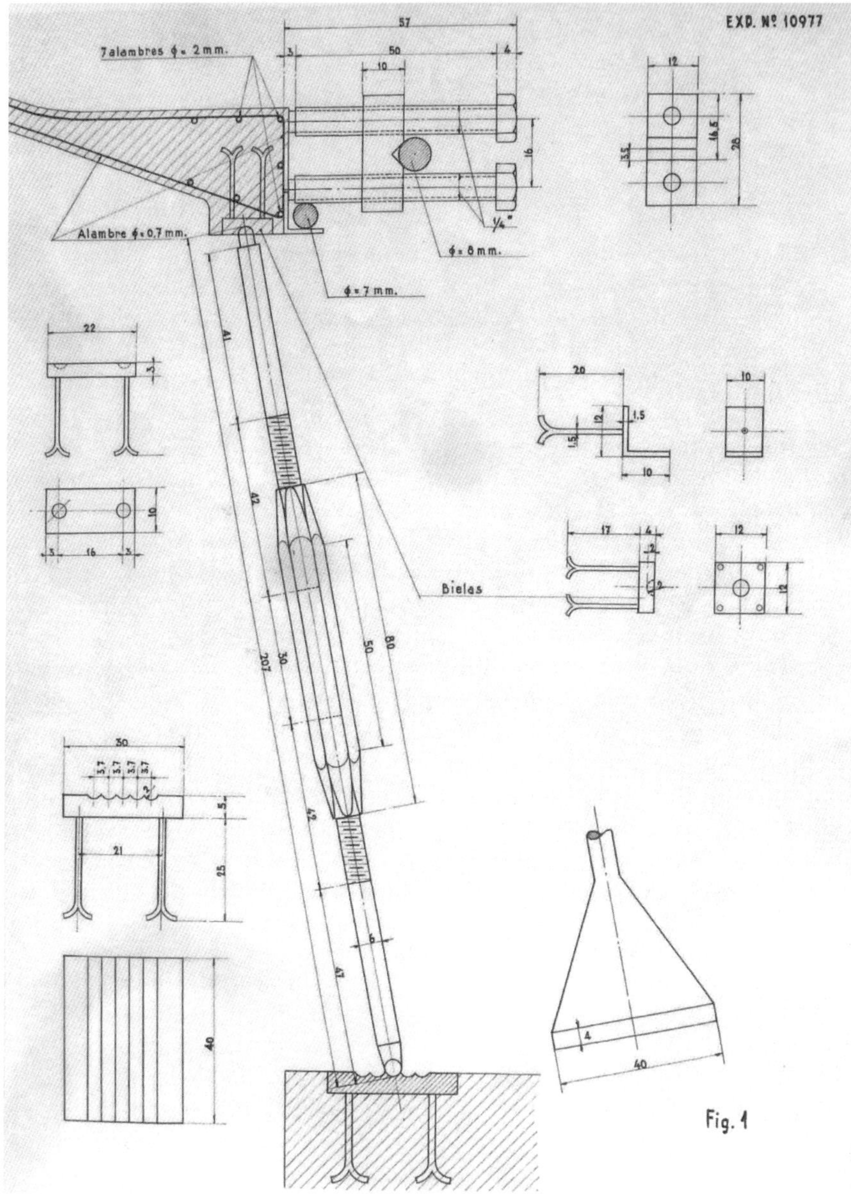

Figure 16.5 Details of a strut supporting the model shell which allowed movement of the perimeter, adjustment of the tie-rod tension and a support plate that allowed the inclination of the strut to be changed. (Source: LCEMC)

Eight loading conditions were studied in which vertical displacements and rotations were measured in both axes and in an oblique plane. Tensions were recorded in several sections of the tie rod, and also the horizontal displacements of the perimeter. One of the three models was used for a test in which the load was increased up to collapse that occurred by buckling at a load of 9.6 kN/m^2. During this test the downward deflection of the crown of the dome was measured.

With the second model, several loading and unloading cycles were carried out up to 4 kN/m^2, verifying the perfect elasticity of the structure. Several further loading and unloading cycles were carried out, with the load acting on only half of the roof, both on the halves divided by the major axis and by the minor axis.

Using the same model, the impact of movements of a support was studied. In order to do this, the shell was loaded with 4 kN/m^2 and one of the supports was lowered. It was observed that the vault followed the descent of the support until it reached 0.5 mm, at which point it ceased to maintain contact with the support and the dome was then supported on only 31 struts. The support was then restored to the point at which contact with the shell was made and then one of the adjacent supports was lowered. When the new descent reached 0.6 mm, radial cracks began to appear near the supports adjacent to the two that had been lowered. At that moment the test was stopped, the position of both supports was restored and the cracks closed up.

The second model was also used to study the impact of the inclination of the support struts. In the previous tests the supports had an inclination of 9° to the vertical. In the new test, the supports aligned with the major axis were placed vertically, those on the minor axis with an inclination of 12°, and the intermediate struts were inclined at angles between 8° and 12°. With this disposition it was verified that the tensions in the various sections of the tie rod were equal.

A further test was undertaken with the shell under no load, in which the force in the tie rod was increased until the crown of the dome was elevated by the same amount that it had fallen when subjected to a load of 1 kN/m^2. Under these conditions, the same load was then re-applied and it was found that the shell recovered its original position. In this way, the force in the tie rod necessary to effect decentring of the dome was obtained. Finally, this second model was progressively loaded up to collapse at 9.9 kN/m^2. From a load of 5 kN/m^2 radial cracks began to appear and, at 9.9 kN/m^2, the entire membrane collapsed due to buckling.

A third model was built with the same characteristics in which the tie rod was pre-tensioned with a stress of 120 N/mm^2. It was loaded up to rupture which occurred at a load of 12.5 kN/m^2. In this case the radial cracks began to appear with a load of 7 kN/m^2 and the final collapse was caused by total buckling of the shell (Figure 16.6). Compared to the collapse test using model one, for the same load, the deformation of the crown of the roof was smaller than in the first test without prestressing [37].

16.4.4 An experimental shell roof

In November of the same year, 1949, the *Instituto Técnico de la Construcción y el Cemento* commissioned the construction of a model of a 'reinforced-mortar shell vault' made up of twelve separate pieces [38]. It was a circular dome with a

Figure 16.6 Collapse of the shell roof in which the change in curvature can be seen. (Image: LCEMC)

central oculus, made up of equal and independent modules supported on twelve columns located at the intersections of the modules. The model was made to scale of 1:5, representing the prototype 14.0 m in diameter with a thickness of 35 mm, giving a self-weight of 1 kN/m^2. The model shell was 7 mm thick over its entire surface except at the edge beams where the section was thicker. The shell was reinforced by a single-layer, diamond mesh of 0.6 mm wires, with a tensile strength of 250 N/mm^2, at 10 mm centres (Figure 16.7a). The perimeter beams were linked by two 2 mm diameter bars, and also incorporated some pins with which the contiguous modules were connected. The surface of the shell module contained a number of holes by which the loads could be suspended (Figure 16.7b). These were made by fixing small nails in the formwork prior to placing the mortar, which protruded by a length equal to the desired thickness

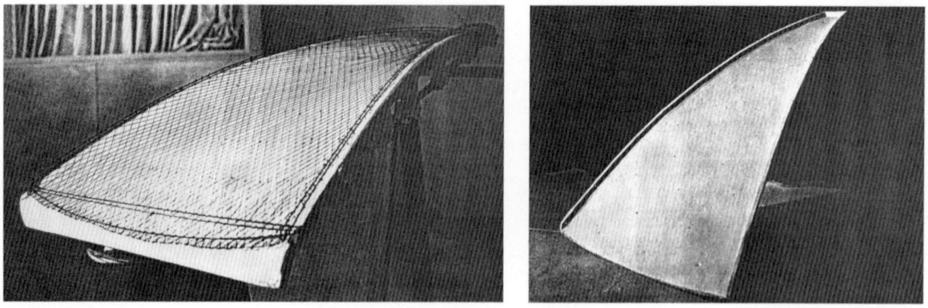

(a) (b)

Figure 16.7 1:5 scale model of an experimental shell roof. (a) A roof module, before concreting showing the arrangement of the reinforcement. (b) The finished module, including stiffening edge fins, in which can be seen the holes through which the loads were suspended. (Images: LCEMC)

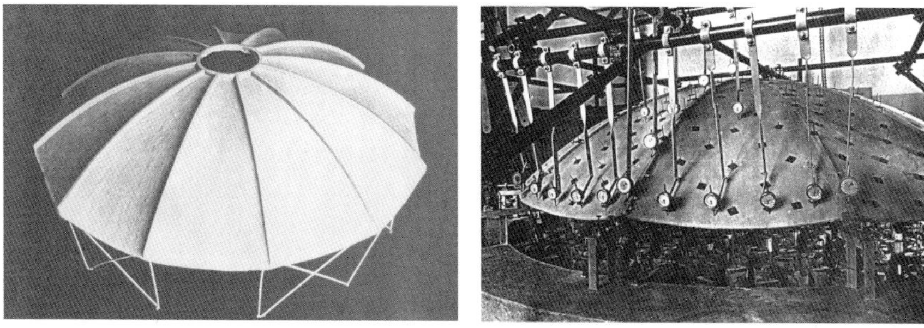

(a) (b)

Figure 16.8 Model of an experimental shell roof. (a) Architectural model of the structure. (b) The 1:5 scale measurement model under load, showing the arrangement of the clock gauges for measuring deflections. (Images: LCEMC)

of the shell. On the one hand these served as a guide to ensure that the thickness of the shell was constant over the entire surface, and on the other, the holes allowed the weights to be anchored when loads were applied to the model shell (Figure 16.8).

The testing routine, developed in previous tests, would become a standard procedure for testing the later reduced-scale models. Steel cylinders were hung from the shell using the small pre-cast holes, and beneath the model was a small pool in which counterweights floated. By changing the water level the load could be increased or decreased at will, in a gradual manner, in order to reproduce a slowly increasing load. In this way, the upper surface of the model shell could be kept uncluttered and able to provide access for the measuring devices. A further advantage of the unobstructed upper face of the shell was that it allowed easy observation of the deformation process, in particular, identifying the instant at which cracks first appeared in the shell. The loading process could then be stopped in order to study the crack formation in detail and then, by slowly increasing the load, follow its evolution up to rupture. Alternatively, the loading process could be interrupted at the right moment to prevent the collapse of the model when it failed, so that it was possible to study precisely how it had occurred.

The shell was tested up to a load of 3 kN/m^2. For different loads during the tests, the displacements along the line of a diameter coinciding with a joint between adjacent modules were recorded; it was noted that one point near the oculus rose while the remaining points all moved downwards. In addition, the horizontal displacements of the twelve vertices of the perimeter were measured and also the vertical displacements of several intermediate points between the supports. The structural form that these tests proved was an entirely innovative way of constructing a lightweight reinforced-concrete shell roof using prefabricated elements; however, there is no evidence that it was ever used at full scale.

16.5 Model studies for concrete dams

In December 1946 Torroja was made president of the first board of directors of the *Empresa Nacional Hidroeléctrica Ribagorzana* (ENHER) and in 1953 the

Eduardo Torroja Technical Office carried out model studies for the same company which was planning the Canelles Dam, near Lleida in north-eastern Spain [39, 40]. To help decide the shape of the arch dam, twenty-three models of different shapes were commissioned and tested [41].

Models 1 to 10 were made at a scale of 1:150 (one metre high) and loaded up to rupture. The mortar with which the model was made was a mixture of sand from the River Manzanares (near Madrid) with foundry sand and various additives to increase its strength. Models 11 to 22 were made at a scale of 1:200 with litargel which could be poured into the suitable mould. It had a density of 17 kN/m^3, a Poisson's ratio of 0.46, a thermal expansion coefficient of 0.0002 and no apparent shrinkage. Model 23 was used to test the stability of the slopes.

The studies for the Canelles Dam provided the opportunity to develop a testing technique for dams that was used many more times in the following years. For models 11 to 22, first of all, a plaster model of the landscape and the dam was built following the instructions of the client. Using this model, a mould was made with which to cast the elastic model. The surface of the moulds was covered with shellac, polished and greased to facilitate removal of the model from the mould and ensure that the surface of the model was completely smooth. One mould was made for the terrain and another for the dam. The mould was fixed to a stiff steel structure and the molten litargel used to make the model was poured at a temperature of 50 °C. At the same time that the model was made, a series of test pieces with the same material was made. These specimens were tested prior to model loading and during measurement. In the documentation kept, the values obtained in the compression tests of the litargel specimens, indicating the temperature, the duration of the load and the value of Young's modulus obtained, are given. The values obtained varied between 10.5 kg/cm^2 at a temperature of 26.6 °C and a Poisson's ratio of 0.392 at 24.2 °C, or 17.7 kg/cm^2 at 19.5 °C and a Poisson's ratio of 0.35 at 16 °C [42].

Finally, the model of the dam was fixed to the model of the terrain and the whole fixed to a steel structure, to which was also fixed a metal plate that limited the void into which the loading liquid would be poured. The surfaces of the upstream and downstream faces of the dam were coated with brittle lacquer that would reveal the surface strains (Section 16.4.1).

With the device built, a first operation was carried out with the aim of obtaining the layout of the isostatics on both faces of the dam. The surfaces of the model were painted with a brittle lacquer ('strain sensitive lacquer No. 105' made by Pinchin Johnson). The model reservoir was then filled to the test height with the sodium sulphide solution. The pressure exerted produced the deformation of the dam wall and, with it, the cracking of the lacquer with which it was covered. Then, the face downstream was impregnated with the same sodium sulphide solution. The cracks that formed in the lacquer coating allowed the directions of the principal stresses to be seen when black lead sulphide was produced by the reaction of sodium sulphide with the lead oxide contained in the litargel. This same method was repeated for all the tests on dams that remain in the archives of the LCEMC. In this way, the isostatic lines could be drawn on both faces of the dam. The model was then unloaded and the 30 mm gauge length of the extensometer was positioned along the direction of the isostatics. Then the model reservoir was refilled with a liquid of known density (in this case Vaseline oil, 8.1 kN/m^3). In order to measure the surface strains in the dam wall, several loading and unloading cycles were carried out, and measurements were taken 15 minutes after the full loading or unloading. The value of the strains was taken as the average value of the observed readings. At the same time that the model was being loaded, the test reference pieces of the litargel were also loaded and their deformation measured. In this way the value of Young's modulus and Poisson's ratio could be measured, at the same temperature as the model tests being made. The displacements were converted to surface stresses in the model and, using the similitude conditions of the model, these were converted to stresses in the full-size structure.

Using this same procedure, twelve further series of tests were carried out between 1953 and 1961. Each series of tests took around ninety days and required about 600 hours of a technical assistant, 400 hours of skilled labour, 200 hours of a specialist technician and about 100 hours of a section leader, although, these times could vary according to the complexity of the project.

Most of these tests consisted of checking a completely defined project, in which the value of the stresses and movements of the dam wall predicted from the model tests could be compared later with those measured in the completed structure. However, in some cases, such as the La Capella Dam, the results of a first test served to modify the designs for the project [43] and the model tests were repeated a few months later with the new form [44].

In 1959 another model study was undertaken with a different, forensic purpose – three models were tested as part of the investigation into the causes of collapse of the Vega de Tera buttress dam which failed in January 1959, destroying the village of Ribadelago in north-western Spain, claiming 144 victims [45].

16.6 Reduced-scale roof models.

16.6.1 The Haas shell roof for a factory in Nadam Havenwerke, Delft

In 1956 Torroja was commissioned by A. M. Haas of the Delft School of Technology to make and test a model of a prestressed, reinforced-concrete roof. The intention was to build a warehouse with twenty-two equal bays, each 80.0 m long by 12.0 m wide, for a factory in Nadam Havenwerke [46]. The element to be tested consisted of a shell that functioned as a continuous beam comprising two sections of 40.0 m long, 12.0 m wide and 2.0 m deep, with a shell 70 mm thick and with stiffening ribs every four metres [47, 48] (Figure 16.9). The model was built at a scale of 1:10 with reinforced mortar and pre-stressing reinforcement (Figure 16.10). Two different types of mortar were used, one for the shell section and another for the lower section and the longitudinal beam that formed the channel. The model was assembled using wires of different diameters arranged in accordance with the layout provided in the proposed design. Both in the channel beam and in the shell, pre-stressed reinforcement of 2 mm diameter Hinderer steel wire were placed in plastic tubes of 3 mm external diameter in order to eliminate adhesion with the mortar. Prestressing was applied using the Barredo system with hydraulic jacks [49].

Figure 16.9 1:10 model of the ribbed shell roof for a factory in the Nadam Havenwerke, Delft, 1956. (Image: LCEMC)

(a)

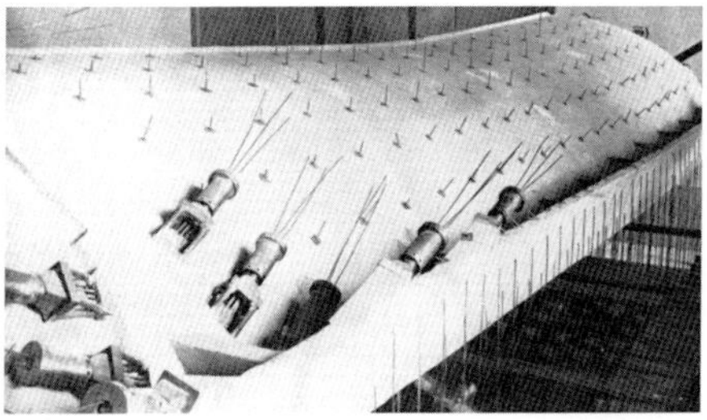

(b)

Figure 16.10 1:10 model of the roof for a factory in the Nadam Havenwerke, Delft. (a) Details of the loading system showing the steel cylinders floating in the water tank beneath the model. (b) Detailed view showing the anchors of the prestressing cables of the Barredo system. (Images: LCEMC)

Seven days after concreting the shell, the formwork was removed and between 13 and 28 days the full prestressing load was applied. The loading system was similar to the previous tests and several loading and unloading cycles were carried out until reaching the load requested by the client of 2.62 kN/m^2 in the shell and 2.98 kN/m^2 in the channel. In this test it was verified that the model behaved elastically – after removal of the imposed load, the original geometry was fully recovered. In order to assess the safety of the beam against collapse, a test was carried out increasing the load above the working load, measuring the displacements up to 150% of the expected peak load. When the load was increased from 125% of peak load to 150%, the displacements observed in the model increased by a factor of 2 in most places, and by 3 and even 5 in some locations. The magnitude of the observed displacements at 150% of peak load and the appearance of cracks in the shell, led to the decision to remove the measurement devices. The load was then increased to 200% of the planned peak load, at which point the structure collapsed (Figure 16.11).

(a)

(b)

Figure 16.11 1:10 model of the roof for a factory in the Nadam Havenwerke, Delft. (a) (b) Collapse of the prestressed model at twice the maximum design load. (Images: LCEMC)

16.6.2 The Labour University of Tarragona

In the same year, 1956, Torroja designed the roof of a building at the *Universidad Laboral de Tarragona* campus, working in collaboration with the architects Antonio de la Vega, Manuel Sierra and Luis Peral and engineers Florencio del Pozo and Alfredo Páez. A 12 cm thick folded-plate structure was proposed to roof over the 57.6 x 20 m warehouse. The roof was made up of triangular-shaped sections joined together along their edges and supported by triangular porticoes located in the façades of the long sides of the building. The folded sections functioned as beams that span the 20.0 m width of the building [50].

To check the capacity of the roof and verify the results expected in the project, a test was carried out on a model built to scale of 1:12. It was made using reinforced and prestressed mortar, reproducing the layout of the reinforcement planned for the actual roof [51]. Three sections of the folded-plate roof were constructed so that the continuity effect could be reproduced. The three sections were loaded with the same load, at the same time, and strains were measured only in the central section. The reinforcement of the shell was made with 0.5 mm diameter wires, and the prestressing reinforcement, consisting of 2 mm diameter wires in PVC tubes, was placed in each plate of the roof. With the reinforcement in place, the mortar was poured without interruption and without joints. After four days, the support plates and tensioning devices were installed. Each prestressing cable was loaded up to 50% of the expected tension. After 28 days, the tension in each cable was raised up to 4.5 kN. Once the loading and measuring devices were in place, several loading and unloading cycles on the roof were carried out up to a value of 3.85 kN/m^2. In the central section, the measurements of the deflections normal to the plane were taken. After the elastic test had been completed, the loading devices were modified to achieve higher loads. In the rupture test, an imposed load of 10.6 kN/m^2 was reached, which meant a safety factor of 3 against collapse (Figure 16.12). The structure was built and is still in service in perfect condition.

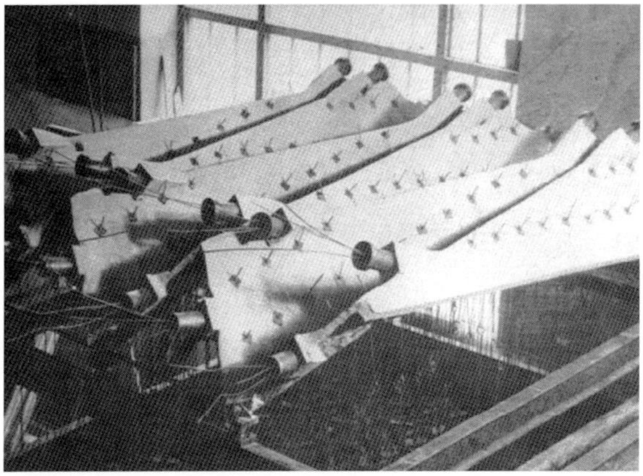

(a)

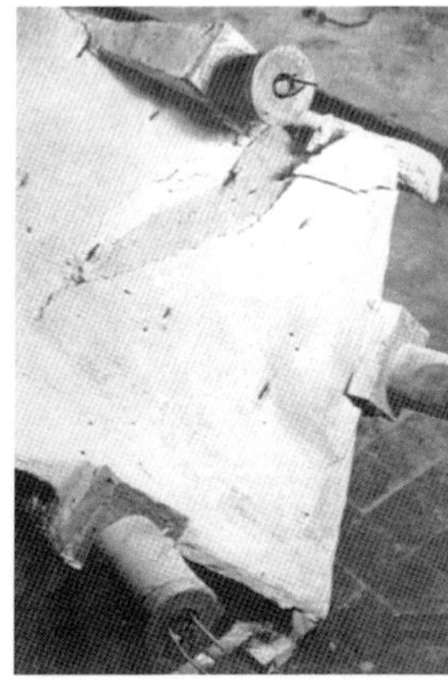

(b)

Figure 16.12 1:12 scale of the folded-plate roof for the Labour University of Tarragona. (a) Model showing the anchors of the prestressing cables and the loading system using steel cylinders floating in the water tank. (b) Detail of fracture of the shell. (Images: LCEMC)

16.6.3 Club Táchira, Caracas, Venezuela

In 1955 Torroja began working on the design of a roof for the Club Táchira complex in Caracas and several early designs were proposed featuring various different forms [52]. After a successive exchange of ideas between the architect José Fructuoso Vivas and Torroja and his collaborators, a final form was chosen in 1957. Its shape was determined by translation of a plane (catenary) curve over a spatial curve (Figure 16.13). The shell roof would cover an area 60 m long by a maximum of 38 m wide.

A 1:12 scale model was made using prestressed, reinforced mortar and tested in the same year [53]. To build the model it was necessary to use a material with the same Young' modulus and Poisson's ratio as the concrete with which the roof would be built. Furthermore, its strength at the same age should be similar. The size of the aggregate used should also allow a shell with a thickness of around 7 mm to be made. In order to find a suitable mortar mix to give a strength of 18 N/mm^2 at 14 days, six different sets of samples were made with different water-cement ratios varying between 0.5 and 1. This led to the selection of a mix with a water-cement ratio of 0.575 (5.95 kg of sand, 2.17 kg of cement and 1.25 litres of water). The model was assembled with wires with a yield strength of 720 N/mm^2 and prestressing wires, tensioned to 4.0 kN in each group of three wires. In some cases, in order to reduce the ultimate tensile strength of the wires used, to bring it closer to that of the steel that would actually be used in the construction, these wires were subjected to heat treatment. As in previous cases, and due to the small size of the elements to be tested, the number of cables and anchors used was reduced, although the planned prestressing force was maintained. The shape of the roof is a translation surface obtained from a catenary moving along a curved directrix. This curve that forms the ridge of the roof is seen Figure 16.13.

For the construction of the model a formwork made of plaster of Paris was used, covered with a layer of shellac and lightly greased to prevent the mortar from

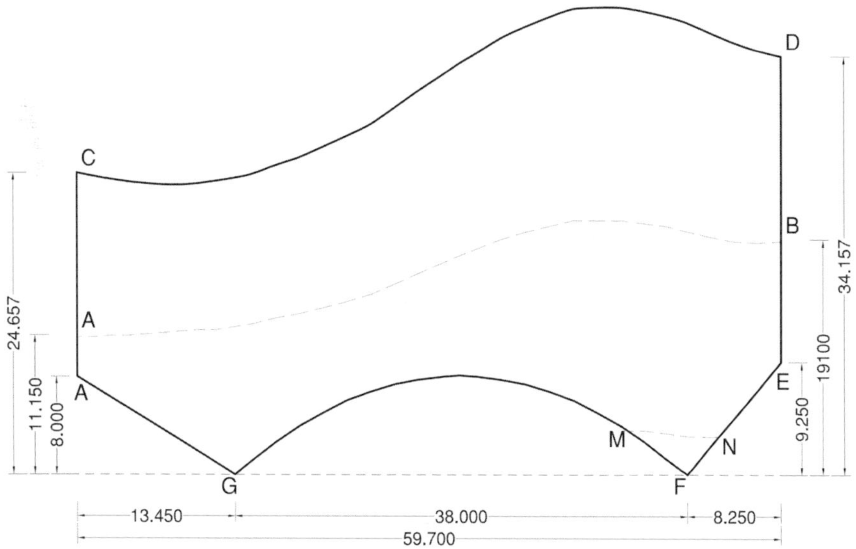

Figure 16.13 Plan view of the proposed roof for the Club Táchira, Caracas, showing the curved line of the ridge. (Image: Joaquín Antuña)

sticking. As was the usual procedure, nails of a length equal to the thickness of the shell were fixed in the formwork in order to ensure the correct thickness of the shell over its entire surface, and to make the holes through which cables were passed to the weights used to load the model (Figure 16.14).

Figure 16.14 1:12 scale model of the proposed roof for the Club Táchira, before placing the measuring devices and applying the loads. The struts supporting the edges of the shell can be seen. (Image: LCEMC)

The asymmetric shape of the shell, the study of various support conditions and the asymmetric effect of wind loads wind made it necessary to carry out a detailed study with many different load conditions. Numerous tests were carried out on the model to simulate different hypotheses. The weight of the prototype shell plus the insulation and waterproofing was assumed to be 2.4 kN/m^2 and, in order to study the effect of the wind, a load of half this amount was applied to one half of the shell.

In the first test the load was gradually increased to 2.4 kN/m^2, until a stable position was reached when the clock gauges showed that movement had ceased. It was then increased to 3.6 kN/m^2 and, after stabilising, the deflection of the shell was measured at 40 different points. Finally the model was unloaded and it was possible to verify the perfect recovery after the deformation that had been experienced. The same procedure was repeated twice more, in each case clock gauges were placed at different points and so that a total of 120 different points were measured (Figures 16.15 and 16.16).

After the first test, two more tests were carried out in which the load of 3.6 kN/m^2 acted on only one of the halves in which the ridge line divided the shell, and the other half was loaded at 2.4 kN/m^2. The load distribution was changed for the following test in order to simulate the asymmetrical effect of the wind action. In all three cases, measurements were taken at 120 points.

Another test was then carried out in which the load was increased to 6.0 kN/m^2, during which no cracks or signs of breakage were observed. This showed that the structure had a safety factor against collapse of at least 2.5. At that time, instead of continuing the test until the collapse was reached, a series of tests were carried out in which the support conditions were modified in a number of ways. First, the position of the horizontal support was fixed to prevent movements in that direction. With this new arrangement, two new tests were carried out with a load of 3.6 kN/m^2 for which new measurements of strain and rotation were taken.

After making several modifications to the supports and repeating the tests, the stiffener located in the MN position (Figure 16.16) which prevented the shell from buckling in that area, was removed. With the new arrangement, the shell was loaded up to 5.1 kN/m^2 and rupture occurred in the area from which

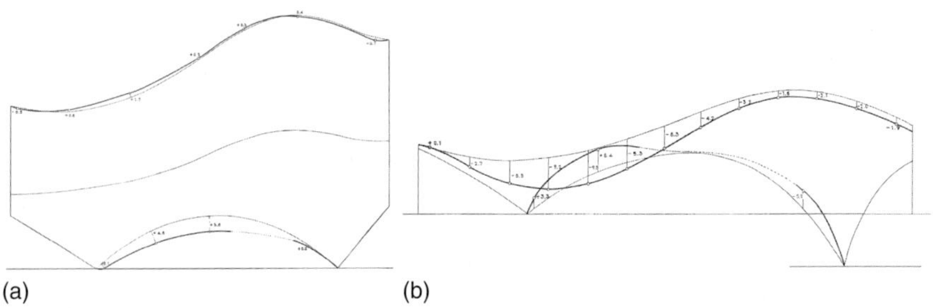

(a) (b)

Figure 16.15 Diagrams showing measured movements of the model shell. (a) Horizontal movements of the horizontal edge supported on the ground (top) and the free edge supported on struts (bottom). (b) Vertical movements of the ridge of the shell and the free edge. (Images: LCEMC)

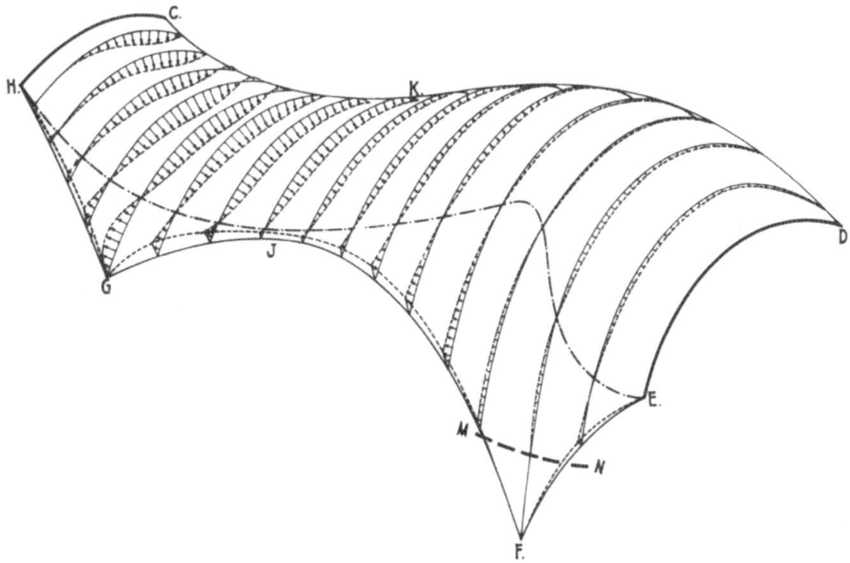

Figure 16.16 Pespective view showing bending moments in the shell. The curved line HE is the line of zero bending. (Image: LCEMC)

Figure 16.17 View of the collapsed model after the loads had been removed. 1 indicates the failure of section M-N (Figure 16.16). 2 and 3 indicate the rupture at the other two supports of the free edge. (Image: LCEMC)

the stiffener had been removed. The buckling and rupture of this zone (1 in Figure 16.17) was followed almost simultaneously in zones 2 and 3 [37]. Thanks to the loading system used, the increase in displacements at the start of the collapse meant that the imposed loads were rapidly reduced, so that both the loading weights and the measuring devices could be removed and the initial stage of the collapse could be studied.

The results obtained in the tests, which were completed in November 1957, demonstrated the safety of the proposed shell structure that had been designed in May 1957. Although the design for the project was completed, it was, in the end, not carried out. (The existing shell roof of the Club Táchira was not designed by Torroja).

The model tests for the Club Táchira roof were probably the most painstaking and thorough of all those carried out in the laboratory while Torroja was its director. The procedure developed for the first of these tests was used frequently until 1961, giving proof of its versatility and the possibility of conducting complex model tests with limited means.

16.6.4 The roof of Bacardi's offices in Havana

During 1959, Mies Van de Rohe produced a design for a new head office for the Bacardi Company in Havana, Cuba. Its roof was a square bi-directional ribbed slab supported by two columns on each side. The length of each side of the building was 54 m and the ribbed slab consisted of two orthogonal families of 19 beams at 3 m centres, varying in depth from 1.4 m to 1.85 m and joined by a slab on their upper edges [54] (Figure 16.18).

The model was made at a scale of 1:15 using reinforced mortar with a strength of 35 N/mm^2. The beams were reinforced with wires between 0.6 mm and 1.7 mm in diameter. All the beams were prestressed with wires of 1.5 mm diameter with a tensile strength of 2300 N/mm^2. These wires were sheathed in PVC tubes, and prestressing jacks of the Barredo system were used. The objective of the test was to determine the movements and rotation in the full-size structure from those measured in the model. Vertical displacements were measured in a triangle corresponding to 1/8 of the deck, to a precision of 0.01 mm, corresponding to 0.15 mm in the actual deck and, at the same points, the rotation was measured in two perpendicular directions to a precision of 0.0003 radians.

The client would be responsible for interpreting the results and obtaining the value of the stresses in the different sections. Because of the difference in Young's modulus between the materials in the full-size structure and the model (20,000 N/mm^2 and 19,700 N/mm^2 respectively) and the scale of the model, the displacements in the prototype would be 14.8 times those measured in the model. The measured rotations would be the same in both the prototype and the model.

In a small model, the dead-weight action must be simulated by applying a part of this load. If the shell were made of reinforced concrete, different types of load could be applied at any time. But in the case of a prestressed roof like this one, before applying all the prestressing it was necessary that part of the test load was acting in order to simulate the self-weight. The model was decentred ten days after concreting, and the distribution plates and anchorage cones were installed after forty days. Up to 25% of the calculated force was then applied to the prestressing wires. The formwork was then stripped and the loading devices were put in place. In order to simulate the uniformly-distributed load, nine point loads were applied to each coffer, making a total 2916 weights that were suspended from the deck and floated in the pool beneath the model. As in previous cases, the loads were carefully controlled by raising or lowering the water level.

75 days after concreting, the load simulating the self-weight began to be applied to enable the full level of prestress to be applied. The initial intention was to reach 3 kN/m^2, but at 1.5 kN/m^2 it was observed that the deflection of the centre of the

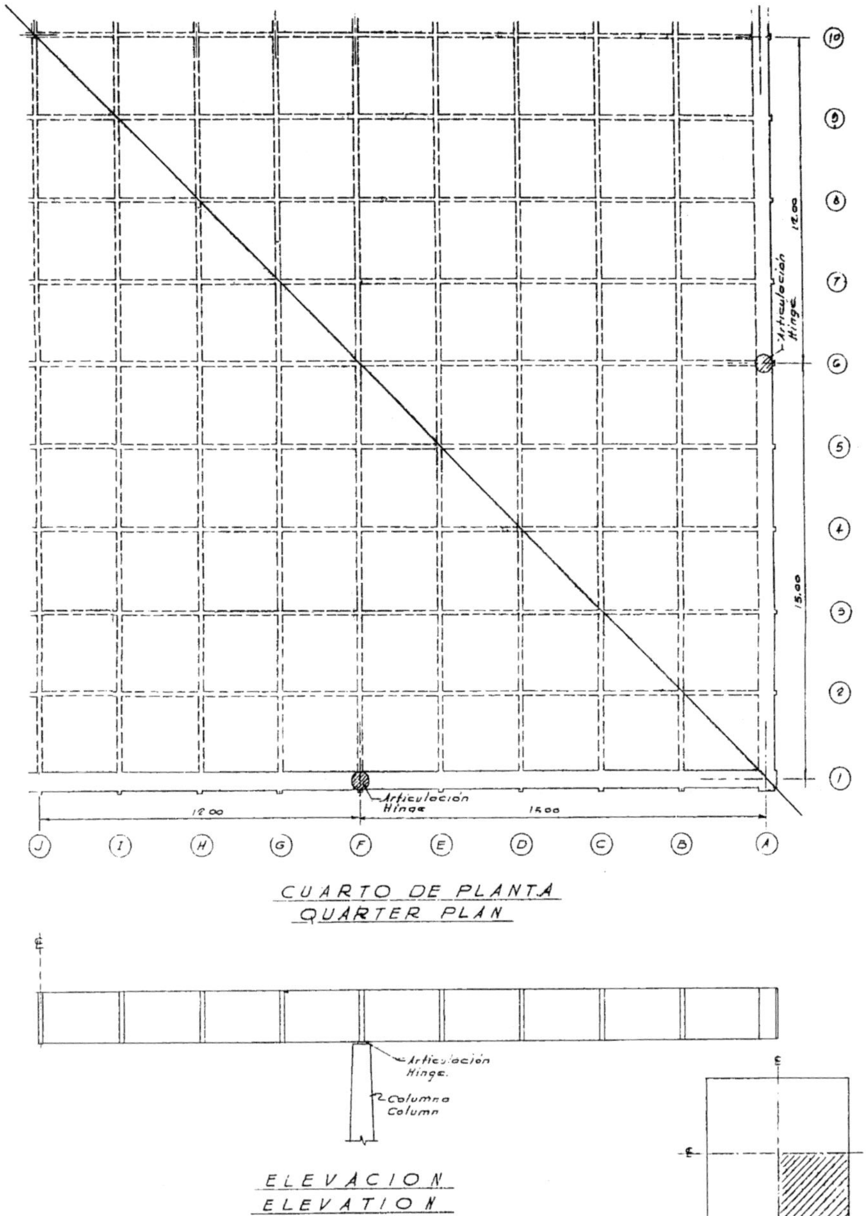

Figure 16.18 Roof proposed for the offices of Bacardi in Havana, October 1958. Plan view of one quarter of the ribbed slab. (Image: AET, CEDEX-CEHOPU)

roof was already 8.2 mm and, considering this to be excessive, it was decided that the full level of prestressing would be applied at this load and that, in the successive tests, it would not be allowed to reduce the load below this value. After two days the force in the cables was increased to 5.5 kN, the level of prestress required by the client.

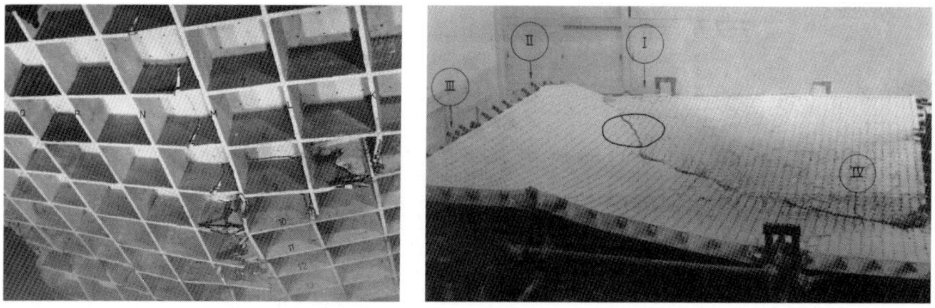

(a) (b)

Figure 16.19 1:15 scale model of the ribbed floor for the offices of Bacardi in Havana. (a) View of the failure in beams 6 to 13 between M and N. (b) View of the collapsed slab after the loads had been removed. (Images: LCEMC)

A number of load trials were then conducted. Starting from 1.5 kN/m^2 the full load of 4.5 kN/m^2 was reached and the loading and unloading cycle repeated several times. In each cycle, deflections and rotations were measured in two directions at various points. Mechanical clock gauges with ranges of 1 cm and 2 cm, with a sensitivity of 0.01 mm were used to measure the deflections and, to measure the rotations, clinometers with a sensitivity of 0.0003 radians were used. The results were averaged from the various loading cycles and scaled up to full scale using the laws of similitude.

After the elastic tests, the load was raised to cause collapse. At 7.5 kN/m^2 no cracks were visible. At 8.0 kN/m^2 intermittent, impact-like noises were heard and a hollow of significant depth was produced. At 8.5 kN/m^2 the loading process was stopped and the deck was inspected. Cracks were found in the lower part of beams 6 to 15 between N and M (Figure 16.19). Shear cracks appeared in the edge beams at an angle of 45°. The fissures were large and constituted real cracks. When the load was reduced the cracks did not close; the load was raised again to 10.0 kN/m^2 and the cracks in the edge beams increased in quantity and intensity. At 11.0 kg/m^2 there was a total collapse due to the abrupt failure of the edge beam at the support 6 of beam A. In the instant prior to the collapse, the cracks in the edge beam had increased rapidly. The collapse of the slab was not symmetrical. From the photographs taken it can be seen that one of the supports seems to have moved because the slab has fallen at that point and the support has come off the edge of the slab. The headquarters building for Barcardi in Havana was not built. However, Mies used the same design, this time built in steel, for the Neue Nationalgalerie in Berlin which was opened in 1968 and was his last building.

16.6.5 The Church of La Paz, Barcelona

The church of La Paz was a project of the architect José Miguel Serra Dalmases which consisted of a rectangular plan 50 x 23 m [55]. The roof of the hall consisted of two cylindrical shells joined by a longitudinal generatrix. Although the project was finally not built, the roof structure was fully designed and tests on a reduced-scale model were completed [56]. The model was built at a scale of 1:9, reproducing the reinforcement layout in the form of two orthogonal meshes on each face and, in between, the prestressing reinforcement was sheathed in

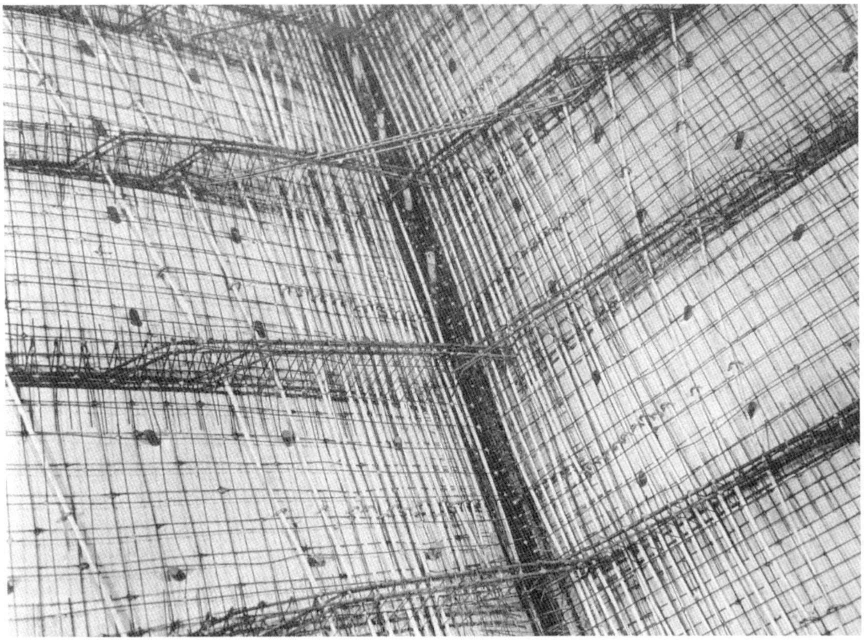

Figure 16.20 1:9 scale model for the roof of the Church of La Paz in Barcelona, 1961. Detail showing reinforcement and PVC tubes for prestressing cables. (Image: LCEMC)

PVC tubes which also served to separate the two meshes (Figure 16.20). In order to load the model, instead of the floating the weights in water which had generally been used up to that time, a different method of loading began to be used. The imposed load, in the form of a sufficient number of equally-distributed point loads, was applied by means of tubular latex strips placed between the slab and a metal frame. The loads were imposed by separating the metal frame from the slab by means of jacks (Figure 16.21). The model was tested up to collapse

Figure 16.21 1:9 scale model of the roof for the Church of La Paz, showing reinforcement and prestressing cables in the slab. (Image: LCEMC)

but no test results seem to have been preserved in the project file in the LCEMC Archive.

16.6.6 The grandstand canopy for the Madrid Canódromo (dog track)

In 1961, a grandstand and canopy was commissioned for the dog track in Madrid, and one of the engineers working on the project was Torroja's son, José Antonio (b.1933). The canopy was a folded-plate cantilever in reinforced concrete, tied down at the rear, a sort of tribute to Torroja's own stadium roof for the Hipódromo de la Zarzuela in Madrid, constructed in the 1930s. The 41.6 m long canopy comprises thirteen identical sections, 3.2 metres wide and 23 m long, of which 17 m was a cantilever over the seating, supported on a column and tied down at the rear [57].

Given the complexity of the structural action, and limits of theoretical analysis, tests on a Plexiglas model were commissioned to determine the stresses, deflections and strength of the cantilever. In addition, several smaller models were made in plaster and tested up to failure. The main (elastic) model of three sections of the roof was built at a scale of 1:25 with Plexiglas sheets of different thicknesses (4, 8 and 10 mm), and bonded with glue [58]. The ties were enhanced with steel rods that were anchored to the steel support frame. Holes were drilled in the plastic sheet to enable 1130, previously pre-stretched, rubber bands to be attached to the roof at one end and, at the other, to a concrete slab that could be moved vertically. The distribution of the points of application of the load was chosen to reflect the distribution of self-weight in the actual structure (Figure 16.22).

Two types of measurement were taken in the model. The bending stresses in the plates were calculated by measuring the curvature between two points separated by 30 mm and assuming that the section stiffness of the sheet was that of the gross section of concrete. For the measurement of vertical and horizontal displacements, a photographic procedure was used which consisted of photographing the model loaded and then unloaded on the same negative. In the double-exposure photograph it was possible to measure the distance between the two images of a white line marked on a piece of matt-black paper stuck to a small piece of Plexiglas. The measurements were made on the negative using a microscope that achieved an accuracy of one hundredth of a millimetre.

The studies on the model of the canopy for the Canódromo heralded the end of the test procedures developed by Torroja in his first years leading the work of the laboratory. For the tests on the model of the roof for the Church of La Paz, a new procedure for loading the model had been introduced; for the Canódromo a new material, Plexiglas, was introduced. The process of applying the loads was similar to that used for the church but, in this case, the loads were applied by means of the displacement of a weights located under the model and also, a new photographic procedure for measuring the displacements in the model was introduced. With this project, a new era in the Laboratory's history began in which Carlos Benito and Florencio del Pozo took a new direction in which numerous tests of great complexity were carried out and in which new measurement

Figure 16.22 Canódromo, Madrid, 1961. 1:25 scale Plexiglas model loaded using 1130 rubber bands. (Image: LCEMC)

techniques were used which led to an increase in the quantity and the precision of the data obtained.

16.6.7 Reflections on model testing by Carlos Benito

Carlos Benito presented the results of a test carried out for the canopy for the Canódromo in a paper [59] in which he also presented a reflection on the work with models that had been done up to that time. On the one hand, the materials with which the models are made had the same breaking strength as those used in the actual construction. The reduced-scale model reproduced the shape of the structure and the reinforcement to the same scale, and in number and position. This was a laborious process that could take about three months: the model was built in six weeks, and acquired the desired mechanical characteristics in a further two weeks; in the third month, the tests and their interpretation were carried out.

In general, the purpose of the test was to check or verify a project whose shape and dimensions were already defined. It was not economically viable to study several original forms and compare the results. Thanks to the test it was possible to conclude that the project had an adequate safety factor and its stiffness was sufficient. But it could also show that the safety factor was not adequate, and the structure would need to be strengthened. In such cases, working with the model often provided information on how the structure could collapse and how its strength could be increased. That shortfall in working with models concerned the staff in the laboratory. Working with models could not be used as a design tool and was, rather, a tool for verifying already-defined designs. The challenge was to avoid studying forms that could be tested analytically using known solutions, but rather to investigate new forms that could only be validated by experimental testing.

In order to use working with models as a design tool, one of the conditions was to be able to build the models more economically. When building models of reinforced-concrete shells, the most expensive part of the process was the fabrication of the reinforcement. If you could build small-scale models more quickly, you would have a fast and economical tool for use on live projects. To build such models it would be necessary to have an economic material available which was easy to model and which had similar strengths in tension and compression. In order to obtain a material with these characteristics, the laboratory had commissioned a study to determine the composition of a material with the following characteristics: tensile and compressive strengths of 0.02 N/mm^2 or greater, and reaching this value after two or three days, and not increasing after that time; linear-elastic behaviour up to fracture and a dry consistency allowing it to be applied with a spatula [60].

No such material was found, but a mixture of 300 parts Paris gypsum, 200 parts water and 0.5% sodium citrate did produce a material that achieved a compressive strength of 10 N/mm^2 and a tensile strength of 5 N/mm^2 after three days (measured in a simple bending test) and a Young's modulus in compression of 1200 N/mm^2. With this material they built a 1:25 scale model of the canopy for the Canódromo de Madrid which was tested until collapse (Figure 16.23). In this way they found a preliminary result that opened the door to using models as a design tool for structures that could not be analysed theoretically – using an economic material with which models could be made in a short time and tested up to collapse. Nevertheless, this line of work was not continued, and the test of the Canódromo grandstand canopy seems to have been the only project that used this new technique.

(a) (b)

Figure 16.23 1:25 scale model of the canopy of the Canódromo, made with an experimental gypsum-based material which accurately modelled the brittle behaviour of concrete. (a) The test rig. (b) The model after collapse. (Image: LCEMC)

16.7 Final remarks

Torroja was director of the *Laboratorio Central de Ensayo de Materiales de Construcción* for 20 years. In the first three years he managed to build a new building in which different laboratories were installed, with a team of technicians and craftsmen who were able to establish procedures for the construction and testing of models of various types, and that gave useful results almost from the beginning. Although it was not an institution of the the size of Arturo Danusso's *Istituto Sperimentale Modeli e Strutture* (ISMES) in Bergamo, in which 45 tests were carried out in 1961 alone [61] (see Chapter 15), some pioneering work was carried out, such as the testing of the cylindrical gridshell and of prefabricated and prestressed shells, carried out between 1948 and 1949. With the techniques acquired in the first few years, a large number of tests were carried out in the following two decades and, in the last of the tests shown, for the Iglesia de la Paz and the Canódromo, we see the intention to introduce new materials and procedures in order to be able to extend the scope of model testing. One of the key objectives was to use them as a tool to investigate new structural forms and new materials. However, that line of work was not continued. Torroja followed up his interest in new structures such as the prefabricated roof modules that he presented at the symposium held at the same Institute in September 1959 under the title 'Non-traditional processes of shell construction' [62]. In this case, the proposed modules were tested at full-scale.

Torroja's work between 1939 and 1961 was extensive and his work in the development of the laboratory was only one part of it. He maintained his activity as a designer, carrying out more than 600 projects in those years; as a professor in the *Escuela de Ingenieros de Caminos* he published several textbooks and, as director of the *Instituto Técnico de la Construcción y el Cemento,* he managed to develop it into an important research centre with wide recognition. In the Laboratory he formed a multidisciplinary team that proved able to adapt to the technical changes that were taking place in the experimental research on structures at the end of the 1950s, and in the following years several complex tests were carried out. The use of tests on reduced-scale models came to an end in 1970 with the testing of the roof of the Palau Blaugrana sports arena in Barcelona (Figure 16.24), a project for which Florencio del Pozo was the engineer. The shell spanning 74.9 m is just 10 cm thick, with ribs 18 cm thick [63]. The Plexiglas model was made in four quadrants, moulded in an inverted orientation. The shell, just 3 mm thick, and the ribs, 5 mm thick, were formed separately by heating the mould to 100°-120° for a period of three days and then cooled slowly over twenty-four hours. The ribs were glued to the shell and then the four quadrants were glued together (Figure 16.25). Loads were applied using blocks of gypsum plaster lowered from the framework above, onto a 4-cm layer of foam rubber to distribute the load onto the shell. Deformations were measured at fifty points and surface stresses measured using sixty-five extensometers and seventy strain-gauge rosettes, giving data for a total of 275 points. After various tests in the elastic range the model was finally loaded to collapse which occurred at 4.5 times the maximum design load.

Figure 16.24 Palau Blaugrana sports arena, Barcelona, 1971. Engineer: Florencio del Pozo. Exterior view of the shell roof spanning 74.9 m. (Image: [64])

(a) (b)

Figure 16.25 1:35 scale Plexiglas model of the dome (2.3 m diameter) for the Palau Blaugrana sports arena, Barcelona. (a) The model in its loading rig. (b) View from beneath showing the rib structure, and the loading framework visible through the transparent shell. (Images: LCEMC)

The tests carried out at the LCEMC, as well as having served as verification of some designs, can now have an important didactic value. These projects for which their form and reinforcement are already known, as well as their response to certain known loads, can be used as an object of study in courses on structural analysis and design. Students can analyse these structures and explain the adequacy or not of the results obtained by theoretical analysis and those observed in the tests on physical models. Undoubtedly, model tests can have new uses.

Further reading

A complete bibliography on Eduardo Torroja can be consulted in:

Antuña, J., (2003). *Las estructuras de edificación de Eduardo Torroja*. PhD Thesis. Universidad Politécnica de Madrid. http://oa.upm.es/1s48/

The Torroja Archive can be consulted at http://www.cehopu.cedex.es/etm/indices/obraindx.htm

Further background on Torroja's works can be found in his main publications:

Torroja, Eduardo (1957) *Razón y Ser de los tipos estructurales*. Madrid: Instituto Técnico de la Construcción y del Cemento.

Torroja, Eduardo (1958) *Philosophy of Structures*. New York: F.W. Dodge Corporation.

Torroja, Eduardo (1958) *The structures of Eduardo Torroja: An autobiography of engineering accomplishment*. New York: F.W. Dodge Corporation.

Torroja, Eduardo (1999) *Las estructuras de Eduardo Torroja*. Madrid: Centro de Estudios y Experimentación de Obras Publicas.

References

1 Eduardo Torroja Archive. http://www.cehopu.cedex.es/etm/etm_index.htm (28/5/2020)

2 AA. VV. (1948) *Cincuentenario del Laboratorio Central de Ensayo de Materiales de Construcción.*

3 Eduardo Torroja Archive. http://www.cehopu.cedex.es/etm/obras/ETM-248.htm (11/9/2019)

4 Eduardo Torroja Archive. File No. 439 *Informe sobre el laboratorio de Ensayo de Materiales*. http://www.cehopu.cedex.es/etm/expt/ETM-248-001.htm (28/5/2020)

5 Eduardo Torroja Archive. File No. 504 *Laboratorio de Modelos Reducidos*. http://www.cehopu.cedex.es/etm/expt/ETM-248-010.htm (28/5/2020)

6 Villena, L. (1943) *Fotoelasticidad*. Madrid: Instituto Técnico de la Construcción y la Edificación.

7 Villena, L. (1946) *El nuevo aparato fotoelástico del Laboratorio*. Monografía 33 del LCEMC.

8 Villena, L. (1947) *Estudios de gelatinas para estudios fotoelásticos*. Monografía 44 del LCEMC.

9 Villena, L. (1947) *Estudios de gelatinas para estudios fotoelásticos*. Monografía 58 del LCEMC.

10 Villena, L. (1949) *Sobre los estados de tensión plana (justificación del ensayo óptico de tensiones)*. Real Academia de Ciencias Exactas, Físicas y Naturales. Serie de Ciencias Físicas. Tomo III. Memoria N° 1. Madrid.

11 Villena, L. (1951) *Estudio fotoelásticos tridimensionales con modelos de gelatina*. Monografías del LCEMC Vol 73.

12 Eduardo Torroja Archive. File No. 530 *Métodos fotoelásticos*.

13 Eduardo Torroja Archive. File No. 531 *Sección de los acueductos de hormigón armado*.

14 Eduardo Torroja Archive. File No. 532 *Investigación sobre la relación de resistencias entre probetas cilíndricas y cúbicas*. Results published in Benito, C. (1946) *Relación entre las resistencias en probetas cilíndricas y cúbicas*. Monografía número 35 del LCEMC.

15 Eduardo Torroja Archive. File No. 533 *Estudio para el establecimiento de un método de determinación de lasŁ tensiones en piezas de hormigón.*
16 Eduardo Torroja Archive. File No. 690 *Modelo reducido ENASA.* http://www.cehopu.cedex.es/etm/expt/ETM-335-002.htm (28/5/2020)
17 The following files are conserved in the archive of the Laboratorio Central: No. 9113 *Investigación fotoelástica sobre modelos en gelatine*; No. 9681 *Estudio de piezas armadas en gelatina fotoelástica*; No. 10709 *Estudios isostáticos de lacas frágiles*; No. 10683 *Estudio en modelos planos*; No. 9565 *Estudio teórico y fotoelástico de emparrillado*; No. 10881 *Determinar la suma de las tensiones principales por analogía eléctrica*; No. 11181 *Construcción y ensayo de rotura de un modelo reducido de una bóveda membrana en mortero armado.*
18 Archive of LCEMC, File No. 9981 *Estudio de piezas armadas con gelatinas fotoelástica.*
19 Archive of LCEMC, File No. 9565 *Estudio teórico y fotoelástico de emparrillados.* The results are published in Torroja, E. & Villena, L. *Estudio teórico y fotoelástico de emparrillados.* Monografía No. 6, LCEMC.
20 Benito, C. (1951) *Estudios fotoelásticos tridimensionales con modelos de gelatina.* Monografía No. 73, LCEMC.
21 Archive of LCEMC, File No. 9983 *Estudio fotoelástico de rigidizadores en vigas de alma estrecha,* 1946; No. 10062 *Estudio de túneles en Modelos Reducidos,* 1947; No. 9616 *Estudio fotoelástico para una presa de contrafuertes,* 1947; No. 10768 *Estudio fotoelástico de las laderas de los embalses,* 1948.
22 Archive of LCEMC, File No. 14381 *Ensayo fotoelástico de un pórtico,* 1954.
23 Archive of LCEMC, File No. 19835 *Ensayo fotoelástico de un modelo reducido de central subterránea,* 1960.
24 Archive of LCEMC, File No. 9980 *Estudio de membrana en paraboloide hiperbólico,* Nov. 1946.
25 Archive of LCEMC, File No. 10173 *Estudio de un viaducto sobre el río Miño en modelo reducido,* 1947.
26 Archive of LCEMC, File No. 10914 *Ensayos con una bóveda de mortero armado en Modelos Reducidos.*
27 Archive of LCEMC, File No. 12882 *Modelos reducidos, láminas.*
28 Benito, C. (1959) Comprobación experimental de cubiertas laminares por medio de modelos reducidos. *Informes de la construcción* 12 (111-112). Archive of *Informes de la construcción* since 9 (87) (1957) available at http://informesdelaconstruccion.revistas.csic.es/index.php/informesdelaconstruccion/issue/archive (12/9/2019)
29 Eduardo Torroja Archive. File No. 791 *Iglesia de Xerallo.* http://www.cehopu.cedex.es/etm/expt/ETM-372-001.htm (28/5/2020)
30 Torroja, E. (1948) Rapport sur les voiles minces construits en Espagne. In: *Proceedings of Third Congress Association Internationale des Ponts et Charpentes, 13-18 Septembre, Liège.*
31 Eduardo Torroja Archive. File No. 665 *Concurso de nave para la fábrica de ENASA.* http://www.cehopu.cedex.es/etm/expt/ETM-335-001.htm (28/5/2020)

32 Torroja, E. (1951) *Laboratorio central de ensayo de materiales de construcción.* Madrid: Instituto Técnico de la Construcción y del Cemento.

33 Del Pozo, Florencio (1956) Voiles minces cylindriques formés par une maille triangulaire. In: *Report of IABSE congress, 25 June – 2 July, Lisboa-Porto.*

34 Pozo, F. (1960) *Cubiertas laminares cilíndricas formadas por una malla triangular de perfiles metálicos.* Madrid: Instituto Técnico de la Construcción y del Cemento.

35 Editorial (1958) Costillares, Instituto Técnico de la Construcción y del Cemento, *Informes de la construcción* 10 (99).

36 Archive of LCEMC, File No. 10977 *Construcción en mortero armado de tres bóvedas membrana a escala 1:10.*

37 Benito, C. (1960) *Experimental testing of thin shells by means of reduced scale models.* North Carolina State College: Student Publications of the School of Design. 3-15.

38 Archive of LCEMC, File No. 11181 *Construcción y ensayo de rotura de un modelo reducido de una bóveda membrana de mortero armado.*

39 Aznar, E. (2015) *Aproximación a la Historia de la Empresa Nacional Hidroeléctrica del Ribagorzana.* ENHER. doi: 10.13140/RG.2.1.4401.0404 (60 pages).

40 Eduardo Torroja Archive. File No. 784 *Presa de Canelles.* http://www.cehopu.cedex.es/etm/expt/ETM-369-001.htm (28/5/2020)

41 Archive of LCEMC, File No. 13354 *Construcción de 23 modelos reducidos de presas en arco.*

42 Archive of LCEMC, File No. 17711 *Determinación de las tensiones principales originadas por el empuje del agua en los paramentos de una presa.*

43 Archive of LCEMC, File No. 14838 *Determinación de las tensiones principales originadas por el empuje del agua en los paramentos de la presa de La Capella.*

44 Archive of LCEMC, File No. 15314 *Determinación de las tensiones principales originadas por el empuje del agua en los paramentos de la presa de La Capella.*

45 Archive of LCEMC, File Nos. 18720, 19073 & 19605. Various model tests to determine the cause of the collapse of the Vega de Terra dam in 1959.

46 García, R. & Valcarce, M.T. (2009) Cylindrical shell construction: the shell roof on the Jamin factory at Oosterhout, Netherlands. In: *Proceedings of the Third International Congress on Construction History, Cottbus.*

47 Archive of LCEMC, File No. 14283 *Construction in reinforced and post-stressed mortar of a model of shell roof in 1:10 scale.*

48 Benito, Carlos (1956) Ensayo, en modelo reducido, de una cubierta laminar de hormigón pretensado. *Informes de la construcción* 9 (81).

49 Barredo, R. (1955) Pretensado 'sistema Barredo'. *Informes de la construcción* 8 (70).

50 Eduardo Torroja Archive. File No. 824 *Cubierta plegada para la Universidad Laboral Francisco Franco en Tarragona.* http://www.cehopu.cedex.es/etm/expt/ETM-392-001.htm (11/9/2019)

51 Archive of LCEMC, File No. 15944 *Construcción en mortero armado y postensado, y ensayo de un modelo de cubierta laminar a escala 1:12.*

52 Eduardo Torroja Archive. File No. 840 *Edificio principal de Club Táchira en Caracas.* http://www.cehopu.cedex.es/etm/expt/ETM-445.htm (28/5/2020)

53 Archive of LCEMC, File No. 16389 *Construcción, en mortero armado y postensado, y ensayo de un modelo de cubierta laminar a escala 1:12.*

54 Archive of LCEMC, File No. 19340 *Construcción en mortero armado postensado y ensayo hasta rotura de un modelo de la cubierta del edificio de Bacardí en Santiago de Cuba.*

55 Eduardo Torroja Archive. File No. 906 *Cubierta de la Iglesia de la Paz.* http://www.cehopu.cedex.es/etm/expt/ETM-471-001.htm (28/5/2020)

56 Granell, J. (1961). Recent tests on reduced shell models carried out in the Laboratorio Central de Ensayo de Materiales de Madrid. In: *World conference on Shell Structures, October 1-4, 1962, San Francisco*, publication No. 1187 (ed. S.J. Medwadowski), 235-244. Washington D.C: National Academy of Sciences/ National Research Council.

57 Anon (1962) Canódromo Madrileño. *Informes de la construcción* 14 (138): 51-68.

58 Archive of LCEMC, File No. 20145 *Ensayo en modelo reducido de una cubierta plegada bajo la acción de su peso propio.*

59 Benito, C. (1961) Essais comparatifs puor le choix d'une forme resistente. In: *Proceedings of the Symposium on Shell Research, Delft August 30- September* 2. 282-286.

60 Archive of LCEMC, File No. 12882 *Estudio de la composición de un material con las características descritas.* Closed in Nov. 1961.

61 Neri, G. (2014) *Capolavori in miniatura. Pier Luigi Nervi e la modellazione strutturale.* Milano: Mendisio Academy Press, Silvana.

62 Astudillo, R. (2014) Eduardo Torroja y la International Association for Shell and Spatial Structures (IASS). *Informes de la construcción* 66 (536): 1-6.

63 del Pozo Frutos, F. (1973) La cúpula del Palau Blau-grana. *Hormigón y Acero.* 24 (107): 27-46.

64 JoJan. Licensed under the Creative Commons Attribution 3.0 Unported license. https://commons.wikimedia.org/wiki/File:F.C._Barcelona_318.JPG (25/10/2019)

17

Scale models for structural testing at the Cement and Concrete Association, UK: 1951-1973

Edwin Trout

17.1 Introduction

'*Throughout history, one can find examples of the use of models for design purposes since, by their use, the designer avoids being restricted by the analyses at his disposal and does not necessarily have to make any assumptions concerning the behaviour of the structure being considered*' (R.E. Rowe in 1965) [1].

During the 1950s and 1960s, concurrent with theoretical advances and the development of computers as a design tool, the Cement and Concrete Association (C&CA) at Wexham Springs in Buckinghamshire, England, earned a significant reputation as 'the home of research using scale models' [2]. There were other British research centres testing models – Imperial College in London for one – and there were similar institutions abroad, for example ISMES at Bergamo in Italy and the University of Illinois. But for a while, the C&CA was at the forefront: 'a pioneer in the UK for this kind of work on structural models' [3].

17.2 The Morice years (1951-1957): establishing a reputation for expertise

The story starts in the immediate post-war years with the formation of a Structures Department at the newly established research station at Wexham Springs. At its head was the young engineer, Peter Morice (b.1926), a recent graduate from the University of Bristol, who joined the Association in May 1948. Within months he was informed that planning permission had been obtained and funding approved to build a new laboratory on the site: the Structures Laboratory. Morice was charged with preparing the necessary specification. His experience of such laboratories had been gained at university where, along with similar facilities at the UK Government Building Research Station near Watford, their role had been limited to testing materials and load testing individual elements such as beams and columns. With the freedom born of a new enterprise, Morice was encouraged to think more widely:

Physical Models: Their historical and current use in civil and building engineering design,
First Edition. Edited by Bill Addis.

> 'What I envisaged … was the possibility of load testing combinations of full-sized structural components, and large-scale models of full structures, since it was difficult to represent the true behaviour of concrete at a microscopic scale' [3].

A visit to the naval dockyard at Rosyth in Scotland confirmed his thoughts of including a large area of strong floor with a regular grid of holding-down sockets, to which a framework could be attached to provide jacking points for loading. Moveable jacks, accurately controlled with high-pressure oil through flexible hoses, were purchased from Tower Brothers, and electrical-resistance strain gauges used in conjunction with a multi-gauge measuring system acquired from the aircraft instrument makers, Savage & Parsons. A demountable mechanical, or 'Demec', gauge was subsequently developed by Morice and Base [4] and used to considerable effect in the Department's work. Construction commenced on this, the first new building at Wexham, and the Structures Laboratory was duly completed in early 1951. The official opening took place in May 1951. It was recorded in that year's annual report, with a comment on its prospects by the Technical Director, Dr A.R. Collins:

> 'The chief development at Wexham Springs during the year has been the bringing into service of the new Structures Laboratory which has allowed the Structures Section to begin practical research on an adequate scale. Judging by the requests for help from consultants and others concerned with the design of structures the Structures Section has an almost unlimited scope for future work; the chief difficulty is one of choosing the most urgent items for attention' [5].

Prestressed concrete and shell structures were to dominate the early years and in the months that followed the laboratory's inauguration, the experimental programme thus facilitated was 'started by tests on large models of interconnected prestressed beams … and testing of large-scale concrete models of shell roofs'. The purpose of such work at this stage was to develop a simplified and improved method of calculation and, indeed, the interplay between test results and generally-applicable theoretical methods would be a constant feature of department's work over the next two decades.

17.2.1 Shell roofs (1951-1953)

In 1952 work started on the testing of a continuous shell roof specimen under the influence of surface loading and of relative settlement of its supports. The preparation for testing of a small anticlastic shell was also started and the shell roof programme continued into 1953:

> 'The tests which were carried out in the Structures Laboratory from 1951 to 1953 on a continuous-shell model indicated that the ultimate strength of the shell was very great, and the results of the experiment greatly impressed the London District Surveyors who visited the Laboratory on 12th May this year' [6].

A symposium on Concrete Shell Roof Construction was held in on 2-4 July 1952, and the proceedings published [7]. This work, while exploring the fundamentals of what, in Britain, was still an innovative method of construction, helped establish an expertise at the C&CA that was made available for testing the design of actual schemes later in the decade. Morice was among the contributors, presenting a paper and participating in the discussion. His paper on research into concrete-shell structures addressed, in part, the use of models for the study of roofs. He referred to tests at the University of London's City and Guilds College where researchers working under Professor A.A.L. Baker – chairman of the C&CA's advisory board – had constructed two cylindrical shells, one of them only 7/8 in. thick (One inch (abb. 'in.') is 25.4 mm). He continued with an account of the C&CA's work on continuous shells, describing both programmes as 'major landmarks in the engineering development of concrete shells for roof structures' [8].

Perhaps the earliest example is one for which there is no reference to models in the literature, but which is recalled by Morice: the Bournemouth bus garage. This was planned in 1948 and its main building, with an extensive cylindrical shell roof in prestressed concrete, was completed in 1952 [9]. Its design prompted a paper in the *Structural Engineer* [10], which referred to sources in the C&CA's library and during the discussion was commented on by Hajnal-Konyi, a shell specialist later involved with the C&CA's testing. Ancillary buildings were added the following year [11]. The consulting engineer in both instances was Alfred Goldstein of Travers Morgan & Partners, a fellow speaker at the 1952 symposium and the promoter of a new method of calculation for shell structures. Two years later he designed the Clifton Bridge (see Section 17.2.3), for which the C&CA was to test a model. It seems that advances in shell design were shared by a limited pool of engineers.

17.2.2 Prestressed road bridges (1953-1956)

Other foundations were laid early on for work on: prestressed flat slabs; continuous prestressed beams and prestressed, interconnected-beam bridge decks; comparisons of prestressed and reinforced-concrete beams; and on friction in prestressing cables. The tests were variously on site and on full-sized elements in the laboratory and, though there is no record of model testing on bridges in 1952, that was soon to change in preparation for the expected announcement of a major government road programme. With road building suspended since the late 1930s, there was considerable pent-up demand for improved communications in the UK and the C&CA saw a major opportunity for concrete and other cement-bound materials. Just as it did for prestressed concrete, it established a development group for road paving. At the overlap of these two interests, was the development of the prestressed road bridge. Here there was scope for the reduced-scale model work:

> 'We have also made good progress in our studies of methods of designing wide bridges of comparatively small span of the type that would be used for motor roads. The analysis of the stresses in these bridges presents considerable difficulties especially when one of the single heavy loads now becoming common on our roads runs close to one side of the bridge. The

> experimental work has been done at Wexham Springs on models but has been checked on full-scale bridges with the help of the bridge engineers of the Ministry of Transport who provided their 120-ton vehicle for loading tests' [12].

Initial experiments directed research towards more practical structural forms, investigating the design characteristics required for T-beam and box-beam structures, and into the effects of skew. Such work would dominate the Structures Section's output in the mid-1950s, once the £50m road programme was finally announced in December 1953.

1954 proved to be a busy year in this particular field, with the Structures Department investigating interconnected beam grillages, beams and box slabs, in close co-operation with the Engineering Section of the Ministry of Transport. As they explained, 'their problem is to be able to carry safely very heavy single abnormal loads of up to 200 tons, on bridges which are also designed for a moderately heavy distributed load. The Association's research enables an assessment to be made of the possible redistribution of these abnormal loads'. The Technical Director went on to declare, 'our work on bridge decks has aroused keen interest among bridge engineers and the meeting of the Institution of Structural Engineers at which Dr Morice and Mr Lewis read a paper on the subject [13] was the occasion of a public appreciation by the Chief Bridge Engineer of the Ministry of Transport' [14, 15].

Theoretical work, and full-scale testing of bridges in situ, was supplemented by laboratory tests on models [16], and a paper on this aspect was prepared for the major conference entitled 'Correlation of Stresses and Displacements by Calculation and/or Laboratory Experiments' [17]. Other output leading up to that included a series of technical reports of restricted circulation, prepared by G. Little: *The Distribution of a Load in a Box-section Bridge from Tests on a Xylonite Model* [18], *Laboratory Tests for Load Distribution in a Model Prestressed Bridge* [19] and, with R.E. Rowe, *Load Distribution in multi-webbed bridge structures from Tests on Plastic Models* [20].

Little's initial experiments to test the theoretical calculations of Guyon and Massonnet were on a Xylonite model constructed to the smallest practicable scale (Xylonite was a celluloid-type material patented in 1869). At a scale of 1:40, the miniature bridge was only 18 in. long and 9 in. wide (45 x 22.5 cm), and consisted of seven main beams and an equal number of diaphragms, uniformly spaced, with top and bottom plates (Figure 17.1). The model structure was placed on a Rolled Steel Joist, which provided a datum for the deflectometers. With Xylonite's modulus of elasticity only 10% of steel, small loads produced quite measurable deformations. These were obtained by the application of 7 lb weights and measured by electrical-resistance strain gauges. 'Transverse deflection profiles were recorded at each line of diaphragms for different positions of a single load concentrated on an area 1 ½ in. x 1 in. The corresponding longitudinal and transverse strains were measured by electrical-resistance strain gauges'. Little observed:

> 'The experimental distribution of deflexions and longitudinal bending moments proved to be a close agreement with the distribution given by a theoretical treatment involving distribution factors. The distribution patterns were found to be virtually unaffected by changes to the longitudinal

Figure 17.1 1:40 scale Xylonite model test of prestressed concrete beams for a road bridge, 1954. (Source: [22] Figure 8)

> position of the load. On the other hand, there were large discrepancies between the experimental and theoretical transverse bending moments. … Nevertheless, it was confirmed that the maximum transverse moment occurs at the centre of the bridge when the load is at its position of minimum eccentricity' [21].

Little's second series of model tests were to confirm or contradict the conclusions drawn from measurements taken from a full-size structure, the Fleet Bridge in Hampshire, using a common-enough method of constructing short-span bridges: 20 precast beams placed side by side and stressed transversely to form a slab, skewed at 15°. Here the load test, with the Ministry of Transport's abnormal loads trailer, had been discontinued at 72 tons when a radical change in distribution occurred and attention was transferred to a scale model. At a scale of 1:4 this model was considerably larger than the Xylonite example (Figure 17.2).

The tests were for a single concentrated load at mid-span. For this the bogie of the MOT abnormal load trailer was also reproduced at ¼ scale. The loads were steadily increased and, while recovery of the model at 7 tons (equivalent to almost 100 tons on the actual bridge) was 'reasonably good', 14 beams bent and cracked at 9 tons (144 tons at full scale). However, as Little acknowledged, 'It would be misleading to scale up these figures to give the ultimate behaviour of the full-size bridge since all of the cables in the model were ungrouted whilst, in addition, the amount of transverse prestress in the model was rather less than in the prototype'. 'Neither cable size and position nor dead weight stresses could be scaled down effectively', he added. Despite these caveats, he felt that 'the desirability of a considerably higher amount of transverse prestress has been demonstrated for full efficiency and economy in behaviour' [23].

Early in 1955, Little and Rowe undertook further experiments on plastic models (both Perspex and Xylonite) to test load distribution in multi-webbed bridge structures against theoretical interpolation formulae. Besides returning to the

Figure 17.2 1:4 scale reinforced-concrete model of Fleet Bridge in Hampshire being tested to confirm measurements made on the full-scale structure. (Image: C&CA Archive)

Xylonite model referred to above, they repeatedly tested a beam-and-slab model in Perspex [24]. The purpose was to find the correct interpolation to be given to the torsional parameter, to be consistent with the assumptions in theoretical analysis of bridge structures intermediate between a no-torsion grillage and a full torsion slab and to establish the validity of a formula based on Massonet and Guyon co-efficients. Starting with a simple T-beam, with a span of 18 in and width of 12 in, more diaphragms were added successively as the test progressed. These ranged from none to three, equally-spaced in the span. To test the load, the slab was simply supported with a hinge at one end and a roller bearing at the other, and the load applied at mid-span. The beams were also subjected to torsion and flexure.

This programme of research into prestressed concrete bridges was summarised in a paper presented at the IABSE congress and published as 'Load Distribution in Right Highway Bridges' in its proceedings in 1956 [22, 25]. It described work undertaken on the one-quarter scale model, a multi-webbed box model in Xylonite and beam-and-slab bridge model in Perspex described above. Recalling the introductory statement in 1954, the authors explained the work had been 'necessitated by the Ministry of Transport's recent introduction of an 'abnormal load' of 180 tons, and therefore an accurate assessment of its effects on new designs for bridges, especially medium-span bridges where the effect of a single vehicle could be greater than the distributed effect of traffic'. The C&CA experiments were compared with analyses based on linear-elastic theory, which were described as 'particularly applicable' to prestressed concrete structures. The authors concluded that 'The agreement between the theoretical and experimental values is very good' and that 'The investigations have shown that the interpolation formula can be used to determine the distribution of load in a uniform bridge structure which is neither a slab nor a simple grillage'. They also argued that the actual flange widths should be used, not the effective flange widths, as laid down in current Codes of Practice [22]. Their work was an implicit call for change.

Besides this fundamental research in association with the Ministry, the practice soon spread to a wider constituency, as consulting engineers approached the

Figure 17.3 A small model in Perspex of interconnected portal frames to correlate with the linear analysis techniques using matrices. (Image: C&CA Archive)

Association to test aspects of their designs for proposed structures. In September 1954 the Technical Director was able to report: 'An important part of our work in structural concrete is the testing done at the request of designers on models, prototype members or complete structures. This work gives us valuable general information and provides a service that cannot be obtained elsewhere'. Of the models prepared in this manner, the one with perhaps the highest profile was the Clifton Bridge in Nottingham, as Collins was keen to highlight: 'One of the models tested has been that of the new bridge over the Trent which will have the largest span (275 ft, 84 m) of any prestressed bridge in the country' [26]. Looking back at this period, Morice also identified the Clifton Bridge as pivotal:

'Towards the late 1950s the expertise of the Structures team in building small-scale models was applied to aid the design of complex structures in a similar way to the work recently established at ISMES in Bergamo. … Elastic small-scale models in Perspex, fully strain gauged and appropriately loaded were built of proposed structures, which included the Clifton Bridge at Nottingham for Travers Morgan & Partners. Very soon, following a careful in-depth study, more ambitious models in micro-concrete, fully prestressed, were built and tested' [27].

'The small-scale model tests', he added, 'provided an opportunity to compare experimental results with those produced by the matrix theory' recently developed by Morice himself, R.S. Jenkins and J.C. de C. Henderson of Imperial College (Figure 17.3).

17.2.3 Clifton Bridge, Nottingham (1954-1955)

As the largest prestressed span in the UK, using both the Magnel-Blaton and Lee-McCall systems, this bridge was bound to attract attention. A new bridge to carry the A52 over the river to the west of Nottingham, and relieve pressure on the established Trent Bridge, had been proposed for some years. Without central funding no progress had been made until, under pressure from the suburban expansion, the city council undertook to finance it, to trunk road standards, out of local funds.

Travers Morgan & Partners had produced a design but, given the scale and novelty of the project, approached the Association to confirm the results of calculation by means of model tests. The designers had concluded that, if a reasonable longitudinal-compressive-stress margin in the tensile zones were obtained by prestressing, and if the reinforcement were carefully detailed to avoid weak planes in cross-section, no additional specific measures need be taken to deal with the skew effects in the cantilever. Model testing would obtain deflexions for load and prestress with which to confirm this analysis. It would also assess the eccentricities of deck reactions at the abutments and piers caused by the skew and incorporate the findings into the design of the prototype.

Accordingly, a model of the north cantilever and end span was made of Perspex to a scale of 1:50 (Figure 17.4). Fillets were omitted and plain butt joints used. The end block was simulated by a rectangular section of laminated Perspex of equivalent moment of inertia. At the abutment end, hardwood was used for this purpose to allow fixings to be attached. For loading, tins of lead shot were used, supported from wire loops that were passed through the top and bottom slabs and over small wooden blocks. As the number of sufficiently accurate gauges was limited, only five deflection readings could be taken for any one loading condition.

Preliminary tests for Young's modulus were carried out and both creep and recovery were found to be acceptable. When the model was loaded, it became apparent that torsional rotation occurred at the pier and abutments, so transverse-rotation calibration tests were performed to enable the correction of further measurements. The model was also tested to ensure no transverse horizontal deflections occurred when loaded.

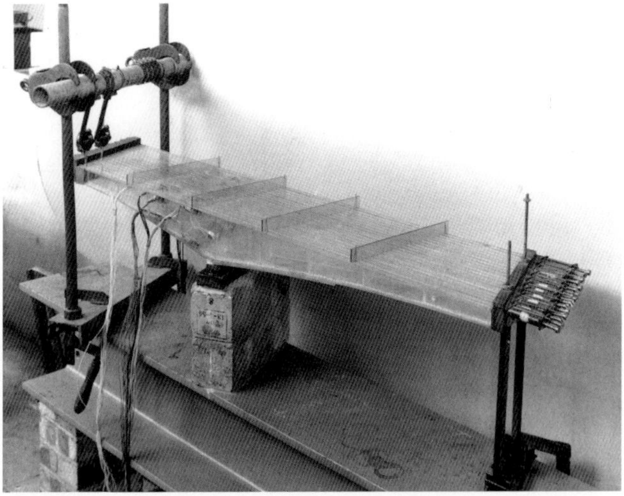

Figure 17.4 1:50 scale Perspex model of Clifton Bridge, 1954. (Image: C&CA Archive)

Load-deflection tests were undertaken in two stages. Firstly, loads were suspended in 5 lb increments, with deflection plotted against load. A load of 25 lb was then selected as most appropriate for the second stage, which consisted of 'taking deflexion readings for all grid points for the test load in turn at all grid points'. To apply the prestressing force, twenty piano wires were positioned over the top slab and a stressing screw turned to compress a calibrated spring. The slack in the wire was taken up at the other end by a grub-screw anchorage, and the stressing screw released. When comparing the early prestressing results with calculations, a discrepancy was discovered and lengthy investigation of the possible causes ensued; the actual load in the wires was established by measuring with a microscope the extension of a known length. Further work followed to determine the effect of the skew on the abutments, which was found to be considerable.

In an extensive account of the testing process [28], it was acknowledged that 'many teething difficulties were encountered in obtaining satisfactory functioning of the model and prestressing equipment' and that, in all, the lab work took four months to accomplish. However, the objectives of the test were obtained and it was concluded that the design adopted was adequate and the assumptions confirmed. That said, the rocker columns and abutment base were increased in size and the degree of prestress raised, directly as a result of the tests.

Once the bridge had been built, in 1956, the Prestressed Concrete Development Group arranged a visit by 348 of its members, followed by a luncheon as guests of the City Corporation. The Perspex model was then displayed jointly by the Association and Travers Morgan at the ICE Conversazione on 20 June 1956. The scheme was finally completed and opened to traffic in March 1958.

17.2.4 1957

1957 was Morice's last year as head of Structures and, while the direction of the Department's work continued much as before, there was a 'large increase in consultation by engineers to local authorities', particularly relating to 'design problems arising with concrete bridges and the abnormal loads'. Experimental work in the laboratory included the construction and testing of a second model of a proposed precast, prestressed concrete box-section bridge, suitable for small spans and capable of sustaining the MOT abnormal vehicle. It was followed by model tests of two small-scale bridges, using prestressed concrete inverted Tee beams and in-situ concrete filling, to compare the use of transverse prestress and transverse mild-steel reinforcement on the distribution of loads in the bridge deck.

A number of ad hoc tests were carried out for, and assistance given to external designers. These ranged from full-scale to micro-structural model tests, and included the construction of ...

- A Perspex beam of an 'Intergrid floor' for the Prestressed Concrete Co.
- A Perspex beam for Stressed Concrete Design Ltd
- A Perspex model of the interconnected portal frame to be incorporated in a boiler house, for Clarke, Nicholls and Marcel of Bristol

... along with assisting the consulting engineer E.W.H. Gifford in his test on a scale-model prestressed concrete beam to be used in the Perth Narrows Bridge, Australia.

Significantly for the department's future direction, R.E. Rowe reported in March 1957 on a visit to the *Istituto Sperimentale Modelli e Strutture* (ISMES) at Bergamo in Italy, one of the world's leading research centres involved in the use of scale models for testing structural designs against theoretical analyses [29]. The object of ISMES was to help engineers assess the actual behaviour of concrete structures as distinct from that given on the basis of certain assumptions and mathematical analyses and, in particular, to provide a means of quickly testing intuitive designs for intractable problems. Tests on models were the obvious means by which assistance could be given since they enable engineers to either confirm or improve intuitive designs relatively quickly. To further its objective the Institute had studied in all its aspects the subject of similitude between models and actual structures and had developed materials which simulate the behaviour of concrete. Rowe's description reveals clear parallels with the emerging pattern of work at the C&CA and is an early indication of conscious policy choices that were to shape its programme in the early 1960s. Morice left the C&CA in December 1957, to take up a chair in engineering at the new University of Southampton, and was succeeded by his former deputy, Roy Rowe.

17.3 The Rowe years I (1958-1966): model testing on a cost repayment basis

The annual report of September 1958 opened with a repeat of the previous year's statement that the Association was '... increasingly consulted by engineers to local authorities before the preparation of bridge construction projects, during the subsequent calculations and detailing and during the construction stage ... This co-operation is encouraged by the Ministry of Transport and especially by the Bridges Section' [30]. The similarity with Bergamo can be seen, though ISMES offered a fully-commercial service. The year's work programme included theoretical work, publication of *Right Bridges* [31], a computer program for beam design and experimental work on three composite slab bridges and also, specifically, scale-model tests.

Although the C&CA was well-funded through a levy on the price of cement, it was not averse to involving other interested parties, and the uptake of consultancy work appears to have suggested a semi-commercial approach, with model testing of structures undertaken by the Association on 'a cost repayment basis'. It was introduced 'to provide facilities for investigating those problems arising during the design of specific structures which could not be solved by normal methods of analysis' [32]. Perhaps this development was an expression of the 'simple cost-display system for research projects' that had been developed in 1953. At the time it was said that its value had 'yet to be determined in use', but five years of research had stimulated a market.

While bridge construction was to be the backbone of the research programme, scale-model testing under the new regime found its earliest application to buildings. According to the annual report, 'The use of small-scale models to determine the [stiffening] effects of cladding and infilling upon the behaviour of framed structures is being studied. Several small frames filled with plaster have been tested and a second series of such frames is now being prepared'. But shell structures were of particular interest: 'During the year a study of the design and analysis of hyperbolic-paraboloid shells was started. A 1:10 scale model of an actual shell of this type was cast and is now under test. Further shells will be tested to facilitate the analysis' [33].

17.3.1 Shell roofs (1958-1961)

This first model of a hyperbolic-paraboloid was for the consulting engineer working on a garage and showroom project for the Lincoln Motor Car Company, designed in 1958 by Sam Scorer of the architectural practice Denis Clarke Hall Scorer & Bright. Realizing the novel roof of the garage was the responsibility of Hungarian émigré engineer Dr K. Hajnal-Konyi. It consisted of four hypar shells, each 2 ½ in. (63 mm) thick, supported on slender columns to provide an unobstructed area 96 ft by 106 ft 9 in. (29.3 x 32.5 m). The lower corners of the slabs were connected by a prestressed tie-beam cast monolithically with the slab. Hajnal-Konyi approached the Association to test a scale model of his proposals. It was commissioned to confirm his assumptions and, particularly, to determine the effect of the tie-beam on the stresses in the slab and the stresses due to the imposed load. For its part, the Association hoped the test might provide additional information that would guide similar designs in the future.

The model was to be at one-tenth scale. It was made of wire-reinforced sand/cement mortar, cast in a timber mould to a thickness of 0.25 in. and cured under hessian. The wires were secured in holes drilled through timber edge strips. Once stripped, the shell was positioned on four corner supports, each with a ball seating that was constrained to move horizontally in the direction of a diagonal. A tie consisting of two Lee-McCall bars connected the lower corners. To simulate a uniformly-distributed load, weights were suspended from 48 tripods mounted on the shell to create a total of 144 point-loads. Vertical deflexions were measured at 25 positions on the shell surface, as well as at the four supports, and strain was measured at various points with 114 electrical-resistance foil gauges (Figure 17.5).

Testing was protracted. Strain readings were possible during a very limited time, allowing only ten readings to be recorded in a five-minute test. Consequently, a minimum of twelve tests had to be carried out for each of the seven load stages. On completion of the tests, in 1959, the shell was finally loaded to destruction. In the event, it performed better than designed, with a load factor of at least 5, and provided data from which the tie-bar force, corner reactions, deflexion and stress distributions in the actual structure were calculated. The findings of the test programme were presented by Hajnal-Konyi at the international colloquium on construction processes for shell structures held in Madrid in September 1959, and recorded by L.L. Jones in a C&CA Technical Report [34]. The actual roof in

(a) (b)

Figure 17.5 1:10 micro-concrete model of the hyperbolic-paraboloid roof for a car showroom in Lincoln, UK, 1958. (a) The test rig. (b) Detail showing means of applying the loads. (Images: C&CA Archive)

Lincoln was built in 1959, becoming the largest of its type in the UK – though it earned the epithet 'a Modernist monstrosity' when listed in 2000.

A similar project followed in 1959, this time for a series of umbrella shells at the Texas Instruments factory, Bedford. Its distinctive roof was to consist of 29 slabs, each 48 ft (14.6 m) square. The client, with a head office in Dallas, wanted the hyperbolic-paraboloid roof to be cyclone-proof and engaged Oscar Faber & Partners as consulting engineer. A one-sixth scale model was commissioned, primarily to determine the deflection under uniform loads and non-uniform loads such as (perhaps more realistically for British weather conditions) snow. Testing established the stresses in both the slab and edge beams and obtained the ultimate load for its designer. The Manton Lane site opened for occupation in December 1960 and was notable not only for its hypar shells, but also the use of the lift-slab method of construction.

A third model, a pseudo-hyperbolic-paraboloid roof for the exhibition hall at the Commonwealth Institute in Kensington, was started in 1959, though the study ran on into 1960. This time the model was much smaller, at a scale of 1:32, and made of the by now familiar Perspex. The eight-month study cost £680 (Figure 17.6). The proposed design by A.J. and J.D. Harris was for a roof 186 ft 6 in. square: a central hyperbolic paraboloid, 93 ft 3 in. square, supported at the two low corners and surrounded by four trapezia, each warped in the opposite way to the adjoining part of the central square. They were designed as rectangular lattices of slender concrete, pin-jointed, and it was proposed that the tension members should be of prestressed concrete, and the compression members of reinforced concrete.

The model agreed on was one of the trapezoidal warps, with the aim of determining deflexion patterns. Load was applied by hanging containers of lead shot from the nodes of the model; the weights were supported on a platform under the model which could be lowered in a single movement to impose the loads simultaneously over the whole shell. As in similar tests, deflexions were measured with dial gauges, and strain by electrical-resistance strain gauges connected to a photographic recorder. Test results indicated the complex nature of the roof's

Figure 17.6 1:32 scale Perspex model for the roof of the Commonwealth Institute, London. (Image: C&CA Archive)

structural behaviour and 'confirmed that reliance on an analysis based on the assumption of pinned joints would have been unwise'. The deflexions were shown to be unacceptable and the type of edge support identified as critical; the design was modified accordingly. The lessons learned were published in a C&CA report by Geoffrey Base [35].

The fourth, and most spectacular model test from this period was for the roof of Smithfield Poultry Market in London. It was to be roofed with an elliptical-paraboloid shell, 68.6 m x 38.7 m in plan, with a rise of only 9.1 m. It was said to be the 'boldest' shell ever constructed, where boldness is defined as span / (rise x thickness). The membrane was to be 75 mm thick, except where it thickened towards the edges, and would be pierced by 24 skylights. It was constrained by a prestressed edge beam, which would tie the corners and would itself rest on 32 perimeter columns. Ove Arup & Partners approached the Association to investigate whether buckling would be a critical factor in the design and to test a scale model of their proposals. A 1:12 scale model was constructed in micro-concrete and reinforced with 1.2 mm diameter wire. 3600 spacers were needed to stabilise the reinforcement while the concrete was poured. Unlike the final design, prestress was applied externally. The dimensions of the model were impressive – 5.72 x 3.23 m and just 6 mm thick (Figure 17.7) [36, 37].

The tests aimed to resolve a number of key questions:

- Establish the failure mode and failure load
- Measure the elastic behaviour to validate the model
- Determine the loads on the perimeter columns
- Determine deformation around the lighting holes
- Compare measured behaviour with the results of elastic analysis
- Verify that post-tensioning of the edge beams would lift the shell off its formwork.

Figure 17.7 1:12 scale model for Smithfield Poultry Market, 5.72 x 3.23 m. Formwork and reinforcement are in place, awaiting the 6 mm layer of micro-concrete. (Image: C&CA Archive)

Deflexions of the shell were measured to an accuracy of one thousandth of an inch (0.025 mm) using 44 clock gauges attached by long wires to the shell in 44 positions. 55 electrical-resistance, foil strain gauge rosettes (12 x 12 mm) were attached to one quarter of the shell – a total of 330 active gauges (Figure 17.8). Mechanical strain gauges were used to measure movement of the edges of the shell and the ring beam.

The load was applied pneumatically using eight air cushions in five different arrangements:

- Uniformly over all four quarters of the shell
- On one half of the shell (divided by the short axis), measuring the behaviour of a loaded quarter
- As above, measuring the behaviour of an unloaded quarter

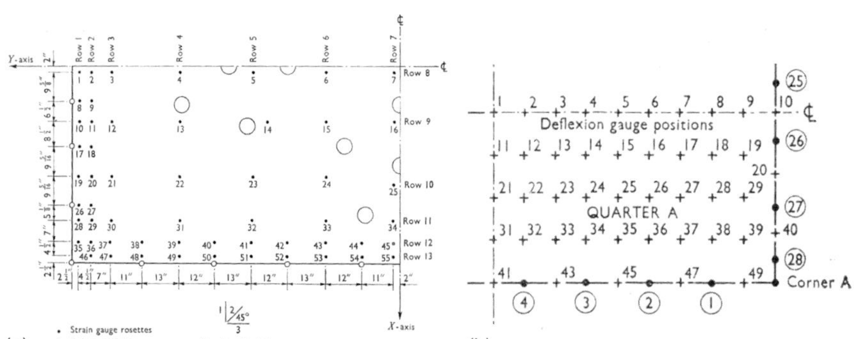

Figure 17.8 Measuring points on the model. (a) Positions of the 55 strain-gauge rosettes (on one quarter of the shell only). (b) Positions of the 44 clock gauges for measuring vertical deflexions of one quarter of the shell. (Source: [37])

- On one half of the shell (divided by the long axis), measuring the behaviour of a loaded quarter
- As above, measuring the behaviour of an unloaded quarter.

The strain-gauge readings were recorded photographically but only 100 could be recorded for each load test. Given the number of loading patterns, the different sizes of loading and the limitations of recording the data, over 200 load tests had to be made, generating some 80,000 strain measurements during the four-month testing programme.

The model was eventually loaded to failure to determine the distribution of load (Figure 17.9). The prestress in the edge beams was annulled when the load was only a little above the design load and so failure of the shell would be caused by the spread of cracks into the membrane. As the designer was more concerned with buckling load, the corners were heavily restrained to prevent failure initiating from the edge beams. Buckling failure was eventually achieved at a total load of 9.1 kN/m^2, though this would have been much lower without the restraint. That accomplished, two undamaged sections of shell were retrieved and loaded to determine the modulus of elasticity.

The conclusions drawn from the series of tests yielded answers to the various questions posed:

- Failure was by buckling (not shear), as expected
- Failure load was about four times the design load
- Direct stresses in the shell were low
- Bending stresses were generally of a similar magnitude, but higher in the region of the edge thickening
- There were no significant stress concentrations around the light holes

Figure 17.9 The model just after failure. Four air bags are still in position on the undamaged half of the shell. The restraining ring beam is also visible. (Image: [C&CA Archive])

- Prestressing had to be raised 40% to prevent cracking of the shell
- Post-tensioning of the edge beams would raise the real shell 9 mm off the formwork to permit decentring.

The project cost £2,400 to complete.

17.3.2 Symposium on Models for Structural Design (1959)

The Department's work on shells was reported to a wider audience in a symposium on Models for Structural Design, organised by the Association and held at the ICE on 29 July 1959 [38]. It attracted 350 delegates. Speakers included Prof. Guido Oberti, Director of ISMES in Bergamo, Prof. Carlos Benito, from the material testing laboratory in Madrid (who later succeeded Torroja as director), Prof R.C. Redshaw (Birmingham), and the architect, Richard Sheppard. Besides accounts of a model dam tested at Bergamo and a model of the new bridge at London Road Station at Manchester, tested at Imperial College [39], six C&CA projects were presented by Roy Rowe. These included the recently-tested models of hypar roofs at Nottingham and Bedford, and looked back at Clifton Bridge and the long-running work on slab bridges used to establish a theory for calculating transverse bending moments. In describing the Medway Viaduct, Rowe also looked forward to a series of multi-span structures – viaducts, approach roads and elevated roadways – that were to dominate the Department's work in the early 1960s.

17.3.3 Bridges and multi-span structures (1959-1962)

The proposed 21-span Medway Viaduct, for which Freeman, Fox & Partners were the consulting engineers, was, at 3,270 ft (997 m), to be the longest bridge constructed in prestressed concrete anywhere in the world. The innovative design provided for end spans 300 ft long, two cantilevers each 200 ft long and a central suspended span 100 ft long. To test the distribution of stresses, a model of the cantilevered beam was constructed in micro-concrete at a scale variously described in the reports as 1:13 or 1:14 (Figure 17.10).

Another novel multi-span structure tested by the C&CA in 1959 was the Hammersmith Flyover. As little was known about the ultimate behaviour of prestressed concrete beams under simultaneous bending, shear and torsion, it was thought advisable to conduct a scale-model test to check the design, especially the ultimate strength of the critical section above the joints. The Association undertook to build and test a realistic model of part of the spine beam, consisting of four beam segments and four cantilever units projecting out to either side of a column. Like the completed structure, the model was made of prestressed concrete, though scaled down to 1:12 (Figure 17.11). It was tested under two loading regimes in which the respective conditions of each carriageway were varied, and the ratios of bending moment, torsion and shear obtained. The model was eventually loaded to destruction and failed only after the more onerous loading requirements had been achieved. The programme took five months and cost £1,000. Construction proceeded in the confidence that the designed strength

Figure 17.10 1:13 (or 14) scale micro-concrete model of Medway viaduct, 1959. (Image: C&CA Archive)

was sufficient for the purpose intended. Indeed, the final structure incorporated an additional upstand that had been deemed necessary for construction of the model, by which its effectiveness had been demonstrated [40].

Other work on models progressed at the Association, and was the subject of a report by Rowe published by the Portland Cement Association in America [41]. One of the models was a quarter-scale model of a standard design for overbridges, also described as 'Shear Connector Bridge' proposed by the British Transport Commission [42]. The year-long testing programme cost £1,450.

17.3.4 Commonwealth projects overseas (1961-1962)

The example of the *Istituto Sperimentale Applicazioni in Calcestruzzo* (ISAC, the predecessor of ISMES) in Bergamo – established in 1946 to test designs for the Piave River dam and partly funded by the hydro-electricity industry – is hinted at by the C&CA's next choice of project. In his report, Rowe had written: 'Hence it is possible to determine the factors of safety or load factors of such structures as large dams, built on foundation strata of varying properties, from tests to destruction on appropriate models' [41]. The C&CA would test its first dam in the summer of 1961, when it prepared a 1:32 scale model of the Mangla Dam for a hydro-electric power scheme in Pakistan (Figure 17.12). This dam consisted of a prestressed cantilever of variable section with a large-diameter aperture for one of the inlet pipes near its base. The test, commissioned by the consulting

Figure 17.11 1:12 scale prestressed, micro-concrete model of precast units for the Hammersmith Flyover, 1959. (Image: C&CA Archive)

engineer Mr H.G. Cousins, was to determine the stress distribution caused by hydrostatic pressure on a section of the cantilever close to the pipe. A model was built in micro-concrete (necessary to establish the three-dimensional state of stress), and tested in a horizontal position in the laboratory over a six-month period. The results obtained were used in detailing the actual structure.

Other Commonwealth projects were undertaken at this time. For five months in late 1961 the Association built a 1:48 scale model of Huntley's Point Overpass, part of the Gladesdale scheme in Australia. This was a hollow spine beam bridge, of varying-section and with a cantilevered deck; notably, it was curved in plan along a radius of 350 ft along its centre line. The prestressed concrete bridge was being designed by G. Maunsell & Partners and so the project was to harness the Association's experience in bridge testing, and to build on an established relationship with the consulting engineer (having worked together on the Hammersmith Flyover, the two parties went on to test designs for the Mancunian Way in 1963). G. Maunsell & Partners required the self-weight deflexions and reactions during early construction; changes caused by jacking or tying down at various supports; and the live-load influence lines for deflexion and reaction in the completed structure. Perspex was chosen as a material suitable for the model for the study of elastic behaviour, concerning only deflexions and reactions.

Figure 17.12 1:32 scale model of the Mangla Dam, Pakistan, 1961. (Image: C&CA Archive)

The tests simulated self-weight and point loading with bags of lead shot, and the distribution of deflexion was established.

It was followed in 1962 by a project of an entirely different type, though also located in Australia: the Sydney Opera House. The Association's involvement in this high-profile project was commissioned by Ove Arup and Partners, and was to test the use of resin as a means of jointing precast-concrete units. Specifically, it was to assess the shear strength of the joint. A series of tests were therefore carried out to simulate the stress conditions existing in the joint, formed in different ways and with various release agents. These enabled the design stresses to be assessed and informed the construction procedures eventually adopted on site. The model tests for the main shells of this building were undertaken at the University of Southampton directed by Professor Morice.

17.3.5 Metropolitan Cathedral at Liverpool (1961-1964)

Unlike the traditional Church of England cathedral also being erected in Liverpool at the time, the Roman Catholic metropolitan Cathedral was of a strikingly novel polygonal design. It was a very large structure: 220 ft tall, excluding the non-structural crown of thorns, with an overall-all diameter of 345 ft. The clear distance inside the frames was almost 200 ft. Its size alone made weight an important consideration and so the design aimed to make the structure as light as possible with minimum cross-sections for the structural members.

A model test was commissioned in 1961 by the consulting engineer, Messrs Lowe and Rodin, to provide information on the interaction between various components under self-weight and wind loading, and the effects of differential shrinkage. Physical confirmation of the design assumptions were deemed all the more necessary when considered against the intended lifespan of 500 years. To achieve that, all possible abnormal loads were accounted for by the engineers, including hurricane-force winds and crashing aircraft cutting through one or more of the ribs. And although the assumptions were conservative, the designers wished to 'add a few more drops to the bucket of knowledge'; they were prepared to cover the cost of a model test. (By the end of the 16-month programme, this eventually amounted to £2,500.)

As the model was needed to study both elastic behaviour and to determine the ultimate strength, a single large model in reinforced micro-concrete was proposed, to a scale of 1:32. To have made the model any larger would have been impractical, as the loading rig, twice the size of the model, already filled the available laboratory space. Changes to the design, however, forced a reduction in scale to 1:29. The eventual model was therefore 7 ft 8 in. (2.34 m) tall with an overall diameter of 12 ft (3.66 m).

The micro-concrete was a continuously-graded mixture of Heston sand and aggregate in a ratio of 3:1. The water/cement ratio was set at 0.55 to achieve the workability required to fill the congested sections. Scaled-down reinforcement of acceptable stress-strain properties was provided in the form of wires of two diameters, grouped together and fabricated into cages on a jig. The task of constructing the model was protracted. The foundation plate or 'reaction platform', was cast in April 1962 (Figure 17.13); the ribs mounted around a polystyrene core former and the ring beam were then added. 32 precast panels were glued into place with Araldite. The shallow dome, which provided the horizontal stiffness, was simulated by a sheet of plywood. The self-weight stresses in the model were only 1:29 of those in the full-size structure and the balance – approximately 9.25 tons – had to be applied by an external rig to 208 point loads distributed to represent the weight of the structure.

Testing started in June 1963, with complete loading and unloading cycles carried out in five minutes and the results averaged. Following the tests to simulate self-weight, a further series was carried out to investigate the effects on the ribs of shrinkage in the ring beam and panels. Wind loading was simulated by applying 128 point loads to the ribs of the model in three different patterns, determined from wind-tunnel tests carried out by the National Physical Laboratory in Teddington (Figure 17.14). Each loading was applied at least three times. The final test – loading to destruction – was carried out on 27 June 1964. With the notable exception of the cooling towers at Ferrybridge power station (see Section 17.3.10), the C&CA's model testing programme

(a) (b)

Figure 17.13 1:29 scale model of Liverpool Metropolitan Cathedral, 1963. (a) Formwork and reinforcement for the base, awaiting micro-concreting. (b) Detail of the ribs, the plywood sheet (representing the shallow concrete dome which provided horizontal stiffening), and the reinforcement for the ring beam awaiting the micro-concrete cover. (Images: C&CA Archive)

throughout the rest of the decade was dominated by viaducts or elevated roadways.

17.3.6 Cumberland Basin scheme, Bristol (1962)

Following on from the Association's earlier bridge work was a project to assist the design of elevated roadways in Bristol's crowded city docks. Here roads were to be supported at different levels on columns arranged in threes across the width of the carriageway, spaced at regular intervals. Where one road was to cross another, especially at a skewed angle, it would become necessary to rearrange the position of the columns from two rows on opposite sides of a rectangular section of slab. The transitional arrangement of slabs would incorporate a skewed parallelogram flanked by two trapezoidal spans, before the columns could again be arranged perpendicular to the line of the road. The designer, working for Freeman, Fox & Partners, required information on column reactions and slab stress, each under two categories of loading.

In discussing a possible model test, the C&CA agreed to build a scale model of the parallelogram and trapezoidal sections, extended to the standard rectangular section to account for end effects. Loading it would provide data on the distribution of moments and reactions in the structure. This would enable more effective placing of reinforcement and efficient sizes of columns and column heads. In addition, the designers asked for information on the effect of a single point load anywhere in the parallelogram to help understand such structures better and

Figure 17.14 1:29 scale model of Liverpool Metropolitan Cathedral, showing the system for applying wind loads to the structure. (Image: C&CA Archive)

contribute to a more knowledgeable approach to designing complicated continuous slabs in future. As all the information needed related to working-load conditions, an elastic test was all that was needed. With the brief agreed, a model was cast to a scale of 1:24, resulting in a fully prestressed concrete slab 10 ft long and 1 in. thick (Figure 17.15). The design was modified for simplicity (and therefore economy) by dispensing with the edge-stiffening beams and widening the slab a little to compensate. In the end, the test took 5 months and cost £500.

The micro-concrete mix consisted of all-in sand and 1/4 in. Mount Sorrel granite in a ratio of 1:2; an aggregate to cement ratio of 3.0; and a water/cement ratio of 0.42. The slab was post-tensioned uniformly in both directions, to 1,500 lb/in^2 (10.3 N/mm^2) longitudinally and 1,000 lb/in^2 (6.9 N/mm^2) transversely. The force in each wire was estimated by measuring the wire's extension to one thousandth of an inch with dial gauges. At 35 days, electrical-resistance strain gauges were attached, bonded to the slab with Araldite. The slab was supported

Figure 17.15 1:24 prestressed micro-concrete model for the Cumberland Basin scheme, Bristol, UK, 1962. (Image: C&CA Archive)

on 20 columns, each comprising a load-measuring device with a ball seating at the top and a roller-bearing at the base. All the load cells were attached by threaded bars passing through the slab and placed under a slight pre-compression.

Loading was applied to provide two types of measurement: column reactions, measured by load cells; and strains, measured with an automatic data-logger. Uniformity of load was simulated by 16 point loads applied to each fifth of the slab, measured with a load cell, and the results superposed. In each loading position a total of 5,000 lb (2,268 kg) was applied in 1,000 lb steps, the step readings enabling the linearity of the load-strain and load-reaction relationships to be checked.

An articulated 1:24 scale model of the standard abnormal heavy-goods vehicle was made and 4,000 lb applied to the model to simulate concentrated loading. Readings were taken every 4 in. as the load was applied successively along the slab, constrained by the position of the centre and kerbs. As many as 228 strain gauges were used on this model, exceeding the capacity of the recorder, and so they were divided into three groups and recorded in turn. Strain gauges were mounted on the upper and lower surfaces of the slab to check for consistency. Moreover, three sets of readings for each load position allowed a check for consistency on the column reactions, as these were read by a different recorder.

Over a testing period of 15 weeks, more than a quarter of a million readings were taken and examined, and the exercise noted as a natural candidate for the computerisation that was then being introduced. The data were plotted as influence lines as the vehicle moved along the established lines of travel. These were presented as contours of maximum stress at each gauge position, allowing the designer to place and proportion the reinforcement to best effect [43, 44, 45].

17.3.7 The Manchester Skyway Bridge / Mancunian Way (1963-1964)

In 1963, G. Maunsell and Partners, the consulting engineers for what was then known as the Manchester Skyway Bridge, asked the Association for a model test to provide fundamental design data for the superstructure, principally the diffusion of stresses induced by post-tensioning cables.

The structure was to form the first section of a network of elevated roads and consisted of a 32-span viaduct. The superstructure was designed as parallel hollow-box spine beams of prestressed concrete, each carrying two cantilevered slabs. The two carriageways were separated throughout and each supported by single tapering columns with two bearings. It was expected that segmental construction would be used, in a manner similar to that employed on the Hammersmith Flyover (also designed by Maunsell). But, unlike Hammersmith, the cantilevers would be cast with the box beam in monolithic sections, and it was necessary to know how much of the prestress would extend into the cantilevers as the cable's profile changed within the web. Other information sought included the stress distribution under working loads; the torsional properties; the effect of point loads on the cantilever and the structure's ultimate strength.

A model of a typical span, with two quarter-span cantilevers to allow an assessment of the effects of continuity, was built in reinforced and prestressed micro-concrete to a scale of 1:12. This scale was selected as it yielded a minimum concrete thickness of ½ in., the smallest practicable size in view of the reinforcement and casting considerations (Figure 17.16). The micro-concrete mix consisted of gap-graded sand, with an aggregate/cement ratio of 2.70 and water/cement ratio of 0.50. The cube strength at 28 days was 7,000-7,500 lb/in^2 (51.7 N/mm^2). To reproduce a single span, with overhangs to simulate support

Figure 17.16 1:12 prestressed micro-concrete model of the Mancunian Way, 1963-64. (Image: C&CA Archive)

conditions at each end, required 27 precast sections. These were mortared together to make a model about 17 ft long. Prestressing force was gradually applied through six 1/4 in. diameter strands, three in each web. Casting and assembling such a model took quite a while, though preparing segments in pairs helped speed the process. However, fabricating the reinforcement cages to the required degree of accuracy was particularly laborious, with special jigs made to ensure correct spacing and cover [46, 47, 48].

Prestress amounting to 39,200 lb was required for the finished model, but this was applied gradually, in stages. Firstly 6,000 lb prestress was applied to enable the model to bear its own weight. Forces and loads were gradually added, according to a carefully-devised programme, to avoid tensile stresses opening up the joints. Sufficient force was applied at one end of a strand until a reading was obtained on the load cell at the other end, enabling total friction losses to be determined. This was then repeated at the other end, enabling an estimate of the prestress at mid-span to be obtained.

The dead load was applied at 12 points along the span by means of a jack and spreader system, while dead load at the overhangs was achieved by suspending concrete blocks beneath. The order of placing these blocks allowed any parasitic effects to be accounted for. Deflexions were measured at all load stages and strains measured at five points along the span, before and after each loading. Other tests subjected the bridge to both eccentric and central lane loading, and also to torsion, in order to measure the distribution of edge shears which would exist should the carriageway structure be connected to its twin, across the central reservation. Finally, in 1964, it was loaded to destruction. The laboratory programme had taken 11 months and cost £1,100.

In many ways this was a landmark test. By using micro-concrete and erecting the model according to the construction method proposed for the real structure, it was more realistic than many of the tests previously undertaken. Because the model had been envisaged from the beginning of the project, relevant information was available to the engineers at the time required. It was also planned from the start that the C&CA (and the Road Research Laboratory) would have access to the actual structure when completed in order to carry out tests, and so enable correlations with the model to be made. This research did take place (see Section 17.4) and the structure went on to win the Concrete Society Award in 1968 [49].

17.3.8 Meeting on *Model Testing*, 1964

The above schemes – Liverpool Metropolitan Cathedral, Cumberland Basin and the Manchester Skyway, along with Smithfield Market – were described in detail and presented to a wider audience at a C&CA meeting entitled *Model Testing* at the Royal Aeronautical Society in London on 17 March 1964 [50, 51]. Their selection for such treatment is partly a reflection of their currency, but also their importance as examples of this medium of structural analysis. Rather like the meeting in 1959, *Model Testing* represents a way mark in the adoption of such techniques by both the Association and the engineering profession generally, as a means of understanding the behaviour of structures and a reliable and practical guide to their design.

In his introductory lecture, Rowe rehearsed the C&CA's credo and set out the alternatives available to the structural engineer faced with design problems, identifying a conventional approach, based on the theory of elasticity and considering only the service load conditions; an ultimate-load approach, based on considerations of the structure's collapse; the application of computers to either approach; or fourthly, as he put it, 'the use of model analysis and testing as a design tool'. This had been the object of attention by civil and structural engineers in recent years, he acknowledged, though he also argued it has been a long-standing practice dating back to the time of Michelangelo.

Rowe described two types of model, both of which had been tested at the C&CA over the years. One type was the 'elastic model' used to test theoretical assumptions of ideal elastic behaviour, and in which the model is, in effect, 'a computer dealing with a specific problem'. The other was the 'realistic model', made of a material similar to that of the actual structure, used to test the actual behaviour of the structure and the difference between theory and practice. He outlined the pros and cons of various materials used in model testing: plastics such as Perspex (the subject of a C&CA study [52]), gypsum plaster, asbestos cement sheet, and cement mortar or micro-concrete. Perspex and micro-concrete were perhaps the principal materials used at the C&CA, while reference was made to the pumice or barites-based mortars used by the Association's counterpart at Bergamo.

Perhaps most importantly, Rowe explained how the C&CA – and indeed other institutions at Illinois and Bergamo – had attempted to establish correlation between the behaviour of a model and that of the actual structure simulated by it. His statement is worth quoting at length:

> 'In its earlier work on the load distribution problem in bridge structures the Association carried out research on small-scale bridges which were effectively models of hypothetical bridges. On the basis of the results obtained, a load-distribution theory was developed which was suitable for the design of actual bridge structures. This theoretical approach was then confirmed by tests on actual bridge structures and, in effect, the confirmation showed indirectly the excellent correlation between the behaviour of a model and the actual structure. In one of these cases, the Fleet Bridge, a 1:4 scale model was subsequently built and tested in the laboratory ... and, to summarize the findings, it may be stated that the correlation as regards both deflexion and strain (or bending moment) was within about 5% in both the linear-elastic and non-linear ranges of behaviour' [53].

Rowe concluded the conference with a brief review of earlier projects to indicate the time taken to construct and test the models involved and the financial costs involved.

17.3.9 Elevated roads (1964-1966)

Following the model tests for the renamed Mancunian Way were those for another elevated section of road, the two-tier Tinsley Viaduct on the M1 motorway near Sheffield. The proposed design of this viaduct, which stretched between Tinsley and Wincobank, was for upper and lower decks of prestressed concrete secured to a trussed- or steel box-girder positioned between them. The requirement was to test the proposed joints at this connection. The joint details and a portion of the girder and deck structure were modelled at one-fifth scale and tested under a system of loads that simulated the anticipated conditions. These tests confirmed the adequacy of the design, and construction commenced in Spring 1965. The Tinsley Viaduct eventually opened in March 1968.

Testing the designs for London's Western Avenue Extension, or West Way, between Paddington and White City, was commissioned in 1965 by the consulting engineers G. Maunsell & Partners. The contract related to Section 5 of the project, the stretch of road over the railway lines near Westbourne Park Station. Other tests relating to Section 6 were carried out at Imperial College.

Initially a Perspex model at a scale of 1:10 was made of the cross section to study the distribution of transverse prestress in the structural elements. Then a 1:16 scale model of one of the 208 ft spans was made in micro-concrete, extended with two cantilever portions to achieve the correct support conditions found in the actual structure (Figure 17.17). Cellular segments were individually cast, transversely and vertically stressed, and mortar-jointed on a jig. The model was then partially stressed longitudinally and lifted onto bearings before the remaining prestress was applied, along with the simulated self-weight. Tests of various design loadings were carried out – including simulated lane loading and abnormal vehicle loading – before the structure was loaded to collapse. The complex bending, shear and torsion stresses were measured by mechanical and

Figure 17.17 1:16 scale model in micro-concrete of the Western Avenue Extension, 1966, showing the data logger used to capture the strain-gauge readings. (Image: C&CA Archive)

electrical-resistance gauges and the many readings, captured by a data logger, were fed directly into the Association's Sirius computer for processing and analysis. The work started in 1965 and was completed in 1966 [54, 55].

17.3.10 CEGB cooling tower (1965)

Much of the C&CA's model testing originated in a brief by the designer at an early stage in the project, with a clear view of the information required from the tests. The outcome might change the design, but the purpose of the test remained largely as envisaged at the outset. Testing the CEGB cooling tower design was a notable exception.

In 1965 the Association become involved with the Central Electricity Generating Board in a general programme of research into the behaviour of cooling towers under wind loading; it was to establish the correlation with earlier CEGB models made of elastic materials (such as Perspex) and tested in a wind tunnel. The C&CA built a 1:25 scale model of a proposed standard design for a hyperboloid cooling tower; made of micro-concrete, it stood 13 ft tall (Figure 17.18).

However, the testing programme was considerably modified in wake of the Ferrybridge disaster, in which three out of eight recently-built cooling towers at

Figure 17.18 1:25 scale micro-concrete model of a standard CEGB cooling tower, 1965. (Image: C&CA Archive)

Ferrybridge power station collapsed in a gale. Each of these was 325 ft tall and 280 ft in diameter at the base, with a wall thickness of 5 in. On 1 November 1965 the first suffered a rippling of the shell at 250 ft, followed by a pulsating movement in a large area of the tower wall, in the middle of which a hole duly appeared, followed by collapse. This was followed by another tower shortly afterwards and a third an hour or so later [56]. The C&CA programme was redirected to aid the investigation into the cause of the failure and reconstruction began on site in April 1966. The effects of self-weight on the stresses were incorporated in the model and the research programme was extended in 1967 to include a study of additional elastic loading effects, including the effects of guying tower cranes to the shell. The study was completed in 1968 when buckling failure was obtained under lateral loading. Final tests were carried out on a section cut from the remains of the model, to determine the material properties of the micro-concrete against which to re-analyse earlier results.

17.4 The Rowe Years II (1966-1973): applying and reporting research

In many ways the cooling-tower project stood at the threshold of another change in direction at the C&CA, because in 1966 Rowe was promoted to the role of Director of Research & Development. Although model testing at the request of consulting engineers continued, albeit at a reduced frequency, the emphasis in the late 1960s tended towards applying the lessons of completed tests to improve best practice in design. And for this it was essential to secure the correlation of theoretical analysis, test results and the performance of structures when built.

In this first year, research on concrete subjected to complex stress systems considered the results of earlier tests on the Mancunian Way and Western Avenue Extension models, in comparison with computer-based theoretical analyses. Then, during construction of the actual Mancunian Way in the following year, observations of the structure were compared with results of the model test and the correlation found to be satisfactory. Finally, in 1968, a comparison was made between the test results and those from the structure itself during stressing and under live loading [57].

Model testing of generic elements was also undertaken. Cracking in full-scale prototype bridge slabs with two-way reinforcement was also modelled and tested at 1:9 and 1:3 scale during 1966 and, a year later, the Shear Study Group was able to report, 'Careful control of the mix design allows the production of small, easy-to-test, models which reasonably represent the behaviour of full-scale beams' [58].

17.4.1 Viaducts (1967-1970)

Proposals for an approach road to a crossing of the River Yarra in Melbourne, Australia, later named the West Gate Bridge, were seen at the time as building on the Association's experience with the Hammersmith Flyover, Mancunian

Way and Western Avenue Extension. Once again the consulting engineer was Maunsell, jointly with Freeman Fox and Partners. Here the problem was a complex connexion between three elements: a precast box beam, cantilever arm and an in-situ deck. An experimental solution was deemed best and so a one-ninth scale model of a transverse section was built in micro-concrete. A six-month programme of testing in 1967 determined the ultimate load factor, and confirmed the design as adequate. It did indicate, however, that careful detailing of the connexions between the spine and cantilever was necessary. The programme was written up and the results published in 1969 [59].

Later in the year, a model of the Gateshead viaduct on the A1 – an elevated roadway of multi-cell box construction designed by Ove Arup & Partners – was built to investigate behaviour under working load conditions in the zone where the elevated section and slip roads merged, where 'stiffnesses are not easy to determine'. The object was to verify the method of analysis used for the design, rather than determine directly how the actual structure would behave, and so a geometrically-simplified model in Perspex was decided upon. The section of the slip-road junction that was selected consisted of five spans, with two end cantilever spans, supported at 18 points. At a scale of 1:48, the resulting model was 10 ft long (Figure 17.19).

Initial experimental work was carried out in 1968 [60, 61]. Analysis and comparison with predicted behaviour followed and the test results were used to confirm the calculations of the structure's performance. Some 450 loading cases were considered and strains measured at approximately 200 positions on the model. A supplementary series provided data on deflexions and strains at a number of other points. In the end, about 500 tests were carried out, each consisting of the application of a point load or a line load and the measurement of the resulting strains at up to 95 different locations: point loads at quarter points of each beam and line loads along them. Readings were recorded on punched tape by

Figure 17.19 1:48 scale Perspex model of the A1 viaduct junction at Gateshead, England, 1968. (Image: C&CA Archive)

data loggers and processed by the Association's Sirius computer [62]. By 1970, the experimental work on the model ended and a report was prepared [63]. The Perspex model was then cut up for generic research into simply-supported spans in box-beam bridges and testing resumed.

The final model test of an infrastructure project in this period was for the Department of the Environment and the consulting engineers W.S. Atkins & Partners. The subject was the elevated section of the M11 London-Cambridge motorway and the aim was to determine the capacity in shear and bending in a connection between in-situ concrete and precast beams – the new MOT/C&CA standard M-beams.

The proposed design was an extension of the established method of multi-span construction – an in-situ deck over simply supported girder beams – combining standard-section precast members with an in-situ capping beam and the joint transversely prestressed. W.S. Atkins had some concerns about the shear capacity if shrinkage or settlement were to cause cracking in the lower half of the joint section so, at the request of the DoE, testing was carried out by the C&CA on beam connections modelled at a scale of 1:5. The work took place in two phases in 1970-71. In the first, tests of five single, longitudinal elements, one beam wide, were stressed to the maximum loads required – and loads, rather than deformations, were the point of interest. In Phase 2, tests were carried out on a single three-beam element to simulate transverse strength and stiffness.

As testing individual elements in Phase 1 was likely to lead to a conservative evaluation of the entire deck, Phase 2 was designed to investigate the combined effects of longitudinal and transverse forces across a broader section. The precast beams were much as in Phase 1, with additional flexural reinforcement. However, continuity steel in the top slab was reduced in line with amendments to the design and ducts were cast in to permit the application of transverse bending stresses with Macalloy stressing bars. After preliminary testing to determine the distribution of transverse prestress in the model and established the effective width to be used, two tests were carried out with different stress distributions. As in Phase 1, longitudinal loading was applied to cause a crack at the connection, then a cantilever load was introduced when the precast beams showed signs of yielding. Laboratory work ended in 1971, after an eight-month programme, and the three-beam unit was tested to failure. The final report was prepared in 1973 [64].

The outcome was considered satisfactory, supporting the proposed design as adequate for the purpose, with the cracked section capable of supporting almost twice the design's maximum sheer stress. The results prompted some modifications of the reinforcement during the programme and some economies were achieved by dispensing with the tapering of the end blocks that were a feature of the prototype precast beams' design.

17.4.2 Three-year research programme supported by CIRIA (1967-1970)

The Association's expertise in bridges was recognised in 1967 with the award of a three-year research programme by the Construction Industry Research and Information Association, on various aspects of bridge design and construction.

Though some would relate to new construction methods, much of it was an extension of the C&CA's earlier work. By 1968 the programme included work on:

- Slab-type bridges – a quarter-scale model of a 22 m span highway bridge, voided and reinforced transversely, was built and tested for information on cellular construction. Experimental work was also conducted on models of in-situ slab bridges to define the most efficient ratio of longitudinal to transverse steel. Models at different scales were used.
- Single-cell boxes with cantilevers – comparing measurements made on the Mancunian Way itself with those of the model.
- Multi-cell boxes with cantilevers – a literature survey was undertaken and, in anticipation of experimental work, methods for the rapid construction of micro-concrete models were developed.

17.4.3 Model techniques (1969-1973)

1969 was the first year in which the C&CA reported not only 'model tests of particular bridges', but 'model techniques' too, much of it derived from the aforementioned CIRIA contract. This reveals the Association to have been involved in a) Studying micro-concrete mix design; b) Investigating methods of adjusting stress-strain characteristics of small-diameter wires; and c) Investigating the accuracy of models in simulating the flexural cracking of reinforced concrete, with 1:3.7 scale models. The annual report added: 'Results applied to the testing of small beams which are models of specimens previously tested'. This work coalesced during the early 1970s into two series of tests that, while reliant on scale models, were of more fundamental and general interest than the project-specific testing of the 1960s.

Experimental work on the 1:3.7 scale models of eight of the slabs used in the general investigation into cracking was completed in 1970. The mechanism, the final report concluded, was 'the same as in the larger structures, crack widths and spaces do not scale directly but are a function of the specimen size, the properties of the micro-concrete and the bond characteristics of the model reinforcement' [65]. However, it was acknowledged that models could help with prediction. Related work on cracking in 1:3.7 scale slabs to investigate the effect of scale and size on crack widths and spacings was published the following year [66]. A further study of size effects on cracking in 1:10 scale models was completed and a report prepared in 1973.

Work on model beams continued into 1970, comparing moment-curvature behaviour in a number of 1:10 scale models with that of full-sized specimens. It concluded that care might be needed in interpreting data resulting from tests on micro-concrete models, as the moment-curvature behaviour does not necessarily scale linearly. A paper was prepared for the Sydney symposium in 1972 and an interim report was prepared [67]. The project was completed in 1973.

The period between 1969 and 1973 saw an increasing tempo of publishing reports and conference papers – a paper on modelling techniques was presented to ASCE national meeting in Cleveland, April 1972 – and this was characteristic of the final phase of model testing at the C&CA. Completing its output on major projects, the Association published the reports on model studies for the West Gate Bridge [68] (presented at the 1st Australian Conference on Structural Models, University of Sydney, May 1972), the Western Avenue Extension [69], the Gateshead viaduct [70] (also presented at the 1972 conference in Sydney) and the Motorway M11 [71]. These reports would prove to be the last ones that the Association published on model testing.

17.5 Final remarks

The C&CA's record in model testing will bear considerable scrutiny. New techniques and expertise were developed in the 1950s; a commercial consultancy practice was established by the end of the decade; the designers of a number of nationally- and internationally-significant structures were supported, right through to the M11 tests; test results were published and fundamental research undertaken for wider benefit. The Association was at the technological frontiers, exploring shell structures and prestressed concrete in the 1950s, and motorways and innovative long-span bridges and viaducts in the 1960s. Yet the programme of project-specific testing petered out in the early 1970s. Was this because demand dried up, or the Association's priorities changed? Had sufficient lessons been learned, or were models starting to be overtaken by computer programs?

What seems certain is that in its heyday, with structures research under Rowe, the C&CA could point with confidence to the importance of scale models as a design tool. In March 1965, Rowe and his deputy, G.D. Base, followed up their lectures at the *Model Testing* meeting of 1964 with a paper for discussion at a meeting of the ICE. While some of the content was a restatement, the authors felt able to conclude as follows:

- 'There is adequate evidence to show the reliability and accuracy of the prediction of structural behaviour by means of model tests.
- Suitable materials exist having the necessary characteristics for the simulation of the principal construction materials at all stages in their behaviours.
- Experimental techniques of considerable refinement have been developed which greatly facilitate measurement and interpretation of measurements in models of all types and in different materials.
- Model analysis and testing, when carried out by skilled personnel, is re-emerging as a design tool in its own right which deserves the same consideration as accorded to the established methods of design when any structural problem is being studied. By its very nature it offers the designer the opportunity to realize imaginative structures with the same facility which at present obtains for conventional, dare we say, stereotyped structures' [72].

And this seems as good a statement as any other, of the C&CA's position in this key period.

Appendix

Cement & Concrete Association: Selected Model Tests undertaken for Clients 1952-1970

	Project	Dates[1)]	Duration (months)	Cost(£)
1952	Bournemouth Bus Garage	c.1952	–	–
1954	Fleet Bridge, Hampshire	-Oct 1954	–	–
	Clifton Bridge, Nottingham	-1954	4	–
1957	Perth Narrows Bridge, Australia	-1957	–	–
	Boiler house, Bristol	-1957	–	–
1958	Lincolnshire Motor Co, Garage	1958-59	–	–
1959	Texas Instruments factory, Bedford	-1959	–	–
	Medway Bridge	-1959	–	–
	Hammersmith Flyover	-1959	5	1,000
	Commonwealth Institute roof	1959-60	8	680
1960	*Two series on bridges hinges*	-1960	–	–
	Third started (incl motorway bridge)	1960-61	–	–
	Smithfield Poultry Market roof	1960-61	9	2,400
1961	*BTC overbridge type*	1961	12	1,450
	Mangla hydro dam, Pakistan	1961	6	300
	Huntley's Point Overpass, Australia	1961-62	5	500
	More hinges started	1961-62	–	–
	Metropolitan Cathedral, Liverpool	1961-64	16	2,500
1962	Cumberland Basin viaduct	1962 (Oct)	5	500
	Four beams for Sir Owen Williams	1962	–	–
	Sydney Opera House panel joints	1962	–	–
1963	Manchester Skyway Bridge	1963-64	11	1,100
1964	Tinsley Viaduct	1964	–	–
	Reservoir structure	1964	–	–
1965	West Way (Western Ave Extension)	1965-66		
	CEGB cooling tower	1965-68	–	–
1967	R Yarra / West Gate Br approaches	1967		
	Gateshead Viaduct slipway junctions	1967-70		
1970	M11 elevated section	1970-71	8	–

1) Project years before 1960 are based on annual reports referring to the 12 months to 30 September, so may include the final quarter of the preceding year.

Cement & Concrete Association: Selected Model Tests undertaken for Clients 1952-1970 (Continued)

Client	Material[2)]	Scale	Notes
[R. Travers Morgan & Partners]?	–	–	see 1952 symposium
[MOT and Hants CC]?	MC	1:4	for MOT abnormal loading
R. Travers Morgan & Partners	P	1:50	
E.W.H. Gifford	MC	–	
Messrs Clarke, Nicholls & Marcel	P	–	an interconnected portal frame
Dr K Hajnal-Konyi	MC	1:10	biggest hypar roofin the UK
Oscar Faber & Partners	MC	1:6	building opened Dec 1960
Freeman, Fox & Partners	MC	1:14	or 1:13
G. Maunsell & Partners	MC	1:12	
AJ and JD Harris	P	1:32	
–	–	–	incl Wentbridge Viaduct
Ove Arup & Partners	MC	1:12	larger / flatter than similar shells
MOT/ Br Transport Commission	–	1:4	? 'Shear Connector Br' (1964)
H.G. Cousins	MC	1:32	
G. Maunsell & Partners	P	1:48	
(the last, for Glamorgan CC/ MOT)	–	–	
Messrs Rowe & Rodin	MC	1:29	cast 62; tests Jun 63 - 27 Jun 64)
Freeman, Fox & Partners	MC	1:24	photo Oct 1962
Sir Owen Williams	–	–	
Ove Arup & Partners	MC	–	
G. Maunsell & Partners	MC	1:12	
–	–	1:5	
–	–	–	cause of structural damage
G. Maunsell & Partners	MC	1:16	+ preliminary 1/10 in Perspex
Central Electricity Generating Board	MC	1:25	
Maunsell + Freeman, Fox & Ptnrs	MC	1:9	
Ove Arup & Partners	P	1:48	
WS Atkins & Partners / DoE	MC	1:5	

[2)] P = Perspex; MC = Micro-concrete, i.e. fine aggregate- or sand-cement mortar scaled appropriately.

References

1 Rowe, R.E. & Base, G.D. (1966) Model analysis and testing as a design tool. *Proc.Instn.Civ.Engrs* 33 (2): 183-199. Discussion: (1967) 36 (3): 677-695.
2 Pomeroy, D. (1999) *History of the Cement & Concrete Association and BCA*. Crowthorne: BCA. 17.
3 Morice, P.B. (2011) A sketch of Wexham Springs history: the early years of the Cement and Concrete Association, 1948-1958. *Construction History* 26: 41-53.
4 Morice, P.B. & Base, G.D. (1953) The design and use of a demountable mechanical strain gauge for concrete structures. *Magazine of Concrete Research* 5 (13): 37-42.
5 Collins, A.R. (1951) *Annual report*. London: C&CA. 3.
6 Collins, A.R. (1953) *Annual report*. London: C&CA. 8.
7 Witt P.J. (ed.) (1954) *Proceedings of a symposium on concrete shell roof construction, 2-4 July 1952*. London: C&CA.
8 Morice, P.B. (1954) Research on concrete shell structures. In: *Proceedings of a symposium on concrete shell roof construction, 24 July 1952* (ed. P.J. Witt), 99-126. London: C&CA.
9 Anon. (1952) 150 ft span prestressed shells. *Concrete Quarterly* 14: 36-38.
10 Kirkland, G.W. & Goldstein, A. (1951) The design and construction of a large prestressed concrete shell roof. *Structural Engineer* 24 (47, April): 107-127; Discussion: 24, Nov: 306-311.
11 Anon. (1953) The extensions to Bournemouth bus garage. *Concrete Quarterly* 19: 12-13.
12 Collins, A.R. (1953) *Annual report*. London: C&CA. 3.
13 Morice, P.B. & Lewis, H.E. (1956) *The strength of prestressed concrete beams and simple plane frames. Strength of Concrete Structures Symposioum*. London: Cement and Concrete Association. 400-477.
14 Collins, A.R. (1951) *Annual report*. London: C&CA. vii.
15 Morice, P.B. & Little, G. (1954) Load distribution in prestressed concrete bridge systems. *The Structural Engineer* 32 (3): 83-111.
16 Morice, P.B. (1957) *The work of the Research and Development Division of the Cement and Concrete Association: Structures Department*. TRA/256. London: C&CA.
17 ICE (1956) *Conference on the correlation between calculated and observed stresses and displacements in structures, London, September 1955*. London: Institution of Civil Engineers.
18 Little, G. (1954) *The distribution of a load in a box-section bridge from tests on a Xylonite model*. TRA/141. London: C&CA. Also in Magazine of Concrete Research 6 (18): 121-132.
19 Little, G. (1954) *Laboratory tests for load distribution in a model prestressed bridge*. TRA/155. London: Cement and Concrete Association.
20 Little, G. & Rowe, R.E. (1955) *Load Distribution in multi-webbed bridge structures from Tests on Plastic Models*. TRA 185. London: Cement and Concrete Association.

21 Little, G. (1954) *The distribution of a load in a box-section bridge from tests on a Xylonite model.* TRA/141. London: C&CA. 1.

22 Little, G., Morice, P.B. & Rowe, R.E. (1956) Load distribution in right highway bridges. In: *Proceedings of the IABSE Congress Lisbon/Oporto,* 1956. 423-450.

23 Little, G. (1954) *Laboratory tests for load distribution in a model prestressed bridge.* TRA/155. London: C&CA. 8.

24 Little, G., Morice, P.B. & Rowe, R.E. (1956) Load distribution in right highway bridges. In: *Proceedings of the IABSE Congress Lisbon/Oporto, 1956,* 423-450. 448, Figure 15.

25 Morice, P.B., Little, G. & Rowe, R.E. (1955) *Design curves for the effects of concentrated loads on concrete bridge decks.* TRB/202. London: Cement and Concrete Association.

26 Collins, A.R. (1954) *Annual report.* London: C&CA. vii.

27 Morice, P.B. (2011) A sketch of Wexham Springs history: the early years of the Cement and Concrete Association, 1948-1958. *Construction History* 26: 41-53. 51.

28 Finch, R.M. & Goldstein, A. (1959) Clifton Bridge, Nottingham: initial design studies and model test. *Proc. ICE* 12 (3): 289-316.

29 Rowe, R.E. (1957) *Istituto Sperimentale Modelli e Strutture, Bergamo, Italy.* TRA/351. London: C&CA.

30 Collins, A.R. (1958) *Annual report.* London: C&CA. 7.

31 Rowe, R.E. (1958) *The design and testing of a type of bridge suitable for medium right spans subjected to abnormal loading.* TRA/292. London: C&CA.

32 Collins, A.R. (1960) *Annual report.* London: C&CA. 23.

33 Collins, A.R. (1958) *Annual report.* London: C&CA. 7-8.

34 Jones, L.L. (1960) *Tests on a one-tenth scale model of a hyperbolic paraboloid shell roof.* TRA/334. London: C&CA.

35 Base, G. (1962) *Tests on a Perspex model anticlastic roof of lattice construction.* Technical Report TRA/358. London: C&CA.

36 Jones, L.L., Base, G.D., Corcoran, B.J. & Somerville, G. (1962) *Tests on a one-twelfth scale model of an elliptical paraboloid shell for Smithfield Market.* Design Note DN/21. London: C&CA.

37 Jones, L.L. & Base, G.D. (1965) Test on a one-twelfth scale model of the dome shell roof for Smithfield Poultry Market. *Proc. ICE* 30 (1): 109-130.

38 Anon. (1960) Testing of model structures. *Concrete & Constructional Engineering.* 55 (7, July): 283-286.

39 C&CA (1959) *Symposium on Models for Structural Design.* London: C&CA.

40 C&CA. (1964) Model testing. *Proceedings of a one-day meeting held at the Royal Aeronautical Society, London, on 17 March 1964 and organized by the Cement and Concrete Association.* London: C&CA.

41 Rowe, R.E. (1960) Work on models at the C&CA. *Jnl. Portland Cement Association R&D Lab* 2 (1): 4-10.

42 C&CA. (1964) Model testing. *Proceedings of a one-day meeting held at the Royal Aeronautical Society, London, on 17 March 1964 and organized by the Cement and Concrete Association.* London: C&CA.

43 Best, B.C. (1964) Certain aspect of model testing for the Cumberland Basin Scheme. In: *Model testing. Proceedings of a one-day meeting held at the Royal*

Aeronautical Society, London, on 17 March 1964 (ed. C&CA), 1-12. London: C&CA.

44 Best, B.C. & West, R. (1965) Tests on a model of a continuous skewed slab supported on columns. *Building Science* 1 (1): 21-32.

45 Rowe R.E. & Base, G.D. (1966) Model analysis and testing as a design tool. *Proc.Instn.Civ.Engrs* 33 (2): 183-199. 189-191.

46 C&CA. (1964) Model testing. *Proceedings of a one-day meeting held at the Royal Aeronautical Society, London, on 17 March 1964 and organized by the Cement and Concrete Association.* London: C&CA.

47 Somerville, G., Roll, F. & Caldwell, J.A.D. (1965) *Tests on a one-twelfth scale model of the Mancunian Way*. TRA/394.

48 C&CA (1966) *Mancunian Way*. Pub'n 91.007. London: C&CA.

49 Somerville, G. (1964) The model test of the Manchester Skyway Bridge. In: *Model testing. Proceedings of a one-day meeting held at the Royal Aeronautical Society, London, on 17 March 1964* (ed. C&CA), 17-24. London: C&CA.

50 C&CA. (1964) Model testing. *Proceedings of a one-day meeting held at the Royal Aeronautical Society, London, on 17 March 1964 and organized by the Cement and Concrete Association.* London: C&CA.

51 Anon. (1964/65) Model analysis and testing as a design tool. *Structural Concrete* 2: 195-196.

52 Schroder H. (1963) *Structural models in Perspex.* Departmental Note DN/30. London: C&CA.

53 Rowe, R.E. (1964) Model analysis and testing – its relation to established methods of design and its accuracy in predicting the behaviour of the actual structure. In: *Model testing. Proceedings of a one-day meeting held at the Royal Aeronautical Society, London, on 17 March 1964* (ed. C&CA), 1-8. London: C&CA.

54 C&CA (1968) *Western Avenue Extension.* Pub'n 91.011. London: C&CA.

55 Swann, R.A. (1970) *The construction and testing of a one-sixteenth scale model for the prestressed concrete superstructure of Section 5, Western Avenue Extension.* Technical Report TRA 441. London: C&CA.

56 Shellard, H.C. (1967) Collapse of cooling towers in a gale, Ferrybridge, 1 November, 1965. *Weather* 22 (6): 232-240.

57 C&CA (1968) *Annual report*. London: C&CA. 7.

58 C&CA (1967) *Annual report*. London: C&CA. 10.

59 Taylor, H.P.J. & Clements, S.W. (1969) *Tests on a one-ninth scale model of a transverse section of the West Gate Bridge, Melbourne.* TRA 425. London: C&CA.

60 C&CA (1968) *Annual report*. London: C&CA. 15.

61 Smyth, B. & Srinivasan (1969) The analysis of the Al viaduct, Gateshead. *The Arup Journal* 4 (3): 16-25.

62 C&CA (1972) *Gateshead Highway.* Pub'n 91.014. London: C&CA.

63 Evans, D.J. (1971) *Tests on a 1/48 scale Perspex model of Gateshead Viaduct.* Departmental Note DN/3008. London: C&CA.

64 Sturrock, R.D. (1974) *Tests on model bridge beams in precast to in situ concrete construction.* Report 42.488. London: C&CA.

65 Beeby, A.W. (1970) *An investigation of cracking in slabs spanning one way.* Report 42.433. London: C&CA.
66 Clark, L.A. (1971) *Crack similitude in 1:3.7 scale models of slabs spanning one way.* Technical Report 42.455. London: C&CA.
67 Evans, D.J. & Odgard, A. (1972) *Moment-curvature behaviour of 1/10 scale model reinforced concrete beams.* Departmental Note DN/3010. London: C&CA.
68 Taylor, H.P.J. & Clements, S.W. (1969) *Tests on a one-ninth scale model of a transverse section of the West Gate Bridge, Melbourne.* TRA 425. London: C&CA.
69 Swann, R.A. (1970) *The construction and testing of a one-sixteenth scale model for the prestressed concrete superstructure of Section 5, Western Avenue Extension.* Technical Report TRA 441. London: C&CA.
70 C&CA (1972) *Gateshead Highway.* Pub'n 91.014. London: C&CA.
71 C&CA (1974) *M11.* Technical Report 42.888. London: C&CA.
72 Rowe R.E. & Base, G.D. (1966) Model analysis and testing as a design tool. *Proc.Instn.Civ.Engrs* 33 (2): 183-199. 197.

18

Heinz Hossdorf: his contribution to the development of physical model testing

Pepa Cassinello

18.1 Introduction

Heinz Hossdorf (1925-2006) was an outstanding and atypical structural engineer. His self-taught personality encompassed cognitive knowledge of the world that surrounds us with the independence of academic disciplines and universities degrees, as if Philosophy had not been subdivided into the different branches that make up the body of Science. His technical and scientific contributions belong to very different fields such as: small-scale physical model tests, electronics, structural engineering, architecture, photography, computer applications, three-dimensional CAD (ITS-10), new materials, heavy machinery, and also revolutionary calculation methods such as the Finite Element Method [1].

He wanted to follow the path of his own intuition. For this reason, as soon as he finished his studies in civil engineering in Zurich and Aachen, the young Heinz Hossdorf founded his own experimental laboratory in Basel in 1953. This path not only allowed him to build innovative architecture and engineering work which he tested using physical model tests, but also to revolutionize and modernize this scientific field. From his first beginnings he was in contact with Eduardo Torroja who was, at that time, the most relevant and renowned expert in this field. The innovative works built by Heinz Hossdorf, like those built by Maillart, Torroja and Nervi, belong to the most important legacy of modern architecture and engineering.

In 1967 Hossdorf was one of the first persons to take advantage of the appearance of the computer to automate the entire process of experimenting with physical models. However, despite the importance of this work, the most important scientific contribution by Hossdorf in this field was the creation of physical "Hybrid Tests", in which he used the computer to integrate two types of reading – one analogue and one digital – to combine the contributions of both [2, 3, 4]. It is not surprising that his contribution in this field is considered as the most important in the continuity of the pioneering development carried out by Eduardo Torroja a generation before.

Physical Models: Their historical and current use in civil and building engineering design,
First Edition. Edited by Bill Addis.

18.2 Tests using physical scale models

Small-scale physical models have been used to check the structural behaviour of new structural forms since the seventeenth century. To build the unbuilt – *to innovate* – has been and always will be the most coveted ambition of the great masters of architecture and engineering. For this reason, it is no surprise that, despite the wide span of time and obvious differences that separate Eduardo Torroja's and Heinz Hossdorf´s respective oeuvres, they shared, among other traits, fidelity to the key principle of innovation: research-experimentation.

From the first moment, Heinz Hossdorf used physical model tests to design his structures. Not in vain, he knew how Eduardo Torroja was able to design new structural forms using physical model tests [5]. He had a very close relationship with the Spanish Technical Institute of Construction and with Eduardo Torroja whom he first met in 1954 [6].

On the other hand, Heinz Hossdorf wanted to build architectural and engineering structures following the model of living beings that make up Nature. For that reason, imitating living Nature, most of the works constructed by Hossdorf were stressed or prestressed. This is an aspect that undoubtedly complicated the realization of his physical model tests, but he overcame this. His works were built using very different structural forms, stressed and prestressed systems, and materials (concrete, plastic, stone, wood, steel). His works include the remarkable cylindrical shell roof for the Gummibandweberei (factory workshop) in Gossau, near St Gallen in Switzerland (1954) [7], the Vischer House in Basel (1960), the VSK Warehouses in Wangen (1961), the library at the University of Basel (1964), the Pavilion for the Swiss National Exhibition in Lausanne (1964), and the Basel Stadttheater (1968-1976). He also designed a prestressed bridge that caught the attention of Eduardo Torroja, but was not built [8].

Given the heterogeneity of resistant forms and materials that Hossdorf used, he had to make very different model tests, always looking into new methods to be able to establish the best similarity between physical-mechanical behaviour of the small-scale model with the prototype (full-size reality). Following Eduardo Torroja, he used physical model tests to check the structural behaviour of the new structural forms that he designed following his intuition. He did not use model tests to generate structural forms, as Antoni Gaudí or Heinz Isler did.

Heinz Hossdorf showed his own point of view on model tests. First of all, he explained that it is very important to reflect the geometric similarity with natural structures. Although, models were built at a smaller size than reality but with geometric similarity, because it is a cheaper and easier way, it is necessary to achieve similitude for other physical characteristics. For this reason, Hossdorf went beyond Galileo's square-cube law and he introduced, in his observations, new physical laws related to the scale effect, such as, surface tension, blood pressure or heart rate. The analysis of a drop of water helped him to verify that the linear dimensions of two similar drops of the same liquid are inversely proportional to the square root of the acceleration due to gravity. Thus, a drop of water on the moon will be related to one on earth if the former is 2.5 times as large as the latter [9]. He also explained how gravity would affect the size of human beings if there were analogous beings on a planet like Jupiter, where they would be only 80 centimetres tall, and they would move and think faster.

Based on his scientific knowledge of the world that surrounds us, he designed his own way to make physical model tests. He always said that if we want to progress, we have to keep in mind that it is necessary to check everything – even those things that others, long before us, previously checked. He thought that walking the same path travelled by others allows us to see new things.

18.3 The construction of scale models

As mentioned above, Heinz Hossdorf wanted to follow the path of his own intuition and he used physical model tests to check the structural behaviour of his innovate designs. Scale models had to be built in such a way that their structural behaviour could be extrapolated to the real structure. For this reason, the physical composition and appropriate groupings of the materials, along with suitable model scaling ensured that the moduli of elasticity and Poisson's ratio would be the same in the model and the actual structure. A further factor also had to be considered: for the model to reproduce the behaviour of a real structure under known loads, both in the elastic stage and when cracking and failing under increasing loads, the stress-strain relationship generated in the model had to be wholly analogous to that of the actual structure [10].

First of all, Heinz Hossdorf looked for the best materials to build his small-scale physical models. And because he was sure that the perfect material did not exist, he explored different new materials such as plastics mixed with Arlite (expanded clay) and aluminium powder. He usually used acrylic resin or translucent epoxy resin because these materials have the advantages of being machinable and transparent, thus making it easy to observe the weaknesses in connections; but he also used other materials such as wood or micro-concrete when appropriate.

18.3.1 Models made with wood

Hossdorf used wood as a material to build physical models of small timber roofs. In this way, it was easier to establish the correspondence of results between the model and reality because they were built in the same material. A typical example was the hyperbolic paraboloid, timber shell roof for the house of Antoinette Vischer in Alsace in 1960 [11, 12] (Figures 18.1 and 18.2). It was built with curved wooden boards, as used to make wine barrels, which were fastened at the edges of hyperbolic paraboloid. The physical model (1:10) was made with the same material using the same process of construction. The architects were Schwartz, Gutmann and Gloor [13].

Another model made of wood was for the folded-plate timber roof of a Pavilion near Basel, designed with Otto Senn in 1959 [14]. This Pavilion has a pentagonal plan with sides of 10 m. Hossdorf optimized the structural form by introducing four thin tensioned wires between each of the five inclined beams. The 1:20 scale model was prestressed in the same way (Figures 18.3 and 18.4). This was an

Figure 18.1 1:10 scale timber model for the Vischer House, 1960. (Image: Heinz Hossdorf)

Figure 18.2 The Vischer House, Basel (1960), under construction. (Image: Heinz Hossdorf)

ingenious structural and architectural solution that allowed natural light to enter through the four large windows under the roof.

18.3.2 Models made with micro-concrete

Hossdorf used micro-concrete to build scale models to check the structural behaviour of his innovate concrete-shell roofs, as Eduardo Torroja had done.

(a) (b)

Figure 18.3 Timber pavilion near Basel, 1959. (a) 1:20 scale prestressed timber model. (b) Pavilion under construction. (Images: Heinz Hossdorf)

Figure 18.4 Bruder Klaus Church in St. Gallen, Switzerland (1957-1958). 1:20 scale micro-concrete model of the shell roof. (Image: Heinz Hossdorf)

One example was the Bruder Klaus Church in St. Gallen, Switzerland, with a shell roof designed with the architects Ernst Brantschen and Alfons Weisser and built in 1957-58 [15]. Being a complex structural geometric form, Hossdorf made a 1:20 scale-model built in micro-concrete to check its structural behaviour; the model shell was just 3.5 mm thick. It was for this test that Hossdorf developed and used the first hydraulic loading device in his laboratory (Figure 18.4).

Hossdorf also used micro-concrete to build the physical model of his famous shell roof for the warehouse of the Verband Schweizerische Konsumvereine in

Figure 18.5 Wangen warehouse, 1959. 1:10 scale micro-concrete model of one bay of the roof under uniform gravity loading. (Image: Heinz Hossdorf)

Wangen, near Olten in Switzerland (1958-1961) [16, 17]. It was one of the first post-tensioned concrete roof structures in the world. The post-tensioned roof was formed by curved-shell beams, spanning 25.20 metres between supports [18, 19]. Each shell beam was built with 18 prefabricated reinforced-concrete elements (8.40 x 1.40 metres), post-tensioned with four cables on their external surface. This new system was called 'aerial post-tensioning', in which the steel armature is not placed inside the concrete mass, thus allowing them to be made very thin (4.5 cm).

Hossdorf carried out different tests in order to determine the complex process of post-tensioning, to check the structural behaviour (tension-deformation), and to check the best system of joining the precast reinforced-concrete elements. The 1:10 scale models were made of micro-concrete and post-tensioned with cables exactly as they would be built at full-scale (Figures 18.5 and 18.6).

18.3.3 Models made with acrylic or epoxy resin

Heinz Hossdorf made most of his small-scale physical models using acrylic resin or translucent epoxy resin because these materials have the advantages of being

Figure 18.6 Wangen warehouse. 1:10 scale micro-concrete model erected to study the interaction between two bays of the roof. (Image: Heinz Hossdorf)

machinable and transparent, making it easy to observe the weaknesses in connections, and their low stiffness meant that deformations would be larger and easy to measure. This was the case for the thin-shell concrete roof of the University Library in Basel (1964) which was designed with the architect Otto Senn [20]. In this complex structure he combined the concrete shell with prestressed concrete beams and triangulated steel girders. He built a small-scale model (1:20) using acrylic resin with high precision and with all the geometric details of the real design (Figures 18.7 and 18.8). In this way, he was able to observe not only the structural behaviour but also all the complex connections between all the structural elements. The model test results also enabled him to establish the most rational sequence of construction.

Heinz Hossdorf used acrylic resin to build the 1:50 scale model of the large prestressed concrete shell roof of the Stadttheater in Basel [21]. This building was designed with the architects Felix Schwarz and Rolf Guttmann (1968-1976) and was the competition winner (Figure 18.9). Undoubtedly, the Hossdorf's considerable experience with models allowed him to build innovative structural shells like this, just as Torroja and Nervi had done years before [22]. The theatre roof is formed by a large structural shell of irregular plan, which spans a distance of 60 m. He reproduced in the model all the details with a high precision and he concentrated measurement points in the most critical areas such as the edges prestressing anchors (Figure 18.10).

Since the behaviour of a complex monolithic concrete structure can only be understood as a close interaction between the very differently deformable elements – the shell roof, stage tower, wall panels and foyer walls – an elastic model test was by far the most efficient means of simulating the true load-bearing behaviour. To observe, measure and evaluate this behaviour, a model made of acrylic resin was used which was firmly glued together and reproduced in every detail. Apart from the statical uncertainties that the model test was expected to clarify, the most important questions were the actual distribution of the shell anchoring forces on the wall sections and enclosing walls, as well as the

Figure 18.7 University Library, Basel, 1964. 1:20 scale model in acrylic resin. (Image: Heinz Hossdorf)

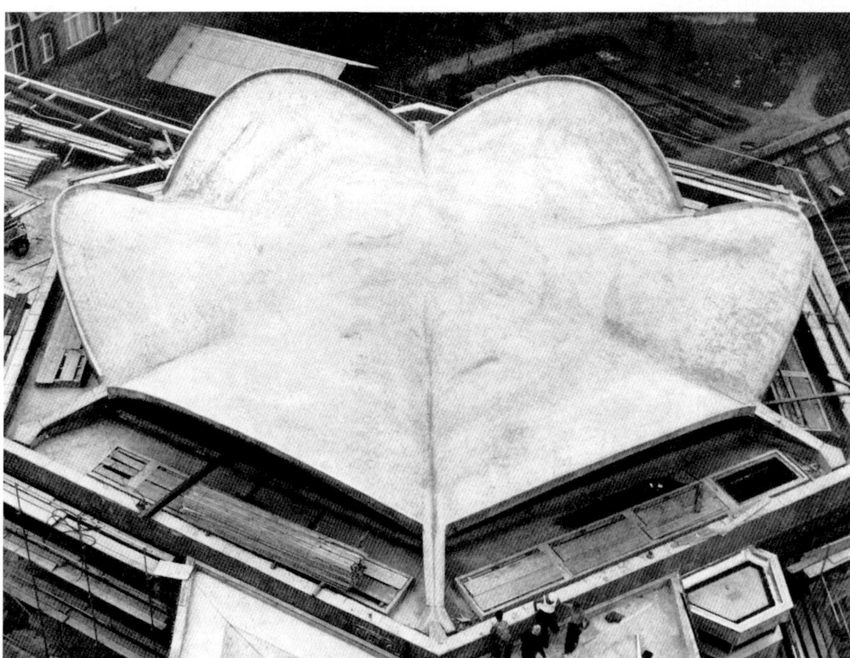

Figure 18.8 University Library, Basel, 1964. (Image: Heinz Hossdorf)

Figure 18.9 Stadttheater, Basel (1968-1976). Architects: Felix Schwartz and Rolf Gutmann. (Image: Heinz Hossdorf)

Figure 18.10 Stadttheater, Basel. 1:50 scale acrylic model. (Image: Heinz Hossdorf)

method of support of the cross beam for the stage tower anchoring. In detail, the following problems were addressed:

- the size and distribution of the shell anchoring forces in the abutment walls and lateral folding walls.
- membrane stresses in the flaccid zones and the bending stress in the cantilever beams there.
- radial thrusts from the side walls.
- introduction of membrane forces via the spatial cross beam to the stage-tower cross walls.

While the classical strain-gauge measuring technique would provide good information about local material stresses, which would provide a picture of the flow of stresses due to the introduction of loads into the shell, this technique hardly reveals anything about the key questions about the shear forces in the wall elements. This special problem prompted the laboratory to develop a completely new experimental concept in view of the key role played by the model experiment in understanding the load-bearing capacity of the theatre roof. This was the basis for the later systematically-developed general experimental method of the 'closed equilibrium systems'.

18.3.4 Models made with aluminium

In 1960 Hossdorf used aluminium sheet to build the model of a gravel silo at Günzgen in Kanton Solothurn to determine the bending stresses in its prefabricated reinforced-concrete walls (Figure 18.11) [23]. It was very expensive to make the complex structural form and, for this reason, he used a model to check walls and bridge decks. As he explained, the aluminium sheet allowed the model to be

Figure 18.11 Gravel silo, Günzgen, Switzerland (1960). 1:20 (est.) scale model made of aluminium sheet to set the bending stiffness of prefabricated reinforced-concrete walls. (Image: Heinz Hossdorf)

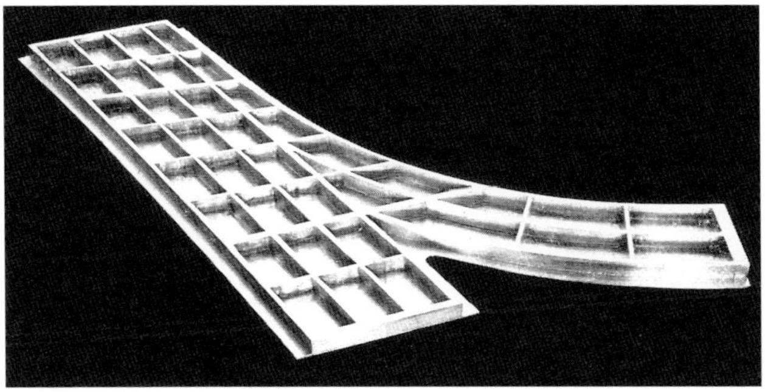

Figure 18.12 Aluminium model of a reinforced-concrete bridge junction on the approaches to the Köhlbrandbrücke suspension bridge in Hamburg. It was milled out of a solid piece. (Image: Heinz Hossdorf)

made with maximum precision and was well suited to check structural behaviour of a simple plate.

A great advantage of aluminium is that it is a very favourable material with which to make monolithic, three-dimensional models, which can be particularly useful for modelling reinforced-concrete structures. It can be machined easily and with great accuracy, and its material properties are precisely known. Hossdorf chose aluminium to make a model of an elevated road junction leading onto a bridge. The model was milled out of a solid piece of the metal (Figure 18.12).

18.3.5 A model made with steel and a polyester membrane

In 1964 a highly innovate work by Hossdorf was featured on the covers of the many of the most important technical journals, including the Journal of the International Association for Shell Structures. It was the Commercial Exchange Pavilion at Expo'64, the National Exhibition of Switzerland in Lausanne [24, 25]. The structure was a prestressed plastic membrane and this was the first time that a plastic material had been used as an architectural material to cover space occupied by people. It was an unforgettable image which, as Hossdorf said, he had designed in imitation of a field of giant transparent tulips (Figure 18.13).

Figure 18.13 Commercial Exchange Pavilion at Expo'64, the National Exhibition of Switzerland. Lausanne. (Image: Heinz Hossdorf)

The pavilion consisted of 24 modules, the prefabricated tulips, each 18.5 m square, covering a total area of 111 x 74 metres (8,124 m^2).

After seeking the best translucent material for the membrane, he finally chose a 3 mm polyester membrane, reinforced with 30% glass fibre. Never before had a hypar shell been stretched as a membrane between slender steel edge reinforcements which also served as compression members in a cantilevered composite structure carrying the roof loads. Although the statical behaviour of the central load-bearing body the 'flower' could be easily understood, and the global stress of its elements could be estimated easily by a simple hand calculation, this would not apply at all to the kind of local interactions between the polyester membrane and steel edge reinforcement.

In contrast to the familiar mode of action of hypar shells under surface load, where the edges are held by pure shear forces, the tensioning of the membrane due to the prestressing mechanism would inevitably result in lateral loading of the edge members. The bending stresses that result are all the more critical as the slim profiles were already likely to buckle due to the very high prestress. Due to the strong adhesive bond between the membrane and the steel edge reinforcement, they were able to work together compositely so that the polyester membrane could serve both to provide stability that would prevent buckling, and also to contribute to carrying the bending load. This complex interaction was impossible to calculate.

First, an accurate 1:6 scale model of a single core element was built, using an estimated cross-section for the edge profile. The glass-fibre reinforced polyester membrane, 0.5 mm thick, was also modelled as realistically as possible with regard to its strength properties. The load was applied by a rope pulling in the direction of the force imposed by the tension tie around the whole flower. The rope was connected via pulleys to the water barrel visible in the background of Figure 18.14. After a satisfactory behaviour of the element subject to the

Figure 18.14 Pavilion, Expo'64. 1:6 scale model test of the core element. (Image: Heinz Hossdorf)

Figure 18.15 Pavilion, Expo'64. 1:6 scale model test of the assembly of the 8 component elements of the flowers. (Image: Heinz Hossdorf)

prestressing loads had been achieved, the dimensions of the edge reinforcement were used as the basis for the construction of the 1:6 model of a complete flower, with all eight component elements (Figure 18.15). This model was used to simulate the effects of all types of loading, in particular the stresses and deformations, which were later compared with measurements on the completed structure. In addition, the natural vibration frequencies of the structure were determined and the long-term creep behaviour of the plastic was observed.

The results of wind-tunnel tests on a 1:500 scale model of the whole Pavilion structure, formed by the 24 membrane modules, demonstrated that wind was the most dangerous load for this type of structure as they are so light and can generate strong turbulence. They confirmed the need to fit bracing and steel anchor cables to the perimeters of the Pavilion structure.

18.4 Evolution of his experimental laboratory

18.4.1 New equipment

Heinz Hossdorf designed his own experimental laboratory and most of the installations, devices and appliances he need to realize his revolutionary small-scale model tests. Step by step he was incorporating new measuring devices, test rigs, materials and all the small details that he needed to improve his tests. He never settled for what he already knew; at all times he looked for new methods that would allow him to get closer and closer to the ultimate knowledge of making models whose structural behaviour was closer to full-scale reality – he opened up a new way of modernising the use of physical model tests. For example, he developed many different and innovative methods for applying loads to models. Previously in elastic tests, it was usual to reproduce the prestressing by incorporating tensile elements analogous to those used at full

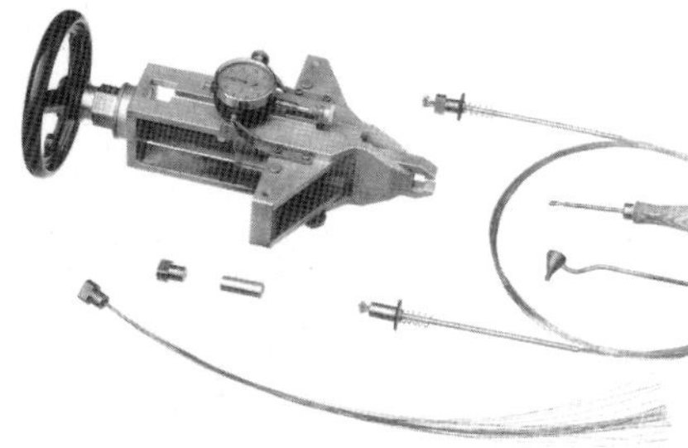

(a)

(b)

Figure 18.16 Testing of presstresed models. (a) Device for applying prestressing forces. (b) Ultimate strength test on a 1:10 micro-concrete model of a V-shaped, prestressed roof beam. (Images: Heinz Hossdorf)

scale. This procedure was very cumbersome and slow since, once the orientation of cables was fixed, the prestressing could be changed only by mechanically varying the model. He developed a new simulation technique for applying external loads to reproduce the effect of any group of prestressing cables, without having to introduce the cables inside the model. The tool designed by Heinz Hossdorf comprised a mechanical device for applying tension with a dynamometer, steel-wire cables, a head for fastening the ends of the wires, and anchoring pieces (Figure 18.16). This system was subsequently used to facilitate the verification of the structural behaviour of bridges.

18.4.2 Computers and the Hybrid Test

In 1967 Hossdorf introduced the use of a computer to automate the entire process of experimenting on physical models. The most important scientific contribution by Hossdorf in the field of physical scale-model testing was the creation of physical 'Hybrid Tests', in which he used the computer to integrate two types of reading – one analogue and one digital – to combine the contributions of both. The digital-analogical converter was very important in the chain of measuring devices between the transducer of the physical quantity to be measured and the computer. As he pointed out, this converter made possible the direct management of the computer using the analogue signals of measurement with which it was then possible to assess the observed phenomena [26].

During the computer revolution of the 1970s, many predicted the end of physical scale-model testing. Hossdorf defended the idea that the best way forward was to integrate the two technologies, obtaining a greater and more powerful technique for designing new structural forms never previously imagined. He carried out several scale-model test in which he compared the results obtained in the tests with those obtained by computer using the Finite Element Method. To do this he literally drew on the model the same division of elements that he had used in his calculations by computer (Figure 18.17).

18.4.3 Hybrid tests with acrylic models

Heinz Hossdorf focused hiss work mainly on architectural structural forms and most of his physical scale-models were made to check his own structural

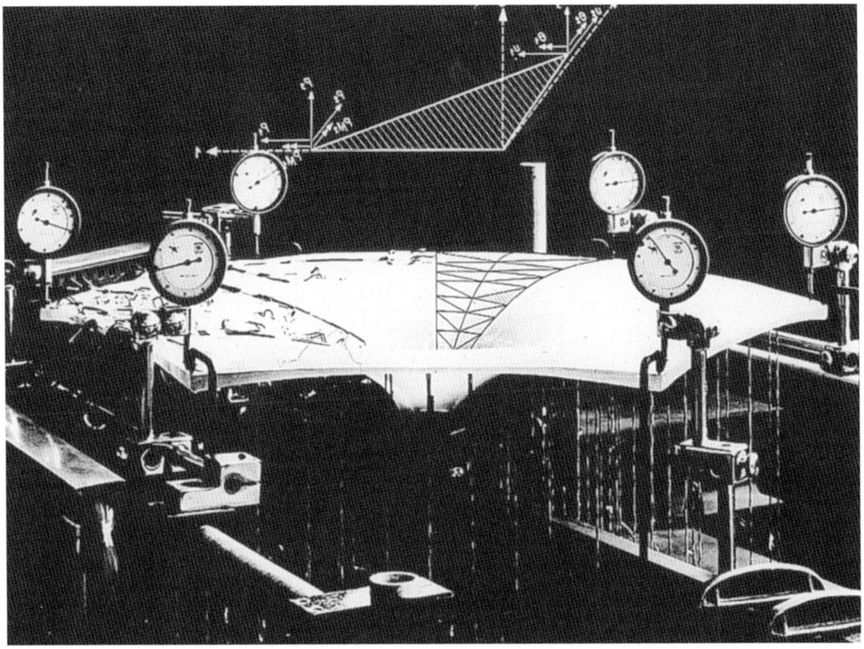

Figure 18.17 Comparison of results from a physical scale model test with results of analysis using the Finite Element Method. (Image: Heinz Hossdorf)

intuition. However, as his reputation for innovation in the field of model testing grew and spread, he was commissioned by many different institutions, companies, engineers and architects to undertake model tests. He undertook tests for the Ministry of Research and Technology in the Federal Republic of Germany. For Schindler Waggon, in 1972, he studied the vibration of railway carriages using a hybrid elastic model built with acrylic resin (Figure 18.18a). For the Realini and Bader Engineering company he carried out a dynamic test for the design of a prestressed concrete canopy to provide protection against rock falls, using a model built with micro-concrete (1980). For Gebr. Engineering in Basel, he tested hybrid elastic models of prestressed bridges using influence-line

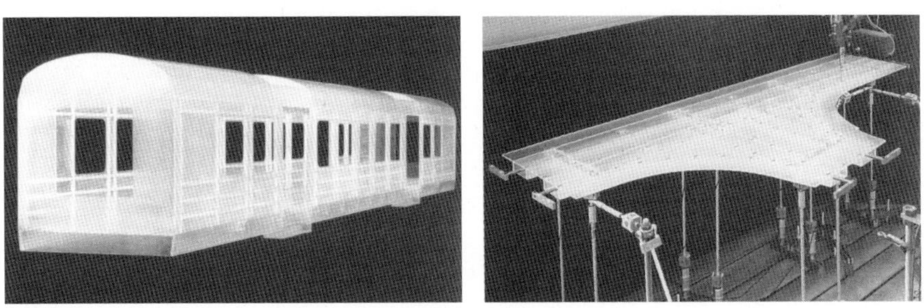

(a) (b)

Figure 18.18 Hybrid elastic models. (a) 1:10 model in acrylic resin for vibration tests on railway carriages. (b) 1:100 scale acrylic resin model for a prestressed concrete bridge. (Images: Heinz Hossdorf)

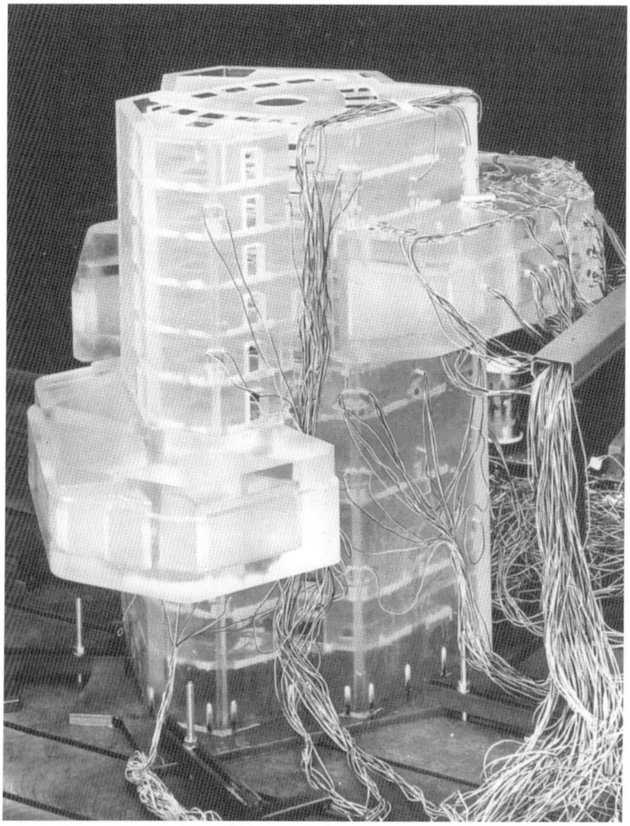

Figure 18.19 1:50 scale acrylic resin model for transfer structures in the National Westminster Tower in London. Image: Heinz Hossdorf)

software (Figure 18.18b). For the contractor Richard Costain, he made tests on a new technique for constructing high-rise buildings, and for consulting engineers Pell Frischmann, he undertook hybrid model tests in 1977 for the design of transfer structures in the innovative 200-metre high, post-tensioned tower for the National Westminster Bank in London. The 1:50 scale acrylic resin model was fitted with 200 foil strain gauges whose output was stored on magnetic tape. This data provided the input for analysis software on a main-frame computer [27] (Figure 18.19).

18.5 Final remarks

Although this chapter has only been able to present a few of Heinz Hossdorf's many hundreds of innovative projects, it has hopefully given an insight into his unique approach. He achieved so much to open new and revolutionary ways in developing the science of model testing at an international level. Details of many more of his projects and his own philosophy of engineering can be found in his published legacy, especially the book compiled a few years before his death in 2006, *Das Erlebnis Ingenieur zu sein* (The experience of being an engineer) [28, 29].

Acknowledgements

All images in this chapter belong to the archive which was donated to me by Heinz Hossdorf. Once again, I want to thank Heinz Hossdorf, known to me as Enrique. He was a very special friend of my family. I will be forever grateful for his endearing affection and his unforgettable lessons.

References

1 Cassinello, P. (2006) En Memoria de Heinz Hossdorf / In Memory of Heinz Hossdorf. *Informes de la Construcción* 58 (502): 63-81. http://informesdelaconstruccion.revistas.csic.es/index.php/informesdelaconstruccion/article/view/389/461 (1/11/2019)
2 Hossdorf, H. (1971) *Modellstatik*. Wiesbaden and Berlin: Bauverlag GMBH.
3 Hossdorf, H. (1972) *Modelos reducidos: metodo de calculo* (trans. Carlos Benito Hernández). Madrid: Instituto Eduardo Torroja.
4 Hossdorf, H. (1974) *Model Analysis of Structures* (trans. C. van Amerongen). New York: Van Nostrand Reinhold.
5 Cassinello, P. (2016) *Eduardo Torroja Museum (Catalogue)*. Madrid: Fundación Eduardo Torroja.
6 Cassinello, P. (1996) Heinz Hossdorf. *Arquitectura (Colegio Oficial de Arquitectos de Madrid)* 305: 108-114.
7 Hossdorf, H. (2003) *Heinz Hossdorf: Das Erlebnis Ingenieur zu sein*. Basel: Birkhäuser Verlag. 18-25.
8 Torroja, E. (1955) Puente pretensado de piedra natural. *Informes de la Construcción* 7 (71).
9 Hossdorf, H. (1974) *Model Analysis of Structures* (trans. C. van Amerongen). New York: Van Nostrand Reinhold. 64.
10 Benito, C. (1959) Comprobación experimental de cubiertas laminares por medio de modelos reducidos. *Informes de la Construcción* 12 (111): 60-80.
11 Hossdorf, H. (2003) *Heinz Hossdorf: Das Erlebnis Ingenieur zu sein*. Basel: Birkhäuser Verlag. 144-147.
12 Cassinello, F. (1973) *Construcción Carpintería*. Madrid: Rueda. 512, 513, 560, 597.
13 Anon. (1969) Vivienda Unifamiliar, Hegenheim, Suiza. *Informes de la Construcción* 22 (216): 15-20. Madrid: Instituto Eduardo Torroja.
14 Hossdorf, H. (2003) *Heinz Hossdorf: Das Erlebnis Ingenieur zu sein*. Basel: Birkhäuser Verlag. 178-180.
15 Hossdorf, H. (2003) *Heinz Hossdorf: Das Erlebnis Ingenieur zu sein*. Basel: Birkhäuser Verlag. 26-29.

16 Hossdorf, H. (2003) *Heinz Hossdorf: Das Erlebnis Ingenieur zu sein*. Basel: Birkhäuser Verlag. 30-39.

17 Joedicke, J. (1962) *Schalenbau: Konstruktion und Gestaltung*. Stuttgart: Karl Krämer Verlag. 72-73.

18 Hossdorf, H. (2003) *Heinz Hossdorf: Das Erlebnis Ingenieur zu sein*. Basel: Birkhäuser Verlag. 30-39.

19 Rühle, H. (1969) *Räumliche Dachtragwerke: Konstruktion und Ausfuhrung*. Berlin: VEB Verlag für Bauwesen. Vol.1 (Beton Holz Keramik): 109.

20 Hossdorf, H. (2003) *Heinz Hossdorf: Das Erlebnis Ingenieur zu sein*. Basel: Birkhäuser Verlag. 78-89.

21 Hossdorf, H. (2003) *Heinz Hossdorf: Das Erlebnis Ingenieur zu sein*. Basel: Birkhäuser Verlag. 96-111.

22 Cassinello, P. & Torroja, J.A. (2018) The pursuit for new structural forms: Eduardo Torroja and Frei Otto. *Journal of the International Association for Shell and Spatial Structures* 59 (1): 34-42.

23 Hossdorf, H. (2003) *Heinz Hossdorf: Das Erlebnis Ingenieur zu sein*. Basel: Birkhäuser Verlag. 46-51.

24 Hossdorf, H. (1964) Design of a Polyester pavilion reinforced with glass fiber for the 64 Swiss Exhibition. *Bulletin of the International Association for Shell Structures. IASS* 19 (Oct.): 17-32.

25 Hossdorf, H. (2003) *Heinz Hossdorf: Das Erlebnis Ingenieur zu sein*. Basel: Birkhäuser Verlag. 58-77.

26 Cassinello, P. (2002) Ciencia y creación en la obra de Heinz Hossdorf. *Arquitectura (Colegio Oficial de Arquitectos de Madrid)* 327: 34-49.

27 Hossdorf, H. (2003) *Heinz Hossdorf: Das Erlebnis Ingenieur zu sein*. Basel: Birkhäuser Verlag. 224-245.

28 Hossdorf, H. (1974) *Model Analysis of Structures* (trans. C. van Amerongen). New York: Van Nostrand Reinhold.

29 Hossdorf, H. (2003) *Heinz Hossdorf: Das Erlebnis Ingenieur zu sein*. Basel: Birkhäuser Verlag.

19

Soap-film and soap-bubble models

Berthold Burkhardt

19.1 Some historical notes

Physical models made with the surfactant films produced from liquids, are part of basic research in architecture, structural design and theoretical mathematics. Liquid soap and soap bubbles have been known since ancient times. Floating bubbles and bubbles on liquids are a natural physical phenomenon. Besides their existence in nature, we know soap bubbles in art, in the performing arts and in entertainment and play. In paintings, soap bubbles often appear as motifs, for example in the painting 'Kinderspiele' by Pieter Bruegel the Elder from the year 1560; and since the early 1980s, the American entertainer Tom Noddy has been inspiring people around the globe with his show 'Bubble Magic'.

Of decisive importance are the experiments with water described by Leonardo da Vinci (1452-1519) in his Codex Leicester, in which he dealt with the surface tension of the 'flying sphere'. Since then, many physicists and mathematicians have studied the phenomena of soap films and bubbles, their structure as well as their appearance and the play of colours. In 1743, the Swiss mathematician Leonhard Euler (1701-1783) proved in his writings that soap films are always minimal surfaces. A lightweight surface structure is of inestimable importance for architecture.

The founding father of the modern science of soap bubbles was the Belgian physicist Joseph Plateau (1801-1883). He established the so-called Plateau's Laws that govern the formation of soap films and bubbles but was not able to see many of his experiments as he became blind in 1843. He published his findings nevertheless, and his collected papers on the subject were published in 1873 [1]. In Britain the subject of soap bubbles was popularised in a series of lectures given to young people in 1888-1889 by the physicist Charles Vernon Boys (1855-1944) published as 'Soap bubbles and the forces which mould them [2].

The best-known question in this context is 'Plateau's problem'. This is the task to find, for a given closed, curved space, all surfaces with the smallest area, i.e. the minimal surfaces. In 1958 the Berlin architect Frei Otto began to experiment with soap films in his development centre for lightweight construction. In 1954 he published his widely-acclaimed dissertation 'Das hängende Dach'

Physical Models: Their historical and current use in civil and building engineering design,
First Edition. Edited by Bill Addis.

(The Hanging Roof) [3], in which models for lightweight, wide-span, load-bearing surface structures, made with fine threads, nets, foils and model gypsum, were shown. Soon after, there were also models made with soap films and bubbles, which were used both for basic research and as design aids for buildings made of nets, membranes and lightweight shells [4, 5]. In the course of this research, in the 1970s, he developed close contact and scientific exchange with the mathematician Stefan Hildebrandt and his research group at the University of Bonn in a special research field of the German Research Foundation 'Non-linear Partial Differential Equations' [6, 7, 8].

By dealing with soap bubbles, Frei Otto and his team gained access to the structure and development of nature's objects. At a memorable conference on 11 February 1973 at the Institute for Lightweight Structures at the University of Stuttgart, biologists, architects and engineers addressed the key sentence 'The Pneu (inflatable) is the physical basis of living nature. The term Pneu, derived from the soap bubble, was further defined as a system in which a surface, exposed only to tensile stress, envelops a medium [9]. The soap bubble phenomenon thus also reached biology as a scientific discipline.

19.2 Manufacture of soap films and bubbles

The experimental production of soap films is relatively simple. Since Plateau, nothing has fundamentally changed in the experiments. Thin wires are still bent into a closed figure, or the edges are created by flexible threads and then immersed in a soap solution. The result is visible when you pull the frame out. We are talking here about self-forming forms in which only the boundary is given. The surface forms itself in the experiment [10].

Surfactant films cannot be produced from normal water. In the short term, small floating bubbles form on water, but these are mainly of interest for scientific studies.

In his 'Experiment of the Month' in October 1982 Jearl Walker described the following recipe according to A. Weiss: 'The simplest soap solution consists of ordinary soap or a detergent and clean, best distilled water. To make the soap film more durable, glycerine can be added. If the soap or detergent is mixed with water and glycerine in a ratio of about 1:3:3, the soap film can last up to three quarters of an hour' [11]. Further simple formulations can be found in publications of the Institute of Lightweight Structures at the University of Stuttgart [12]. All soap-film experiments carried out at Frei Otto's institute used the soap solution with the brand name Pustefix, developed by Tübingen chemist Dr. Rolf Hein in the 1950s. The degree of possible dilution depends on the experiment.

A soap film consists of three layers namely, two outer layers with orientated surfactant anions and, between them, a liquid layer with still-unordered surfactant anions. The films appear in spectral colours that depend on the thickness of the film. Due to gravity, the liquid flows downwards during an experiment, the film becomes thinner until it bursts (in a bubble) or tears (in a film) [13]. Shortly before the destruction, the soap film is no longer visible, a so-called black hole develops.

In this state, the thickness of a surfactant film is smaller than the wavelength of the light, i.e. not visible.

The surface tension of a surfactant film can be measured with a simple device developed by Frei Otto in the 1980s. If an open wire bow is pulled out of a soap solution to form a soap film, the pull-out height can be converted into a force by applying a corresponding previous calibration.

Soap film models for the preliminary design of lightweight surface structures made of membranes or nets are about the size of a hand or marginally larger. The influence of the self-weight of the soap film is negligibly small with this size of the model. Size is not only related to consistency and durability, but also to the difficulty of capturing or recording the geometry by photographic and graphical representations.

In order to explore the limitations of size, a 4.5 m long soap film with a width of 10 cm, and a soap film in a flat circular ring with a diameter of 1.0 m, were produced at the Institute for Lightweight Structures using the usual soap solution (Pustefix). The soap film in the flat ring showed very clearly a sag due to the influence of its own weight.

Another possibility for a longer 'shelf life' for soap films is the use of thicker soap liquids which, however, make it more difficult to create particularly complicated shapes, or change the shape when they are produced, due to unequal stresses in the model. Frozen soap films and bubbles are also feasible, but they are also unsuitable for experiments on architectural models, as they cannot adapt to changing edge conditions in the design process. When freezing soap bubbles, the internal temperature of the bubble must be kept constant and equal to the external temperature, otherwise they will implode.

Due to their physical and chemical consistency, soap films have a short life, usually only a few seconds. This disadvantage of a short-term experiment is compensated in that the soap film can be created again and again in the same form within the given boundary frame after renewed immersions. In multi-symmetric configurations, e.g. cubes, the formed shape of the soap films can also be oriented to the other axis of symmetry (Figure 19.1).

A longer-life bubble can only be achieved in very high air humidity which prevents evaporation, by replenishing the run-off liquid with new soap solution. In order to improve the durability and precision of the geometric recording of soap-film models, a so-called soap-film machine was built at the Institute (IL) in 1973 (refs. 326-331). A climatic chamber was made, with a steel frame and transparent walls on all sides made of glass and acrylic (Figures 19.2 and 19.3). In this climatic chamber a 100% cooled air humidity can be produced which achieved a life span of several hours for the soap film. By atomising the soap solution, the soap films receive sufficient new liquid to stabilise the film. In order to study the behaviour of minimal-surface soap films under the action of wind, a small wind tunnel could be attached to the soap-film machine to blow a gentle stream of air onto the model.

A vertical, height-adjustable spindle leads into the climatic chamber. At the lower end there is a 360° rotatable ring or base plate for fixing the model edges. The model can now be immersed in a swing-in container of soap solution. On two rails, which run through the climate chamber, there is an optical bench.

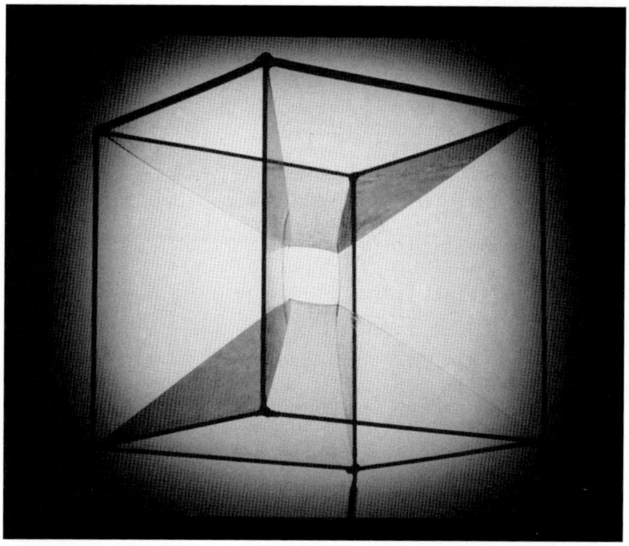

Figure 19.1 A minimum-area soap film created within a rigid cube boundary. (Image: Berthold Burkhardt)

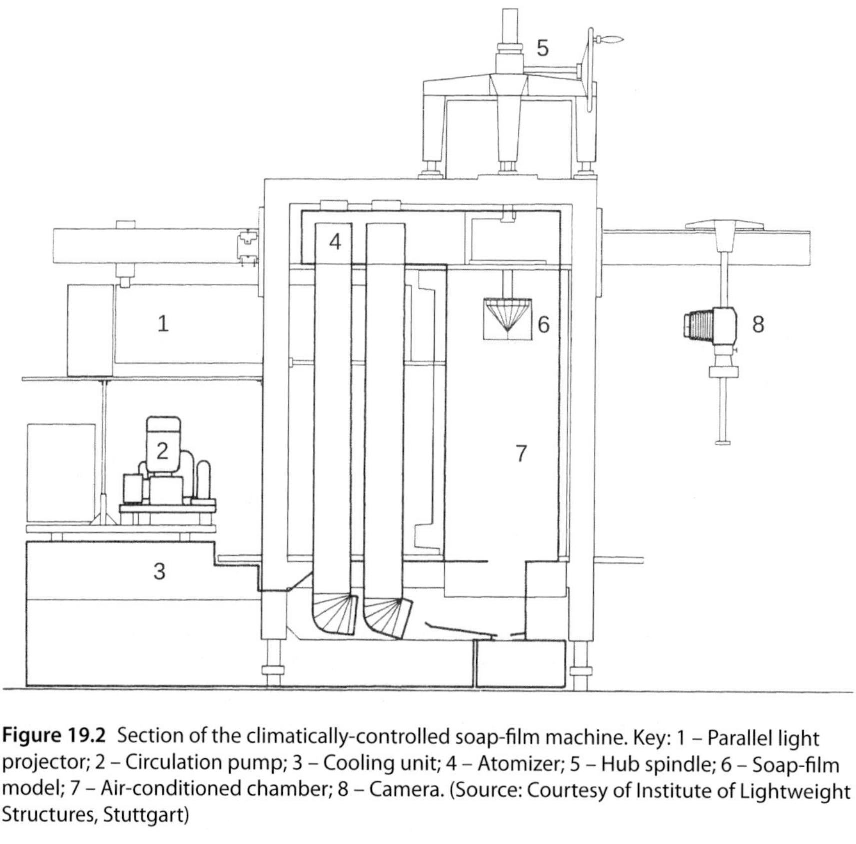

Figure 19.2 Section of the climatically-controlled soap-film machine. Key: 1 – Parallel light projector; 2 – Circulation pump; 3 – Cooling unit; 4 – Atomizer; 5 – Hub spindle; 6 – Soap-film model; 7 – Air-conditioned chamber; 8 – Camera. (Source: Courtesy of Institute of Lightweight Structures, Stuttgart)

Figure 19.3 The climate-controlled soap-film machine at IL. (Image: Berthold Burkhardt)

By means of parallel light projection installed on it, the contours of the model can be photographed and recorded against the background of the measuring grid, by rotating it step by step. A small wind tunnel is available as a demountable addition to this apparatus. Blowing and suction fans allow the observation of soap bubbles and soap films under the action of wind pressure and suction.

Klaus Linkwitz and his colleague Wolfgang Faig at the Institute for Geodesy and Photogrammetry at the *Technische Hochschule* in Stuttgart began to develop close-range photogrammetry for recording and measuring models of lightweight structures during the design and development of the German Pavilion at EXPO '67 in Montreal at the Institute for Lightweight Structures under the direction of Frei Otto. However, the techniques developed successfully for models made of thin wires and nets like the Montreal model, were extremely difficult to reproduce for soap film models. Close-range photogrammetry needs fixed points on the surface as a reference, which could hardly be produced on soap films, even using the finest particles such as powdery flower seeds. A solution was found in projecting a raster line onto the soap film to show the spatial deformation more effectively (Figure 19.4). The colour patterns on soap films, which appear as interference fringes, also help to make the soap film visible. However, due to the internal liquid flow within the film, these colour stripes change constantly and these methods do not provide help with the photogrammetric surveying.

Other techniques were also tried. The *moiré* method was tested on solidified soap films to determine geometric changes, but with limited success. Also,

Figure 19.4 Grid lines Projection on soap film. (Image: Courtesy of Institute of Lightweight Structures, Stuttgart)

blown smoke was introduced into single or multi-chambered soap bubbles and foams, which made the bubbles easier to observe, especially the junctions between bubbles.

19.3 Creating soap film surfaces

19.3.1 The soap film as a minimal surface area

The minimal surface is the smallest surface between arbitrarily formed, closed lines or surfaces, as already known by Leonhard Euler. It is characterised by the fact that, at each point on the surface, the sum of all positive and negative radii of curvature is zero. Membranes or networks whose stresses are the same in all directions always assume the shape of a minimal surface.

Flat minimal surfaces are only formed between polylines that lie in a plane. Between all other spatial lines, anticlastic (saddle-shaped) minimal surfaces are always formed. In soap bubbles, which form under internal pressure, the shape is always dome-shaped, i.e. synclastic. Such single-sided curved surfaces are never minimum surfaces. Saddle surfaces also exceed the minimum area if the tension, and thus the curvature, is changed in one direction.

Besides the long-known catenoids, which can be formed between two rings, the minimum surface area between skewed square frames is also known. The boundaries of these minimal surfaces, also called four-point surfaces, can be formed by rigid lines or by soft lines (thread edges). In three-dimensional edge configurations, minimal surfaces are formed from several partial surfaces (Figure 19.5). These surfaces meet in three planes at angles of 120° to each other, as a condition of uniform surface tension in surfactant films. In addition to those created between linear boundaries, minimal surfaces can also be created with surfaces as boundaries.

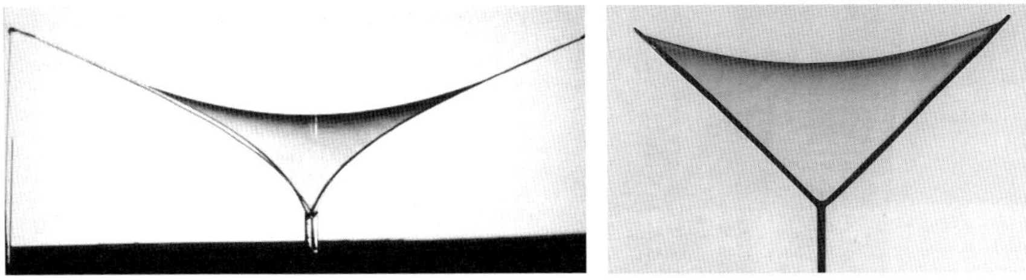

(a) (b)

Figure 19.5 (a) Soap film in a four-point frame with thread edges. (b) Soap film in a four-point frame with rigid edges. (Images: Courtesy of Institute of Lightweight Structures, Stuttgart)

19.3.2 Support of surfaces

It is not possible to apply point loads to the surface of soap films. A needle can penetrate through a soap film without causing deformation; it either passes through the film or bursts it. Only flat elements such as a ring, a circular garland, a disc or a ball are able to raise (high point) or lower (low point) a flat film. In this way, the forces are transferred over a larger distance or surface, and the single-point support spreads to the 'inner' edge.

For the size and shape of the deflection, the same relationships of diameter to height apply as for the catenoid, the minimum surface area between two rings. Since the self-weight of soap bubbles is very low in relation to the tension, high and low points can be treated equally.

Just as with soap films, textile membrane structures or nets must also be supported from single-point supports in the surface by extending individual loads over the largest possible transfer lengths or surfaces. The constructional possibilities to transfer these forces at high and low points include garlands, rosettes, rings, rope loops or branched suspensions [14]. This principle of single-point support is also found in all structural assemblies made of different materials, where a single support has to carry a beam or a plate.

Four means of supporting the edges of full-scale membrane structures were developed from countless experiments using soap films – rope loops, ridges, ridges and valleys, and arches.

Using the constructionally simple means of a rope loop, the so-called eye, it is possible in tent construction to deform membranes through high and low points without a significant increase in tension. Here, too, the ground plan area, loop size and degree of deformation are related. First, a rope loop of constant curvature (circular) lies within a flat film of soap solution. When lifting or lowering, the thread forms a loop which causes the same surface tension in all directions to form a space curve of constant curvature (Figure 19.6), (see Figure 19.11a).

Membrane surfaces can be supported by ridges that run between the edge and a peak. The total area is divided by the ridges into individual areas, which can each be regarded as minimum areas. Ridges or valleys act like flexible rope edges. With an increasing number of ridges, sharper, i.e. taller, tent shapes are possible. A three-ridge tent with tightly-stretched ridges over an equilateral triangular ground plan, corresponds to the shape of a tetrahedron with a spatial

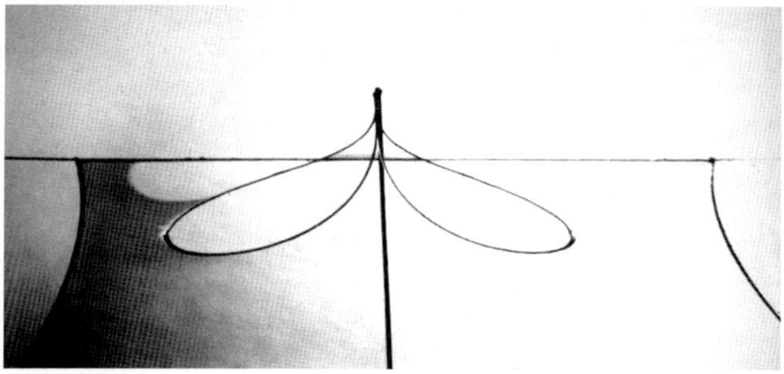

Figure 19.6 Rope loops used to convey the tension in a soap film to a single point. (Image: Courtesy of Institute of Lightweight Structures, Stuttgart)

angle at the peak of 109° 28' 16". If the angle becomes more acute, intermediate film surfaces are formed according to the number of ridges, as predicted by Plateau's laws of bubble formation. The more ridges that are introduced, and the more sag they have, the smaller the area of the intermediate films, even at acute angles. If the peak of a form made of ridges is resolved using circular rings or garlands, the same laws of formation apply in principle.

Membranes made of soap films can be stretched in spatially-deformed frames to form not only saddle-shapes but also wave-shapes. With the help of alternating ridge and valley ropes between alternating high and low points, a flat surface becomes a wave surface. As a rule, these are minimal surfaces with non-rigid edges (threads in the model) as warped 4-point surfaces which can be multiplied in parallel, radial or crosswise configurations. If the high and low points, and the associated ridges and valleys, are arranged in parallel rows, parallel waves occur (Figure 19.7); if two ridges and valleys intersect at right angles, they are referred to as a cross wave. If the high and low points lie on a circular line, this is a star waveform.

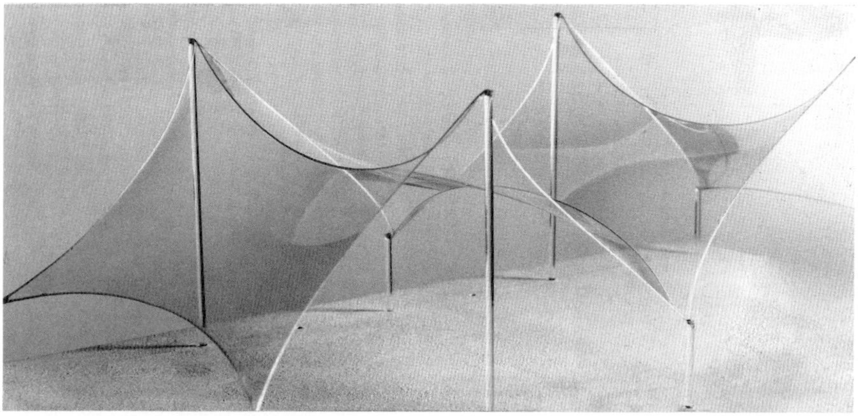

Figure 19.7 Longitudinal wave with ridge and valley (Image: Courtesy of Institute of Lightweight Structures, Stuttgart)

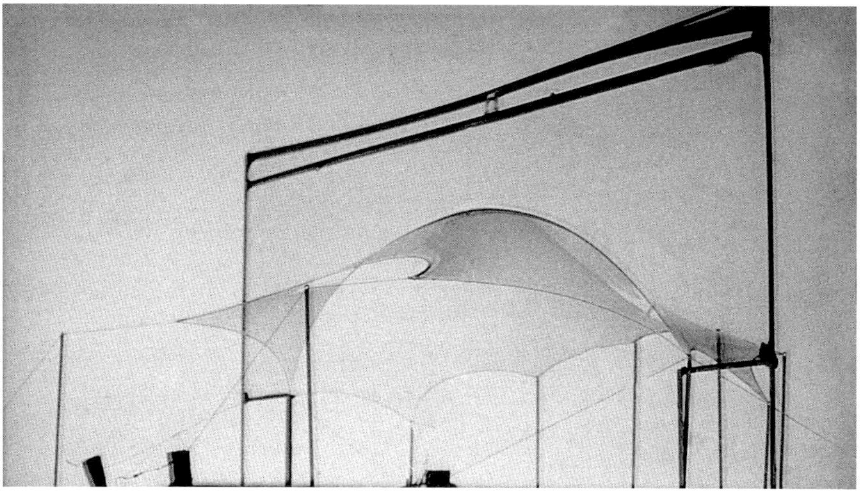

Figure 19.8 Arch structure formed by three lamellae – one vertical in a rigid frame, and two side films tied to point supports by means of thread boundaries. (Image: Courtesy of Institute of Lightweight Structures, Stuttgart)

In complex frames, soap films form systems which, as already described for nodes in minimal net systems or free edges, adjoin each other. If, for example, the minimum surface of a cube is found using a soap bubble model, the surfaces meet in shared Y-shaped edges at an angle of 120°. In the soap bubble model for an arched tent, the edges for the two adjacent surfaces are usually formed from threads. The third (vertical) soap-film surface, with which the arch interfaces, is produced by a vertical wire frame (Figure 19.8). This subframe has no influence on the self-forming minimum surfaces or the arch itself. Several arches in succession are thus feasible when linked by stretched membrane surfaces. Such supporting arches should only be designed for a normal force, if the minimum surface area of the tent membrane prevents it from buckling sideways.

19.4 Other forms of film and bubble

19.4.1 Minimal net surfaces

By means of two-dimensional soap films and their one-dimensional projection, path systems with the smallest total length can be found experimentally and distributed for any number of points in a plane. The prerequisite is that the plane of the system is the same everywhere, that all points have the same properties, and that all connections are of equal value.

In a test arrangement, flat soap films are formed between horizontal, parallel planar surfaces and vertical linear elements, which are perpendicular to the surfaces. If several soap films are connected to form a communicating system, a minimum surface configuration is created. An arrangement of nodes and branches forms with three arms and three 120° angles. If this system is projected into the

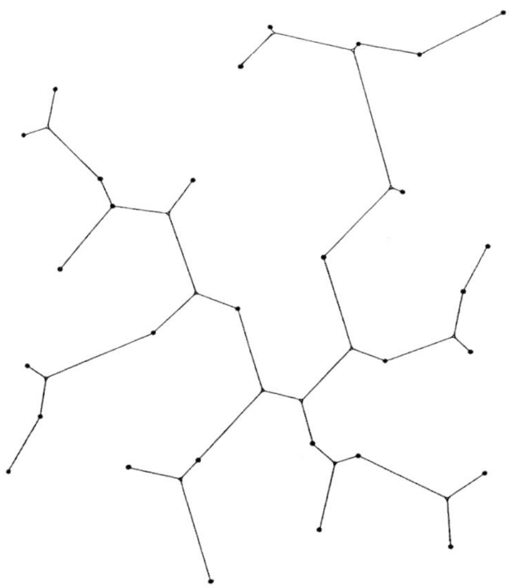

Figure 19.9 Two-dimensional minimum mesh. (Image: Berthold Burkhardt)

plane of the horizontal surfaces, one dimension of the film is eliminated. The spatial arrangement of surfaces becomes a flat arrangement of lines, and minimal surfaces become minimal meshes or networks. This method in the soap film model can also be used to generate minimal networks of unequal connections or ground plan topographies [15] (Figure 19.9).

19.5 Air-inflated structures – the pneus

Soap bubbles always take the form of the smallest area. They are therefore also minimal surfaces, enveloping a maximum volume with a minimum surface area. The shape and size of the bubbles are directly dependent on the ratio of the surrounding and internal air pressures, as well as the surface tension of the soap film.

In contrast to the open, saddle-shaped minimal surfaces between edges, the closed bubbles have synclastic surface curvatures. Soap bubbles float freely and, ideally, assume a round ball shape. Since smaller bubbles have higher internal pressures than larger ones, they are more precisely formed and more dimensionally stable.

On fixed, rigid edges of circular form, spherical segments that are smaller, equal to or larger than hemispheres can be formed pneumatically with soap bubbles (Figure 19.10). With fixed edges, it is also possible to form dome-shaped bubbles over triangular, quadrilateral or polygonal ground plans, as well as over free spatial edge configurations. Studies with such soap bubbles on fixed edges are the basis for finding the shape of air halls [16]. If the ground plan is circular, the doubly-curved bubbles are synclastic at all points. In the case of angular or constricted ground plan shapes, areas with bell-shaped, anticlastic surfaces can also form in the corners (gussets). Very flat bubble shapes can be produced pneumatically over fixed outer edges and inner low points.

Figure 19.10 Multi-chamber pneumatic soap-film test. (Image: Berthold Burkhardt)

19.6 Bubbles with free edges

With the help of a so-called free edge, the surface curvatures of large, extensive air halls with inner low points can be determined. The fixed acrylic glass walls form a 60° or 90° grid of clonable areas around a single low point of a cut-out model. Between the fixed edge of a low point and the boundary surface of the model, the liquid edges of a soap film adjust themselves freely. The soap films are positioned normal to the walls at each point of contact. The resulting boundary curves correspond to the surface curvatures of an air hall membrane in the axis of symmetry of the cross-section. A module derived from the shape of this soap membrane can be added repeatedly on all sides to form a so-called infinite surface.

Bubbles floating on a liquid surface or a wetted surface always have an exact hemispherical shape. The liquid edges with their circular free edges are adjusted in such a way that a maximum volume is enclosed with a minimum surface area. The edges of the synclastic hemisphere are perpendicular with their contact surfaces to the support plane.

Floating bubbles on a surface can be moved as desired. If they collide, they arrange themselves by self-forming into the configuration of lowest energy level. The internal pressure can be inferred from the respective bubble size. Smaller bubbles have larger internal pressures than larger ones. The form of the shared boundary walls between bubbles also depend on the size and, hence, the internal pressure; only if the bubbles are the same size and have the same internal pressure, are the shared walls flat [17].

19.7 Soap bubbles with nets

Membrane surfaces as well as pneumatic constructions can be reinforced by nets, both in the soap-bubble model and in the full-size realisation. This can increase stability and reduce the deformations under wind load. The area formed by the nets may differ from a minimum area since the soap film forms within the individual cells of the mesh or net.

19.8 Using soap films in the design of structures

Tension-stressed surface structures such as tents or rope-net constructions can be created and planned in the preliminary design stages using soap-film models. The big advantage is, first of all, the uniform surface tension of the film, which then corresponds to the pretension and the uniform material thickness of membranes and nets. However, buildings must be able to withstand wind loads and, in certain cases, snow loads. These aspects need to be taken into account in determining the final form and in detail-design calculations. Soap films suitable for buildings always have an anticlastic curvature. If the area of the ground plan is enlarged, the high points rise to a height that has architectural and functional limits. Support and the addition of single surfaces become sensible considerations (Figure 19.11).

Once the basic shape for a membrane or cable net construction has been found with a model study, or a partial study made of soap films, the transfer to a larger-scale textile model can take place. Suitable materials for this are fine fabrics (e.g. women's stockings) or fine-meshed nets (e.g. tulle or net-curtain fabric). If fabrics with hexagonal meshes are used, the distribution of the surface stresses can be read directly from the deformation of the individual cells of the mesh. A particularly good example of this process was the King Abdulaziz

(a) (b)

Figure 19.11 IL Pavilion at the University of Stuttgart (a) Soap-film model using a single rope loop. (b) The cablenet construction. (Images: Berthold Burkhardt)

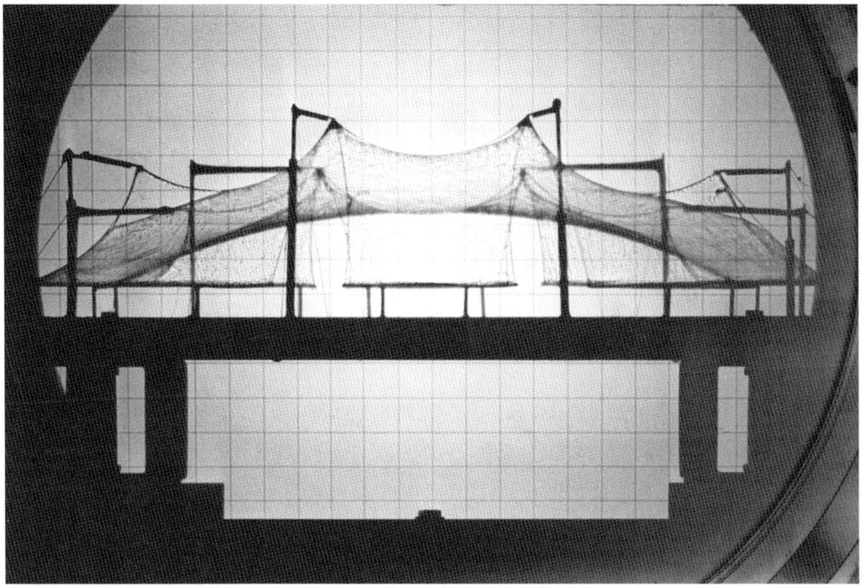

Figure 19.12 Soap-film model for the sports hall, Jeddah (end view, using parallel light). (Image: Courtesy of Institute of Lightweight Structures, Stuttgart)

Figure 19.13 Soap-film model for the sports hall, Jeddah (side view, using backlight). (Image: Courtesy of Institute of Lightweight Structures, Stuttgart)

sports hall for the University of Jeddah, designed by architect Rolf Gutbrod with Frei Otto, and engineers Ove Arup and Partners and Buro Happold in the mid-1970s. On the basis of form studies using soap-film models (Figures 19.12 and 19.13) a suspended-chain model was constructed (Figure 19.14). After further optimisation of the form using the chain model, a fine polyester fabric was used to create the final form of the sports hall (Figures 19.15 and 19.16).

Figure 19.14 Suspended-chain model for the sports hall, Jeddah. (Image: Courtesy of Institute of Lightweight Structures, Stuttgart)

Figure 19.15 Final model of the sports hall, Jeddah in fine polyester fabric. (Image: Courtesy of Institute of Lightweight Structures, Stuttgart)

19.9 Concluding remarks

Research and experiment with these sensitive but highly-accurate surfactant films, the soap bubbles and films lead to optimizations of lightweight surface structures made with membranes and nets. In the experimental testing, they allow a multiplicity of variations which, following the principle of self-formation, can be achieved only by changing the boundary conditions. Preliminary studies of entire membranes or net constructions, or even only parts of them, led to a wide range of optimised surface structures such as the Sternwellenzelt (Star wave canopy) at the Tanzbrunnen in Cologne (Figures 19.17 and 19.18), the stadium roof of the Olympic facilities in Munich in 1972 (see Section 14.3.4), and, for the (un-executed) design for a 2-km diameter pneumatic roof over a city in the Arctic [18], all by Frei Otto.

Figure 19.16 King Abdulaziz sports hall for the University of Jeddah, 1981. Designed by architect Rolf Gutbrod with Frei Otto, and engineers Over Arup and Partners and Buro Happold. The largest dimensions of the hall are 120 x 90 x 18 m high, and it covers an area of 7,500 m^2. (Image: Courtesy of Institute of Lightweight Structures, Stuttgart)

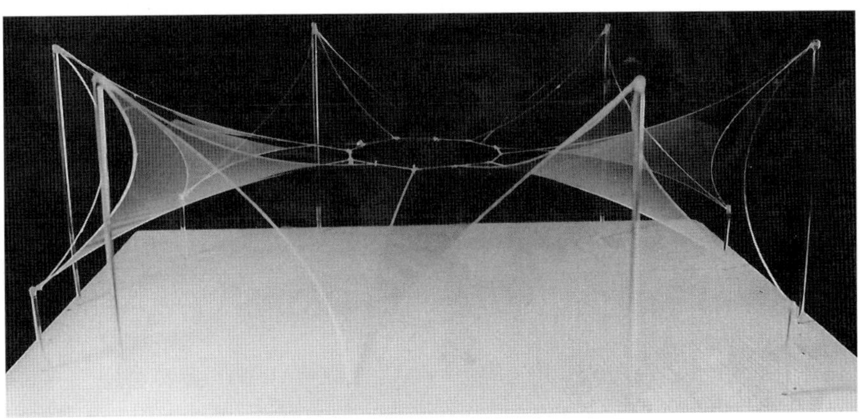

Figure 19.17 Soap-film experiment for the Sternwellenzelt, Tanzbrunnen, Cologne. (Image: Courtesy of Institute of Lightweight Structures, Stuttgart)

Figure 19.18 Sternwellenzelt at the Tanzbrunnen in Cologne, 1957. (Image: Berthold Burkhardt)

The experiments with soap bubbles and minimal surfaces also made a decisive contribution to the investigation of the formation and growth processes in living and non-living nature through the Biology and Construction Working Group headed by Frei Otto, Johann Gerhard Helmcke and myself. In many technical fields, even including space research, the soap bubble is used as a comparison for spherical accumulations of stars and comets, possibly physically as well as theoretically.

References

1 Plateau, J.A.F. (1873) *Statique expérimentale et théorique des liquides soumis aux seules forces moléculaires* (2 Vols). Paris: Gauthier Villars.

2 Boys, C.V. (1896) *Soap bubbles and the forces which mould them.* London: SPCK. (Reprinted 2012, as *Soap bubbles: their colors and forces which mould them.* New York: Dover Books.)

3 Otto, Frei (1954) *Das hängende Dach. Gestalt und Struktur.* Berlin: Ullstein.

4 Otto, Frei & Trostel, Rudolf (1962) *Zugbeanspruchte Konstruktionen.* Vol. 1 of 2 vols. Frankfurt, Berlin: Ullstein.

5 Otto, Frei, & Schleyer Friedrich-Karl (1966) *Zugbeanspruchte Konstruktionen.* Vol. 2 of 2 vols. Frankfurt, Berlin: Ullstein.

6 Hildebrandt, Stefan & Tromba, Anthony (1987) *Panoptimum – Mathematische Grundmuster des Vollkommenen.* Heidelberg: Spektrum der Wissenschaft.

7 Hildebrandt, Stefan & Tromba, Anthony (1995) *The Parsimonious Universe: Shape and Form in the Natural World.* New York: Copernicus/Springer Verlag.

8 Hildebrandt, Stefan & Tromba, Anthony (1996) *Kugel, Kreis und Seifenblasen: Optimale Formen in Geometrie und Natur.* Basel, Boston, Berlin: Birkhäuser. (with comprehensive bibliography)

9 Otto, Frei, Bach, K., Helmcke, J.-G. et al. (1977) *Pneus in Natur und Technik: Pneus in Nature and Technics.* Mitteilungen des Instituts für leichte Flächentragwerke. IL 9. Universität Stuttgart.

10 Gaß, Siegfried (1990) *Experimente: Experiments.* Mitteilungen des Instituts für leichte Flächentragwerke. IL 25. Universität Stuttgart.

11 Walker, Jearl (1982) Experiment des Monats – Farbmuster auf Seifenfilmen. In: *Spektrum der Wissenschaft* 11: 124-129. Heidelberg.

12 Bach, Klaus, Burkhardt, Berthold & Otto, Frei (1987) *Seifenblasen: Forming bubbles.* Mitteilungen des Instituts für leichte Flächentragwerke. IL 18. Universität Stuttgart.

13 Faig, Wolfgang (1969) *Vermessung dünner Seifenlamellen mit Hilfe der Nahbereichsphotogrammetrie.* (Diss.) München: Verlag der Bayerischen Akademie der Wissenschaften.

14 Bubner, Ewald (1997) *Membrankonstruktionen – Verbindungstechniken.* Essen: Wehlmann. (Publication of Doctoral Thesis, Essen, 1969).

15 Minke, G., Burkhardt, B. & Schöfl, G. (1969) *Minimalnetze: Minimal nets.* Mitteilungen des Instituts für leichte Flächentragwerke. IL 1. Universität Stuttgart.

16 Otto, Frei, Dickson, Michael, Burkhardt, Berthold, et al. (1983). *Lufthallenhandbuch: Air Hall Handbook*. Mitteilungen des Instituts für leichte Flächentragwerke. IL 15. Universität Stuttgart.
17 Sloterdijk, Peter (1998-2004) *Sphären*. 3 Bde. (Bd. 1 *Mikrosphärologie. Blasen*; Bd. 2 *Makrosphärologie. Globen*; Bd. 3. *Schäume*). Frankfurt am Main: Suhrkamp.
18 Otto, Frei, Tange, Kenzo et al. (1971) *Stadt in der Arktis: City in the Arctic*. Mitteilungen des Instituts für leichte Flächentragwerke. IL 2. Universität Stuttgart.

20

The model as a concept: the origins of the design methods of Sergio Musmeci

Lukas Ingold

20.1 The interdependence of form and method

The multifaceted work of the Italian structural designer Sergio Musmeci (1926-1981) was essentially driven by various design methods originating in the natural sciences. His interest in mathematics led him to explore physical models made of soap films or rubber membranes for the purpose of design. Such experiments enabled him to derive novel forms directly from the configuration of the forces. This particular approach to form-finding was developed in a genealogy over more than a decade and resulted in the emergence of a completely new spectrum of shell structures in reinforced concrete. However, many of the designs remained unbuilt, except for his main work, the *Ponte sul Basento* at Potenza in Southern Italy (1967-1975). This bridge exemplifies his fundamental research on the form of structures, which sought the inversion of basic principles of structural mechanics: 'There is no reason why the unknown factors should always be the internal stresses and not, for example, the geometric parameters that define the form itself […]' [1]. The genesis of the form of the bridge was inseparably related to the development of the design methods, a circumstance that can be observed throughout the whole professional career of Musmeci. His curiosity to search for new forms led him to new design methods, and the application of these new methods changed the appearance of his designs. This chapter aims to trace this interdependence of form and method, where the physical model represents not just a simple tool but rather a conceptual instrument which links natural sciences and physical reality.

In order to grasp Musmeci's intentions, the chapter follows a reverse reading, starting from the built object. The first part illustrates how the design for the Ponte sul Basento reflects his thinking as a structural designer, which was shaped by the specific conditions of the post-war period in Italy. The second part focuses on the design method for the bridge, in which both physical and mathematical models became important instruments for deriving its form, describing its geometry, and analysing its structural behaviour. The third part exemplifies how the conception of the bridge did not arise suddenly or accidentally. Instead, the design

Physical Models: Their historical and current use in civil and building engineering design,
First Edition. Edited by Bill Addis.

should be considered the product of a continuous development, starting in the mid-1950s. The fourth part aims to contextualize Musmeci's work, as his idealistic approach addressed possibilities and problems very similar to those of such contemporaries as Frei Otto (1925-2015). Their endeavour is embedded in a long history of ideas that can be traced back to the mathematical sciences of the eighteenth century. In retrospect, Musmeci's work demonstrates a radical application of scientific ideas to architecture and engineering, as the final part shows.

20.2 The quest for the form

The structural designs of Sergio Musmeci were driven by a very basic quest for the form. The Italian engineer aimed to derive the form of structures in an immediate way from the laws of nature or simply from the conditions of the force flow. The Ponte sul Basento in Potenza (1967-1975) serves as an exceptional example of this approach and may be deemed his most significant contribution. The bridge paradigmatically illustrates his design method, in which physical models became important instruments for the whole planning process. Musmeci described the genesis of the project, from the first ideas to the construction process, in an issue of the periodical *L'industria italiana del cemento* [2], a valuable source for studying his design intentions and methods. However, Musmeci began his manifold explorations of novel forms for structures almost two decades before he designed the bridge in Potenza. His endeavour sprang from the particular constraints on Italy in the aftermath of World War II.

20.2.1 Scarcity as potential

Musmeci's professional career started directly after the war, when Italy underwent deep political, economic and social changes. It was a time of glaring contrasts. On the one hand, this period of reconstruction was characterized by poverty, the provision of employment and the restriction of resources. On the other hand, this was the beginning of the *Miracolo economico* (ca. 1945-1968), which saw sprawling urban development, the initiation of large infrastructure projects, such as the *Autostrada del Sole*, and the heyday of cultural production in architecture, art, literature and cinema [3]. Despite the evolving building activity, the early post-war period was afflicted by a scarcity of resources: construction materials were limited and expensive, while labour was readily available and cheap. Consequently, the demand for an economical construction method corresponded with the characteristics of reinforced concrete, which became the primary construction material at that time [4]. In structural engineering, reinforced concrete represented an important research subject, leading to a range of technological developments such as prefabrication and prestressing. From an aesthetic point of view, the materiality expressed by reinforced concrete resulted in the creation of a new, post-fascist architectural identity [5].

These multifaceted conditions framed the environment of Musmeci's education and early career. He graduated from the University of Rome in both civil engineering (1948) and aeronautical engineering (1953), and he worked for the two most highly-reputed structural engineers in Italy at that time, Pier Luigi Nervi (1891-1979) and Riccardo Morandi (1902-1989), before founding his own practice in 1954 together with his wife, the architect Zenaide Zanini (1926-2015) [6]. Furthermore, Musmeci had the opportunity to collaborate on numerous occasions with some of the best-known Italian architects of his time, such as Adalberto Libera (1903-1963), Giuseppe Vaccaro (1896-1960), Carlo Mollino (1905-1973) and Bruno Zevi (1918-2000). Although he was surrounded by an influential circle of architects and engineers, his deepest interests were related to a completely different field, the natural sciences.

Since his childhood, Musmeci had pursued his curiosity in astronomy. Later, he developed interests in such natural sciences as physics, chemistry and, above all, mathematics. His designs for structures found their genesis in his intellectual world of science, the *pensiero scientifico* [7]. Driven by his explorative curiosity, his work comprised a broad spectrum of mostly new structural forms, which creatively addressed the concept of material efficiency as a direct response to the limited resources. Consequently, the constraint of scarcity was understood as potential. Musmeci's early works in the 1950s ranged from experimental operations with folded-plate structures to the first attempts with simple vaults. Subsequently, at the end of the 1950s and the beginning of the 1960s, various forms of more sophisticated shells and gridshells became a major aspect of his research. Physical models played a crucial role in developing such conceptions. Soap-film and rubber-membrane models offered possibilities for the creation of complex surfaces with a high degree of double curvature. But Musmeci used physical models in a range of scales and materials for both the design and the verification of structures, as he most impressively demonstrated with the example of the Ponte sul Basento. In the later projects of the 1970s, the physical models were increasingly replaced by electronic calculations [8]. Early computers became a key instrument for his designs of spatial meshes and spatial lattice structures. Such lightweight principles led to a higher degree of redundancy. The very regular and systematic order followed a definition of points in space, similar to atoms in crystals. These investigations represent another impressive example of the profound impact of scientific thinking on Musmeci's imagination [9]. Unfortunately, these spatial structures were only developed in a prototype state, due to his early death at an age of only 54 years.

20.2.2 Continuity of force and form

One of the most spectacular buildings of the post-war period was constructed at the approach to city of Potenza. The Ponte sul Basento has an exceptional presence, and the design marks a completely new and non-canonical direction from the perspectives of both architecture and engineering (Figure 20.1). The particular form of the bridge expresses an extreme purity, and the very slender deck is supported by only four spatially-deformed vaults. The thin shell represents a structure arising from its own internal logic, and it impressively demonstrates the ability of the designer to give a shape to inscrutable forces [10]. Musmeci aimed to explore the 'relation between the form of the structure and its architectural

Figure 20.1 The Ponte sul Basento, Potenza, Italy, 1967-75. The deck is supported by a thin reinforced-concrete shell. Structural design: Sergio Musmeci. (Image: [12])

expression' in a very immediate way, and he was fascinated by 'the perception of statical conditions which become manifested through their form' [11].

The 286 m long bridge connects the motorway with the elevated city centre of Potenza by crossing the *Basento* river and a railway. During the development of the project, Musmeci was assisted by the technical engineer for the local authorities, Aldo Livadiotti, the engineer Emanuele Filiberto Radogna, and the architect Zenaide Zanini. Their spectacular design replaced a preliminary bridge scheme, envisaged as prefabricated beams resting on columns, and evoked the elation of Gino Viggiani, the president of the client's consortium for the industrial development of the city [13]. The project, consisting of only two basic elements, the bridge deck and the substructure in the form of the shell, was constructed entirely of in-situ reinforced concrete. The slender bridge deck has a width of 16 m and a maximum thickness of only 1.3 m. Both sides of the deck are slightly curved, and the edges are also smoothly rounded. The divisions of the deck, expansion joints based on the Gerber-hinge principle, remain almost invisible. The thin shell below supports the bridge deck at a few points across a distance of 17 m. The shell itself, a continuous surface with double curvature, comprises four connected spans of around 58 m. The general thickness of the shell is only 30 cm, but along the edges, its thickness increases up to more than a metre. The edges above the supports are particularly remarkable for their shape which follows a spiral line (Figure 20.2). The highly-complex geometry of the shell, particularly visible along these curved edges, demonstrates an unexpected architectural expressiveness. Hence, the object embodies a membrane-like, almost 'fluid' appearance [14]. Between the deck and the continuously-developing surface, an exceptional architectural space serves as an area accessible to pedestrians.

In the context of Italian architectural discourse of the post-war period, the bridge design exemplified the *architettura organica* [16]. This term was introduced by the architectural historian Bruno Zevi, a friend and mentor of

Figure 20.2 Ponte sul Basento. The twisting edges correspond to the manipulation of boundary conditions in the form-finding studies using a soap-film model. (Image: [15])

Musmeci, and signified a tendency following the school of Frank Lloyd Wright (1867-1959) [17]. Zevi had great respect for Musmeci's unique approach, and saw in the work a response to his call for 'the involvement of all architectural elements for the orchestration of the statics' [18]. By investigating the basic constitution of the form, inseparably related to the distribution of its inner forces, Musmeci's endeavour ideally reflected the ideas of Zevi. The resulting designs proved to be very efficient in their use of material. Consequently, he aimed to focus on 'spatial configurations, which require the minimum compatible volume of material under a given statical assignment'.

For the design of the Ponte sul Basento, Musmeci intended to create exactly such a spatial structural scheme:

> 'The three-dimensional space offers an enormous enrichment of possibilities. Every surface with double curvature can potentially be used as a membrane without any bending, if certain conditions of the borders are fulfilled [...]. If we add the condition that the compression-only forces

> have to be uniform, every part of the membrane starts to contribute in the same way and the material is used with maximum efficiency in every point and in every direction. The inner compression forces of the membrane are resisted by the borders, and when they are free from any bending, they follow the design of pure spatial arches' [19].

The surface evolves continuously between the spatially-curved edges, which define the boundary conditions. In fact, the surface and the edges find an equilibrium in their stress state with each other. Due to the smoothness of the surface, the forces can be distributed equally and concentrations of stresses can be avoided. Consequently, the continuity of forces finds its expression through the continuity of the form.

20.2.3 Between economic construction and material efficiency

Musmeci's forms and methods represented an unparalleled approach in the post-war period in Italy. His structural designs were far more radical and incisive than those of most of his contemporaries, such as Pier Luigi Nervi, Riccardo Morandi, Silvano Zorzi (1921-1994), Aldo Favini (1916-2013), and Angelo Mangiarotti (1921-2012). While these structural designers tended to focus on the economy of the construction process and the integration of such technological achievements as prefabrication and prestressing, Musmeci aimed to derive his forms directly from the conditions of the force flow [20].

The design of the Ponte sul Basento is based on an extremely idealized structural system in which the aspect of efficiency seems to be completely restricted to the minimum use of material and does not reflect the construction process at all [21]. The bridge was built with rather outdated methods, and the construction took more than four years [22]. In contrast to the continuity of the geometry, the shell was not cast in one single phase but was divided into several shorter segments. First, the segments directly above the foundations were built as solid transverse arches between the four point supports. Afterwards, the four longitudinal spans were executed one after the other. A very elaborate but already outdated system was used to scaffold the centring [23]. Cambered timber girders, which followed the transverse guiding curves of the geometry, allowed the complex formwork to be controlled. However, for the almost vertical parts of the shell, a counteracting formwork had to be moulded. The circuitous and protracted construction process followed the rather conventional construction method proposed by the contractor because Musmeci did not develop a construction technique that better reflected the form's geometrical complexity. This might have occurred due to an overreaching confidence that any kind of structure could be built, but the conception of the shell neglected numerous questions about its 'constructability'.

Instead of placing importance on the method of construction, Musmeci informed the design entirely from the constitution of the inner forces within the structure. As a consequence, the quest for the form rendered the quest for the method salient for his designs. Model studies enabled the evolution of the explorative approach. They were not only used for the verification of structures; they were also deployed for the purpose of design. In this regard, Musmeci was the first in Italy to enter the field of form-finding. The Ponte sul Basento provides an exemplary survey of the potentials and problems of an emerging design methodology that marks a distinct tendency in the history of architecture and engineering in the twentieth century.

20.3 The physical model as instrument

Thinking through models became a main driver of Sergio Musmeci's work. Both physical and mathematical models are always abstractions. But they do not merely represent an isolated state of the design or a particular simulation of the reality. Instead, models must be understood as alternating 'media of translation' in the step-by-step evolution of a project. Consequently, the approach was based on an interdependency of various physical and mathematical models with the aim of taking advantage of both. Where mathematical models became difficult to describe and series of equations too circuitous to solve, simple physical models started to play a crucial role. Simplification was fundamental to Musmeci's experimental approach: 'In fact, it often happens that the most important and significant aspects are the qualitative, and they may be understood and experimented with using simply-constructed models and with very rough calculations' [24].

In a larger context, Musmeci's approach to physical models was embedded in the empirical culture of structural engineering in Italy. During the post-war period, it was very common to use physical models in test laboratories to verify large-scale structures in reinforced concrete. Even at the beginning of the 1930s, the engineer Arturo Danusso (1880-1968) had started to carry out physical model tests at the *Politecnico* in Milan and became an influential figure on the Italian engineering scene (see Chapter 10). In 1951, together with his assistant Guido Oberti (1907-2003), he founded the *Istituto Sperimentale Modelli e Strutture* (*ISMES*) in Bergamo (see Chapter 15). The laboratory was the most renowned experimental facility in Italy and allowed the rapid and systematic testing of structural conceptions whenever intractable problems arose in mathematical analysis. Numerous physical models of bridges, towers, halls and dams were tested in Bergamo, and Pier Luigi Nervi also analysed many of his structural designs in this laboratory. While physical models were extensively used for structural analysis, it was rather uncommon to use them in the early design process. Apart from the funicular chain model, a method for determining the thrust line that was widely known from the rule of Robert Hooke (1635-1703) and the studies of Antoni Gaudí (1852-1926) and others (see Chapter 2), physical models had rarely been used for design purposes.

But for Musmeci, the physical model became a vital instrument for deriving new kinds of forms for his structures. The design of the Ponte sul Basento was initiated with such experimentation, and the example of the bridge illustrates how physical models were used throughout the design process, from the early stage of the design to the verification of the elaborated structure. Altogether, four physical models were made for various purposes, in different materials and at several different scales [25, 26]. As already mentioned, the use of physical models was inseparably connected to the mathematical definitions of the form.

20.3.1 Visualizing mathematics: the soap-film model

The membrane-like appearance of the Ponte sul Basento constitutes its spatial presence, but it also immediately reveals the design method of the structure. The basic 'motif' of the shell was derived from studies with a soap-film model (Figure 20.3). In such an experiment, a closed frame with a predefined form is dipped into a soap solution. While the frame provides the boundary conditions, the soap film is allowed to 'find the form' of the surface in the space within the frame. The resulting surface represents the smallest possible surface between the given boundary curve (see Chapter 19). The stresses within the liquid are as low as possible, because they are distributed equally and in all directions. Consequently, stress peaks in the surface are avoided, or in the case of the soap-film model, simply result in the collapse of the thin film.

Musmeci's approach was based on a broad theoretical corpus from the mathematical sciences. The form of the Ponte sul Basento was mainly informed by the theory of *minimal surfaces*. Such surfaces can be developed either with a mathematical description or with a physical model such as a soap film, as described. Minimal surfaces evoke a special interest in the design of structures, because they can be applied to both membrane and shell structures, in which only pure tension or pure compression stresses occur. The specific conditions arise from the geometric description of the surface. In mathematical terms, minimal surfaces consist of *zero mean curvature*. On every minimal surface, two opposite directions of curvature can be observed. At every point, these two principal curvatures have the same degree of curvature, but the sign is inverted, meaning one has a positive curvature, the other negative. Musmeci took this condition as a self-constraint for his design: 'generating a form of a surface, which is able to function with uniform compression forces in all directions, is a mathematical topic antecedent to the design of the form' [28]. The experiment with the soap-film model represented a very early step in the design process for the Ponte sul Basento. The model had

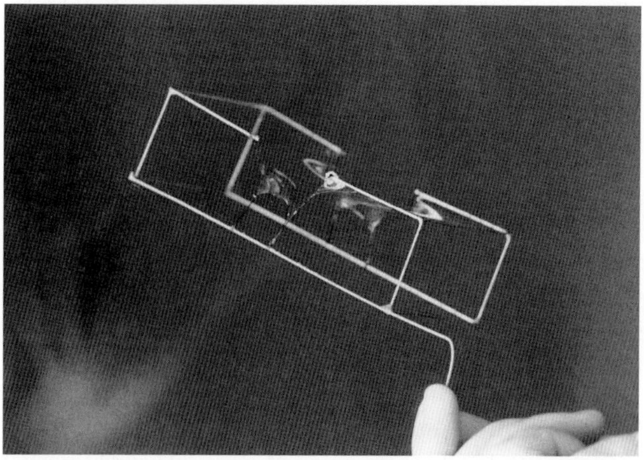

Figure 20.3 The soap film represents the smallest possible surface within the given boundary curve. The hand in the photograph is that of Musmeci's daughter. (Image: [27])

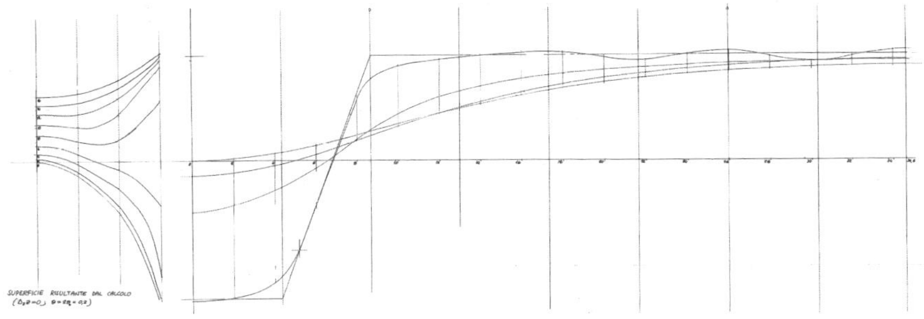

Figure 20.4 Calculated minimal surface based on a straight boundary line. (Image: [29])

the purpose of translating the abstract mathematical theory of minimal surfaces into a physical reality. In fact, the physical model enabled Musmeci to visualize the mathematics as a real form.

Nonetheless, the conditions of the experiment were less strict than they might appear, and they allowed Musmeci to manipulate the result in various ways. First, the boundary curve was not fully predefined, because it consisted of a stiff, yet open, timber frame and a flexible cotton thread within the space. The partly-flexible boundary curve enabled not only the surface to self-generate but also, to some extent, its boundary line. Literally, the surface of the soap film pulled the thread into its spiral form, which was used to display the spirally-curved edges of the shell. Indeed, this was important, because Musmeci was able to describe the general shape of the surface mathematically (Figure 20.4) but not the exact disposition of the edge line. A second manipulation of the soap-film model involved the general distortion of the form. Since the compression forces in the transverse direction were foreseen as notably smaller than in the longitudinal direction, the transverse curvature was tripled. As a consequence, the physical model was built three times wider in the transverse direction. Such manipulations indicate Musmeci's distinctive intuition: while the general form was directly related to a mathematical theory, the result was consciously adapted. It seems that the design approach constantly oscillated between the directness of the theoretical application and the experimental intuition of the designer. As the design evolved further, a similar procedure of manipulation can be observed for all four physical models.

20.3.2 Measuring the geometry: the rubber-membrane model

The soap-film model was mainly used to derive the form, but it was difficult to capture the geometry of that form. The convergence towards a determinable geometry was enabled by the second step, the rubber membrane model (Figure 20.5). This physical model was an elastic surface strongly distorted to simulate the configuration of uniformly-distributed stresses. Within a frame, metal clamps held the rubber membrane and represented the points of the external forces from the deck and the foundations. The clamps were connected through strings so that the external loads were applied equally. The edges of the membrane were also reinforced with rubber cords to represent the strengthened

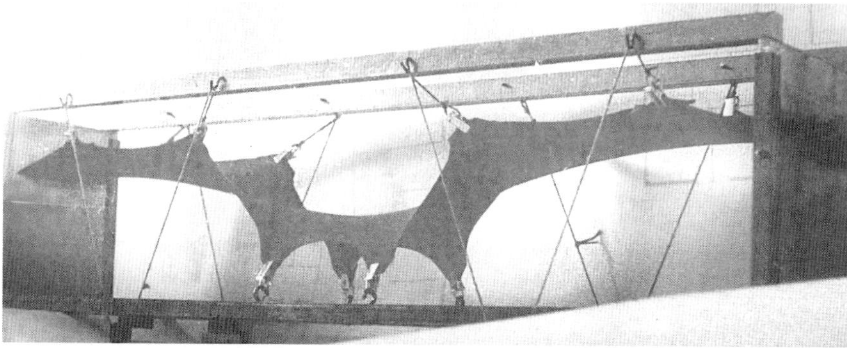

Figure 20.5 The rubber-membrane model (probably 1:50 scale), enabled the geometry to be measured. (Source: [46] 25)

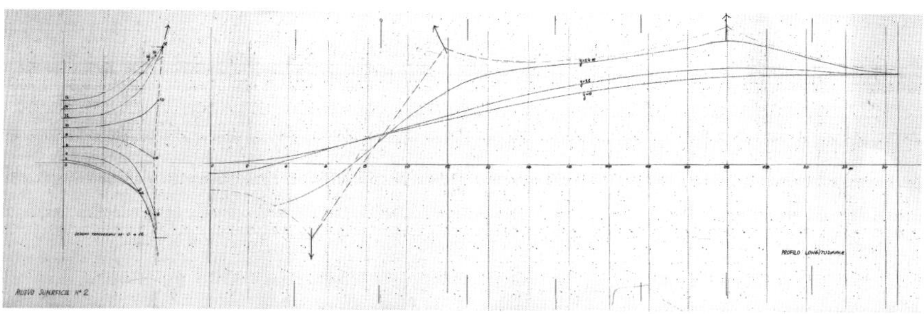

Figure 20.6 Contour lines measured from the rubber-membrane model. (Image: [33])

borders. Even though the model represented a tension-only stress state, it could easily be translated into the reciprocal condition of pure compression.

The main purpose of the model was to measure the geometry of the form. The rubber was marked with a grid pattern on its upper surface to facilitate surveying the geometry in space. This allowed Musmeci to trace the main contour lines of the form (Figure 20.6). Even though this procedure represented only an approximation of the intended form, the measured relief of the physical model proved to be very similar to the calculated geometry. The mathematical model referred to theories concerning minimal surfaces by Rudolf Trostel (1928-2016) [30], which had been published in Frei Otto's book *Tensile Structures* [31, 32]. Based on these calculations, the geometry of the surface was defined for the subsequent development of the design. Furthermore, the rubber-membrane model enabled the intuitive analysis of the structural behaviour. With this simple instrument, he could observe the deformations under various load cases and test variations on the structural system. For instance, it was again possible to manipulate the form consciously in order to reduce the stresses in the transverse direction.

The case of the Ponte sul Basento exemplifies how minimal surfaces can be applied for the design of thin shells, as they can for gridshells, membranes and cable nets. Such very idealized conceptions bear inherent structural potential, but they also bear challenges in their realization. Their limited geometrical tolerance and lower structural redundancy made the application complex and circumstantial. For example, structures with negligible self-weight, such as tensile membranes, are ideally represented with a minimal surface. But shell structures in reinforced concrete have to deal with higher dead loads. Moreover, in the case of a bridge, the live load might also become a governing factor that was not conceived in the original form. The superposition of these effects can defeat the initial simplicity of the form and make the analysis of the structure rather difficult.

20.3.3 Gauging stresses and strains: the methacrylate model

While the first experiments with physical models were used for the design, the purpose of the later studies was the verification of the structure. For the third study, a physical model of two spans of the bridge was built to a scale of 1:100. The model was made of methacrylate, a transparent plastic material often referred to by its trade name, *Perspex* (Figure 20.7). In order to produce the complex form, the plastic material was heated and pressed between two moulds, which were composed of a block of hardwood and a counter-block cast in aluminium [34].

The physical model study allowed Musmeci to measure precisely the stresses and strains under a range of load conditions. The measurements were conducted in the *Laboratorio di Ricerche su Modelli* at the Faculty of Engineering of the University of Rome using electrical-resistance strain gauges. In the case of the bridge, it was especially useful to verify the strains occurring in asymmetric load cases.

The project was developed further by determining the shell definitively through the calculation of a series of finite differences. This numerical method was used to approximate geometries that are difficult to describe otherwise. In this mathematical model, a segment between the foundation and the longitudinal vertex was projected onto the plane. This surface was subdivided in an orthogonal grid, which showed the longitudinal and transverse axes. The borders were represented as funiculars in order to remain in equilibrium with the forces accumulating in the surface. This step made it possible to gain a precise understanding of the geometry of the form (Figure 20.8). At the same time, the

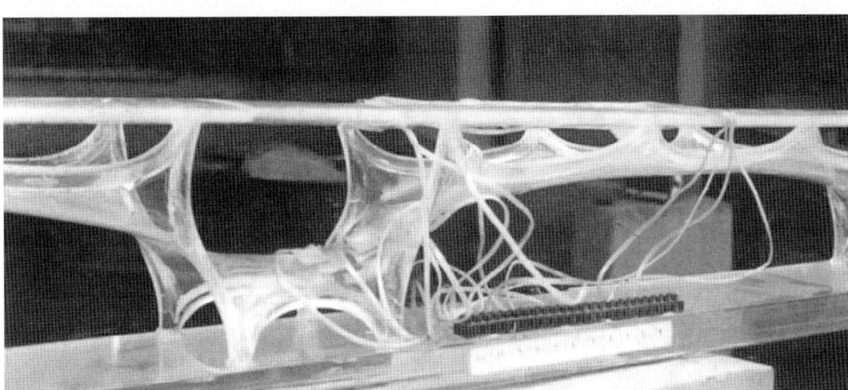

Figure 20.7 1:100 scale methacrylate model used to determine stresses and strains with electrical-resistance strain gauges. (Image: [35])

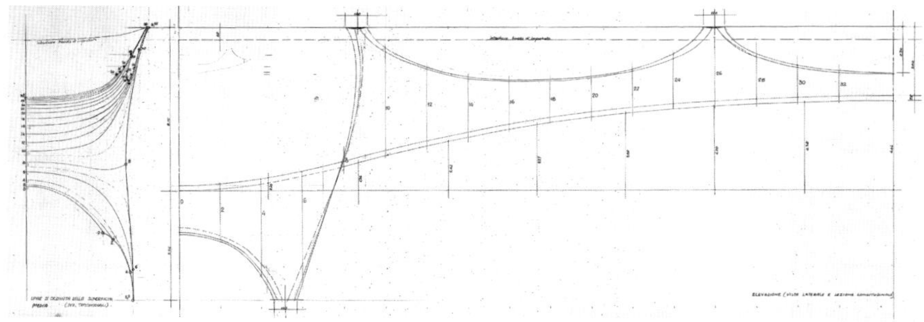

Figure 20.8 Translation of the surface into a shell, with a general thickness of only 30 cm. (Image: [36])

strain-gauge analysis provided a comprehensive understanding of its structural behaviour. Nevertheless, the production of a further physical model became necessary. The fourth and final physical model was built at the request by the public authorities to gain an additional and enhanced verification of the structure.

20.3.4 Analysing the material behaviour: the reinforced micro-concrete model

The final physical model test was conducted at a scale of 1:10 in the *Istituto Sperimentale Modelli e Strutture* (*ISMES*) in Bergamo. The design of the Ponte sul Basento was materialized with an enormous effort: The model was made using its 'real' material, a reinforced micro-concrete (Figure 20.9, see Figure 15.23). Again comprising two spans of the bridge, the structural model had a length of around 14 m and a shell thickness of 3 cm. Not only were the dimensions proportionally represented in the model, but also both the compressive strength of the micro-concrete and the tensile strength of the reinforcement were adapted

Figure 20.9 1:10 scale reinforced micro-concrete model at the *Istituto Sperimentale Modelli e Strutture* in Bergamo. (Image: ISMES Historical Archive)

in proportion to the full-scale project. This model was produced by pouring the micro-concrete between the two halves of a plaster mould.

The study enabled Musmeci to analyze the material behaviour of the structure comprehensively. After some preliminary tests, a series of elastic experiments allowed the stresses to be measured at 200 points and under five loading conditions. The loads were applied with 224 hydraulic jacks, and the deformations were recorded with clock gauges and electrical strain gauges. The measurements of bending effects resulting from asymmetric load cases were particularly useful, since these could not be considered in the calculations based on membrane theory. After the extensive series of experiments, the structure was loaded to collapse. While the deck and the ultra-thin vault withstood the applied loads, the foundation blocks, not part of the structural model, eventually failed and caused the immediate collapse of the entire structure. Accumulating the loads, the deformations and the scaled material strengths, the model of the bridge was loaded 2.9 times higher than the calculated ultimate load and 7.3 times higher than the design live load. Amusingly, Musmeci described the extensive physical model tests of the ultra-thin vault as 'an eggshell in a bench vice' [37].

When the tests in Bergamo were concluded in March 1971, construction on-site should already have been in progress. However, the structural model in north Italy had a direct link to the construction site in the south. The moulds of the model were reused to prepare a plaster model of a quarter of the vault. The model was provided to the contractor as a physical reference in addition to the calculated geometry of the mould. This fact is indicative of the challenges presented by the complexity of the form; it resulted in considerable difficulties in the construction process that appeared almost insurmountable and required more than four years to overcome. While the physical models enabled the rapid simulation of specific conditions at a reduced scale during all phases of the planning, the translation of the form into a real, built object became very circuitous. The models served as instruments of simplification throughout the whole process from design to verification, but at the real scale of the construction, this methodical approach represented an alienation from reality. However, the modelling process can also be considered from another critical perspective: during the whole planning process, the form of the bridge was translated repeatedly. Every geometric translation presented the possibilities both to refine the conception and to lose some of the original qualities of the form. In the case of the bridge at Potenza, such a shift can be observed in the increase of the shell thickness along the edges, which adversely affected the elegant appearance [38]. The built work shows a distinct contrast to the original modelled form, which was absolutely expressive from an aesthetic and structural point of view through its pure and fluid shape.

20.4 The development of the method

By the time Sergio Musmeci came to develop his design for the Ponte sul Basento, he had already been experimenting with similar structural forms for many years. Consequently, the bridge design must be understood as the result of a long evolution [39, 40], a genealogy of conceptions. The development started directly after World War II. Musmeci's particular interest in material-efficient forms can be traced back to his time as a student. He graduated in 1948 with a thesis project on thin shells in reinforced concrete [41], at that time an emerging structural typology that was introduced during the interwar period from Germany [42]. Such shells were used as structures for large spans with a low grade of reinforcement. Hence, shells became a main reference for Musmeci's early works. However, he was searching for conceptions beyond both the known structural typologies and the canonical aesthetics, an intention informed by his deep curiosity about the natural sciences. This interest also brought him to new design methods. In this regard, the genealogical development did not concern the form alone; it was strongly related to the evolution of the design methods behind the genesis of the form.

20.4.1 Genealogy of conceptions

Musmeci's first projects were designed as single-curved vaults or as folded plate structures, another emerging structural typology, which he studied through many creative endeavours [43, 44]. These projects featured spans of 12-35 m and comprised sports halls, foyers, cinemas and workshops. During the same period, the proposal for the *Ponte sull'Astico* in Caltrano (1956) was designed in collaboration with the architects Sergio Ortolani and Antonio Cattaneo. The segments of the arched bridge were configured as thin and slightly-curved plates and integrated into a statical system with four hinges, which would have been stiffened through smaller lateral parts. The design method still referred to a classical mechanical analysis based on influence lines [45], but the project indicated Musmeci's first investigations of doubly-curved surfaces. However, the double curvature was only used for local self-stiffening and not as a general principle for the overall form of the arched bridge. In later projects, the appearance of the structures converged towards more continuous forms. An early example is the design for the *Ponte sull'Aglio* in Barberino di Mugello (1957), often only known as a bridge prototype for the *Autostrada del Sole* [46]. The design consists of several stacked arches. While the project was derived graphically, it was represented as a physical model made of branching threads, which was produced later for didactic reasons [47].

In further development converging towards the design of the Ponte sul Basento, Musmeci began to explore in greater depth the structural potential of doubly-curved surfaces. The most important project in this evolution was the proposal for the *Ponte di Tor di Quinto* in Rome (1958), a competition project that he designed together with the engineer Ugo Luccichenti. The long deck of this bridge is supported by six spatially-deformed shells. The basic form of these abutments was derived from a rubber-membrane model, one of the first occasions when Musmeci used physical form-finding models (Figure 20.10). By gripping the elastic sheet at four points, a surface can be generated with double curvature. During the design process, Musmeci experimented with many variations, testing diverse sizes and orientations of the abutments, options with holes

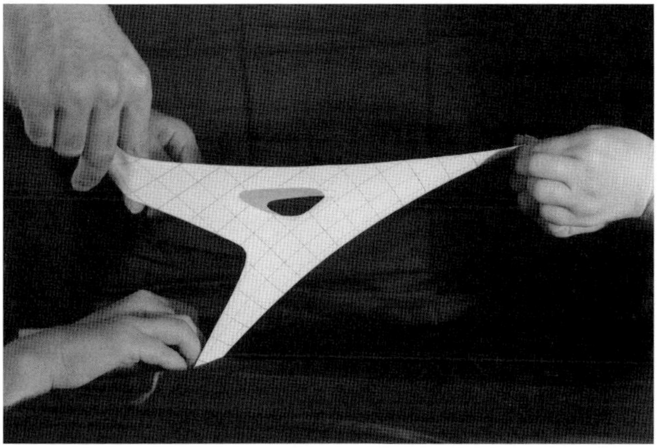

Figure 20.10 Form-finding experiment for the abutment of the Ponte di Tor di Quinto in Rome, 1958. The method is similar to that used for the design of the Ponte sul Basento. (Image: [49])

in the surface, and others with a higher degree of curvature along the borders of the shell. The project showed many similarities to the Ponte sul Basento in both its structural principle and method of design adopted. Musmeci conceived the form 'as a structure in space' and referred in the technical description of the competition entry to the theory of minimal surfaces and the possibility of deriving such a form with a soap-film model: 'The surface resulting from uniform compression forces is also the surface with a minimal size. Therefore, it shows only a minimal need of material [...] for its maximal utilization' [48]. Even though no depictions of these very first soap-film studies are known to have survived, it seems remarkable that Musmeci had already designed with minimal surfaces in 1958.

20.4.2 Emergence of a motif

After Musmeci had developed new design methods for the first bridge conceptions during the late 1950s, he continued to use these methods for the exploration of similar forms in the following years. These ideas were applied to various designs for diverse uses, at a range of scales, and in several geometric configurations, both continuous and discretized. The basic motif of the Ponte sul Basento, a saddle-like surface with a high degree of inverse curvature, was seen for the first time in the competition project for the *Palazzo del Lavoro* in Turin (1959), which was developed in collaboration with the well-known architect Carlo Mollino and his colleague Carlo Alberto Bordogna. For the festival hall of the *Italia '61* exhibition, which was held for the centenary of the unification of Italy, the three designers produced a broad spectrum of ideas but were unable to condense them into one single concept. Therefore, they submitted three different proposals, all enriched through an explorative design approach resulting in strong expressiveness of the structures. The proposal preferred by Musmeci spanned the large hall with a vaulted structure. Featuring an array of openings,

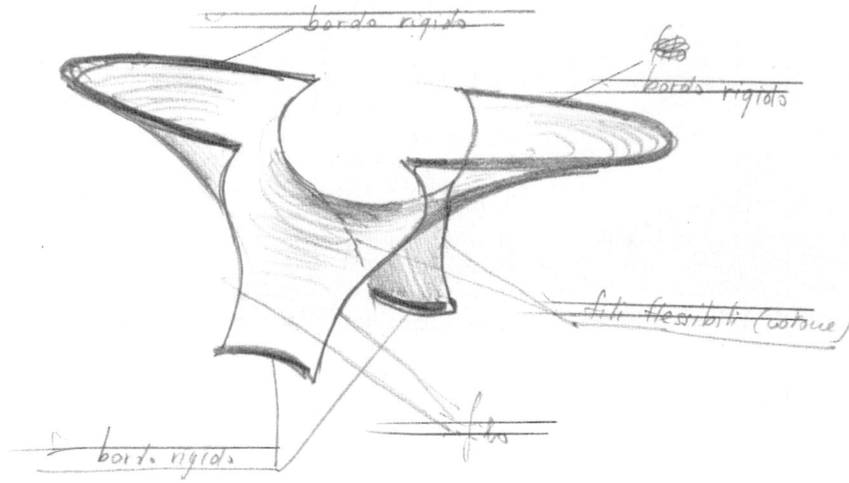

Figure 20.11 Design for the Palazzo del Lavoro in Turin. Drawing of a column whose form was derived using a soap-film model. (Image: [51])

the thin shell consisted of a very complex geometry. However, the columns of the structure were even more expressive. They showed exactly the same motif, termed an *anticlastic geometry* (Figure 20.11) as used later in the Ponte sul Basento. And clearly, this highly-complex form was derived from physical model studies with soap-film experiments [50], which Musmeci had discovered shortly before.

The same series of projects for large-span halls included the proposals for the *Mercati Generali* in Rome (1960), designed with the architects Annibale Vitellozzi, Massimo Castellazzi and Tullio Dall'Anese, and the *Stabilimento Ursus Gomma* in Cassino (1961), developed with Zenaide Zanini. Both designs remained unbuilt but were based on the experience of the project in Turin [52]. In these two examples, Musmeci aimed to integrate all structural elements of the design in one richly-shaped surface. It no longer represented a simple vault but rather an elaborate, spatially-developed continuous form incorporating at once the roof, the supporting columns and the openings for lighting the interior space. The two designs are particularly captivating because of the large, round openings around the slanted, 'chalice-shaped columns' (Figure 20.12). All parts are seamlessly integrated into the overall structure, so that no geometric segmentation is visible. For these designs, it became impossible to distinguish between a 'local' and 'global' understanding of the structure, because every single piece of the thin shell is an integral part of the whole structure. Even though the form was derived from mathematics, in the design process Musmeci relied mainly on his intuition to define its general characteristics.

He applied similar principles even for projects on a larger scale. The competition entry for the *Ponte sul Lao* in Laino Borgo (1964), envisaged as the longest viaduct of the southern extension of the *Autostrada del Sole*, comprised seven

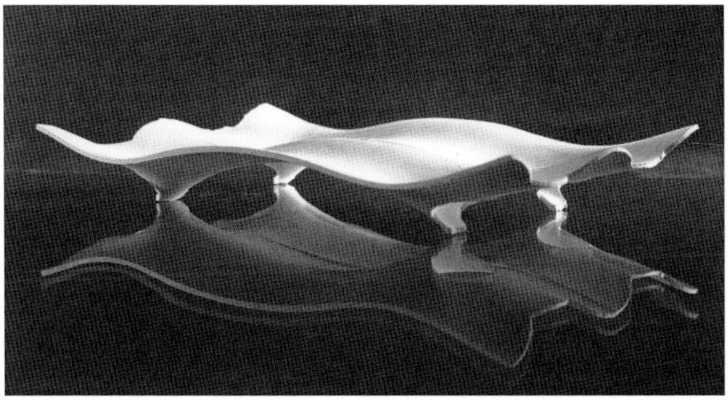

Figure 20.12 Roof proposed for the Mercati Generali in Rome. The reinforced-concrete shell is supported by chalice-shaped columns, placed below the openings. (Image: [53])

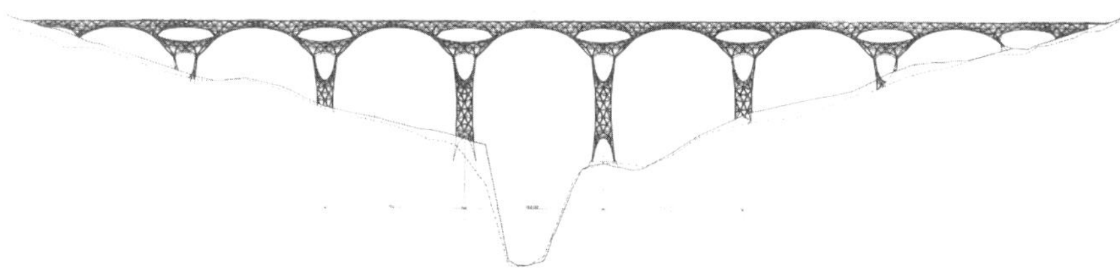

Figure 20.13 The competition design for *Ponte sul Lao* in Laino Borgo (1964) was an antecedent design for the Ponte sul Basento. Due to the large scale, the surface was translated into a gridshell. (Source: [46] 20).

spans of 125 m. The proposal showed many similarities to the Ponte sul Basento, because the basic motif of a spatial arch was the same, but due to the large scale, the doubly-curved surface was conceived as a gridshell (Figure 20.13). The structure would have been constructed with hollow reinforced-concrete elements, which would have been filled with concrete. In his design process, Musmeci followed the same approach as for the other projects. He derived the basic form by operating with soap-film models [54], but the precise order of the grid was defined through a mathematical model (Figure 20.14). For this procedure, the spatial arch was divided into quarters.

Throughout the whole genealogy of conceptions, the design method was driven by both physical and mathematical models. As the discussion of the development process has shown, the same modelling techniques were used by Musmeci on various occasions and resulted in similar forms, often even in the same motif. However, he creatively varied the elements in his approach to adapt the designs to the specific demands and scales. Unfortunately, all these projects of the late 1950s and early 1960s remained unbuilt. But during this time, his forms and methods developed continuously with every project, until the commission for the Ponte sul Basento gave him the opportunity to express his conception in a real built object.

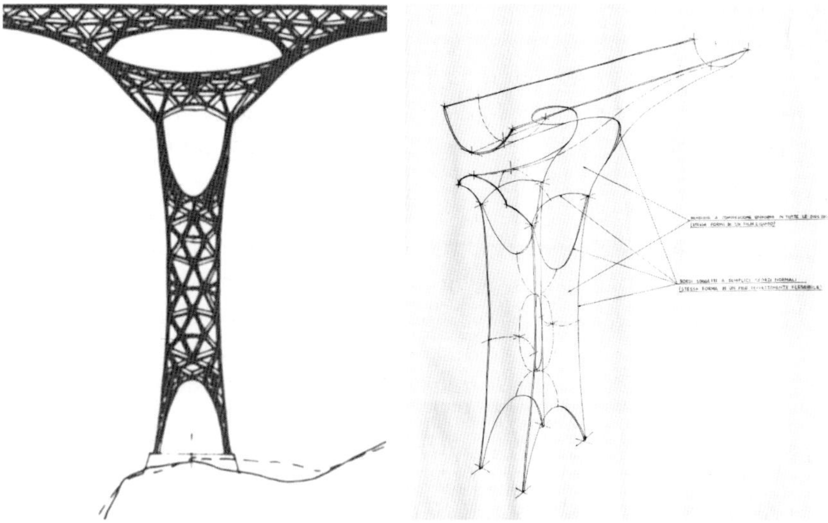

(a) (b)

Figure 20.14 The *Ponte sul Lao* in Laino Borgo. (a) Detail of one of the piers. (Image: [55]). (b) Sketch of the soap-model experiment. (Source: [46] 20).

20.5 The origin of the method

The idealism expressed in Sergio Musmeci's experiments represented a uniqueness during the post-war period in Italy. But his approach can be embedded in both a larger geographical context and a wider historical setting [56]. His research resembled very similar design methods to those of other experimental architects and engineers of the period, such as Frei Otto, Heinz Isler (1926-2009), Robert Le Ricolais (1894-1977) and Lech Tomaszewski (1926-1982) [57]. They all used physical modelling techniques, such as hanging textiles and stretched or inflated membranes, or fluids to discover new forms of structures. Even though the spectrum of these investigations was very broad, most of the structural designers began to develop a particular interest in soap bubbles during the late 1950s, as did Musmeci. Such experiments had an inestimable effect on their work. They searched in a very experimental manner and with plenty of variation for spatially-developed surfaces and volumes. These almost invisible and fluid objects addressed a novel vocabulary of forms far beyond the prevalent canonical understanding. This kind of 'fluidification of form' marked a new aesthetic notion common to these architects and engineers. Moreover, it mirrored the common fascination for the purity of forms found in nature. In this respect, their research referred not only to the same methods and aesthetic ideas. Their imagination was driven by an intellectual world that was based on the natural sciences and aimed at understanding problems in a highly-integrated manner.

20.5.1 Form-finding in a history of ideas

The common curiosity in the natural sciences also reveals the origin of the experimental methods. Since none of the structural designers really invented their instruments of form-finding, these were also adapted from the fields of physics and mathematics. The use of soap-film models belongs to a long history of ideas that trace the evolution of mathematical theory on minimal surfaces. The development started around 200 years before its application to architecture and engineering, as the German mathematician Stefan Hildebrandt (1936-2015) has shown [58].

During the second half of the eighteenth century, several mathematicians researched the field of differential geometry in order to describe surfaces of minimal area. An important step was the formulation of a minimal-surface equation, which was published around 1760 by the Italian mathematician Joseph-Louis Lagrange (1736-1813). During the early-nineteenth century, theories on capillarity led to the insight that these minimal surfaces can be displayed by soap films. The Belgian physicist Joseph Antoine Ferdinand Plateau (1801-1883, see Section 19.1) formulated what is called *Plateau's problem*, which asks whether a minimal surface can be inscribed in any closed spatial boundary curve. Studies to solve the problem and continuing various investigations into minimal surfaces were carried out by numerous renowned mathematicians in several academic institutions. The University of Göttingen in particular developed a long tradition of mathematical sciences and the best reputation in that field. Among other professors at Göttingen, Hermann Amandus Schwarz (1843-1921) was able to formulate important contributions. He found a way to create additions to several minimal surfaces without interrupting the continuity of those surfaces. Other important figures at Göttingen were David Hilbert (1862-1943) and the American, Jesse Douglas (1897-1965), who solved Plateau's problem around 1930 at almost the same time as the Hungarian Tibor Radó (1895-1965). Furthermore, the University of Göttingen amassed a large collection of physical models to exemplify the mathematical theories. Among the collection were metal frames, which allowed soap films to be created and which were discussed by Hilbert in his work *Anschauliche Geometrie* (1932) [59]. In fact, Musmeci referred to the Italian version of Hilbert's book [60], where the geometric description of minimal surfaces was introduced and the use of soap-film models was illustrated.

The long history of ideas of mathematical theories related to minimal surfaces has also been described by D'Arcy Wentworth Thompson (1860-1948) in his book *On growth and form* (1917). The phenomenon of surface tension for liquids was introduced in correlation with the mathematical theory in the chapter 'The forms of cells' [61]. Thompson also explained how to test the theory with physical models: 'we may bend a wire into a closed curve, plane or not plane, we may always, under appropriate precautions, fill the entire area with an unbroken film' [62]. Thompson's treatise served both as a broad corpus of scientific theories and as an approach to the question of beauty in nature. During the twentieth century, the publication became widely known and established as both a theoretical framework for natural scientists and an important source of inspiration for a whole cultural milieu of architects, engineers, designers and artists. Consequently, the book was a key reference in the field of form-finding. Musmeci for instance, studied Thompson's work with his students at the University of Rome [63].

20.5.2 Between theory and empiricism

In order to outline the origins of the methods of form-finding, it is important to trace how knowledge was transferred. In this regard, referencing plays a crucial role in the early methodical development of the different structural designers, particularly in the case of Sergio Musmeci and Frei Otto. Through his curiosity in mathematics, Musmeci found the theories of David Hilbert. Consequently, the physical model served rather as evidence for the mathematical laws than as an empirical instrument. In contrast, Otto developed his early studies in relation to the published lectures of the English physicist Charles Vernon Boys (1855-1944) [64], which comprised imaginative explorations of a series of empirical experiments with soap bubbles [65]. Hence, Otto's use of physical models reflected the empiricism promoted by Boys. Nevertheless, the mathematical base for his experiments was developed through collaborations with the engineer Rudolf Trostel and the mathematician Stefan Hildebrandt [66]. Generally, Otto's studies can be described as rather empirical, and the physical model was made for the 'observation' of a large variation of forms, while Musmeci's work was rather theoretically influenced, and the physical model was used as the 'visualization' of the mathematical model.

In this sense, Musmeci and Otto approached the same methodology from different sides. Even though they produced similar results, their work found application in very different fields: while Musmeci designed shell structures in reinforced concrete, Otto built lightweight tents. Also, their research framework was completely different: Otto conducted fifteen years of intense research on design with soap-film models. For him, the physical model was not a simple tool, rather it became an 'apparatus' [67]. In his laboratory, he developed a technically very refined device for the creation of soap-film models, called the *Minimalflächengerät*. This machine was equipped to regulate the temperature, and the humidity of the air and with all instruments necessary for precise geometric measurement (see Figure 19.3). A whole team worked on the experiments at the *Institut für leichte Flächentragwerke (IL)* at the University of Stuttgart. Similar to Otto's design approach, the extreme variability of the soap-film models and the evanescence of their frothy and fluid condition expressed his ideal of an adaptable architecture [68]. In Musmeci's case, the facilities for his physical-model studies were rather basic. He created the soap-film models at home with the help of his family. The hand visible on the historic photographs is that of his daughter Livia (Figure 20.3) [69]. But this simplicity mainly characterized the early design phase. During the project's development, the physical model tests became ever more sophisticated, in particular when stresses and strains were measured and the material behaviour was analyzed. The large-scale test in the laboratory in Bergamo was professionally documented, unlike the studies during the early design stage.

Even though the purpose and the application of the physical model studies show decisive differences, the forms derived by Musmeci and Otto have much in common. Interestingly, both started to experiment with soap-film models in around 1958. Otto was interested in soap films for his pneumatic studies and his tent designs, and Musmeci referred to minimal surfaces for the design of the Ponte di Tor di Quinto. Musmeci became aware of Otto's work only many years later [70], probably first through periodical articles and later, in-depth, through the English edition of Otto's book *Tensile Structures* (1967). In fact, Musmeci and Otto met in 1977 in Rome for a panel discussion, where Riccardo Morandi and Bruno Zevi were also involved [71]. An interesting comparison concerns the emergence of a very particular form, the chalice-shaped column. In Musmeci's project for the Mercati Generali in Rome (1960), the vaulted form connects the roof seamlessly to the columns and also has an opening inscribed. For a long time, Musmeci was not aware that Otto had developed a very similar geometry for his tensile structures only shortly afterwards [72], because exactly the same motif can be found in Otto's soap film studies from 1964 onwards [73]. Together with the architect Rolf Gutbrod, Otto implemented this geometrical principle in the tensile roof structure of the *German Pavilion* at the *Expo '67* in Montréal. The arc of the main cable formed a large drop-like opening, called a *Seilauge*. Aside from the vertical inversion, the motif appeared exactly the same in both works.

While Otto gained the theoretical background through collaborations with experts, Musmeci started to work on a comprehensive theory of statics on his own. By formulating a theorem of the structural minimum, called *Il minimo strutturale* [74], he researched solutions using a minimum volume of material. In this calculation process, he first set an equilibrated system of external forces and then aimed to determine the form of a structure with a minimal amount of material, depending on its resistance. Based on this definition, he derived a specific value, the product of the length of each structural member and their corresponding inner force. This procedure illustrates that Musmeci tried to approach the question of form-finding from different sides: through the use of both physical models and theoretical description.

20.6 The scientification of the design

Sergio Musmeci aimed to inform architecture and engineering through the theories and the methods of the natural sciences. His lifelong curiosity in the fields of mathematics, physics and chemistry enabled him to gain a broad corpus of scientific knowledge, which provided the foundation for his entire work. From his extensive explorations emerged a full spectrum of structures in which the form correlates directly with the inner forces. Such structural designs proved to use an impressively low quantity of material. In fact, this material efficiency could be seen both as a direct response to the scarcity of resources during the post-war period in Italy, and as a guiding idea for his basic research on structures, where the conception of the form became even more relevant than the built object itself. As a consequence of this radical approach, practical questions about the construction technology were increasingly disregarded, which was quite unusual in Italy at that time. But Musmeci consciously focused only on scientific questions: 'Technology is made through countless applications, which do not directly depend on each other. Instead, science comprises a few great ideas coordinated in a system. Technology means practice and science means theory. The first has an influence on the chronicle, the latter on the history' [75]. Musmeci's understanding marked a thorough scientification of the design. The idealism expressed in this approach became characteristic for a whole generation of structural designers, where the conciseness of the idea was considered more important than its implementation. Hence, the original constitution of the form, essentially emerging through experiments with physical models, marks the key moment of the design process. For example, the soap-film model clearly exemplifies in its fluidity the beauty of its appearance, and also reveals the inherent structural potential of minimal surfaces. Consequently, this physical model can be understood as the physical artefact closest to the original imagination. As a spatial visualization of the scientifically-derived form, the model becomes the driver of the design.

Acknowledgements

The author would like to thank Joseph Schwartz, Mario Rinke, Pierluigi D'Acunto and Simon Milligan for their constructive review of this text.

References

1 Musmeci, Sergio (1980) Struttura ed architettura. *L'industria italiana del cemento* L (10): 771-786. 784. Translated by the author.
2 Musmeci, Sergio (1977) Ponte sul Basento a Potenza. *L'industria italiana del cemento* XLVII (2): 77-98.
3 Kirk, Terry (2005) *The architecture of modern Italy. Volume II: Visions of utopia. 1900-present*. New York: Princeton Architectural Press. 143-206.
4 Poretti, Sergio (2013) *Italian modernisms: architecture and construction in the twentieth century*. Rome: Gangemi. 209-235.
5 Forty, Adrian (2012) *Concrete and culture: a material history*. London: Reaktion Books. 89-99, 208-212.
6 Giovannardi, Fausto (2012) Modernes Universalgenie. *Tec 21*. 138 (18): 18-19.
7 Musmeci, Sergio (1979) Architettura e pensiero scientifico. *Parametro* 80: 34-47.
8 Musmeci, Sergio (1972) Il calcolo elettronico e la creazione di nuove forme strutturali. In: *Architettura & computer. Il computer nella progettazione* (ed. Maria Zevi), 147-166. Rome: Bulzoni.
9 Musmeci, Sergio (1979) La genesi della forma nelle strutture spaziali. *Parametro* 80: 12-33.
10 Jones, Victor J. (2016) *Un pont à part. A distant bridge. Sergio Musmeci & Zenaide Zanini, Potenza 1966-1976*. Geneva: Mētis Presses. 38, 47.
11 Musmeci, Sergio (1977) Ponte sul Basento a Potenza. *L'industria italiana del cemento* XLVII (2): 77-98. 77-78. Translated by the author.
12 MAXXI Rome, Archive of Sergio Musmeci and Zenaide Zanini (F19088).
13 Capomolla, Rinaldo (2005-2006) Il ponte sul Basento, ovvero l'invenzione di una forma 'ancora senza nome' *Casabella* LXX (739-740): 12-19. 15.
14 Nicoletti, Manfredi (1999) *Sergio Musmeci. Organicità di forme e forze nello spazio*. Turin: Testo & Immagine. 25, 60.
15 MAXXI Rome, Archive of Sergio Musmeci and Zenaide Zanini (F18349).

16 Nicoletti, Manfredi (1999) *Sergio Musmeci. Organicità di forme e forze nello spazio.* Turin: Testo & Immagine. 5-6.
17 Ghirardo, Diane (2013) *Italy. Modern architectures in history.* London: Reaktion Books. 151-152.
18 Zevi, Bruno (1973) *Il linguaggio moderno dell'architettura: guida al codice anticlassico.* Turin: Einaudi. 43-47. Translated by the author.
19 Musmeci, Sergio (1977) Ponte sul Basento a Potenza. *L'industria italiana del cemento* XLVII (2): 77-98. 80. Translated by the author.
20 Iori, Tullia & Poretti, Sergio (2015) Il linguaggio delle strutture. In: *SIXXI 2. Storia dell'ingegneria strutturale in Italia* (eds Tullia Iori & Sergio Poretti), 7-21. Rome: Gangemi.
21 Iori, Tullia & Poretti, Sergio (2015) Il linguaggio delle strutture. In: *SIXXI 2. Storia dell'ingegneria strutturale in Italia* (eds Tullia Iori & Sergio Poretti), 7-21. Rome: Gangemi. 18.
22 Capomolla, Rinaldo (2005-2006) Il ponte sul Basento, ovvero l'invenzione di una forma 'ancora senza nome' *Casabella* LXX (739-740): 12-19. 14-15.
23 Giannetti, Ilaria (2017) *Il tubo Innocenti. Protagonista invisibile della scuola italiana di ingegneria.* Rome: Gangemi. 117-123.
24 Musmeci, Sergio (1980) Struttura ed architettura. *L'industria italiana del cemento* L (10): 771-786. 784. Translated by the author.
25 Abruzzese, Donato & Tursi, Alessandro (2004) L'ottimizzazione della progettazione strutturale nella ricerca di Musmeci. In: *Il ponte e la città. Sergio Musmeci a Potenza* (ed. Margherita Guccione), 82-93. Rome: Gangemi.
26 Adriaenssens, Sigrid, et al. (2015) Early form finding techniques of Sergio Musmeci revisited: the Basento Viaduct project. In: *Proc. International Association for Shell and Spatial Structures. Annual symposium.* 1-12. Amsterdam: IASS.
27 MAXXI Rome, Archive of Sergio Musmeci and Zenaide Zanini (F54465).
28 Musmeci. Sergio (1980) Cited in Nicoletti, Manfredi, Intervista a Sergio Musmeci 2/3. 15 pages, 6. Unpublished. MAXXI Rome. Archive of Sergio Musmeci and Zenaide Zanini. Translated by the author.
29 MAXXI Rome, Archive of Sergio Musmeci and Zenaide Zanini (27943).
30 Capomolla, Rinaldo (2011) Structure and form: The theory of 'minimal surfaces' and the bridge over the Basento river by Sergio Musmeci. *Docomomo Journal* 45: 46-51. 50.
31 Trostel, Rudolf (1962) Berechnung der Membranen. In: *Zugbeanspruchte Konstruktionen. Gestalt, Struktur und Berechnung von Bauten aus Seilen, Netzen und Membranen. Band 1* (ed. Frei Otto), 170-296. Frankfurt-Berlin: Ullstein Fachverlag.
32 Trostel, Rudolf (1967) Calculation of membranes. In: *Tensile Structures: design, structure and calculation of buildings of cables, nets and membranes. Volume 1* (ed. Frei Otto), 169-296. Cambridge, Mass.: MIT Press.
33 MAXXI Rome, Archive of Sergio Musmeci and Zenaide Zanini (27942).
34 Capomolla, Rinaldo (2005-2006) Il ponte sul Basento, ovvero l'invenzione di una forma 'ancora senza nome' *Casabella* LXX (739-740): 12-19. 14-15.
35 MAXXI Rome, Archive of Sergio Musmeci and Zenaide Zanini (F19107-1).
36 MAXXI Rome, Archive of Sergio Musmeci and Zenaide Zanini (28229).

37 Musmeci, Sergio (1977) Ponte sul Basento a Potenza. *L'industria italiana del cemento* XLVII (2): 77-98. 86. Translated by the author.

38 Capomolla, Rinaldo (2005-2006) Il ponte sul Basento, ovvero l'invenzione di una forma 'ancora senza nome' *Casabella* LXX (739-740): 12-19. 15.

39 La Torre, Carlo (2004) Il ponte sul Basento a Potenza: un percorso di ricerca. In: *Il ponte e la città. Sergio Musmeci a Potenza* (ed. Margherita Guccione), 94-107. Rome: Gangemi.

40 Capomolla, Rinaldo (2007) Le 'forme organiche strutturali'. Materia e spazio nelle opere di Sergio Musmeci. *Rassegna di architettura e urbanistica* XLI (121-122): 135-148.

41 Giovannardi, Fausto (2012) Modernes Universalgenie. *Tec 21.* 138 (18): 18-19.

42 Herzog, Max (1999) *150 Jahre Stahlbeton (1848-1998).* Berlin: Ernst & Sohn. 3-8, 13-15.

43 D'Acunto, Pierluigi & Ingold, Lukas (2016) The approach of Sergio Musmeci to structural folding. *Proc. International Association for Shell and Spatial Structures. Annual symposium* 1-10. Tokyo: IASS.

44 Ingold, Lukas & D'Acunto, Pierluigi (2017) Structural folding as a source of research for Sergio Musmeci. *Proc. International Association for Shell and Spatial Structures. Annual symposium.* 1-10. Hamburg: IASS.

45 MAXXI Rome. Archive of Sergio Musmeci and Zenaide Zanini. Plan documents 28348, 28349, 28351.

46 Musmeci, Sergio (1979) *Sergio Musmeci.* Pordenone: C.E.S.I.C.A. 4.

47 La Torre, Carlo (former student and collaborator), interviewed by the author on 12.07.2018. 58 pp., 52. Unpublished.

48 Musmeci, Sergio (1958) 110d Relazione, 2, 5. MAXXI Rome. Archive of Sergio Musmeci and Zenaide Zanini. Folder P12/18. Translated by the author.

49 MAXXI Rome, Archive of Sergio Musmeci and Zenaide Zanini (F54403).

50 Musmeci, Sergio (1959) Notes and sketches. MAXXI Rome. Archive of Sergio Musmeci and Zenaide Zanini. Folder 13/3.

51 MAXXI Rome, Archive of Sergio Musmeci and Zenaide Zanini (74054).

52 Musmeci, Sergio (1980) Cited in La Torre, Carlo, Intervista di Sergio Musmeci, 7.10.1980. 47 pages. 23-24. Unpublished. MAXXI Rome. Archive of Sergio Musmeci and Zenaide Zanini.

53 MAXXI Rome, Archive of Sergio Musmeci and Zenaide Zanini (F54596).

54 Pedio, Renato (1967) Viadotto in Calabria. *L'achitettura, cronache e storia* XII (138): 792-793.

55 MAXXI Rome, Archive of Sergio Musmeci and Zenaide Zanini (28475).

56 Tursi, Alessandro (2004) Il viadotto Musmeci, sintesi dell'architettura strutturale del '900. In: *Il ponte e la città. Sergio Musmeci a Potenza* (ed. Margherita Guccione), 70-81. Rome: Gangemi.

57 Escher, Cornelia (2017) Model – experiment – environment. In: *Frei Otto. Thinking by modeling* (eds Georg Vrachliotis, Joachim Kleinmanns, Martin Kunz & Philip Kurz), 53-60. Leipzig: Spector Books.

58 Hildebrandt, Stefan & Tromba, Anthony (1996) *Kugel, Kreis und Seifenblasen. Optimale Formen in Geometrie und Natur.* Basel: Birkhäuser. 128-181.

59 Hilbert, David & Cohn-Vossen, Stephan (1932) *Anschauliche Geometrie.* Berlin: Springer. 161-190.

60 Musmeci, Sergio (1977) Ponte sul Basento a Potenza. *L'industria italiana del cemento* XLVII (2): 77-98. 81, 98.
61 Thompson, D'Arcy Wentworth (1917) *On growth and form*. Cambridge: University Press. 201-231.
62 Thompson, D'Arcy Wentworth (1917) *On growth and form*. Cambridge: University Press. 230.
63 La Torre, Carlo (former student and collaborator), interviewed by the author on 12.07.2018. Unpublished.
64 Otto, Frei (1962) Pneumatische Konstruktionen. In: *Zugbeanspruchte Konstruktionen. Gestalt, Struktur und Berechnung von Bauten aus Seilen, Netzen und Membranen. Band 1* (ed. Frei Otto), 1-169. Frankfurt-Berlin: Ullstein Fachverlag. 11.
65 Boys, Charles Vernon (1896) *Soap bubbles and the forces which mould them*. London: Society for Promoting Christian Knowledge.
66 Otto, Frei (1987) Introduction. In: *Seifenblasen: Forming bubbles*. Mitteilungen des Instituts für leichte Flächentragwerke. IL 18. (Eds Bach, Klaus, Burkhardt, Berthold & Otto, Frei), 310-319. Universität Stuttgart.
67 Vrachliotis, Georg (2017) Thinking in models. Architecture and operative aesthetics in the work of Frei Otto. In: *Frei Otto. Thinking by modeling* (eds Georg Vrachliotis, Joachim Kleinmanns, Martin Kunz & Philip Kurz), 20-33. Leipzig: Spector Books. 29-30.
68 Escher, Cornelia (2017) Model – experiment – environment. In: *Frei Otto. Thinking by modeling* (eds Georg Vrachliotis, Joachim Kleinmanns, Martin Kunz & Philip Kurz), 53-60. Leipzig: Spector Books. 57.
69 Musmeci, Alberto (brother) and Musmeci, Livia (daughter), interviewed by the author on 26.10.2017. Unpublished.
70 Musmeci, Sergio (1979) Le tensioni non sono incognite. *Parametro* 80: 34-47. 44.
71 Zevi, Bruno (1977) Una casa che va a vela. *L'espresso* 27 March: 73-75.
72 Musmeci, Sergio (1980) Cited in La Torre, Carlo, Intervista di Sergio Musmeci, 7.10.1980. 47 pages. 23-24. Unpublished. MAXXI Rome. Archive of Sergio Musmeci and Zenaide Zanini.
73 Otto, Frei (1987) Introduction. In: *Seifenblasen: Forming bubbles*. Mitteilungen des Instituts für leichte Flächentragwerke. IL 18. (eds Bach, Klaus, Burkhardt, Berthold & Otto, Frei), 10-21. Universität Stuttgart.
74 Musmeci, Sergio (1971) *La statica e le strutture*. Rome: Cremonese. 79-132.
75 Musmeci, Sergio (1979) Architettura e pensiero scientifico. *Parametro* 80: 34-47. 35-36. Translated by the author.

21

Heinz Isler and his use of physical models

John Chilton

21.1 Introduction

The Swiss engineer Heinz Isler (1926-2009) is recognised as one of the great designers of thin reinforced-concrete shells of the twentieth century. His first freely-shaped shell was built at the Hotel Kreuz, Langenthal, in 1954-55 [1, 2], and the last four were completed at Pfäffikon, near Zürich, in 2008-9 [3]. The form of the surfaces of Isler's shells were predominantly found using physical models. This was a departure from the more common, mathematically-determined, surfaces that were used by other shell builders in the mid-twentieth century – for example, Eduardo Torroja (1899-1961), Félix Candela (1910-1997) and Pier Luigi Nervi (1891-1979).

Throughout his career Heinz Isler was a fervent advocate of the use of physical models in engineering design, being of the opinion that engineers build physical objects, in a physical world, subject to physical laws. He believed that a precisely-made physical model reveals both its three-dimensional statical behaviour and the impact of the physical properties of the materials of which it is made. When touched, a physical model gives some indication of the stiffness of the structure. It also allows the designer to use all their physical senses, for instance, to compare proportions or hear and see evidence of structural distress (creaking and cracking) when a model is loaded [4].

Isler's interest in models was profoundly influenced by his professor and mentor at Eidgenössische Technische Hochschule (ETH) Zürich, Professor Pierre Lardy (1903-1958) who, in turn, had been motivated by Professor Eduardo Torroja's design development through scale models at his laboratory in Madrid. Lardy established his own laboratory in Zürich and, on graduation, Isler worked as one of his assistants until May 1953. This gave him a rigorous apprenticeship in building accurate physical, statical models with appropriate boundary conditions, in the application of strain gauges and loading systems, and in their use to predict stresses, deformations and stability of the full-scale structure. He also learnt how to compare these predictions with tests made on full-scale buildings [5].

Physical Models: Their historical and current use in civil and building engineering design,
First Edition. Edited by Bill Addis.

21.2 Modelling techniques

In the late 1950s, thin reinforced-concrete shells were almost exclusively mathematically-described surfaces e.g. spherical and parabolic domes, barrel vaults, conoids, hyperboloids and hyperbolic paraboloids (commonly named 'hypars') [6, 7]. At the First Congress of the International Association for Shell Structures (IASS, now the International Association for Shell and Spatial Structures) in 1959, which was organized in Madrid by Eduardo Torroja, Heinz Isler proposed alternative modelling methods for form-finding of thin concrete shells in his short but highly-influential paper *New Shapes for Shells* [8]. In his paper Isler introduced three physical form-finding modelling techniques, which he referred to as:

- the freely-shaped hill
- the membrane under pressure
- the hanging cloth reversed (inverted) [9, 10].

The last figure in this paper consisted of a 4 x 10 matrix that was populated with thirty-nine sketches of potential shell forms [11], all drawn by Isler, with the remaining space provocatively filled with the word 'etc.' to suggest the infinite variety of shell forms that could be modelled using his proposed methods [12, 13, 14, 15].

21.2.1 The freely-shaped hill

Isler did not apply the freely-shaped hill method widely. However, he did use this technique to form a small nuclear bomb shelter where soil dug from a roughly circular trench of around 4 m diameter was heaped in the middle to create a small 'hill' on which a 200-mm thick reinforced-concrete dome shelter was cast [16]. The soil inside was then removed and heaped on the shell to give further protection from nuclear blast. A similar method was used to produce formwork, moulded in sand in the grounds of Isler's office, for 6-m span, glass-reinforced polyester (GRP) shell vaults [17]. In the inverted form, he made a swimming-pool liner only a few millimetres thick, also of GRP [18].

21.2.2 Membrane under pressure

Isler related that his membrane-under-pressure technique had been inspired in 1954 by his observation of the double-curved form of his pillow [19]. He emulated the effect of the stuffing in the pillow by inflating a rubber membrane between a base board and a fixed frame, having a 'cut-out' with scaled dimensions of the proposed roof plan (Figure 21.1a). Introduction of air under pressure between the base board and elastic membrane produces a rounded surface in tension which, on reversal of the load, would be an appropriate shape for a thin shell in pure compression (Figure 21.1b).

Photo 21 in Isler's paper *Generating Shell Shapes by Physical Experiments* [4] which describes in some detail his form-finding methods, shows him measuring

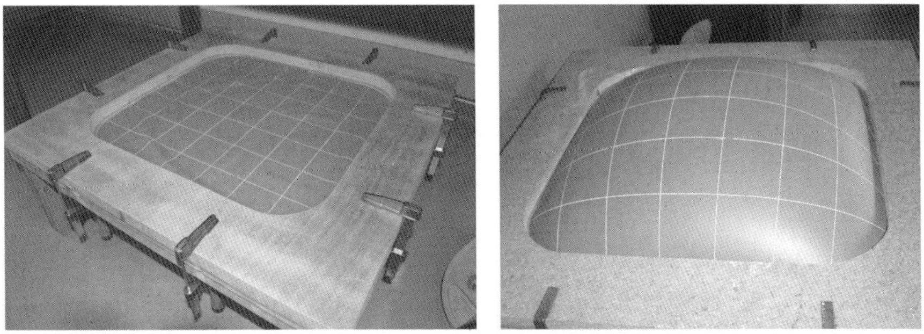

(a) (b)

Figure 21.1 (a) Apparatus made by Isler for form-finding by membrane under pressure. (Image: © John Chilton). (b) Apparatus with membrane inflated. (Image: © Ekkehard Ramm)

the three-dimensional form of the inflated membrane directly – which he did to an accuracy of 0.02 mm. However, he comments that sometimes he noticed that the membrane rose during the measuring process, when one might have expected it to sink if pressure was being lost through leakage. He concluded that this was due to expansion of the air under the membrane caused by his own body heat! His remedy was to wear thick clothing and fingerless gloves whilst undertaking this critical procedure [20]. In later demonstrations he showed how, by covering the membrane with a thin layer of gypsum plaster before inflation, the form after inflation could be captured on hardening. On removal, this cast could be used directly as a preliminary model for presentation, or processed further to obtain a more detailed geometry of the surface.

Nonetheless, Isler was well aware of the statical limitations of this technique – the air pressure on the membrane acts perpendicular to the surface while the major loads on the real concrete shell, self-weight and snow load, act vertically under gravity. Therefore, he generally used this technique for shells with relatively low rise/span ratio, where the difference between the two load-cases was less significant and could be accommodated in the detailed design of the surface.

21.2.3 Hanging cloth reversed

Isler's 'hanging cloth reversed' technique (the more familiar term 'inverted' is used in this chapter) was also inspired by his observation of physical form, in this case wet hessian, as used to protect freshly-cast concrete during the curing process, which he saw draped over a reinforcement mesh. A suspended rope or chain has negligible bending stiffness and will take on an appropriate profile to resist any applied loads in pure tension. If the load distribution is changed, the profile of the rope will also change. The hanging cloth, net or membrane behaves similarly in three dimensions. Hence, under load it will change shape until the surface is in pure tension, (Figure 21.2a). Under the same loading the inverted structure will be stressed in pure compression [21] (Figure 21.2b).

The method consists of three stages. A thin elastic sheet resting on a flat surface is coated with a layer of wet plastic material of consistent thickness. When

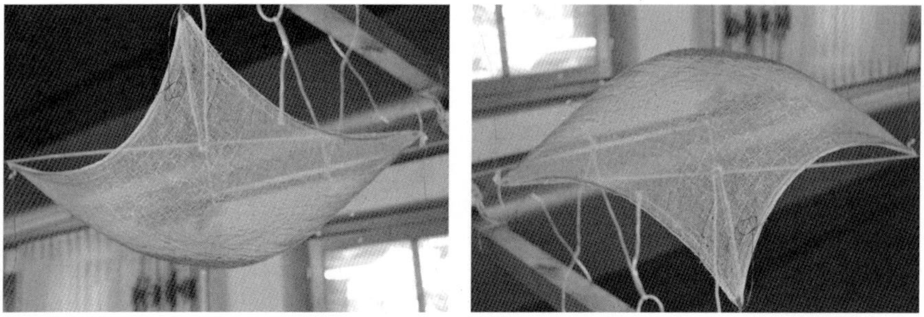

(a) (b)

Figure 21.2 (a) One of the many hanging cloth models made by Isler and (b) the same model shown inverted for use as a compression structure. (Image: © John Chilton)

this mix approaches the appropriate consistency – not too liquid to slide off a curved surface and not so set that it will crack when curved – the membrane is carefully suspended between supports in the same relative position, to scale, as the supports of the full-scale shell. (On cold winter nights Isler also sprayed water onto objects, including fabric and nets, to freeze and create solid sculptural ice shell forms). Once set, the cast is inverted, the membrane is removed and the surface is revealed [22]. Finally, its three-dimensional form is measured precisely. Isler accomplished this final stage using a purpose made jig (Figure 21.3a) to find the x, y and z co-ordinates of a grid of points on the cast surface, to an accuracy of 0.01 mm. To ensure that the form was accurately captured, the grid was less dense where curvature was less pronounced (in the centre of the shell) (Figure 21.3b) and denser in areas of higher curvature (such as near supports) (Figure 21.3c).

Although this method would appear simple to apply, several factors influence the resulting form. If the membrane, when flat, exactly fits the area encompassed

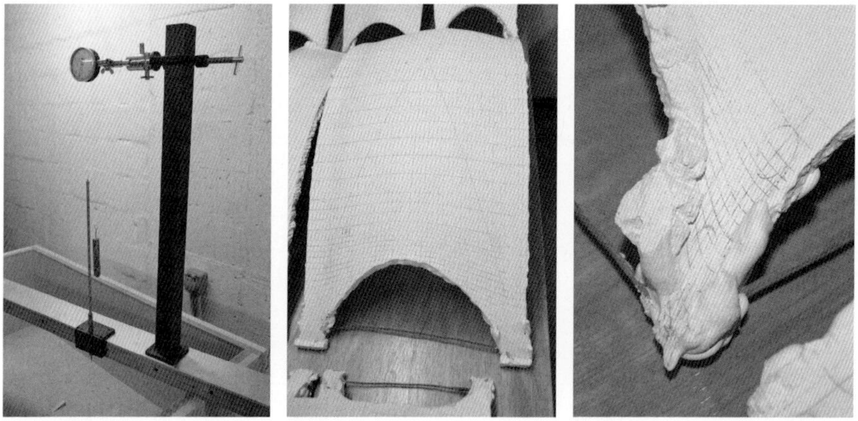

(a) (b) (c)

Figure 21.3 (a) The jig used by Isler to measure the three-dimensional form of his inverted hanging membrane models. (b) Typical overall distribution of measuring points on cast shell. (c) Example showing the dense measurement grid near a support. (Images: © John Chilton) by the supports, unless it is very thin and flexible, the form generated is likely to be relatively flat. Such forms are generally of little practical use due to their low rise/span ratio and the relatively high stresses in the shell. If, on the other hand, the initial area of membrane is much larger than the area encompassed by the supports, the resulting form is likely to have wrinkles (undesirable in a smooth shell surface) and a high rise/span ratio, which would require more material. The skill of Isler as the designer was to find an architecturally, structurally and economically acceptable compromise among these competing factors.

This method of form-finding is eminently suitable for determining efficient contours for thin reinforced-concrete shells. Although reinforced concrete is not normally thought of as a flexible material, when its thickness is small in relation to the span of the shell, its bending stiffness is low relative to the span and it can be considered to act as a thin membrane. On the other hand, the self-weight of the shell is relatively high – around 240 kg/m^2 for a 100 mm thick shell – and this is usually the dominant load case. Hence, if the shell form is determined for the self-weight condition, theoretically it will be in compression most of the time. Environmental loads such as wind and snow may result in asymmetrical loading which would, in theory, require modification of the surface profile to maintain the shell in pure compression. However, their intensity is usually relatively low compared to the self-weight, and the shell is designed to have sufficient stiffness to resist this small, induced bending.

A key difference between the hanging cloth or membrane, in tension, and the inverted form when used as a thin reinforced-concrete shell in compression, is that the latter is constrained by instability due to buckling, especially at the edges. Isler's form-finding method results in a double-curved edge profile that averts the need for edge thickening. Nevertheless, Isler was well aware of the potential for buckling and determination of the critical buckling load was the objective of many of the tests he carried out on the structural/verification models described below in Section 21.3.

The engineer Ekkehard Ramm at the University of Stuttgart has listed the advantages of physical modelling of shells, namely: the ease of assessing their aesthetic qualities; easy adaptability to design constraints; physical detection of weak points in the surface; and easy measurement for subsequent analysis. However, he also alludes to potential problems: a lack of compatibility between the model and real building material; the possibility of unwanted wrinkling; validity for just one load case; buckling and non-linear response [23]. He also points out that, for a woven fabric or net, the resulting hanging form is dependent on the orientation of the warp and weft relative to the plan [24]. For this reason, when determining the final shell form Isler usually used a high quality latex rubber membrane.

Ramm believed that '…[this] method has been brought to perfection for RC roof shells by Heinz Isler…' [25] and it produced shells that has earned Isler the accolade 'structural artist' [26]. Examples include shells for a motorway service station Deitingen Süd (1968) and the Sicli SA factory in Geneva (1969), as well as many swimming pool and sports hall roofs.

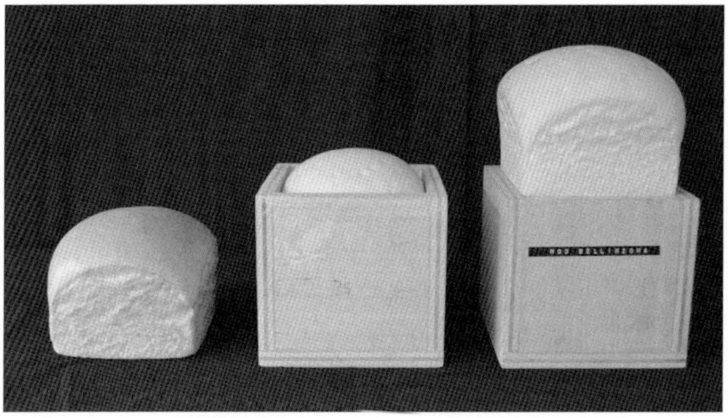

Figure 21.4 Expansion form by Isler. (Image: © John Chilton)

21.2.4 Expansion forms

Although not included in his paper *New Shapes for Shells*, Isler also explored a further modelling technique, the expansion of foam restricted by a solid boundary – much like bread or cake rising in a baking tin (Figure 21.4). This provided the inspiration for shell roofs for a supermarket in Bellinzona (1964), the Kilcher Factory, Recherswil, near Solothurn (1965) and the Bürgi Garden Centre, Camorino (1973).

21.2.5 Modelling rules and need for precision

Isler always stressed the importance of three main rules in his physical modelling. Firstly, the full-scale structure and model must be in direct geometric proportion. The second is that, whatever material is used for the model, the stress-strain diagram should be of similar form to that of the material used in the full-size shell; he notes, for example, that Plexiglas and concrete are reasonably compatible. Finally, the application of the same load per square metre on both model and real structure will generate the same stresses in both [27]. In terms of modelling technique, he emphasized the need for absolute precision, pointing out that any errors in the modelling by hand will be scaled up. In the case of Isler's shells, this scaling factor was between twenty and fifty.

21.3 The application of physical models in Isler's design process

Heinz Isler had originally intended to follow a career as an artist having exhibited his sketches and watercolours in 1944 whilst still at school [28]. He was, therefore, a meticulous observer of the natural world, a skill that he carried over into his ultimate career as an engineer. He used physical models at all stages of his design process. Visitors to Heinz Isler's offices at Lyssachschachen, near Burgdorf, Switzerland, were immediately struck by the number and variety of physical models on display in the offices, basement and even in the surrounding

Figure 21.5 Heinz Isler enjoyed observing nature, as evidenced by the shell-covered bird table outside the office meeting room window. (Image: © John Chilton)

grounds – including a plastic shell as cover for the bird table outside the office meeting room window (Figure 21.5). These ranged from simple exploratory models, designed to aid a qualitative understanding of structural behaviour and/or the architectural qualities of his shells, to precise scale models used for their structural verification.

21.3.1 Exploratory models

Isler used exploratory models in diverse ways, for instance, to propose the provisional architectural and structural form of the Heilig Geist Kirche (Holy Spirit Church) Lommiswil, near Solothurn, Switzerland (1967) (Figure 21.6), which was then adopted by the project architect Roland Hanselmann. Here, with a few sticks and some thick knitting wool, he was able to convey the concept of wall sections cantilevering from the foundations and supporting a free-form plan hyperbolic-paraboloid roof.

More specifically structural exploratory models were used when Isler designed a 961 m^2 shell roof for a supermarket in Bellinzona in 1964; it was based on an expansion form and rested on four point supports. Here, because of the slenderness, just 80 mm, and geometrical constraints of the proposed shell, he was concerned about it buckling outwards near the supports, due to high compressive forces and because the line of thrust might deviate outside the shell under certain load conditions. Rather than adopt a more conventional response – to further increase the thickness of the shell in the area of concern – Isler explored the possibility of introducing a horizontal ring beam before finally selecting his innovative alternative solution in which vertical restraints are connected to the mass of the basement and concealed in the mullions of the façade glazing [29]. He demonstrated the feasibility of this system with a simple physical model in which two alternatives can be seen. Application of downward pressure on the model

Figure 21.6 Exploratory model for the Heilig Geist Kirche (Holy Spirit Church) Lommiswil, near Solothurn, Switzerland (1967). (Image: © John Chilton)

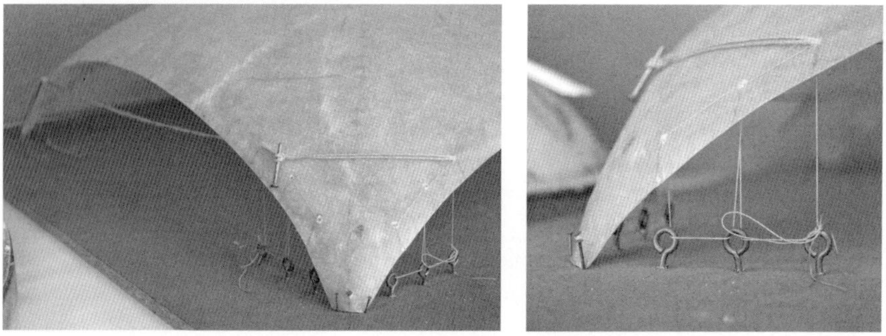

(a) (b)

Figure 21.7 (a) 1:50 scale model exploring the use of restraints concealed in façade glazing to avoid potential buckling of shell in Bellinzona (1964). (b) Detail of buckling restraint. (Images: © John Chilton)

shell induces an outward movement of its surface near the supports, which is resisted by tension in the 'cables' connected to the base (Figure 21.7).

In addition, to reassure himself about its stability, Isler built a 1:10 scale model of the Bellinzona shell in the grounds surrounding his office (Figure 21.8). Made of micro-concrete just 8 mm thick with two layers of steel wire reinforcement [30], the scale model does not have the buckling restraints that were installed

(a) (b)

Figure 21.8 (a) and (b) 1:10 scale model, 3.2 x 3.2 m, of the Bellinzona shell in the grounds of Isler's office in 2009. (Image: © John Chilton)

in the full-size shell. At the time the photographs were taken, the scale model had been exposed to snow and wind loads comparable to those experienced by the real shell for 45 years, and had survived with no visible sign of buckling or distress.

Another example of Isler's use of exploratory models is the latex casts, shown slightly squashed in Figure 21.9a, that he made to assist his appreciation of the buckling behaviour of free-form shells for the Steinkirche, at Cazis in Graubunden (1996) [31]. Here the shell forms, which were intended to symbolize large rounded pebbles extracted from a mountain stream, were produced by the architect Werner Schmidt [32] for his competition-winning design. By casting the forms in latex rubber and removing the proposed lenticular window openings, Isler was able to prod and pull the surface to see directly how the incised shell deformed under load. His tactile exploration of the problem of force transfer across the openings led Isler to introduce lacing bars that act in

(a) (b)

Figure 21.9 Steinkirche, Cazis by architect Werner Schmidt (1996). (a) Latex casts of ovoid shells used to explore buckling behaviour qualitatively. (b) Styrofoam model to determine construction geometry, with shell contractor Heinz Bösiger. (Images: © John Chilton) tension or compression, as required. Further physical exploratory models were also made from Styrofoam (Figure 21.9b) to assist the contracting firm Willi Bösiger AG in determining the detailed geometry for construction of the shells.

21.3.2 Form-finding models

To engineers working with conventional framed structures, composed of beams and columns, the notion of having to ascertain the three-dimensional form of one's structure, before being able to analyse its performance, may be somewhat surprising. However, for thin concrete shells, the imposition of a specific geometry, with little or no regard to determining an efficient form, can lead to heavier structures requiring massive stiffening edge beams or alternative means of edge support, as can be seen in the shell designed by Eero Saarinen for the Kresge Auditorium, Boston, USA (1955) [33]. To avoid such difficulties, Isler employed the form-finding modelling techniques described in Section 21.2 to determine precisely the three-dimensional geometry of efficient doubly-curved shell surfaces, primarily stressed in pure compression under their own weight. As implied in his paper *New Shapes for Shells*, there exists an infinite number of valid solutions, and these need to be assessed against appropriate design criteria. In the discussion of his paper, where his ideas were challenged by some of the most eminent engineers of the day [34], it is reported that he listed five design criteria (functional, shaping, architectural or artistic expression, statics, and others – acoustics, light); however, photographic evidence shows that he actually listed slightly different criteria on a blackboard during his presentation – functional, shaping, artistic expression, statics, construction and cost [35].

Isler first used the inflated membrane technique to establish the geometry of three shell roofs in a factory for the firm Glas Trösch AG, Bützberg (1955), designed by architect W. Grünig. A 1:40 scale frame was made for the 21.7 x 14 m shells (Figure 21.10a) in which the edge profiles of the frame were curved in elevation (Figure 21.10b).

Subsequently, similar frames were used to form-find his industrial 'bubble' shells. These shells were generally square or rectangular in plan, bounded by post-tensioned edge beams and supported only at the four corners. Hundreds of these shells were constructed, typically just 80 to 100 mm thick, with spans up to 40 m. The largest, for the COOP warehouse at Wangen bei Olten, was 54.6 x 58.8 m and up to 150 mm thick (1960). On the same site, Heinz Hossdorf built a multibay, north-light, post-tensioned shell roof over the 100 x 193 m warehouse (4 x 23 bays) (see Section 18.3.2). Through his physical modelling of the square/rectangular plan, Isler observed that reverse curvature develops in the surface close to the corners. Consequently, he modified the boundary of his bubble shells to avoid this. By providing curved fillets between adjacent edge beams, he maintained a synclastic curvature over the whole shell surface, whilst simultaneously providing additional space for pre-stressing anchorages at the corners.

Although the greatest number of Isler shells were form-found using the inflated membrane method, the contours of Isler's most innovative and elegant shells were obtained using his hanging cloth (or membrane) technique. The first built

(a) (b)

Figure 21.10 (a) Original 1:40 scale frame and rubber membrane used to determine the form of 21.7 m x 14 m shell roofs for Glas Trösch AG, Bützberg. (b) Edge profile detail. (Images: © John Chilton)

example using this technique was a shell roof for Gips Union, Bex, (1968), which has a trapezoidal plan, 31.5 m in length and tapering from 25.0 m to 17.4 m in width, on four corner supports [36, 37] (Figure 21.11). Isler described how, through iterative modifications of the hanging model, its performance was gradually enhanced [38].

In that same year, Isler employed the same modelling technique to determine the shape of a matching pair of triangular shells for a motorway service station at Deitingen Süd, Switzerland. These demonstrated how a shape determined through the force of gravity acting on a flexible physical model, can produce a highly efficient and durable thin shell surface of 31.6 x 26.0 m without edge-stiffening beams. Yet, perhaps the most complex of Isler's shells form-found by this method is that for the former Sicli SA factory (now the Pavilion Sicli cultural centre) in Geneva. Constructed in 1968-69, this consists of two interlinked free-form surfaces on seven supports. Working with local architect Constantin Hilberer, Isler proposed several alternative solutions for the roof, including different combinations of shell and membrane structures, before finally selecting the single surface shell [39] (see Figure 21.17).

The most-numerous examples of Isler shells form-found by hanging membrane models are for swimming pools (generally of square plan of around 30 x 30 m) and sports halls (typically of 48 m span and 17.8 m or 18.6 m in width) for example the Norwich Sports Village in Britain [40, 41]. These models were also used to assess the aesthetic properties of the shells and relative efficiency of the surfaces. According to Isler, the most efficient was a modified version of the sports hall (51.7 x 18.6 m), constructed for an air museum at Dübendorf near Zürich, which exhibited almost no deflection at mid-span under uniform loading. The structural verification model for this project is shown in Figure 21.12.

Figure 21.11 Development models and plan template (top right) for Gips Union, Bex, (1968). (Image: © John Chilton)

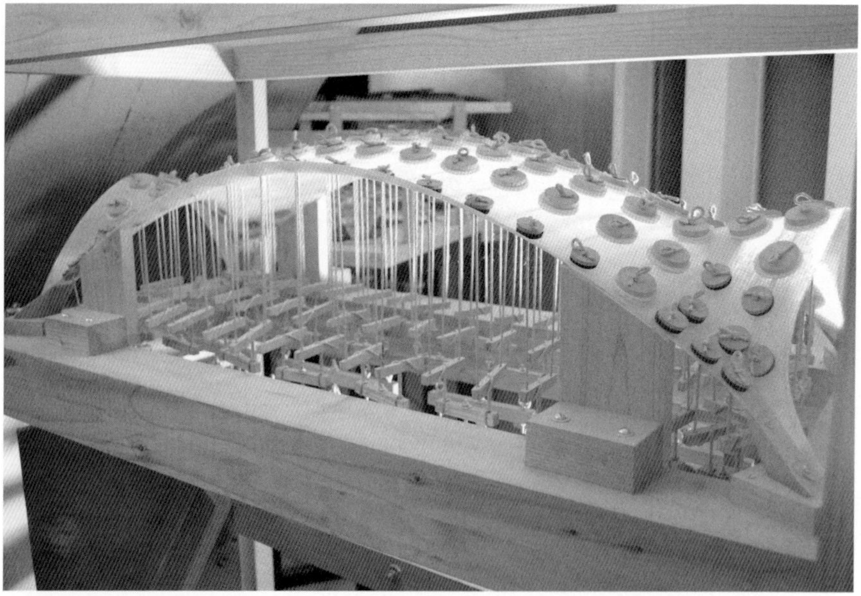

Figure 21.12 1:50 scale model of the modified version of the sports hall (51.7m x 18.6m), constructed for an air museum at Dübendorf near Zürich. (Image: © John Chilton)

21.3.3 Structural verification/validation models

Following selection of the most appropriate form, and accurate measurement of its surface, the next and most critical step in Isler's design process was structural verification/validation, using Plexiglas or reinforced polyester models at a larger scale. These he used to determine stresses in the shell surface under different load and support conditions, including application of pre-stress and the effect of movement of supports. It should be noted here that a key stage of Isler's shell construction method was the post-tensioning of edge beams in his 'bubble' shells, and of tie beams between supports for his point-supported shells. This post-tensioning ensured that the shell surface was maintained in compression under all load conditions. A convenient side effect of the post-tensioning was to lift the shell after casting and thus to separate it from the formwork. The same models were also used to assess buckling behaviour.

The earliest example of Isler using a physical model to verify the structural behaviour of a freely-shaped, doubly-curved shell on a rectangular plan was for the 24.5 by 14.5 m roof of a concert hall at the Hotel Kreuz, Langenthal, (Architect H. Sohm, Utzenstorf), in 1954-55 whilst he was working at Weder & Prim in Burgdorf. Realising that a fully double-curved shell would have improved buckling resistance compared to a cylindrical shell, a 1:30 scale hardwood mould of the proposed shell geometry was produced. This was then used as a template to form a 3.4 mm thick acrylic (Plexiglas) sheet, which replicated the original mould geometry to within 0.1 mm [42, 43]. After adding straight acrylic edge beams, the model was tested in a purpose-made rig, (Figure 21.13). A balanced beam system was used to apply load through rubber pads at sixty-four points on the surface to simulate a uniformly-distributed load, strains were measured at seventy-four points by electrical-resistance strain gauges connected to a data logger and pre-stress was applied by pistons using oil pressure to tension scaled

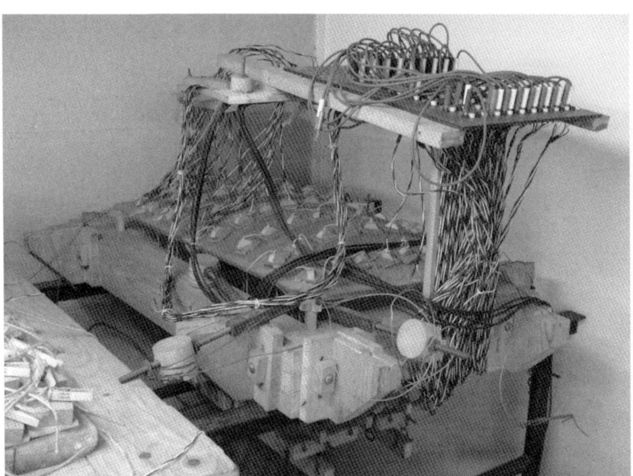

Figure 21.13 Remains of the original test rig for the Hotel Kreuz shell in the basement of Isler's offices in 2011. (Image: © Ekkehard Ramm)

cables passing through channels in the edge beams. In the following weeks of experimental work twelve load cases were investigated to determine normal forces, moments and buckling behaviour, to inform the appropriate dimensioning of the shell [44]. In particular, the effect of pre-stress and support conditions was established. It was found that these factors significantly influenced the stresses in the shell, which outperformed the equivalent cylindrical shell.

Initially Isler had assumed that load would be distributed reasonably evenly to the perimeter beams. However, his experiments showed that they carried only about 10% of the total applied load with the remaining 90% being concentrated towards the corners. This led him to develop his corner supported shells [45]. A more refined version of this test rig was used to verify the performance of the standard, normally square, 'bubble' shells which were built in many sizes from 15 x 15 m up to 54.6 x 58.8 m (Figure 21.14a). Isler soon established his own design office where he employed similar models to assess the structural characteristics of progressively more-sophisticated shells, for example, a shell constructed in 1965 for F. Kilcher at Recherswil, near Solothurn, which is based on an expansion form (Figure 21.14b).

The Heilig Geist Kirche (Holy Spirit Church) Lommiswil, near Solothurn, Switzerland (1967) (Figure 21.15), is one of the few examples of an Isler shell based on hyperbolic-paraboloid geometry [46]. This complex structure has a free-form plan. Its bounding walls of individual precast units vary in height and are contiguous with the shell surface, which is post-tensioned to eliminate tensile stresses. Isler used a physical model to assess the interaction between the shell and the supporting structures, and the stresses induced in the surface by the pre-stress (Figures 21.16a and 21.16b).

Of the several structural verification models on display in the Isler office, before they were moved to the Isler archive at the ETH, Zürich in 2012, was the first built example of a shell designed using a hanging membrane – a synclastically-curved roof for Gips Union in Bex, Switzerland, 31.5 m long x 25.0 m wide, tapering down to 17.4 m. Isler was particularly concerned by the relatively-limited double curvature of the surface along the free edges, and used a physical model to investigate the shell's resistance to bucking instability (Figure 21.17). The model

(a) (b)

Figure 21.14 More-sophisticated test rigs for (a) the standard square-plan 'bubble' shells and (b) the 1:50 scale verification model for a shell based on an expansion form. (Images: © John Chilton)

Figure 21.15 Heilig Geist Kirche (Holy Spirit Church) Lommiswil, near Solothurn, Switzerland (1967). (Image: © John Chilton)

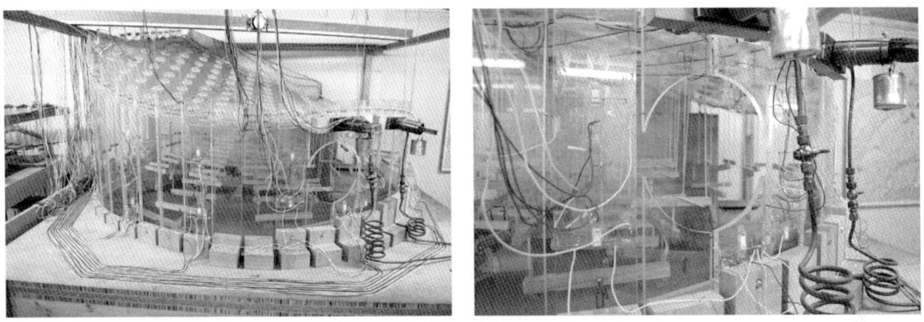

(a) (b)

Figure 21.16 1:20 scale model of the Heilig Geist Kirche (Holy Spirit Church) Lommiswil. (a) The test rig. (b) Detail showing small hydraulic jacks used to apply pre-stress to the shell surface, and the disposition of strain gauges around the circular window opening in the precast wall panels. (Images: © John Chilton)

shell is pin-supported on wooden blocks, hinged at their base and connected diagonally across the plan by steel bars. This arrangement allowed Isler to pull the bases inwards to simulate the pre-stress that would be applied to the full-scale shell. Through successive testing and slight modifications to the original proposed form, in particular the profile of the free edges, the critical buckling load of the shell was increased by 30%, whilst maintaining the same shell thickness [47]. A smaller shell of similar trapezoidal plan form (24.4 m long and tapering from 19.6 m to 11.76 m) but with more-pronounced curvature, was constructed for an architect's office in Sargans, Switzerland, also in 1968 [48].

Just as the most impressive of Isler's form-finding models were for the Sicli SA factory shell roof in Geneva (1968), this project also required the most-impressive structural-verification model, due to both its physical scale and the complexity of the surface (Figure 21.18). This roof comprised two interlinked shells – one of roughly rectangular plan (35 x 30.0-31.5 m) and the other, a more

Figure 21.17 1:50 scale structural verification model for Gips Union, Bex (1968). (Image: © John Chilton)

(a) (b)

Figure 21.18 1:50 scale structural verification model for the Sicli SA factory shell roof (1968). (a) The test rig. (b) Detail of modelling of the foundations. (Images: © John Chilton)

freely-shaped plan form (35 x 22.5 m) – that share one support where the two shell surfaces merge. This complicated the modelling of the pre-stressing which connected diagonally opposed foundations across each surface (Figure 21.18b).

21.3.4 Is enlargement of models allowed?

An important question that Isler himself raised was 'Is it allowed to enlarge results of models to objects of real size?' [49]. He answered by saying that, provided scaling and similitude rules are followed, and models are made precisely with quality materials, the technique is, indeed, applicable. However, he emphasised that the shape-finding material must be completely bending-free at the instant the shape is created, so that it is subject to only in-plane membrane stresses. If, on the other hand, the material has some degree of bending stiffness, the shape will not be a pure tension surface. When inverted, the areas subject to bending would require greatly increased shell thickness to provide the scaled-up bending stiffness. He concluded that '...absolute fluidity of the shape-making model is of utmost importance' [50].

21.4 Isler's use of models in teaching student engineers and architects

Heinz Isler was well-aware of the power of models in communicating his ideas to students of all ages. In his office basement he frequently lectured to students, school children and fellow professionals, both architects and engineers. In the later years of his career, as new commissions diminished, he displayed his research test rigs and physical models like museum exhibits (Figures 21.19a and 21.19b).

As an introduction, visitors were confronted by a simple model demonstrating the relative structural efficiency of double-curved shells compared to more conventional linear beam structures. The model beam was a wide, thin sheet of plastic spanning between two supports. When loaded with a small weight at the centre of the span, it deflected by approximately 25% of the span! At its side there was a double-curved, tied-arch shell of the same span and made from a plastic

(a) (b)

Figure 21.19 Physical models displayed in the Isler offices (a) A shell in a test rig in 2009. (Image: © John Chilton). (b) Exploratory physical models in 2003. (Image: © Ekkehard Ramm) sheet of similar width and thickness. This was loaded at the mid-span with 30 times the load applied to the flat beam, yet showed negligible deflection [51].

Isler also passed on his enthusiasm for double-curved forms in extended student seminars in universities, for example at Darmstadt, Karlsruhe, Princeton (for two-weeks in April 1983 [52]) and Berlin. Using materials such as clay, moulding sand, plaster and plastics, students were encouraged to play with curved forms and rediscover the inherent natural creativity of their childhood, which Isler believed was often stifled by more conventional higher education [53].

21.5 Dissemination of Isler's modelling techniques to engineers and architects

Isler also lectured widely. He was very happy to talk about his work and to show the results of his modelling techniques, but was rather reluctant to reveal fine details of the design process. However, he did on occasion give some insight. As an example, he gave a spectacular demonstration – observed by the author – of his inflated membrane and hanging membrane form-finding techniques on stage during the 2nd meeting of IASS Working Group 15: Structural Morphology, held in Stuttgart, in 1994. This inspiring exposition took place under an Isler shell, form-found using the inverted hanging technique – the Musicalsaal, Naturtheater, Stetten auf den Fildern (architect Michael Balz).

Starting with the inflated membrane, Isler highlighted the factors that need to be satisfactorily controlled to ensure a successful result [54]. These include: the recipe and consistency of the gypsum plaster (Isler said he had his own special mix); the evenness of the thickness of the plaster layer – in his demonstration this was achieved by Isler drumming on the frame with his fists to spread the liquid plaster; the setting time of the plaster – inflate too early and the plaster will run off the membrane, inflate too late and it will crack! Using a paper cup to form the opening for a central skylight, the plaster shell was formed by inflation in a few seconds (Figure 21.20).

Turning to the hanging models, Isler demonstrated this using a piece of open-weave fabric, which he dipped into a bath of liquid polyester. The shell shape appeared instantly on removal from the resin (Figure 21.21), and was left to set rapidly under an intense ultraviolet lamp. While he was enthusiastically explaining what we had just seen, disaster was only just averted when the audience drew Isler's attention to the fumes emanating from the hanging model, which were about to be ignited by the hot lamp! Sadly, this caused Isler to resolve never to repeat this demonstration in public again.

In the associated paper entitled *Creating Shells*, Isler emphasised the high degree of accuracy that he normally required from his modelling [55] and noted that this would not be apparent in the live demonstrations.

Since Isler died in 2009 there has been increasing interest in his shells from engineers and architects, and a desire to better understand his physical modelling techniques. For instance, for a 'Bend and Break' course at Delft University

Figure 21.20 Shell model formed by an inflated membrane during Structural Morphology seminar, Stuttgart, 1994. (Image: © John Chilton)

Figure 21.21 Inverted hanging-membrane formed during Structural Morphology seminar, Stuttgart, 1994. (Image: © John Chilton)

of Technology (TU Delft) in 2017 and 2018, students used Isler's inverted membrane technique to produce reinforced-gypsum shells (Figure 21.22). Form-finding software was used to replicate the form of each physical model and FEM used to predict its failure load. The twenty-three models were subsequently load tested under surface pressure applied by air extraction from a vacuum bag, exhibiting behaviour similar to reinforced concrete [56]. It was remarked that the load test revealed that 'Lots of new phenomena of shell structures occur, which cannot be simulated by computer software' [57].

Figure 21.22 Inverted hanging-membrane gypsum shell models produced in a student workshop at Delft University of Technology (TU Delft). (Image: © Andrew Borgart)

21.6 Isler's contribution to the use of models in structural design

Isler was a staunch defender of the effectiveness of his physical modelling techniques used at all stages of his design process. Nevertheless, he was also well aware that they were only an approximation of the real structure. He was adamant that the most important model was the one constructed at full scale and considered these to be further experiments. Accepting that, realistically, it is very difficult (and expensive) to monitor stresses and strains in a full-scale shell, he chose to closely monitor their shape at fixed points on the surface. This recording of the shell outline occurred immediately after striking the formwork until the form of the shell had stabilised, sometimes many years after completion [58]. For example, he used the inverted hanging membrane to find the shape of sports halls: then, through feedback from measurement of the constructed shells, the longitudinal section of the form was gradually modified (first by 3 cm then by 6 cm) until an optimum solution was approached. Similarly, through post-construction monitoring of deflections, he detected small inaccuracies in the form-finding in two areas of the Sicli shell [59].

David Billington (1927-2018), of Princeton University, championed the study of structural art. He has commented that Heinz Isler's shells demonstrate the required qualities of efficiency, economy and elegance [60], adding that those derived through his use of physical models are exemplars that should be followed by engineers and architects in the twenty-first century.

In 2003, in conversation with the author and Ekkehard Ramm at his office near Burgdorf, Heinz Isler remarked that he thought he had been successful '...about a dozen times...'. He was not more specific but it is reasonable to assume that he was mainly referring to the shells form-found with the inverted membrane technique, which he said always gave the best results. He went on to say that he had confidence that, once he was gone, his design methods and philosophy would be taken up by others in the future '...in their own way...' [61, 62].

Today, it is more likely that engineers and architects would work with digital modelling for both form-finding and structural verification, rather than with physical models. However, Isler's argument that, when designing three-dimensional reinforced-concrete shells, a flat drawing board does not give a true representation of the finished article [63, 64, 65] still holds true – the two-dimensional drawing board has just been replaced by a two-dimensional computer screen. On the other hand, if exploratory physical models are produced, it is now possible to scan the geometry rapidly for input into finite element analysis software. Also, if digital form-finding is used, rapid-prototyping and 3D-printing technologies can be used to produce physical manifestations, with potential for assessment of architectural and aesthetic qualities and even physical testing [66, 67].

Over a career of more than 50 years, Heinz Isler was responsible for the engineering (and in some cases, architectural) design of hundreds of elegant thin reinforced-concrete shells realised through his use of physical models for their form-finding and structural verification. He demonstrated what can be achieved by precision 'hands-on' design methods that still have a place in supplementing the new digital technologies. The blending of physical and digital modelling affords engineers and architects the opportunity to fulfil the potential of the infinite continuum of solutions suggested in the final figure of his 1961 paper *New Shapes for Shells* [8, 11].

References

1 Weder, A. & Isler, H. (1955). Die 'Buckelschale', ein interessantes Bauelement [The 'Hump shell', an interesting building element]. *Schweizerische Bauzeitung* 73 (23): 346-349. http://dx.doi.org/10.5169/seals-61927

2 Isler, Heinz (1957). Discussion on shell research. In: *Proceedings of the Second Symposium on Concrete Shell Roof Construction, 1-3 July 1957*, 327-328. Oslo: Teknisk Ukeblad.

3 Bösiger, Heinz (2011). The Building of Isler Shells. *Journal of the International Association for Shell and Spatial Structures*, Madrid, IASS, 52 (3) No. 169: 161-172.

4 Isler, Heinz (1993). Generating Shell Shapes by Physical Experiments. *Bulletin of the International Association for Shell and Spatial Structures*, Madrid, IASS, 34 (1) No. 111: 53-63. 53. [Lecture presented at the IASS Symposium, Copenhagen, 1991]

5 Isler, Heinz (1993). Generating Shell Shapes by Physical Experiments. *Bulletin of the International Association for Shell and Spatial Structures*, Madrid, IASS, 34 (1) No. 111: 53-63. 54.

6 Ramm, Ekkehard (2004). Shape Finding of Concrete Shell Roofs. *Journal of the International Association for Shell and Spatial Structures*, Madrid, IASS, 45 (1) No. 144: 29-39.

7 Joedicke, Jürgen (1963). *Shell Architecture.* New York: Rheinold Publishing. 22-28.

8 Isler, Heinz (1961). New Shapes for Shells. *Bulletin of the International Association for Shell Structures.* Madrid, IASS, Paper C-3, No. 8.

9 Isler, Heinz (1961). New Shapes for Shells. *Bulletin of the International Association for Shell Structures.* Madrid, IASS, Paper C-3, No. 8.

10 Chilton, John (2009). 39 etc... : Heinz Isler's infinite spectrum of new shapes for shells. In: *Proceedings of the International Association for Shell and Spatial Structures (IASS) Symposium 2009, Valencia: Evolution and Trends in Design, Analysis and Construction of Shell and Spatial Structures* (eds C. Lázaro & A. Domingo), 51-62. Valencia: Editorial de la UPV (Universidad Politécnica de Valencia). http://hdl.handle.net/10251/6465. 52-54.

11 Chilton, John (2009). 39 etc... : Heinz Isler's infinite spectrum of new shapes for shells. In: *Proceedings of the International Association for Shell and Spatial Structures (IASS) Symposium 2009, Valencia: Evolution and Trends in Design, Analysis and Construction of Shell and Spatial Structures* (eds C. Lázaro & A. Domingo), 51-62. Valencia: Editorial de la UPV (Universidad Politécnica de Valencia). http://hdl.handle.net/10251/6465. 55.

12 Isler, Heinz (1961). New Shapes for Shells. *Bulletin of the International Association for Shell Structures.* Madrid, IASS, Paper C-3, No. 8.

13 Chilton, John (2000). *Heinz Isler: The Engineer's Contribution to Contemporary Architecture.* London: Thomas Telford. 24.

14 Ramm, Ekkehard & Schunk, Eberhard (2002). *Heinz Isler Schalen.* Zürich: Hochschulverlag an der ETH. 80.

15 Chilton, John (2010). Heinz Isler's Infinite Spectrum: Form-Finding, in Design. *Architectural Design. Special issue: The New Structuralism: Design Engineering and Architectural Technologies* 80 (4): 64-71. 66.

16 Isler, Heinz (1956). Feldmässiger Unterstand in Beton (Atom-Schutzbunker). *Technische Mitteilungen für Sappeure, Pontoniere und Mineure,* No. 4, Zürich: 158-179.

17 Chilton, John (2000). *Heinz Isler: The Engineer's Contribution to Contemporary Architecture.* London: Thomas Telford. 144.

18 Chilton, John (2000). *Heinz Isler: The Engineer's Contribution to Contemporary Architecture.* London: Thomas Telford. 20.

19 Chilton, John (2000). *Heinz Isler: The Engineer's Contribution to Contemporary Architecture.* London: Thomas Telford. 50.

20 Isler, Heinz (1993). Generating Shell Shapes by Physical Experiments. *Bulletin of the International Association for Shell and Spatial Structures,* Madrid, IASS, 34 (1) No. 111: 53-63. 61.

21 Chilton, John (2010). Heinz Isler's Infinite Spectrum: Form-Finding, in Design. *Architectural Design. Special issue: The New Structuralism: Design Engineering and Architectural Technologies* 80 (4): 64-71. 67.

22 Billington, David (2003). Heinz Isler: Structural art in thin-shell concrete. In: *The Art of Structural Design: A Swiss Legacy,* 128-162. Princeton University Art Museum / Yale University Press. 146.

23 Ramm, Ekkehard (2004). Shape Finding of Concrete Shell Roofs. *Journal of the International Association for Shell and Spatial Structures*, Madrid, IASS, 45 (1) No. 144: 29-39. 31.
24 Ramm, Ekkehard (2004). Shape Finding of Concrete Shell Roofs. *Journal of the International Association for Shell and Spatial Structures*, Madrid, IASS, 45 (1) No. 144: 29-39. 32-33.
25 Ramm, Ekkehard (2011) Heinz Isler Shells – The Priority of Form. *Journal of the International Association for Shell and Spatial Structures*, Madrid, IASS, 52 (3) No. 169: 143-154. 150.
26 Billington, David (2003). Heinz Isler: Structural art in thin-shell concrete. In: *The Art of Structural Design: A Swiss Legacy* (ed. David Billington), 128-162. Princeton University Art Museum / Yale University Press.
27 Isler, Heinz (1993). Generating Shell Shapes by Physical Experiments. *Bulletin of the International Association for Shell and Spatial Structures*, Madrid, IASS, 34 (1) No. 111: 53-63. 54-55.
28 Chilton, John (2000). *Heinz Isler: The Engineer's Contribution to Contemporary Architecture*. London: Thomas Telford. 13-14.
29 Chilton, John (2000). *Heinz Isler: The Engineer's Contribution to Contemporary Architecture*. London: Thomas Telford. 79.
30 Chilton, John (2000). *Heinz Isler: The Engineer's Contribution to Contemporary Architecture*. London: Thomas Telford. 77.
31 Steinkirche Cazis, https://de.wikipedia.org/wiki/Steinkirche_Cazis (Accessed 7th January 2019)
32 Chilton, John (2011). Shells for Two Churches. *Journal of the International Association for Shell and Spatial Structures*, Madrid, IASS, 52 (3) No. 169: 173-183.
33 Ramm, Ekkehard (2004). Shape Finding of Concrete Shell Roofs. *Journal of the International Association for Shell and Spatial Structures*, Madrid, IASS, 45 (1) No. 144: 29-39. 36.
34 Isler, Heinz (1961). New Shapes for Shells. *Bulletin of the International Association for Shell Structures*. Madrid, IASS, Paper C-3, No. 8.
35 Chilton, John & Chuang, Chu-Chun (2017). Rooted in nature: aesthetics, geometry and structure in the shells of Heinz Isler. *Nexus Network Journal* 19: 763-785. https://doi.org/10.1007/s00004-017-0357-5. 767-768.
36 Gallenmüller, M. (1999). *Numerische Formfindung von Flächentragwerken am Beispiel einer Schale von Heinz Isler*. Diplom Thesis, Institute for Structural Mechanics, University of Stuttgart.
37 Längst, P., Michalski, A. & Lienhard, J. (2017). Integrated Design Approach for Shell Structures Using Isogeometric Analysis. *Nexus Network Journal* 19 (3): 629-643. https://doi.org/10.1007/s00004-016-0320-x
38 Isler, Heinz (1993). Generating Shell Shapes by Physical Experiments. *Bulletin of the International Association for Shell and Spatial Structures*, Madrid, IASS, 34 (1) No. 111: 53-63. 57.
39 Chuang, Chu-Chun and Chilton, John (2016) Design and Modelling of Heinz Isler's Sicli Shell. In: *Proceedings of the IASS Annual Symposium 2016, 'Spatial Structures in the 21st Century'* (eds. K. Kawaguchi, M. Ohsaki, & T. Takeuchi), 10 pages. Tokyo: IASS.

40 Chilton, John (1992). Shell comeback. *Concrete Quarterly* 173: 22-24.
41 Addis, Bill (2001). *Creativity and Innovation: The structural engineer's contribution to design*. Oxford: Architectural Press. 44-45.
42 Weder, A. & Isler, H. (1955). Die 'Buckelschale', ein interessantes Bauelement [The 'Hump shell', an interesting building element]. *Schweizerische Bauzeitung* 73 (23): 346-349. 347. http://dx.doi.org/10.5169/seals-61927
43 Beckh, Matthias & Schützeichel, Rainer (2017). Paths to Form Finding: The formative years of Heinz Isler. In: *Proceedings of the IASS Annual Symposium 2017 'Interfaces: architecture. engineering. science'* (eds Annette Bögle & Manfred Grohmann) Paper 10337. Hamburg: IASS.
44 Isler, Heinz (1957). Discussion on shell research. In: *Proceedings of the Second Symposium on Concrete Shell Roof Construction, 1-3 July 1957;* 327-328. Oslo: Teknisk Ukeblad. 327.
45 Weder, A. & Isler, H. (1955). Die 'Buckelschale', ein interessantes Bauelement [The 'Hump shell', an interesting building element]. *Schweizerische Bauzeitung* 73 (23): 346-349. 347. http://dx.doi.org/10.5169/seals-61927
46 Chilton, John (2011). Shells for Two Churches. *Journal of the International Association for Shell and Spatial Structures,* Madrid, IASS, 52 (3) No. 169: 173-183. 173-178.
47 Isler, Heinz (1993). Generating Shell Shapes by Physical Experiments. *Bulletin of the International Association for Shell and Spatial Structures,* Madrid, IASS, 34 (1) No. 111: 53-63. 57.
48 Isler, Heinz (1979). New Shapes for Shells – Twenty Years After. *Bulletin of the International Association for Shell and Spatial Structures,* Madrid, IASS 20 (3) and 21 (1) Nos 71 and 72: 9-26. 19.
49 Isler, Heinz (1994). Creating Models. In: *Proceedings of the 2nd International Seminar on Structural Morphology, University of Stuttgart, 1994. Application of Structural Morphology to Architecture.* 221-230.
50 Isler, Heinz (1994). Creating Models. In: *Proceedings of the 2nd International Seminar on Structural Morphology, University of Stuttgart, 1994. Application of Structural Morphology to Architecture.* 221-230. 230.
51 Chilton, John (2000). *Heinz Isler: The Engineer's Contribution to Contemporary Architecture.* London: Thomas Telford. 33.
52 Billington, David (2011) Heinz Isler: From Delft to Princeton and Beyond, *Journal of the International Association for Shell and Spatial Structures,* Madrid, IASS, 52 (3) No. 169: 135-141. 140.
53 Ramm, Ekkehard & Schunk, Eberhard (eds) (2002). *Heinz Isler – Schalen: Katalog zur Ausstellung*. Zürich: Hochschulverlag an der ETH. 81.
54 Isler, Heinz (1994). Creating Models. In: *Proceedings of the 2nd International Seminar on Structural Morphology, University of Stuttgart, 1994. Application of Structural Morphology to Architecture.* 221-230. 223.
55 Isler, Heinz (1994). Creating Models. In: *Proceedings of the 2nd International Seminar on Structural Morphology, University of Stuttgart, 1994. Application of Structural Morphology to Architecture.* 221-230. 226.
56 Borgart, Andrew, Li Qingpeng & Eigenramm, Peter (2018). Manufacture and loading test of reinforced gypsum shells generated from hanging models. In: *Creativity in Structural Design: Proceedings of the Annual Symposium*

of the International Association for Shell and Spatial Structures (eds Sigrid Adriaenssens & Caitlin Mueller), Paper 432, 8 pages.

57 Borgart, Andrew, Li Qingpeng & Eigenramm, Peter (2018). Manufacture and loading test of reinforced gypsum shells generated from hanging models. In: *Creativity in Structural Design: Proceedings of the Annual Symposium of the International Association for Shell and Spatial Structures* (eds Sigrid Adriaenssens & Caitlin Mueller), Paper 432, 8 pages. 8.

58 Isler Heinz, (2008). Shell Structures: Candela in America and What We Did in Europe. In: *Seven Structural Engineers – The Felix Candela Lectures* (ed. Guy Nordenson), 86-101. New York: Museum of Modern Art. 99-100.

59 Isler, Heinz (1979). New Shapes for Shells – Twenty Years After. *Bulletin of the International Association for Shell and Spatial Structures*, Madrid, IASS 20 (3) and 21 (1) Nos 71 and 72: 9-26. 25.

60 Billington, David (2003). Heinz Isler: Structural art in thin-shell concrete. In: *The Art of Structural Design: A Swiss Legacy*, 128-162. Princeton University Art Museum / Yale University Press. 159.

61 Isler, Heinz (2003). *Conversations with Heinz Isler, by John Chilton and Professor Ekkehard Ramm, recorded on 18 March 2003, at Buro Isler*, Lyssachschachen, Burgdorf, Switzerland.

62 Chilton, John (2010). Heinz Isler's shells – a lasting legacy. In: *Structures and Architecture* (ed. P.J.S. Cruz), 795-802. Leiden: CRC Press/Balkema.

63 Isler, Heinz (1961). New Shapes for Shells. *Bulletin of the International Association for Shell Structures*. Madrid, IASS, Paper C-3, No. 8. [Remarks by Isler reported in the Discussion, when responding to Torroja's comments on his presentation].

64 Isler, Heinz (1997). Is the Physical Model Dead? In: *Structural Morphology: Towards the New Millennium, Proceedings of the IASS International Colloquium on Structural Morphology at the University of Nottingham* (eds. J.C. Chilton, B.S. Choo, W.J. Lewis & O. Popovic), 270-274.

65 Chilton, John (2009). 39 etc... : Heinz Isler's infinite spectrum of new shapes for shells. In: *Proceedings of the International Association for Shell and Spatial Structures (IASS) Symposium 2009, Valencia: Evolution and Trends in Design, Analysis and Construction of Shell and Spatial Structures* (eds C. Lázaro & A. Domingo), 51-62. Valencia: Editorial de la UPV (Universidad Politécnica de Valencia). http://hdl.handle.net/10251/6465. 56.

66 Chilton, John (2010). Potential unrealised? The shells Heinz Isler might have built ... . In: *Proceedings of the International Symposium of the International Association for Shell and Spatial Structures (IASS), Shanghai. Spatial Structures – Temporary and Permanent* (eds Q. Zhang, L. Yang & Y. Hu), 3155-3162. Beijing: Architecture & Building Press.

67 Chilton, John & Chuang, Chu-Chun (2017). Rooted in nature: aesthetics, geometry and structure in the shells of Heinz Isler. *Nexus Network Journal* 19: 763-785. https://doi.org/10.1007/s00004-017-0357-5. 768.

22

Models for the design development, engineering and construction of the Multihalle for the 1975 Bundesgartenschau in Mannheim

Ian Liddell

22.1 Introduction

The Mannheim Multihalle was a landmark structure in the development of lightweight shell structures [1]. The specialist engineers for the roof were Ove Arup & partners who were appointed late into the design and construction programme. Six models were used in the design and engineering process.

1) A simple wire-mesh model as part of the scheme design.
2) A hanging-chain net model that defined the geometry of the grid shell.
3) A vacuum-formed rigid plastic model used for wind-tunnel testing to establish the design wind loads.
4) A Perspex strip loading model of a trial gridshell, to validate the modelling technique.
5) A Perspex strip loading model of the larger gridshell at Mannheim to investigate buckling loads.
6) A weighted wire-mesh model to study the behaviour of the grid during lifting and finishing.

22.2 Early gridshells

After exploring the potential for tensile structures in the 1950s [2], Frei Otto (1925-2015) became interested in light-weight shells which could be form-found using the Hookean principle of inverting a funicular structure. Robert Hooke used this method to show Christopher Wren how the Dome of St Paul's might work (see Section 3.2.1). This form-finding method also suggests a construction method using an equal-mesh square grid of timber laths or steel rods, thin enough to be readily bent into shape. A square grid can be moulded to a doubly curved surface by the deformation of the grid squares into rhombi. Such a structure Otto described as a gridshell (*Gitterschale*).

In 1962, Otto built, with some students at the University of California in Berkeley, a trial structure in the form of a dome using steel rods, standing on

Physical Models: Their historical and current use in civil and building engineering design,
First Edition. Edited by Bill Addis.

Figure 22.1 Trial gridshell structure at Essen 1962. (Source: Courtesy of the Institute for Lightweight Structures, Stuttgart)

four points. Later that same year he made a trial timber structure at Essen on a 15 x 15 m super-elliptical plan (Figure 22.1). In spite of incurring a number of breakages of the laths, this structure was successfully completed and clad with transparent plastic foil.

Inside the cablenet tent structure for the Montreal Expo '67, designed by Frei Otto, two small auditoria were required and these were made using grid-shell construction similar in size to the Essen shell. The timber meshes were prefabricated in Germany and sent to Canada folded into bundles, where they were opened up and installed on site. The grids were clad with thin plywood sheets to form the enclosures.

Between May 1971 and May 1973 a research project was undertaken at Frei Otto's Institute for Lightweight Structures (IL) at Stuttgart University. The work was carried out by a Japanese/German team called STI, and funded by the Seibu Company (Japan), Kenzo Tange + Urtec (Architects) and the Institut für Leichte Flächentragwerke, University of Stuttgart.

Their work included a large number of study models and investigations into the optimum methods of making chain nets for hanging-chain models, and their output is all recorded in *Gitterschalen*, published in 1974 [3].

22.3 Initial design for the Mannheim shells

In January 1970, it was decided to hold the 1975 Bundesgartenschau in Mannheim. Carl Mutschler & Partners of Mannheim were selected as architects and Heinz H Eckebrecht of Frankfurt as landscape architect for the Herzogenried Park. The multipurpose hall building was to be the central feature arranged alongside a through route, with a restaurant on the other side. A feature of the

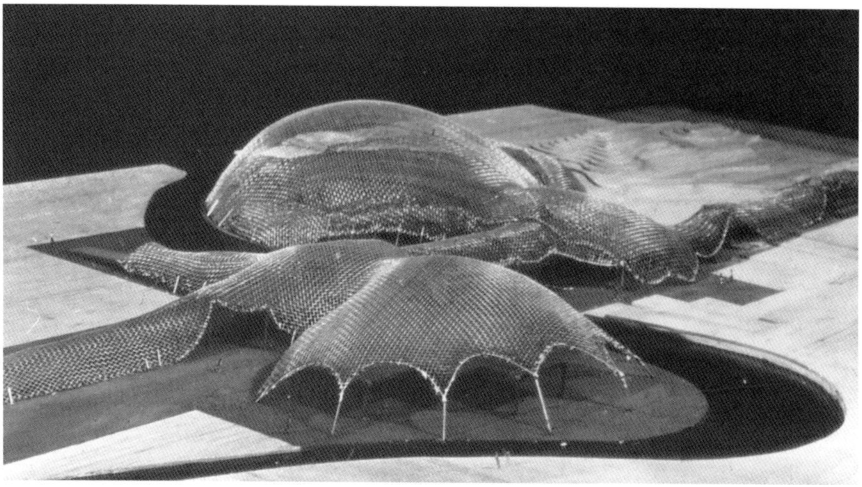

Figure 22.2 Wire-mesh 'sketch' model. (Image: Courtesy of the Institute for Lightweight Structures, Stuttgart)

site in the flat plain of the Rhine was a large mound formed from demolition material left from the war, and the landscape architect wanted a continuation of the hilly form by architectural means. After some initial crazy ideas, the architects remembered Otto's grid shell trials and went to meet with him at Atelier Warmbronn. After some preliminary shapes were made with wire mesh, the architects and the collaborators at the Atelier Warmbronn, particularly Evald Bubner, developed a 1:500 wire mesh model with two large dome structures and the linking covered walkways (Figure 22.2). The mesh, which could be moulded by hand, was similar to that used for normal kitchen sieves. This was effectively the sketch design.

22.3.1 The hanging-chain model

The research group STI had studied the effects of different net geometries. A triangular net can be fitted onto any curved surface but one with all the links of equal length can only exist on a flat surface. A hexagonal net has three links meeting at each node so as a hanging net it can always be in equilibrium, with all the links in tension (Figure 22.3a).

A four-sided net with equal-length links can be drawn on a doubly-curved surface starting from two central lines and using a compass to define the node points, but it will not necessarily hang in the same shape as the original with all the links in tension (Figure 22.3b). If chain links are used, any links in 'compression' will go slack; with rod links, the links could resist some compression but the slackness could be lost in the loose node rings and this would not be apparent (Figure 22.4).

A square grid has the advantage that a scaled-up version of a model can be built with a grid of flexible laths of equal length. However, there are also less advantageous properties; the distribution of forces between the two directions is not as equal as it would appear. Small changes in length of the chains result in large

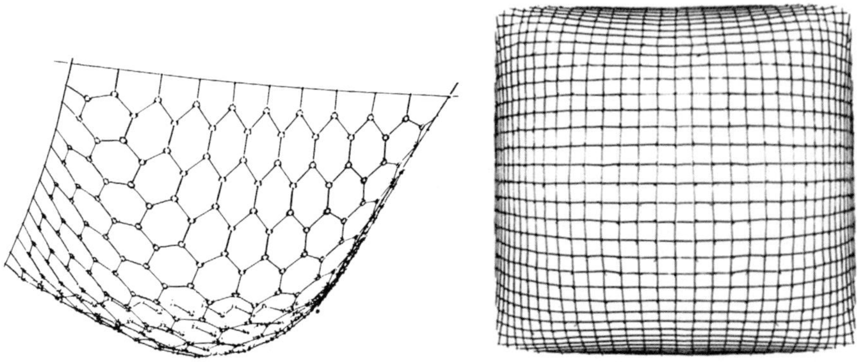

(a) (b)

Figure 22.3 (a) A hexagonal net. (b) A hanging model made with a four-sided net with equal meshes. (Images: Courtesy of the Institute for Lightweight Structures, Stuttgart)

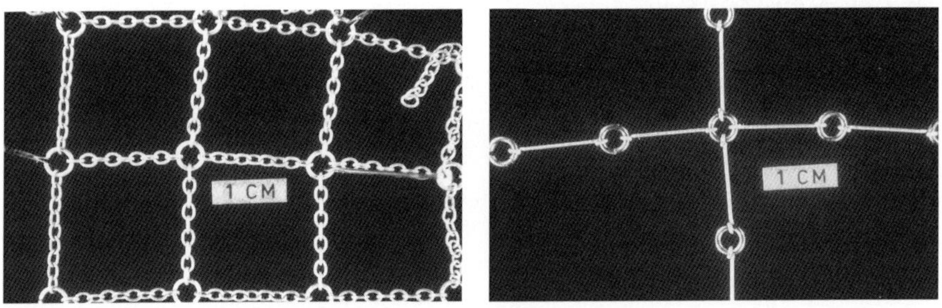

(a) (b)

Figure 22.4 Details of the chain model of the shell. (a) Chain links. (b) Hooked-rod links with rings. (Images: Courtesy of the Institute for Lightweight Structures, Stuttgart)

changes in the distribution of the forces. To try and smooth this out, the links in the final model were joined to the boundaries with soft springs (Figure 22.5).

The model was measured by stereo photography and the photos processed to generate the coordinates of the nodes (Figure 22.6). The form was then processed by engineers in the *Institut für Anwendungen der Geodäsie im Bauwesen* (IAGB) at the University of Stuttgart, led by Professor Linkwitz. To correct the nodes and smooth out the forces, they used the analytical method of force-densities that had recently been developed in that department.

In August 1973, drawings and specifications were prepared from the measured model. These were sent out to a few companies to get prices for the ground works, the timber shells and the fabric coverings as three separate packages. Three companies were selected that would form an *Arbeitsgemeinschaft* (working team) for the construction.

The Mannheim engineers Bräuer und Späh were entrusted with the statical calculations. In the middle of 1973, they realised that the stability of the shell was not accessible to purely mathematical calculations and, on the suggestion of Frei Otto, the London firm of Ove Arup & Partners, represented by Ted Happold and myself, were engaged in September 1973 while Brauer und Späh continued with

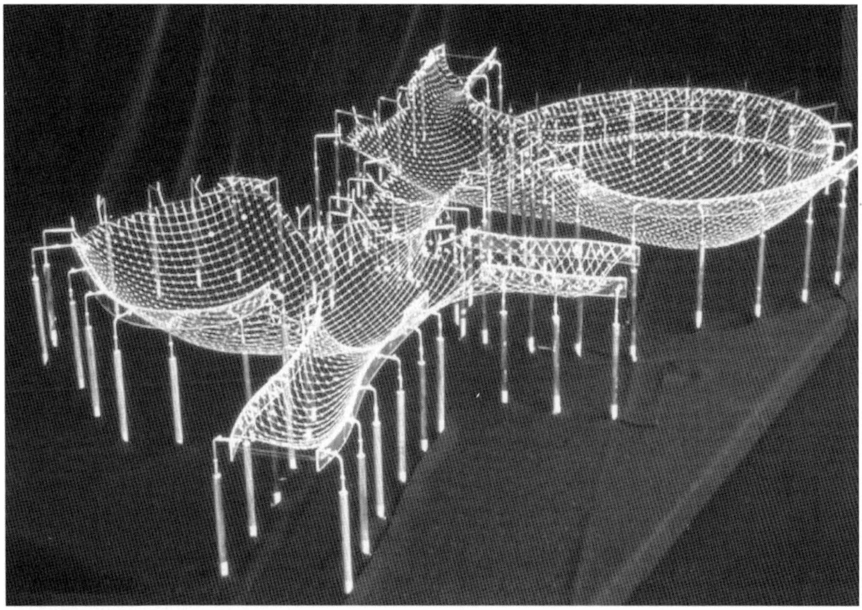

Figure 22.5 The final hanging-chain model for the structure was made with hooked-rod links connected with rings at the nodes. (Image: Courtesy of the Institute for Lightweight Structures, Stuttgart)

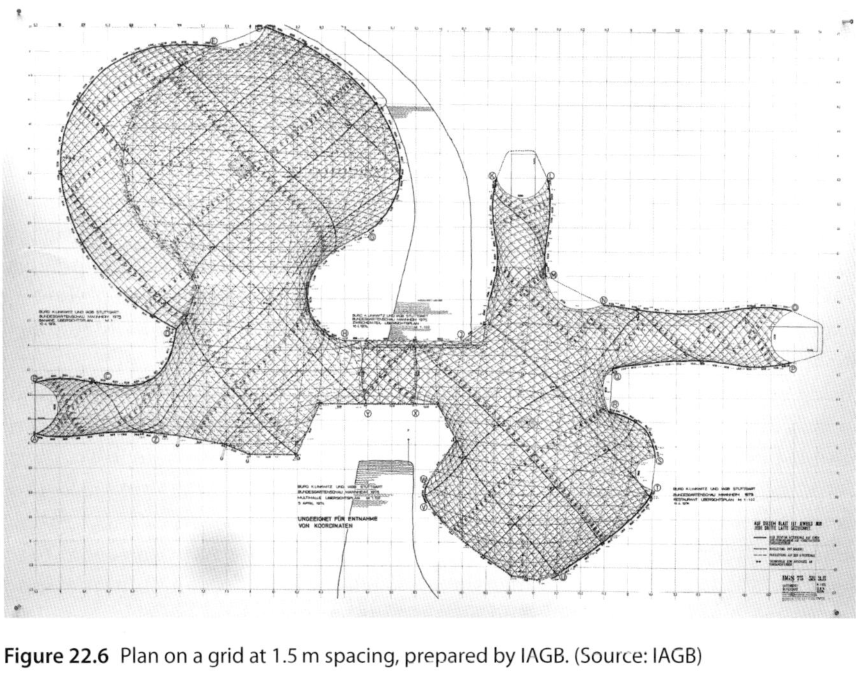

Figure 22.6 Plan on a grid at 1.5 m spacing, prepared by IAGB. (Source: IAGB)

the ground and base works. The exhibition was due to open on 1 May 1975 and the start of construction on site was scheduled for December 1973.

22.3.2 Initial engineering work

The engineers were faced with an absence of knowledge as to how the grid structure would behave and were required to confirm in four weeks that they thought it could be safely built [4, 5]. They made an immediate start on four fronts:

a) Wind tunnel investigations into the design environmental loads (wind, snow)
b) Hand calculations to get approximate values for the member forces
c) Desk studies and hand calculations on shell buckling
d) Model testing to investigate the buckling loads.

Hand calculations were based on a paper by D.T. Wright on buckling in reticulated shells [6]. This paper gave a simple formula for the buckling load of a fully-triangulated doubly-curved shell with pin-jointed members. This arrangement is not the same as that for grid shells, but the formula did indicate that, to get sufficient out-of-plane stiffness, 100 x 100 mm laths would be required, rather than 50 x 50 mm that had initially been proposed. This told us that detailed studies on the buckling in this unusual shell structure would be required.

22.4 Wind loads

22.4.1 Wind pressure and scale modelling

Wind pressures on structures are commonly related to the wind-stream stagnation pressure, $(\rho V^2)/2$, by means of the non-dimensional pressure coefficient, *Cp*. For conventional structures, values of *Cp* over the surface are well tabulated in Codes of Practice. In general, these have been calculated from wind tunnel tests. Little information was available for unusual shapes, and so it was decided to carry out wind-tunnel tests to establish the pressure distribution on the Mannheim shells. They were carried out by the British Hydrodynamics Research Association (BHRA) in their low-speed wind tunnel at the Cranfield Institute, which has a working section 1.8 m square. 1 in 20 year wind speeds were calculated using data supplied by Wetterdienst, Mannheim, using the statistical theory of extreme values as set out by Gumbel [7].

For true modelling of a prototype structure, all parameters should be modelled. To achieve this, the non-dimensional ratios describing the structure and the wind should be constant, and the important parameters should be represented as well. A compromise has to be reached for conventional low-speed wind tunnels in that the dimensionless number, Reynolds Number, *Re*, cannot be scaled.

The dynamic behaviour of lattice shells is uncertain, but it was considered that, because of the large length dimension and high structural damping, the wind response would be similar to the effect of an equivalent static pressure. This same simplification eased the modelling and allowed the use of a static rather than an aeroelastic model to determine the *Cp* distribution. It also applied to the modelling of the wind and reduced the need to scale the turbulence intensity. An approximation to the velocity distribution near the ground was made to allow a single reference velocity, *Vz*, to be used for evaluation of *Cp* values.

Velocity scaling is, then, not dependent on the model scale, but on the measuring equipment and other physical limitations in the tunnel. While it is not possible to scale *Re*, this parameter primarily controls the flow pattern around the structure and, thus, is of importance. For flow in a fully-developed turbulent boundary layer, such as atmospheric winds, flow pattern is independent of *Re*. This type of flow is often ensured by the normal sharp-edged wind-tunnel models used. Devices such as trip wires or surface roughening are necessary for small rounded models such as the Mannheim lattice shell. By this means, flow conditions may be well scaled even though *Re* is not identical. These considerations led to the choice of a static rigid hollow model to determine *Cp* values for wind-loading calculations

22.4.2 Wind-tunnel tests

For convenience, the length scale of the model was chosen as 1:200 and the area of surrounding land was determined by the size of the wind tunnel turntable (Figure 22.7). A rigid model of urethane foam was made by Fairey Surveys from a contour drawing supplied by Frei Otto. From this, several clear vacuum-formed models were made with a shell thickness of approximately 2 mm. This produced

Figure 22.7 Wind-tunnel model of the Multihalle set in the surrounding park (Image: Ove Arup & Partners)

Figure 22.8 Moulded-plastic model with flow visualisation over the surface using paraffin smoke. (Image: Ove Arup & Partners)

a rigid model which was transparent, and which had openings around the sides, as required. The model was firmly attached to a base board and about 150 pressure taps were connected to measure relative pressure over the outside and inside surfaces of the model. A large roughened fairing was provided up-stream of the model to produce a turbulent boundary layer.

With regard to flow verification and visualization, to verify the assumption that *Re* scaling was not important, a series of runs were made in the tunnel with smoke injected upstream of the model. It was hoped to gain some appreciation of the pattern of air flow over the structure, and to discover any problems of high velocity in the surrounding areas (Figure 22.8). As was expected, the flow pattern showed no visible change with tunnel speeds of reasonable order and above, and it was concluded that the trips and sharp-edge site features were sufficient to ensure that the flow pattern was turbulent and independent of *Re*.

While no quantitative information was to be gained from smoke tests, the following was observed:

- Very small flows occurred inside the building. This may underestimate the real condition due to scale effects of small holes; however, it would appear that flow rates inside the building will be low and unlikely to cause discomfort.
- The flow around the southern area of the structure appeared to have a high turbulence component from the many trees in the area
- The large tree in the Biergarten near the middle of the structure appeared to reinforce a standing vortex in this area for most wind directions. Velocities

here are unlikely to be greater than the natural wind velocity, but on days of high wind speed, some discomfort may be experienced in this area.

22.4.3 Determination of pressures.

The model was instrumented with 150 pressure taps over the surface, including several on the inside. The total pressure at each tap location was measured by manometer. Prior to any readings, a traverse with a calibrated National Physical Laboratory pitot-static tube was carried out upstream and downstream of the model to determine the boundary-layer profile. The banks of manometers were photographed for each run (Figure 22.9).

The processing of manometer readings into *Cp* values was done by BHRA. These were then calculated for wind directions each 30° from North ($\phi = 0°$) with extra readings around the apparently critical wind direction up the Rhine Valley of $\phi = 210°$. Each wind direction has an associated relative probability which may be used to predict the final wind loading on the structure. In order to present the mass of information in a more useful form, a computer contouring technique was used to produce *Cp* contours for each wind direction. The data values were also plotted at their measuring locations. The computer plot for *Cp* contours was 'smoothed' by hand to remove the unavoidable anomalies produced by the irregular structural shape. The contours and the data values for wind at 210° are shown in Figure 22.10.

Figure 22.9 Bank of 30 alcohol manometers that were recorded photographically and processed to get external and internal pressure coefficients. (Image: Ove Arup & Partners)

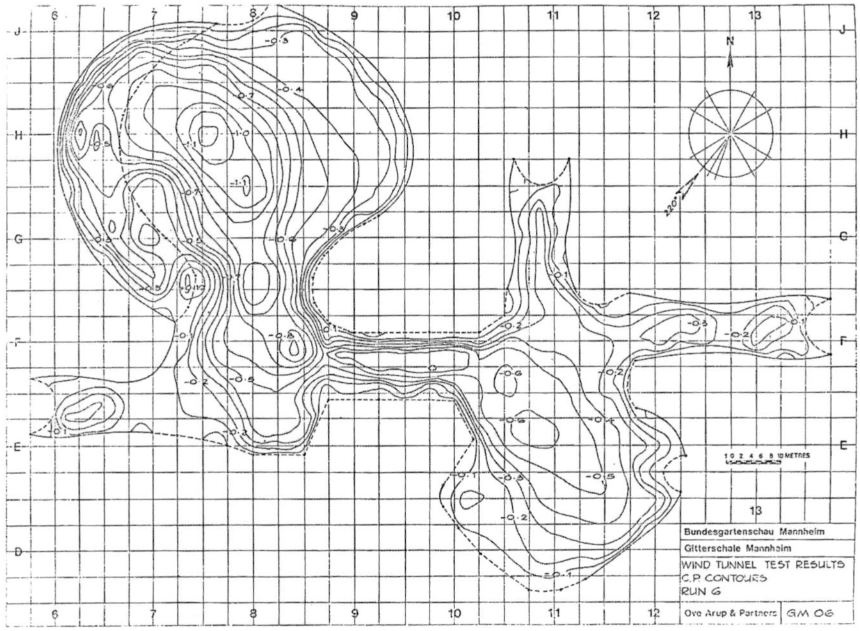

Figure 22.10 Contour plot of external surface pressure coefficients for wind direction 210°. (Image: Ove Arup & Partners)

22.5 Structural model testing: the Essen model

As soon as structural design was started, it was decided to make a 1:16 scale model of the 15-m span Essen dome to help gain some understanding of the behaviour of lattice shells. It was also decided to make the laths as thin and flexible as possible so that the structural properties resembled those of the Mannheim Multihalle shell. The most suitable material for the model laths was Perspex strip which had a Young's modulus one quarter that of timber. It was also easily worked and jointed.

The basic lattice for the model consisted of 3 x 1.7 mm laths at 50 mm centres. The dimensions of the model were 93 x 93 x 30.5 cm high. The joints were made by drilling small holes in the members at 50 mm centres and then passing small pins through the holes. The pins were then bent over to complete the joint. The laths were cut to length using the original cutting pattern obtained from Atelier Warmbronn, and were tied down with adhesive tape to a preformed base board (Figure 22.11). The model was loaded by hanging bundled 100 mm nails, with an average weight of 12.5 g, at the nodes.

Each load case consisted of a uniformly-distributed load plus a point load at one of three points denoted as centre, side and corner. With the UDL constant, the point load was increased and its deflection measured. This enabled the failing load to be determined. The UDL was then increased and the same procedure repeated. The failing point loads were then plotted against the UDL to establish the critical UDL. Four sets of tests were carried out with different conditions of diagonal restraint:

A - pinned joints;

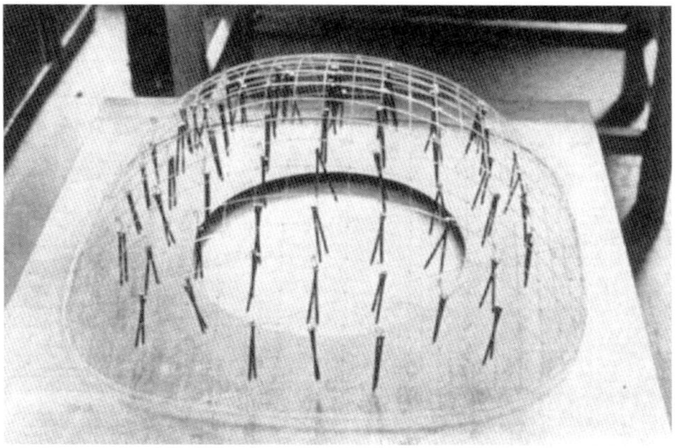

Figure 22.11 Model load test on the Essen gridshell. (Image: Courtesy of the Institute for Lightweight Structures, Stuttgart)

B - glued joints, i.e. no rotation at the joints;
C - pinned with loose ties;
D - glued with nylon ties at all nodes.

For test case A, the 'corner' point gave the lowest failing load (Figure 22.12). With this case the geometry of the structure allowed large sway movements to take place which greatly reduced the failing load. This mode of failure was prevented by the addition of some diagonal stiffness, and with ties, this point became the strongest. The addition of diagonal stiffness increased the collapse loads and decreased the deflections prior to collapse, thus making the collapse far more sudden. The centre point gave the most consistent failing loads. In the graph in Figure 22.12 the maximum centre point load is plotted against UDL to predict the critical UDL for each of the four sets of tests.

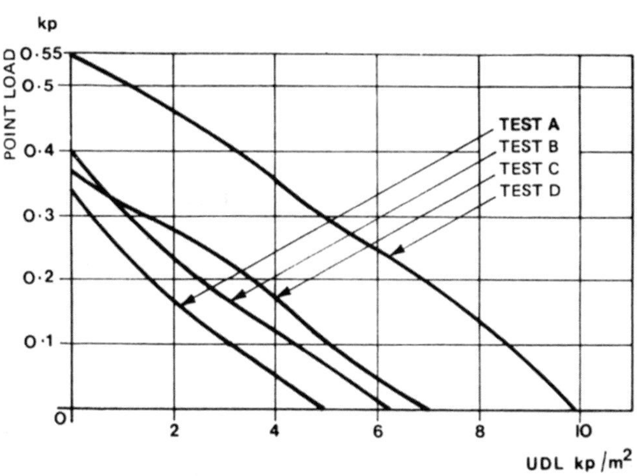

Figure 22.12 Essen gridshell model tests. Graph of centre point load against uniformly-distributed load. (Source: Courtesy of the Institute for Lightweight Structures, Stuttgart)

22.6 Predicting failure loads from model tests

22.6.1 Scale factors

Buckingham's method of dimensional analysis [8] (see Section 13.2.2) was used to compare the properties and behaviour of the Essen model to that of the real Essen shell, and also to try to predict the failure loads of the Mannheim shells. (The scaling issues and structural analysis are discussed more fully elsewhere [9]). The properties of the structure which define and control the behaviour are as follows:

S	is the span. If the model is geometrically scaled then its size can be represented by a typical dimension, say the span.
$\frac{EI_{xx}}{a}$	is the out-of-plane bending stiffness of the surface, per unit length. (a = spacing of laths).
$\frac{EI_{yy}}{a^3}$	is proportional to the contribution of the timber members to diagonal stiffness, if the joints between timber members are rigid.
$\frac{EA}{a}$	is the axial stiffness along the timber members, per unit length.
$\frac{E'A'}{ka}$	is proportional to the contribution of the ties to the diagonal stiffness (A = area of ties, ka = tie spacing)

The collapse load, q_{cr}, is defined as a function of:

$$S, \frac{EI_{xx}}{a}, \frac{EI_{yy}}{a^3}, \frac{EA}{a}, \frac{E'A'}{ka}$$

As there are six terms involved, including q_{cr} and two dimensions, force and length, four independent non-dimensional groups can be formed, such as:

$$\text{I} = \frac{q_{cr}}{\dfrac{EI_{xx}}{aS^3}} = \frac{q_{cr}}{\dfrac{EI_{xx}}{a} \times \dfrac{1}{S^3}}$$

$$\text{II} = \frac{S^2 I_{yy}}{a^2 I_{xx}} = S^2 \times \frac{EI_{yy}}{a^3} \times \frac{1}{\dfrac{EI_{xx}}{a}}$$

$$\text{III} = \frac{S^2 A}{I_{xx}} = S^2 \times \frac{EA}{a} \times \frac{1}{\dfrac{EI_{xx}}{a}}$$

$$\text{IV} = \frac{S^2 E'A'}{kEI_{xx}} = S^2 \times \frac{E'A'}{ka} \times \frac{1}{\dfrac{EI_{xx}}{a}}$$

It is possible to form other non-dimensional groups, such as $\left(\frac{a^2A}{I_{yy}}\right)$, but these further groups are only combinations of equations I to IV, above. Hence:

$$\frac{q_{cr}}{\frac{EI_{xx}}{aS^3}} = f\left[\frac{S^2I_{yy}}{a^2I_{xx}}, \frac{S^2A}{I_{xx}}, \frac{S^2E'A'}{kEI_{xx}}\right] \tag{22.1}$$

Equation (22.1) shows that non-dimensional group I is a function of II, III and IV only. Hence, if the model is constructed so that the numerical values of the groups II, III and IV are the same as in the full-size structure, then the group I will also have the same value on the model and full size structure. If group I has the same value for the model and the full-size structure, then the collapse load on the structure can be found from the collapse load on the model, since $\left(\frac{EI_{xx}}{aS^3}\right)$ is known for both.

The deformation of the grid shell is mainly due to out-of- plane bending and diagonal distortions of the grid squares. If the diagonal stiffness is much less than the axial stiffness then the influence of the group III properties can be ignored. The diagonal stiffness is made up of the in-plane bending stiffness, group II and the tie stiffness, group IV. If the ties have a greater stiffness than the in-plane bending then group IV will overshadow group II. In this case, group I will be the same for the model and full scale structure, provided that the tie stiffness is correctly scaled.

Therefore, to scale up the critical load from structure 1 (the model) to structure 2 (the prototype), this expression was used:

$$\frac{q_{cr(1)}}{q_{cr(2)}} = \frac{\left[\frac{EI_{xx}}{aS^3}\right]_{(1)}}{\left[\frac{EI_{xx}}{aS^3}\right]_{(2)}}$$

22.6.2 Prediction of collapse loads

The structure was varied by altering the in-plane diagonal stiffness by adding ties and gluing the joints. Buckingham's method was used to evaluate the scale factors in the areas of largest radii of curvature so the results could be applied to the largest of the Mannheim shells.

In Table 22.1 the results from the Essen model testing are scaled up to predict failing loads for the Essen shell and for the Mannheim shells with both double and single grids. The Mannheim shells are not geometrically similar, and so an average value of the radius over the large area of the dome was taken as the controlling dimension (S) and compared with a similar dimension for the Essen model. The predicted collapse load for the Multihalle shell with single layer was 3.8 kg/m^2, with a double layer it was 100 kg/m^2 and with ties it was 160 kg/m^2. This information gave the design team the confidence to proceed with the project using a double-layer grid.

Table 22.1 Collapse load for the Essen gridshell, based on model tests, and predicted collapse loads for the Multihalle gridshell.

	Collapse load including structure own weight (kgf/m^2)	
	Rigid joints	Rigid joints with ties
Essen shell (single layer)	150	- a)
Mannheim Multihalle (single layer)	3.8	- a)
Mannheim Multihalle (double layer)	100	160
Mannheim restaurant shell (single layer)	5.8	- a)
Mannheim restaurant shell (double layer)	150	240

a) Model results cannot be meaningfully scaled as the ratio of tie stiffness to out of plane bending is not correct.

The results demonstrated quite clearly the low buckling capacity of a single layer grid for the Mannheim shells. They also indicated the advantages of having a double-layer grid with ties, and on the basis of these tests the decision was taken to double the grid.

Unfortunately, the model tests for a single layer grid cannot accurately predict the behaviour of a double layer grid, as the out-of-plane shear stiffness becomes important and cannot be modelled. The importance of this effect was not fully realized until after a Multihalle model had been made and tested, and the results compared with the computer model. This is discussed in Section 27.7.3, below.

22.7 Structural model testing: the Multihalle model

22.7.1 The Perspex model

A model of the Multihalle was constructed in a similar way to the Essen model described above (Figure 22.13). It was to a scale of 1:60 and consisted of 1.4 mm

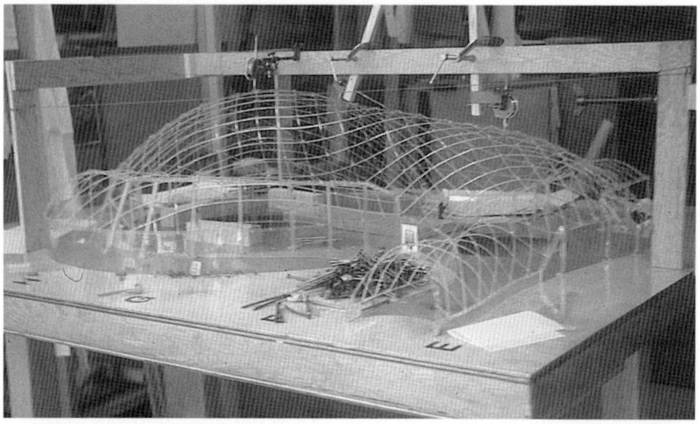

Figure 22.13 1:60 scale Perspex load model of the Multihalle shell. (Image: Ove Arup & Partners)

deep by 2.6 mm wide Perspex members in a single layer, with a joint spacing of 50 mm. Each Perspex member therefore represents 60 x (5/50) = 6 double-layer members on the full scale structure.

The joints were not glued since the tests of the full-scale joint showed that a single bolt connection did not provide sufficient stiffness to prevent relative rotation of the members in the plane of the lattice. Furthermore, the tests on the Essen model showed that the effect of gluing the joints is slight. This effect is further reduced if ties are present.

The model was built on a timber base board with Perspex rod and sheet to represent the supporting columns, beams, arches and concrete walls. No attempt was made to scale the stiffness of the supporting structure except for the arch in the Multihalle. This is because the deflections of the boundaries would not be large enough significantly to affect the behaviour of the shell.

Some of the tests were performed with diagonal ties consisting of stout linen thread added to the model. The ties were at a spacing corresponding to $3\sqrt{2}$ m on the full-scale structure, and were attached to the nodes by wrapping the thread around the pins and then adding a drop of glue.

The model was tested using a similar procedure to that employed for testing the Essen model. A distributed load was first applied to the model, and then a point load at various locations in turn. At each location the deflection of the structure was recorded as the point load was increased. The test locations are shown in Figure 22.14.

This procedure was adopted since it is possible to predict the collapse load from the load-deflection curves without actually collapsing the model. There was reluctance to collapse the model since, once started, the collapse is difficult to stop and would result in considerable damage. This was especially so when the model has ties due to the higher load carrying capacity, and this was demonstrated in the final Essen model test.

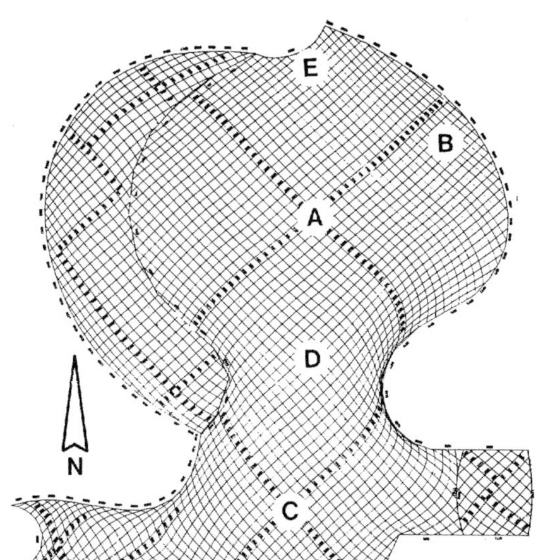

Figure 22.14 Plan of Multihalle grid showing the five test point-load points. (Image: Ove Arup & Partners)

The load was again applied using 100 mm nails each weighing 12.5 g. The deflection of the structure under a point load was measured, using a dial gauge for the first experiments. Later, a lever arrangement was fabricated out of fine wire to avoid the friction inherent in a dial gauge, and this enabled readings to be obtained using smaller point loads. This meant that the stiffness of the structure could be evaluated without affecting it by gross local deformations.

The load/deflection curves for each point, without and with ties, were plotted and extrapolated to give the collapse UDL [9]. In each case the curve which predicted the lowest collapse UDL was used. The collapse UDL for the model was thus found to be 2.8 kgf/m^2 with no ties and 12.5 kgf/m^2 with ties.

As with the Essen model tests, the results were scaled using the non-dimensional group $\left(\dfrac{q_{cr}}{\left(\dfrac{I_{xx}}{aS^3}\right)}\right)$. Scaled up in this way, the collapse load of the Multihalle shell was predicted as 63 kgf/m^2 with no ties and 280 kgf/m^2 with ties and consequently ties were adopted for the real structure.

22.7.2 Testing of the timber joint details

In addition to the testing of reduced-scale models, it was recognised that the structural behaviour of the spliced laths of Western hemlock, and the nodes was so crucial to the stability of the gridshell, that some full-scale testing would be needed. The timber testing included load/rotation tests on the grid intersections and creep bending tests of the laths, the aim being to get better data on the member and connection stiffness to use in the computer modelling. The capacity of the boundary connections was also tested. Of particular concern was the transmission of shear force between the upper and lower members of a pair of laths (Figure 22.15). The structure relied on a pair of laths acting as a beam with the theoretical stiffness being increased 26 times compared to that of a single lath. The compound stiffness relies of the rigidity of the shear connection between the laths which is formed of single 50-mm square laths crossing at right angles held by a single through bolt. Changes in moisture content will cause the timber to expand and contract. If the bolt is done up tight, the expansion will increase the stress in the timber causing it or the bolt to creep or yield, so that, when it dries out and shrinks, it will become loose. To prevent this, it was decided to use disc springs to ensure that the compression was maintained. Tests were carried out with varying amounts of spring compression and this indicated a reasonable amount of joint stiffness until the bolts were completely slack.

22.7.3 Comparison with computer results

1973 was still early days for computational analysis which had only begun in the early 1960s. The modelling for Mannheim was done on a Univac 1108 with augmented core storage and magnetic tape drives for input and output. The solution technique was the Newton-Raphson iteration method in which successive approximations are made and terminated when a sufficient degree of accuracy is obtained. Since the model had to be re-set for each iteration this could be a

Figure 22.15 Testing a section of the grid for shear stiffness. (Image: Ove Arup & Partners)

slow process, so care had to be taken in setting up the computer model to make it as efficient as possible. A standard analysis program was selected, but it had to be modified to take into account that one straight member represented a number of curved members and the fact that shear deflections, which would reduce the out-of-plane bending stiffness, could not be ignored and that some stiffness would be reduced by axial force.

The ties were modelled as timber sections, hence the wire area was increased by the modular ratio. The ties in the model could take compression as well as tension, the problem of one tie going slack was ignored since the stiffness was quite low.

The Multihalle model had 192 nodes, 297 members and 254 ties. A total of fourteen computer runs was carried out for various load cases and member properties. The first four of these runs were invalid as it was discovered that the program was not able to accept members with very low shear stiffness. The program was altered before the fifth run to include the member stiffness equations given previously. The ninth run with a UDL of 105 kgf/m^2 failed to converge but, in a way, that indicated that only a small reduction in load was necessary.

The buckling collapse load for the structure of 105 kg/m^2, was considered satisfactory for the heated dome. However further testing on the shear stiffness generated by the cross nodes was actually 40% of that used in the computer model. Computer runs with this value of shear stiffness had indicated that failure would occur. Hand calculations indicated that members with high axial forces would fail in compression, and so it was decided that the shear stiffness had to be increased in these areas. The load test on the completed structure validated the accuracy of the computational model.

The computer model, with additional out-of-plane shear introduced in highly-stressed areas, predicted a collapse load of slightly more than 100 kgf/m^2 under uniform loading. The difference between this and the 280 kgf/m^2 predicted by the physical model tests occurred because the Perspex model did not scale the shear deformation of the double layer members of the full-scale structure. This deformation is largely controlled by the slip/unit force, η, of the individual node joints. A suitable non-dimensional group, including η, could then be added to those in Table 22.1, above, which would be $\left(\frac{\eta S^2}{ab^2E}\right)$. It would not be possible to construct a model in which both this group and group III $\left(\frac{S^2A}{I_{xx}}\right)$ are scaled, if η is low and varying, using a single-layer model. The Perspex model corresponds to a full-scale structure with a very high value of joint stiffness and therefore predicts a higher collapse load than the computer model in which realistic values of η were used. In spite of this drawback, the model was useful in providing qualitative data showing the type of buckling failure to be expected and the critical areas of the structure.

22.7.4 Modelling the installation on site

The installation on site began with the contractors laying out the laths for the Multihalle shell on the ground, without having a plan for lifting it. Their idea was to lift it with cranes but they realised that the shell would have to be supported in its correct position while all the nodes were set to the correct level, the node bolts tightened and the all boundaries fixed. The additional shear-block pieces and diagonal ties would also have to be installed before the supports could be removed. To assist the contractors with understanding the problems involved, the engineers made a model using wire mesh weighted so that the bending stiffness

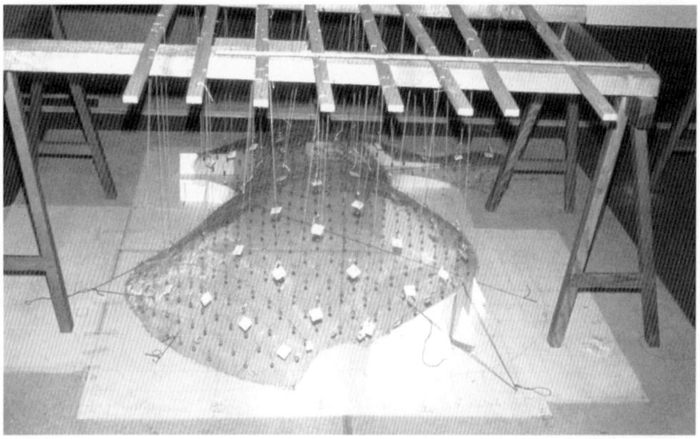

Figure 22.16 Weighted mesh model to test out support points. (Image: Ove Arup & Partners)

Figure 22.17 Method of supporting and lifting the grid. (Image: Ove Arup & Partners)

and self-weight were modelled (Figure 22.16). This indicated that the supports needed to be provided on a 9 x 9 m spacing. The contractors then proposed to use a scaffold tower system with timber spreaders to distribute the loads. The towers were jacked up in 33 cm stages with extra sections being added in (Figure 22.17). When the grid was lifted to give sufficient headroom, forklift trucks were used both to jack up the towers and to move the bases to keep the towers vertical. The installation and lifting model had been quickly made at a time when the contractor needed guidance on how to lift the grid. It broke the log-jam at that moment and enabled the on-site work to be completed on time.

For structural reasons, as well as aesthetic, it was most important that the completed shells should have a smooth surface, closely following the original design. When lifted, the grid was found to deflect significantly between the towers, and so a procedure was developed to deal with the problem. This involved over-lifting the towers or using intermediate push-ups to raise the areas between to the correct heights, and then tightening the node bolts on those lines; the towers were then lowered again to return those areas back to their correct levels. With this procedure the shells were eventually manoeuvred into the correct shape and all the stiffening work was completed. Once the shell was finished it was covered with a specially-made PVC-coated fabric that was partially translucent and allowed some directional sunlight in, creating shadows inside (Figure 22.18).

22.8 Concluding remarks

Physical modelling was widely used in the 1970s, especially by Frei Otto at his Institute and at his Atelier. For the engineers, the Perspex loading models gave some initial information about the buckling capacity of the shells and the direction to take to refine the design. However, to validate the structures it was necessary to use the still quite-limited computational modelling available at the time. I think the greatest value of the physical modelling was to help the engineers, who were working in a different language in a different country, to

Figure 22.18 The completed Multihalle. (Image: Ove Arup & Partners)

communicate to the rest of the design team, the proof engineer and the client, all the additional details required to ensure the design life of the structure.

Opened in 1975, the Multihalle was intended as a temporary building with a design life of just twenty years. It was given listed-building status in 1998. The building was closed for public events in 2011 following the discovery of decay in some of the laths and some sagging at the crown of the gridshell. As it now approaches the age of 45 years, the local authority in Mannheim is committed to saving the building and discussions are under way as to how this can be achieved with minimal impact on the original structure.

References

1 Burkhardt, Berthold (ed.) (1978) *Multihalle, Mannheim.* Mitteilungen des Instituts für leichte Flächentragwerke. IL 13. Universität Stuttgart.

2 Otto, Frei (1967) *Tensile Structures* (2 Vols). Cambridge, Mass.: MIT Press.

3 Burkhardt, Berthold & Hennicke, Jurgen (eds) (1974) *Gitterschalen: Gridshells.* Mitteilungen des Instituts für leichte Flächentragwerke. IL 10. Universität Stuttgart.

4 Happold, E., Liddell, I. & Dickson, M. (1976) Design towards convergence: a discussion. *Architectural Design* 46 (7): 430-435.

5 Addis, Bill (1998) Design methodology. In: *Happold: The Confidence to Build* (ed. Derek Walker & Bill Addis), 86-95. London: Taylor and Francis.

6 Wright, D.T. (1965) Membrane forces and buckling in reticulated shells. *Journal of the Structural Division, ASCE* 91 (1): 173-202.

7 Gumbel, E.J. (1967) *Statistics of extremes.* New York: Columbia University Press.

8 Buckingham, E. (1914) On Physically similar systems: illustrations of the use of dimensional equations. *Physical Review* 4 (4): 345-376.

9 Happold, E & Liddell, W.I. (1975) Timber Roof for the Mannheim Bundesgartenschau. *The Structural Engineer* 53 (3): 99-135.

Section D

Physical models used in non-structural engineering disciplines

Although Leonardo hints in his notebooks about building a model canal to determine the flow rate in a full-scale canal that he proposed to link the Arno valley to Pisa, we know nothing more of what he did. After the behaviour of structures, the flow of water was the next engineering discipline that attracted the attention of scientists – Bélidor began writing his *Architecture Hydraulique* in 1737. The 'art of conducting, raising and managing water' was of great military and social importance and soon many scientists were conducting experiments to determine the flow of water in channels and pipes and over weirs, as well as investigating the movement of sand in river beds and estuaries in order to prevent navigable channels from silting up. The first measurement models of estuaries were built in the 1870s, for the Rivers Garonne, near Bordeaux, and the Seine near Le Havre, and for the Mersey estuary near Liverpool. Many hydraulic laboratories were built in the first two decades of the twentieth century and attention was soon turning also to modelling the action of waves on harbour walls and flood defences and to find ways of preventing erosion on beaches.

As with water, making calculations about the movement of air and its impact on buildings was possible only when the flow was streamlined – a situation that is never encountered in the built environment. The transition from conducting experiments to determine the force of wind on geometric shapes such as a square or circular plate, to measuring the forces on models of buildings began in the 1890s. The techniques of using measurement models were developed rapidly in the 1910s and 1920s for studying whole aircraft and their wings and propellers and by the 1930s wind-tunnel tests were yielding results of use to the designers of buildings such as the Empire State Building in New York. The breakthrough for using wind tunnels in civil and building engineering came in 1965 when Alan Davenport perfected his design for a Boundary-Layer Wind Tunnel which was able to measure loads on buildings in the atmospheric boundary layer in which the airflow is turbulent rather than streamlined.

The end of the nineteenth century also marked the start of seeking ways to model the behaviour of buildings and other structures under seismic loads, and the behaviour of sound in rooms such as lecture theatres and concert halls. By the 1920s the equipment had developed to a degree that enabled engineers to use results for their designs, and the 1930s saw a growing interest in these techniques in both engineering disciplines.

The final chapter in this section deals with a relatively-modern engineering discipline – the use of centrifuges to generate high g-forces to accelerate and accentuate the behaviour of soil under load, for example in embankments, retained earthworks and the interaction between soil and the foundations of superimposed structures. An interesting note to this story is that the idea of centrifuge testing was first proposed in 1869 when Édouard Phillips, a French engineer, criticised the model tests undertaken for the design of Fairbairn and Stephenson's Britannia Bridge opened in 1850, which had been carried out using point loads on the bridge. He proposed applying a uniformly-distributed load by rotating a model of the bridge about an axis parallel to the girders. The first centrifuges were built in Russia and the USA in the 1930s and, from the start, were used to investigate soil behaviour.

23

The historical use of physical model testing in free-surface hydraulic engineering

Bill Addis

23.1 Introduction

Since ancient times mankind has sought to reap the benefits that water brings, such as irrigation, hygiene and transport, and also to minimise the tremendous damage it can inflict, such as flooding and erosion of river banks as well damage from the effects of waves in the sea. Hydraulic engineering, alongside bridges and fortifications, formed the body of all the early engineering treatises. Hydraulics was also the first engineering subject that received the full attention of the earliest modern scientific studies in the Age of Enlightenment, especially in France. As well as construction details for structures relating to hydraulics, such as bridge piles, locks and water channels, the classic work *Architecture hydraulique* (1737-53) by Bernard Forest de Bélidor (1698-1761) contains many detailed accounts of experiments such as determining the flow rates of water through orifices and pipes, over weirs and in channels, as well as a great many tables of results [1]. These results were useful to a design engineer, not directly as design rules but rather as results of scientific experiments to determine fundamental principles and, in particular, to find general equations that would allow the results to be applied to actual engineering structures. Bélidor's book was the first to make use of calculus for engineering purposes (to calculate water discharge from vessels and pipes) and he established a scientific approach that was followed by many other, especially French scientists and engineers. Henri Pitot (1695-1771) invented his eponymous tube in 1732 to measure the speed of the River Seine in Paris. For the first time it was possible to measure the speed of flow across the whole cross-section of a channel [2] (Figure 23.1). It was developed to take its modern form by Henri Darcy in 1868 [3]. It also established a basis for the first hydraulic research facilities that were founded, albeit on a small scale, towards the end of the eighteenth century, both in Europe and the USA [4, 5, 6]. Nevertheless, during this period, there is no mention of reduced-scale models being used for any design purposes.

In fact, the earliest written evidence of hydraulic modelling is found in the notebooks of Leonardo da Vinci who, during his work as a military and civil engineer in around 1500, planned at least one canal, to link Florence with the

Physical Models: Their historical and current use in civil and building engineering design,
First Edition. Edited by Bill Addis.

(a) (b)

Figure 23.1 (a) Operators holding a Pitot tube to measure the speed of the water over the whole cross section of a test water channel, c.1815. (b) Detail of the Pitot Tube. (Source: [2] Plate IV)

Mediterranean [7], and it seems likely that he may have made a model to help determine the flow rate:

> 'First I shall make some experiments before proceeding; my purpose being to conduct the experiment first and then, having determined the cause, to show why the experiment was forced to work out as it did. And this should be the correct rule which investigators of Nature's ways should follow; even though Nature may begin with the cause and end with the experiment, we must pursue the opposite plan, that is, begin, as stated, with the experiment and investigate the cause with its aid' [8, 9].

The movement of free-surface water poses great problems for engineering scientists trying to understand its behaviour because its movement is frequently turbulent and relatively unconstrained, and it usually interacts with the bed of a river or the sea beneath, and with the wind above. Since the mid-nineteenth century, the work of hydraulic engineers has focused on two main fields in the study of natural water movement. The first is the flow of water in rivers and estuaries which causes the bed to move, with the result that navigable waterways can silt up and become too shallow to use; similar problems arise from the tidal movement of water in harbours. The second problem is the erosion of river banks and coastlines arising from both the action of flowing water and also the action of waves. Understanding how these processes work underlies the design of measures taken to manage the effects of water movement such as sea walls, river banks, groynes, breakwaters, harbour walls and quays.

In addition to these examples of natural water flow, since the late-nineteenth century engineers have needed to understand how to manage the flow of water through spillways and penstocks in dams, as it responds to rainfall and changing heights of water in reservoirs, and also to study the effects of introducing man-made structures such as bridges, jetties and piers in rivers or tidal waters. Physical models have played a major role in developing an understanding of the various phenomena and also testing the effectiveness of proposed engineering works before they are built, often leading to modifications of designs.

23.2 The nineteenth-century pioneers

23.2.1 Louis J. Fargue

The earliest known tests on a movable-bed river model were conducted by the French engineer Louis J. Fargue in 1875. As a young military engineer his first secondment in 1848 had been to analyse the considerable amount of data that had been collected about the navigable channel in the Garonne River between Bordeaux and the sea, a distance of some 30 km. The river meanders and even within the width of the river, the channel meanders relative to the centre line. Fargue found that none of the current ideas were of help in explaining this, some claiming that the water always tends to flow in a straight line, others that it follows the curves of the centre line, and others that the route of the river was irrelevant and that what was needed was to ensure that the width of the river bed should be made to increase steadily towards the sea. These different ideas about the water flow gave rise to two conflicting approaches to managing the channel – to dredge it regularly, or to modify the river banks to create water flow that did not lead to silting up. Fargue's task was to find the best strategy. Fargue soon noticed that there was a relationship between the curvature of the river banks and the depth of the channel and he began to develop ideas about why. This led him, in 1875, to build a reduced-scale model of the river, parallel to the actual river [10]. The 'little river' contained four bends of known radius, some of which were made sharper for later tests (Figures 23.2 and 23.3). The channel was about 60 metres long and 1 m wide. The flat bottom was covered with a 30 cm layer of sand taken from the main river. The linear scale in plan was 1:100 and, in the vertical direction, 1:20, and the water speed was reduced by 1:1.6.

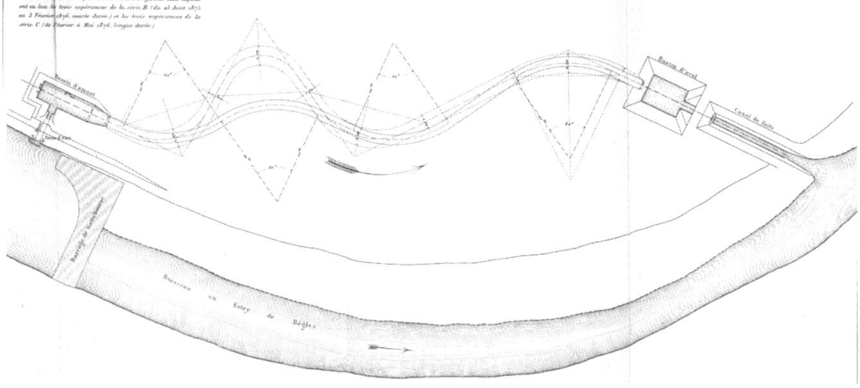

Figure 23.2 1:100 horizontal scale model river built by Louis Fargue, 1875. (Source: [10] Figure 12)

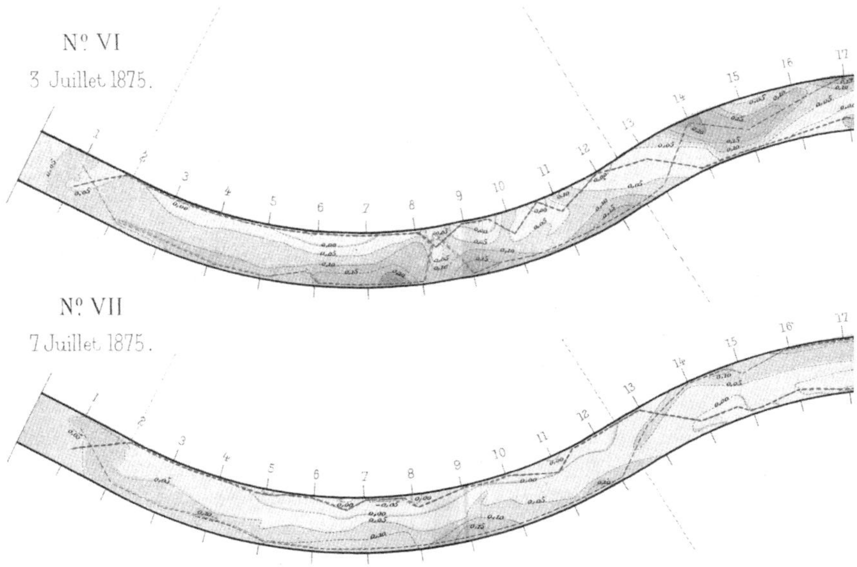

Figure 23.3 Survey of the bed profile of the model river after two tests. (Source: [10] Figure 12)

His twenty-one experiments were carried out over a period of a year and ranged in duration from 45 minutes to 160 hours. The flow rate was regulated between 150 and 250 litres/sec using an adjustable weir at the outlet of the main storage tank, and the water flowed at 0.75-1 m/s. He noted the effects on the river bed of the turbulent flow in the curves, and the build-up of sand on the inside of the bend, which suggested that water was flowing transversely across the river section as well as along the main axis of flow. He also noted that the flow meandered from one side of the river to the other in the straight portions – in other words, the flow seemed to be naturally unstable between the banks. Furthermore, he found regular peaks and troughs in the river bed, respectively at intervals of 3.7 and 2.1 metres along the axis of flow, which did not quite agree with the locations he had found in the actual river during his surveys of a decade earlier. In later experiments he achieved a greater degree of agreement between the model and the actual river, and an overall conclusion was that the river bed had fewer peaks and troughs when transition curves in the river were more gradual. Fargue went on to make a model of the River Seine between Rouen and the English Channel, and the River Scheldt at Antwerp, and also undertook theoretical studies to help understand and interpret his experimental findings [11].

23.2.2 Osborne Reynolds

The British engineering scientist Osborne Reynolds (1842-1912) began a series of tests in 1885, using a movable-bed model of the estuary of the River Mersey to the west of Manchester. This opportunity came about as the result of a commission by the promoters of the Manchester Ship Canal who were seeking confirmation that no changes would be produced in the main channels of the estuary by the canal works which had been designed to modify, very slightly, the line of the Cheshire shore. Like Fargue, Reynolds was intrigued by the different depths of water in the channel and the fact that the main channel lay on the convex (northern) side of the curved estuary, rather than the concave south side that would be usual in a river bend. A key question was how the water entering the estuary after a low tide interacted with the water still flowing out to sea from the inland end of the estuary.

To resolve these questions, he built a model of the estuary at the University of Manchester. The horizontal scale was 1:31,800 and the vertical scale, 1:960 [12]. By trial and error, initially generating the tides by hand, he found that he could replicate the observed filling and emptying of the estuary basin using a tidal period of about 40 seconds, a time he also showed using non-dimensional similarity between the wave velocities and heights, in the ratio $\sqrt{v/h}$. He subjected the model to 2000 tides and succeeded in replicating the location and the depth of the main channel in the Mersey (Figure 23.4). To study the possible influence of the size of the model, Reynolds built a second model at 1:10,800 horizontal scale and 1:396 vertical scale, for which the tidal period was 80 seconds. His description of these experiments is a model of scientific literature, and the key part is reproduced in Appendix 7. Apart from satisfying his curiosity about the unusual behaviour of the Mersey estuary he was also able to conclude that 'this method of experimenting seems to afford a ready means of investigating and determining beforehand the effects of any proposed estuary or harbour works; a means which, after what I have seen, I should feel it madness to neglect before entering upon any costly undertaking' [13].

Very soon, Reynolds joined a 'Committee appointed to investigate the Action of Waves and Currents on the Beds and Foreshores of Estuaries by means of Working Models'. Reynolds was the key investigator and the work occupied him for over three years, resulting in three exemplary reports, totalling 59 pages of text and 40 plates, describing the apparatus, experiments undertaken and the results [14, 15, 16]. The aim of the first experiments was 'to determine in what respects, and to what extent, the distribution of sand in the beds of model estuaries of similar lateral configuration is affected by the horizontal and vertical dimensions, and the relation which these bear to one another and to the tide period so as to place the laws of similarity on which the practical applications of the method depend, on as firm an experimental basis as possible' [17].

Being concerned to test the effectiveness of the modelling process, it was decided at the outset to construct two models – one as large as could be accommodated in the laboratory and the other at half that scale. Any unanticipated scale effects could thus be exposed by comparing the results from the two models. The working trough of the main model was 4.8 x 1.15 m and the tidal range was 51 mm and tidal period 30 seconds (Figure 23.5). This was equivalent to a scale of about 1:20,000, representing a full-scale area of about 70 x 30 km. To enable the sand contours to be plotted, the study area was covered with sheets of glass, under which was a grid of black cotton threads at 15 cm spacing, and the depth of the sand bed was measured using a grid of tiny vertical brass scales that did not interfere with the movement of the sand particles. The tides were created using an electric motor, geared down by 1:6,000 to operate the water pump. The apparatus was tested first using pea-sized balls of paraffin wax in two colours,

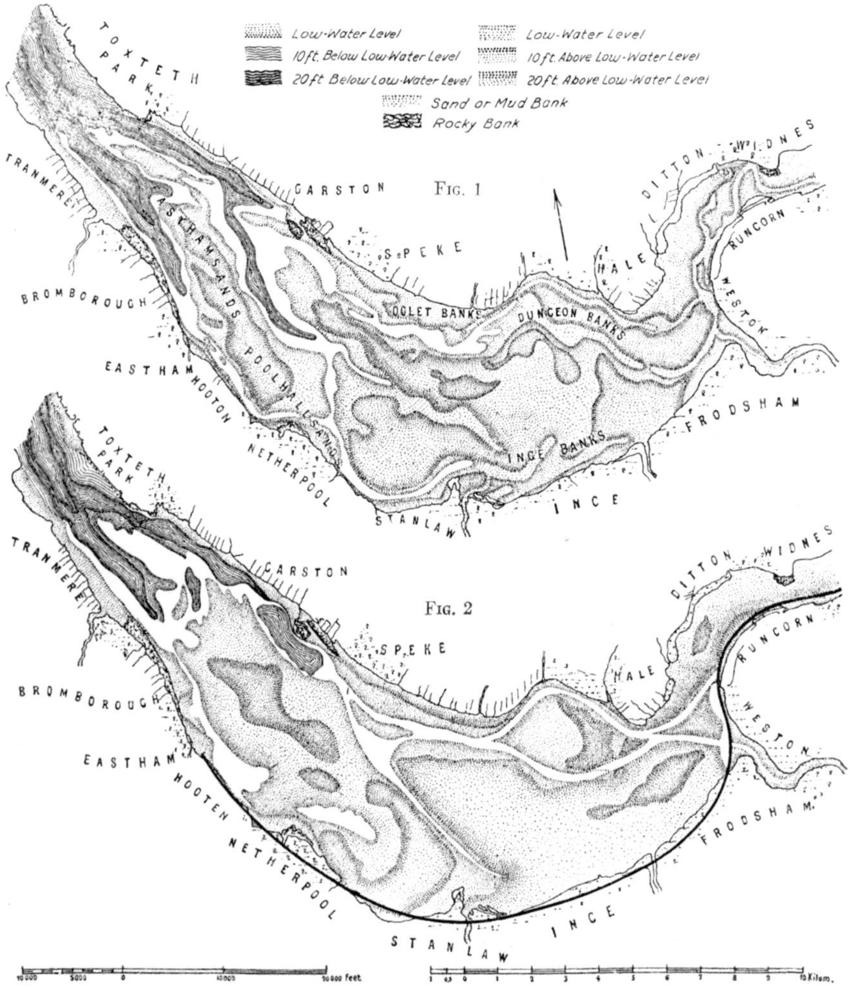

Figure 23.4 1: 31,800 horizontal scale model of the estuary of the River Mersey. Above: Map of the bed of the estuary. Below: Bed of the estuary reproduced in the model after 2000 tides. The black line on the south side of the estuary is the line of the proposed Manchester Ship Canal for which the studies were made. (Source: [13])

one of which just floated, and the other rested on the bottom. They were placed in lines perpendicular to the tide and their movement observed over many tidal cycles. It was found that the surface balls were greatly affected by 'scum' on the water which, although invisible, prevented them moving freely with the water. Water for the experiments was well cleaned. Two different types of sand were used for the experiments: one was a shell sand from Huna Bay on the north coast of Iceland, the other, fine beach sand from Calais. The experiments typically had durations between five and twenty thousand tides – full scale periods of between about 7 and 30 years (Figure 23.6).

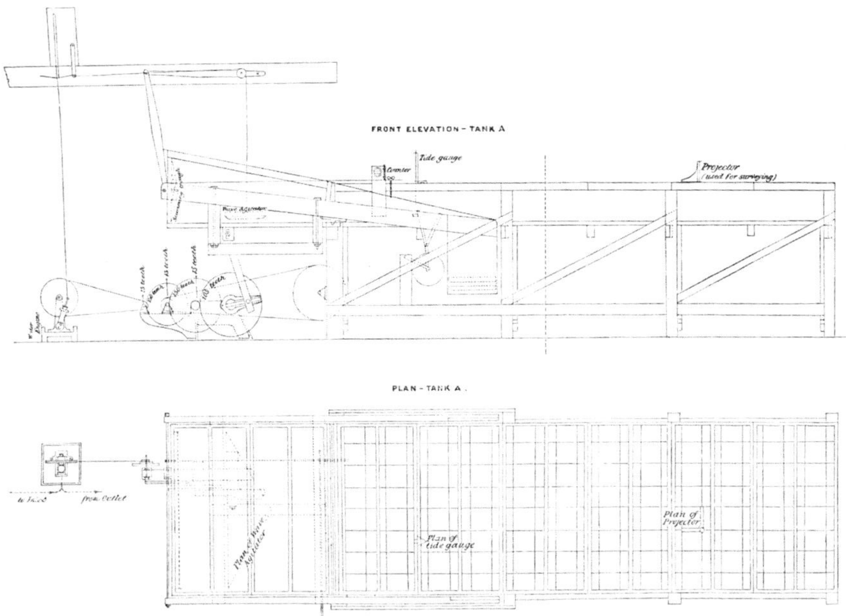

Figure 23.5 Reynolds' apparatus for creating tides acting on a sandy beach. The trough was 1.15 m wide and 4.8 m long. The tidal range was 51 mm. (Source: [14])

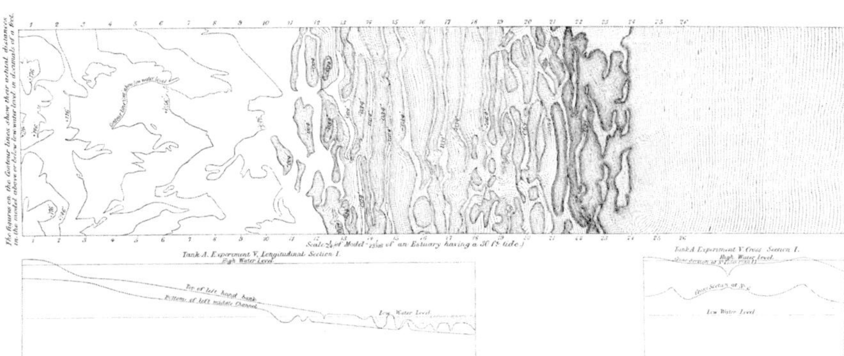

Figure 23.6 Plan and sections of the beach / sea bed after 12,697 tides with a period of 53.4 seconds and 3,589 tides of 49.3 seconds. (Source: [14])

Reynolds' aim was to discover if there 'laws' that explained the slope of the 'beach', how the sand moved and how the characteristic ripples in the sand were formed. The tidal range in the model and their frequency were varied and, in some tests, additional water was introduced from the land side. In fact, although he replicated the full-scale behaviour in this first series of experiments – the formation of a smooth beach that later became grooved and rippled – he discovered no laws. Nevertheless, he planned and carried out several further series of experiments that looked at tidal behaviour in a V-shaped estuary.

In striving for greater similitude between model and full-scale behaviour, he investigated the effects of the size and density of the sand grains, the exaggeration of the vertical scale (necessary to make the depths measurable), and the introduction of land water. To these ends he doubled the length of his model and improved several of the measuring techniques and varied the tidal period and generally increased the number of tides in each study. To these aims, in the final year of experimentation, Reynolds added a study of the effect of introducing spring and neap tides, rather than using uniform tides. In his conclusion in 1891 he noted that the laws of similarity had been determined and verified across a wide range of conditions likely to affect an estuary. In the final two years he had studied the effects of over 1,200,000 tides – representing the experience of 2000 years. The conditions for similarity had been verified for models with horizontal scales between 1:10,000 and 1:40,000, and a vertical exaggeration of between 60 and 100. He concluded, finally, that 'It thus appears that this system of investigation has been tested over a great portion of the ground it is likely to cover, and that most of the difficulties that are likely to occur have been met and the necessary precautions found. It would seem, therefore, by carefully observing these precautions, the method may now be applied with confidence to practical problems' [18].

Reyonlds' contribution to free-surface water modelling was definitive and paved the way for others to have confidence in using a reliable experimental technique.

23.2.3 Vernon Harcourt

Unlike Reynolds, Leveson Vemon-Harcourt (1839-1907) was a consulting civil engineer and keen to apply the technique developed by Reynolds to live problems. His first opportunity seems to have been to follow up on an interest that had been shown in Reynolds' experimental techniques in Manchester by some French engineers responsible for managing the River Seine between Rouen and the English Channel. Vernon-Harcourt undertook a study of this estuary in order to compare a number of alternative strategies that were being proposed for maintaining the navigable channel. Works to train the channel had begun in 1848, but were stopped in 1870 when it was found that the channel was becoming badly silted up. Vernon-Harcourt knew of fourteen schemes that had been proposed for resolving the problem and their great variety illustrated 'very forcibly the great uncertainty which exists, even in a special case where the conditions have been long studied, as to the principles which should be followed in designing training works. It is evident that no reasoning from analogy could prevail amongst such very conflicting views' [19]. He built a model of the estuary nearly 3 metres long in his London office, with a horizontal scale of 1:40,000 and vertical scale of 1:400, giving a model tidal range of just 18 mm. He began by using silver sand in his model, but found that it became covered with a film that inhibited its mobility. In seeking an alternative he tried powdered pumice, flower of sulphur, pounded coke, violet-powder, fuller's earth, lupin seed, the grains of coffee grounds, fine sawdust from boxwood and lignum vitae, and powder from Bath brick. All were found to be unsuitable, being too course or dirty or sticky. Finally, in July 1887, he found a suitable fine-grain sand, with a small admixture of peat, on Chobham Common, belonging to the Bagshot beds.

The model was made to correspond to detailed survey results from 1834 and then the model set in action for a period equivalent to the intervening 53 years. The resulting shape of the model estuary bed was found to correspond closely with the results of a more recent survey. The second stage was to introduce into the model some tin sheets to represent the training walls that had been constructed and the model was worked for a further period to form the accretion that had actually occurred in the estuary. This stage occupied Vernon-Harcourt and his assistant during all his spare time for eighteen months and provided the foundations for the third stage – to model the effects of introducing more training walls, corresponding to what he considered to be five of the most promising of the various alternatives that had been proposed (Figure 23.7). Two of the arrangements of training walls were tested over 6000 tides and the other three

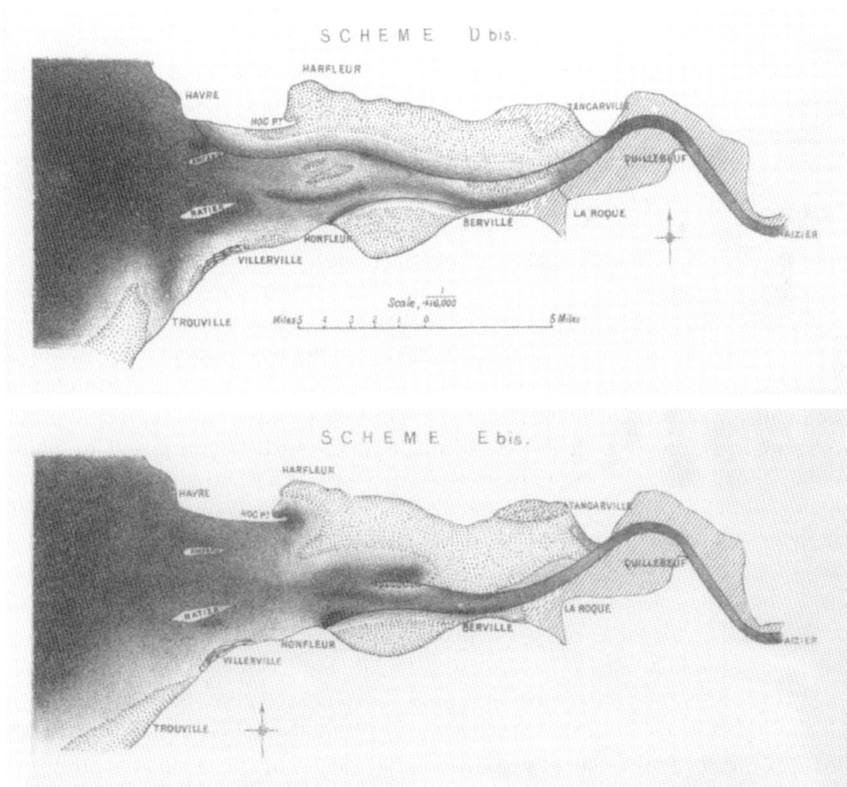

Figure 23.7 Model results by Vernon-Harcourt of two alternative arrangements of training walls aiming to create a permanent navigable channel in the estuary of the River Seine, France, 1888. (Source: [19])

were modified after 3,500-5,000 tides to make changes intended to avoid excessive accretion. Vernon-Harcourt presented the results to the French authorities and they began their own model studies before carrying out the chosen intervention. The current course of the river, by the way, is to the south of the estuary and the entire north western part is now occupied by a huge dock development.

Vernon-Harcourt went on to carry out a study of the Mersey estuary, following up on Reynolds' work, to determine how best to link the Manchester Ship Canal to the main navigation channel on the other side of the estuary. He made a working model of the estuary in his London office, with a horizontal scale of 1:30,000 and a vertical scale of 1:500 with a bed formed of fine Bagshot sand. The experiments were directed to the solution of two problems: the influence of training walls in the wide upper estuary on the channel below Liverpool and across the bar; and the effects of training works in the lower estuary on the channel across the bar. His test demonstrated that 'training walls in the upper estuary would be injurious, owing to the resulting accretion, [while] training walls in the lower estuary would improve the depth of the outlet channel' and that 'such training walls, combined with dredging, offer the best prospect of forming a direct stable, and deepened channel across the bar' [20].

Like Reynolds, Vernon-Harcourt made a final positive comment about the usefulness of this type of tidal model: 'If I succeed in demonstrating with the model that the originally existing conditions can be reproduced typically; and if, moreover, by placing regulating works in the model, the same changes can be reproduced that were brought about by the training works actually built, then I am sure that I can take the third, most important step, namely, of investigating, with every promise of success, the probable effect of the projects that have been proposed for extending the training dikes towards the mouth of the river' [21].

23.3 The first hydraulics laboratories 1900-1930

23.3.1 Hubert Engels

The first hydraulics laboratory dedicated to free-surface hydraulic modelling was opened at the *Technische Hochschule* in Dresden in 1898. It was the brainchild of the hydraulic engineer Hubert Heinrich Engels (1854-1945) and has led to him sometimes being called the founder of modern hydraulics research [22]. From 1877 Engels had worked as a site engineer for several harbour, river and canal projects and finally for a bridge over the Nordelbe at Hamburg. This very practical experience moulded his approach to hydraulics during his academic career that began in 1887 at the *Technische Hochschule* in Braunschweig and continued when he moved to Dresden in 1890. At Dresden, Engels had access to a small hydraulics laboratory that had been built by the professor of thermodynamics, Gustav Zeuner (1828-1907). The apparatus was contained in a zinc tray 6 m long by

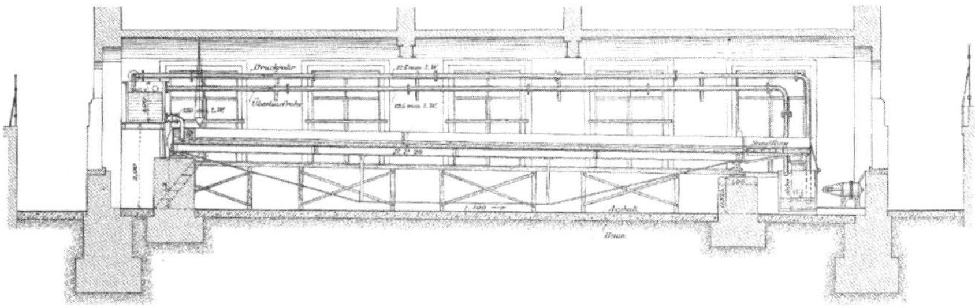

Figure 23.8 Tiltable channel at the river engineering laboratory at Dresden-Übigau experimental station, 1898. (Source: [23])

40 cm wide and 10 cm deep. It could have a bed of sand but its flow capacity was just 3 litres per second. In 1893, Engels used this apparatus for studying the protection of bridge piers against scouring, the first time such tests had been made at small scale. An important aim of the tests was to prove the validity of drawing useful conclusions from such small-scale tests; he also used them to supplement his lectures on hydraulics. Engels raised his confidence in model tests when determining the resistance to movement of model ships at the 63 x 5 m towing tank at the ship-building experimental station at Dresden-Übigau. He applied successfully for funding to construct a river-engineering laboratory and it opened in 1898. Limited by the size of the basement room available, the tiltable channel was 13.4 x 2 m and 40 cm deep (Figure 23.8). It had a capacity of 30 litres per second [23].

Engels undertook many fruitful studies of the effects of constructing structures such as groynes, spur dykes and bridge piers in rivers. He also studied the creation of sandbanks in straight and meandering channels and used sands of two different colours to demonstrate the movement of the sand [24] (Figure 23.9). While many engineers found his work useful and reliable, he found others who were highly sceptical about the validity of using small-scale models, sometimes based on curious arguments – a vet would hardly be entrusted to operate on an elephant if he had gained his knowledge of anatomy from a mouse.

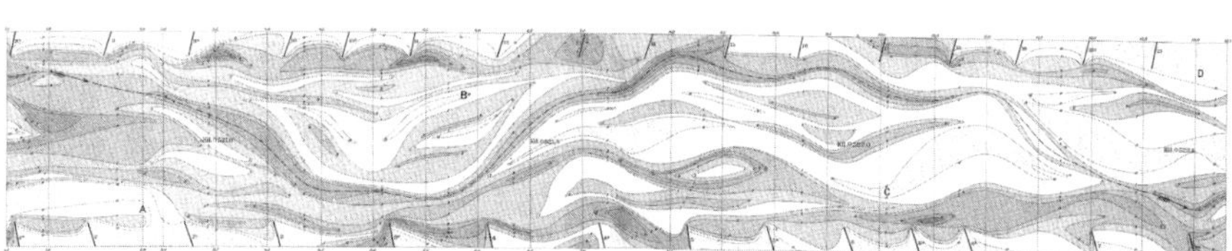

Figure 23.9 1:161 model test of a two-kilometre section of the river Elbe with spur dykes, using sand with an average diameter of 0.44 mm (compared to 0.78 mm in the actual river). (Source: [23])

The size of this first facility was soon too small for new engineering problems for which solutions were sought, for example the hydraulics of dams, sluices and reservoirs. A new 'river engineering laboratory', later named after Engels, was installed in the basement (63 x 9.9 x 3.3 m) of the new civil engineering building at Dresden and opened in 1913. It had a capacity of up to 200 litres per second.

23.3.2 Theodore Rehbock

Theodore Rehbock (1864-1950) was appointed as chair of hydraulics at the *Technische Hochschule* in Karlsruhe in 1899 after several years working on major bridge and hydraulics projects. Within two years he had established a large river-hydraulics laboratory, initially with a single flume 17.8 long by 2 m wide and 40 cm deep with a capacity of 60 litres per second. A second flume was also constructed, with glass sides to enable the water to be studied, for example, as it flowed over a weir or sluice [25, 26, 27]. These facilities were immediately used to study a large range of phenomena including: changes to river beds due to groynes and jetties, the effect on river beds of bridge piers and abutments, the silting up of harbour entrances, water flow over weirs, and erosion by water from spillways (Figures 23.10 and 23.11).

Most of the early work was intended mainly to develop this new modelling technique, understand the new test facility and to gain a good understanding of the behaviour of river beds, rather than undertaking studies for particular projects. This changed in around 1909 when Rehbock began to use his facilities for various types of study at the design stage of proposed hydraulic works [28] (Figures 23.12a and 23.12b). In 1925 a new hydraulic laboratory was built at Karlsruhe with a main experimenting room 74 m long and 11.5 m wide, with two smaller rooms

Figure 23.10 The river-hydraulics laboratory at the *Technische Hochschule* in Karlsruhe built by Theodore Rehbock and opened in 1901. (Source: [26] Plate 20)

Figure 23.11 Apparatus for drawing contour lines positioned over a river bed created by the artificial river. (Source: [27] Plate 33)

on upper floors. A drawing of the layout in 1927 shows the layout of 36 experimental models (Figure 23.13). At the time, this was probably the largest hydraulic testing laboratory in the world. Rehbock continued working until in his seventies on major projects such as the exit lock to the Zuiderzee in the Netherlands and the remodelling of spillways at the Ricobayo Dam near Zamora in western Spain to reduce erosion. In 1935, at the age of 71, he realised his dream of creating an international organisation for research in hydraulics, and was one of the founder members of the International Association of Hydraulic Research (IAHR) now named the International Association for Hydro-Environmental Engineering and Research.

23.3.3 Model testing in the 1920s

Taking their lead from the laboratories established by Engels and Rehbock, many other universities in Europe established new hydraulics laboratories. In many cases these built on the work of hydraulics engineers, both in industry and the academic world, who had been conducting various experiments on the flow of water for many decades, especially in connection with canals, rivers, weirs, locks, dams, pumps, pipes and turbines. The laboratories of Engels and Rehbock particularly impressed an American hydraulics engineer, John Freeman (1855-1932), who undertook a study tour of European hydraulics laboratories in 1924. He proposed to some engineers he had met that a book should be produced, giving details of the many facilities he had visited or heard about. The first version of the book, in German, was published in 1926 [29] and an expanded English version appeared in 1929 [30]. The result is one of the most

Nº AND DATE	DESIGNATION OF EXPERIMENTS	CONSTRUCTION LABORATORY
1 1913-1914	Flood water structure for Pegnitz River across the city of Nürnberg.	Experiments were made with models of the intake structure as follows: 9 models at 1/100 scale, 2 at 1/50 scale, 1 complete model at 1/100 and 1 half model at 1/50 with glass walls. Also 3 combination models of the inlet and outlet works at 1/50.
2 1914-1921	Experiments on the effect of dock walls along the Main River at Würzburg on the discharge and on the condition of the bed.	Model of Main River with vertical scale exaggerated 5 times. Horizontal scale 1/500. Vertical scale 1/100. Studies of eddies and directions of flow in the unobstructed channel and after its having been made narrower. Accurate determination of the longitudinal river profile.
3 1916-1917	Devising a suitable form of overfall weir for the River Sihl at Zürich to serve as a tunnel for the left-shore Zürich rail-road.	Models Studied 1.- Official design. Overfall weir with steep drop-off and step 2.- Overfall weir with sloping deck 3.- Three designs, I, II, and III, with sloping decks and depressed tumble-bays. The designs show various forms of river walls.
4 1919	Experiments for designing the intake structure for the feeder tunnels of the Murg development at Raumünzach dam.	Plaster-of-Paris model of intake structure set in a wooden box. Glass discharge shaft for making observations on entrained air.
5 1920	Experiments on the most suitable manner of rebuilding the dam of the Wolfshek Mfg. Plant in the Murg valley, which had been damaged by a flood wave.	Models Studied 1.- Original form of dam. 2.- Dam profile having a crest circular in outline with radius = 2.3 ft. 3.- Dam profile conforming to the lower surface of a freely aerated nappe.

Figure 23.12a Experiments carried out at the old river-hydraulics laboratory of the Karlsruhe Enginering College 1901-1921: Summary of main experiments carried out for projects. (Source: [28] Plate II)

AND DESCRIPTION OF APPLIANCES USED	RESULTS OF EXPERIMENTS
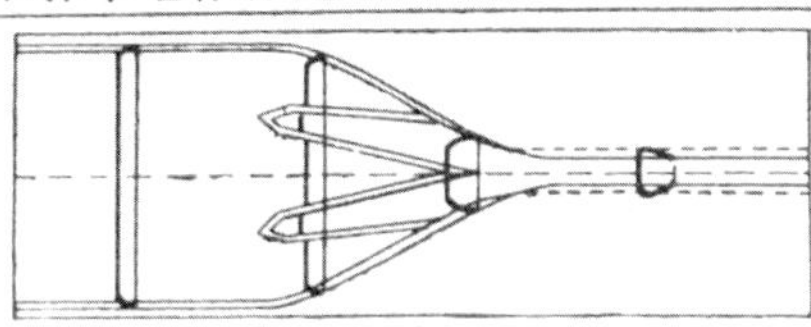 FORM OF INTAKE FINALLY ADOPTED	An intake structure was evolved that would take care of discharge velocities as high as 17.4 ft. per sec. without giving rise to waves. Practically complete conversion of kinetic energy into potential energy was accomplished by raising the 1¼ mile canal connecting with the tunnel by about 1 meter (3.28 ft.)
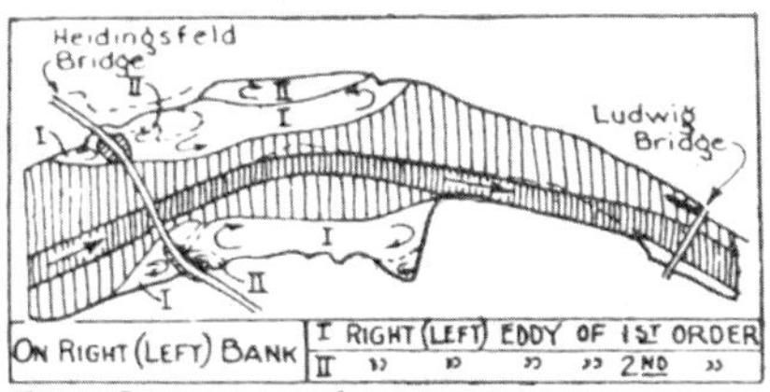 FLOW DIAGRAM OF IMPROVED RIVER-BED	Existence shown of large shore eddies in the unobstructed river. Backwater above bridge was studied showing that because of oblique currents the bridge opening had practically no effect on the discharge. Final answer: The construction of dock walls will not be harmful to the river channel or bridge.
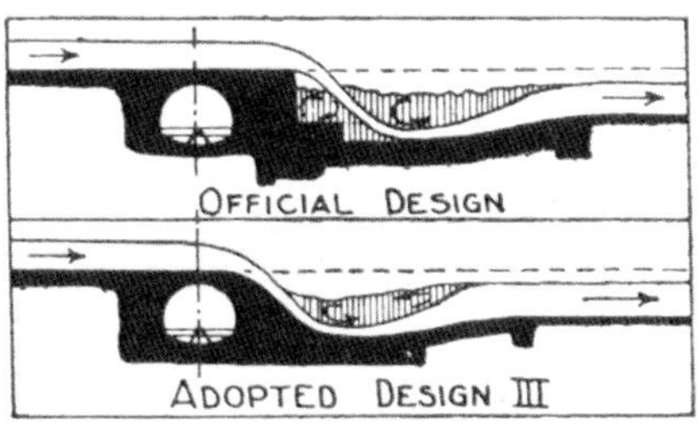	Determination of the most suitable form of overfall weir and tumble-bay with reference to stability, discharge capacity, and debris transported. Recording of flow lines. Study of the form and stableness of the surface backroll and of means for absorbing its energy.
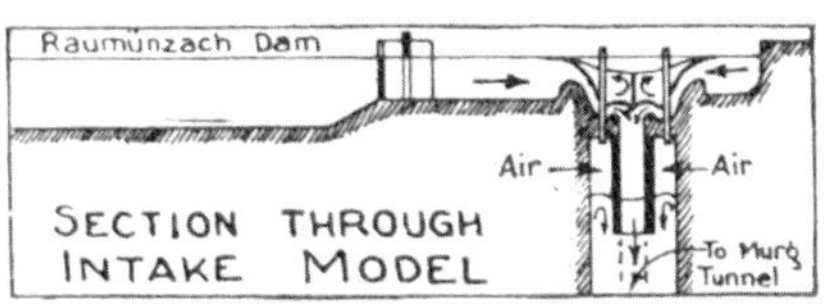	Determination of best type of intake shaft for preventing the entraining of air into the Murg tunnels. Installing a forebay which prevented air from being entrained proved to be the solution for insuring definite discharges.
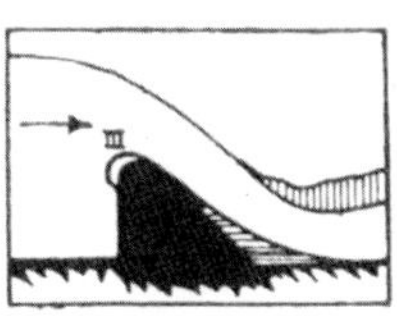 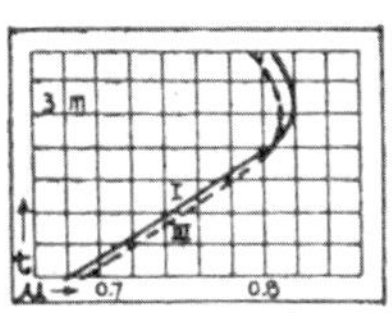 DAM TYPES I AND III VALUES OF μ FOR I AND III	Investigation of causes of damage to dam. Determination of a suitable profile for rebuilding the dam of reinforced concrete.

Figure 23.12a (*Continued*)

6 1921	Working out the design for the main spillway of the Friedland water power plant of the East Prussian Power Co with a maximum spillway capacity of 14,100 cu.ft. per sec.	Models Studied (1) Original design with 3 baffle sills and 3 aprons. After repeated experimentation: (2) Spillway without aprons but with dentated sill added in 1923. In all, twelve distinct models were experimented with.
7 1921	Working out the design for the waste weir on the river Dhünn for the proposed river harbor at Wiesdorf on Rhine below Cologne.	Models Studied (1) Proposed waste weir with extended apron. (2) Modification: waste weir with apron shortened (3) Modification: drop with: (a) clinging nappe; (b) free nappe. (4) Modification: inclined plane provided with obstructions.
8 1921	Calibration of a measuring gate in the tailrace of the low-head power development of the Murg project, which serves as a rating station for the water utilized in the Murg project	Models Studied (1) Exact half model of rating station at 1:10 reproduction ratio. Width of model = 1.8 ft. Middle pier for half its width fastened to one flume wall. Gate openings determined by means of a measuring wedge. (2) Part model at 1:5 ratio. A section of the overall width was placed between glass walls in order to observe flow phenomena under the gate.
9 1921	Study of flow in surge tank of Itter power development and determination of size of tumble-bay	Models Studied (1) Proposed arrangement: Length of overflow = 8.2 ft.; siphon 6.6 ft. wide and 1.6 ft. deep. Tumble-bay 49 ft. long by 10 ft. wide. (2) Modification: Length of overflow 8.2 ft.; siphon 4.1 ft. wide and 2.6 ft deep. Tumble-bay 26 ft. long and 6.5 ft. wide. (3) Proposed design: Length of overflow 19.7 ft.; no siphon; tumble-bay 26 ft. by 6.6 ft. wide.
10 1921	Experiments on the scouring action of the circular piers of a skew railroad bridge across the Wisent River for the Nürnberg railroad board.	Models Studied (1) Bridge piers circular in section set in natural river bed. (2) Bridge piers with elliptical noses set in natural river bed. (3) Massive pier footings circular in section. (4) Massive pier footings bounded by straight lines. (5) Massive pier footings elliptical in section.

Figure 23.12b (As Figure 21b) (Source: [28] Plate II cont.)

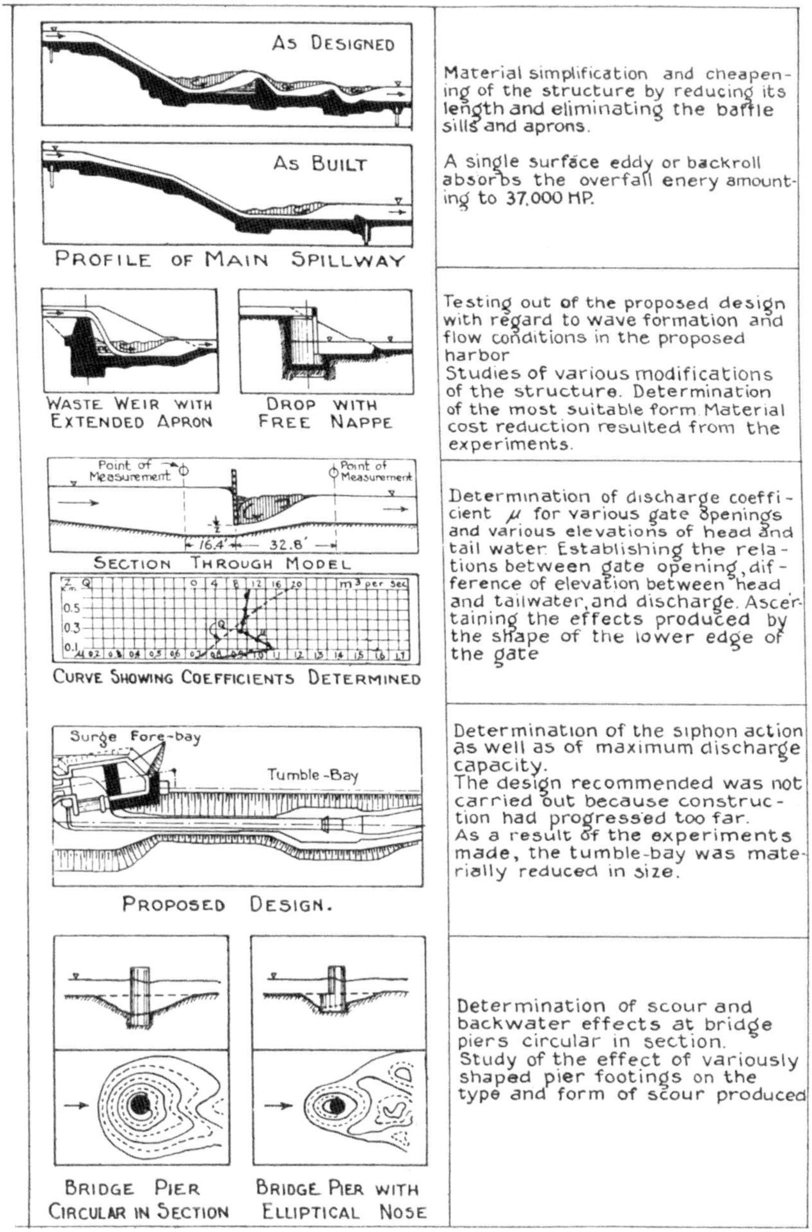

Figure 23.12b (*Continued*)

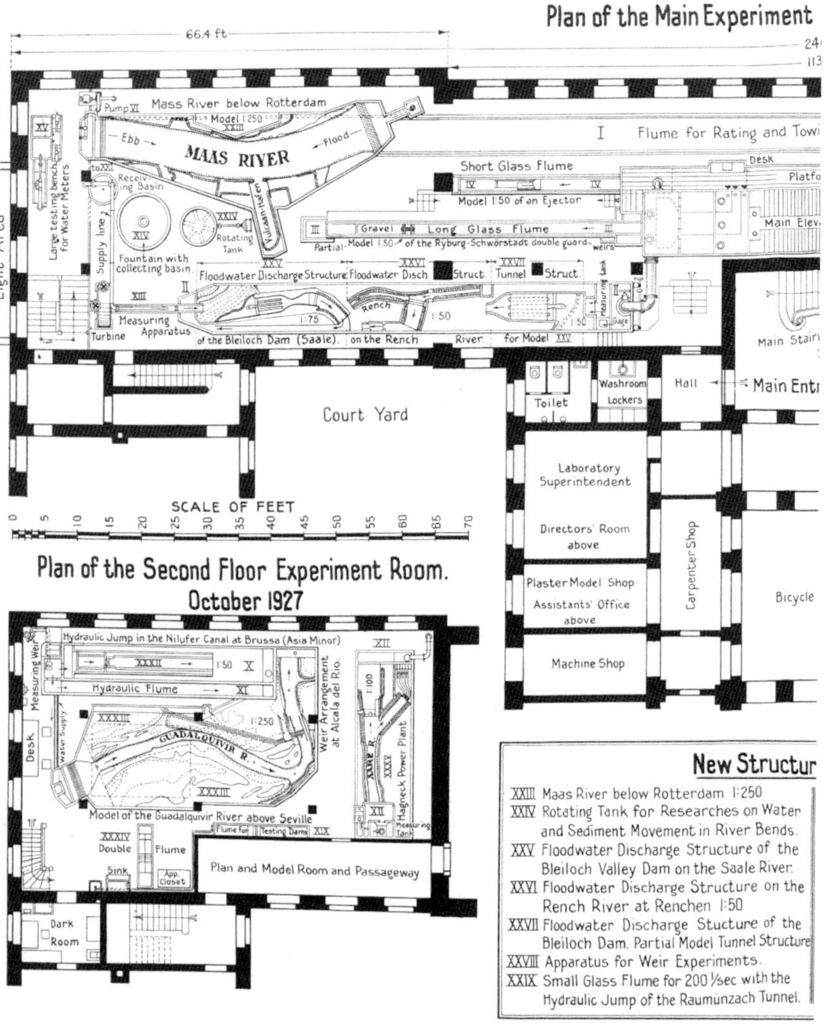

Figure 23.13 New hydraulic laboratory of the Technische Hochschule, Karlsruhe 1927. (Source: [28] Plate IV)

comprehensive studies of the research and model testing facilities in any branch of engineering. The English version runs to 868 A4 pages with nearly 1000 figures, most of which are photographs.

The book gives details of eighteen facilities, and the experiments undertaken in them, in cities in North and East Europe – Dresden, Karlsruhe, Braunschweig Danzig, Gratz, Darmstadt, St Petersburg, Wilhelmshaven, Berlin, Vienna, Munich, Budapest, Brünn in Czechoslovakia, Stockholm, Zurich, Hanover, Delft and Trondheim in Norway; five in Italy – Milan, Padua, Stra near Venice, Pisa and Rome; and two in France – Bauvert near Grenoble and Toulouse. John Freeman also gave an overview of more than a dozen laboratories in the USA, some of which had been experimenting since the 1840s, including Lowell

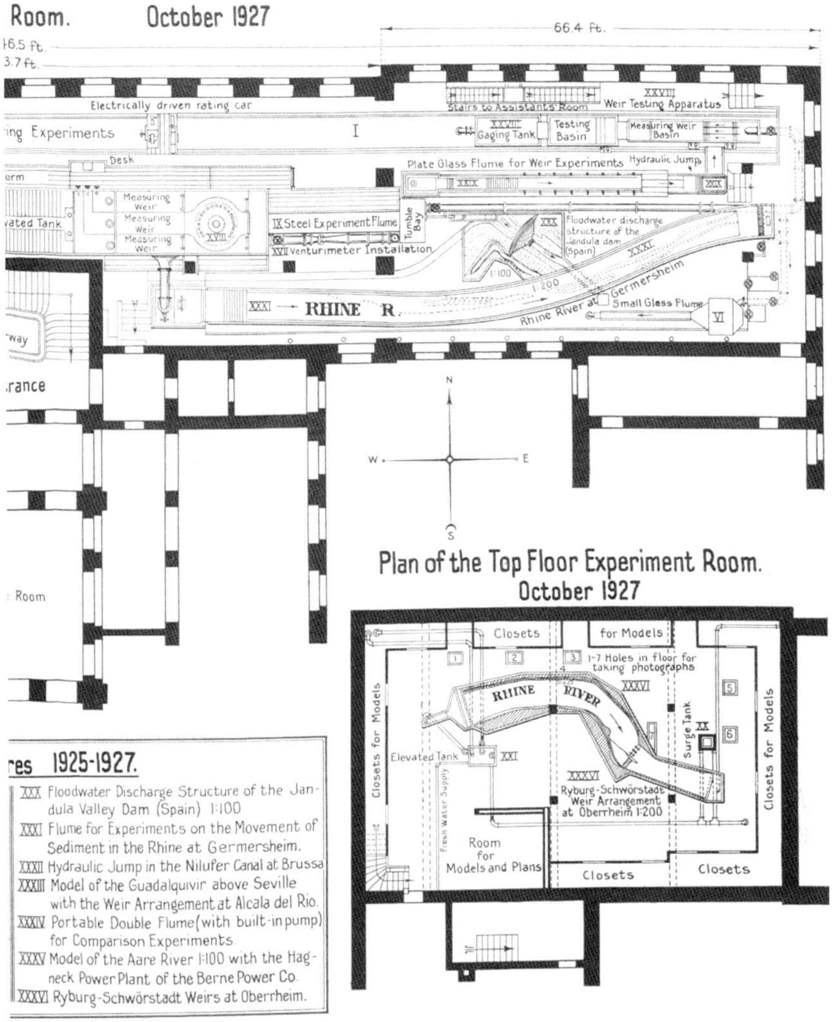

Figure 23.13 *(Continued)*

Massachusetts, Boston Water Works, Cornell University, the State University of Iowa, Worcester Polytechnic Institute and Birmingham Alabama.

The book gives an excellent state-of-the-art review of hydraulic research in the 1920s embracing moveable bed models as well as more traditional models for determining the flow rates of weirs, syphons, spillways and stilling basins. Hydraulic modelling had reached maturity.

Many engineering schools had small fluids laboratories for teaching purposes and these could be adapted to conduct experiments for a live-project. This was the case for the Giza School of Engineering in Cairo and the construction of the Aswan dam, some 900 km to the south. When it opened in 1902, the old Aswan Dam on the River Nile in Egypt was the largest masonry dam ever constructed and retained one of the greatest rivers in the world. The original dam was 22 m high and was increased to 27 m in 1909-1912 and new, larger sluices were needed. There was no precedent for building sluices to deal with such large quantities of water, and little hydraulic theory of use. The use of models would have been familiar to many who had worked on the project from their use by Benjamin Baker (see Section 7.3), but in the middle of Egypt, the engineers Henry Hurst (1880-1978) and David Watt had no access to any papers on the use of models. Nevertheless, they embarked on a series of model tests with the help of the School of Engineering at Giza to determine the best shape for the inlet to the sluices to achieve maximum discharge. After the early tests, one of the engineers return to Britain on leave and took the opportunity to find some literature on model testing, both in aerodynamics and Reynolds' work in Manchester, gathering some fifteen papers to help them plan and carry out the remainder of their model tests [31].

The school's experimental flow channel was modified to accommodate the models of the sluice. Six models were made, at scales between 1:67 and 1:33. The first (1:67) was of oiled and varnished pine, with a sluice gate of brass, but swelling of the wood caused difficulties. The second (1:33) was made of mahogany, painted with oil paint, with a channel cross-section of 10.5 x 6 cm. Generally the model was made to an accuracy of a few tenths of a millimetre, although some swelling of the wood occurred, and a 1% correction factor was introduced. The third model (1:50) was made with a brass inlet to an accuracy of 0.1 mm. The fourth model had a longer section made with brass plate, and the fifth was made entirely of brass with a wooden frame to represent the dam wall. The final model (1:50) was also made of brass with a cast-iron plate to represent the upstream face of the dam. Altogether about 1500 measurements of discharge flow rate were made, for different heads and with the sluice open by different degrees. For a geometric scale of n, the theoretical scale factor for flow rate was $n^{5/2}$, and this was born out by the experiments over a wide range of conditions. A number of the measurements were compared with measurements of discharge in the actual sluice in the dam to verify and calibrate the measurements made on the models. The agreement was generally within 1%. The authors concluded that models could be used to determine the discharge rates for large sluices with an average accuracy as good as obtained using a current-meter to measure full-scale rates, and that they provided a cost-effective way of studying many hydraulic problems at a manageable scale, echoing Reynolds' comment that it would be 'madness to neglect [model tests] before entering upon any costly undertaking' (Section 23.2.2).

There were, of course, still many engineers working on live projects who had to improvise, making models on site to test their design ideas and to corroborate their theoretical calculations, especially when far from the universities and polytechnics of Europe or North America. For example, in 1918, Henry Dare (1867-1949) was resident engineer during the construction of the Burrunjuck Dam near Canberra in Australia. The narrow gorge meant that it was necessary to build the spillways perpendicular to the line of the dam and the engineers were not able to find any design guidance for such a situation. The engineers had built a 1:32 scale model from which some measurements of discharge rates were obtained, but the model was destroyed by a flood before further data could be gathered. During construction, however, the river flooded to a level 60%

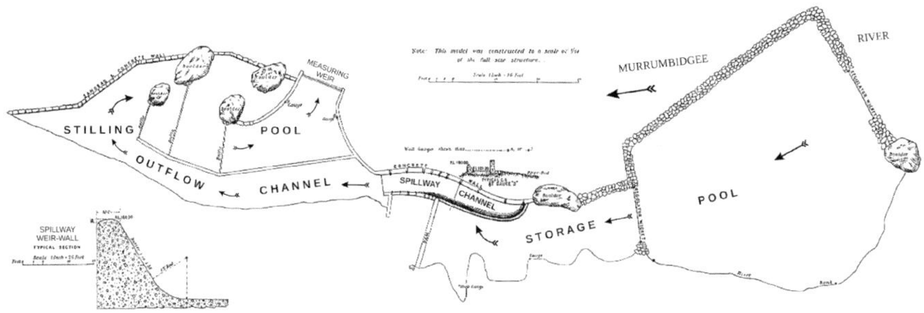

Figure 23.14 1:24 scale model of the spillway for the Burrunjuck Dam, Australia, 1918. (Source: [32] re-labelled by Bill Addis)

higher than previously recorded and it was realised the spillway would need to be redesigned [32]. Dare built a new model at a scale of 1:24 alongside the main river about 200 metres downstream of the dam (Figure 23.14). It comprised miniature versions of the dam wall, the spillway weir-wall and channel, a storage pool, wicket regulator, a stilling pool and a measuring weir. The total length of the model was about 60 m. The main aims of the tests were to determine the flow conditions in the spillway channels and to determine how the existing designs for the north and south spillways could be improved, and how the spillways would perform with the equivalent of a 3 m surcharge in the reservoir. Even before measurements were taken it was found that the storage pool was too small and it was enlarged. The discharge rates were found to be too low because of the great turbulence caused by the water flowing over the spillway weir perpendicular to the main direction of the channel. The profile and line of the spillway weir and the line of the training wall forming the opposite wall of the channel were modified to generate a smoother flow transition.

23.3.4 Hydraulics Research Laboratory, Poona, India, 1916

In the early-twentieth century, India was faced with diverse problems of irrigation, ranging from the alluvial rivers in the northwest of the country to the impermeable black cotton soils overlying disintegrated basalt in the Deccan Plateau to the south and east. Attempts to migrate the experience with irrigation canal systems in the north to the south led to major difficulties as the agricultural crops, the pattern of rainfall and the natural drainage were entirely different. The land quality soon deteriorated due to waterlogging and efflorescence of sodium salts. In 1916 the regional government in Bombay (Mumbai) created the Special Irrigation Cell in Poona (Pune) to research into more-effective irrigation methods to suit the local geology, climate and agriculture. The scope of research soon grew to include the hydraulics of rivers and, in 1919, the Irish engineer Claude Cavendish Inglis (1883-1974), then serving in India, was appointed to undertake research in hydraulics and flooding [33, 34]. In 1925 the research centre moved to larger premises at Khadakwalsa Dam near Poona, and Inglis was appointed its director. Under his directorship it was so successful that similar laboratories were soon set up in the other states in India. In 1936 it was taken over by the

Government of India and its name was changed to the Central Hydrodynamic Research Station. By 1938 the laboratory had undertaken over a hundred major model studies and was one of the first hydraulics research establishments, if not the first, to use large working models with erodible boundaries. Inglis' experience led him to identify eight markedly different types of hydraulic model:

1. Models which are geometrically similar in shape to the prototype and give geometrically similar results
2. Models which are geometrically similar in shape; but do not give geometrically similar results
3. Regime models in which conditions of flow are maintained constant, with complete freedom as regards silting and scouring.
4. Combined erodible channel and rigid models
5. Rigid and semi-rigid, vertically-exaggerated models in which conditions are imposed, e g tidal models
6. Models with mobile protection – falling aprons, spurs, protection around piers, etc
7. Meandering river models, and
8. Combined rigid, mobile and meandering river models.

He also concluded that that much larger models than had generally been used in hydraulic laboratories in the past were required to give accurate results, especially in the case of erodible models, and that most flow problems – even investigations into the movement of silt – which had hitherto generally been investigated in laboratories in parallel sided flumes, as though they were 2-dimensional questions, are essentially 3-dimensional problems [35, 36]. Among many examples, he described a 52-m long model of the River Indus at the Sukkur Barrage Canals to determine whether still-pond regulation or open-flow regulation would be more effective at keeping the canals free from silt; his conclusion that still-pond regulation would be more effective was validated when the canals opened. He also predicted, however, that a silt bank would build up on one bank unless two vanes were constructed upstream to divert the water. This recommendation was not implemented, the authorities preferring to 'wait and see'. Several years later such a silk bank did form during flooding and one of the canals silted up. Since the vane solution could no longer be implemented, new studies had to be conducted to find what would be a much more expensive solution [36]. Inglis remained director of the Research Station until 1945, and when he returned to Britain he was invited to set up a similar research laboratory. This was first located at the National Physical Laboratory in Teddington, near London and opened in 1947. After four years, it relocated to its own premises in Wallingford near Oxford where it is today known as HR Wallingford (see Section 23.6.2).

23.4 Hydraulic modelling in the USA in the 1930s

23.4.1 Iowa Institute of Hydraulic Research, 1920-

Perhaps the foremost hydraulics research facility in the USA during the 1920s had been the Iowa Institute of Hydraulic Research in Iowa City, under the leadership of Floyd Nagler (1892-1933) between 1920 and 1933 [37]. The size and number of model tests were boosted in 1932 when its new building provided fifty times more floor area than the previous one. The University dam provided river water under a three-metre head in two large river channels – one 3 m wide and 90 m long, the other 4.8 m wide and 30 m long, for testing models and structures requiring high flows. The new laboratory also provided two steel flumes for studying river models, one tiltable and 3 m wide, the other fixed and 7.5 m wide, three glass-sided flumes and six timber flumes of various lengths [38]. The type of projects being studied reflected the massive building programme in 1930s-America related to flood management and the construction of dams. Nagler was particularly instrumental in promoting the use of physical hydraulic models to design the Mississippi River locks and dams, such as the Hastings Dam on the Mississippi River [39]. The Iowa Hydraulics Laboratory became a major modelling site for federal projects in the Eastern United States and Nagler's successors, including Hunter Rouse (1906-1996) [6], have maintained the international pre-eminence of the Iowa Institute of Hydraulic Research (IIHR) to this day.

23.4.2 The U.S. Army Waterways Experiment Station, 1929-1998

Although there had existed many hydraulic laboratories in the USA in universities and water supply companies since the mid-nineteenth century [40, 41], after visiting the many European hydraulic research laboratories, John Freeman returned to the USA with a zealous determination to change things. He persuaded the U.S. Army corps of engineers of the national significance of hydraulic engineering, especially for flood control and coastline erosion. The Waterways Experiment Station (WES) was founded in 1929 and initially occupied an area of about 60 hectares in Vicksburgh, Mississippi which incorporated a 16-hectare supply reservoir. At that time physical modelling was not widely accepted:

> 'In 1930, […] most hydraulic engineers thought of models as mere toys for the youngsters of the profession. A very large group was openly sceptical and many voiced their disdain. This attitude was shared by many officers and civilian employees of the Corps, and a few held ill-concealed resentment of the new-fangled approach. Some high officers actually feared that model testing imposed a threat to sound engineering based on theory and computations. … As a result of proof provided by the station as to the validity of models, engineers do not build large or expensive hydraulic structures today until they are first tested on a small scale. This is a complete revolution in engineering concepts' [42].

The first project undertaken at WES was to build a model of the Illinois River to establish the limit of backwater from the Mississippi (Figure 23.15). But results were needed in thirty days. Templates were cut from sheet steel and fitted to the ground which was carved out with a grapefruit knife to form a miniature river channel. A weir box was constructed, and the weir calibrated. However, the engineers had only one chance at getting the test right – the soil in the grounds

Figure 23.15 The first model test at the U.S. Army Waterways Experiment Station in 1931. (Source: [43] p. 1)

of the WES was a loess which could be easily carved out but, once disturbed, the soil became dust and could not be carved again.

Soon a fixed-bed model of about 160 km of the Mississippi River was proposed. Initially a scale of 1:2000 was proposed, to keep the size (80 m) and cost down, although this was much smaller than the scales used in Europe, typically about 1:300. In fact, the model was finally built at 1:2400 because graph paper was available only with subdivisions in multiples of 12 – no metric graph paper could be found. A vertical distortion of 1:20 was used and the entire model was made in concrete (Figure 23.16). At the time of its completion, it was the largest flood model in the world:

Figure 23.16 1:2400 scale model of the Mississippi River, 1931-35. This section of the model was used to prevent capture of the Mississippi River by the Atchafalaya River. (Source: [43] p. 15)

> 'Everything was there in miniature – levees, tributaries, bridges, swamps, and forests, even the willow thickets. The model required intensive study and research and when completed, every sandbar, every miniature clump of willows and every prominent terrain feature had its place and its reason for being there. Forests and thickets were simulated by galvanized screen, turned upright in an exact pattern determined from aerial photographs. It may have looked like a toy to visitors, but the model was a valuable tool for hydraulic engineers. The flood of 1927 was duplicated and, by taking the river readings for any given day in the past, the model could be activated to represent that particular time. A crew of eighteen men was needed to operate the model and read its 210 gauges. A day's rise and fall on the river could be reproduced in 5 ½ minutes, and the 1927 flood that sent the river rampaging out of its banks for five months could be recreated in 14 hours. The practical purpose of the model was to test the effects of possible improvements along the river. Would a levee of a certain height be sufficient to protect an area from a flood such as occurred in 1927, assuming that all levees held? If a cutoff were made in a certain bend, what would be the effects on flood heights and flood duration? Would the backwaters be raised or lowered? It would be much cheaper to make an experimental cutoff on the model for a few dollars than to spend a half million on the prototype without knowing what the results would be' [43].

By the middle of 1933 some sixty-five fixed-bed models had been built for various projects, both outdoors and under cover. Techniques for conducting shoaling studies in fixed-bed models through the use of lightweight granular materials injected into the model, were developed in about 1937. From 1936 moveable-bed models were being made for navigation studies of rivers, using crushed coal for the river bed.

The first hydraulic model at the WES that simulated tides and tidal currents was constructed in 1933. The rise and fall of the tide and the reversing tidal currents were reproduced approximately by manual operation of valves installed in pumped inflow and gravity outflow lines connected to the two ends of the model. In 1937, mechanized submerged weirs were developed for control of tidal elevations and the directions and velocities of tidal currents; these weirs were programmed by an electromechanical control system that significantly improved the accuracy with which tidal phenomena could be reproduced. In 1939, the mechanized weirs were replaced by mechanized rising-stem valves resulting in simplified control of tidal phenomena and a more faithful reproduction of real tides. The significance of density currents on hydraulic and shoaling phenomena in estuaries was recognized in the late 1930s, and techniques for operating estuary models with both salt and fresh water were developed in the early 1940s [44]. By the end of 1938 nearly 250 model studies of all types had been completed.

In the early days of the war, the WES was asked by the US Navy to model the harbour of New York. The model study was undertaken to find a more accessible route to the Brooklyn Navy Yard. Only two routes provided access to/from the sea: one was usable at all times, but the second route, through the East River and Long Island Sound, was accessible to large ships only for short periods during

Figure 23.17 1:1000 scale model of New York harbour. Manhattan and Brooklyn Bridges can be seen in front of the person, and the Statue of Liberty at the bottom. (Source: [43] p. 36)

high tide. Military necessity required the availability of both routes and the WES studies resulted in recommendations that enabled it to be opened up for use at all times [45] (Figure 23.17). The WES continued its work under the same name until 1998 when it was incorporated into the U.S. Army Engineer Research and Development Center (ERDC).

23.4.3 The Beach Erosion Board, U.S. Army Corps of Engineers, 1930-1963

During the late 1920s, it was recognised, especially in the State of New Jersey, that the erosion of beaches by wave action needed to be better understood. There were three main reasons: the first was a growth in seashore recreation by the increasingly affluent residents of New York and Philadelphia, which made new demands on the nature of the shore and the increasing value of properties and land adjacent to the sea; the second was the need to explain why coastal erosion had been observed to be becoming more severe in recent years; and the third was the realisation that erosion was a problem that had to be tackled on a very large scale, by government bodies, not by the owners of individual properties.

The tool for investigating beach erosion was the wave tank, in which waves of different heights and frequencies could be generated and their action on model beaches could be observed, as well as the effectiveness of measures to mitigate the effects. The Beach Erosion Board (BEB) was formed in 1930 and its first laboratory wave tank was constructed in 1932 at Fort Humphreys (now Fort Belvoir), Virginia. It was 7.3 m long, 3.6 m wide and 45 cm deep. After only a few model studies it was realised that a larger installation was needed and in 1937 a tank 26 x 4.3 m with a depth of 1.2 m was built [46].

The work of the BEB at this time served both to inform the public authorities on how they might best invest money in safeguarding sections of the coastline and also to establish the scientific study of wave action. The BEB was active until 1963 when it was replaced by the Coastal Engineering Research Center (CERC) [47, 48].

23.4.4 Current Hydraulic Laboratory Research Reports, 1933-1942

In 1933 a series of reports was begun by the US Bureau of Standards in response to a proposal made at the meeting of the World Power Conference in Berlin in 1930. The proposal was made by Friedrich Schaffernak, a hydraulics engineering professor at the Technische Hochschule in Vienna that 'an organization be established for the purpose of interchanging, in the field of hydraulic engineering, the results of laboratory research and of observations of natural conditions corresponding thereto' [49]. In the first report, the laboratory facilities of thirty-one research organisations including universities, private research institutes and government establishments such as the WES were included, from which eleven reported their current research projects – a total of ninety-nine, of which the WES alone reported on forty-three, all of which were for real projects (rather than scientific research). The projects ranged from undergraduate and post-graduate projects to large-scale flooding studies, and included model studies relating to hydraulic machinery, and even the holding ability of anchors, as well as studies relating to dams, tidal estuaries, rivers and beaches.

The reports were generally published twice a year (three in 1933 and annually from 1938) providing an excellent picture of hydraulic research in the USA during the 1930s. From 1934 references to papers and projects in many European countries were also included.

23.5 Hydraulic modelling in other countries in the 1930s

Of course, research in countries apart from the USA continued to grow during the 1930s, but there is no need for duplication here. The remarkable achievements of the Central Hydrodynamic Research Station in Poona, under Claude Inglis, have already been described (see Section 23.3.3). Two further projects will illustrate typical developments in model testing during this period.

23.5.1 The Severn Estuary barrage

Probably the largest tidal-model experiments made in Britain before WWII were to investigate the potential for, and impact of a barrage across the Severn Estuary, where it is about 3.4 km wide, to generate tidal energy. The Severn Estuary has a very large tidal range and is, from a technical point of view, a natural candidate for a tidal energy scheme. However, the potential impact on about a hundred kilometres of coastline and the complex network of sandbanks would be enormous. The study, undertaken by A.H. Gibson (1878-1959), a former pupil of Osborne

Figure 23.18 The model of the Severn Estuary at the University of Manchester, 1927. (Image: Courtesy of University of Manchester)

Reynolds, was commissioned in 1926 to determine the effect of a barrage on siltation and sandbanks, on the tidal range, the currents above and below the barrage and, of course, the magnitude of the head of water to drive the turbines. A horizontal scale of 1:8500 and vertical scale of 1:100 were chosen, giving a model 13.5 m long, representing about 130 km of deep-water channel in the river and estuary, with a maximum tidal range of 12.5 cm in the model [50] (Figure 23.18).

The bed of the estuary was moulded in accordance with detailed surveys made in 1849. The model was then operated for a period corresponding to seventy-eight years to create the bed for the estuary in 1927, which was then compared with data from a survey undertaken in 1927. A good agreement was found, which effectively validated the model. The model barrage was then introduced and the model operated for a period of up to sixty years into the future, to determine the various effects of the barrage. It was found that the 1:100 vertical scale led to slopes that were too steep for the model sand and so the model was reconstructed at a scale of vertical scale of 1:200, and the seventy-eight-year period of tides was repeated. Great attention had to be paid to the introduction of silt from the five rivers feeding into the estuary, and the salinity of the water. Silt coagulates in sea water and settles more quickly than in fresh water. It was found that it would be necessary to add 227 kg of sea salt to the model each day to produce the right effect, which would be cumbersome and costly. It was eventually found that adding 1 kg per day of alum achieved the required properties of the model 'seawater'. Surveying the depths of the river bed also proved to be a challenge. Overhead photography and observing via a mirror at 45° both proved unsuitable because the boundaries between sand, mud and water were not clear. Finally, sounding rods were used to measure the depth of the bed below a frame covering the whole model. Each survey of the whole estuary took about two days.

Considerable efforts were taken to find the best sand for the model. Sand moves along a river bed in five different ways which depend on the size and weight of the grains as well as the water speed – sliding, rolling, jumping, in suspension and collective movement in sand-waves. Gibson studied the work of Reynolds and Vernon-Harcourt and carried out tests on ten different sands to find the ones which best achieved shapes corresponding to the surveyed contours of the bed. Three grades of pumice were also tested. Finally a single sand, with a mean grain diameter of 0.18 mm was chosen and used for all the tests.

The model was operated many times from the 1849 baseline situation, and up to sixty years after the 1927 date of the experiments, with accurate representation of spring and neap tides, with and without the barrage, with and without the introduction of additional silt, in order to raise confidence that the model was representing accurately the behaviour of the real estuary (Figure 23.19). The effect of water temperature in the model were also tested. The conclusions were that the introduction of the barrage would not have an adverse effect on navigation below the barrage and would improve navigation above the barrage, especially at low water. It was also found that the drainage of sewage into the river and estuary would not be affected adversely, and inundations due to floods in the River Severn would be reduced.

The second series of experiments beginning in 1929 looked at six designs for the barrage, with different alignments, different lengths of sluices, different numbers

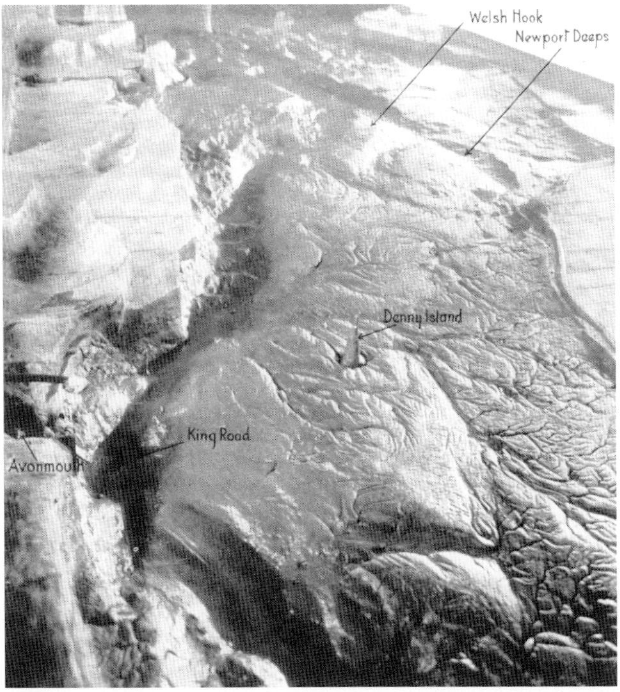

Figure 23.19 Model of the bed of the estuary after the equivalent of sixty years operation with the barrage. (Image: Courtesy of University of Manchester)

of turbines, and different heights of the sill of the sluice gates. Taken all together, the model studies for the barrage across the Severn Estuary was probably the most sophisticated study of free-surface water behaviour carried out up to that time. In the final design of the barrage, it was calculated that the energy that could be generated from the sixty-seven Kaplan turbines had a maximum output of 685 MW and an output of 3190 MW-hours per tide. Gibson went on to give the Vernon-Harcourt lecture at the Institution of Civil Engineers in 1936, giving an overview of his work on tidal and river models, including the River Severn and, fittingly, ended his paper citing Vernon-Harcourt's comment about model testing, that it would be 'madness to neglect [model testing] before entering upon any costly undertaking' [51] (see Section 23.2.3).

23.5.2 An underwater breakwater for the port of Leixões, Portugal

In 1936, the hydraulics laboratory at the Swiss Federal Institute of Technology at Lausanne was asked to study the effectiveness of a submerged breakwater that was planned to shelter the port of Leixões in northern Portugal. The aim of the study was to determine the best slope angles on the seaward and harbour sides of the breakwater, both in terms of the calming effect of the breakwater and its robustness under that action of Atlantic storms, at both low and high tides, and also the best form of construction combining concrete blocks and stone rubble [52]. The initial proposal was for a symmetrical section with both slopes at 5:4, made up of concrete blocks weighing 54 tonnes placed on a base of stones up to 4 tonnes on a core of stones up to 2 tonnes (Figure 23.20). Breaking waves of various heights and wavelengths were filmed to enable detailed study of the calming effect of the breakwater (Figures 23.21 and 23.22). Regarding the robustness of the breakwater against storm damage, tests were made with larger concrete blocks, and with modified use of the stone rubble beneath and between the concrete (Figure 23.23).

Based on the results obtained from the tests, a number of changes were made to the original design proposal, of which the most significant were: the layer of small stones up to 2 tonnes was omitted; the weight of the concrete blocks was increased from 54 to 78 tonnes; the slope angle was changed on the seaward face from 5:4 to 2:1, and on the harbour face from 5:4 to 3:2; the seaward toe berm was extended. (Figures 23.24 and 23.25).

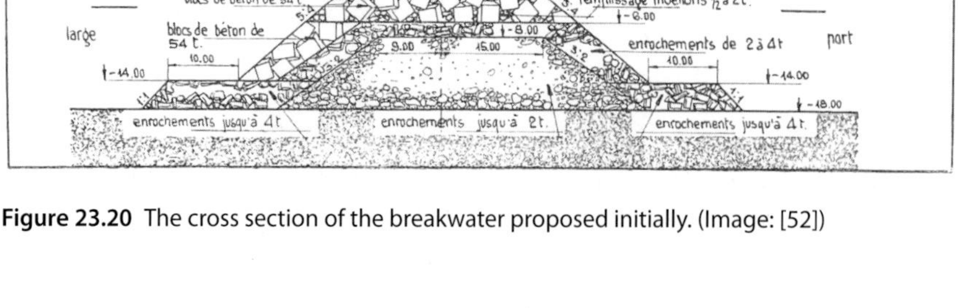

Figure 23.20 The cross section of the breakwater proposed initially. (Image: [52])

Figure 23.21 1:70 scale breakwater cross-section in position in the 15 m x 30 cm wave tank. (Image: [52])

Figure 23.22 A photo showing the calming effect of the breakwater, for a wave height of 6 m and wavelength of 120 m. (Image: [52])

23.6 Some post-war developments in free-surface water modelling

23.6.1 The BEB and WES post-war

After the war the activities of both the WES and BEB grew with increased vigour. At the BEB, the original interest in beach erosion widened considerably

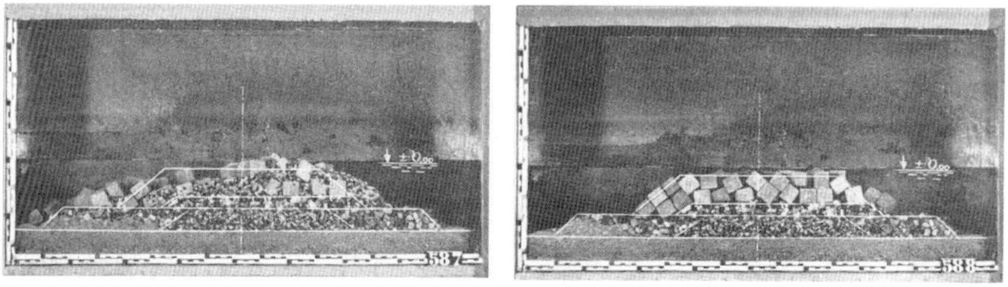

(a) (b)

Figure 23.23 Photos showing the damage caused by a 10 hour storm with a wave height of 9 m and wavelength of 140 m. (a) With the original construction. (b) Without the small stone rubble. (Images: [52])

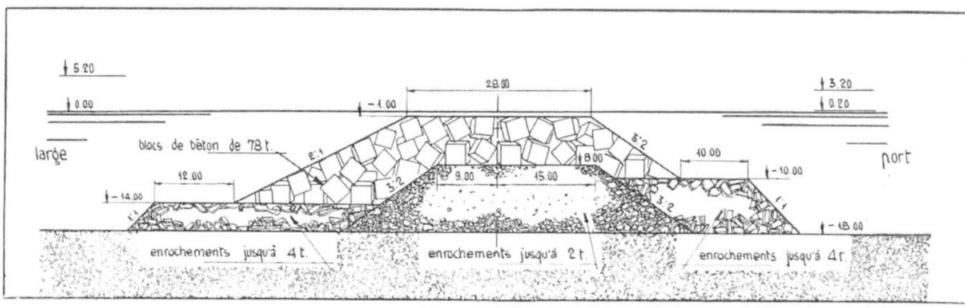

Figure 23.24 The modified cross section of the breakwater. (Image: [52])

Figure 23.25 Photo showing the damage caused to the modified design for the breakwater by a 12 hour storm with a wave height of 9 m and wavelength of 220 m. (Image: [52])

to include: waves in deep water; waves in shallow water; currents in shallow water; factors affecting supply and movement of beach material to the littoral zone; the significance of natural formations; beach processes; functional design and the effects of man-made structures. To support these activities three new facilitates were constructed. In 1947 an indoor wave tank, 29 m by 45 cm wide and 36 cm deep, was opened. In 1953, an outdoor, shore-processes test basin, 92 m x 46 m and 90 cm deep, with ten moveable wave generators, opened for studies related to floating breakwaters and to beach processes, including the use of movable bed materials. And in 1955 a much larger, open-air wave tank, 194 metres long, 4.5 m wide and 6 m deep, was opened. It held 3,750 m^3 of water and 2 m high breakers could be generated. Early studies included beach erosion and the overtopping of levees [53]. In parallel with these developments, a new field of 'coastal engineering' was established, drawing academics and practising engineers together by means of an annual conference, first held in 1950 [54] and continuing to the present day, whose proceedings provide thorough coverage of the use of physical models.

In 1943 the WES began construction of what would become the largest and most complex model of a river system ever built – the Mississippi Basin Model (MBM). The Mississippi catchment area covers about one third of the area of mainland America, extending about 2,700 km west-east, and the same north-south. At a horizontal scale of 1:2000, the model would cover an area of about 85 hectares, extending about 1.35 x 1.35 km. At a vertical scale of 1:100, the height differential on the model was 15 m. The total length of rivers in the model was about 13 km [55, 56] (Figure 23.26). Construction of the model began in August 1943, using German prisoners of war to prepare the whole site ready for building the model itself. The number of prisoners employed as labourers, surveymen, draftsmen, calculators, engineering assistants, inspectors, mechanics, cooks and clerks, peaked at 1,800, many of whom remained until May 1946 when they were repatriated. The river channels and the overbank flood areas (the area used for testing) were moulded in concrete, and the remaining ground was stabilised by compression. Precast-concrete sections of the rivers were made off-site, brought in on trucks and set in their final location on 3 m long piles. Particular care had to be taken to giving the river channels and

Figure 23.26 The Mississippi Basin Model. At a horizontal scale of 1:2000 it covered an area of 85 hectares. View looking upstream on the Ohio River from Evansville, Indiana. Tennessee and Cumberland Rivers are in the foreground, showing the site of the Kentucky and Barkley Dams. The Tradewater and Green Rivers are shown in the centre. (Image: [57])

overbank flood areas appropriate surface roughness, sometimes by treating the surface of the concrete and sometimes by fixing layers of expanded metal. Instrumentation and control of the water during tests were major challenges. Initially about 600 people were needed to run a large test, but this was reduced to about sixty after the development and introduction of bespoke automated electrical systems for water control and logging the thousands of measurements taken across the huge site. The model was final completed in 1965, but tests using the partially-completed model began in 1949. Between 1951 and 1973, when the MBM was retired, it was used to undertake over seventy-five major studies of levees, floodways, cutoffs, reservoirs and other interrelated flood control mechanisms.

Fortunately the Mississippi section of the model was fully operational by 1952. In April of that year a major flood threatened to reach a crest discharge of almost twice the maximum flood of record at Sioux City, Iowa. The Missouri River Division (MRD) requested the WES to operate the MBM 24 hours a day to assist in predicting crest stages and discharges. During a critical 16-day period, WES and MRD personnel maintained almost constant contact by telephone. The model tests first indicated that certain levees were not high enough to contain the crest of the flood. The MRD evacuated the affected areas and raised the levees within five days of the WES prediction. Flood crests exceeded the original levels of the levees but were contained by the new additions, saving two cities from flooding. Many other remedial actions were taken based on predictions from this spectacular use of the model, and MRD claimed it was a prominent factor in the success of the flood fight which prevented damages estimated at $65 million. From 1959, when the model was substantially complete, it was used to replicate the historic floods of 1937, 1943, 1945 and 1952 and also three hypothetical floods caused by possible events in spring, summer and winter. Each flood was repeated six times for different scenarios and the results used to assess the operation of existing and projected reservoirs in flood management. The success of the MBM was well-publicised and, from 1964, it was opened to tourists, attracting some 5000 visitors a year and even today is well-featured on the Internet.

Another model built at the WES and popular with visitors was that of the Niagara Falls. It was built to determine what effect, if any, a huge hydraulic power-plant might have on the famous tourist attraction. The objective at the real falls was to generate nearly eight billion more kilowatt hours of electricity without sacrificing the falls' grandeur and flow of water that were defined in a treaty between Canada and the United States in 1950 [58]. The flow of the river is nearly constant throughout the year, at about 5,660 m^3/sec; the treaty stipulated a minimum flow over the Falls of 2,830 m^3/sec during the day, and 1,415 m^3/sec at night, and any excess over these quantities could be diverted to generate electricity. Water in the model flowed from the international boundary, over rock-studded rapids and dropped into a mist-shrouded chasm spanned by a tiny bridge – every significant feature of the Niagara River from Lake Erie to beyond the rapids was reproduced (Figure 23.27). Built to a horizontal scale of 1:360 and vertical scale of 1:60, the model compressed 41 km of the river's length into a bed that was 79 m long; the model covered about the area of a football field. The river bed was formed of precast concrete shells 5 cm thick, with some removable

Figure 23.27 Model of the Niagara Falls at the WES. Horizontal scale 1:360, vertical scale 1:60. The model covered about the area of a football field. (Image: [43] p. 72)

sections to allow alternative arrangements of the proposed hydroelectric scheme to be studied. The model tests studied the effects on the flow of water over the falls of different design options for the intakes for the turbines while maintaining the minimum flows stipulated in the treaty. The engineers even floated ice cubes across the cascades to reproduce winter conditions. The study was able to determine the effects on the Niagara River and Falls of the proposed additional diversions of water for power, and also the extent and nature of remedial works required for maintenance of adequate flow conditions in the river and at the Falls. This remedial work involved excavations in the cascades to redistribute water more evenly around the horseshoe to maintain the spectacle with the reduced flow rate (2,830 m^3/sec) during peak visitor times.

The movement of water in models was generally measured using small current meters but these do not give a good overview of flow patterns. In 1955 a model was made at WES to find a way of reducing the erosion of Point Chehalis at the mouth of Grays Harbor on the Pacific coast, west of Seattle (Figure 23.28). To help maintain the channel into the harbour, two jetties had been constructed at the turn of the century, and reconstructed in 1940 after storm damage. These repairs led to an increased erosion of Point Chehalis which, it was predicted, would soon disappear unless corrective measures were found and implemented. To make visible the currents that were causing this erosion a photographic technique was used to compare the baseline situation with new flow patterns generated by the introduction of various groynes and breakwaters. This was achieved by taking fixed exposure pictures of confetti mixed in with the water, enabling both the direction and speed of the water to be determined [59] (Figures 23.29 and 23.30).

An addition to WES in 1978 was the construction of a model for visitors; it was a small-scale model of a river flowing to the ocean (Figure 23.31). The model showed the effects of rainfall on a stream where flood controls are used. Dams on

Figure 23.28 Solid-bed model of Grays Harbor and Point Chehalis at WES, 1955. Horizontal scale 1:800, vertical scale 1:80. (Image: [59])

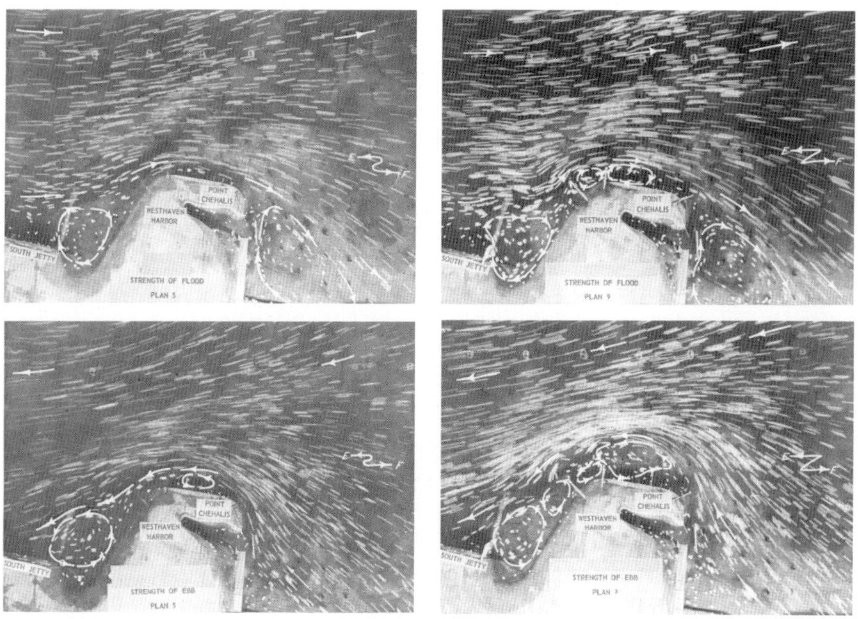

Figure 23.29 Point Chehalis. Flow vectors in flood and ebb tides. Left: Baseline study. Right: Scheme with final arrangement of groynes and breakwater as shown in Figure 23.30. (Image: [59])

the river automatically released water through spillway gates, and a small-scale model towboat demonstrated the use of locks. The area surrounding the river and harbour was carefully landscaped to scale with tiny towns and cities, trucks, cars, houses, industries – even a tennis court. Downstream, the effects of tides and waves in a harbour were demonstrated to explain various methods used by the Corps of Engineers for harbour protection. At the push of a button, a visitor could start the rain in the mountains and watch the many ways it affected everyday life

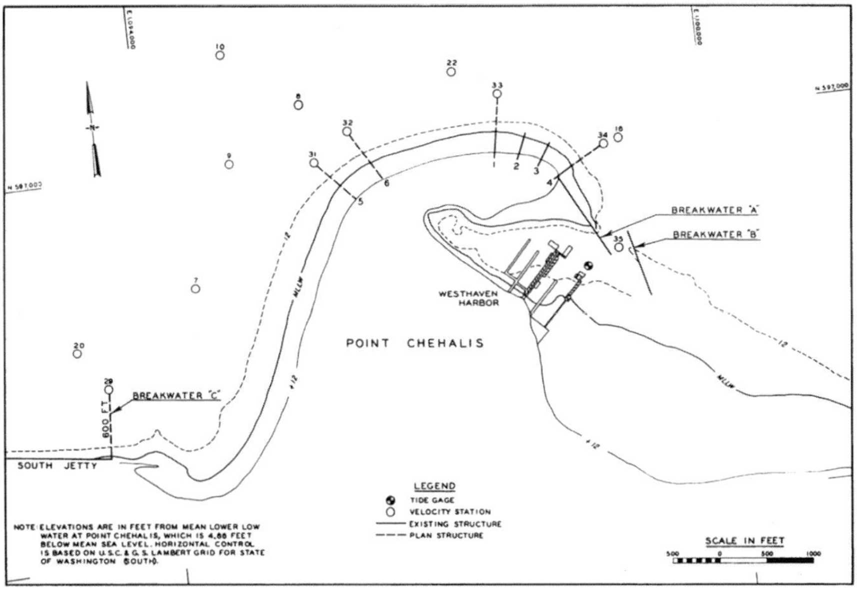

Figure 23.30 Point Chehalis. Final arrangement of groynes and breakwater adopted after studies of many different arrangements. (Image: [59])

Figure 23.31 Automated water resources model demonstrating to visitors the various activities of the WES, 1978. (Image: [43] p. 180)

until it eventually emptied into the ocean. The model included flood control, river channel stabilization and navigation, hydroelectric power production, harbour protection, conservation, recreation, and watershed management – a succinct summary of the work of the WES over the previous five decades [60].

23.6.2 Some developments in Europe

In Europe there was perhaps a greater diversity and wider scope in model testing than in the USA. This led to the establishment of many international model testing laboratories which survive to the present day, including the Hydraulic Research Station, Wallingford (now HR Wallingford) in the UK [61, 62], Delft Hydraulics Laboratory in the Netherlands, the Danish Hydraulic Institute in Copenhagen [63], and SOGREAH in Grenoble, France [64]. On the occasion of the 10th Congress of IAHR in 1963 in England, the US Bureau of Reclamation compiled and circulated a report of the visits by delegates to six hydraulic laboratories in Britain and France which illustrated some of the differences between activities in Europe and the USA [65].

In the immediate post-war period, the leading UK exponent of physical hydraulic models was Professor Jack Allen (1905-1984), appointed Head of the Engineering Department at the University of Aberdeen in 1946 [66]. He had been taught at Manchester University by A.H. Gibson, who was himself a pupil of Osborne Reynolds. Allen dedicated a lot of time and laboratory space to studies of Aberdeen Harbour and the shoaling of the River Tay estuary, with a very large model of the Tay in the premises of the Dundee Harbour Board. He examined the difficulty of scaling sediment movements in a physical model, and the application of dimensional analysis to practical engineering problems [67]. Others have subsequently addressed the difficulty of scaling sediment movements in physical models, using a wide variety of materials, but with little success. The most successful has been the limited use of coal dust as a representative beach material in coastal models.

Sir Claude Inglis, returning from India, was tasked by the UK government with setting up a UK Hydraulics Research Station (HRS), which became located at Wallingford, Oxfordshire. A purpose-designed experimental model hall, with an area of 13,000 m^2 and large roof-spans between columns, was constructed for model investigations, together with four large covered wave-basins totalling 3900 m^2, and a number of flumes up to 122 m long and 3.6 m wide. During this period there was also ongoing development of new and improved techniques for making the water flow pattern and velocity more amenable to measurement and representative of the parameters intended to be measured, for example using tracer dies in the water, or floating small lighted candles on the surface – both more reliable than using sawdust or confetti that would sink and clog the pumps. An in-house team designed and manufactured the machinery and instrumentation to control the models and take measurements in the laboratory facilities. The commercial engineering studies were underpinned by a government-funded programme of fundamental research into hydraulics processes. In 1979 a new wave basin, 35 x 30 m with a centre depth of 4 m, was constructed in which a real sea could be modelled over a large area for the testing of offshore structures and moorings. A computer-controlled 80-paddle wave generator could produce a random sea pattern at angles up to 89°, and wind and tidal currents could also be superimposed. The HRS laboratory undertook – and, as HR Wallingford, still undertakes (see Chapter 31) – a wide range of model studies of hydraulic structures, rivers, estuaries, coasts and offshore installations for projects in the

Figure 23.32 1:50 scale model of the Thames Surge barrier to study the effectiveness of measures for protecting the river bed. (Image: Courtesy of HR Wallingford)

UK and worldwide, including collaboration with other governments to establish hydraulic models and modelling capabilities overseas, for example in Australia, Argentina and Hong Kong. Between 1968 and 1978 the HRS built six models for the outer Thames Estuary at scales varying from 1:25, for the barrier gate, to 1:1000, one of which was 115 m long, to support studies for the feasibility, design and construction of the Thames Tidal Surge Barrier, London [68, 69] (Figure 23.32). The apogee of physical model studies at Wallingford was probably the construction, in a purpose-built building, 75 x 32 m and column free, of a vast tidal model of the Thames estuary and the southern North Sea for a feasibility study of an airport in the outer Thames estuary [70] (Figure 23.33).

23.6.3 The starry sky

One particularly noteworthy development in free-surface modelling in the 1950s was by the French engineer Émile-Georges Barrillon (1879-1967). He devised a technique known as *le ciel étoilé* – the 'starred sky' or 'starry sky' in English. Its main purpose was to be able to quantify wave disturbance over a wide area, without using measurement instruments that interfered with the water surface. The principle is as simple as it is ingenious. A regular array of small lights is mounted just below the ceiling of the room containing the model (Figure 23.34). A camera mounted in the ceiling takes a photograph of the water surface, with an exposure time of at least one wave period (for regular waves), and the photograph records the movement of the reflection of each of the lights in the ceiling. The size, orientation, shape and degree of distortion of the movement of the reflection of an individual light, indicates the slope of the waves, from which can be calculated the heights and directions of the waves. These data can then be used to estimate forces exerted by the waves on ships or harbour walls.

Conventional measurement techniques for waves in hydraulic laboratories employ point measurements using either resistance- or capacitance-type wave

Figure 23.33 1:1000 horizontal scale and 1:100 vertical scale model of the outer Thames estuary for studying the impacts of an airport proposed to be built on the Maplin Sands. The 75 x 32 metre building is column free. The dozens of small cylinders on the model sea bed were used to introduce turbulence into the flow. (Image: Courtesy of HR Wallingford)

probes. Their disadvantage is that they record only point measurements and information about a large two-dimensional field depends on having very large numbers of probes, which is costly. It is also possible that the wave probes are not located at the critical locations. The main advantage of the starry-sky technique is that it gives a record of an entire wave field [71, 72].

In 1974, the Hydraulics Research Station at Wallingford, England, undertook a study of some proposed improvements to the harbour at Wick in Caithness, Scotland (Figure 23.35). The model was constructed at a scale of 1:100, in which, using Froude's laws of dynamic similarity, wave periods and velocities would be one tenth of those at full scale [73].

The studies identified causes of problems due to reflections off the piers and harbour walls (Figure 23.36). Three principal measures were adopted to reduce the effect of the reflections – a spending beach (to absorb wave energy), an extension to the main pier, and the construction of a stone groyne against the existing piers. The starry-sky model was able to quantify the reduction in wave height (by 20-40%), especially in the vicinity of the harbour mouth, achieved by extending the existing pier, constructing groynes or by building a new breakwater (Figure 23.37).

Figure 23.34 A 'starry sky' fitted to the ceiling of the experimental laboratory. (Image: Courtesy of HR Wallingford)

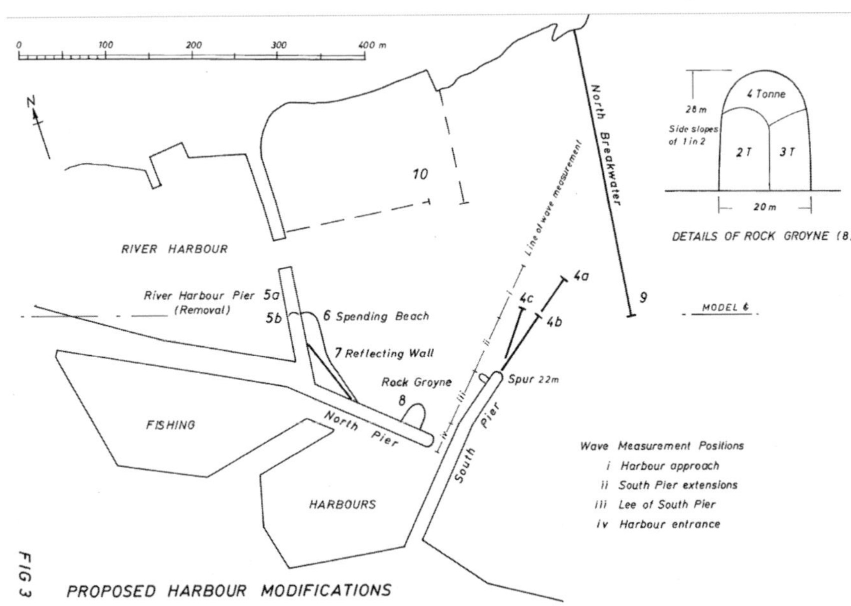

Figure 23.35 Diagram of various modifications proposed for the Wick harbour. (Image: Courtesy of HR Wallingford)

Figure 23.36 Study of problems caused by reflections of waves off the harbour walls and piers. (Image: Courtesy of HR Wallingford)

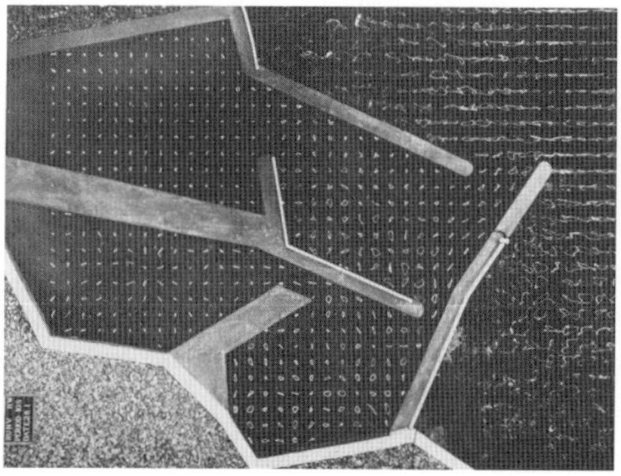

(a)

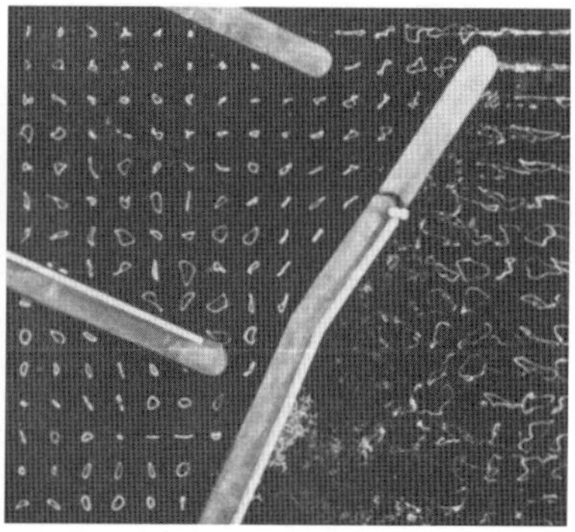

(b)

Figure 23.37 Photographs showing distorted reflections of the 'starry-sky' light grid, indicating the agitation of the water surface and, hence, wave heights. (Image: Courtesy of HR Wallingford)

23.7 Conclusion

The use of fixed-bed and moveable-bed models for studying the behaviour of rivers, estuaries and beaches developed quickly from the late-nineteenth to the mid-twentieth century. Wave tanks too developed during the same period from being an instrument of scientific investigation to a direct aid to design. They have all become major tools in the armoury of engineers seeking guidance on the most effective ways of preventing the blocking of navigation channels and erosion of land-water boundaries, managing floods and designing canal locks, dams and barrages. The inherent difficulty of creating mathematical models of water movement has ensured that physical models continued to develop throughout the twentieth century.

During the 1970s and 1980s, computer modelling began to supplant many forms of physical model. Initially, computer modelling was constrained by the capacity and speed of even the largest computers, and the limitations of presenting results as visual graphics. Early computer models crudely averaged three-dimensional flow patterns and provided none of the detailed visual information that could readily be seen in a physical model. Nevertheless, in 1978 a new feasibility study of tidal power generation in the Severn Estuary elected to use faster and cheaper computer modelling rather than the usual large physical model. As computing power has rapidly increased and more-sophisticated computer model techniques have evolved, the large physical hydraulic model has generally been supplanted by computer modelling. The advantages of computer models are their relative cheapness and flexibility, being easily altered to test ideas and variations. They are particularly suited to the study of dispersion of hot or polluting discharges, and the inclusion of sediment transport, which are difficult if not impossible to reproduce in a physical model.

However, computer models need the physical processes to be described precisely in mathematical terms. In complex situations, particularly those involving waves and/or flows in and around structures, the physical hydraulic model is still the preferred and, often, the only model tool fit for the purpose. Physical models also offer a more holistic overview of a proposed scheme, in which interactions and fine details can be visually identified. They thus have a public role in the presentation of proposed projects to clients and committees who may need to understand and be convinced of project proposals. At the beginning of the twenty-first century, physical and computer models are used in tandem, each being chosen as appropriate to the complexity and required results of the project, as we shall see in Chapter 31.

Acknowledgements

I am grateful to Mike Thorn, formerly of HR Wallingford, for providing valuable information and helping to improve this chapter.

References

1 Bélidor, Bernard Forest de, (1737-1753) *Architecture hydraulique, ou, L'art de conduire, d'elever et de menager les eaux pour les différens besoins de la vie.* (4 Vols). Paris: Jombert.

2 Dubuat, P.L.G. (1816) *Principes d'hydraulique et de pyrodynamique, verifié par un grand nombre d'expériences.* (New Edition, 3 vols). Paris: Firmin Didot.

3 Darcy, H. & Bazin, H. (1885) *Recherches hydrauliques.* Paris: Imprimerie impériale. 2 Pts. 2 vols. in 1 and atlas of plates.

4 Rouse, H. & Ince, S. (1957) *History of Hydraulics.* Iowa: Iowa Institute of Hydraulics Research, State University of Iowa.

5 Freeman, John R. (1929) Historical development of hydraulic researches and some noteworthy hydraulic laboratories in America. In: *Hydraulic laboratory practice* (ed. John R. Freeman), 1-32. New York: American Society of Mechanical Engineers.

6 Rouse, H. (1976) *Hydraulics in the United States, 1776-1976*. Iowa: Iowa Institute of Hydraulics Research, State University of Iowa.
7 Parsons, W.B. (1939) *Engineers and Engineering in the Renaissance*. 2nd Edn (1968). Cambridge, Mass.: MIT Press. 323-344.
8 Leonardo da Vinci. MS. E. FOL. 55 f. Bibliothèque de l'Insitut de France
9 Thierry, G. de (1929) A Review of European Hydraulic Laboratory Practice. In: *Hydraulic laboratory practice* (ed. John R. Freeman), 579-587. New York: American Society of Mechanical Engineers. 581.
10 Fargue, Louis (1894) Expériences relatives à l'action de l'eau courante sur un fond de sable. *Annales des Ponts et Chaussées, Mémoirs et documents*, 7. Série, Vol. VII, 1. Semester: 426-466, Plates 11-19.
11 Fargue, Louis (1908) *Hydraulique fluviale. La forme du lit des rivières à fond mobile*. Paris: Gauthier.
12 Ippen, A.T. (1970) Hydraulic Scale Models. In: *Osborne Reynolds and Engineering Science Today* (eds J.D. Jackson & D.M. McDowell), 199-224. Manchester: The University Press.
13 Reynolds, Osborne (1888) On certain laws relating to the regime of rivers and estuaries, and on the possibility of experiments on a small scale. In: *Report of the 57th Meeting of the British Association for the Advancement of Science held in Manchester in 1887*, 555-562. London: John Murray. 562.
14 Reynolds, Osborne (1890) Report of the Committee appointed to investigate the Action of Waves and Currents on the Beds and Foreshores of Estuaries by means of Working Models: On Model Estuaries. *Report of the 59th Meeting of the British Association for the Advancement of Science, Newcastle-upon-Tyne, Cardiff, September 1889*, 327-343, Plates I-IX. London: Murray.
15 Reynolds, Osborne (1891) Second Report of the Committee appointed to investigate the Action of Waves and Currents on the Beds and Foreshores of Estuaries by means of Working Models: On Model Estuaries. *Report of the 60th Meeting of the British Association for the Advancement of Science, Leeds, September 1890*, 512-534, Plates I-XVIII. London: Murray.
16 Reynolds, Osborne (1892) Third Report of the Committee appointed to investigate the Action of Waves and Currents on the Beds and Foreshores of Estuaries by means of Working Models: On Model Estuaries. *Report of the 61st Meeting of the British Association for the Advancement of Science, Cardiff, August 1891*, 386-404, Plates II-XIV. London: Murray.
17 Reynolds, Osborne (1890) Report of the Committee appointed to investigate the Action of Waves and Currents on the Beds and Foreshores of Estuaries by means of Working Models: On Model Estuaries. *Report of the 59th Meeting of the British Association for the Advancement of Science, Newcastle-upon-Tyne, Cardiff, September 1889*, 327-343. London: Murray. 327-328.
18 Reynolds, Osborne (1892) Third Report of the Committee appointed to investigate the Action of Waves and Currents on the Beds and Foreshores of Estuaries by means of Working Models: On Model Estuaries. *Report of the 61st Meeting of the British Association for the Advancement of Science, Cardiff, August 1891*, 386-404. London: Murray. 393.
19 Vernon-Harcourt, L. (1888). The Principles of Training Rivers through Tidal Estuaries, as Illustrated by Investigations into the Methods of Improving the

Navigation Channels of the Estuary of the Seine. *Proceedings of the Royal Society of London* 45: 504-524.

20 Vernon-Harcourt, L. F. (1890) Investigations into the Effects of Training Walls in an Estuary like the Mersey. *Proc. Roy. Soc.* 47: 142-144. 144.

21 Ippen, A.T. (1970) Hydraulic Scale Models. In: *Osborne Reynolds and Engineering Science Today* (eds J.D. Jackson & D.M. McDowell), 199-224. Manchester: The University Press. 204.

22 Bollrich, G. (1987) Julius Weibach's and Hubert Engels' contributions to hydraulics and hydraulic research development. In: *Hydraulics and hydraulic research: a historical review* (ed. Gunther Garbrecht), 275-280. Rotterdam: Balkema.

23 Engels, H. (1900) Das Flußbau-Laboratorium der Königlich Technischen Hochschule in Dresden. *Zeitschrift für Bauwesen* 50 (7-9): 343-360.

24 Engels, H. (1929) The hydraulic laboratory at the Technical University of Dresden. In: *Hydraulic laboratory practice* (ed., John R. Freeman), 83-107. New York: American Society of Mechanical Engineers. (Translation and revised edition of Thierry, George Henry de & Matschoss, Conrad (eds) (1926) *Die Wasserbaulaboratorien Europas; Entwicklung, Aufgaben, Ziele.* Berlin: Verein Deutscher Ingenieure.)

25 Larsen, A.P. & Bernhart, H.H. (1987) Theodor Rehbock (1864-1950). In: *Hydraulics and hydraulic research: a historical review* (ed. Gunther Garbrecht), 281-291. Rotterdam: Balkema.

26 Rehbock, Theodor (1903) Das Flussbaulaboratorium der Großherzoglichen Technische Hochschule, Fridericiana' zu Karlsruhe. *Zeitschrift für Bauwesen* 53: 103/104-135/136.

27 Rehbock, Theodor (1910) Neueinrichtungen des Karlsruher Flussbaulaboratoriums. *Zeitschrift für Bauwesen* 60: 313/314-327/328.

28 Rehbock, Theodor (1929) The river-hydraulic laboratory of the Technical University at Karlsruhe. In: *Hydraulic laboratory practice* (ed., John R. Freeman), 111-242. New York: American Society of Mechanical Engineers.

29 Thierry, George Henry de & Matschoss, Conrad (eds) (1926) *Die Wasserbaulaboratorien Europas; Entwicklung, Aufgaben, Ziele.* Berlin: Verein Deutscher Ingenieure.

30 Freeman, John R. (ed) (1929) *Hydraulic laboratory practice.* New York: American Society of Mechanical Engineers. (Translation and expanded edition of *Die Wasserbaulaboratorien Europas*).

31 Hurst, H.E. & Watt, D.A.F. (1924) The similarity of motion of water through sluices and through scale models: Experiments with models of sluices of the Assuan Dam. *Min. Proc. Inst. Civ. Engs* 218: 72-112.

32 Dare, H.H. (1922) The spillways of Burrinjuck Dam. *Min. Proc. Inst. Civ. Engs* 214: 333-345.

33 http://cwprs.gov.in/ Also http://cwprs.gov.in/Page/Introduction.aspx (15/9/2019)

34 Thomas, A.R. & Paton, A. (1975) Claude Cavendish Inglis. 3 March 1883--29 August 1974. *Biographical Memoirs of Fellows of the Royal Society* 21: 366-388.

35 Inglis, C.C. (1938) The use of models for elucidating flow problems based on experience gained in carrying out model experiments at the Hydrodynamic Research Station, Poona. *Proc. National Institute of Sciences of India* IV (4): 219-239.

36 Inglis, C.C. (1949) *The behaviour and control of rivers and canals (with the aid of models).* Poona: Central Waterpower Irrigation and Navigation Research Station.

37 Mutel, C.F. (1998) *Flowing Through Time. A History of the Iowa Institute of Hydraulic Research.* Iowa: IIHR – Hydroscience & Engineering, University of Iowa.

38 Lane, E.W. & Soucek, E. (1938) *The road map of hydraulic engineering in Iowa.* Studies in Engineering: Bulletin 15. Iowa: University of Iowa. https://pdfs.semanticscholar.org/da81/f3c56690a345879b0599d589887d91411fb4.pdf (15/9/2019)

39 Nelson, M.E. (1932) *Laboratory tests on hydraulic models of the Hastings Dam.* Studies in Engineering: Bulletin 2. Iowa: University of Iowa. https://pdfs.semanticscholar.org/2d67/c6bdad234bd1e2600ad8e08aee91354807bc.pdf (15/9/2019)

40 Freeman, John R. (1929) Historical development of hydraulic researches and some noteworthy hydraulic laboratories in America. In: *Hydraulic laboratory practice* (ed. John R. Freeman), 1-32. New York: American Society of Mechanical Engineers.

41 Macagno, E.O., Moran, D.D. & McDougall, D.W. (1971) *The first half century of hydraulic research at the University of Iowa.* Iowa City: Iowa Institute of Hydraulic Research.

42 Tiffany, J.B. (ed.) (1968) *History of the Waterways Experiment Station.* Vicksburg, Miss.: U.S. Army Engineer Waterways Experiment Station. App.I-1.

43 Cotton, Gordon A. (1979) *A history of the Waterways Experimental Station 1929-1979.* Vicksburg, Miss.: U.S. Army Engineer Waterways Experimental Station. 16.

44 Tiffany, J.B. (ed.) (1968) *History of the Waterways Experiment Station.* Vicksburg, Miss.: U.S. Army Engineer Waterways Experiment Station. App.V-9.

45 Cotton, Gordon A. (1979) *A history of the Waterways Experimental Station 1929-1979.* Vicksburg, Miss.: U.S. Army Engineer Waterways Experimental Station. 35-37.

46 Quinn, M-L. (1977). *The History of the Beach Erosion Board, U.S. Army, Corps of Engineers, 1930-63.* MR 77-9, U.S. Army, Corps of Engineers. Fort Belvoir, Va.: Coastal Engineering Research Center. 27.

47 Hudson R.Y., Hermann F.A. et al. (1979) *Coastal hydraulic models. Special Report 5.* US Army Corps of Engineers, Coastal Engineering Research Center, Vicksberg, Mississippi.

48 Moore, Jamie W. & Moore, Dorothy P. (1991) *History of the Coastal Engineering Research Center 1963-1983.* Vicksburg, Miss.: U.S. Army Engineer Waterways Experiment Station.

49 Bureau of Standards (1933) *Report on Current Hydraulic Laboratory Research in the United States. Report I-1 (April).* Washington: U. S. Department of Commerce. 1.

50 Gibson, A.H. (1933) *Construction and operation of a tidal model of the Severn Estuary*. London: HMSO.

51 Gibson, A.H. (1936) Tidal and river models. *Jnl Instn Civ. Engrs* Vol 3 No.8: 699-722.

52 Stucky, A. & Bonnard, D. (1937) Contribution à l'étude expérimentale des digues maritimes en enrochements. *Bulletin Technique de la suisse romande* 63 (18): 232-237.

53 Quinn, M.L. (1977). *The History of the Beach Erosion Board, U.S. Army, Corps of Engineers, 1930-63*. MR 77-9, U.S. Army, Corps of Engineers. Fort Belvoir, Va.: Coastal Engineering Research Center. 77-78.

54 Johnson, J.W. (ed.) (1951) *Proceedings of the First Conference on Coastal Engineering, Long Beach, California, November 1950*. Berkeley, California: Council on Wave Research.

55 Foster, J. E. (1971) *History and Description of the Mississippi Basin Model*. Vicksburg: WES (MBMR 1-6).

56 Fatherree, Ben H. (2004) *The first 75 years: History of hydraulics engineering at the Waterways Experiment Station*. Vicksburg: US Army Corps of Engineers, Engineer Research and Development Center. 73-75, 89-95.

57 https://commons.wikimedia.org/wiki/File:Miss-Basin-Model2-s.jpg Public domain.

58 Corp of Engineers, U.S. Army (1955) *Preservation and enhancement of Niagara Falls: Hydraulic model investigation*. Technical Memorandum 2-4111. Vicksburg: Waterways Experiment Station.

59 Corp of Engineers, U.S. Army (1955) *Plans for the improvement of Grays harbour and Point Chehalis: Hydraulic model investigation*. Technical Memorandum 2-417. Vicksburg: Waterways Experiment Station

60 Cotton, Gordon A. (1979) *A history of the Waterways Experimental Station 1929-1979*. Vicksburg, Miss.: U.S. Army Engineer Waterways Experimental Station. 180.

61 http://www.hrwallingford.com/about/70years/celebrating-70-years-of-working-with-water (15/9/2019)

62 http://www.hrwallingford.com/celebrating-70-years-working-with-water/from-hro-to-hr-wallingford (15/9/2019)

63 https://www.dhigroup.com/global/references/emea/overview/supporting-the-design-of-a-major-new-coastal-highway (15/9/2019)

64 https://www.arteliagroup.com/en/group/who-we-are/history/sogreah (15/9/2019)

65 Peterka, Alvin J. (1963) *Report of attendance at the Tenth Congress of IAHR, England, September 1963*. Paper 194. United States Department of the Interior, Bureau of Reclamation

66 Willets, B. & McEwan, I. (2006) *A historical sketch of engineering fluid mechanics in Aberdeen, 1946-2006*. University of Aberdeen. https://www.abdn.ac.uk/engineering/documents/historical-sketch-fluid-mechanics.pdf (16/9/2019)

67 Allen, J. (1952) *Scale Models in Hydraulic Engineering*. London: Longman.

68 Institution of Civil Engineers (1977) *Thames Barrier design.* Proceedings of the conference held in London, 5 October 1977. London: Institution of Civil Engineers.
69 http://www.hrwallingford.com/celebrating-70-years-working-with-water/modelling-the-tidal-thames (16/9/2019)
70 http://www.hrwallingford.com/celebrating-70-years-working-with-water/working-on-the-maplin-model (16/9/2019)
71 Laurent, J. (1948) Hydraulique. Etude des ports sur modèles réduits. *La revue maritime* Part I - 26: 657-671. Part II - 27: 798-811.
72 Valembois, J. (1952). Methods used at the national hydraulic laboratory of Chatou (France) for measuring and recording gravity waves in models. In: *Proceedings of the NBS Semicentennial Symposium on Gravity Waves, NBS, June 18-20, 1951. Circular 521,* 255-263. Washington: National Bureau of Standards.
73 Hydraulics Research Station (1975) *Wick Harbour, Caithness. Hydraulic model studies of proposals for the improvement of conditions at the fishing harbour.* Report EX706. Wallingford: Hydraulic Research Station.

24

The historical use of physical model testing in wind engineering

Bill Addis

24.1 The scientific study of wind

24.1.1 The force of the wind

The first scientific tests to determine the force exerted by the wind were undertaken by John Smeaton (1724-1792) in the 1750s during his experiments to improve the efficiency of windmills [1]. He created a near-steady wind speed by rotating an arm with windmill sails mounted at the end (Figure 24.1). His measurements, which were in close agreement to those obtained by a colleague, Mr Rouse, using a different apparatus, showed that the wind force per unit area was proportional to the square of the wind speed – P (in pounds per square foot) = 0.005 V^2 (V in miles per hour) i.e. P (N/m^2) = 1.2 V^2 (m/s). The Smeaton-Rouse formula was used by many engineers in the nineteenth century to estimate wind loads on structures.

During the nineteenth century, many scientists and engineers, including Benjamin Baker (1840-1907), engineer of the Forth Bridge and Gustav Eiffel (1832-1923), conducted full-scale experiments to determine both maximum wind speeds experienced in storms, and the force that wind at different speeds exerts on structures. The results varied considerably although the general relationship $P = k.V^2$ was always confirmed – only the value of k was disputed. By the 1870s more empirical data had been collected and instruments for measuring air speed had improved. The Wind Pressure (Railway structures) Commission was established to investigate wind loads after the collapse of the Tay Bridge in 1879 recommended using the formula P (N/m^2) = 2.4 V^2 (m/s) up to a maximum value of 2.68 kN/m^2 to be used for design calculations for railway bridge and viaducts, together with a factor of safety of four when calculating stresses in structural ironwork, which could be reduced to two when calculating resistance to overturning against gravity forces alone [2, 3, 4].

Interest in the magnitude of wind loads on structures grew throughout the nineteenth century, especially with the growing number of very large bridges in exposed locations and, from the late 1880s in Chicago and New York, tall buildings with small footprints. The incorporation of bespoke wind-bracing structural systems in tall buildings became standard from around 1890.

Physical Models: Their historical and current use in civil and building engineering design,
First Edition. Edited by Bill Addis.

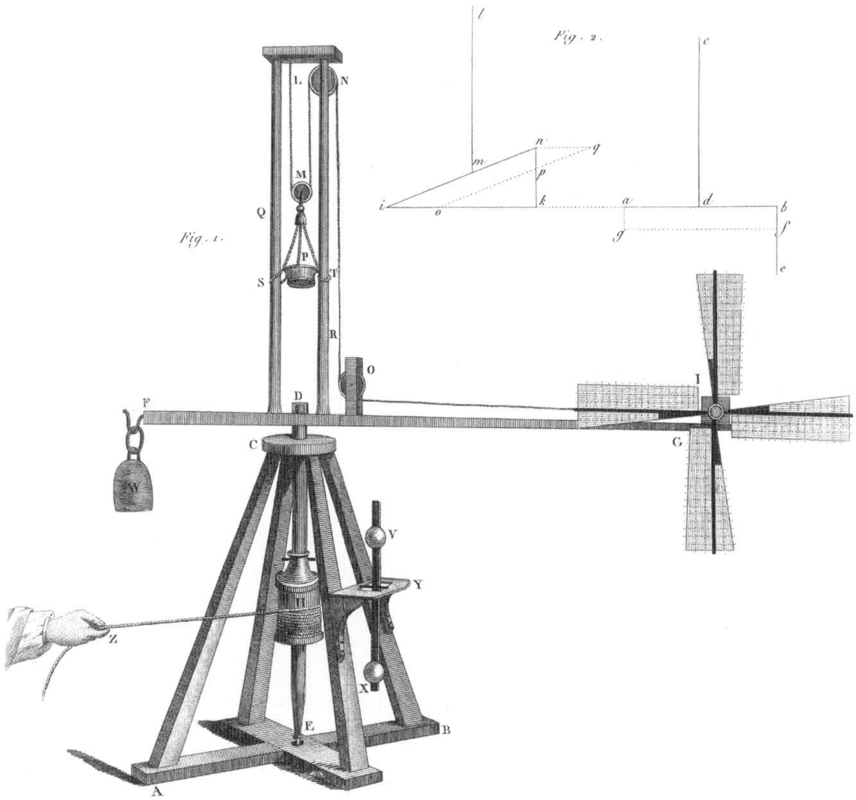

Figure 24.1 John Smeaton's model windmill for determining the effectiveness of the sails to convert the force of the wind to a force that raised a weight. (Source: [1])

However, the need for more information about wind loads did not grow out of scientific curiosity. Rather, there was a feeling that the engineering design community was in a state of confusion or, indeed, near chaos. William Kernot, Professor of Engineering at the University of Melbourne, described the situation he saw in Australia in around 1890.

> 'With regard to the horizontal forces due to that ubiquitous agent, the wind, and which in multitudes of cases are the only important horizontal forces the structures are called upon to endure, the present state of knowledge is most unsatisfactory and discreditable. Publications professing to be scientific authorities give arbitrary, inconsistent, and occasionally unintelligible directions, while an examination of engineering practice as exhibited in large and costly structures in Australia and elsewhere reveals the most extraordinary inconsistencies. Structures in exposed positions and offering large surface to the wind are not infrequently absolutely devoid of wind bracing, while others placed in sheltered localities and offering but little surface are filled with costly lateral members having no possible utility except as aids against wind pressure. ....
>
> Recently a Royal Commission reported on certain iron bridges on a railway in Tasmania and condemned them as weak under wind pressure.

The designer of the bridges protested, and further advice was sought: The matter being referred to the author of this paper, he calculated the resistance and found that it would take no less than 200 pounds per square foot to overturn the girders! And yet the Royal Commission condemned them, but never dreamt of condemning the railway carriages that passed over them, and which a similar calculation showed would infallibly capsize under a pressure of only 25 [pounds per square foot]. Under these circumstances of conflicting authorities and wildly inconsistent practice it appeared to the writer that a paper on wind pressure, comparing and discussing the data and rules given in the books and detailing results of his own inquiries, calculations, and experiments, could not fail to be of use and interest' [5].

24.1.2 Early wind tunnels

Interest in the fluid mechanics of air grew in the 1870s as the study of fluid dynamics as a scientific discipline was developing. The first wind tunnel for measuring lift and drag on various potential shapes for aircraft wings was designed by Frank Wenham (1824-1908) and constructed in 1871 by John Browning (c.1831-1925), two members of the Aeronautical Society of Great Britain. It had a 45 x 45 cm cross-section and was 3.7 m long. Air was propelled at speeds up to 65 kph by a fan driven by a steam engine. Although the airflow was unsteady and not streamlined, it was able to demonstrate that the lift-to-drag ratio was unexpectedly high for surfaces at a low angle of incidence and that, for a given surface area, a long narrow wing was more effective than a short wide one. Confidence in the possibility of powered flight was raised significantly. Another wind tunnel, built in the early 1880s by Horatio Phillips (1845-1924), an English aviation pioneer, was unusual in that the gas flow was steam rather than air. He studied various aerofoil shapes and demonstrated the benefit of wings with a cambered upper surface [6]. And in 1887, the German physicist Ernst Mach (1838-1916) photographed a bullet moving faster than the speed of sound (not in a wind tunnel), using the *schlieren* technique to make the shock waves visible. All this was before Otto Lilienthal's successful glider flights in the 1890s, and the Wright brothers' first controlled, sustained, powered flight in 1903.

Wind tunnels were built in many countries during the first decade of the twentieth century, including the UK, the USA, Italy, Germany, France and Russia [7]. These were of two basic types – the open-circuit type, usually named after the National Physical Laboratory (NPL) in Teddington near London, where wind tunnels of increasing size were built in 1903 (60 cm diameter), 1910 (1.2 x 1.2 m) (Figure 24.2) and 1912 (2.1 x 2.1 m), or after Gustav Eiffel who built his similar wind tunnel in 1909.

The second type of wind tunnel was the closed-circuit type, first designed and built in 1908 by Ludwig Prandtl (1875-1953) at the University of Göttingen in Germany. Prandtl seems first to have had the idea for a closed-circuit when he made a water channel on this principle for studying turbulence around objects in a fluid stream, with water serving as an analogue for air [8] (Figure 24.3). Using a camera mounted above the channel, he recorded examples of turbulence and eddy formation (Figure 24.4).

Figure 24.2 The 1.2 x 1.2 m 'wind channel' built at the National Physical Laboratory, Teddington, UK, 1910. (Image: National Physical Laboratory)

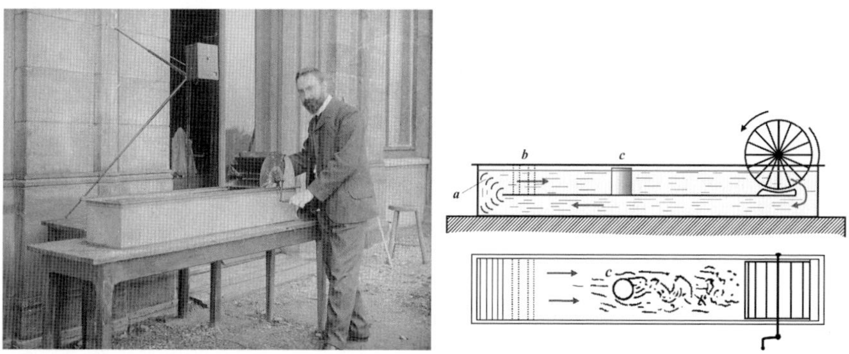

(a) (b)

Figure 24.3 (a) Ludwig Prandtl with his closed-circuit water channel, 1904. A camera is positioned above the channel (Image: *Max-Planck-Institut für Dynamik und Selbstorganisation*, Göttingen). (b) The closed-circuit water channel. (Source: [8], redrawn Bill Addis)

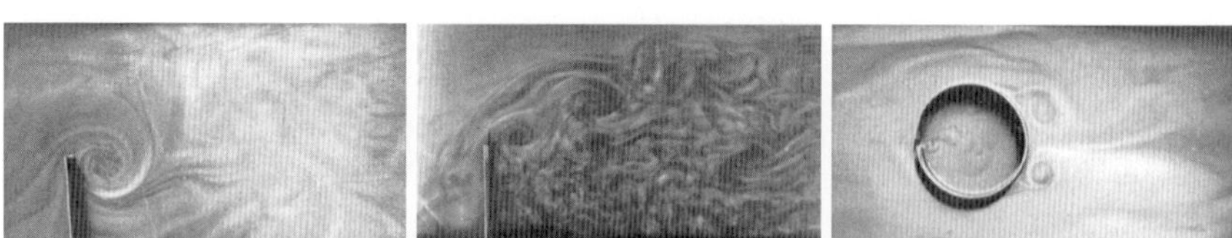

(a) (b) (c)

Figure 24.4 Photographs taken by Planck showing turbulence and formation of eddies in his water channel, 1904. (Source: [8])

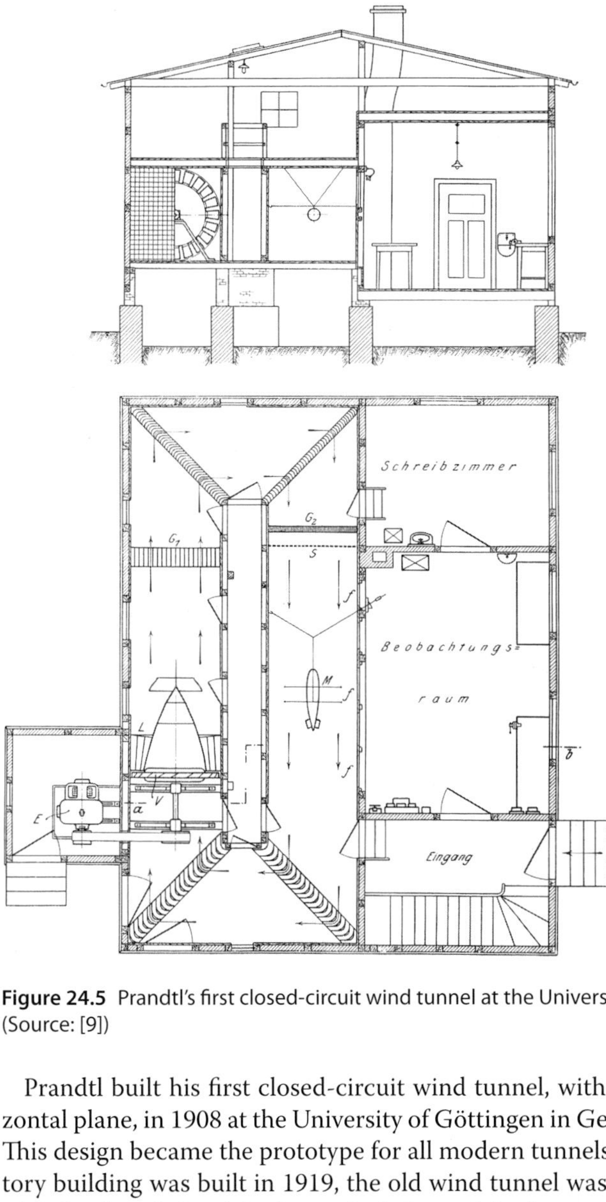

Figure 24.5 Prandtl's first closed-circuit wind tunnel at the University of Göttingen, 1908. (Source: [9])

Prandtl built his first closed-circuit wind tunnel, with the circuit in the horizontal plane, in 1908 at the University of Göttingen in Germany [9] (Figure 24.5). This design became the prototype for all modern tunnels. When a larger laboratory building was built in 1919, the old wind tunnel was dismantled and rebuilt with the closed circuit in the vertical plane, giving better access to the test section of the tunnel (Figure 24.6).

Prandtl's particular contribution to using wind tunnels to study buildings and bridges is that he discovered that the air in contact with a surface parallel to the airflow is stationary, and its speed increases to the ambient wind speed across a thin layer of air that he called the *boundary layer*, the thickness of which is highly sensitive to the roughness of the surface [8]. This was important for civil and building engineers, as all structures are located in this layer of the atmosphere, and themselves contribute to the roughness of the surface of the earth. This is

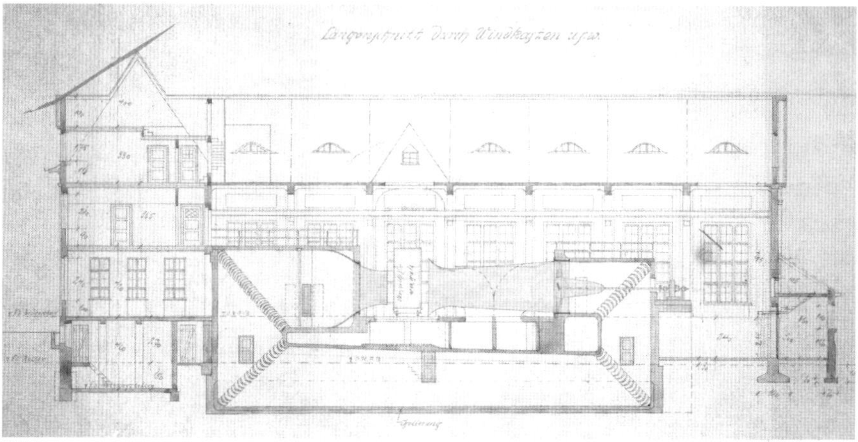

Figure 24.6 Prandtl's second closed-circuit wind tunnel, 1919, with the circuit in the vertical plane. (Image: Max-Planck-Institut für Dynamik und Selbstorganisation, Göttingen)

the basis of boundary-layer wind tunnels that were developed in the 1960s (see Section 24.6).

24.2 The first measurement of wind-pressure loads on model buildings

24.2.1 William Charles Kernot, 1893

William Kernot (1845-1909) constructed his 'blowing machine' at the University of Melbourne to resolve the uncertainties about wind loads that were still rife in the 1890s, for which he needed 'an artificial wind, a current of air of uniform or nearly-uniform velocity and direction' [10]. His apparatus consisted of a fan, 71 cm in diameter with a pitch of 102 cm, driven by an Otto gas engine and including two flywheels in the power train to minimise fluctuations in speed, blowing air through a duct onto a model mounted on a table beyond the end of the duct (Figure 24.7). The first section of the duct was a cylinder, 76 cm in diameter and 91 cm long. Kernot discovered that the propeller did not deliver a uniform current of air. Rather, it produced a cylindrical shell of air, about 15 cm thick, moving helically along the duct, surrounding a central core of dead or motionless air, about 46 cm in diameter. Air was taken out of the cylinder of moving air into a rectangular duct, in a direction tangential to the helical path of the air in the cylinder (about 60° to the axis). This funnel-shaped section diverged to form a rectangular cross-section 30 x 25 cm delivering a uniform current of air. (The illustration is interpreted from Kernot's rather imprecise description).

The model under test was mounted on a three-wheeled carriage with gunmetal wheels rolling on a heavy engineer's surface plate in order to reduce vibrations from the moving machinery. The surface plate and the carriage were encased in a box to protect them from the wind. The models being tested were mounted on a stem that projected upwards from the carriage through a slot in the casing.

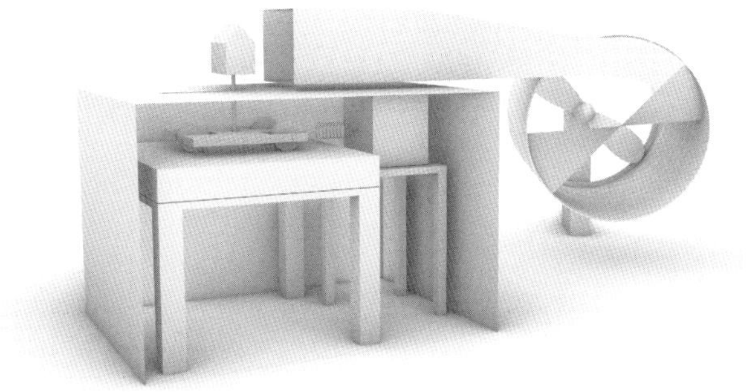

Figure 24.7 William Kernot's 'blowing machine' for creating a uniform current of air directed onto various models, 1893. (Image: Eike Schling)

The force on the model was measured by a delicate, calibrated spring balance. The flat models he tested were made of cardboard and the standard three-dimensional shapes (e.g. a cylinder) were wooden geometrical models borrowed from the university's store of teaching equipment.

Kernot's first series of experiments was to determine the law connecting wind pressure and velocity. He tested various discs, spheres, cubes and rectangular plates. The air speed was calculated knowing the speed of the gas engine, but only two engine speeds could be set using the governor. In one set of tests the air speeds were calculated at 7.1 and 3.7 m/s, and in the second, 6.5 and 2.7 m/s. These data were felt to be insufficient to settle the important matter of the velocity-pressure law and so a different experiment was devised. A plate 30 x 30 cm was fixed to a bicycle that was propelled at speeds from 0.5 to 6.7 m/s. The pressures proved too small to measure reliably so the plate was increased to 58 x 58 cm. Despite the crudeness of this test, Kernot was happy with the consistency of the results and settled on the relationship (in imperial units) of $P = 0.0033\ V^2$, compared to Smeaton's formula, $P = 0.005\ V^2$.

Kernot's next series of experiments, using the blowing machine was to determine the relationship between the pressure on solids of various forms and on thin plates corresponding in area to the projection of each solid with the objective of obtaining a 'modulus' (today, the drag coefficient) for each shape. The different shapes he tested included a cube, rectangular block, pyramid, cylinder, octagonal prism, cone, sphere and hemispherical cup (as used in an anemometer). He also conducted tests on a number of engineering structures – roofs (on a simple rectangular building), 'girder bridges' and 'lattice work'. Kernot selected these structures because they were specifically addressed in the standard structural engineers' design book of the time that he referred to as 'Stoney on stresses' [11]. He tested roofs of different pitches – 30°, 45° and 60° – noting the uplift especially on the 30° pitch. He tested bridges made with two parallel, solid girders, noting the shading effect of the windward girder which varied with different spacings between the girders. Finally, he tested pairs of lattice girders, using a card model 15 x 20 cm, at different spacings, noting the pressure on the leeward lattice was a minimum when the spacing was equal to the height of the girder.

Kernot's assistant also noted the shelter provided to a small surface when a larger one was place *behind* it, on the leeward side.

Kernot conclusions were succinct and practical. Designers should assume 'a maximum wind pressure of not more than 20 pounds per square foot in exposed locations [and] in very sheltered locations half the above values may be taken'. 'Square, round and octagonal towers, chimneys, spires, railway carriages, girder bridges and roofs are to be reduced to their equivalent normal plane surfaces by means of the moduli and rules' that he had established. And, finally, that 'factors of safety of not less than two for stability and three for strength should be used' [12]. It is not known how trustworthy Kernot's guidance was considered.

It is interesting that, when Kernot continued his experiments to determine wind pressures after 1893, he abandoned his blowing machine and reverted to a whirling machine like Smeaton's which was easier to control, allowing him to measure wind pressures continuously during the tests, and to calculate wind velocities more precisely. He admitted that his blowing machine had been unsatisfactory because of the difficulty of measuring air speed, and the use of the bicycle to establish the velocity-pressure law was unsatisfactory because it had only been possible to estimate the speed of the bicycle, not to measure it, and the measurements of pressure were unsatisfactory as they were taken out of doors 'where the atmosphere itself was always to a greater or lesser extent in motion' [13].

24.2.2 Johann Irminger, 1894

Johann Otto Valdemar Irminger (1848-1938) was introduced to questions about air flow around curved objects by a friend and marine engineer, H.C. Vogt (1848-1928), who believed that a significant proportion of the lift on an aerofoil, propeller or sail arose from a *negative* pressure on one side, adding to the positive pressure on the other. Irminger was the manager of the Copenhagen Gas Works, and designed a wind tunnel that used the low pressure inside the gas works' 30-metre tall, 5-metre diameter chimney to draw air in through a rectangular opening 28 x 16.5 cm in the chimney wall. A duct with polished internal surfaces, 23 x 11.5 x 102 cm long was inserted into the hole (Figure 24.8). The flow rate

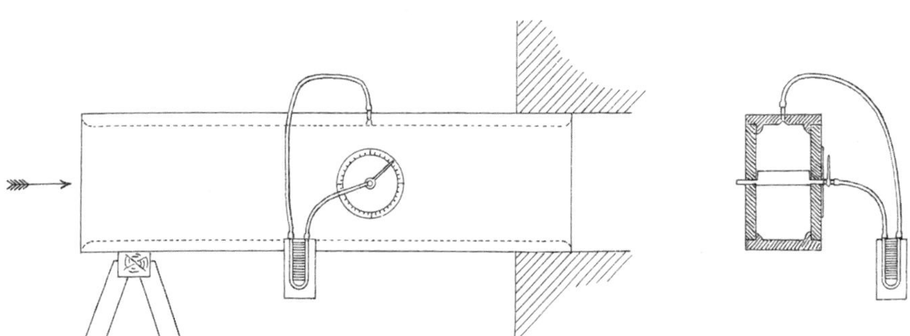

Figure 24.8 Irminger's wind tunnel, 1894. The air flow was created by suction from the 30 m tall chimney of the Copenhagen Gas Works. (Source: [15])

could be easily controlled between about 7.3 and 14.6 m/s. The test models had holes in their surface which were connected by a flexible hose to a manometer to measure the local (total) air pressure relative to the static pressure. The pressure at any hole was measured by closing off the other holes with sticking plaster. The models could also be rotated to present a different orientation to the air flow [14, 15, 16].

Irminger first tested a simple, hollow, rectangular 'aerofoil' made of sheet metal 38.1 x 2.2 mm and was able to verify his friend's hypothesis about the important role of negative pressures in generating lift on aerofoils, especially at an angle of attack of 3-5° (Figure 24.9). This revelation opened the way to a new understanding of air flow and air pressure which greatly influenced the approach to designing aerofoils.

Irminger went on to make measurements on a wide range of other shapes, including buildings with pitched and curved roofs and a gasholder (Figure 24.10). His aim was to check the many inconsistent formulae used by engineers to calculate wind loads. He found great discrepancies and, in particular, all the experiments demonstrated the large and often unexpected negative pressures acting

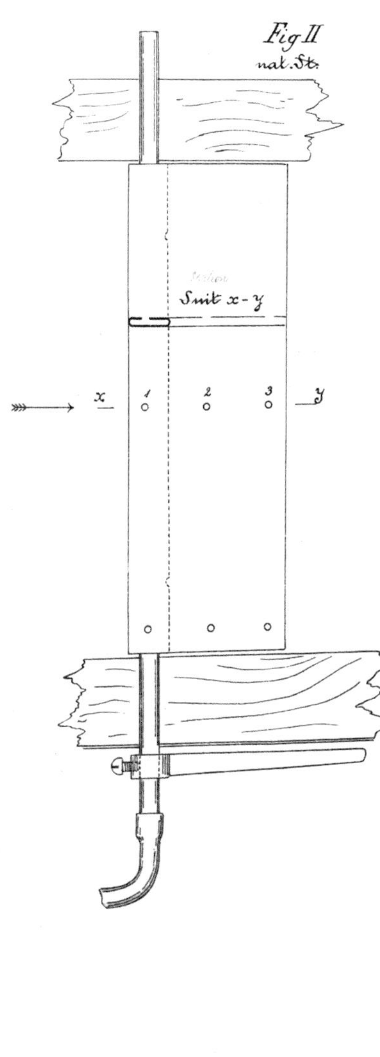

Figure 24.9 Irminger's flat-plate test model, made of iron sheet, had six holes in the surface to measure the pressure at several locations. Its orientation to the air flow could be rotated. (Source: [14])

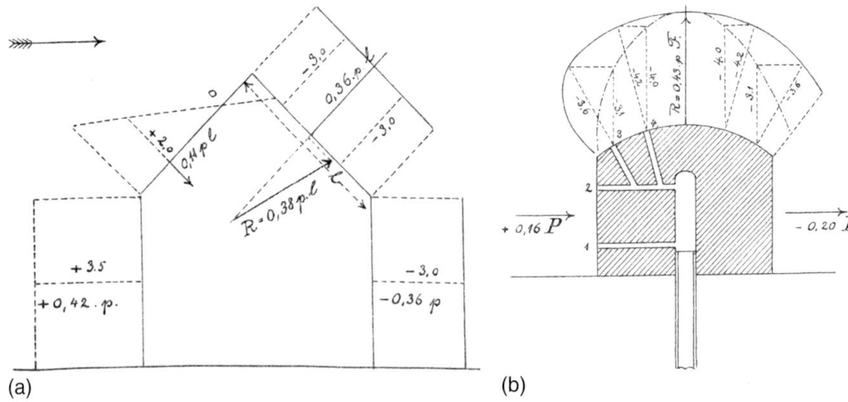

Figure 24.10 Diagrams showing pressures on two building forms measured by Irminger. (a) A building with a pitched roof; the hollow model was made with sheet metal. (b) A wooden model of a gas holder including a base plate to represent the ground surface. It had four pressure taps (holes) and could be rotated about a vertical axis. (Source: [16])

on building surfaces. This helped in finding reasons behind unexplained damage caused by the wind to a variety of building and bridges. He cited the example of a span of the Louisville-Jeffersonville Bridge in the USA that was said to have been lifted vertically off its supports in a high wind in 1894, and fell again 9 metres downwind in an upright position. His general conclusion was that 'the wind acts in a direction quite different from that assumed and usually provided for in the design of bridges, roofs and other structures; the dead weight may not be sufficient to resist such action, and our attention is, therefore, called to the necessity of securing our structures more firmly to their supports' [17].

24.2.3 National Physical Laboratory, Teddington, UK, 1903-08

The National Physical Laboratory at Teddington in south-west London, was established in 1900 and the first laboratories opened in 1902 under the direction of Thomas Stanton (1865-1931). One of the first projects undertaken by Stanton, at the suggestion of Sir Benjamin Baker, had the specific aim of determining the wind loads that act on structures by carrying out tests on small scale models in a 'wind channel', as it was called at the time. This wind channel, completed in 1903, was of circular cross-section, 60 cm in diameter, with a vertical axis. Air was sucked downwards through the test section which made it easier to measure the force on test specimens around 10 cm across using a simple balance (Figure 24.11) [18]. The accuracy with which measurements of force and air pressure were made was extraordinary. The forces were measured to an accuracy of 0.0045 grams. The air speeds achieved, mainly around 6-7 m/s (22-26 kph), were measured using a Pitot tube; using a sophisticated inclined manometer, the air pressures were measured to an accuracy of 0.014 N/m^2.

In addition to the total wind load on a surface, measurements were also taken to determine the distribution of pressure across the faces of the plates, both on the windward and leeward sides. Other tests were undertaken on a single square bay

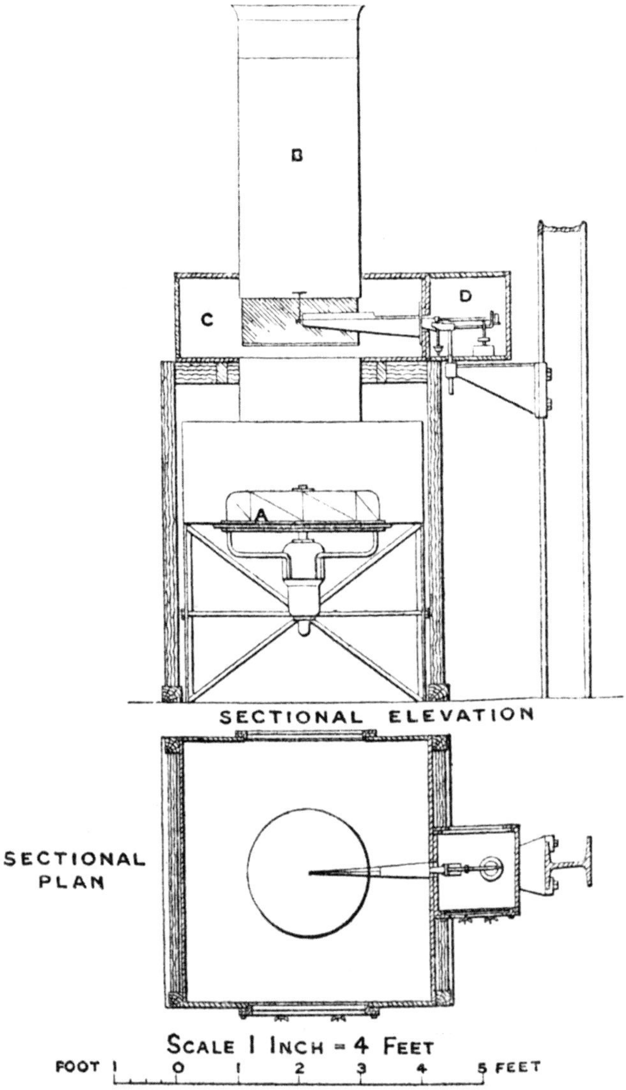

Figure 24.11 The first 'wind channel', built by Stanton in 1903, at the National Physical Laboratory, Teddington, UK. The channel was 60 cm in diameter and 1.35 m long. (Source: [18]).

of a lattice girder, and pairs of such bays, one behind the other, as would be found in a lattice-girder bridge, to determine the shielding effect of the windward lattice. In various tests on a section of lattice about 7.5 cm square, different parameters were altered – the spacing between girders, the size of the models (100%, 75% and 50%), and the width of the lattice members (3.2 and 4.8 mm).

Stanton concluded that 'it appears that the resistance of any structure in a uniform current of air may be predicted from experiments on models of the structure in an apparatus of the kind here described' [19]. He went on to test a 2.5 cm wide section of a simple rectangular building with walls 7.5 cm high, and the roof at

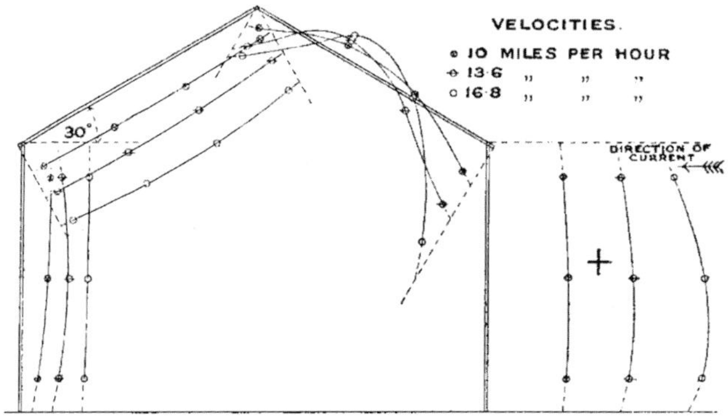

Figure 24.12 Normal pressures, positive and negative, measured on the external surfaces of a model house, with walls 7.5 cm high, due to winds from the right at 4.5, 6.1 and 7.5 m/s. (Source: [18])

three different pitches – 30°, 45° and 60° (Figure 24.12). He also tested a model of a lattice bridge girder, 17.5 x 2.5 cm, consisting of seven square bays – a scale of about 1:100. The tests, undertaken with a wind speed of 6.2 m/s, included a single girder, two parallel girders tested both with and without a roadway, and with spacings of 1, 1.5 and 2 times the height, and also orientated both normal and inclined to the wind direction. In each case the result was a wind load per unit area, at standard atmospheric temperature and pressure, that a designer could use.

Stanton's work was well-received and highly praised by most engineers who made their comments known, welcoming a return (sic) to using small-scale model testing. However, there were also those who voiced their scepticism about using such small-scale models, especially given the highly variable and often turbulent behaviour of wind, both in the test apparatus and at full size. Mindful of these comments, and the extraordinary accuracy that had to be achieved in taking measurements, Stanton stopped working with his wind channel and began a series of tests in the open air, mounting the models on a 16-metre tall tower in the grounds of the National Physical Laboratory [20]. He repeated many tests that he had previously made in the wind channel, including forces on flat plates, pitched roofs and a lattice girder, 8.8 x 1.1 m, at a scale of about 1:2, fifty times larger than the model he tested in the wind channel. The main conclusion of these tests was that the ratio of wind load on a structure such as a braced girder, to that on a square plate with the same area, is the same for models of any scale and for the full-size structures – what we call today the ratio of drag coefficients. After these tests were completed, in 1907, the NPL redirected its wind-channel research towards aviation [21].

24.3 Making visible the dynamic behaviour of fluids

24.3.1 Schlieren photography

The *schlieren* technique is based on the phenomenon observed when air is seen rising from a hot road surface. (*Schlieren* is the German word for 'streaks'). The varying density of the air causes light to be refracted to different degrees and create a distorted image. When set up as a photographic technique, light refracted in two different directions passes through filters of different hues or colours, revealing areas of high or low pressure. It was invented by the German physicist August Toepler (1836-1912) in 1864 to study supersonic motion, and used by Ernst Mach in 1887 to photograph the shock waves of a supersonic bullet. It is still widely used in supersonic wind-tunnel research. In building engineering it was used by Wallace Sabine and Franz Osswald to make visible the air waves in their acoustic model studies (see Section 26.1.1). Today it is used in water-bath model studies of air movement in buildings (see Chapter 36).

24.3.2 Visualisation of flow lines

In the mid-1890s Henry Hele-Shaw (1854-1941) devised a means for making visible the streamlines in flowing water that was based on the fact that a fluid adjacent to a boundary surface cannot become turbulent. His apparatus consisted of two sheets of thick glass separated by as small a distance as was practical – typically about 0.3 mm, sealed on two opposite sides to create a channel and allowing water to flow between the other pair of opposite sides [22, 23]. A large number of coloured streams were introduced at one side, which flowed past an object and out at the other side. In one such apparatus used at the National Physical Laboratory in Teddington, the die used was Nestlé condensed milk [24]. The technique could demonstrate both streamline flow, recorded photographically (Figure 24.13a), or changing and turbulent flow such as eddies in the wake of an object, recorded using film (Figure 24.13b). These eddies were later called von Kármán vortices, after the Hungarian aerodynamicist Theodore von Kármán (1881-1963), which were so significant in helping to explain the collapse of the Tacoma Narrows Bridge in 1940 (see Section 24.5). The Hele-Shaw apparatus was used mainly as a demonstration with relevance to water flow around ships and propellers, and air flow around aircraft components. While potentially of interest to civil engineers, there is no evidence they made use of the technique at this time.

At about the same time as the work of Hele-Shaw, the French photography pioneer Étienne-Jules Marey (1830-1904) developed a similar technique using smoke trails in streamline airflow. He injected fifty-seven lines of smoke into the airflow,

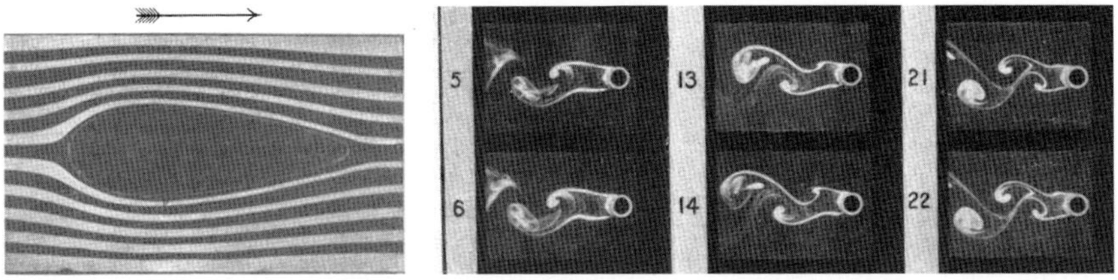

(a) (b)

Figure 24.13 Hele-Shaw visualisation of water flow. (a) Streamline flow past an aerofoil (Source: [23]); (b) Turbulent flow with eddies being shed in the wake of a cylindrical object. (Image: National Physical Laboratory)

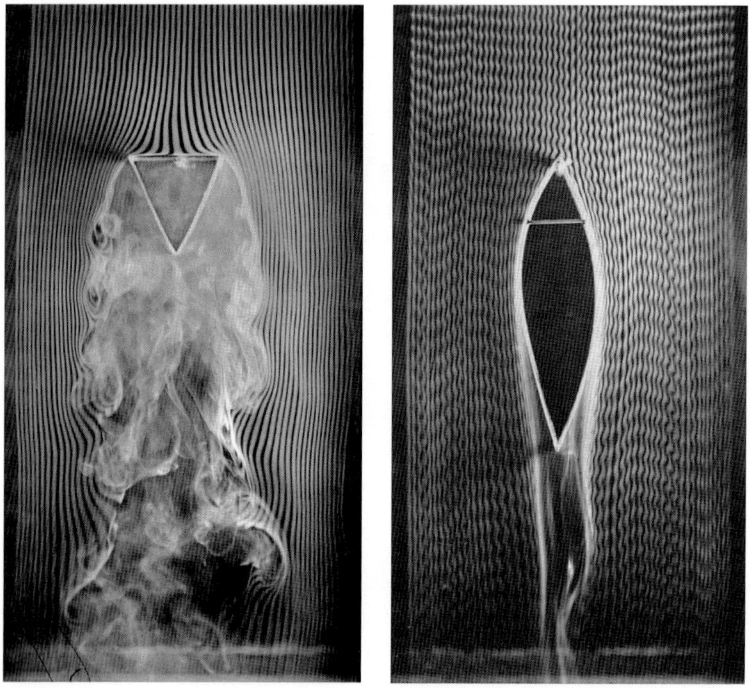

(a) (b)

Figure 24.14 Photographs of smoke trails in air created by Étienne-Jules Marey c.1890. (a) Turbulent flow behind a bluff object. (b) Flow lines with pulsed smoke trails to enable local velocities to be measured. (Source: [26])

upstream of an object being studied. Although probably not the first time this idea was used, his photographs make his contribution significant. They revealed streamline flow around some objects, giving also an indication of speed, as the streamlines become more closely spaced when speed increases [25, 26]. They also show turbulent flow in the leeward side of bluff objects, including the shedding of eddies, alternately from each side of an object (Figure 24.14a). Marey went further and created timelines in the smoke streams by vibrating the smoke injector sideways to produce regular ripples [27, 28]. The spacing of the undulations along a given streamline indicated the local velocity of the fluid which could be measured on photographs using a ruler in the field of view (Figure 24.14b). Despite the quality of the photographs and the content, it is not known whether his work influenced civil and building engineers.

A more precise means of estimating local velocities of air was developed by several experimenters, using water as the model material and timed photographic exposures to record the movement of particles on the surface of the flowing water. At the NPL a technique using droplets of oil on the water surface was developed. The water surface was filmed and then a series film frames, taken at the same point in the repeating cycle of shedding eddies, was projected onto a screen. The lines made by the moving droplets in each image were drawn on the screen and a composite picture of the streamlines was built up.

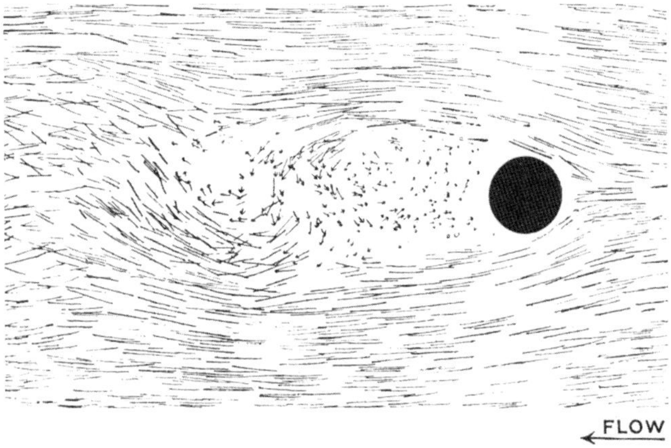

Figure 24.15 Model test for measuring flow velocities. The flow of water, as a model for airflow, around a cylindrical obstacle was made visible using droplets of oil, c.1915. (Image: National Physical Laboratory)

The image shown in Figure 24.15 was built up by the superposition of drawings from eighty separate photographs [29]. The length and directions of individual lines gave absolute velocities. If the technique had been perfect, the streamlines would not have crossed but, nevertheless, the image makes the flow pattern very clear. As with the visualisation of streamlines by Hele-Shaw and Marey, it is not known if civil engineers made use of these techniques in their work as designers. Nevertheless, they became well-known at least as qualitative models for the behaviour of air passing objects such as bridges, towers and radio masts.

In the late 1920s, Otto Flachsbart also used water channels to model the flow of air around structures, making the flow visible by sprinkling aluminium powder on the water surface and photographing the results to enable flow speeds and directions to be measured (see Section 24.4.1).

24.4 Wind-tunnel model studies in the 1920s and 1930s

The use of wind tunnels grew rapidly during the 1920s and 1930s mainly, of course, for testing aeroplane designs and also for various studies to streamline the profiles of trains, cars and ships. Wind tunnels grew in size and experimental techniques improved, although they were still firmly based on the approaches established in the decade following Irminger's work pioneering work in the mid-1890s. The use of wind tunnels for building and civil engineering projects benefited from these improvements, but there were still no tunnels built exclusively for these branches of engineering – wind-tunnel time had to be rented from the aerodynamicists. All major countries expanded their wind-tunnel test facilities in the 1920s and 1930s but here is not the place to report these developments. For our purposes, this era can be represented by the work in three countries – Otto Flachsbart at Göttingen and Hannover in Germany, Johann

Irminger and Christian Nøkkentved in Denmark, and Hugh Dryden and George Hill at the United States Bureau of Standards.

24.4.1 Model studies in Germany

The German engineer Otto Flachsbart (1898-1957) began his lifelong career in fluid mechanics as a student of Ludwig Prandtl at Göttingen, and gained his doctorate in 1928 with a historical study of the early seventeenth-century hydraulic systems at Goslar in the mining region of the Harz Mountains, which formed the world's most important pre-industrial power generation and supply system. Working with Prandtl and his closed-circuit wind tunnel at Göttingen, he conducted various studies concerning wind loads on structures. After moving to the Leibniz University in Hannover, initially as Professor of hydro-and aerodynamics and, from 1932, as Professor of mechanics, he constructed a wind tunnel and continued his research using physical models to provide a foundation for understanding wind loads.

Flachsbart began his study of wind loads on buildings under Ludwig Prandtl in 1926. His early series of tests were on the pressures exerted by wind on simple building forms of different heights, widths and roof forms, similar to those studied earlier by Irminger and at the NPL. These tests were extended in 1927 to include his second series of tests studying wind pressures acting on pairs of simple building forms, spaced at different distances, and a series of five such buildings, again at different spacings. The models were 16 x 8 x 8 cm high with a roof pitch of 40° and each was tested at three different orientations to the wind, 0°, 45° and 90°. The pressures were measured at thirty-six points on all external surfaces [30].

Flachsbart also tested models of a large racecourse grandstand with a flat roof. The model was 32 x 16 x 8 cm high, open on the windward side, and both external and internal pressures were measured at forty-two points. The model was tested both with and without side walls, as well as a further test with a 2 cm horizontal slot along the full width of the back wall of the grandstand. (Figure 24.16). Flachsbart conducted these tests under two different conditions – in the normal airflow in the wind tunnel with a uniform wind-speed profile, and also with the models on a flat surface creating a boundary layer with a wind-speed profile similar to that actually experienced above the ground. In other words, it corresponded to the wind-speed distribution in the Prandtl atmospheric boundary

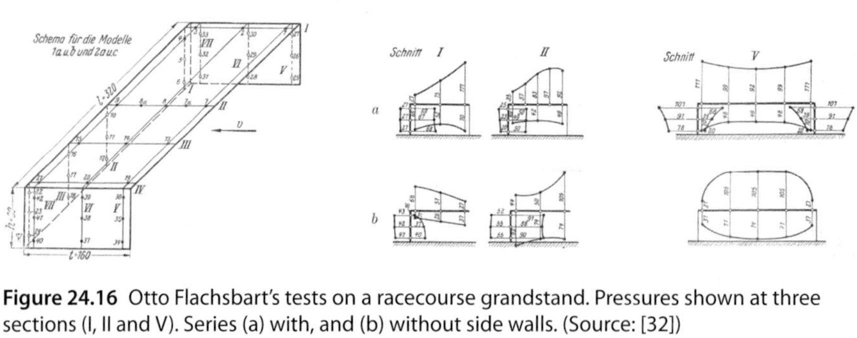

Figure 24.16 Otto Flachsbart's tests on a racecourse grandstand. Pressures shown at three sections (I, II and V). Series (a) with, and (b) without side walls. (Source: [32])

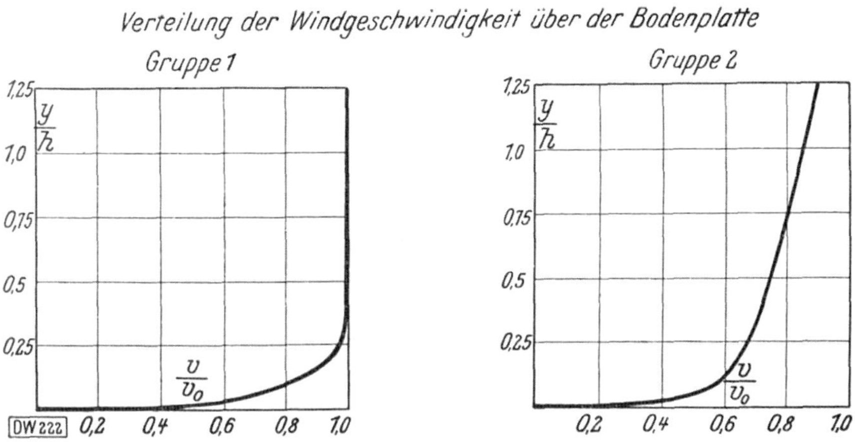

Figure 24.17 The two profiles of wind speed used by Flachsbart. (a) Normal, uniform profile. (b) Non-uniform profile corresponding to the wind-speed distribution in the Prandtl atmospheric boundary layer. (Source: [32])

layer (Figure 24.17). Finally Flachsbart conducted several tests varying only the airspeed in the wind tunnel, from 10 to 30 m/s, and demonstrated that the pressure coefficients for his model were largely independent of the Reynolds Number.

To investigate the nature of the airflow around the grandstand, he used a water channel, 60 x 60 cm in section, originally made for towing models of ship hulls. He made the flow around the model visible using aluminium powered sprinkled on the water surface. The model was mounted on a carriage, together with the film camera. The carriage was moved along at a constant speed with the model partly immersed in the water, with a plate fixed to the lower part of the model to ensure the flow was two dimensional. He filmed the generation of eddies as air flowed over several shapes of the roof, and determined the local velocities by measuring the image in a single frame of the film [31] (Figure 24.18).

Flachsbart began publishing papers after he moved to Hannover where he was able to use his new wind tunnel. Some of the papers were of a general nature, emphasising the importance of considering both the precise nature of the wind and its gusts, and also the aerodynamic behaviour of air impacting on structures and the dynamic response of these structures [32, 33, 34, 35]. He also used his wind-tunnel tests on models to provide the fundamental basis needed for the design of structures, as well as the application of aerodynamic theory to analysis of the problems and interpretation of the experimental results. He provided the results of his investigations and specific guidance on designing a wide range of

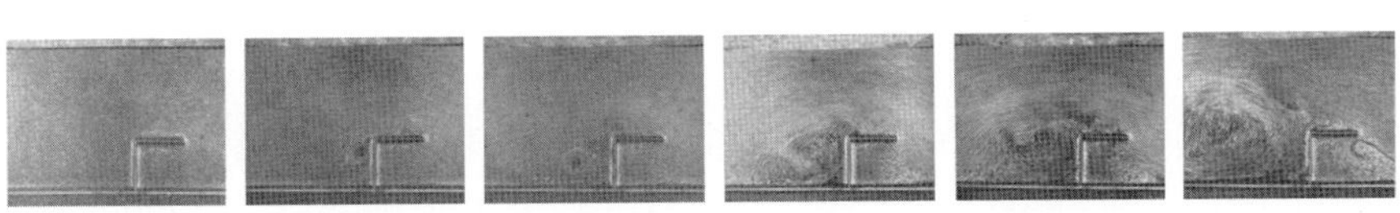

Figure 24.18 Single frames from film of a model of a racetrack grandstand, showing the build-up of a trailing vortex. The model was towed through the water and aluminium powder was used to make the flow visible. Otto Flachsbart, 1928. (Source: [31])

Figure 24.19 Models of various buildings tested by Irminger. (Source: [41])

structures, including chimneys [36], closed and open buildings [37], gas holders [38], two-dimensional [39] and three-dimensional frame and lattice structures [40].

24.4.2 Model studies in Denmark

Johann Irminger, who had built the first wind tunnel for testing buildings in 1894, renewed his work using models to determine wind loads in 1927 when he began, at the age of 80, collaborating with Christian Nøkkentved (1892-1945) at the University of Copenhagen. Now using a standard closed-circuit tunnel, they developed Irminger's earlier work by increasing the number and range of building shapes, with sheet-brass models typically 5 cm high up to eaves level (Figure 24.19). They produced two comprehensive reports with a multitude of graphs of pressure distribution [41] (Figure 24.20). These results provided a basis for wind-load calculations in Europe for several decades.

In parallel with this work, Nøkkentved made model tests to study turbulence in the wake of buildings. For this he used the water-channel equipment that Flachsbart had made at Göttingen. He demonstrated the formation of a large, stagnant vortex in the wake of a building which effectively formed a streamline shape past which the moving fluid could flow easily. The existence of this vortex gave a physical explanation for the negative pressures he had measured in the wind tunnel [42]. He also demonstrated the formation of a small vortex at the foot of the windward face of the building in some tests. Nevertheless, Nøkkentved found that a region of suction can change into a region of pressure on a building, depending on the height of the boundary layer in relation to the height of the building. He observed very differing pressure distributions, depending on the ratio of these two heights, which could be varied by systematically changing the surface roughness of the wind tunnel floor, ranging from lacquered to corrugated cardboard. Different suction patterns were observed for each roughness, leaving the researchers puzzled and needing to make more measurements in the real world at full size [43].

24.4.3 Model studies in the USA

Having already obtained his PhD at the age of 20, Hugh Dryden (1898-1965) was appointed the director of the newly-created Aerodynamics Division of the National Bureau of Standards in the USA. Although the work focused mainly on aircraft studies, he also undertook studies related to wind loads on building. His criticism of previous work was that the models were often simple geometric

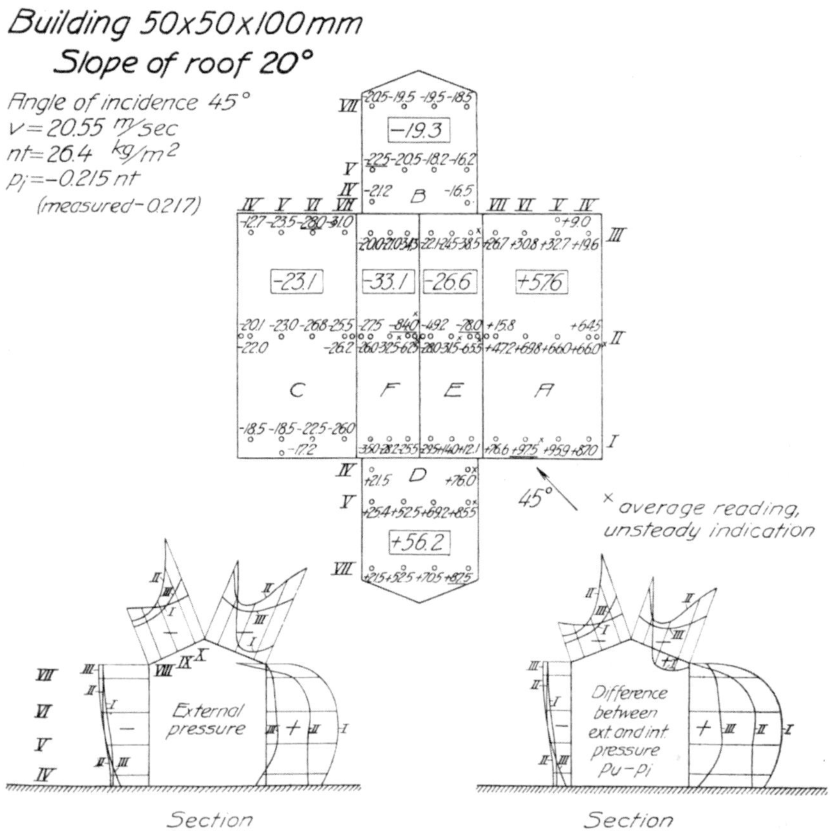

Figure 24.20 Pressures measured on a model building tested by Irminger. (Source: [41])

shapes or very crude models of buildings. He advocated the use of models that represented the buildings being studied much more faithfully.

The first model he chose to study was a small, square, high-rise building [44]. The wind tunnel was of the NPL type, 28 m long, with a working section 15.3 m long and 3 m in diameter (Figure 24.21). The model was made of brass sheet 6 mm thick, and was 20 x 20 x 62 cm tall. Seven columns of ten pressure taps, each 1 mm in diameter, were drilled in one face of the model; a further 16 holes were drilled in one quadrant of the top face, which could be placed in any orientation by rotating the model about its vertical axis (Figure 24.22). For each hole a nipple was fixed on the inside to allow a rubber tube to connect it to the manometer (Figure 24.23). Two different techniques were used to represent the ground on which the building sat and for each, an accurate plot of wind speed was made to enable corrections to be made to pressure measurements. In one, the model rested on the floor of the tunnel, for which there was a great variation of speed across the width of the model; in the other, the model was mounted on a flat board in the centre of the tunnel and the air speed was found to be almost uniform. Measurements of pressure were taken at orientations of the model to the airflow at 15° intervals, and for air speeds of approximately 6, 12, 18 and 24 m/s.

Figure 24.21 The NPL-type wind tunnel at the US Bureau of Standards, 1925. (Source: [44])

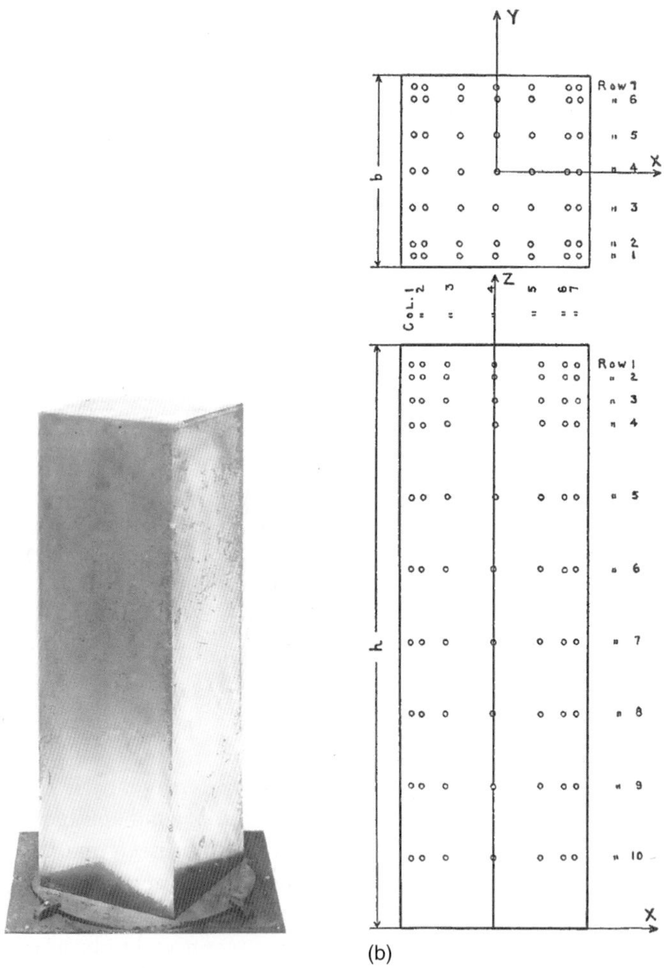

(a) (b)

Figure 24.22 (a) The model skyscraper, made of 6-mm thick brass sheet. (b) The locations of the pressure taps. (Source: [44])

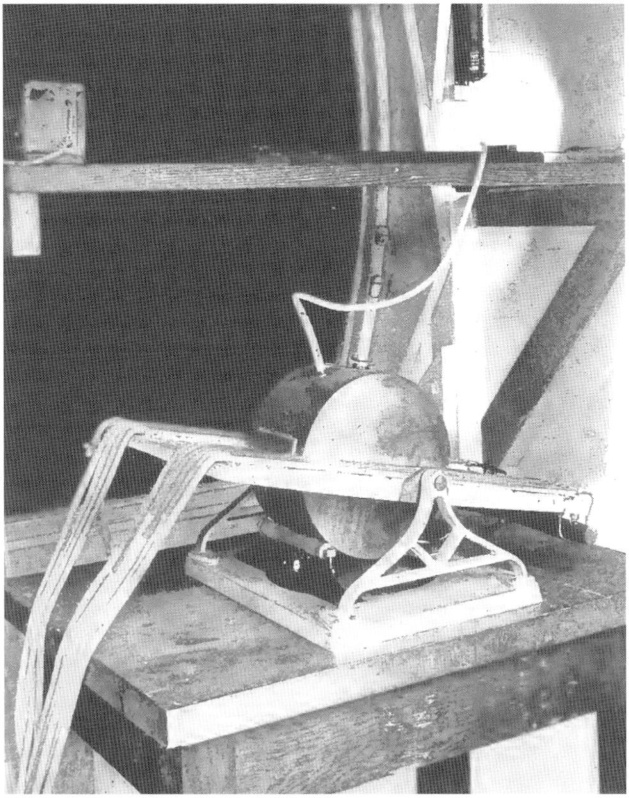

Figure 24.23 An array of ten inclined manometers connected by rubber tube to ten pressure taps. (Source: [44])

The many tests generated a huge number of quantitative data which were summarised in a series of pressure distribution maps to enabled the worst-case scenarios to be identified, including the greatest positive and negative pressures on the faces of the building, and the greatest negative pressure on the top of the building (Figure 24.24). Although not discussed in detail in their study, it was noted that individual data would be valuable for the designers of individual structural elements and the façade systems and windows in a high-rise building. The key conclusions of the study were more general: the greatest overturning moment on the building, and the greatest torsion about a vertical axis.

Dryden and Hill went on to study the wind loads on tall chimneys, comparing the results from tests on models – an experimental chimney made with a timber structure and lightweight cladding, and a real chimney – demonstrating and discussing certain consistencies and inconsistencies among the results [45]. They were aware of the characteristic vortex shedding from such structures, but did not study the phenomenon. They then went on to undertake model tests on a 'mill building' which was essentially a low-rise rectangular building with a pitched roof [46].

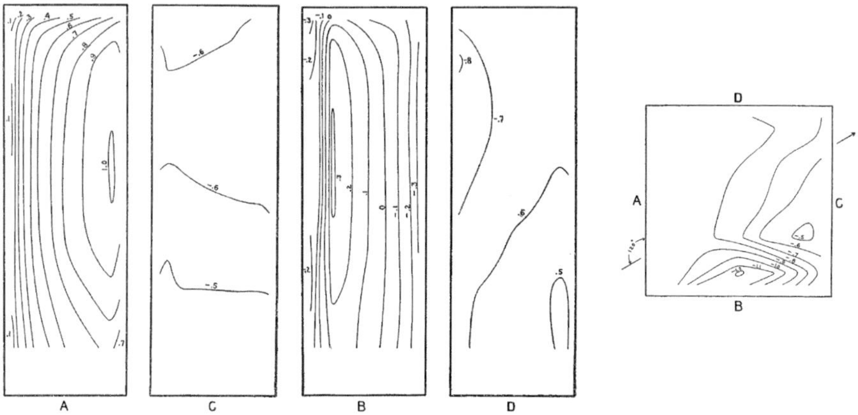

Figure 24.24 Equal-pressure contours showing the pressure distribution over the four faces and the roof of the building when face A was orientated at 120° to the wind direction. (Source: [44])

Figure 24.25 1:250 scale model of the Empire State Building in the US Bureau of Standards wind tunnel, looking downstream. (Source: [49])

Probably the best known of the building study made by Dryden and Hill was for the design of the Empire State Building in New York in 1932 [47]. The 1:250 scale model of the building, 1.53 m tall, was made of rolled aluminium plates, 6 mm thick, except for the top tower, which was constructed of wood. The test procedure followed that developed in their 1926 study on an idealised high-rise building [48]. There were thirty-four pressure taps around the perimeter of the model at three different heights, and a number of additional taps in the building edges at various heights (Figure 24.25). The results consisted of values for the maximum positive and negative pressures and it was noted that the areas of negative pressure were of far greater extent than had been expected. It was also demonstrated, by attaching a fine thread to the model, that the air passing over the top of the building then flowed downwards and formed a large eddy that extended a metre downstream. They also noted that, when the wind was normal to the face of the building, the pressures were greatest at the centre, reducing towards the corners. This led them to recommend not placing the entire wind-bracing structural system in the end panels of the building, and to include some in the centre.

Given the non-uniform geometry of the building, it was decided to make an independent measurement of the overturning moment on the building. This was achieved by placing the base of the model on a fulcrum, and fixing wires to the model and the wind-tunnel walls so that the out-of-balance force in the wire fixed near the top of the model could be measured using a spring balance (Figure 24.26).

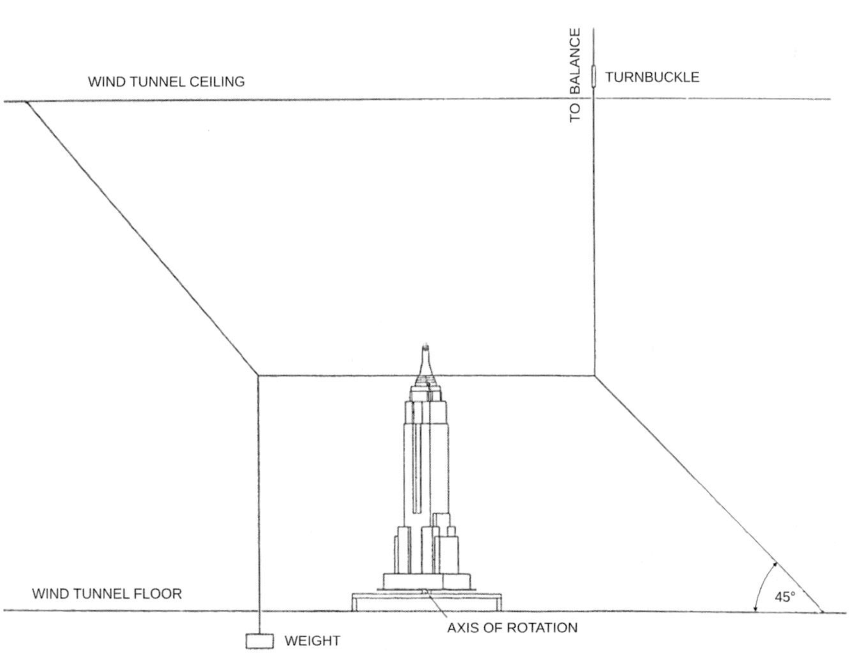

Figure 24.26 Diagram of the arrangement for measuring the overturning moment on the model of the Empire State Building. (Source: [47] (Relabelled))

24.5 The Tacoma Narrows Bridge collapse, November 1940

The spectacular collapse of the Tacoma Narrows suspension bridge near Seattle, on 7 November 1940, in a wind of 19 m/s while twisting at a frequency of 36 cycles per minute, was recorded in some of the most spectacular minutes of film ever shot [50]. One outcome of this collapse was a 14-year programme of research into the behaviour of suspension bridges, which included a considerable use of wind tunnels. It led to a revolution in the way that suspension bridges were designed, introducing, for the first time, wind-tunnel model tests as an essential stage in the design procedure [51, 52].

24.5.1 The static model used for the design of the bridge

Following the normal practice for the design of large suspension bridges at the time, Frederick Farquharson (1923-1961) had built and tested a 1:100 static model at the University of Washington in 1938 (Figure 24.27) [53, 54]. The purpose of this was to study the deflections of the towers and cables arising from alternative sequences of constructing the main span. The model was highly simplified, with only one suspension cable and single weights representing the sections of the bridge deck. The main advantage of using the model was speed – a mathematical computation of a new load case would occupy two men for 1-2 days, whereas the same results could be achieved using the physical model in ten minutes.

24.5.2 Dynamic model studies undertaken before the collapse

Problems with the Tacoma Narrows bridge were clear for all to see as soon as the roadway was in place in early 1940. In relatively light winds, the deck

Figure 24.27 1:100 scale model used to undertake static tests to help plan the construction sequence. (Source: [56] Figure 25) (Image: 1949 Farq I Fig.25 p. 35)

Figure 24.28 1:100 scale model used to undertake dynamic tests and determine the natural frequencies for horizontal and vertical vibrations. (Source: [56] Figure 30) (Image: 1949 Farq I Fig.30, p. 46)

of both the side spans and the main span of the bridge oscillated in several ways – horizontally, vertically and in a twisting fashion. It already seemed likely that this problem might be linked to its unprecedented slenderness – the span-to-width ratio for the deck – which gave a very approximate indication of its stiffness for all three modes of oscillation. In the early 1930s, a span-to-width ratio of forty was typical; in the Golden Gate Bridge (1933-37) the ratio was forty-eight; in the Deer Island Bridge (1938) it was fifty-four, and for the Tacoma Narrows Bridge it was seventy.

To study the benefits of various remedial options, Farquharson built a dynamic model of the bridge, also at 1:100 scale giving 'full dynamic similitude' with the full-size bridge. This model had two suspension cables and separate weights to represent the weights and inertia of the cable and sections of the deck (Figure 24.28). A horizontal stiffening truss in the roadway was also included. The dynamic load, equivalent to a wind speed of 45 m/s, was applied horizontally to the deck of the model using 100 electromagnets, able to impose a maximum lateral deflection of 61 mm at mid-span. By applying the point loads in various different sequences, and by tuning the stiffness and the damping effect of the towers, two natural frequencies were determined: for horizontal vibration of the deck it was found to be 42 cycles per minute, a period of 1.43 seconds, equivalent to 14.3 seconds in the full size structure. The natural frequency in a vertical direction was found to be 0.789 s (7.89 s at full scale) [53, 55, 56]. No results were taken for torsional vibration, not least because the loading technique employed could not excite such movement. A study was also undertaken to determine the increase in damping that would be achieved by introducing stays between the centre of the side spans and ground anchors in the estuary bank. The most-effective locations for the hold-down stays were found to be 106 m and 82 m respectively from the east and west ends of the bridge which increased the logarithmic damping by 27%; they were installed in April 1940. A large number of model studies were also made to determine the benefits of stays from the top of each tower to the road deck, in both the main and side spans, but these were not implemented for a number of practical reasons, including their different behaviour in different ambient temperatures. All this work was finished by spring 1840, before the bridge had been completed.

During this same period, in early 1940, the effects of wind on the road decks of suspension bridges were being investigated. Static wind-tunnel tests were undertaken by Farquharson at the University of Washington on drag and lift forces on a slightly-inclined deck, including ways of reducing drag, either by cutting holes in the deck girders, and by adding a fairing on the outer faces of the girder [57, 58, 59].

After these tests it was decided in September-October 1940 to modify the nearly-complete bridge by adding fairings to the deck girders but the bridge collapsed before work could begin. Around this time some model tests were also undertaken at the Case School of Applied Sciences using a water channel to investigate the generation of eddies in the wake of the flow around the I-shaped section of the deck [60]. Although these confirmed the production of eddies, no evidence of *alternating* eddies was observed and it was concluded, therefore, that the generation of eddies did not contributed to exciting periodic torsional loads on the bridge deck – a conclusion that was later proved to be incorrect.

24.5.3 Model studies begun immediately after the collapse

After the collapse of the bridge in November 1940, a further series of model tests were undertaken in the wind tunnel of the Guggenheim Aeronautical Laboratory at the California Institute of Technology and were published in the report of the Board of Engineers in March 1941 [61]. The director of the laboratory was the eminent Hungarian aerodynamicist Theodore von Kármán who was also one of the Board of Engineers who conducted the inquiry into the collapse. He was a former pupil of Ludwig Prandtl in Göttingen and had first identified the periodic, alternate shedding of vortices in the wake of a cylinder in 1911. Disturbed by political developments in Germany, he accepted the invitation to become director of the Guggenheim Aeronautical Laboratory in 1930. The 1:80 scale model, 3 x 20 x 109 cm, made of brass, represented an 87 m long section of the bridge deck and was suspended from a support rig, developed by some trial and error, to provide the appropriate damping and resistance to torsion (Figure 24.29). The tests were able to replicate the torsional instability of the deck at certain low wind speeds. Several tests were carried out on similar models of the bridge deck with truss stiffening rather than solid plates; torsional instability could not be demonstrated in these models, giving the first clear evidence that eddies might, indeed, have been a significant cause of the oscillations [62].

Based on the outcome of these tests on models of short sections of the bridge deck, a full model of the bridge, at a scale of 1:234, was tested at the California Institute of Technology [63]. It was 5.7 m long with a road deck made of a hard rubber extrusion 5 cm wide; the main span was 3.6 m. With the towers in the model pin-jointed at their foot, it was found that there was too much damping to reproduce the torsional oscillations. By changing to a rigid-foot connection and with a spring connection between the cables and the towers, the damping was reduced and torsional oscillations were reproduced (Figure 24.30).

Figure 24.29 1:80 scale model of a section of the road deck, tested in the wind tunnel at the Guggenheim Aeronautical Laboratory in the California Institute of Technology. (Source: [55] Figure VIII-23) (Image: Amman et al Fig VIII-23)

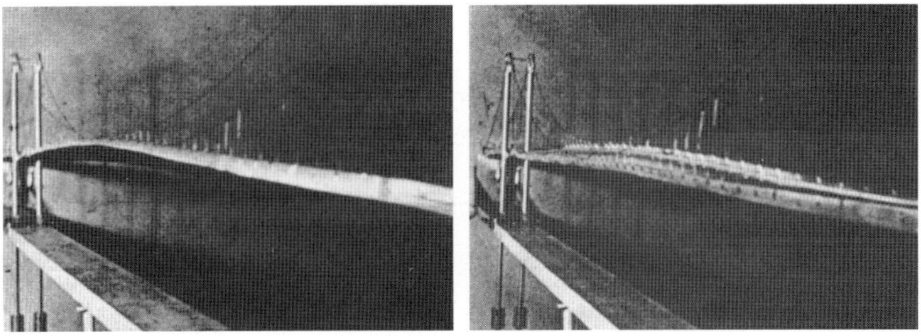

Figure 24.30 1:234 scale model in the wind tunnel at California Institute of Technology displaying torsional oscillation in a 2.55 m/s air flow. (Source: [55] Figure VIII-26)

Finally in this period of the research programme, some further tests were undertaken at the Massachusetts Institute of Technology and the Washington State College in summer 1941 to try and demonstrate the production of periodic, alternating vortices in the wake of the bridge deck section [64]. These tests used an unusual application of birefringence (or double refraction) observed using a polariscope. It had been discovered that a dilute colloidal suspension of California bentonite displayed a degree of birefringence proportional to the rate of shear in the fluid. This had the effect of making the fluid flow visible in a manner similar to that achieved using schlieren photography or, more crudely, aluminium powder of the surface of water. A 1:100 scale model of the bridge deck section, 122 x 25 mm, was placed in a stream of bentonite 9.5 mm deep, and the flow for various angles of attack to the stream were observed using a polariscope [65] (Figure 24.31). The results for the bridge section demonstrated the generation of eddies in the wake of the deck, but not periodic, alternating eddies. However, such eddies were demonstrated for a rectangular box section at

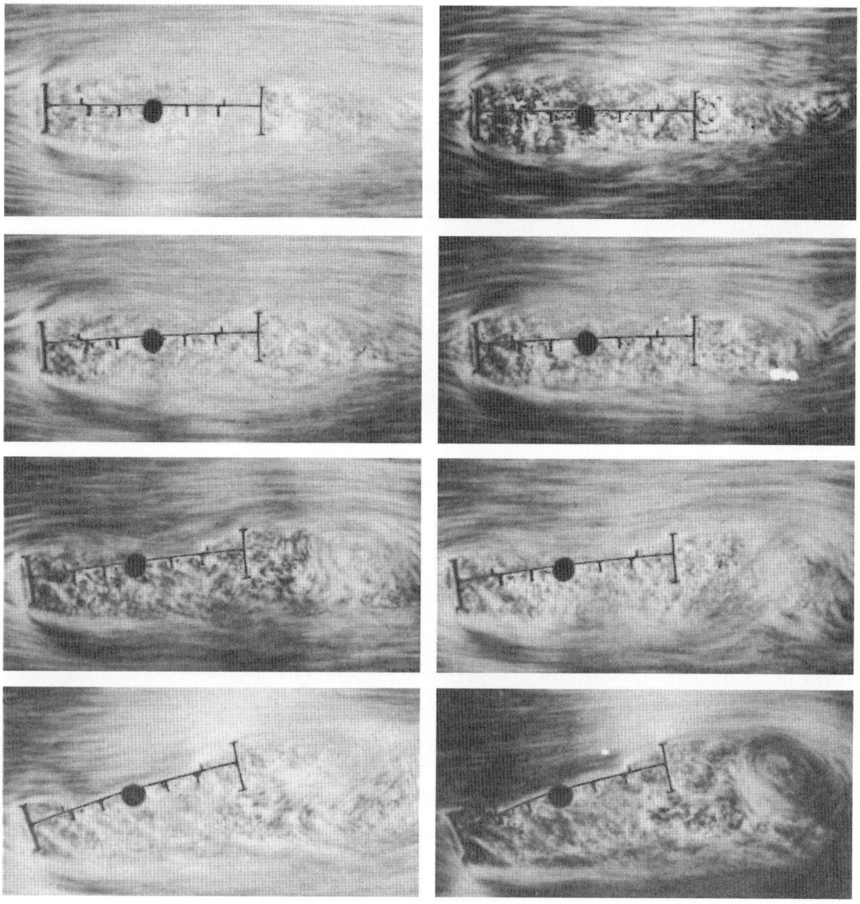

Figure 24.31 1:100 scale model in a flow channel, using a birefringent colloid bentonite, and made visible using a polariscope. Speed: left 1.95 m/s; right 2.44. Angle: top to bottom 0°, 3°, 7°, 15°. (Source: [65])

certain fluid velocities and angles of incidence. Although interesting, this work did not influence the main Tacoma Narrows investigations.

24.5.4 Studies using the full aerodynamic model of the original bridge, at 1:50 scale

In order to test a full 1:50 scale model of the entire bridge, it was necessary to design and build a bespoke wind tunnel, large enough to take the 30-metre long model, with a main span 17 metres long [66]. A closed-circuit tunnel was built in 1942 at the University of Washington in which the curved timber roof, 9.5 metres high, also served as the return duct for the air (Figure 24.32). The air was blown around the circuit using 10 propellers in a line, each 2.29 m in diameter. These pressurised a plenum chamber from which the air was forced through damping screens into a 4:1 reduction in the height of the air jet. The tunnel could deliver

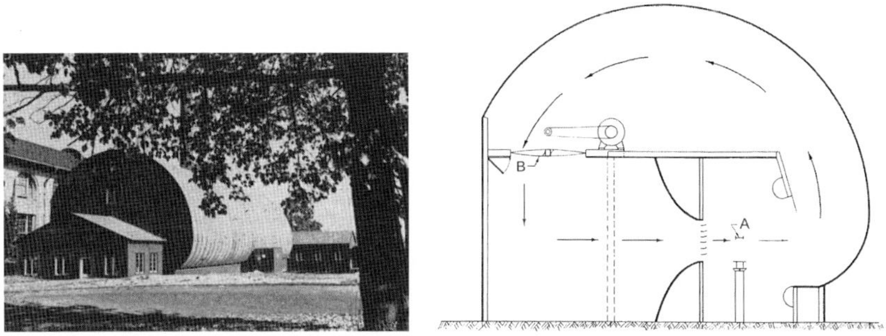

(a) (b)

Figure 24.32 (a) The suspension bridge laboratory, University of Washington. (b) Cross-section of the wind tunnel. Model suspended at A; propellers at B. (Source: [66] Figures 2 and 10)

a smooth wind at speeds from 0.22 m/s up to 8.1 m/s which was equivalent to a wind of 200 kph at full scale. The throat delivering the air was 1.2 m high and about 30 m long. The first task of these tests was to reproduce the behaviour of the original bridge that had been observed. This served to prove the wind tunnel and the testing technique (Figure 24.33).

Following the proving of the wind tunnel and the full 1:50 scale model of the original design, tests were undertaken to study the effect on the oscillation of the bridge of the stiffness provided by the backstays and the side spans of the bridge (by removing them, in turn). The bulk of the following tests were undertaken on short sections of the bridge deck, mainly about 1.5 m long, as such models were quicker and cheaper to make (Figure 24.34). A crucial aspect of this work was developing the mountings for the section models that would allow dynamic movement and provide the appropriate restraint and damping due to the suspension cables and adjacent lengths of the bridge deck. Various mountings were used for models at scales of 1:50, 1:31.6, 1:47 and 1:90. The tests were made to determine the effectiveness of various modifications to the deck in achieving a more streamline section and to reduce vortex shedding. These modifications included curved fairings on the outer faces of the solid deck girder, different arrangements of the pedestrian walkways, and slots in the deck to allow air to pass through when the deck was inclined to the wind direction. The effectiveness of the slots was also studied using streams of titanium tetrachloride smoke to make the airflow visible [50, 67] (Figure 24.35).

In parallel with these tests at Washington University, Theodore von Kármán, at the California Institute of Technology, undertook a thorough investigation using section models at 1:90 and 1:30 scale to determine the lift, drag and moment characteristics of different cross-sections, especially their tendency to create and shed vortices. The tests proved to von Kármán that the oscillation of the original bridge had indeed been due to vortex shedding, although a slightly different explanation is sometimes given today [68]. A key conclusion of these various tests on section models was that truss stiffening of the bridge deck was preferable to a solid plate girder, both in terms of the stiffness it could provide and the reduced generation of vortices.

Figure 24.33 Full 1:50 scale model of the original Tacoma Narrows bridge, in a wind of 1.16 m/s, oscillating in vertical mode at a frequency of 1.97 cycles per second. (Source: [66] Frontispiece)

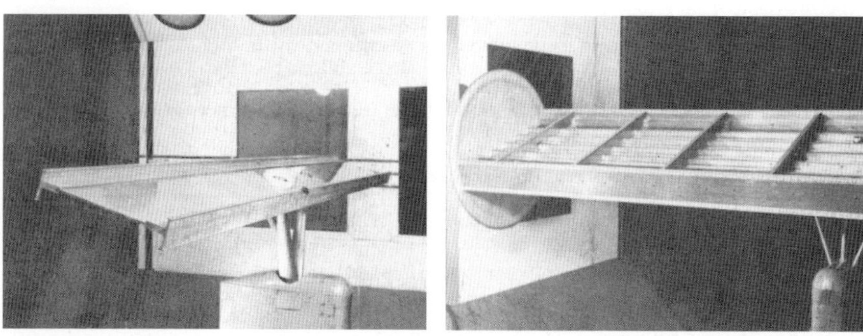

Figure 24.34 Tests on 1:50 scale sections about 1.5 m long, of the road deck to study the effects of various slotted decks and arrangements of the pedestrian walkways, on reducing turbulence and wind resistance. Left: the original deck section. Right: Slotted deck. (Source: [67] Figures 51 and 52)

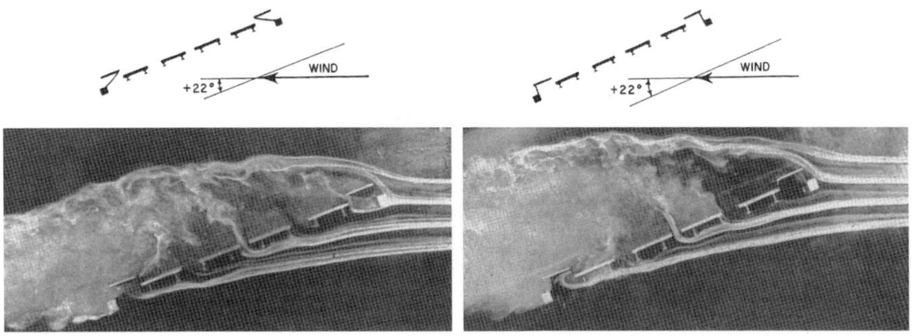

Figure 24.35 Tests on 1:50 scale sections, 75 mm long, of the road deck to study the effects of various slotted decks and arrangements of the pedestrian walkways, using streams of smoke to make the airflow visible. (Source: [67] Figure A-11)

24.5.5 Studies using a 1:50 scale model for the design of the new Tacoma Narrows Bridge

Following the extensive model studies to determine the cause of the collapse of the original Tacoma Narrows bridge, the engineers turned their attention to the design of the replacement bridge, making use of the various model-testing techniques they had developed [69]. Following the conclusions of the model studies on the first bridge design, the road deck was stiffened using deep trusses rather than solid plate girders. A full model of the proposed new bridge, at a scale of 1:50 and 30.5 metres in length, was constructed for testing in the facility built to test the full model of the first bridge (Figure 24.36). It was in regular use during 1943-1945, and used occasionally until 1950, during which time the model was modified many times. However, the cost of each modification was considerable – if a change was made to the road deck, each of the 162 sections of deck had to be modified. To accelerate the test programme, extensive studies were also undertaken by von Kármán at the California Institute of Technology, on 1.54 m long sections of the deck at 1:50 scale, mounted on sprung supports. The section model tests were used to compare the effectiveness of various ways of reducing lift and the formation of vortices, including a number of different arrangements of spoiler mounted on the deck truss, streamlining the deck using pedestrian walkways outside the main hangers, and introducing slots into the deck. Different arrangements of slots were also studied using smoke to make the air currents visible, as had been done in the studies of the first bridge. The results of the section tests were used to define a final design for the bridge for which a new 1:50 scale model was then built. Apart from verifying the overall adequacy of the design, these tests were also used to correlate the results obtained using section models and the full model. This provided a valuable demonstration of the accuracy of section model tests as a means of predicting the dynamic behaviour of the full-size bridge.

Figure 24.36 1:50 model of the preliminary design for the new Tacoma Narrows Bridge (wind from right to left). Beneath the full model is a section model. Key: A - adjustable pitot tube; B - pitch meter used in settting angle of attack; C - one of 21 pitch meters used in the full model; D - SR-4 strain gauge mounted on torsion spring; E - torsion spring (one mounted at each end of model). (Source: [67] Fig 35)

24.5.6 The legacy of the Tacoma Narrows bridge collapse

The fourteen-year research programme following the collapse of the Tacoma Narrows Bridge established three particular conclusions concerning the form and stiffness of the decks of suspension bridges and the benefit of introducing damping:

- truss-stiffening of decks was preferable to girder stiffening since it substantially increase the wind speed at which oscillations would occur;
- the torsional stiffness of a truss-stiffened deck was usefully improved by incorporating lateral bracing at both the top and bottom of the trusses (so forming the box section which was achieved, literally, using welded steel plates, in the Severn Bridge some 25 years later) [70]
- the overall tendency for the bridge to oscillate could be significantly reduced by connecting the deck and main cables longitudinally at centre span, by ties incorporating damping action.

Of wider significance was the establishing of a radically new approach to designing suspension bridges which, for the first time, involved the use of wind tunnels. The report of the investigations concluded that the use of the section-model tests and the associated mathematical analysis would yield 'adequate knowledge of the aerodynamic characteristics of a proposed suspension bridge'. This became the established method for designing suspension bridges for the following thirty years.

24.6 The boundary-layer wind tunnel

The significance of the boundary layer in fluid flow adjacent to a fixed boundary was first recognised by Ludwig Prandtl in 1904 [8]. It seems first to have been taken into account in the early 1930s in wind tunnel tests on model buildings by Otto Flachsbart, a pupil of Prandtl's (see Section 24.4.1). He made comparisons between pressures observed in model buildings using a uniform velocity profile and one with a significant boundary layer arising from the roughness of the 'ground' around the model. In particular, he noted that both negative and positive pressures were, in contrast to what might have been expected, reduced for the case of rough ground [71].

The 'atmospheric boundary layer', as it came to be known, is the height above ground level that is affected by the roughness of the earth's surface. It is typically around 50 metres high, but can extend to several hundred metres depending on the nature of the surface roughness, ranging from the open sea, to open countryside, trees, low-rise buildings and high-rise buildings. The key characteristic of wind in this boundary layer is that it is turbulent and, with a portion of total energy embodied in this turbulence, the average wind speed is lower than at higher altitude. While Flachsbart had taken into account the reduction of wind speed in the boundary layer, he did not address the issue of turbulent flow.

24.6.1 Martin Jensen

What we might call the modern approach to wind engineering was, to a large extent, initiated by the Danish engineer Martin Jensen (1914-1990), who had been a student of Nøkkentved, himself a disciple and collaborator of Irminger [72, 73]. Martin took over Nøkkentved's work on the sheltering effect of vegetation in the late 1940s and completed his doctorate on the same subject, for which he also studied the drifting of snow and dispersion of smoke from chimneys which, of course, directly revealed the intensity of turbulence in the atmospheric boundary layer. Accepting the difficulty of conducting experiments at full scale, Jensen concentrated on the use of reduced-scale models and wrote his seminal paper on model-scale tests in turbulent wind in 1958 [74]. With his collaborator Niels Franck he published a fuller treatment of the subject in 1963-1965 [75, 76]. In essence, he established that similitude is achieved if the ratio of the model scale to the roughness height of the boundary surface, embodying both the depth of the boundary layer and the degree of turbulence, is maintained constant. Since 1986 this ratio has been named the Jensen Number, *Je*, in recognition of its creator.

24.6.2 Alan Davenport

Alan Davenport (1932-2009) is widely recognised as the father of the modern boundary-layer wind tunnel [77]. He began his study of wind at the University

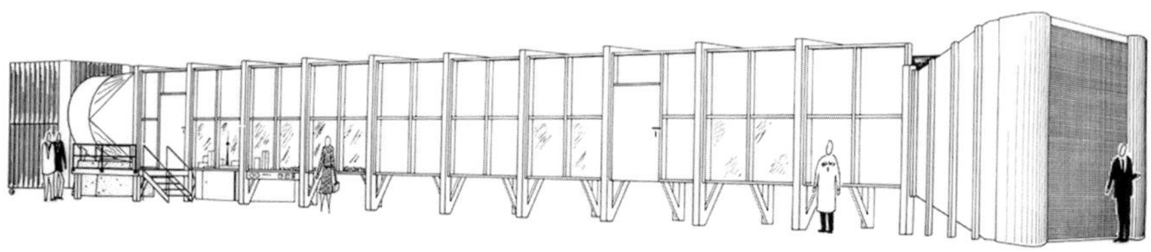

Figure 24.37 Davenport's first, open-circuit Boundary-Layer Wind Tunnel at the University of Ontario, opened in 1965. (Image: Courtesy of the Boundary-Layer Wind Tunnel Laboratory, UWO)

of Bristol in England under the guidance of Alfred Pugsley (1903-1998), who revolutionised our understanding of the safety of structures, and from whom he acquired an understanding of the statistical nature of many live loads on structures, particularly on long bridges and tall buildings. After completing his doctorate in 1961, Davenport was recruited to the University of Western Ontario where he raised the necessary support to construct the first boundary-layer wind tunnel dedicated to testing models of buildings and bridges [78] (Figure 24.37). It was an open-circuit tunnel with a test section measuring 2.4 m wide, 1.5-2.1 m high and 33 m long. It delivers wind speeds up to 9.7 m/s. The second, larger BLWT was a closed-circuit type and opened in 1985. It had two test sections – a low-speed section 5 x 4 m high and 52 m long, with wind speeds up to 10 m/s, and a high-speed section 3.4 x 2.5 m high and 39 m long, with wind speeds up to 28 m/s (Figure 24.38). Both wind tunnels include space upstream of the test model for small obstacles – often upturned plastic water cups in the early days – to be fixed to create the appropriate roughness of the ground.

An early study undertaken in the first BLWT was for the design of the A. Murray McKay Bridge at Halifax, Nova Scotia [79]. The first tests were on 1:320 and 1:40 models of short sections of the truss-stiffened bridge deck, both during construction and when complete – a procedure that had become standard following the outcome of the Tacoma Narrows studies; the results from these tests were in

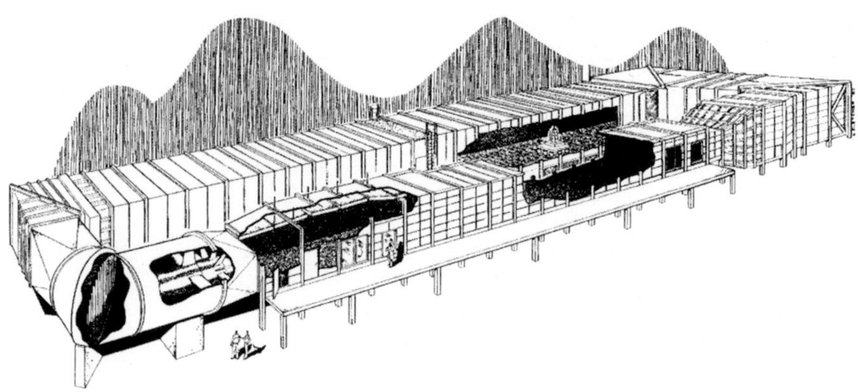

Figure 24.38 Davenport's second, closed-circuit Boundary-Layer Wind Tunnel at the University of Ontario, opened in 1984. (Image: Courtesy of the Boundary-Layer Wind Tunnel Laboratory, UWO)

close agreement. When an aeroelastic model of the full bridge, at a scale of 1:320, was tested in uniform airflow it was found to be stable, but when tested in turbulent boundary layer flows, the model displayed random oscillations in both the vertical and horizontal directions which raised doubts about the dependability of relying only on tests using short sections of the deck. After careful analysis of the results, Davenport proposed the use of a 'taut-strip' model which introduced some prestress into the bridge deck to reproduce more accurately the elastic, dynamic response of the deck. This technique was tested first on a model of the A. Murray McKay Bridge and, later, the well-studied Golden Gate and Tacoma Narrows bridges, and has since become an accepted tool in wind-tunnel testing [80].

In the early years of the BLWT Davenport also began testing the dynamic response of tall buildings by constructing a model frame ('stick') structure behind the exposed façade of a model building, to reproduce its overall stiffness. He had, in fact, already tested such a model of the New York World Trade Centre in the Colorado State University Meteorological Wind Tunnel which, in the early 1960s, was the only one in which a turbulent boundary-layer wind could be generated. These tests enabled suitable tuned-mass dampers to be installed in the buildings to reduced oscillations that would be uncomfortable for the buildings' occupants. In 1970 Davenport began similar dynamic studies for the Sears Tower in Chicago which, at the time, was the tallest building in the world (Figure 24.39). This aeroelastic model of the tower had twenty-one degrees of freedom and provided predictions of the overall peak structural wind loads and the wind-induced deflections and accelerations of the tower. The BLWT studies also included a detailed assessment of air currents at ground level to investigate possible discomfort for pedestrians walking near the base of the building [79].

(a) (b)

Figure 24.39 Aeroelastic wind-tunnel model of the Sears Tower, Chicago. (a) Internal stick structure enabling the model to respond dynamically to winds. (b) In situ with surrounding buildings. (Images: Courtesy of the Boundary-Layer Wind Tunnel Laboratory, UWO)

Davenport's contribution to wind engineering was not, however, limited to building the BLWT. He established a new and clear logical train of thought with regard to wind loading on structures. Rather than the earlier approach based on determining an appropriate wind load, in N/m^2, and an effective area over which this load acted, he advocated a chain of thought which recognised that the wind loading experienced by a particular structure is determined by: (a) the combined effects of the local wind climate, described in statistical terms; (b) the local exposure to the wind, which is influenced by the topography and terrain roughness; (c) the aerodynamic characteristics of the structure; and (d) the potential for increased loads arising from wind-induced resonant vibrations. This approach, now generally referred to as Davenport's wind-loading chain, crucially includes also an assessment of the importance of the effects of wind loads on both the structure and the people using the structure, for example their ability to walk in outdoor areas, and their tolerance of wind-induced motions of a building.

24.7 Final remarks

The development of model testing in currents of air developed along several strands from the 1890s – measuring the forces on models and correlating these with measurements on their full-size equivalents; likewise, measuring air pressure directly, and making the flow of air visible using water as a model and using smoke trails in air. After the Tacoma Narrows Bridge collapse in November 1940, wind-tunnel studies of section models became the standard means of testing suspension bridge designs for the effects of dynamic forces, especially due to vortex shedding. Finally, the importance of turbulent airflow in the atmospheric boundary layer was recognised and the boundary-layer wind tunnel, the standard test facility used today, was born. This enabled the turbulence in the model boundary layer to reflect the actual site of a bridge or building, just as the wind-tunnel model represented the structure itself.

References

1 Smeaton, J. (1759) An experimental Enquiry concerning the natural Powers of Water and Wind to turn Mills, and other Machines, depending on a circular Motion. *Phil. Trans.* 51: 100-174.

2 HMSO (1881) *Report of the Wind Pressure (Railway structures) Commission.* London: HMSO.

3 Fleming, R. (1915) *Six monographs on wind pressures (Pressure factors, specification requirements, mill-building stresses, rigid-joint wind bracing for office buildings).* New York: Engineering News/Hill.

4 HMSO (1880) *Report of the Court of Inquiry and Report of Mr Rothery upon the Circumstances attending the Fall of a Portion of the Tay Bridge on 28 December 1879.* London: HMSO.

5 Kernot, W.C. (1893) Wind pressure. *Proceedings, Australasian Society for the Advancement of Science* V: 573-581. 573-574.

6 Anderson, John D. (1997) *A History of Aerodynamics*. Cambridge: Cambridge University Press.

7 Randers-Pehrson, N.H. (1935) Pioneer wind tunnels. *Smithsonian Miscellaneous Collections* 93 (4). 20 pp. 4 Plates.

8 Prandtl, L. (1904) Über Flüssigkeitsbewegung bei sehr kleiner Reibung. *Verhandl. III. Intern. Math. Kongr. Heidelberg* 2: 484-491.

9 Prandtl, L. (1909) Die Bedeutung von Modellversuchen für die Luftschifffahrt und Flugtechnik und die Einrichtungen für solche Versuche in Göttingen. *Zeitschrift des Vereins Deutscher Ingenieure* 53: 1711-1719.

10 Kernot, W.C. (1893) Wind pressure. *Proceedings, Australasian Society for the Advancement of Science.* V: 573-581. 577.

11 Stoney, B.B. (1873) Theory of Strains in Girders and similar Structures. (2nd ed.) London: Longman. 411-415, 524.

12 Kernot, W.C. (1893) Wind pressure. *Proceedings, Australasian Society for the Advancement of Science.* V: 573-581. 581.

13 Kernot, W.C. (1895) Addendum to paper on wind pressure. *Proceedings, Australasian Society for the Advancement of Science.* VI: 741-745.

14 Irminger, J.O.V. (1894) Nogle Forsøg over Trykforholdene paa Planer og Legemer paavirkede af Luftstrømninger (Some experiments on pressure conditions on planes and bodies, affected by air currents). *Ingeniøren* 3 (17) (28 April): 101-105, Plan I-II, (18) (5 May): 107-109, Plan III-IV.

15 Irminger, J.O.V. (1894) Some experiments respecting the conditions of pressure on planes and bodies generated by currents of air. (book at ICE)

16 Irminger, J.O.V. (1899) Wind pressure on buildings and other structures. *Proc. Institution of Gas Engineers,* London: 187-202.

17 Irminger, J.O.V. (1895) Recent Experiments on Wind Pressure. (Abstract of Irminger 1894). *Engineering News and American Railway Journal* (New York) 33 (7): 109-110. 110.

18 Stanton, T.E. (1904) On the resistance of plane surfaces in a uniform current of air. *Minutes of the Proceedings of the Institution of Civil Engineers* 156: 78-112; Discussion 113-126.

19 Stanton, T.E. (1904) On the resistance of plane surfaces in a uniform current of air. *Minutes of the Proceedings of the Institution of Civil Engineers* 156: 78-112. 104.

20 Stanton, T.E. (1908) Experiments on wind pressure. *Minutes of the Proceedings of the Institution of Civil Engineers* 171: 175-200; Discussion, 201-214; Correspondence, 214-226.

21 Zahm, A.F. (1914) Report on European aeronautical laboratories. *Smithsonian Miscellaneous Collections* 62 (3). 23 pp. 11 Plates.

22 Heli-Shaw, H.S. (1898) Experiments on the nature of the surface resistance in pipes and on ships. *Transactions of the Institution of Naval Architects* 39: 145-152; Discussion 153-156. Plates XX-XXIII.

23 Heli-Shaw, H.S. (1898) Investigation of the Nature of Surface Resistance of Water and of Stream-line Motion Under Certain Experimental Conditions.

Transactions of the Institution of Naval Architects 40: 21-40; Discussion 40-46. Plates VII-XVII.

24 Bairstow, Leonard (1920). *Applied aerodynamics.* London: Longmans. 344.

25 Marey, E.-J. (1900) Des mouvements de l'air lorsqu'il rencontre des surfaces de différentes formes. *Comptes rendus de l'Académie des Sciences, Paris* 131 (Jul-Dec): 160-163.

26 Marey. E.-J. (1902) Le mouvement de l'air étudié par la chronophotographie. *J. Phys. Theor. Appl.* 1 (1): 129-135.

27 Marey, E.-J. (1901) Changements de direction et de vitesse d'un courant d'air qui rencontre des formes diverses. *Comptes rendus de l'Académie des Sciences, Paris* 132 (Jan-Jun): 1291-1296.

28 Didi-Huberman , G. & Mannoni, L. (2004) *Mouvements de l'air. Étienne-Jules Marey, photographe des fluides.* Paris: Gallimard.

29 Bairstow, Leonard (1920) *Applied aerodynamics.* London: Longmans. 345.

30 VDI (1928) Beitrag zur Frage der Berücksichtigung des Windes im Bauwesen. In: *11. Jahrbuch der Deutschen Gesellschaft für Bauingenieurwesen,* 160-169. Berlin: Verein Deutscher Ingenieure (VDI).

31 VDI (1928) Beitrag zur Frage der Berücksichtigung des Windes im Bauwesen. In: *11. Jahrbuch der Deutschen Gesellschaft für Bauingenieurwesen,* 160-169. Berlin: Verein Deutscher Ingenieure (VDI).

32 Flachsbart, O. (1930) Winddruck auf Bauwerke. *Die Naturwissenschaften* 18: 475-479.

33 Flachsbart, O. (1931) Der gegenwärtige Stand der Winddruckforschung. *Jahrbuch, 1930, der Deutschen Gesellschaft für Bauingenieurwesen* 6: 108.

34 Flachsbart, O. (1932) Grundsätzliches zur Frage des Winddrucks auf Gebäude. *Bauwelt* 27: 660-664; 28: 692.

35 Flachsbart, O. (1932) Winddruck auf vollwandige Bauwerke und Gitterfachwerke *Proceedings of IABSE Congress, Zurich.* 153-172.

36 Flachsbart, O. (1931) Winddruck auf Schornsteine. *Naturwissenschaften* 19: 759-760.

37 Flachsbart, O. (1932) Winddruck auf geschlossene und offene Gebäude. In: *Ergebnisse der aerodynamischen Versuchanstalt zu Göttingen, IV. Lieferung* (ed. L. Prandtl & A.Betz) 128-134. Munich & Berlin: Verlag von R. Oldenbourg.

38 Flachsbart, O. (1932) Winddruck auf Gasbehälter. In: *Ergebnisse der aerodynamischen Versuchanstalt zu Göttingen, IV. Lieferung* (ed. L. Prandtl & A.Betz) 134-138. Munich & Berlin: Verlag von R. Oldenbourg.

39 Flachsbart, O. (1934) Modellversuche über die Belastung von Gitterfachwerke durch Windkräfte. 1. Teil: Einzelne ebene Gitterträger. *Der Stahlbau* 7 (9): 65-69; 7 (10): 73-79.

40 Flachsbart, O. (1935) Modellversuche über die Belastung von Gitterfachwerke durch Windkräfte. 2. Teil: Räumliche Gitterfachwerke. *Der Stahlbau* 8 (9): 57-63; 8 (9): 65-69; 8 (10): 73-77.

41 Irminger, J.O.V. & Nøkkentved, C. (1930 /1936) *Wind pressure on buildings: experimental researches.* (First and second series.) (Tr from the Danish by A.C. Jarvis & O. Brodsgaard). Kobenhavn: Danmarks Naturvidenksabelige Samfund.

42 Nøkkentved, Chr. & Flachsbart, O. (1932) Wind pressure on buildings. *Proceedings of IABSE Congress, Zurich.* 365-378.
43 Larosea, G.L. and Franck, N. (1997) Early wind engineering experiments in Denmark. *Journal of Wind Engineering & Industrial Aerodynamics* 72: 493-499.
44 Dryden, H.L. & Hill, G.C. (1926). Wind Pressures on Structures. *Scientific Papers of the Bureau of Standards.* No. S523. 20: 697-732.
45 Dryden, H.L. & Hill, G.C. (1930). Wind Pressure on Circular Cylinders and Chimneys. *Bureau of Standards Journal of Research* (RP221) 5 (3): 563-693.
46 Dryden, H.L. & Hill, G.C. (1931). Wind *Pressure on a Model of a Mill Building,* Bureau of Standards Journal of Research (RP301) 6 (4): 735-755.
47 Dryden, H.L. & Hill, G.C. (1933). Wind pressure on a model of the Empire State Building. *Bureau of Standards Journal of Research,* (RP 545) 10 (4): 493-523.
48 Dryden, H.L. & Hill, G.C. (1926). Wind Pressures on Structures. *Sci. Papers of the Bureau of Standards.* No. 523, Bull. 20: 697-732.
49 Addis, B. (2007) *Building: 3000 years of design engineering and construction.* New York/London: Phaidon.
50 https://archive.org/details/uwlibraries_oclc892993126_tacoma (26/06/2019)
51 Addis, W. (1994) Design Revolutions in the History of Tension Structures. *Structural Engineering Review* 6 (1): 1-10.
52 Addis, W. (1990) *Structural Engineering: the Nature of Theory and Design.* Chichester, UK: Ellis Horwood. 169-173.
53 Farquharson, F.B. (1940) *A dynamic model for the Tacoma-Narrows suspension bridge.* Bulletin of the University of Washington Engineering Experiment Station. Technical Notes, No.1 (June). Seattle: University of Washington.
54 http://archiveswest.orbiscascade.org/ark:/80444/xv77728
55 Ammann, O.H., von Kármán, T. & Woodruff, G.B. (1941) *The failure of the Tacoma Narrows bridge. A Report to the Hon. J.M. Carmody, Administrator, Federal Works Agency, Washington.* Pasadena: Board of Engineers. Appendix VII. https://babel.hathitrust.org/cgi/pt?id=mdp.39015021064251&view=1up&seq=13 (18/9/2019)
56 Farquharson, F.B. (1949) *Aerodynamic stability of suspension bridges with special reference to the Tacoma Narrows Bridge. Part I: Investigations prior to October, 1941.* Seattle: University of Washington Engineering Experiment Station. 41-54. https://catalog.hathitrust.org/Record/001613008 (18/9/2019)
57 Anon (1940) Model tests showed aerodynamic instability of Tacoma Narrows bridge. *Engineering News Record* 125 (21st November): 674-677.
58 Ammann, O.H., von Kármán, T. & Woodruff, G.B. (1941) *The failure of the Tacoma Narrows bridge. A Report to the Hon. J.M. Carmody, Administrator, Federal Works Agency, Washington.* Pasadena: Board of Engineers. Chapter IV (96-126).
59 Farquharson, F.B. (1949) *Aerodynamic stability of suspension bridges with special reference to the Tacoma Narrows Bridge. Part I: Investigations prior to October, 1941.* Seattle: University of Washington Engineering Experiment Station. 55-74.

60 Ammann, O.H., von Kármán, T. & Woodruff, G.B. (1941) *The failure of the Tacoma Narrows bridge. A Report to the Hon. J.M. Carmody, Administrator, Federal Works Agency, Washington.* Pasadena: Board of Engineers. 104-106.

61 Ammann, O.H., von Kármán, T. & Woodruff, G.B. (1941) *The failure of the Tacoma Narrows bridge. A Report to the Hon. J.M. Carmody, Administrator, Federal Works Agency, Washington.* Pasadena: Board of Engineers.

62 Ammann, O.H., von Kármán, T. & Woodruff, G.B. (1941) *The failure of the Tacoma Narrows bridge. A Report to the Hon. J.M. Carmody, Administrator, Federal Works Agency, Washington.* Pasadena: Board of Engineers. VIII-2 - VIII-16.

63 Ammann, O.H., von Kármán, T. & Woodruff, G.B. (1941) *The failure of the Tacoma Narrows bridge. A Report to the Hon. J.M. Carmody, Administrator, Federal Works Agency, Washington.* Pasadena: Board of Engineers. VIII-17 – VIII-20.

64 Farquharson, F.B. (1949) *Aerodynamic stability of suspension bridges with special reference to the Tacoma Narrows Bridge. Part I: Investigations prior to October, 1941.* Seattle: University of Washington Engineering Experiment Station. 75-77.

65 Farquharson, F.B. (1947) Early M.I.T. tests of Tacoma bridge failure. *Civil Engineering* 401 (July): 43-44.

66 Farquharson, F.B. (1952) *Aerodynamic stability of suspension bridges with special reference to the Tacoma Narrows Bridge. Part III: The investigation of models of the original Tacoma Narrows Bridge under the action of wind.* Seattle: University of Washington Engineering Experiment Station.

67 Farquharson, F.B. (1954) *Aerodynamic stability of suspension bridges with special reference to the Tacoma Narrows Bridge. Part IV: Model investigations which influenced the design of the new Tacoma Narrows Bridge.* Seattle: University of Washington Engineering Experiment Station. Appendix III.

68 Steven Ross, et al. (1984) *Construction Disasters: Design Failures, Causes and Prevention.* New York: McGraw Hill Education. 216-239.

69 Farquharson, F.B. (1954) *Aerodynamic stability of suspension bridges with special reference to the Tacoma Narrows Bridge. Part IV: Model investigations which influenced the design of the new Tacoma Narrows Bridge.* Seattle: University of Washington Engineering Experiment Station.

70 Walshe, D.E. (1967) A resume of the aerodynamic investigations for the Forth Road and the Severn Bridges. *Proceedings of the Institution of Civil Engineers* 37 (1): 87-108; Discussion, 38 (4): 781-783.

71 Flachsbart, O. (1932) Winddruck auf geschlossene und offene Gebäude. In: *Ergebnisse der aerodynamischen Versuchanstalt zu Göttingen, IV. Lieferung* (ed. L. Prandtl & A. Betz) 128-134. Munich & Berlin: Verlag von R. Oldenbourg.

72 Larose, G.L. and Franck, N. (1997) Early wind engineering experiments in Denmark. *Journal of Wind Engineering & Industrial Aerodynamics* 72: 493-499.

73 Davenport, A.G. (1992). Martin Jensen: an appreciation. *Journal of Wind Engineering and Industrial Aerodynamics* 41 (1-3): 15-22.

74 Jensen, M. (1958) The model-law for phenomena in natural wind. *Ingeniøren* (International Edition) 2 (4): 121-128.
75 Jensen, M. & Franck, N. (1963) *Model-scale tests in turbulent wind. Part 1: Phenomena dependent on the wind speed, shelter at houses, dispersal of smoke.* Copenhagen: Danish Technical Press.
76 Jensen, M. & Franck, N. (1965) *Model-scale tests in turbulent wind. Part 2: Phenomena dependent on the velocity pressure, wind loads on buildings.* Copenhagen: Danish Technical Press.
77 Surry, D., Galsworthy, J. Hangan, H., King, P. & Kopp, G. (2003) Engineering Symposium to honour Alan G. Davenport for his 40 years of contributions. *Journal of Wind Engineering and Industrial Aerodynamics* 91 (12-15): 1301-1906.
78 Roberts, Siobhan (2013). *Wind Wizard: Alan G. Davenport and the Art of Wind Engineering*. Princeton: Princeton University Press.
79 Isyumov, N. (2012) Alan G. Davenport's mark on wind engineering. *Journal of Wind Engineering and Industrial Aerodynamics* 104-106 (May-July): 12-24.
80 Davenport, A.G., Isyumov, N. & Miyata, T. (1971) The experimental determination of the response of suspension bridges to turbulent wind. In: *Proc. Third Int. Conf. on wind effects on buildings and structures, Tokyo.* 1207-1220. Tokyo: Saikon.

25

The historical use of physical model testing in earthquake engineering

Bill Addis

25.1 Early shaking tables

25.1.1 Seismology and seismographs in the nineteenth century

As with many scientific disciplines, modern seismology began during the Enlightenment with the observation of events, such as the Lisbon earthquake in 1755, and descriptions of what happened and the type of damage that was caused. By the late-eighteenth century, several scientists began to discuss the possible causes and mechanisms of earthquakes, for example Thomas Young was the first, in around 1810, to propose that earthquakes seemed to be elastic waves passing through the earth [1]. During the first half of the nineteenth century several scientists in Britain and a number of earthquake-prone countries, especially Italy and Japan, developed instruments for measuring the ground movement in earthquakes, as well as recording the movement on paper. The first reliable, modern seismograph was developed in the early 1880s by the Scottish physicist and engineer, James Alfred Ewing (1855-1935) [2]. At the invitation the Imperial University of Tokyo, Lord Kelvin recommended a number of Scottish scientists and engineers to establish new departments. Ewing, then aged 23, was recommended as Professor of Mechanical Engineering and Physics, a post he held from 1878 to 1883 [3]. Kelvin also recommended Thomas Gray (1850-1908) for the position of Demonstrator of Physics and Instructor of Telegraphy at the Imperial College of Engineering in Tokyo, which he held from 1878 to 1881. A little earlier, in 1876, the English engineer John Milne (1850-1913) had been appointed Professor of Geology and Mining at the Imperial College of Engineering in Tokyo. All three of them had their interest in seismology awakened by a major earthquake in Tokyo in 1880 [4] and Milne, Ewing and Gray founded the Seismological Society of Japan in the same year.

Ewing's seismograph was able to record, as three separate traces on a rotating disc, the movement of the earth in the x, y and z directions. Sekei Sekiya (1854-1896), who had been one of Ewing's first researchers, used the traces from

Physical Models: Their historical and current use in civil and building engineering design,
First Edition. Edited by Bill Addis.

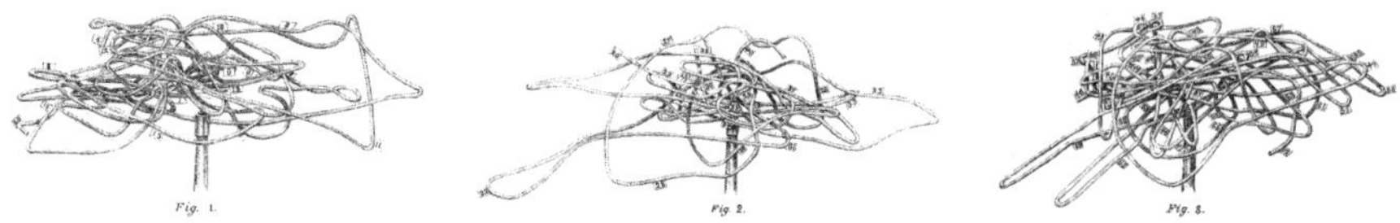

Figure 25.1 A wire model showing the motion of an earth particle during the earthquake of 15 January 1887 in Tokyo, preserved in the Whipple Museum in Cambridge. (Source: [5] Plate XXVII)

an earthquake on 15 January, 1887 to make a three-dimensional model with copper wire showing the movement of a hypothetical point during a period of 72 seconds. The wire model was made fifty times larger than the actual movements, and split into three periods of 20, 20 and 32 seconds. Each model is about 80 mm across and each second is marked by a small metal tag fixed to the wire [5] (Figure 25.1). Sekiya was appointed as the world's first Professor of Seismology in 1886.

25.1.2 The Milne-Omori shaking table c.1890

In around 1890, John Milne and Fusakichi Omori (1868-1923), a young researcher working under Sekiya conducted the first seismic-loading experiments on buildings, simulating the movement of an earthquake by pulling on a building with ropes and letting go suddenly. They went on to build the first shaking table to simulate the movement of an earthquake in order to test structural elements, such as a brick column (Figure 25.2). It consisted of a table, freely suspended or mounted on wheels that could be moved back and forth in one direction using a hand crank. The main purpose of the early shaking-table research was not to seek how to construct buildings to resist earthquakes; rather it was to try to calibrate ground-motion intensity with acceleration by observing the overturning of stone or masonry structures. Nevertheless, apart from Omori, little local interest in seismology developed and by 1892 activity in the Seismological Society had almost ceased due to a lack of interested members. Milne returned to Britain in 1895 after his laboratory was destroyed by fire and research in Japan reverted to the direct observation of earthquakes in order

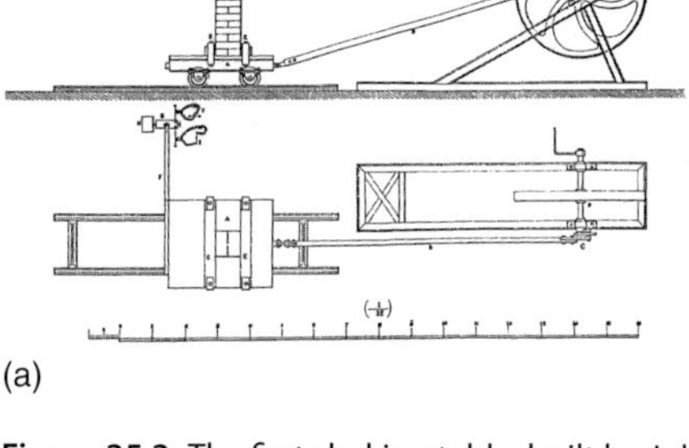
(a)

(b)

Figure 25.2 The first shaking table, built by John Milne and Fusakichi Omori in Tokyo, 1893. (a) (Source: [6]). (b) (Source: [7])

to determine exactly what movements occurred during earthquakes and their aftershocks.

25.1.3 F.J. Rogers at Stanford University, 1908

As had happened in 1880 in Tokyo, interest in simulating the behaviour of an earthquake was aroused by another large earthquake – this time the 1906 earthquake in California. One outcome of studies of the damage caused was the major contribution made by the weak soil on which many buildings were constructed. This was contrary to the belief of many that buildings built on rock foundations would be damaged more because of the rigid connection between the earth and the building. A programme of testing was undertaken in which the behaviour of model buildings founded on soft soils was studied. Professor F.J. Rogers at Stanford University built a more sophisticated shaking table than that built by Milne and Omori. It was powered by an electric motor linked by a crank to a board about 50 x 75 cm. The first experiments were with heaps of sands with different water contents. When the table was moved back and forth by 2-3 cm, it was found that the peak of a mound of wet sand moved by a greater distance than the movement of the table – an entirely unexpected result. This provided the stimulus to build a larger table which could move a box 100 x 85 and 30 cm deep, mounted on rollers. The box could be shaken at different frequencies and amplitudes up to 10 cm (Figure 25.3a). These experiments revealed that when the sand had less than 10% water, its movement mirrored the box, while wet sand moved more than the box. At intermediate wetness, the surface of the sand broke up, forming fissures or crevasses that opened and closed periodically. The motion of the box and a block embedded in the sand were recorded on a revolving drum. The results for one test with wet sand show the block moving more than the box, and at almost constant speed, rather than the sinusoidal movement of the box [8] (Figure 25.3b). Further tests were carried out with various types of sand, gravel and stones, and different degrees of wetness, but few useful conclusions were drawn. Perhaps because Rogers was a physicist rather than a ground engineer, the experiments mainly concerned a physical phenomenon, rather than a search for guidance for construction engineers.

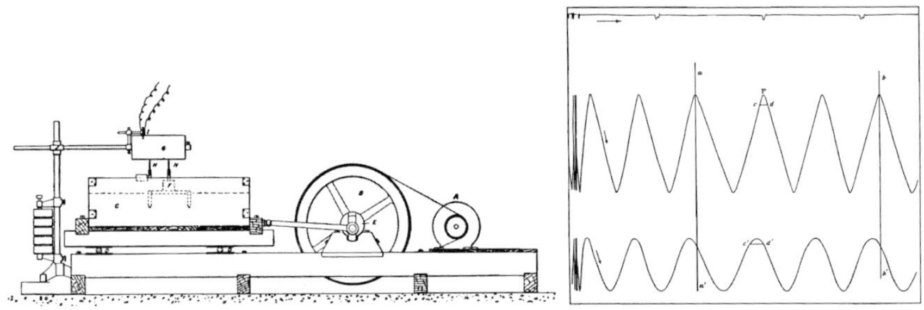

(a) (b)

Figure 25.3 (a) The shaking table built by Rogers at Stanford University in 1908. (b) Graphical record of the movement of the box (bottom) and a block in the soil (centre), and the clock pulse (top). (Source: [8] Figures 60 & 61)

25.2 Shaking tables in the 1930s

25.2.1 Lydik Jacobsen at Stanford University

After completing his doctoral research in 1927, Lydik Jacobsen (1897-1976) founded a Vibration Laboratory at Stanford University and soon became interested in the work of Rogers carried out 20 years earlier at the same University. He built a shaking table to study, like Rogers, the behaviour of soils under seismic loading. At 4 x 3 m, it was larger and more robust than that of Rogers and it allowed the possibility of applying loads either by the impact of a 405 kg, 7-m long pendulum, or an unbalanced flywheel mounted on the table itself (Figure 25.4). Jacobsen shook soil in a box 4.8 x 1.05 x 0.75 m high, and reported that he observed liquefaction of the sand at a water content of 13-14% [9, 10]. Jacobsen also used his shaking table in the first systematic research into earthquake-resistant timber housing which began in 1929. He tested a number of timber wall panels, 2.7 m square, to determine whether it was structurally more effective to lay the facing planks diagonally rather than horizontally, as was usual [11].

During this early work Jacobsen recognised the need for more sophisticated models and methods of measuring the forces and deflections of the model during a test. In 1930-31, he adapted his original shaking table for this new study (Figure 25.5) and undertook the first model tests on a multi-storey structure – a proposed seventeen-storey building for the Olympic Club in San Francisco. The model was a mechanical one with springs providing the effective stiffness of the columns and walls. The mass of each floor was provided by steel plates. This model had no vertical freedom and was able to display only a shear distortion, in which the floors remained level. Experiments were conducted for both a conventional first storey and one having considerable flexibility – the 'flexible first storey' concept. The model was loaded by horizontal vibrations, impact from the pendulum, or excited by random motions. The results were recorded on film for later analysis.

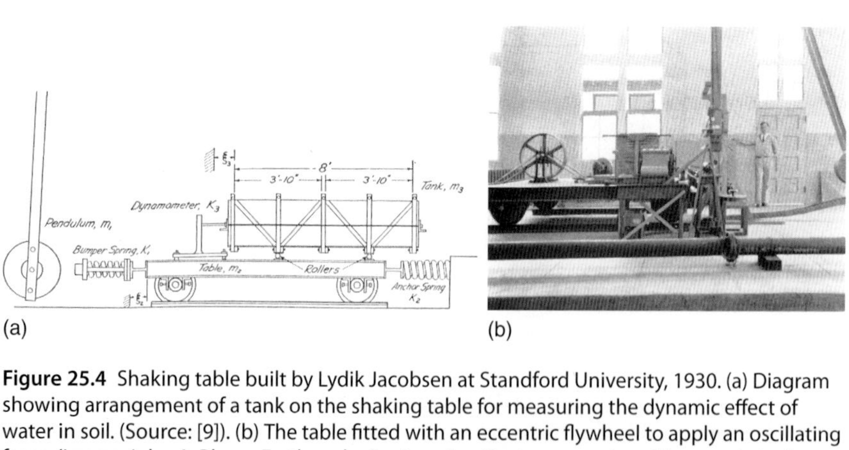

(a) (b)

Figure 25.4 Shaking table built by Lydik Jacobsen at Standford University, 1930. (a) Diagram showing arrangement of a tank on the shaking table for measuring the dynamic effect of water in soil. (Source: [9]). (b) The table fitted with an eccentric flywheel to apply an oscillating force. (Image: John A. Blume Earthquake Engineering Center, reproduced by permission)

Figure 25.5 Lydik Jacobsen's shaking table at Stanford University, during testing of an unknown 16-storey building, c.1933. (Image: John A. Blume Earthquake Engineering Center, reproduced by permission)

This work led to the ground-breaking studies undertaken by Jacobsen and his students John Blume (1909-2002) and Harry Hesselmeyer. The study was undertaken on a 1:30 scale model of the fifteen-storey Alexander Building in San Francisco, for which Blume and Hesselmeyer were able to obtain a full set of drawings. This model was much more sophisticated than the earlier one, and very difficult to construct (Figure 25.6). Each storey had five degrees of freedom – two horizontal, one vertical and two rotational, about horizontal axes. It was a mass of aluminium and steel plates and tubes, steel springs and steel ball bearings which allowed each floor to roll. The bending of the steel plate provided the flexibility for overall flexure of the structure. The model also included an additional basement storey to allow soil-structure interaction to be simulated. At that time there were no electrical strain gauges and the motion was recorded using mechanical gauges mounted between each pair of floors (Figure 25.7). These amplified the movement which was recorded on film and later analysed, frame by frame (Figure 25.8) [12].

The analysis undertaken in Blume and Hesselmeyer's joint thesis finally raised the understanding of the effects of earthquakes on high-rise buildings to a level that allowed structural engineers to consider loads due to earthquakes when designing their buildings, and hence provide structures that would safely resist these loads [13]. Blume went on to build an unbalanced-flywheel machine for applying loads at different frequencies to actual structures – over a hundred high-rise buildings, including the real Alexander building, and a number of dams – to determine their resonant frequencies, and compare these results with theoretical calculations.

Figure 25.6 Model, 1:30 scale, of the 15-storey Alexander Building, made by John Blume and Harry Hesselmeyer, 1932-33. (Image: John A. Blume Earthquake Engineering Center, reproduced by permission)

Figure 25.7 Detail of the model of the 15-storey Alexander Building. (Image: John A. Blume Earthquake Engineering Center, reproduced by permission)

Figure 25.8 A still from the film of a test on the model of the Alexander Building. (Image: John A. Blume Earthquake Engineering Center, reproduced by permission)

During the following quarter century, Jacobsen and Blume made many improvements to their table and to the means of measuring and recording the test results. He devised dynamometers to measure deflections of 1-2 mm in a time interval of a twentieth of a second in order to compute forces and moments, including electromagnetic sensors that could feed a signal into an oscilloscope. Jacobsen and Blume always undertook careful theoretical analyses to inform and interpret the experimental results, despite the difficulty of solving up to fifteen simultaneous equations by hand, aided only by a Marchant electric calculating machine. Jacobsen commented:

> 'In order to make inferences from model experiments reliable, a considerable acquaintance with the theory underlying the similitude problem must be acquired; a mere experimentation without a conscious theoretical foundation cannot be expected to yield results on which an engineer is willing to risk the safety of human beings and invested capital' [14].

25.2.2 Nahago Mononobe in Tokyo

Dr Nahago Mononobe at the University of Tokyo built a number of shaking tables from the late 1920s. He used tables of the hanging-box type to study the pressure exerted by sand on the walls of a container, as a model for earth acting on a retaining wall. One such box was 2.64 x 1.33 x 1.33 m high and weighed 8 tonnes; another was 110 x 43 x 40 cm high and weighed 560 kg. Loads were applied by means of an electromagnet vibrator and pressures against the walls were measured with piezo-electric pressure gauges. The output of the gauge was fed to a 'delicate string electrometer' whose displacement was magnified five time by a telescope and this image recorded on a photographic plate [15].

He also used a larger table, 1.8 x 7.3 m, which was actuated by three electric motors to give rectilinear motion along the three major axes – an early example of a table with seismic simulation in three degrees of freedom. He referred also to shaking-table experiments in which piezoelectric crystals were embedded in soil samples to measure soil pressures as they occurred during testing [16, 17].

Mononobe used a shaking table of the Jacobsen type, driven by eccentric-mass vibrators, to study the behaviour of earth dams subject to seismic loads. Displacements of 1:100 scale model dams were measured using optical methods including 'motion pictures' [18, 19].

25.2.3 Arthur Ruge at MIT

During the 1933 earthquake at Long Beach, California a number of elevated water tanks had collapsed which meant there was a lack of water to deal with fires caused by the earthquake. A year later, Arthur Ruge (1905-2000), a Research Associate in seismology at Massachusetts Institute of Technology commenced work on studying the stability of water towers and began by constructing a new type of shaking table. The table, about 1 m square, was suspended from above by piano wires and prevented from twisting by stabilising wires attached to the floor [20]. Initially Ruge used a rotating eccentric mass to give the table, which was restrained by springs, a sinusoidal motion. He soon realised that such motion was a poor representation of the actual movement of an earthquake. He devised a means of driving the table by means of an oil-filled piston whose position was changed by pumping oil into one side of the actuator or the other. This was an important conceptual leap as it effectively applied a strain to the system, not a force, just like a real earthquake. Ruge was also able to use the trace of an actual earthquake, recorded as a shadowgraph, to drive the motion of the piston [21]. An essential component of this mechanism was to provide an instant measure of the position of the table to provide the essential positive feedback in the control system. He achieved this using an 'electric eye' which provided a signal to feed into an amplifier which controlled the electromagnetic drive of the hydraulic actuator. The light beam was interrupted by a rotating disc of wood whose circumference had been cut to replicate the shadowgraph recording of the earthquake of 1933 [22]. As the disc rotated, so the light beam was obscured in proportion to the shape of the circumference to generate an electrical signal. To avoid unstable hunting in the control system, he also used a signal proportional to the speed of the control valve to apply a damping input to the motion of the valve. For its time, this was a highly-sophisticated application of control engineering (Figure 25.9). To ensure that the results of his model tests were reliably interpreted at full scale, Ruge also took account of the similitude requirements of the experiments.

As well as creating the first modern shaking table, Ruge was also one of two engineering credited with the invention of the electrical-resistance strain gauge (see Section 13.4.5). In 1937-1938, he was seeking a way of measuring the minute deformations of the reduced-scale model of the metal water tank as it was being shaken. He had the idea of gluing a thin sheet of metal to the model that would

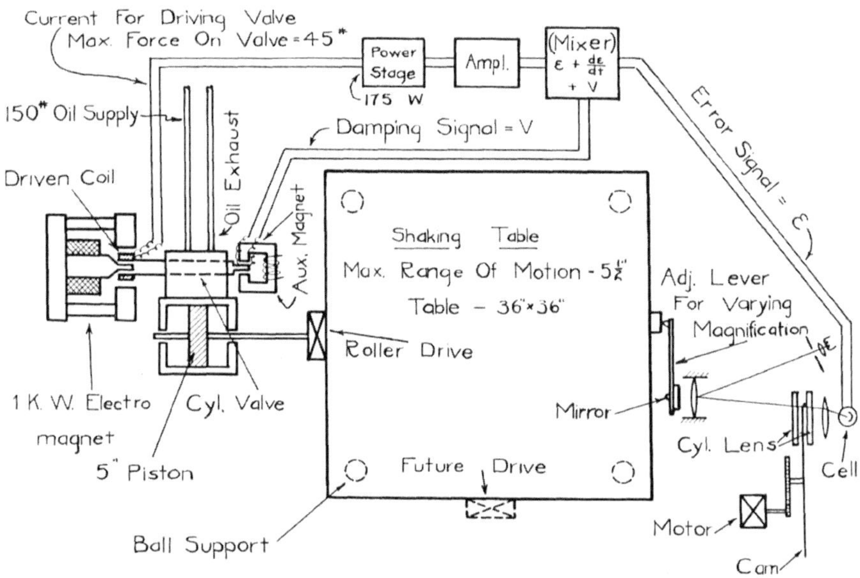

Figure 25.9 Schematic diagram of the Massachusetts Institute of Technology shaking table. (Source: [21] Figure 1)

be stretched when the model deformed. As it changed length, so its electrical resistance changed – a measure of this resistance thus gave a measure of the surface strain of the model. He developed this idea, with the other co-inventor, Edward Simmons from California Institute of Technology, to produce the first bonded-wire strain gauge, the SR-4 (incorporating both their initials), measuring about 20 x 10 mm. This invention revolutionised the activity of experimental stress analysis.

25.3 Post-war developments in shaking tables

25.3.1 More, bigger, better

After the war the numbers of shaking-table facilities grew steadily as countries recovered from the hiatus caused by the Second World War. In 1932 there were only five or six shaking tables in operation in the world, only one of which, in Japan, had three electric motor actuators giving three-degrees-of-freedom seismic simulation [23]; by the end of the century there were over fifty. The main drivers for shaking-table development in this period were (a) to provide a more and more faithful replication of the wave form of real earthquakes, and to achieve this in more than a single direction (degree of freedom), and (b) to increase the size and power of the shaking table itself [24, 25]. This was sometimes achieved using shaking tables of the Jacobsen type, for example at ISMES. In 1956, at the First World Congress on Earthquake Engineering, Guido Oberti presented the most recent addition to their range of model-testing apparatus to test the dynamic behaviour of buildings and dams, typically at a scale of 1:100. A steel platform, 3 x 4.5 m, suspended from a rigid frame, could be excited in three ways – a composite system of pendulum and springs to produce shock waves; an eccentric-mass machine producing either horizontal or vertical motion at 2 to 25 Hz: and four synchronized electromagnetic vibrators producing unidirectional and harmonic motion [26].

Technical developments during the war brought many benefits to experimental seismology. There were many new instruments for measuring displacements (e.g. the linear variable differential transformer) and accelerations. Hydraulic power systems and their control systems had been developed for military use and, during the 1950s, hydraulic actuators for shaking tables became more common. The growing sophistication of the control systems allowed the generation of specified waveforms, accelerations, velocities and displacements. Finally there arrived the computer – perhaps the first use of a computer to control a shaking table was at the University of Illinois in 1966 [27].

25.3.2 Shaking tables with six degrees of freedom

The growth of the nuclear-power industry in the 1950s and 1960s was a direct stimulus to the development of shaking tables. Given the horrific consequences of damage to a nuclear reactor, it was necessary to subject models to a greater range and magnitude of seismic loads, and also to the possibility of bomb damage. To meet these aims, the first tables with full six degrees of freedom were developed – linear movement along three orthogonal axes, and rotation about each of these axes. The most effective solution was found to be to use eight actuators – four attached to the mid-points of a square table to generate one directional or rotational movements of the table and to control yaw, and four vertical actuators at the corners of the table to generate vertical movement and to control pitch and roll (Figure 25.10). The first of these was built in Germany in the late 1970s, but dismantled when Germany decided to abandon nuclear energy. Construction

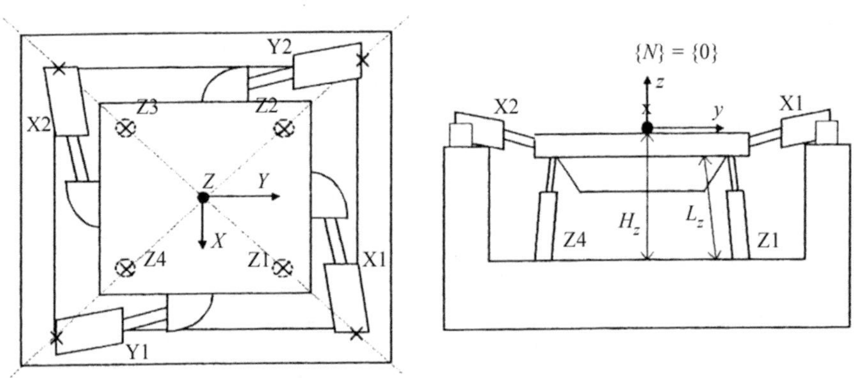

Figure 25.10 The six-degrees-of-freedom shaking table at the Earthquake Engineering Research Centre, Bristol University, UK. Diagrams showing the usual arrangement of the eight actuators. (Source: [25] Figure 3)

of several tables with six DOF finally began in Europe in the early 1980s, most of which have aluminium tables, to reduce the inertia: a 3 x 3 m table at Bristol in Britain, a 4 x 4 m table in 1984 at ISMES in Bergamo and another (steel) one in 1986, a 4 x 4 m table at the Technical University of Athens in 1986, and another at ENEA (Italian National Agency for New Technologies, Energy and Sustainable Economic Development). In the 1990s four tables were built for the CEA (*Commissariat à l'Energie Atomique*) near Paris. As may be imagined, the growing complexity of the mechanical features of the table designs was matched by more and more sophisticated systems for controlling the tables and measuring the responses of the models under test [24].

25.4 Final remarks

During their first hundred years, shaking tables have probably been used more by engineering scientists conducting experiments to study the response of soils and major structures such as dams, high-rise buildings and bridges to dynamic loading, than by engineers working on particular projects. However, the understanding gained by seismologists, for example regarding soil pressures, soil liquefaction, and the resonant frequencies, ultimate strength and safety of large structures, have quickly been fed into design codes of practice.

References

1 Musson R.M.W. (2013) A history of British seismology. *Bulletin of Earthquake Engineering* 11: 715-861. 742.

2 Muir-Wood, R. (1988) Robert Mallet and John Milne – earthquakes incorporated in Victorian Britain. *J. Earthquake Eng. and Struct. Dyn.* 17(1): 107-142.

3 Latimer, C. (2008) Kelvin and the Development of Science in Meiji Japan. In: *Kelvin: Life, Labours and Legacy.* (eds R. Flood, M. McCartney & A. Whitaker), 201-223. Oxford: University Press.

4 Musson R.M.W. (2013) A history of British seismology. *Bulletin of Earthquake Engineering* 11: 715-861. 791-808.

5 Sekiya, S. (1887) A model showing the motion of an earth-particle during an earthquake. *Journal of the College of Science (Imperial University, Japan)* 1: 359-363.

6 Muir-Wood, R. (1988) Robert Mallet and John Milne – earthquakes incorporated in Victorian Britain. *Journal of Earthquake Engineering and Structural Dynamics* 17 (1): 107-142.

7 Masciari-Genoese, F. (1915) *Trattato di costruzioni antisismiche preceduto da un corso di sismologia.* Milan: Ulrico Hoepli. 262.

8 Rogers, F.J. (1908) Experiments with a shaking machine. In: *The California Earthquake of April 18, 1906: Report of the California State Earthquake Investigation Commission* Pub. No. 87 (eds A.C. Lawson et al.), Vol. I, Part 2: 326-335. Washington D.C.: Carnegie Institution of Washington. Reprinted in *Bulletin of the Seismological Society of America* (1930) 20 (3): 147-160.

9 Jacobsen, L.S. (1930) Motion of soils subjected to simple harmonic motion vibration. *Bulletin of the Seismological Society of America* 20 (3): 160-195.

10 Jacobsen, L.S. (1976) Earthquake engineering research at Stanford in the period of 1906-1960. *Earthquake Engineering Research Institute Newsletter* 10 (3): 1-10.

11 Jacobsen L.S. (1930) Experimental study of the dynamic behaviour of models of timber walls. *Bulletin of the Seismological Society of America* 20 (3): 115-146.

12 Blume, J.A. (1994) *The EERI Oral History Series: John A. Blume.* (Interviewer Stanley Scott.) Oakland (CA): Earthquake Engineering Research Institute.

13 Blume, J.A. & Hesselmeyer, H.L. (1934) *The reconciliation of the computed and observed periods of vibration of a fifteen-storey building.* Post-graduate thesis. Stanford University.

14 Severn, R.T. (2011) The development of shaking tables – a historical note. *Earthquake engineering and structural dynamics* 40: 195-213. 201.

15 Mononobe, N. & Matsuo, H. (1932) *Experimental investigation of lateral earth pressures during earthquakes. Bulletin of the Earthquake Research Institute (Tokyo)* X (Part 4): 884-902.

16 Mononobe, N. (1930) Earthquake-proof construction of masonry dams. *Proc., 1930 Meeting of the Eastern Section of the Seismological Society of America.* Washington: U.S. Bureau of Standards. 18.

17 Reitherman, Robert (2012) *Earthquakes and engineers: an international history.* Reston VA, USA: American Society Civil Engineers. 267.

18 Mononobe, N., Takata, A. & Matumura, H. (1936) Seismic stability of earth dams. *Second World Congress on Large Dams,* Washington DC.

19 Severn, R.T. (2011) The development of shaking tables – a historical note. *Earthquake engineering and structural dynamics* 40: 195-213. 201.

20 Ruge, A.C. (1934) The determination of earthquake stresses in elastic structures by means of models. *Bulletin of the Seismological Society of America* 24 (3): 169-230.

21 Ruge, A.C. (1936) A machine for producing earthquake motion from a shadowgraph of the Earthquake. *Bulletin of the Seismological Society of America* 26 (3): 201-205.

22 https://ascelibrary.org/action/showImage?doi=10.1061/9780784410714&iName=master.img-003.jpg&w=360&h=450 (17/10/2019)

23 Reitherman, Robert (2012) *Earthquakes and engineers: an international history.* Reston VA, USA: American Society Civil Engineers. 251-252.

24 Severn, R.T. (2011) The development of shaking tables – a historical note. *Earthquake engineering and structural dynamics* 40: 195-213.

25 Severn, R.T. & Stoten, D.P. & Tagawa, Y. (2012) The contribution of shaking tables to earthquake engineering. *Proceedings of 15th World Conference on Earthquake Engineering,* Lisbon.

26 Oberti, G. & Tonini, D. (1956) Development of aseismic design and construction in Italy by means of research on large model tests. *Proceedings of First World Conference on Earthquake Engineering*, Berkeley, CA.

27 Sozen, M.A., Otani, S., Gulkan, P. & Nielsen, N.N. (1969) The University of Illinois earthquake simulator. *4th World Conf. on E'quake. Eng.*, Santiago, Chile, Vol. 2, Session B5, 140-149.

26

The historical use of models in the acoustic design of buildings

Raf Orlowski

26.1 Early twentieth century

26.1.1 Model studies using the sound-pulse method

Wallace Clement Sabine (1868-1919) was the father of modern acoustics. As a young physicist working at Harvard University, he became involved in room acoustics when he was invited, in 1895, to improve the acoustics of the recently-built Fogg Lecture Hall which had very poor intelligibility. This led him to develop an experimental approach to studying reverberation, sound intensity and room acoustics. He used an organ pipe as the sound source and, working before it was possible measure sounds electrically or to record them, he relied on his ears to judge the intensity and quality of the sound [1]. In 1913, he was the first to study the reflection characteristics of auditoria by using a recently-developed technique for photographing the propagation of a sound pulse within a model of a section of a theatre [2]. The method relied on the technique of *schlieren* photography invented in 1864 by August Toepler (1836-1912) which makes visible any changes in air density [3]. A sound pulse was generated in the model and, as the sound wave and its reflections passed through the model section, they refracted the light and the image was recorded on a photographic plate [3, 4] (Figure 26.1).

Sabine tested models of the New Theatre in New York, in both elevation and plan, to verify that the sound spread uniformly through the theatre and also to demonstrate the effects of fitting a canopy in the main auditorium to try to improve its acoustics (Figure 26.2). He later collaborated with the architect C.H. Blackall while he was designing the Scollay Square Theatre in Boston, and used his sound-pulse technique to demonstrate the acoustic improvement achieved in the final design, compared to an earlier scheme.

Sabine's method was further developed, to be more practical, at the National Physical Laboratory (NPL) in Britain in around 1922 [5, 6]. A large Wimshurst machine, capable of generating a potential of 100 kV, was used to charge up a number of capacitors. Using trigger-sparks, the capacitors were discharged through two spark-gaps marked 'sound spark' and 'light spark', both within a

Physical Models: Their historical and current use in civil and building engineering design,
First Edition. Edited by Bill Addis.

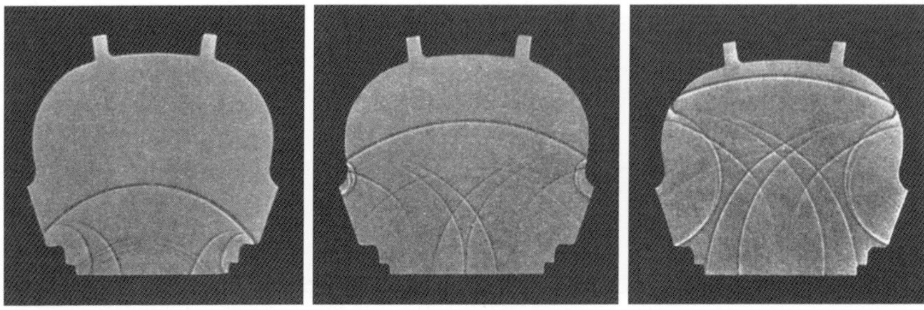

Figure 26.1 Sound-pulse photographs by Sabine showing the progress of a single wave through a plan of a theatre. (Source: [4] Figures 22-24)

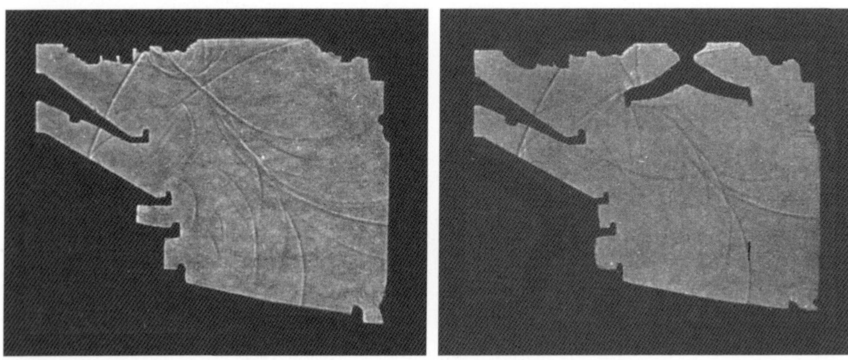

Figure 26.2 Sound-pulse photographs by Sabine showing the effect of modifying a theatre ceiling. (Source: [4] Figures 16, 19)

light-tight box. The 'sound spark' generated a sound pulse which was illuminated a fraction of a second later by light from the delayed 'light spark'. The light from the 'light spark' cast a shadow of the sound-pulse on a viewing screen or photographic plate at the end of the box. The model section of the building whose reflection characteristics were required was located so that the gap of the 'sound spark' was within the interior of the section and at right angles to its plane. The model was made of ebonite or hardwood, usually at a scale of 1 inch to 32 feet (scale factor 1:384) (Figures 26.3 and 26.4). This apparatus was capable of making photographs showing the progress of waves in a time-interval corresponding to about 125 milliseconds in a full-size building.

The sound-pulse technique was used to study the acoustics of the Royal Institution Lecture Theatre in London [5] (Figure 26.5). The photographs show that, in general, the sequence of sound reflections and their delays were favourable for good speech intelligibility, although the sound reflected from the upper part of the wall behind the speaker, which finally returned to the floor after a second reflection from the ceiling, contributed to an echo effect perceived in the front seats. It proved possible to eliminate this slight defect by mounting some sound-absorbent material on the upper part of the wall behind the lecturer.

The sound-pulse method was further developed in Switzerland in the 1930s by the Swiss engineer Franz Max Osswald (1879-1944). He was Switzerland's first

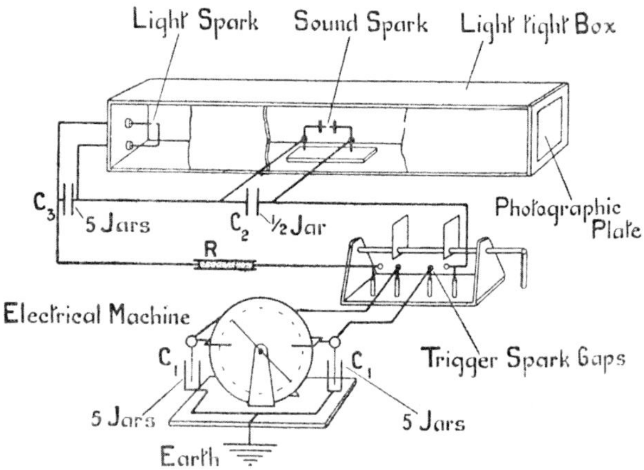

Figure 26.3 Diagram of apparatus for sound-pulse photography developed at the NPL. (Source: [5] Figure 16)

Figure 26.4 NPL apparatus for sound-pulse photography. (Image: Courtesy of National Physical Laboratory)

specialist in architectural acoustics and founder of the first applied acoustics laboratory at the ETH, in Zurich. He set up his laboratory in 1930 with an apparatus very similar in concept to that of Sabine and at the NPL (Figure 26.6a) [3, 7, 8]. Osswald undertook many studies similar in kind to those of Sabine and the quality of the photography and data that could be extracted, reflected the technical progress that Osswald had made. He undertook many studies for cinemas and concert halls, for example, investigating the effects of different profiles for the walls and ceilings [9] (Figure 26.6b). One of his first studies was for an auditorium with a movable ceiling that would enable the volume to be decreased from 8750 to 6100 m^3, and the reverberation time from 1.6, suitable for large-scale classical music, to 1.3 seconds, suitable for more intimate music and the spoken word [10] (Figure 26.7).

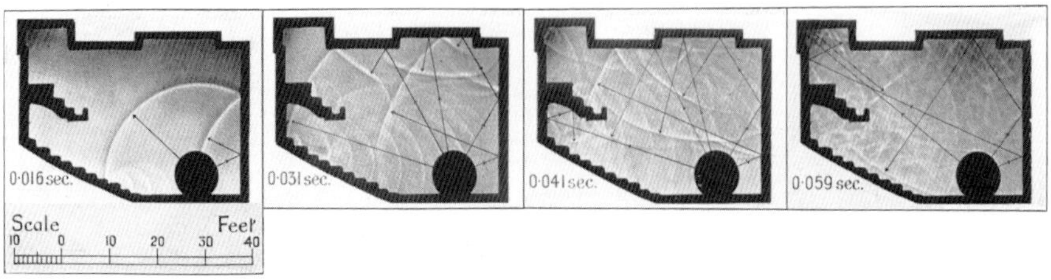

Figure 26.5 Sequence of sound-pulse photographs taken at the NPL of the Royal Institution Lecture Theatre. (Image: Courtesy of National Physical Laboratory)

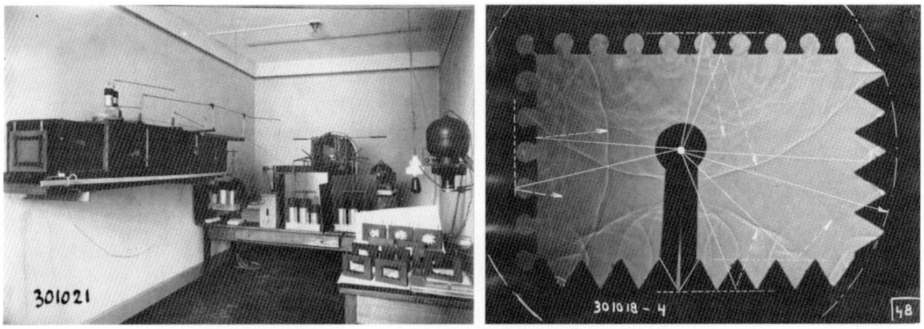

(a) (b)

Figure 26.6 (a) Apparatus room at Osswald's applied acoustics laboratory at ETH Zurich, 1930. (b) Study of wall profiles in cinemas, 1930. (Images: Image Archive, ETH Library Zurich)

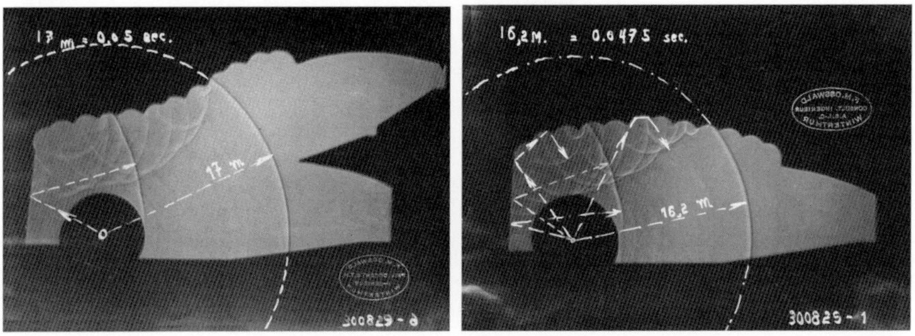

(a) (b)

Figure 26.7 Sound-pulse study of an auditorium with variable volume, 1930. (a) Ceiling raised. (b) Ceiling lowered. (Images: Image Archive, ETH Library Zurich)

26.1.2 Model studies using the ripple-tank method

The ripple-tank method was based on the fact that ripples in a small tank of water are suitable for modelling sound waves in a space; wavelengths were typically comparable with the size of the models being tested. The method had is origin in the early nineteenth century when Thomas Young used a ripple tank to model the behaviour of light waves and to demonstrate interference. The technique was adapted by the NPL in the 1920s to provide a visual demonstration of

Figure 26.8 Ripple tank at the National Physical Laboratory, 1920s. (Source: [5] Figure 31)

the propagation of sound waves. A great advantage of this technique, especially compared to the sound-pulse method, was that it was easy to carry out and quick to modify the cross-section of a room to see the effect of changes on the acoustic behaviour [5] (Figure 26.8).

A typical ripple tank arrangement involved a model of a longitudinal section of a hall at a scale factor of about 1:50 placed on the glass bottom of a shallow tank of water. The reflection characteristics of the model section were observed by the behaviour of ripples produced by a plunger just dipping into the water at the location of the sound source in the hall. To help view the progress of the waves, light was shone upwards through the glass bottom of the ripple tank to cast a shadow of the waves on a screen placed above. By taking a series of photographs at intervals, the impact of the waves on selected features of the room could be shown. In one such sequence of photographs of waves in an auditorium section, the distance from crest to crest between successive ripples corresponded to sound waves of length from 0.5 metres to 1.4 metres in the full-scale building, i.e. sounds in the important frequency range 250-700 Hz (Figure 26.9). Rays showing the direction of the wave fronts were added to a diagram and the auditorium section judged to be satisfactory. With regard to possible echoes, the dimensions of the hall indicate that most sounds reach listeners after one reflection within one-fifteenth of a second of the direct sound. However, reflections from the backs of the galleries are directed upwards towards the ceiling and on reflection arrive at the floor too late to reinforce the direct sound [4].

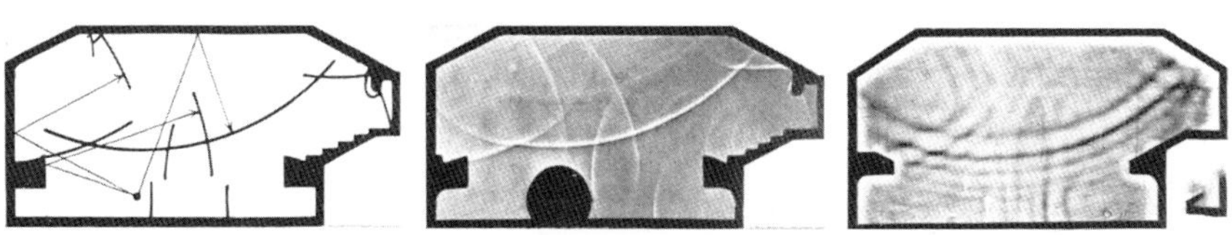

(a) (b) (c)

Figure 26.9 The progress of ripples in a sectional model of an auditorium. (a) Geometric construction. (b) Sound pulse image. (c) Ripple tank image. (Source: [5] Plate VII)

26.1.3 Three-dimensional model studies using light rays

A fundamental advantage of three-dimensional models is that they enabled complex three-dimensional sound propagation to be explored. A simple but useful way of doing this is to use light rays rather than sound sources, since light rays can be assumed to behave like sound waves at high acoustic frequencies. We find in Vitruvius the idea of imagining sound as travelling in straight lines like rays of light and this formed the basis of guidance to theatre designers from the early-nineteenth century when the first books on acoustics and theatre design were written [11]. Using a narrow light beam, the paths of individual reflections can be traced. Reflective surfaces can be modelled by mirrors and absorbent surfaces by matt black materials.

The Japanese acoustician/architect Takeo Satow (1899-1972) described in 1929 his own version of such a modelling apparatus which he had been using fruitfully for several years [12]. A polished metallic model of the cross section of an auditorium was placed in a box with a glass lid. Light rays radiated from the position of the stage and were made visible by means of smoke introduced into the box (Figures 26.10 and 26.11). A small cylindrical prism could be placed in the auditorium where a listener might be. The prism reflected light arriving from all directions perpendicularly to the incident rays so that the intensity of light (i.e. sound) arriving from different directions could be made visible (Figure 26.12). Satow considered his method particularly advantageous in allowing different auditorium sections to be made and modified easily, since it required little specialist training to use, and could also be used to convince people with little knowledge of acoustics.

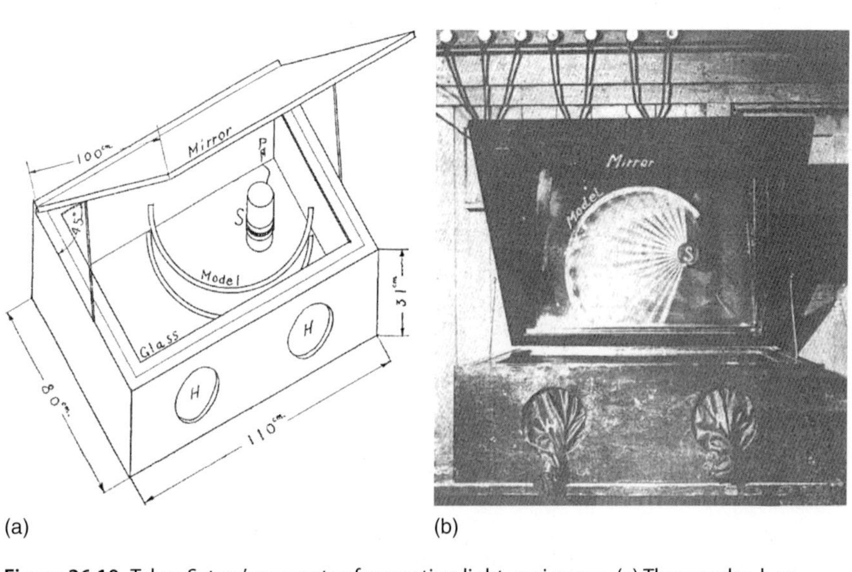

(a) (b)

Figure 26.10 Takeo Satow's apparatus for creating light-ray images. (a) The wooden box showing the source of light rays, position of the model, glass top to the box and mirror used to view the rays. (b) The experimenter's view of a model under study. The holes at the front of the box provide hand access to adjust the model. (Source: [12])

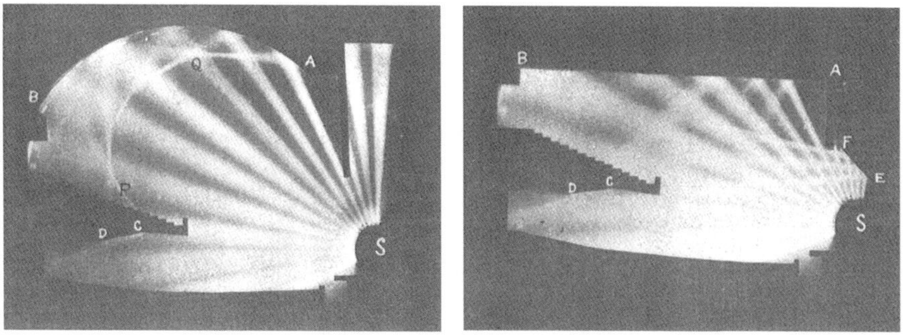

(a) (b)

Figure 26.11 Light-ray model to study the acoustical performance of an auditorium. (a) An auditorium with a curved ceiling, showing multiple reflections to the gallery seating. (b) An auditorium with a flat ceiling (AB) and reflector (EF) above the stage, showing fewer long-path reflections. (Source: [12])

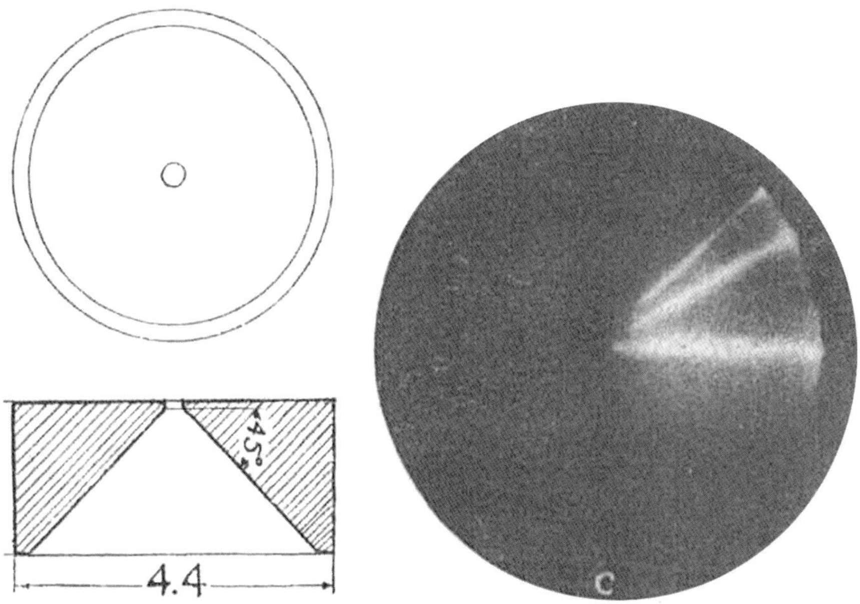

(a) (b)

Figure 26.12 Cylindrical prism used to show the light/sound intensity arriving from different directions. (Source: [12])

Another technique using the principle of ray tracing was developed in the 1920s at the NPL in Teddington, England, and was especially useful for studying the acoustic performance of a complex three-dimensional space for which a sound-pulse or ripple-tank model would have been inadequate. In order to trace the path of a 'sound ray' in a model auditorium, a pencil beam of light was directed from the stage in the model towards various parts of the auditorium, and a number of small mirrors were mounted on the model to reflect the

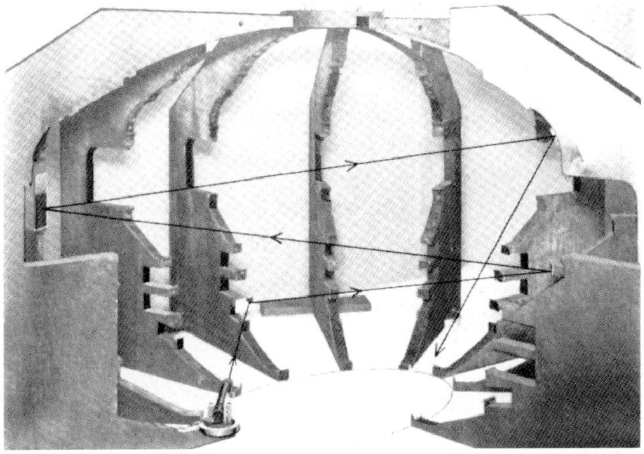

Figure 26.13 Skeleton three-dimensional model of a theatre auditorium, National Physical Laboratory, Teddington, 1920s. A light beam is used to represent the path of a sound wave. (Source: [5] Plate VII)

light onwards, and so on for 4 or 5 times (Figure 26.13): 'A small light 'gun', mounted upon bearings graduated in altitude and azimuth, is placed in the model auditorium at the position of the speaker, and directs light in any desired direction. By dividing the hemisphere above the speaker's head into 100 equally distributed regions, it is possible to trace out all probable echoes and repeated reflections in the hall' [13]. It also enabled the mean free path of a sound wave to be estimated and thus give an indication of the reverberation time of a hall. The technique is still used today where readily-available laser pointers can provide a suitably narrow and intense light beam which can be shone in various directions into a scale model.

Another form of model analysis using light was developed in 1935 by engineers at Philips in Eindhoven who were working on the reconstruction of the firm's theatre [14]. The existing Philips Theatre had suffered from problems of intelligibility, especially at the back of the hall. The ceiling with flat surfaces would be replaced by a curved ceiling; the hall would also be shortened and a balcony added to maintain the number of seats (Figure 26.14).

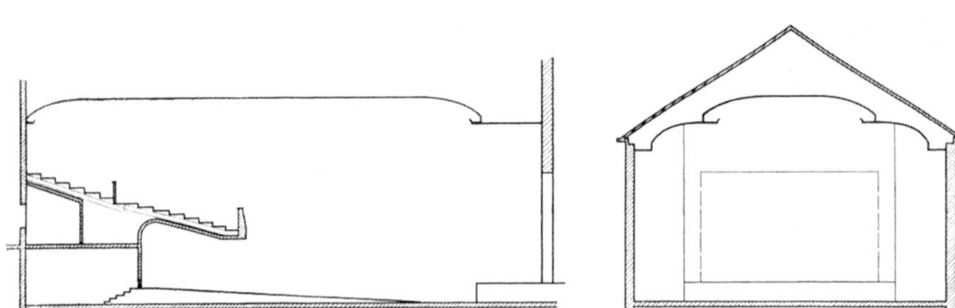

Figure 26.14 Design for the proposed reconstruction of the original Philips Theatre. (Source: [14])

(a) (b)

Figure 26.15 1:10 scale models in aluminium sheet of (a) the original Philips Theatre and (b) the new design for the theatre after reconstruction. The opal glass floor of the model has been removed to show the interior. The location of the lamp representing the 'sound source' can be seen on the stage of the model auditorium. (Source: [14])

Two models were made of aluminium sheet which reflects about 50% of incident light (Figure 26.15); the light intensity would thus be less than 10% after four reflections. A lamp was placed at the performing area and the light intensity in the seating area was measured. This was done by placing a ground glass screen in the seating area and recording the light intensity photographically.

The test on the model of the original theatre showed a significant fall in light intensity from the stage towards the back of the auditorium (Figure 26.16a). In the redesigned theatre, shorter and with a curved ceiling, the reduction in light

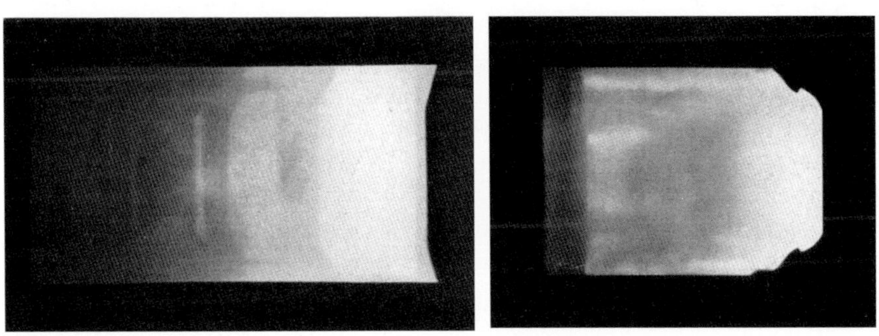

(a) (b)

Figure 26.16 (a) Photograph of light intensity in the model of the original theatre. (b) Photograph of light intensity in the model of the redesigned theatre, with a balcony. (Source: [14])

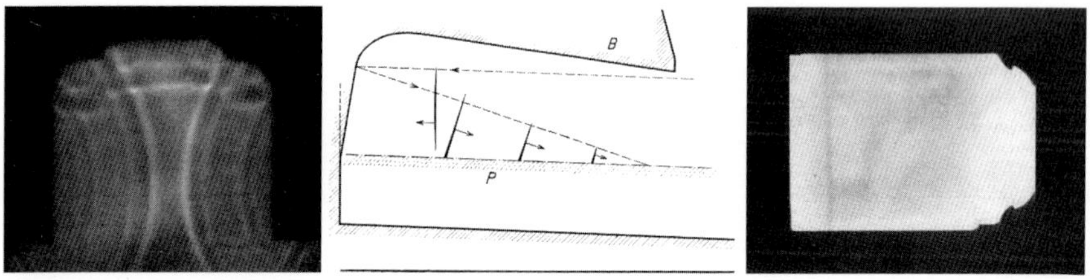

(a) (b) (c)

Figure 26.17 (a) Ripple-tank model of the cross section of the hall, showing reflections between the walls. (b) Drawing made from the ripple-tank study of the benefit gained by inclining the rear wall of the auditorium beneath the balcony. (c) Photograph of the even light intensity in the model of the redesigned theatre, with the rear wall under the balcony inclined. (Source: [14])

(sound) intensity towards the rear is shown to be less serious, but low-intensity light beneath the balcony can be identified (Figure 26.16b).

A ripple-tank model study was also undertaken to study the sound reflecting between the parallel walls. It showed that the sound continued to reflect back and forth without being directed down to the audience, thereby increasing the reverberation time of the hall (Figure 26.17a). It was thus decided to fix wooden panelling to the walls to absorb some sound and reduce the amount by which these reflections would increase the reverberation. Another ripple-tank study was carried out to resolve the problem of the low sound intensity beneath the balcony. The model showed sound being reflected off the rear wall back into the auditorium. By inclining the rear wall forward slightly, the ripple-tank model demonstrated the reflection of this incident sound downwards to the audience (Figure 26.17b). The effectiveness of this solution was demonstrated using the light model, rebuilt with the inclined wall, showing a more even sound intensity across the whole of the ground floor audience seating (Figure 26.17c).

26.1.4 Three-dimensional model studies using sound

Physical scale models were first used to measure and analyse the acoustic performance of auditoria by the German physicist Friedrich Spandöck (1904-1966) in the early 1930s, and developed while he worked at the laboratory for sound equipment at Siemens & Halske in Berlin [15]. Using simple dimensional analysis to find suitable dimensionless constants, he showed that the acoustic behaviour of a room varies inversely with the model scale – for a one-fifth scale model, the sound frequencies used for testing need to be five times higher than normal frequencies. He also recognised the need to ensure the temperature, pressure and density of the air in the model room were identical to a real room, since these affect the speed of different frequencies of sound differently. Spandöck made his models at a scale of 1:5 using 3 cm thick, tongue-and-groove plasterboard supported by wooden battens. The test frequencies were raised by the linear scale factor, from 500 and 800 Hz to 2500 and 4000 Hz, respectively. He built three models – a simple box shape 4 x 3 x 2.6 m high (Model I), the same with a sloping floor (Model II), and another with a semi-circular wall at one end (Model III).

His first test method was a qualitative assessment of the room acoustic. He played a gramophone record of some speech and music at five times the correct speed, and then used a second phonograph, located at various places in the room where a listener would be, to record the received sound on disc. He then replayed this disc at one fifth of the recording speed. Although he acknowledged its limitations, he concluded: 'this method was suitable for testing the actual intelligibility of speech and audibility of a projected space, which is ultimately only determined by the ear, and to convey an immediate impression of the acoustics to be expected in the various places before its execution'.

In his second method, Spandöck used a loudspeaker as the sound source and a microphone connected to an oscilloscope to display the sound intensity in different places in the model rooms. As the microphone traversed the room the amplitude was recorded photographically. His first concern was to demonstrate that tests on a model were indeed a reliable representation of the acoustic behaviour of a similar full-size room. He then studied the effect of the three different room shapes on the reverberation time – the simple box shape, the room with sloping floor (shorter) and with a semi-circular wall behind the stage (longest). He also compared the reverberation time with bare wall surfaces and then with the surfaces covered in felt to absorb sound. In Model III he made a traverse from the front to the back of the hall, recording the sound intensity, first with bare surface, then with felt-covered surfaces. This showed that, when the surfaces were covered in felt, the intensity is reduced and also that the variation in intensity over the length of the hall was almost constant, except at the two ends (Figure 26.18).

Spandöck also photographed the oscilloscope screen when a regular pulsed sound was recorded in all three models. These showed a relatively pure sound in Model II, a slightly distorted sound in Model I, and a rather more distorted sound in Model III (Figure 26.19).

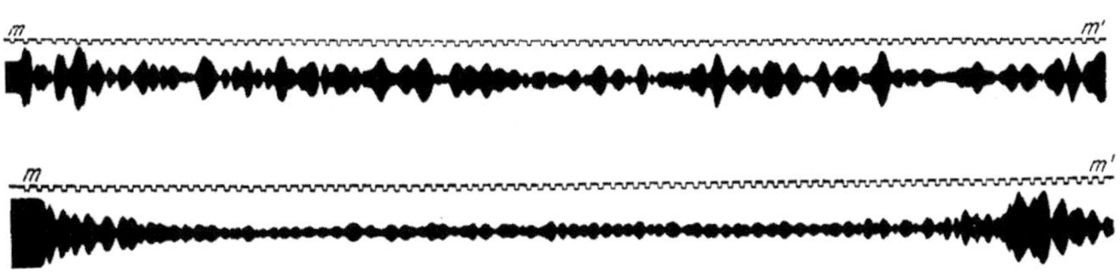

Figure 26.18 Variation of sound intensity from front to back of the room. With bare surfaces (above) and felt-covered surfaces (below). (Source: [15])

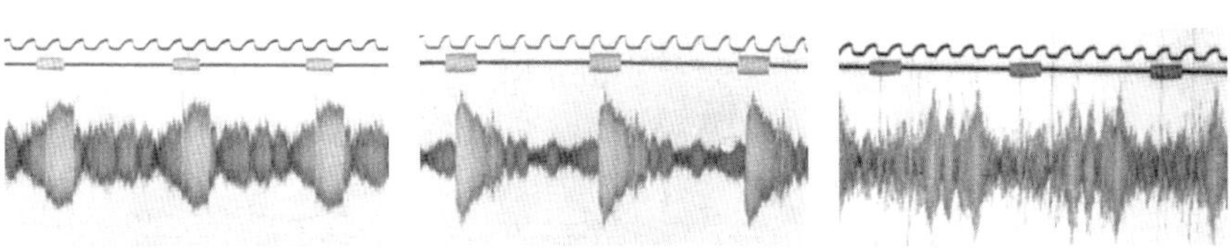

Figure 26.19 Wave form of pulsed sound recorded in Models I, II and III. (Source: [15])

Spandöck's main conclusion from his studies was that tests made on a 1:5 model could be used to predict the acoustic behaviour of a full-size room. He also concluded that the shape of a room was the most important influence on the quality of the room acoustic, especially the reverberation time, which was different from the conclusion to be drawn from using Wallace Sabine's simple formula for reverberation time.

Similar studies were undertaken in the late 1930s in a number of university physics departments which all demonstrated the possibility of using models during the design of an auditorium. In the 1940s, the Danish acoustician Vilhelm Jordan improved the technique by using a magnetic tape recorder on which he recorded samples of speech. Using this method he could demonstrate the intelligibility of speech in rooms with differing amounts of sound absorption [16]. Nevertheless, the use of acoustic scale modelling did not enjoy widespread use by building designers until after the war.

26.2 Model testing in the 1960s and 1970s

26.2.1 Developing new modelling techniques

The acoustic modelling technique that Spandöck had initiated in the 1930s was limited technically by the transducers, electronics and recording equipment available to him. The Danish acoustician Vilhelm Jordan had developed the technique into a useful design tool but a more substantial breakthrough came in the 1960s due mainly to the research work of Brebeck, Bucklein and Krauth who were guided by Spandöck himself [17]. They thoroughly reviewed the principles of acoustic scale modelling and developed instrumentation, equipment and techniques to optimise the procedure.

In essence, the principles of acoustic scale modelling are straightforward and based on the fundamental equation:

$$\text{Speed} = \text{Distance}/\text{Time} = \text{Frequency x Wavelength}$$

The speed of sound is constant for a given medium, so that if air is used in the model, when distance is reduced by a scale factor, the time is reduced by the same factor. Considering the second part of the equation, the ratio of the wavelength of sound to the dimensions of the room must remain the same in the model as at full scale, so that reducing the room dimensions by the scale factor means that the wavelength needs to be reduced by the same factor. This means that the frequency of sound must be increased by the same factor – for a model at a scale of 1:10, the frequency must by increased 10 times.

A complication arises by increasing the frequency of sound because this also increases the absorption of sound as it propagates through the air. For perfect modelling, the air absorption should increase by the scale factor but in fact it increases much more than this. Therefore, the air absorption must be reduced or a correction applied.

The other aspect which has to be matched in an acoustic model is the sound absorption of the room surfaces. This requires that the sound absorption coefficients of the surfaces at full-scale frequencies have the same value as model materials at model frequencies.

The final complication of acoustic scale modelling is the characteristics of the sound sources and microphones. In the case of sound sources, they need to operate predominantly at ultrasonic frequencies and this presents challenges in reproducing appropriate directional properties. For microphones, their size needs to be small to preserve the correct directionality which means they have low sensitivity and therefore, poor signal-to-noise ratios.

All the above issues were thoroughly addressed by Brebeck and appropriate solutions were developed. To model air absorption, they proposed drying the air to a low value of relative humidity, around 2%, which reduces the absorption and provides a very good match at scale factors of around 1:8 and 1:10. Smaller models would have a less good match.

To identify suitable model materials, a model reverberation chamber was constructed, together with air drying equipment, to test the absorption coefficients of promising samples.

For objective measurements of, for example, reverberation time, an electric spark was developed which is an effective point source with omnidirectional characteristics. For subjective measurements, an electrostatic loudspeaker was employed which can operate at frequencies up to 100 kHz. Using a tape recorder with a speed change of 1:10, recordings of speech or music could be made up to a full-scale equivalent frequency of 10 kHz which was sufficient to provide good quality sound for comparison with full-scale recordings. Recordings of this kind were made in a 1:10 scale model of the Meistersinger Hall in Nuremberg and also in the actual hall. These recordings were presented to a panel of listeners who reported a close similarity between the model and the actual venue.

This major advance by Brebeck and his colleagues was the starting point for an acoustic modelling programme adopted by the British Broadcasting Corporation (BBC) in 1968 when it was planning to build a number of new orchestral studios and needed a reliable method of ensuring that the acoustics would be suitable.

A feasibility study was carried out by Harwood and Burd [18] who proceeded to build a 1:8 scale model of an existing BBC orchestral studio (Maida Vale) as a proving experiment [19]. There is little difference between using a 1:8 scale model and 1:10 scale model although a 1:8 scale model operates at frequencies which are three octaves above full-scale frequencies and this was a little easier for the instrumentation. The proving experiment was successful and the BBC then made major advances in developing model loudspeakers and microphone systems so that they could make recordings up to the equivalent full-scale frequency of 12.5 kHz with a very good signal-to noise ratio of around 50 dB. The BBC went on to use this modelling technique to design the acoustics of a large new orchestral studio in Manchester [20].

26.2.2 Sydney Opera House

For the design of the unusual interior spaces of the Sydney Opera house, the Danish acoustician Vilhelm Jordan (1909-1982), a pioneer in the use of scale models in acoustic design, was appointed to advise on how to achieve a satisfactory acoustic for the many different uses, including concerts, opera and speech that were proposed. Since these different uses required different acoustic characteristics, one early plan for the major hall was to provide a large moveable ceiling that could be adjusted to create the most suitable reverberation time. So radical were the proposals for both the concert hall and the opera theatre, and different from established forms, that Jordan advised it would be necessary to make acoustic tests using scale models. Over five or six years, during which Jørn Utzon, the original architect, resigned and new architects were appointed, models of several different proposals for both auditoria were made at 1:10 scale (Figure 26.20). Made of wood, they included models of the audience with bodies made from neoprene and cardboard heads [21]. The tests soon established that it would not be possible to create an acoustic that could serve the original idea of a multi-purpose auditorium. The curved form of the walls also caused problems in achieving a satisfactory distribution of reflected sound in all parts of the auditoria.

The model tests demonstrated the benefits of introducing flat side walls to the concert hall to enhance lateral reflections, and to add reflective interiors to the boxes. The model testing also showed that the original ceiling of the auditorium, consisting of large catenaries hung longitudinally from the external shell, left the orchestra area suffering from a deficiency of early-reflected sound, making it difficult to play their instruments. Adding suspended reflectors over the stage did not substantially change this situation and it was decided to reduce the distance between the side walls and raise the auditorium ceiling substantially.

Figure 26.20 Vilhelm Jordan inside his 1:10 scale model of the 2800-seat concert hall of Sydney Opera House used to test different acoustic treatments of the walls and ceiling. (Image: Jordan Acoustics)

Many different shapes and positions of reflectors suspended above the orchestra were studied throughout the design stages and the final arrangement used small toroidal discs (slightly convex on the underside) as reflectors, suspended from the ceiling by adjustable cables.

The acoustics of both finished auditoria were finally assessed in a number of test performances with real orchestras and audiences. Magnetic-tape recorders were used to record the decay of sound after firing a gun on stage (with a blank cartridge), and getting the conductor to halt the orchestra abruptly while playing Beethoven at full power. The results of these tests were used to recalibrate the model tests and to identify a number of adjustments that could be made to the various moveable reflectors to achieve the best acoustic. Jordan's work on the Sydney Opera House was a milestone in the development of acoustics.

26.3 Model studies at one-eighth scale in the late-twentieth century

26.3.1 Music studios and auditoria

With advances in transducers, recording equipment and electronics, it became possible to carry out the technique originated by Spandöck with much greater authenticity. A system similar to the one developed by the BBC was employed in an acoustic modelling laboratory at the University of Cambridge, set up in 1975 to develop modelling techniques. In particular, the laboratory was used to investigate the design of two new major auditoria being planned for London in the 1970s. One was to be the Olivier auditorium, part of the new National Theatre, and the other was the Barbican concert hall, the first permanent home of the London Symphony orchestra. Models of these two auditoria were built at one-eighth scale.

There were three special requirements for the modelling laboratory, in addition to a large-enough space for housing one-eighth scale models: first, instrumentation for observing room-acoustical phenomena at model scale; secondly, suitable equipment for scaling air absorption; and thirdly, facilities for modelling the absorptive properties of boundary surfaces.

Standard acoustic laboratory instrumentation was generally suitable for modelling work, since much of it could operate at model frequencies, namely up to 100 kHz; a special item was a tape recorder with an eight-to-one speed change facility. Transducers, however, were a challenge and considerable effort was devoted to designing spark sources, air jets and transducers with specific directional characteristics.

As already mentioned, air absorption, and its rapid increase with frequency, is a complicating factor in acoustic modelling when air is used as the medium. For perfect modelling, the air absorption coefficient in the model should be increased by the scale factor. In ambient conditions, the increase with frequency is much greater than this. A satisfactory solution was developed that involved reducing the molecular component of air absorption, which is due to the interaction between water-vapour and oxygen molecules, by removing water-vapour.

An air-drying system using silica gel at the absorbing agent was able to maintain the relative humidity in a model at 2%. This value gives an air absorption coefficient for a one-eighth scale model which closely matches the value required for perfect modelling of full-scale ambient conditions.

With regard to modelling absorption characteristics, all model surfaces should ideally match, at scale frequencies, the impedance and variation in impedance with angle of incidence of their full-scale counterparts at normal frequencies. In practice, however, it proved satisfactory, at least in auditoria, to model only the absorption coefficient at random incidence and materials were selected using a one-eighth scale model reverberation chamber. Typical materials used were polyether foam to model seating upholstery and furnishing velvet to model carpeting. The models themselves were constructed of wood and the interiors were painted with gloss varnish, which gave a good match for acoustically-hard and reflective surfaces such as concrete, brick and plasterwork.

26.3.2 The Olivier Auditorium, National Theatre, London

A one-eighth scale model of the Olivier Auditorium in London was built in the mid-1970s as the building itself was nearing completion and, although it was not used in the design process, it proved very important in correcting some acoustical deficiencies that became apparent when the theatre opened.

The model was built out of plywood and a thick coat of varnish was applied to internal surfaces to model the exposed concrete interior. The seating was made of wood and lined with polyether foam to model the upholstery. The model was tested by scaling the reverberation time and then making both subjective assessments and objective measurements. In both cases a sound source was required with the same directionality as a human speaker but operating at model frequencies. A special transducer was developed for this which consisted of a high-frequency loudspeaker loaded by a tapering horn so that the sound was emitted from a 6 mm aperture well-separated from the magnet assembly of the driver [22].

For subjective recordings, this source was used in combination with a simple dummy head which consisted of two 6-mm omnidirectional microphones separated by a model head. (This crude model dummy head was substantially improved later by Japanese acousticians who were able to accurately reproduce the outer ears on a model head). The test procedure involved replaying anechoic speech at eight times normal speed through the model and recording the signal received by the microphones. The recording was then replayed at normal speed over headphones for subjective assessment. The quality of the recordings was sufficiently authentic to enable differences to be heard between different seats in the auditorium, and the technique was used to improve the speech intelligibility in certain seats by adjusting panels in the ceiling and installing acoustically-absorbent panels at strategic locations.

Objective measurements required a criterion which correlates well with speech intelligibility and is simple to measure at model scale. The measure adopted, which is based on the impulse-response, was the 50-ms energy fraction. This measures, at a given listening position, the proportion of total energy arriving within 50 ms of the direct impulse. For satisfactory intelligibility it was considered that half the total energy should arrive within 50 ms of the direct impulse, giving a minimum value for the fraction of 0.5. Thus, the aim of the model adjustments, in objective terms, was to achieve values in excess of 0.5 throughout all the seats.

26.3.3 The Barbican Concert Hall, London

The design of the Barbican Concert Hall in London was unconventional in many respects which made the studies on its model particularly interesting. Like the Olivier Theatre model, the Barbican model, although tested in the late 1970s, was not influential in the design but proved useful in examining remedial measures [23, 24].

The dimensions of the hall reveal that it is relatively short for a 2000 seat venue (32 m from the stage front to the rear) whilst, at 43 m, it is very wide. The ceiling is characterised by two pairs of very deep beams, with a depth of 3.7 m, running along and across the ceiling to provide structural support. The ceiling has a coffered structure but to promote additional diffusion and avoid shadowing effects due to the roof beams, an intricate array of 2000 spheres was suspended from the ceiling. The unconventional nature of this design and, in particular, the impossibility of predicting the behaviour of the suspended spheres, was the principal benefit of testing a one-eighth scale model (Figure 26.21).

Subjective tests were carried out, as for the theatre model, except that the sound source now consisted of two one-eighth scale loudspeaker arrays which formed a stereo pair after a design by the BBC. Music recorded through the model sounded convincing with adequate signal-to-noise ratio; variations could

Figure 26.21 1:8 scale model of the Barbican Concert Hall, with the author inside. (Image: Raf Orlowski)

be heard between different seat positions which corresponded with expectations from objective results. The subjective testing system proved a useful tool in helping to make comparative assessments of acoustic quality with changes in design.

A full range of objective measurements was made in the model including Reverberation Time, Early Decay Time, Clarity Index, Loudness and Early Lateral Energy Fraction. A key finding of these measurements was that the spheres were important in increasing the diffusion in the hall and also increasing the spatial impression or the sensation of being enveloped by sound, a desirable feature in concert halls.

When the Barbican Concert Hall was completed in 1982 and all 2000 spheres were in position, it was unfortunately found that the Reverberation Time was rather short, probably due to excessive acoustic absorption of the seating. This had not been observed in the model as the seating had been modelled on typical upholstered seating rather than the actual seating for which no data was available. In an attempt to increase the Reverberation Time in the hall, the spheres were removed but this did not provide the uplift intended and probably decreased the valued attribute of envelopment which had been demonstrated in the model. Further modelling of the Hall was not pursued at this stage and work was focused on measuring the sound absorption of full-scale elements such as the seating with a view to decreasing absorption and increasing reverberation.

26.3.4 Assessment of model studies at one-eighth scale

Measurements on the Barbican and National Theatre models enabled an overall assessment of the value of one-eighth scale models in auditorium design. The results indicated that the important characteristics of a full-size auditorium could be reproduced in a model although some limitations are imposed by the transducers. Some of these limitations have been resolved more recently by developments in transducer design. Therefore, by testing a one-eighth scale model both objectively and subjectively, the risks of acoustic inadequacies in the actual building are substantially reduced and the technique continues to be used in the twenty-first century for the most prestigious auditoria. The disadvantages of modelling at this scale are cost, space requirements and time taken to construct and possibly modify the model. These limit the use of such models to a largely confirmatory role with only limited scope for comparing different options.

26.4 Model studies at one-fiftieth scale

Much of the difficulty of changing one-eighth scale models to reflect changes in a design can be overcome by using much smaller models at a scale of, say, 1:50. The great advantage with smaller models, in addition to their reduced cost and size, is that they could be built early in the design process and permit several different design options to be tested and evaluated. Thus, a 1:50 scale model had the potential for a much greater influence on the overall design and, thereby, enable a more controlled approach to achieving the desired acoustic conditions. The disadvantage, however, is that objective measurements are limited to the 1 kHz octave at the equivalent full-scale frequency because the absorption of sound energy by air becomes excessive. At the 2 kHz octave (100 kHz at model frequency), for example, the air absorption is nearly four times greater than needed to achieve perfect modelling. There are also limitations in the frequency range of transducers such as microphones. Also, subjective assessments using real speech or music were not viable. Nevertheless, the potential limitations were overcome, and the feasibility of 1:50 scale modelling was validated by Michael Barron at the University of Cambridge by comparing full-scale and model measurements, and the technique was used extensively during the following 30 years [25, 26].

26.4.1 Glyndebourne Opera House and some concert halls

The techniques of acoustic scale modelling at 1:50 scale became very popular in acoustic consultancy in the 1990s and 2000s. The firm of Arup, which had developed a large portfolio of performing arts projects from the 1980s onwards, employed acoustic models at 1:50 scale in almost every one of its auditorium designs, including the Copenhagen Opera House and the Oslo Opera House.

An early example was the design of Glyndebourne Opera House which was acoustically challenging because of its circular form (Figure 26.22). The 1:50 scale modelling was particularly beneficial in helping to develop a successful design for Glyndebourne as the method accurately reproduces the reflection effects of curved surfaces, while computer modelling, at that time, was unable to model the scattering of sound from curved surfaces accurately, a shortcoming

Figure 26.22 1:50 scale model of Glyndebourne Opera House. (Image: Over Arup & Partners)

Figure 26.23 1:50 scale model of Bridgewater Hall, Manchester, UK. (Image: Over Arup & Partners)

that was partially resolved in early part of the twenty-first century. Key elements that were designed using the model were the convex surfaces that were formed on the walls and in the ceiling area in order to prevent focusing of sound by the circular brick drum.

A 1:50 model was also used in the design of concert halls, for example the Bridgewater Hall in Manchester which is a major concert space with 2400 seats (Figure 26.23). In this case, as well as testing the whole auditorium model, it was useful to test certain individual surfaces of the space such as the ceiling. The original design of the ceiling comprised a series of uniform steps made of concrete and there was concern that this could lead to a comb filtering effect caused by repetitions of reflections with a slight delay between each; this would have led to a distortion of sound. Testing the ceiling alone demonstrated that this effect could occur and the design was modified by angling the steps alternately so that reflections were scattered. An additional non-acoustic use of a 1:50 acoustic scale model was to be able to measure the volume of the auditorium which is, of course, important for calculating the reverberation time. This was done by filling the model with small polystyrene balls which were then emptied out and measured in a measuring cylinder.

26.4.2 Factory buildings

Following their work on auditorium modelling, the Cambridge team were tasked in the late 1970s with an investigation of other building types, in particular, factories. The initial aim was to develop a model of an existing factory to establish how accurately the sound field can be reproduced at model scale. If successful, the model would be used for predicting the effect of various noise-control measures and as a research tool for the study of factory acoustics [27, 28, 29].

The factory selected for the modelling exercise had a characteristic factory shape namely, a large flat building with a relatively small height (10 m) compared to its length (120 m) and width (45 m). Inside the factory were twelve main production lines and ancillary equipment for manufacturing domestic lightbulbs. A scale factor of 1:16 was chosen for the model on the basis that this was the largest size practicable for testing in the laboratory environment. Using this scale factor, measurements could be made up to the equivalent of the 4 kHz octave full scale which is a reasonable upper limit for measurements in dB(A).

It was not necessary to reproduce in the scale model the actual noise levels in the factory with all the machines running, since the acoustics of a factory enclosure can be adequately characterised by the reverberation time and the sound propagation with distance from an omnidirectional source. If the sound power of the machines is also known, the sound pressure level at any point in a factory can be evaluated. Sound propagation measurements in the model were made using a crossing, air-jet source in conjunction with a 6 mm or 3 mm diameter capacitor microphone. The air-jet was calibrated in terms of sound power at model frequencies so the sound propagation curves could be expressed in absolute terms, namely sound pressure level minus sound power level. This permits evaluation of noise levels in the factory if the sound power levels of machines is known.

The process of matching the model to the real factory gave considerable insight into the acoustical behaviour of a factory enclosure. For example, a dominant feature was the low-frequency absorption and transmission of sound by the lightweight roof. This effect, coupled with the scattering effect of the machines, dominated the reverberation time and sound propagation at low frequencies. Modelling the sound absorption and transmission of the lightweight roof proved difficult and it was not possible to reproduce the variation of absorption with angle of incidence which was considered to be relevant; rather, the random incidence absorption coefficient was modelled. The scattering and absorption effects of the machinery were modelled by various wooden blocks and metal cylinders.

The measured results showed a good match with the full-scale factory in terms of reverberation time and sound propagation. The model reverberation time was within 10% of the full-scale value at all except the lowest frequencies. The sound propagation in dB(A) was modelled to an average accuracy of about 1 dB. In the six individual octave bands from 125 Hz to 4 kHz, the accuracy of modelling sound propagation was typically within 2 dB. This match was considered to be sufficiently good for investigations into noise control.

Two popular measures for reducing noise in factories were tested in the model namely, suspended absorbers and barriers. Suspended absorbers produced a modest reduction in level – at a typical density of one absorber per square metre of floor area, the overall reduction in level was 3 dB(A). Barrier performance was, as expected, limited by ceiling reflections but it could be usefully enhanced by adding suspended absorbers. A combination of multiple absorbent barriers and suspended absorbers produced an overall reduction in level of 8 dB(A).

The factory model was not only useful for testing various noise control measures but proved a useful research tool for studying theoretical models. Murray Hodgson developed this application by constructing a multi-purpose model at 1:50 scale with which he was able to make comparisons with theoretical models for predicting reverberation time and sound propagation.

26.4.3 Underground railway stations

A commercial application of the modelling techniques developed for factories was adopted by the firm Arup in the 1990s for an investigation into possible acoustic treatments in London Underground stations with the overall aim of reducing train noise and increasing the intelligibility of the public address system. Detailed investigations at full scale were not possible because of access restrictions. Moreover, calculation methods in such spaces were not well developed and not proven. Acoustic scale modelling therefore was the best option to evaluate the effect of various treatments.

A scale factor of 1:16 was selected enabling model measurements up to the 4 kHz octave full-scale which was considered necessary for investigating speech intelligibility. Models were constructed of existing 'deep tube' stations which have a circular cross-section and these were calibrated successfully against full-scale measurements. Accurately-scaled sound absorptive materials were introduced into the models in a variety of locations and forms, and key acoustic-response characteristics measured, in particular the variation of sound propagation with the distance from a point source. These indicated that, of the chosen test arrangements, the most effective location for absorptive material was a zone immediately below the tunnel crown. The optimal shape of this absorption was in the form of an elliptical tube suspended centrally and running along the crown.

With the model absorption in place, model loudspeakers of known directional characteristics were installed at the equivalent of 4 m centres on the platform wall. Measurements of Speech Transmission Index at various platform locations gave values between 0.6 and 0.8 (i.e. significantly above the target of generally accepted 0.5). The model elliptical tube was constructed at full-scale and installed in St John's Wood underground station in London for full-scale trials. Although the trial was successful, the device was not immediately adopted by London Underground but shelved for possible future use.

Scale modelling of underground stations was extended to geometries which were rectangular in section and eventually a design guide was drawn up stipulating the location and amount of acoustic treatment to install in stations to provide sufficient speech intelligibility from the public address and voice alarm systems.

A particularly interesting example of scale modelling of an underground railway station was the new station at Canary Wharf in London designed by architects Foster and Partners. The main challenge of this project was that the architects envisaged a cathedral-like space for the concourse where the surface finishes were predominantly of concrete. With a large volume and hard surface finishes, the likelihood of achieving a Speech Transmission Index of 0.5 was very low, although this is a statuary requirement by the Fire Brigade.

A model of the station was built at 1:50 scale and model-scale loudspeakers for public address were specially developed to have the same directional characteristics as a full-scale column loudspeaker (Figure 26.24). Clusters of these model loudspeakers were deployed throughout the concourse in a typical layout which permitted measurements of Speech Transmission Index at various locations throughout the concourse. Sound absorptive treatment was applied on

Figure 26.24 1:50 scale model of Canary Wharf station (Image: Over Arup & Partners)

various surfaces in the model to determine the amount required and the optimum locations for it to achieve the required Sound Transmission Index. The most suitable (and available) places proved to be the walls at high level and the ceiling. Applying sound absorptive treatment to the ceiling proved to be problematic for the architects as it detracted from the pure lines of the exposed concrete ribs. However, a solution was reached by developing an array of stainless-steel tubes backed by acoustic absorbent which fitted neatly between the ribs. The Canary Wharf exercise showed how scale modelling can help to determine a specification to meet a key acoustic requirement as well as satisfactorily resolving an architectural design issue.

26.4.4 Corporate and Government Buildings

An early application of 1:50 modelling for a corporate building was in the design of the Lloyd's of London Building by Richard Rogers Architects in the mid-1980s. The design comprised extensive open-plan offices and a very tall atrium at the top of which the Lutine bell was hung – this bell is rung when a ship is lost at sea. The aim of the modelling was to ensure that the acoustics of the open-plan offices would provide a good balance between privacy and good communication with nearby colleagues, and also that the sound in the atrium was well-controlled. Recommendations from the model tests passed into the design of the full-scale building which turned out to be acoustically successful.

A particularly-interesting modelling exercise was in the design of the New Parliamentary Building (Portcullis House) in Westminster, London by Michael Hopkins Architects in around 2000 (Figures 26.25 and 26.26). The building design contained a large concourse with a roof made entirely of glass comprising a number of large, concave surfaces. 1:50 scale modelling of the space revealed

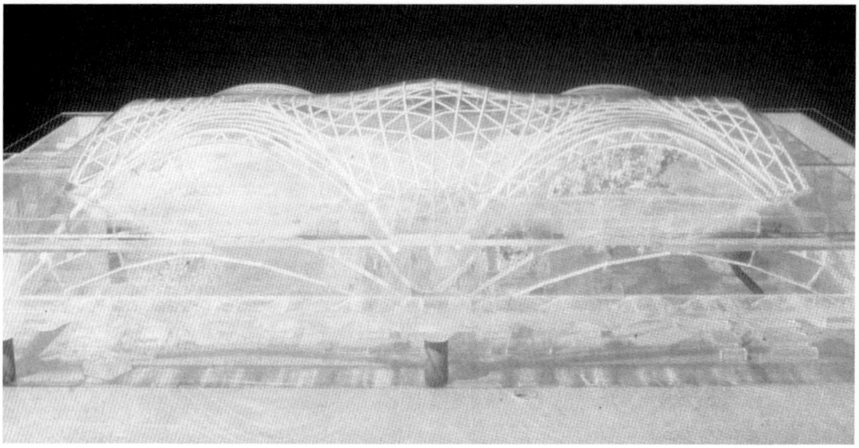

Figure 26.25 1:50 scale model of the concourse at the New Parliamentary Building – Portcullis House (exterior view) (Image: Over Arup & Partners)

Figure 26.26 1:50 scale model of the interior of the concourse at Portcullis House with model trees which served to moderate the internal acoustics. (Image: Over Arup & Partners)

strong focusing of sound and a long reverberation time. The architects were reluctant to apply any acoustically absorbent material to the underside of the glass roof but confirmed that the space would be fitted out with large trees. Further model tests with scale-model trees indicated that the trees had enough acoustical effect, particularly in terms of sound diffusion, to control the overall sound in the space. The trees have now been in place for nearly two decades and are successfully controlling the acoustics.

26.5 Physical modelling of acoustics – the first hundred years

The idea of investigating the acoustic properties of a space using physical models is now over 100 years old and many ingenious techniques were developed with the available technology of the day. As electronics, acoustic transducers and recording mechanisms developed, acoustic scale modelling took advantage of them and provided increasingly-accurate objective measurements and subjective recordings. The result of the development work has led to the extensive use of scale models in designing new auditoria and correcting the deficiencies of existing ones.

As computer modelling of the acoustics of spaces became more accurate, there has been a shift in the use of acoustic scale models towards the use of acoustic computer models. However, as acoustic computer models still have shortcomings in the way they deal with the diffraction of sound (at least in a reasonable time scale), acoustic scale models have remained an important resource in that they can model the interaction of sound waves with surfaces and obstacles perfectly. It is for this reason that acoustic scale models are still in use in the twenty-first century.

References

1 Sabine, Wallace C. (1922) *Collected papers on acoustics.* Cambridge. Mass.: Harvard University Press.

2 Foley Arthur L. & Souder, Wilmer H. (1912) A New Method of Photographing Sound Waves. *Physical Review* 35, no. 5: 373-386, Plates I-V.

3 von Fischer, Sabine (2017) A visual imprint of moving air: methods, models, and media in architectural sound photography, ca. 1930. *Journal of the Society of Architectural Historians* 76 (3) Sept.: 326-348. 329-330.

4 Sabine, Wallace C. (1913) Theatre Acoustics. *American Architect* 104: 257-279. Reprinted in Sabine, W.C. (1922) *Collected papers on acoustics.* Cambridge. Mass.: Harvard University Press.

5 Davis, A.H. & Kaye, G.W.C. (1927) *The acoustics of buildings.* London: G. Bell & Sons.

6 Davis, A.H. (1934) *Modern acoustics.* London: G. Bell & Sons.

7 von Fischer, Sabine (2013) *Hellhörige Häuser: Akustik als Funktion der Architektur, 1920-1970.* PhD Thesis. ETH Zurich. Permanent Link: https://doi.org/10.3929/ethz-a-010150274 (17/10/2019)

8 von Fischer, Sabine (2019) *Das akustische Argument: Wissenschaft und Hörerfahrung in der Architektur des 20. Jahrhunderts.* Zurich: GTA Verlag.

9 Osswald, F.M. (1936) Raumakustik in geometrischer Betrachtung. *Zeitschrift für technische Physik* 17 (12): 562-563.

10 Osswald, F.M. (1930) Akustischer Konzert- und Vortragssaal mit veränderlichem Volumen. *Schweizerische Bauzeitung* 96 (18): 224-225.

11 Addis, Bill (2009) A Brief History of Design Methods for Building Acoustics. *Proceedings of the Third International Congress on Construction History, Cottbus, Germany, 2009.* 1-10.

12 Satow, Takeo (1929) Acoustics of auditorium ascertained by optical treatment in models. *Proceedings of the World Engineering Congress. Tokyo.* Paper 118.

13 Davis, A.H. (1934) *Modern acoustics.* London: G. Bell & Sons. 282.

14 Vermeulen, R. & de Boer, J. (1936) Optical model experiments for studying the acoustics of theatres. *Philips Technical Review* 1 (2): 46-52.

15 Spandöck, F. (1934) Akustische Modellversuche. *Annalen der Physik* 20: 345-360.

16 Jordan, V.L. (1941) *Elektroakustiske undersøgelser af materialer og modeller.* (In Danish). Doctoral Thesis, Copenhagen.

17 Brebeck, D., Bucklein, R., Krauth, E. & Spandöck, F. (1967) Akustisch ähnliche Modelle als Hilfsmittel für die Raumakustik. *Acustica* 18 (4): 213-226.

18 Harwood, H.D. & Burd, A.N. (1970) *Acoustic scaling: General outline.* Report 1970-13. BBC. https://www.bbc.co.uk/rd/publications/rdreport_1970_13 (17/10/2019)

19 Harwood, H.D., Burd, A.N., Spring, N.F., Randall, K.E. & Smith M.K.E. (1972) *Acoustic scaling: An evaluation of the proving experiment.* Report 1972-03. BBC. https://www.bbc.co.uk/rd/publications/rdreport_1972_03 (17/10/2019)

20 Harwood, H.D., Lansdowne, K.F.L. & Randall, K.E. (1975) *Acoustic scaling: The design of a large music studio for Manchester: final report.* Report 1975-35. BBC. https://www.bbc.co.uk/rd/publications/rdreport_1975_35 (17/10/2019)

21 Jordan, V.L. (1973) Acoustical Design Considerations of the Sydney Opera House. *Journal and Proceedings of the Royal Society of New South Wales* 106 (1&2): 33-53. https://royalsoc.org.au/images/pdf/journal/106_12_jordan .pdf (25/6/2020)

22 Orlowski, R.J. (1981) An eighth-scale model speech source for subjective assessments in acoustic models. *Journal of Sound and Vibration* 77 (4, August): 551-559.

23 Barron, M. (1997) Acoustic scale model testing over 21 years. *Acoustics Bulletin* 22 (3): 5-12.

24 Barron, M. (2010) *Auditorium Acoustics and Architectural Design.* (2nd Ed.) London: Spon Press. 180-185.

25 Barron, M. & Chinoy, C.B. (1979) 1:50 scale acoustic models for objective testing of auditoria. *Applied Acoustics* 12 (5, September): 361-375.

26 Barron, M. (2010) *Auditorium Acoustics and Architectural Design.* (2nd Ed.) London: Spon Press.

27 Hodgson, M.R. & Orlowski, R.J. (1987) Acoustic scale modelling of factories. Part I: Principles, instrumentation and techniques. *Journal of Sound and Vibration* 113 (1, February): 29-46.

28 Hodgson, M.R. & Orlowski, R.J. (1987) Acoustic scale modelling of factories. Part II: 1:50 scale model investigations of factory sound fields. *Journal of Sound and Vibration* 113 (2, March): 257-271.

29 Orlowski, R.J. & Hodgson, M.R. (1988) Acoustic scale modelling of factories. Part III: The potential of scale models for predicting the effect of noise control measures. *Journal of Sound and Vibration* 121 (3, March): 525-545.

27

Geotechnical centrifuge models – a history of their role in pre-construction design

William H. Craig

27.1 Introduction

While physical model testing as an adjunct to engineering design may not have been as common in ground engineering as in other disciplines, it is none the less an ongoing activity which has gained in prominence in the last fifty years with the development of the techniques of modelling aboard large centrifuges; and this despite the advances in numerical modelling which have generally had the opposite effect on physical modelling in many other engineering disciplines. This is not the place to give full details of the techniques used, or even the considerations of similarity which lead to and follow the choice of the centrifuge technique. A good general introduction can be found in the various chapters of Taylor [1]. In the specific context of 'design', in 1985 the Author edited a volume entitled 'Application of Centrifuge Modelling to Geotechnical Design' which arose from a symposium in Manchester the previous year under the auspices of the International Society for Soil Mechanics and Foundation Engineering (ISSMFE) and its Technical Committee on Centrifuge Testing [2]. This TC was set up in 1982 with initial responsibility placed in the hands of the British Geotechnical Society, under the chairmanship of Professor Andrew Schofield at Cambridge University. The title for the symposium was carefully considered. The centrifuge technique was seen as having a role in the design of specific real structures at the pre-construction stage as well as in the development of broader generic guidance. This present chapter concentrates on the former approach and is largely based on experience at the University of Manchester where Professor Peter Rowe built a very large centrifuge on this basic premise in 1970-71 (see Section 27.3.2). Other UK university groups have since used machines of smaller capacity over the decades – pre-eminently at Cambridge, but the philosophy of their use has been more focused to retrospective analysis of failures and the development of design in a generic sense with a longer-term impact and not usually related to site-specific applications.

This chapter will concentrate on personal experience but also covers work on prototypes in a number of countries. After initial experiments in the USA and USSR in the 1930s and in the post WWII era, centrifuge modelling is now carried

Physical Models: Their historical and current use in civil and building engineering design, First Edition. Edited by Bill Addis.

out extensively in many nations and became increasingly developed following publications from Japan, the UK (Cambridge) and the USSR at the ISSMFE International Conference in 1969 in Japan. There is now a huge international literature with contributions from many more countries.

This approach to modelling was first mooted in the nineteenth century at a time when structural modelling was becoming a significant tool for the engineer [3, 4]. Most textbooks on the theory of modelling and on dimensional analysis are rooted in the field of fluid mechanics and, while they may pay brief attention to the geotechnical world, it is peripheral to the main thrust. The lack of apparent interest relates to the inherent difficulties of satisfactorily representing the ground and its behaviour within a model. Geotechnical materials are complex – often three phase, particulate (except in the field of solid geology), with behaviour which is dependent on effective stress levels and/or temperature and gravity-generated loads which play a significant role in behaviour in almost every instance. This may be in terms of driving/active forces for flows and possible instabilities, but also as a reactive agent in the generation of mobilised soil strength. The underlying principle of centrifuge modelling is that if a small model at 1:N linear scaling is subjected to accelerations aboard a centrifuge generating Ng, then the self-weight induced stresses of the soil and rock and of any other suitably-scaled structural components will be as in the prototype under the action of earth's normal gravitational acceleration 'g'. This is a simplistic introduction and there are many caveats and limitations to modelling with real and replacement materials. Drainage and flows of ground fluids can be modelled and static, cyclic and dynamic events, e.g. blast, impact, seismic, can be simulated to such an acceptable degree of similitude so as to be useful. From models, projections can be made for both generic and site-specific design studies, within limitations that need to be recognised – but this is equally true of extrapolations from all other types of model. Because of the complexity of the behaviour of ground materials their incorporation in real physical models represents a benefit that may be missing in generalised conceptual and computational models, hence the extent of the uptake of the technology in recent decades. Without the simulation of self-weight effects, geotechnical models with linear scaling on the laboratory floor or bench top are rarely able to generate realistic results.

27.2 Historical review

27.2.1 The beginnings of centrifuge testing of physical models

Curiously, the origins of centrifuge testing can be traced back indirectly to the Britannia Bridge across the Menai Straight between Anglesey and the Welsh mainland. Working with William Fairbairn and Eaton Hodgkinson, Robert Stephenson arranged for a substantial programme of experiments which demonstrated buckling failures in simple section models – a form of failure not previously known in cast iron bridges. The models evolved to include more complex sections with improvements and additions, together with a decision to have the final tubes continuous over four spans. In the absence of calculations to give guidance on crushing and buckling the model tests were crucial in determining structural form.

Following a fire on the bridge in 1970 it was the subject of a major reconstruction and redesign, in the course of which new model tests were carried out in relation to wind loading on the superstructure (see Husband & Husband [5], Wardle & Lucas [6]). In discussion of these papers on the bridge repair, the Author drew attention [7] to the published thoughts of Édouard Phillips (1821-1889) [3]. Phillips first proposed the use of a centrifuge in 1869 for structural model testing, with specific reference to the Britannia Bridge which had been designed twenty years earlier and completed in 1850 [4] (see Section 6.7). Phillips advocated the use of a centrifuge with a model at 1:50 scale, rotating about a vertical axis at a radius of 2 m and a speed of 2½ rev/s, generating accelerations of 50g. The proposed model of 8.6 m overall length would have simulated the entire bridge structure, and if subjected to speeds of 3½ or 4⅓ revs/s could have been overstressed by a factor of 2 or 3. Phillips also deduced that, for dynamic loads, inertial time-scaling in the centrifuge is the same as linear scaling, a fact familiar to geotechnical modellers today.

It seems that none of these ideas came to fruition and there is no record of anyone in the nineteenth century using a centrifuge for modelling elastic, or any other, structures. Phillips' ideas were many years ahead of their time and it appears that no centrifuge model tests on any sort of engineering structure were carried out until the independent development of such techniques in the USA and USSR in the 1930s.

The idea of using centrifuges for modelling various processes was developed by Philip Bucky (1899-1952) in 1931 at the University of Columbia [8] and independently by Nikolai Davidenkov (1879-1957) [9] and Georgi Pokrovsky (1901-1979) [10] in 1933 who, with his colleague Il'ya Fedorov (1903-1986) initiated a stream of modelling work in the USSR which continues today [11].

Philip Bucky's interest lay in deep mining and strata mechanics and he first mentions centrifuge modelling in a paper in 1931 [8]. He studied models of salt domes, which may experience prototype movement rates of the order of 1 cm/year, using centrifuges running at 1000g+ initially using a machine with a working radius of around 220 mm and later with one of 0.6 m radius. Bucky's work in mining continued for several years and was applied specifically to design [12].

In 1929, Nikolai Davidenkov initially generated increased body forces in models by utilising the sudden deceleration of bodies from free fall. Model structures involving soils were placed in a box and dropped onto springs, which effected controlled decelerations up to 100g over small distances. He argued that when models survived this loading the equivalent prototype construction would not fail. The perceived drawback was that such model loadings were of short duration, effectively instantaneous, while gravity loads on prototypes endure for indeterminate time periods which were not modelled. This technique is described in the later monograph of Pokrovsky and Fedorov [13] and was demonstrated in Manchester, by Professor Schofield in the 1970s, during a short course given to promote centrifuge technology and the new centrifuge at the University of Manchester Institute of Science and Technology (UMIST). Schofield dropped model containers, designed for use in the centrifuge, from a height of some 4–5 m onto simply supported steel bars, which yielded plastically giving computable decelerations of high magnitude.

Davidenkov recognised the possibilities of rotating a model about a fixed axis and utilising centrifugal forces, although he himself did not continue along this route. It seems that his 1933 paper relating to the use of models in a centrifuge was something of a diversion from his mainstream activity and, while he proposed the use of a centrifuge, he gave no details of any machinery and reported no experimental results [9]. Pokrovsky and Fedorov [13] suggest that his aspirations were for a centrifuge some 10 m in diameter, which was too large to gain the support of others at that time.

Georgi Pokrovsky recognised the possibility of centrifuge use in July 1932, formally publishing his ideas in the following year [14]. In December of 1932 the first centrifuge model tests, looking at earth embankment slope stability using sandy soils were carried out by I.I. Kyrenshikov at the soil mechanics laboratory of the Moscow Institute of Hydraulic Engineering and Hydrology. The centrifuge there was subsequently used for a number of studies – notably the stability of slopes on the channel between the Ysa and Volga rivers. Subsequently, the first centrifuge in the USSR specifically for geotechnical work was built in Moscow at the soil mechanics laboratory of the Institute of Hydro-techniques and Hydrogeology. This allowed the modelling of various processes in soils and rocks and a dedicated facility was created for the use of the method of centrifugal modelling that allowed the examination in small models of the stress–strain state of soil constructions in an artificially created gravitational field.

Il'ya Fedorov first designed a mini-centrifuge for laboratory studies, and a much larger machine to his design was installed at Fundamentstroi in 1936 – at that time the most powerful machine in the USSR [15]. He cooperated effectively with Pokrovsky for many years and, over the years of their collaboration, they published seven books in Russian on centrifuge modelling and blasting in mining construction and hydraulic engineering.

Bucky and others continued to publish outputs in English from experiments in North America looking at a number of geology/rock mechanics problems using a combination of centrifuge and photoelastic modelling [16, 17, 18, 19] – see Clark [20] for a wider discussion. Hoek reported some photoelastic work on a centrifuge he built in South Africa in the 1960s [21]. He sought to study the three-dimensional stress distribution around large mining excavations using the technique of stress freezing (see Section 12.3.1). The model was subjected to a gravity field of 100 g at a temperature of 160°C. While still under load the model was cooled to lock in stresses and strains and was later cut into 3-mm slices and the stress distributions analysed using a polariscope. However this work was not pursued after Hoek moved to the United Kingdom later in the decade. More recently, Allersma [22] has used photoelastic particles in soils-related experiments in a centrifuge. However, all these studies, while useful and enlightening, are generic rather than site-specific.

The theory of modelling and the principles of similitude had been thoroughly studied since the 1930s for many applications outside geoengineering mainly in other engineering disciplines. In the USA, Henry Langhaar (1909-1992), who had a background as a seismographer and later as a structural research engineer and in hydrodynamic research at the Waterways Experiment Station in Vicksburg, Mississippi, deals at length with 'Dimensional Analysis and Theory of Models' without any reference to soil models, although he does consider similarity and model testing specifically in relation to design [23]. He gives examples of various fields of activity in which models have proved invaluable: hydraulic structures, rivers and harbours, hydraulic machines, airplanes, structures and ships – but nothing geotechnical. He writes that 'a model study furnishes useful qualitative indications of the characteristics of the prototype. Usually quantitative information is also sought'. In relation to geological phenomena he notes that 'exact laws of similarity cannot be deduced' and that only 'empirical and semi-empirical interpretations of model data are employed'. In a chapter looking at particular examples applied to problems of stress and strain, he considers, in a subsection on small deflections in elastic structures, (i) the stresses in an arched dam, loaded in the model by a heavy fluid (mercury) to replace water in order to increase gravity-induced stresses, and (ii) the stresses in an airport pavement, considering Westergaard's elastic theory. No model tests are reported or referenced and he makes no reference to centrifuges. In the USSR, Sedov covers similar ground to Langhaar, also without addressing geotechnics, although he does consider specifically situations where the weight of a structure is an essential characteristic parameter [24].

In a general text on 'Experimental Modelling in Engineering' in 1982, David and Nolle consider the role, in many situations, of the gravitational field, which they note should be considered as a variable and not a universal constant since it may apply to any system regardless of its field strength [25]. They address prototype design problems, particularly where mathematical theory is too complex or even non-existent and refer to the possibility 'to change the strength of the gravitational force field, as would be the case of a satellite in orbit'. They do not appear to be familiar with the centrifuge model concept but it is of interest here to note work reported elsewhere in reduced acceleration conditions such as that of Cherkasov et al. [26]. Geotechnical problems that David and Nolle address are those of the transfer of granular materials and the traction dynamics of all-terrain vehicles, but with nothing site-specific.

Once again concentrating on fluid mechanics Ivicsics [27], writing in Hungary, also picked up the possibilities of the centrifuge, citing three papers by Pokrovsky. In the UK, Schofield [28] picked up references to Pokrovsky's work when translating a text by Sokolovski [29]. Later twentieth-century work in several countries (from the 1960s in Japan, UK, USSR, following smaller initiatives in South Africa [30] and the USA) is described in the 1988 volume 'Centrifuges in Soil Mechanics' [31] and by the Author in 2001 [32]. It is of interest to note that, almost fifty years after Hoek built his machine, a new centrifuge was commissioned in 2012 at the University of Pretoria in South Africa. A description of this new development commented that access to geotechnical centrifuges on the African continent had previously been limited to one in Egypt, since 2002; no mention was made to Hoek's machine built in the 1960s [33].

Professor Andrew Schofield is the doyen of centrifuge modellers, in the English-speaking world at least, and his post-retirement book 'Disturbed Soil Properties and Geotechnical Design' covers much ground and, in the present context, deals quite succinctly with his discovery of the centrifuge technique and his approach to the use of models in the design context [28]. Much of this lies in the retrospective use of models to help understand events that have occurred in the field in order to develop better, generally plastic, design approaches. Although he first became involved in centrifuge modelling in Cambridge after translating Sokolovski's book [29] he spent some years at UMIST in Manchester where he built a large machine commissioned before that of Rowe, which was in a laboratory a kilometre or so away, in the Department of Engineering in the Faculty of Science of the Victoria University of Manchester. The two institutions have since merged into the present University of Manchester. While at UMIST, Schofield experimented with the drop-box technique mentioned above, though this was never more than a demonstration. The first large centrifuge in Cambridge was built just before his return there. The growth and dominance of the University of Cambridge modelling group built up throughout his career and has continued beyond his retirement, guided subsequently by Professors Bolton, Madabhushi, Mair and Soga. Centrifuge modelling at Cambridge has focused on the development of engineering science rather than its use on live construction projects.

In Japan Professor Mikasa of Osaka City University developed theories of consolidation for soft clay and utilized the centrifuge in these studies. His work continued for many years and, like Rowe in Manchester, Mikasa was involved in site-specific design of large dams [34]. The widespread use of the centrifuge today in the Japanese construction industry has its roots in the early work of Mikasa and his successors, notably Professor Kimura. The different national culture of Japan may be partly responsible for the substantially different centrifuge infrastructure that has developed there. Elsewhere in the world, centrifuges remain predominantly in academic and government agency hands, while in Japan, a large part of the centrifuge activity lies much closer to the principal designers and constructors of new works.

27.2.2 Early use of physical modelling in geology

This and the following section review some general modelling similarity/dimensional analysis texts before looking at the perceived role of various model types specifically in geotechnical engineering and geology at various scales. I would prefer the term geo-engineering, but that has recently become associated with engineering on a planetary scale that is beyond our present scope. Although geotechnical/geological situations are not well covered in the general modelling literature, it will be instructive to review the perceived role of geotechnical physical models expressed in a number of more recent publications focused in this area.

The classic work on models used in the study of geological structures was written in 1937 by Marion Hubbert (1903-1989) [35]. Following on from this, the interesting work aboard a centrifuge relating to models in the field of geology at a regional scale stems largely from Hans Ramberg (1917-1998) working in Sweden in the 1950s and 1960s [36]. He used replacement/analogue viscous materials within quite small models to replicate endogenic processes, treating rocks as continua having macroscopic material characteristics such as a yield point, ultimate strength, effective viscosity, rigidity etc. He recognised that anisotropy was the rule, rather than the exception within crystalline rocks and was unable to account for the consequences. Yoxall [37] also uses the term 'analogue' when substitute substances are incorporated in models but refers to models as 'iconic' when prototype substances are used. Zelikson [38, 39] provides more recent examples of this sort of work but, as with Ramberg, it is not associated with any particular site or with design in any sense. He also considers a wide range of problems that can/might be simulated including geophysical and man-made high-energy events. Future possibilities will arise to make use of the largest centrifuges now available to investigate design aspects in problems relating to deep underground caverns and to deep repositories for radioactive waste, though these are likely to continue to use analogue, rather than site-specific, rock materials.

27.2.3 The growth of centrifuge modelling in geotechnical engineering

Writing a major text on geotechnical modelling early in the present century, David Muir Wood covers the whole gamut of geotechnical models and modelling (empirical, theoretical, numerical, constitutive, physical etc.) and devotes a whole chapter to centrifuge modelling covering both hardware and modelling techniques but giving little detail of site-specific modelling for design purposes [40]. While acknowledging the present Author's expressed view that 'If centrifuge work is to continue, it should have a positive role beyond phenomenological studies and the development of design rules by parametric variation in idealised, non-specific models' [41], Muir Wood notes that, in practice, the technique has been used more for generic problems, indicating some of the limitations of these models. In a closing section of the monograph he writes that 'modelling to support design may be performed to provide reassurance. Modelling to further geotechnical understanding should deliberately set out to surprise'. This seems a limited view. Models can not only support design but can lead it, precisely because they allow comparison between options at an early stage of a project as well as an indication of quantitative behaviour of a subsequent prototype, within limitations which must always be recognised explicitly.

It is also useful to look at the contribution that reports of physical modelling have made to the premier geotechnical journal *Géotechnique* in the sixty years from its inception to 2008. In a historical review of geotechnical papers in all publications in the years 1836-1948 [42], there is no reference to models in geotechnical engineering: there was simply nothing of significance in the discipline at a time when modelling in hydraulic and structural engineering was increasingly prominent. During the years 1948-2008 *Géotechnique* published around 200 papers primarily related to physical modelling but only around ten of these were before 1968. Over the next forty years some ninety papers were devoted specifically to centrifuge modelling and there have been many more since, as summarised by White [43]. In looking forward to the next sixty years from 2008 Simpson and Tatsuoka include a short paragraph on physical model testing:

> ‘Centrifuge testing is now well established as a means to simulate a given prototype stress state, and its use will continue as a means of studying mechanisms of behaviour. The testing method has developed with time in terms of data accuracy, model scale and applicable loading history (including stage construction, monotonic and cyclic loading). Further improvements can be expected, such as better modelling of construction sequences, geotechnical processes, and inclusions such as geosynthetics and piles. However, it is not certain whether centrifuge testing will ever be able to tackle issues such as viscous behaviour and ageing, and issues of particle size remain problematic’ [44].

It is an indication of the apparently counter-intuitive rise of the physical model in this discipline that a specialist journal devoted to the topic has been published since 2001. The *International Journal of Physical Modelling in Geotechnics* is published by the Institution of Civil Engineers in London and covers conventional as well as high-g modelling. In the first issue, Randolph and House assess the ‘Complementary Roles of Physical and Computational Modelling’, looking explicitly at reasons why physical modelling might be undertaken either for research or ‘in a design situation’ [45]. In some respects they cover the same ground as Mitchell, writing a decade earlier in Canada, who considers the role of centrifuge modelling as a consulting tool [46]. Also in the IJPMG, in 2015, Gaudin gives a brief overview in an editorial article of the state-of-modelling within the discipline, ending with his belief in ‘the recognition of physical modelling as a fundamental contributor to geotechnical science’ [47]. The continuing success of this publication is testimony to the health of physical modelling in the geotechnical disciplines.

Another excellent review source is the ICE Manual of Geotechnical Engineering (2012). In this multi-author work, Burland notes that structural and geotechnical engineers have to interact and that there can be differences in approach and difficulties in communication [48]. With regard to modelling, he defines the so-called geotechnical triangle, of which one corner is the ‘appropriate model’, which may be conceptual, physical or analytical. Whatever the approach, he notes that it is frequently the case that what is being sought is an understanding of the basic mechanisms of behaviour rather than quantitative detail. This assumes that the geometry of the structural form is, to an extent, well defined and can be idealised. This is not always the case in design where novelty in structural form may need to be assessed and comparisons made between options, as with Fairbairn and Stephenson’s models for the Britannia Bridge in the 1840s. As will be demonstrated below, with examples principally from offshore and coastal geotechnics, there is a particular role for the physical model here.

27.3 Physical model testing for site-specific prototypes

27.3.1 Excavations

In 1973, the Author introduced a discussion at the 8th International Conference on Soil Mechanics and Foundation Engineering, giving three examples where centrifuge models had been used in relation to prototype structures [49]. One, an excavation collapse in stiff fissured clay, was modelled post-failure using a block of soil retrieved from the site (670 x 545 x 368 mm). The second was a simulation of a circular multi-panel diaphragm wall, modelled in micro-concrete, which was being designed for a soft clay site in the Netherlands, using a site sample 1 m square [50]. As a contribution to a design study the test demonstrated that the diaphragm wall, with potential weakness at the panel joints, could perform in ring-compression without internal propping. Even when bending cracks occurred around the diaphragm under axisymmetric loading, the integrity of the wall could be maintained. Eventual collapse of the model was brought about by significant asymmetric loading. Although the structure under consideration at the design stage was not built, this was as a result of wider factors and the design-stage modelling was not invalidated. The third example related to design-stage studies of an embankment dam [51, 52].

27.3.2 Dams and embankments

As had been the case with several new centrifuges, the first uses of the machine that Rowe built in Manchester in 1970-71 were related to slope stability assessments and, indeed, these had dictated the design of this particular centrifuge with a long model-axis parallel to the rotational axis, avoiding exaggerated curvature effects in 2-D, and plane-strain models placed in the plane of machine rotation. The machine had a symmetrical, fixed-arm rotor with a radius to the face plates of 3.2 m. and a rotor-arm height of 2.6 m (Figure 27.1). Designed to run at a maximum acceleration of 200g it attained accelerations in excess of 130g with the 200 HP motor installed in 1971. The maximum model size was 2.0 x 1.0 m in plan, 0.6 m deep and the maximum payload mass of model plus its structural container and all loading/monitoring equipment was 3.5 tonnes.

Figure 27.1 The University of Manchester centrifuge and the Author, with a model in place on the rotor arm. (Image: University of Manchester)

Figure 27.2 Post-test, side view of a 1:100 scale model of Grimwith reservoir embankment, with a complex failure mode. (Image: University of Manchester)

In order to increase the size of a Victorian reservoir at Grimwith in the English Pennine hills in the 1970s, it was necessary to raise the old embankment, built in the 1860s, by some 20 metres. The vertical clay-core dam was to be raised by placing fill on the upstream face with a new inclined core supported by a stiffer clay-fill shoulder with sand drainage blankets to help dissipate construction pore pressures. Using a simple 'gravity turn-on' approach various models were brought to failure only by overloading and by incorporating clay at strengths well below the specified prototype range (Figure 27.2). An impressive full-size photographic print of the model, 2 m long and 600 mm high, has been prominent on the laboratory wall for over forty years, clearly showing a complex failure mode, even in 2-D – a tension crack behind the slope crest, plastic failure in the soft-clay core, and brittle rupture in the centre of one of the stiffer clay layers within the supporting embankment moving into the sand drainage blanket only at a point when reduced overburden stress rendered this material weaker than the clay on either side. Despite the short-comings of the model stress-path this was a powerful demonstration of the capability of the models and contributed to the acceptance of this particular design.

In a parallel study at the same time for the proposed Bewl Bridge reservoir being built in Kent, in South-East England, large blocks of heavily-fissured, weathered and un-weathered, Wadhurst Clay recovered, undisturbed, from the site, were cut and built into the foundations of models with a vertical central clay core linked to an unusual horizontal cut-off layer and sealing trench near the slope toe on the upstream side. The models, brought to failure in the same manner as for Grimwith, showed non-circular slip mechanisms which were associated with slumping of the soft central core and transfer of vertical load, via the stiffer compacted siltstone supporting shoulders to induce rapid pre-collapse increases in pore pressures in the soft-clay cut-off with drops in the pore pressure ratio under the centre line (Figure 27.3).

Both these earth dams were duly completed and have stood for several decades – full details of the modelling can be found in PhD theses on work carried out under Rowe's supervision [52, 53]. It should be noted that these two embankment studies each involved a small number of models (single figures) with limited variations in slope geometry and material properties. Malushitsky reports on activity on an altogether bigger scale describing the outcomes from

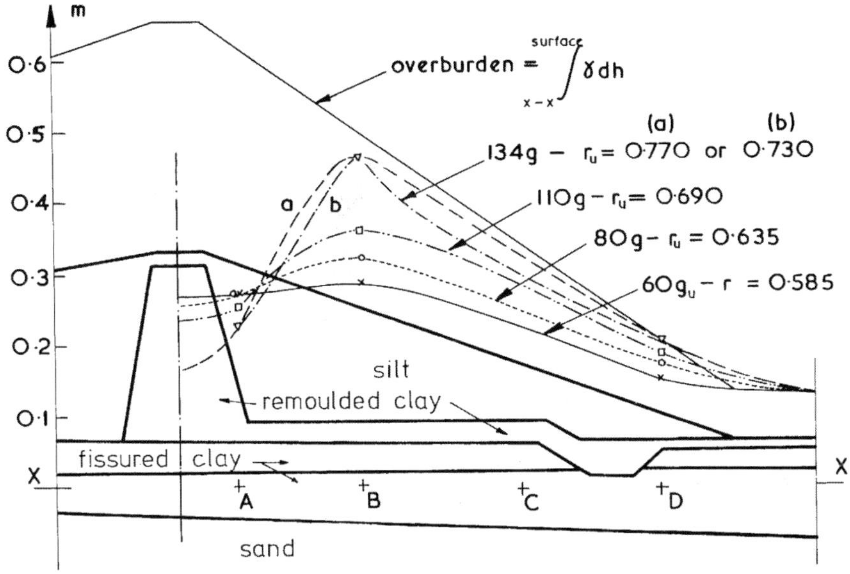

Figure 27.3 Pore-water pressure distribution in the foundation below the cut-off blanket of Bewl Bridge reservoir model. Overburden stresses are shown as an equivalent head of water for purposes of comparison with the pore pressure heads plotted. (Image: University of Manchester)

255 models of waste-heap embankments in the context of the Ukrainian mining industries [54]. Some of this monumental piece of research was site-specific, other parts were more generic. Notably, he estimated the economic benefits of the design changes put into practice as a result of the work.

27.3.3 Offshore and marine structures

Although the centrifuge that Rowe built was partially funded by the Water Resources Board, the momentum in modelling shifted elsewhere in the late-1970s with the lack of major embankment projects in the UK and the growing interest in offshore oil and gas industries in the North Sea and elsewhere. The Manchester group was widely involved. In the late 1970s, offshore piling capability and design methods developed rapidly and pile lengths in use routinely exceeded any that could be modelled realistically in the UK or elsewhere. Although extensive model programmes were carried out, they were seldom related to particular sites at the design stage. In relation to the small number of first-generation gravity platforms, placed in the North Sea in water depths up to 100 m, centrifuge modelling was significant [55]. These platforms with enormous footprints and only shallow sub-surface penetrating skirts, if any, were novel. The height of the notional 'design wave' which induced large cyclic horizontal loads on the structure and its foundation was increasing by the month, and the industry was unsure how shallow foundations in any soils, often layered, would respond, or to what depths failure mechanisms might penetrate. At a time when any insights were invaluable, information gleaned from physical models being tested late into the night in Manchester were communicated to a client overseas the following morning and were common knowledge in London later the same day. In the words of an old adage, 'In the world of the blind, the one-eyed man is king' and any useful input was valuable to a new technology. Model studies, both generic and site-specific, indicated that, under cyclic loading, potential foundation failure modes involved localised weakening of soils close to the base of the structure and quite shallow sliding mechanisms [55, 56].

Some years later, a second generation of gravity platforms evolved and model studies carried out in Manchester between 1983 and 1987 contributed to the design process for one such platform in the Troll Field of the North Sea in around 300 m water depth, where the seabed consists of considerable depths of soft clay. In a collaborative study involving an oil company, designers, construction contractor, independent consultant and the university, a programme of ten major model tests was carried out over a four-year period. It is pertinent to note that, while the initial contract for modelling indicated a detailed programme in advance, the reality was that this was modified in the light of results arising from each previous test and of other inputs from many sources – a very flexible approach is needed in such cases, as innovative design cannot be defined in advance and even apparently straightforward research programmes often wander from the line of initial intention.

In the case of the Troll site, initial plans for a tripod platform, with quite short peripheral and internal ground-penetrating skirts at each foot and additional piled support, moved towards a deeply-penetrating monopod foundation with multiple internal cells nearly 40 m deep below the mudline (Figure 27.4). Even the somewhat reduced footprint of the monopod structure was too large to be modelled with full similarity in the largest centrifuges available. Models were tested at 1:300 linear scale under accelerations of only 100g [57], using applied vertical stresses at the mudline and soil-strength profiles as for the prototype. In this instance collapse failure in terms of rapid settlement or movement off station

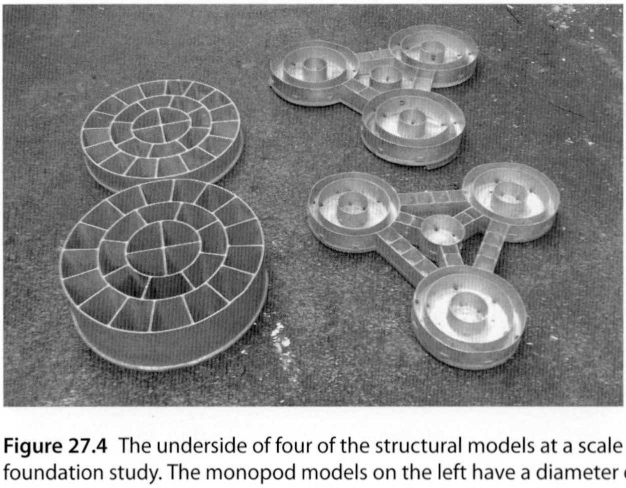

Figure 27.4 The underside of four of the structural models at a scale of 1:300 for the Troll Field foundation study. The monopod models on the left have a diameter of 0.515 m. (Image: University of Manchester)

under loading was not a major issue and models were taken to the point where rotational movements and lateral displacements at the level of the sea surface or the superstructure above were unacceptable in terms of equipment serviceability and human factors. In all cases the models used clay retrieved from the site and reconsolidated to an appropriate effective stress level to reproduce an undrained strength profile at 1:300 scale. Figure 27.5 shows the model set-up in the centrifuge just before the testing container was lowered into the test position on the rigid, fixed-arm rotor of the centrifuge shown in Figure 27.1. Many centrifuges elsewhere utilise model containers which are supported on hinges at the end of otherwise rigid rotating arms – given the dimensions and orientation of the Manchester machine this was not a viable option when the machine was designed with slope studies in mind. As was the case with most of the truly 3-D models in the various projects reported here, little can be seen of the soil and foundation when loading rigs, mounted above the foundation models, obscured the view.

Figure 27.5 Troll Field platform model ready for testing in centrifuge. (Image: University of Manchester)

In the early model tests on offshore structures subjected to simulated wave loads, the loading system employed semi-automated switching of valves to feed compressed air into the chambers of double-acting pneumatic cylinders. These were later replaced by hydraulic actuators supplied with oil at high pressure from an off-machine pump, controlled by servo-valves which, we were advised, had previously only been used under the acceleration levels associated with missile systems (less than 20g). This seemed like a leap in the dark and something of an act of faith at the time, but experience over a human generation of activity indicated not a single servo-system malfunction, even under long-sustained accelerations of 100g. Similar systems are now quite common on larger centrifuges. At that time, the multiple monitored outputs in response to cyclic loading were recorded initially in analogue form and digitised later. Today, all monitoring and possibly all loading control of the model would be digital, though a capacity for manual intervention in real time would be maintained.

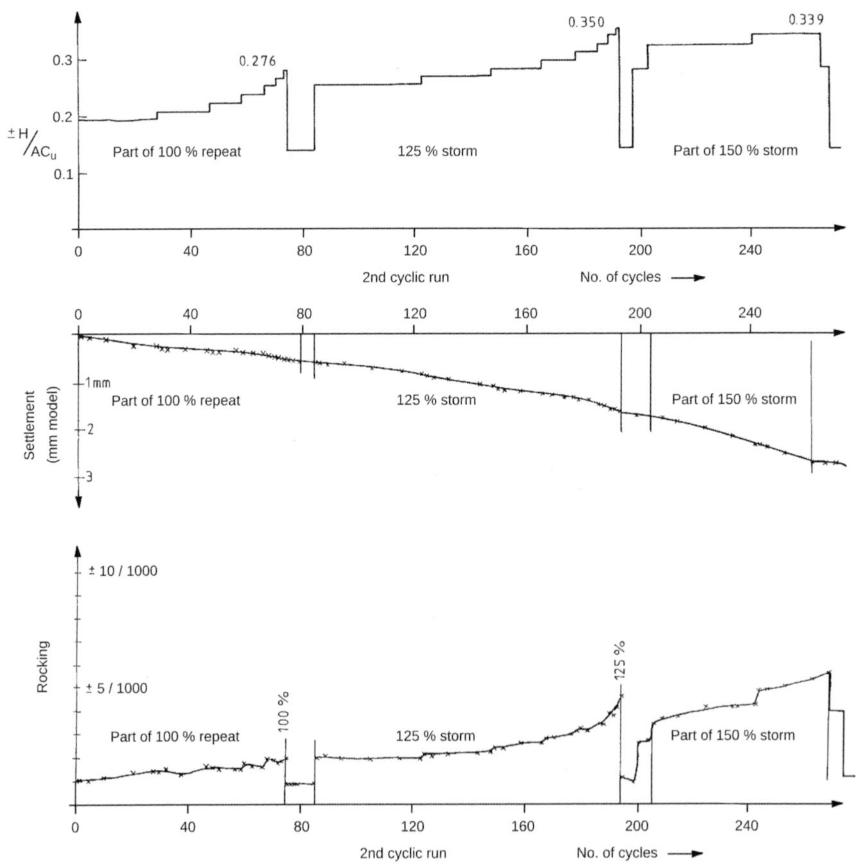

Figure 27.6 Output from one Troll Field model taken beyond design storm loading.
H = Horizontal load applied to model at an appropriate height above mudline.
A = Plan contact area between model structure and seabed.
Cu = Undrained shear strength of seabed soil at lowest point of structural foundation.
(Image: University of Manchester)

The Troll Field model was first subjected to increasing acceleration to the required 100g before cyclic horizontal loading was applied in increasing 'parcels' of simulated waves of increasing height to the 100% design storm pattern. Subsequent further loading sequences up to 125% of the design load envelope, and beyond, were applied (Figure 27.6). At that time (mid-1980s) no attempts were made to simulate the random nature of the wave pattern as this was not the approach adopted elsewhere in other parallel, theoretical design inputs. This could have been relatively-simply programmed and monitored even then, but was not thought to be appropriate.

This four-year programme of physical model testing had a significant effect on the evolution of structural form and prediction of likely prototype performance of an offshore platform of unprecedented size. The platform was placed in its final location more than a decade after modelling was commenced. In 1995, when the structure was completed and floated to its final seabed location, it was the largest structure ever moved across the face of the earth. The initial design was conceived as a conventional, piled tripod structure, governed by settlement considerations. The final design, with deep foundation cells in a monopod configuration, was more-critically dependent on structural dynamics and cyclic displacements. More general descriptions of the project and the structure as built and installed are available – see an article in the British Geotechnical Society journal, *Ground Engineering*, at the time of float out in 1995 [58]. The centrifuge contributed significantly to acceptance of the shift in design.

27.3.4 Transmission line pylon foundation

Another interesting use of a single model within a site-specific design relates to a high-voltage transmission line structure built across the Jamuna River in Bangladesh [59]. As part of the design process a single simplified centrifuge model of one of the foundation caissons was tested. This was not a commissioned work but rather an opportunistic add-on to a broader ongoing programme of modelling the lateral deflections of offshore piles. On this occasion no material from the actual site was used and, as in all other examples cited, this pre-construction model was but one input to the design decision data on a major project.

27.3.5 Netherlands national security – coastal defence

The protection of the Netherlands from a repetition of flooding from the sea became a national priority after the disasters of 1953. Work progressed for decades to improve coastal defences and build barriers across various sea inlets. The closure across the Oosterschelde was the final piece of the project and involved engineers from many countries and inputs from many sources. The barrier, which contains multiple piers and gates which are closed only when flooding threatens, was completed in 1986. Centrifuge modelling, which took place in two stages, made significant contributions to the final design. The initial stage [60] attempted to model field tests with a caisson foundation on natural and densified sand at the site. Simulation of behaviour on the densified sand was generally satisfactory but simulation of uniform foundation sand at 50% relative

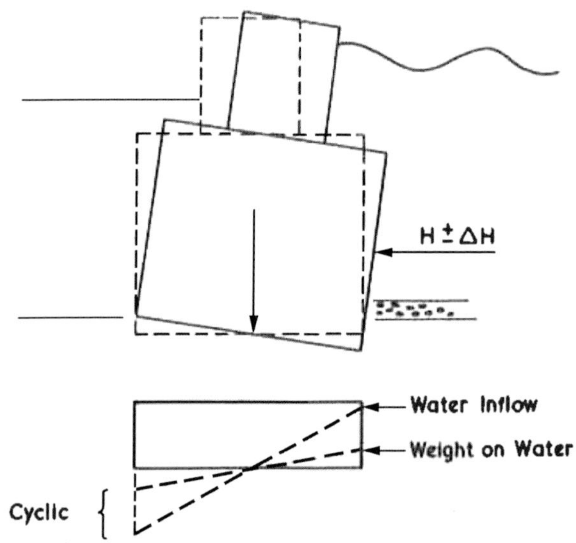

Figure 27.7 Backward tilt of Oosterschelde caisson structures under wave loading. [56]

density failed to capture the observed field performance. Only by including a pattern of very loose zones within model foundation strata were models able to replicate the field in retrospect. Subsequently the scheme design evolved to use more-deeply-embedded foundations and a lengthy modelling sequence was carried out as one element in the optimisation process to find minimum deflections of the piers and gate sills. Many models were tested with varying forms and burial depths for the piers [61] and the entire modelling project of the 1970s is summarised by Rowe [56]. A surface caisson might tilt backwards towards the wave-loaded seaward-side due to softening of the foundation under the heel (Figure 27.7). For a particular pier foundation, there is a critical depth of burial which minimises the sill deflection (Figure 27.8). The models were one component only in a very complex design project which resulted in successful completion of the *Oostercheldekering* in 1986.

27.4 Centrifuge model testing for more-general geotechnical problems

27.4.1 Pipeline interactions

While problems relating to many forms of offshore structure were addressed from the mid-1970s, the use of centrifuge modelling applied to sub-sea pipeline problems arose fairly late and at linear scales typically between 1:10 and 1:40, reflecting the modest diameter of the prototype pipes. Two particular scenarios have attracted attention, both related to buckling under cycles of expansion/contraction associated with intermittent passage of hot fluids, interspersed with periods when the pipe is at ambient sub-sea temperatures.

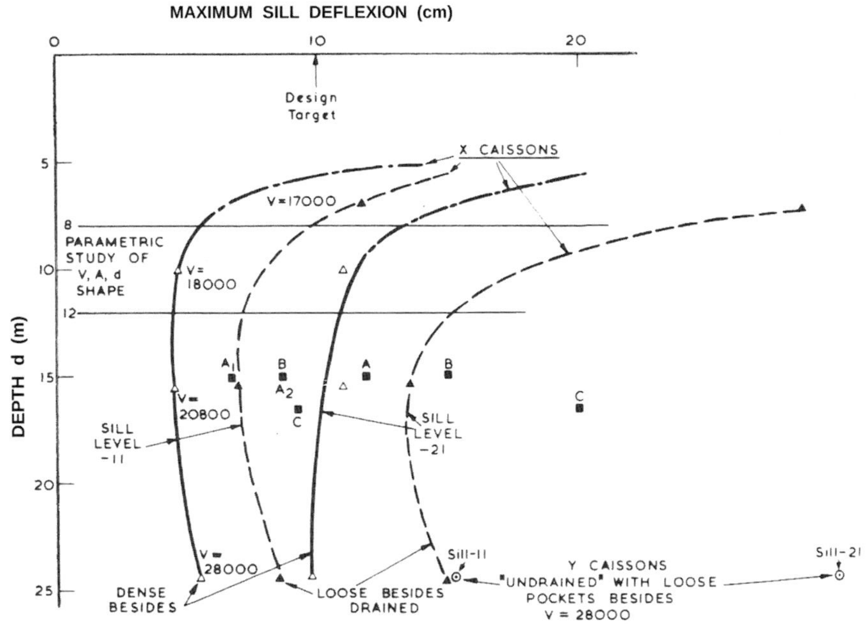

MONOLITHS

	Shape	Area A	Weight V	(Gravel)
A_1	Rectangular Square Ends	1120 m^2	29600t	3640
A_2	" " "	780	25500	2250
B	" Rounded "	1034	29600	3340
C	Sloping Faces	1000	25000	7500
Caissons	3 Cylinders	600	20800	0

Figure 27.8 Critical depth of foundation to limit sill deflection at Oosterschelde site. [56]

Depending on the geometry of the pipeline installation, e.g. depth of burial and/or coverage by rock-dumping, either localised or continuous, buckling can occur in the vertical or horizontal plane. Uncontrolled rapid vertical movements are associated with so-called upheaval buckling which may occur when a pipe expansion, initially suppressed by the build-up of axial compression stresses, overcomes the restraining forces in the cover materials. In the horizontal plane slower, more-controlled movements can occur where the pipeline can 'walk' across the seabed, displacing seabed material and forming berms which must be pushed aside if further excursion from the initial line is to occur. Both scenarios have been extensively simulated using pipeline element models in 2-D, subjected usually to displacement-controlled movements in one plane or the other. Studies of this type have been both generic and site-specific in a number of laboratories around the world.

For one site-specific project modelled in Manchester, material was retrieved in vibrocore tubes from the seafloor and built in to the model in a remoulded condition as the cover material over the pipeline crown, to a depth of around two diameters. The stiff pipeline element, around fifteen diameters in length, was used in the centrifuge model to produce a simple load/displacement plot (Figure 27.9).

While a simple load/displacement plot can be useful to the various contributors to a pipeline project (SI contractor, pipeline owner/operator, oil/gas company,

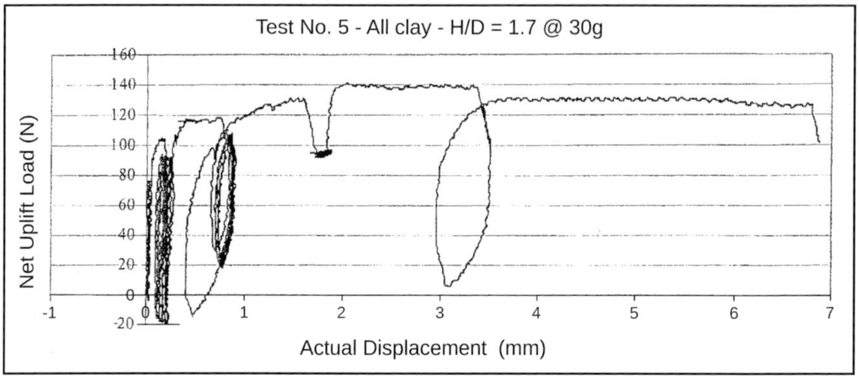

Figure 27.9 Sub-sea pipeline model study – typical uplift performance of a pipe element. (H = Depth of backfill cover above crown of pipeline of diameter D). (Image: University of Manchester)

designer, consultant – all of whom have at one time or another commissioned such tests), additional information can be obtained using relatively short 2-D plane section modelling with transparent glass or polycarbonate side walls, coupled with modern observation/instrumentation capabilities including Particle Image Velocimetry (PIV) techniques.

With regard to pipeline studies, two notable difficulties are outstanding and these may typify some of the limitations which apply to the use of physical (in our case) but also other forms of modelling.

- Offshore pipelines are not simply wished into position. Typically, they are placed in a ploughed trench which is later back-filled with the broken-up or remoulded displaced material and possibly further rock-dump or similar to increase the confining stresses. Alternatively, they may be jetted into place using a process which involves violent remoulding of the parent soil, again with possible later addition of backfill.
- The true buckling phenomenon in a pipeline is difficult to model due to constraints on model geometry that limit the ratios of pipeline length:diameter. Such a study was undertaken on the Manchester centrifuge using a highly-miniaturised pipe, at a much-higher scale ratio, subjected to fluid flow under increasing temperatures in an idealised soil scenario and not on a site-specific basis [62]. This study made use of the full 2-m straight dimension available in Manchester, parallel to the rotational axis of the centrifuge. True buckling models under conventional gravity have also been studied [63].

27.4.2 Simulating construction processes

Modelling of construction processes can often be intractable. Although the realistic installation/backfill process for pipelines is difficult to simulate, various approaches can be taken to simulate backfill reinstatement in idealised centrifuge models, using materials recovered from a range of site-specific vibrocore and other samples retrieved from the seabed [64]. None is entirely satisfactory but, in the future, there will be improved on-centrifuge robotics installed on larger machines that will facilitate better simulation of a number of construction processes as yet poorly replicated. Realistic simulation of any soil excavation procedure currently remains difficult, as does realistic placement and compaction and densification of cohesive fills. Some piling procedures can be reasonably simulated but others less so. Simulation of grouting has also been shown to be possible. The future promises considerable advances in capability but there is much to be developed.

27.4.3 Control and testing modes

In a similar vein, it is sometimes pertinent to assess the choice between modelling situations either under load/stress control or under displacement/strain control. The pipeline buckling problem is a case in point. In real life, loads and stresses build up in a pipe which moves, often quite rapidly, in response i.e. the situation is load controlled. The pipe element tests, such as that shown in Figure 27.9, are most frequently carried out under conditions of controlled displacement and displacement rate, allowing more-thorough monitoring and data capture. This benefit is also found in much laboratory testing under displacement control of models and material elements when 'post-peak' behaviour can be followed by brittle failure situations. G.B. Clark added a perspective to the overall centrifuge/self-weight position:

> 'A unique characteristic of the centrifuge is that models continue to be loaded by the same forces after initial failure. This type of loading is not possible in other types of loading device. In many geotechnical situations, with both soil and rock, the post-initial failure behaviour of a mass is as critical as the conditions which lead to initial failure' [20].

In relation to the assessment of the bearing capacity of offshore jack-up rig foundations, there are two common scenarios that have been assessed by centrifuge modelling by several groups around the world;

- Controlled ongoing plastic failure as footings penetrate deeply into beds of soft clay.
- Uncontrolled, punch-through failures when the bearing capacity in a stronger upper soil layer is overcome and the footing can plunge rapidly downwards through a softer underlying layer, with potentially very serious consequences for the structure above.

In practice both these situations in the field are load controlled, with no possibility of rapid release of the total load on the structure as a whole, and only responsive redistribution of that load in the short term. Model tests, usually on single footings, have been displacement-controlled often, but not always, with constraints against rotational displacements. Such tests allow detailed observation of the whole load/displacement behaviour and assessment of the large-strain displacement patterns around the footings as they penetrate, and also as they are subsequently uplifted and removed. The effects of the gross displacements on nearby piles can also be modelled. Early, simple 3-D tests [65, 66] have been followed more recently by much-greater and more-thorough series of studies, notably in Australia and Singapore [67, 68].

27.4.4 Visual observations

In the early days of modelling on the centrifuge, real-time visual observation was often provided (1960s onwards) by a tube-based television camera mounted at a suitable orientation close to the centrifuge axis. Still images for measurement were taken by statically-mounted cameras in various positions, with high-speed flash illumination. Attempts were also made to use specialist high-g movie cameras mounted on the rotor. Over the years these have been replaced by miniature solid-state CCD television cameras which can be mounted, in or alongside models, at the end of the centrifuge rotor arm, as well as near the centre of rotation – what works in the stumps at televised cricket matches at unit gravity also works in centrifuge systems around the world. Later, solid-state still cameras were installed close to models to capture images for measurement purposes.

27.5 Concluding remarks

This chapter has reviewed the background to physical and, in particular, centrifuge modelling in geotechnics and also some of the previous review assessments of the role of these 'hard' models in this technical area. Authors' views differ of course, partly dependent on their own intellectual closeness to particular approaches and to physical closeness to modelling facilities of sufficient size to be significant in a site-specific context. Smaller models are of diminishing use due to the need to include suitable soil/particulate materials and other construction details. The question as to whether or not they can be used as an integral part of a pre-construction, site-specific design process depends, to some extent, on location of facilities (which may be in-house with a contractor or design house, or in universities, or in public-sector research institutions, as in several countries but not in the UK) as well as the overall attitudes of those who control them who may or may not allow access to outside parties.

Drawing largely on personal experience in the University of Manchester group, where the guiding light who built and ran the UK's largest centrifuge, in terms of payload capacity if not of radius, was a renowned international consultant, it has been demonstrated that these models do have a potential role as one component among several in an interactive design process on major projects. In the examples cited, some of the model studies have supported the progress of such a process which might have reached similar conclusions had they not taken place. In others, they have clearly provided insights and raised questions that have led to changes in the design path. On occasion, the models were part of a process which did not come to fruition – though that does not in any way imply that the effort has been wasted. Even a negative result, proving a particular design option is non-viable, is valuable. Over a career spent in teaching as well as in research, the Author has learned that one of the hardest tasks is to persuade students to look at several design options and pursue them all until a judgement on preference can be rationalised and particular avenues can be closed off.

Great artists make many sketches and not all lead to finished 'works'. One should not look at physical modelling in any sense as a single strand or 'monochromatic' exercise. The recognised 'old master' artists drew sketches and cartoons in a single colour, or at best in a restricted palette, before fully realising their multi-coloured designs in a different medium on wood, plaster or canvas, using assistants to cover parts of the surface. Alternatively, working in 3-D, the great sculptors used models and *maquettes* in different materials in order to develop their ideas and determine perspectives, in any sense, and relative dimensions. One of the greatest sculptures of the Western world is Michelangelo's masterful David. There is a copy (a retrospective model?) on public view in the *Piazza della Signoria* in Florence but the original is held in the *Galleria dell'Accademia* in the same city, for protection. Even there, it remains vulnerable and centrifuge model studies have recently been carried out to help assess its vulnerability to overstress and possible tensile cracking and failure in the lower limbs [69]. Other cultural assets with a more geotechnical component that have been the subject of centrifuge modelling are the tower at Pisa in Italy [70, 71] and stone pagoda structures in Korea [72].

Architects have long taken the model approach on board and still do so, both in hard physical models and in virtual reality, usually working alongside other professionals with complementary skill bases. Engineers should not be reticent in this respect – physical models still have a role today as they had in earlier decades and centuries. Taken with explicit recognition of limitations and approximations they can still surprise designers, point to otherwise unseen ways forward and also provide warnings of potential difficulties and dead-ends – all outcomes of true value. Even in these days of increasing computational hegemony, the final output of almost all construction industry design has a physical reality to which the physical model can contribute.

At the turn of the present century, a meeting was held at Ascona, hosted by the geotechnical staff of ETH in Zurich, looking at the twin roles of both constitutive and centrifuge modelling. In contributing to a debate under the title 'Centrifuge tests are a sufficient means of solution to geotechnical design problems' the Author interpreted the word 'sufficient' in the mathematical sense as meaning 'necessary and sufficient' for a particular purpose:

> 'I think there is not now, and never has been, an engineer to whom the centrifuge modelling technique has seemed so versatile and ubiquitous that no other tools or approaches to design are necessary. ... That the centrifuge will play a part in design in its broadest sense is not in doubt. This will not be sufficient for full design solutions, but will nonetheless be an integral part of many specific projects' [73].

This remains a sustainable view.

References

1 Taylor, R.N. (Editor) (1995) *Geotechnical Centrifuge Technology*. London: Blackie Academic & Professional.

2 Craig, W.H. (Editor) (1985) *Proc. Symp. Application of Centrifuge Modelling to Geotechnical Design*. Rotterdam: Balkema.

3 Phillips, E. (1869) De l'equilibre des solides elastiques semblables. *CR Acad. Sci., Paris* 68: 75-79.

4 Craig, W.H. (1989) Édouard Phillips (1821-89) and the idea of centrifuge modelling. *Géotechnique* 39: 697-700.

5 Husband, H.C. & Husband R.W. (1975) Reconstruction of the Britannia Bridge: I Design, II Construction. *Proc. Inst. Civ. Engrs*, Part 1, 58: 26-66.

6 Wardle, J.B. & Lucas, J.C. (1975) Britannia Bridge: stress investigation before and after the fire. *Proc. Inst. Civ. Engrs*, Part 1, 58: 177-193.

7 Craig, W.H. (1975) Discussion on Wardle & Lucas (1975), Britannia Bridge: stress investigation before and after the fire. *Proc. Inst. Civ. Engrs*, Part 1, 58: 673.

8 Bucky, P. B. (1931) The use of models for the study of mining problems. *Tech. Pub. Am. Inst. Min. Engrs.*, No. 425, 28 pages.

9 Davidenkov, N.N. (1933) The new method of the application of models to the study of the equilibrium of soils (in Russian). *J. Tech. Phys., Moscow* 3: 131-136.

10 Pokrovsky, G.I. (1933) On the application of centrifugal force for modelling earthworks in clay (in Russian). *J. Tech. Phys., Moscow* 3: 537-539.

11 Craig, W.H., Vinogradov, V.V., Frolovsky, Y.K. & Zaytsev, A.A. (2015) Pioneers of Centrifuge modelling. *Int. J. Physical Modelling in Geotechnics* 15(1): 3-18.

12 Bucky P.B. & Fentress A.L. (1934) Application of principles of similitude to design of mine workings. *Trans. Am. Inst. Min. & Met. Eng.* 1: 25-50.

13 Pokrovsky, G.I. & Fedorov, I.S. (1969) *Centrifugal Modelling in the Mining Industry*. Moscow: Niedra Publishing House. (In Russian; draft translation in English (1975) by UK Building Research Establishment library translation service.)

14 Pokrovsky, G.I. (1933) On the use of a centrifuge in the study of models of soil structures. *Zeitschrift für Technische Physik* 14 (4): 160–162 (in German); also *Technical Physics Magazine* 3 (4): 537–539.

15 Pokrovsky, G.I. & Fedorov, I.S. (1968) *Centrifugal Modelling in Civil Engineering*. Moscow: Gosstroyizdat Publishers. (Sometimes translated as 'in Construction', or 'in the Building Art'). Draft translation in English (1975) by UK Building Research Establishment library translation service.)

16 Bucky, P.B., Solakian, A.G. & Baldin, L.S. (1935) Centrifugal method of testing models. *Civil Engineering* 5: 287-290.

17 Sinclair, D & Bucky, P.B. (1940) Photoelasticity and its applications to mine pillar and tunnel problems. *Trans. Am. Inst. Min. & Met. Eng.* 139: 225-252.

18 Polivka, J.J. (1942) Analysis of gravity load stresses by photoelastic methods. *Proc. 16th Semi-annual Eastern Photoelasticity Conf.* 40-46.

19 Wang, C.S., Boshkov, S.H. & Wane, M.T. (1968) The application of barodynamic photostress techniques to the study of the behaviour of rock beams loaded by their own weight. *Proc. 9th Symp. On Rock Mechanics, Am. Inst. Metallurgical Engrs.*

20 Clark, G.B. (1988) Centrifugal testing in rock mechanics. In: *Centrifuges in Soil Mechanics* (ed. W.H. Craig, R.G. James & A.N. Schofield), 187-198. Rotterdam: Balkema.

21 Hoek, E. (1965) The design of a centrifuge for the simulation of gravitational force fields in mine models. *J. South African Institute of Mining & Metallurgy* 65 (9): 455-487.

22 Allersma, H.G.B. (1998) Stress analysis on photoelastic particles in centrifuge tests. *Proc. Conf. Centrifuge* 98, 1: 61-66, Rotterdam: Balkema.

23 Langhaar, H.L. (1951) *Dimensional analysis and theory of models.* New York: John Wiley & Sons.

24 Sedov, L.I. (1959) *Similarity and dimensional methods in mechanics.* London: Infosearch.

25 David, F.W. & Nolle, H. (1982) *Experimental Modelling in Engineering*. London: Butterworths.

26 Cherkasov, I.I., Mikheev, V.V., Petrukhin, V.P., Palukin, V.A., Snarskii, A.S. & Shvarov, V.V. (1970) Influence of gravity on the mechanical properties of soils. *Soil Mechanics & Foundation Engineering* 7 (1): 25-31.

27 Ivicsics, L. (1975) *Hydraulic models.* Budapest: Research Institute for Water Resources Development.

28 Schofield, A.N. (2005) *Disturbed soil properties and geotechnical design.* London: Thomas Telford Publishing.

29 Sokolovski, V.V. (1960) *Statics of Soil Media.* London: Butterworths. (Based on a second Russian edition of 1954).

30 Hoek, E. (1963) Experimental study of rock-stress problems in deep-level mining. In: *Experimental mechanics. Proceedings of the first International Congress on Experimental Mechanics held in New York, November 1-3, 1961* (ed. B.E. Rossi), 177-194. Oxford: Pergamon Press.

31 Craig, W.H., James, R.G. & Schofield A.N. (Editors) (1988) *Centrifuges in Soil Mechanics.* Rotterdam: Balkema.

32 Craig, W.H. (2001) The seven ages of centrifuge modelling. *Proc. Int. Workshop on Constitutive and Centrifuge Modelling: Two Extremes,* 165-174. Lisse: Balkema.

33 Jacobsz, S. W., Kearsley, E. P. & Kock, J. H. L. (2014) The geotechnical centrifuge facility at the University of Pretoria. *Proc. 8th Int. Conf. Physical Modelling in Geotechnics* 1: 169-174.

34 Mikasa M., Takada N. & Yamada K. (1969) Centrifugal model test of a rockfill dam. *Proc. 7th Int. Conf. Soil Mechanics & Foundation Engineering* 2: 325-333. México: Sociedad Mexicana de Mecánica de Suelos.

35 Hubbert, M.K. (1937) Theory of scale models as applied to the study of geologic structures. *Bull. Geol. Soc. Am.* 48: 1459-1519.

36 Ramberg, H. (1967) *Gravity, deformation and the earth's crust*. London: Academic Press.

37 Yoxall, W.H. (1983) *Dynamic models in earth-science instruction.* Cambridge: Cambridge University Press.

38 Zelikson, A. (1985) Hydro-geotechnical modelling on large centrifuges. *Proc. Symp. The application of Centrifuge Modelling to Geotechnical Design,* 155-170. Rotterdam: Balkema.

39 Zelikson, A. (1991) Large centrifuge study of rock folding and fracture. *Proc. Conf. Centrifuge 91,* 599-606. Rotterdam: Balkema.

40 Muir Wood, D. (2004) *Geotechnical modelling.* London: Spon Press.

41 Craig, W.H. (1985) Centrifuge modelling for site-specific prototypes. *Proc. Symp. Application of Centrifuge Modelling to Geotechnical Design.* 485-501. Rotterdam: Balkema.

42 Chrimes, M. (2008) Geotechnical publications before Géotechnique. In: *The Essence of Geotechnical Engineering: 60 years of Géotechnique* (ed. J. H. Atkinson), 31-48. London: Thomas Telford.

43 White, D.J. (2008) Contributions to Géotechnique 1948-2008: physical modelling. In: *The Essence of Geotechnical Engineering: 60 years of Géotechnique* (ed. J. H. Atkinson), 49-60. London: Thomas Telford.

44 Simpson, B. & Tatsuoka, F. (2008) Geotechnics: the next 60 years. In: *The Essence of Geotechnical Engineering: 60 years of Géotechnique* (ed. J. H. Atkinson), 105-113. London: Thomas Telford.

45 Randolph, M.F. & House, A.R. (2001) The complementary roles of physical and computational modelling. *Int J. Physical Modelling in Geotechnics* 1: 1-8.

46 Mitchell, R. J. (1991) Centrifuge modelling as a consulting tool. *Can. Geotech. J.* 28: 162-167.

47 Gaudin, C. (2015) Editorial. *Int. J. Physical Modelling in Geotechnics* 15 (1): 1-2.

48 Burland, J. (2012) Structural and geotechnical modelling. In: *ICE Manual of Geotechnical Engineering* (eds. Burland et al.), Vol. 1, 27-34. ICE Publishing, London.

49 Craig, W.H. (1974) Discussion. *Proc. 8th Int. Conf. Soil Mechanics and Foundation Engineering,* 4.2: 15-16.

50 Rigden, W.J. & Rowe, P.W. (1975) Model performance of an unreinforced diaphragm wall. *Proc. Conf. on Diaphragm walls and anchorages,* 63-67. London: Inst. Civil Engrs.

51 Rowe, P.W. (1972) The relevance of soil fabric to site investigation practice. *Géotechnique* 22: 193-300.

52 Pavlakis, G. (1976) *An investigation into the performance of a model earth dam.* PhD. Thesis, University of Manchester.

53 Craig, W.H. (1974) *Model studies of the stability of clay slopes.* PhD thesis, University of Manchester.

54 Malushitsky, Yu. N. (1981) *The Centrifugal Model Testing of Waste-heap Embankments.* Cambridge: Cambridge University Press. (Translation of 1975 Russian Edition.)

55 Rowe, P.W. (1975) Displacement and failure modes of model offshore gravity platforms founded on clay. *Conf. Offshore Europe '75,* 218: 1-17.

56 Rowe, P.W. (1983) Use of large centrifugal models for offshore and nearshore works. *Symp. Geotechnical Aspects of Coastal and Offshore Structures,* Bangkok. Rotterdam, Balkema. 21-33.

57 Craig, W.H. (1993) Partial similarity in centrifuge models of offshore platforms. *Proc. 4th Canadian Conf. Marine Geotechnical Engineering* 3: 1034-1042.

58 Anon (1985) Dwarfing all others. *Ground Engineering* 28(5): 24-26.

59 Hinch, L.W., McDowell, D.M. & Rowe, P. W. (1984) Jamuna River 230kV crossing – Bangladesh, l. Design of foundations. *Proc. Inst. Civ. Engrs,* Part 1, 76: 927-949. (This is one of a linked set of four papers on the project with later discussion in 80: 731-753).

60 Rowe, P.W. & Craig, W.H. (1976) Studies of offshore caissons founded on Oosterschelde sand. In: *Design and Construction of Offshore Structures: Proceedings of the conference held on 27-27 October 1976* (eds J.P Blanc & M. Monro), 49-55. London: Thomas Telford.

61 Rowe, P.W. & Craig, W.H. (1978) Prediction of caisson and pier performance by dynamically loaded centrifugal models. *Symp. on Foundation Aspects of Coastal Structures,* Vol. 2, Paper IV.3. 16 pages. Delft.

62 Moradi, M. & Craig, W.H. (1998) Observations of upheaval buckling of buried pipelines. *Proc. Conf. Centrifuge 98.* Rotterdam: Balkema. 693-698.

63 Taylor, N. & Tran, V. (1996) Experimental and theoretical studies in sub-sea pipeline buckling. *Marine Structures* 9: 211-257.

64 Craig, W.H. (2014) Simulating cohesive backfills in centrifuge models of subsea pipelines. *Proc. 8th Int. Conf. Physical modelling in Geotechnics* 1: 459-464.

65 Craig, W.H. & Chua, K. (1990) Deep penetration of spud-can foundations on sand and clay. *Géotechnique* 40 (4): 541-556.

66 Craig, W.H. (1998) Spud-can foundations: installation with deep penetration and subsequent removal. *Proc. Inst. Civ. Engrs, Geotechnical Engineering* 131: 146-151.

67 Hossain, M.S., Hu. Y, Randolph, M.F. & White, D.J. (2005) Limiting cavity depth for spudcan foundations penetrating clay. *Géotechnique* 55: 679-690.

68 Purwana, O.A., Leung, C.F., Chow, Y.K. & Foo, K.S. (2005) Influence of base suction on extraction of jack-up spudcans. *Géotechnique* 55: 709-719.

69 Corti, G., Costagliola, P., Bonini, M., Benvenuti, M., Pecchioni, C., Vaiani, A. & Landucci. F. (2015) Modelling the failure mechanisms of Michelangelo's David through small-scale centrifuge experiments. *J. Cultural Heritage* 16: 26-31.

70 Cheney, J.A., Abghari, A. & Kutter, B.L. (1991) Leaning instability of tall structures. *ASCE J. Geotechnical Engineering* 117(2): 297-318.

71 Viggiani, C. (2002) The Tower of Pisa is back to the future. *Proc. Int. Workshop on Constitutive and Centrifuge Modelling: Two Extremes* (ed. S. Springman), 3-34. Lisse: Balkema.

72 Kim. D-S. & Park, H-J. (2014) Centrifuge modelling on seismic behaviour of stone architectural heritages. *Proc. 8th Int. Conf. Physical modelling in Geotechnics,* 1: 47-60.

73 Craig, W.H. (2002) The seven ages of centrifuge modelling. *Proc. Int. Workshop on Constitutive and Centrifuge Modelling: Two Extremes* (ed. S. Springman), 165-174. Lisse: Balkema.

Section E

Physical modelling in the twenty-first century

After an American newspaper prematurely printed an obituary of the author Mark Twain, he is said to have responded 'The reports of my death are greatly exaggerated'. Something similar might be said about the use of physical models in civil and building engineering. Despite remarkable progress in the power of computers and analytical software since the 1970s, the countless ways that they have benefited engineering, and the regular claims that the testing of physical models is a thing of the past, the use of physical models in the present century is alive and well.

The final section of the book has chapters reporting on current practice in the use of measurement models in all the disciplines covered in earlier chapters, and a few more. Their content is clear from the list of contents for the book.

The book concludes with three chapters that do not follow the pattern of earlier ones. Jan Knippers, from the University of Stuttgart, describes work that has been undertaken in the field of biomimetics in which the behaviour of biological entities – all manner of living organisms – can be used as models for different systems within modern buildings, ranging from climatic control, to fibre reinforcement and adaptive geometries. The final technical chapter is by Mamoru Kawaguchi, the eminent Japanese structural engineer who, like many Japanese engineers, has been a great believer in and advocate of using physical models for the whole of his professional life (sadly he died, aged 87, in May 2019). In this chapter he describes how he put his understanding of model analysis to good use in designing a 100 metre-long *Koinobori* (a Japanese carp-shaped windsock) some twenty or thirty times larger than the normal ones flown by children on 5th May each year – what better example could there be of engineering ingenuity put to good use.

The final chapter of the book looks at where we might go from here. There are very few examples of engineers' measurement models on display in museums. This can surely not be right for objects that have played such a fundamental role in the progress of civil and building engineering over the last three centuries, through their use on many major projects as well as in the engineering scientists' laboratory. Dirk Bühler and Christiane Weber present a number of proof-of-concept models and measurement models that we know do exist. They then make a case for saving some significant models, conserving them as necessary, and putting them on display in museums to ensure that their contribution to engineering progress can be recognised, celebrated and carried forward into the future.

28

Physical models as powerful weapons in structural design

Mamoru Kawaguchi

28.1 Introduction

Today we use computers very often, or too often we might say, and resort to *analytical models* using software installed in them. Another type of models are *physical models* which are always real and sufficiently reliable if they are made carefully and used correctly. The most attractive examples of this type may be form-finding funicular models attempted for masonry and concrete structures by Antoni Gaudí and Heinz Isler. An important feature of physical models is that we can see them, touch them and feel them, and when we load them to fracture, we can expect that the same thing may happen to their full-size prototypes. In the present chapter the author exemplifies the usefulness of physical models by showing several models which he has utilized to make various decisions at the different stages of structural design.

28.2 Aesthetic models

Aesthetic models are the simplest and most direct type of models which we can use effectively in our design process. They are manufactured in different sizes according to the purposes of the models. Small models are used to examine the possible forms of the structures or buildings. Materials are seldom important for this size of model. We can use drawing paper, polystyrene board, balsa wood and so on, which are cheap and easy to work with. With a small-scale model, say 20-50 cm, we examine the whole shape of the structure that we are going to design, its silhouette and visual atmosphere, the general arrangement and assembly of the structural components, etc.

Models of medium-scale, say 1-3 m, are used to examine more detailed characteristics of structures. With these models we study the three-dimensional constitution of the structure, the relationship of individual components to the whole structure and the effects of the colour and texture of materials. We carefully select

Physical Models: Their historical and current use in civil and building engineering design,
First Edition. Edited by Bill Addis.

the materials for the models of medium size so that we can duly study the above features.

Large models, usually of parts of a structure and often as big as the prototype, are used to check the important details of the structure. We use this type of model to discover visual information for making the final decisions about the structural design. So we pay the maximum attention to select correct materials not only in terms of dimensions but also of colour and texture. In Figures 28.1 to 28.5 various

Figure 28.1 Small-scale model (1:50) used to study the overall 'atmosphere' of the Inachus Bridge in its context. (Image: Mamoru Kawaguchi)

Figure 28.2 The completed Inachus Bridge, Oita, Japan, 1994. Span 35 m. (Image: Mamoru Kawaguchi)

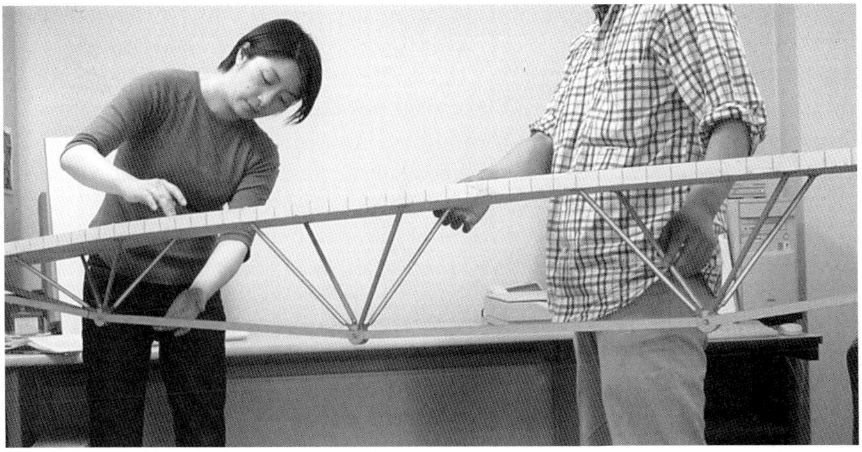

Figure 28.3 Medium-scale (1:10) model used to study the proportions of the various structural elements and the textures of the stone and steel components. (Image: Mamoru Kawaguchi)

Figure 28.4 Large-scale (1:1) partial model to study the forms and proportions of the elements in a principal node connection. (Image: Mamoru Kawaguchi)

Figure 28.5 The node in the completed structure. (Source: Mamoru Kawaguchi)

sizes of models which the author used for the design of Inachus Bridge at Oita in Japan [1] are shown in comparison with their prototypes, to illustrate the ways in which those models were utilized. A small model was made of paper, bamboo strip and polystyrene board, and was used to examine the total atmosphere of the 'incomplete truss', the proportions of the stone deck and steel chain and the visual effect of the handrails. Different designs of handrail were incorporated into the model to find the most suitable one for the bridge (Figure 28.1).

A medium-scale model was made of Styrofoam, balsa plates and wood bars, and was used to find the most suitable proportions and dimensions of constituent members and important details. Textures of stone blocks and steel members were also studied in this phase of design study. The method of prestressing the stone deck was also examined with this model (Figure 28.3).

A large model was made for the important steel node in the lattice to study the problems of connecting the lattice and chain members and the overall appearance of the node. The methods of welding in the factory and on site were also examined using this model. The model was made of polystyrene products in the shapes of plates and cylinders and were sprayed with silver paint to give a metallic appearance (Figures 28.4 and 28.5).

28.3 Mechanism models

Mechanism models are used to check a structural design when the structure is designed to change its shape to a large degree, either during or after its construction. The purpose of this model is to see if every part of a structure can move freely without interference with other parts of the structure. We made more than one mechanical model almost every time we have designed a structure using the Pantadome principle [2, 3]. The purpose of using a mechanism model in connection with the Pantadome system is threefold: to confirm that the movement is unimpeded with no interference among different components of the structure; to ensure that the structure is stable at every stage during the movement; and to check if the structure is sufficiently stiff and strong to resist lateral loads. While the first of these purposes can nowadays be achieved using a three-dimensional CAD model, the other two issues can only be conveniently confirmed using the physical mechanism model (Figure 28.6).

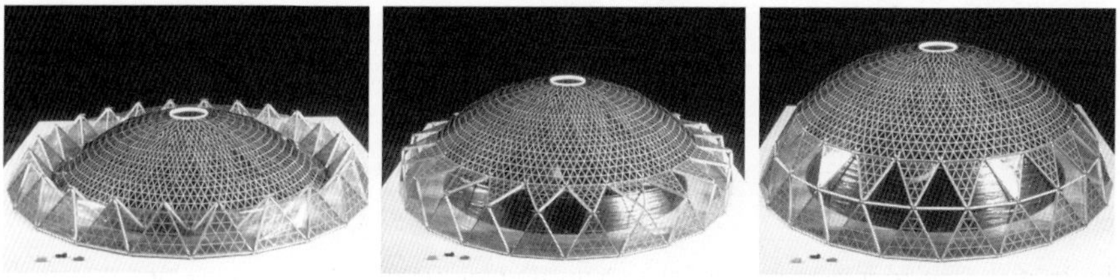

Figure 28.6 Small-scale generic model (1:100) to study the mechanical mechanism and erection procedure for the Pantadome. (Image: Mamoru Kawaguchi)

28.4 'Touch and feel' models

One of the most advantageous features of physical models is that we can touch them, push them, pull them and directly feel their behaviour. For touching and feeling purposes, the models are made with rather deformable materials. The model is loaded using hands or fingers so that the person can feel the intensity of the applied loads. The deformation due to the loading is felt by the person, and affirmed by the eyes. This kind of model is particularly suitable for the structural designer in order to develop an intuitive understanding of the 'force flow' in the structure he/she is going to create. Such models can also be used effectively for educational purposes.

An example is shown here in a paper 'touch and feel' model that was made in order to understand the general behaviour of a singly-curved, semi-rigid tensile roof. The model was made of drawing paper and was loaded using a whippletree by which we could apply uniform loading over the whole or half area of the roof.

Since paper has a low value of Young's modulus, the roof produces perceptible deformation when it is loaded by a finger. Since each part of the roof frame is made of paper, it is easy to tell whether a certain member is in tension or compression. If a member is in tension, a member with slight curvature created in the hand-making process tends to straighten, and when it is in compression, it tends to become more curved (Figures 28.7 and 28.8).

Another example of a 'touch and feel' model is the one that we made with wooden rods and thin wire to help our design staff intuitively understand the behaviour of a 32 m communications tower located in a large science park which includes a large synchrotron radiation facility called SPring-8 which can produce X-rays of 8 giga-electron volts. With these facilities scientists are examining the structure of protein and we began to design the tower at the time when the human genome was first fully decoded. We thus thought that it would be meaningful to adopt the double helix as the motif of our tower. The location, near Kobe in Hyōgo

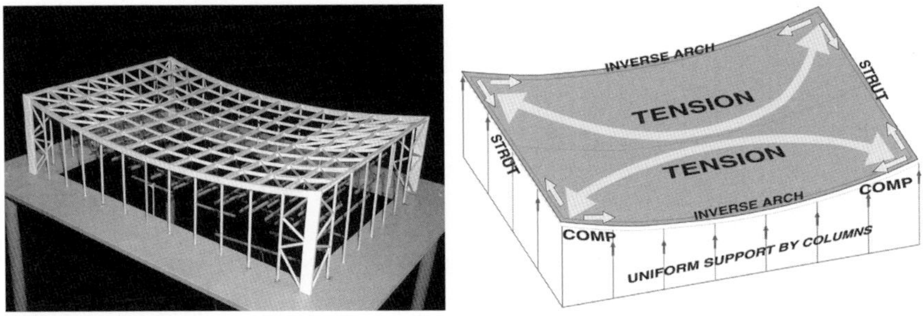

(a) (b)

Figure 28.7 Steel roof structure for Osakaya distribution centre in Kagoshima city. (a) Small-scale 'touch and feel' paper model (49.5 x 27 cm). (b) Diagram showing the approximate flow of forces derived from using the paper 'touch and feel' model. (Images: Mamoru Kawaguchi)

Figure 28.8 The full-scale steel roof structure for Osakaya distribution centre in Kagoshima city. (Image: Mamoru Kawaguchi)

(a) (b)

Figure 28.9 The Genome Tower, Hyōgo, Japan, 2002. (a) The 'Touch and feel' model. (b) The completed tower. (Images: Mamoru Kawaguchi)

Prefecture, is subject to earthquakes and so the principal function of the structure was to transmit the lateral loads due to wind and earthquakes safely to the foundations. We wanted the double helix, made with steel tubes and cables, to be as structurally efficient as possible.

The structure consists of a central compression tube, 400 mm in diameter, from which are suspended eight 38-mm diameter rods. To provide shear resistance we wanted to avoid a Schukov-type lattice arrangement, preferring to use the 140-mm diameter tubes which form the double helix itself, together with an additional structural system to link the two helices. The 'touch and feel' model was extremely useful in determining the requirements of this additional system and comparing alternative arrangements of its structural members (Figure 28.9). The final structure was confirmed by computer analysis.

28.5 Structural behaviour models

28.5.1 Full-size structural models

Full-size models are made at the final phase of design, and are often used to confirm the appropriateness of a design. They are usually tested to grasp the true behaviour of the prototypes or their parts up to their fracture. Full-size models are normally expensive to make, and the testing of such models is also a costly process. Normally, analytical study and experiments with smaller models are carried out before we proceed with full-size experiments. However, when a certain structural component is to be produced in large numbers, such as the main node of a space frame, it is likely to be highly-effective to conduct a series of experiments with full-size models.

When we designed the grand roof of Festival Plaza for Expo '70 in Osaka, we had to design a huge space-frame roof covering around 292 x 108 metres, supported on only six columns. This consisted of two bays of 108 x 75.6 m, with cantilevers of 16.2 m and 37.8 m outside the column grid. Furthermore, the roof was very heavily loaded, as it was not a mere roof, but an elevated exhibition venue accommodating more than 4000 people wandering within the depth of the roof. Moreover, at the time of Expo '70, cast steel was still a relatively new structural material [4, 5].

Since there were more than 600 space-frame nodes in the roof, design of these cast-steel components formed a part of utmost importance in the whole design of the grand roof structure. After thorough study of the roof by analytical methods for all the possible loading, including shell analysis of the cast steel joints, we carried out a series of tests on the full-size joints (Figure 28.10a). These results were compared to those calculated using the analytical model (Figure 28.10b). Reduced-scale models were also made to help enhance the aesthetic appearance of the node (Figure 28.11).

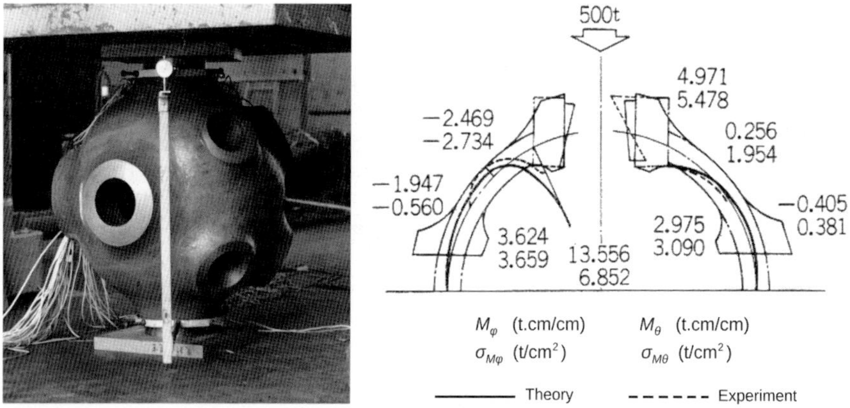

(a) (b)

Figure 28.10 Node for the grand roof of the Festival Plaza, Expo '70, Osaka, Japan. (a) A full-scale prototype of the steel node, 0.92 m in diameter, being tested in tension. (b) Diagram showing a comparison of results obtained with the analytical model and those obtained using the physical model. (Images: Mamoru Kawaguchi)

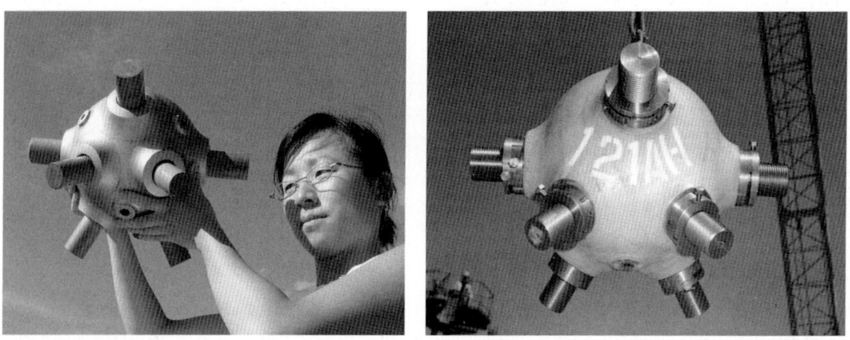

(a) (b)

Figure 28.11 Node for the grand roof of the Festival Plaza, Expo '70, Osaka, Japan. (a) The large-scale (1:4) aesthetic model of the node. (b) The completed node. (Images: Mamoru Kawaguchi)

28.5.2 Reduced-scale structural models

The most common experimental testing used in the development of structural designs makes use of reduced-scale models. Model theory, based on the use of dimensional analysis, is one of the most fundamental and reliable weapons that can be used in structural design. Beside classical ways of undertaking experimental studies, model theory can be used to predict the behaviour of huge, new creations. If the similitude requirements between the model and prototype are carefully observed, the behaviour of the prototype, not only in elastic regions but also in elasto-plastic regions and even its fracture behaviour, can be well-predicted from the information obtained from reduced-scale model tests.

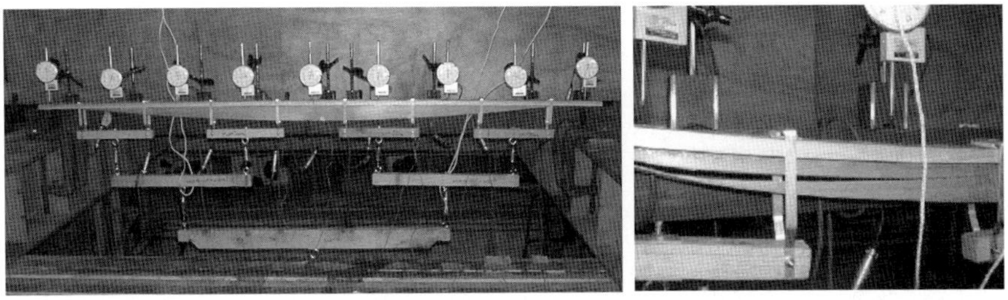

Figure 28.12 Onishi Hall, 2004. (a) 1:10 scale model of the steel-reinforced timber beam, under uniformly distributed loading, applied by means of a whippletree. (b) Fracture of the scale model under load. (Images: Mamoru Kawaguchi)

One project in which we used small-scale model testing was a small gymnasium for a town north of Tokyo called Onishi. To reflect the town's important forestry industry, one of the competition requirements was to use locally produced, high-quality cedar for the gymnasium. Unfortunately, while this timber is excellent as a finishing material, it is not well suited to structural applications. An additional challenge was that the architect with whom we collaborated had a particular preference for slender structural members. To meet these challenges, we designed composite beams consisting of a thin cedar beam sandwiched between two steel plates. To prove this construction we undertook tests on reduced-scale models, using model theory to scale up the results for the full-scale beams spanning 20 metres and 70 cm deep at mid-span (Figure 28.12).

28.6 Concluding remarks

Reduced-scale and full-size physical models continue to have an important role in structural design alongside the common use of computers to study analytical models. They can be used to provide both quantitative data and a qualitative feel for structural behaviour. They can also help develop an aesthetic appreciation of proportions, dimensions and appearance. Models of many types are also helpful in developing the understanding of structures in students of both engineering and architecture. Unlike to the output of computer programmes, physical models have the capacity to impart feeling and understanding directly into the heart and soul of the human designer.

References

1 Kawaguchi, M. (1996) Granite Pedestrian Bridge, Beppu, Japan. *Structural Engineering International* 6 (3): 148-149.

2 Kawaguchi, M. (1994) Application of Pantadome System to Long-Span Roof Structures. Symposium (British Group): Places of assembly and long-span building structures, Birmingham. *IABSE Report* No.71: 91-100.

3 Kawaguchi, M. (1998) Neuartige Tragwerke und ihre Formen – Beispiele des Pantadome-Systems. *Bautechnik* 75 (11): 948-954.

4 Tsuboi, Yoshikatsu & Kawaguchi, Mamoru (1972) The space frame for the symbol zone of Expo '70. In: *Proceedings 1971 IASS Pacific Symposium Part II, on Tension Structures and Space Frames, Tokyo and Kyoto,* 893-904. Tokyo: Architectural Institute of Japan.
5 Chilton, John (2011) *Space Grid Structures.* Oxford: Architectural Press. 61-64.

29

Physical modelling of structures for contemporary building design

Bruce Martin

29.1 Introduction

When we design familiar forms to be constructed using established materials, we can often develop the concept, complete the analysis and communicate the proposals using a combination of hand sketches, computer models, drawings and reports. We draw on previous experience of physical objects, ranging from our observations of the world around us to the many laboratory tests that inform the design guides and codes that we reference to develop and test our designs.

When our designs extend into the realm of unusual forms or materials, we may go beyond the references and guidance that underlie normal design practice. In these cases physical models may play an essential part in developing our understanding of, and achieving confidence in, the behaviour of the final constructed object.

The Stavros Niarchos Foundation Cultural Center (SNFCC) opened in June 2016 to provide new homes for the Greek National Library and the Greek National Opera, and a large, much needed, Public Park in Athens [1]. For the bulk of the project, the need was for good-quality conventional engineering solutions. However, we recognised that there would be some 'special areas' in which particular challenges would require unusual engineering solutions. For some of these 'special areas' physical models played a key part in developing and verifying the engineering designs. Each model served a particular purpose and the models were correspondingly varied, ranging from a rough-and-ready Meccano thought piece, through functioning 3D-prints of spring and bearing assemblies, to full-strength, house-sized parts intended for testing to destruction. This chapter contains a description of why and how physical models were used to inform the structural-engineering design of the SNFCC.

29.2 The canopy – tensegrity then ferrocement

A 100 x 100 m canopy is supported on the opera house and shades the upper part of the park which continues over the library and opera house. The canopy

Physical Models: Their historical and current use in civil and building engineering design,
First Edition. Edited by Bill Addis.

Figure 29.1 The canopy above the opera house. (Image: Expedition Engineering)

supports a one hectare array of photo-voltaics which provides the base power supply for the day-to-day operation of the SNFCC (Figure 29.1).

The initial architectural concept was for a 'mist-like' structure supporting photo-voltaic 'leaves' which would cast dappled shade on the upper part of the park. Many frame arrangements were investigated and simple physical models made of straws, thread and plastic sheet were used to help the team to visualise some of the options. The team closed in on a tensegrity system which would minimise the visual bulk of the structure (Figure 29.2).

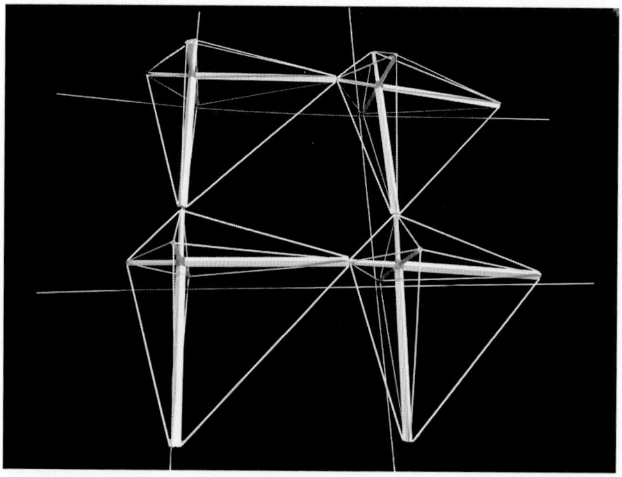

Figure 29.2 Concept development model – working towards a tensegrity canopy. (Image: Expedition Engineering)

Figure 29.3 Half-scale model prototype of the tensegrity canopy. (Image: Expedition Engineering)

Achieving the correct prestress in the tensegrity structure would be critical to the performance of the canopy. We developed what appeared to be a simple method for achieving the required pre-stress with relatively few stressing points. Since the team was not aware of close precedents, it was decided to build a physical prototype to test the proposed stressing system and to demonstrate that the structure would perform in accordance with the analysis model. A half-scale model of a part of the canopy was commissioned, fabricated and assembled in the Department of Engineering at Cambridge University (Figure 29.3).

Meanwhile the requirements for access walkways, cables, inverters and transformers to service the photo-voltaics had been developed. The extent of these services adversely transformed the visual bulk of the canopy and led to a reopening of the architectural concept. Ioannis Mitsos went on to use the tensegrity prototype to inform his PhD, along the way confirming the stressing system and structural performance. However, the tensegrity prototype played no further part in the development of the design for the canopy.

In the revised concept Renzo Piano, the architect, wanted to achieve a very thin profile on elevation, and a joint-free soffit with a finish 'like a yacht'. The revised design would need to enclose the services and provide a simplified access system to the photo-voltaics. The client required the scheme to be constructible by at least five contractors so that all aspects could be competitively tendered. Structurally, we had to limit any increase in the load that would be applied to the opera house for which the detailed design was well underway and could not be substantially changed.

Figure 29.4 A ferrocement sample cut from the prototype. (Image: Expedition Engineering)

The new design requirements led the team to undertake a short, but intense, study which identified ferrocement as the material best suited to the new requirements. Ferrocement is a type of ultra-thin reinforced concrete, invented by Joseph Lambot in 1848 and developed by Pier Luigi Nervi in the twentieth century. Ferrocement combines common materials and techniques in a non-standard manner to form unusually thin reinforced concrete – typically 15 to 60 mm thick [2]. In the new design, curved ferrocement skins would form the top and bottom surfaces of the canopy with steel bracing connecting the skins to create a single structure capable of spanning 55 x 75 m over the central part of the opera house. To limit the weight and visual bulk of the canopy, the ferrocement would be painted, but not over-clad (Figure 29.4).

We quickly confirmed that the canopy would be a larger, more-engineered ferrocement structure than any of the precedents, and that it would fall outside the usual range of the ferrocement and reinforced-concrete design codes. We identified that we would need to use physical models to:

- demonstrate that the ferrocement could be built using locally-available labour with no prior knowledge of ferrocement,
- confirm that the ferrocement did perform in accordance with theory extrapolated beyond the usual ranges of the codes, and
- demonstrate to the design team and the companies tendering for the project that the ferrocement canopy was constructible.

The project manager, Neil Clemson of Faithful and Gould, supported us and the client agreed.

The first stage of developing the ferrocement design was for our team of engineers to gain hands-on experience so that we understood what we were asking other people to build. Paul Nedwell of the University of Manchester arranged a one-day workshop for us to build ferrocement samples. Although the samples that we created were rudimentary, the workshop was a quick and effective method of making us aware of the principles of ferrocement construction and arrangements that were likely to be more or less challenging to construct, particularly in relation to achieving full mortar penetration through the multiple layers of mesh reinforcement.

There were three interacting strands in the development of the canopy geometry: a finite-element analysis model, 3D renders, and multiple sanded wood samples which allowed a tactile appreciation of the form and an assessment of the adjustments introduced by modifications to stiffen the analysis model.

The next stage of physical modelling was to demonstrate that locally-available labour unfamiliar with ferrocement could construct high-quality ferrocement. Samples were required to establish that mortar penetration could be achieved in areas with particularly congested reinforcement and that a high quality of finish could be achieved. Panels were also needed for load tests. Before starting to make ferrocement, preliminary physical trials were required to establish methods for fabricating and fixing the many 3-mm spacers which held the reinforcement clear of the formwork, and to test a range of geometries for the end fittings to pump hoses to achieve the most-effective mortar penetration through the mesh. Preliminary trials of formwork types were also required (Figure 29.5).

An impromptu decision to video Dr Pete Winslow, the project engineer, jumping on one of the early ferrocement test panels turned that jump into one of the most useful physical models of the project. Although entirely unquantified and unscientific, the video was extremely effective in giving the team confidence that the 'wafer-thin' 'reinforced concrete' was strong and robust (Figure 29.6).

The Jump Test was supported by many physical loading tests completed by Professor Kourkoulis and his team in the Department of Mechanics at the National Technical University of Athens (NTUA). The purpose of the tests was not to establish the strength of the ferrocement panels, rather the purpose was

Figure 29.5 Ferrocement trial panel cast on Perspex – too smooth! (Image: Expedition Engineering)

Figure 29.6 'Jump test' on a ferrocement trial panel. (Image: Expedition Engineering)

Figure 29.7 Testing a ferrocement panel at NTUA. (Image: Expedition Engineering)

to demonstrate that the panels were consistently stronger than was predicted by our, intentionally-conservative, analysis model. To complete the tests NTUA built a new test rig which was able to apply large axial compression loads to full-scale ferrocement panels along two orthogonal axes (Figure 29.7).

The connection between the internal steel-bracing members and the ferrocement skins was unusual. The ferrocement was too thin for us to cast steel plates in the section, and we had to limit thickenings to avoid too great an increase in the load on the opera house. A steel spider arrangement was developed which

Figure 29.8 Fit test of the steel-to-ferrocement connection before the mortar was applied. (Image: Expedition Engineering)

pushes against a slight thickening in the ferrocement to transfer loads between the steel bracing and the ferrocement. The ferrocement thickening is reinforced with very many 1.6-mm diameter steel shear links. This arrangement falls well outside the usual scope of the guidance in the design codes for the arrangement of shear reinforcement. However, load tests on physical models confirmed that the Eurocode guidance for design of shear links in reinforced concrete held good, even at this small scale (Figure 29.8).

Once we had confidence in the component parts we constructed a 16 x 6 x 5 m, full-scale physical prototype of a part of the canopy. The primary aim of this prototype was to demonstrate that the ferrocement canopy was constructible using locally-available labour and materials. A secondary benefit was that we would be able to load the prototype to provide further verification of the analysis model. The intention was that the prototype would remain in place as a reference, while the main canopy was built. Therefore, the initial load test completed soon after the prototype was complete was limited to the characteristic loads. This allowed us to confirm stiffness and deflection without risking damage to the prototype. In due course, when the time came to remove the prototype, we carried out what was intended to be a load test to destruction. In the event we substantially exceeded the theoretical failure load, but failed to damage the prototype. The test did at least demonstrate that our analysis model was conservative (Figure 29.9).

The final part of physical modelling of the ferrocement canopy came in planning the construction sequence. Two smaller ferrocement canopies (17.5 x 17.5 m and 22.5 x 22.5 m) were designed for the roofs of two of the project outbuildings (now cafés). The construction sequence was planned so that these canopies would be built before the main canopy above the opera house, allowing the contractor to demonstrate capability in ferrocement construction before starting on the main canopy where there would be little scope to rectify errors.

Figure 29.9 Load-testing the ferrocement prototype of part of the canopy. (Image: Expedition Engineering)

29.3 Column heads with bearings, springs and dampers

The canopy is supported on 30 columns which bear on the roof of the opera house. The connection between the head of the columns and the canopy is special – each connection includes an arrangement of vertical bearings, springs and dampers which soften the connection. This allows the column load to remain approximately constant as the canopy moves vertically relative to the column. The arrangement enables the column-head connections to accommodate three unusual requirements:

i) The top surface of the 100 x 100-metre canopy is exposed to the sun – particularly before the PVs are installed, and when the PVs are removed for maintenance. The bottom surface of the canopy is shaded. When the top surface is warmed by the sun it expands causing the canopy to 'banana' with the centre moving upwards and the edges descending. The effect is significant, since construction, real in-service daily thermal movements of up to 20 mm have been recorded between the moving canopy and the 'static' column heads. Traditional rigid connections between the columns and the canopy would have introduced very large axial loads into the columns as the opera house acted to resist the vertical movements. These column loads would, in turn, have resulted in very large forces within the canopy and opera house structures.
ii) Some of the columns are supported on long-span reinforced-concrete beams. In relation to the opera house beams, the double-skinned canopy is stiff but weak. If we had adopted traditional rigid connections between the columns and the canopy, creep deflection of the long-span beams would have changed

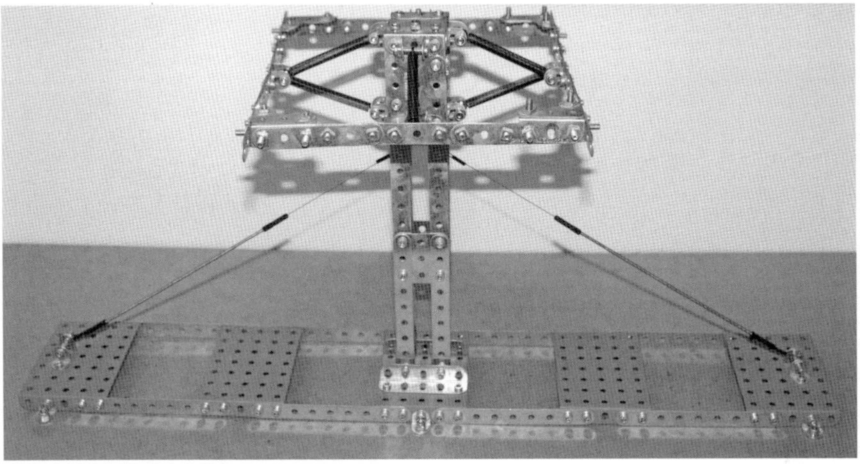

Figure 29.10 A rough-and-ready Meccano concept model of the sprung column heads. (Image: Expedition Engineering)

the structural action so that some columns became hangers suspending the opera house beams from the canopy – until the canopy broke.

iii) The opera house design was well progressed when we developed the revised canopy concept, and we were therefore limited in the maximum load that each canopy column could apply to the opera house. Wind passing over the canopy results in local pressure concentrations around the edges and, particularly, near the corners of the canopy. The 'softened' column heads allow the canopy to deflect and transfer some of the peak wind loads to adjacent columns.

The use of springs to deliver vertical loads onto columns is unusual. A rough-and-ready Meccano model was used to quickly communicate the idea within the team. The tactile response of the model to probing hands appeared to help the team to accept the concept (Figure 29.10).

We understood that we were asking the contractors to build something unusual, and that the requirement could easily be missed in a sea of tender documents. To avoid this we took a working 3D-printed model, complete with springs and bearing surfaces, to the mid-tender interviews and made a short presentation about the column-head detail, what we required it to do, and why (Figure 29.11). The successful contractor responded some months later with their own working 3D-print of the design as shown on their fabrication drawings.

Because the column-head detail is so unusual a full-size column head was tested at Aristotle University of Thessaloniki. As well as demonstrating the general structural performance of the complete assembly, the laboratory tests provided information on the real dynamic properties of the complete assembly. This data was used to tune the specification for the dampers which limit potential resonance under wind loading (Figures 29.12 and 29.13).

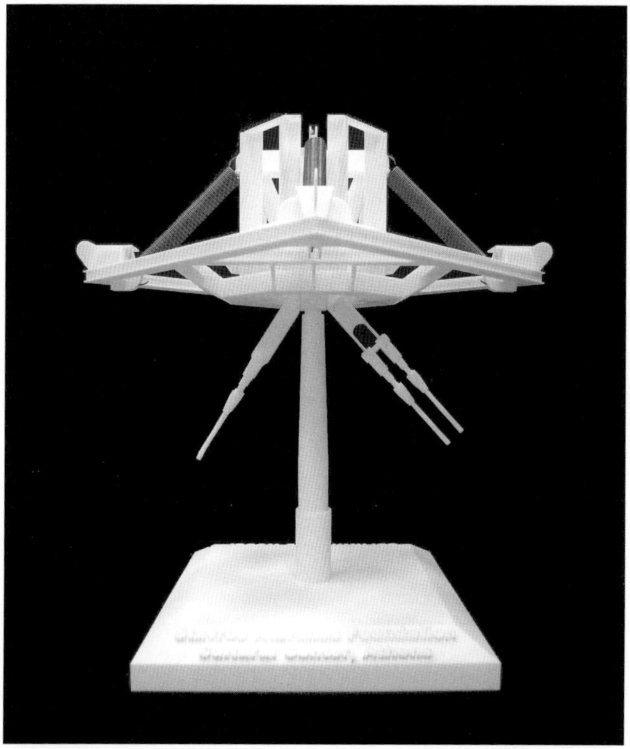

Figure 29.11 The 3D-printed model of sprung column-head arrangement used in the mid-tender interviews. (Image: Expedition Engineering)

Figure 29.12 A local detail during testing of a full-size column-head assembly. (Image: Expedition Engineering)

Figure 29.13 Column head installed inside the partially complete canopy. (Image: Expedition Engineering)

29.4 Seismic base isolation

Although seismic base isolation is unusual in Athens, the library and opera-house buildings are supported on seismic base isolators. As well as protecting the building fabric, the base isolation is intended to ensure that these public buildings will be available for community support after an earthquake.

There are several types of base isolation, each with particular advantages. At the SNFCC, the decision was largely driven by the desire to minimise the cost of future maintenance which would be paid for by the Greek state. Pendulum base isolators were selected. However, these can be difficult to understand. In principle, at each support point, the building is supported on a smooth curved surface. During an earthquake, the building slides back and forth on the curved surfaces, with the end result being that the building behaves as though it were suspended on a pendulum.

As we were exploring the merits of different types of isolator, the firm Maurer provided a desk-top-friendly working sample. We built a very simple model to complement the sample, which demonstrates the different behaviour of a rigid platform and a pendulum-supported platform when exposed to manual shaking. Together the sample and model were very effective in explaining to the client and other non-technical members of the team what the seismic base-isolation does, and how it works (Figure 29.14). Laboratory testing of the dynamic performance of the base isolators at EUCENTRE, Pavia and the University of California, San Diego formed an important part of the quality-control process.

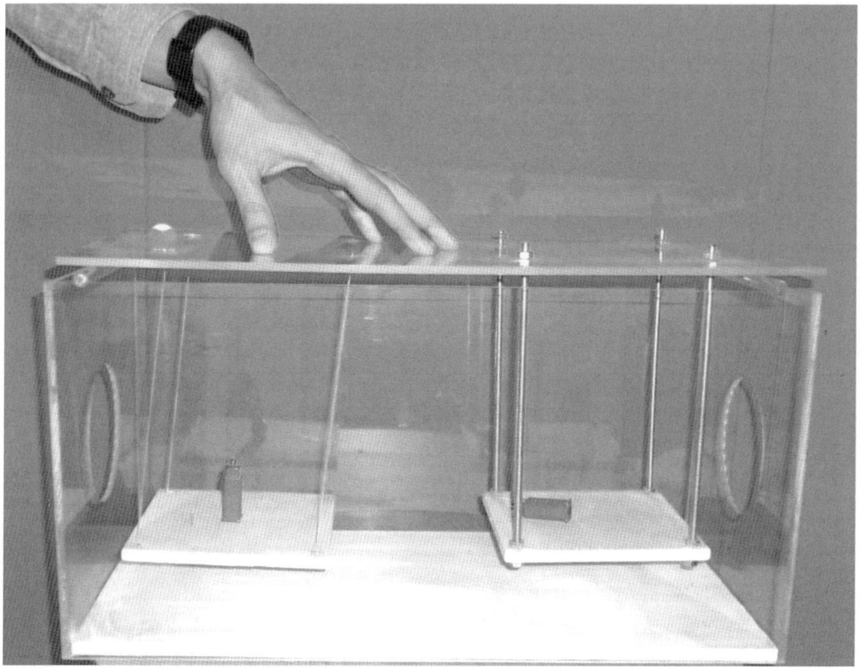

Figure 29.14 Model to demonstrate the performance of pendulum base-isolation. (Image: Expedition Engineering)

Figure 29.15 The model in the wind tunnel. (Image: Expedition Engineering)

29.5 The wind-tunnel model

Computer models do not yet model the real behaviour of the wind as accurately as physical models can. As is common for larger projects, a physical model was built and tested in the wind tunnel at the CSTB laboratory in Nantes to establish the wind loads acting on the canopy (Figure 29.15).

29.6 The flexible mast

Above the canopy is a 40 m feature mast which is designed to flex visibly in the normal daily winds. It is difficult to make a mast sufficiently flexible to deform noticeably in normal winds while also being strong enough to withstand the extreme wind event for which a building must be designed.

A not-very-successful precedent was an important reference for the project. A visit and video clip taken on a windy day helped to communicate the care that would be required to deliver a mast that really does flex visibly in normal winds.

The mast was fabricated by Maxspar at their workshop in Genoa docks. Before completion, the mast was erected in the docks on a light-wind day, with a rope attached to the tip. The team assembled, the mast was watched, and the rope was pulled. While these tests were unscientific, they did demonstrate that the mast would move in light winds and that a relatively small increase in load (applied using the rope) caused significant additional deflection. The trial erection formed an important part in the acceptance procedure for the mast.

29.7 The future use of physical models in structural design

Physical models still inform us in ways that computer models cannot. In fact, physical models continue to do what they have always done – they:

- demonstrate how structures will behave when our designs extend beyond the references and guidance that underlie normal design practice,
- provide information on types of behaviour that are difficult, or currently impossible, to model mathematically,
- help to communicate ideas that are hard to understand, and
- build confidence, particularly among non-technical members of the team, more effectively than abstract mathematical or computer models do.

During the development of the design of the SNFCC, physical models were tested in a laboratory to confirm that ferrocement does perform in accordance with theory extrapolated by an order of magnitude beyond the usual scale of reinforced concrete design. Similarly, the full-scale physical tests of the seismic base-isolators, column-head assemblies and ferrocement prototype each demonstrated the behaviour of aspects of the design that extended beyond the scope of normal design practice and methods.

We used wind-tunnel tests to establish the design wind loads. Physical trials of different mortar mixes and placing methods were used to select working methods that ensured full penetration of mortar through multiple layers of steel meshes. Sample panels identified formwork and trowelling systems that consistently delivered the required finish on the ferrocement. Each of these tests was used to provide information on types of behaviour that are difficult, or currently impossible, to model mathematically.

The various physical models of the column heads, and the pendulum base-isolation system, were used to communicate ideas that were difficult to understand.

The "Jump Test" on a ferrocement panel, the ferrocement prototype, and the trial erection of the flexible mast were each fundamental to achieving confidence across the team that the structures would perform as intended.

Mathematical models, and computer models in particular, enable designers to complete many complex analyses quickly. However, abstract models are limited by the physical references on which they are based. There will always be designs which, to satisfy project requirements, extend beyond the physical references that underlie normal design practice. Designs such as these will often be based on ideas that are novel and, initially, difficult to grasp. In such cases, physical models will continue to play an essential part in developing, testing and communicating structural designs.

Acknowledgements

Thanks are due to the many companies involved in making and testing the physical models used in the SNFCC project. Where Joint Ventures are listed, the name of the sub-consultant or sub-contractor is in italic font.

Client for design and construction: The Stavros Niarchos Foundation. Beneficiary and owner of the completed cultural centre: The Greek Nation. Architect: Renzo Piano Building Workshop + *BETAPLAN*. Project manager: Faithful + Gould. Structural engineer: Expedition Engineering + *OMETE*. Services engineer: Arup + *LDK*. Contractor: Salini Impregilo + *Terna*. Structural engineer to the contractor: Penelis Consulting Engineers S.A. Tensegrity canopy prototype: Cambridge University, Department of Engineering. Ferrocement hands-on workshop: The University of Manchester, School of Mechanical, Aerospace and Civil Engineering. Construction of ferrocement test panels and prototype: Consulteam S.A. Ferrocement laboratory testing: National Technical University of Athens (NTUA), Department of Mechanics. Ferrocement prototype testing: Consulteam S.A. / Salini Impregilo + *Terna*. Column head laboratory testing: Aristotle University of Thessaloniki. Seismic base isolation concept model: Maurer SE. Seismic base isolation laboratory testing: EUCENTER, Fondazione Centro Europeo di Formazione e Ricerca in Ingegneria Sismica, Pavia & University of California, San Diego Department of Structural Engineering. Flexible mast: Maxspar Srl, Genova.

References

1 Jones, K., Martin, B. & Winslow, P. (2017) Innovation in structural engineering – the art of the possible. *The Structural Engineer* 95 (1): 14-21.

2 Iori, T. & Poretti, S. (2013) Pier Luigi Nervi: His Construction System for Shell and Spatial Structures. *Journal of the International Association for Shell and Spatial Structures* 54 (176 & 177): 117-126.

30

Models in the design of complex masonry structures

David Wendland

30.1 Introduction

Masonry, as an assembly of rectangular blocks, at first glance seems to be most appropriate for creating plane vertical walls, which today is its normal employment. However, it also has great potential for building curved surfaces, as it can be seen in many fascinating examples of curved and undulating walls as well as vaulted ceilings and domes, throughout the history of architecture in buildings of sculptural and spatial quality, for instance in the work of Francesco Borromini and Guarino Guarini (Figure 30.1), or the vaults designed by Balthasar Neumann and Antoni Gaudí (Figure 30.2) as well as in innumerable vernacular constructions made of brick or stone masonry.

Curved surfaces are essential in vaults built in masonry, which until the nineteenth century were the most-widely used construction system for durable and fire-proof roofs and ceilings. As these structures work solely in compression, vaults are stable by virtue of their shape and therefore must be curved (Figure 30.3). Depending on the boundary conditions, the surfaces may have a single or double curvature and, following the equilibrium diagram, in many cases their shape has a complex geometry for structural reasons.

The possibility of 'forming' curved surfaces in masonry with a regular bond pattern depends on some geometric rules of the assembly [1, 2]. For instance, the possible radius of curvature is related to the direction of the courses in the masonry. It may therefore be necessary to arrange the bond pattern according to the desired curvature of the surface. The courses are always inscribed in parallel planes: complex surfaces are sometimes divided into smaller portions with different directions of the courses, providing a force-fit connection between these parts. There is also some geometric relation between the direction of such seams and the direction of the courses, which makes it necessary to devise an integral planning procedure where all these aspects can be correlated.

Vaults can be constructed on a formwork that supports the masonry during construction, but there are also traditional techniques that enable bricklaying without such support [1, 3, 4]. In free-handed vault construction (i.e. formwork-free), the possibilities of applying double curvature to the masonry

Physical Models: Their historical and current use in civil and building engineering design,
First Edition. Edited by Bill Addis.

Figure 30.1 Turin, Palazzo Carignano, by Guarino Guarini. The curved surfaces of the façade are built entirely in brick masonry with a regular bond pattern. (Image: David Wendland)

Figure 30.2 Detail of the church in Parque Güell, by Antoni Gaudí. The shaping of the masonry surface by means of curvature and shifting of the courses is clearly visible. (Image: David Wendland)

Figure 30.3 Brick masonry vault at Lierop (NL). The surface of this nineteenth century vault is doubly-curved with a complex geometry. (Image: David Wendland)

shell is not limited by any restrictions from a supporting structure: the surface of the vault can assume doubly-curved shapes that would go beyond the possibilities of, for instance, constructions supported on wooden boards. On the other hand, this procedure imposes some constraints on the shape of the shell: it can be performed only if the masonry is stable at every moment, which is usually achieved by enabling every single course of the masonry to work as an arch in a direction parallel to the beds. This leads to some characteristic consequences for the shape of the vault surface.

In designing complex shapes in masonry, and in particular the shape of vaults built without formwork, it is therefore essential to control the geometric rules for the assembly of blocks and joints, while also meeting the requirements of the self-supporting masonry texture. For performing this task, the use of physical models turns out to be highly advantageous: models that reproduce the masonry texture, facilitate the control of the spatial arrangement of the courses, and even simulate the process of bricklaying. By the latter, it is possible to anticipate the arrangement of self-supporting courses in free-handed vault construction and perform the self-formation process that shapes the vault surface – a concept that was developed in the 1980s by Frei Otto and his colleagues at the University of Stuttgart [5].

Beyond the specific issue of masonry, physical models in general have great benefits in designing architecture with complex shapes. The use of models is an interactive process of working on an object that is located in our environment and therefore immediately accessible to our spatial perception [6], as we shall discuss later. Hence, physical models offer excellent possibilities for clarifying spatial relations, and for designing, inventing and controlling spatial structures with complex shapes.

In historical and traditional vault construction, the interrelation between the layout of the masonry texture and the possible curvature, as well as the need for stable courses during construction without formwork, is only rarely described in the historical technical literature. Nevertheless, it is essential for reproducing or even understanding these structures: today, only a few specialists are still familiar with the art of vault construction. It is extremely difficult, if not impossible, to explain by means of architectural drawings the principles underlying the shape of vaults in regular masonry bond and with self-supporting courses, given the complex geometrical relations and also to the properties of free-form shapes (Figure 30.4). On the other hand, the geometric principles of curved masonry surfaces can be easily described in physical models that simulate the masonry texture and also reproduce the construction process. Such models have turned out also to be very useful in research on historical and traditional vault construction.

Physical models are also of use in the case of Gothic vaults, where the geometric description of the shell surface is rather complex. This is due to the fundamental

Figure 30.4 A nineteenth-century traditional brick vault built without formwork, Geldrop (NL) The shape can assume high geometric complexity and the free-form is the outcome of a 'self-formation' governed also by the construction procedure. (Image: David Wendland)

principle of Gothic vault design, according to which the curves of the arches and groins do not result from the intersection of defined surfaces, but are predetermined as a primary feature – as autonomous curves with a very simple geometric definition, usually describing circle segments inscribed in vertical planes. The surfaces of the webs must adapt to these given curves, and in consequence their shapes cannot be defined in terms of elementary geometric objects.

In addition to these boundary constraints, the shape of the surface between the arches is conditioned by the curvature within the masonry texture. Along the courses, a certain curvature is necessary to provide the stability of the single courses during bricklaying; in this direction, it is possible to produce a rather small radius of curvature. Whereas, in the direction perpendicular to the courses, the possible curvature is reduced: the courses are inscribed in parallel planes due to the uniform height of the blocks, so that the curvature is limited by the possibility of shifting the courses relative to each other. Radial rotation of the courses is possible only if the courses are straight or have a very small rise. Therefore, in doubly-curved masonry surfaces, rotating bed joints can be achieved only in very narrow sections, for instance near the springing of the vault. Furthermore, in order to create a continuous bond over a large portion of the vault, especially in the areas where forces concentrate, it is advisable to arrange the courses more or less perpendicular to the plane of the cross arch.

In simulations carried out using physical models (Figure 30.5), the arches of the ribs and groins could be easily represented by cardboard planes. In between, the geometrical arrangements of the masonry bond are reproduced by means of balsa blocks glued together. In these models, made to a scale of approximately 1:20, the blocks representing the units are often larger, as the exact size of the units has little influence on the principal character of the shape of the shell and on the process

Figure 30.5 1:20 scale model simulation of the masonry texture in a vault built without formwork (free-handed). The characteristic shape, visible in the curvature of the ridge, is reproduced as a direct consequence of the geometrical rules according to which the bricks (each 10 x 5 x 2 mm) were laid. (Image: María José Ventas Sierra)

of self-generation. The bricklaying can be recreated, course by course, from the base upwards, assessing the effects of the rules mentioned above according to the chosen inclination of the bed-joint planes. Alternatively, in many cases it is more useful to assume at the outset the spatial position of the courses in an upper portion, for instance where the longest courses occur, and then extrapolate downwards to clarify which is the best inclination of the courses at the beginning – such bricklaying from the top to the bottom is obviously possible only in the model. In this way, it is possible to verify whether, with the attempted solution for the portion above the mid-point, the masonry texture can be extended in a regular manner to the upper and lower ends of the shell surface without exceeding the limits of possible curvature across the courses.

This same procedure could also be used to better understand the masonry texture and shape of other complex masonry structures, such as the late-Gothic 'diamond vaults' which were quite common during the fifteenth and sixteenth centuries in some regions of central Europe (Figure 30.6). These structures are characterized by the procedures of free-handed vaulting and the texture of rather thick masonry surfaces. Their construction procedure was not documented and could be investigated only by means of surveys of existing structures. The evidence from the built objects, and the procedures with which they were created, could be understood only by reproducing their structural texture in models and, after that, in full-scale experiments [7].

Hence, in the attempt at recovering historical technical knowledge, simulation models can play a key role in the research. By their use, it is possible to reproduce and understand the formation process of the masonry surfaces, and to develop geometric or parametric descriptions of the masonry shell that can also be used in design processes and for numerical modelling.

Figure 30.6 1:20 scale model simulation of the masonry texture of a late-Gothic 'diamond vault'. (Image: María José Ventas Sierra)

30.2 Modelling traditional Iranian vaults

One type of vault which is common in vernacular Persian architecture is built on a rectangular plan with arched courses (squinches) from the four corners, without using any formwork or centring – the bricklaying is carried out completely free-handed [8, 9] (Figure 30.7). In the absence of any geometrically determined supporting elements, the shape of these vaults is completely dependent on the inherent geometrical rules of the masonry fabric, and predominantly determined by the requirement of stability in every single course during the construction process.

The geometry of the shape is more complex that it seems, but the self-formation process that generates the shape can be easily simulated in physical models (Figure 30.8). In such models that reproduce the geometrical rules of the masonry fabric and the procedure of building with stable courses, it has been possible also to identify the boundary conditions that determine the final shape, such as the inclination chosen for the planes in which the single courses of the masonry are inscribed. By this simulation, it is possible to better understand the relationship between the construction system and the architectural shape of these buildings, and precisely describe the building procedure which we can consider an intangible aspect of cultural heritage. Moreover, it can be applied directly in the conservation of the built heritage.

After the disastrous earthquake in 2003 that destroyed the historical buildings in Bam Citadel, Iran, a number of studies were carried out into how these traditional vaults could be reconstructed using the original procedure, while at the same time predicated on modern structural design methods based on numerical

Figure 30.7 Construction of a traditional Iranian vault built by extending squinches to meet in the centre. (Source: [10] By permission of the Oxford University Exploration Society)

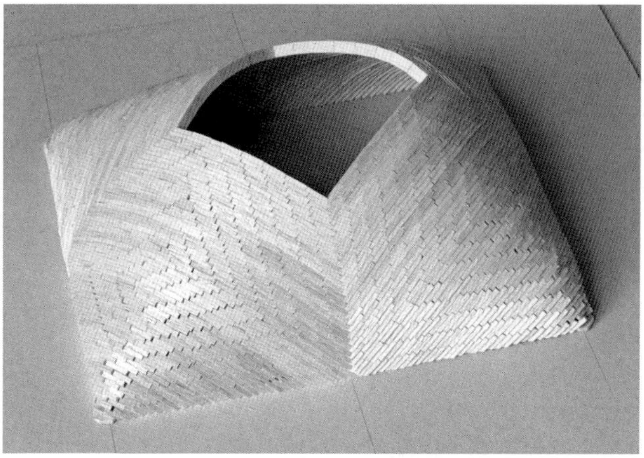

Figure 30.8 1:20 scale model simulation of a traditional Iranian dome built with self-supporting courses. (Image: María José Ventas Sierra)

modelling of the mechanics of the vault performed before the execution. Hence, the shape of the vaults had to be anticipated, predicting the outcome of what the masons would do spontaneously in the vernacular building practice without following any design procedure. This shape had to be described in a way that could be processed within the software environment for Finite Element Modelling.

This task was carried out by developing a procedural surface model using physical models and thus, by reverse geometric engineering – forming geometric features from a survey of the digital model – creating a parametrically-defined surface (Figure 30.9). Although the geometry is complex, because the curved shape has a rectangular plan at the base while the horizontal sections in the upper part are round (causing edges that disappear), the CAD surface based on NURBS (non-uniform rational base-splines) geometry can be easily handled, and

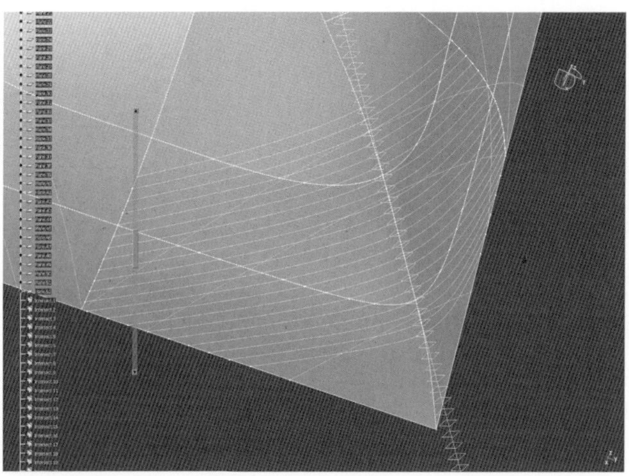

Figure 30.9 Parametric surface model based on the reverse geometric engineering of the physical model, and on descriptions of the building process. (Image: David Wendland)

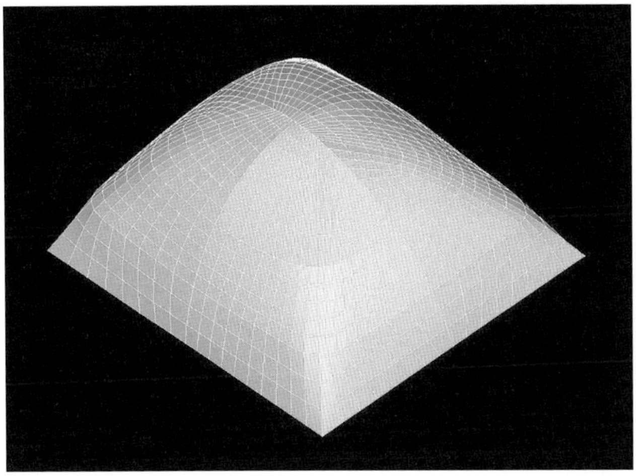

Figure 30.10 Meshed model of the vault for FEM. (Image: Mahmoud Nejati)

the mesh applied with great freedom by the operator (Figure 30.10), optimizing the possibilities of numerical modelling.

A similar procedure was carried out for other vaults in the same building at Bam Citadel, which can be described as barrel vaults between transverse arches. The arches are designed as pointed arches defined by several circle segments according to historical design procedures. They were constructed on formwork that predefines their geometry, either *in situ* or by making a very slender layer of the arch on the floor, using gypsum to bond the bricks, and then raising this arch into position where it served as permanent formwork for the remaining courses of brickwork. The shells between the arches are built free-handed with courses that are stable due to their shape, forming single arches that are inscribed between parallel inclined planes (Figures 30.11 and 30.12). Hence, these surfaces cannot be described in elementary geometric terms, but as free-form surfaces determined by the assembly rules of the brick masonry and the procedures of spontaneous bricklaying. For the architectural and structural design and, in particular, for obtaining the surfaces to be captured by numerical modelling, the surfaces were reproduced in model simulations, the result of which was then modelled through reverse geometric engineering (Figures 30.13 and 30.14).

30.3 Experimental construction of a free-form shell structure in masonry

In order to demonstrate once again the fascinating 'other side' of masonry – beyond the plane wall – an experimental free-form shell structure was developed and built in 2004-2007 at the *Technische Universität* Dresden [11]: the structure was entirely designed, planned and built by participating students at the School of Architecture (Figures 30.15 and 30.16). The aim of this experiment was to investigate the feasibility of a shape of high geometric complexity in masonry, using

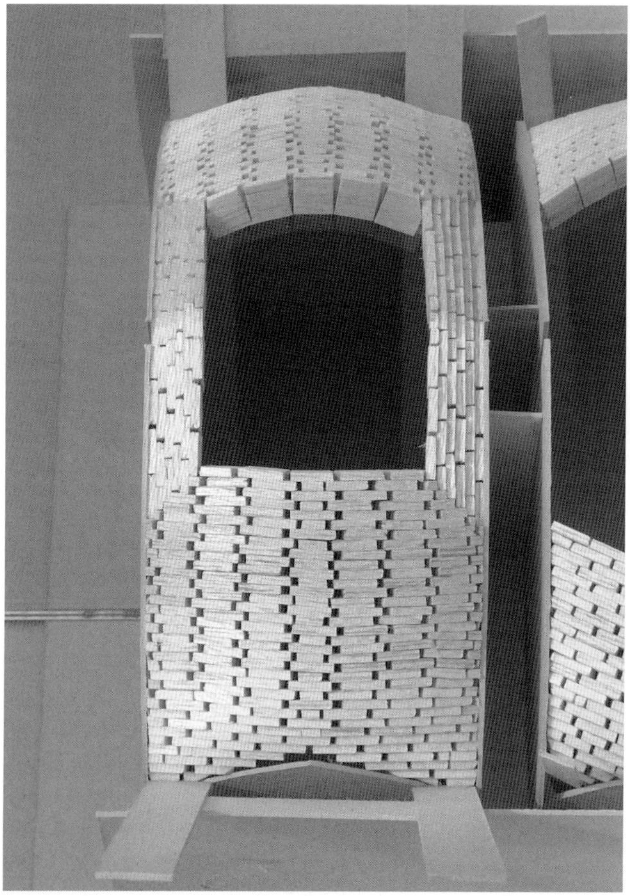

Figure 30.11 Model for a second vaulted ceiling at Bam Citadel, with flat shells between transverse arches. (Image: David Wendland)

current materials while remaining within the scope of the building codes. For realization of the sculptural shape, no formwork was needed because the brick masonry could be built with self-supporting courses: here, for the first time, traditional vault construction was implemented for an architecture project in the digital age.

In this project, the potentials of both physical and digital modelling were combined in an integrated design process. Numerical modelling enabled the structural analysis of shapes with complex geometry; information technology provides the means of realizing such surfaces with high fidelity. The development of the shape, on the other hand, was predominantly carried out using physical models. The design of the shape is the outcome of a sculptural process in interaction with the material (Figure 30.17); and the solution for the curved masonry surface and the layout of the masonry bond was performed with physical models (Figure 30.18). A core issue of this design process was therefore the integration of the different media in a continuous design process, which is essential for ensuring

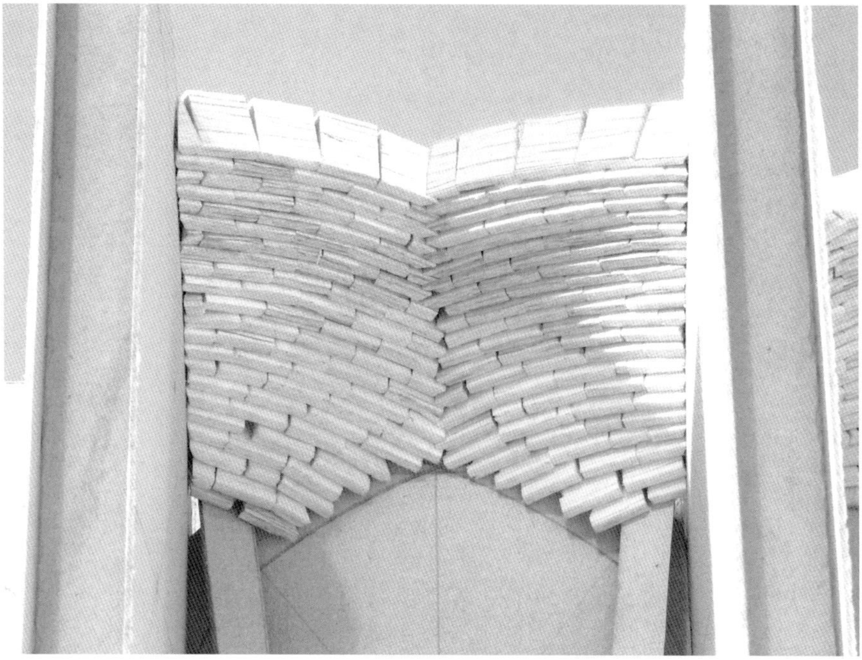

Figure 30.12 Alternative configuration of bricks proposed for the vault shown in Figure 30.11, which facilitates construction at the squinches and can easily be built free-handed. (Image: David Wendland)

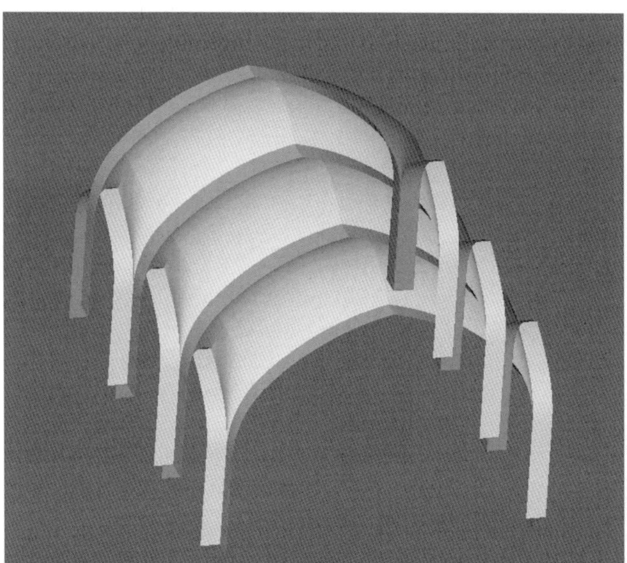

Figure 30.13 CAD model created as the basis for numerical modelling. (Image: David Wendland)

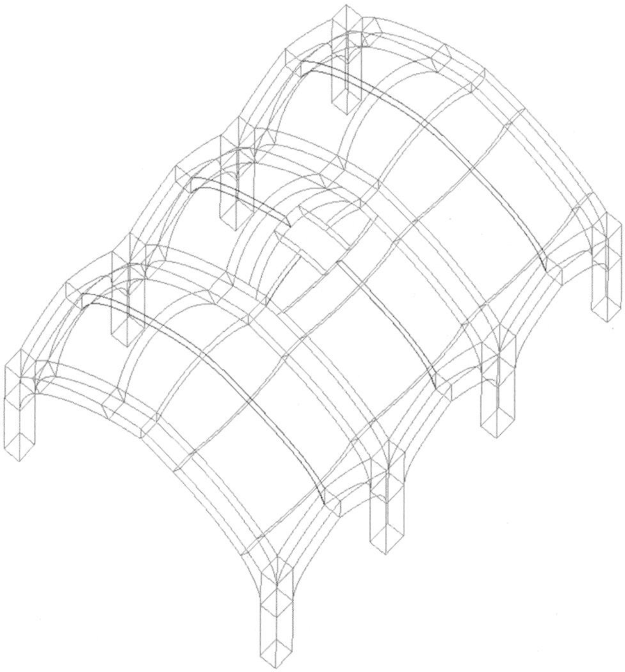

Figure 30.14 View of the CAD model of the vault from above, showing only the edges of the single NURBS surfaces. (Image: David Wendland)

Figure 30.15 Free-form shell structure made of unreinforced brick masonry, TU Dresden, 2016. (Image: María José Ventas Sierra)

safety, quality and serviceability of a building, while maintaining the aesthetic qualities of the original design.

The richness of shapes that can arise from the sculptural working process on physical models is beyond comparison and can result from any kind of forming process. This can involve human interaction with the material, as has always been

Figure 30.16 The masonry texture is arranged according to the geometric properties of curved masonry with a regular bond. (Image: María José Ventas Sierra)

Figure 30.17 The shape of the building is developed in a sculptural working process using Plasticine models. (Image: David Wendland)

the case for sculptors working with clay, plaster or other materials to be deformed by hand or with tools; or it can involve aleatoric processes, natural and physical processes such as melting and solidifying or erosion. It can also involve the formation of minimal surfaces and equilibrium processes which have often been undertaken for shape-optimization of structures [5] and which, in the

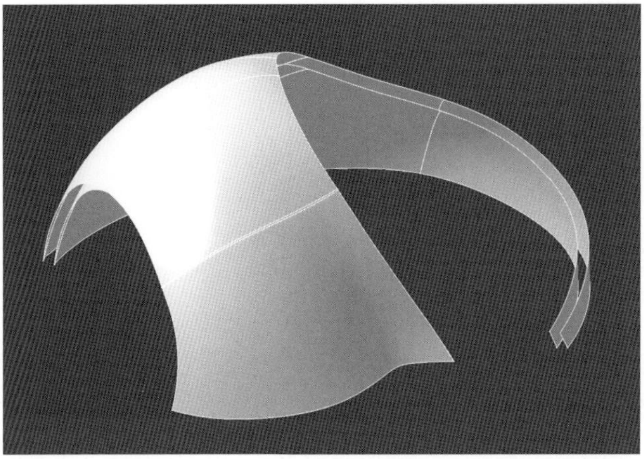

Figure 30.18 CAD model obtained by reverse geometric engineering from a survey of the physical models. The NURBS surfaces correspond well with the designed shape and can be transferred to the software tools for numerical modelling. (Image: David Wendland)

context of contemporary artistic concepts, we can also consider to be sculptural processes [12].

The other outstanding feature of physical models is the intuitive control of the shape – the immediate and total perception which may also combine with the possibility of interaction. This notion has been expressed by many designers including Antonio Gaudí, Heinz Isler and Frank Gehry. It also corresponds with empirical studies on the psychology of perception [6] which have shown the fundamental difference between the perception of an image, in terms of projection, and the direct perception of an object in our environment. The perception of our surroundings and of the objects in it, is not at all a passive attitude – it is an activity, both mental and physical, that couples the movement and our notion of it with the flow of the changing optical array, extracting the invariants by processing the information about both the environment and the self: Gibson speaks of 'kinaesthetic' and of 'ambulatory vision' [6]. We must therefore consider any optical representation of an object and its shape, even if dynamically generated by the interaction of the operator like in a CAD interface, as not being part of our surrounding environment, but rather, as a codification of reality which must again be decoded. Only the real object within our immediate reach is accessible to our mind through our activity of perception. Moreover, the real object gives the opportunity of connecting haptic and visual sensing, touching the surface, perhaps feeling with the hand the changing gradients of its curvature. Physical models can therefore be considered as an optimal tool, especially for reconnecting to the original intention of the design.

In order to exploit the advantages of physical models in the design process, it is essential to overcome their deficiency in not providing the exact description of the shape needed for reproduction at large scale such as a building [13, 14]. In fact, such a model cannot be scaled without transforming it; for instance, any error or inaccuracy will be also enlarged by the same scale factor, and the change of the object size will alter the relation between shape and texture [6, 11]. The

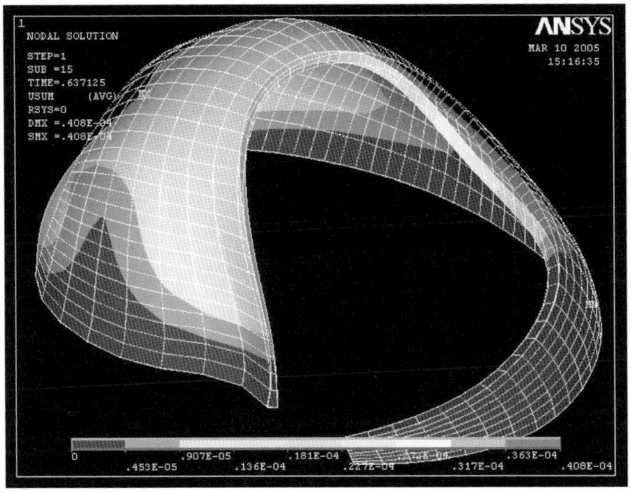

Figure 30.19 Numerical modelling (FEM) of the structural behaviour, showing the structural problems in the initial design. (Image: Mahmoud Nejati, David Wendland)

production of a digital model, therefore, cannot be conceived of as an automated process or merely as a completion. Rather, reverse geometric engineering is a translation process that is both intelligent and creative: going beyond the geometric survey of the physical model on which it is based, it consists in a reinterpretation of the shape.

In this process of reinterpretation or re-modelling, there is also room for considering the function of the digital model – for instance structural modelling. In the particular case of the brick shell, this regarded the disposition of smooth NURBS surfaces (Figure 30.19) that could be easily handled within the software environment for numerical modelling (Figure 30.20), which imposed little or no constraint for applying meshes, and above all, could easily be modified. In one phase of the design process, the shape of the shell was gradually modified to improve its mechanical behaviour, changing the shape carefully and assessing the effects of these changes by repeated trials of numerical modelling (Figure 30.21).

In spite of being conceived of as a shell structure built in masonry, subjected only to compression forces, in which the structural behaviour is therefore intrinsically connected with the shape, the shape of the building in this case was not determined *a priori* by means of a shape in statical equilibrium. The aim was to find a 'third way', between the conception of optimized shell shapes where the shape is determined by the structural behaviour, and the common approach in which structural stiffness is used to relegate the geometric form to having only secondary importance in determining the structural behaviour. It was therefore attempted to develop a shape according to several criteria, in which structural form was not the primary driver, but was still an influence on obtaining a reasonable shell structure – seeking to exploit the range of possibilities for the design of shell structures, within the limits given by the intrinsic interrelation of shape and structural behaviour in this type of structure. The sculptural process in the first stage was therefore accompanied by building simple funicular models,

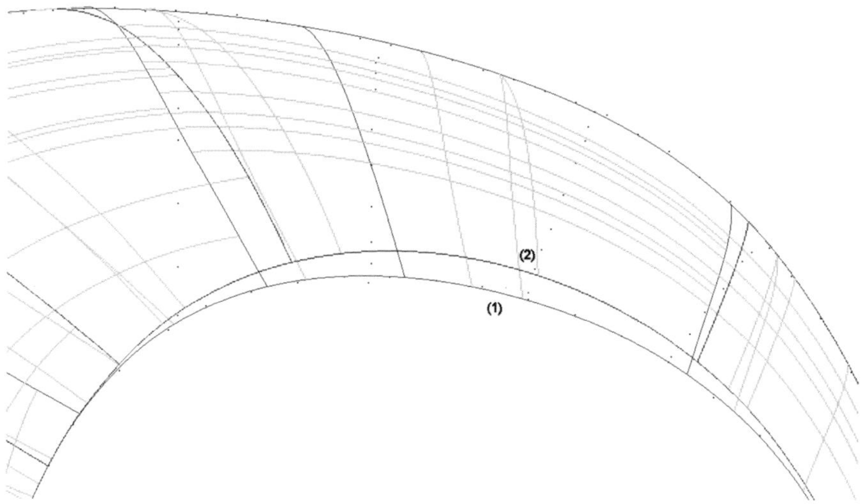

Figure 30.20 Alterations of the shape to improve the structural behaviour were carried out first on the CAD model, and their effect immediately verified in the FEM environment. (Image: David Wendland)

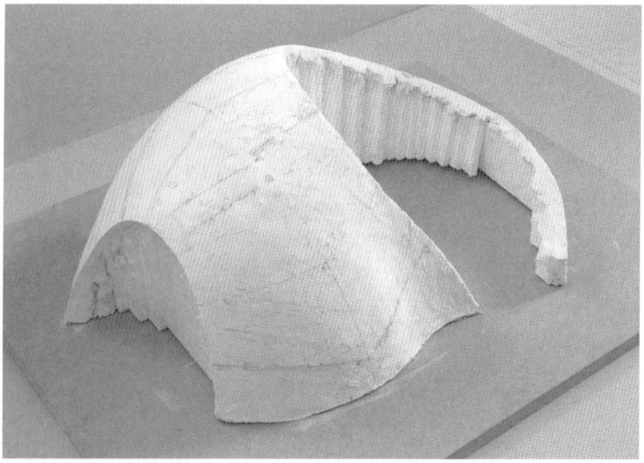

Figure 30.21 Physical model in expanded polystyrene with a plaster coating, produced from the CAD data after the modifications suggested by the results of the numerical modelling. The final decisions and changes regarding the shape were carried out using this model. (Image: David Wendland)

while developing a convincing design solution for the shape from the aesthetic point of view – rigorously working on physical models that allowed immediate interaction with the material.

After modifying the shape step-by-step in the CAD model, based on the outcome of numerical modelling of the mechanical behaviour, the modified shape was once again translated into a physical model (Figure 30.22). This gave the opportunity for a final assessment of the alterations that took place throughout

Figure 30.22 1:20 scale model for determining the best solution for the masonry texture (Image: David Wendland).

Figure 30.23 Execution of the shell, showing details of the centring, the geometric control during bricklaying, and the preparation of the seam. (Image: David Wendland)

the process. Hence, the last word on the shape of the building was spoken by means of a physical model.

The bonding of the brickwork was also developed using a physical model, studying possible solutions in correspondence with the geometrical rules of the masonry texture. It would have been impossible to provide a uniform pattern of the courses able to follow the curvature throughout the entire surface. In order to realize the curvature of the shell, the surface had to be divided into smaller portions, where the courses ran in different directions according to the degree of local curvature in the surface. A force-fit connection between the different portions was provided by seams in the masonry; this detailing is common in traditional vault construction and can be seen especially in brick vaults from the nineteenth century [1, 4] (Figures 30.16, 30.23 and 30.24) (cf. Figure 30.3). In the execution, a centring consisting of two-dimensional arches was built, based on the information from the CAD model. As in traditional vault construction, no continuous support for the shell surfaces was needed. The courses are stable and the shape was interpolated by means of flexible bars. For developing the spatial development of the vault, organizing the seams in a reasonable pattern, and assessing the relationship between bond pattern and curvature in every portion of the shell, the physical model built with balsa blocks turned out to be a perfect design tool. The experimental building produced in a university teaching context demonstrated both the potential of physical models integrated in a design process and their role in creating new curved masonry surfaces.

Figure 30.24 Execution of the shell: bricklaying in the portions with inverse double curvature. (Image: David Wendland)

30.4 Reconstruction of the vault in the Chapel of Dresden Castle: a masonry structure with complex geometry

The experience gained in creating the experimental brick shell, concerning the possibilities of forming curved surfaces in masonry with regular bond patterns, as well as the potential of physical models, was further exploited in the design of a large vaulted ceiling built in brick masonry, for the reconstruction of the vault in the chapel of Dresden Castle which was completed in 2013 [15] (Figure 30.25).

The chapel is located in the northern wing of the palace and was built in the mid-sixteenth century as one of the first protestant churches in Germany. As far as is known, it originally had a vaulted ceiling with looping ribs: as in other examples in Saxony and Bohemia, the stone ribs of the vault described curves both in plan and elevation. In the eighteenth century, the chapel was closed after the court had converted to the Catholic faith. The building was then subject to major alterations and the vault was demolished. After the partial destruction of the palace in World War II, the more stable parts of the perimeter walls were preserved, the pillars consolidated with concrete and, in the 1980s, the external envelope of the building was reconstructed. The works carried out in 2011-2013 were aimed at reconstructing the vaulted ceiling of the former chapel and rehabilitating the parade rooms located above on the second floor.

The original ceiling is only poorly documented: some preserved small fragments of the ribs and an interior view of the chapel in an etching from the

Figure 30.25 The reconstructed brick vault in the chapel of Dresden Castle, 2013. (Image: María José Ventas Sierra)

seventeenth century give little evidence about the vault itself, but confirm that it had curvilinear ribs. More information on how the original vault must have looked and especially on the structure and construction details, was obtained from the study of comparable buildings [16, 17]. The ceiling of the reconstructed chapel cannot therefore be considered as a replica of the original vault, but rather as a new construction that recreates the design process and the construction system, using original materials including sandstone ribs, hand-made bricks and lime mortar. Like the original examples, it is conceived as a self-supporting shell structure in unreinforced masonry [18].

In vaults with looping ribs, no major structural role can be assigned to the ribs which, because of their spatial curvature, cannot take imposed normal forces. The main load-bearing element of such constructions is therefore the masonry shell above the ribs. In the historical examples, and also in the new vault in Dresden Castle, the shape of this masonry shell has a complex geometry.

On the one hand, the shape of the vault surface is conditioned by the curves of the ribs. In general, in late-Gothic vaults, the curves of the ribs are the primary feature of the architectural design and were defined as autonomous curves to which the vault has to adapt. On the other hand, the shape is also determined by structural requirements, as the stability of the shell is due solely to its shape. This second condition sometimes leads to a separation between the vault surface and the curves of the ribs. As in much historical and traditional vault construction, the realization of the complex geometry of the masonry surface was made possible by avoiding the need for formwork, by using free-handed vaulting.

In the design process for the vault reconstruction, the spatial curve system of the ribs was determined first, according to historical design schemes [14]. On this basis, the construction system of the masonry structure was developed. At first, the possibilities of creating membrane surfaces generated to fit the given curve of the ribs were explored.

In the design of the masonry shell, the shape and brickwork bonding texture had to be developed as an intrinsic interrelation. Moreover, the arrangement of the brick courses was conditioned by the local curvature of the vault surface, the requirement of free-handed vaulting and, in this case, also by the detailing of the connections between the vault masonry and the stone ribs. All this was achieved by working with scale models where the masonry texture could be simulated and the complex geometrical boundary conditions could be inspected (Figure 30.26). In the model, different solutions were proposed and discussed with the architects and the structural designers (Figure 30.27). In some cases, the spatial position of the courses had to be determined in a higher part of the vault, and the lower portions developed from there (Figure 30.28). Hence, beyond simulation, the model served also as a direct means of communication, enabling immediate visual control. During intensive sessions gathered around the model, the different solutions were verified against the requirements for the stability of the masonry shell, and the appearance and aesthetic qualities in regard to the design preferences of the architectural project. The materials used – cardboard and balsa blocks assembled using non-permanent glue – enabled rapid changes and the development of trials for alternative solutions.

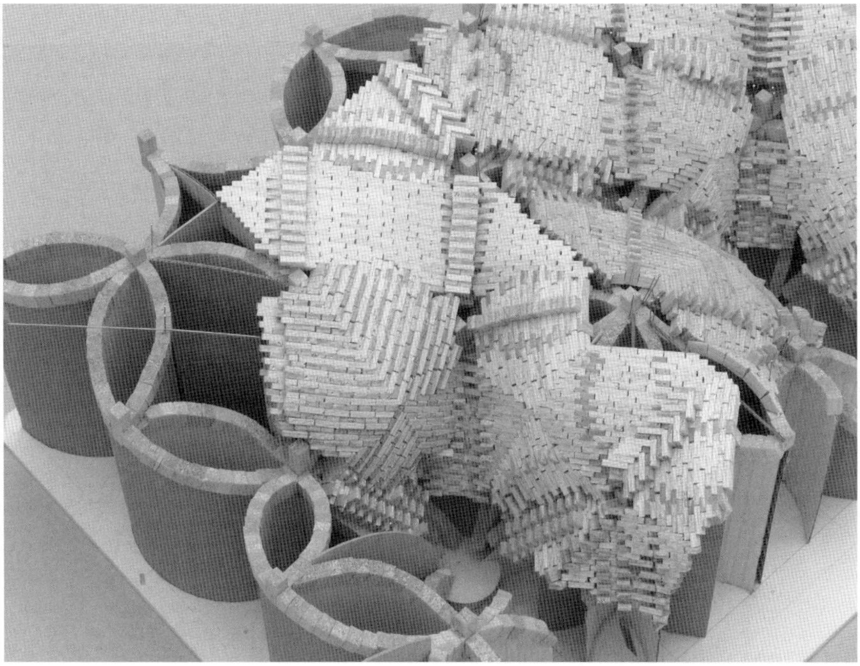

Figure 30.26 1:20 scale model simulating the masonry texture in the shell structure of the vault. (Image: María José Ventas Sierra)

Figure 30.27 In some cases, the spatial position of the courses must be determined in a higher portion of the vault, and the texture of the lower portions derived from there. (Image: María José Ventas Sierra)

Figure 30.28 Detail of the model, showing thin wires used to control the continuity of the surface flow. (Image: María José Ventas Sierra)

Figure 30.29 Detail of the model in an advanced stage of the simulation of the masonry. (Image: María José Ventas Sierra)

For structural verification and for the elaboration of the building specifications, the outcome of the process carried out on the model also had to be translated into a digital model. This could be done by using the curves of the ribs that had been modelled in CAD from the beginning, and according to the shape control used in the last stage of the model simulation, which consisted of wires to ensure a smooth surface flow from the summit to the abutment (Figure 30.29). The latter was also used for orientation in creating Bezier surfaces. Hence, a digital surface model was created that could be compared with membrane surfaces in order to assess the stability, and which helped identification of plane sections that could be used in order to produce centring arches.

During the execution of the vault, the physical model was also used to explain the complex masonry texture to the bricklayers (Figures 30.30 and 30.31). As the

Figure 30.30 Realization of the brick masonry: detail of the connection between a thick portion of the shell, stone rib that remain below the shell surface, and reinforcing masonry ribs on the extrados of the shell. (Image: David Wendland)

Figure 30.31 View from below onto the shell masonry, with all centring arches still in place. (Image: David Wendland)

vault was built without formwork, the shape had to be carefully controlled during execution, in order to maintain the validity of the structural design. For this, centring arches were mounted (Figures 30.32 and 30.33), and between them the same geometric guidance system was used as had been developed when building the physical model. In place of wires, the bricklayers used thin wooden rods. This was further confirmation of the usefulness of simulating the building process at reduced scale in such constructions, for instance, to explain the detailing of the masonry in the vertical diaphragms between the lower parts of the ribs and the shell, and their force-fit connection with the load-bearing shell (Figure 30.34). Overall, the success of this approach was due to the integration of the physical model into the design process, based upon the techniques developed for the experimental brick shell structure at the TU Dresden.

Figure 30.32 Free-handed vaulting over single centrings, the curves of which were obtained from the CAD model. (Image: David Wendland)

Figure 30.33 Geometric control during bricklaying, in accordance to the wires in the physical model. (Image: David Wendland)

30.5 Conclusion

In many aspects of architectural and engineering design, physical models have today been replaced by numerical modelling and three-dimensional CAD modelling – digital tools that have great benefits in providing direct information flow across all stages of the design process. Nevertheless, physical models maintain some outstanding qualities that make them ideally-complementary to other design tools – and which, also thanks to information technology, can be integrated into the design process in order to exploit their potentials fully. The unique quality of physical models lies in their immediate accessibility for our spatial perception and imagination. This is significant in sculptural working processes, and of great benefit for the communication between the people

Figure 30.34 The vertical diaphragms in the lower portions of the vault connect the shell with the lower ribs by means of the masonry bond. (Image: María José Ventas Sierra)

involved in the design; in the case studies presented, this was especially evident in the discussions between architects and engineers, and in the instruction of the bricklayers. Beyond that, the potential of physical modelling can be exploited in design tasks that require total mastery of spatial interrelations. This can be the case in many traditional and historical structures that we wish to maintain by repair and integration, and whose design processes also constitute a valuable asset as intangible cultural heritage. This applies in particular to masonry structures with complex geometry which, as we have learnt, represents masonry construction at its best.

Acknowledgements

The models simulating the masonry texture of historical vaults, as well as the model of the masonry shell for the reconstructed vault in Dresden Castle, were built by María José Ventas Sierra and the author.

The reconstruction of the shape of traditional Iranian barrel vaults at Bam Citadel (Iran) was carried out by the author in the preparation of the pilot project 'Sistani House'. At Wolfram Jäger Consulting Engineers, the FEM modelling of the Iranian squinch dome was by Tammam Bakeer.

The experimental shell structure at the TU Dresden was developed and realized in 2004-06, based on a concept by the author within several courses in the Faculty of Architecture, the Chair of Building Construction and Conservation (Prof. C. Schulten) and the Chair of Structural Design (Prof. Dr.-Ing. W. Jäger). The courses were held by Florian Schneider, Wolfgang Kurtz, and the author. The FEM within the iterative design process was carried out by Mahmoud Nejati.

The reconstruction of the chapel in the Dresden Castle is a public construction project performed by *Staatsbetrieb Sächsisches Immobilien- und Baumanagement* (SIB), Dresden (Germany). The architectural design of the reconstructed vault, completed in 2013, is by J.-U. Anwand Architekten, the structural design by KUS Ingenieurbüro für Baustatik und Tragwerksplanung. The author was involved in this project as scientific consultant and collaborated in the conception and design of the shell structure in masonry and in the supervision of its construction.

References

1 Wendland, D. (2008) *Lassaulx und der Gewölbebau mit selbsttragenden Mauerschichten. Neumittelalterliche Architektur um 1825-1848.* Petersberg: Michael Imhof.

2 Wendland, D. (2010) Zur Herstellung gekrümmter Flächen aus Mauerwerk. In: *Mauerwerksbau Aktuell – Praxishandbuch 2011* (eds K.-J. Schneider, G. Sahner & R. Rast), 110-124. Berlin: Bauwerk Verlag.

3 Wendland, D. (2004) Some considerations on the shape of the caps of vaults. In: *Proceedings 4th International Seminar 'Structural analysis of historical structures', Padova* (ed. C. Modena, P. Lourenço & Roca), 111-120. Leiden: Balkema.

4 Wendland, D. (2007) Traditional vault construction without formwork: Masonry pattern and vault shape in the historical technical literature and in experimental studies. *International Journal of Architectural Heritage* 1 (4): 311-365.

5 Gaß, S. (1990) Experimente / Experiments: Form, Kraft, Masse, 5. *Mitteilungen des Instituts für Leichte Flächentragwerke, Vol. 25.* Stuttgart: Krämer.

6 Gibson, J.J. (1979) *The Ecological Approach to Visual Perception.* Boston: Houghton Mifflin.

7 Wendland, D. (2014) Reverse Engineering und Experimentelle Archäologie – Forschungen zu Bau, Planungsprinzipien und Entwurfskriterien spätgotischer Zellengewölbe. In: K. Schröck and D. Wendland (ed.): *Traces of Making. Entwurfsprinzipien von spätgotischen Gewölben / Shape, Design and Construction of Late Gothic Vaults,* Petersberg, 10-37.

8 Besenval, R. (1984) *Technologie de la voûte dans l'orient ancien.* Paris, Éditions Recherche sur les Civilisations.

9 Zamarshidi, H. (1998) *Mi'mari-i Iran: ijra-yi sakhtiman ba masalih-i sunnati.* Tehran: Intisharat-i Zumurrud.

10 Ward-Perkins, J.B. (1958) Notes on the structure and building methods of early Byzantine Architecture. In: *The Great Palace of the Byzantine emperors. Second Report.* (Ed. D. Talbot Rice), 52-104. Edinburgh: Walker Trust.

11 Wendland, D. (2009) Experimental Construction of a Free-Form Shell Structure in Masonry. *International Journal of Space Structures* 24 (1): 1-11.

12 Wendland, D. (2003) Model-based form-finding processes: 'Free forms' in structural and architectural design. In: *Eduardo Torroja – From the philosophy of structure to the art and science of building* (eds F. Levi, M. Chiorino & C. Bertolini), 104-119. Milano: Franco Angeli. Also available at http://elib.uni-stuttgart.de/opus/volltexte/2001/761/

13 Burkhardt, Berthold (ed.) (1978) *Multihalle Mannheim*. Mitteilungen des Instituts für leichte Flächentragwerke. IL 13. Universität Stuttgart.

14 Linkwitz, K. (1995) Two examples of integrated formfinding and numerically controlled premanufacturing: The timber shells of Bad Neuenahr 1993 and Maulbronn 1995. In: *Spatial Structures: Heritage, Present and Future. Proceedings of the IASS International Symposium*, Milano 1995, 1397-1407.

15 Wendland, D. & Ventas Sierra, M. J. (2013) Designing a Masonry Shell in the Reconstructed Vault of the Palace Chapel at Dresden – An Attempt of Recovering the Forgotten Art of Late Gothic Vault Construction. *Informes de la construcción* 65 (No. Extra 2: *Construcción Histórica en Piedra / Historic Stone Construction*) 49-63.

16 Wendland, D. (2012) Arches and Spirals – The Geometrical Concept of the Curvilinear Rib Vault in the Albrechtsburg at Meissen and Some Considerations on the Construction of Late-Gothic Vaults with Double-Curved Ribs. In: *Nuts & Bolts of Construction History: Culture, Technology and Society* (eds by R. Carvais et al.), 351-357. Paris: Picard.

17 Wendland, D. (2019) *Steinerne Ranken, wunderbare Maschinen: Entwurf und Planung spätgotischer Gewölbe und ihrer Einzelteile.* With contributions by M. J. Ventas Sierra, M. Aranda Alonso, and A. Kobe. Petersberg: Michael Imhof.

18 Wendland, D., Anwand, J.-U. & Kröning, M. (2013) Das Kappenmauerwerk: Entwicklung einer geometrisch komplexen gemauerten Schalenkonstruktion für ein spätgotisches Schlingrippengewölbe In: *Das Schlingrippengewölbe der Schlosskapelle Dresden* (ed. Sälchsisches Staatsministerium der Finanzen), 108-118. Altenburg: Klaus-Jürgen Kamprad.

31

Physical modelling of free surface water – current practice

James Sutherland

31.1 Introduction

Laboratory-based physical modelling has been the most widely-accepted method of undertaking hydraulic modelling for many years [1]. The steady rise of computing power has meant that this position is changing [2]. 2-D (depth-averaged) numerical flow models are being replaced by more detailed 3-D flow models or even computational fluid dynamics (CFD) models.

For example, back in the 1970s the modelling of tides and sediment transport in an estuary or large river would normally have been undertaken using a physical model. Today, such studies are almost always conducted using numerical models – with the exception of some very large-scale models. As another example, wave propagation studies in shallow water are now also undertaken using numerical models, where previously they were done using physical models.

However, physical modelling has been evolving alongside computational modelling. The main reason for this is that Moore's law (which states that computer power doubles about every 18 months) also applies to the equipment used to generate the input conditions to a physical model and to measure the resulting flows, sediment transport and bathymetry. The increases in computer power, the development of Ethernet technologies for transferring data and the shrinking of electronics mean that more and more detailed measurements at finer resolution and higher frequencies can be measured in physical models and in the field [3, 4]. This has allowed more detailed investigations of increasingly complex phenomena to be undertaken. This trend is expected to continue and new measuring techniques are being developed, allowing physical measurements at a greater spatial density [2, 5]. Knowledge of model scaling is also developing [1, 6, 7, 8] which improves our understanding of the applicability of model results.

As a result, there has been a trend for computer models taking over the simulation of flow and wave conditions that are linear, or close to linear, while physical models remain dominant in areas with strong non-linearity. There has also been an increase in the combined use of physical and numerical models, known as hybrid or composite modelling, to address problems in both structural design and environmental fluid mechanics [9, 10].

Physical Models: Their historical and current use in civil and building engineering design,
First Edition. Edited by Bill Addis.

Physical models can also play an important role in studies relating to climate change. Numerical models used for climate-change impact analysis cannot be validated against field measurements for conditions beyond the range of the current climate. Since we have no field data truly reflecting future conditions, a key challenge in climate-change adaptation is to rigorously test models using proxies of future conditions. Physical modelling offers the opportunity to test present-day situations and possible climate-change adaptations against a range of hydraulic boundary conditions, such as sea-water level, wave height and flow speed that exceed present-day conditions and are compatible with climate-change projections.

This chapter shows how physical models are being used today to address practical problems in the design and operation of various developments and infrastructures. The following sections describe:

- the main areas of physical testing undertaken in commercial European laboratories;
- the capabilities of some of the modern measurement techniques;
- how physical models are increasingly used with numerical models to address practical concerns.

It does not cover academic research or the more detailed, specialist equipment that is generally confined to research projects.

31.2 Physical model testing

Physical model testing plays an important part in the development and validation of the design of many coastal, maritime and freshwater hydraulic structures. Some of the most common forms of structure tested in hydraulic laboratories include breakwaters and floating structures. These may, of course, be combined in the design of a new harbour or marina, or could be entirely separate. Some examples of the different phenomena tested are given in the following sections.

31.2.1 Wave overtopping of structures

Waves passing over the top of a structure can cause damage and disruption to the structure itself and to assets being protected by that structure. Different types of activity and structure have different acceptable overtopping rates, commonly expressed as the average volume of water to pass a metre length of the structure in each second. A range of conditions is likely to be tested, from those occurring several times a year – to determine safe operational conditions – to extreme design conditions and commonly, an overload condition.

An example of a wave-overtopping test carried out in a wave flume, with random waves approaching straight onto the structure is shown in Figure 31.1. This type of test is known as a two-dimensional (2-D) test. Three-dimensional (3-D) tests can also be undertaken in wave basins. Often a 2-D test at a larger scale (such as 1:20) is carried out in a flume, while a 3-D test at a scale of 1:40 to 1:60 is carried

Figure 31.1 A 1:30 scale model test showing wave overtopping of a seawall. (Image: © HR Wallingford)

out in a wave basin. Modern wave paddles, driven by electric motors, can be used to generate different standard wave spectra, such Pierson-Moskowitz or JONSWAP spectra, or specialised spectra, such as combinations of swell (long-period waves generated a long distance away) and wind-sea (with shorter period waves, generated more locally). In the case of three-dimensional tests, the wave conditions can be long-crested (coming from a single direction) or short-crested (coming from a range of directions). Wave conditions can even be varied along the face of a wave-maker to represent spatial variations in incident wave conditions.

31.2.2 Breakwater stability

Another common type of hydraulic test is used to study the stability of a breakwater. These can be constructed from scaled rock or scaled concrete armour units, commonly hired from the patent holder. Rock sizes are scaled to take account of variations in the density of water and rock armour, while filter-layer scaling has to take account of permeability as well. An example of a model breakwater is shown in Figure 31.2. Templates are normally used to achieve the correct cross-section of every layer, while concrete armour units often have to be placed in a specialised way to create the correct packing density.

The movement of armour can be measured using a range of methods. Traditionally, the elevation across a number of sections through a model has been measured using a point gauge. Increasingly, terrestrial laser scanners or other optical techniques, such as photogrammetry or structure-from-motion, have been used to create a three-dimensional digital elevation model (DEM) of the entire structure (Figure 31.3). It is possible to get much more detailed assessments of damage suffered using a DEM than using a few profiles.

31.2.3 Loads on structures

Loads on structures can be measured using a whole-body force table or pressure sensors (Figure 31.4). Whole body force tables can be used to measure forces and

Figure 31.2 Waves breaking on a physical 1:48 scale model of a breakwater. (Image: © HR Wallingford)

Figure 31.3 Laser scan of a 1:50 scale model breakwater used to reveal damage suffered during a test. (Image: © HR Wallingford)

(a) (b)

Figure 31.4 (a) Caisson with whole-body force measurement. (b) Detail of model showing pressure transducers (insert) in the front face. (Image: © HR Wallingford)

moment applied to a stiff model of the structure. A small degree of movement of the structure must be allowed, in order for forces to be registered.

Pressure measurements are conducted using an array of pressure transducers placed in the middle of the structure where the maximum pressures are commonly assumed to occur. Due to advances in computers, data acquisition and sensor technology in recent years, the number of pressure sensors typically used has increased, with deployments of sixteen to thirty-two becoming more common. Total forces are calculated by integrating the point measurements, assuming uniform pressure distribution between the sensors. For more complex studies the pressure distribution is significantly less well known and the calculated resultant force may vary quite considerably from the actual force [11, 12]. Hence, there is a requirement to develop pressure measurement systems with high spatial and temporal resolution. This is already occurring through the application to hydraulics of matrix-based tactile sensors [13].

31.2.4 Motion of ships and other floating structures

Motions of floating bodies may be measured using different technologies, such as gyroscopes, accelerometers, potentiometers, laser, optical systems or mechanical systems. Mechanical measurement of motion, for instance with magnetostrictive sensors or weights and pulleys, induces mass, stiffness and damping effects that corrupt the phenomenon being studied. Therefore, measurements should be performed by robust and non-contact systems, to avoid any interference with the physical model response.

Six Degree of Freedom (6-DOF) motions (i.e. surge, sway, yaw, heave, roll and pitch) of free-floating or moored vessels or structures can be determined by laser

Figure 31.5 1:63 scale model of a harbour for studying the motion of a moored ship showing the camera (a) and white circular targets on the ship for the motion-detecting system (b). (Image: © HR Wallingford)

displacement sensors or a real-time optical tracking system (Figure 31.5). The core components of an optical tracking system are:

- two or more high-resolution and high-speed infrared video cameras;
- a number of small, lightweight reflective marks rigidly fixed to the model;
- a high-speed data network; and
- a data-acquisition and processing unit.

Also shown in Figure 31.5 are:

(c) metal chutes and storage tanks to collect water overtopping the front of the breakwater, so that an average overtopping rate can be calculated;
(d) mooring lines and mooring-line load measurement systems
(e) tripods with capacitive wave gauges to measure time series of surface elevation at a point. The line of gauges in front of the breakwater is used to check the spatial distribution of wave energy, while those inside the harbour are used to check the level of wave energy entering the harbour.
(f) scaled rock armour that is used to check the level of damage that the breakwater will suffer under design storm conditions (often including a 20% overload condition).

The principle of operation consists in exposing the reflective marks to the infrared light emitted by the cameras, and detecting the light reflected; it is also possible to use active marks, able to emit infrared light. Each camera measures a 2-D position of the reflective marks. The system combines processed data from two or more calibrated cameras to calculate the 3-D position of the markers. If several markers are attached to the model (ship or floating structure) its 6-DOF motions may be calculated in real-time. The accuracy of the system within a measuring area of 10×10 m can be better than ±1 mm for translational motions and $\pm 0.1°$ for rotational movements.

Inertial measurement units (IMUs) featuring solid-state accelerometers and gyroscopes, as used in drones, gliders and industrial machinery, are becoming increasingly accurate and can be used in physical models. Speed (rotational or linear) and displacement can be determined from single and double integration of acceleration with considerable accuracy, allowing 6-DOF motions to be determined. These units can be installed within structures and / or where optical access is difficult.

Measurements of water-surface elevation can be made using a range of wave gauges, including resistive, capacitive, servo motor, laser and ultrasonic technologies.

31.2.5 Scour around structures

The presence of a structure causes flow speeds (whether caused by waves or currents) to increase locally. This can lead to increased rates of sediment transport and scour – the local lowering of the bed level around the structure. It is important for the design of many structures that the potential depth of scour is known. In cases where deep scour cannot be tolerated, scour protection can be provided in the form of rock armour, concrete units, concrete mattresses, rock-filled bags, geotextile containers filled with sand, rock-filled gabions or frond mats.

The depth of scour is often investigated using a physical model, with these studies often including the stability of scour protection designs. In recent years there has been increasing investment in marine renewable-energy devices and this has led to a large number of physical model tests of the foundations of these devices – particularly of offshore wind turbines. The original foundations for these devices were almost always monopoles, but as wind farms move into deeper water, a wider array of foundations, such as steel jackets and concrete gravity-base structures are being tested.

About fifteen years ago the depth of scour and the movement of scour protection were normally measured in model tests using a mechanical bed profiler, which moved along a beam making regular measurements of the depth to the seabed or armour. The spatial variation of scour around a foundation could be measured by moving the beam around the structure to survey along different radial lines. Alternative approaches involved burying thin, regularly-marked rods in the seabed and counting how many marks were exposed, or lowering the water level and recording the shoreline using an overhead camera. In recent years, technologies have been developed which collect much more data, often with considerably less human intervention. These techniques include the use of terrestrial laser scanners or underwater laser scanners. An example of an underwater laser scanner is the ULS200s shown in Figure 31.6. A fan of laser light illuminates the seabed and the point of contact with the bed is recorded by the digital camera. This scanner uses a triangulation method to determine the seabed location.

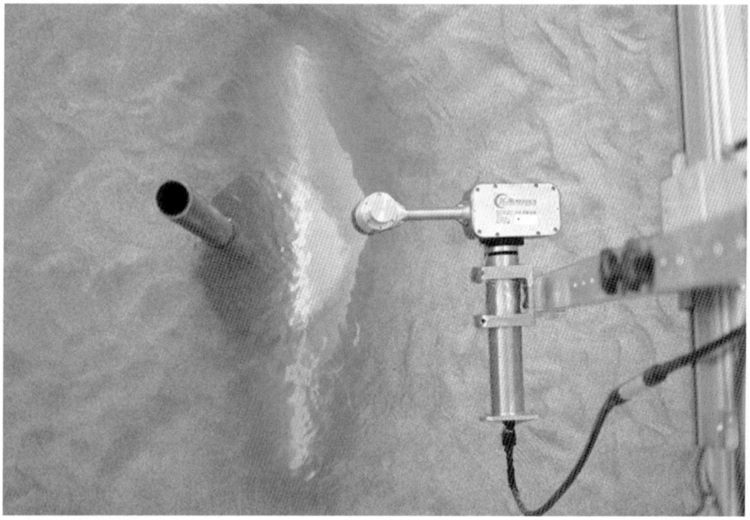

Figure 31.6 Underwater laser scanner surveying the bathymetry around a monopole foundation. (Image: © HR Wallingford)

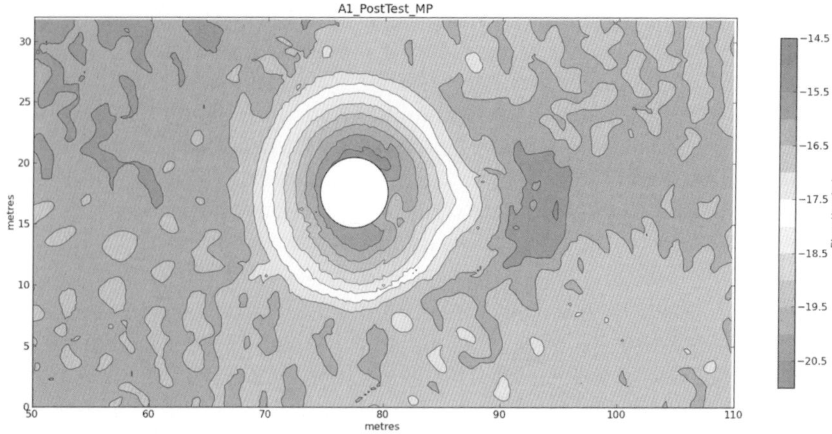

Figure 31.7 Bathymetry changes around a monopole during a test at full scale. (Image: © HR Wallingford)

The scanner can be moved over the seabed on a 1-D beam or a 2-D profiler to build up a detailed bathymetry, or rotated about its axis to scan over a swath of bathymetry. An example of a bathymetry built up using a 2-dimensional profiler and scanner is shown in Figure 31.7. An area of approximately 4 m by 4 m can be surveyed using this system with a spatial resolution of about 4 mm, in about 20 minutes. Other techniques, such as photogrammetry, structure from motion [14], terrestrial laser scanning and point probes can also be used to measure scour around a structure [15]. These techniques generally require the basin to be drained before scanning, although some photogrammetry has been done above and below water level in a still basin.

Figure 31.8 Waves approaching a detached offshore breakwater from the top of the picture and diffracting round it to form a tombolo in its lee. (Image: © HR Wallingford)

31.2.6 Other model types

Physical models are also used today for: beach optimisation, seawater intakes, spillways, drop shafts and other hydraulic structures. Figure 31.8 shows waves creating a tombolo, or mound, inshore from a detached, offshore breakwater in a wave basin at HR Wallingford. The relative length of the breakwater and relative distance from the shoreline are important parameters in determining the response of the beach that can be tested in a physical model.

Figure 31.9 shows a physical model of a pumping-station intake. As the efficiency of the pumps is affected by the vorticity of the flow entering the mouths of the pumps at the ends of the six pipes at the bottom of the figure, models are used to test different configurations of chambers and channels. Today these studies can also be done using CFD models, and comparisons between the results of tests using physical models and CFD models are useful in establishing the validity of CFD solutions.

31.3 Combined use of physical and digital models

Physical and numerical models both have limitations that can restrict their independent use. Composite modelling – the combined use of physical and digital models – can lead to different forms of improvement. For example, being

Figure 31.9 1:10 scale model of a pumping station intake. (Image: © HR Wallingford)

able to model problems that cannot be modelled by either physical or numerical modelling alone, increasing quality at the same cost, obtaining the same quality at reduced cost or reducing uncertainty at the same cost [9, 10]. The following sections illustrate common ways that numerical and physical models can be used together in design.

31.3.1 Model nesting

Model nesting is the most commonly applied and traditional method of linking numerical and physical models. A global or regional numerical model provides the external boundary conditions for a smaller area, or more-detailed model, such as a physical model.

31.3.2 Use of a numerical model to design a physical model

Many physical models are designed with the assistance of numerical models, for example:

- a coastal-area numerical model, set up to model a wave basin to optimise the positions of wavemakers and wave guides;
- the numerical modelling of the approach geometry for a model dam has been used to minimise the head box dimensions, while maintaining the correct approach flow. This reduced construction costs of the model, while increasing the model size and reducing Reynolds number scale effects.

31.3.3 Physical model of one component of a system

The more straightforward (less nonlinear) aspects of a hydraulic system may be modelled adequately using numerical models, yet there may be elements of particular complexity or unusual arrangement that cannot be modelled with confidence using a standard modelling approach. The composite modelling approach to such a problem involves the following stages:

1. Initial modelling of the system using one or more numerical models. This stage is used to optimize the layout as much as possible;
2. Physical modelling of the most complex part of the system with appropriate measurements to characterize the results;
3. Incorporation of the physical model results into the next round of numerical modelling, either by parameterization of them or through their use in calibrating the numerical model.

31.3.4 Model training and calibration

A physical model can be used to calibrate a numerical model, which can then be run for different cases, or used to obtain data where there were no physical model measurements. Alternatively, the physical model can be used to supply the data that goes into an artificial intelligence model, such as a neural network [16]. Possibly the most common model use in design of coastal structures is employing a neural network to train a simulator of wave overtopping rates [17].

31.4 Conclusion

The use of physical hydraulic models has evolved, and will continue to evolve, as computer power increases. Increasing emphasis is being given to complex, turbulent flows, with more-linear problems being almost exclusively studied using numerical models. The increasing capability of computational fluid dynamics (CFD) models is the latest challenge to physical models. There are now so many CFD solvers available, and methods to set up and run CFD models that a wide range of solutions can be produced. Physical models have a history of over a century of successful development and can now produce more detailed results than ever before. The continued development of numerical and physical models in parallel – to compare and complement each other – is therefore expected to continue.

Acknowledgements

This work has received funding from the European Union's Horizon 2020 research and innovation programme under grant agreement No 654110, HYDRALAB+.

References

1 Hughes, S. A. (1993) *Physical Models and Laboratory Techniques in Coastal Engineering*. (Advanced Series on Ocean Engineering No.7). Singapore: World Scientific.

2 Sutherland, J. & Evers, K.-U. (2013) *Foresight study on the laboratory modelling of wave and ice loads on coastal and marine structures*. HYDRALAB Deliverable D2.3.

3 Aberle, J., Rennie, C.D., Admiraal, D.M. & Muste, M. (eds) (2018) *Experimental Hydraulics: Methods, Instrumentation, Data Processing and Management*. IAHR Monograph (2 Volumes). Leiden: CRC Press / Balkema.

4 Frostick, L.E., McLelland, S.J. & Mercer, T.G. (eds.) (2011) *Users guide to physical modelling and experimentation. IAHR design manual*. Leiden: CRC Press/Balkema.

5 Sutherland, J. & Evers, K.-U. (2013) Foresight study on the physical modelling of wave and ice loads on marine structures. *Proceedings of the 35th IAHR World Congress, Chengdu, China*.

6 van Rijn, L.C., Tonnon, P.K., Sánchez-Arcilla, A., Cáceres, I. & Grüne, J. (2011) Scaling laws for beach and dune erosion processes. *Coastal Engineering* 58 (7): 623-636.

7 Heller, V. (2011) Scale effects in physical hydraulic engineering models. *J. Hydraul. Res.* 49 (3): 293-306.

8 Heller, V. (2017) Self-similarity and Reynolds number invariance in Froude modelling. *J. Hydraul. Res.* 55 (3): 293-309.

9 Sutherland, J. & Barfuss, S. (2011) Composite Modelling, combining physical and numerical models. *Proc 34th IAHR World Congress, Brisbane, Australia*, pp 4505-4512. CD-ROM.

10 Gerritsen, H. & Sutherland, J. (2011) Composite Modelling. In: *Hydralab. A user's guide to hydraulic modelling and experimentation* (eds J. Kirkegaard, G. Wolters, J. Sutherland, R. Soulsby, L. Frostick, S. McLelland, T. Mercer & H. Gerritsen), 171-219. Leiden, Netherlands: CRC Press / Balkema.

11 Alderson, J.S & Allsop, N.W.H. (2007) Comparisons of observed pressure integrated and directly measured global force measurements on caisson breakwater models. In: *Proceedings of the 5th Coastal Structures International Conference, CSt07, Venice, Italy, 2-4 July 2007*, 249-258. World Scientific.

12 Cuomo, G., Allsop, W. & Takahashi, S. (2010) Scaling wave impact pressures on vertical walls. *Coastal Engineering* 57 (6): 604-609.

13 Stagonas, D., Marzeddu, A., Gironella, X, Sanchez-Arcilla, A. & Muller, G. (2016) Measuring wave impact induced pressures with a pressure mapping system. *Coastal Engineering* 112 (June): 44-56.

14 Westoby, M.J., Brasington, J., Glasser, N.F., Hambrey, M.J. & Reynolds, J.M. (2012) 'Structure-from-Motion' photogrammetry: A low-cost, effective tool for geoscience applications. *Geomorphology* 179 (15 Dec): 300-314.

15 Todd, D., Sutherland, J., Crossouard, N., Rankine, R., Rigden, T. & Whitehouse, R. (2016) Comparison between methods for creating DEMs

of physical models. *Proc 6th Int Conf on the Application of Physical Modelling in Coastal and Port Engineering and Science (Coastlab '16).*

16 Aamir, M. & Ahmad, Z. (2019) Estimation of maximum scour depth downstream of an apron under submerged wall jets. *Journal of Hydroinformatics* 18 (5).

17 Van der Meer, J.W., Allsop, N.W.H., Bruce, T., De Rouck, J., Kortenhaus, A., Pullen, T., Schüttrumpf, H., Troch, P. & Zanuttigh, B. (2018) *EurOtop: Manual on wave overtopping of sea defences and related structures. An overtopping manual largely based on European research, but for worldwide application.* www.overtopping-manual.com (3/8/2019)

32

Boundary layer wind tunnel model testing – current practice

Francesco Dorigatti

32.1 Introduction

Considerable progress has been made since the 1970s in wind-tunnel model testing of building and civil engineering structures. On the one hand, there has been a rapid and steady development of technologies which have translated into the introduction of new equipment and innovative and ever-more-accurate experimental techniques. On the other hand, this progress was nurtured by the steadily growing demand for wind-tunnel testing by building and civil engineering designers, as well as from the construction industry.

Today, the role of wind-tunnel testing in building and civil engineering is different from the past. While it is still crucial in supporting the design of the world's most exceptional and unique structures, it is now no longer carried out only on a restricted number of special projects. In the last few decades wind-tunnel tests have been conducted to inform the design of the vast majority of new tall buildings, long-span bridges and other types of structure exposed to wind, as well as to help retrofitting existing structures vulnerable to wind. Furthermore, they have been widely used to study the local wind conditions in urban environments, such as central areas of large cities, either at ground level to assess the safety and comfort for people, or at height to investigate the dispersion of pollutants. Wind-tunnel testing has thus been transformed from being a bespoke and often research-oriented approach to become a standard and routinely-used methodology. It is now recognised and sometimes recommended in a number of international building codes and design guides [1, 2, 3] and plays a key role in best-practice design of buildings, civil engineering structures and urban environments.

32.2 Recent developments in test facilities

32.2.1 Principles of boundary layer wind tunnel modelling

A boundary layer wind tunnel (BLWT) is a particular type of test facility that simulates, at reduced scale, the 'stationary' (i.e. stable) gusty natural winds

Physical Models: Their historical and current use in civil and building engineering design,
First Edition. Edited by Bill Addis.

typical of the lower atmospheric boundary layer [4, 5]. By generating a calibrated turbulent onset flow and measuring its effects on scale models of buildings and civil engineering structures, the ultimate aim of these facilities is to predict what happens at full-scale.

At full-scale, the characteristics of stationary winds at a given site depend on the topography and surroundings in its immediate proximity [6] as well as on the type of terrain at a distance of up to several kilometres from the site itself [7]. Both these aspects require modelling at reduced scale for a natural wind to be simulated properly. Therefore, to develop a calibrated turbulent onset flow in a BLWT, a 'surrounding model' is used to reproduce in detail the buildings and structures in the immediate proximity of a given site, while 'flow-conditioning devices' are employed to simulate the effect of the terrain roughness at further distances upwind.

BLWT surrounding models replicate the massing and location of every neighbouring building and structure, as well as any significant topography and elevation change near the study site. They are generally less detailed than the 'target models', i.e. the models of the buildings or structures under study, and reproduce an area of at least 250-300 m in radius from the study site at full-scale [8, 9]. However, their exact extent varies from one project to another, as it mostly depends on the type of test, model scale and size of the wind tunnel.

BLWT flow-conditioning devices are generally installed in a 'straight fetch' upstream of the working section, which is typically 10 to 20 m long. These devices include grids, fence barriers and spires, which are generally set at the fetch inlet, as well as roughness elements distributed on the floor along the entire fetch [10, 11] (Figure 32.1). Multiple types, sizes and combinations of flow-conditioning devices can be used in the same fetch to simulate different types of turbulent flow, which gives the flexibility to reproduce a variety of onset wind conditions in the same wind tunnel.

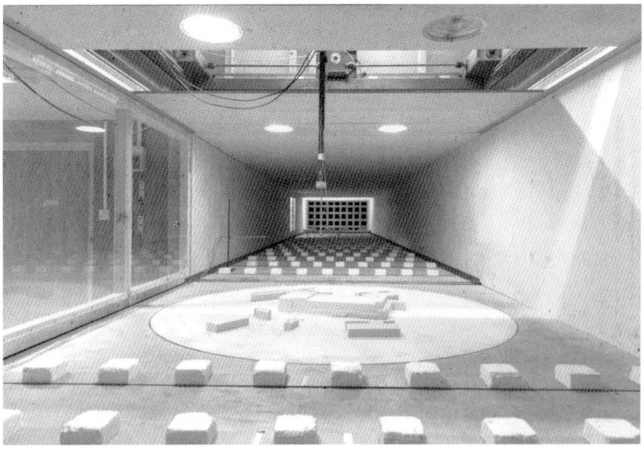

Figure 32.1 Flow-conditioning grid, fence and floor roughness in the 8 x 4 feet Atmospheric boundary layer wind tunnel at Cranfield University. 1:400 scale model. (Image: Courtesy of Cranfield University)

32.2.2 Modern boundary layer wind tunnels

A variety of BLWTs have been built during the last few decades to address the increasing demand for modelling and experimental study of wind effects on building and civil engineering structures. As it happened historically, many of these facilities belong to Universities or research institutions and are devoted mainly to carrying out fundamental research. In addition, though, several modern BLWTs are now privately owned. Generally operated by wind engineering consultancy firms, these are 'industrial' – rather than research – facilities and are used almost exclusively for commercial purposes, i.e. to conduct tests that support the design of specific engineering projects.

On the one hand, the design and working principles of the modern BLWTs build on the legacy of the facilities first introduced in the 1960s by Alan Davenport at the University of Western Ontario (see Section 24.6.2). As such, most facilities still have a floor turntable to accommodate both the surrounding and target model together, so that they can easily be rotated to facilitate testing for 360° wind directions, and use passive flow-conditioning devices on a straight long fetch upstream of the working section to develop a turbulent onset flow simulation.

On the other hand, since the 1970s, a variety of novel instrumentation has progressively been employed in BLWTs, which heavily exploits modern electronics and computer systems and considerably enhances the measuring capabilities of today's facilities compared to the past [4, 12]. Automation has been introduced to improve operational efficiency. Furthermore, active turbulence generation methods have been developed, which rely on using multiple electronically-controlled fans or actively controlled flow-conditioning devices to simulate a turbulent onset flow in short-fetch wind tunnels [13, 14, 15, 16].

The BLWTs built over the last 50 years present a wide variety of layouts. They have open or closed-return wind circuits, employ different types of flow-conditioning devices, may have open or closed working sections, and use either axial fans or centrifugal blowers to supply the required airflow. Also, they have been built at different sizes and can operate at different wind speeds, as they are optimized for testing different types of structure. Generally, modern BLWTs have working sections whose width goes from 2 to 3 m up to more than 10 m, and operate at airflow speeds ranging from 10 to 25 m/s.

In current practice, 'standard-size BLWTs' have working sections of 2 to 3 m wide and are mostly used to test models built at scales of 1:200 to 1:500 [5, 17, 18]. Experiments carried out in facilities of this size generally fulfil the requirements for accuracy and reliability imposed by most design codes and building standards. They tend to enable testing with blockage ratios in the order of 10% or less and Reynolds numbers in the order of 10^4 to 10^5, which ease the scaling-up of the wind tunnel results to full-scale. Furthermore, using relatively small facilities and small-scale models limits both the operational costs and the time required for building the models. As such, standard-size BLWTs are well suited for industrial use, i.e. for testing commercial projects, because they entail lower costs and allow for shorter timescales while providing sufficiently accurate simulations.

Figure 32.2 GVPM closed-circuit wind tunnel, Politecnico di Milano. 13,8 m wide x 3,8 m high x 35 m long Boundary-Layer Test Section with flow-conditioning triangular spires and floor roughness blocks. (Image: Courtesy of Politecnico di Milano)

'Large-size' BLWTs have working sections around 10 m wide or more (Figure 32.2), which offer the opportunity to accommodate either large-scale models or extended surrounding models at scales of 1:200 to 1:500. Compared to standard-size facilities, these entail higher costs and longer timescales relating both to their operation and to the manufacture of larger wind-tunnel models. As a result, large-size BLWTs are mostly used for research, while their commercial use is limited to projects that can justify the extra costs. For example, to test at high Reynolds numbers or to model accurately small-size geometrical details by using large-scale models, or to investigate comprehensive models of extremely large structures such as long-span bridges and urban environments.

The use of automation in modern BLWT aims to expedite testing and, in turn, increase the capacity of the facility, which is the ability of carrying out more tests in a given timeframe. As such, automation is widely employed particularly in industrial BLWTs, which target the highest utilization rates. Motorised turntables operated remotely from a control room are adopted to minimise the testing time. In addition, two or more interchangeable turntables can be used so that while a model mounted on a first turntable is being tested in the wind tunnel, another model and its instrumentation would be either set up or dismantled on a different turntable placed outside of the working section. This would reduce considerably the downtime between tests. Also, automation of the passive flow-conditioning devices can be implemented to almost eliminate the test downtime associated with changing the fetch setup manually, which is often required to simulate different onset wind profiles for different wind directions during the same test. For example, electric or hydraulic actuators controlled by computers can be employed to change the depth and pattern of the floor roughness elements and to modify the size and shape of the spires and grids (Figure 32.3).

Figure 32.3 Motorized turntable, automated floor roughness and automated flow-conditioning spires in the Irwin Wind Tunnel at RWDI. 1:230 scale full-aeroelastic model of the Gordie Howe International Bridge in Detroit, USA and Windsor, Canada. (Image: Courtesy of RWDI)

32.2.3 Novel test facilities

A small number of novel test facilities has been built in recent years to complement and expand the capabilities of conventional BLWTs. These facilities implement new approaches to experimental testing and consist mainly of full-scale or large-scale laboratories, and of transient-wind simulators.

Full-scale and large-scale laboratories enable experiments on full-scale or large-scale models subject to stationary (stable) gusty winds. Compared to undertaking measurements in the field, using real structures subject to natural winds, these facilities provide the significant advantage of creating controlled and repeatable wind conditions which reduces considerably the uncertainty and scatter of results. By minimising or removing scaling effects, they can also overcome many of the challenges related to using traditional atmospheric BLWTs, such as Reynolds number effects [5], and the need for accurate modelling of complex architectural features and small-size elements (e.g., fins, louvres, porous screens, cavities). Furthermore, by subjecting real and complex buildings or structures to controlled winds in such facilities, tests can be carried out to investigate the mechanisms of wind-induced failure for some of their individual components [19]. Among others, these laboratories include the Research Center of the Insurance Institute for Business & Home Safety (IBHS) [20] and the Wall of Wind at Florida International University [21], both in USA, the Wind Engineering Energy and Environment (WindEEE) Dome at University of Western Ontario, in Canada [22], the Jules Verne Climatic Wind Tunnel at the Centre Scientifique et Technique du Bâtiment, in France [23] and the proposed Full-Scale Storm Simulator (FSSS) in Japan [24].

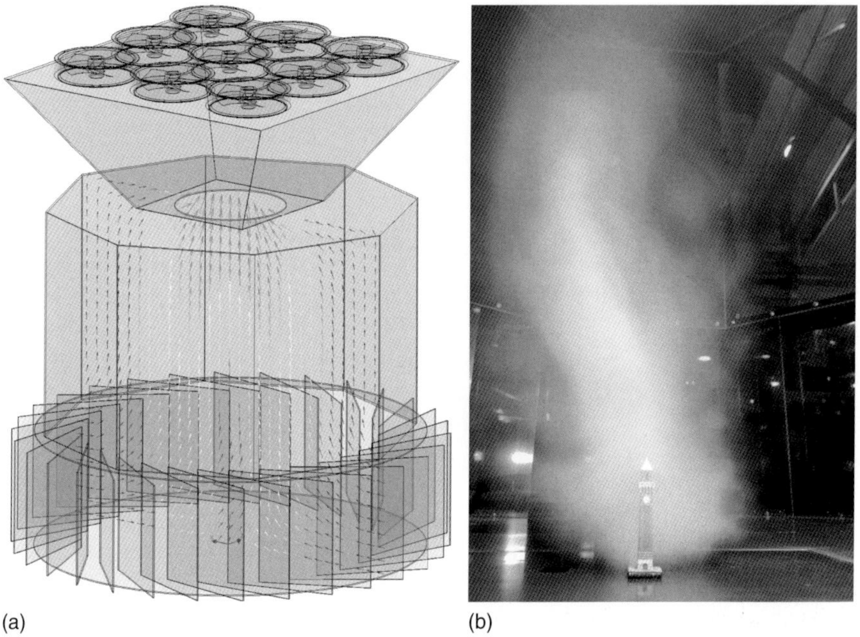

Figure 32.4 Tornado Generator at the University of Birmingham. (a) Diagram and flow field of the working apparatus. (b) Flow visualization of a tornado. (Image: Courtesy of University of Birmingham)

Transient-wind simulators are designed to reproduce at reduced scale non-stationary (i.e. non-stable) extreme wind events such as thunderstorm downbursts and tornadoes, and to investigate their effects on buildings and structures. Transient in their nature, these wind systems are characterised by flow patterns considerably more complex than stationary winds, which historically have made it difficult to simulate them physically at reduced scales in a controlled environment [25, 26]. However, renewed impetus to develop dedicated facilities capable of studying downbursts and tornadoes arose out of research that has argued that the localized high wind speeds generated in these wind events can be relevant – and perhaps critical – for the design of buildings and other exposed structures [27, 28, 29]. This led to several transient-wind simulators being built, including the WindEEE Dome at the University of Western Ontario in Canada, [22, 30], the VorTECH simulator at Texas Tech University [31] and the Tornado Simulator at Iowa State University in the USA [32] and both the Downburst Simulator and Tornado Generator at the University of Birmingham in the UK [33] (Figure 32.4).

32.3 Recent developments in measuring techniques

A variety of measurements can be collected during BLWT tests. Depending on the aim of a wind tunnel study, different quantities can be measured during testing, which require dedicated measuring equipment and may also require the use of specific types of physical models. The three types of measurement most often made are wind velocities, wind-induced surface pressures and wind-induced overall forces and moments. In addition, measurements of accelerations, displacements, deflections and local wind-induced stresses experienced by a model can be made, as well as measurements of flow dispersion by monitoring the concentration of a tracer substance [17, 34, 35].

As most wind-induced phenomena are unsteady and fluctuating, their assessment requires not only mean values to be measured, i.e. time-averages, but also, and more importantly, peak values. As a results, the vast majority of the measurements collected during BLWT tests involve acquiring the time-series of a given signal by sampling at a sufficiently high rate and for a relatively long period of time. This enables both the time-average and dynamic fluctuation of such a signal over time to be captured and, in turn, both mean and peak values to be calculated [36]. For example, gust velocities determine the safety and comfort of people in outdoor spaces of urban areas, while peak pressures and peak forces and moments induced by the wind govern the design of cladding and overall structural systems.

In recent years, BLWT testing has benefited both from the development of new and more advanced instrumentation and from the introduction of novel measuring techniques [4, 12]. Quite naturally, such innovations came not only from dedicated and applied research and development, but also extensively leveraged new technology from other fields, particularly electronics and computer systems. Given the continuous progress in these fields, it is therefore anticipated that BLWT measurements will also see further and fast-paced developments in the years to come.

32.3.1 Velocity measurements

Measurements of flow velocity in the proximity of a body subject to an air flow are essential in order to understand the interaction between the fluid and the body and the mechanisms that govern that interaction. For building and civil engineering projects, measuring the characteristics of a wind flow in the proximity of a building or structure is necessary for two main reasons: to assess the comfort and safety of people in that area; and to understand the flow mechanisms that determine the wind-induced pressure on the surface of such a building or structure and, in turn, the overall aerodynamic loads acting on it.

Pitot and Pitot-static tubes are still heavily used in modern BLWTs. They are employed in combination with a variety of differential pressure transducers and manometers, mostly to characterise the wind tunnel boundary-layer flow and to provide measurement of the reference mean wind velocity during testing. However, the use of Pitot tubes becomes challenging when measuring local wind conditions either in close proximity to a body or within narrow and partially enclosed spaces at small scale, where there is no clear prevailing flow direction with which to align the probe. Furthermore, their ability to resolve accurately the short-term velocity fluctuations is limited by their relatively low frequency response. Therefore, a number of alternative velocity-measurement techniques and instrumentation are currently employed to complement them in BLWTs.

32.3.1.1 Hot-wire and hot-film anemometers (HWA, HFA)

Hot-wire and hot-film anemometers (HWA, HFA) are the instruments that provide the best frequency response, up to 100 kHz if needed, and have the capability of resolving all three components of the velocity [12, 37] (Figure 32.5a). They ensure excellent performance in terms of accuracy and resolution. Nevertheless, they find their main limitation in the fact that the wire probe is prone to break if hit by small particles carried in the airflow. Unfortunately, breakage of HWA probes is not uncommon in test environments such as a BLWT in which the airflow is not immune from small-particle contamination [35]. The fragility of HWA probes can therefore become a challenge, aggravated by the complexity of the process to repair and recalibrate them.

32.3.1.2 Multi-hole pressure probes

An alternative to the HWA and HFA consists of multi-hole pressure probes, for instance Cobra probes [38, 39] (Figure 32.5b). These work on the principle of simultaneously measuring multiple pressures at small pressure taps installed with different orientations on the probe head and correlating them to resolve the three components of the local flow speed, as well as the static pressure. Their frequency response can reach 2 kHz, which is not as good as for HWAs and HFAs, but suffices for most BLWT applications. The main advantage of these probes, which has contributed to their widespread use, is that they can deliver accurate measurements and are often found to be more robust and easier to handle in a BLWT than HWAs and HFAs. However, they do not replace the latter entirely. The measurement ranges for multi-hole pressure probes tend to be narrower than those of HWAs and HFAs, which requires different probes with different ranges to carry out measurements at low and high velocities. Also, although they resolve the three components of the fluctuating velocity, these

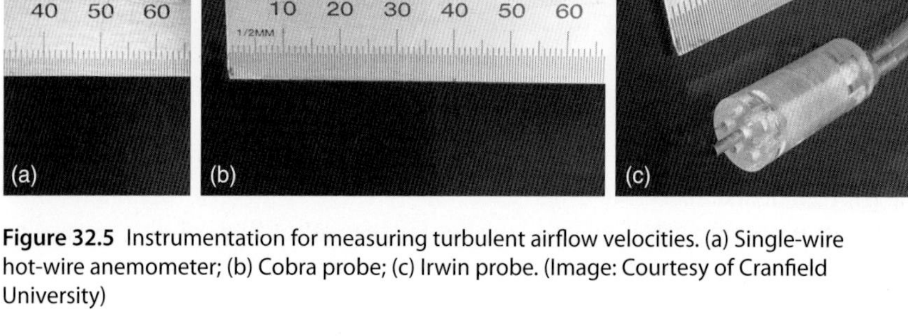

Figure 32.5 Instrumentation for measuring turbulent airflow velocities. (a) Single-wire hot-wire anemometer; (b) Cobra probe; (c) Irwin probe. (Image: Courtesy of Cranfield University)

probes generally have a 'cone of acceptance' for the direction of the mean flow, which prevents them from resolving any reverse flows and can limit their use when the direction of the mean flow is not known beforehand.

32.3.1.3 Irwin probes and surface pressure sensors

Irwin probes are widely employed in BLWTs in combination with differential pressure transducers to measure the magnitude of the mean and peak velocities in proximity to a surface. For example, they are routinely used to measure wind velocities at ground level in wind tunnel studies of pedestrian safety and comfort.

The Irwin probe present an extremely simple and compact design and work by detecting the instantaneous pressure difference between the top of the sensor tube, at the height of interest, and the surface in proximity to where the measurement is taken [40] (Figure 32.5c). The time series of this pressure difference is then converted into a velocity time series by applying a calibration function and is processed to give the mean and peak wind speeds at the height of interest. Irwin probes are omnidirectional in the sense that they detect and resolve the magnitude of the velocity vector regardless the direction of the oncoming mean flow. However, they do not resolve the direction of the velocity vector. This ability was added later to the original Irwin probe in a number of directional surface sensors that have been developed in recent years [41, 42]. These sensors are more complicated in their design and generally require measuring multiple differential pressures, instead of just one.

Irwin probes are very simple in their design, which makes them easy to manufacture and relatively inexpensive. They are robust, easy to calibrate, install and use and deliver results accurate enough for most BLWT applications. The ability of Irwin probes to resolve the time fluctuations of the velocity depends on the pressure transducers used in combination with them, and is affected by the time lag and attenuation generated by the tubing system [43]. Although this results in a lower frequency response than for HWAs and Cobra Probes, it is generally adequate to ensure peak-velocity measurements of similar accuracy to a HWA [40, 44].

32.3.1.4 Non-intrusive measurements

Laser Doppler Velocimetry (LDV) and Particle Image Velocimetry (PIV) are non-intrusive velocity-measuring techniques. Although they are two different techniques, they rely on similar working principles as they both employ high-power lasers, optical systems, high-speed light detectors or cameras, and micro-metre size flow-seeding particles [35, 37, 45].

LDV and PIV resolve the velocity field in a certain portion of the flow field by monitoring the movement of extremely small particles (diameter 1 to 10 μm) immersed in the fluid, in the assumption that their speed is the same as the flow itself. High-powered lasers are employed to illuminate the portion of the flow field under study and detect the velocity of the particles by measuring over time the portion of laser light that they reflect.

LDV can resolve up to three components of the velocity with a frequency response in the order of a few kHz and it is generally used to investigate limited portions of a flow field. Standard PIV can measure only two velocity components, which lie on a defined plane oriented through the fluid domain. Although the extension of such a plane is larger than the study area for LDV, the time resolution of basic PIV measurements is usually limited to a few Hz. More-advanced and complex versions of PIV are available, such as Stereo-PIV, which measures three velocity components, Tomographic PIV, which investigates a volume – rather than a 2D-plane – of the fluid domain and time-resolved PIV, which enable sampling at a high frequency [45, 46].

LDV and PIV are particularly useful when the use of a probe would introduce notable perturbations and interferences in the flow. This generally happens when the size of the probe is large compared to the extent of the flow area being studied. For example, when small-scale models are used and velocity measurements need to be taken in street canyons or in confined spaces which are part of a building or bounded between two or more structures. However, the cost of the equipment and the complexity of the experimental setup and procedure are such that LDV and PIV have had limited use in BLWT studies until now and these were mainly for research applications [35].

32.3.2 Pressure measurements

Pressure measurements are carried out in modern BLWTs to provide a variety of information. They enable indirect measurements of flow velocity using techniques and instrumentations such as multi-pressure probes and Irwin probes. They can also detect and quantify local wind-induced loads on the surface of buildings and structures, which are then processed to predict design wind loads for cladding and secondary structural elements. Furthermore, if simultaneous pressure measurements are integrated over the envelope of a building or structure, they can also provide predictions of overall wind-loading and wind-induced motion for a building or other structure [4, 17, 35]. The standard methodology for measuring wind-induced surface pressures in a BLWT is to install pressure taps on the surface of scale models and connect them to remotely positioned pressure scanners using flexible PVC tubing [12].

Reduced-scale physical models for pressure testing in wind tunnels are nowadays mostly produced using 3-D printing technologies. These rigid models are hollow by design to allow for hosting the flexible pressure tubing in the inside space. This tubing connects directly to the pressure taps from the inside of the model shell and then run through the bottom of the model to reach the pressure scanner, which is often placed underneath the wind-tunnel working section. Pressure-test models present a relatively thick shell compared, for example, to force balance and aeroelastic models, which eases the installation of taps and also provides sufficient stiffness to prevent them from flexing or vibrating noticeably when subject to wind during testing. Finally, to facilitate access to the inner side for tap installation, pressure models can also be 3-D printed in multiple pieces and assembled only at the end, after the pressure taps have been installed on each piece.

Pressure taps, typically several hundreds, are installed on wind-tunnel models by drilling small holes (typically ~1 mm diameter) through their outer shell (Figure 32.6). The ensemble of pressure taps on a model is called 'tap layout'.

Figure 32.6 1:400 scale wind-tunnel model for surface pressure measurements. (a) Details of the pressure tubing inside the model. (b) Model mounted in the wind tunnel for testing. (Image: Courtesy of RWDI)

The number and positions of the pressure taps that form a tap layout are bespoke to the model geometry and to the type of pressure test (e.g., cladding or HFPI test). In general, the more pressure taps that are monitored, the better the resolution of the measured data. Therefore, as many taps as possible tend to be installed on wind-tunnel models for pressure testing. Limitations to the maximum number of pressure taps come primarily from the size of the scale model, which limits how many taps can fit into its shell, as well as the number of flexible tubes that can be hosted in its internal cavity. The number of channels available in the pressure scanner and acquisition system can also limit how many pressure taps can be monitored during one test. If there are more pressure taps than channels available, successive tests can be carried out, each monitoring a different groups of taps; however, this leads to longer testing times and higher costs.

Pressure taps are mainly placed in locations on the model envelope where the peak pressures, either positive or negative, are expected to occur. Also, multiple and closely spaced taps are positioned where sharp variations of pressure are anticipated, to achieve a better resolution in those critical locations. In practice, this translates into having many closely spaced taps near the edges of a building or structure, either on the façade or on the roof where large negative pressures and sharp pressure gradients are expected to occur due to flow separation. Pressure taps are also placed in the middle area of each façade. Their aim here is to capture the peak positive pressures, but they tend to have a relatively larger spacing compared to near the edges because the pressure in the stagnation regions tends to be fairly uniform. Finally, dedicated pressure taps can be placed either on or near to architectural details and appendages such as fins, louvres, balconies and parapets.

On scale models of tall buildings with fairly uniform cross-sections, pressure taps tend to be organised in 'rings', i.e. they are placed around the perimeter of the building at fixed heights. In this case, multiple rings of taps are distributed to map the variations of wind-induced pressure with height. As vertical pressure gradients are generally much smoother than the horizontal ones, such as in proximity to edges, the vertical spacing between the rings is larger than the distance between taps on the same ring. Also, the vertical spacing is smaller at the bottom and top of the building compared to the middle portion in order to better capture any three-dimensional flow effects.

Modern pressure scanners employed for BLWT testing are high-frequency multi-channel Scanivalve-type systems. These are stand-alone units that employ multiple piezo-resistive transducers combined with electronically-controlled pneumatic manifolds to monitor multiple signals simultaneously at high sampling rates [12]. If needed, multiple units can be linked and operated together to enable several hundreds, or even more than a thousand signals to be recorded synchronously and data can be sampled at 500 Hz and higher.

High sampling rates of 500 Hz or higher are required to resolve the fluctuations of the wind-induced pressure with time and, in turn, to enable calculation of their peak values over averaging times as short as half a second at full-scale. Also, to preserve the high-frequency response of the measurements despite using remote-mounted pressure scanners and relatively long tubing, either tube restrictors are employed or Tube Transfer Functions (TTF) are applied for data processing [43].

32.3.3 Measurements of overall forces and moments

The aerodynamic overall forces and moments acting on a wind-tunnel scale model of a building or structure can be measured directly using force-balance systems. Force balances are mostly used in combination with rigid models to detect the mean and background aerodynamic loads [47]. However, if employed for aeroelastic model testing, they resolve also the wind-induced resonant loading.

Different types of force balance provide direct measurements of overall loads in BLWT testing. They rely mostly on strain gauges or piezoelectric sensors to resolve six components of the wind-induced loads (i.e. three base moments and three shears) and are generally accommodated either inside the model or below it, under the wind-tunnel floor (Figure 32.7). To resolve the full range of wind-induced fluctuations in the loading, it is crucial that a force-measuring system ensures a good dynamic response at high frequency. This depends primarily on the balance itself that needs to be designed to combine appropriate sensitivity and resolution with relatively high stiffness, which enables it to resolve high frequency fluctuations. Furthermore, the force-balance system needs to be rigidly anchored onto the wind-tunnel turntable and insulation needs to be provided to prevent signal contamination due to mechanical vibrations or electromagnetic noise coming from external sources. Finally, the frequency response depends also on the model used for testing. Force-balance test models are generally both extremely stiff and very light, so that their natural frequency of vibration once mounted on the balance is in the order of 100 Hz. This is above the

Figure 32.7 Rigid wind-tunnel model of a tall building mounted on a High Frequency Force Balance (HFFB) system for measurement of overall wind forces and moments. (Image: Courtesy of RWDI)

typical frequency range of interest for wind-induced excitations on wind-tunnel models and therefore would not interfere with any aerodynamic measurements.

Wind-tunnel scale models for force-balance tests can be either made out of light foam, 3-D printed or hand-built. Foam models are generally sting-mounted, for example on underfloor balances for testing tall buildings. This setup works quite well because, even if foam is not particularly stiff, the internal sting secured on the balance provides sufficient rigidity to the whole system. 3-D printed models, which are slightly heavier and considerably stiffer than foam models, can be mounted directly onto the force balance system. 3-D printed and foam models produced by computer-numerically-controlled (CNC) machines can replicate highly complex geometries accurately at small scale. Also, they can be produced quickly compared to hand-built models. Hand-built models, made of light materials like balsa wood, are still used in certain cases to reduce the model mass to less than an equivalent foam or 3-D printed model. In certain instances, hybrid models can be used where some parts are 3-D printed, some made out of foam, and others hand-built.

32.3.4 Measurements of gas dispersion

Measuring the dispersion of a given gas injected into an oncoming wind flow is required in BLWT studies that investigate how an exhaust gas or pollutant spreads in an urban environment driven by the wind and, in turn, how it might affect the health or comfort of people. In these experiments the wind-driven dispersion of a given exhaust gas is quantified by measuring the concentration of such a gas in the wind tunnel at multiple locations. To this end, a viable measuring technique relies on seeding the exhaust gas mixture with a carbon tracer gas, for example methane, and then collecting air samples at any location of interest on the scale model to analyse their carbon concentration using a flame ionisation detector (FID). In this technique, the concentration of the exhaust gas at any sample locations, and therefore its dispersion rate, can be obtained from the concentration of the tracer gas measured in proportion to its concentration at the exhaust [34, 48].

32.4 Wind-tunnel tests on buildings and urban environments

Wind-tunnel tests on buildings and urban environments represent one of the major applications of modern BLWTs and consist mostly of three types of study:

1) predictions of wind-induced pressures on building envelopes;
2) analyses of overall wind-induced responses of tall buildings;
3) assessments of pedestrian-level winds in urban environments.

Nowadays, provisions and guidelines for wind design of buildings are provided in a number of national and international building codes and design manuals [1, 2, 3]. Nevertheless, these provisions are inherently conservative and therefore do not enable design optimization. Furthermore, because they are relatively simple and quite general in order to ease and to widen their applicability, they are not appropriate for use in complex cases. Hence, BLWT testing is often acknowledged within such codes as the appropriate methodology for use in such cases. For instance, wind-tunnel tests might be recommended by codes when the height or shape of a building is unusual or unprecedented, because of the complexity of the façade systems, or in the presence of challenging surroundings. Also, these tests might be deemed necessary considering the high likelihood of strong winds at a certain site, due to the wind climate, to its exposure, or both.

In current practice the use of code-based calculations is largely limited to predicting wind loading on buildings at a preliminary design stage. BLWT testing, however, is an effective tool both to assess and to optimize the wind effects on tall buildings and urban environments at all design stages. BLWT testing is, therefore, effectively the current industry standard and represents the reference methodology to inform the final designs for wind.

32.4.1 Wind-induced pressures on building envelopes

The envelope of a building is the interface between its interior space and the outside environment, which typically includes both the façades and the roof. Beyond aesthetics and architectural considerations, wind is one of the most important environmental factors to consider in the design of a building envelope.

The effects of a wind blowing onto a building are to generate a pressure field over its external surface and, indirectly, to affect the pressure of the internal air volume. In certain conditions, strong winds can produce extremely large localized wind loads at a number of critical positions of either the façade or the roof of the building. Therefore, it is necessary to predict the magnitudes of such forces to inform the selection and sizing of the cladding and also of its framing system. The wind loads used for cladding design are the *peak net pressures*, i.e. loads per unit area that result from the largest difference between simultaneous wind-induced actions on the two sides of a cladding element. They include both positive and negative pressures, which are directed towards the inside and outside of the building, respectively. On cladding elements that separate the building inner space from the outside, the peak net loads are given by the difference between the wind-induced *external pressures* and the *internal pressures*. For cladding mounted onto the building exterior and exposed to winds on both sides (such as parapets, louvres, fins, canopies, etc.), the net wind loads are the differences between the wind-induced external pressures on the two sides of each cladding element. Generally, as individual cladding elements are relatively stiff, no significant contribution to their wind loads comes from inertial forces. However, such forces sometimes affect cladding elements like long and slender fins or louvres which are flexible enough to experience wind-induced vibrations.

The main objective of a cladding wind-tunnel study is to provide the designer with an accurate prediction of the local peak net pressures induced by the wind on the envelope of a given building. To this end, the local wind-induced external pressures on the scale model are measured directly during the wind tunnel test and then processed to predict their peak values at full-scale. From these, then, the final peak net wind pressures to be used for cladding design are obtained including an allowance for internal pressure and applying the *wind climate data*, which account for likelihood of the design wind speed to occur from any specific wind direction. Internal pressures are not measured directly during the tests, as the interior of a building is neither modelled at scale nor instrumented. In most cases they are calculated from code provisions, which assume pressure equalization in the building interior governed by uniform leakage [1, 2, 3]. Alternatively, they can be estimated from the measured external pressure to account for the effects of openings in the building envelope associated either with the presence of operable cladding or with breaches caused by the impact of flying debris [49, 50].

Model scales in the range of 1:200 to 1:500 are used for most cladding tests. Typical tap layouts comprise an average of 500-1000 taps in scale models of most mid- and high-rise buildings. Surrounding models have radii of 300-500 m at full-scale, and testing is usually carried out to investigate thirty-six wind directions, i.e. every 10°. High-frequency pressure measurements are collected employing remotely-mounted multi-channel pressure scanners (Section 32.3.2)

The results of a cladding wind-tunnel study can, in theory, be reported presenting the peak-positive and peak-negative pressures obtained at every measuring location and for any of the thirty-six wind directions. This, however, would be impractical as it would entail presenting 36 x 2 sets of wind-induced pressures, i.e. positive and negative peak pressure for each wind direction. To summarize and facilitate the use of the results for design, only the envelope of the positive and negative worst-cases are often presented which consists of the largest peak-positive and peak-negative pressures obtained at each location for

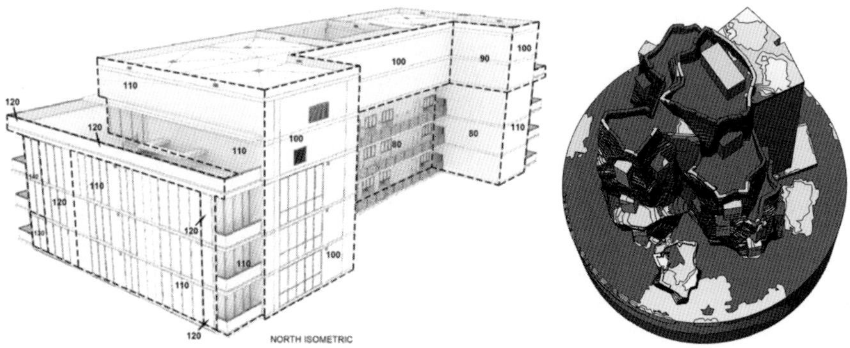

(a) (b)

Figure 32.8 Examples of cladding wind-tunnel study results. (a) Specified pressures on each area of a building envelope. (b) 3-D colour contour pressure coding. (Image: Courtesy of RWDI)

all wind directions. Also, to ease further the interpretation of the results, the peak wind pressures obtained at the pressure tap locations can be interpolated over the entire building envelope. To this end, different interpolation methods can be used, which then produce different types of deliverable, including three-dimensional drawings and contour plots (Figure 32.8).

32.4.2 Overall wind-induced responses of buildings

32.4.2.1 Structural wind loading and wind-induced motion

The study of the overall responses of a building to wind involves the assessment of two main aspects: the overall structural wind loading and the wind-induced building motion. The analysis of the overall wind loads informs the design of the primary structural systems of a building, including its foundations, for strength and serviceability. The assessment of the building motion, on the other hand, examines the accelerations and torsional velocities induced by the wind, to ensure that such a building also provides an appropriate level of comfort for the occupants.

For low-rise buildings the determination of the overall wind loading represents the main focus. This is because they are sufficiently rigid laterally to experience hardly any wind-induced movement. However, since the lateral stiffness progressively decreases with height and slenderness (i.e. the ratio of the building height to its narrowest side), the predictions of both wind loading and wind-induced motions are of paramount importance for tall buildings [18].

32.4.2.1.1 Structural wind loading The overall forces and moments induced by the wind on a building are sustained by its primary structural system and conducted to the ground through the foundations. These loads result from the combination of the mean loading with the background and resonant loading [18]. The mean wind loading is constant in time and proportional to the onset mean wind speed. The background loading is the dynamic quasi-steady loading produced by the fluctuations in the onset wind velocity integrated over the entire building envelope. The resonant loading is also a dynamic loading and consists of the inertial forces and moments that develop as a result of the building mass being in motion and oscillating at its own natural modes of vibration under the action of the wind. The magnitudes of resonant loads depend on the stiffness, damping and mass distribution within the building. They tend to be small for low-rise buildings, but increase to become of the same order, or even larger than background loads on tall and slender buildings.

The overall wind-induced responses of a tall building can also be categorized either as along-wind or across-wind depending on their direction relative to the mean wind direction. The overall mean wind loading and static deflection are intrinsically along-wind. The dynamic wind loads, however, as well as the resulting wind-induced building oscillations can be either along-wind or across-wind. Dynamic along-wind responses are caused primarily by the gustiness in the onset wind, while across-wind responses are generally associated with vortex shedding – the formation and detachment of alternate vortices from the trailing edge of a bluff body, which induce a fluctuating aerodynamic force leading to oscillations of the body perpendicular to the mean wind direction. The frequency at which these vortices shed depends on both the shape and size of the body, and the onset wind speed. In a tall building, if vortex shedding occurs at a frequency similar to one of its natural frequencies of vibration, it can lead to considerable across-wind loads and accelerations, up to several times larger than the along-wind ones. The onset of vortex shedding can thus threaten either the building strength or its serviceability, in which case mitigation measures need to be adopted to control the building response [18].

32.4.2.1.2 Wind-induced building motion Several studies concerning the human perception of wind-induced building motion have been conducted in recent decades [51]. These were based either on surveys from occupants of real buildings or on research conducted using motion simulators. Results indicate that human perception of building motion, and therefore the occupants' comfort, are related to two main parameters: wind-induced accelerations, primarily, and torsional velocity, secondly. Human perception of wind-induced building motion depends not only on the magnitude of peak accelerations, but also on the frequency at which the building oscillates and, interestingly, on the primary activity of a building occupant. However, it was also found in these studies that human perception of wind-induced building movement is highly subjective. Therefore, it is not straightforward to define either a unified set of acceleration thresholds associated with the level of comfort, displeasure or discomfort perceived by the building occupants, or a series of motion limits that should not be exceeded in order to prevent a given building from causing occupants to complain. Table 32.1 summarises an approximate relation between the magnitudes of peak wind-induced accelerations and their effects in terms of building motion and occupants' perceptions [52, 53].

Design criteria for occupant comfort are provided in the ISO 10137: 2007 [54] which indicates upper limits for peak acceleration at the one-year return period as functions of the building frequency of vibration for both residential and office occupancies. Guidance based on practice in North America is also

Table 32.1 Occupants' perception of wind-induced accelerations in tall buildings (Source: [52, 53])

Acceleration (milli-g)	Effect
< 5	Humans cannot perceive motion.
5 to 10	Sensitive people can perceive motion. Hanging objects may move slightly.
10 to 25	Majority of people perceive motion. Motion level may affect deskwork. Long-term exposure may produce motion sickness.
25 to 40	Desk work becomes difficult or almost impossible. Ambulation still possible.
40 to 50	People strongly perceive motion. It is difficult to walk naturally and standing people may lose balance.
50 to 60	Most people cannot tolerate motion and are unable to walk naturally.
60 to 70	People cannot walk or tolerate motion.
> 85	Objects begin to fall and people may be injured.

included in the NBCC 2010 [55] which indicates 15-25 milli-g as an acceptable limit for ten-year return period peak accelerations. Instead of a threshold, the AIJ-GBV-2004 [56] provides guidelines on the probability of wind-induced acceleration perceptions being reported by building occupants, as a function of frequency of vibrations. Finally, guidelines for torsional velocities have been suggested by Isyumov [57] who proposed upper limits of 1.5 and 3 milli-radians/s for the one-year and ten-year return periods, respectively.

32.4.2.1.3 Structural wind-tunnel studies Structural wind-tunnel studies tend to be conducted mostly on tall buildings over 80-100 m in height for which they provide a greater benefit in terms of increased accuracy and reduced conservatism compared to using a code-based analytical approach. These studies can be either 'confirmatory studies' or 'aerodynamic optimization studies' and they can be carried out using two main methodologies: 'rigid model testing' and 'aeroelastic model testing'.

Confirmatory wind-tunnel studies are undertaken to validate and refine an existing preliminary structural design which is typically based on code-based calculations. Furthermore, they assess the wind-induced building motion against the applicable comfort criteria in order to minimize the risk of future complaints from the building occupants. Although they are generally instructed in a late phase of the design process (i.e. detail design), these studies enable structural design optimization and, in turn, significant cost saving. Nevertheless, they rarely lead to any aerodynamic optimization of the building shape, because any modification of the building architecture would be highly problematic and costly at such a late design stage.

Wind tunnel aerodynamic optimization studies are carried out for buildings whose wind-induced responses are deemed to be so severe as to threaten their own feasibility. For example, buildings of unprecedented height and slenderness or having certain critical shapes, or buildings subject to extremely strong winds due to either a hostile wind climate or unfavourable exposure and surroundings. In these cases, wind-tunnel studies are commissioned which consist of a series of iterative tests conducted throughout the design process. At first, an initial test establishes the baseline wind-induced behaviour of the building which is driven by its original architecture and structural design. After that, if severe wind responses are found, one or more optimisation workshops are carried out to investigate what changes in the building geometry can mitigate such responses. If this process starts at an early stage of the project (i.e. concept or scheme design), then minor adjustments to the building architecture are still possible. Therefore, a number of building geometries can be selected to investigate through discussion between wind consultants, structural engineers and architects and, based on the test results, an optimum design can be identified that resolves the issues, both in terms of building architecture and on wind-related structural reliability and occupant comfort.

32.4.2.2 Rigid-model tests

Rigid wind-tunnel models reproduce at reduced scale only the shape of the building and not its dynamic properties. Tests that employ these models do not simulate the building responses to wind directly in the wind tunnel, but only measure the mean and background wind loads. From these, the resonant responses are calculated analytically through computation of the generalized forces based the building dynamic properties. Mean, background and resonant loading are then combined and processed in consideration of the wind climate to predict the overall peak wind loads, accelerations and torsional velocities for the building at full-scale [18, 47, 58].

Rigid model testing represents a simplified approach and their results can be marginally less accurate compared to aeroelastic testing, particularly when examining extremely flexible buildings. Nevertheless, a simpler methodology offers a number of advantages. Rigid models are easier, quicker and therefore cheaper to manufacture, and also to instrument and test in the wind tunnel. They offer the opportunity of processing the wind-tunnel data obtained from a single test with multiple analytical versions of the building dynamic properties, with no need to repeat the test as long as there is no change in the building geometry. Furthermore, rigid model tests are effective for investigating structural and aerodynamic interactions of multiple buildings connected to each other, such as two towers linked via a skybridge or through a common podium [59].

Rigid wind-tunnel models are used for two types of structural study of tall buildings: high-frequency force balance (HFFB) and high-frequency pressure integration (HFPI) tests. HFFB tests employ rigid models mounted on high-frequency force-balance systems to measure directly the overall loads induced by the wind on model. These systems can be underfloor balances, mounted below the wind tunnel floor and on which wind tunnel scale model shells are secured [60], or they can be bespoke balances partly accommodated

Figure 32.9 1:400 scale rigid wind tunnel model mounted on a bespoke HFFB system partially accommodated inside the model. 'Ping An' International Financial Centre in Shenzhen, China. (Image: Courtesy of RWDI)

inside the building model [18] (Figure 32.9). Any of these systems need to measure the five key components of the loading: two horizontal shears, two overturning moments and the torsional moment (determining the lift force is not of primary importance for wind-induced responses of tall buildings) [47].

Typical scales for HFFB models range from 1:200 to 1:500, with surrounding models that extend between 300 and 500 m from the project site, which is similar to cladding model tests. HFFB tests are carried out changing the model orientation by 10° at each step to investigate thirty-six different wind directions in total. Unlike in aeroelastic tests, a single wind tunnel run is carried out for each wind direction because variations of the onset wind speeds and structural damping ratios at full-scale are accounted for analytically during data processing. For analysis, the wind-tunnel data are combined with the structural properties of the building, as provided by the structural engineer, to predict the overall peak wind-induced responses for each of the wind directions tested. Then, the wind climate data are applied to account for wind directionality and to predict the final peak wind responses to be used for design. The test results include a series of static equivalent peak base wind loads, the corresponding floor-by-floor wind load distribution and a series of load combination factors that define how these should be applied to obtain the critical wind-loading scenarios for design. This information is provided to the structural engineer to validate the building design for strength and to carry out deflection checks. In addition, to assess the expected occupant comfort, predictions of wind-induced peak accelerations and torsional velocities are also provided and presented alongside their associated design criteria.

HFPI tests are fundamentally different from HFFB because they do not measure the overall wind loading directly. HFPI tests use a rigid scale model equipped with pressure taps and the overall wind loading on such a model is obtained by integrating wind-induced pressures measured simultaneously over its entire envelope [61]. However, HFPI and HFFB tests are consistent with each other in that only mean and background loading are obtained directly from the wind-tunnel measurements, while the resonant responses are computed analytically and accounted for in the analysis. Within this process, the main difference between HFPI and HFFB is that HFPI tests use distributions of wind loading with height obtained through integration of measured pressures, while HFFB tests extrapolate such distributions from base loading. HFPI tests rely on the same type of rigid wind tunnel models, instrumentation and test procedure as used for cladding tests. Synchronous pressure measurements collected during cladding wind-tunnel tests can thus also be used to carry out HFPI studies.

Previous studies have shown that there is good agreement between the overall wind responses predicted using HFFB and HFPI studies [62]. HFPI studies offer the advantage that they can be carried out in combination with cladding tests, thus reducing costs and testing times compared to doing the cladding and structural studies separately. On the other hand, HFPI tests might require longer preparation before testing compared to HFFB because they require installation of pressure taps on the model. Also, HFPI tests are not well suited to investigate buildings with irregular geometries, complex façades or many balconies, because such instances would require an excessive number of pressure taps to ensure a sufficient resolution for pressure integration.

Although marginally more approximate compared to aeroelastic model tests, both HFFB and HFPI tests are sufficiently accurate to predict the wind-induced responses of the vast majority of modern tall buildings. Furthermore, as they are relatively simple methodologies, they are cheaper, faster and provide notably more flexibility than aeroelastic testing. As a result, both HFFB and HFPI tests are the current standard for confirmatory structural wind-tunnel studies on buildings [18]. In addition, by using modular rigid scale models, HFFB tests are also routinely used to conduct aerodynamic optimization tests.

32.4.2.3 Aeroelastic model tests

Aeroelastic model tests employ wind tunnel models that reproduce at reduced scale both the aerodynamic and the dynamic structural behaviour of a full-scale building. They can thus simulate at scale the wind-induced oscillations of the building and enable the overall wind load and accelerations to be measured directly on the model. To achieve this, an aeroelastic model is designed to replicate not only the geometry, but also the stiffness, structural damping, mass distribution and, in turn, the natural modes of vibrations predicted for such a building at full-scale. Depending on which degrees of freedom are critical and warrant investigation in a certain building, then different types of aeroelastic models can be used for testing [63].

A *full-aerolastic building model* has multiple degrees of freedom and is used to analyse the building response to wind in both sway and torsion. Such a model is challenging to design and build, and leads to long timescales and high project costs. For this reason and also considering that torsional wind-induced responses are rarely critical for modern tall buildings, full-aeroelastic models are seldom employed for commercial projects.

A *spine aeroelastic building model* can be used to study the building response to wind only in lateral directions. This type of model can reproduce multiple sway modes of vibration, including second and higher order modes, but not torsional modes. The main design feature of spine aeroelastic models is to decouple and address separately the modelling of the stiffness from that of mass and geometry, and of damping. They comprise a metal flexible spine that runs through their centre and a number of shell segments that are mounted discretely on the spine (Figure 32.10). The spine delivers the required bending stiffness and modal

Figure 32.10 1:500 scale *spine-aeroelastic building model* showing the internal metal spine, internal model instrumentation and shell segments. Burj Khalifa in Dubai, UAE. (Image: Courtesy of RWDI)

deformations at scale, while the shell segments replicate the building's outer shape and provide the required mass distribution. Small segments separated by tiny gaps are used for the outer shell to prevent them from contributing to the model stiffness or damping. Magnetic or friction devices are used to introduce and control the damping of the model.

A *stick aeroelastic building model* can be employed to investigate the building response associated only with its first order sway modes of vibration. This type of model is a simplified version of a spine model that uses a rigid, rather than flexible, spine, i.e. a 'stick', which is hinged at or slightly below ground level. Spring systems and damping devices positioned under the wind tunnel floor are connected to the rigid spine to provide the stiffness and damping that control the model oscillations about the pivot point. An outer model shell mounted on the spine is still used for these models to reproduce the building geometry and the mass distribution. However, in this case, a single shell can be used, with no need for articulation of multiple segments.

As an aeroelastic model reproduces at scale deflections and oscillations of a given building, the selection of its geometrical scale relates directly to velocity and time scales and, in turn, dictates the wind speed for testing and influences the design of the model stiffening system [5, 63]. Typical scales for aeroelastic building models are in the range of 1:300 to 1:600. Similar to rigid models, aeroelastic wind tunnel models are tested with 300-600 m radius surroundings for thirty-six wind directions, in 10° steps. However, unlike HFFB and HFPI models, they require multiple wind-tunnel runs to examine a range of onset wind speeds and different levels of structural damping for every wind direction. Their main instrumentation includes strain gauges and accelerometers installed on the models, laser displacement transducers and, sometimes, a base-mounted balance, all of which are used to record time-series sampling at high frequencies.

After testing, the building responses measured in the wind tunnel are converted from model to full-scale and then processed in combination with the wind climate data to predict the wind-induced peak responses. The format and content of the final results are similar to those from HFFB tests and include the base wind loads, the floor-by-floor loads, the corresponding load combination factors, and an assessment of the accelerations and torsional velocities.

Wind tunnel tests on aeroelastic building models provide the most accurate and comprehensive simulation at reduced scale of the wind-structure interaction. By reproducing the building movement directly, they can capture the aeroelastic effects, such as aerodynamic damping, which are proportional to the amplitude, frequency and velocity at which a building oscillates in the wind. Nevertheless, these tests require longer timescales and higher costs compared to rigid model tests, due to the complexity of the models, instrumentation and testing protocol. Furthermore, they offer only limited flexibility, if any, for accommodating changes to the building dynamic properties. An aeroelastic model is built to replicate both the outer geometry and dynamic behaviour of a full-scale building. Therefore, if variations in the structural design were to be implemented after the first wind-tunnel test which change the building dynamic behaviour, then a new model and a new test would be required to examine the effects of such changes on the wind-induced responses, even if the building shape were to remain the same.

Aeroelastic tests on tall buildings tend to be instructed only for projects whose challenges relating to wind-induced responses demand a high accuracy, which incurs high costs. For example, these tests are often performed on highly-flexible buildings which, due to their height, slenderness, or structural design, are likely to experience aeroelastic effects such as negative aerodynamic damping [18]. Even in these cases, though, aeroelastic tests are usually performed only as final confirmatory tests and typically follow earlier wind-tunnel studies conducted using rigid models.

32.4.2.4 A case study: 432 Park Avenue, New York

432 Park Avenue is a 96-storey residential tower located in Manhattan, just southeast of Central Park. Completed in 2015 with a height of 425.5 m, it is currently the second tallest building in New York City, and the tallest residential building in the Western hemisphere. The building is mostly known for its narrowness, with a square floor-plan only 28.5 m wide that remains uniform over its entire height. Such a narrow footprint combined with a considerable height leads to a slenderness ratio of 1:15, which was unprecedented for a residential building and posed considerable challenges relating to its ability to withstand the loads and oscillations induced by the wind [64, 65, 66].

Having a square floor plan, entirely uniform in height and with sharp corners made the original design proposed for 432 Park Avenue highly vulnerable to vortex shedding. The initial prediction of a period of sway vibration was around ten seconds, which meant that the vortex shedding excitation would have caused very high across-wind accelerations on the top floors and compromised the human comfort. An optimization of the design was therefore carried out to control and reduce the building response to wind and to make sure that an adequate level of comfort was provided to the occupants.

First, a bespoke assessment was conducted using a motion simulator to establish what level of wind-induced accelerations would be acceptable for comfort. Although motion simulators are used to conduct research on wind-induced accelerations [67], their employment relating to a specific building design is not common. This is because sufficient guidance is normally provided by the acceleration criteria and guidelines available in a number of building codes. The predicted vibration for 432 Park Avenue, however, fell outside the range covered by the building codes and so motion simulation tests had to be carried out. Once an acceptable level of wind-induced accelerations was set, in the range of 8-10 milli-g [64], three major strategies were adopted to optimise the building design and mitigate its response to wind: a bespoke structural design; the use of a supplementary damping system; and an aerodynamic optimization of the building shape [66].

The bespoke primary structural system developed for 432 Park Avenue consists of a reinforced-concrete dual tube-in-tube system [66]. This employs an inner shear-wall core, an external frame on the tower perimeter formed by beams and columns and five outriggers that connect these two subsystems every twelve floors. This structural design contributed to controlling the wind-induced responses by providing the building with high lateral stiffness. In addition, the

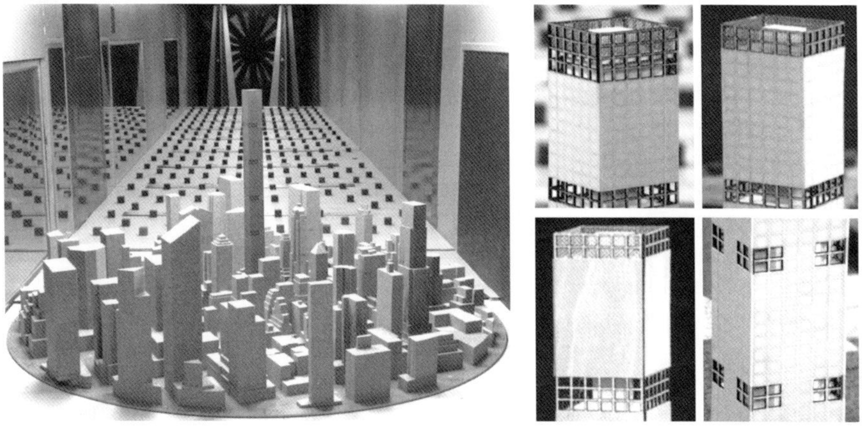

(a) (b)

Figure 32.11 1:400 scale wind-tunnel model of 432 Park Avenue in New York: (a) HFFB model with surroundings. (b) Selected options for openings in the façade investigated during aerodynamic optimization. (Image: Courtesy of RWDI)

large mass associated with using concrete, and the thickening of the slabs on the upper stories, also helped to reduce the wind-induced accelerations.

The supplementary damping system installed at 432 Park Avenue comprises of two opposed pendulum Tuned Mass Dampers (TMD) each with a mass of 600-tonnes [68]. These are mounted on the 84th floor, on opposite sides of the core, and reduce the wind-induced building accelerations by up to 15% [65].

The aerodynamic optimization of 432 Park Avenue was developed in a wind-tunnel test workshop using a 1:400 scale HFFB model (Figure 32.11a). The outcome was a mitigation strategy that consisted of introducing openings in the building façades at discrete levels distributed over the building height. This solution was welcomed because it required only minor changes to the building shape and therefore preserved the original architectural concept. On the other hand, in aerodynamic terms, the presence of the openings allowed for the wind to flow through the building at certain levels and, in so doing, to break the correlation between the vortices shedding on the leeward side at different heights. This, in turn, reduced the magnitude of the fluctuating forces generated on the building across-wind which considerably mitigated the wind-induced accelerations. During the wind-tunnel optimization workshop, a variety of options relating to these openings were tested to predict the associated reduction of wind-induced accelerations. These studies included exploring the number of adjacent floors to be opened up, their height in the building, and what portion of the façade should be open at each level (Figure 32.11b). The design finally adopted comprises six double-floor openings, every twelve storeys in the building, which also provide space that could be used for plant rooms. Confirmatory wind-tunnel tests were then carried out to confirm the viability of the final design, which included HFFB and aeroelastic structural wind tunnel tests, as well as cladding tests.

32.4.3 Pedestrian level winds in urban environments

Amongst the effects of atmospheric winds blowing on urban areas are not only the wind loads on the buildings, but also the wind conditions that develop around them. Understanding and controlling the wind flows that develop in the proximity of and in between buildings is extremely important, particularly at ground level, where the local wind conditions may affect the people who stand, walk or cycle there [4, 69]. First, and most importantly, strong and gusty winds near the ground can pose a safety threat. They could cause people to lose balance or, in extreme cases, be blown over, fall and potentially get injured. Secondly, even when they are not strong enough to cause safety concerns, the local winds on the ground can still affect people's comfort because they can make it difficult or unpleasant for certain activities to be carried out in certain areas.

32.4.3.1 Urban wind flows in proximity to the ground

Experiencing strong and gusty winds locally on the ground in urban areas is not uncommon. For example, downdrafts, created by tall buildings that deflect the oncoming wind downwards, can result in high local wind speeds at pedestrian level when they hit the ground and are diverted horizontally. Or wind gusts may develop downwind of the sharp corners of a building as the flow from the windward façade accelerates around those corners and then suddenly detaches from the building to form vortices. Or again, high local wind speeds are caused by funnelling, i.e. flow acceleration through openings in a building or narrow gaps along streets between adjacent tall buildings [4, 36]. Interestingly, whether or not any of these conditions occur depend not only on the strength and direction of the overall wind, but also on the geometry, size, height and location of the buildings in a particular area, which determine how the flow is diverted and directed locally near the ground. Furthermore, how often they occur during a year depends on the wind climate.

32.4.3.2 Lawson criteria for pedestrian safety and comfort

The safety of pedestrians in wind in an urban area is assessed based on the local wind speed near the ground and its likelihood of occurrence. The perception of comfort, on the other hand, is evaluated by relating the local wind speed and its probability of occurrence with the type of activity in which people are engaged.

A number of criteria have been proposed and adopted over the years to assess the safety and comfort of pedestrians in urban environments [4, 69, 70, 71]. These include the criteria proposed by Lawson which are currently widely adopted both in the UK and other countries. To assess safety, Lawson introduced a 'distress criterion' that sets the limit of acceptability at Beaufort Force 6 wind speed (i.e. an hourly average wind speed of 10.95 to 14.1 m/s), with a maximum frequency of occurrence of one hour per year (i.e. a probability of exceedance of 0.01%). To assess comfort, the Lawson criteria categorize six different human activities and, for each of these, set an upper threshold consisting of a wind-speed range and its highest frequency of occurrence. The wind speed adopted in the Lawson criteria is taken at heights representative of pedestrians, such as head height, and the wind-speed ranges associated with each comfort category relate to the mariners' Beaufort Wind Scale adapted for use on land (Table 32.2). No

Table 32.2 Lawson criteria for pedestrian comfort in wind and description of human activities. (Sources: [4, 71])

Comfort Category	Upper threshold: Maximum frequency of occurrence	Beaufort Force	Hourly average wind speed (m/s)	Description
Roads and Car Parks	6%	5	8.25 – 10.95	Winds of this magnitude are considered a nuisance for most activities, and wind mitigation is required
People around buildings	2%	5	8.25 – 10.95	High speeds that can be tolerated if one's only objective is to walk, run or cycle without lingering
Pedestrian walk-through	4%	4	5.60 – 8.25	Breezes that would be appropriate for window shopping and strolling along a city/town centre street, plaza or park
Pedestrian standing	6%	3	3.35 – 5.60	Gentle breezes acceptable for main building entrances, pickup/drop-off points and bus stops.
Entrance doors	6%	3	3.35 – 5.60	
Sitting	1%	3	3.35 – 5.60	Light breezes desired for outdoor restaurants and seating areas where one can read a paper or sit for long periods.

provisions are included in these criteria to account for the effect of any other environmental agent.

It can be seen from the definitions of the comfort categories proposed by Lawson that the more frequently a given wind-speed threshold is exceeded, the lower the level of comfort. Also, the more active the people are, the more they accept stronger winds and higher frequencies of occurrence as comfortable. As the Lawson criteria relate to the Beaufort Scale, the wind-speed ranges associated with each comfort category are expressed in terms of hourly averages. However, it is recognized that human comfort in wind depends not only on the mean, i.e. the hourly average wind speed, but also on the size and strength of the wind gusts. Therefore, an 'equivalent hourly-average gust speed' was introduced to account for the contribution of wind turbulence which can be calculated from the local gust wind speed and, together with the corresponding hourly average, compares directly to the thresholds set for each category to determine the comfort level.

32.4.3.3 Mitigation measurements

If wind conditions on the ground are found to be either unsafe or uncomfortable for the intended use it is usually advisable for them to be mitigated [71]. Mitigation can be achieved by modifications to the building geometry, such as changes

Figure 32.12 1:500 scale wind-tunnel model for Pedestrian Level Wind (PLW) testing including landscaping. Burj Khalifa in Dubai, UAE. (Image: Courtesy of RWDI)

in massing, addition of fins, sunshades and porous features on the façades, or addition of canopies and screens, particularly in proximity to entrances. Alternatively, landscaping on the ground can divert the oncoming flow and shelter locations deemed critical for wind. Trees, hedges and other vegetation can be used for landscaping together with or as an alternative to solid and porous screens or artistic features such as public sculptures (Figure 32.12).

32.4.3.4 Pedestrian-level wind tests

Experimental measurements on scale models in BLWTs have been carried out to assess the pedestrian level winds (PLW) in built environments for many years. Recently however, they have been in greatly-increased demand as a consequence of the ever-growing number of mid- to high-rise buildings that have been built in many cities around the world. The higher the buildings, and the more closely spaced they are in urban areas, the more complex and potentially severe become the wind-induced flow conditions on the ground. Furthermore, the more that pedestrians use and transit through those areas, the higher the risk that adverse local wind conditions lead to widespread public complaints or, even worse, to consequences that might compromise people's safety. To address this high demand and to prevent wind-related issue being raised, it has now become routine that the pedestrian comfort and safety in wind in urban areas are assessed using BLWT tests at a very early design stage of a new development. If required, it is also common to implement counter measures that mitigate any risk of having unsafe or uncomfortable wind conditions for the intended use of the spaces. In some cases this is now a formal requirement posed by the local planning authorities that grant permission to build [71].

Most PLW wind-tunnel models are of similar size and scale to those used for cladding and structural studies. Model scales are in the range of 1:200 to 1:500, while surrounding models reproduce a 300- 600 m radius around the site of interest. Exceptions occur when masterplans that extend over large areas are analysed which might require using larger surrounding models. It is common for PLW tests to investigate multiple configurations of surroundings in proximity of a given target building or development to analyse the effects of existing and future scenarios. In addition, to provide a benchmark for the local wind conditions, a baseline configuration is also generally tested that consists of all buildings and structures at and in proximity to the project site as they are before any of the proposed development is built. The wind directions tested can be thirty-six in total, one every 10°, as for cladding and structural tests. However, for PLW tests this requirement is sometimes relaxed and only sixteen wind directions, one every 22.5°, are tested to carry out an assessment [69].

The current standard for PLW wind-tunnel tests is to employ Irwin probes or similar types of surface velocity sensors. This type of instrument is best suited for such tests because it detects the magnitude of the time-average and gust flow speed in proximity to the ground. Furthermore, compared to more sophisticated options such as HWA or multi-hole pressure probes, Irwin probes are cheaper, easier to calibrate and install, and extremely robust. If detailed investigations are also required to resolve the local direction of the flow, then they can be supplemented by using directional surface sensors, HWA or multi-hole pressure probes. A tap layout is prepared before testing to establish the measuring locations which are set to capture the most critical wind conditions expected within a given urban environment. Irwin probes are installed directly on the ground plane of the wind-tunnel scale model prior to testing and then connected to a remote pressure scanning system. In addition, it has become common to measure the local wind conditions also at height in any outdoor space accessible to people that is part of the proposed development. For example, in areas like balconies, terraces, outdoor pool decks and viewing decks. Irwin probes or similar surface pressure sensors embedded into the building model are widely used to carry out these measurements in parallel with those monitoring at ground level (Figure 32.13).

The number of measuring locations in a PLW test depends greatly on the specific circumstances of the projects. Generally, several hundred locations are monitored during each test. In principle, the number of channels available in the pressure scanner poses a limit to how many probes can be monitored simultaneously during the same test. However, if the number of measuring locations exceed the capacity of the pressure scanner, multiple tests can be carried out to monitor different groups of probes during successive runs. After completion of testing, the wind-tunnel results are combined with the wind climate data for the site to predict the local wind speeds at the measuring locations together with their probability of occurrence. This enables wind-related safety to be assessed at each location, both on the ground and on building outside-spaces at height, as well as a comfort category to be defined according to the adopted assessment criteria (Figure 32.14).

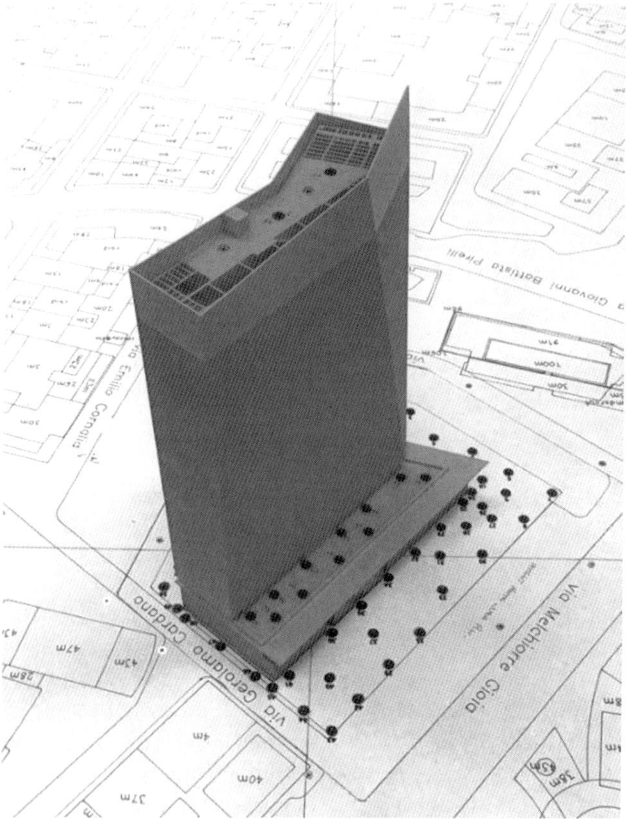

Figure 32.13 1:300 scale wind-tunnel model for PLW testing including Irwin probes installed on the ground, podium and roof terraces. Porta Nuova Gioia in Milan, Italy. (Image: Courtesy of RWDI)

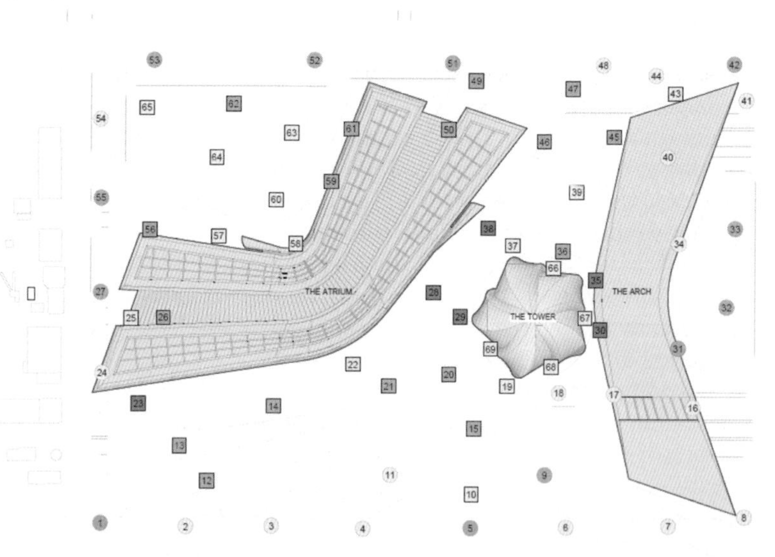

Figure 32.14 Example of wind-tunnel test results for Pedestrian Level Wind studies using colour-coding to indicate different comfort categories. (Image: Courtesy of RWDI)

32.5 Wind-tunnel tests on bridges

Together with investigations on buildings and urban environments, experiments on scale-model bridges represent one of the most common uses of modern BLWTs. Over the last fifty years, the progress made in construction techniques, materials and technology has been pushing back the boundaries of bridge engineering. Bridges have become longer and longer, with a world-record main span of 1991 m reached by the Akashi Kaikyo Bridge, in Japan [72]. Moreover, modern bridges can be extremely slender, have low structural damping and be considerably lighter than in the past. All these aspects increase their vulnerability to wind-induced excitations and expose them to the risk of experiencing aerodynamic instabilities whatever their span. It has therefore become more and more important, as well as being required by codes of practice and design standards, that appropriate wind-tunnel studies are conducted – not only to inform the design for wind of new structures, but also to support the retrofitting of already existing bridges [2, 73, 74].

32.5.1 Principles of bridge aerodynamics

Like buildings, bridges are surrounded by complex flow fields generated by the gustiness of the onset winds and by the turbulence that develops when such winds impinge on the bridges. However, in the case of bridges, such a complexity is aggravated by the contribution of aeroelastic effects, i.e. the fact that the large oscillations which a bridge undergoes when subject to wind have, in turn, an effect on the wind field around it [5]. As a result, a bridge can experience a series of steady and unsteady aerodynamic effects, which include:

- Static and dynamic wind loading
- Vortex-shedding
- Static and dynamic aeroelastic instabilities, such as torsional divergence, galloping and flutter.

The envelope of the responses to wind of a given bridge is determined by a combination of these effects the contribution of which is expressed through dedicated sets of steady and unsteady non-dimensional aerodynamic coefficients [75]. These parameters include *static force and moment coefficients, nonstationary aerodynamic (flutter) derivatives* and *admittance functions* which can be extracted from wind-tunnel tests of scale model bridges that usually investigate the deck, towers and cables of the bridge separately [5, 76].

Building on the research and engineering work that was carried out in the middle of the last century, the current methodology for analysing the wind-induced responses of bridges rely on the combined use of wind-tunnel testing and analytical / numerical methods [75, 77, 78]. Wind-tunnel tests on reduced-scale models of bridges are conducted for many different purposes. By replicating at reduced scale the airflow-structure interaction and then measuring the bridge responses to wind, these experiments investigate the aerodynamic stability either of the entire bridge, or of the deck, tower and cables separately. Furthermore, wind-tunnel tests are also performed to measure the steady and unsteady aerodynamic forces acting on the deck, towers and cables individually and to extract the corresponding non-dimensional aerodynamic and aeroelastic coefficients. These coefficients serve as the input to the numerical simulations of the full bridge which are performed to predict its response to wind using computer software.

Numerical simulations that predict the bridge responses are generally based on Finite Element Models (FEM) which reproduce the characteristics of a given bridge in terms of mass, stiffness and damping to simulate its dynamic behaviour. As part of the simulations, the onset turbulent wind conditions specific to the site are recreated by producing time-series of the fluctuating wind-velocity field that impinges on the bridge and from these, the steady and unsteady aerodynamic excitations are calculated using the aerodynamic non-dimensional coefficients obtained experimentally. By applying such aerodynamic excitations to the FEM, the wind-induced overall loading, localized stresses, deflections and accelerations of the bridge can be predicted, which inform the structural design [79].

32.5.2 Full-aeroelastic bridge model tests

Full-aeroelastic models replicate at reduced scale the geometry, mass, stiffness and structural damping of full-size bridges. These models simulate in a wind tunnel not only the aerodynamic forces acting on a bridge but also its natural frequencies and modes of vibration and, in turn, its wind-induced responses [80]. Full-aeroelastic models are built and tested during the design phase of a bridge to predict the wind-induced responses and wind loading in its final configuration. A model is made of either the whole bridge or only its main span, which tends to be the portion most vulnerable to wind. Models can also be made of a partially-constructed bridge to examine its wind-induced behaviour during the construction stages when a bridge may be particularly vulnerable [80].

The design approach to full-aeroelastic bridge models relies on modelling the stiffness, mass and shape of the structure using separate features, which is similar to, but more complex than, a spine aeroelastic building model. A full-aeroelastic bridge model is made of multiple spines and cables linked one to another which form its internal supporting structure and provide the required stiffness. Onto this spine and cable system are mounted a number of shell segments which form the external shape and contribute to the mass of the model. Each shell segment is rigidly attached to the spines only in one point and small gaps of 1 to 2 mm are allowed between adjacent segments, so that they do not come into contact and therefore do not contribute to either the model stiffness or damping. Elements such as springs, sliding or hinge joints, weak flexures and lump masses are employed to connect different spine elements, to anchor the model into the wind tunnel working section, or to provide it with appropriate localised stiffness or mass. For example, if the side spans of a bridge are not entirely modelled in the wind tunnel, weak flexures are often fitted at both ends of the bridge deck to replicate at scale their contribution to the dynamic behaviour of the main span. The structural damping of the model depends mainly on the inherent damping of the spine and cable systems and aims to be in the range of 0.5% to 1.5% of critical damping.

Figure 32.15 1:300 scale full-aeroelastic model. Messina Strait Suspension Bridge, Italy. (Image: Courtesy of Politecnico di Milano)

Internal spines for full-aeroelastic bridge models are mostly made out of metal, for example aluminium or brass, and can be machined to size to match the required stiffness at model scale. Cables can be made out of piano wire of different diameters. These wires are normally used in combination with a series of discrete cylinders having larger diameter than the wire which are clamped onto it and evenly distributed along its length to replicate at scale the mass and aerodynamic drag predicted for the cables at full-scale (Figure 32.15). Also, calibrated springs are often mounted in series with the piano wire to reduce its axial stiffness and match the target value obtained through scaling of the properties for the full-scale cable. Shell segments for the deck and the towers can be hand-built out of balsa wood, or machined using foam or resin materials offering different strengths and densities, or nowadays, often 3-D printed. Other elements of the model, including springs, weak flexures and links between spines and cable tend to be bespoke and custom-built depending on the individual application. The mounting of a full-aeroelastic bridge model on the floor of the wind-tunnel working section is often made as rigid as possible, so that the model anchoring does not interfere with the dynamic behaviour of the model itself. However, there are instances when the soil conditions at the bridge site at full-scale are so weak that they significantly affect the stiffness and, in turn, the dynamic behaviour of the whole structure. In these cases, it becomes necessary that the stiffness of such foundations is also reproduced at scale as part of the aeroelastic model and therefore bespoke floor-support systems are designed and employed in the model mounting inside the wind tunnel.

Typical scales for full-aeroelastic bridge models are in the range of 1:100 to 1:300 [8, 73, 80]. These scales are larger than those commonly used for other aeroelastic models, for example of tall buildings, as they result from scaling criteria that are more complicated than for other structures. When suspension bridges are studied, for which gravity provides a significant contribution to the overall stiffness of the main cables, the Froude number similitude needs to be considered [81]. This dictates a relationship between the model scale and the scaling ratio for the velocity that leads to extremely low wind speeds for testing if excessively small model scales are adopted. Therefore, using small scale models tends to be avoided to allow for reasonably high test speeds which, amongst other benefits, mitigate the issues consequent on having a low Reynolds number for testing. On the other hand, because the length of bridges at full-scale can be several hundred metres or even more than a kilometre, modelling them at 1:100 to 1:300 scale requires the use of a large BLWT for testing, with a working section 10 m or more wide.

Full-bridge areoelastic models are often tested in isolation. However, if buildings or other structures of significant size are present in the vicinity, they can sometimes be included as surroundings in the wind tunnel model to provide an accurate representation of their aerodynamic interferences on the bridge. Complex topography is also modelled to reproduce at scale its effects on the wind impinging on the bridge. Finally, if two parallel bridges are to be built side by side in close proximity, then two full-aeroelastic models are tested together in a wind tunnel to capture in full their mutual aerodynamic interference (Figure 32.16).

Wind-tunnel testing of full-aeroelastic bridge models is conducted in turbulent flow, which is calibrated to provide an adequate simulation at scale of

Figure 32.16 1:211 scale full-aeroelastic models of two parallel bridges in close proximity. Tacoma Narrows Bridges, Washington State. (Image: Courtesy of RWDI)

the gusty onset winds expected at the bridge site. Tests in smooth flow can also be performed for theoretical and validation purposes, and to assess the effect of the onset turbulence, although they are not representative of any actual onset wind condition at full-scale. As long as the whole wind-tunnel model is accommodated on a turntable, a number of wind directions can easily and quickly be investigated by rotating the model. However, provided that they include any of the prevailing winds, the most critical wind directions for bridge stability and overall wind-induced responses are usually those perpendicular to the bridge axis. The range of flow velocities for testing in the wind tunnel is determined by the wind speeds predicted at the bridge site at full-scale by applying appropriate scaling factors. Tests are conducted to investigate whether any vortex shedding or divergent aeroelastic instability such as flutter, occur and to measure the wind-induced loads, displacements and accelerations. Instrumentation employed during full-aeroelastic bridge testing include miniaturised accelerometers mounted on the model and laser displacement transducers, both placed at carefully chosen positions on the deck, towers and cables to monitor the bridge responses. In addition, strain gauges and miniaturised load cells can be integrated into the spine and cable system to measure stresses and forces at critical locations, such as the tower bases [80, 82].

The main benefit of full-aeroelastic bridge model testing is the ability to measure directly at scale the wind-induced displacements and accelerations of a bridge model and so to assess the onset of any vortex shedding and diverging instabilities. Based on these tests, the wind-induced behaviour of the bridge at full-scale can be predicted. Moreover, the results can also be used to validate the numerical simulations from which the wind-induced loads to be used when designing the bridge are obtained [83]. Nevertheless, the design of full-aeroelastic bridge models is far from being straightforward and the models are extremely complex, costly and time consuming to build. Furthermore, testing often requires the use of large BLWT facilities which are not easily available, expensive to operate and face challenges when simulating at reduced scale the large integral length scales and spatial correlation of turbulence typical of atmospheric winds. Finally, by reproducing the whole bridge, full-aeroelastic models do not offer the flexibility to investigate subcomponents of a bridge individually. They are therefore not suitable for obtaining the aerodynamic and aeroelastic non-dimensional coefficients associated with the deck, towers or cables [82].

32.5.3 Sectional bridge model tests

Wind-tunnel tests on sectional models can be performed to investigate the aerodynamic and aeroelastic behaviour of individual sub-components of a bridge. They are carried out primarily on bridge decks, but can also be conducted to investigate sections of the towers or the cables [5, 8, 84].

Sectional bridge-deck models consist of rigid models that reproduce at relatively large scales a short length of the bridge deck. They can be tested either statically or in a dynamic configuration. Static sectional models do not move or vibrate subject to wind. They are mounted on force balances or load cells and are used to measure forces and moments to obtain the static aerodynamic

Figure 32.17 1:14 scale sectional model of Tintagel Footbridge, UK. (Image: Courtesy of RWDI)

coefficients. Dynamic sectional models are supported at both ends by moving rigs which allow them to vibrate at calibrated frequencies and also control their structural damping. These models are used for dual purposes: to extract the unsteady aerodynamic coefficients such as aerodynamic derivatives and admittance functions; and to investigate the onset of vortex shedding and aeroelastic instabilities.

In their most common setup, sectional bridge-deck models are designed to span the entire width of the wind tunnel and are mounted at both ends onto bridge rigs that are placed outside of the working section (Figure 32.17). However, models consisting of a live mid-portion mounted in between two dummy sections, with endplates at both ends, can also be used and set up within wide wind-tunnel test sections and supported either from beneath or suspended on tensioned cables (Figure 32.18).

Bridge sectional model rigs are often mounted on the side of the wind tunnel working section and generally consist of a series of rails that allow model oscillations in two or three degrees of freedom: vertical, torsional around the longitudinal axis, and lateral along-wind. These rigs employ multiple series of calibrated springs to control the frequencies of vibration of the model which are set in each degree of freedom independently. Furthermore, they also include suitable bearings to minimise the inherent damping of the system when it moves, so that a controlled level of structural damping can be deployed during testing using magnetic or friction-based devices. Both the frequencies of vibrations and the mass of a sectional deck model are determined through scaling of the full-scale bridge properties. In this process, to enable appropriate investigation of two-degree-of-freedom dynamic instabilities, it is essential that the frequency ratio between the corresponding vertical and torsional modes of vibrations is

Figure 32.18 1:30 scale sectional model of the Adige cable-stayed bridge near Padua in Italy. (Image: Courtesy of Politecnico di Milano)

maintained between full size and model scale [81]. Alternative types of sectional model rigs have also been developed that rely on hydraulic actuators to control and impose the motions [85].

Sectional bridge-deck models are built at considerably larger scales compared to full-aeroelastic models. Typical scales are in the range 1:10 to 1:100 [8, 73], but can be larger for relatively short spans in highly-flexible bridges such as many modern footbridges. The use of such large scales enables a better Reynolds number similitude to be obtained for testing [81]. It also allows these models to reproduce in detail not only the shape of the deck itself, but also many of the appendages mounted on the deck such as guard rails and traffic barriers, light posts and wind fences. This increases the accuracy of the simulation because these features have the potential to affect the aerodynamic performance of the deck significantly [86]. In terms of size, guidelines suggests that sectional scale models should replicate a portion of span at least four times longer than the width [8]. In practice, most of these models are generally no more than 2-3 m in span to make sure they are sufficiently rigid and do not bend or twist excessively during testing.

Sectional bridge-deck models are tested over a range of wind speeds. Measurements of static forces and moments are carried out at speeds in the upper bound of wind-tunnel capabilities, typically between 15 and 30 m/s, in order to maximize the Reynolds number for testing and obtain stronger signals from the force-measuring systems. Dynamic tests are performed over a range of flow speeds, generally between 1 and 30 m/s. These are selected depending on the expected wind speed at full-scale and on the scaling ratio adopted for velocity which, in turn, follows from the model scale and the scaling ratio for frequency [81].

Static tests and dynamic tests undertaken to check stability are usually carried out both in smooth and turbulent flow. The results from smooth flow tests provide a theoretical baseline for the aerodynamic behaviour of the bridge. Any full-scale prediction is based on the tests with turbulent onset flow, which is calibrated to provide at scale a partial turbulence simulation of a natural gusty wind experienced at full-scale [8]. Dynamic tests for extracting the flutter derivatives are carried out in smooth flow, which is necessary to remove any contribution from the onset turbulence and to isolate the effects of the self-excited aerodynamic forces to which such coefficients apply [5]. Measurements aimed at identifying aerodynamic admittance functions are carried out with turbulent onset flow [85, 87]. Different wind-tunnel test techniques can be adopted to obtain these unsteady aerodynamic coefficients, including both the free-vibration and forced-vibration methods. Instrumentation for static sectional model bridge testing consists essentially of the load-measuring systems integrated at both ends of the scale model. For dynamic testing the arrangement also includes accelerometers and laser displacement transducers. Multiple flow-speed measurements are also collected upstream and at the model location in the working section during these tests to enable blockage corrections to be applied, if necessary.

Tests on reduced-scale sectional bridge models represent a simplified methodology compared to full-aeroelastic bridge model testing. This means that these tests can neither resolve the three-dimensionality of the wind-bridge interaction, nor provide a direct simulation of the wind-induced bridge responses in full. However, relying on a simplified approach also gives a number of advantages compared to more comprehensive methodologies. First, once a sectional bridge rig has been developed, it can be used to test many different bridge models. Secondly, sectional bridge models are considerably easier to design and build than full-aeroelastic models. They are therefore cheaper and can be constructed and tested in shorter timeframes. Furthermore, they give large flexibility in terms of testing: for example, by adjusting the springs and damping devices of the bridge rig, they enable modes of vibration of a different order to be investigated. Also, when required, the same sectional model can be used to carry out a dynamic optimisation of the deck which consists of implementing minor changes to the deck geometry and repeating the tests to assess the effects on both the stability and the aerodynamic loading. Finally, sectional model tests complement well the numerical simulations of the bridge responses to wind, because they can provide the full set of steady and unsteady aerodynamic non-dimensional coefficients that are required as an input to such simulations.

Overall, scale sectional model tests represent the current standard methodology to investigate a bridge's susceptibility to wind-induced instabilities and to measure its non-dimensional steady and unsteady aerodynamic coefficients. As such, they are routinely undertaken both to inform the design of new bridges and to support the retrofitting of existing ones, such as recently undertaken for the Golden Gate Bridge in San Francisco (Figure 32.19).

32.5.4 Other wind-tunnel tests on bridges

In order to reproduce at scale the geometry and the vertical and torsional sinusoidal modes of vibration of a bridge deck, taut-strip models can be used [88]. These are flexible models used to examine the steady and unsteady wind-induced

Figure 32.19 1:60 scale sectional model of the Golden Gate Bridge in San Francisco. (Image: Courtesy of RWDI)

responses of a bridge deck subject to a three-dimensional turbulent flow, with no inference from with other elements of the bridge. The geometry of the deck and its mass are appropriately scaled to replicate the full-size bridge, and its vertical and torsional stiffness are provided by means of two or more tensioned wires or tubes, placed inside the model and running along the its entire span. The level of pretension is chosen to achieve the appropriate frequencies of vibration for the given geometry and mass of the model deck. Taut-strip model testing represents a simplified methodology compared to full-scale aeroelastic models. However, using flexible models subject to a 3-D turbulent flow simulation is more accurate compared to sectional model tests, which instead use rigid models suspended on moving rigs subject to a 2-D flow. Although the complexity, timescale and costs related to building and testing a taut-strip model exceed those associated with conducting a sectional model experiment, they are considerably less than those associated with performing full-aeroelastic model testing. As such, taut-strip model testing gives the opportunity to refine the predictions of sectional model testing without having to afford a full-aeroelastic test (see Section 24.6.2).

Wind-tunnel tests are also often carried out on scale models of free-standing bridge towers, with no cables or deck. These are undertaken to investigate the stage in construction before the towers are fully connected with the rest of the bridge infrastructure. In this scenario the stiffness of a bridge tower is likely to be significantly lower than when the bridge is fully constructed. Its wind-induced responses and particularly its susceptibility to vortex shedding can thus exceed those experienced after the completion of the bridge, making this stage critical in the structural design [75, 89]. In principle, a simplified investigation of the responses of a bridge tower to wind could be carried out using a sectional model. These tests would include static force measurements to obtain the aerodynamic load coefficients used in wind-loading calculations, as well as dynamic tests to examine susceptibility to vortex shedding. However, since these tests are undertaken using a 2-D flow simulation, they cannot replicate the three-dimensional flow effects that develop both at the top of the tower and around any crossbeams in the tower, leading to less accurate results. For this reason, 3-D flow tests on a free-standing aeroelastic tower model may be undertaken to obtain predictions of behaviour that are sufficiently accurate to inform the design (Figure 32.20).

Figure 32.20 1:100 scale aeroelastic model of a free-standing bridge tower. Messina Strait Suspension Bridge, Sicily. (Image: Courtesy of RWDI)

32.6 Wind-tunnel tests on other structures

In addition to buildings and bridges, a variety of other civil engineering structures have emerged over the years as being vulnerable to significant wind-induced excitations. Amongst others, these include long-span roofs, highly-flexible building appendages such as spires, antennae and louvres, as well as various types of artwork. Wind-tunnel tests on scale models of these structures are now commonly undertaken in modern BLWTs in order to support their design. BLWT tests are also often conducted to investigate wind effects on urban environments other than pedestrian level winds, such as wind-driven dispersion of plumes and pollutants [4, 8].

32.6.1 Long-span roofs

Long-span roofs are employed on a variety of modern structures, from sport stadia, to airport or railway terminals and to low-rise and large footprint commercial buildings such as shopping malls. Being extremely flexible structures of large extent, these roofs often have low natural frequencies of vibration, which make them highly prone to wind-induced oscillations of large amplitude and, in turn, high resonant wind loads [90, 91]. As a result, long-span roofs nowadays often undergo BLWT testing as part of their design.

The most common methodology adopted for BLWT testing of long-span roofs relies on carrying out high-frequency simultaneous pressure measurements of the same type as those conducted on buildings (Figure 32.21). From these measurements, predictions of the localised peak-pressures can be obtained to inform the design of the cladding elements and their fixings. In addition, by performing an HFPI study, the local pressure measurements can be integrated over the roof surface and combined with the dynamic properties of the structure to predict the overall wind loads which inform the design of the primary and secondary structural elements of the roof [91, 92]. As an alternative, aeroelastic scale models can be built and tested to investigate extremely flexible long-span roofs susceptible to wind-induced oscillations of large amplitude [93, 94]. However, these tests are

(a)

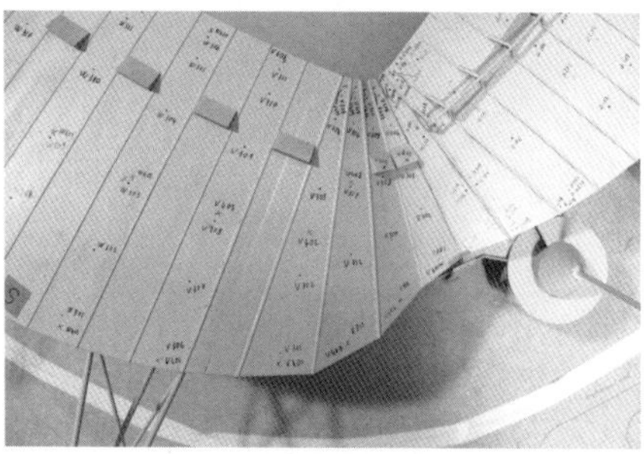

(b)

Figure 32.21 1:250 scale model of a long-span stadium roof for the Etihad Stadium in Manchester, UK. (a) Wind-tunnel model surrounded by adjacent buildings. (b) Detail showing pressure taps for cladding and HFPI studies. (Images: Courtesy of RWDI) rarely conducted because of their considerable complexity, long timescales and high costs.

The experimental methodology employed to carry out pressure testing on scale models of sport stadia and other long-span roof structures is very similar to those used for pressure testing on buildings. Model scales in the range 1:200 to 1:400 and surrounding models with 300-600 m radius are used. Tests are conducted for thirty-six wind directions, i.e. every 10°. Generally, between several hundreds and a thousand pressure taps are installed on scale models of long-span roofs, the exact number of which depends on the size and complexity of the model. For studies that investigate roofs of enclosed buildings, pressure taps are installed only on the upper surface of such roofs and the tubing is hosted inside the building model. No pressure measurements are required from the underside of these roofs as they are not exposed to wind. However, cantilevered roofs typical of sport stadia require pressure taps on both the upper and lower surfaces, as both are directly exposed to wind. Modelling of these types of roof thus poses notable challenges for the installation of taps and running the pressure tubing without interfering with the wind flow in proximity of the model.

32.6.2 Building appendages and superstructures

A wide variety of appendages and superstructures are commonly fixed to either the top or the façades of modern buildings and, in some instances, they are sufficiently flexible by design to become vulnerable to wind. These include, amongst others: building spires and antennae; architectural features such as louvres, fins and sunshades; and sometimes forms of artwork [8, 18].

In many cases, analytical assessments based on information available from the literature are appropriate to predict both the wind loads and the aeroelastic behaviour of these elements. Otherwise, educated estimates can also be obtained based on local pressure measurements collected in proximity of such elements through a cladding wind-tunnel test of the entire building. However, there are circumstances where these approaches do not suffice, either due to the unconventional shape of these appendages and superstructures, or to their large size or extreme flexibility. In these cases, dedicated BLWT tests can be carried out to investigate the wind-induced behaviour of these features.

Most of the wind tunnel tests on building appendages or superstructures employ aeroelastic models, which are necessary in order to obtain an accurate simulation in consideration of the high flexibility of such building features. Also, given their relatively small size at full-scale, large model scales in the range 1:1 to 1:20 are typically used in these tests. It is important that a calibrated turbulent flow simulation consistent with such a large scale is produced in the wind tunnel for testing so that the effects associated with the gustiness of the onset wind can be properly accounted for.

32.6.3 Plumes and pollutant dispersion in urban environments

Due to wind-driven dispersion, exhaust gases, smoke plumes and other pollutants can significantly affect the air quality in outdoor urban areas at great distances from the source of emission. Furthermore, if they re-enter a building, they obviously have an impact also on the internal conditions and either lead to unpleasant environmental conditions or, in some cases, can pose risks to human health [95].

To prevent serious issues in relation to plumes and pollutant dispersions, particularly in urban areas, design guidelines are provided in codes and standards for exhaust stack design [96]. However, the complexity of the wind flow in urban environments is such that bespoke BLWT studies are often required, for example as part of an environmental impact assessment – first to assess the effect of exhaust-gas dispersion on the air quality and secondly, if serious issues are identified, to help develop suitable mitigation strategies.

Wind-tunnel measurements of gas dispersion are typically carried out by modelling urban areas within full-scale radii of 300-900 m, at model scales of 1:200 to 1:500. Tests are conducted in turbulent onset flow simulations, which are calibrated to replicate at scale the gustiness of the natural winds and to capture its effects on the gas dispersion. To reproduce the effect of buoyancy, it is essential for wind dispersion studies to maintain the Froude similitude between model-scale and full-scale. By relating the flow-velocity scaling to the model scale, the Froude similitude imposes relatively low wind-tunnel test speeds, typically in order of 0.5-2.5 m/s. It is also necessary to select an exhaust gas for wind-tunnel testing that maintains the same density ratio with respect to air as at full-scale. Plumes at full-scale are generally lighter than air due to their high temperature, but such high temperature cannot be directly simulated in the wind tunnel. The solution is to use light gas mixtures at ambient temperature in the wind tunnel; gases typically used in these mixes include ethane, helium and nitrogen mixed in a proportion that is calibrated to obtain the required density ratio [34, 48, 97].

During a test, the selected mixture of gas is released at a calibrated flow rate at the exhaust location under study. To model the interaction between the plume and the onset wind accurately, it is important that a consistent scaling is adopted in the wind-tunnel experiment between the plume exhaust flow velocity and the onset wind velocity. To monitor the exhaust dispersion the concentration of a carbon tracer gas mixed in the exhaust, is measured using a flame ionisation detector (FID). Generally, multiple wind directions and different onset wind speeds are tested in wind-tunnel plume-dispersion studies. Furthermore, not only is the concentration of exhaust gases monitored in multiple locations by collecting a number of samples, but also multiple exhaust locations can be examined. As a result, it is not uncommon for these studies to require a large number of wind tunnel runs to be carried out and, in consequence, become time consuming and expensive to commission.

32.7 The future of the BLWT

In recent decades, digital models of the movement of air have been created using Computational Fluid Dynamics (CFD). CFD describes a wide range of techniques for studying fluid flows by modelling them numerically, rather than experimentally. It does not rely on conducting tests on physical models, but exploits computational resources to perform virtual simulations.

Two of the most common CFD techniques currently adopted to study turbulent wind flows around buildings and civil engineering structures are RANS and LES [98, 99]. The former, which is the simplest, solves the Reynolds-averaged Navier-Stokes (RANS) equations using a given empirical model to reproduce all turbulence numerically. By contrast, software using Large Eddy Simulation (LES) is more advanced and resolves the large-scale turbulence in a flow directly while relying on empirical models only to reproduce the turbulence at small scale. Other CFD techniques currently in use include Unsteady-RANS (URANS), Detached Eddy Simulations (DES), Very Large Eddy Simulations (VLES) [100], as well as Smoothed-particle hydrodynamics (SPH) and Lattice Boltzman Methods (LBM) [101].

In principle, CFD simulations can be used to undertake numerical modelling in investigations that have been carried out historically using BLWT testing. These include assessing wind loading and wind-induced responses of buildings and civil engineering structures, as well as studying pedestrian level winds and wind-driven pollutant dispersion. However, the results obtained through CFD are not always reliable and their accuracy varies considerably depending on the application.

There is now some consensus that good-quality RANS simulations are capable of predicting pedestrian level winds in urban environments with an accuracy similar to BLWT tests. Nevertheless, there are still significant concerns on the ability of RANS and LES to predict with sufficient accuracy any wind-induced local peak pressure, overall wind loading or wind-induced response of buildings and civil engineering structures [101]. Despite continuing research to develop CFD techniques suitable for these studies, their full validation against the current standard represented by BLWT testing has not yet been achieved. Therefore, the use of CFD for these purposes is currently not recognised in most international wind codes and building standards [100].

In the future, the rapid increase anticipated in computational power suggests that CFD techniques will improve further and be able to produce accurate results over a wider number of wind engineering studies. At present, though, they do not yet have sufficient reliability to replace physical modelling and BLWT testing. Rather, CFD and BLWT testing are currently considered two complementary methodologies. They can, and should, be used together at different stages and for different purposes to support the design of buildings, civil engineering structures and urban environments [100, 101].

References

1 ASCE (2016) *Minimum Design Loads and Associated Criteria for Buildings and Other Structures.* ASCE 7-16. Reston, VA: American Society of Civil Engineers.

2 EN (2010) *Eurocode 1: Actions on structures-Part 1-4: General actions-wind actions*. EN 1991-1-4:2005+A1:2010. European Committee for Standardization (CEN).

3 Standards Australia (2011) *Structural design actions. Part 2: Wind Actions*. Australian/New Zealand Standard, AS/NZS 1170.2:2011. AS/NZS 117.

4 Lawson, T.V. (2001) *Building Aerodynamics*. London: Imperial College Press.

5 Simiu, E. & Scanlan, R.H. (1996) *Wind effects on structures: fundamentals and applications to design*. 3rd ed. New York: Wiley.

6 ESDU (2000) *Mean wind speeds over hills and other topography*. Data item 91043. London: Engineering Sciences Data Unit International.

7 ESDU (2010) *Computer program for wind speeds and turbulence properties: flat or hilly sites in terrain with roughness changes*. Data item 01008. London: Engineering Sciences Data Unit International.

8 ASCE (2012) *Wind tunnel testing for Buildings and Other Structures*. ASCE 49-12. Reston, VA: American Society of Civil Engineers.

9 AWES (2019) *Quality Assurance Manual: Wind Engineering Studies of Buildings*. (AWES-QAM-1-2019). Australasian Wind Engineering Society (AWES).

10 Cook, N.J. (1978) Wind-tunnel simulation of the adiabatic atmospheric boundary layer by roughness, barrier and mixing-device methods. *Journal of Wind Engineering and Industrial Aerodynamics* 3 (2-3): 157-176.

11 Irwin, P.A. (1981) The design of spires for wind simulation. *Journal of Wind Engineering and Industrial Aerodynamics* 7 (3): 361-366.

12 Barlow, J.B., Rae, W.H. & Pope, A. (1999) *Low-speed wind tunnel testing*. New York: Wiley.

13 Bienkiewicz, B., Cermak, J.E., Peterka, J.A. & Scanlan, R.H. (1983) Active modeling of large-scale turbulence. *Journal of Wind Engineering and Industrial Aerodynamics* 13 (1): 465-475.

14 Nishi, A., Kikugawa, H., Matsuda, Y. & Tashiro, D. (1999) Active control of turbulence for an atmospheric boundary layer model in a wind tunnel. *Journal of Wind Engineering and Industrial Aerodynamics* 83 (1): 409-419.

15 Kröger, L., Frederik, J., Wingerden, J. van, Peinke, J. & Holling, M. (2018) Generation of user defined turbulent inflow conditions by an active grid for validation experiments. *Journal of Physics: Conference Series*, 1037. doi: 10.1088/1742-6596/1037/5/052002.

16 Smith, J.T., Masters, F.J., Liu, Z. & Reinhold, T.A. (2012) A simplified approach to simulate prescribed boundary layer flow conditions in a multiple controlled fan wind tunnel. *Journal of Wind Engineering and Industrial Aerodynamics* 109: 79-88.

17 BLWTL (2007) *Wind tunnel testing: A General Outline*. Edited by Alan G Davenport Wind Engineering Group. London, Ontario: Boundary Layer Wind Tunnel Laboratory, University of Western Ontario. http://www.blwtl.uwo.ca/User/Doc/Wind%20Tunnel%20Outline-Web.pdf (4/9/2019)

18 Irwin, P., Denoon, R. & Scott, D. (2013) *Wind tunnel testing of High-Rise Buildings*. London: Routledge.

19 Kopp, G.A. (ed.) (2018) *Large-Scale and Full-Scale Laboratory Test Methods for Examining Wind Effects on Buildings*. Lausanne: Frontiers in Built Environment.

20 Standohar-Alfano, C.D., Estes, H., Johnston, T., Morrison, M.J. & Brown-Giammanco, T. M. (2017) Reducing Losses from Wind-Related Natural Perils: Research at the IBHS Research Center. *Frontiers in Built Environment* 3: 9. doi: 10.3389/fbuil.2017.00009.

21 Fu, T.-C., Aly, A.M., Bitsuamlak, G., Chowdhury, A. & Simiu, E. (2010) Flow simulation in 12-fan Wall of Wind testing facility. In: *The 2nd American Association for Wind Engineering Workshop (2AAWE).* Marco Island, Florida.

22 Hangan, H., Refan, M., Jubayer, C., Romanic, D., Parvu, D., LoTufo, J. & Costache, A. (2017) Novel techniques in wind engineering. *Journal of Wind Engineering and Industrial Aerodynamics* 171: 12-33.

23 Gandemer, J. (1992) La soufflerie climatique 'Jules Verne'. *Journal of Wind Engineering and Industrial Aerodynamics* 41 (1): 43-54.

24 Tamura, Y., Nishijima, K., Matsui, M., Phuc, P. Van & Yang, Q. (2017) From Load Estimation to Performance Estimation – From Model-Scale Test to Full-Scale Test: With Special Interest in Asian Region. *Frontiers in Built Environment.* https://www.frontiersin.org/article/10.3389/fbuil.2017.00008 (4/9/2019).

25 Oreskovic, C., Savory, E., Porto, J. and Orf, L.G. (2018) Evolution and scaling of a simulated downburst-producing thunderstorm outflow. *Wind and Structures* 26 (3): 147-161.

26 Refan, M., Hangan, H. & Wurman, J. (2014) Reproducing tornadoes in laboratory using proper scaling. *Journal of Wind Engineering and Industrial Aerodynamics* 135: 136-148.

27 Letchford, C.W., Mans, C. & Chay, M.T. (2002) Thunderstorms – their importance in wind engineering (a case for the next generation wind tunnel). *Journal of Wind Engineering and Industrial Aerodynamics* 90 (12): 1415-1433.

28 Jesson, M., Sterling, M., Letchford, C. & Baker, C. (2015) Aerodynamic forces on the roofs of low-, mid- and high-rise buildings subject to transient winds. *Journal of Wind Engineering and Industrial Aerodynamics* 143: 42-49.

29 Haan, F.L., Balaramudu, V.K. & Sarkar, P.P. (2010) Tornado-Induced Wind Loads on a Low-Rise Building. *Journal of Structural Engineering* (American Society of Civil Engineers) 136 (1): 106-116.

30 Hangan, H. (2014) The Wind Engineering Energy and Environment (WindEEE) Dome at Western University, Canada. *Wind Engineers, JAWE* 39 (4): 350-351. doi: 10.5359/jawe.39.350.

31 Mayer, L.J. (2009) *Development of a large-scale simulator*. Texas Tech University, MSc Thesis. https://ttu-ir.tdl.org/handle/2346/20590 (4/9/2019).

32 Haan, F.L., Sarkar, P.P. & Gallus, W.A. (2008) Design, construction and performance of a large tornado simulator for wind engineering applications. *Engineering Structures* 30 (4): 1146-1159.

33 Gillmeier, S., Sterling, M. & Hemida, H. (2017) A parametric study of the effect of a tornado generator's geometric design on the flow field. In: *The 7th European and African Conference on Wind Engineering*. Liège, Belgium.

34 Cermak, J.E. (1975) Applications of Fluid Mechanics to Wind Engineering – a Freeman Scholar Lecture. *Journal of Fluids Engineering, ASME* 97 (1): 9-38.

35 Cochran, L. (2006) State of the Art Review of Wind Tunnels and Physical Modelling to Obtain Structural Loads and Cladding Pressures. *Architectural Science Review (Taylor & Francis)* 49 (1): 7-16.

36 Cook, N.J. (1985) *The designer's guide to wind loading of building structures. Part1: Background, damage, wind data and structural classification.* Garston: Building Research Establishment; London: Butterworths.

37 Tropea, C., Yarin, A. & Foss, J.F. (2007) *Springer Handbook of Experimental Fluid Mechanics.* Berlin: Springer-Verlag.

38 Hooper, J.D. & Musgrove, A.R. (1991) Multi-hole pressure probes for the determination of the total velocity vector in turbulent single-phase flow. In: *4th International Symposium on Transport Phenomena in Heat and Mass Transfer*. Sydney, Australia. 1364-1374.

39 Watkins, S., Mousley, P. & Vino, G. (2004) The Development and Use of Dynamic Pressure Probes With Extended Cones of Acceptance (ECA). In: *15th Australasian Fluid Mechanics Conference.* Sydney: The University of Sydney.

40 Irwin, H.P.A.H. (1981) A simple omnidirectional sensor for wind tunnel studies of pedestrian-level winds. *Journal of Wind Engineering and Industrial Aerodynamics* 7 (3): 219-239.

41 Palombi, E. & Cook, N.J. (2010) Development and Application of a Directional Irwin Sensor. In: *9th UK Conference on Wind Engineering*. 25-28. Bristol: Wind Engineering Society (WES).

42 Pearce, W., Wu, H. & Kilpatrick, J. (2014) Development of a directional probe for wind speed and comfort assessment. In: *11th UK Conference on Wind Engineering*. Birmingham: Wind Engineering Society (WES).

43 Irwin, H.P.A.H., Cooper, K.R. & Girard, R. (1979) Correction of distortion effects caused by tubing systems in measurements of fluctuating pressures. *Journal of Wind Engineering and Industrial Aerodynamics* 5 (1): 93-107.

44 Wu, H. & Stathopoulos, T. (1994) Further experiments on Irwin's surface wind sensor. *Journal of Wind Engineering and Industrial Aerodynamics* 53 (3): 441-452.

45 Raffel, M., Willert, C.E., Wereley, S. & Kompenhans, J. (2007) *Particle Image Velocimetry – A Practical Guide (2nd Edition).* Berlin: Springer-Verlag.

46 Vogel, A. & Lauterborn, W. (1988) Time resolved particle image velocimetry. *Optics and Lasers in Engineering* 9 (3-4): 277-294.

47 Tschanz, T. & Davenport, A.G. (1983) The base balance technique for the determination of dynamic wind loads. *Journal of Wind Engineering and Industrial Aerodynamics* 13 (1): 429-439.

48 Garry, K.P. (1988) A simulation technique for the study of buoyant plume dispersion of offshore platforms. In: *International Seminar on Problems of Simulation in Wind Tunnels,* 229-239.

49 Dale, J., Irwin, P., Kumar, S., Kilpatrick, J. & Galsworthy, J. (2013) Probabilistic Approach to Determining Internal Pressures Based on Wind Tunnel Measurements. In: *Proceedings of the ATC and SEI Conference on Advances in Hurricane Engineering 2012: Learning From Our Past, October 24-26, 2012, Miami, Florida* (eds C.P. Jones & L.G. Griffis), 1027-1034. Reston, VA: American Society of Civil Engineers.

50 Irwin, P.A. & Sifton, V.L. (1998) Risk considerations for internal pressures. *Journal of Wind Engineering and Industrial Aerodynamics* 77-78: 715-723.

51 Kwok, K.C.S., Burton, M.D. & Abdelrazaq, A.K. (2015) *Wind-Induced Motion of Tall Buildings – Designing for Habitability*. Reston, VA: American Society of Civil Engineers.

52 Gunawardena, T., Fernando, S., Mendis, P., Waduge, B. & Hettiarachchi, D. (2017) Wind analysis and Design of Tall Buildings: the State of the Art. In: *8th International Conference on Structural Engineering and Construction Management (ICSECM)*. Sydney, Australia.

53 Yamada, M. & Goto, T. (1975) The criteria to motions in tall buildings. In: *Pan Pacific Tall Buildings Conference proceedings: tall buildings and the growth of cities. January 26-29, 1975, Honolulu, Hawaii*. 233-244.

54 ISO (2007) *Bases for design of structures – Serviceability of buildings and walkways against vibrations* (ISO 10137:2007). International Organization for Standardization (ISO).

55 NRC (2010) *National Building Code of Canada (NBCC 2010)*. Ottawa, Ontario: National Research Council of Canada (NRC).

56 AIJ (2004) *Guidelines for the evaluation of habitability of building vibration* (AIJ-GBV-2004). Architectural Institute of Japan (AIJ).

57 Isyumov, N. (1993) Criteria for acceptable wind-induced motions of tall buildings. In: *International Conference on Tall Buildings*. Rio de Janeiro: Council on Tall Buildings and Urban Habitat (CTBUH).

58 Davenport, A.G. (1961) The application of statistical concepts to the wind loading of structures. *Proceedings of the Institution of Civil Engineers* 19 (4): 449-472.

59 Xie, J. & Irwin, P.A. (1998) Application of the force balance technique to a building complex. *Journal of Wind Engineering and Industrial Aerodynamics* 77-78: 579-590.

60 Cook, N.J. (1983) A sensitive 6-component high-frequency-range balance for building aerodynamics. *Journal of Physics E: Scientific Instruments* (IOP Publishing) 16 (5): 390-393.

61 Steckley, A., Accardo, M., Gamble, S.L. & Irwin, P.A. (1992) The use of integrated pressures to determine overall wind- induced response. *Journal of Wind Engineering and Industrial Aerodynamics* 42 (1-3): 1023-1034.

62 Dragoiescu, C., Garber, J. & Suresh Kumar, K. (2019) A Comparison of Force Balance and Pressure Integration Techniques for Predicting Wind-Induced Responses of Tall Buildings. In: *Proceedings of Structures Congress, 2006*. 1-10. Reston, VA: American Society of Civil Engineers.

63 Isyumov, N. (1982) The Aeroelastic Modelling of Tall Buildings'. In: *International Workshop on Wind Tunnel Modeling Criteria and Techniques in Civil Engineering Applications. Gaithersburg, Maryland, USA*. 373-407. Cambridge: Cambridge University Press.

64 Macklowe, H. (2015) The Complex Path to Simple Elegance: The Story of 432 Park Avenue. In: *Proceedings of the CTBUH 2015 New York Conference*. 48-53. New York. http://global.ctbuh.org/resources/papers/download/2437-the-complex-path-to-simple-elegance-the-story-of-432-park-avenue.pdf (4/9/2019).

65 Pearson, A. (2015) Manhattan Slimline. *Tall Buildings: Concrete Quarterly Special Issue* No.1: 12-13. https://www.concretecentre.com/getmedia/3d3e1259-ccc4-46b3-987e-eac1a207e3fe/CQ_253_Autumn2015_Tall_Buildings_Special.pdf.aspx (4/9/2019)

66 Marcus, S., Mena, H., Yalniz, F. & Shirley, C. (2018) 432 Park. *STRUCTURE magazine* July: 32-33. https://www.structuremag.org/?p=13350 (4/9/2019)

67 Burton, M.D., Kwok, K.C. & Abdelrazaq, A. (2015) Wind-Induced Motion of Tall Buildings: Designing for Occupant Comfort. *International Journal of High-Rise Buildings* 4 (1): 1-8.

68 Haskett, T. & Smith, A. (2017) Feeling at Home in the Clouds. *STRUCTURE magazine* December: 14-17. https://www.structuremag.org/?p=12403 (4/9/2019).

69 ASCE (2004) *Outdoor human comfort and its assessment: state of the art.* Edited by Task Committee on Outdoor Human Comfort. Reston, VA: American Society of Civil Engineers.

70 AWES (2014) *Guidelines for Pedestrian Wind Effects Criteria.* Australasian Wind Engineering Society (AWES).

71 RWDI and City of London (2017) *Wind Effects and Tall Buildings. Planning Advice Note. Guidelines and best practice for assessing wind effects and tall buildings in the City of London.* City of London Corporation.

72 Miyata, T. & Yamaguchi, K. (1993) Aerodynamics of wind effects on the Akashi Kaikyo Bridge. *Journal of Wind Engineering and Industrial Aerodynamics* 48 (2-3): 287-315.

73 Highways Agency (2001) Design Rules for Aerodynamic Effects on Bridges. In: *Design Manual for Roads and Bridges.* Vol. 1, Sect. 3, Part 3. BD 49/01. London.

74 AASHTO (2017) *AASHTO LRFD Bridge Design Specifications.* 8th edn. American Association of State Highway and Transportation Officials (AASHTO).

75 Diana, G., Rocchi, D. and Belloli, M. (2015) Wind tunnel: a fundamental tool for long-span bridge design. *Structure and Infrastructure Engineering (Taylor & Francis)* 11 (4): 533-555.

76 Davenport, A.G. (1962) The response of slender, line-like structures to a gusty wind. *Proceedings of the Institution of Civil Engineers.* 23 (3): 389-408.

77 Ostenfeld, K.H. & Larsen, A. (1992) Bridge engineering and aerodynamics. In: *Aerodynamics of Large Bridges: Proceedings of the First International Symposium on Aerodynamics of Large Bridges, Copenhagen, Denmark, 19-21 February 1992* (ed. A. Larsen), 3-22. Rotterdam: Balkema.

78 Scanlan, R.H. (1992) Wind dynamics of long-span bridges. In: *Aerodynamics of Large Bridges: Proceedings of the First International Symposium on Aerodynamics of Large Bridges, Copenhagen, Denmark, 19-21 February 1992* (ed. A. Larsen), 47-57. Rotterdam: Balkema.

79 Stoyanoff, S. & Dallaire, P.-O. (2013) A direct method for calculation of wind loads on LongSpan bridges. In: *12th Americas Conference on Wind Engineering 2013, ACWE 2013.* 1316-1325. Seattle, WA, USA.

80 Irwin, P. (1992) Full aeroelastic model tests. In: *Aerodynamics of Large Bridges: Proceedings of the First International Symposium on Aerodynamics of*

Large Bridges, Copenhagen, Denmark, 19-21 February 1992 (ed. A. Larsen), 125-135. Rotterdam: Balkema.

81 Tanaka, H. (1990) Similitude and modelling in wind tunnel testing of bridges. *Journal of Wind Engineering and Industrial Aerodynamics* 33 (1-2): 283-300.

82 King, J.P.C. (2003) 'The foundation and the future of wind engineering of long span bridges – the contributions of Alan Davenport', *Journal of Wind Engineering and Industrial Aerodynamics.* doi:10.1016/j.jweia.2003.09.007.

83 Zasso, A., Stoyanoff, S., Diana, G., Vullo, E., Khazem, D., Serzan, K., Pagani, A., Argentini, T., Rosa, L. and Dallaire, P.O. (2013) 'Validation analyses of integrated procedures for evaluation of stability, buffeting response and wind loads on the Messina Bridge', *Journal of Wind Engineering and Industrial Aerodynamics.* doi: 10.1016/j.jweia.2013.07.013.

84 Hjorth-Hansen, E. (1992) Sectional model tests. In: *Aerodynamics of Large Bridges: Proceedings of the First International Symposium on Aerodynamics of Large Bridges, Copenhagen, Denmark, 19-21 February 1992* (ed. A. Larsen), 95-112. Rotterdam: Balkema.

85 Diana, G., Resta, F., Zasso, A., Belloli, M. and Rocchi, D. (2004) 'Forced motion and free motion aeroelastic tests on a new concept dynamometric section model of the Messina suspension bridge', *Journal of Wind Engineering and Industrial Aerodynamics.* doi: 10.1016/j.jweia.2004.01.005.

86 Wyatt, T.A. (1992) Recent British developments: Windshielding of bridges for traffic. In: *Aerodynamics of Large Bridges: Proceedings of the First International Symposium on Aerodynamics of Large Bridges, Copenhagen, Denmark, 19-21 February 1992* (ed. A. Larsen), 159-170. Rotterdam: Balkema.

87 Larose, G.L. (1999) Experimental determination of the aerodynamic admittance of a bridge deck segment. *Journal of Fluids and Structures* 13 (7-8): 1029-1040.

88 Davenport, A.G., King, J.P.C. & Larose, G.L. (1992) Taut strip model tests. In: *Aerodynamics of Large Bridges: Proceedings of the First International Symposium on Aerodynamics of Large Bridges, Copenhagen, Denmark, 19-21 February 1992* (ed. A. Larsen), 113-124. Rotterdam: Balkema.

89 Larose, G.L., Falco, M. & Cigada, A. (1995) Aeroelastic response of the towers for the proposed bridge over Stretto di Messina. *Journal of Wind Engineering and Industrial Aerodynamics* 57 (2-3): 363-373.

90 Borri, C. & Costa, C. (2007) Bridge Aerodynamics and Aeroelastic Phenomena. In: *Wind Effects on Buildings and Design of Wind-Sensitive Structures.* (eds T. Stathopoulos & C.C. Baniotopoulos), 167-200. Vienna: Springer.

91 Holmes, J.D. (2012) Modern techniques for effective wind load distributions on large roofs. In: *The 2012 World Congress on Advances in Civil, Environmental and Materials Research (ACEM '12).* Seoul, Korea.

92 Holmes, J.D. (2016) Effective wind load distribution for large roofs, with incorporation of resonant components. In: *12th UK Conference on Wind Engineering*. 39-46. Nottingham: Wind Engineering Society (WES).

93 Irwin, H.P.A.H. & Wardlaw, R.L. (1980) A wind tunnel investigation of a retractable fabric roof for the Montreal Olympic Stadium. In: *Proceedings of the Fifth International Conference, USA,* 1979. 925-938.

94 Kawakita, S., Bienkiewicz, B. & Cermak, J.E. (1992) Aeroelastic model study of suspended cable roof. *Journal of Wind Engineering and Industrial Aerodynamics* 42 (1-3): 1459-1470.

95 Petersen, R., Cochran, B.C. & Carter, J. (1999) Evaluating and Designing Exhaust Systems to Avoid Fume Reentry and Adverse Health Effects. In: International Society for Pharmaceutical Engineering – 1999 Interphex Conferenc and Exibition. New York. 169-185.

96 *2011 ASHRAE Handbook – Heating, Ventilating, and Air-Conditioning Applications (I-P Edition).* Atlanta, GA: American Society of Heating, Refrigerating and Air-Conditioning Engineers.

97 Petersen, R.L., Cochran, B.C. & LeCompte, J.W. (2002) Specifying Exhaust Systems that Avoid Fume Reentry and Adverse Health Effects. In: *ASHRAE Transactions 2002, June 22, 2002 -June 26, 2002, Honolulu,* 1039-1049 (Paper HI-02-15-3).

98 Blocken, B. (2014) 50 years of Computational Wind Engineering: Past, present and future. *Journal of Wind Engineering and Industrial Aerodynamics* 129: 69-102.

99 Blocken, B. (2015) Computational Fluid Dynamics for urban physics: Importance, scales, possibilities, limitations and ten tips and tricks towards accurate and reliable simulations. *Building and Environment* 91: 219-245.

100 Thordal, M.S., Bennetsen, J.C. & Koss, H.H.H. (2019) Review for practical application of CFD for the determination of wind load on high-rise buildings. *Journal of Wind Engineering and Industrial Aerodynamics* 186: 155-168.

101 Phillips, D.A. & Soligo, M.J. (2019) Will CFD ever Replace Wind Tunnels for Building Wind Simulations? *International Journal of High-Rise Buildings* 8 (2): 107-116.

33

Model testing using shake tables – current practice

Amarnath Kasalanati

33.1 Introduction

Structures are subjected to a variety of loads and environmental conditions during their lifetime. Many of the loads such as live loads from people, traffic, thermal loads, wind and snow are frequent; others are rare, such as high wind load or peak snow load. A few, such as earthquake loads, are very rare. In load conditions such as wind, snow, braking and centrifugal loads (in bridges), a direct force is applied on the structure. These conditions can be simulated in the laboratory settings by applying a force through actuators and studying the response of the structure under the effect of those loads. On the other hand, earthquakes do not apply a force – rather, they impose a movement at the base of the structure. The dynamic characteristics of the structure (mass, stiffness, natural frequency, damping, nonlinear behaviour, etc.) generate the forces in that structure. Therefore, study of the structural behaviour for earthquake actions involves two components: generating the ground motion at the base, and capturing the dynamic characteristics of the structure. Thus, attention needs to be paid to the faithful replication of ground motion as well as dynamic characteristics of the model, especially when nonlinearity is expected in the model as a result of earthquake loading. The main focus of this chapter is on the means by which ground motion is applied at the base of the structural model.

The ground motion is replicated on a scale model of the structure placed on a platform or platen that moves in a manner similar to the ground in an earthquake. These platforms are called by several names: 'shake tables', 'shaking tables' or 'earthquake simulators'. Of these, the term 'shake table' is used slightly more often than others. Also, this phrase is used interchangeably to mean the whole system or just the platen itself. Very early shake tables utilized springs, rotating wheels or pendulum strikes to simulate lateral motion, which mainly produced sinusoidal motion. More realistic earthquake motions were later simulated using electric motors and then hydraulic actuators, first in one direction and subsequently in two and three directions.

In 1995, there were about twenty-five shake table facilities around world with a platen area greater than 10 m^2 and a payload of over 20 tonnes [1]. Today there are

Physical Models: Their historical and current use in civil and building engineering design,
First Edition. Edited by Bill Addis.

over 120 shake tables in the world of notable size and capacity that are capable of producing earthquake motions, installed in universities, research institutions and some private organizations [2, 3, 4]. These range from 1 to 300 m^2 in area, with a payload of a few tonnes up to 2000 tonnes. The range of dynamic performance goes up to 2 m displacement, 2 m/sec velocity and 10g acceleration, for different tables. The working frequency is usually under 20 Hz for larger tables and up to 200 Hz for smaller tables. The cost of these tables ranges from tens of thousands of dollars for a small table to hundreds of millions of dollars for large tables.

The following sections cover various aspects of testing specific to shake tables, in particular the importance of physical testing for seismic cases, the key components of the shake table, a few important shake tables around the world and some notable projects tested on them. Further details can be found elsewhere [5, 6, 7, 8, 9].

33.2 The need for physical testing

There have been tremendous advances in analytical simulation techniques over the past few decades. The analytical models are more detailed, the analysis engines are sophisticated and computers are a lot more powerful. With these advances, modern performance-based design methods rely on analytically simulating nonlinear behaviour of a complex system with three-dimensional interaction under dynamic conditions. Such analytical simulations require experimental validation under dynamic, nonlinear and three-dimensional conditions. Ironically, these analytical advances have led to an *increase* in the need for good-quality physical testing in the field of earthquake engineering. Moreover, the size and range of civil engineering structures does not allow easy, broad conclusions to be drawn from a few prototype tests.

Many important developments in earthquake engineering are due to 'learning from past earthquake damage'. These 'natural experiments' of strong earthquakes acting on a specific type of structure are too infrequent for a quick design revision. Unfortunately, these learning opportunities are usually accompanied by widespread destruction. Shake tables provide the opportunity to 'learn from simulated earthquakes acting on model structures' without the damage and destruction having an impact on communities. To interpret the results better, or to develop more accurate understanding, the models need to be 'full-scale' or 'realistically-large-scale', the ground motions need to reproduce actual recorded earthquake motions (or as close as possible) and the models need to represent entire structures, rather than elements or sub-assemblages. While reduced-scale testing provides obvious advantage of economy and feasibility, many structural phenomenon require testing at close to full-scale. This is mainly because there are certain minimum sizes below which the characteristics of reinforced concrete, masonry and wood cannot be accurately replicated. Similarly, while tests on sub-assemblies or elements provide valuable information, they do not accurately represent certain constraints and boundary conditions. Finally, the dynamic effects associated with inertia, damping, rate-dependent response, rocking, pounding and contact problems can be accurately studied by physical testing in appropriate dynamic conditions. This highlights the need for physical testing for seismic cases, and doing so at as large a scale as physically and economically feasible.

Despite the demonstrated need for seismic testing, most of the shake tables around world have less than 50% utilization [10]. The costs and complexities of tests are part of the problem. Also, shake tables are perceived more as a research tool or an academic toy. In the past, large earthquakes with a new type of failure have increased the shake table testing until some solution strategies for the specific problem were developed. Recently, the focus has shifted to testing non-structural components, specifically critical equipment, whose failure renders a building non-operational for long durations. Some of the regulatory agencies have required shake-table testing for critical non-structural elements to obtain seismic certification for their use in essential facilities such as hospitals. This has increased usage of shake tables. In these tests, the actual unit (full-scale) is placed on the shake table and tested.

33.3 Key components of a shake table

A shake table consists of a platen driven by hydraulic actuators, with the model of a structure attached to the platform. Actuators, controlled by servo-valves, apply the earthquake motion to the platform. A servo-control system allows an input signal, such as a digitized record of an earthquake motion, to cause the actuators to move the shake table and ensures good fidelity or tracking between input and table motion through a closed loop. Range of motion could be as simple as linear (one degree of freedom, or 1-DOF), or bilinear (two degrees of freedom, or 2-DOF) to full three-dimensional control (three linear and three rotational degrees of freedom, or 6-DOF). Modern shake tables use a combination of mechanical, hydraulic and electric/electronic subsystems, acting together to produce the desired movement (Figures 33.1 and 33.2). Capabilities of earlier shake

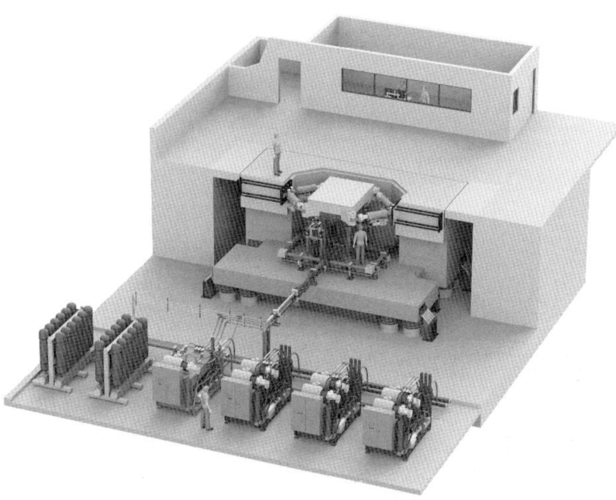

Figure 33.1 Illustration showing the general arrangement of shake-table systems. (Image: Courtesy of MTS Systems Corporation)

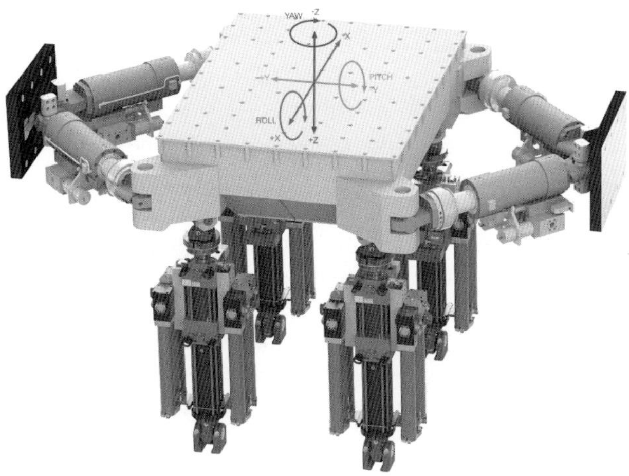

Figure 33.2 The platen and actuators of a 6-DOF shake table. (Image: Courtesy of MTS Systems Corporation)

tables were limited mainly by the capacities of servo-valves, hydraulic actuators and control systems available at that time. These systems have reached a higher level of sophistication in recent years and most new shake tables in the world have limitations mainly due to economic constraints. Different systems of the shake table are described in detail in the following sections.

33.3.1 Mechanical aspects of the shake table

The mechanical aspects of the system include the reaction block, the physical layout of the actuators, the platen, the test-model anchoring arrangements, the test specimen, and restrainers / guides / hydrostatic bearings to limit some degrees of freedom.

The reaction block is usually a large concrete structure around the platen with the hydraulic actuators located between the two. The absolute movements of the platen are actually slightly less than the movements between actuator ends because of the movement of the reaction block. Large mass and stiffness of the reaction block are necessary to keep its displacements to a fraction of a millimetre during peak actuator forces. This results in the actuator displacements and platen displacements being much closer to each other. A larger mass of the reaction block results in better control of the platen and reduced transfer of vibrations to the surrounding soil. However, massive reaction blocks are very expensive and have low resonant frequencies. Depending on surrounding soil stiffness and radiation damping (the energy dissipated from the foundation to the surrounding soil) a reasonable reaction-block-to-driven-mass ratio is usually chosen between 10 (for example, UC Berkeley or UC San Diego shake tables) and 100 (such as E-Defense table). This choice is usually made with future additions / improvements in mind because it is very difficult to change the reaction block mass later on.

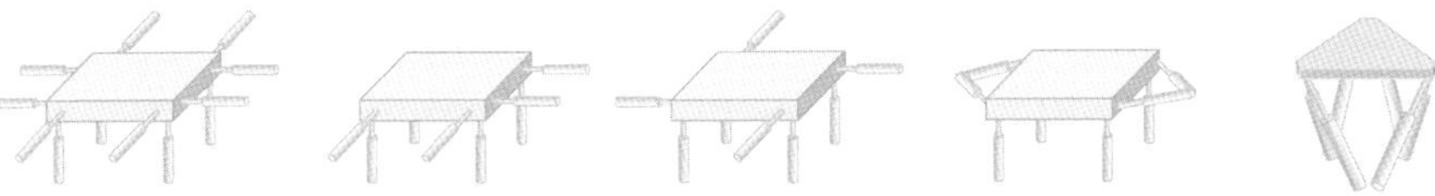

Figure 33.3 6-DOF Shake Table Configurations: (a) Balanced Orthogonal, (b) Unbalanced Orthogonal, (c) Pinwheel, (d) Vee, and (e) Stewart Platform. (Image: Amarnath Kasalanati)

The actuator configuration is usually determined by the number of degrees of freedom, physical constraints around the table such as available floor space, ease of access to the top of the table (floor level or elevated tables), and the ability to combine multiple tables. Six degree-of-freedom of movement can be achieved with as few as six actuators, arranged in a configuration called 'Gough-Stewart platform' or simply 'Stewart Platform'. These tables are used widely in the automobile and aerospace industries, where the rotational and translational movements are equally important. In seismic testing, two horizontal displacements and the vertical displacement are much more important than rotations. In general, rotational control in seismic simulators is needed mainly to keep the rotations close to zero. As a result, a minimum of four vertical actuators are used to control pitch and roll motions and a minimum of four horizontal actuators are used to control yaw motion. These actuators can be arranged in different configurations (Figure 33.3) which include Balanced orthogonal (the CGS Shake table in Algeria), Unbalanced orthogonal (the E-Defense table in Japan), Pinwheel (the EQUALS table at the University of Bristol and the UC Berkeley table). Horizontal actuators can also be arranged in a Vee shape (the University at Buffalo), and only six actuators are needed for the Stewart Platform.

All these configurations have some level of static and dynamic cross-coupling of different degrees of freedom that affect the table control, in particular during larger displacements and higher overturning moments. A recent study shows that the horizontal Vee arrangement of actuators has the least dynamic cross-coupling while the Stewart Platforms have the most dynamic cross-coupling [11].

The platen, or the flat platform or the *shake table* can be made of concrete or steel and must be rigid enough to transfer the actuator loads to the test specimen. The platen is designed to be stiff to ensure less interference with the test specimen over the whole frequency range of testing. The platen is provided with holes or an anchoring mechanism in a specified pattern to attach the test specimens to the platen. When fewer than 6-DOF is implemented, low friction sliders, guides or hydrostatic bearings are used to restrain those degrees of freedom.

33.3.2 Hydraulic system

A hydraulic system consists of a pump that pressurizes the hydraulic fluid, valves that control the flow direction or limit the pressure, and hydraulic cylinders also known as actuators. Hydraulic systems are widely used to apply force or move various things. Simple directional valves will allow the hydraulic cylinder to apply the full force in one direction or the other or stay at a set point. Such directional valves can only produce a triangular wave and cannot reproduce earthquake motion. To achieve more variable or nuanced motion of the actuator, servo valves are used.

Figure 33.4 A three-stage servo valve. (Image: Amarnath Kasalanati)

A servo valve accepts an electrical signal (a command), and directs a supply of fluid to the actuator at a pressure proportional to the electrical signal it receives. Servo valves can be single-stage or multi-stage, and can continuously vary the pressure they supply from zero to the peak pressure provided by the pump. With feedback control, servo valves can control the position, velocity, pressure or force of an actuator very accurately. Multi-stage servo valves progressively increase the pressure or actuation in the system and can control high-powered actuators with a small electrical signal (Figure 33.4). Advances in servo-controlled hydraulics has been a key development in achieving earthquake motions in modern shake tables.

33.3.3 Control systems of a shake table

A controller provides the electrical signal to the servo valve. This signal is known as the 'target', or 'command', or 'reference' input. In *feedback control,* the position or the force of the actuator is measured and fed back into the controller that sends the electrical signal to the servo valve (Figure 33.5). The difference between the target (or reference or command) and the actually-achieved position (feedback) is the error signal. The controller modifies the input based on the error. If the applied correction at the controller is proportional to the error, then the procedure is called *proportional control.* The feedback is measured at a sampling frequency of several hundred times per second, thereby making it possible to take advantage of trends or changes in the error. For example, an *integral* of the error calculates the cumulative error over a time and can help close the gap in case of a stagnant error. On the other hand, the *differential* of the error helps read the current trend of the error (decreasing or increasing) and make adjustments accordingly. To enhance the control and minimize the

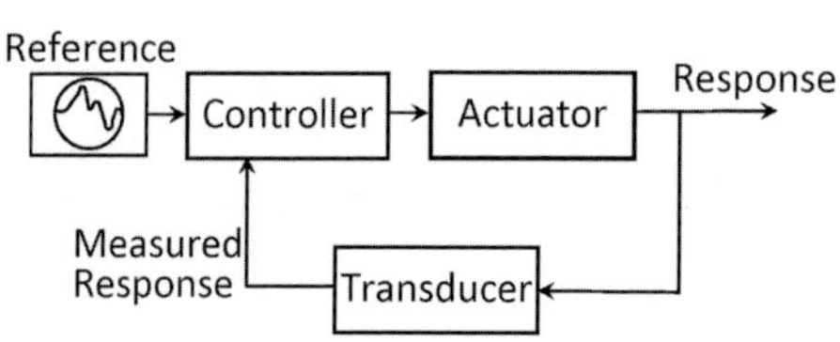

Figure 33.5 Block diagram illustrating feedback control. (Image: Amarnath Kasalanati)

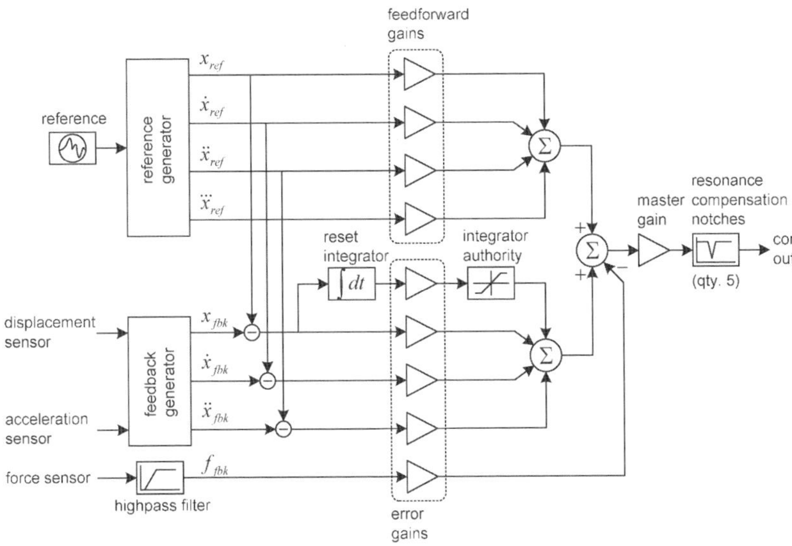

Figure 33.6 Block diagram of a three-variable control system. (Image: MTS Systems Corporation)

error quickly, the integral and the differential of the error are used in addition to the proportional part. Using a weighted combination of these three is known as Proportional-Integral-Differential (PID) control.

In general, shake tables are displacement controlled. That is, the displacement of actuators is controlled through the feedback loop, thereby controlling the movement of the platen. However, using only the displacement feedback results in poor acceleration tracking at high frequencies. To improve this, the displacement feedback is augmented by other parameters, such as the velocity and acceleration, giving what is called *three-variable control* (Figure 33.6). MTS Systems, the global leader in seismic simulator manufacturing, adopts three-variable control to improve the system bandwidth over a wide range of frequencies [12]. Various control algorithms are used, or proposed by researchers, to achieve specific objectives [13, 14, 15].

33.3.4 Shake table performance limitations

The displacement capacity of a table is limited by the displacement stroke of the actuators which limits the low-frequency or long-period range of the table's operation (Figure 33.7). A table's velocity is limited by the flow rate of the valves, usually measured in litres/min. The velocity limit usually governs in the mid-frequency range of table operation. A table's acceleration is limited by the force applied by the actuators and the combined weight of the table and the test specimen. This limit is in the high frequency zone of the table's operation. It should be noted that high fidelity in signal reproduction is not required in all shake table tests. In many cases, comparison between the experimentally-observed behaviour and analytically-predicted structural response can be done afterwards using the achieved table motions as input to

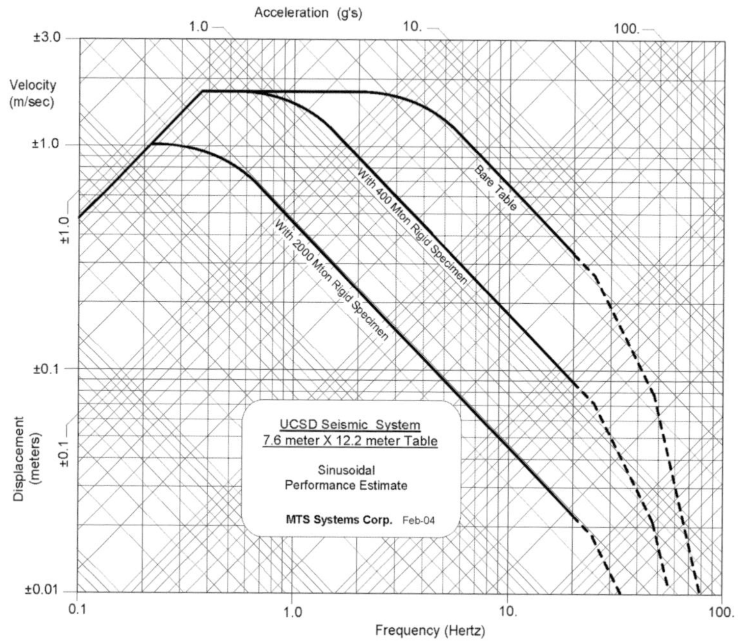

Figure 33.7 Displacement, velocity and acceleration limits of a shake table at University of California San Diego. (Image: MTS Systems Corporation)

the analytical model. In such cases, it is usually important to understand of the sources of errors that occur in reproducing an earthquake motion. Nevertheless, there are cases, such as hybrid simulation and geographically-distributed testing, where accurate reproduction of a specific signal is important.

The performance of a shake table may thus be defined by several capacity parameters: displacement capacity is determined by the stroke of the actuators (direct actuator stroke in an orthogonal arrangement or its component in case of Vee or Stewart arrangements); the velocity limit is driven by the oil flow rate, which in turn is limited by the hydraulic power; force limit is based on the actuator piston area and the hydraulic pressure; acceleration limit is based on the force limit and combined mass of the test article and platen; overturning limit is based on the vertical actuator forces and their plan layout (individual vertical actuator forces and distances between them). As the displacement capacity increases, the interaction between different degrees of freedom increases. That is, the actuators apply the load at an angle at the peak displacement and thereby reduce the system limits in other directions to a lower number than combination of individual actuators. Moreover, the maximum limits in velocity may not be achieved at the same time due to flow-rate limitations. Further, while resisting a high overturning moment, the vertical actuators cannot apply the full vertical accelerations because some of them will be acting in the opposite direction to resist the overturning [16, 17, 18].

As with any experiment, technical challenges in shake-table testing range from choice of model to interpretation of results. Specifically, these include: the choice of scale for the test specimen, based on the dimensions & capabilities of the shake table; accuracy of reproduction of the desired table motion, interaction of movement in one direction with movements in other directions; the shake table's own internal nonlinearities; interaction of the shake table with the specimen; interaction of the shake table with the reaction mass; collection of pertinent data from the tests; and finally, the interpretation of collected information. Extensive research is available on these topics and is beyond the scope of this chapter.

33.4 Notable modern shake tables in the USA

33.4.1 The University of California Berkeley shake table

In 1964, the Great Alaska earthquake and the Niigata earthquake in Japan caused widespread damage and led to considerable research (funding) in the field of earthquake engineering. In the 1960s, researchers from UC Berkeley, led by Professors Joseph Penzien and Ray Clough pursued several large-scale, ambitious plans with far-reaching impact. Some of these endure to date and have contributed immensely to earthquake engineering research and practice. These include: founding of the Earthquake Engineering Research Center (EERC) and the National Information Service for Earthquake Engineering (NISEE), construction of a 6.1 x 6.1 m shake table, and writing the first book on Structural Dynamics.

These researchers are credited with the development of the first modern servo-controlled shake table of significant size which led to development of other large shake tables in Japan, Mexico, Yugoslavia, China and elsewhere [19]. The original table control system produced motions in the vertical and one horizontal direction. In the 1990s, this table was modified to provide motions in the vertical and two horizontal directions with control of three rotational motions as well (a 6-DOF table). This table continues to provide testing for research and commercial projects and is operated by the Pacific Earthquake Engineering Research Center (PEER).

Interestingly, the original proposal by this research team was even more ambitious: a 30.5 x 30.5 m, 6-DOF shake table with fifty-two servo-controlled hydraulic actuators, with the ability to test full-scale structures weighing up to 1800 tonnes and accelerations up to 0.67g [20]. Detailed feasibility studies estimated the cost of such a table to be nearly $20 million and annual operating costs of $3 million (approximately $150 million and $22.5 million, respectively, at 2019 values). A two-year research program was initiated on feedback control using the most advanced electro-hydraulic systems at that time, with involvement from MTS Corporation. In the end, it was determined that the technical challenges involved in a 30.5 x 30.5 m, 6-DOF shake table were too great and that plan was abandoned. Instead, the researchers settled for a *medium-sized* 2-DOF shake table with better operating characteristics which would be a good facility in its own right. Without a secure source of funding for the full project, the team proceeded to build the facility as economically as possible, with hydraulic equipment donated by the US military. In Professor Penzien's own words '...we really bootstrapped the whole project. We believed in what we were doing, but did not have the funding from start to finish. We had go ahead with the confidence that funding would be there when the next phase of construction came' [21]. After nearly five decades, this *medium-sized* facility, with its improvements made in the 1990s, continues to be the largest 6-DOF shake table in the United States.

In the current configuration, the platen is driven horizontally by eight 75,000 lb (330 kN) actuators arranged in a pinwheel configuration, and vertically by four 75,000 lb (330 kN) actuators located at the corners. An MTS 469 controller commands the movement of the shake table. The controller provides for closed loop control of motion in translation, and rotation about the three principal axes. The platen and attachments weigh close to 45 tonnes and the surrounding reaction block weighs 717 tonnes. Test specimens weighing up to 68 tonnes have been tested on this table. Actuators are located in the chamber underneath the table, thus allowing the table surface to be at the same level as the laboratory floor. Another unique feature of this table is the presence of vinyl-covered nylon fabric around the platen, connecting the platen and the reaction block. This *air bag,* the inflated tube around the square platen in Figure 33.8, allows the chamber below the table to be pressurized to 1.5 psi (10 kPa). This air pressure supports the weight of the platen and the test specimen, thereby allowing the vertical actuators to apply similar forces up and down.

One of the side benefits of table initiation exercises was the unintended study of site response. During the early tuning of the shake table, several high-intensity harmonic motions were used, gradually increasing from low frequency and to a higher frequency. The main focus of this tuning was on the table's response to input commands and how well the motion was replicated. However, it also provided an excellent opportunity to study ground motions induced by the actuators and reaction block on the surrounding soil. When the frequency of the vibrating foundation matched a characteristic site frequency, a large amount of energy

Figure 33.8 The University of California Berkeley shake table. (Image: Amarnath Kasalanati)

was transmitted to the surrounding soil. This happens only during the harmonic motions and when the frequency is gradually changed. The unintended effect led to research on how the energy would be transmitted away from the source.

33.4.2 The NEES shake tables in the USA

The US National Science Foundation (NSF) invested over $200 million in 2000 on an unprecedented initiative called the George E. Brown Network for Earthquake Engineering Simulation (NEES), combining the experimental facilities of fifteen US Universities distributed throughout the country. This initiative funded both the construction of new facilities and significant upgrading of existing ones. It also integrated them electronically with a unique IT infrastructure, supported access to them by other research institutions, and shared the data by disseminating the research to a much wider audience. These facilities included geotechnical centrifuges, large-scale structural-testing laboratories, a tsunami wave basin, field and mobile facilities for field experimentation, and three high-performance shake tables. These shake tables were built at the State University of New York at Buffalo (UB), University of Nevada at Reno (UNR), and University of California, San Diego (UCSD). The shake tables at UB and UNR are medium-sized and the UCSD shake table is substantially larger. Each facility has some unique features and complements the other facilities, thereby providing a versatile range of options for NEES researchers [22].

Since the mid-1980s, the State University of New York at Buffalo has operated a 3.7 x 3.7 m shake table funded by NSF. This 5-DOF table (one horizontal, one vertical and three rotational controls) with a load capacity of 20 tonnes was the centre of earthquake engineering research activities in the eastern United States [23, 24, 25]. As part of the NEES program in 2004, UB developed two high-performance shake tables to complement the original table. The unique feature of the UB NEES facility is two 6-DOF tables that are relocatable in a long trench. These tables can be controlled either independently or as a single synchronized table. The tables can be spaced as far apart as 40 metres in the trench to simulate seismic loading on a model long-span bridge. Each table is capable of a load of 50 tonnes. In addition, a large reaction wall, located adjacent to the tables, is equipped with dynamic actuators to enable real-time dynamic hybrid testing.

The key feature of the facility at the University of Nevada at Reno is the ability to relocate four shake tables anywhere on the strong floor of the laboratory to accommodate large and uniquely-shaped test specimens. It currently has three biaxial shake tables each measuring 4.25 x 4.25 m with at 50-tonne load capacity, and a fourth 6-DOF table measuring 2.8 x 2.8 m. Biaxial tables have custom-designed hydrostatic bearings to prevent pitch, roll, yaw and uplift without requiring a trench, which enables the tables to be located anywhere on the strong floor. The shake tables can be operated independently, either in phase or differently from each other. In 2011, all four tables were used as supports for testing a 1:2.5 scale model of a three-span curved bridge, 44 m long with a 24 m radius, a unique test that used nearly 1000 m^2 of laboratory space (see Section 33.7.3).

At the University of California, San Diego the Large High Performance Outdoor Shake Table (LHPOST) is a facility that can produce severe near-source earthquake ground motions on large-scale structures. With a steel platen 12.2 x 7.6 m, maximum displacement of ±0.75 m and velocity of 1.8 m/sec, it is the largest, single-axis shake table in the United States. Being an outdoor facility, it can accommodate tall specimens, such as a full-scale wind turbine or multi-storey buildings. Working in conjunction with the nearby laminar soil shear-box and refillable soil pits, this shake table has the unique capability of Soil-Foundation-Structure Interaction. Over the last 15 years, this facility has tested several 'landmark experiments' that are known for their completeness, realistic scale and realistic seismic input. This table is currently undergoing an upgrade to become the largest 6-DOF table in the US by 2021 [26, 27, 28].

33.5 Shake tables in Japan

Japan has a long history of constructing and using shake tables. Large shake tables in Japan are found not only in universities, but also in large private companies and research institutions. Many medium-sized shake tables have been built in Japan since the servo-hydraulic developments in 1960s and 1970s. The 1964 Niigata earthquake and associated liquefaction spurred interest in foundation research and large shake tables to provide suitable simulations of seismic motion. One of the early large-scale shake tables was a 15 x 15 m table, at the Mitsubishi Industries National Centre for Disaster Prevention in Tsukuba Science City (north of Tokyo) which is capable of testing a 500 tonne test article horizontally or a 200 tonne test piece vertically [29]. Later, the Public Works Research Institute in Tsukuba City constructed a 6 x 8 m table with a 100 tonne load capacity and acceleration capacity of 1.7g, and four tables of 2 x 2 m to study the multiple support excitation of long span bridges. Several private companies in Japan built their own shake tables in 1980s but details of the performance of these tables have not been publicised.

33.5.1 E-Defense shake table

During the great Hanshin-Awaji earthquake, more widely known as Kobe earthquake, in 1995, it was evident that the very strong ground motion near the fault ('near-field effects') caused severe structural damage well beyond what would have been expected. In the aftermath of this earthquake, the National Research Institute for Earth Science and Disaster Prevention (NIED) and the Science and Technology Agency of the Japanese Government (STA) built a three-dimensional, full-scale earthquake testing facility. This facility in Miki City, near Kobe, Japan, known as the E-Defense Shake Table, short for Earth-Defense, is currently the largest shake table in the world.

The design team set out criteria for the facility based on lessons from the Kobe earthquake and experience from other large testing facilities. The basic requirements for the facility were to reproduce realistic earthquake motion, equal to or larger than the Kobe earthquake, on full-scale structures. The velocity and displacement performances are emphasized as much as, if not more than the acceleration performance. The motion recorded at the JR Takatori Station during Kobe earthquake had a peak velocity of 1.7 m/sec and peak displacement of 0.5 m. Therefore, the requirements for the facility were set to a maximum velocity of 2 m/sec and maximum displacement of ±1 m. Since most structures are rectangular, designers chose a rectangular shape for the table to use the available area most effectively. The platen was designed to be larger than existing tables – dimensions of 15 x 20 m were chosen as a good balance between size and control. Movements in both horizontal directions were designed to have the same performance because structures could experience maximum input in any direction.

The maximum test-specimen weight limit was set to 1200 tonnes based on a four-storey reinforced-concrete structure or a liquid tank with a large pier to simulate liquefaction. To facilitate testing of tall and realistic structures, the table was designed for an overturning capacity of 150 MNm. The data acquisition system can record 900 channels to monitor structural and non-structural elements during the testing.

The facility resulting from these requirements is an impressive engineering achievement in its own right. As noted in Section 33.3.5, high displacement and velocity capacities require longer actuators and high flow rates, respectively. To achieve a displacement capacity of ±1 m (2 m total stroke), the horizontal actuators needed to be nearly 9 m long, with a cylinder diameter of nearly 1.5 m and an overall weight of 116 tonnes. The pistons are 830 mm in diameter and machined to a precision of 0.2 mm or better. The length and weight of the actuator required separate development of self-tightening high-pressure seals, pressure-balanced piston rings and spherical hydrostatic bearings to prevent seizure under self-weight at high velocities [30].

Four 3.175 MW gas engines are located in a nearby building to power the hydraulics. A total of ten horizontal actuators are arranged in unbalanced orthogonal configuration and fourteen vertical actuators support the platen. Linkage rods with specifically-designed spherical bearings connect the platen and the actuators and allow simultaneous, large three-dimensional movement of the platen. The mass of the platen is 750 tonnes and consists of thirty-two welded steel blocks. The individual block size was limited to about 5.5 x 3.3 x 3 m to allow transportation to the site. The reaction block has a weight of 200,000 tonnes – over 100 times the combined masses of the platen and largest test specimen. In addition, the reaction block was anchored to the bedrock directly. Three servo valves are used for each horizontal actuator to provide a total flow rate of 45,000 l/min, to achieve a velocity of 2 m/sec. One servo valve with 15,000 l/min flowrate is used in each vertical actuator, to achieve a vertical velocity of 0.7 m/sec [31, 32, 33].

This shake table can produce three-dimensional motion in excess of Kobe earthquake motion on a 1200 tonne test specimen as well as being able to test full-scale multi-storey buildings to collapse. By 2017, eighty-four experiments had been conducted at the facility, an average of eight experiments per year, without any accidents or injuries [34].

33.6 Shake tables in Europe and Asia

There are more than ten shake tables of significant size in Europe. The largest square table is the Azalee table at Saclay in France. This 6-DOF table is 6 x 6 m in plan and has a payload of 100 tonnes. The largest rectangular table is at the Laboratory for Training and Research in Earthquake Engineering and Seismology (TREES Lab) of EUCENTRE in Pavia, Italy. This uniaxial (1-DOF) table is 5.6 x 7 m in plan and has a payload of 140 tonnes. There are also many large shake tables in China, with over a dozen tables in excess of 10 m^2 plan area. The National Center for Research in Earthquake Engineering (NCREE) in Tainin City in Taiwan, ROC, recently opened a 6-DOF high performance 8 x 8 m shake table [35]. A few notable tables in Europe and China are described in the following sections.

33.6.1 LNEC 3D Earthquake Simulator, Lisbon

The Structural Dynamics Laboratory in *Laboratorio Nacional de Engenharia Civil* (LNEC) houses a shake table with three translational degrees of freedom. The rotational degrees of freedom are minimized by torque tube systems, one for each of the pitch, roll and yaw axes. The table measures 5.6 x 4.6 m in plan and has a payload of 40 tonnes [36]. Translational couplings are minimized by a system of linking rods connecting the actuators. In addition, the weights of the table and test specimen are compensated by a passive nitrogen hydraulic system. Each of the translations can also be mechanically blocked.

33.6.2 Dynamic Testing Laboratory at IZIIS, Skopje

The Institute of Earthquake Engineering and Engineering Seismology (IZIIS) was founded in 1965 after the 1963 Skopje earthquake in the Republic of Macedonia (formerly Yugoslavia). It was the first institution of its kind in Europe and one of the oldest of this type in the world. The dynamic testing laboratory at IZIIS has three shake tables: the largest is 5 x 5 m with a 40 tonne payload and is capable of applying vertical and one horizontal motion. Over a period of 30 years, IZIIS has performed numerous shake-table tests on historic structures (see Section 33.7.2).

33.6.3 EQUALS Laboratory, University of Bristol

The Earthquake and Large Structures Laboratory (EQUALS) is a part of Bristol Laboratories for Advanced Dynamics Engineering (BLADE) at the University of Bristol, UK. It houses a 15-tonne capacity, 6-DOF shake table. The table is 3 x 3 m, driven by 8 actuators with 4 horizontal actuators in a pinwheel arrangement. It has a displacement capacity of $\pm$0.15 m and peak velocities of 1 m/s. Some of the unique capabilities of this facility are a 15 m tall, strong wall adjacent to the table and the availability of 40 actuators, providing a highly-adaptive dynamic test facility for a variety of earthquake tests.

33.6.4 Multi-Function Shake Table (MFST) Array, Tongji University, Shanghai

The MFST Array facility in Tongji University is part of the International Joint Research Laboratory of Earthquake Engineering (ILEE). This facility consists of four 6 x 4 m shake tables and a reaction wall. Each table has two horizontal translational degrees of freedom. The tables can be placed anywhere in two parallel trenches 70 m and 30 m long. Two tables have a payload of 30 tonnes and the other two, 70 tonnes. In each horizontal direction, these tables can move ±0.5 m, with a peak velocity of 1 m/s and acceleration up to 1.5 g. They can be placed to test multi-span bridges, long-span bridges up to 70 m span, or combined together to test a large structure. A nearby strong wall provides the opportunity for real-time hybrid simulation by connecting the specimen to the wall by an actuator.

33.7 Some notable projects tested with shake tables

Shake table tests have been performed on a wide variety of reduced-scale or full-size structures such as historic buildings, residential buildings, multi-storey buildings, parking structures, single-span bridges, multi-span bridges, suspension bridges, cable-stayed bridges, water tanks, wind turbines, underground tunnels, shell structures, nuclear power plant equipment, base-isolated structures and structures with supplemental damping. These models were constructed with a range of materials such as concrete, steel, masonry, timber and composite materials. In addition, a wide range of non-structural components have been tested including power distribution equipment, pipelines, ceiling fixtures, air-handling units, generators, stacked wine racks, frozen storage units, super computers and medical equipment. The tests were performed for a range of purposes: to validate analysis results, to examine the feasibility of a design or to advance the understanding of structural behaviour to develop improved design procedures. A few notable examples are presented here.

33.7.1 Testing of concrete-dam models at UC Berkeley

The 1967 Koynanagar earthquake near Mumbai in India, and subsequent damage to the Koyna earth dam, highlighted dam-safety concerns in seismically-active regions of the world due to the potential hazard to downstream populations. Several nonlinearities can be expected in the response of a concrete dam such as the opening of vertical joints between concrete monoliths, horizontal cracking of monoliths and reservoir cavitation. These non-linearities necessitated experimental verification of analysis methods available at that time. Two 1:150 scale models were used for shake-table tests at UC Berkeley in 1980: an arch-rib model of Techi arch dam in Taiwan, to study the joint-opening mechanism, and a gravity dam section model of Koyna Dam, India, to study dynamic cracking and the cavitation mechanism [37] (Figure 33.9). To meet similitude requirements, the model

Figure 33.9 1:150 scale model of Koyna Dam tested on the UC Berkeley shake table. (Image: By permission of the PEER-NISEE Library)

materials needed to have a strength and modulus of 1/150 of the prototype material, i.e. a modulus of 186 MPa and compressive strength of 0.186 MPa. A mix of casting plaster, Celite, sand and water were used to achieve appropriate modulus and strength. Lead powder was added to achieve the required density. The arch-rib model was progressively tested to failure with joint openings observed along the way. The gravity dam section model cracked in a manner very similar to that observed at the Koyna Dam during the 1967 earthquake. Cracking of the top part of gravity dam and subsequent post-cracking tests demonstrated that a cracked gravity dam can still retain the reservoir and that cracking does not necessarily represent failure of such structures. The cavitation mechanism was observed experimentally leading to further study of this phenomenon. Subsequent tests on a 1:50 scale model of Koyna dam by US Bureau of Reclamation resulted in similar findings and confirmed the results for UC Berkeley testing of 1980 [38] (Figure 33.10).

(a) (b)

Figure 33.10 (a) 1:50 scale model of Koyna Dam. (b) Connection between the actuator and the platen. (Image: US Bureau of Reclamation)

(a) (b)

Figure 33.11 (a) 1:2.75 model of the church of St Nikita, Banjani, Macedonia mounted on the shake table. (b) Detail showing damage to the model church. (Images: Courtesy of IZIIS)

33.7.2 Testing of models of historic buildings, at IZIIS, Macedonia

Historic structures usually suffer severe damage in strong earthquakes due to their stiff and brittle structural components. Any retrofit to improve the ductility needs to be minimally invasive to preserve the historic fabric of the structures. For over half a century, IZIIS has performed numerous shake-table tests on historic structures. These include: testing a model of a historic masonry house in 1986; testing 1:2.75 models of Byzantine Churches in the 1990s (Figure 33.11); testing a 1:6 model of the ancient Marcus Aurelius column in Rome; testing in 2006-07 of a 1:6 scale model of the Mustafa Pasha Mosque in Skopje, built in 1942; and testing of 1:5.5 scale model of a twelfth-century Gothic Cathedral. Great care was taken to make the models true replicas of the original structures by using materials and construction methods similar to original construction (stone, brick and lime mortar). Crack patterns and dislocation of blocks observed in shake-table tests were similar to observations of damage on the original structures in previous earthquakes. In addition, effective retrofit strategies were developed and tested on the models. Shake-table tests on these models were invaluable in assessing the seismic stability of these structures, investigation of retrofit techniques and refinement of numerical models [39].

33.7.3 Testing of a curved bridge at University of Nevada, Reno

A significant number of highway bridges (entry/exit ramps, freeway interchanges) are horizontally curved, yet there have been few experimental studies of such bridges. The I-5/SR-14 freeway interchange in California suffered extensive damage during the 1971 San Fernando (or Sylmar) earthquake and highlighted the need for closer investigation of curved bridges. There followed

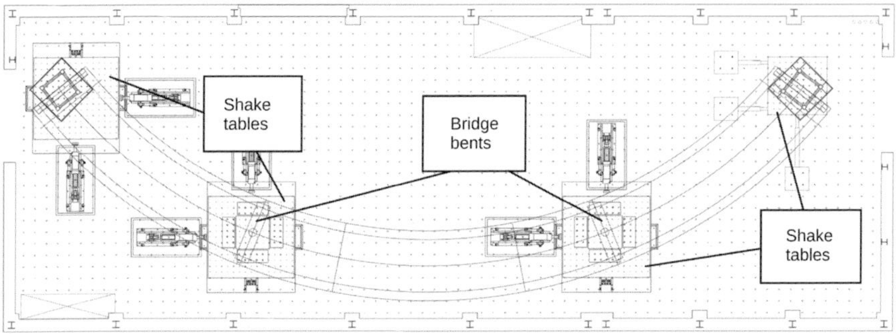

Figure 33.12 Test setup at the Nevada Seismological Laboratory (University of Nevada, Reno) for study of a 1:2.5 scale model of a curved road bridge. (Image: Courtesy of UNR)

tests on reduced-scale models of two such structures in California – the 1:200 scale model of Ruck-A-Chucky Bridge [40, 41] and the 1:30 scale mode of the I-5/SR-14 freeway [42]. These scale factors required the models to be distorted because there was no practical way to obtain full similitude. By increasing the model scale closer to the full-size structure (1:2.5), the testing of a multi-span curved bridge at UNR in 2011 was able to meet the similitude requirements by adding weights to the structure. The model was 44 m long, with a 24 m radius and a 3.6 m wide concrete deck, supported on steel-plate girders. The test set up, in which the size of the model dwarfs the size of the tables, took advantage of four relocatable shake tables in a 1000 m^2 laboratory (Figures 33.12 and 33.13). Several configurations were examined to study various components: conventional and rocking-column design; abutment design with and without backfill to simulate soil-structure interaction; full and partial seismic isolation; and the effect of live load simulated by trucks on the superstructure. It was an extensive study of curved-bridge performance using a physical model and the results of the research were used to develop new design procedures for horizontally-curved bridges [43, 44].

Figure 33.13 The curved-bridge model at UNR (Image: Amarnath Kasalanati)

33.7.4 Testing of a wind turbine at UC San Diego

The past twenty years have seen significant growth in wind power in earthquake-prone areas, leading to an increased need to properly account for seismic hazard in the design of wind turbines. While the design to meet the actions of wind forces on turbines is a well-developed field, seismic design practices at that time varied greatly, leading to significant under- or over-design. Operability of wind farms immediately following an earthquake ensures power for rescue and recovery efforts. This leads to designs where turbines need to behave elastically in earthquakes. Moreover, the homogeneity of turbines in a wind farm makes most of them vulnerable to a single earthquake with unfavourable characteristics. Experimental investigation of a large (full-scale or near full-scale) wind turbine had not been performed before 2004. The outdoor shake table at UC San Diego was uniquely suited for testing a full-scale unit measuring nearly 30 m in height. The 65 kW unit had a thin-walled, tubular-steel tower 22 m tall, a rotor diameter of 16 m and total weight of 10.5 tonnes. The testing program consisted of applying from 50% to 200% of one component of the 1992 Landers earthquake in California. Due to the elastic design of the turbine, results showed nearly linear behaviour over the range of different seismic tests. Shake-table testing revealed rotational and torsional motion in addition to the longitudinal motion, due to the mass distribution of the rotor and nacelle, highlighting the need to consider these effects in analytical modelling. Furthermore, shake-table test results showed that the measured damping of the full-scale unit was closer to 0.6%, significantly lower than the 5% damping assumed by building design codes. This experimental study provided a valuable basis for refining the analytical models and developing improved design procedures for wind turbines [45].

33.7.5 Testing of an isolated five-storey building at E-Defense

Seismic isolation is an effective method to mitigate the earthquake effects on structures and non-structural components. Shake-table testing of isolated structures has unique challenges because of the long vibration period of the structure and shake table's ability to replicate the motion in that range. The first two decades of modern shake-table testing was on reduced-scale models and the time reduction required by the similitude laws enabled the models to be tested on shake tables. Advances in the size and capabilities of shake tables in the last two decades has provided an opportunity to test full-scale or near full-scale isolated structures. A prominent case study is the large-scale testing program conducted collaboratively between NIED, NSF and NEES in 2011 using the E-Defense shake table in Japan. A full-scale, five-storey steel moment frame weighing 530 tonnes with plan dimensions of 10 x 12 m and a height of 15.8 m was used in this study (Figure 33.14). Testing consisted of three configurations: fixed-base, base isolation with triple friction pendulum bearings, and base isolation with lead-rubber isolation bearings in combination with cross linear bearings. The main objectives of the program were: to demonstrate the effectiveness of base isolation on a full-scale shake table test; to study

Figure 33.14 Testing an isolated structure at the E-Defense Facility in Miki City, near Kobe, Japan. (Image: Amarnath Kasalanati)

the effect of vertical acceleration on structural performance; and to study the response of non-structural components in a realistic configuration. Overall, the effectiveness of base isolation to reduce structural demands and lower the accelerations was clearly demonstrated. An interesting finding of the study was that the non-structural component disruption was closely correlated to the vertical ground motion intensity. With peak table accelerations in excess of 1 g, several ceiling panels were dislodged. In addition, during large lateral displacements, load transfer between the vertically stiff and soft components of the isolation system was observed. This study highlighted the need for considering horizontal-vertical coupling in the design of structures with continued functionality objectives [46].

33.8 Concluding remarks

Shake tables have been an essential part of many advances made in earthquake engineering in the past century. Modern shake tables are a combination of mechanical, hydraulic and electric/electronic subsystems, and have reached significant sophistication in recent years. Some of the key improvements in the last few decades include movement with six degrees of freedom, better servo-controlled hydraulic systems, larger and high-performance tables that test full-scale or near full-scale structures. A wide range of buildings, bridges and special types of structures have been tested on shake tables. Results were used to validate analysis results, examine feasibility of a design or advance the understanding of the behaviour of structures. As with any experimental study, there are technical challenges in shake-table testing due to size limitations, boundary conditions and table control. Nevertheless, accurate reproductions are achieved by most modern tables with ever-improving control algorithms.

The number of shake tables of significant size has grown to over 120 around the world. Despite their capabilities, the utilization of many shake tables remains less than 50%. Reasons for the underutilization include the complexity and costs of tests, and the perception that shake tables are only a research tool. In reality, shake tables are tools to improve reliability and safety of structural designs, to analyse the causes of damage in earthquakes and to develop new techniques of extending the life of existing structures in seismic areas. Recently, the focus has shifted to testing non-structural components, specifically critical equipment and the regulations requiring that shake-table testing is undertaken for critical non-structural elements are starting to increase usage of shake tables for such purposes.

Acknowledgements

The author is grateful to many mentors, colleagues and collaborators, who have provided a great deal of information and insight about shake-table testing over the past 25 years. In particular, interactions with the following individuals are gratefully acknowledged: Michael Constantinou, Andrei Reinhorn, Andrew Whittaker and Mark Pitman of UB; Ian Buckle, Gokhan Pekcan and Patrick Laplace of UNR; Masayoshi Nakashima, Taichiro Okazaki and Tomohiro Sasaki of E-Defense; Gianmario Benzoni, Danny Innamorato and Joel Conte of UC San Diego; J.S. Hwang and K.C. Chang of NCREE, Taiwan; Stephen Mahin, Jack Moehle, Khalid Mosalam and Clement Barthes of UC Berkeley. Christina Bodnar-Anderson, NISEE-PEER Librarian, provided information about UC Berkeley Shake Table. Any omission of facilities in this chapter is due to space limitations and is sincerely regretted.

References

1 Seible, F. & Shing, B. (1995) Worldwide Survey of Earthquake Engineering Testing Facilities. *Assessment of Earthquake Engineering Research & Testing Capabilities in the United States – Proceedings*. EERI Publication No. WP-01A. Oakland, CA: Earthquake Engineering Research Institute. Tables 5a & 5b; A-30-31.

2 Fardis, M.N. & Rakicevic, Z.T. (Eds.) (2012) *Role of Seismic Testing Facilities in Performance-Based Earthquake Engineering, SERIES Workshop*. (Geotechnical, Geological and Earthquake Engineering, No. 22). Dordrecht: Springer.

3 Yao, J., Dietz, M., Rui, H.Y., Xiao, W.T. & Yue, D. (2016) An overview of control schemes for hydraulic shaking tables. *Journal of Vibration and Control* 22 (12): 2807-2823.

4 Nuclear Energy Agency (2004) *Experimental Facilities for Earthquake Engineering Simulation Worldwide*. NEA/CSNI/R(2004)10. Paris: OECD.

5 Severn, R.T., Stoten, D.P. & Tagawa, Y. (2012) The Contribution of Shaking Tables to Earthquake Engineering. In: *Proceedings 15th World Conference on Earthquake Engineering, Lisbon, 2012.* Vol. 3: 1044-1053.
6 Seible, F. & Shing, B. (1995) Worldwide Survey of Earthquake Engineering Testing Facilities. *Assessment of Earthquake Engineering Research & Testing Capabilities in the United States – Proceedings.* EERI Publication No. WP-01A. Oakland, CA: Earthquake Engineering Research Institute.
7 Severn, R.T. (2011) The Development of Shaking Tables – A Historical Note, Earthquake Engineering and *Structural Dynamics* 40: 195-213.
8 Penzien, J., Bouwkamp, J.G., Clough, R.W. & Rea, D. (1967) *Feasibility study large earthquake simulator.* Report No. EERC-67-1. University of California, Berkeley: Earthquake Engineering Research Center.
9 Rea, D. & Penzien, J. (1973) Dynamic response of a 20 x 20 ft. shaking table. *Proceedings of Fifth WCEE, Rome* Vol. 2: 1447-1456.
10 Seible, F. & Shing, B. (1995) Worldwide Survey of Earthquake Engineering Testing Facilities. *Assessment of Earthquake Engineering Research & Testing Capabilities in the United States – Proceedings.* EERI Publication No. WP-01A. Oakland, CA: Earthquake Engineering Research Institute. Table 6, A-32.
11 Thoen, B.K. (2017) Exploring Inherent Cross-coupling Among Various Seismic Shake Table Configurations, Brad Thoen, *MTS. Force and Motion* 31, January 2017.
12 Nowak, R.F., Kusner, D.A., Larson, R.L. & Thoen, B.K. (2000) Utilizing modern digital signal processing for improvement of large scale shaking table performance. *Proceedings of the 12th World Conference on Earthquake Engineering, Auckland, New Zealand, Jan. 30-Feb. 4*: 2035-2042.
13 Yao, J., Dietz, M., Rui, H.Y., Xiao, W.T. & Yue, D. (2016) An overview of control schemes for hydraulic shaking tables. *Journal of Vibration and Control* 22 (12): 2807-2823.
14 Stehman, M. & Nakata, N. (2013) Direct acceleration feedback control of shake tables with force stabilization. *Journal of Earthquake Engineering* 17 (5): 736-749.
15 Stoten, D.P. & Shimizu, N. (2007) The feedforward minimal control synthesis algorithm and its application to the control of shaking-tables. *Proceedings of the Institution of Mechanical Engineers, Part I: Journal of Systems and Control Engineering* 221 (3): 423-444.
16 Luco, J.E., Ozcelik, O. & Conte, J.P. (2010) Acceleration tracking performance of the UCSD-NEES Shake Table. *Journal of Structural Engineering, ASCE* 36 (5): 481-490.
17 Trombetti, T.L. & Conte, J.P. (2002) Shaking table dynamics: results from a test-analysis comparison study. *Journal of Earthquake Engineering* 6 (4): 513-551.
18 Thoen, B.K. (2017) Exploring inherent cross-coupling among various seismic shake table configurations. *MTS. Force and Motion: Geo/Civil/Seismic Testing* 31 (January). (On line)

19 EERI (2004) *Connections: EERI Oral History Series. No.11 Joseph Penzien.* Oakland CA: Earthquake Engineering Research Institute. https://www.eeri.org/monographs-and-oral-histories/oral-history-series/ (2/9/2019)

20 Penzien, J., Bouwkamp, J.G., Clough, R.W. & Rea, D. (1967) *Feasibility study large earthquake simulator.* Report No. EERC-67-1. University of California, Berkeley: Earthquake Engineering Research Center. 13-14.

21 EERI (2004) *Connections: EERI Oral History Series. No.11 Joseph Penzien.* Oakland CA: Earthquake Engineering Research Institute. https://www.eeri.org/monographs-and-oral-histories/oral-history-series/ (2/9/2019)

22 NEES (2014) *NEES: 2004-2014. A Decade of Earthquake Engineering Research.* West Lafayette, IN: Purdue University.

23 Ketter, R.L. (1983/1984) The UB Seismic Simulator. *Engineering Progress of Western New York* 3 (2, Winter).

24 Reinhorn, A.M. & Prawel, S.P. (1986) Ferrocement in a Large Shaking Table. *Journal of Structural Engineering (ASCE)* 112 (2, Feb): 401-416.

25 Hwang, J.S., Chang, K.C. & Lee, G.C. (1987) *The System Characteristics and Performance of a Shaking Table.* Technical Report NCEER 87-0004. Buffalo, NY: National Center for Earthquake Engineering Research.

26 Van Den Einde, L, Restrepo, J., Conte, J.P., Luco, E., Seible, F., Filiatrault, A., Clark, A., Johnson, A., Gram, M., Kusner, D. & Thoen, B. (2004) Development of the George E. Brown Jr. Network for Earthquake Engineering Simulation (NEES) Large High Performance Outdoor Shake Table at the University of California, San Diego. *Proceedings 13th World Conference on Earthquake Engineering*, Vancouver, BC Canada: Paper No. 3281.

27 Luco, J.E., Ozcelik, O. & Conte, J.P. (2010) Acceleration Tracking Performance of the UCSD-NEES Shake Table. *Journal of Structural Engineering, ASCE* 36 (5): 481-490.

28 Luco, J.E., Ozcelik, O., Conte, J.P. & Mendoza, L.H. (2011) Experimental study of the dynamic interaction between the foundation of the NEES/UCSD Shake Table and the surrounding soil: Reaction block response. *Soil Dynamics and Earthquake Engineering* 31 (7): 954-973.

29 Sawada, K. et al. (1970) Strength and vibration characteristics of large type shaking table. *Technical Review* (Mitsubishi Heavy Industries Ltd.) 7 (3).

30 Ohtani, K., Ogawa, N., Katayama, T. & Shibata, H. (2004) Construction of E-Defense (3-D full-scale earthquake testing facility) *Proceedings 13th World Conference on Earthquake Engineering Vancouver, B.C., Canada, August 1-6, 2004.* Paper No. 189.

31 Sawada, K. et al. (1970) Strength and vibration characteristics of large type shaking table. *Technical Review* (Mitsubishi Heavy Industries Ltd.) 7 (3).

32 Ohtani, K., Ogawa, N., Katayama, T. & Shibata, H. (2003) World's largest shaking table takes shape in Japan (3rd Report). *Transactions 17th International Conference on Structural Mechanics in Reactor Technology (SMiRT 17), Prague, Czech Republic, August 17-22, 2003.* Paper K12-1.

33 Ogawa, N., Ohtani, K., Katayama, T. & Shibata, H. (2001) Construction of a three-dimensional, large-scale shaking table and development of core technology. *Bulletin of Phil. Trans. R. Soc. London* Series A, 359 (1786 – Dynamic Testing of Structures): 1725-1751.

34 Nakashima, M., Nagae, T., Enokida, R. & Kajiwara, K. (2018) Experiences, accomplishments, lessons, and challenges of E-defense – Tests using world's largest shaking table. *Japan Architectural Review* 1 (1): 4-17.

35 Gao, Chun-hua & Yuan, Xiao-bo (2019) Development of the shaking table and array system technology in China. *Advances in Civil Engineering* Article ID 8167684.

36 Emílio, F.T., Duarte, R.T., Carvalhal, F.J., Costa, C.O., Vaz, C.T. & Corrêa, M.R. (1989) *The New LNEC Shaking Table for Earthquake Resistance Testing*. Memoire LNEC 757. Lisbon: LNEC.

37 Niwa, A. & Clough, R.W. (1980) *Shaking table research on concrete dam models*. Earthquake Engineering Research Center, Report UCB/EERC-80/05. Berkeley, CA: University of California.

38 Harris, D., Snorteland, N., Dolen, T. & Travers, F. (1999) *Shaking table 2-D models of a concrete gravity dam for computer code validation*. DSO-98-13. Denver, CO: Dam Safety Office, US Department of Interior.

39 Shendova, V., Rakicevic, Z.T., Krstevska, L., Tashkov, L. & Gavrilovic, P. (2012) Shaking table testing of models of historic buildings and monuments: IZIIS' experience. In: *Role of Seismic Testing Facilities in Performance-Based Earthquake Engineering, SERIES Workshop*. (Geotechnical, Geological and Earthquake Engineering, No. 22). (eds M.N. Fardis & Z.T. Rakicevic), 221-245. Dordrecht: Springer.

40 Godden, W.G. (1977) Seismic Model Studies of the Ruck-A-Chucky Bridge. Report prepared for T.Y. Lin International, under terms of their Solicitation No. DS 7159. Denver, CO: United States Department of the Interior, Bureau of Reclamation.

41 Godden, W.G. & Asham, M. (1978) Dynamic model studies of Ruck-A-Chucky Bridge *Journal of Structural Division, ASCE ST12* 104: 1827-1845.

42 William, D. & Godden, W. (1979) Seismic Response of Long Curved Bridge Structures: Experimental Model Studies. *Earthquake Engineering and Structural Dynamics* 7: 107-128.

43 Levi, M.J., Sanders, D.H. & Buckle, I.G. (2012) Seismic Response of Columns in Curved Bridges. In: *Proc. Structures Congress 2012, Chicago, IL, March 29-31, 2012* (Eds J. Carrato & J.G. Burns). Reston, VA: American Society of Civil Engineers.

44 Monzon, E.V., Wei, C., Buckle, I.G. & Itani, A.M. (2012) Seismic Response of Full and Hybrid Isolated Curved Bridges. In: *Proc. Structures Congress 2012, Chicago, IL, March 29-31, 2012* (Eds J. Carrato & J.G. Burns). Reston, VA: American Society of Civil Engineers.

45 Prowell, I., Veletsos, M. & Elgamal, A. (2008) Full Scale Testing for Investigation of Wind Turbine Seismic Response. *Proceedings of the 7th World Wind Energy Conference, Kingston, Ontario, Canada*.

46 Sasaki, T., Sato, E., Ryan, K.L., Okazaki, T., Mahin, S. & Kajiwara, K. (2012) NEES/E-Defense base isolation tests: effectiveness of friction pendulum and lead-rubber bearing systems. *Proceedings 15th World Conference on Earthquake Engineering, Lisbon, 2012*. Vol. 15: 12097-12106.

34

Geotechnical centrifuge modelling – current practice

David White

34.1 The role of the geotechnical centrifuge

Geotechnical centrifuge modelling is a technique in which small-scale models of geotechnical structures are subjected to enhanced gravitational forces within a centrifuge, while events or processes are simulated. The technique is motivated by the need to respect the influence of stress level on the strength and stiffness of geomaterials and the influence of body forces on geotechnical stability.

Small-scale models of geotechnical systems are widely used for teaching [1] and for research, and many instructive studies are produced that shed light on fundamental aspects of geotechnical behaviour. However, when contemplating the response of a full-scale geotechnical system, a small-scale model has limitations. Firstly, the stress:strain response of soil or its strength and stiffness, depend on the effective or intergranular stress level, which is the total stress minus the pore-water pressure. Therefore small models with low effective stresses, do not directly replicate field conditions. Secondly, self-weight body forces within the model are lower than the larger equivalent, so deformations and failure mechanisms driven by these forces are not correctly simulated, unless the strength and stiffness properties are modified accordingly, which is problematic with natural soils.

Geotechnical centrifuge modelling provides a technique that overcomes both these limitations and brings collateral benefits. The centrifuge establishes a modelling environment in which stresses and strains match a full-size equivalent that is larger than the model by a factor N, where N is the ratio between the centrifuge acceleration and Earth's gravity. In this way, the correct stress-strain response operates within the model and body forces are scaled correctly. An additional benefit is that time-dependent processes happen more quickly than in the full-size prototype, by a factor of N^2. This means that the centrifuge provides accelerated testing of field-scale processes. It is possible to perform 'whole life' simulations in which many years of loading and environmental exposure and the resulting effect on the geotechnical system, can be modelled in a practical timescale. For example, at a typical scaling factor of $N = 100$, a 24-hour centrifuge test allows 27 years of behaviour to be simulated.

Physical Models: Their historical and current use in civil and building engineering design,
First Edition. Edited by Bill Addis.

The centrifuge therefore supports small-scale geotechnical modelling in a similar way to how wind-tunnel testing allows small-scale fluids experiments to replicate larger-scale behaviour faithfully. Both approaches follow scaling principles that are underpinned by the relevant dimensionless group [2].

The historical development of geotechnical centrifuge modelling set out the genesis of this technique through the twentieth century especially the industry-focused achievements of the Manchester group under Professor P.K. Rowe (see Chapter 27). As the twentieth century closed, new geotechnical centrifuge facilities were being established throughout the world, primarily hosted by academic institutions. The number of operating geotechnical centrifuge facilities worldwide is now higher than ever, contrasting with the general trend sometimes associated with physical modelling laboratories. However, the application of centrifuge modelling to live projects has perhaps not grown in direct proportion to the spread of centrifuge facilities.

This chapter illustrates the recent evolution and use of geotechnical centrifuge modelling incorporating a description of modern facilities and the technologies and capabilities that they offer, as well as some examples illustrating how centrifuge modelling is applied to projects for ground improvement, foundation design and the simulation of whole-life system behaviour.

34.2 State-of-the-art geotechnical centrifuge facilities

34.2.1 Centrifuge machines

The state-of-the-art of geotechnical centrifuge facilities is defined both by the centrifuge machines themselves and the equipment and models that are flown within them. Centrifuge machines specifically for geotechnical modelling were first built around 1970 [3, 4, 5]. They are usually characterised by their diameter or by their maximum capacity given by the product of the model weight and the acceleration level in g-tonnes.

We are therefore reaching a transitional time as the early machines approach the end of their working life and, for the first time, organisations are contemplating rebuilding and replacing their machines or reducing or even removing the space previously dedicated to geotechnical centrifuge modelling.

At Cambridge, where I spent my formative years, the Turner beam centrifuge has operated for over 40 years. This 9-m diameter, 150-g-tonne machine remains busy today, fully booked by new PhD students. Meanwhile, at Manchester, the original 6.4 m diameter, 250 g-tonne, Rowe centrifuge has been decommissioned and the space repurposed. The geotechnical group has instead taken ownership of a more compact 1.2 m diameter, 6-g-tonne beam centrifuge located elsewhere on campus.

The University of Western Australia, where I later undertook research, hosts the busiest geotechnical centrifuge facility worldwide, with three machines each spinning for more than 200 days per year. In 2011 work began on building a new 10-m diameter, 240-g-tonne beam centrifuge [6] (Figure 34.1), which now

Figure 34.1 Modern geotechnical centrifuge facilities. Actidyn C72 10-m diameter, 240-g-tonne beam centrifuge at the University of Western Australia. (Image: Courtesy of University of Western Australia)

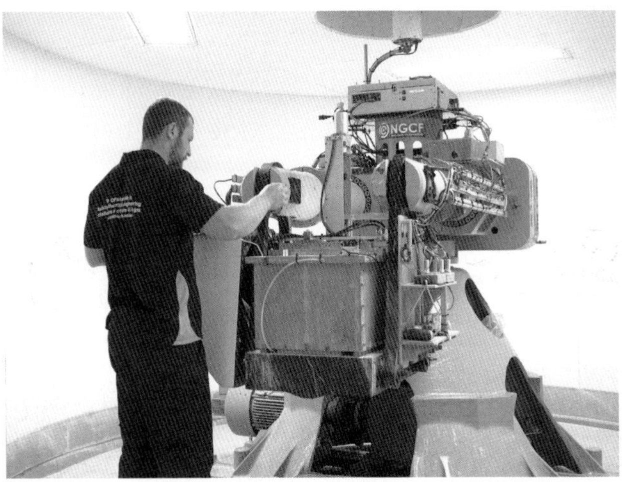

Figure 34.2 Modern geotechnical centrifuge facilities. Actidyn C61 3.6-m diameter, 40-g-tonne beam centrifuge at the University of Western Australia. (Image: Courtesy of University of Western Australia)

complements the 30-year-old 3.6-m diameter, 40-g-tonne beam centrifuge [7] (Figure 34.2) and a smaller, 20-year-old 1.2-m diameter, 290-g-tonne drum centrifuge [8]. The 40-g-tonne machine was one of the first geotechnical centrifuges capable of running unattended, with automatic monitoring and shutdown systems. As a result, projects on this machine have pioneered long-term tests with multiple days of continuous centrifuge running, allowing 'whole life' behaviour to be observed. The overall usage of the UWA 40-g-tonne machine since installation in 1989, typically with N = 100, corresponds to more than 100,000 years of geotechnical history.

Now at the University of Southampton, I am supporting the establishment of a 6-m diameter, 150-g-tonne beam centrifuge as part of the UK government's investment in a National Infrastructure Laboratory. This new facility joins seven other active geotechnical centrifuge facilities at UK universities in Brighton, Cambridge, City (London), Dundee, Manchester, Nottingham and Sheffield.

Meanwhile, in China, more than fifteen new geotechnical centrifuges have been established in the past twenty years [9]. This population of machines will be crowned by a new facility at Zhejiang University – referred to as CHIEF, the Centrifugal Hypergravity Interdisciplinary Experimental Facility expected to enter operation in 2023. The CHIEF will comprise two gigantic centrifuges with a capacity of 1500 g-tonnes, one with the capability to fly small models at accelerations up to 1500 g, and one with the capability to test models up to 30 tonnes in weight.

Diversification is therefore the trend in geotechnical centrifuge machines – there is no evolution towards a single, ideal scale. Instead, complementary types of facility are being constructed, offering a balance of capacity, usability and operability appropriate to their host.

34.2.2 Modelling technologies

The models and equipment flown on geotechnical centrifuges have evolved hugely since the late twentieth century, reflecting new technologies for sensing, data acquisition, control and miniaturisation, as well as improved methods of creating model ground [10, 11].

Early experiments used either large block samples taken from the field or material reconstituted from dried sediment, by means of a slurry for clays or deposition through air or water for sands. It is now possible to reconstitute geomaterials through a well-controlled stress history so that particular profiles of strength and stiffness can be 'dialled-up' through a computer-controlled consolidation press which replicates the desired geological overburden history. There have even been attempts to 3D-print a heterogenous bed [12] and individual granular particles [13].

Once the sediment sample has been prepared, miniature representations of real world civil engineering processes can be installed above the sample and brought to life once the model is spinning at the representative g-level. Many ingenious devices have been developed to replicate construction activity at small scale. These robotic devices operate while the centrifuge is 'in-flight', representing processes such as volume loss during tunnel construction [14], excavation [15] and casting of diaphragm walls [16]. Other machines replicate ground improvement processes through dynamic and vibratory compaction [17, 18], the action of a vibratory roller [19], or through deep mixing [20]. Sometimes scale-model robots represent their real-life equivalent, for example a 1:10 scale excavator, 16 cm long, crawling up and down an earth embankment to study its stability and safety against overturning on slopes of different angles [21] (Figure 34.3).

If needed, the centrifuge model can be exposed to realistic environmental conditions. The forces of nature can be replicated at small scale in environmental

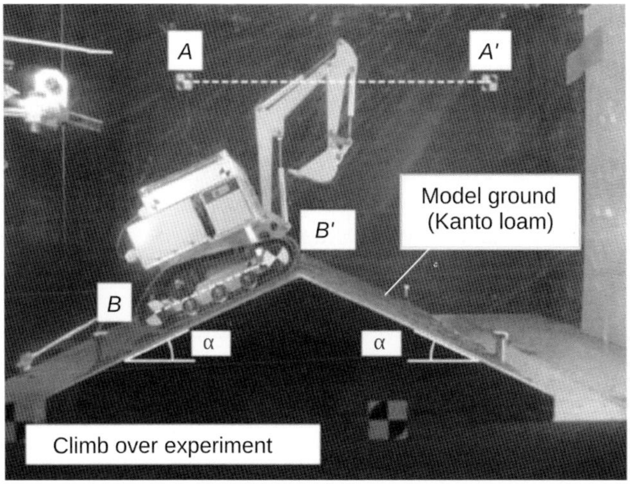

Figure 34.3 Radio-controlled 1:10 scale model excavator in tests to study its stability while climbing and descending an earth embankment. (Source: [21])

chambers. These sealed chambers allow control of boundary conditions associated with infra-red radiation, wind, temperature, humidity and water table which may cycle according to diurnal, seasonal or long term patterns. For example, the seasonal wetting and drying cycles of clay soils can be mimicked by subjecting the model to controlled cycles of hot and cold weather, or evaporation and rainfall – with the water droplets being appropriately scaled [22, 23, 24, 25]. The accelerated scaling of time processes in the geotechnical centrifuge allows these environmental simulations to span multiple decades. These types of test can reveal the resilience of earthworks, landfill systems and coastal defences [26, 27].

The model can also be subjected to short-term natural hazards. Earthquake loading can be imparted using sophisticated servo-controlled shaking tables [28] and tsunamis or landslides can be generated via miniature flume arrangements [29, 30]. Counter measures to protect structures can be mimicked with faithful adherence to scaling laws using miniaturised reinforced concrete [31] and structural dampers [32]. The control offered by 3D printing allows detailed modelling of building stiffness, for example to explore the interaction between complex masonry structures and their foundations, such as nineteenth-century buildings affected by new infrastructure tunnelling under cities [33]. Natural structures such as plant roots can even been incorporated in the model, using either real juvenile plants [34] or 3D-printed replicas [35].

The soil model can be investigated by small-scale versions of the site investigation techniques used in reality – so centrifuge laboratories are equipped with miniaturised versions of penetrometer and pressuremeter devices [36, 37, 38, 39]. Non-intrusive measurements of the deformation of samples can also be made using image analysis. Modern photogrammetric and image correlation technologies such as Particle Image Velocimetry (PIV) can detect displacements in 2D and 3D to an accuracy of around a micron [40, 41, 42, 43]. To compare this to conventional surveying, consider a centrifuge model at an acceleration of 100g using a

1 m square strongbox. The micron-level precision in the centrifuge is equivalent to a surveying system with 0.1 mm precision acting across a test region that is approximately the size of a football pitch.

As a centrifuge test progresses, it may be interacting with parallel numerical simulations, through so-called real-time, hybrid sub-structure testing. In common with other forms of physical testing, this approach allows the behaviour of a more extensive system beyond the physical model to be analysed, with a numerical approach being adopted for the remainder. Rapid and continuous feedback between the two can be achieved with modern software and hardware [44].

To illustrate how some of these technologies have been brought to bear on different types of projects, three examples are described below, highlighting particular case studies where a 'live' project has relied directly on centrifuge models to guide and verify the design process.

34.3 Complex ground improvement – the Rion-Antirion Bridge

The Rion-Antirion bridge has connected the town of Rio on the Peloponnese peninsula with the mainland of Greece since its opening in 2004. It is one of the world's longest multi-span cable-stayed bridges and sits in a seismic region. The bridge has three main spans, each 560 m long, crossing water that is 65 m deep (Figure 34.4). The foundation soils are soft alluvial sediments with layering of silty sand, sandy clay and clay, with gravel close to the surface. The dead weight carried by each pier is 800 MN and an additional load equivalent to 600 MN acting horizontally at a height of 33 m above the seabed is applied during the design earthquake [45, 46, 47].

Figure 34.4 The Rion-Antirion Bridge, completed 2004. (Image: [48])

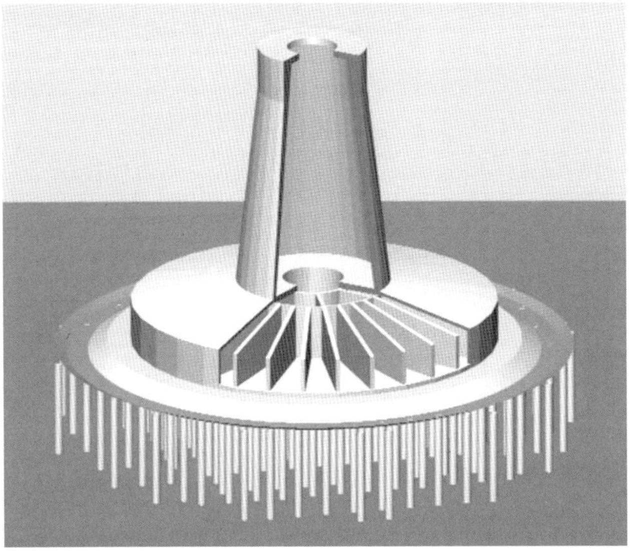

Figure 34.5 Rion-Antirion bridge. Illustration of bridge pier concept. (Image: Institution of Civil Engineers)

The soft seabed soils initially appeared to favour foundation solutions that transmit the load to deeper material via piles or a deep caisson. However, these solutions were ruled out due to challenges associated with the construction or installation processes.

The design soon focused on a surface foundation but the strength of the underlying soil remained a challenge. A novel reinforcement approach was conceived, using pipe piles that are not connected to the pier – these are referred to as inclusions to distinguish them from conventional piles (Figure 34.5). This lack of structural connectivity allows the pier to slide sideways under the design seismic load while rotation is prevented by the strengthening of the inclusions – a powerful application of the principles of performance-based seismic design [49, 50].

In order to provide sufficient confidence in this solution an extensive programme of centrifuge modelling was performed to provide the necessary validation and optimisation of this novel construction solution. A programme of testing was performed at the *Laboratoire de Ponts et Chaussées* in Nantes, France, (now known as IFFSTAR) which faithfully modelled the *in-situ* soil strength profile and the grid of almost 200 inclusions beneath the foundation (Figure 34.6a). The centrifuge tests indicated the ultimate capacity of the composite foundation system which aligned to within 15% of theoretical predictions and also provided information that could not easily be derived by mathematical analysis. A key observation was that the foundation system showed no degradation of stiffness even under several cycles of loading up to 75% of the failure load. This resilience provides an indication of the system reliability under a wider range of design conditions including multiple design events beyond the critical loads alone. The model test arrangement is shown in Figure 34.6b. A cross-sectional view after failure of the foundation under an elevated load is shown in Figure 34.7.

(a) (b)

Figure 34.6 Centrifuge model of ground improvement at Rion-Antirion bridge. (a) Centrifuge model of foundation inclusions. (b) Centrifuge model of foundation with system for applying cycles of overturning load. (Image: Jacques Garnier)

Figure 34.7 Failure mechanism beneath a pier of the Rion-Antirion bridge. (Image: Jacques Garnier)

In this example, the centrifuge modelling programme provided performance data for a foundation solution that could not be reliably generated using numerical techniques and which were necessary to provide confirmation that the design met the performance-based acceptance criteria. The particular combination of a gravel layer, unconnected exclusions and intense cyclic loading brought together too many complex interactions. Conventional calculations or numerical simulations would not have provided adequate certainty.

The project programme for the Rion-Antirion Bridge allowed the physical modelling activity to be carried out within the design programme without increasing the overall project duration. Two campaigns of physical modelling were performed. In the initial phase, the layout and geometry of the inclusions were varied to investigate the optimal design configuration. In the second phase, particular attention was focused on the effect of cyclic loading and a smaller number of more-highly instrumented tests were performed.

Construction proceeded as planned, with the bridge opening ahead of schedule in August 2004. The project chief, Jean-Paul Teyssandier, concluded that 'in the event of an earthquake in the region, everything will be destroyed except the bridge' [51].

34.4 Offshore oil and gas platforms – the Yolla and Maari projects

Geotechnical centrifuge modelling has been widely applied to offshore engineering challenges, including for project-specific studies that explore novel design solutions, for example, the modelling studies prompted by the early oil and gas developments (see Section 27.3.3). These were driven by the need to verify designs for huge offshore structures that were far beyond the scale and loading experienced for onshore civil engineering structures. This trend has continued throughout the development of the North Sea and, as centrifuge technology spread worldwide, it has been adopted by the US and Australasian oil and gas industries [52, 53, 54].

Two of the most recent project examples where centrifuge modelling played a key role in optimising and verifying the design are the Yolla A and Maari platforms. Yolla A is in Australia's Bass Strait, approximately 140 km offshore from Kilcunda, Victoria and was installed in 2004. Maari is located 80 km off the coast of South Taranaki, New Zealand and was installed four years later. These platforms were two of the first deployments of the Arup partnership's proprietary ACE design – the 'Arup Concept Elevator'. The ACE concept is a self-installing, steel gravity-based platform (Figures 34.8a and 34.8b). The entire structure, including both the foundation and the topsides facilities for oil and gas

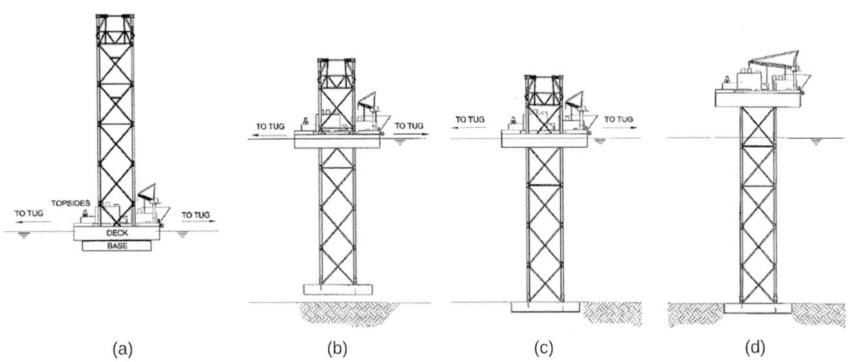

Figure 34.8 The ACE platform-skirted foundation. Installation sequence: (a) Start of base lowering. (b) Base above seabed. (c) Skirts penetrate into seabed. (d) Deck lifted clear of water. (Image: Phil Watson)

production, is assembled at an onshore yard. The facility is then floated to site, and can be installed without the need for heavy lift operations. Installation of the platform takes place by lowering the foundation on steel legs to the sea-floor. The skirts of eight foundation compartments are then embedded into the seabed by a combination of self-weight – by lifting the deck slightly out of the water – and suction applied within the compartments (Figures 34.9 and 34.10). Once the foundation is fully installed, the platform is then jacked up above the highest waves, before welding the platform-leg connections.

Centrifuge modelling was used in a variety of ways to optimise and validate the novel aspects of this design concept during the two projects. The installation process relies on a suction system to reduce the pressure in the water beneath the foundation base plate, in order to create a net downwards force on the foundation to assist installation. In addition, for the Yolla project, the suction system was also used to dewater a layer of silt within the soil plug over the month following installation. This preloading process was necessary to minimise the subsequent settlements of the structure that would occur after the installation was complete and the topsides were welded into position. This silt layer was beneath a less permeable clay layer, so the suction system included piping down the skirts to nozzles close to the skirt tip (Figure 34.9). It was identified that the suction process could lead to ingress of silt through these nozzles, damaging the soil plug and potentially also affecting the ability to apply suction. Centrifuge modelling was used to mimic the specific behaviour close to the nozzles, trialling different types of filter material [55]. In this unusual case, the centrifuge environment was not used to test a small-scale model, but simply as a convenient means of matching the stress and flow conditions around prototype filter elements. This study informed the choice of filter design which worked effectively during the Yolla installation and preloading [56].

Figure 34.9 Yolla platform A; The ACE platform skirted foundation. Interior view of the prototype skirted foundation. (Image: Phil Watson)

At Maari, a different consideration, related to the suction installation process, was examined using centrifuge model tests. The soil layering at Maari is composed of interbedded sands and silts. At the design stage it was uncertain whether the foundation could be successfully installed through the layers or whether the suction process would instead lift one or more soil layers up into the skirts, leaving the foundation proud of the seabed and lacking the required capacity.

Centrifuge modelling was used to simulate the installation process using a model of a single foundation compartment. This model was installed into a layered soil model using different combinations of suction and static self-weight. A relationship was derived showing combinations of suction and self-weight that led to skirt penetration for the Maari soil conditions (Figure 34.10). It was shown that the plug would remain intact and not lift away from the sea floor. The results from the centrifuge modelling programme were used to finalise the arrangements of the foundation design. They also provided a basis for controlling and forecasting the installation response [57].

Finally, a third centrifuge study was performed to validate the stability design of the Maari platform. The layered soil created uncertainty regarding the overturning capacity of the foundation system. The ACE platform design is a relatively light structure, while Maari is a harsh ocean environment where waves higher than 25 metres were predicted. These loads cause the trailing side of the foundation to be subjected to net uplift. This creates tension in the soil plug which must be sustained in order to preserve the overturning capacity.

For this study, a centrifuge model with appropriate layering was created and a model of the full Maari foundation system was installed and subjected to cyclic loading at and above the design conditions (Figure 34.11). Despite the uplift loading, the overturning capacity remained well above the design load and differential

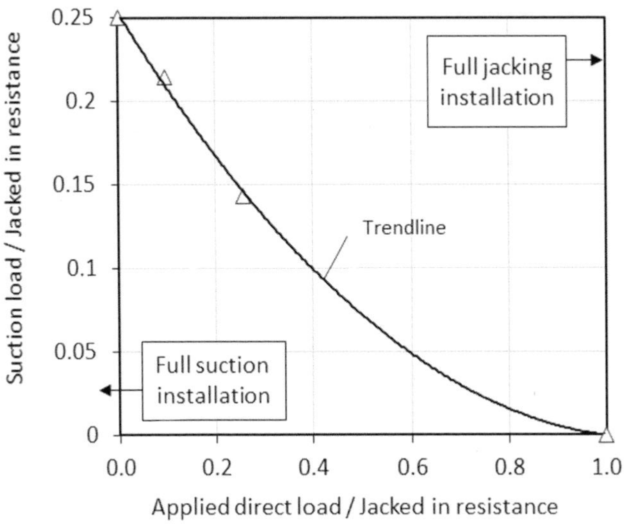

Figure 34.10 Centrifuge-based relationship between direct load and suction requirement for Maari platform installation. (Image: Phil Watson)

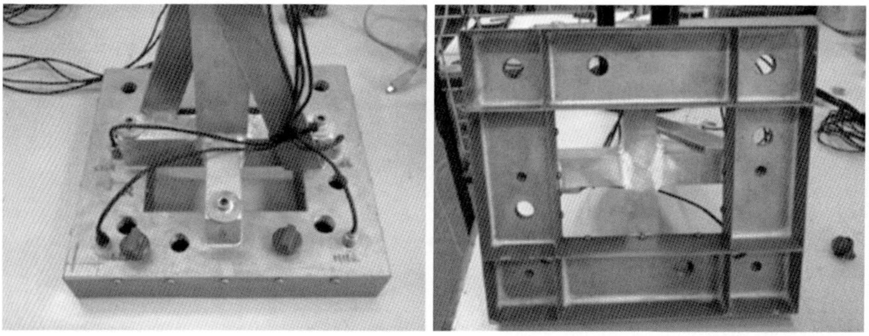

Figure 34.11 The ACE platform skirted foundation. Centrifuge model used for Maari foundation stability testing (a) View from above. (b) View from beneath. (Image: Phil Watson)

settlements were acceptable with both the leading and trailing sides of the foundation undergoing net settlement in the long term. These results gave confidence in the analytical techniques used to assess the foundation capacity, including the reliance on short term uplift capacity.

The cost benefits associated with the novel ACE system came at the expense of an installation process that had limited flexibility if operations did not go exactly as planned. Nevertheless, the model tests confirmed that the proposed approach would be reliable. Furthermore, the Bass Strait and offshore New Zealand are extremely onerous ocean environments with challenging soil conditions. Centrifuge modelling played a key role in assuring the design of these two novel structures which have now both been operating for more than a decade.

34.5 Sub-sea pipelines – modelling of 'whole life' system behaviour

A second example of centrifuge modelling being used in support of offshore projects is the modelling of seabed pipelines that are subjected to thermal expansion during operation. Indeed, an early use of centrifuge models was to assess the behaviour of buried pipelines when they expand and can emerge from their protective backfill cover – a process known as upheaval buckling (see Section 27.4.1).

For pipelines installed in waters where trawl fishing is not performed, the pipelines need not be buried for protection and it is simpler and more cost effective to leave the installed pipelines on the seafloor. Thermal expansion and contraction still occurs each time the pipeline is started up or shut down which changes the temperature of the contents. If the pipeline is laid on the seafloor this expansion tends to be accommodated in buckles that form by the pipe moving laterally on the seabed.

The formation and behaviour of these buckles must be estimated during design to ensure that the pipeline strength is adequate and fatigue will not occur. It is

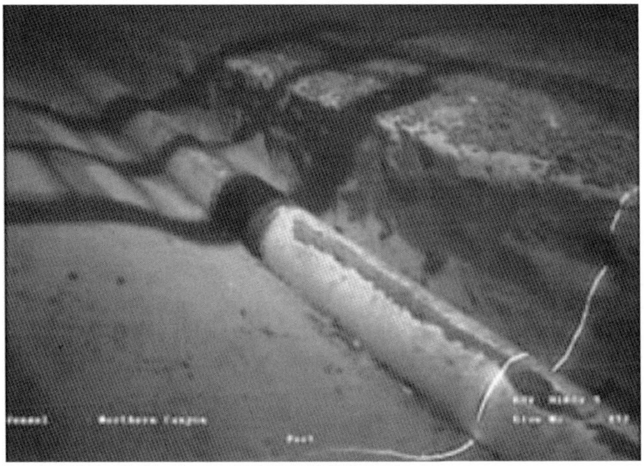

Figure 34.12 Example of a seabed trench created by a laterally-buckling pipeline. (Image: Dave White)

often necessary to introduce engineered buckles at specific locations along the pipeline, but the number of additional structures to trigger these needs to be minimised. A view within the buckle of a real pipeline on the seabed illustrates the deep trench that can form as a result of the movement of the pipe (Figure 34.12).

The influence of friction between the pipe and the seabed can be critical in the design of these pipelines. However, this lateral resistance is difficult to predict, particularly through the many hundreds of cycles of expansion and contraction that can occur during a typical operating life. Theoretical and numerical models are currently inadequate because pipeline-seabed interaction involves gross changes in the local bathymetry of the surrounding seabed and also changes in soil properties. As the pipe is laid on the seafloor it moves dynamically while hanging from the lay vessel which remoulds the soil, causing a reduction of strength by a factor of up to 10. The strength will then recover, at least partially, particularly if the pipe is weighted down with water to provide stability while testing is undertaken prior to operation [58, 59]. Once the pipe enters operation further changes in local bathymetry and soil strength occur as the pipe cycles back and forth eroding a trench. The behaviour varies between surface sliding and bulldozing over soil with varying strength. However, structural analysis of the pipeline behaviour requires this complexity to be distilled into simpler representations such as spring-sliders that can be incorporated into a model of the full length of the pipeline.

Centrifuge modelling techniques have been developed to replicate these processes, in order to provide the necessary design inputs. With appropriate actuation and control systems, a model of a pipe element, representing perhaps a 5 m length, can be subjected to motion and loading that represents the full sequence of dynamic laying, water-filling and then operational cycles of movement across the seabed [60, 61]. The aftermath of such a model test is shown in Figure 34.13,

Figure 34.13 Seabed deformation after centrifuge model test of pipe element. (Image: Dave White)

illustrating the formation of trenches and berms within the model seabed similar to the field example.

Over the past decade many offshore projects have relied on bespoke centrifuge modelling programmes to characterise this lateral pipeline-seabed interaction during lateral buckling. These projects include BP-operated ventures around the world in Angola, Azerbaijan and Egypt, as well as most of the projects that have underpinned Australia's booming liquid natural gas industry including Gorgon, Pluto, Ichthys, Wheatstone and Browse.

These physical model tests have been used to select an appropriate system of structures and pipeline routing to mitigate expansion-induced movements. The model-testing programmes use soil samples recovered from the seabed that are reconstituted to mimic the seabed conditions. The full sequence of pipeline-seabed interaction is modelled, from the installation of the pipeline onto the seabed, through commissioning (when the pipe becomes water-filled and therefore heavier for a period) and subsequent operational cycles.

A typical set of results from this type of experiment is shown in Figure 34.14 The separate sequences that represent the lay process, the initial breakout when started up and the subsequent cycles back and forth across the seabed are highlighted. These elegant patterns show the gross changes in apparent friction and the complexity of the behaviour. There is a progressive increase in the constraint on the pipeline as larger and larger soil berms are created on each side of the deepening trench. A typical modelling programme involves 10-20 similar tests that explore the relevant parameter space. These usually include variations in the sequence of movements, such as the amplitude of the cycles and the progressive drift of their limits and the vigour of the laying process. The programme may be used to design multiple pipelines and so variations in weight and diameter would be included. The full set of results are then used to calibrate the relevant pipe-soil interaction model that is used in the structural modelling of the system [62].

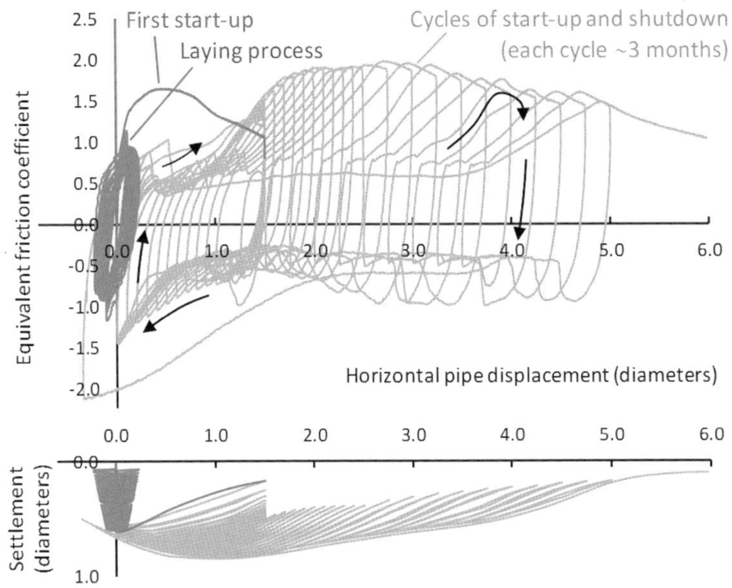

Figure 34.14 'Whole life' modelling of the settlement and lateral restraint of a pipe element within a lateral buckle. (Image: Dave White)

34.6 Concluding remarks – Why use geotechnical centrifuge modelling?

The examples above illustrate cases where geotechnical centrifuge modelling is of particular value during design, with the greatest ability to influence and improve the final structural form. The value of centrifuge modelling is generally greatest when the geotechnical system involves:

- a new design concept or ground improvement process for which full-scale experience and performance data is lacking
- an unusual soil type or stratigraphy
- evolving soil properties linked to complex loading or environmental boundary conditions.

All these cases can be challenging to assess using conventional analysis methods or numerical models and additional confidence can be provided by centrifuge modelling results.

The latter point – long term evolution of the system – is addressed by 'whole life' long-term centrifuge modelling. This approach takes advantage of the accelerated consolidation processes and modern miniaturised technologies that can impose realistic loading and changing environmental conditions. This type of centrifuge modelling provides performance data that are otherwise unobtainable within a project timescale.

A further benefit of centrifuge modelling that has been found to add confidence and understanding as project engineers work together towards an optimum design is simply that centrifuge models are real physical events that project owners and designers can watch and inspect. They can observe the state of the ground around their planned design after it has been subjected to loading up to and beyond the worst conditions that are expected. The failure mechanisms, whether dramatic plastic shearing or an accumulation of movement and cracking, can be seen in a way that is always more compelling than results from a computer simulation. As we walk into the centrifuge chamber after a successful test and peer into the strongbox, we could be flying in a helicopter over the planned structure fifty years in the future.

References

1 Craig, W. (1989) The Use of a Centrifuge in Geotechnical Engineering Education. *Geotechnical Testing Journal* 12 (4): 288-291.
2 Garnier, J., Gaudin, C., Springman, S.M., Culligan, P.J., Good-ings, D., Konig, D., Kutter, B., Phillips, R., Randolph, M.F., & Thorel, L. (2007). Catalogue of Scaling Laws and Similitude Questions in Geotechnical Centrifuge Modeling. *Int. J. Phys.Modell. Geotech.* 7 (3): 1-24.
3 Roscoe, K.H. (1970) The influence of strains in soil mechanics. *Géotechnique* 20 (2): 129-170.
4 Rowe, P.W. (1972) The relevance of soil fabric to site investigation practice. *Géotechnique* 22 (2): 195-300.
5 Schofield, A.N. (1980) Cambridge geotechnical centrifuge operations. *Géotechnique* 30 (3): 227-268.
6 Gaudin, C., O'Loughlin, C.D. & Breen, J. (2018) A new 240 g-tonne geotechnical centrifuge at the University of Western Australia. In: *9th Int. Conf. Phys. Modelling in Geotechnics, London,* 501-506.
7 Randolph, M.F., Jewell, R.J., Stone, K.J.L. & Brown, T.A. (1991) Establishing a new centrifuge facility. In: *International conference on centrifuge modelling, Centrifuge 91, Boulder, Colorado, USA,* 3-9.
8 Stewart, D.P., Boyle, R.S. & Randolph, M. (1998) Experience with a new drum centrifuge. In: *Centrifuge 98, Tokyo, Japan* (eds T. Kimura, O. Kusakabe, & J. Takemura), Vol. 1, 35-40. Rotterdam, Netherlands: CRC Press/Balkema.
9 Hou Y.J. (2018) Development of geotechnical centrifuges and facilities in China. In: *9th Int. Conf. Phys. Modelling in Geotechnics, London,* 77-85.
10 White D.J. (2008) Contributions to Géotechnique 1948-2008: Physical Modelling. *Géotechnique* 58 (5): 413-421.
11 Take W.A. (2018) Current and emerging physical modelling techniques. In: *9th Int. Conf. Phys. Modelling in Geotechnics, London,* 101-109.
12 Pua, L.M., Caicedo, B., Castillo, D. & Caro, S. (2018) Development of a 3D clay printer for the preparation of heterogeneous models. In: *9th Int. Conf. Phys. Modelling in Geotechnics, London,* 155-160.
13 Adamidis, O., Alber, S. & Anastasopoulos, I. (2018) Investigation into 3D printing of granular media. In: *9th Int. Conf. Phys. Modelling in Geotechnics, London,* 113-118.

14 Song, G., Marshall, A.M. & Heron, C.M. (2018) A mechanical displacement control model tunnel for simulating eccentric ground loss in the centrifuge. In: *9th Int. Conf. Phys. Modelling in Geotechnics, London,* 455-460.
15 Ma, X.F. & Xu, J.W. (2018) Centrifuge model tests on excavation in Shanghai clay using in-flight excavation tools. In: *9th Int. Conf. Phys. Modelling in Geotechnics, London,* 419-424.
16 Choy, C.K., Standing J.R. & Mair, R.J. (2007) Stability of a loaded pile adjacent to a slurry-supported trench. *Géotechnique* 57 (10): 807-819.
17 Kundu, S. & Viswanadham, B.V.S. (2018). Influence of tamper shape on dynamic compaction of granular soil. In: *9th Int. Conf. Phys. Modelling in Geotechnics, London,* 1205-1209.
18 Nagula, S., Mayanja, P. & Grabe, J. (2018) Deep vibration compaction of sand using mini-vibrator. In: *9th Int. Conf. Phys. Modelling in Geotechnics, London,* 1229-1233.
19 Caicedo, B. & Escobar, J. (2018) Centrifuge modelling of continuous compaction control (CCC). In: *9th Int. Conf. Phys. Modelling in Geotechnics, London,* 131-136.
20 Lee, F.H., Lee, C.H. & Dasari, G.R. (2006) Centrifuge modelling of wet deep mixing processes in soft clay. *Géotechnique* 56 (10): 677-691.
21 Hori, T. & Tamate, S. (2018) Centrifuge model test on the instability of an excavator descending a slope. In: *9th Int. Conf. Phys. Modelling in Geotechnics, London,* 383-388.
22 Take, W.A. & Bolton, M.D. (2002) An atmospheric chamber for the investigation of the effect of seasonal moisture changes on clay slopes. In: *Proc. Int. Conf. Phys. Modelling Geotech., St John's,* 765-770. Rotterdam: Balkema.
23 Khan, I.U., Al-Fergani, M. & Black, J.A. (2018) Development of a rainfall simulator for climate modelling. In: *9th Int. Conf. Phys. Modelling in Geotechnics, London,* 507-512.
24 Bhattacherjee, E. & Viswanadham, B.V.S. (2018) Development of a rainfall simulator in centrifuge using modified Mariotte's principle. In: *9th Int. Conf. Phys. Modelling in Geotechnics, London,* 337-342.
25 Archer, A. & Ng, C.W.W. (2018) A new environmental chamber for the HKUST centrifuge facility. In: *9th Int. Conf. Phys. Modelling in Geotechnics, London,* 489-494.
26 Take, W.A. & Bolton, M.D. (2011) Seasonal ratcheting and softening in clay slopes, leading to first-time failure. *Geotechnique* 61 (9): 757-769.
27 Laporte, S., Siemens, G.A. & Beddoe R.A. (2018) Physical modelling of roads in expansive clay subjected to wetting-drying cycles. In: *9th Int. Conf. Phys. Modelling in Geotechnics, London,* 175-178.
28 Brennan, A.J., Knappett, J.A., Bertalot, D., Loli, M., Anastasopoulos, I. & Brown, M.J. (2014) Dynamic centrifuge modelling facilities at the University of Dundee and their application to studying seismic case histories. In: *8th International Conference on Physical Modelling in Geotechnics 2014, Perth, Australia,* Vol.1: 227-233.
29 Boylan, N., Gaudin, C., White D.J. & Randolph M.F. (2010) Modelling of submarine slides in a geotechnical centrifuge. In: *Int. Conf. on Physical Modelling in Geotechnics. Zurich,* 1095-1100.

30 Exton, M.C., Harry, S., Mason, H.B., Yeh, H. & Kutter, B.L. (2018) Novel experimental device to simulate tsunami loading in a geotechnical centrifuge. In: *9th Int. Conf. Phys. Modelling in Geotechnics, London,* 371-375.

31 Knappett, J.A., Brown, M.J., Shields, L., Al-Defae, A.H. & Loli, M. (2018) Variability of small-scale model reinforced concrete and implications for geotechnical centrifuge testing. In: *9th Int. Conf. Phys. Modelling in Geotechnics, London,* 241-246.

32 Boksmati, J. & Madabhushi, S.P.G. (2018) Development of model structural dampers for dynamic centrifuge testing. In: *9th Int. Conf. Phys. Modelling in Geotechnics, London,* 343-348.

33 Ritter, S., DeJong, M.J., Giardina., G. & Mair, R.J. (2018) 3D printing of masonry structures for centrifuge modelling. In: *9th Int. Conf. Phys. Modelling Geotechnics, London,* 449-454.

34 Liang, T., Bengough, A., Knappett, J., Muir Wood, D., Loades, K.W., Hallett, P.D., Boldrin, D., Leung, A. & Meijer, G. (2017) Scaling of the reinforcement of soil slopes by living plants in a geotechnical centrifuge. *Ecological Engineering* 109 (Pt B): 207-227.

35 Liang, T., Knappett, J.A., Bengough, A.G. & Ke, Y.X. (2017) Small-scale modelling of plant root systems using 3D printing, with applications to investigate the role of vegetation on earthquake-induced landslides. *Landslides* 14 (5): 1747-1765.

36 Stewart, D.P. & Randolph, M.F. (1991) A new site investigation tool for the centrifuge. In: *Proc. Int. Conf. on Centrifuge Modelling – Centrifuge* '91 (eds H.Y. Ko & F.G. McLean), 531-538. Rotterdam: Balkema.

37 Thorel, L., Gaudin, C. & Garnier, J. (2007) A cone pressuremeter for soil characterisation in the centrifuge. *International Journal of Physical Modelling in Geotechnics* 7 (1): 23-30.

38 BagBag, A., Doherty, J. & Lehane, B. (2016) Stress-strain response of fine silica sand using miniature pressuremeter. In: *Proc. 5th Int. Conf. on Geotechnical and Geophysical Site Characterisation, ISC'5* (eds B.M. Lehane, H.E. Acosta-Martinez & R. Kelly), 749-754. Queensland: Australian Geomechanics Society.

39 Carey, T., Gavras, A., Kutter, B., Haigh, S.K., Madabhushi, S.P.G., Okamura, M., Kim, D.S., Ueda, K., Hung, W.Y., Zhou, Y-G., Liu, K., Chen, Y-M., Zeghal, M., & Manzari, M. (2018) A new shared miniature cone penetrometer for centrifuge testing. In: *9th Int. Conf. Phys. Modelling in Geotechnics, London,* 293-298.

40 White, D.J., Take, W.A. & Bolton, M.D. (2003) Soil deformation measurement using Particle Image Velocimetry (PIV) and photogrammetry. *Géotechnique* 53 (7): 619-631.

41 Stanier, S.A., Blaber, J., Take, W.A. & White, D.J. (2016) Improved image-based deformation measurement for geotechnical applications. *Canadian Geotechnical Journal* 53: 1-13.

42 Take, W.A. (2015) Thirty-sixth Canadian Geotechnical Colloquium: Advances in Visualization of Geotechnical Processes through Digital Image Correlation. *Canadian Geotechnical Journal* 52 (9): 1199-1220.

43 Le, B.T., Nadimi, S., Goodey, R.J. & Taylor, R.N. (2016) System to measure three-dimensional movements in physical models. *Géotechnique Letters* 6 (4): 1-7.

44 Idinyang, S., Franza, A., Heron, C.M., & Marshall, A.M. (2018) Millisecond interfacing of physical models with ABAQUS. In: *9th Int. Conf. Phys. Modelling in Geotechnics, London,* 209-214.

45 Pecker, A. & Teyssandier, J.P. (1998) Seismic design for the foundation of the Rion-Antirion Bridge. *Proc. I.C.E., Geotechnical Engineering* 131 (1): 4-11.

46 Pecker, A. & Garnier, J. (1999) Use of Centrifuge Tests for the Validation of Innovative Concepts in Foundation Engineering. In: *Proc. 2nd Int. Conf. on Earthq. Geotech. Eng., Lisbon,* 433-439.

47 Pecker, A. (2006) Enhanced seismic design of shallow foundations: example of the Rion Antirion Bridge. *4th Athenian Lecture on Geotechnical Engineering*. Athens, Greece. https://www.gefyra.gr/images/user/Publications/4th%20Athenian%20Lecture.pdf (20/10/2019)

48 This work has been released into the public domain by its author, Kampini. https://commons.wikimedia.org/wiki/File:Rio_Antirrio_bridge.jpg (20/10/2019)

49 SEAOC (1995) *Vision 2000: Performance-based seismic engineering of buildings.* Sacramento (CA): Structural Engineers Association of California.

50 ISO (2005) *ISO 23469 Bases for design of structures – Seismic actions for designing geotechnical works.* Geneva: International Organization for Standardization.

51 https://www.vinci-construction-projets.com/en/realisations/rion-antirion-bridge/ (20/10/2019)

52 Murff, J.D. (1996) The geotechnical centrifuge in offshore engineering. *Proc. Offshore Technology Conference, 6-9 May, Houston, Texas.* Paper OTC 8265.

53 Watson, P.G. & Randolph, M. (1998) Skirted foundations in calcareous soil. *Proc.I.C.E., Geotechnical Engineering* 131 (3): 171-179.

54 Gaudin, C., Clukey, E.C., Garnier, J. & Phillips, R. (2011) New frontiers for centrifuge modelling in offshore geotechnics. In: *Proceedings of 2nd International Symposium on Frontiers in Offshore Geotechnics.* Vol. CD: 155-188. The Netherlands: CRC Press/Balkema.

55 Gaudin, C., Watson, P.G. & Randolph, M.R. (2006) Centrifuge Testing of Offshore Filters. *Geotechnical Testing Journal* 29 (4): 348-354.

56 Watson, P.G. & Humperson, C. (2007) Foundation design and installation of the Yolla A Platform. *Paper presented at 6th International Conference on Offshore Site Investigation and Geotechnics – London, 2005.*

57 Watson, P., Jackson, G. & Humpheson, C. (2019) The Maari WHP platform – design, installation and performance. In: *Proc. 13th Australia New Zealand Conference on Geomechanics. Perth, Australia 1-3 April 2019,* 635-640.

58 Gaudin, C. & White, D.J. (2009) New centrifuge modelling techniques for investigating seabed pipeline behaviour. In: *Proc. 17th Int. Conf. on Soil Mechanics & Geotechnical Engineering. Alexandria, Egypt.* (eds M. Hamza, M. Shahien & Y. El-Mossallamy), Vol.1. 448-451. Amsterdam: IOS Press.

59 Cocjin, M., White, D.J. & Gourvenec, S.M. (2018) Softening and consolidation around seabed pipelines: centrifuge modelling. *Géotechnique* 68 (2): 118-132.

60 White, D.J., & Cathie, D.N. (2010) Geotechnics for subsea pipelines. In: *Proc. 2nd Int. Symp. on Frontiers in Offshore Geotechnics, Perth.* (eds S. Gourvenec & D.J. White), 87-123. Boca Raton: CRC Press/Balkema.

61 Gaudin, C. & White, D.J. (2012) Centrifuge modelling to support the design of subsea pipelines. In: *Proc. GeoCongress 2012, San Francisco,* 2402-2411. ASCE STP225.

62 Hill, A. & White, D.J. (2015) Pipe-soil interaction: Recent and future improvements in practice. In: *Proc. 3rd Int. Symp. on Frontiers in Offshore Geotechnics, Oslo* (ed. V. Meyer), 387-392. Boca Raton: CRC Press/Balkema.

35

The use of physical models in acoustic design – current practice

Raf Orlowski

35.1 Introduction

In the twenty-first century, there has been relatively little development in modelling techniques in acoustic engineering, largely because fewer people are now using them, for example, the transducers with an electric spark providing a good omnidirectional source remain the same as two decades ago. Acoustic scale modelling of internal spaces such as auditoria has now largely been overtaken by acoustic computer modelling.

However, acoustic computer modelling is still unable to reproduce, within a reasonable time scale, the complex interaction between sound waves and modulated surfaces such as those found in most concert halls. In other words, acoustic computer models find it difficult to simulate the diffusion of sound – they have to do it by adopting various types of approximation. By contrast, acoustic scale models are able to reproduce authentically the diffusion of sound from irregular surfaces because the interaction between the sound wave and the surfaces in the model is exactly the same as at full scale – only the wavelength and scale have been changed. A recent comparison between acoustic scale modelling and computer modelling by Jens Holger Rindel [1] has concluded that both methods can contribute in different ways to the design of an auditorium.

The authenticity of acoustic scale models means that they still have real value although, nowadays, they tend to be used only for the most prestigious projects as the cost of scale modelling is high compared with computer modelling. Examples of some recent prestigious projects making use of scale models are described below.

35.2 Elisabeth Murdoch Hall, Melbourne

The brief for the design of a new 1000-seat recital hall in Melbourne in Australia required a traditional shoebox shape but interpreted in a modern idiom. The hall is part of the Melbourne Recital Centre and has been named the Elisabeth Murdoch Hall. The traditional shoebox halls, such as the Musikvereinssaal in Vienna,

Physical Models: Their historical and current use in civil and building engineering design,
First Edition. Edited by Bill Addis.

Figure 35.1 1:50 scale acoustic model for Elisabeth Murdoch Hall, Melbourne. View of the stage. (Image: Raf Orlowski)

completed in 1870, contain a lot of ornamentation which scatters and diffuses sound that is incident on these modulated surfaces. To reproduce this scattering effect in a modern hall, surface finishes need also to be modulated but need to follow the aesthetic intentions of the architect.

The architects for the Elisabeth Murdoch Hall, Ashton, Raggatt and McDougall, were presented by the acoustician with a set of rules to follow for designing the modulated walls and ceiling. These rules included such parameters such as the depth of modulation, the dimensions of modulating elements and their locations throughout the auditorium. Once the architect had developed the designs, which turned out to be a pseudo-random surface pattern on rectangular panels which were offset from each other, it was necessary to test their effect acoustically because the diffusion of sound in a space can have a significant influence on the reverberation. An acoustic scale model was the only effective method of testing the effect of the diffusing surfaces and a very accurate model was constructed at 1:50 scale where full-scale dimensions down to 50 mm or less were reproduced in the model (Figure 35.1).

The model testing involved the measurement of all the key acoustic parameters including Reverberation Time, Early Decay Time, Clarity Index and Strength. The results showed an ample Reverberation Time of around 2 seconds, with all the other parameters giving very good values [2]. The design of the diffusing surfaces was reproduced in the full-scale hall and the venue has become one of the most favoured recital halls in Australia.

35.3 Concert hall, Krakow Congress Centre

The 'vineyard terrace' type of concert hall was pioneered by the German architect, Hans Scharoun, in the 1960s in his design for the new Philharmonic Hall

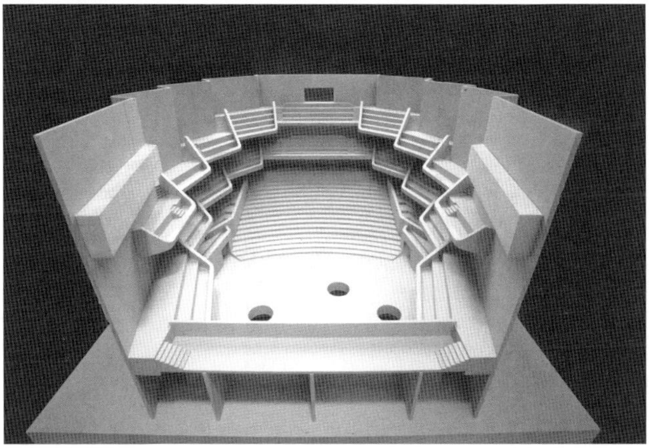

Figure 35.2 1:50 scale acoustic model for the concert hall in the Krakow Congress Centre, Poland. (Image: Ingarden and Ewy Architects)

for Berlin. This design has proved very successful, not only for the Berlin hall but for many other halls around the world which have adopted the same concept. The design of vineyard-terrace halls does not have the same degree of acoustical certainty as shoebox-shaped halls. Therefore, testing an acoustic scale model is important to demonstrate that the design will be successful.

The new concert hall in the Krakow Congress Centre in Poland is a modification of the traditional vineyard terrace as the stage is at one end rather than towards the centre. To test the acoustic performance of this hall, a 1:50 scale model was tested to demonstrate that the specified acoustic parameters could be achieved (Figure 35.2). The holes in the stage are for different positions of the sound source. The construction of the concert hall followed the model design closely and the measured acoustic parameters on completion confirmed the results predicted from the model [3].

35.4 An acoustic scale model on wheels

The Royal Festival Hall in London is one of the most prestigious concert halls in the international world of music. During its refurbishment between 2005 and 2007 it was important to ensure that modifications influencing the acoustics would achieve the best possible outcome. Several major changes were proposed for the refurbishment of the Hall including replacing the heavy timber reflectors above the orchestra with lightweight fabric ones, and changing the geometry of the walls around the stage. To help with this, an acoustic scale model was built at 1:20 scale. This scale factor has the advantage of producing results up to the 4 kHz octave at full-scale which is usually the highest frequency of interest in acoustic measurements. However, this scale factor results in quite a large model – nearly 3 metres long in the case of the Royal Festival Hall. To facilitate ease of transportation and also to permit good access, the model was built in

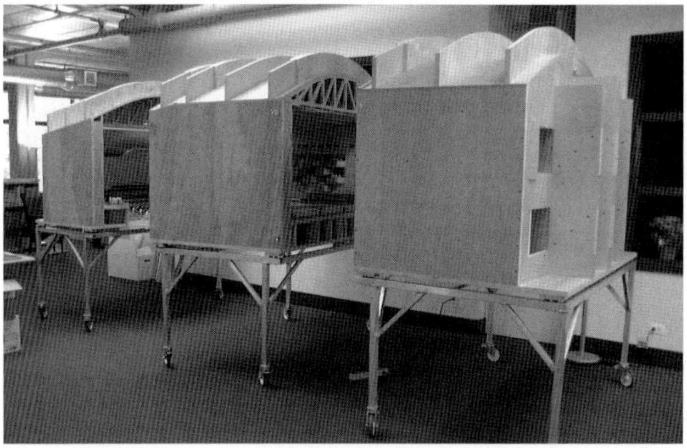

Figure 35.3 1:20 scale acoustic model or the Royal Festival Hall, London. (Image: Kirkegaard and Associates)

three sections, mounted on wheels [4] (Figure 35.3). The model was essential to help reach decisions about which of the several proposed changes should be implemented.

35.5 The ongoing popularity of acoustic models at scale factors of 1:10 or similar

A number of acoustic consultancies that developed an extensive use of large acoustic models in the late twentieth century, have continued using them in the twenty-first century because they believe that the results obtained from them give the best insight into the acoustic characteristics of the space under investigation.

One such firm, Peutz, based in the Netherlands, has consistently used large models at scale factors of 1:10 or 1:12 for their work on auditorium acoustics. An example is the recent refurbishment of the Staatsoper in Berlin, built in 1742, and rebuilt twice since, after its destruction. The consultants at Peutz wanted to be sure that raising the ceiling by 5 metres, a very costly exercise, would achieve the desired acoustical improvements. A 1:10 scale model was built which even included LED lights fitted to the balconies – not strictly necessary for acoustic measurements, but useful in giving the client a better impression of the auditorium (Figure 35.4). Members of the audience and a few musicians on stage are represented by pyramids of foam rubber which replicate the sound absorption of a person. The two black posts mounted on light circular bases, stage left, are the sound sources; the sound is emitted from a tiny opening at the top of the black cone. The three pairs of metallic rods on stage (and also in the audience areas) are the microphones. The small condenser microphones are mounted at the top and the preamplifiers fitted in the thicker part at the bottom. The sources and microphones are at standard heights – corresponding to 1.5 m for sources and 1.2 m for receivers at a scale of 1:10. A model listener, incorporating small microphones, can be placed in any seat in the auditorium to enable the received sound to be recorded (Figure 35.5). The model tests confirmed that raising the

Figure 35.4 1:10 scale acoustic model of the Staatsoper in Berlin. (Image: Peutz Associates)

(a) (b)

Figure 35.5 (a) 1:10 scale acoustic model of the Staatsoper in Berlin. The acoustician is holding a model listener whose ears are microphones for recording the sound at different places in the auditorium. (b) A model listener. (Images: Peutz Associates)

ceiling to increase the volume of the hall by about 50% would make a substantial improvement to the acoustics of the opera house and, when this modification was implemented, a very successful outcome was reported by opera goers [5].

35.6 Final remarks

Acoustic scale models have more than one function. As well as providing a means for testing the acoustics of a space during the design stage, they provide an opportunity for the client to get a visual impression of what the auditorium will look like. Acoustic scale models at 1:50 scale which are very flexible during the design process (see Section 26.4), can give can give a good visual impression but perhaps lack visual impact. By contrast, a 1:10 scale acoustic model, which is typically the size of a normal living room, can have a high visual impact on the client, particularly if he or she can climb into it.

Although such large models are costly and unwieldy as tools in developing the acoustic design, they can have an enormous impact on the client and can provide confidence in the project and help to secure funding. Large acoustic scale models can thus provide not only useful data about the acoustic characteristics of an auditorium, but can also help in the delivery of the whole project. For this reason, it is very likely that such models will continue to be used well into the future, at least for major performance spaces.

References

1 Rindel, J.H. (2011) Room acoustic modelling techniques: a comparison of a scale model and a computer model for a new opera house. *Building Acoustics* 18 (3-4): 258-280.

2 Orlowski, R. (2008) The acoustic design of the Elisabeth Murdoch Hall, Melbourne, Australia. In: *Proceedings of Acoustics -08, 29 June – 4 July 2008, Paris.* Paris: Société Française d'Acoustique. 340-344.

3 Orlowski, Raf (2015) The acoustic design of a vineyard hall and a shoebox theatre at the new Krakow Congress Centre. In: *Proceedings of 9th International Conference on Auditorium Acoustics, 29-31 October 2015, Paris.* See also: https://aasarchitecture.com/2014/06/ice-krakow-ingarden-ewy-architekci-2.html/ (12/9/19)

4 http://www.acoustics.salford.ac.uk/acoustics_info/concert_hall_acoustics/?content=testing (20/10/2019)

5 Vercammen, Martijn & Lautenbach, Margriet (2018) Staatsoper Unter den Linden Berlin. *Proceedings of the Institute of Acoustics* 40 (3): 64-72.

36

Water-bath modelling – small-scale simulation of natural ventilation flows

Owen Connick

36.1 Introduction

Buildings are generally large, expensive and often architecturally unique. This means that almost every building is, in effect, a one-of-a-kind prototype. Considering the costs involved in planning, constructing and operating a full-scale building which is the first of its kind, designers are always seeking to improve design-stage modelling and prediction of how the building will perform when occupied. Furthermore, designers and architects are constantly having to convey and justify their grand plans to an end client who ultimately foots the bill.

Fluid mechanics is one of the most complicated and far-ranging areas of science. When considering airflows into, out of and within buildings, the length scales that must be resolved range from the sub-millimetre to the multi-kilometre.

Computational Fluid Dynamic simulations have improved vastly in recent years. With advances in computing power enabling large simulations to be run on standard desktop machines. For simulating airflows into and out of buildings or rooms, networked or zonal pressure models, sometimes referred to as Dynamic Thermal Models/Simulations (DTM/S) have become the default tool of choice.

What all computational simulations lack, however, is a tangible physical effect that can be observed, measured, tweaked in real time and, thereby, innately understood. It is this understanding and this realism which physical modelling will always offer and which computer models will always struggle to provide.

36.2 Historical perspective

The use of small-scale models to simulate airflows is ancient. In the fifteenth century, Leonardo Da Vinci probably tested small-scale models of his flying machine inventions. In the eighteenth century, John Smeaton made experiments with windmills to determine the power they could generate (Section 24.1.1) In the late-nineteenth century, researchers at the Royal Aeronautical Society

Physical Models: Their historical and current use in civil and building engineering design,
First Edition. Edited by Bill Addis.

developed the technique of using small-scale models in wind-tunnels to simulate full-scale aerodynamic behaviour. Later, the work of Reynolds and others enabled the results of these small-scale experiments to be related directly back to the full-scale flows using the laws of similitude.

The use of small-scale models to study airflows within buildings can be traced back to work done in the 1960s on exploring the ventilation of hot gases from fires in large, industrial buildings [1]. The model consisted of a box made of asbestos wood, at a scale of about 1:100, with one side open and a vent in the roof, and a small gas flame to represent the fire. Laws of similitude were applied and a number of nomograms (graphical calculation tools) were produced to aid designers of venting systems in factories for different situations according to the geometry of the space and size of the fire. In the 1970s, research conducted at the Fire Research Station (now part of Building Research Establishment in the UK) on the ventilation of smoke from a multi-level shopping mall used 1:10 scale models with air as the working medium [2]. Since that time, the use of physical models to study the behaviour of fire has grown rapidly, for both safety and forensic purposes [3].

Fire and smoke ventilation are one regime of building ventilation, an extreme application we hope never to experience in the real building. Ventilation in general, however, and in particular natural ventilation, is a regime we expect to see throughout the building's occupation, and which we need to understand at the design stage in order to predict confidently what the airflows will be in the full-scale building post-completion. There arise a number of issues with the use of air to model ventilation airflows at small scale. In both cases described above, the objective of the study was to observe the depth of a smoke layer which formed and to link this to the effectiveness of the ventilation. If we try to extend the same method to study buoyancy-driven ventilation, the driving forces are extremely small and making the airflow visible is difficult. Seeding the airflow with artificial smoke is possible but fraught with issues. And, crucially, buoyancy-driven flows of air at small scale are dynamically dissimilar to the buoyancy-driven flows of air at full scale which they are seeking to replicate.

36.3 Fundamentals and theory – developing the technique

To overcome some of the problems associated with using air as the working fluid, water can be used instead. Ventilation airflows in buildings are characterised by high Reynolds (Re) and Peclet (Pe) numbers, indicating that viscous and diffusive effects are of secondary importance in a flow which is dominated by advection (i.e. transport by bulk motion). In air, ventilation airflows will typically have $Re > \sim 10^3$ and $Pe > \sim 10^3$. Although it is not necessary to match these values exactly in the small-scale model, similar orders of magnitude are required to ensure dynamical similarity/similitude. An added bonus of this dynamic similarity is that results measured at small scale, appropriately non-dimensionalised, can be related directly back to the full-scale building, by applying suitable scaling factors.

For many years, the University of Cambridge has been a centre of development for intellectual thought on buoyancy-driven fluid mechanics – the movement of fluids driven by differences in density. The fluid mechanics of plume behaviour, in particular the entrainment into plumes and changes in the average density of fluid within a plume with height, had been known for some time [4]. During the 1980s and 1990s, a research team led by Paul Linden at the University of Cambridge sought to apply this understanding of plume dynamics to develop simplified theoretical models for natural displacement ventilation in buildings (rooms) with localised sources of heat [5]. These theoretical models were elegant because they could easily be used by engineers to predict the steady state ventilation rates and temperatures in buildings as a function of various factors such as heat load and the location and area of vents, as well as the transient development of conditions in the internal space. In order to help verify the proposed mathematical modelling technique, the researchers used dimensional analysis to develop a suite of analogous water-bath model experiments and compared the experimental findings with their theoretical predictions. Although the paper by Linden was highly mathematical, it heralded the first use of water-bath experiments to enhance our understanding of natural airflows in buildings. The field of natural ventilation research and the application and extension of the mathematical modelling approach, had started, and is now being applied to many aspects of air-flow in buildings, such as ventilation in hospitals [6].

36.4 Water-bath modelling

Buoyancy-driven fluid flows can be generated in a water-bath model by changing the density of portions of the water. This can be achieved in one of two ways:

- Heating part of the water to reduce its density and cause it to rise (Figure 36.1a) (It is also possible, though rare, to introduce cooling to specific zones using iced water.)
- Increasing the salinity of part of the water to increase its density and cause it to fall (Figure 36.1b).

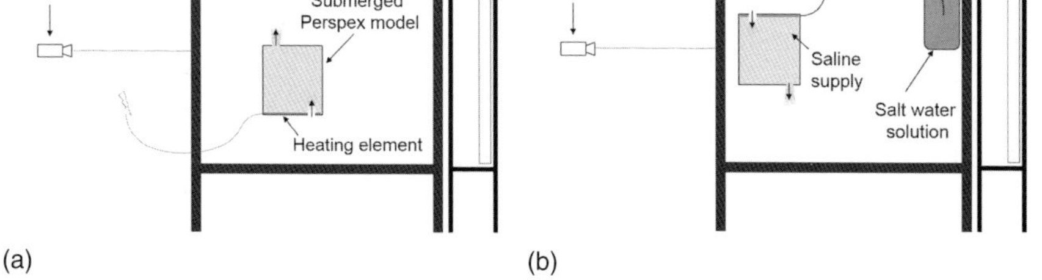

Figure 36.1 Water-bath modelling techniques. (a) Electrically-heated wires. (b) Salt solution. (Images: Owen Connick)

In each case the flow of the lower/higher density fluid is made visible by lighting the water bath from behind and can be enhanced by colour-dyeing the fluid in specific regions of interest.

36.4.1 Point sources of buoyancy

Much of the early work done in this field focused on a single source of buoyancy in a space with two vents, one at high level and one at low level. This was soon developed by various researchers to consider more complex factors such as multiple sources of buoyancy, sources of buoyancy of different strengths and transient (as opposed to steady-state) flows [7, 8, 9].

36.4.2 Distributed heat loads

In the early 2000s researchers in the BP Institute at the University of Cambridge further developed the water-bath modelling technique to examine the impact of distributed heat loads (buoyancy sources) rather than a limited number of discrete, point sources. The initial attempts involved circulating hot water through a heat exchanger in a model tank, but this technique was soon superseded by electrical heating elements which made it much easier to quantify the energy input to the model. This style of physical model was first used to verify a new suite of simplified mathematical models for airflow in buildings.

The technique was quickly applied to real building projects, where heat sources at different levels were being considered. The first project to benefit from the technique was Chapelfield Shopping Mall in Norwich [10]. Within the shopping centre, air-conditioned shops open onto a naturally-ventilated central mall. The water-bath modelling technique was developed to help take account of the air-conditioned spaces by means of iced-water supplied to the relevant interior areas of the model. The relative strength of the air-conditioning from adjacent spaces and the heat loads in the naturally-ventilated mall were varied to explore the consequences. This revealed both the macro-scale up/down flow ventilation patterns and also the circulation patterns that would be experienced.

36.4.3 Wind and buoyancy

Also during the 2000s the water-bath modelling technique was further developed by combining the buoyancy-driven displacement ventilation with air flow driven by the wind. Water-bath tanks were set up using a flume arrangement to provide a background flow. The experiments undertaken using this configuration initially focused only on verification of some theoretical models developed at the BP Institute. In particular, Ben Lishman and Andy Woods had identified theoretically that multiple steady states might exist for the same set of boundary conditions [11]. Experiments were undertaken to ascertain whether these multiple steady states could be observed physically; in essence, the experiments were undertaken to help establish whether both modes were in fact stable. Although this work spawned a number of subsequent research studies, the technique has not been widely used in actual building projects – the model set-up is rather complex and the circumstances required for these particular multiple steady states to exist are quite rare. In practice, as was revealed by the water-bath modelling studies, if there are openings at high level on more than one side of a building/space, then the likelihood of multiple steady states is very low.

36.4.4 Current and future work

The most intriguing recent research using water-bath modelling focuses on situations more commonly encountered in real buildings. Examples include the prediction and impact of bi-directional flow through an opening and the hybrid effect of a mechanical supply or extract on a space which is predominantly naturally-ventilated.

Under certain conditions, a heated space with multiple horizontal openings at the same height can exhibit oscillatory flow behaviour. The oscillations occur as the flow pattern migrates from uni-directional flow through each of the vents to bi-directional flow. If this effect were to be demonstrated with mathematical or computational modelling, a discerning audience would always wonder whether an instability (if observed at all) was actually a physical one or a numerical one. The use of physical modelling meant that no such questions arose as the instability was clearly observable in the (dynamically similar) physical model.

Many modern buildings in reality exhibit some form of 'hybrid' or 'mixed-mode' ventilation in which mechanical equipment, governed by a computer-controlled building management system, is used to assist or manage a naturally-driven airflow. Such hybrid ventilation flows are still not fully understood, and the use of complementary theoretical and small-scale experimental modelling seeks to improve the modelling and control of such airflows by enhancing our understanding of the underlying fluid mechanics.

36.5 Some applications of water-bath modelling in design

Engineers have quickly learned that the water-bath modelling technique could be used to model airflows specific to a particular building. Although the primary application of water-bath modelling had been to help verify mathematical model results which, in turn, enabled ventilation rates and temperatures to be predicted, it was the visualisation in three dimensions of fluid flow, evolving over time, which has had the greatest impact amongst architects, designers and – indeed – end clients. This improvement in the clarity of communication of how air can move in a building has transformed the understanding by non-specialists of a proposed ventilation strategy – particularly for naturally-ventilated spaces. A number of examples are laid out below to give the reader some idea of the range in scale and type of buildings for which water-bath modelling has been successfully applied.

36.5.1 Breathing Space, Rotherham

Breathing Space in Rotherham, UK, is a National Health Service facility which offers specialist medical treatment for people suffering from respiratory conditions. Constructed in 2007, the building has a low-energy natural-ventilation strategy incorporating a 'thermal mass labyrinth'. In summer, when the outside air is warm, a series of concrete-lined tunnels underneath the main building are used to cool the fresh incoming air. During the mid-seasons of spring and autumn, the same labyrinth is used to preheat fresh air before it is supplied to the building. In both modes, air is exhausted through high-level windows atop the central atrium, with fans being used to assist the natural ventilation flow if necessary. In winter, the underground labyrinth is closed off, so that air both enters and exits at atrium roof level, in a mixing ventilation mode [12].

During the pre-construction design stage, a small-scale water-bath model (approximately 1:80 scale) was constructed to test the effectiveness of different ventilation-stack designs (stack height, size of openings, location(s), control of vents, etc.), and of different locations and sizes of the ventilation openings elsewhere in the building.

The Perspex model was 30 x 20 x 20 cm high, with six 10-mm diameter vents at the base, nine in the roof and three in each vertical façade (Figure 36.2). Thermocouple temperature sensors can be seen, positioned at several heights in the model to enable measurement of spatial variations in temperature and stratification.

It is useful to note that the model geometry does not precisely replicate the real building geometry, but seeks to represent the features relevant to this fluid-dynamic study. For example, in the real building the atrium roof has a curved saw-tooth form. For the purposes of the small-scale model, however, it would be unnecessarily complicated to reconstruct this curved form when straight saw-tooth sections will behave similarly with regard to simulating the natural ventilation flow. The hot-wire heating elements are also visible on the floor of the model. The heat input supplied to these can be manipulated in order to model a range of internal-external temperature differences – simulating operation of the building under different external weather conditions.

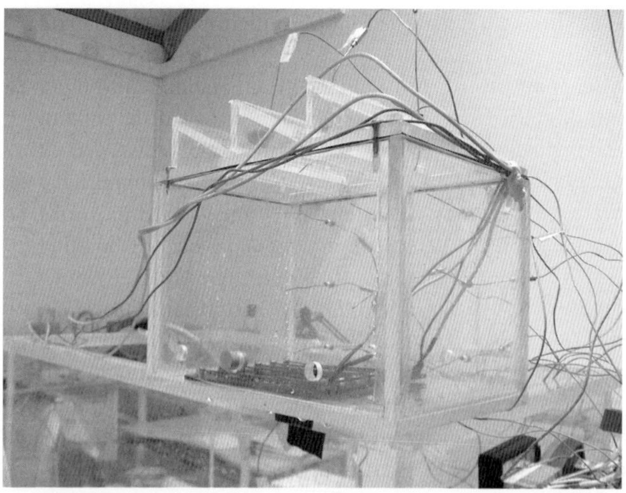

Figure 36.2 1:80 scale Perspex model of the atrium space, instrumented and ready to be submerged within the water-bath tank. (Source: Breathing Buildings)

36.5.2 Aldwyck Housing Group HQ, Houghton Regis, UK

This three-storey headquarters building, located near Luton, features open-plan offices either side of a central triple-height atrium and was completed in 2003. The building was designed with a low-energy heating and ventilation strategy. Fresh air would be drawn in through the façade on each storey, across the open-plan office, where it would be warmed by heat gains from occupants and equipment. This warm, stale air would then rise through the central atrium, and exhaust through stack vents atop the atrium roof. At night in summer the exposed concrete floor slabs were cooled by night air and the coolth radiated during the following day (Figure 36.3). The intention of this strategy was to create similar, environmentally-comfortable conditions on all storeys.

Soon after the building was occupied, however, it was noticed that the internal environmental conditions were not the same on each of the three floors. The building managers were interested in understanding more about how the natural-ventilation system was actually functioning, in order to determine whether there could be ways of reducing the differences in internal temperature between floors, especially in summer. A set of water-bath model experiments provided the key to understanding and remedying these issues. The two images below show the resulting steady 'end-state' of water-bath experiments with different initial conditions. Different coloured dyes were used for visualisation purposes only and do not necessarily indicate different temperatures between the two storeys.

Figure 36.4(a) shows the intended ventilation strategy in operation. Fresh water (simulating air) enters through openings on the lower left side of each storey, is warmed by heating elements across the floor and exhausts from the room into the atrium, wherefrom it rises under buoyancy and ventilates out of the atrium high-level vents. This steady state was achieved by introducing the heat load to both the ground floor and first floors at the same time and allowing the system to evolve naturally. Figure 36.4(b), in contrast, shows an entirely different, but

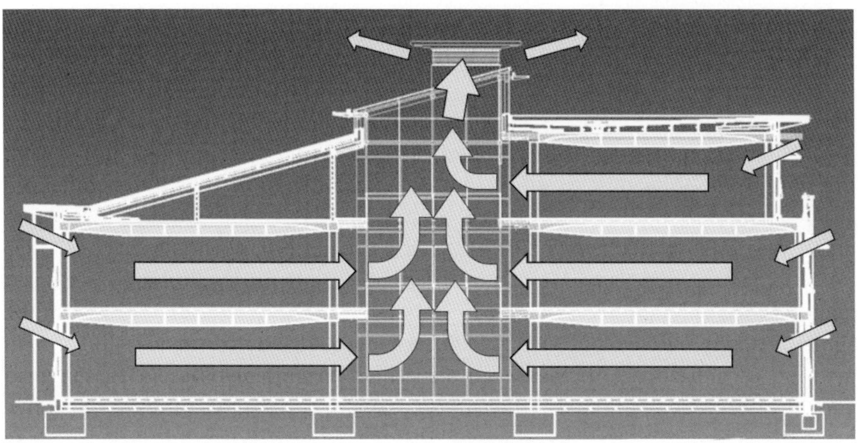

Figure 36.3 Section of the building showing the intended air flow to achieve natural ventilation. (Image: Bill Addis)

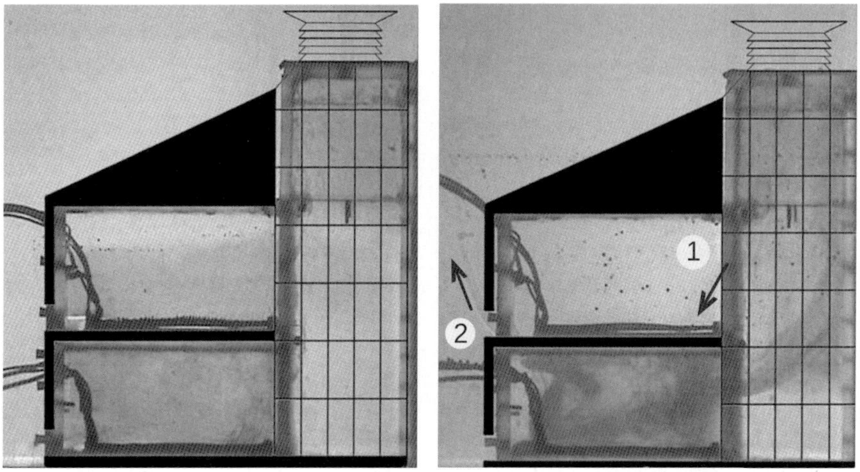

(a) (b)

Figure 36.4 The model under test. (a) The intended ventilation strategy in operation, with air entering the room through vents in the façade and flowing through the room into the atrium. (b) The unintended steady state found using the water-bath model, with air entering the room from the atrium (1) and exiting through the vent (2). (Images: Breathing Buildings)

also stable steady state. In this case, the heat input on the first floor was activated sooner than that on the ground floor, resulting in the first floor being initially hotter than the ground floor. Air, and in this case water, will always take the path the of least resistance, which in this situation is for air to be drawn into the first floor from the atrium and to be exhausted through the low-level façade vent on that storey – the opposite direction from what was intended. This could be made visible using different coloured water to show the flows (Figure 36.4):

1) Coloured water moving from the atrium into the first-floor room.
2) A plume of warm air escaping through the low-level vent in the external facade of the room.

Water-bath model experiments enabled both the effect and the cause of that effect to be identified. It was found that, whilst the ventilation strategy involved purging both floors with fresh air overnight, this 'night cooling' was more effective on the ground floor – resulting in greater cooling of the concrete slab on the ground floor when compared with that on the first floor. At the beginning of each day, therefore, the first-floor slab was invariably hotter than the ground floor which made it difficult for the desired ventilation flow regime to become established [13]. The matter was resolved by improving the effectiveness of the first-floor night cooling.

36.5.3 Bloomberg European HQ, London

Bloomberg's new European headquarters in London, UK took ten years to design and is one of the of the most environmentally-friendly office buildings ever built (Figure 36.5). Designed by Foster + Partners and building services

Figure 36.5 The Bloomberg Building, London. (Image: [14])

engineers Sweco, it was winner of the 2018 RIBA Stirling prize for architecture. The building features nine floors of open-plan office space. The lowest two floors and the top floor are sealed and mechanically ventilated. The intermediate six stories, however, are naturally ventilated whenever the conditions are suitable. Fresh air is brought in through automated vents in the façade and exhausted through vents atop a central atrium; the atrium doubles as an architecturally unique helical staircase, flooding the centre of the building with natural daylight and creating a connection between floors.

The design of this building combined all the most-advanced airflow modelling techniques available. Full dynamic thermal-mass simulations were performed on multiple ventilation strategies and building use scenarios. Computational Fluid Dynamic simulations were used to assess external wind effects on the building façade; and water-bath modelling was used to examine air flow into and through the complicated internal layout of the proposed floorplate/atrium structure. At 1.6 x 1.6 m and 0.8 m high, this is the probably largest water-bath model ever constructed for such a purpose (Figures 36.6 and 36.7).

The model used electrically-heated wires to induce a buoyancy-driven flow within the model (Figure 36.8); coloured dyes were then introduced to enable visualisation of those 'airflows' in real time, in 3-dimensions. More than 10 kW of electrical power was used to drive the simulation – an engineering challenge in itself, requiring significant health and safety checks to be performed prior to setting up or running the experiment. The tests were filmed to allow detailed analysis after the tests were completed.

Figure 36.6 1:35 scale Perspex model of the Bloomberg Building. View from above. (Image: Breathing Buildings)

Figure 36.7 1:35 scale Perspex model of the Bloomberg Building. View through the facade. (Image: Breathing Buildings)

Figure 36.8 The model ready for the test to be filmed. The cables entering the water-bath tank are connected to the electrically-heated wires to drive the water movement. (Image: Breathing Buildings)

The water-bath model tests revealed that on the uppermost naturally-ventilated level, plume outflows were clearly visible from perimeter vents, as well as plume outflows from the top of the central atrium. This indicated that, whilst cross-flow ventilation was occurring on this level, there was also a substantial element of single-sided ventilation at the perimeter due to air entering and leaving via the same façade inlets. This was probably caused by the relatively limited buoyancy head on this floor which meant that the driving force for ventilation was weakest on that level. This was one of many useful insights gleaned from the water-bath modelling exercise which the architects used to help inform changes in the design and treatment of the upper storey [15, 16].

The visual and qualitative evidence resulting from the water-bath experiments were both key elements in demonstrating to the client the validity and efficacy of the proposed natural/hybrid ventilation strategy which is now fully-implemented in the occupied building. A testament to the power of this technique is that the model occupied pride of place in the foyer of Foster+Partners' London office for some time.

36.6 Experimental techniques and flow visualisation

The case studies described above predominantly make use of basic flow visualisation where coloured dyes injected at strategic times and areas of the experimental model can be easily observed. Such techniques are ideal for extracting qualitative conclusions such as demonstrating the relative impact of changing the heat source or the effect of adjusting the size or location of vents.

When there is a need for quantitative results which might include, for example, measuring plume velocities, inferring the steady-state temperature which results from a particular experimental set up, or tracking the evolution of the stratification with time, more sophisticated experimental techniques can be used.

36.6.1 Dye attenuation

When working with saline solutions in water, the dye attenuation technique enables easy visualisation of the inflow and outflow volume fluxes which can readily be recorded for later analysis using shadowgraph visualisation and a video camera. In this approach, saline solution which has been coloured with a dye for which the absorption spectrum is known is introduced to the experimental model at a measured density and volume flux – which are variable and easily controllable. This enables precise control of the input buoyancy flux, equivalent to the heat input to a building. Images taken at regular intervals can be compared and cross-referenced with discrete density sample measurements to construct a time series of how the buoyancy (analogous to temperature) evolves through the course of an experiment.

36.6.2 Schlieren imaging

Schlieren imaging is an optical imaging technique which can be used to observe density differences, in particular density gradients. The technique works by projecting parallel light through the experimental tank and onto a viewing screen. Different densities within the model refract the parallel light rays by varying degrees, projecting an image that can be filmed for later analysis.

36.6.3 PIV, PTV and LIF

Over the last ten years, the rapid development of advanced imaging techniques, combined with computational analysis, has further expanded what is possible with water-bath modelling. For the measurement of velocities and for tracking the evolution of a flow with time, techniques such as Particle Image Velocimetry (PIV), Particle-Tracking Velocimetry (PTV) and Laser-Induced Fluorescence (LIF) are now bringing additional insight to the field of water-bath modelling.

36.7 Concluding remarks

The future for physical water-bath modelling is bright. There have been a number of recent examples where major projects have benefited from water-bath modelling. These include the Apple Campus building in Cupertino, California for which water-bath modelling was used alongside dynamic thermal modelling and detailed computational fluid dynamics, as part of the design stage efforts to understand how this complicated building would behave once occupied. Apple Campus is one of the largest buildings in the world to have been designed with the support of water-bath modelling [17].

An important feature of water-bath modelling is that the visualisation of flows can help every member of the design team and client team alike to understand the potential flow patterns which can thus enhance the overall design process. There is no other technique which enables us to physically observe time-varying fluid flow in real time in three dimensions.

As the construction industry has become more aware of the importance of occupant health and well-being, so the attention paid to internal environments is greater now than it ever has been. To this must be added the heightened focus on the impacts of climate change and the need to reduce energy consumption in the built environment. Water-bath modelling has the potential to offer valuable insights in both of these areas and is likely to become a very cost-effective tool for building design engineers.

References

1 Thomas, P.H., Hinkley, P.L., Theobald, C.R. & Simms, D.L. (1963) *Investigations into the flow of hot gases in roof venting*. Fire Research Technical Paper No.7. London: Her Majesty's Stationery Office.

2 Morgan, H.P., Marshall, N.R. & Goldstone, B.M. (1976) *Smoke hazards in multi-level shopping malls: some studies using a model two-storey mall*. Report CP 45/76. Boreham Wood (UK): Fire Research Station, Building Research Establishment.

3 Saito, Kozo (ed.) (2008) *Progress in Scale Modeling: Summary of the First International Symposium on Scale Modeling (ISSM I in 1988) and Selected Papers from Subsequent Symposia (ISSM II in 1997 through ISSM V in 2006)*. New York: Springer Publishing.

4 Turner, J.S. (1973) *Buoyancy Effects in Fluids*. Cambridge: Cambridge University Press.

5 Linden, P.F. (1999) The fluid mechanics of natural ventilation. *Annual Review of Fluid Mechanics* 31 (1): 201-238.

6 Todd, S.P., Adamu, Z.A., Cook M.J. & Price A.D.F. (2014) Natural personalised ventilation for hospital wards: Experimental validation. *CIBSE ASHRAE Technical Symposium, Dublin, Ireland, 3-4 April 2014*. 15pp.

7 Linden, P.F., Lane-Serff, G.F. & Smeed, D.A. (1990) Emptying filling boxes: the fluid mechanics of natural ventilation. *Journal of Fluid Mechanics* 212: 309-335.

8 Hunt, G.R., Holford, J.M. & Linden, P.F. (2001) Natural ventilation by the competing effects of localised & distributed heat sources. In: *Proc. 14th Australasian fluid mechanics conference, Adelaide, Australia*: 545-548. Adelaide: Causal Productions.

9 Gladstone, C. & Woods, A.W. (2001) On buoyancy-driven natural ventilation of a room with a heated floor. *Journal of Fluid Mechanics* 441: 293-314.

10 Livermore, S. (2006) *Aspects of buoyancy-driven natural ventilation.* PhD Thesis, University of Cambridge.

11 Lisham, B. & Woods, A.W. (2006) The control of naturally ventilated buildings subject to wind and buoyancy. *Journal of Fluid Mechanics* 557: 451-471.

12 Department of Health (2013) *Health Technical Memorandum 07-07 Sustainable health and social care buildings: Planning, design, construction and refurbishment*. 54-56. London: Department of Health.

13 Livermore, S. (2006) *Aspects of buoyancy-driven natural ventilation.* PhD Thesis, University of Cambridge.

14 David Holt. Licensed under the Creative Commons Attribution 2.0 Generic license. https://commons.wikimedia.org/wiki/File:Bloomberg_European_Headquarters,_London.jpg (21/10/2019)

15 Astbury, J. (2017) Building study: Foster ramps it up at Bloomberg. *Architects Journal* 244 (22), 15 Nov.: 20-36.

16 Pearson, A. (2017) Designing a natural ventilation strategy for Bloomberg's central London HQ. *CIBSE Journal on-line.* https://www.cibsejournal.com/case-studies/designing-a-natural-ventilation-strategy-for-bloombergs-central-london-hq/ (Consulted 21/10/2019)

17 https://src.lafargeholcim-foundation.org/dnl/7df1948e-3035-4e4f-a437-6f15ab4c4510/A18LabMusma-BostonMentalWellness.pdf (16/11/2019)

37

The use of biological models for building engineering design

Jan Knippers

37.1 Historical background

'All machinery is derived from nature' wrote Vitruvius in his *Ten Books on Architecture* [1] in the first-century B.C., which indicates that humankind has always looked at nature as a model for building and construction. In the history of architecture there have been countless attempts to transfer nature's richness into built form, such as the Corinthian capitals and in Gothic vaults.

Modern civil engineering was formed towards the end of the eighteenth century and soon referred to nature. John Smeaton (1724-1792), the 'father of civil engineering' described a tree trunk: '... every section of the tree is nearly of an equal strength in proportion to what it has to resist' both where the tree rises from its root bole and where a branch meets the main trunk [2]. He used this observation to justify the profile of his Eddystone lighthouse from 1759 (Figure 37.1).

Later, during the period of iron construction in the nineteenth century, Carl Culmann (1821-1881), the pioneer of graphical methods of analysis in engineering suggested to the anatomist Georg Hermann von Meyer, that trabecular patterns in bones are aligned along lines of principal stress according to functional loading. Culmann drew an analogy between the patterns inside a femur (the human thighbone) and the stress trajectories of a cantilevered curved crane by William Fairbairn [3]. The published papers from this communication caught the attention of the anatomist Julius Wolff (1836-1902) and stimulated his development of the trajectorial hypothesis of bone adaptation – now called 'Wolff's Law' [4, 5, 6] (Figure 37.2).

However, the typical trussed girders and arches of the nineteenth century with hinges and pinned joints that were designed by Carl Culmann and his contemporaries, are a human invention and cannot be found in nature. They were developed to enable a quick and reliable calculation of deformations, forces and loading of the individual members using a simple set of formulae. A natural structure such as the femur places material where it is needed during growth. This leads to framed and Vierendeel-like systems with rigid and thickened joints displaying a very different type of structural behaviour, making use of bending at the nodes. It is therefore not surprising that there is no evidence to justify the frequent reading

Physical Models: Their historical and current use in civil and building engineering design,
First Edition. Edited by Bill Addis.

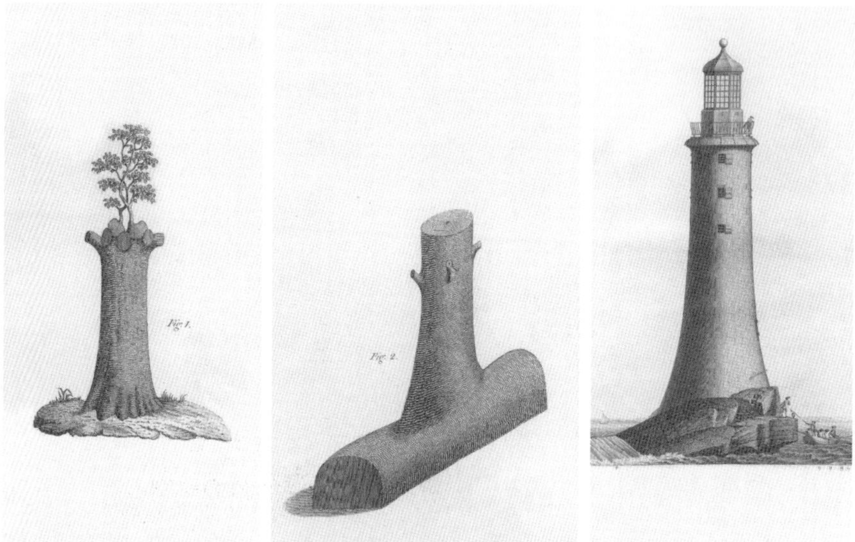

Figure 37.1 John Smeaton's appeal to biological models to underpin the form of his Eddystone Lighthouse, 1759. (Source: [2])

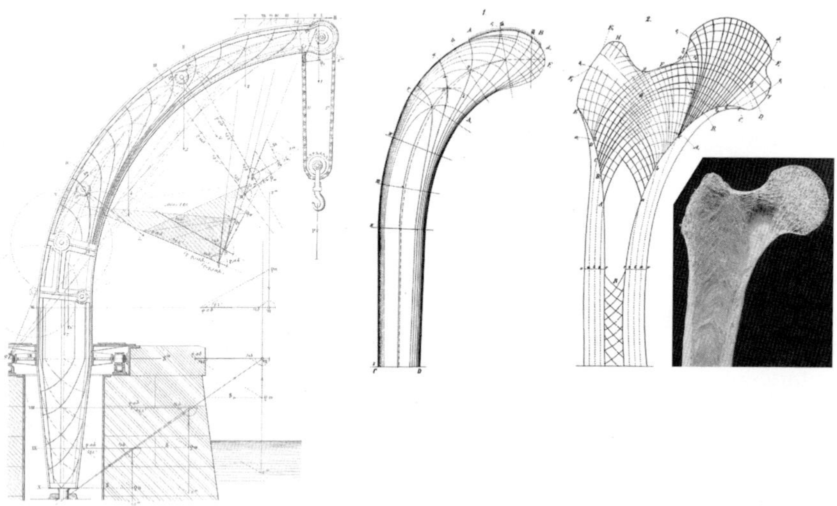

Figure 37.2 Comparison of Culmann's diagram of bending in a Fairbairn crane (Source: [3] Plate 11) with Julius Wolff's diagram comparing the stress trajectories in an idealised crane (based on Culmann's drawings) and a femur (Source: [4] Plate XVIII), and the internal structure of a femur head that reflects the stress trajectories. (Source: [7])

in popular literature that Gustave Eiffel and his engineers Maurice Koechlin and Émile Nouguier were inspired by the femur in the construction of his tower.

Another claim that we can often find in books is that the ridge-and-furrow roof system that Joseph Paxton developed for green houses and, later, for the Crystal Palace, was inspired by the huge leaves of the waterlily *Victoria amazonica*, the strength of which Paxton tested by setting his daughter Annie afloat on one of them. Again, no written evidence can be found that an analysis of the relevant plant morphology led to the design of this greenhouse structures.

These examples may illustrate that there was an early interest by architects and engineers in natural constructions, but a transfer of functions from nature to technology did not take place.

Around the turn of the twentieth century, general interest in natural sciences was raised considerably by some key publications. One of them was Ernst Haeckel's beautifully illustrated 'Kunstformen der Natur' (Art forms of nature) from 1904 [8, 9] that gained enormous popularity and could be found in every bourgeois household. Drawings of radiolarians from this and other books of Ernst Haeckel (1834-1919), a zoologist in Jena, were an inspiration for Walther Bauersfeld (1879-1959) when developing the first 'geodesic' dome that formed the structure of the planetarium in the same city, Jena [10] (Figure 37.3). However, as with the other examples given above, the link is not a direct one, since, unlike the flexible, hexagonal cells of the radiolarian, Bauersfeld's dome is fully triangulated and, hence, very stiff. A few years later, the Scottish biologist D'Arcy Wentworth Thompson (1860-1948) published his studies of analogies and correlations between biological forms and mechanical phenomena, including a drawing of the radiolarian *Aulonia hexagona*. The influence of his book *On Growth and Form* from 1917 [11] on the development of natural science, mathematics and arts cannot be overestimated. Both these publications also attracted a lot of interest, of course, among the engineers, but they were of limited relevance for engineering design.

This situation continued throughout the entire twentieth century. Frei Otto became interested in natural constructions from the early stages of his career in the 1960s, inspired by his contact with the zoologist Johann-Gerhard Helmcke (1908-1993). Otto pursued the investigation of natural constructions parallel to his developments in lightweight structures throughout his entire professional life.

(a) (b)

Figure 37.3 (a) Drawings of Radiolarians by Ernst Haeckel. (Source [8] Plate IX). (b) Framework for the experimental dome for a Planetarium in Jena. (Source: [12])

Nevertheless, he did not aim for a knowledge transfer from biology to technology; rather, he saw it more the other way round: as an expert in lightweight structures he supported the biologists in understanding the form-defining mechanical principles behind natural constructions.

Buckminster Fuller had a similar approach. After he became famous for the construction of geodesic domes, natural scientists provided him with many examples of geodesic structures on the micro-level. He regarded this as a validation of his own technical achievements and supported the scientists in helping them understand the mechanical formation principles in nature.

37.2 A new way of thinking in the twenty-first century

Until the end of the twentieth century, the morphological differentiation of form and material in systems of nature was far beyond the scope of design and fabrication. Therefore, the transfer from nature to technology was not aimed for. Nevertheless, reference to nature was important for many pioneers in architecture and engineering, in order to illustrate the aims of their efforts including the beauty, elegance, lightness and efficiency of structures in nature.

This state-of-affairs changed fundamentally around the turn to the twenty-first century with the introduction of computational methods in all fields of science and technology.

In biology, microcomputer tomography and other digital imaging methods have generated data models that can be directly transferred to Finite Element Methods in engineering. Simulation technologies allow for a quantitative analysis of highly-differentiated structures. They provide scientific gain in both directions: they support biologists in their understanding of the functional morphology of living beings and they facilitate the abstraction of functional principles of natural role models for transfer to fabrication. Additive manufacturing enables a differentiation of chemical and physical properties, similar to what we find in nature.

Hence, the advent of computational methods has enabled a direct exchange of knowledge between fields of research that were previously widely separated and allowed a redefinition of the relationship between nature and technology.

The most successful examples of biomimetic research in recent decades have been located in the realm of structured surfaces – self-cleaning, friction-reducing, anti-fouling, self-adherent, etc... One reason for these success stories in science is that such technical coatings are structured on a similar scale to that of their biological models, i.e. the lotus leaf, shark skin, gecko feet and others. The situation is more complex for load-bearing structures and building envelopes for which the challenge of scaling from a smaller biological model to a typically larger building structure arises. In this context, the issue of scaling addresses not only size, but also loading, life expectancy, etc... Thus, for building structures a linear and direct transfer of functional principles from a biological model to a building structure cannot be expected.

However, this does not mean that the study of biological models is not useful for today's structural engineers. Research and development in engineering is typically directed towards the incremental improvement of well-established building systems. In contrast, biomimetic research can be seen as an invitation or even an instruction to think outside the box because natural structures follow completely different design principles. They are composed of only a limited number of chemical elements and basic molecular components. Through self-organization, living organisms form multi-layered, hierarchically-organized, finely-tuned and highly-differentiated structures that exhibit multiple networked functions. This differentiation of form and material is a key aspect for their performance and resource efficiency, achieving 'less material' through 'more form'. In these respects, biological structures differ fundamentally from building constructions because the latter follow the paradigm of arranging as few different parts as possible, in as simple a way as possible. Hence, biomimetic research has the potential to unlock innovation through the reconsideration of design and construction principles in engineering, even though a direct transfer of functional principles from biology to building structures is achievable only in a very few cases [13]. In the following sections, two examples illustrate this point.

37.3 The Elytra Filament Pavilion

The Institute for Computational Design (ICD, Professor Achim Menges) and the Institute of Building Structures and Structural Design (ITKE, Professor Jan Knippers) at the University of Stuttgart developed an alternative approach to fibre-reinforced polymer (FRP) constructions from 2012 [14]. The aim was to make use of the structural potential of anisotropic fibre-based material systems and, at the same time, avoid the high costs and large quantities of waste of conventional fabrication methods by using moulds.

Studies of natural lightweight fibre structures accompanied these developments from the beginning. The Elytra, the protective shell of the abdomen and wings of the beetle, provided a suitable model. The Elytra is made of a chitin fibre composite material that forms a double-layered shell and is coupled by trabeculae, i.e. internal bracing structures. Seamless structural integration of the upper and lower shells is achieved through continuous arrangements of fibres within each trabecula. Morphological differentiation addresses the geometric articulation of each trabecula and their arrangement within the global structural system (Figure 37.4).

The genetically-controlled self-organization during the growth of the Elytra has been transferred to a robotic additive winding process. This requires a high level of abstraction which, in the present case, becomes visible through a modular construction in contrast to the continuous seamless Elytra. The aim is to replace the costly and toxic waste-causing mould of conventional FRP constructions with a light, discrete and resource-efficient steel frame that holds the resin-soaked fibres in place during the additive layering process and allows for easy adaption to various geometries. After the curing of the resin, the steel frame is dismantled and a pure fibre composite structure remains.

This approach was used for an installation at the John Madejski Garden of the Victoria and Albert Museum in summer 2016, called the Elytra Filament Pavilion

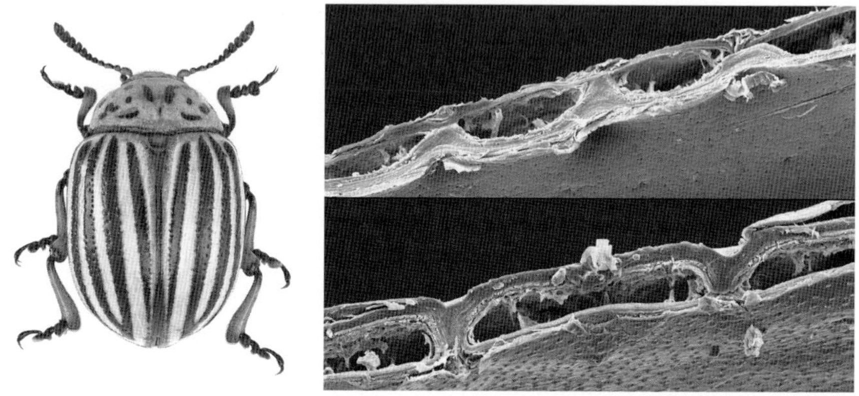

(a) (b)

Figure 37.4 (a) Colorado Potato Beetle (*Leptinotarsa decimlineata*) (Image: Udo Schmid). (b) SEM scans of *Leptinotarsa decimlineata* elytron scanned by Prof. Oliver Betz at University of Tuebingen. (Image: Oliver Betz)

(a) (b)

Figure 37.5 (a) Robotic winding process (Image: ICD/ITKE, University of Stuttgart). (b) Elytra Filament Pavilion – Victoria and Albert Museum, London, 2016. (Image: NAARO)

[15] (Figure 37.5). The roof structure was configured as a planar, slightly inclined canopy supported by seven columns and covered by a polycarbonate sheet to provide rain coverage and drainage. The canopy was a modular structure, initially made up of thirty-six hexagonal roof modules covering a total area of approximately 200 m^2 with a weight of just 9 kg per m^2 (the weight of glass and carbon fibres in the roof cells and columns). The outer dimensions of each component are identical: a depth of 40 cm and diameter of 2.40 m. After winding, each module was tempered in an oven for about six hours at 80 °C. The bolts and sleeves at the component joints and the column bases are the only metal parts in the structure. The installation includes an on-site robotic fabrication unit to facilitate growth and adaption over time. Four further components were manufactured on site during public events and subsequently installed within the structure.

As this structure was to be exposed to thousands of museum visitors, structural integrity needed to be verified very carefully. For the simulation, a reduced

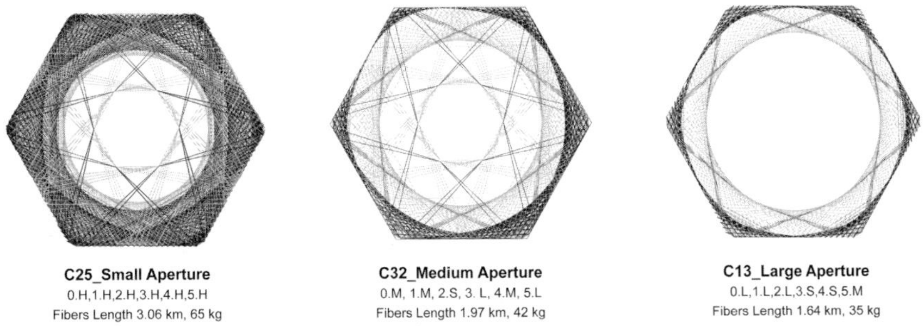

Figure 37.6 Components with load-adapted fibre layout. (Image: ICD/ITKE, University of Stuttgart)

FE-Model was used. The woven mesh was represented as a continuous surface and the woven edge connectors were idealized as lines. The primary purpose of this reduced analysis was to guide the load-adapted design of the components: the diameter of the central aperture and the reinforcement with carbon fibres, along the edges and on the hyperbolic surface, were differentiated according to the loading of the respective component in the reduced simulation (Figure 37.6).

The simulation was not sufficient fully to prove the integrity of the system, as relevant mechanical phenomena of the fibre-mesh were not considered: in particular, the local buckling of compressed fibres, spanning free between contact points, and the capacity of the contact points to resist local shear and tension. Therefore, a selection of roof modules and columns was tested physically [16].

The structural system of the Elytra Filament Pavilion was reliable and easy to handle during fabrication and installation. The modular components can easily be adapted to larger spans, higher loads and different geometries. Larger buildings, based on the same material and fabrication concept, are currently being designed and will soon follow.

37.4 Plant movements as concept generators for adaptive building systems

Biomimetic research is repeatedly shifting from classical topics, such as structural efficiency, to design principles such as adaptivity or multi-functionality, just as are found in biological structures to meet several and sometimes also contradictory requirements in parallel.

Research on kinetic and adaptive building envelopes illustrates this aspect. In contrast to the automotive industry and other fields of technology, kinetic building structures or envelopes are typically unique items, often planned and built without reference projects or prototypes. This leads to the key question of how complexity can be reduced to a minimum. Botany provides a radical answer that stands in total contradiction to typical technical solutions based on rigid-body motion. Many plant organs move without any specific mechanical elements by means of the locally-adapted and adaptive flexibility of their components.

Figure 37.7 Bird of Paradise Flower (*Strelitzia reginae*). (Image: Ruslou Koorts)

Whether and how such plant movements can be used in artificial systems has been investigated at the Institute of Building Structures and Structural Design (ITKE, Professor Jan Knippers) at the University of Stuttgart and the Plant Biomechanics Group (PBG, Professor Thomas Speck) at the University of Freiburg since 2008.

A screening of various plants identified the movements of the bird of paradise flower (*Strelitzia reginae*) as a suitable kinematic principle and biological model. The Strelitzia's flower features two adnate petals that form a perch for pollinating birds (Figure 37.7). When a bird sits on this perch to sip the nectar, its weight causes the perch to bend down. In a simultaneous movement, the petal lamina exposes the anthers that are otherwise hidden and kept safe. This bending actuates a lateral unfolding of the wings of the flower. A gradual abstraction transforms the elastic kinematics into a simple mechanism that consists of a thin-shell element attached to a beam. The equilibrium path of the shell element is a non-symmetrical bending motion, triggered by the torsional buckling that is induced by the uniaxial bending of the attached beam. This torsional buckling phenomenon is well-known to engineers. However, in structural design it is usually considered as a failure mode that needs to be prevented. Nature, however, exploits this principle and uses it actively to initiate large deformations by small actuation forces.

Based on the first physical models and numerical simulations, a prototype façade-shading system called Flectofin was developed [17] (Figures 37.8 and 37.9).

An opportunity for introducing such systems on a larger scale was presented at the Thematic Pavilion at EXPO 2012 in Yeosu, Korea. A kinematic media façade with 108 individually controllable fins was planned and built. The façade can adapt to light conditions and physical building conditions and allows the artistic staging of special lighting effects. It has a total length of 140 m and a height of

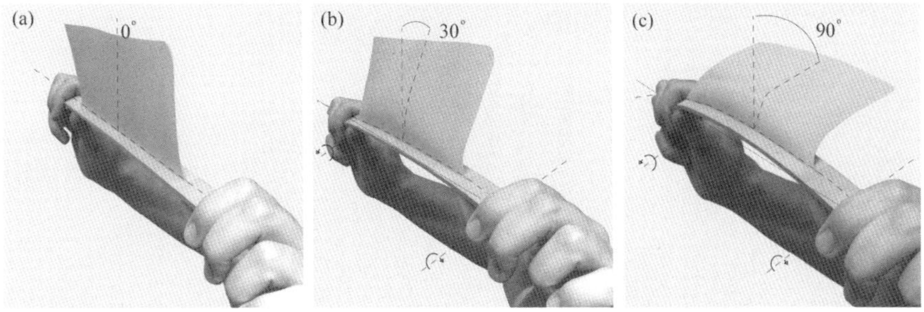

Figure 37.8 Mechanism of Flectofin. (Image: ITKE, University of Stuttgart)

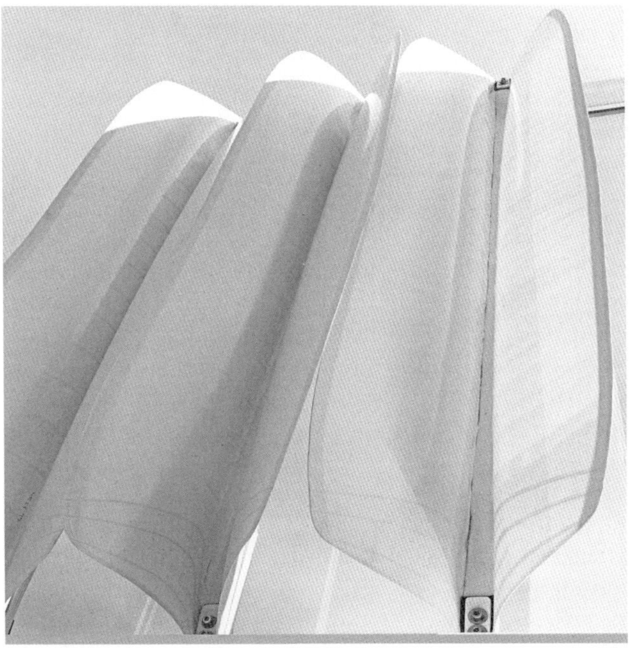

Figure 37.9 Compliant mechanism for the façade shading system called Flectofin. (Image: J. Lienhard)

between 3 and 14 m, and is designed to withstand the very high wind speeds on the Korean coast [18] (Figure 37.10).

We initially attempted to scale Flectofin to the size of the façade. However, this proved to be difficult in its original configuration. On the one hand, it did not fulfil all aspects of the architectural design; on the other hand, without additional structural reinforcement, it did not offer sufficient stability to withstand the high wind loads. Inspired by our research on plant movements, another kinetic system was developed. The façade was made with slightly-curved plates supported by two hinged corners at the top and the bottom. At the other two corners, a small compressive force was applied in the plane of the fin, leading to controlled buckling. This principle showed locally smaller strains than the Flectofin original

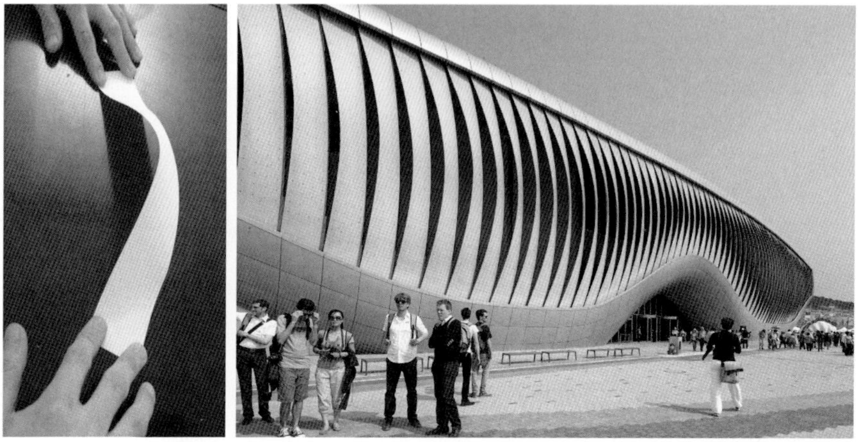

Figure 37.10 Thematic Pavilion, Yeosou, South Korea, 2012. Architect: Soma Architecture Vienna, Engineer Kinetic Façade: Knippers Helbig Engineering, Stuttgart. (a) Compliant mechanism. (b) Built structure. (Images: ITKE, University of Stuttgart)

and does not open completely. It perfectly matched the initial design intentions of the architects and offered a favourable ratio of structural stability and actuation energy. The elastically-deformable fins were made of glassfibre-reinforced polymers. They are up to 14 m high, 1.25 m wide and only 9 mm thick, with an additional stiffener on the side with less-elastic deformation. When open, the curved geometry, together with the residual stress state, results in a very strong system that deforms only a few millimetres under high wind loads. In its closed state, the adjacent plates are clamped together so that the facade can withstand even the strongest storms without damage.

Both the Flectofin movements and the Yeosu shading system are made possible by means of elastic deformations that are associated with correspondingly large strains. Both cases show nonlinear deformation that, from a structural engineer's perspective, is considered to be a stability failure and usually needs to be prevented by bracing or reinforcement. Even though the principle of the Yeosu façade does not directly follow the abstraction of plant movement, the underlying idea was inspired and derived from the observation and analysis of natural role models. This encouraged the search for solutions lying outside the traditional methods of design and construction and going beyond preconceptions, such as avoiding disproportionately growing deformations and stability failure modes.

37.5 Concluding remarks

These examples show that a linear understanding of biomimetics in the sense of a direct transfer of concepts and strategies from biology into architecture, might only be sufficient if the focus is on the abstraction of single functions as found in technical products such as the shading system Flectofin. In architectural tasks, such as the design of the façade in Yeosu or the Elytra Pavilion in London, however, a variety of functions have to be met simultaneously within a given set of constraints. In addition, the biological models very often have to be scaled up to a large size, leading to increasing difficulties to meet such challenges. In the context of architecture and building construction, a linear and direct abstraction and transfer are therefore difficult to maintain. Instead, an expanded definition of biomimetics is required: the analysis of natural form and function principles has the potential to stimulate architects and engineers to use fundamentally new strategies in architectural design and technical implementation, leading us to solutions beyond existing typologies.

References

1 Vitruvius (c.30-15 BC) *De Architectura.* Book X Chapter 1, para 4.
2 Smeaton, J. (1791) *A narrative of the building and a description of the construction of the Edystone lighthouse with stone …* London: Hughs. 42.
3 Culmann, K. (1866) *Die Graphische Statik.* Zurich: Verlag Von Meyer & Zeller.
4 Wolff, J. (1870) Über die innere Architektur der Knochen und ihre Bedeutung für die Frage vom Knochenwachstum. *Arch. Pathol. Anat. Physiol. Klin. Med.* 50: 389-453.
5 Wolff, J. (1892) *Das Gesetz der Transformation der Knochen.* Berlin: Hirschwald.
6 Skedros, J.G. & Baucomb, S.L. (2007) Mathematical analysis of trabecular 'trajectories' in apparent trajectorial structures: The unfortunate historical emphasis on the human proximal femur. *Journal of Theoretical Biology* 244: 15-45.
7 Nascher, I.L. (1914) *Geriatrics: the diseases of old age and their treatment.* Philadelphia: Blakiston's Son & Co. Plate following p.24.
8 Haeckel, Ernst Heinrich (1862) *Die Radiolarien.* Berlin: Georg Reimer.
9 Haeckel, Ernst Heinrich (1904) *Kunstformen der Natur.* Leipzig: Bibliographisches Institut.
10 Bauersfeld, Walther (1957) Projection Planetarium and Shell Construction. *Proc.Inst.Mech.Eng.* 171 (1): 75-80.
11 Thompson, D.W. (1917) *On Growth and Form.* Cambridge: Cambridge University Press.
12 Photo (n.d.) Universitätsarchiv der TU Berlin, 423 Nachlass Franz Dischinger, Nr.118.
13 Knippers, J. & Speck, T. (2012) Design and construction principles in nature and architecture. *Bioinspiration & Biomimetics* 7 (1): 015002.
14 Menges, A. & Knippers, J. (2015) Fibrous Tectonics. *Architectural Design* 85 (5): 40-47.
15 Prado, M., Dörstelmann, M., Menges, A., Solly, J. & Knippers, J. (2017) Elytra Filament Pavilion: Robotic Filament Winding for Structural Composite Building Systems. In: *Fabricate: Rethinking Design and Construction* (eds. A. Menges, B. Sheil, R. Glynn, & M. Skavara), 224-231. London: UCL Press.

16 Knippers, J. (2017) The Limits of Simulation: Towards a New Culture of Architectural Engineering. *Technology, Architecture + Design* 1 (2): 155-162.

17 Lienhard, J., Schleicher, S., Poppinga, S., Masselter, T., Milwich, M., Speck, T. & Knippers, J. (2011) Flectofin: a hingeless flapping mechanism inspired by nature. *Bioinspiration & Biomimetics* 6 (4): 045001.

18 Knippers, J., Scheible, F., Oppe, M. & Jungjohann, H. (2012) Bio-inspired Kinetic GFRP-façade for the Thematic Pavilion of the EXPO 2012 in Yeosu. *Proceedings of the IASS-APCS-Symposium on 21-24 May 2012 in Seoul, Korea.*

38

Flying a 100-metre long Jumbo Koinobori

Mamoru Kawaguchi

38.1 Introduction

The flying of fabric carps called *Koinobori* has been a traditional event in Japan for several centuries (Figure 38.1). They are flown in the breeze by large numbers of Japanese families with their children early in May every year, or more specifically, on 5th May, Children's Day, to celebrate the growth of their children and to pray for their health and success in the future. There is a town to the north of Tokyo called Kazo which has been producing Koinoboris for a long time mainly using cotton as the material for their carps. Early in 1988 a group of volunteers from Kazo had the innocent idea of making a fabric carp 100 m in length to advertise their town. They enlarged the pattern of a domestic-size carp, typically a few metres long, up to a length of 100 m. They cut the fabric according to the pattern, sewed up the cut fragments of fabric to form a Jumbo carp and painted it to finish off the fabric fish.

When the carp was completed, the volunteers wanted to show it to the public and, since they thought that the carp would be too large to fly normally in the sky, they attempted to show off the carp in a somewhat more modest manner. They put a wire rope through its stomach and raised the rope using three cranes, like laundry being hung on a washing line. As soon as the hoist started and the carp had left the ground, it was caught by the breeze and suddenly, accompanied by a screaming noise, its stomach skin was badly ripped apart (Figure 38.2). The reporters and film crews from several TV companies, who had come to the site expecting to witness the successful display of the carp, were all disappointed and left.

The volunteers from Kazo who had devoted themselves to making the Jumbo Koinobori were very much discouraged by the failure but they soon tried to recover from the shock. They mended the torn Koinobori and began searching for an expert who might be able to advise them how they could display the carp more safely and, hopefully, to fly it in the air. They visited various authorities in the fields of fluid mechanics, physics and applied mathematics, but all in

Physical Models: Their historical and current use in civil and building engineering design,
First Edition. Edited by Bill Addis.

Figure 38.1 Koinoboris flying on Children's Day at Chizu, Tottori prefecture, Japan. (Image: [1])

Figure 38.2 The hoisted Jumbo Koinobori ripped apart by the wind, as drawn by a child who witnessed the accident. (Source: Mamoru Kawaguchi)

vain. All the experts said they could not suggest ways to calculate how such an enormous carp could be safely constructed.

Finally, volunteers from Kazo came to us to ask for our advice. While the volunteers did not really expect that the carp might fly after experiencing the miserable failure which had occurred when they simply tried to hoist it up for display, they still put to us their innocent and enthusiastic question – would it be at all possible that the Koinobori could swim normally in the air?

38.2 Model theory and dimensional analysis

We thought that this sort of question might be answered most clearly using model theory or, more specifically, dimensional analysis, since we Japanese know very well that the carps of normal size (2-5 m in length) celebrated at home can swim in a mild breeze without problems. The only difference between the Jumbo carp and domestic carps is their size; their shapes are mathematically similar to each other and this situation is one to which we know that dimensional analysis can be conveniently applied. The following shows how we found a rational way to predict the behaviour of a jumbo carp of 100 m in length by means of model theory.

From the Buckingham's Pi theorem (see Section 13.2.2) the following relations are reduced among the participating quantities, where ***v, p, L, σ*** are, respectively, the wind speed necessary for the carp to fly, air pressure, linear dimensions, and stresses in the fabric. The subscripts ***p*** and ***m*** indicate prototype and model, and λ is the scale factor, being the ratio of linear dimensions between the prototype and the model. The important thing in this case was that the jumbo carp was not in perfect similarity to the carps of ordinary size, since the jumbo carp was made with the same fabrics as for an ordinary carp. So the theory had to be modified to take account of the imperfect similarity. It was concluded that the jumbo carp flies in the same wind speed as an ordinary carp, it swims very slowly and produces stresses λ times as large as those in ordinary ones. An outline of the derivation of the above conclusions using model theory is as follows:

For perfect similarity:

$$v_p = \sqrt{\frac{p_p}{p_m}}.v_m = \sqrt{\frac{L_p}{L_m}}.v_m = \sqrt{\lambda}.v_m \quad \cdots\cdots\cdots \quad (1)$$

with the stresses in the fabric $\sigma_p = \sigma_m$ .... (2)

For imperfect similarity (when the same fabric is used for prototype and model):

Eq. (1) becomes $v_p = v_m$ .... (1')

and the stresses in the fabric are $\sigma_p = \lambda\,\sigma_m$ .... (2')

Through this application of dimensional analysis, we could derive the following results (for full details of the analysis, see [2]):

- The Jumbo Koinobori swims in a breeze of the same speed as for the domestic carps, i.e. it begins to swim at a wind speed of 2-3 m/sec.
- The tensile stresses in the skin of the flying Jumbo Koinobori are λ times as large as those for the domestic carps. (If a domestic carp is 5 m long, then $\lambda = 100/5 = 20$)
- The Jumbo Carp swims very slowly, taking λ times longer for the same action as a domestic carp.

The first result from the analysis above was very welcome to those involved in the project. The second one, however, highlighted the need to consider carefully the strength of the skin (fabric) of the Jumbo Koinobori in which stresses 20-50 times higher than those in domestic carps may occur. The third result provided simple visual information about the speed of movement of the Jumbo Carp.

38.3 Other technical issues

Among other technical problems to be solved before the Jumbo Koinobori could be launched into the sky, the most important was the level of stresses that might occur in various parts of its body. The following three kinds of force that might arise during the flight were identified as the most significant for design:

- The circumferential forces that may occur in the skin of the carp due to the stomach air pressure.
- The longitudinal forces that may occur in the skin due to the stomach air pressure and the wind flow along the carp.
- The local forces that may be caused by the mouthpiece ring.

The first of the above forces are very essential and are the most significant for design. A Koinobori is designed and manufactured in such a way that the diameter of the stomach is larger than that of its mouth for the purpose of inflating the stomach when swimming in the wind. The air stream that separates from the main external flow of the wind at the mouth goes through the stomach of the carp and joins the external stream of the wind again at the tail. The speed of the airflow in the stomach is thus lower than that at the mouth and outside of the carp, since the cross-sectional area of the stomach is larger than that of the mouth. According to the Bernoulli's theorem, the air pressure in the stomach is therefore larger than at the mouth which is the same as the external air pressure. This means that the air pressure inside the stomach is larger than outside the carp. This was confirmed through a series of wind-tunnel tests on 10 m long model carp (Figures 38.3 and 38.4). This mechanism inflates the stomach of a carp when it is swimming in the air and the skin of the carp needs to withstand the stresses due to this pressure.

Assuming the carp is in the shape of a cylinder, both the circumferential and the longitudinal stresses are found easily using membrane analysis. The calculations revealed that the fabric used for Kazo Jumbo Koinobori was strong enough

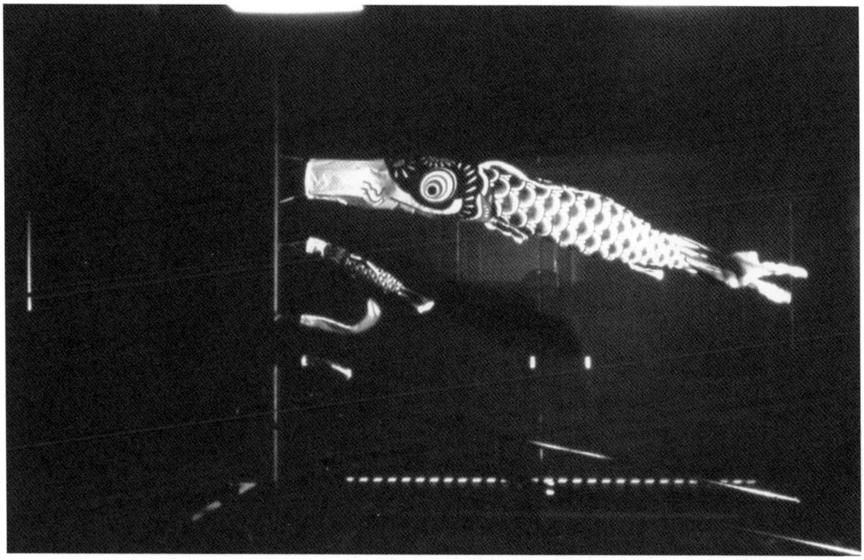

Figure 38.3 1:10 scale model Koinobori (10 m long) in the wind tunnel. (Source: Mamoru Kawaguchi)

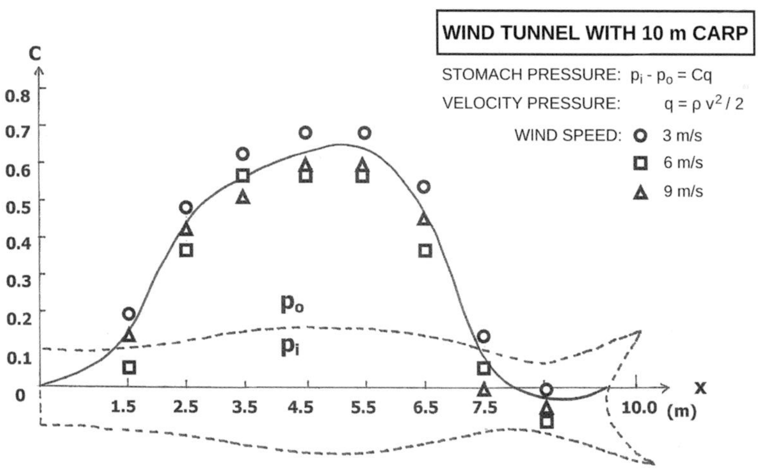

Figure 38.4 Distribution of stomach air pressure coefficients. (Source: Mamoru Kawaguchi)

to withstand the stresses described above but the strength of the seams was seriously lacking. The seams of the jumbo Koinobori were re-sewn and finished using portable professional sewing machines that were loaned at no cost to the volunteers by a tent manufacturer.

The third kind of force which occur around the mouthpiece was also important. The mouthpiece is indispensable for a carp to introduce the air stream smoothly into the stomach. For family-size carps, mouthpieces of bamboo are often used. The ring for the mouthpiece is subjected to a large compressive force while the

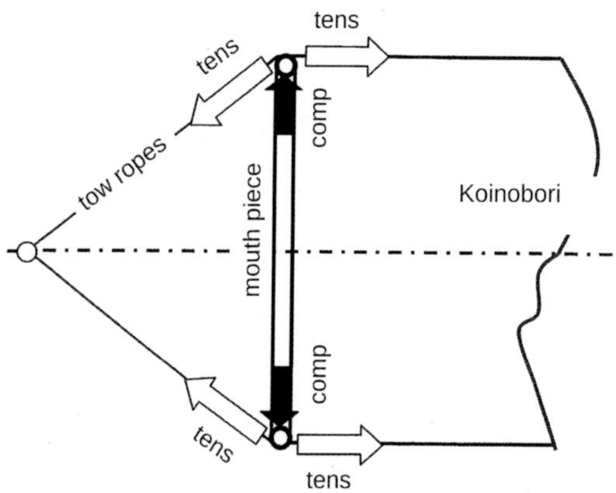

Figure 38.5 Forces in the mouthpiece of a Koinobori. (Source: Mamoru Kawaguchi)

Figure 38.6 The aluminium mouthpiece set out on the ground. (Source: Mamoru Kawaguchi)

carp is swimming (Figure 38.5). If the ring were designed to avoid the danger of buckling due to the compression, a steel tube would be needed weighing some 300 kg, which is nearly half the total weight of the Koinobori; this would have made the whole project unfeasible.

This problem was solved by adding twelve ropes arranged around the circumference of the mouthpiece, and connected to the ring in a similar way to the spokes of a bicycle wheel. With this arrangement the ring could be made with aluminium tube 50 mm in diameter, having a total weight of just 30 kg (Figure 38.6).

38.4 Launching the Jumbo Koinobori

For a Koinobori to fly in the air it should normally be hoisted to a height that is approximately the same as its length. And so every family which celebrates the

Figure 38.7 The Jumbo Koinobori laid out on the ground. (Source: Mamoru Kawaguchi)

flying of domestic Koinoboris has a pole which is temporarily built each May for hoisting the carps.

For the Jumbo Koinobori of Kazo we decided not to design a pole on which to hoist it, but rather we hired a crane that could suspend the carp from a height of 100 m. It did not prove difficult to find a crane that had a boom of sufficient length, capable of carrying the weight of the carp and withstanding the lateral forces due to the wind.

On a fine day towards the end of April in 1988, the Jumbo Koinobori was laid out in the field beside the Tone River in Kazo (Figure 38.7). The weight of the carp was some 700 kg and, with a length of 100 m it was about 1.5 times as long as a 747 Jumbo jet plane.

The weather was fine but we had no wind in the morning. We had to wait for several hours before a gentle breeze began to blow in the field. When the speed of the breeze reached 3 m/sec, the crane was set in motion, slowly raising the mouthpiece of the carp. The carp, which had been lying flat on the field as if it had been sticking to the ground, began to inhale the air and its stomach slowly inflated to form a three-dimensional shape. The crane continued to raise the mouthpiece and finally, when the tail of the Koinobori lifted off from the field, the huge carp began to fly elegantly in the air (Figure 38.8). The spectators sitting on the field and waiting for the flight of the Koinobori became elated at the sight of its graceful swim, high above in the sky, and a great clapping of hands broke out.

Figure 38.8 The Jumbo Koinobori soaring in the breeze. (Source: Mamoru Kawaguchi)

38.5 Concluding remarks

Flying the Jumbo Koinobori was established thereafter as one of the most important annual events of the town of Kazo. It has been celebrated on the third of May, Constitution Day, every year since, and the recorded flight of the Jumbo Koinobori has been broadcast on 5th May, the Children's Day. The people of Kazo are very proud to show off each year the graceful flight of the largest flying object in the world [3]. The story of how the dream of flying the Jumbo Koinobori came true was told in a 45-minute TV programme, and is included in the archives of NHK, Japan Broadcasting Corporation, so that it is publically accessible to everyone. The Jumbo Koinobori has also travelled abroad to perform its graceful flight – twice in Hawaii and once in Kaiserslautern, Germany [4].

References

1 663highland. Licensed under the Creative Commons Attribution-Share Alike 3.0 Unported license. https://commons.wikimedia.org/wiki/File:150425_Koinobori_Chizu_Tottori_pref_Japan01bs.jpg (6/11/2019))

2 Kawaguchi, M. (1996) Flying a Jumbo Carp in the Breeze. *Journal of Wind Engineering* 66: 53-58. (in Japanese)

3 https://www.youtube.com/watch?v=Uisl-fNap3M (6/11/2019)

4 Kawaguchi, M. (2006) Flying the Jumbo Koinobori in the German Sky. *Wind Engineers (JAWE)* 31 (4): 205-11. (in Japanese)

39

Epilogue: A future for models from the past

Dirk Bühler and Christiane Weber

39.1 Introduction

After their validation and celebration in the present book, there cannot remain any doubt about the necessity and usefulness of engineering models, and the central role they have played in the development of civil and building engineering, alongside materials technology, engineering theory and engineering drawings. Models were used both as an essential tool in engineering design and in the development of engineering science, to test and verify new theories. As tools they were and are able to serve, sometimes as a 'proof of concept' (e.g. Fontana models, Grubemann's marketing models, Augsburg models), at other times to communicate how something works or should be built, and on many occasions to raise confidence in a proposed design of method of construction. In the case of model-testing, two main types have generally been used in engineering. First there are the mechanical or assembly models, which are not necessarily intended to behave as the full-size engineering work but to help understand, analyse, improve and verify an engineering design. Secondly there are measurement models (*Messmodelle* in German) which enable measurements of engineering behaviour, in miniature, usually to predict the behaviour of a full-size work, a method closely linked to engineering calculations, material technology and construction methods.

Probably the most common purposes of original models until the early twentieth century were what we might call 'proof of concept' today – to test ideas, such as new types of machines or structures, and to work out precisely how they should be constructed. The need for proof of concept led to the creation of many mechanical models to test and demonstrate the working of mechanical machines, for example the hydraulic machinery in Augsburg (see Section 1.4.1). Other models served to convey engineering designs to the people who would build them – three-dimensional drawings, so to speak. They could demonstrate details of the assembly of large timber structures, especially their complex joints.

It was also found that traditional physical models had two other vital functions. They could form a vital part of the training of young constructors during their apprenticeships: countless masterpiece models have been produced as a

Physical Models: Their historical and current use in civil and building engineering design,
First Edition. Edited by Bill Addis.

mandatory examination task at the end of apprenticeships – for example roofs, bridges and complex spiral staircases for carpenters and joiners. Models were also used to publicise innovative designs, either to visitors, such as the models of pumps and mills in Augsburg, or to potential clients, as with the models of timber bridges made by the Grubenmann brothers.

Finally, of course, apart from the measurement models that are the main focus of this book, many models were made for educational purposes – to demonstrate structural or other engineering principles, such as the behaviour under load of an arch or truss or tensile structure or soil, or the flow of water through channels and the effects of wind or earthquakes.

The important question is now, what happens to these models after they have been built and used? Unfortunately they are seldom show-pieces eligible for art galleries because they have usually suffered from their usage or even been destroyed and thus are often considered objects unsuitable for collections. This is why, measured by their significance, too few of them have been conserved and, unfortunately, even fewer presented to the public. Their significance is not an aesthetic one, but an intangible one based, in most cases, in their experimental setup which can help us to retrace the ideas of the engineer and the implementation of the results for a structural or other engineering design. To improve this situation a new validation of these objects is necessary with regard to their universal usefulness: they are not just historical records of past progress, now obsolete; on the contrary, they help us to understand the provenance of the knowledge we have today and the epistemological routes by which we achieved it. It is possible still to learn from them and to question and improve them constantly. In addition, these models are, if exhibited, enormously useful for familiarizing the public in general with the work of civil engineers and at the same time to inspire and recruit students. Though rather few in number, some important examples of physical engineering models in museum collections can be found and may be useful to those engaged in engineering education or historical research.

39.2 Historical study of engineering models

39.2.1 Museum collections

The numbers of models related to building and civil engineering on display in museums today are few compared to the number of architectural models. Many of these are recently-built models intended to educate the visitor about engineering in the past, such as the construction of iconic buildings and bridges [1]. Only a very small proportion of museum models are original, dating from the time of construction, and even fewer are known to have been used by the engineers working on the projects. These few examples are, therefore, worthy of particular attention.

The engineering models found in museum collections today may have served any of the functions of models, and perhaps more than one – often we shall never know the true answer. Collections of models related to building and civil engineering can be found in a wide variety of different locations, including museums of architecture, technical museums, regional and national museums, university collections and many city institutions that collected models of their own engineering works. Sadly, however, even when they exist, many of these collections are no longer on public view, and remain locked away in storage rooms, accessible only through museum inventories.

Perhaps the oldest models related to civil engineering are in Germany, on display in the collections in the *Maximilian Museum* in Augsburg [2] and the industrial museum in Nuremberg [3] (see Sections 1.2.5 and 1.3). The collection in Augsburg is particularly remarkable because it dates back to the establishment of the model chamber in the city of Augsburg, built up from the beginning of the sixteenth century with models of the Perlach-Tower (proposals dated 1503 and 1514), the Lueginsland-Tower (1514), the roof framework of the Sankt Servatius Church Tower (1543) – all of them scattered in different city depositories until Elias Holl dedicated the room above the Golden Hall in the new city hall in 1620 to be the model chamber. The models are now – in part – on display in the *Maximilian Museum.* This collection contains engineering models made and used by Elias Holl (1573-1646), one of the best-known renaissance architects and engineers. A century later the hydraulic engineering models of Caspar Walter (1701-1769) became part of the collection and have contributed to establishing his fame. The largest collection of scientific instruments in Germany, donated by the Bavarian Academy of Science and including many engineering models from the nineteenth and twentieth centuries are kept in the *Deutsches Museum* in Munich [4].

The *Musée des Arts et Métiers* in Paris has three galleries with permanent exhibitions dedicated to construction and engineering, showing original engineering models dating mainly from the nineteenth century; many more are kept in storage [5]. They represent bridges (Figure 39.1), canals and sluices, roof trusses and scaffolds, several construction sites and a representative collection of cranes and pile

Figure 39.1 Model of the pier and swing bridge over the Rapids Canal (Vietnam). Manufactured by Digeon J. et Fils aîné for the World Fair 1900 in Paris to demonstrate their capabilities. Scale: 1:25. (*Musée des Arts et Métiers*, Paris, Inv.-Nr.13298. Image: Dirk Bühler)

drivers. Many of these models were clearly masterworks made by apprentices, but would also have served to teach younger craftsmen.

Other models in the *Musée des Arts et Métiers* were made with the specific aim of demonstrating and explaining engineering principles, many of which were made by Digeon J. et Fils aîné, ingénieur constructeur of Paris, in in the 1880s. One example is an apparatus for applying loads to the individual voussoirs of an arch, measuring the thrust of the arch on the foundations and demonstrating the failure of the arch by formation of hinges between voussoirs (Figure 39.2). A great many such models were made in the *École des ponts et chaussées*, the civil engineering school founded in 1747 and the *Ècole d'arts et métiers*, founded in 1780 (and, of course, other educational institutions throughout the world).

The largest and oldest collections of models relating to engineering in Great Britain are kept in the *Science Museum* in London [6] however, only a handful of these are models used for engineering investigations. One remarkable example, from around 1772, is the model of a proposed timber arch bridge spanning 275 m across the River Foyle in Derry, Ireland [7] (see Section 1.5.2). The model is about 5.8 m long, at a scale of 1:48. This design by Johann Konrad Altherr and Johann Konrad Langenegger was an inspiration for Ivan Kulibin's design for his own bridge of similar size across the Neva at St Petersburg (see Section 4.2.2). Another model for a bridge across the Neva, proposed by the Swiss bridge builder named Aldon and dating from 1762, is now displayed at the *Central Museum of Railways of the Russian Federation* in St. Petersburg (see Figures 4.9 and 4.10). There also exists in the Hermitage Museum in St Petersburg a model, most probably made by Kulibin himself in around 1770, of one of the early designs for the bridge (Figure 39.3). The bridge has an upper deck for pedestrians, between the two

Figure 39.2 Apparatus demonstrating the working of a voussoir arch, made by Digeon J. et Fils aîné, around 1880. (*Musée des Arts et Métiers*, Paris, Inv.-Nr.11265. Image: Dirk Bühler)

Figure 39.3 A model made of lime wood for a bridge over the River Neva, c.1770, most probably by Kulibin. The model arch has a span of 235 cm, indicating a scale of approximately 1:125. (Image: Inna Regentova. © The State Hermitage Museum, St. Petersburg)

main arches, and a lower deck for carriages suspended from the arches, including a drawbridge section in the centre to allow the passage of sailing vessels on the river with tall masts. The model is waiting some conservation work and then due to go on public display.

Also in London's *Science Museum* there are two rubber models and a collection of diagrams, photographs and scientific papers relating to the measurement of stresses in masonry dams, made by J. S. Wilson and W. Gore at the City and Guilds College in 1905 [8] (see Figure 7.13). There are also several micro-concrete models and photo-elastic models, including a section of the Westway elevated-road, London, used in model tests made by the Cement and Concrete Association near London (see Section 17.3.9), as well as some typical models utilised in photoelastic investigations in the Structural Engineering Laboratory of Imperial College, London. Of particular interest to the development of experiments using reduced-scale models, there are also the original models of ships used by William Froude to measure the resistance of ships' hulls [9] (see Figure 13.1).

The *Soane Museum*, also located in London, owns a total of 402 models, some of them of engineering relevance, such as models of timber roofs, a model of an engine for driving piles, a model for a brick construction method and some models showing structural details of Sir John Soane's own buildings [10]. Of particular note are models made by Soane in 1828 when he was responsible for repairing the roof of Inigo Jones's Banqueting House in London. Three models were produced to show the original structure of the roof, together with the means he adopted to strengthen the trusses [11]. The collection of the University of Oxford at the *Ashmolean Museum* owns several significant objects of civil engineering [12]. The original museum was built between 1679 and 1683 in order to host the collection of Elias Ashmole (1617-1692) and developed constantly to a considerable size, mainly after 1714 when demonstrations began to accompany the lectures.

A catalogue from 1790 lists a model of a machine for breaking pieces of wood, models of cranes, pile drivers, pumps and mills and even 'Smeaton's pulley with a three-legged stand' [13]. However, many of them are currently not on display. The *Whipple Museum of the history of Science* in Cambridge has the original model made by Professor Seikei Sekiya in 1887 showing the movement of a soil particle in an earthquake that occurred in Tokyo [14] (see Figure 25.1). The *National Museum of Scotland* in Edinburgh has several models made by William Jessop during design of Smeaton's Eddystone lighthouse in the winter of 1756-1757 [15] (see Section 1.5.3).

In Switzerland, the *Grubenmann Museum* at Teufen in Canton Appenzell has been open to the public since 2012 and displays many original and recent models of wooden bridges and roof trusses from the eighteenth century [16]. The Museum has only one original model by Grubenmann, the Trogen Bridge (c. 1745-1755) [17, 18], on permanent loan from the Kantonsschule, Trogen (see Figure 1.9). The original model of the Grubenmann brothers' Schaffhausen Bridge (c.1755) is exhibited at the *Museum Allerheiligen* in Schaffhausen [19, 20]. Other original Grubenmann models that have survived include the bridge at Wettingen (around 1765), kept in the Tiefbauamt (Civil engineering department) at Aarau in Canton Aargau [21, 22], and a bridge at Stein-am-Rhein [23, 24, 25]. An original model of a small timber bridge (about 10 m span) over the Hemishoferbach in Schaffhausen, Switzerland, survives in the Gemeindekanzlei (council offices) at Hemishofen and has recently been restored [26]. It was made in 1792 by Hans Conrad Stadler (1752-1819), who often worked with Johannes Grubenmann [27], and the bridge itself was completed in the following year by another builder.

Among the many experiments carried out by engineer-architects in the nineteenth century, a number of models by Georg Ludwig Friedrich Laves (1788-1864) are held in the *Historisches Museum* in Hannover, Germany. He was particularly noted for developing a fish-belly or lenticular girder in around 1839 and ten of his models of such girders that he tested have survived (Figure 39.4). A model of his Stadtgraben Bridge (1835) in Hannover, used by the engineer to test the bearing capacity of the Laves timber girder [28], is preserved and on display in the *Historisches Museum* in Hannover [29].

Italy does not have large national museums of engineering and technology like the other countries mentioned. However, engineering models can occasionally be found in various small, specialized collections. Some of them show physical and mechanical experiments useful in engineering, while other engineering models related to the construction of the buildings are displayed near the buildings themselves. For example, the *Museo Galileo* in Florence [30]), Padua (Natural and medical sciences) [31], the *Palladio Museum* in Vicenza, including many models of his buildings and bridges (but no original models) [32], the *Museo Leonardo da Vinci* in Rome including nineteenth- and twentieth-century models of his many engineering designs [33], and the Archaeological Museum in Naples [34]. In Milan, the *Museo Nazionale Scienza e Tecnologia Leonardo da Vinci* has a remarkable collection of models based on Leonardo's manuscripts that have been made since the 1950s [35]. These technical models (without established scale) represent 104 significant examples from the whole range of Leonardo's ideas on

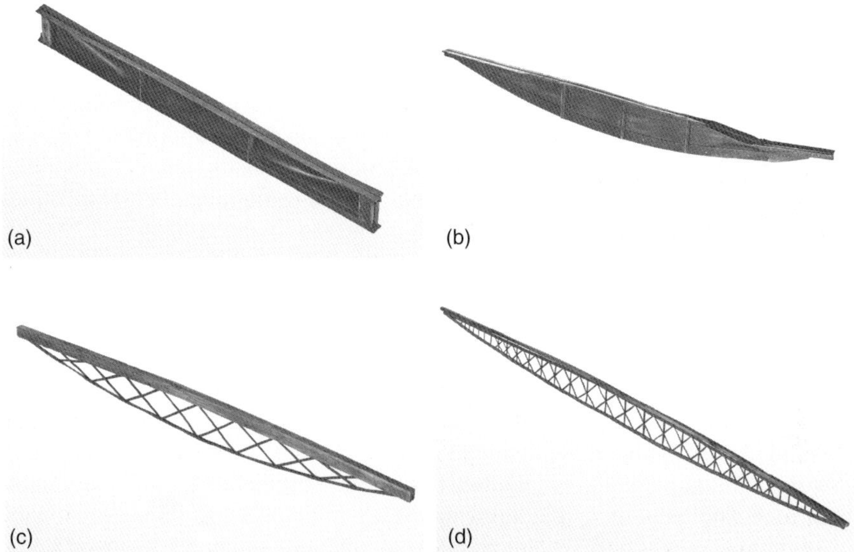

Figure 39.4 Models of Georg Laves preserved in the Hannover *Historisches Museum*. (a) I-beam made of sheet metal and wood, 27 cm long. (b) Fish-belly girder made of sheet metal, 27 cm long. (c) Fish-belly truss girder made of iron and wood, 33 cm long. (d) Fish-belly truss girder made of sheet metal, wood and iron wire, 59 cm long. (Images: Courtesy of Hannover Historisches Museum)

science and technology. As well as conveying Leonardo's ideas, these models are, effectively, also tests of the viability of Leonard's designs [36]. The difficulties encountered when making these real, three-dimensional models after his superb drawings have exposed many practical problems and confirm the suggestion that Leonardo probably did not make many models himself.

39.2.2 Exhibitions

In 1997, the *Centre Pompidou* in Paris presented an exhibition of unprecedented and unequalled size and comprehensiveness – *L'art de l'ingénieur. Constructeur, entrepreneur, inventeur*. The exhibition featured many models as well as original artefacts. The lengthy book that accompanied the exhibition mentions many of the models and includes a catalogue of the exhibits [37].

In 2012, the *Deutsches Architekturmuseum* (DAM) in Frankfurt [38] devoted an exhibition to the subject of architectural models – *Das Architekturmodell: Werkzeug, Fetisch, Kleine Utopie* (The Architectural Model: Tool, Fetish, Small Utopia). The exhibition featured 1240 models by 419 architects from twenty-five countries. Of these, only nine form-finding models by Frei Otto had any engineering purpose. The catalogue, which focuses on many aspects of making, collecting and exhibiting architectural models, contains an essay about Otto's model-making [39].

In 2016, the *Südwestdeutsches Archiv für Architektur und Ingenieurbau* (saai) mounted an exhibition at the Karlsruhe Institute for Technology, displaying most of the models of Frei Otto which were donated to the archive some years earlier. The exhibition, *Denken in Modellen* (Thinking by modelling), was illustrated by over 200 models and tools and instruments associated with his model making, as well as more than 1000 photographs, drawings, sketches and films, much of which was captured in the accompanying book [40]. Most of the exhibits illustrated Otto's experimental approach to form-finding using a variety of techniques that have been discussed elsewhere in this book (see Chapters 3, 14 and 19).

For an exhibition in 2017, *Maquetas y modelos históricos: ingeniería y construcción*, held at the *Fundación Juanelo Turriano* in Madrid, thirty-three models related to historical engineering and construction in Spain, some of them dating back to the eighteenth century, were put on display for the very first time. They had been found in twenty-three different locations in Spain including the technical universities of Madrid, Barcelona, Valencia, Sevilla and some smaller cities [41]. Unfortunately there were no measurement models – a task for further research – but the exhibition demonstrated that there are probably still many more models of engineering interest which are not well-known and whose historical importance is perhaps not recognised [42, 43].

39.2.3 Conferences

Physical models have begun to receive some scholarly attention in recent years. A series of three conferences began in 2009 at the Technical University in Munich under the thematic topic *Models and Architecture* [44]. Of twenty-one papers by delegates from Germany, Austria, Switzerland, France, Italy and Spain, only one of the papers addressed engineering models – by Rainer Graefe on Gaudì's form-finding models (see Section 3.5.3). The second conference, in 2015 in Paris, was entitled *La maquette – Un outil au service du projet architectural* [45]. As in Munich, of twenty papers only one dealt with engineering models – by Christiane Weber reviewing structural model testing at the University of Stuttgart (see Chapter 14). The third meeting in 2016 was in Rome at the *Accademia nazionale di Santa Luca*, entitled *Il modello architettonico. Funzione ed evoluzione di uno strumento di concezione e di realizzazione.* Of the six papers presented, only one addressed engineering models – the net- and soap-bubble models of Frei Otto (see Chapter 19) [46]. The conclusion from these conferences is, then, that the role of engineering models in construction projects of the past, and the contributions made by the various actors involved, has not yet attracted the attention of the wider academic community.

39.3 Some surviving models from the twentieth century

Many engineering models do not survive, of course, because they are destroyed during testing. Many models, such as acoustic and hydraulic models, were simply dismantled and some of the components were salvaged for making new models. However, an equally serious threat to the survival of models is when an institution closes or refocuses its interest towards new activities; entire collections of models have been lost in this way. Nevertheless, some models have survived and it is to be hoped that these models continue to survive and, when necessary, are conserved, and also that models that are at present simply 'lying around' somewhere may be recognised for their value and saved, either in the original institution or in a museum. The projects mentioned in this section represent a few examples known to various chapter authors of this book.

39.3.1 Models for designing concrete shells

Dyckerhoff & Widmann was one of the firms whose own testing laboratories finally closed in the 1960s, no doubt with the loss of many models. Franz Dischinger (1887-1953) had been a pioneer of model testing from the 1920s (see Chapter 9). One experimental model shell does remain, known for the iconic photograph of it supporting 39 employees, including Dischinger himself (see Figure 9.5). The Biebrich shell is now relocated in the factory grounds of Dyckerhoff GmbH in Wiesbaden-Amöneburg [47].

Only one model remains from the *Institut für Modellstatik* at the University of Stuttgart – the 1:27 scale model in Perspex of the hyperbolic-paraboloid roof of the Alster swimming baths in Hamburg. It is now mounted in the entrance hall of the Department of Civil Engineering at the university, standing nearly two metres high (see Section 14.2.2). Fortunately its destruction was prevented in 2009, but now its condition is deteriorating as some of the glued joints have become brittle and parts have suffered damage from impact. Until now, its large size has prevented it from becoming part of a collection [48].

A few models remain from ISMES in Bergamo, though none of the iconic structures by Nervi. A 1:100 scale wind tunnel model of the Norfolk Scope Arena, Virginia (Arena built 1965-1971), produced at ISMES and tested at the *Politecnico di Torino* 1964, still exists, and has been conserved by *Centro Museo e Documentazione Storica* at the *Politecnico di Torino* (CEMED) in Turin [49] (see Figure 15.19a). Two models used during the design of St. Mary's Cathedral, San Francisco by Nervi still exist. One is a 1:100 scale wind-tunnel model, produced at ISMES in 1964 and tested at the *Politecnico di Torino*, and now preserved at CEMED (see Figure 15.16). The other is a 1:36.89 scale model of St. Mary's Cathedral made of resin that was used for static and dynamic tests at ISMES in 1965, and is now located at the *Dipartimento architettura rilievo disegno urbanistica storia* (DARDUS) at the *Università Politecnica delle Marche*, in Ancona, Italy (Figure 39.5). The same department owns another model from ISMES, a 1:50 scale model used for static and dynamic tests in 1960 during the design by Nervi of the Leverone Field House in Hannover (Figure 39.6).

The contents of Heinz Isler's test laboratory were donated after he died to the *Institut für Geschichte und Theorie der Architektur* (Institute for the History and Theory of Architecture) at ETH in Zurich [50]. This archive consists of over 500

Figure 39.5 A surviving model of St. Mary's Cathedral, San Francisco designed by Nervi, 1964-1965. 1:36.89 scale model made from resin for elastic and dynamic tests. (Image: Gabriele Neri)

(a) (b)

Figure 39.6 Model for the Leverone Field House, Hannover. (a) Model under test in 1971. (Image ISMES). (b) Model now located at the *Dipartimento architettura rilievo disegno urbanistica storia at Università Politecnica delle Marche*, in Ancona. (Image: Gabriele Neri)

models of various sizes and types, of which thirty-seven are measurement models and seven or eight of these are still mounted in their original test apparatus (Figure 39.7).

Sadly no original physical models remain of the many structures designed by Eduardo Torroja (see Chapters 11 & 16) or Heinz Hossdorf (see Chapter 18).

Figure 39.7 Model of one of Heinz Isler's characteristic shell roofs in its original testing apparatus. Now located in the Isler Archive at the ETH, Zurich. (Image: Bill Addis)

39.3.2 Frei Otto models

Frei Otto retired in 1990, and in 2011 the Southwest German Architecture and Engineering Archive acquired most of the models built in Otto's office in Warmbronn and exhibited them in 2017 at the *Zentrum für Kunst und Medien* (ZKM, Centre for Art and Media) [51]. The inventory embraces over 400 models, and is thus among the most comprehensive collections of models from twentieth-century Germany that exists. Among these are many form-finding models but scarcely any measurement models. A small measurement model of Otto's IL pavilion at Stuttgart, donated by Bertold Burkhardt to the *Deutsches Architekturmuseum,* (DAM), Frankfurt on the occasion of the exhibition *Das Architekturmodell: Werkzeug, Fetisch, Kleine Utopie,* bears witness to the first experiments with tensile structures and their design methods (Figure 39.8). The IL Pavilion effectively served as a large-scale experiment for Otto's much larger German Pavilion at the Montreal Expo in 1967. As the 1:75 measurement model of the Montreal Pavilion no longer exists, the small model of the IL Pavilion is the only representative of the unique experiments using models for the design of the German Pavilion [52]. The model now complements the DAM's collection of approximately twenty of Frei Otto's form-finding models. The founding director of the *Deutsches Architekturmuseum,* Heinrich Klotz (1935-1999), first presented Frei Otto's designs and models at the DAM exhibition *Vision der Moderne* in 1986 [53].

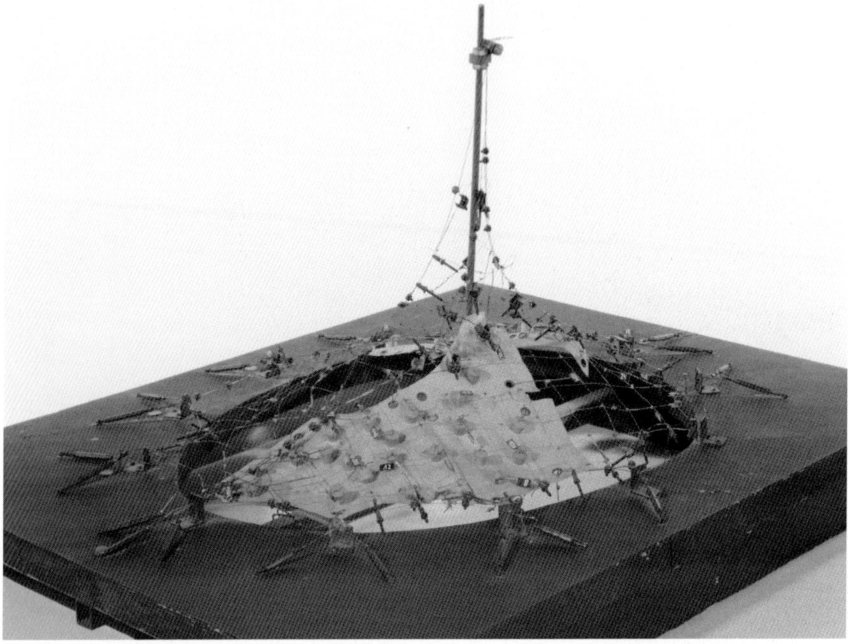

Figure 39.8 Measurement model of Frei Otto's IL Pavilion at the University of Stuttgart. (Image: Architekturmuseum München)

In Otto's original IL Pavilion, which now houses the *Institut für Leichtbau Entwerfen und Konstruieren* (ILEK), there are still several form-finding models for the Munich Olympic Stadium roof and a measurement model that was used to define the low point of the cablenet roof over the Olympic swimming pool at a scale of 1:25 [54]. This model is still located on its original marble slab which serves as a rigid base. However, despite having a cover, the model is slightly damaged and is in need of conservation. The IL Pavilion also still houses the unique apparatus constructed to measure and photograph soap-film models (see Figure 19.3). This apparatus is still cared for and is surely a candidate for perseveration in a museum.

Fortunately, there also remains the large measurement model (about 2 x 1.5 m) that was used by Otto in developing his design for the large sports hall for the Munich Olympic Games in 1972 (Figure 39.9). It is on public display in the Information Centre at the Olympia Park in Munich.

39.3.3 Miscellaneous structural models

For the design of the suspension bridge across the Lillebælt strait in Denmark in 1964, the firm C. Ostenfeld & W. Jønson Consulting Engineers (now COWI) commissioned tests to be undertaken on a 6-m long model. Fortunately, the model was rescued from destruction by Professor Guido Morgenthal at the Bauhaus University in Weimar, where it is now on display [55, 56] (Figure 39.10).

Figure 39.9 1:125 scale measurement model for the Olympic Sports Hall at the Munich Olympics, 1972. Now located in the Information Centre at the Olympia Park in Munich. (Image: Christiane Weber)

Figure 39.10 Measurement model of the suspension bridge across the Lillebælt designed by C. Ostenfeld & W. Jønson Consulting Engineers. Probably dating from 1964, it is displayed today at the Bauhaus University, Weimar. (Image: Guido Morgenthal)

Many consulting engineers probably kept, and still keep models used during the design stages of their major projects, as souvenirs if nothing else. Sadly most are finally discarded. Mamoru Kawaguchi is one of the (probably many) exceptions. In the offices of his firm in Japan many of their models are kept and used to educate young engineers. A kinetic model of a Pantadome, 1.5 x 1 x 1 m, (see Figure 28.6) has recently been moved to Mukogawa Women's University at Nishinomiya, near Osaka in Japan, where it will be installed so that the students can appreciate it.

39.3.4 Replica structural models

A few special structural models have become so well-known from photographs that they have been reconstructed long after the original was lost, partly to understand better how they were built and used, partly as a tribute to their remarkable qualities, and also to put on display.

One such example was one of the models made by Antonio Gaudí (1852-1926), who used hanging models rom 1898 to 1908 for developing the design of a church to be built in the workers' settlement in the Colonia Güell near Barcelona. After this model had been lost, the *Kunsthaus Zürich* commissioned a reconstruction of the former model in 1982. It was manufactured at Frei Otto's Institute in Stuttgart in collaboration with the Gaudí-Group from the Technical University of Delft in the Netherlands (see Figure 3.38). The model is well documented [57] and is now on display at the museum of the Basilica of the Sagrada Família in Barcelona (Figure 39.11).

Figure 39.11 Reproduction of Gaudi's hanging model for Colònia Güell, Barcelona, in the museum of the Sagrada Família in Barcelona. (Image: [58])

Figure 39.12 Reduced-scale reproduction of Torroja's 1:30 scale model for the *Frontón Recoletos* in Madrid. Now at the *Musée Nationale d'Art Moderne* in Paris. (Image: Bill Addis)

Among many engineering models at the exhibition *L'art de l'ingénieur. Constructeur, entrepreneur, inventeur* at the *Centre Pompidou* in Paris in 1997, the highlight was a replica of Torroja's model of the *Frontón Recoletos* in Madrid (Figure 39.12) (see Section 11.3.3). It remains in the collection of the *Musée Nationale d'Art Moderne* in Paris catalogued as an 'architectural model'.

39.3.5 Models from non-structural engineering disciplines

In the early days of using shake tables to study the dynamic behaviour of buildings in response to seismic actions, Lydik Jacobsen, working at the John A. Blume Earthquake Engineering Center, commissioned a 1:30 scale model of extraordinary sophistication to investigate the dynamic behaviour of the fifteen-storey Alexander Building in San Francisco. Each storey had five degrees of freedom – two horizontal, one vertical and two rotational, about horizontal axes (see Section 25.2.1). The model, about 2 metres high, is on display in the Blume Center at Stanford University.

As previously mentioned, many consulting engineering retain models they have used, subject to the limitations of making space for large models, for example, acoustic models of concert halls, perhaps 2 x 2 x 1.5 m high! In the field

Figure 39.13 A number of wind-tunnel models stored by consulting engineers Wacker Ingenieure. (Image: Wacker Ingenieure)

of wind engineering there is, in some countries, an obligation to keep models for the purpose of satisfying the requirements of professional insurance. Many such consultancies, such as Wacker Ingenieure – Wind Engineering, in Birkenfeld, Germany, are likely, by arrangement, to allow visits to view wind-tunnel models [59] (Figure 39.13).

A model of the Sydney Opera House used for wind-tunnel testing in 1961 at the National Physical Laboratory in Teddington, UK is now in the *Museum of Applied Arts and Sciences* in Sydney [60] and a charming video of the wind-tunnel test is viewable on the Internet [61].

Although better-known for structural testing, ISMES in Bergamo Italy produced wind-tunnel models (at 1:100 scale) for the Norfolk Scope Arena, Virginia (Arena built 1965-1971) and St. Mary's Cathedral, San Francisco (church built in 1963-1971). They were tested at the Politecnico di Torino in 1964 and have been conserved and put on display at the *Centro Museo e Documentazione Storica at Politecnico di Torino* in Italy [62] (see Figures 15.16 and 15.19a).

In the *Deutsches Museum* in Munich there is a truly remarkable 'proof-of-concept' working model, at 1:30 scale, devised and built by Konrad Zuse (1910-1995), of a machine to produce a tall tower, for example to support a wind turbine. The *Helixturm* extrudes the building frame vertically by inserting panels sequentially around the circumference of the building while simultaneously jacking it up [63, 64]. In a curious way, it seems to be a mechanical antecedent of 3D-printing. First constructed in 1993, it was meticulously restored to working order later in the 1990s [65].

Finally, in this short review of some existing engineering models there is one that survives in a good state of repair, but will never find a place in a museum. It is the 1:2,400 scale model of the Mississippi river basin, near Jackson MS, which occupies about 80 hectares (see Figures 23.16 and 23.26). It is currently looked after by an enthusiastic group that helps with its maintenance and encourages visits by the public [66].

39.4 A plan for the future

It is in the nature of civil and building engineering that the majority of models made to assist with the design and construction processes have not survived; and this is as true in the twentieth and twenty-first centuries as it was in the Renaissance, the Age of Enlightenment or the nineteenth century. It is clear, however, that few of those models which have survived are on display today: most are now in storage in museums, often because they are in collections dealing with fields much broader than building and civil engineering. Others, perhaps, are simply 'lying around', attracting little attention or appreciation of their engineering-historical value. This highlights the need to celebrate more formally and systematically the importance of models in building and civil engineering.

All these models, be they already present in exhibitions, just kept in museums' depositories or even simply stored somewhere in academic institutions, require a new evaluation in terms of their significance for the history of civil engineering. Abandoning such models would mean devaluing and forgetting the importance of our technical achievements throughout history and to lose them as an inspiration and example for future research in engineering.

By comparison, the value of architectural models in design and planning and the necessity to conserve them has been established long ago by architects. This is largely because it is generally clear to architects that a rigorous study and knowledge of the entire history of their profession is fundamental to the development of new ideas and designs. The principle that rules this idea is: without history, there is no future, or, in other words: an ahistorical performance of our work would leave us lost in time. If we make ourselves aware that our presence is not just to have arrived somewhere, but is the result of many steps of development throughout history, it is obvious that this history is indispensable for a future to be built up. This basic idea, axiomatic in architecture, has not yet been established in the convictions, the education and the everyday practice of civil and building engineers.

In this context the models we have been talking about in this book, play an essential role, because they teach us to make comprehensible the thought processes for the development of new engineering designs and checking the feasibility of applying theories in construction practice. This is the main reason why we surely have an obligation to preserve this irretrievable legacy from the past. Furthermore, we have to put some of these models on display as a vital component of engineering history, for specialists, of course, but with probably even more emphasis for the public in general, in order to make people aware of the extraordinary and varied achievements of building and civil engineers in our history, and thus to anticipate the challenges of the future. This responsibility for the past must include the requirement to explain the models within their historical context, not only during the formation but also permanently during the course of professional life. One of the best possible means to give these models a meaning is to celebrate their value in exhibitions and museums.

In order to reach this goal, the role of museums and their exhibitions in collecting and promoting engineering models could be more widely publicised, and inventories and catalogues of the relevant objects in collections could be made more useful by ensuring that building and civil engineering models can be found by search engines. Ideally, a comprehensive inventory of all available engineering models should be prepared, annotated with explanations and evaluations of their engineering historical significance. Not least, of course, museums could more-actively seek and acquire engineering models of historic significance.

An important final step in validating the importance of such models must be conducting new research into their historical intention, their realization and their meaning in order to establish their significance for the cognitive process in engineering design. Only in recent times have engineering models received some scholarly attention. Future doctoral research projects could focus on the many different fields of building and civil engineering, and study their use of physical models.

Finally, of paramount importance when keeping artefacts from the past, is the question of their conservation and, when appropriate, their restoration. This is an accepted practice for buildings and bridges, alongside paintings and furniture, but has not yet attracted the attention of specialists in the field of engineering models.

It is to be hoped that this book can serve as a stimulus to all these processes.

Acknowledgements

Thanks are due to Sebastian Hoyer at Braunschweig Technical University for information about Georg Ludwig Friedrich Laves.

References

1 Bühler, Dirk (2019) La maqueta del puente de Neuilly, en el Deutsches Museum. In: *Proceedings of the IX Congreso Nacional de Historia de la Construcción, Soria, 9-12 de octubre de 2019*. 169-180.

2 http://kunstsammlungen-museen.augsburg.de/maximilianmuseum (16/8/2019)

3 http://museen.nuernberg.de/museum-industriekultur/dauerausstellung/industrie-im-wandel/ (16/8/2019)

4 http://www.deutsches-museum.de/sammlungen/ (16.8.2019)

5 https://www.arts-et-metiers.net/musee/recherche-sur-les-collections (17/8/2019)

6 https://collection.sciencemuseum.org.uk/search (17/8/2019)
7 https://www.scienceandsociety.co.uk/results.asp?source=1923-0016 (17/8/2019)
8 Wilson J. S. & Gore W. (1905) Rubber cross-sectional model with sharp corners relating to the measurement of stress in masonry dams (Object No. 1973-218/1) Rubber cross-sectional model with rounded corners relating to the measurement of stress in masonry dams (Object No. 1973-218/3). City and Guilds College, London. Science Museum, London.
9 https://collection.sciencemuseumgroup.org.uk/objects/co8040652/one-of-four-original-swan-and-raven-models-used-by-froude-study-models-hydrodynamics-hydrodynamic-resistance (18/8/2019)
10 https://www.soane.org/collections/ (17/8/2019)
11 http://collections.soane.org/object-m1215 (17/8/2019)
12 http://collections.ashmolean.org/collection/about-the-online-collection (16.8.2019)
13 Fox, Celina (2009) *The Arts of Industry in the Age of Enlightenment*. New Haven & London: Yale University Press. 154.
14 http://www.sites.hps.cam.ac.uk/whipple/explore/models/ajapaneseearthquakemodel/ (7/11/2019)
15 https://www.nms.ac.uk/explore-our-collections/collection-search-results/?item_id=180908 (17/8/2019)
16 http://www.zeughausteufen.ch/das-grubenmann-museum/die-ausstellung/ (18/05/2018)
17 Killer, Joseph (1942) *Die Werke der Baumeister Grubenmann*. Doctoral thesis, ETH. Zürich: Gebr. Leemann & Co. 22. https://www.research-collection.ethz.ch/handle/20.500.11850/133364 (18/8/2019)
18 Maggi, Angelo & Navone, Nicola (eds.) (2003) *John Soane and the wooden bridges of Switzerland. Architecture and the culture of technology from Palladio to the Grubenmanns*. Mendrisio: Archivo del Moderno, Accademia di architettura & London: Sir John Soane's Museum. 51 (Cat. No. 2).
19 Killer, Joseph (1942) *Die Werke der Baumeister Grubenmann*. Doctoral thesis, ETH. Zürich: Gebr. Leemann & Co. 24.
20 http://www.stadtarchiv-schaffhausen.ch/online_archivp/daten/bildgross/153946.jpg
21 Killer, Joseph (1942) *Die Werke der Baumeister Grubenmann*. Doctoral thesis, ETH. Zürich: Gebr. Leemann & Co. 45.
22 Maggi, Angelo & Navone, Nicola (eds.) (2003) *John Soane and the wooden bridges of Switzerland. Architecture and the culture of technology from Palladio to the Grubenmanns*. Mendrisio: Archivo del Moderno, Accademia di architettura & London: Sir John Soane's Museum. 34, 51 (Cat. No. 1).
23 Steinmann, Eugen (1982) Baumeister Hans Ulrich Grubenmann (1709-1783) von Teufen. *Appenzellische Jahrbücher* 110: 3-57. 25, Plate 16.
24 Killer, Joseph (1942) *Die Werke der Baumeister Grubenmann*. Doctoral thesis, ETH. Zürich: Gebr. Leemann & Co. 48.

25 Maggi, Angelo & Navone, Nicola (eds.) (2003) *John Soane and the wooden bridges of Switzerland. Architecture and the culture of technology from Palladio to the Grubenmanns*. Mendrisio: Archivo del Moderno, Accademia di architettura & London: Sir John Soane's Museum. 51 (Kat.-Nr. 3).

26 Tanner, Hermann (2014) *Ein Brückenmodell und seine Geschichte*. Gemeinde Hemishofen, Switzerland. http://www.hemishofen.ch/fileadmin/Redaktoren/Dokumente/News/Ein_Brueckenmodell_und_seine_Geschichte.pdf (18/8/2019)

27 Steinmann, Eugen & Witschi, Peter (1987) Johannes Grubenmann der Jüngere von Teufen und Appenzell: Brückenbauer und Klosterarchitekt. *Appenzellische Jahrbücher* 115: 3-28. http://doi.org/10.5169/seals-283328 (18/8/2019)

28 Weber, Helmut (1964) Georg Ludwig Friedrich Laves als Bauingenieur. In: *Georg Ludwig Friedrich Laves*, Georg Hoeltje, 201-252. Hannover: Steinbock Verlag. 204-208, Figures 2-8.

29 https://www.hannover.de/Kultur-Freizeit/Museen-Ausstellungen/Museumsf%C3%BChrer/Museen-von-A-bis-Z/Historisches-Museum-Hannover (7/11/2019)

30 https://www.museogalileo.it/en/ (18/8/2019)

31 https://www.musme.it/en/homepage_musme/ (10.06.2019)

32 https://www.palladiomuseum.org/models/ (18/8/2019)

33 https://www.leonardodavincimuseo.com/en/the-exhibition/ (18/8/2019)

34 https://www.museoarcheologiconapoli.it/en/collections/ (18/8/2019)

35 http://www.museoscienza.org/ (18/8/2019)

36 Giorgione, Claudio (2009) *Leonardo da Vinci. The models collection*. Milan: Museo Nazionale Della Scienza e Della Tecnologia Leonardo da Vinci.

37 Picon, Antoine (1997) L'art de l'ingénieur. Constructeur, entrepreneur, inventeur. Paris.

38 https://dam-online.de/en/ (16/8/2019)

39 Weber, Christiane (2012) Frei Otto – Experimentelle Modelle. In: (2012) *Das Architekturmodell: Werkzeug, Fetisch, Kleine Utopie*. (eds, Oliver Elser & Peter Cachola Schmal), 45-50. Zürich: Scheidegger & Spiess.

40 Vrachliotis, Georg, et al. (eds) (2017) *Frei Otto: Denken in Modellen*. Leipzig: Spector Books.

41 Fundación Juanelo Turriano (ed.) (2017) *Maquetas y modelos históricos – Ingeniería y construcción. Catálogo de la exposición del 7 de junio al 17 de septiembre de 2017 en el Centro Conde Duque en Madrid*. Madrid: Fundación Juanelo Turriano.

42 https://www.juaneloturriano.com/en/exhibitions/exhibition/2017/05/29/historic-mock-ups-and-scale-models.-engineering-and-construction (16/8/2019)

43 https://www.juaneloturriano.com/en/news/2014/03/17/animaci%C3%B3n-en-3d-del-artificio-de-toledo-de-juanelo-turriano (16/8/2019)

44 Schuller, Manfred (ed.) (2009) *Modelle und Architektur: Les Maquettes d´architecture: Models and Architecture*. Munich: Technische Universität München: Lehrstuhl für Baugeschichte, Historische Bauforschung und Denkmalpflege; Architekturmuseum. http://www.sabinefrommel.eu/Booklet_Kolloquium_ModelleUndArchitektur.pdf (18/8/2019) https://www.dbz.de/

download/152583/2610_einladung_modelle.pdf (18/8/2019) https://www.dbz.de/download/152583/2610-Flyer_Modelle_und_Architektur.pdf (18/8/2019)

45 Amsellem, Guy & Grubert, Mireille (eds) (2015) *La Maquette. Un outil au service du projet architectural. Actes du colloque qui s'est entue les 20-21 mai 2011 à la Cité de l'architecture & du patrimoine.* Paris: Éditions des cendres.

46 https://www.accademiasanluca.eu/2016.04.12_comunicato_1.pdf (18/8/2019)

47 May, Roland; Stegmann, Knut; Engelberg, Meinrad von (2017). Das Runde muss in das Eckige. Die Biebricher Dywidag-Versuchsschale von 1931. In: *Ingenieurbaukunst 2018 – Made in Germany. Hg. von der Bundesingenieurkammer.* Berlin: Ernst & Sohn: 152-161.

48 Weber, Christiane (2015). The last witnesses – physical models in architecture and structural design, taking the Technical University in Stuttgart as an example. In: *Proceedings of the 5th International Congress on Construction History, Chicago* (eds Brian Bowen, Donald Friedman, Thomas Leslie & John Ochsendorf) Vol. 3: 569-576.

49 http://www.comune.torino.it/museiscuola/propostemusei/toeprov/centro-museo-e-documentazione-storica---politecnic.shtml (16/8/2019) Or? MAP-MuseoArchivioPolitecnico https://areeweb.polito.it/strutture/cemed/001/Index2.htm (16/8/2019)

50 http://www.schwartz.arch.ethz.ch/Forschung/islerarchiv.php?lan=en (16/8/2019)

51 https://www.saai.kit.edu/?glossary=otto-frei (16/8/2019)

52 Leonhardt, Fritz, Egger, Harald & Haug, E. (1968) Der deutsche Pavillon auf der Expo '67 in Montréal – eine vorgespannte Seilkonstruktion. *Der Stahlbau* 4: 97-106; 5: 138-145.

53 Klotz, Heinrich (1986) *Vision Der Moderne.* Munich: Prestel.

54 Weber, Christiane (2015). The last witnesses – physical models in architecture and structural design, taking the Technical University in Stuttgart as an example. In: *Proceedings of the 5th International Congress on Construction History, Chicago* (eds Brian Bowen, Donald Friedman, Thomas Leslie & John Ochsendorf) Vol. 3: 569-576, Figs. 7 and 8.

55 Letter from Guido Morgenthal to Christiane Weber, August 24 2011.

56 Ostenfeld, C., Haas, G. & Frandsen, A.G. (1970) Motorway Bridge across Lillebælt. Publication XI: Model Tests for the Superstructure of the Suspension Bridge. *Bygningsstatiske Meddelelser* 41 (2): 99-112.

57 Tomlow, Jos (1989) *Das Modell. Antoni Gaudis Hängemodell und seine Rekonstruktion: neue Erkenntnisse zum Entwurf für die Kirche der Colònia Güell.* IL34. Stuttgart: Institut für Leichte Flächentragwerke.

58 Rüdiger Marmulla. Licensed under the Creative Commons Attribution-Share Alike 3.0. https://commons.wikimedia.org/wiki/File:SagradaFamiliaStatikmodell.jpg (22/10/2019)

59 http://www.wacker-ingenieure.com/ (19/8/2019)

60 https://collection.maas.museum/object/12041 (19/8/2019)

61 https://www.youtube.com/watch?v=XLKEPDIMtm0 (19/8/2019)

62 http://www.comune.torino.it/museiscuola/propostemusei/toeprov/centro-museo-e-documentazione-storica---politecnic.shtml (16/8/2019)

63 https://www.deutsches-museum.de/fileadmin/Content/data/020_Dokumente/040_KuT_Artikel/2009/33-4-50.pdf (8/9/2019)

64 http://www.deutsches-museum.de/ausstellungen/kommunikation/informatik/filme/ (8/9/2019)

65 https://www.ar.tum.de/fileadmin/w00bfl/rkk/content_uploads/Eibisch_Helixturm_Dipl-Arb-TUM_2009.pdf (8/9/2019)

66 https://friendsofmrbm.org/ (18/8/2019)

Appendices

With the exception of the descriptions by William Fairbairn and Edwin Clarke of the model tests made for the Britannia Bridge in the 1840s (see Chapter 6), very little was written about the use of physical models by engineers until the beginning of the twentieth century. The following appendices present seven of these early texts, ranging from Vitruvius to Osborne Reynolds.

Appendix A.1	Extract from Vitruvius (c.30-15 BC)
Appendix A.2	Extract from Galileo (1638)
Appendix A.3	Leonhard Euler (1766) A simple rule to determine the strength of a bridge or similar structure, on the basis of the known strength of a model
Appendix A.4	Extract from report: Telford's design for new London Bridge (1801)
Appendix A.5	Experimental Models, *The Builder* (1846)
Appendix A.6	On model experiments, *The Civil Engineer and Architects Journal* (1847)
Appendix A.7	Osborne Reynolds (1888) Extract from paper: On certain laws relating to the regime of rivers and estuaries.

Appendix A.1

Extract from Vitruvius (c.30-15 BC)

Marcus Vitruvius Pollio, (c.30-15 BC) *De architectura.* Book 10, Chapter XVI, Paragraphs 3-7.

'3. Diognetus was a Rhodian architect [engineer], to whom, as an honour, was granted out of the public treasury a fixed annual payment commensurate with the dignity of his art. At this time an architect [engineer] from Aradus, Callias by name, coming to Rhodes, gave a public lecture, and showed a model of a wall, over which he set a machine on a revolving crane with which he seized an *helepolis* [siege tower] as it approached the fortifications, and brought it inside the wall. The Rhodians, when they had seen this model, filled with admiration, took from Diognetus the yearly grant and transferred this honour to Callias.

4. Meanwhile, king Demetrius [...] making war on Rhodes, brought with him a famous Athenian architect [engineer] named Epimachus. He constructed at enormous expense, with the utmost care and exertion, an *helepolis* one hundred and thirty-five feet high and sixty feet broad. He strengthened it with hair and rawhide so that it could withstand the blow of a stone weighing three hundred and sixty pounds shot from a *ballista*; the machine itself weighed three hundred and sixty thousand pounds. When Callias was asked by the Rhodians to construct a machine to resist this *helepolis*, and to bring it within the wall as he had promised, he said that it was impossible.

5. For not all things are practicable on identical principles, but there are some things which, when enlarged in imitation of small models, are effective, others cannot have models, but are constructed independently of them, while there are some which appear feasible in models, but when they have begun to increase in size are impracticable, as we can observe in the following instance. A half inch, inch, or inch and a half hole is bored with an auger, but if we should wish, in the same manner, to bore a hole a quarter of a foot in breadth, it is impracticable, while one of half a foot or more seems not even conceivable.

Physical Models: Their historical and current use in civil and building engineering design,
First Edition. Edited by Bill Addis.

6. So too, in some models it is seen how they appear practicable on the smallest scale and likewise on a larger. And so the Rhodians, in the same manner, deceived by the same reasoning, inflicted injury and insult on Diognetus. Therefore, when they saw the enemy stubbornly hostile, slavery threatening them because of the machine which had been built to take the city, and that they must look forward to the destruction of their state, they fell at the feet of Diognetus, begging him to come to the aid of the fatherland. He at first refused.

7. But after free-born maidens and young men came with the priests to implore him, he promised to do it on the condition that if he took the machine it should be his property…'.

Appendix A.2

Extract from Galileo (1638)

Galileo Galilei (1638) *Two new sciences.* (Translated Henry Crew & Alfonso de Salvio, 1914) New York: Dover Publications. 130-132.

'(Salviati) From what has already been demonstrated, you can plainly see the impossibility of increasing the size of structures to vast dimensions either in art or in nature; likewise the impossibility of building ships, palaces or temples of enormous size in such a way that their oars, yards, beams, iron bolts and, in short, all their other parts will hold together; nor can nature produce trees of extraordinary size because the branches would break down under their own weight; so also it would be impossible to build up bony structures of men, horses or other animals so as to hold together and perform their normal functions if these animals were to be increased enormously in height; for this increase in height can be accomplished only by employing a material which is harder and stronger than usual, or by enlarging the size of the bones, thus changing their shape until the form and appearance of the animals suggest a monstrosity. …[1]

To illustrate briefly, I have sketched a bone whose natural length has been increased three times and whose thickness has been multiplied until, for a correspondingly large animal, it would perform the same function which the small bone performs for its small animal (Figure 27). From the figures here shown you can see how out of proportion the enlarged bone appears. Clearly then if one wishes to maintain in a great giant the same proportion of limb as that found in an ordinary man he must either find a harder and stronger material for making the bones, or he must admit a diminution of strength in comparison with men of medium stature; for if his height be increased inordinately he will fall and be crushed under his own weight. Whereas,

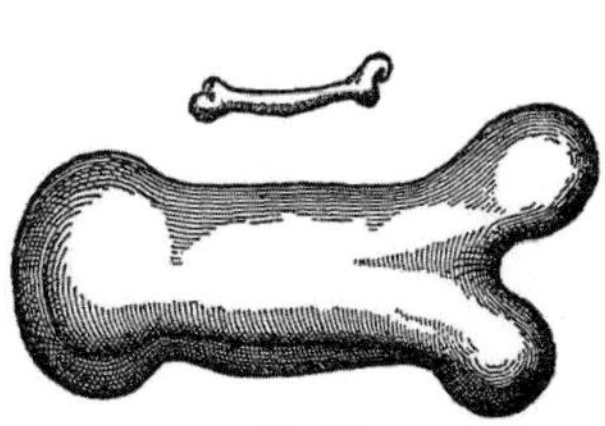

Fig. 27

1 The subject of magnitude and similitude in the natural world is discussed at length in Chapter 2 of *On Growth and Form* by D'Arcy Thompson (abridged version) (1961) Cambridge: Cambridge University Press.

Physical Models: Their historical and current use in civil and building engineering design,
First Edition. Edited by Bill Addis.

if the size of a body be diminished, the strength of that body is not diminished in the same proportion; indeed the smaller the body the greater its relative strength. Thus a small dog could probably carry on his back two or three dogs of his own size; but I believe that a horse could not carry even one of his own size.

(Simplicio) This may be so; but I am led to doubt it on account of the enormous size reached by certain fish, such as the whale which; I understand, is ten times as large as an elephant; yet they all support themselves.

(Salviati) Your question, Simplicio, suggests another principle, one which had hitherto escaped my attention and which enables giants and other animals of vast size to support themselves and to move about as well as smaller animals do. This result may be secured either by increasing the strength of the bones and other parts intended to carry not only their weight but also the superincumbent load; or, keeping the proportions of the bony structure constant, the skeleton will hold together in the same manner or even more easily provided one diminishes, in the proper proportion, the weight of the bony material, of the flesh, and of anything else which the skeleton has to carry. It is this second principle which is employed by nature in the structure of fish, making their bones and muscles not merely light but entirely devoid of weight [due to buoyancy].'

Appendix A.3

Leonhard Euler (1766) A simple rule to determine the strength of a bridge or similar structure, on the basis of the known strength of a model

Euler (1776). Regula facilis pro diiudicanda firmitate pontis aliusve corporis similis excognita firmitate moduli. *Novi Commentarii academiae scientiarum Petropolitanae* 20: 271-285, Plate II Figs 7-9 (see Section 4.3). [1]

This question has recently come up in the context of a permanent bridge to be built across the river Neva. There have been several approaches to this enterprise, and to this end models have been made, according to whose structure the real bridge would be built. Most people have held the opinion that the real bridge would have sufficient strength if the model had proved a certain resistance. They believed obviously that if the model had carried a load proportional to that to be carried by the real bridge, then there could be no doubt that the real bridge, built accordingly to the scale model, would have sufficient strength. However, this conclusion is erroneous, as is evident from the fact that such a bridge certainly cannot be built over an arbitrarily long span, such as one or several miles, since it would collapse due to its own weight, independently of the strength exhibited by the model. From this it is clear that the strength of the real bridge cannot be deduced from the strength of the model according to the principle of direct proportionality. Therefore, I am going to inquire in more detail how the strength of the bridge can be determined by the strength of the model.

1 Several facsimile versions are available via the Internet:
https://www.biodiversitylibrary.org/item/38593 (includes Figures 7, 8 and 9) (25/9/2019)
http://eulerarchive.maa.org//docs/originals/E480.pdf (25/9/2019) (No figures)
https://babel.hathitrust.org/cgi/pt?id=mdp.39015018393705;view=1up;seq=7 (25/9/2019) (Figures 8 and 9 on p. 733; no Figure 7).

Physical Models: Their historical and current use in civil and building engineering design,
First Edition. Edited by Bill Addis.

§2 The strength of both the bridge and the model depends on the strength of its individual parts and the way they are joined together, and has to be derived from their failure loads. However, the loads which such a machine has to carry act in two different ways, namely in tension and in bending. The first loading type occurs when ropes are applied to hold the whole machine, whereas the other loading type is predominant in the case of a machine composed of beams which resist the bending to such an extent that they do not break. Both types have to be studied in detail if we want to determine the strength with which the individual parts resist to rupture and breaking.

§3 With respect to the first type, we consider a rope with a given thickness, and we determine which amount of tension it can carry without rupture. Firstly, it is evident that this strength depends on the tenacity of the individual fibres of the rope, for it is clear that the strength of similar ropes which are composed by a similar arrangement of fibres is the higher, the bigger the section of the rope is. So if we call the section of such a rope $= cc$, then the force which it will resist is proportional to cc. Therefore, we can assume a certain length L so that the product Lcc characterizes the force which the rope will sustain. Here, it is evident that the quantity L depends on the tenacity of the fibres. Therefore, if the product Lcc would represent the weight of a rope of length L, then this would indicate the maximum length which the rope could reach if suspended vertically and carrying only its own weight.

§4 Now this length L depends on the force with which the smallest particles of the individual fibres are attached to each other, or rather, compressed to each other by some external force; if this compression results exclusively from the atmospheric pressure, which we know from two polished marble pieces which are so smooth that no air can enter the contact surface, and which are pressed together only by the barometric pressure $=k$, then we find, with the section $=cc$, that the cohesion can be overcome by a force corresponding to a mercury column of basis cc and height $=k$. Therefore, if mercury is n times heavier than the material of the rope, we obtain $\mathrm{L} = nk$, that is $=28$ inches. However, experience shows that most ropes resist a much higher load, whence it results that the reason for the cohesion is not to be sought in the atmospheric pressure, but rather in the elastic force of the aether which compresses all bodies in contact if the aether is totally excluded from the contact surface. Many experiments show that this force of the aether must be almost a thousand times stronger than the atmospheric pressure; so one can easily understand that the length L will reach a thousand feet or even more.

§5 Whatever may be the reason of the strength or cohesion of solid bodies, it is not strictly necessary for the present investigation to know the length L with absolute accuracy, since we do not consider the absolute resistance of some material from which the machine is made; rather, the present question touches only on the comparison of the strength of different ropes made of the identical material. Therefore, whatever that length L may be, if we compare two ropes which are made in the same way, one of which has the section cc, and the other $=\mathrm{CC}$, then the respective strengths will be Lcc and LCC, or their relation will be as cc:CC, that is to say the ratio of the section areas, and this relation suffices to establish the comparison of the strengths of different ropes provided that they are made in a similar way and that the forces pulling at these ropes are applied in a similar way, which we may safely assume since in such models of machines normally everything is made to perfect similarity.

§6 This said, we now consider two ropes *acb* and ACB (Plate II, Figures 7 and 8), which are not only made from similar material, but also tensioned in the same way. Let the section of the thinner one be $=cc$, and that of the thicker one $=\mathrm{CC}$, and let a denote the distance ab of the smaller one, and $\mathrm{AB}=\mathrm{A}$ the distance of the thicker one. Now let v be the force which the thinner rope carries before rupture, and V the force which the thicker one can take. Provided that the loads are applied in a similar way, it is clear that $v : \mathrm{V} = cc : \mathrm{CC}$, or $\mathrm{V} = \frac{\mathrm{CC}}{cc} v$, so that the dimensions a and A enter the computation only in the way they follow the ratio of the thicknesses.

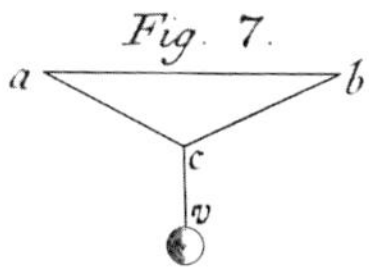

Fig. 7.

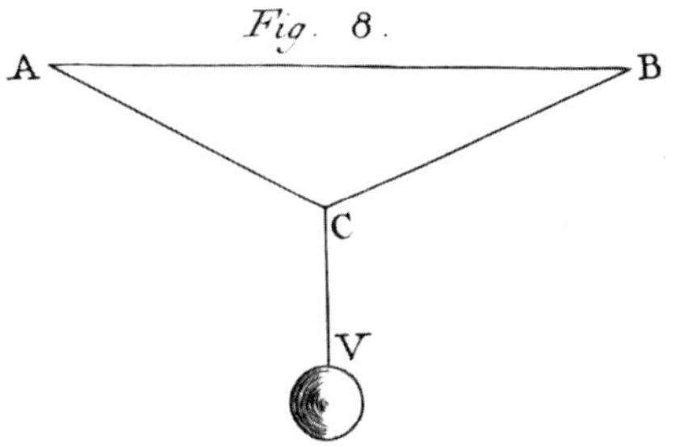

Fig. 8.

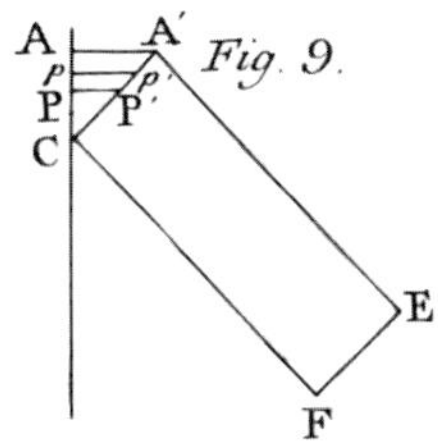

Fig. 9.

§7 The forces v and V can be split into two parts each, one of which corresponds to the weight of the rope, including the parts to which the load is attached, whereas the other part corresponds to the load to be carried by either rope, and which will be imposed or attached to it, so that we can tell the strength of the thicker one if we know the strength of the thinner one. Let the weight of the smaller rope be denoted by p, and that of the bigger one by P, then it follows from similarity that $p : \mathrm{P} = abc : \mathrm{ABC}$, or $\mathrm{P} = \frac{\mathrm{ABC}}{abc} p$. Now, let the load attached to the smaller rope at rupture be $=q$, and let us define Q to be the ultimate load of the bigger rope.

§8 Now, we have to consider both forces v and V to be composed by the weight of the rope and the load: We will have $v = p + q$ and $\mathrm{V} = \mathrm{P} + \mathrm{Q}$, for we have to add the weight of each rope and the respective load. Now, since $\mathrm{V} = \frac{\mathrm{BC}}{bc} v$, we find $\mathrm{P} + \mathrm{Q} = \frac{\mathrm{BC}}{bc}(p + q)$. Therefore, since $\mathrm{P} = \frac{\mathrm{ABC}}{abc}$, we find the load that will be carried by the bigger rope as $\mathrm{Q} = \frac{\mathrm{BC}}{bc}(p + q) - \frac{\mathrm{ABC}}{abc} p = \frac{\mathrm{BC}}{bc}(p + q) - \frac{\mathrm{A}}{a} p$. From this, it follows directly that unless $(p + q) > \frac{\mathrm{A}}{a} p$ or $q > \frac{(\mathrm{A} - a)}{a} p$, the bigger rope will not be able to carry any load at all, but will rather break under its own weight, even though the smaller rope will be able to carry a considerable load. From this, it is already evident that the machine cannot be scaled from the model by an arbitrary factor, but only to the size A, which will be $A = \frac{(p+q)}{p}a$ and beyond this the machine cannot stand.

§9 Now, it is no longer difficult to determine whether it is possible to construct a bridge with ropes across the river Neva, which is around 1000 Rhenanian feet wide [c.314 m]. To this end, a model has to be built according to some scale, say 20 feet long, which has to be perfectly similar to the real bridge, so that $\frac{A}{a} = 50$. Unless this model will carry a load which is more than 49 times bigger than its own weight, the bridge will not stand. And if this should happen to work, then it is nevertheless clear that this bridge will not work if the width of the river is bigger. Furthermore, it is not at all sufficient that the bridge will carry its own weight, but it is also necessary that it will support considerable loads, so that the value Q from our formula attains the required value.

§10 Having dealt with the strength of ropes, we now move on to investigate how beams or other rigid body resist rupture. To this end, we consider the beam ACEF (Plate II, Figure 9), which is already bent to such a degree that it barely does not break in AC. Therefore, the angle ACA′ is the maximum inflexion which it can carry, so that the extreme fibres AA′ cannot be extended further. Let this maximum elongation AA′ be denoted by l, and let us denote by L the force which holds them together; it is clear that this force L is the same that we have before considered as the ultimate force of ropes; therefore, if the whole beam would be pulled by a force acting in the direction AE, then the whole would not be pulled apart if the force would =L.

§ 11. Now, we denote the thickness of the beam AC=c, while the width of the beam, when it starts to break away around the point C, will be denoted by =b. Now, we denote the indeterminate abscissa from A in direction to C by CP=x, where the elongation of the fibres will be $PP' = \frac{lx}{c}$; the contracting force of these fibres will be smaller than in the extreme fibre AA′, so that this force will be $\frac{Lx}{c}$. Multiplying this by the element Pp=dx, we obtain the elementary force $\frac{L\,x\,dx}{c}$, which we have to multiply by the width of the beam, so that we get $\frac{L\,b\,x\,dx}{c}$ [2]. However, since the breaking acts around the axis C, the moment of this elementary force with respect to the axis is $\frac{L\,b\,x\,x\,dx}{c}$, yielding the integral $\frac{Lbx^3}{3c}$. Inserting the total height of the beam, x=c, we obtain the total moment of the forces to which the beam will resist, which is equal to ⅓Lbcc. This expression contains four dimensions, three of which describe a volume whose weight equals to a breaking force, while the fourth dimension corresponds to the length of a lever at whose end the force is applied. Therefore, the beam will certainly break if the moment of all the applied forces with respect to the point C exceeds the value ⅓Lbcc from the formula.

2 dx is the differential of x.

§12. Now, if in the model the dimensions *b* and *c* refer to the single small beams whose strength has to resist the rupture, then we will assume that the same letters B and C will denote the corresponding dimensions of the beams from which the real bridge is made. The moments of the forces which the model, or the bridge, respectively, will resist will be as *bcc* to BCC. Here, it has to be noted that the forces are acting in the same sense along the thickness AC. By contrast, if the forces act on the width of the beam, which we denote by =*b*, then the moments of the forces will be as *bbc* to BBC.

§13. Now, let the length of the model =*a*, of the bridge =A. Then *v* denotes the sum of all forces which the model is able to carry without breaking, while V denotes the sum of all forces whose action [the bridge] will withstand [3]. In both cases we assume that these forces are distributed over the length of the model and the bridge in the same manner. Hence it is clear that the moments of all these forces in the model and in the bridge follow the ratio *av*:AV, and we conclude from all this, that it must hold that

$$av : \mathrm{AV} = bcc : \mathrm{BCC}$$

and hence that $$v : \mathrm{V} = \left(\frac{bcc}{a}\right) : \left(\frac{\mathrm{BCC}}{A}\right)$$

§14. Next, we consider the vertical forces which the bridge has to carry, and which are due to the weight of the bridge, and the loads which it will withstand; we denote the thickness of the beams which resist the vertical loads by the letters *c* and C, while the other dimension *b* describes their width. Firstly, we state that the weight of the model which has to be carried is =*p*, while the additional payload is =q, which may easily be determined by experiments by distributing along its length increasing loads, until it begins to cede. For the bridge itself the letter P will denote the weight, while Q is the sum of all payloads which the bridge has to support; with these definitions, we arrive at

$$v = p + q \quad \text{and} \quad \mathrm{V} = \mathrm{P} + \mathrm{Q}$$

so that we obtain,

$$(p + q) : (P + Q) = \left(\frac{bcc}{a}\right) : \left(\frac{\mathrm{BCC}}{A}\right)$$

and for the bridge itself,

$$(\mathrm{P} + \mathrm{Q}) = \left(\frac{a\mathrm{BCC}}{\mathrm{A}bcc}\right)(p + q)$$

From this equation we can determine the sum of all loads which the bridge will support; if it is smaller than required, than we can boldly state that the bridge constructed according to the model will not have sufficient strength.

§ 15. Therefore, if we assume that the bridge itself is made according to the model, in a similar structure and from a similar material, its weight P as compared to the weight of the model will be as ABC to *abc*. Therefore, from the known weight *p*

3 In the original, this half-sentence lacks the predicate 'the bridge'.

of the model, we obtain the weight of the bridge itself as $P = \left(\frac{\text{ABC}}{abc}\right) p$; from this it follows that $\text{Q} = \left(\frac{a\text{BCC}}{\text{A}bcc}\right)(p+q) - \left(\frac{\text{ABC}}{abc}\right)p$. This rule permits us to derive the total load Q which the bridge itself will be able to sustain, from the strength of the model, which has to be determined by experiments. From what we have said before, it follows that we have to take care that the negative part $\left(\frac{\text{ABC}}{abc}\right)p$ does not exceed the positive part $\left(\frac{a\text{BCC}}{\text{A}bcc}\right)(p+q)$, since otherwise the bridge simply will not stand, but collapse under its own weight. For this not to happen at all, it is required that $p + q > \left(\frac{\text{A}ac}{\text{aaC}}\right)p$ or $q > \left(\frac{\text{A}ac}{\text{aaC}} - 1\right)p$.

§ 16. So far, we have assumed that the bridge is in each and every dimension similar to the model. (For it is easy to understand that our formula even holds if the three major dimensions A, B, C do not hold the same ratio with each other as the three minor dimensions a, b, c as long as the difference is not too big). So we state that the individual dimensions of the model with respect to those of the bridge itself obey the ratio 1:n, so that $\text{A} = na$, $\text{B} = nb$, and $\text{C} = nc$. Therefore, we obtain the following equation for the load, $Q: \ \text{Q} = n^2(p+q) - n^3 p$ or $\text{Q} = nn(p + q - np)$. From this it follows that a necessary condition for the bridge to stand is $q > (n-1)p$, since otherwise it will collapse due to its own weight. Furthermore, it is required that the load q which is carried by the model exceeds this limit $(n-1)p$ the more the higher the load Q is which the real bridge is supposed to carry. For example, if $n = 30$, it is required that $q > 29p$. If $q = 30p$, then $\text{Q} = 900p$; if, however, $q = 31p$, then we would get $\text{Q} = 1800p$; and if $q = 32p$, then $\text{Q} = 2700p$, and so on. From which one may understand that if q is somewhat bigger than $29p$, then the bridge will already carry a significant load; namely, for each increment the value increases by $900p$.

§ 17. However, if the bridge constructed in perfect similarity to the model does not provide the required strength, then we can at least remedy this defect by increasing the thickness C of the beams beyond the ratio 1:n. Even though the similarity to the model does not persist in this case, the conclusions which have led to our formulae still hold without restriction, at least as long as the dissimilarity is not enormous and the loads are still distributed evenly along the length of the bridge. No matter how we compare the bridge with the model, it is not required that all the three dimensions a, b, c, and A, B, C all hold the same ratio between each other.

§ 18. Therefore, the length ratio $a : \text{A} = 1 : n$, or $\text{A} = na$ is maintained, and also the width ratio $\text{B} = nb$, whereas we fix the height ratio at $C = zc$, then our formula for the payload (§ 15) then assumes the following form: $\text{Q} = zz(p+q) - nnzp$. From this equation we can determine the unknown ratio z if the payload Q to be carried by the bridge is given, then $z = \dfrac{nnp + \sqrt{n^4 pp + 4(p+q)\text{Q}}}{2(p+q)}$. Therefore, if the construction of the bridge permits increasing the beam thickness to this extent, then the bridge will have sufficient strength.

§ 19. We are going to illustrate this formula with some examples. We assume that the width of the river is thirty times as big as the span of the model, that is n=30. We assume, in addition, that the bridge has to have such a strength that it carries the load $Q = 3600p$; however, the model has been found to carry only a load $q = 30p$, so that the bridge built in perfect similarity to the model would just carry Q = 900p. Therefore, we set, in our formula, $q = 30p$ and $Q = 3600p$, and we find

$$z = \frac{900 + \sqrt{900^2 + 4.31.3600}}{2.31}, \text{ which reduces to}$$

$$z = \frac{900 + 30\sqrt{900 + 16.31}}{2.31}, \text{ whose value is } z = 32.595.$$

Therefore it becomes evident that in this case the bridge will gain sufficient strength if the thickness of the beams is $32\frac{3}{5}$ times as big as in the model, whereas the two remaining dimensions only have to be thirty times as big; in this case, the deviation from the similarity will not arouse any concern.

§ 20. However, this remedy can also be applied if the payload q determined by the experiments is found to be considerably less than 29p. To show this in general, we assume that $q = (\lambda - 1)p$, while the required payload remains at $Q = 3600p$. Therefore, since $p + q = \lambda p$, we get

$$z = \frac{900 + 30\sqrt{900 + 16\lambda}}{2\lambda} = \frac{15}{\lambda}(30 + \sqrt{900 + 16\lambda}).$$

In the following table we show the values of z resulting from this formula, beginning with the value $\lambda = 33$, and continually decreasing:

λ	$q = (\lambda - 1)$	$z = \frac{15}{\lambda}(30 + \sqrt{900 + 16\lambda})$
33	32p	30.81
32	31p	31.68
31	30p	32.85
30	29p	33.58
29	28p	34.62
28	27p	35.74
27	26p	36.61
26	25p	38.23
25	24p	39.63
24	23p	41.14
23	22p	42.79
22	21p	44.58
21	20p	46.54
20	19p	48.70

§ 21. From this table we understand that if the strength of the model attains only $q = 20p$, in which case the construction of the bridge would seem to be rejected, it suffices to give it sufficient strength by just increasing the thickness of the beams according to the ratio 1:48.70. Similarly, one can easily determine also in other cases when bridges are proposed to be built in one or some other way, if they will have sufficient strength or not, and whether, in the latter case, their strength can be increased arbitrarily.

§ 22. However, since bridges also have to withstand sufficiently to the impact of the winds, it is also necessary to expose the model to an extremely strong wind. For if it will withstand this impact, then we can also be sure that the bridge itself will not suffer any damage from such a wind. Namely, the wind load is proportional to the surface area, so that it holds the ratio $1{:}nn$, and since the resistance to the wind increases in the same ratio, it is clear that the bridge itself will sustain the same wind loads as the model.

§ 23. However, if it should turn out that it has to be feared that the bridge will not have enough strength, then its resistance to wind loads can easily be increased, if a larger width of the beams can be accepted. Furthermore, it would be useful to give the bridge a greater width at both ends than towards the middle, so that it will act against the wind so to speak like a horizontal arch, which will be much more stable towards the wind. It is also clear that nothing has to be feared if we add additional width to a bridge which has already been found to have sufficient strength before.

Translation by Stefan Holzer.

Appendix A.4

Extract from report: Telford's design for new London Bridge (1801)

Editor's note.

One of many subjects addressed by the Select Committee considering further Measures for Improvement of The Port of LONDON was the need for a new bridge to replace old London Bridge which was a hindrance to shipping. In 1800, designs were submitted by several engineers including a scheme proposed by Thomas Telford and mechanical engineer James Douglas for a cast-iron arch with a single span of 600 feet across the River Thames (see Figure 2.23). Such was the boldness of this proposal that the Select Committee was charged with investigating its viability.

Seventeen eminent mathematicians, scientists and engineers were invited to answer twenty-one questions to help the Select Committee make their recommendation. Of particular interest for this book are two questions relating to the use of models to test the feasibility of the bridge. Below are reproduced the responses of the experts to these two questions. After these responses, is reproduced the responses to these same two questions given by Thomas Young as part of his article 'Bridge' in a supplement to the Encyclopaedia Britannica published in 1824.

The responses are of particular interest because they constitute a very rare opportunity to see contemporary opinions of scientists and engineers about the use of models in 1800. They are mainly illuminating because of their lack of scientific content and clarity, as indeed Young himself noted: 'The results of these inquiries are not a little humiliating to the admirers of abstract reasoning and of geometrical evidence; and it would be difficult to find a greater discordance in the most heterodox professions of faith, or in the most capricious variations of taste, than is exhibited in the responses of our most celebrated professors, on almost every point submitted to their consideration'.

Physical Models: Their historical and current use in civil and building engineering design,
First Edition. Edited by Bill Addis.

Nevertheless, we do find in some responses, especially of Charles Hutton and Thomas Young, some comments about the viability of using some aspects of the behaviour of a small-scale model structure to be a reliable predictor of behaviour at full scale, and others not. Also, that some behaviour was independent of size, some could be scaled up linearly, and some according to more complex relationships. However, except in the case of Young, these ideas went little further than Galileo's square-cube law. In 1800, civil and building engineers were still a very long way from reaching a good understanding of the effects of scale.

Bill Addis

Report (1801) from Select Committee upon further Measures for Improvement of The Port of LONDON. In: *Reports from the Committees of the House of Commons,* Vol.XIV (1803), 604-635. London: HMSO. Appendix, 606-635.

'The obvious Advantages which would be obtained if the Communication could be effected by Means of a Single Arch, as well as the Magnificence of the proposed Structure, appeared to give the latter Design a particular Claim to the Notice of Your Committee; yet the Attempt was of so novel a Nature, that they thought it absolutely necessary for their own Information, as well as for the Purpose of affording some Grounds upon which the House might hereafter form their Judgment as to its Expediency, to request the Opinions of some of the Persons most eminent in Great Britain for their Theoretic as well as Practical Knowledge of such Subjects [seventeen in number]. For this Purpose, the following [twenty one] Queries were drawn up and transmitted, together with the engraved Design of Messrs. Telford and Douglass, and the Explanatory Drawings, annexed to this Report, to such persons as were supposed to be most capable of affording them Information. …

VIII. Is it necessary or advisable to have a model made of the proposed bridge, or any part of it, in cast iron? If so, what are the objects to which the experiments should be directed; to the equilibration only, or to the cohesion of the several parts, or to both united, as they will occur in the intended bridge?

IX. Of what size ought the model to be made, and what relative proportions will experiments, made on the model, bear to the bridge when executed?

The responses of the invited experts to the two questions concerning models were as follows:

Dr. Nevil Maskelyne, F.R.S., Astronomer Royal, Greenwich.
VIII. I think it might be useful to have an exact model of the present bridge, as soon as the plan of it shall be in some measure settled, or of some part of it, made of cast iron, for trying experiments on; and that the objects to which the experiments should be directed, should be the cohesion of the parts, and the capacity of it to sustain a shock from a body in motion falling perpendicularly against the arch.

IX. – If the model could be made use of afterwards in some of the roadways about the bridge, it might then be worth-while to make it of a considerable size, with its linear dimensions 1/10th or 1/12th of those of the intended bridge, or the model may be made at pleasure with the dimensions, suppose 1/30th or 1/50th. The model will be stronger than the bridge, relative to supporting its own weight, in the same proportion in which the linear dimensions of the bridge are greater than those of the model. For example, if the linear dimensions of the bridge are ten times those of the model, and the bridge may be just broken by its own weight, the model will require ten times its own weight to break it. As to the effect of a weight, suppose a loaded waggon going over the bridge, taking that to be 6 or 8 tons, it will be only 1/1000th of that of the bridge, and its stress on the middle of the bridge will be 1/1000th part of the stress of the whole weight of the bridge there; but it will be easy to make the bridge strong enough to bear 1/1000th more than its own weight.

Rev. A[braham] Robertson, Savilian Professor of Geometry, Oxford.
No response.

Mr. [John] Playfair, Professor of Mathematics, Edinburgh.
VIII. Though a great deal of benefit might accrue to the theory of arches in general, from a well-constructed model of an iron bridge, adapted to the making of experiments, yet I cannot say that I think such a model indispensable with respect to the present bridge. If it were so contrived indeed as to afford the means of measuring the pressure on the abutments, or on other sections of the bridge, it might lead to very important discoveries, and enable us to compare the conclusions of theory with matter of fact. This advantage is, however, too remote and contingent to authorize the recommendation of such experiments as a *necessary* preparation for executing the intended bridge over the Thames. Should they be undertaken, the particular objects of them, and the manner in which the model ought to be constructed, will require to be carefully considered, and would lead to longer discussions than can find a place in this Report.

Mr. J[ohn] Robeson [Robison], Professor of Natural Philosophy, Edinburgh.
VIII. Were I to make a model, it should be of plaister of Paris. I think that such a model would be instructive. One rib is enough, giving it considerable dimensions transversely; I would cast it in one piece, that it may be uniform in its texture. I would set it up between two planks, thus [Fig], and I would pour small shot on it, till it begins to crush, and would notice the progress of failure. I would cut another cast into sections, by passing a fine spring saw along certain radial joints. I would set up this model, by putting slips of card paper in the joints, to make up for what the saw takes out. I would try this model in the same manner, and I should expect useful information from the way in which they failed in the different parts. I apprehend that it will appear, that equilibration is very little concerned, in any other way than what I mentioned at first. A profile might also be true like the sketch B [missing], where the haunches consist of uprights and braces, all acting like the braces of a truss roof. A friable model like this will shew the manner of failing at the crown which I dread, from compression, or from yielding of the abutments; and, by making a middle piece with circular joints, it would be seen how far my proposed remedy is effective.

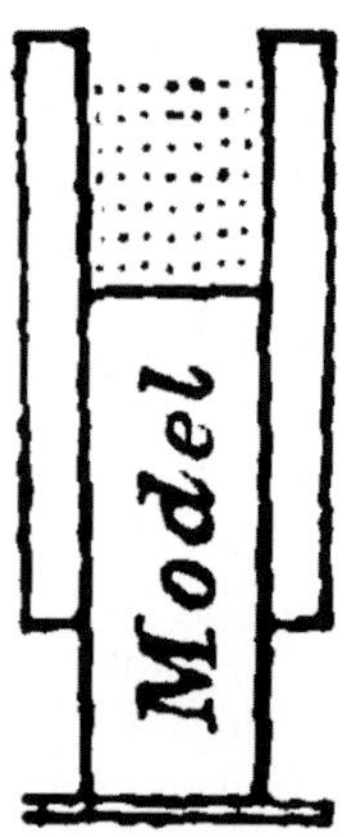

Dr. [Isaac] Milner.
[Mathematician and scientist; one-time Jacksonian Professor of Natural Philosophy at University of Cambridge]
Dr. Milner has been informed, that there is in London a model of the intended bridge; and if so, a careful inspection of it may afford much more light than can possibly be obtained from drawings; but he is still of opinion, that in a matter of

such difficulty and importance, no help of that kind, nor indeed any help whatever, can supersede the necessity of a consultation.

Dr. Charles Hutton, Royal Military Academy, Woolwich.
[Professor of mathematics]
VIII. It appears to be very advisable to have a model made of the whole of the proposed bridge, in cast iron, as well for the greater safety and satisfaction, as for the benefits and improvements to be derived from the experiments to be made with it, and from the experience and knowledge derived from the casting and making it. The objects to which the experiments should be directed, might be, the equilibrium of the whole, the cohesion and fitting of the several parts, the effects of a vertical load on every part separately, and the effects of a horizontal blow or shock against every part in the side of the arch. Also what weight the model frames require to break or to crush them.

IX. The greater the size of the model, the more satisfactory the experiments and conclusions will be. For this purpose, it seems advisable, that the model be not less than the twentieth part [of] dimensions of the bridge, that is, of thirty feet in length. Now, as the solid contents of similar bodies are in the same proportion as the cubes of their linear dimensions, such a model would require only the eight-thousandth part of the weight [of] metal in the bridge, because the cube of 20 is 8,000. So that, as it is estimated, the bridge will require 6,500 tons of metal, it follows, that about three quarters of a ton weight of metal will suffice for the model of thirty feet in length. As to the relative proportions of experiments made with the model; those relating to the equilibrium will be in the same direct proportion with the masses of the model and bridge, as well as those relating to loads or shocks. But the strength of any particular bar or frame will be only as the square of the scantling, while the stress upon it is in the same proportion as the length.

Mr [George] Atwood, of Sloane-street, Knightsbridge.
[Eminent mathematician and physicist, and Fellow of the Royal Society]
VIII. It would be very satisfactory to verify experimentally the principle of adjusting the equilibrium of the arch, and to estimate the degree of force which the fastenings would be able to resist; but if these experiments should be made on the same model, it might be difficult to distinguish sufficiently the effects arising from the adjustment of equilibrium, and any defect thereof, from those which are produced by the fastenings; for which, and other reasons, it appears to me, that more satisfactory results would be the consequence of making experiments for verifying the principle of equilibrium, and on the strength of the fastenings, independently of each other. Every weight which is occasionally brought to press on any part of an arch, gives an impulse to the sections toward a position different from those which they occupy, and causes a stress on the fastenings, applied in order to prevent any such change of place. A similar stress on the fastenings would be the consequence of a want of exact equilibrium among the sections, with this difference; the stress which is caused by occasional weight is temporary, continuing only while the additional weight presses upon the sections; but the stress arising from the want of equipoise in the original construction acts incessantly, and in a course of time may produce a change in the form of the arch,

which would resist without injury the impulses of a greater force that acts upon it during a small interval of time.

IX – No response

Colonel [William] Twiss of Woolwich.
[Eminent military engineer]
VIII. It appears advisable to have a model made of the proposed arch in cast-iron, and the great object will be to ascertain the best mode of casting and putting together the several parts; it will also be a proof of its equilibration, though of that I should have no doubt if calculated by Dr. Hutton.

IX. I should suppose a model fifty feet long, that is, on a scale of one inch to a foot, would answer; and also, that if the real arch is executed equally as well as the model, the proportion of the weight they will bear, will be as the experiment made with timber of the same proportional dimensions. Should the model be so constructed that it could easily be widened to ten or twelve feet, it might sell, as a permanent bridge, for nearly what it cost.

Mr William Jessop, of Newark, Nottinghamshire.
[Eminent canal engineer, bridge builder and early pioneer of railways]
I do not think that any satisfactory results can be deduced from experiments made on a model.

Mr J[ohn] Rennie of Stamford Street, Surrey Road.
[Eminent canal engineer and bridge builder]
VIII. & IX. As the durability of bridges depends not only on the perfection of their construction, but also on the tenacity of the materials of which they are composed, I am of opinion no very satisfactory information can be had from a model, unless made on a large scale. In a work of this magnitude and importance, it should be at least 1/20th part of the size at large, and would be better if made much larger. The experiments should be directed both to the equilibration and cohesion.

Mr James Watt, of Heathfield, near Birmingham.
[Engineer and pioneer of the steam engine]
No response.

Mr John Southern, Engineer, Soho near Birmingham.
[Engineer at Boulton and Watt's works, later becoming a partner of the firm.]
VIII. It will not be possible to make a model of cast iron precisely similar to the bridge, but upon a scale which will be deemed too large and expensive, because the strength of that metal cannot be judged of comparatively, if less than an inch square, which would call for a model whose span was 75 or 100 feet, perhaps; but one might be made in such a manner as to satisfy the Committee what load would break up the bridge, if the opinions which they may receive upon this subject should leave a doubt in their minds. This I would propose to effect by putting the metal which belongs to all the ribs into one, or at most two ribs of the model.

I am decided in the opinion, that the arch should be so constructed as to render it impossible for any weight that can actually be brought upon the bridge to bring on any degree of *tension* upon the parallel limbs of the frames, and therefore cohesion is not an object of the model, so far as tension is concerned; neither is the equilibration to be ascertained by a model, which at the same time shall be capable of determining obviously the largest extra load the bridge will sustain, or just be broken up by.

In fact, the object for which any model should be made, depends upon the doubts which the Committee may entertain on any point, and it should be constructed with the view of determining the question of doubt.

I think it an object worthy the attention of the Committee, to ascertain what weight pressing end-wise upon a rod of cast iron (say of an inch square and a foot long) will crush it or destroy its cohesion, if it does not appear to them that the point has already been decided.

IX—The model should not have a span of less than 30 feet, if made of two ribs, for ascertaining the force that will break up the bridge; and the force in large will be to that in small, as the whole weight of the bridge is to the whole weight of the model; the altitude of the arch frames being proportional to the span, and the bridge and model loaded similarly and proportionally previously to the extra load being applied.

Mr. William Reynolds, of Colebrook Dale, Shropshire.
[Iron master and one-time partner of Abraham Darby]
VIII & IX. I would by all means advise a model to be made of cast iron, which it is very likely would throw some light on many points necessary to be known, previous to the construction of so great a work; as to the size of the model, that must be determined by the engineers.

Mr. John Wilkinson, of Bradley, near Wednesbury, Staffordshire.
[Iron master, known as John 'Iron-Mad' Wilkinson, and prime mover in constructing Ironbridge, completed in 1779]
VIII & IX. It appears to me, that in an undertaking like the present, it is not only advisable, but absolutely necessary, that a model should be made, which would point out more clearly than can possibly be done by theory, if any and what alterations would be requisite. If one upon a scale of 3/4 of an inch to the foot [1:16] was put up, which would be a span of 37 1/2 feet, it would answer every purpose, and an opinion could then be formed, and a determination made with a degree of certainty, which cannot now be done.

Mr Charles Bage, Shrewsbury.
[Designer of the Ditherington flax mill at Shrewesbury, the first cast-iron frame building, 1796]
No response.

General [Samuel] Mr Jeremy Bentham, Inspector General of the Naval Works of the Admiralty.
No response.

Mr T[homas] Wilson.
[Bridge engineer; co-designer of Wearmouth Bridge, Sunderland, 1796)]
No response.'

Young, Th. (1824). Article on 'Bridge'. In: *Supplement to the fourth, fifth and sixth editions of the Encyclopaedia Britannica*. Vol. 2: 497-520, plates. 42-44. (Written in 1816). Edinburgh: Archibald Constable. Republished in part in: Peacock, G. (ed.) (1855). *Miscellaneous Works of the late Thomas Young*. London: John Murray. Vol. 2: 194-247.

'But the results of these inquiries are not a little humiliating to the admirers of abstract reasoning and of geometrical evidence; and it would be difficult to find a greater discordance in the most heterodox professions of faith, or in the most capricious variations of taste, than is exhibited in the responses of our most celebrated professors, on almost every point submitted to their consideration. It would be useless to dwell on the numerous errors with which many of the answers abound; but the questions will afford us a very convenient clue for directing our attention to such subjects of deliberation as are really likely to occur in a multiplicity of cases; and it will perhaps be possible to find such answers for all of them, as will tend to remove the greater number of the difficulties which have hitherto embarrassed the subject.' [Young 1824: 508]

'Question 8.
Experiments on the equilibration of the arch would be easy and conclusive; on the cohesion or connexion of the parts, extremely uncertain; the form and proportion of the joints could scarcely be imitated with sufficient accuracy; and since the strength of some of the parts concerned, would vary as the thickness simply, and that of others as the square or cube of the thickness, it would be more difficult to argue from the strength of the model upon that of the bridge, than to calculate the whole from still more elementary experiments. Some such experiments ought, however, to be made, on the force required to crush a block of the substance employed; and the form calculated to afford the proper equilibrium, might be very precisely and elegantly determined, by means of the method first suggested by Dr Hooke, that of substituting for the blocks, resting on each other and on the abutments, as many similar pieces forming a chain, and suspended at the extremities. It would, however, be important to make one alteration in the common mode of performing this experiment, without which it would be of little or no value; the parts corresponding to the blocks of the arch, should be formed of their proper thickness and length, and connected with each other and with the abutments by a short joint or hinge in the middle of each, allowing room for a slight degree of angular motion only; and every other part of the structure should be represented in its proper form and proportion and connexion, that form being previously determined as nearly as possible by calculation; and then, if the curve underwent no material alteration by the suspension, we should be sure that the calculation was sufficiently correct; or, if otherwise, the arrangement of the materials might be altered, until the required curve should be obtained; and the investigation might be facilitated by allowing the joints or hinges, connecting the block, to slide a little along their surfaces, within such limits as would be allowable, without too great a reduction of the powers of resistance of the blocks.

Question 9.
The size is of little importance, and it would be unsafe to calculate the strength of the bridge from any general comparison with that of the model. There is an Essay of Euler in the Nexo Commentaries of the Royal Academy of Petersburgh (Vol. XX. p. 271) (see Chapter 4 and Appendix A3) relating expressly to the mode of judging of the strength of a bridge from a model; but it contains only an elementary calculation, applicable to ropes and simple levers, and by no means comprehending all the circumstances that require to be considered in the structure of an arch.' [Young 1824: 513-514]

Appendix A.5

Experimental Models, *The Builder* (1846)

Anon. (1846) Experimental models. *The Builder* 4 (193, 17th Oct.): 500.

EXPERIMENTAL MODELS

'Parvis componere magna' [1]

The strength and weight of the building materials at our command being known, the proportions of the parts of a proposed structure, necessary to ensure its stability under the greatest probable stress to which at any time it may be liable, as with regard to a bridge, from a load upon it, may be calculated, sometimes with mathematical nicety [2].

If the design be unusual and complicated, and change of form or motion is likely to take place from causes extraneous of the structure, as in a suspension-bridge from the action of a load passing over it, or of the wind, it is well to use a model, as by its means the changes and vibrations of the proposed structure may be more satisfactorily inferred than from the most laborious calculations.

1 [Editor's comment] From Virgil, *Eclogue 1. Sic canibus catulos similes, sic matribus haedos noram, sic* parvis componere magna *solebam.* Just as I know puppies resemble dogs, and kid goats their mothers, so I used *to compare small things with large ones.*
2 A practical illustration of this is, that an arch constructed of loose cards, packed vertically between upright abutments, may be just made to stand without any change of form taking place when the centre is removed. This occurs when the curve of equilibrium cuts the middle of the depth of the arch at the vertex, and the extrados at the abutments. The force by which the arch is sustained is the elasticity of the cards producing sufficient pressure against the abutments to prevent the arch from sliding down them. The amount of force necessary to effect this must be first ascertained by finding the coefficient of friction of the cards and the rise and span of the arch must be adjusted accordingly. The resistance of the cards is just equal to the elastic force by which they are compressed; the friction generated prevents their slipping. If an arch constructed with the same span but greater rise than this will stand, nevertheless, as soon as the centre is removed, it will change form; the cards will be compressed at the extrados, loose at the intrados towards the abutments, and the reverse at the crown; thus showing partial grains, which in construction should be avoided as much as possible.

Physical Models: Their historical and current use in civil and building engineering design,
First Edition. Edited by Bill Addis.

The model may be made of materials the same as – or different from, if more convenient – those intended to be used on the larger scale; but in either case regard must be had to the relative strengths and weights of the materials when put together in the proposed manner, if we wish to make a just comparison of the little with the great structure.

In masses of similar forms and materials, but of dissimilar dimensions, the strengths are as the squares, but the weights (and of course the strains depending upon the weights) are as the cubes of the respective dimensions. For instance, with respect to two rectangular beams laid horizontally, the first 1 foot deep, 1 foot wide, and 20 feet long between the supports, the other 2 feet deep, 2 feet broad, and 40 feet long, the dimensions are as 1 to 2; the strengths are as the number of parts in the cross section which oppose extension, that is, as the areas which are as the squares of 1 and 2 or 1 to 4; the cubical contents, and, therefore, the weights are to each other as the cubes of 1 and 2 or 1 to 8. It thus appears, that while the resisting power and weight (or straining force) of the smaller beam, are to each other as 1 to 1, the resisting force and weight of the greater beam are to each other as 4 to 8, or as 1 to 2. If we suppose that the latter is just so strong as not to bend with its own weight alone acting upon it, the former would bear a load equal to its own weight, uniformly distributed over it, without bending; for then the resisting power and straining force of each would have the same proportion, viz. 1 to 2. In experimenting with a model, care must be taken that this condition is fulfilled.

If we proposed to construct a suspension bridge, or say for the novelty, a tubular bridge, of 450 feet span, with 500 tons iron, and an experimental model of 45 feet span, the dimensions being as 1 to 10, the cross sections of the great tube would be 100 times greater than those of the model; its weight would be 1,000 times greater, which numbers also are as 1 to 10. To put the model, therefore, upon equal terms with the projected bridge, it would have to be loaded with 9 times its own weight, or with $500/1000 \times 9 = 4\frac{1}{2}$ tons; and the same allowance would have to be made for all external disturbing causes, as the weight of carriages, &c.

If we were to form our model of tin, as more manageable on a small scale, we must proceed in the same way to reduce the strengths and strains to the same ratio. The relative strengths of tin and iron to resist extension, are nearly as 1 to 4, and their specific gravities nearly equal. Making the respective dimensions as 1 to x, the strengths are as 1^2 to $4x^2$, and the strains as 1^3 to x^3.

$$1^2 : 1^3 :: 4x^2 : x^3$$

$$x^3 = 4x^2$$

$$x = 4$$

Whence it appears that if the dimensions were as 1 to 4, a bridge of tin of 450/4, or 112.5 feet span, would be nearly upon the same terms with regard to strength and strain as an iron one of 450 feet span. If we were to make our model of 112.5/3, or 37.5 feet span, we should have to add in our trials only twice its own weight, &c.

If we were to construct the model of lead, the strengths of lead and iron to resist extension being nearly as 1 to 10, and their specific gravities as 11 to 7.

$$1^2 \; : \; 11 \text{ x } 1^3 \quad :: \quad 10x^2 \; : \; 7x^3$$

$$7x^3 = 11 \text{ x } 10x^2$$

$$x = 110/7 = 15.7$$

The dimensions in this case are as 1 to 15.7. The span of the model being made 450/15.7, or 29 feet nearly, no extra loading would be required.

The resistance which bodies offer to extension has been alone referred to; the same reasoning, however, applies to the resistance which they offer to compression. A bridge built of chalk might be made to indicate the strength of one built of granite, if the qualities of the two materials were taken into account.

Appendix A.6

On model experiments, *The Civil Engineer and Architect's Journal* (1847)

Anon. (1847) On model experiments. *The Civil Engineer and Architect's Journal* 10 (November): 352-353; (December): 398.

ON MODEL EXPERIMENTS

Until this present era of the 'railway and the steam-ship and the thoughts that shake mankind', the studies of the engineer, like those of the lawyer, were confined to the acquiring of details of precedent, while the knowledge of the scientific principles of his profession was neglected as of comparatively little importance. Thus, although the recognised modes of construction were numerous, the laws of structural equilibrium were few and but imperfectly developed; the builder was satisfied if the edifice he was about to raise were similar in character and magnitude to others which had been raised before; – this fact at least he knew – they had been found to stand, perhaps for ages; and the most ordinary, *pariter paribus*, style of reasoning might be sufficient to assure him that his own work would be no exception to the general rule. But with the railway arose a new epoch in the history of engineering: works were required to be constructed of unprecedented magnitude and solidity, and for the execution of which a higher amount of mechanical science and a wider range of experience were required. To supply the latter of these two desiderata, numerous experiments have been conducted of late years on the strength of materials, and on models of the whole or most important component parts of proposed structures. It is to these last class of experiments we would now direct the reader's attention, particularly with reference to the difference of amount of thrusts and strains in the model and its original. We shall first consider the case of a simple horizontal girder, composed of a web and an upper and lower flange, and loaded with a given weight; – to this case may be referred almost all the cast iron railway bridges now completed, as well as the proposed tubular Menai Bridge.

Physical Models: Their historical and current use in civil and building engineering design,
First Edition. Edited by Bill Addis.

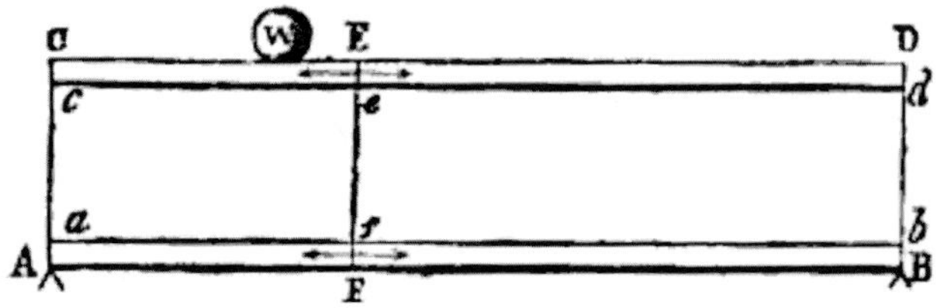

Let AB be the girder, supported at A and B, and composed of the lower flange AB*ab*, the upper flange CD*cd*, and the web *cabd*. Let AB = *au*; AC = *bu*; A*a* = *cu*; C*c* = *du*; C*w* = *hu*.

Let the weight at $w = wu^2$ (because the weight varies as the cube of the scale u); the weight of beam and girder = $w'u^2$; CE = *lu*; R and R' the reactions at A and B.

Then first considering the equilibrium of the whole girder, we shall have

$$R + R' = (w + w').\, u^3$$

$$R'.\, a\, u = \left(w\, h\, u + \frac{w' a\, u}{2}\right).\, u^3;$$

the girder being supposed uniform and symmetrical throughout its mass.

$$\therefore R = \left\{ w\left(\frac{a-h}{a}\right) + \frac{w'}{2} \right\}.\, u^3$$

For the equilibrium of the portion CF, if T be the tension and the thrust of the lower and upper flanges, Y the vertical force at F, *z* the distance between the points of application of the thrust and tension, we have

$$T = T;\quad Y + R = \left(w + \frac{l}{a} w'\right).\, u^3 \text{ and}$$

$$Y\, l\, u = \left(w\, h\, u + \frac{w'.l^2 u}{2a}\right).\, u^3 - Tz$$

$$\text{Whence } Tz = \left\{ \left(\frac{a-l}{a}\right).hw + \left(\frac{a-l}{2a}\right).\, lw' \right\}.\, u^4$$

Now, in applying the results of experiments upon a model girder to its original, all we have to do is to vary the scale u from the scale of the model to the scale of the original. Consequently, we find that T*z* varies as the fourth power of the scale or dimensions of the girder. If the web of the girder be very thin compared with the breadth of the flanges and their vertical depth, – and if their vertical depth, *uc*, *ud*, be small compared with *ub*, – and if *uk*, *uk'* be the width of the upper and lower flanges respectively, – t and t' their thrusts and tensions per square inch respectively – then we shall have

$$T = t\, k\, c\, u^2 = t'k'd\, u^2 \text{ nearly;} \quad \text{and } z = a\, u \text{ nearly.}$$

$$\therefore Tz = t\, a\, k\, c\, u^3 = t'k'a\, d\, u^3 \text{ nearly}$$

:. t and t́ both vary as u nearly; that is, approximately, the tension per square inch on the upper lower flange, and the thrust per square inch on the upper flange, of all similar and similarly loaded girders varies as their scale of linear dimension. This we consider so important a fact, that we shall endeavour to give a proof of it in popular language.

Suppose a vertical section made of a loaded girder at EF; then supposing F the fulcrum about which the mass AE is turned, – AE will be prevented from turning about F by the opposite action of the tension at F and reaction at A, and the weight w and thrust at E, and the weight of AE collected at the centre of gravity of CF. Now the weight w, the weight of CF, and the reaction at A, will all vary as the cube of the dimensions of the girder, if we suppose the girder loaded proportionally to its mass. And the leverage of these forces varies as the linear dimensions of the girder; consequently, their moment about F varies as the fourth power of the dimensions of the girder; therefore, the moments of the tension and thrust at F and E, vary as the fourth power of the dimensions of the girder; therefore, if we suppose Aa and Cc small, the tension and thrust vary as the cube of the scale; but as the tension and thrust are composed of the sum of all the tensions and thrusts per square inch at a vertical section of the flanges, and as the area of this vertical section varies as the square of the scale, in order to make up the fourth power, we must have the tension per square inch varying as the scale u.

We next propose to determine the amount of the load Vu^2 which can be supported at the centre of a girder of the dimensions u, in order that t and $t́$ at the centre may be the same as in a girder of the dimensions $u = 1$ supporting a load w at its centre. We have proved $Tz = Ctu^3$, where C is some constant independent of u.

Making $l = h = a/2$ we have, therefore,

$$C\,t\,u^3 = \left(\frac{Va}{2} + \frac{\acute{w}a}{4}\right)u^4;$$

$$\text{and } Ct = \left(\frac{wa}{2} + \frac{\acute{w}a}{4}\right);$$

$$\therefore \frac{wa}{2} + \frac{\acute{w}a}{4} = \left(\frac{Va}{2} + \frac{\acute{w}a}{4}\right)u$$

$$\therefore 2V = \frac{2w - (u-1).\acute{w}}{w}$$

$$2Vu^3 = (2w - (u-1).w).\,u^2$$

If $(u\text{-}1)\ \acute{w}$ = or exceed $2w$, it follows that the girder in scale u will not be able to support any weight at its centre, without the tension and thrust per square inch being increased.

Example. – A model girder, length 80 feet, weight 10 tons, breaks with a weight of 30 tons in the middle: what weight will break a similar girder 480 feet long?

$$Vu^3 = \frac{(2w - (u-1)\acute{w})u^2}{2} = \frac{(60-50).36}{2} = 180\ \textit{tons}. \quad \textit{Ans.}$$

We have been especially induced to call our readers' attention to the subject of model experiments, from the fact that the proposed tubular bridge over the Menai Straits is to be constructed, as to its dimensions, according to laws developed in a series of experiments, conducted by Mr. Hodgkinson at Blackwall. We much fear that the enormous width to be crossed without a support will prove too much even for the known ingenuity of Mr. Stephenson. If, however, the talents of that illustrious engineer should prove equal to the magnitude of his conception, none will feel more satisfaction than ourselves at his success; and with mingled wonder and pleasure shall we witness – at a respectful distance – his aerial tunnel quivering and bending beneath a load of 'moral agents', happily unconscious of the laws of equilibrium and of the depth of the cold dark waters above which they are being whirled.

(To be continued)

ON MODEL EXPERIMENTS

(Continued from page 353)

In the last number of the Journal we obtained formulae for the comparison of the weights capable of being sustained by similar girders; we now proceed to apply our formulae to the experiments so ably conducted by Mr. Hodgkinson, with reference to the proposed bridge over the Menai Straits. A report of these experiments will be found in the May number of the Journal for 1846, from which we extract the following table and explanation: -

"To obtain the strength of tubes, precisely similar to other tubes fixed on; – but proportionately less than the former in all their dimensions, as length, breadth, depth, and thickness, in order to enable us to reason as to strength from one size to another, with more certainty than hitherto, as mentioned before. Another object, not far pursued, was to seek for the proper proportion of metal in the top and bottom of the tube. Much more is required in this direction. In the three series of experiments made, the tubes were rectangular, and the dimensions and other values are given below.

Length.			Distance between supports	Weight.	Thickness of Plates.	Last observed Deflection.	Corresponding Weight.	Breaking Weight.	Value of f, for crushing Strain.
ft. in.	In.	In.	ft. in.	cwt. qr.	Inch.	Inch.	Tons.	Tons.	Tons.
31 6	24	16	30 0	44 3	·525	3·03	56·3	57·5	19·17
31 6	24	16	30 0	24 1	·272	1·53	20·3	22·75	14·47
31 6	24	16	30 0	10 1	·124	1·20	5·04	5·53	7·74
				lb. oz.			lb.	lb.	
8 2	6	4	7 6	78 13	·139	·46	9,416	9,976	23·17
8 2	6	4	7 6	38 11	·065	·32	2,696	3,156	15·81
8 2	6	4	7 6	..	..	..	..	..	..
4 2½	3	2	3 9	10 12	·061	·485	2,464	2,464	24·56
4 3¼	3	2	3 9	4 15	·03	·13	660	672	13·42

The tube placed first in each series, is intended to be proportional in every leading dimension, as distance between supports, breadth, depth, and thick ness of metal, and any variations, are allowed for in the computation. Thus the three first tubes of each series are intended to be similar; and in the same manner of the other tubes, &c."

Here it will be observed that in the first set the dimensions are four times the dimensions in the second set nearly, and the dimensions in the second set are very nearly twice those in the first. Comparing the first of the second and third sets, we find in the first of the second set the weight of the tube 78 lb. 13 oz., and the breaking weight 9,976 lb.; and in the first of the third set, the weight of the tube 10 lb. 12 oz., and the breaking weight 2,464 lb. Now, by the formula deduced in our last paper, if w be the imposed breaking weight, w' the weight of a girder in scale 1, the breaking weight of a similar girder in scale u will be $\left(\frac{2w-(u-1)w'}{2}\right)u^2$.

$$\text{Here } (u-1) = 1; \quad w' = 10 \text{ lb.} 12 \text{ oz.}; \quad u = 2 \text{ lb.}; w = 2{,}464 \text{ lb.};$$

$$2w - (u-1)w' = 4{,}918 \text{ lb. nearly}$$

$$\therefore \left(\frac{2w-(u-1)w'}{2}\right)u^2 = 9{,}836$$

This, as will be seen, is only 140 lb. less than the breaking weight as found by experiment. It will be observed, moreover, that the tube in the second set is rather more than twice as thick as that in the first, which sufficiently accounts for the slight discrepancy. Comparing now the first experiment of the first and second sets, we shall find the value of the breaking weight in the first, deduced by the formula from the second – too great:

$$\text{putting } u = 4 \text{ lb.}, \quad w' = 78 \text{ lb.} 13 \text{ oz.}, \quad w = 9{,}976 \text{ lb.},$$

the breaking weight were = 70 tons, – an excess over experiment of nearly 13 tons. The reason for this difference is obvious: our formula supposes that the breaking tension per square inch, in all the models compared by it, is constant. This, no doubt, would have been the case in the model tubes of Mr. Hodgkinson, had they been constructed of one uniform plate of metal, and not riveted. The necessity of riveting is one great cause of the mishaps which are constantly occurring in iron bridges, as we endeavoured to show a few numbers back, in a paper on the cause of the breakage of the Dee Bridge [...]: the longer the bridge, the wider and more numerous must be the joints, and the greater the chance of irregular strains. The effect of riveting, we observe, has been to reduce the breaking tension per square inch one-fifth, in the first of the first set of experiments as compared with the first of the last set; – how much that effect would be increased in a tube 460 feet long, or fifteen times the dimensions of the longest experimented on by Mr. Hodgkinson – we leave our readers to judge.

We trust, in the meanwhile, that Mr. Hodgkinson will continue his labours on tubes of still greater length, by which means only can we hope for anything like an approximation to a law for the mean effect of a number of riveted joints. However, as we have often observed, the mischief most to be apprehended is not of a statical but dynamical character, – the constant recurrence of vibrations, tending to loosen the joints, perhaps to impair the mature of the iron, and ultimately so to weaken the structure, that a slight jar or strain may be sufficient to produce sudden disrupture of the whole.

Appendix A.7

Osborne Reynolds (1888) Extract from paper: On certain laws relating to the regime of rivers and estuaries

Reynolds, Osborne (1888) On certain laws relating to the regime of rivers and estuaries, and on the possibility of experiments on a small scale. In: *Report of the 57th Meeting of the British Association for the Advancement of Science held in Manchester in 1887*, 555-562. London: John Murray.

'That this prevalence of the deepest water on the concave side [of the Mersey Estuary] must be the result of the momentum left in the water by the flood at once seemed to me probable; for if the bottom were level or deepest on the Lancashire side the effect of the curved shape would be to cause the flood entering at the northern horn to follow the south-eastern or Cheshire shore, and the momentum of this water would tend to carry it round the head of the estuary and back along the Lancashire side; would, in fact, tend to set up a circulation before the top of the flood was reached; so that on the Lancashire side the water would be moving down the estuary before the ebb commenced; whence, considering that the flood tends to raise the bottom and the ebb to lower it (for the reasons already pointed out), it seems that the stronger flood on the Cheshire side would raise this side, While the stronger ebb on the Lancashire side would lower this. This is supposing the bottom to be level.

In order to verify these conclusions a vessel was constructed having a flat bottom and a vertical boundary of the same shape as the high-tide line of the inner estuary from the rock to the same distance above Runcorn. The horizontal scale was 2′ to a mile (1:31,800), and the vertical scale 1 inch to 80 feet (1:960).

A shallow tin pan was hinged on to the otherwise open channel at the rock, by raising and lowering which, when full of water, the motion of the tide could be produced throughout the model through the narrows; the true form of the bed of the channel was given to the model by means of paraffin. And in order to obtain approximately the proportional depth in the inner estuary, sand was placed level on the bottom so that the high-tide depth was reduced to the equivalent of about twenty feet. The idea in making this model was not so much to obtain a shifting of the sand as to show the circulation of the water as resulting from the flood tide

Physical Models: Their historical and current use in civil and building engineering design,
First Edition. Edited by Bill Addis.

with a level bottom. In the first instance the tide pan was raised and lowered by hand, but as at the first trial it became evident that the model was not only going to show the expected circulation, but was also capable of showing, by the change in the position of the sand, the effect of this circulation on the configuration of the estuary and other important effects, it was arranged that the model should be worked from a continuously running shaft. The working of the model by hand at once showed that there was only one period of working at which the motion of the water in the model would imitate the motions of the actual tide in the Mersey, which period was found to be about forty seconds; a result that might have been foreseen from the theory of wave motions, since the scale of velocities varies as the square roots of the scales of wave heights, so that the velocities in the model which would correspond to the velocities in the channel would be as the square roots of the vertical scales – about 1/33 – and the ratios of the periods would be the ratio of horizontal scales divided by this ratio of velocities, or 33/31,800 = 1/950.

Hence, taking 11.25 hours (40,700 seconds) as the tidal period, the period of the model = 40,700/950 = 42 seconds (about).

This period was adopted for working the model from the shaft. It was then found that the circulation at the top of the flood, which was very evident while the bottom was flat, caused a general rise of the sand on the Cheshire side and lowering on the Lancashire, which went on for about 2,000 tides. That during this time, owing to the increase of flood up the Lancashire side and the diminution of that on the Cheshire side which followed from the deepening of the one and the shoaling of the other, the circulation steadily diminished until its character was so changed that it could no longer be called a general circulation, and that after this, although there were further changes in detail going on in the estuary, the two sides maintained a steady condition as regards depth for low tides.

During this time banks were formed and low-tide channels, which resembled in all the principal features those actually in the Mersey; the eastern bank, with the deep sloynes on the Cheshire side, the Devil's Bank and the Garston Channel, the Ellesmere Channel and the deep water in Dungeon Bay and at Dingle Point – all these were very marked in character and closely approximate in scale.

And, what is as important, the causes of these as well as all minor features could be distinctly seen in the model.

The eastern and Devil's Bank are seen during the process of their formation to be simply an internal bar formed by carrying the sand brought down by the ebb out of the narrows and Sloyne [an anchorage in the Mersey], until debouching into the broad estuary; its velocity is so far diminished that it can no longer carry its charge, just as happens at the mouth of every river. The peculiar configuration of these banks is explained by the existence of two lines of eddies from about half-tide to the top of the flood: the first of these is caused by the sharp corner at Dingle, and lies between Dingle and Garston, the eddies having their centres over the Devil's Bank; and the second, caused by the divergence of the Cheshire Bank towards Eastham, having the lines of centres over the Eastham Bank. These eddies, which during the most rapid part of the flood only effect a diminution of the velocity of the flood, cause, as the velocity slackens toward the top of the flood, back water to set in along both shores, which back waters, starting the ebb, cause this to be strongest over the Garston and Eastham Channels, which are thus kept open.

The lateral configuration of the shores at Dungeon Bay and at Ellesmere is seen to cause back waters to exist in these bays during the whole of the flood in the latter, and from one to two hours before the top of the flood in the former, which fully accounts for the deep water at these points. The existence of these back waters in the actual channel has been verified. There are many other circumstances brought to light by this model, which it is impossible for me here to notice without unduly extending the length of this paper, if, indeed, I have not already done so. I will therefore only remark that a second start was made with the sand flat in this second model, and that the result obtained was the same as regards the general features of the estuary. So interesting were these results that it was decided to try a larger scale. A model, having a horizontal scale of 6 inches to a mile, and a vertical scale of 33 feet to an inch, was therefore made, and the tide produced as before. The calculated period of this model is 80 seconds, and experiment bears this out, any variation leading to some tidal phenomena, such as bonos or standing waves, which are not observed ir the estuary.

The disadvantage of the larger model is the time occupied–a little more than a minute a tide – which means about 300 tides a day, or 2,000 tides a week. On one occasion the model was kept going for 6,000 tides, and a survey was then made of the state of the sand. And this will be seen to present a remarkable resemblance in the general features to the charts of the Mersey, of which three – 1861, 1871, 1881 – are shown [omitted]; in fact the survey from the model presents as great a resemblance to any one of these as they do to each other.

It is impossible for me to enter upon all the points of agreement. Taking into account that in both the estuary and the model there are always changes going on within certain limits, and these changes do affect the currents to a certain extent, it is not to be supposed that there will be exact agreement between the currents at all points and at all states of the tides on the model and estuary. Still there is a general agreement, and in the few verifications I have made I have found that the current found in the model at a particular point and state of tide is also to be found in the estuary.

In one respect the great difference between the model and the estuary calls for remark: this is the much greater depth of the model as compared with its length and breadth. The vertical scale being 33 feet to an inch, and the horizontal scale 880 feet to an inch, so that the vertical heights are nearly twenty-seven times greater than the horizontal distances, such a difference is necessary to get any results at all with such small scale models; and it is only natural to suppose that it would materially affect the action. As a matter of fact, however, it does not seem to do so. And, further, it would seem that, notwithstanding the general resemblance on the regime of the beds of large and small streams running over sand, there is in these a similar difference in vertical scale, the smaller streams not only having a greater slope, but also having greater depth as compared with their breadth and steeper banks. So far as the theory of hydrodynamics will apply, it seems that in the model the effects of the momentum of the water would be greater as compared with the bottom resistances than in the estuary, and I think that they are. But the effects of momentum in the estuary greatly preponderate on the resistances, as shown by the fact that the tide at the top of the flood rises some 2 to 3 feet higher at high spring tides than it does at the rock; nor does it do much more than this in the model. In the model it certainly seems that the general regime is determined by the momentum effects, and from the almost exact resemblance which this regime bears to that of the estuary, it would seem that, although the momentum effects may be diminished by the greater resistance on the bottom, they are still the prevailing influence in determining the configuration of the banks.

Further investigation will doubtless explain this, and also determine the best proportional depths. From my present experience, in constructing another model, I should adopt a somewhat greater exaggeration of the vertical scale. In the meantime I have called attention to these results, because this method of experimenting seems to afford a ready means of investigating and determining beforehand the effects of any proposed estuary or harbour works; a means which, after what I have seen, I should feel it madness to neglect before entering upon any costly undertaking.'

Author biographies

Bill Addis worked as a design engineer with Rolls Royce Aeroengines, where his passion for photoelastic stress analysis and physical modelling was born. He later joined Reading University lecturing on materials science and structures, where he completed his doctorate in the history and philosophy of engineering. After more than fifteen years in the academic world, he worked for a similar time with Buro Happold as a specialist in sustainable construction. He has followed his interest in the history of civil and building engineering for over forty years and written many publications in the field. Based on his PhD, *Structural Engineering: the Nature of Theory and Design* was published in 1990 and in 1999 he edited *Structural and Civil Engineering Design*, Vol. 12 of the series *Studies in the History of Civil Engineering*. His most recent book is *Building: 3000 years of engineering, design and construction*. During the last few years he has enjoyed visiting professorships at six universities in continental Europe.

Joaquín Antuña Bernardo is an architect and professor in the Department of Structures and Building Physics at the Technical School of Architecture in the Technical University of Madrid where he is secretary of the PhD program in Building Structures. His PhD in 2003 gained the Edoardo Benvenuto award. He is a member of the scientific committee for the commemorative exhibition of the centenary of the birth of Eduardo Torroja. Since 1989, he has worked as an architect and structural consultant, carrying out projects such as the Baracaldo stadium, the rehabilitation of the *Teatro de la Comedia* in Madrid, and the Santa Eulalia del Rio Congress Centre in Ibiza.

Dirk Bühler studied and gained a PhD in architecture at the RWTH Aachen. 1979-1992 he carried out research and teaching in the History of Latin-American Architecture in Mexico D.F. and Puebla (Mexico). From 1993 he was senior curator and director of the construction collection at the *Deutsches Museum* in Munich where, in 1998 he established the bridge- and hydraulic engineering exhibitions. Since 2016 he was a senior researcher at the *Deutsches Museum* focusing on the history of bridge building, concrete and engineering models. His other research themes include the study of urbanism, civil architecture and engineering in Latin America during the viceregal and independence periods (1492-1914).

Bertholdt Burkhart is an architect and Professor for structural design at the University of Braunschweig. He worked with Peter Poelzig in Berlin and Frei Otto at the University of Stuttgart on the German Pavilion Montreal 1967 and the

Physical Models: Their historical and current use in civil and building engineering design,
First Edition. Edited by Bill Addis.

membrane roofs for the Munich Olympics 1972. His research interests include nature and technology, the history of lightweight structures and tents, and conservation of twentieth century buildings and structures. In his architectural practice his work has included the *Kanzlerbungalow* in Bonn by Sep Ruf, the former employment office in Dessau by Walter Gropius, various tent structures and historic bridges. He is a member of the Construction History Society, ICOMOS, Docomomo, the Alvar Aalto Society, and others.

Pepa Cassinello is an architect, Director of the Eduardo Torroja Foundation and Titular Professor at the Polytechnic University of Madrid. She specialises in construction and structural design of historic and avant-garde buildings. She has given lectures at the Royal Institute of British Architects, London, *École Polytechnique Féderalé*, Lausanne, *Sapienza Universitá* di Roma, *Universitá degli Studi di Roma Tor Vergata, Technische Universität* Berlin, and the *Deutsches Museum* Munich. She has written and edited books on the work of Eduardo Torroja, Felix Candela, Ildefonso Sánchez del Río, and lightweight structures. She has received awards from the International Association for Shell and Spatial Structures and XIII BEAU Spanish Biennial of Architecture and Urbanism.

John Chilton is Emeritus Professor of Architecture and Tectonics at the University of Nottingham, UK. He is an Executive Council member of the International Association for Shell and Spatial Structures (IASS). With a special interest in the history, design and construction of innovative and non-conventional structures he actively participates in IASS working groups on shell structures, spatial timber structures and structural morphology. He has published books on Space Grid Structures (1999), the Swiss reinforced-concrete shell builder Heinz Isler: The Engineer's Contribution to Contemporary Architecture (2000) and Timber Gridshells: architecture, structure and craft (2016).

Mario Alberto Chiorino is Professor Emeritus of Mechanics of Solids and Structures at the *Politecnico di Torino*, Italy, where he also served as Professor of Theory and Design of Structures and of Structural Analysis of Masonry and Monumental Structures. He is member of the Turin Academy of Sciences and of the Russian Academy of Engineering, an honorary member of ACI American Concrete institute and life member of fib International Federation for Structural Concrete. His interest for structural models developed especially during his studies on the relationship between the art and science of building in the works of Eduardo Torroja and Pier Luigi Nervi, and on the role of experimentation in the conception and assessment of their projects.

Mike Chrimes worked at the Institution of Civil Engineers, London for 37 years, leading the Library and Archives team, and retiring in 2014 as Director of Engineering Policy and Innovation. He has written many articles on information science and construction history, presenting papers at events and conferences in Europe and North America as well as the UK. He is author and editor of a number of books on the history of civil engineering. He is co-author of *Consulting Engineers: The British consulting engineers who created the world's infrastructure*, published in 2019, and edited the proceedings of the Second International Conference on Early Main Line Railways published in 2019.

Owen Connick is Sustainability Associate at Hoare Lea. Prior to joining Hoare Lea, Owen was Technical Director at Breathing Buildings, where he led the design and consultancy teams. After a first degree in Aerospace Engineering at Imperial College London, he completed his PhD on the fluid mechanics of hybrid natural ventilation. His experience and expertise include the use of simple analytical models, and the use of water-bath modelling to simulate buoyancy-driven flows in a laboratory environment. At Breathing Buildings he developed significant expertise in the application of Dynamic Thermal Modelling to simulate thermal comfort and air quality in real buildings, with a particular focus on schools and UK design regulations. Owen has presented papers at numerous CIBSE/ASHRAE seminars, and is keen on many aspects of sustainability in the built environment.

Bill Craig retired from his post as Reader in Geotechnical Engineering in the University of Manchester in 2015, having been on the staff there for forty five years. He is a former Chairman of the British Geotechnical Society (1997-99) and was a senior Fulbright Scholar at the University of Colorado (1989-90). He first became involved in centrifuge modelling as a research student, working with Professor Peter Rowe in 1969, and continued using this technique throughout his career. Alongside technical advances in the field he has always had in interest in the history and development of centrifuge technologies and their applications around the world.

Francesco Dorigatti is a Research Fellow and Wind Tunnel Laboratory Manager at Cranfield University, in the UK. Specialising in wind engineering and experimental aerodynamics, he conducts research and commercial wind tunnel tests on civil engineering structures, ground vehicles and aeronautical applications. A former senior engineer and associate at RWDI, he has led and contributed to over a hundred wind-tunnel studies on tall and super-tall buildings, roofs for sports stadia, and long-span bridges. While at the wind tunnel of *Politecnico di Milano*, in Italy, he was part of the team that designed and tested the 13-metre long full-aeroelastic model of the proposed Messina Strait suspension bridge which, at 3.3 km, would be the world's longest main span.

Bernard Espion is Professor of Structural Engineering at the *Université Libre de Bruxelles* (ULB) where he is also the Director of the Civil Engineering Laboratory. He is a founding member of Construction Histories Brussels, the ULB-VUB joint research group. His main research interests in Construction History lie in structures (reinforced and prestressed concrete, composite steel-concrete, cable roofs, thin concrete shells, bridges), materials and construction processes (cement and concretes, iron and steel, connections), historical evolution of structural theory and design concepts, especially by Belgian engineers, contractors and companies in the nineteenth and twentieth centuries. He contributes to the promotion of Belgian civil and building engineering heritage by publications, lectures and contributions to exhibitions.

Rainer Graefe was born in 1941 in Berlin. He gained his Magister Artium in 1969 for his thesis *Idee des Bauhaus-Theaters*, and his D.Phil. on *Roman theatre vela* in 1976. From 1969 he worked under Frei Otto at the Institute of Lightweight Structures, Stuttgart, establishing in 1984 the research project *History of Constructing* funded by the *Deutsche Forschungsgemeinschaft*. From 1991 he was Chair of the Institute of Architectural History at the University of Innsbruck. In 2005 he founded the *Archiv für Baukunst* in Innsbruck.

In 2008 he created the permanent exhibition *The First Church of Gaudí* at the Diocesan Museum in Barcelona. Since 2010 he has worked on the D-A-CH project *Shoukov's strategies for economical steel structures* in cooperation with ETH Zürich and TU München. In 2014 he founded the *LindenbaumMuseum* in Neudrossenfeld, Germany. His main research interest is the development of historical building construction.

Stefan Holzer studied Civil Engineering at the Technical University in Munich, gaining a doctorate on numerical mechanics in 1992. After two years of professional practice he was appointed associate professor at the University of Stuttgart from 1995-2001, and a Professor of Engineering Mathematics at the University of the Federal Armed Forces, Munich from 2001-2016. Since 2016 he is Professor of Building Archaeology and Construction History at ETH Zurich. His main interest is in the analysis and preservation of historic monuments on which subject he has published numerous national and international papers as well as five books. His research has focused on Early Modern and nineteenth-century construction history, with particular attention to timber and masonry construction. His work as a consulting engineer for conservation has included many major monuments. Since 2017 he has been Chairman of the *Gesellschaft für Bautechnikgeschichte* (German Society for Construction History).

Santiago Huerta worked as an engineer and architect in the 1980s, before completing his doctoral studies at the Polytechnic University of Madrid (UPM) in 1990. Since 1992 he has been Professor of Structural Design at the School of Architecture at UPM where his main interest has been masonry structures. He lectures on Structural Design, Construction History and Historic Structures. During the last thirty years he has worked as a structural consultant for more than 35 historical constructions including the monastery of San Juan de los Reyes, the cathedrals of Tudela, Palma de Mallorca and Santiago de Compostela, and the mediaeval masonry bridge at Navea. In 1997 he was a co-founder of the Spanish Society of Construction History (SEdHC), serving as its Secretary until 2003 and as President since that date.

Lukas Ingold is a lecturer at the Institute of Technology in Architecture at ETH Zurich. He received his master's degree in architecture from ETH Zurich in 2012. After working for the architect Christian Kerez in Zurich, he has been teaching at ETH Zurich at the Chair of Structural Design of Joseph Schwartz. His research focuses on the history of architecture and engineering of the twentieth century and particularly on the work of the Italian structural designer Sergio Musmeci. He has been awarded research scholarships from the Swiss National Science Foundation and the *Istituto Svizzero di Roma*.

Andreas Kahlow studied mechanical engineering at the University of Rostock (Germany) 1971-1975 and worked on the mechanics of solid bodies as an *Assistent* in the Faculty of Shipbuilding at the University of Rostock, 1975-1978. He undertook postgraduate studies at the Humboldt University in Berlin and completed his doctoral thesis in 1981 on the development of engineering mechanics in the nineteenth century. He worked as a technical journalist before beginning research in 1984 in History of Technology at the Berlin Academy of Sciences. From 1992 he was Professor for History of Construction at the faculty of civil engineering at the University of Applied Sciences in Potsdam. He has published many papers and several books on the history of construction and mechanical engineering. He retired in 2018.

Amarnath Kasalanati is the Associate Director at the Pacific Earthquake Engineering Research Center (PEER) in the University of California, Berkeley, where he currently oversees the operations of the 6-DOF shake table. Prior to joining PEER, he was with Dynamic Isolation Systems (DIS) for 19 years as the Director of Engineering. During that time, he was involved in experimental research at SUNY at Buffalo, University of Nevada, Reno, UC Berkeley, UC San Diego, NCREE facility in Taiwan and the E-Defense facility in Japan. Amarnath's research interests include structural control, experimental investigation, and analysis and modelling of seismic isolation and energy dissipation devices. He served as a member of ASCE and AASHTO Code Committees for Seismic Isolation Design.

Mamoru Kawaguchi (1932-2019) was a Japanese structural engineer who began his career working in the office of Professor Yoshikatsu Tsuboi, where he contributed to the suspended roof designed by Kenzo Tange for the 1964 Tokyo Olympics. He founded his own practice Kawaguchi & Engineers in 1964 and designed many innovative and daring membrane structures in Japan, including air-supported structures for Expo'70 held in Osaka and his Pantadome System. His design work has included many special spatial structures: pneumatic and membrane structures, timber structures, seismic isolation systems, bridges and towers, many of which have relied on the use of physical models. In 1972 he became a Professor at Hosei University and was President of the IASS in 1990-91 and 2000-06. He is the recipient of numerous prestigious awards. Sadly, Mamuro Kawaguchi died in May 2019.

Jan Knippers is a practising consulting engineer and since 2000 head of the Institute for Building Structures and Structural Design (ITKE) at the University of Stuttgart. His interest is in innovative and resource-efficient structures created at the intersection of research and development and practice. From 2014 to 2019 he was the coordinator of the DFG collaborative research centre 'Biological Design and Integrative Structures'. Since 2019 he is Deputy Executive Director of the Cluster of Excellence 'Integrative Computational Design and Construction for Architecture' and Vice-rector for Research at the University of Stuttgart. He is author of several books, numerous scientific publications, and member of various advisory boards.

Ian Liddell is a retired structural engineer who was formerly a partner of the engineering design firm, Buro Happold and a visiting professor at Cambridge University Engineering Department. At the start of his career he worked on the roof of the Sydney Opera House. Ten years later he was the engineer in charge of the work in the Arup office on the Mannheim Multihalle gridshell roof, and at the end of his career he was the design engineer of the Millennium Dome in London. He has several awards for his work including the IABSE International Award of Merit and the Gold Medal of the Institution of Structural Engineers.

Roland May read architecture and town planning at the Technical University, Darmstadt. From 2002 to 2008 he was research and teaching associate and 2008/09 lecturer in the Department of History and Theory of Architecture at the same institution. Since 2009 he is research associate at the Chair for Construction History at BTU Cottbus-Senftenberg. He carried out major research projects on the architect Paul Bonatz, the civil engineer Franz Dischinger and the relationship between civil engineers and architects in modern architecture. His publications deal with topics such as the interrelations between architecture and engineering in the 20th century, the cultural impact of civil engineering, and traditionalistic tendencies in modern architecture.

Bruce Martin is a structural engineer who is passionate about creating the best cross-discipline solution to the particular needs of every project. He cares about elegant and practical engineering. He believes in monitoring and minimising the environmental impact of our engineering. Sometimes the best solution requires novel approaches. In such cases Bruce leads the process of identifying and completing the steps required to demonstrate the performance of novel engineering designs, and then deliver an economic and timely project completion. Bruce works in London as an Associate Director with Expedition Engineering.

Raf Orlowski gained a PhD in acoustics, and was then appointed as a research associate at the University of Cambridge working with Peter Parkin and Mike Barron on the development of acoustic scale models for the design of auditoria and other large spaces. He later helped set up acoustic consultancy offices, first for Arup and then for Ramboll where he worked on a number of prestigious architectural projects for which he used acoustic scale models to aid the design. He was formerly a visiting professor at the University of Sheffield and is currently a visiting professor at the Bartlett School of Architecture at University College, London.

Gabriele Neri is an architect who gained his PhD in History of Architecture and Urban Planning. He has been Adjunct Professor of History of Architecture at the Polytechnic of Milan (Italy) since 2011. Since 2012, he is researcher at the Mendrisio Academy of Architecture (Switzerland), since 2012. In 2018 he was appointed member of the Board of Directors of the Triennale Design Museum, Milano, Italy. On the work of Pier Luigi Nervi, he published the volume *Capolavori in miniatura. Pier Luigi Nervi e la modellazione strutturale* (2014) and several essays and in 2014 he edited an Anthology with Nervi's writings and the reissue of his famous book *Scienza o arte del costruire?* His other publications include: *Caricature architettoniche. Satira e critica del progetto moderno* (2015), *Umberto Riva. Interni e allestimenti* (2017), and *Louis Kahn and Venezia* (with E. Barizza, 2018).

Denis Smith (1930-2017) was a historian of engineering, a founding member of the Greater London Industrial Archaeology Society, a long-standing member of the Newcomen Society, and the Institution of Civil Engineers' Panel for Historical Engineering Works. His doctorate on the use of models in bridge design in the nineteenth century is a classic reference work. He wrote many papers on engineering history and industrial archaeology, and compiled the Civil Engineering Heritage guide to London and the Thames Valley. He gave courses in engineering, industrial archaeology and engineering history at many educational establishments in London. He was also a noted player of the accordion in English folk music.

James Sutherland has been a Technical Director at HR Wallingford, near Oxford, UK for over twenty years, specialising in studies related to coastlines and estuaries and working on projects in over 20 countries in Europe, the

Middle East, Asia-Pacific, Africa and the Americas. From 2012-2016 he was HR Wallingford's R&D coordinator. He has published many papers on nearshore hydrodynamics, sediment transport (including scour around structures and coastal erosion) beach management and wave forces on maritime structures.

Edwin Trout joined the British Cement Association in 1995, transferring to The Concrete Society in 2006. He also acts as Executive Officer for the Institute of Concrete Technology. With responsibility for the Society's library – established in 1937 – Edwin developed an interest in early concrete construction and the history of the cement and concrete industries, on which he has given talks and written articles for periodicals, including Construction History. He has written two books, *Some Writers on Concrete: the Literature of Reinforced Concrete, 1897-1935 (*2013) and *Traditional Milling Technology in the English Cement Industry, 1796-1899* (2015). He also writes a monthly column in *Concrete* magazine and has contributed chapters to several books.

David White is Professor of Infrastructure Geotechnics at the University of Southampton, involved in the UK's National Infrastructure Laboratory and the associated geotechnical centrifuge modelling facilities. His research and teaching has focused on geotechnics within the broader themes of offshore and onshore infrastructure for energy and transport. He has pioneered various physical modelling technologies related to the geotechnical centrifuge and 'water tunnel' sediment flumes. He served as Secretary of the ISSMGE Technical Committee for Physical Modelling in Geotechnics for 12 years. He has published more than 300 papers, many related to physical modelling, and is a Fellow of the Royal Academy of Engineering and the Institution of Civil Engineers.

Christiane Weber is a historian of the architectural built environment. She holds a diploma in architecture and an MA in art history from the Technical University of Karlsruhe. She studied in Karlsruhe, the *École Nationale Supérieure d'Architecture de Paris-Belleville* and the *Université de Strasbourg*, and received her doctorate from the Braunschweig Technical University. Since 2013 she has been lecturing and undertaking research at the Franz Leopold University of Innsbruck. Her research interests lie in construction history, phenomena of cultural and technological transfer between France and Germany and the history of the German technical education system. Since 2014 she has been a Feodor-Lynen fellow of the Alexander von Humboldt Foundation.

David Wendland is specialized in design processes and construction of buildings with complex shape, in particular shells and vaults, developing an interdisciplinary approach that integrates architectural design, engineering and the hermeneutic methods from the Humanities. He studied Architecture in Darmstadt, Venice and Stuttgart, where he completed his PhD on vault construction without formwork and the definition of this technique in the Gothic Revival. His scientific consultancy includes the reconstruction of the Late-Gothic vault in Dresden Castle. At the TU Dresden, he carried out research on the construction and design principles of Late-Gothic vaults, for which he was awarded with an ERC Starting Grant. Since 2019 he is professor of Construction History at the Brandenburg University of Technology Cottbus-Senftenberg.

About the series editors

Karl-Eugen Kurrer was born in 1952 in Heilbronn, Germany. Following his degree in civil engineering at Stuttgart University of Applied Sciences, he worked as a structural timber engineer in Heilbronn. He then returned to university to study civil engineering, history of technology and physical engineering sciences at TU Berlin. His dissertation on the development of vault theory from the eighteenth century to 1980 was completed in 1981, and that was followed by the award of a doctorate by TU Berlin in 1986. Between 1989 and 1995, Dr. Kurrer worked for Telefunken Sendertechnik GmbH in Berlin as a designer of antenna systems.

Since 1996, Dr. Kurrer has chaired the Working Group on the History of Technology at the VDI (Association of German Engineers) in Berlin. Between 1996 and February 2018, he was chief editor of *Stahlbau* and (from 2008) *Steel Construction – Design and Research*, journals published by Ernst & Sohn. For more than 40 years, Dr. Kurrer has carried out research on the subject of construction history with special emphasis on theory of structures and structural mechanics. He has published more than 190 papers and several monographs.

Werner Lorenz, born in 1953, graduated from TU Berlin in 1980 with a degree in structural engineering. After his first practical experience in an engineering firm in Berlin in 1984, he returned to TU Berlin, where he gave his first seminars on construction history. During 1988 he spent a research period as a visiting professor at the École Nationale des Ponts et Chaussées in Paris before, in 1992, gaining a doctorate with a thesis about the early history of building with iron in Berlin and Potsdam. He was appointed to the newly created Chair of Construction History at BTU Cottbus in 1993. Since his retirement in 2018, he has continued to work there as honorary professor for construction history and, since 2019, as coordinator of 'Cultural Heritage Construction', a nationwide interdisciplinary priority programme dedicated to the period of high modernism. Most of his historic research is concentrated on the history of construction shaped by industry during the eighteenth, nineteenth and twentieth centuries, and in terms of materials focuses on iron, steel and reinforced concrete. Concurrent with his work at the university, he founded a structural engineering consultancy in 1996 which specializes in the structural refurbishment of historic buildings and bridges.

Physical Models: Their historical and current use in civil and building engineering design,
First Edition. Edited by Bill Addis.

About this series

The *Construction History Series/Edition Bautechnikgeschichte* gives the new discipline of construction history a home for the publication of important works reflecting the full diversity of this subject. The scope of the series ranges from overviews to monographs on individual aspects or structures to biographies of prominent engineers. The two main orientations in construction history – either focusing more on the history of design or more on the history of theory – both receive due consideration.

Editors

Karl-Eugen Kurrer (Berlin) and Werner Lorenz (Cottbus)

Scientific Advisory Board

Bill Addis (Cambridge)
Antonio Becchi (Berlin/London)
Robert Carvais (Paris)
Bernard Espion (Brussels)
Stefan M. Holzer (Zurich)
Santiago Huerta (Madrid)
Andreas Kahlow (Berlin)
Roland May (Cottbus)

Physical Models: Their historical and current use in civil and building engineering design,
First Edition. Edited by Bill Addis.

Index

Physical Models: Their historical and current use in civil and building engineering design,
First Edition. Edited by Bill Addis.

g

n

o